SEAFOOD AND FRESHWATER TOXINS

FOOD SCIENCE AND TECHNOLOGY

A Series of Monographs, Textbooks, and Reference Books

1. Flavor Research: Principles and Techniques, *R. Teranishi, I. Hornstein, P. Issenberg, and E. L. Wick*
2. Principles of Enzymology for the Food Sciences, *John R. Whitaker*
3. Low-Temperature Preservation of Foods and Living Matter, *Owen R. Fennema, William D. Powrie, and Elmer H. Marth*
4. Principles of Food Science
   Part I: Food Chemistry, *edited by Owen R. Fennema*
   Part II: Physical Methods of Food Preservation, *Marcus Karel, Owen R. Fennema, and Daryl B. Lund*
5. Food Emulsions, *edited by Stig E. Friberg*
6. Nutritional and Safety Aspects of Food Processing, *edited by Steven R. Tannenbaum*
7. Flavor Research: Recent Advances, *edited by R. Teranishi, Robert A. Flath, and Hiroshi Sugisawa*
8. Computer-Aided Techniques in Food Technology, *edited by Israel Saguy*
9. Handbook of Tropical Foods, *edited by Harvey T. Chan*
10. Antimicrobials in Foods, *edited by Alfred Larry Branen and P. Michael Davidson*

11. Food Constituents and Food Residues: Their Chromatographic Determination, *edited by James F. Lawrence*
12. Aspartame: Physiology and Biochemistry, *edited by Lewis D. Stegink and L. J. Filer, Jr.*
13. Handbook of Vitamins: Nutritional, Biochemical, and Clinical Aspects, *edited by Lawrence J. Machlin*
14. Starch Conversion Technology, *edited by G. M. A. van Beynum and J. A. Roels*
15. Food Chemistry: Second Edition, Revised and Expanded, *edited by Owen R. Fennema*
16. Sensory Evaluation of Food: Statistical Methods and Procedures, *Michael O'Mahony*
17. Alternative Sweeteners, *edited by Lyn O'Brien Nabors and Robert C. Gelardi*
18. Citrus Fruits and Their Products: Analysis and Technology, *S. V. Ting and Russell L. Rouseff*
19. Engineering Properties of Foods, *edited by M. A. Rao and S. S. H. Rizvi*
20. Umami: A Basic Taste, *edited by Yojiro Kawamura and Morley R. Kare*
21. Food Biotechnology, *edited by Dietrich Knorr*
22. Food Texture: Instrumental and Sensory Measurement, *edited by Howard R. Moskowitz*
23. Seafoods and Fish Oils in Human Health and Disease, *John E. Kinsella*
24. Postharvest Physiology of Vegetables, *edited by J. Weichmann*
25. Handbook of Dietary Fiber: An Applied Approach, *Mark L. Dreher*
26. Food Toxicology, Parts A and B, *Jose M. Concon*
27. Modern Carbohydrate Chemistry, *Roger W. Binkley*
28. Trace Minerals in Foods, *edited by Kenneth T. Smith*
29. Protein Quality and the Effects of Processing, *edited by R. Dixon Phillips and John W. Finley*
30. Adulteration of Fruit Juice Beverages, *edited by Steven Nagy, John A. Attaway, and Martha E. Rhodes*
31. Foodborne Bacterial Pathogens, *edited by Michael P. Doyle*
32. Legumes: Chemistry, Technology, and Human Nutrition, *edited by Ruth H. Matthews*
33. Industrialization of Indigenous Fermented Foods, *edited by Keith H. Steinkraus*
34. International Food Regulation Handbook: Policy • Science • Law, *edited by Roger D. Middlekauff and Philippe Shubik*
35. Food Additives, *edited by A. Larry Branen, P. Michael Davidson, and Seppo Salminen*
36. Safety of Irradiated Foods, *J. F. Diehl*
37. Omega-3 Fatty Acids in Health and Disease, *edited by Robert S. Lees and Marcus Karel*
38. Food Emulsions: Second Edition, Revised and Expanded, *edited by Kåre Larsson and Stig E. Friberg*
39. Seafood: Effects of Technology on Nutrition, *George M. Pigott and Barbee W. Tucker*
40. Handbook of Vitamins: Second Edition, Revised and Expanded, *edited by Lawrence J. Machlin*

41. Handbook of Cereal Science and Technology, *Klaus J. Lorenz and Karel Kulp*
42. Food Processing Operations and Scale-Up, *Kenneth J. Valentas, Leon Levine, and J. Peter Clark*
43. Fish Quality Control by Computer Vision, *edited by L. F. Pau and R. Olafsson*
44. Volatile Compounds in Foods and Beverages, *edited by Henk Maarse*
45. Instrumental Methods for Quality Assurance in Foods, *edited by Daniel Y. C. Fung and Richard F. Matthews*
46. *Listeria*, Listeriosis, and Food Safety, *Elliot T. Ryser and Elmer H. Marth*
47. Acesulfame-K, *edited by D. G. Mayer and F. H. Kemper*
48. Alternative Sweeteners: Second Edition, Revised and Expanded, *edited by Lyn O'Brien Nabors and Robert C. Gelardi*
49. Food Extrusion Science and Technology, *edited by Jozef L. Kokini, Chi-Tang Ho, and Mukund V. Karwe*
50. Surimi Technology, *edited by Tyre C. Lanier and Chong M. Lee*
51. Handbook of Food Engineering, *edited by Dennis R. Heldman and Daryl B. Lund*
52. Food Analysis by HPLC, *edited by Leo M. L. Nollet*
53. Fatty Acids in Foods and Their Health Implications, *edited by Ching Kuang Chow*
54. *Clostridium botulinum*: Ecology and Control in Foods, *edited by Andreas H. W. Hauschild and Karen L. Dodds*
55. Cereals in Breadmaking: A Molecular Colloidal Approach, *Ann-Charlotte Eliasson and Kåre Larsson*
56. Low-Calorie Foods Handbook, *edited by Aaron M. Altschul*
57. Antimicrobials in Foods: Second Edition, Revised and Expanded, *edited by P. Michael Davidson and Alfred Larry Branen*
58. Lactic Acid Bacteria, *edited by Seppo Salminen and Atte von Wright*
59. Rice Science and Technology, *edited by Wayne E. Marshall and James I. Wadsworth*
60. Food Biosensor Analysis, *edited by Gabriele Wagner and George G. Guilbault*
61. Principles of Enzymology for the Food Sciences: Second Edition, *John R. Whitaker*
62. Carbohydrate Polyesters as Fat Substitutes, *edited by Casimir C. Akoh and Barry G. Swanson*
63. Engineering Properties of Foods: Second Edition, Revised and Expanded, *edited by M. A. Rao and S. S. H. Rizvi*
64. Handbook of Brewing, *edited by William A. Hardwick*
65. Analyzing Food for Nutrition Labeling and Hazardous Contaminants, *edited by Ike J. Jeon and William G. Ikins*
66. Ingredient Interactions: Effects on Food Quality, *edited by Anilkumar G. Gaonkar*
67. Food Polysaccharides and Their Applications, *edited by Alistair M. Stephen*
68. Safety of Irradiated Foods: Second Edition, Revised and Expanded, *J. F. Diehl*
69. Nutrition Labeling Handbook, *edited by Ralph Shapiro*

70. Handbook of Fruit Science and Technology: Production, Composition, Storage, and Processing, *edited by D. K. Salunkhe and S. S. Kadam*
71. Food Antioxidants: Technological, Toxicological, and Health Perspectives, *edited by D. L. Madhavi, S. S. Deshpande, and D. K. Salunkhe*
72. Freezing Effects on Food Quality, *edited by Lester E. Jeremiah*
73. Handbook of Indigenous Fermented Foods: Second Edition, Revised and Expanded, *edited by Keith H. Steinkraus*
74. Carbohydrates in Food, *edited by Ann-Charlotte Eliasson*
75. Baked Goods Freshness: Technology, Evaluation, and Inhibition of Staling, *edited by Ronald E. Hebeda and Henry F. Zobel*
76. Food Chemistry: Third Edition, *edited by Owen R. Fennema*
77. Handbook of Food Analysis: Volumes 1 and 2, *edited by Leo M. L. Nollet*
78. Computerized Control Systems in the Food Industry, *edited by Gauri S. Mittal*
79. Techniques for Analyzing Food Aroma, *edited by Ray Marsili*
80. Food Proteins and Their Applications, *edited by Srinivasan Damodaran and Alain Paraf*
81. Food Emulsions: Third Edition, Revised and Expanded, *edited by Stig E. Friberg and Kåre Larsson*
82. Nonthermal Preservation of Foods, *Gustavo V. Barbosa-Cánovas, Usha R. Pothakamury, Enrique Palou, and Barry G. Swanson*
83. Milk and Dairy Product Technology, *Edgar Spreer*
84. Applied Dairy Microbiology, *edited by Elmer H. Marth and James L. Steele*
85. Lactic Acid Bacteria: Microbiology and Functional Aspects: Second Edition, Revised and Expanded, *edited by Seppo Salminen and Atte von Wright*
86. Handbook of Vegetable Science and Technology: Production, Composition, Storage, and Processing, *edited by D. K. Salunkhe and S. S. Kadam*
87. Polysaccharide Association Structures in Food, *edited by Reginald H. Walter*
88. Food Lipids: Chemistry, Nutrition, and Biotechnology, *edited by Casimir C. Akoh and David B. Min*
89. Spice Science and Technology, *Kenji Hirasa and Mitsuo Takemasa*
90. Dairy Technology: Principles of Milk Properties and Processes, *P. Walstra, T. J. Geurts, A. Noomen, A. Jellema, and M. A. J. S. van Boekel*
91. Coloring of Food, Drugs, and Cosmetics, *Gisbert Otterstätter*
92. *Listeria*, Listeriosis, and Food Safety: Second Edition, Revised and Expanded, *edited by Elliot T. Ryser and Elmer H. Marth*
93. Complex Carbohydrates in Foods, *edited by Susan Sungsoo Cho, Leon Prosky, and Mark Dreher*
94. Handbook of Food Preservation, *edited by M. Shafiur Rahman*
95. International Food Safety Handbook: Science, International Regulation, and Control, *edited by Kees van der Heijden, Maged Younes, Lawrence Fishbein, and Sanford Miller*
96. Fatty Acids in Foods and Their Health Implications: Second Edition, Revised and Expanded, *edited by Ching Kuang Chow*

97. Seafood Enzymes: Utilization and Influence on Postharvest Seafood Quality, *edited by Norman F. Haard and Benjamin K. Simpson*
98. Safe Handling of Foods, *edited by Jeffrey M. Farber and Ewen C. D. Todd*
99. Handbook of Cereal Science and Technology: Second Edition, Revised and Expanded, *edited by Karel Kulp and Joseph G. Ponte, Jr.*
100. Food Analysis by HPLC: Second Edition, Revised and Expanded, *edited by Leo M. L. Nollet*
101. Surimi and Surimi Seafood, *edited by Jae W. Park*
102. Handbook of Water Analysis, *edited by Leo M. L. Nollet*
103. Seafood and Freshwater Toxins: Pharmacology, Physiology, and Detection, *edited by Luis M. Botana*
104. Handbook of Nutrition and Diet, *Babasaheb B. Desai*

Additional Volumes in Preparation

Nondestructive Food Evaluation: Techniques to Analyze Properties and Quality, *edited by Sundaram Gunasekaran*

Drug Residues in Foods: Pharmacology, Food Safety, and Analysis, *Nickos A. Botsoglou and Dimitrios J. Fletouris*

Alternative Sweeteners: Third Edition, Revised and Expanded, *edited by Lyn O'Brien Nabors*

Handbook of Dietary Fiber, *edited by Susan Sungsoo Cho and Mark Dreher*

Green Tea: Health Benefits and Applications, *Yukihiko Hara*

Food Processing Operations Modeling: Design and Analysis, *edited by Joseph Irudayaraj*

Handbook of Microwave Technology for Food Applications, edited by Ashim Datta and R.C. (Swamy) Anantheswaran

SEAFOOD AND FRESHWATER TOXINS

PHARMACOLOGY, PHYSIOLOGY, AND DETECTION

edited by
Luis M. Botana
University of Santiago de Compostela
Lugo, Spain

MARCEL DEKKER, INC. NEW YORK • BASEL

Library of Congress Cataloging-in-Publication Data

Seafood and freshwater toxins : pharmacology, physiology, and detection / edited by
Luis M. Botana.
p. cm. — (Food science and technology)
Includes index.
ISBN 0-8247-8956-3 (alk. paper)
1. Marine toxins. 2. Poisonous shellfish. I. Botana, Luis M. II. Series.

QP632.M37 S43 2000
615.9′45—dc21

00-040480

This book is printed on acid-free paper.

Headquarters
Marcel Dekker, Inc.
270 Madison Avenue, New York, NY 10016
tel: 212-696-9000; fax: 212-685-4540

Eastern Hemisphere Distribution
Marcel Dekker AG
Hutgasse 4, Postfach 812, CH-4001 Basel, Switzerland
tel: 41-61-261-8482; fax: 41-61-261-8896

World Wide Web
http://www.dekker.com

The publisher offers discounts on this book when ordered in bulk quantities. For more information, write to Special Sales/Professional Marketing at the headquarters address above.

Current printing (last digit):
10 9 8 7 6 5 4 3 2 1

PRINTED IN THE UNITED STATES OF AMERICA

Preface

The field of toxins is fascinating for many reasons, but perhaps the most intriguing question is why toxins exist in nature. In some cases, such as insects or reptiles, the reason for the existence of toxins is clear; for dinoflagellates or cyanobacteria, we are far from understanding the ecological purpose for the extraordinary diversity of toxins. But whatever the explanation, the toxins show exquisite selectivity to some molecular targets in mammals, making them a real sanitary and economic threat.

Researchers use seafood toxins as elegant tools to study how biological models function. But anyone working in this field knows how difficult it is to integrate this kind of information, since the groups that study seafood toxicity vary in areas of expertise from analytical chemistry, to molecular biology, ecology, physiology, pharmacology, epidemiology, and food technology.

There are numerous well-studied seafood toxins, and it is suspected that a great number are still unknown. Many people are already devoted to trying to isolate them; a clear example is the *Pfiesteria* case in Chesapeake Bay. This book is needed because of the scientific heterogeneity of the disciplines necessary to cope with seafood toxins, and because there is very little integrated information. The book is an excellent reference for anyone in need of integrated information about seafood or freshwater toxins, from their natural genesis to their deeper known mechanism of action.

At first, the book was thought of as a seafood toxin study, but it ended up including cyanobacterial, or freshwater, toxins because of the many similarities between both kinds of toxins. Also, freshwater toxins are a major sanitary concern, a sad example being the 54 people who died recently in Brazil from hemodialysis due to the presence of microcystin in the water. Another reason to include seafood and freshwater toxins in a single volume is that, more and more frequently, seafood living close to estuaries can be contaminated with toxins generated upstream.

Throughout the book, the structure of the chapters is the same for each group of toxins, with chapters reviewing their ecobiology, chemistry, and detection (both chemical and biological), pharmacology, physiology, and toxicology. The classification follows criteria in some cases still under debate, as when including yessotoxin as a diarrheic toxin. Chapters 3 and 4 cover the epidemiology of marine toxins, and Chapter 5 reviews the toxin action of calcium channels, since most dinoflagellate toxins are functionally driven preferentially to the sodium channel. The final chapters review topics of special interest: Chapter 32 explains the new toxins we should expect, Chapter 33 gives examples of the potential of toxins as a source for new drugs, and Chapters 34 and 35 deal with economic aspects of dinoflagellate blooms.

This book is intended as a solid reference, and it should be useful for Ph.D. students and specialists in many fields, such as pharmacologists, physiologists, biochemists, medical doctors, chemists, biologists, and food technologists.

As with any book of this kind, its quality must be attributed to the contributors. They are all leading scientists in their fields, and I thank them for their great effort, since this thankless task was superimposed on their other duties. I am deeply grateful to them. Also, the guidance, encouragement, and work of the editorial team at Marcel Dekker, Inc. is greatly acknowledged.

This work is dedicated to my parents and my wife.

Luis M. Botana

Contents

Contributors

David J. Adams, Ph.D. Professor and Head, Department of Physiology and Pharmacology, The University of Queensland, Brisbane, Australia

A. Alfonso, Ph.D. Professor, Department of Physiology, Campus de Lugo, University of Santiago de Compostela, Lugo, Spain

Darío P. Andrinolo, M.Sc. Biologist, Department of Physiology and Biophysics, Faculty of Medicine, University of Chile, Santiago, Chile

Daniel G. Baden, Ph.D. Professor of Chemistry and Director, Center for Marine Science Research, University of North Carolina at Wilmington, Wilmington, North Carolina

Alan G. Bishop, Ph.D. Researcher, Ecotoxicology Research Unit, Chemistry Department, Cork Institute of Technology, Bishopstown, Cork, Ireland

Luis M. Botana, Ph.D. Full Professor, Department of Pharmacology, University of Santiago de Compostela, Lugo, Spain

A. G. Cabado, Ph.D. Professor, Department of Physiology, Campus de Lugo, University of Santiago de Compostela, Lugo, Spain

Eoin P. Carmody, Ph.D. Researcher, Ecotoxicology Research Unit, Chemistry Department, Cork Institute of Technology, Bishopstown, Cork, Ireland

Fun S. Chu, Ph.D. Professor, Food Research Institute, University of Wisconsin, Madison, Wisconsin

José M. Cotos, Ph.D. Associate Professor, Department of Electronics and Computing Science, University of Santiago de Compostela, Santiago de Compostela, Spain

Marcia Craig, Ph.D. Postdoctoral Fellow, Department of Biochemistry, University of Alberta, Edmonton, Alberta, Canada

David J. Craik, Ph.D. Professor, Centre for Drug Design and Development, The University of Queensland, St. Lucia, Brisbane, Australia

Adam Doble, Ph.D. Director, CNS and Endocrinology Department, Rhône-Poulenc S.A., Antony, France

T. A. Doucette, B.A. Ph.D. Candidate, Department of Anatomy and Physiology, Atlantic Veterinary College, University of Prince Edward Island, Charlottetown, Prince Edward Island, Canada

Rosa Draisci, Ph.D. Director, Department of Residues in Foods of Animal Origin, Veterinary Medicine Laboratory, Italian National Institute of Health, Rome, Italy

Mark Estacion, Ph.D. Senior Researcher, Department of Research, MetroHealth Medical Center, Cleveland, Ohio

M. T. Fernández-Sánchez, Ph.D. Associate Professor, Biochemistry and Molecular Biology, University of Oviedo, Oviedo, Spain

Ambrose Furey, Ph.D. Researcher, Ecotoxicology Research Unit, Chemistry Department, Cork Institute of Technology, Bishopstown, Cork, Ireland

Luis Gandía, Ph.D. Instituto de Farmacología Teófilo Hernando, Facultad de Medicina, Universidad Autónoma de Madrid, Madrid, Spain

Antonio G. García, M.D., Ph.D. Full Professor, Instituto de Farmacología Teófilo Hernando, Universidad Autonoma de Madrid, Hospital Universitario de la Princesa, Madrid, Spain

Bradford David Gessner, M.D., M.P.H. Medical Epidemiologist, Alaska Department of Health and Social Services, Division of Public Health, Anchorage, Alaska

Juan Jesús Gestal-Otero, M.D. Professor of Preventive Medicine, Public Health Department, University of Santiago de Compostela, and Head of Preventive Medicine, Public Health Department, University Hospital of Santiago de Compostela, Santiago de Compostela, Spain

Sonia E. Guzmán-Pérez, Ph.D. Food Scientist, Department of Food Science, Louisiana State University, Baton Rouge, Louisiana

Ken-ichi Harada, Ph.D. Associate Professor, Faculty of Pharmacy, Meijo University, Nagoya, Japan

Carlos Hernández, Ph.D. Full Professor, Department of Electronics and Computer Science, University of Santiago de Compostela, Santiago de Compostela, Spain

Charles F. B. Holmes, BSc, MSc, Ph.D. Associate Professor, Department of Biochemistry, University of Alberta, Edmonton, Alberta, Canada

Kevin J. James, Ph.D. Director, Ecotoxicology Research Unit, Chemistry Department, Cork Institute of Technology, Bishopstown, Cork, Ireland

Seán S. Kelly, Ph.D. Researcher, Ecotoxicology Research Unit, Chemistry Department, Cork Institute of Technology, Bishopstown, Cork, Ireland

Masaaki Kodama, Ph.D. Professor, School of Fisheries Sciences, Kitasato University, Sanriku, Iwate, Japan

Néstor W. Lagos, Ph.D. Biochemist, Department of Physiology and Biophysics, Faculty of Medicine, University of Chile, Santiago, Chile

Ching-Ong Lau, Ph.D. Forensic Scientist, Narcotics Laboratory, Department of Scientific Services, Institute of Science and Forensic Medicine, National University of Singapore, Singapore

Richard J. Lewis, Ph.D. Principal Scientist, Queensland Agricultural Biotechnology Centre, The University of Queensland, Brisbane, Australia

Manuela G. López, Ph.D. Instituto de Farmacología Teófilo Hernando, Facultad de Medicina, Universidad Autónoma de Madrid, Madrid, Spain

M. C. Louzao, Ph.D. Professor, Department of Pharmacology, Campus de Lugo, University of Santiago de Compostela, Lugo, Spain

Luca Lucentini, Ph.D. Researcher, Italian National Institute of Health, Rome, Italy

Bernd Luckas, Ph.D. Professor of Food Chemistry, Faculty of Biology and Pharmacy, Institute of Nutrition, University of Jena, Jena, Germany

Alessandro Mascioni, B.Sc. Researcher, Italian National Institute of Health, Rome, Italy

Jordi Molgó, D.D.S., Ph.D., D.Sci. Director of Research, Laboratoire de Neurobiologie Cellulaire et Moléculaire, CNRS, Gif sur Yvette, France

Daniel T. Monaghan, Ph.D. Associate Professor, Department of Pharmacology, University of Nebraska Medical Center, Omaha, Nebraska

Carmen Montiel, Ph.D. Instituto de Farmacología Teófilo Hernando, Facultad de Medicina, Universidad Autónoma de Madrid, Madrid, Spain

Mohinder S. Nijjar, Ph.D. Professor of Toxicology, Department of Anatomy and Physiology, Atlantic Veterinary College, University of Prince Edward Island, Charlottetown, Prince Edward Island, Canada

Satnam S. Nijjar, B.Sc. (Honors) Professor, Faculty of Medicine, University of Toronto, Toronto, Ontario, Canada

Antonello Novelli, Ph.D. Associate Professor, Department of Psychology, University of Oviedo, Oviedo, Spain

Douglas L. Park, Ph.D. Professor and Head, Department of Food Science, Louisiana State University, Baton Rouge, Louisiana

Gian Paolo Rossini, D.Sc. Professor, Department of Biochemistry, University of Modena and Reggio Emilia, Modena, Italy

Francisco Leira Sanmartin, Ph.D. Head of Toxicological Area, Research and Development, National Technical Center for Fish Products (CECOPESCA), Vigo, Spain

Martin J. Scanlon, Ph.D. Research Fellow, Centre for Drug Design and Development, The University of Queensland, St. Lucia, Brisbane, Australia

Yuzuru Shimizu, Ph.D. Professor, Department of Biomedical Sciences, College of Pharmacy, University of Rhode Island, Kingston, Rhode Island

Kaarina Sivonen, Ph.D. Senior Scientist of the Academy of Finland, Department of Applied Chemistry and Microbiology, Biocenter Viikki, University of Helsinki, Helsinki, Finland

Donald A. Skifter, Ph.D. Staff, Department of Pharmacology, University of Nebraska Medical Center, Omaha, Nebraska

Paul T. Smith, Ph.D. Senior Lecturer, Department of Biological Sciences, University of Western Sydney (Macarthur Campus), Campbelltown, New South Wales, Australia

Ignacio Sordo, M.D. Researcher, Electronics and Computing Science, University of Santiago de Compostela, Santiago de Compostela, Spain

Benjamín A. Suárez-Isla, Ph.D. Chemist, Professor, and Chairman, Department of Physiology and Biophysics, Faculty of Medicine, Institute of Biomedical Sciences, University of Chile, Santiago, Chile

Akira Takai, M.D., D.M.Sc. Associate Professor, Division of Biomolecular Dynamics, Department of Cell Physiology, Graduate School of Medicine, Nagoya University, Nagoya, Japan

Chee-Hong Tan, Ph.D. Associate Professor, Department of Biochemistry, Faculty of Medicine, National University of Singapore, Singapore

R. A. R. Tasker, Ph.D. Professor of Pharmacology, Department of Anatomy and Physiology, Atlantic Veterinary College, University of Prince Edward Island, Charlottetown, Prince Edward Island, Canada

Kiyoshi Terao, M.D, Ph.D. Professor, Yamawaki-Gakuen Junior College, Department of Food Sciences, and Professor Emeritus, Chiba University, Tokyo, Japan

Mark P. Thomas, Ph.D. Assistant Professor, Department of Pharmacology, University of Nebraska Medical Center, Omaha, Nebraska

Magdalena T. Tosteson, Ph.D. Lecturer on Biophysics, Department of Cell Biology, Harvard Medical School, Boston, Massachusetts

Joaquín A. Triñanes, Ph.D. Researcher, Department of Electronics and Computer Science, University of Santiago de Compostela, Santiago de Compostela, Spain

Frances M. Van Dolah, Ph.D. Research Biochemist, Center for Coastal Environmental Health and Biomolecular Research, NOAA/National Ocean Service, Charleston, South Carolina

Patricio Vélez, Ph.D. Biologist, Department of Physiology and Biophysics, Faculty of Science, University of Valparaíso, Valparaíso, Chile

Juan M. Vieites, Ph.D. Director, R+D Unit, National Technical Center for Fish Products (CECOPESCA), Vigo, Spain

Mercedes R. Vieytes, Ph.D. Professor, Department of Physiology, Campus de Lugo, University of Santiago de Compostela, Lugo, Spain

Mariyo F. Watanabe, Ph.D. Professor, Faculty of Geo-environmental Sciences, Rissho University, Magechi, Kumagaya City, Saitama, Japan

Takeshi Yasumoto, Ph.D. Professor Emeritus, Tohoku University, Sendai, and Technical Advisor, Tama Laboratory, Japan Food Research Laboratories, Tokyo, Japan

1

Historic Considerations Regarding Seafood Safety

Takeshi Yasumoto
Tohoku University, Sendai, Japan
Japan Food Research Laboratories, Nagayama, Tama, Tokyo, Japan

I. INTRODUCTION

Worldwide various kinds of seafood are being enjoyed as delicacy and nutritionally healthy food. As demands for and production of seafood increase, it became important to pay more attention to the seafood safety. Problems previously regarded as localized and trivial phenomena may spread to other regions as the result of increasing world trade or proliferation of toxic organisms triggered by environmental changes in the sea. Various types of seafood poisoning so far known are listed in Table 1. Some others of very rare occurrence or only poorly characterized are not included in the list. For reasons beyond public health problems, seafood toxins attract attention of scientists because of their potent bioactivity and unique chemical structures. Chemically the toxins make the most challenging targets because of their structural complexity and extremely limited availability. To pharmacologists and biochemists marine toxins are attractive agents because of their potent and specific action, as exemplified by the use of tetrodotoxin and saxitoxin in sodium channel studies and that of okadaic acid in biochemical studies. Identification of toxin sources and elucidation of their food chain transmission are also interesting from the ecological and environmental point of view. Owing to the difficulty of covering all aspects of seafood toxins, in this chapter, emphasis will be on introducing historical events and anecdotes related to etiological and structural aspects. Among various types of poisoning listed in Table 1, topics related to paralytic shellfish poisoning (PSP), neurotoxic shellfish poisoning (NSP), amnesic shellfish poisoning (ASP), and palytoxin are not included or only briefly mentioned in this chapter, because they will be discussed elsewhere in this book.

II. TETRODOTOXIN POISONING (PUFFERFISH POISONING)

A. Occurrence and Etiology

Pufferfish are a highly esteemed delicacy in Japan despite the presence of a fatal toxin, tetrodotoxin (TTX), in the gonad, liver, and skin. In order to protect consumers, the government imposes a strict regulation defining the edible species and edible tissues, and requests cooks who prepare pufferfish dishes in restaurants to take a training course for a license. Nevertheless, pufferfish poisoning has stayed as the primary cause of the death caused by food poisoning, because there

Table 1 List of Types of Seafood Poisoning

Type of poisoning*	Type of seafood	Toxins	Source of toxins
Tetrodotoxin poisoning	Pufferfish	Tetrodotoxin	Bacteria
Ciguatera	Tropical fish	Ciguatoxin	*Gambierdiscus toxicus*[a]
Palytoxin poisoning	Fish	Palytoxin	*Ostreopsis siamensis*[a]
Crab poisoning	Coral reef crabs	Saxitoxin, tetrodotoxin	Bacteria
		Palytoxin	Zoanthids (?)
PSP	Shellfish	Saxitoxin	*Alexandrium* spp.[a]
DSP	Shellfish	Okadaic acid	*Dinophysis* spp.[a]
		Pectenotoxin	*Dinophysis* spp.[a]
NSP	Shellfish	Brevetoxin	*Gymnodinium breve*[a]
ASP	Shellfish	Domoic acid	*Pseudonitzschia* spp.[b]
AZP	Shellfish	Azaspiracid	Unknown
Seaweed poisoning	Gracilaria	Aplysiatoxin	*Lyngbya majuscula*[c]
		Polycavernoside	Unknown
Turtle poisoning	Marine turtles	Lyngbyatoxin-a	*Lyngbya majuscula*[c]

* PSP, paralytic shellfish poisoning; DSP, diarrheic shellfish poisoning; NSP, neurotoxic shellfish poisoning; ASP, amnesic shellfish poisoning; AZP, azaspiracid poisoning.
[a] Dinoflagellate.
[b] Diatom.
[c] Cyanobacterium.

are people who knowingly eat the liver of the pufferfish and boast that they enjoy the tingling sensation in the month caused by the toxin. The reason for such a suicidal act lies in the fact that pufferfish toxicity markedly varies individually, regionally, and seasonally. Unless a large piece of this fish is eaten, the gourmand has a good chance, but not guaranteed for escaping from intoxication. Because such a marked variation in toxicity suggested the exogenous origin of the toxin, we constructed a fluorometric high-performance liquid chromatographic (HPLC)

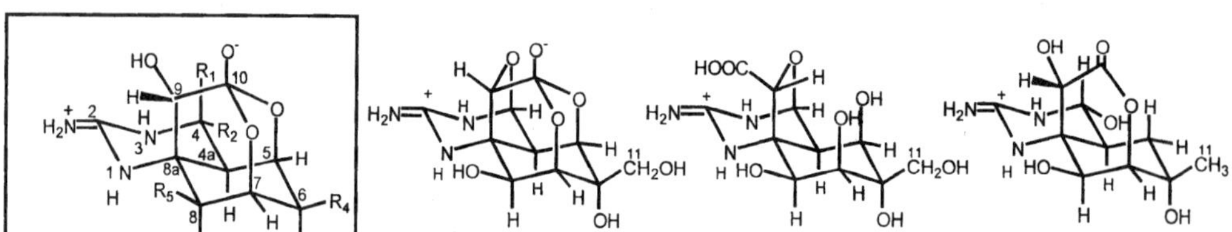

12 4,9-anhydroTTX **13** tetrodonic acid **14** 5,6,11-trideoxyTTX

		R_1	R_2	R_3	R4	R5
1	TTX	H	OH	OH	11-CH_2OH	OH
2	4-*epi*TTX	OH	H	OH	11-CH_2OH	OH
3	6-*epi*TTX	H	OH	11-CH_2OH	OH	OH
4	11-deoxyTTX	H	OH	OH	11-CH3	OH
5	11-oxoTTX	H	OH	OH	11-CHO	OH
6	TTX-11-carboxylic acid	H	OH	OH	11-COOH	OH
7	11-norTTX-6,6-diol	H	OH	OH	OH	OH
8	11-norTTX-6(*S*)-ol	H	OH	OH	H	OH
9	11-norTTX-6(*R*)-ol	H	OH	H	OH	
10	chiriquitoxin	H	OH	OH	11-CH(OH)CH(NH_2)COOH (12 *R*, 13 *S*)	OH
11	TTX-8-*O*-hemisuccinate	H	OH	OH	11-CH_2OH	$OOC(CH_2)_2COOH$ (1' 2' 3' 4')

Figure 1 Structure of tetrodotoxin and its analogues (6 and 11 were prepared by chemical reactions).

analyzer for TTX (1–3) and pursued the food chain. A bacterium initially assigned to *Pseudomonas* sp. and later amended as *Shwanella alga* was confirmed to produce TTX and one of its analogues, 4,9-anhydroTTX. The toxins were isolated and unambiguously identified by thin layer chromatography (TLC), HPLC, fluorometric HPLC, fast atom bombardment-mass spectrometry (FAB-MS), dose-survival time response in mouse, and alkali degradation to 2-amino-6-hydroxymethyl-8-hydroxyquinazoline (4,5).

B. Toxin Chemistry

The structure of TTX was established in 1964 (6–8), but little was known about the occurrence of analogues. The fluorometric HPLC analyzer we constructed was not only superior to the conventional mouse bioassays in its rapidity, sensitivity, and specificity but also was useful in detecting a number of TTX analogues in pufferfish, newts, and frogs. These analogues were isolated and their structures were determined, as shown in Figure 1. These analogues made good tools to probe the interaction between the toxins and the binding site on the voltage-gated sodium channels (9). They also gave a new insight into the biosynthesis of this intriguing molecule. Using the HPLC method, we also detected TTX in fish other than pufferfish, crabs, mollusks, polychate, and an alga (5). More details of TTX studies are described in our review article (10).

III. CIGUATERA

A. History and Etiology

In tropical regions many species of fish may become toxic and cause a combination of neurological and gastroenteric disturbances called ciguatera. Although fatal cases are very rare, the mortality rate for the illness is high, ranging from 2 to 554 cases per 100,000 population in South Pacific countries in 1977. As toxicity in individual fish varies very much, and the occurrence of toxic fish widely differs from place to place, it was easily seen that the toxin is of exogenous origin. However, the true source of the toxin and its chemical structure remained unknown for a long time. The causative toxin was first isolated in a pure form in Hawaii and named ciguatoxin (CTX) (11). However, subsequent attempts to elucidate the structure was unsuccessful because of the inadequate spectroscopic methods available at that time.

The first breakthrough in toxin etiology was made by the discovery of an epiphytic dinoflagellate, *Gambierdiscus toxicus*. In order to test the food chain theory, small herbivorous fish, *Ctenochaetus striatus*, were chosen. Because their teeth are small and loosely attached, they feed on detrital material and sediments by suction and grazing, thereby making them a good tool to test benthic or epiphytic microorganisms. The stomach and gut contents of *C. striatus* were confirmed to contain a CTX-like toxin and a polar toxin which was named maitotoxin (MTX) after the Tahitian name, ''maito,'' for the fish (12). Furthermore, disk-shaped dinoflagellates were abundantly observed in the guts of fish from endemic areas, whereas they were scarce or absent in fish from nonendemic areas (13). The dinoflagellate was initially assigned as *Diplopsalis* sp., but in a later study, a new genus and a new species were created to name it *Gambierdiscus toxicus*. The dinoflagellates were collected from the surface of dead corals in the Gambier Islands, where ciguatera was highly prevalent, and proved to contain toxins closely resembling CTX and MTX (14). When a strain of *G. toxicus* was cultured, however, the organism produced MTX but did not produce CTX in any sufficient level for chemical characterization. Thus, questions were raised about the toxigenicity of the organism. Years later another expedition was made, and several strains of *G. toxicus* were isolated and cultured to be tested for intraspecies variation. Production of two ciguatoxin analogues (CTX4A, CTX3C) by one of the strains was confirmed by isolating the toxins and determining their structures by chemical and spectroscopic methods (15–17). The results of these structural studies finally put an end to the long-persisting arguments about the origin of ciguatera toxins.

B. Chemistry

The first breakthrough in chemical studies on CTX was made in a collaborative study between two research groups in Tahiti and Japan. Four tons of moray eels were collected from endemic areas and 124 kg of the viscera was extracted to yield 0.35 mg of pure CTX (18). By working on this small sample, the structure of CTX was successfully determined (19). The absolute configuration of the toxin was also established recently, as shown in Figure 2 (20). The toxin is one of the most potent natural toxins, having an LD50 of 350 ng/kg (18). Analysis of fish samples implicated in human intoxication revealed that as little as 70 ng of the toxin was enough to cause ciguatera symptoms by oral intake. The toxin in the wild specimen of *G. toxicus* was initially named gambiertoxin but was renamed CTX4B. It has a less oxidized structure indicating that oxidation to CTX takes place during the transmission in the food chain. From *G. toxicus* cultures, a stereoisomer of CTX4B, coded CTX4A, was isolated together with a new analogue, CTX3 (16,17). Chromatographic properties of CTX4A agreed well with those of a toxin previously detected in parrotfish and tentatively named scaritoxin. Therefore, we propose that the name scaritoxin should be replaced with CTX4A, which is structurally better defined. Other CTX analogues with defined chemical structures are shown in Figure 2. Interestingly, CTX in Caribbean fish (C-CTX-1, -2) has a slightly different structural skeleton (Figure 2).

MTX was the major toxin in the viscera of herbivorous fish such as surgeonfish and parrotfish, but its content was insignificant in the flesh, suggesting a limited role of MTX played in ciguatera. However, a crude MTX sample prepared from surgeonfish caused severe vomiting

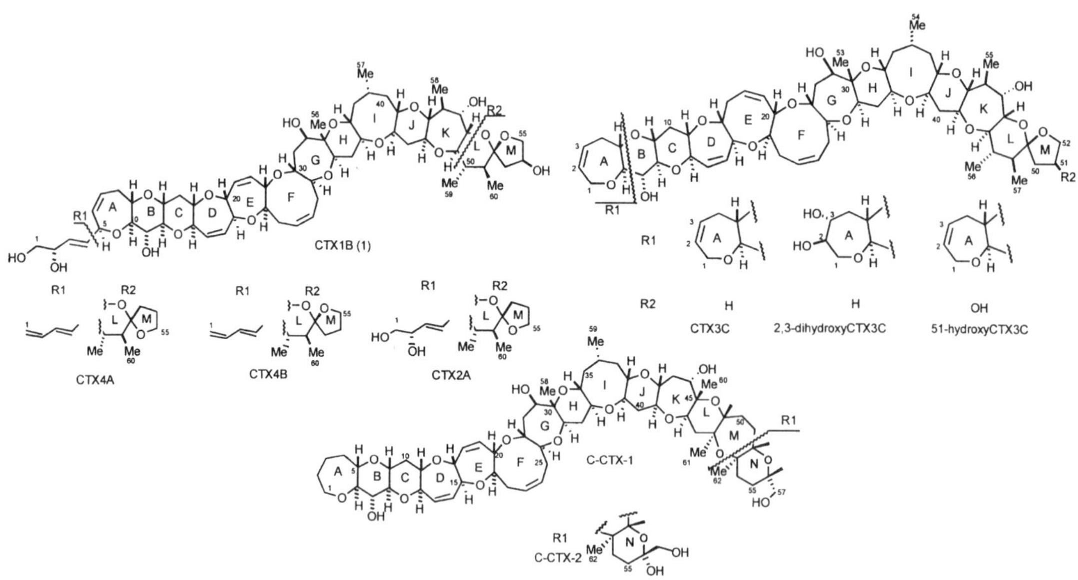

Figure 2 Structures of ciguatoxins.

and diarrhea when given to cats orally (21). As eating small fish without eliminating their viscera is a common practice in many places, the possible contribution of MTX should not be disregarded. Besides mouse lethality, MTX shows potent hemolytic and ichthyotoxic activities (21). Subsequent structural studies unraveled an extraordinary large and complex structure of the molecule. The planar structure (22) and relative stereostructure (23,24) were determined by spectroscopic studies, and the absolute configuration (Figure 3) was determined by partial synthesis (25). MTX is one of the most potent natural toxins (LD50 500 ng/kg, i.p., mice), being only exceeded by a few proteinous toxins of bacterial origins (26). MTX enhances the influx of extracellular calcium ion into cells and also stimulates phosphoinositides breakdown. For further information, we refer readers a review article by Gusovsky and Daly (27), and references therein.

C. Assay or Determination Method for CTX

Screening of toxic fish is usually carried out by intraperitoneally (i.p.) by injecting a lipid fraction of fish into mice and observing them for 24 h. As the concentrations of CTX in fish flesh are very low (0.5 ~ 1 ppb), preparation of the lipid samples involves tedious purification steps. As CTX shares with brevetoxin the same binding site on sodium channel protein, competitive binding assays using radiolabeled brevetoxin is possible (28). Tetrazolium-based cell bioassay for CTX was proposed based on the combined effect of the sodium channel activator veratridine in the presence of ouabain, an inhibitor of Na^+/K^+ATPase, to block sodium efflux (29). Determination by liquid chromatography/mass spectrometry (LC/MC) may provide sensitive and accurate results (30), but the expensive cost of the instrument coupled with the absence of an adequate supply of reference toxins limits wide use of the method. Probably developing an enzyme immunoassay is the most rational approach to screen a large number of individual fish. Owing to the extremely limited availability of CTX for preparation of protein-conjugated antigen, an antibody cross reacting with CTX was prepared by immunizing mice with an okadaic acid–protein conjugate. By using the antibody, an enzyme immunoassay method was proposed (31). Validity of the method is reportedly being tested.

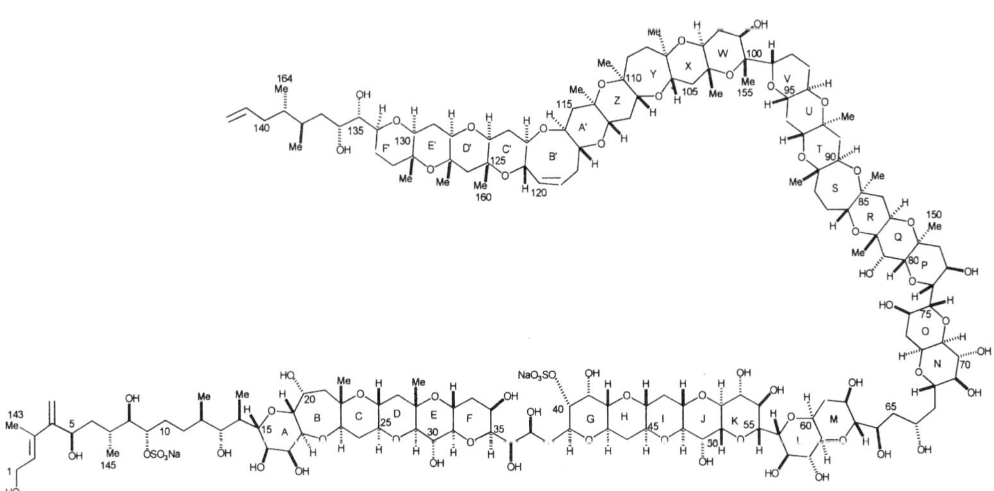

Figure 3 Structures of maitotoxin.

IV. PALYTOXIN POISONING

Among fishermen in tropical regions, filefish and triggerfish have been known to cause severe or even fatal poisoning when the liver was eaten. Palytoxin was identified as the cause in the two species (32,33). Similarly, the toxin was isolated from three species of coral reef crabs, *Lophozozymus pictor*, *Demania alcalai* (34), and *D. reynaudii* (35) as the cause of fatal intoxication. Possibly, those fish and crabs accumulate the toxin by feeding on *Palythoa* spp., the zoanthids from which the toxin was first isolated. More recently, the toxin or its analogue was detected in a tropical sardine, *Herklotsicthys quadrimaculatus*, which was implicated in a fatal intoxication (36). The fish and related species occasionally cause severe or fatal intoxication in various parts of tropical seas, and the poisoning has been distinguished from ciguatera by the name ''clupeotoxism.'' As the dinoflagellate *Ostreopsis siamensis* produces palytoxin and its analogues (37), the probable source of the toxin in the planktivorous sardines was suggested to be *O. siamensis*.

V. SEAFOOD POISONING CAUSED BY MARINE CYANOBACTERIAL TOXINS

Gracilaria seaweeds are commonly eaten in many parts of the Pacific. Yet, occasional poisoning incidents are reported (Table 2). Aplysiatoxin (Figure 4), which had been known as a metabolite of a marine cyanobacterium, *Lyngbya majuscula*, was isolated from the algal sample involved in intoxication in Hawaii (38). From algal samples collected at the time of another toxic incident in Guam, polycavernoside A (Figure 4) and its analogues were isolated and their structures determined (39–41). Although the origin of the polycavernosides have not yet been identified, their structural resemblance with aplysiatoxin and contamination of the algae by cyanobacteria as revealed by microscopic observation were suggestive of a cyanobacterial origin of the toxins.

Another example of a marine cyanobacterial toxin is the occurrence of lyngbyatoxin-a (Figure 4) in a turtle implicated in fatal poisoning in Madagascar (42). Lyngbyatoxin-a was previously isolated as the cause of skin inflamation in swimmers from *Lyngbya majuscula.* As this cyanobacterium is frequently seen in sea grass (turtle grass) fields, it is highly likely that the turtle had accumulated the toxin by eating cyanobacteria-contaminated sea grass. Although seaweeds and marine turtles may sound unusual and trivial foods, it should be borne in mind that these cyanobacterial toxins strongly promote tumor formation through activating protein kinase C (43). The possibility of their accumulation in herbivorous fish should be explored.

Table 2 Gracilaria Poisoning Incidents

Year	Location	Death/Patients	Toxin
1980	Yamagata, Japan	1/4	Unknown
1981	Ehime, Japan	1/4	Unknown
1991	Guam	3/13	Polycavernoside
1992	California	0/3	Unknown
1993	Kanagawa, Japan	1/2	Unknown
1994	Hawaii	0/6	Aplysiatoxin
1996	Guam	0/1	Polycavernoside

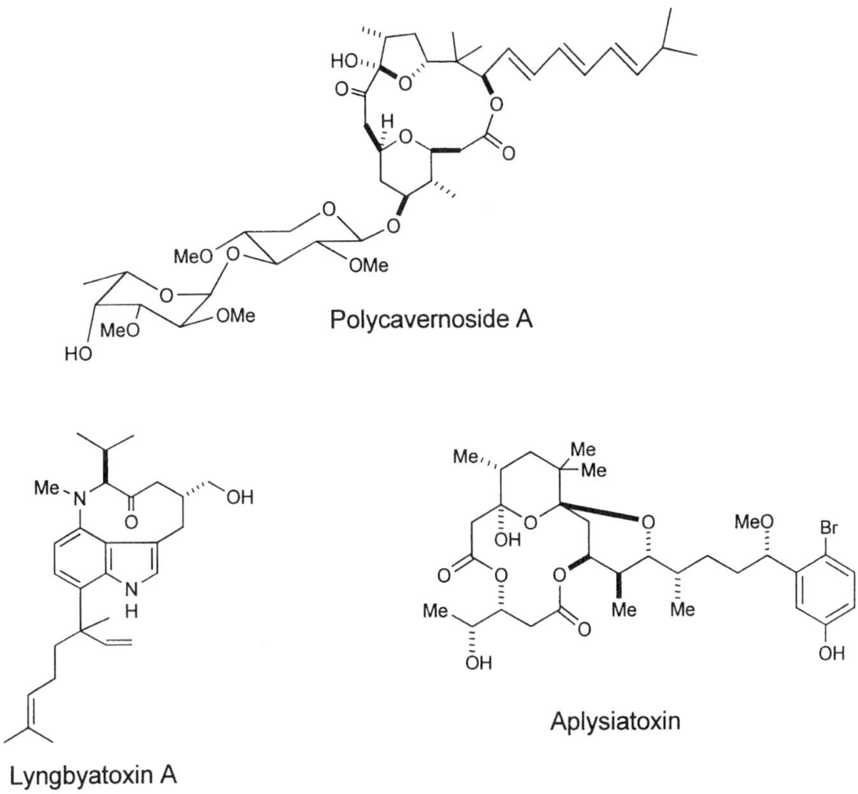

Figure 4 Cyanobacterial toxins implicated in seafood poisoning.

VI. DIARRHETIC SHELLFISH POISONING

A. History and Etiology

In the summer of 1987, a series of food poisonings occurred in the northeastern part of Japan after ingestion of mussels (44). Laboratory examination revealed no pathogenic bacteria, but a lipid toxin lethal to mice by i.p. was found. The toxin concentration was highest in the hepatopancreas but was insignificant in other tissues. Gastrointestinal disturbances such as diarrhea (92%), nausea (80%), vomiting (79%), and abdominal pain (52%) were the major complaints from the patients. The onset time of symptoms from consumption of shellfish ranged from 30 min to 7 h, and patients recovered within 3 days. The illness was named diarrhetic shellfish poisoning (DSP) (45). Although no official record existed, people must have been aware of the poisoning for a long time, as a widely known legend in the region warned that mussels might become toxic during the season of paulownia flowers (June). During the following years, DSP incidents increased both geographically and in frequency. The occurrence of DSP in Europe was first reported in the Netherlands almost at the same time as that in Japan, and the toxic agent was identified as okadaic acid (46).

Microscopic examination of the gut contents of mussels and scallops revealed the presence of many cells of the dinoflagellate *Dinophysis fortii* in toxic shellfish but not in nontoxic ones. Plankton was collected by towing nets and fractionated with sieves of different mesh sizes. The fraction containing *D. fortii* was the most (50–100 μm) toxic to mice. Furthermore, good correlation was observed between the number of *D. fortii* cells and mouse toxicity. Based on these results, *D. fortii* was assigned as the source of the DSP toxins (47). Because none of *Dinophysis* species is cultivable in laboratories, it was necessary to apply a sensitive as well as specific analytical method to verify the toxigenicity of these organisms. Various species of *Dinophysis* were collected under the microscope by manipulating a capillary, and the homogeneous *Dinophysis* samples thus obtained were analyzed by a fluorometric HPLC method. DSP toxins (okadaic acid or dinophysistoxin-1 or both) were confirmed in *D. fortii, D. acuta, D. norvegica, D. acuminata, D. mitra, D. rotundata*, and *D. tripos*, giving solid evidence that these species are the sources of the toxins (48). Remarkably, toxin concentration in shellfish reached an intoxication level with as low *D. fortii* population density as 200 cells per liter in the sea. A benthic dinoflagellate, *Prorocentrum lima*, also produces the same toxins (49) and can be found in shellfish farming areas; involvement of this benthic species in toxicification of shellfish has not been confirmed.

B. Toxin Structures

The toxin first detected in the hepatopancreas of toxic mussels was named dinophysistoxin-1 (DTX1) after its biogenetic origin. Subsequent purification and spectrochemical analyses revealed that DTX1 is a methyl derivative of okadaic acid (OA) which had been isolated first from a sponge, *Halichondria okadai* (50), and then from *P. lima* (49). DTX2 is a regio isomer of OA later isolated from Irish mussels and is produced by *D. acuta* (51,52). DTX3 refers to a group of homologues in which 7-OH in DTX1 is esterified with different kinds of a fatty acid (7-*O*-acyl-DTX1) (53). The most common fatty acid in DTX3 is palmitic acid. As DTX3 is absent in plankton samples, acylation of 7-OH in DTX1 should occur in shellfish. Structurally related 7-*O*-acyl-OA is often included in DTX3 for convenience. Structures of the OA class toxins are shown in Figure 5.

The second group of toxins was named pectenotoxins (PTXs) (Figure 6) after the genus of the scallop *Patinopecten yessoensis* (53). Structurally, PTXs resemble OA in molecular weight and in having cyclic ethers and a carboxyl group in the molecule. Unlike in OA, however, the carboxyl moiety in PTX is in a form of macrocyclic lactone (macrolide). PTX2 is produced by *D. fortii*, and a methyl group at C43 is oxidized in shellfish to alcohol (PTX1), then to aldehyde (PTX3), and finally to carboxyl acid (PTX6). Furthermore, the spiroketal ring system composed of the rings A and B easily undergoes rearrangement and/or epimerization under an acidic condition to produce PTX4 and PTX7 through PTX10 (54). Interestingly, the lactone ring in PTX2 is opened to give a PTX2–seco acid in *D. acuta* (*seco* means a bond cleavage in a ring structure) (55). Analogous with PTX2, epimerization and rearrangement of the rings A and B yield seco acid isomers (Figure 6). PTX2–seco acid showed no significant toxicity against KB cells.

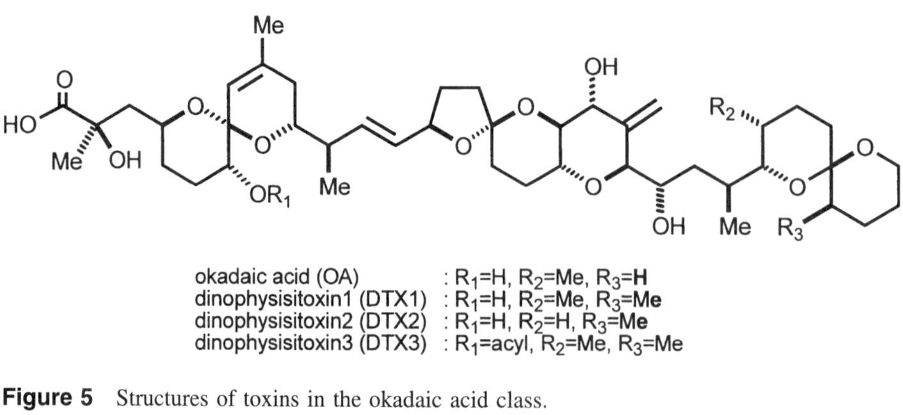

okadaic acid (OA) : R1=H, R2=Me, R3=H
dinophysisitoxin1 (DTX1) : R1=H, R2=Me, R3=Me
dinophysisitoxin2 (DTX2) : R1=H, R2=H, R3=Me
dinophysisitoxin3 (DTX3) : R1=acyl, R2=Me, R3=Me

Figure 5 Structures of toxins in the okadaic acid class.

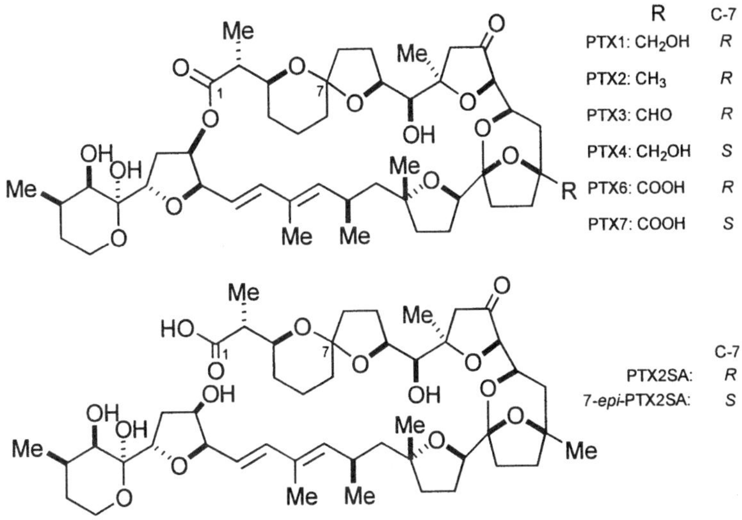

Figure 6 Structures of pectenotoxins (PTXs) and PTX2–seco acids (PTX2SA).

C. Biological Properties of Toxins

OA was first found as a cytotoxic agent by two research groups searching for antitumor drugs (50). Thus, the OA class toxins are toxic to eukaryotic cells but not to the prokaryotics (56). The toxins are diarrhetic (57,58) and tumorigenic (59). The mouse lethality by oral administration is nearly the same as that by i.p. (60). The mechanism of action underlying these activities is explained mainly by their potent inhibitory action against protein phosphatase 2A, 1, and 2B (61). DTX3 does not inhibit the enzyme (62) but is easily hydrolyzed to DTX1 by digestive enzymes such as lipase. The inhibitory effect of OA derivatives on protein phosphatase 2A have been studied in detail (62,63).

Oral administration of PTX2 (250 ~ 1000 μg/kg) to mice caused severe mucosal injuries and fluid accumulation in the small intestine and diarrhea, indicating that it also took part in inducing the DSP symptoms (64). Severe damage to the liver was also observed. When administered by i.p. injection, histopathological changes were prominent in the liver but not in the digestive tract (58,65). PTXs have a potent cytotoxicity (57,66,67) and probably inhibit actin polymerization (68).

D. Determination Methods

The worldwide occurrence of DSP toxins was monitored by mouse bioassays. In order to avoid the use of mice and improve sensitivity and specificity, development of alternative methods is desired. Carboxylic acid toxins easily produce fluorescent esters by reaction with 9-anthryldiazomethane and thereby can be determined with high sensitivity and accuracy (69). Depending on the availability of fluorescent reagents and analytical columns, many modifications of the method evolved. The absence of a carboxylic acid group in PTXs, except for PTX6, make it difficult to attach a fluorophore. Attempts are being made to determine PTXs after conjugating a fluorigenic reagent with the conjugated double bond in PTX molecules. Liquid chromatography using a mass spectrometer as a detector (LC/MS) offers the best solution to determine all the toxins. Sample preparations are simple and the sensitivities are high (70,71). The obstacle in the wide use of the method is the high price of the instrument and the limited availability of reference toxins.

As toxins in the OA class, except for DTX3, strongly inhibit protein phosphatase 2A, enzyme inhibition assays were developed by using either ultraviolet (UV)–absorbing (72) or fluorescent substrates (73). DTX3 has to be hydrolyzed to be detected by these methods. Antibodies to detect OA and DTX1, but not DTX3, were prepared and enzyme-linked immunosorbent assay (ELISA) kits were commercialized (74,75). Reportedly, an antibody suitable for detecting OA, DTX1, and DTX3 was developed (76), but the product is not on the market yet.

VII. TOXINS FOUND IN ASSOCIATION WITH DSP

The low specificity of the mouse bioassay for DSP is a drawback on the one hand but, on the other hand, can be regarded as advantageous, because it detects all kinds of toxins that are soluble in organic solvents. Such toxins are briefly described.

A ladder-shaped polycyclic ether toxin was isolated from the hepatopancreas of the scallop *Patinopecten yessoensis* and named yessotoxin (YTX) (77). Because of its frequent coexistence with DTXs and PTXs, YTX was tentatively included in the DSP category. However, YTX does not cause diarrhea and is less toxic by oral administration than OAs and PTXs (60). YTX also differs from DSP toxins in etiology. YTX is produced by *Protoceratium reticulatum* (78) and its analogue, homoYTX, by *Lingulodinium polyedrum* (79). Both toxins are oxidized to 45-hydroxy derivatives by shellfish. Structures of YTX and its analogues are shown in Figure 7 (80–82). A recently accepted view regarding YTXs is that they should no longer be included in the category of DSP.

In November 1995, at least eight people in the Netherlands became ill after eating mussels produced in Ireland. Although human symptoms were reminiscent of those of DSP, contamination of DSP toxins was very low. The major toxin has many unique structural features and is named azaspiracid because of its unusual azaspiro ring assemblies (Figure 8) (83). Similar poisoning incidents were reported in the following years, and the authors' group proposed the name azaspiracid poisoning (AZP). Because of the probable salt formation between the positively charged amino group and the negatively charged carboxyl group, the reactivity of the two functional groups is low, making it difficult to introduce a fluorescent group. Determination by LC/MS is being developed for azaspiroacid and its analogues.

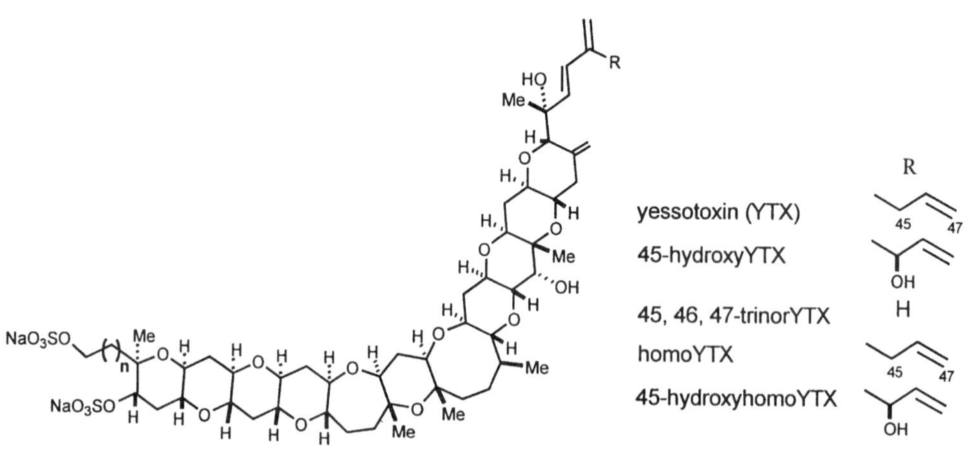

Figure 7 Structures of yessotoxins (YTXs).

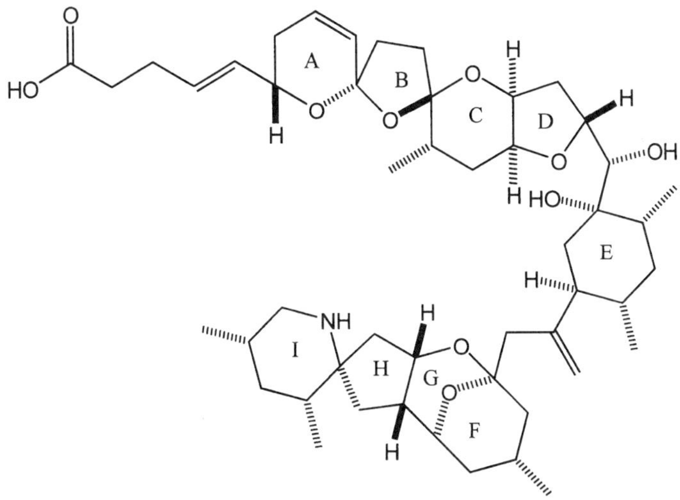

Figure 8 Azaspiracid ($C_{47}H_{71}NO_{12}$, MW 841).

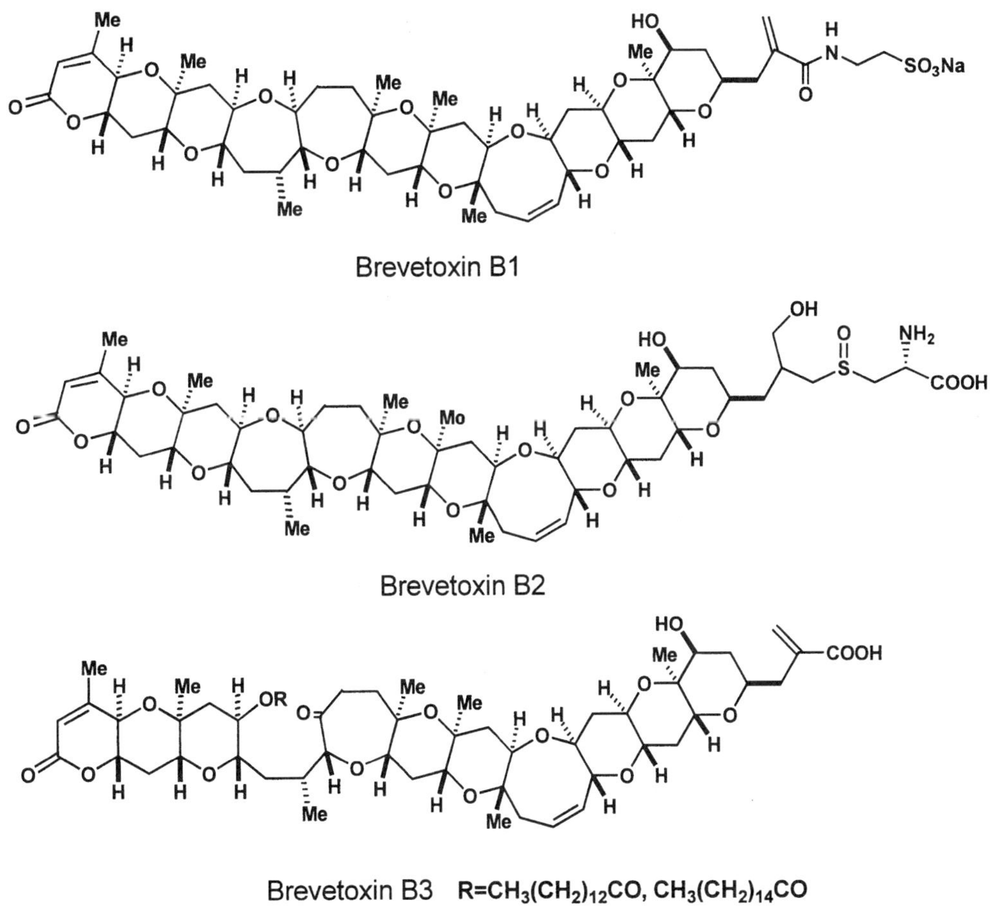

Figure 9 Structures of the brevetoxins isolated from bivalve shellfish.

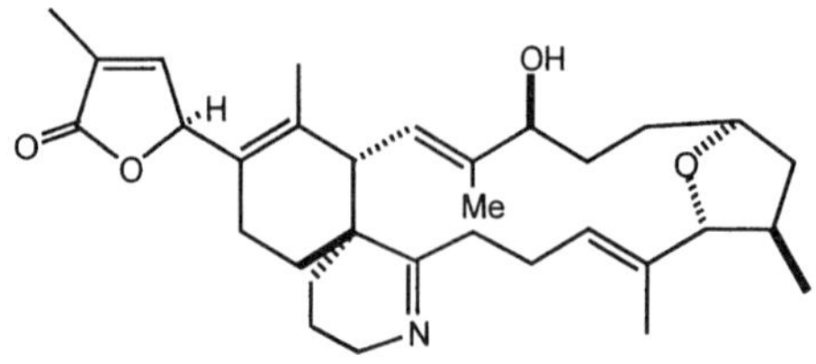

Gymnodimine

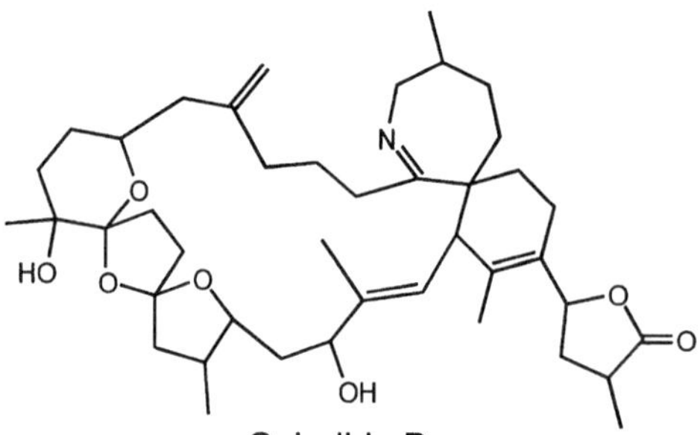

Spirolide B

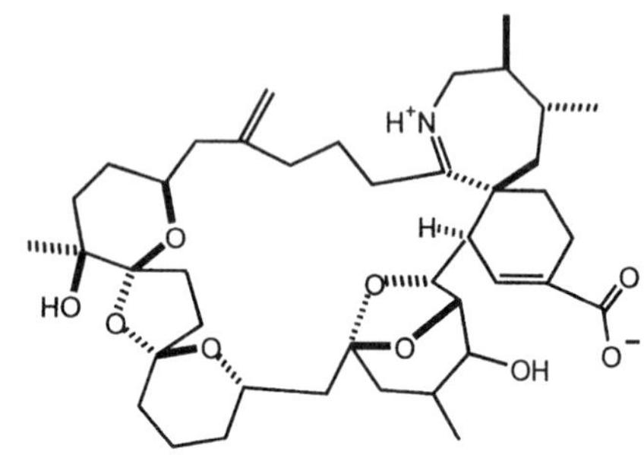

Pinnatoxin A

Figure 10 Cyclic imine toxins isolated from shellfish.

Neurotoxic shellfish poisoning (NSP) is also monitored by mouse bioassays. Structures of toxins isolated from shellfish revealed that the toxins produced by the dinoflagellate *Gymnodinium breve* were biochemically modified when accumulated in shellfish (Figure 9) (84–86).

Three toxins characterized by a cyclic imine structure were isolated from shellfish (Figure 10). Gymnodimine was confirmed to be produced by *Gymnodinium* sp (87,88). The origins of the other two toxins, spirolides (89) and pinnatoxin (90,91), are not known. Involvement of these imine toxins in human intoxication has not yet been confirmed.

Table 3 Marine Dinoflagellate Toxins Containing No Nitrogen Atom

Toxin	Formula	MW	LD50 (μg/kg)[a]	Origin
Okadaic acid	$C_{44}H_{68}O_{13}$	804	200	*Dinophysis* spp.
Dinophysistoxin 1	$C_{45}H_{70}O_{13}$	818	160	*Dinophysis* spp.
Dinophysistoxin 2	$C_{44}H_{68}O_{13}$	804	—	*Dinophysis* spp.
Dinophysistoxin 3	$C_{61}H_{100}O_{14}$	1056	250	*Dinophysis* spp.
Pectenotoxin 1	$C_{47}H_{70}O_{15}$	874	250	*Dinophysis fortii*
Pectenotoxin 2	$C_{47}H_{70}O_{14}$	858	230	*Dinophysis fortii*
Pectenotoxin 6	$C_{47}H_{68}O_{16}$	888	500	*Dinophysis fortii*
Yessotoxin	$C_{55}H_{80}O_{21}S_2Na_2$	1186	100	*Protoceratium reticulatum*
Brevetoxin B[b]	$C_{50}H_{70}O_{14}$	894	60 (rat)	*Gymnodinium breve*
42-Dihydrobrevetoxin B[b] (=PbTx-3)	$C_{50}H_{72}O_{14}$	896	—	*Gymnodinium breve*
Brevetoxin B3	$C_{64}H_{100}O_{17}$	1164	>300	*Gymnodinium breve*
Ciguatoxin	$C_{60}H_{86}O_{19}$	1110	0.35	*Gambierdiscus toxicus*
Maitotoxin	$C_{164}H_{256}O_{68}S_2Na_2$	3422	0.05	*Gambierdiscus toxicus*
Aplysiatoxin	$C_{32}H_{44}BrO_{10}$	668	250	*Lyngbya majuscula*
Polycavernoside A	$C_{43}H_{68}O_{15}$	824	200–400	Cyanobacterium suspected

[a] Determined by i.p. injection into mice.
[b] Not found in shellfish.

Table 4 Nitrogen-Containing Marine Microalgal Toxins

Toxin	Formula	MW	LD50 (μg/kg)[a]	Origin
Saxitoxin	$C_{10}H_{15}N_7O_4$	299	10	*Alexandrium* spp.
Gymnodimine	$C_{32}H_{45}NO_4$	507	450	*Gymnodinium* sp.
Pinnatoxin A	$C_{41}H_{61}NO_9$	711	180	Unknown
Spirolide B	$C_{42}H_{63}NO_7$	693	200	Unknown
Azaspiracid	$C_{47}H_{71}NO_{12}$	841	150	Unknown
Brevetoxin B1	$C_{52}H_{74}NO_{17}SNa$	1039	50	*Gymnodinium breve*
Brevetoxin B2	$C_{53}H_{79}NO_{17}S$	1033	300	*Gymnodinium breve*
Palytoxin	$C_{129}H_{223}N_3O_{54}$	2677	0.5	*Ostreopsis siamensis*
Prymnesin 2	$C_{96}H_{136}Cl_3NO_{35}$	1967	80	*Prymnesium parvum*
Lyngbyatoxin-a	$C_{28}H_{40}N_2O_2$	436	—	*Lyngbya majuscula*

[a] Determined by i.p. injection into mice.

VIII. SUMMARY OF MARINE TOXINS

For the convenience of the readers, the representative seafood toxins are listed in Tables 2 and 3 for their molecular formulae, molecular weights, mouse lethalities by i.p. injection, and biogenetic origins. It should be borne in mind that many of the toxins listed in the Tables 3 and 4 are the metabolically modified products of the precursor toxins produced by the microorganisms.

REFERENCES

1. T Yasumoto, M Nakamura, Y Oshima, J Takahata. Construction of a continuous tetrodotoxin analyzer. Bull Jpn Soc Sci Fish 48:1481–1483, 1982.
2. T Yasumoto, T Michishita. Fluorometric determination of tetrodotoxin by high performance liquid chromatography. Agric Biol Chem 49:3077–3080, 1985.
3. M Yotsu, A Endo, T Yasumoto. An improved tetrodotoxin analyzer. Agric Biol Chem 53:893–865, 1989.
4. T Yasumoto, D Yasumura, M Yotsu, T Michishita, T Endo, Y Kotaki. Bacterial production of tetrodotoxin and anhydrotetrodotoxin. Agric Biol Chem 50:793–795, 1986.
5. T Yasumoto, H Nagai, D Yasumura, T Michishita, A Endo, M Yotsu, Y Kotaki. Interspecies distribution and possible origin of tetrodotoxin. Ann NY Acad Sci 479:44–51, 1986.
6. K Tsuda, S Ikuma, M Kawamura, R Tachikawa, K Sakai, C Tamura, D Amakasu. Tetrodotoxin. VII. On the structure of tetrodotoxin and its derivatives. Chem Pharm Bull 12:1357–1374, 1964.
7. RB Woodward. The structure of tetrodotoxin. Pure Appl Chem 9:49–74, 1964.
8. T Goto, Y Kishi, S Takahashi, Y Hirata. Tetrodotoxin. Tetrahedron 21:2059–2088, 1965.
9. M Yotsu-Yamashita, A Sugimoto, A Takai, T Yasumoto. Effects of specific modifications of several hydroxyls of tetrodotoxin on its affinity to rat brain membrane. J Pharm Exp Ther: 289:1688–1696, 1999,
10. T Yasumoto, M Yotsu-Yamashita. Chemical and etiological studies on tetrodotoxin and its analogs. J Toxicol Toxin Rev 15:81–90, 1996.
11. K Tachibana. Structural studies of marine toxins. PhD dissertation, University of Hawaii, Honolulu, 1980.
12. T Yasumoto, Y Hashimoto, R Bagnis, JE Randall, AH Banner. Toxicity of the surgeonfishes. Bull Jpn Soc Sci Fish 37:724–734, 1971.
13. T Yasumoto, R Bagnis, S Thevenin, M Garcon. A survey of comparative toxicity in the food chain of ciguatera. Bull Jpn Soc Sci Fish 43:1015–1019, 1977.
14. T Yasumoto, I Nakajima, R Bagnis, R Adachi. Finding of a dinoflgaellate as a likely culprit of ciguatera. Bull Jpn Soc Sci Fish 43:1021–1026, 1977.

15. M Satake, M Murata, T Yasumoto. Gambierol: A new toxic polyether compound isolated from the marine dinoflagellate *Gambierdiscus toxicus*. J Am Chem Soc 115:361–362, 1993.
16. M Satake, M Murata, T Yasumoto. The structure of CTX3C, a ciguatoxin congener isolated from cultured *Gambierdiscus toxicus*. Tetrahedron Lett 34:1975–1978, 1993.
17. M Satake, Y Ishibashi, A-M Legrand, T Yasumoto. Isolation and structure of ciguatoxin-4A, a new ciguatoxin precursor, from cultures of dinoflgaellate *Gambierdiscus toxicus* and parrotfish *Scarus gibbus*. Biosci Biotech Biochem 60:2103–2105, 1997.
18. A-M Legrand, M Litaudon, JN Genthon, R Bagnis, T Yasumoto. Isolation and some properties of ciguatoxin. J Appl Phycol 1:183–188, 1989.
19. M Murata, A-M Legrand, Y Ishibashi, T Yasumoto. Structures of ciguatoxin and its congener. J Am Chem Soc 111:8929–8931, 1989.
20. M Satake, A Morohashi, H Oguri, T Oishi, M Hirama, N Harada, T Yasumoto. The absolute configuration of ciguatoxin. J Am Chem Soc 119:11325–11326, 1997.
21. T Yasumoto, R Bagnis, JP Vernoux. Toxicity of the surgeonfishes—II Properties of the principal water-soluble toxin. Bull Jpn Soc Sci Fish 42:359–365, 1976.
22. M Murata, H Naoki, T Iwashita, S Matsunaga, M Sasaki, A Yokoyama, T Yasumoto. Structure of maitotoxin. J Am Chem Soc 115:2060–2062, 1993.
23. M Murata, H Naoki, S Matsunaga, M Satake, T Yasumoto. Structure and partial stereochemical assignments for maitotoxin, the most toxic and largest natural non-biopolymer. J Am Chem Soc 116: 7098–7107, 1994.
24. M Satake, S Ishida, T Yasumoto, M Murata, H Utsumi, T Hinomoto. Structural confirmation of matotoxin based on complete ^{13}C NMR assignments and the three-dimensional PFG NOESY-HMQC spectrum. J Am Chem Soc 117:7019–7020, 1995.
25. M Sasaki, N Matsumori, T Maruyama, T Nomura, M Murata, K Tachibana, T Yasumoto. The complete structure of maitotoxin, II: Stereochemistry of the C135-C142 side chain and absolute configuration of the entire molecule. Angew Chem Int Ed Engl 35:1675–1678, 1996.
26. A Yokoyama, M Murata, Y Oshima, T Iwashita, T Yasumoto. Some chemical properties of maitoxin, a putative calcium channel agonist isolated from a marine dinoflagellate. J Biochem 104:184–187, 1988.
27. F Gusovsky, JW Daly. Maitotoxin: a unique pharmacological tool for research on calcium-dependent mechanisms. Biochem Pharmacol 39:1633–1639, 1990.
28. T Yasumoto, M Fukui, K Sasaki, K Sugiyama. Determinations of marine toxins in food JAOAC Intern 78:574–582, 1995.
29. RL Manger, LS Lega, SY Lee, JM Hungerford, MM Wekell. Tetrazolium-based cell bioassay for neurotoxins active on voltage-sensitive sodium channels: semiautomated assay for saxitoxins, brevetoxins, and ciguatoxins. Anal Biochem 214:190–194, 1993.
30. RL Lewis, A Jones, JP Vernoux. HPLC/tandem electrospray mass spectrometry for the determination of sub-ppb levels of Pacific and Caribbean ciguatoxins in crude extracts of fish. Anal Chem 71:247–250, 1999.
31. Y Hokama, MA Abad, LH Kimura. A rapid enzyme-immunoassay for the detection of ciguatoxin in contaminated fish tissues. Toxicon 21:817–824, 1993.
32. Y Hashimoto, N Fusetani, S. Kimura. Aluterin: a toxin of filefish, *Alutera scripta*, probably originated from a zoantharian, *Palythoa tuberculosa*. Bull Jpn Soc Sci Fish 35:1086–1093, 1967.
33. M Fukui, M Murata, A Inoue, M Gawel, T Yasumoto. Occurrence of palytoxin in the trigger fish *Melichthys vidua*. Toxicon 25:1121–1124, 1987.
34. T Yasumoto, D Yasumura, Y Ohizumi. Palytoxin in two species of xanthid crab from the Philippines. Agric Biol Chem 50:593–598, 1986.
35. AC Alcala, LC Alcala, JS Garth, D Yasumura, T Yasumoto. Human fatality due to ingestion of the crab *Demania reynaudii* that contained a palytoxin-like toxin. Toxicon 26:105–107, 1988.
36. Y Onuma, M Satake, T Ukena, J Roux, S Chanteau, N Rasolofonirina, M Ratsimaloto, H Naoki, T Yasumoto. Identification of putative palytoxin as the cause of clupeotoxism. Toxicon 37:55–65, 1999.
37. N Usami, M Satake, S Ishida, A Inoue, Y Kan, T Yasumoto. Palytoxin analogs from the dinoflagellate *Ostreopsis siamensis*. J Am Chem Soc 117:5389–5390, 1995.

38. H Nagai, T Yasumoto, Y Hokama. Aplysiatoxin and debromoaplysiatoxin as the causative agent of a red alga *Gracilaria coronopifolia* poisoning in Hawaii. Toxicon 34:743–752, 1996.
39. M Yotsu-Yamashita, RL Haddock, T Yasumoto. Polycaversoside A: a novel glycosidic macrolide from the red alga *Polycavernosa tsudai* (Gracilaria edulis). J Am Chem Soc 115:1147–1148, 1993.
40. M Yotsu-Yamashita, T Seki, FJ Paul, H Naoki, T Yasumoto. Four new analogs of polycavernoside A. Tetrahedron Lett 36:5563–5566, 1995.
41. K Fujiwara, A Murai, M Yotsu-Yamashita, T Yasumoto. Total synthesis and absolute configuration of polycavernoside A. J Am Chem Soc 120:10770–10771, 1998.
42. T Yasumoto. Fish poisoning due to toxins of microalgal origins in the Pacific. Toxicon 36:1515–1518, 1998.
43. H Fujiki, T Sugimura. New classes of tumor promoters: teleocidin, aplysiatoxin, and palytoxin. Adv Cancer Res 46:223–264, 1987.
44. T Yasumoto, Y Oshima, M Yamaguchi. Occurrence of a new type of shellfish poisoning in the Tohoku District. Bull Jpn Soc Sci Fish 44:1249–1255, 1978.
45. M Murata, M Shimatani, H Sugitani, Y Oshima, T Yasumoto. Isolation and structural elucidation of the causative toxin of the diarrhetic shellfish poisoning. Bull Jpn Soc Sci Fish 48:549–552, 1982.
46. M Kumagai, T Yanagi, M Murata, T Yasumoto, M Kat, P Lassus, JA Rodriguez-Vazquez. Okadaic acid as the causative toxin of diarrhetic shellfish poisoning in Europe. Agri Biol Chem 50:2853–2857, 1986.
47. T Yasumoto, Y Oshima, W Sugawara, Y Fukuyo, H Oguri, T Igarashi, N Fujita. Identification of *Dinophysis fortii* as the causative organism of diarrhetic shellfish poisoning. Bull Jpn Soc Sci Fish 46:327–331, 1980.
48. JS Lee, T Igarashi, S Fraga, E Dahl, P Hovgaad, T Yasumoto. Determination of diarrhetic shellfish toxins in various dinoflagellate species. J Appl Phycol 1:147–152, 1989.
49. Y Murakami, Y Oshima, T Yasumoto. Identification of okadaic acid as a toxic component of a marine dinoflagellate *Prorocentrum lima*. Bull Jpn Soc Sci Fish 48:69–72, 1982.
50. K Tachibana, PJ Scheuer, Y Tsukitani, H Kikuchi, DV Engen, J Clardy, Y Gopichand, FJ Schmitz. Okadaic acid, a cytotoxic polyether from two marine sponges of the genus *Halichondria*. J Am Chem Soc 103:2469–2471, 1981.
51. T Hu, J Doyle, D Jackson, J Marr, E Nixon, S Pleasance, MA Quilliam, JA Walter, JLC Wright. Isolation of a new diarrhetic shellfish poison from Irish mussels. J Chem Soc Chem Commun: 39–41, 1992.
52. R Draisci, L Giannetti, L Iucentini, C Marchiafava, KJ James, AG Bishop, BM Healy, SS Kelly. Isolation of a new okadaic acid analogue from phytoplankton implicated in diarrhetic shellfish poisoning. J Chromatogr A 798:137–145, 1998.
53. T Yasumoto, M Murata, Y Oshima, M Sano, GK Matsumoto, J Clardy. Diarrhetic shellfish toxins. Tetrahedron 41:1019–1025, 1985.
54. K Sasaki, JLC Wright, T Yasumoto. Identification and characterization of pectenotoxin (PTX) 4 and PTX7 as spiroketal stereoisomers of two previously reported pectenotoxins. J Org Chem 63:2475–2480, 1998.
55. M Daiguji, M Satake, KJ James, A Bishop, L MacKenzie, H Naoki, T Yasumoto. Structures of pectenotoxin analogs, pectenotoxin-2 seco acid and 7-epi-pectenotoxin-2 seco acid, isolated from a dinoflagellate and greenshell mussels. Chem Lett 653–654, 1998.
56. H Nagai, M Satake, T Yasumoto. Antimicrobial activity of polyether compounds of dinoflagellate origins. J Appl Phycol 2:305–308, 1990.
57. Y Hamano, Y Kinoshita, T Yasumoto. Enteropathogenicity of diarrhetic shellfish toxins in intestinal models. J Food Hyg Soc Jpn 27:375–379, 1986.
58. K Terao, E Ito, T Yanagi, T Yasumoto. Ultrastructural changes in the small intestine and liver of suckling mice induced by dinophysistoxin-1 and pectenotoxin-1. Toxicon 24:1141–1151, 1986.
59. H Fujiki, M Suganuma. Tumor promotion by inhibitors of protein phosphatase 1 and 2A: The okadaic acid class of compounds. Adv Cancer Res 61:226–277, 1993.
60. H Ogino, M Kumagai, T Yasumoto. Toxicologic evaluation of yessotoxin. Nat Toxins 5:255–259, 1997.

61. C Bialojan, A Takai. Inhibitory effect of a marine-sponge toxin, okadaic acid, on protein phosphatases. Biochem J 256:283–290, 1988.
62. A Takai, M Murata, K Torigoe, M Isobe, G Mieskes, T Yasumoto. Inhibitory effect of okadaic acid derivatives on protein phosphatase. Biochem J 284:539–544, 1992.
63. K Sasaki, M Murata, T Yasumoto, G Mieskes, A Takai. Affinity of okadaic acid to type-1 and type-2A protein phosphatases is markedly reduced by oxidation of its 27-hydroxyl group. Biochem J 298: 259–262, 1994.
64. M Ishige, N Satoh, T Yasumoto. Pathological studies on the mice administrated with the causative agent of diarrhetic shellfish poisoning (okadaic acid and pectenotoxin-2). Bull Hokkaido Inst Public Health 38:15–18, 1988.
65. ZH Zhou, M Komiyama, K Terao, Y Shimada. Effects of pectenotoxin-1 on liver cells in vitro. Nat Toxins 2:132–135, 1994.
66. T Aune. Toxicity of marine and freshwater algal biotoxins towards freshly prepared hepatocytes. In: S Natori, K Hashimoto, Y Ueno, eds. Mycotoxins and Phycotoxins '88. Amsterdam: Elsevier, 1989, pp 461–468.
67. JH Jung, CJ Sim, CO Lee. Cytotoxic compounds from a two-sponge association. J Nat Prod 58: 1722–1726, 1995.
68. L Spector, NR Shcher, MR Bubb. Actin binding marine natural products as investigative tools and anticancer agents. Proceedings of 9th International Symposium on Marine Natural Products, Townsville, Australia, 1988, OR-25.
69. JS Lee, T Yanagi, R Kenma, T Yasumoto. Fluorometric determination of diarrhetic shellfish toxins by high-performance liquid chromatography. Agric Biol Chem 51:877–881, 1987.
70. S Pleasance, MA Quilliam. Ionspray mass spectrometry of marine toxins. IV. Determination of diarrhetic shellfish poisoning toxins in mussel tissue by liquid chromatography/mass spectrometry. Rapid Commun in Mass Spectrom 6:121–127, 1992.
71. R Draisci, L Lucentini, L Giannetti, P Boria, A Stacchini. Detection of diarrhoetic shellfish toxins in mussels from Italy by ionspray liquid chromatography–mass spectromety. Toxicon 33:1591–1603, 1995.
72. A Tubaro, C Florio, E Luxich, S Sosa, R Loggia, T Yasumoto. A protein phosphatase 2A inhibition assay for a fast and sensitive assessment of okadaic acid contamination in mussels. Toxicon 34:743–752, 1996.
73. MR Vieytes, OI Fontal, F Leira, JMV Baptista de Sousa, LM Botana. A fluorescent microplate assay for diarrheic shellfish toxins. Anal Biochem 248:258–264, 1997.
74. T Usagawa, M Nishimura, Y Itoh, T Uda, T Yasumoto. Preparation of monoclonal antibodies against okadaic acid prepared from the sponge *Halichondria okadai*. Toxicon 27:1323–1330, 1989.
75. WS Shestowsky, MA Quilliam, HM Sikorska. An idiotypic-anti-idiotypic competitive immunoassay for quantitation of okadaic acid. Toxicon 30:1441–1448, 1992.
76. H Hamano. Studies on diarrhetic shellfish toxins, particularly okadaic acid. Bull Med School Osaka Univ 43:67–81, 1991.
77. M Murata, M Kumagai, JS Lee, T Yasumoto. Isolation and structure of yessotoxin, a novel polyether compound implicated in diarrhetic shellfish poisoning. Tetrahedron Lett 28:5869–5872, 1987.
78. M Satake, L MacKenzie, T Yasumoto. Identification of *Protoceratium reticulatum* as the biogenetic origin of yessotoxin. Nat Toxins 5:164–167, 1997.
79. A Tubaro, L Sidari, RD Loggia, T Yasumoto. Occurrence of yessotoxin-like toxins in phytoplankton and mussels from northern Adriatic Sea. In: B Reguera, J Blanco, ML Fernandez, T Wyatt eds. Harmful Algae. Paris, IOC/UNESCO, 1998, pp 470–472.
80. M Satake, K Terasawa, Y Kadowaki, T Yasumoto. Relative configuration of yessotoxin and isolation of two new analogs from toxic scallop. Tetrahedron Lett 37:5955–5958, 1996.
81. H Takahashi, T Kusumi, Y Kan, M Satake, T Yasumoto. Determination of the absolute configuration of yessotoxin, a polyether compound implicated in diarrhetic shellfish poisoning, by NMR spectroscopic method using a chiral anisotropic reagent, methoxy-(2-naphthyl)acetic acid. Tetrahedron Lett 37:7087–7090, 1996.
82. M Daiguji, M Satake, H Ramstad, T Aune, H Naoki, T Yasumoto. Structure and fluorometric HPLC determination of 1-desulfoyessotoxin, a new yessotoxin analog isolated from mussels from Norway. Nat Toxins, 6:235–239, 1998.

83. M Satake, K Ofuji, H Naoki, KJ James, A Furey, T McMahon, J Silke, T Yasumoto. Azaspiracid, a new toxin having unique spiro ring assemblies, isolated from Irish mussels, *Mytilus edulis*. J Am Chem Soc 120:9967–9968, 1998.
84. H Ishida, A Nozawa, K Totoribe, N Muramatsu, H Nukaya, K Tsuji, K Yamaguchi, T Yasumoto, H Kasper, N Berkett, T Kosuge. Brevetoxin B1, A new polyether marine toxin from the New Zealand shellfish, *Austrovenus stutchburyi*. Tetrahedron Lett 36:725–728, 1995.
85. A Morohashi, M Satake, K Murata, H Naoki, HF Kaspar, T Yasumoto. Brevetoxin B3, a new brevetoxin analog isolated from the greenshell mussel *Perna canaliculus* involved in neurotoxic shellfish poisoning in New Zealand. Tetrahedron Lett 36:8995–8998, 1995.
86. K Murata, M Satake, H Naoki, HF Kaspar, T Yasumoto. Isolation and structure of a new brevetoxin analog, brevetoxin B2, from greenshell mussels from New Zealand. Tetrahedron 54:735–742, 1998.
87. T Seki, M Satake, L MacKenzie, HF Kaspar, T Yasumoto. Gymnodimine, a new marine toxin of unprecedented structure isolated from New Zealand oysters and the dinoflagellate, *Gymnodinium* sp. Tetrahedron Lett 36:8995–8998, 1995.
88. M Stewart, JW Blunt, MHG Munro, WT Robinson, DJ Hannh. The absolute stereochemistry of the New Zealand shellfish toxin gymnodimine. Tetrahedron Lett 38:4889–4890, 1997.
89. T Hu, JM Curtis, Y Oshima, MA Quilliam, JA Walter, WM Watson-Wright, JLC Wright. Spirolides B and D, two novel macrocycles isolated from the digestive glands of shellfish. J Chem Soc Chem Commun: 2159–2161, 1995.
90. D Uemura, T Chou, T Haino, A Nagatsu, S Fukuzawa, SZ Zheng, HS Chen. Pinnatoxin A: a toxic amphoteric macrocycle from the Okinawan bivalve *Pinna muricata*. J Am Chem Soc 117:1155–1156, 1995.
91. JA McCauley, K Nagasawa, PA Lander, SG Mischke, MA Semones, Y Kishi. Total synthesis of pinnatoxin A. J Am Chem Soc 120:7647–7648, 1998.

2
Diversity of Marine and Freshwater Algal Toxins

Frances M. Van Dolah
Center for Coastal Environmental Health and Biomolecular Research, NOAA/National Ocean Service, Charleston, South Carolina

I. INTRODUCTION

Microalgae are primary producers that make up the base of both marine and freshwater food webs. Many microalgae also produce novel compounds that exhibit potent biological activities. These are generally considered to be secondary metabolites, that is, compounds which are not essential to the basic metabolism and growth of the organism, and which are present in restricted taxonomic groups. The biosynthesis of secondary metabolites is common in bacteria, as well as in eukaryotic microbes and plants. The roles played by secondary metabolites in the life history of the producing organism are much debated. Many are considered to be chemical defenses, which confer a competitive advantage over other microbes or discourage predation by higher trophic level organisms. Others may serve roles in chemical signaling, and yet others have been proposed to be evolutionary relics. The elaborate biosynthetic pathways required to synthesize these compounds, and the observation that genes involved in biosynthesis of a given secondary metabolite are often clustered within the genome, are consistent with the notion that these metabolites serve a specific function for the organism.

Among the many algal secondary metabolites that have been identified, a number are potent toxins responsible for a wide array of human illnesses, marine mammal and bird morbidity and mortality, and extensive fish kills. Toxins of human health significance originate primarily from three classes of unicellular algae: dinoflagellates, diatoms, and cyanobacteria. Dinoflagellates are responsible for the widest array of toxins. However, of the several thousand species of dinoflagellates known, only a few dozen species appear to be toxigenic. Among the diatoms, a single genus, *Pseudo-nitzschia*, produces a toxin impacting human health. In contrast, among the cyanobacteria, selected species and certain isolated strains of all common planktonic cyanobacterial genera have been identified to produce toxins.

Algal toxins are structurally and functionally diverse, and many are derived from unique biosynthetic pathways. Dinoflagellates, for example, synthesize a variety of polyether toxins, which fall into two structural groups: linear (okadaic acid family and macrocycles) or fused (ladder-like). Dinoflagellate polyethers are synthesized via a novel polyketide synthase pathway which, like other polyketide synthases, builds the compound by the sequential addition of acetates, but also utilizes two glycolates as substrates—a novel mechanism not previously observed (1). Within the dinoflagellates the expression of polyethers does not correlate directly with phylogenetic groupings (2), leaving their origins and purpose open to speculation. Other toxins are clearly polyphyletic in origin or may be derived from lateral gene transfer. For example, the heterocyclic guanidine saxitoxins are produced by certain species of marine gonyaulacoid dinoflagellates, but are also produced by freshwater cyanobacteria. However, in both the dinoflagellates and the cyanobacteria, isolated strains within known toxin-producing species may be nontoxic. It is not yet clear whether the toxin-producing genes are absent from these strains or if toxin expression is environmentally regulated.

Algal toxins that impact human health may be functionally categorized as neurotoxins or hepatotoxins. Neurotoxicity of algal toxins is mediated by diverse, highly specific interactions with ion channels involved in neurotransmission. Such specificity may reflect their role in antipredation; alternatively, it may suggest the presence of conserved structures on primitive targets present in eukaryotic microbes, or it may support the hypothesis of primitive regulatory roles for these compounds. Okadaic acid and microcystins are examples of structurally unrelated compounds which are elaborated by phylogenetically unrelated groups of dinoflagellates (okadaic acid) and cyanobacteria (microcystins) to inhibit a common target, ser/thr protein phosphatases.

Dinoflagellate and diatom toxins impact human health primarily through the consumption of seafood. Filter-feeding shellfish, zooplankton, and herbivorous fishes serve as vectors either directly (shellfish) or through further food web transfer of sequestered toxin to higher trophic levels. Five major human seafood poisoning syndromes result: paralytic shellfish poisoning (PSP), neurotoxic shellfish poisoning (NSP), amnesic shellfish poisoning (ASP), diarrhetic shellfish poisoning (DSP), and ciguatera fish poisoning (CFP). Cyanobacterial toxins primarily impact human health directly through drinking water contamination and result in either neurotoxicity or hepatotoxicity. This chapter provides an overview of the diversity of algal toxins that impact human health, their mechanisms of action, causative organisms, and the geographical distribution of human outbreaks due to these toxin classes. Each of these topics will be considered in further detail in later chapters in this book.

II. PARALYTIC SHELLFISH TOXINS

Paralytic shellfish poisoning (PSP) is the most widespread algae-derived shellfish poisoning on a worldwide basis (Figure 1). The toxins responsible for PSP are a suite of heterocyclic guanidines collectively called saxitoxins (Figure 2), of which there are currently over 21 known congeners. The crystal structure of the parent compound, saxitoxin, was first described by Schantz et al. (3). The structures of saxitoxin congeners vary by differing combinations of hydroxyl and sulfate substitutions at four sites on the molecule (R1–4). Based on substitutions at R4, the saxitoxins can be subdivided into four groups, the carbamate toxins, sulfocarbamoyl toxins, decarbamoyl, and deoxydecarbamoyl toxins. Substitutions at R4 result in substantial changes in toxicity, with the carbamate toxins being the most potent (892–2483 mouse units [MU]/μmol), the decarbamoyl toxins being intermediate in potency (1274–1872 MU/μmol), and the sulfo–carbamoyl toxins generally being the least potent (15–239 MU/μmol) (4).

Paralytic shellfish toxins are produced by a number of genera of gonyaulacoid or gymnodinioid dinoflagellates, including *Alexandrium, Gymnodinium*, and *Pyrodinium*, and they have also been found in freshwater cyanobacteria (considered later in this chapter). Several studies suggest that saxitoxins can be produced autonomously by bacteria isolated from cultures of PSP-producing dinoflagellates (5–7), although chemical identity of the toxic activity present in these bacteria has not been unambiguously confirmed. In addition, endosymbiotic or cell associated bacteria may play a role in the production of paralytic shellfish toxins by dinoflagellates

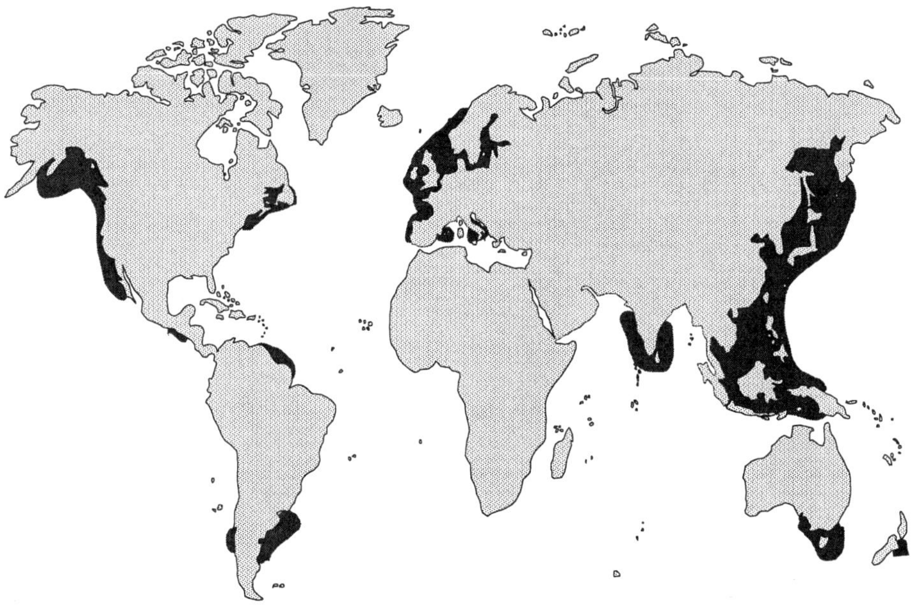

Figure 1 Worldwide incidence of paralytic shellfish poisoning.

(for review, see ref. 8). Paralytic shellfish toxins are produced in varying proportions by different dinoflagellate species and even by different isolates within a species. All congeners are not found in any one species. Metabolic pathways responsible for saxitoxin (STX) biosynthesis have been identified by radio tracer studies (4), which indicate that positions 1–12 of the STX molecule (Figure 2) are formed by the condensation of acetate and arginine. The most prevalent congeners in dinoflagellates, on a molar basis, are the sulfocarbamoyl derivatives. This is of particular interest, because the occurrence of N-sulfated groups is rare among natural products (9), and it may represent another novel dinoflagellate biosynthetic mechanism. N-sulfotransferase and N-oxidase activities have been identified in STX-producing dinoflagellates (10).

The toxin composition in bivalve tissues can differ markedly from that found in the dinoflagellates ingested. The capacity to metabolize PSP toxins varies substantially between shellfish species (for review, see ref. 11). In most shellfish, PSP toxins with a hydroxysulfate at the C11 position undergo epimerization from the β-epimers (GTX3,4 and C2,4) to the α-epimers (GTX1,2 and C1,2) (9). Of public health significance, the N-sulfocarbamoyl derivatives, which account for the majority of toxin in some dinoflagellates, may be metabolically converted to the more potent decarbamoyl congeners when metabolized in some shellfish species.

The role toxicity plays in the life history of the dinoflagellate species which produce them is not clear. It has been suggested that saxitoxins may play a role in nitrogen metabolism, since some strains accumulate toxin up to 60 pg/cell, or about 0.2% of total wet weight. However, the occurrence of healthy nontoxic strains would suggest that the saxitoxins are secondary metabolites which are not essential for dinoflagellate growth (12).

Saxitoxin binds with high affinity ($K_d \sim 2$ nM) to site 1 on the voltage-dependent sodium channel, inhibiting channel opening. The binding affinity of saxitoxin congeners to site 1 varies proportionally with their toxicity in mice (13). The voltage-dependent sodium channel plays a critical role in neurotransmission at both the neuronal synapses and neuromuscular junctions. The polarity of the STX molecule largely excludes it from traversing the blood-brain barrier;

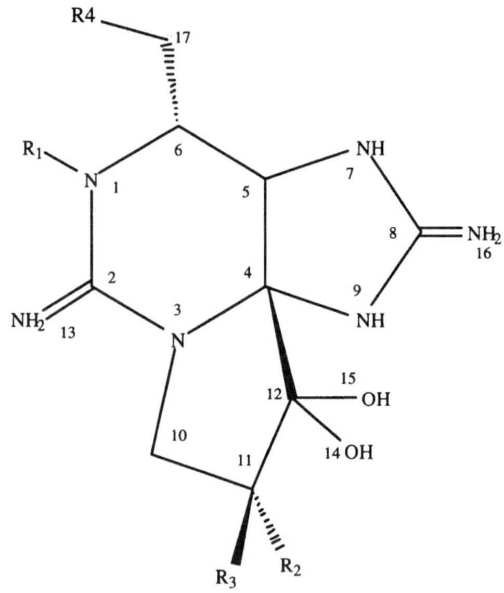

		R1	R2	R3	R4	MU/μmol[1]
Carbamate	STX	H	H	H	OCONH2	2483
	Neo STX	OH	H	H	OCONH2	2295
	GTX1	OH	OSO3-	H	OCONH2	2468
	GTX2	H	OSO3-	H	OCONH2	892
	GTX3	H	H	OSO3-	OCONH2	1584
	GTX4	OH	H	OSO3-	OCONH2	1803
Sulfocarbamoyl	GTX5 (B1)	H	H	H	OCONHSO3-	160
	GTX6 (B2)	OH	H	H	OCONHSO3-	-
	C1	H	OSO3-	H	OCONHSO3-	15
	C2	H	H	OSO3-	OCONHSO3-	239
	C3	OH	OSO3-	H	OCONHSO3-	33
	C4	OH	H	OSO3-	OCONHSO3-	143
Decarbamoyl	dcSTX	H	H	H	OH	1274
	dcNeoSTX	OH	H	H	OH	-
	dcGTX1	OH	OSO3-	H	OH	-
	dcGTX2	H	OSO3-	H	OH	1617
	dcGTX3	H	H	OSO3-	OH	1872
	dcGTX4	OH	H	OSO3-	OH	-
Deoxydecarbamoyl	doSTX	H	H	H	H	-
	doGTX2	H	H	OSO3-	H	-
	doGTX3	H	OSO3-	H	H	-

[1]Oshima, 1995

Figure 2 Structures and relative toxicity of the paralytic shellfish toxins. Toxicity values in mouse units (MU), where 1 MU unit is the amount of toxin required to kill a mouse in 20 min.

therefore, the primary site of STX action in humans is most likely at the neuromuscular junction. This is consistent with the rapid onset (less than 1 h) of symptoms which are classic for paralytic shellfish poisoning, including tingling and numbness of the perioral area and extremities, loss of motor control, drowsiness, incoherence, and, in the case of high doses, respiratory paralysis. The lethal dose in humans is 1–4 mg STX equivalents (14). Clinical symptoms of PSP in humans occurs when approximately 2000 MU (0.72 mg) is ingested, and serious cases generally involve ingestion of 5000–20,000 MU toxin (0.9–3.6 mg) (15). In a study of clinical samples from a PSP outbreak in Alaska in 1994 (16), the clearance of saxitoxins from the blood in humans was found to be less than 24 h, even in patients who had experienced respiratory paralysis and were maintained on life support. Clearance was largely via the urine.

There is currently no widely available antidote for PSP. Anti-STX monoclonal antibodies tested in vitro and in situ show some protection against STX binding and STX-induced reduction of peripheral nerve action potential in rat tibial nerve, suggesting that antibodies may potentially provide useful reagents for protection against toxicity in vivo (17,18). In addition, the potassium channel blocker 4-aminopurine has recently been shown significantly to reverse the effects of STX toxicosis in rats, suggesting that it may be useful as an antidote for PSP (19).

III. NEUROTOXIC SHELLFISH TOXINS

The occurrence of neurotoxic shellfish poisoning (NSP) has historically been limited to the west coast of Florida (Figure 3), where blooms of the dinoflagellate *Gymnodinium breve* initiate offshore and are subsequently carried inshore by wind and current conditions (20). Gulf of Mexico *G. breve* blooms are occasionally carried around the base of Florida by the Loop Current

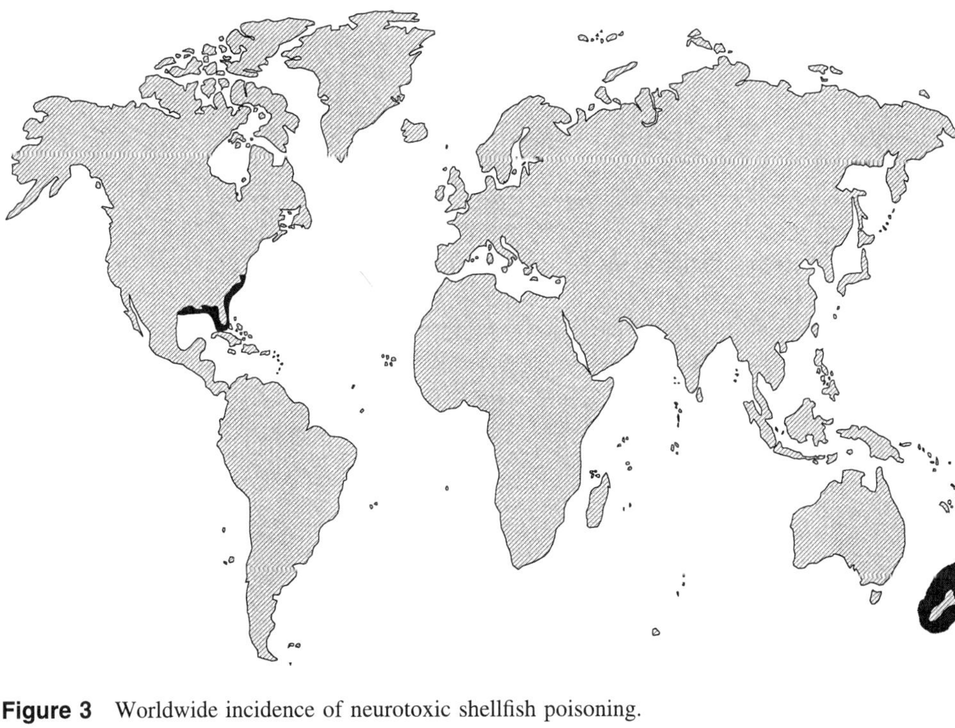

Figure 3 Worldwide incidence of neurotoxic shellfish poisoning.

and northward by the Gulf Stream, resulting in red tides on the east coast of Florida and, in a single incident in 1987, as far north as North Carolina (21). In 1993, an unprecedented outbreak of shellfish toxicity in New Zealand resulted in the identification of additional *Gymnodinium* species (referred to as *Gymnodinium* cf. *breve*) which produce NSP-like toxins, (22). Recently, other fish-killing flagellate species, *Chatonella marina, C. antiqua, Fibrocapsa japonica*, and *Heterosigma akashiwo*, have also been reported to produce this class of polyether toxins (23–25).

The toxins responsible for NSP are a suite of ladder-like polycyclic ether toxins collectively called brevetoxins (PbTx, for *Ptychodiscus brevis*, more recently restored to its earlier taxonomic designation, *Gymnodinium breve*) (Figure 4). Brevetoxin congeners fall into two types based on backbone structure, the brevetoxin B backbone (type 1) and brevetoxin A backbone (type 2). The type 1 congeners are the most abundant in nature, with PbTx-2 and PbTx-3 being the most prevelant in *G. breve*. Although the ring systems in the middle of the molecules differ somewhat, type 1 and type 2 toxins share a lactone in the A ring (''head'') and a conserved structure on the ''tail'' ring, both of which are required for their toxicity (26). Type 2 congeners are more flexible (38 rotatable bonds) than those with the type 1 backbone (31 rotatable bonds), which may play a role in their generally greater potency (27).

Brevetoxins bind with high affinity (K_d 1–50 nM) to site 5 on the voltage-dependent sodium channel (28). Binding to this site both alters the voltage sensitivity of the channel, resulting in inappropriate opening of the channel under conditions in which it is normally closed, and inhibits channel inactivation, resulting in persistent activation or prolonged channel opening. The toxic potency of brevetoxin congeners correlates well with their relative binding to the sodium channel (29). Backbone flexibility may determine the relative ease with which the toxin can intercalate between transmembrane domains of the sodium channel to interact with both the voltage sensor, near the ''outside'' of the channel and the inactivation gate on the intracellular side (30). The symptoms of NSP include nausea, tingling and numbness of the perioral area, loss of motor control, and severe muscular ache. Unlike PSP, NSP has not been known to be a fatal intoxication, with symptoms generally resolving within a few days. Like PSP, there is presently no antidote for NSP.

Unlike most other dinoflagellates responsible for seafood poisonings, *G. breve* is an unarmored dinoflagellate, which is easily lysed in turbulent water. *G. breve* red tides are frequently associated with massive fish kills. The extreme sensivity of fish to the Florida red tide may result from lysis of cells passing through the gills. One route of human intoxication results from the consumption of shellfish that have accumulated brevetoxins by filter feeding. Recent studies in the greenshell mussel demonstrate that brevetoxins can be metabolized by shellfish to yield novel derivatives (31).

An additional route of human exposure to brevetoxins is through respiration of aerosolized toxin, which is the result of cells breaking as a result of wave action. A common symptom associated with exposure to aerosolized brevetoxin is irritation and burning of the throat and upper respiratory tract. In 1996, at least 149 manatees died during an unprecedented epozootic in Florida concurrent with a persistent red tide. Immunohistochemical staining of tissues from affected animals revealed brevetoxin immunoreactivity in lymphocytes and macrophages associated with inflammatory lesions of the respiratory tract and lymphoid tissues (32). Molecular modeling studies have implicated brevetoxin as an inhibitor of a class of lysosomal proteases, the cysteine cathepsins, which are important in antigen presentation (33). The demonstration of brevetoxin immunoreactivity in lymphoid tissue of the manatees raises the possibility of immunosuppression as a second mode by which brevetoxin exposure may affect human health, particularly in individuals with chronic exposure to aerosolized toxin during prolonged red tide incidents.

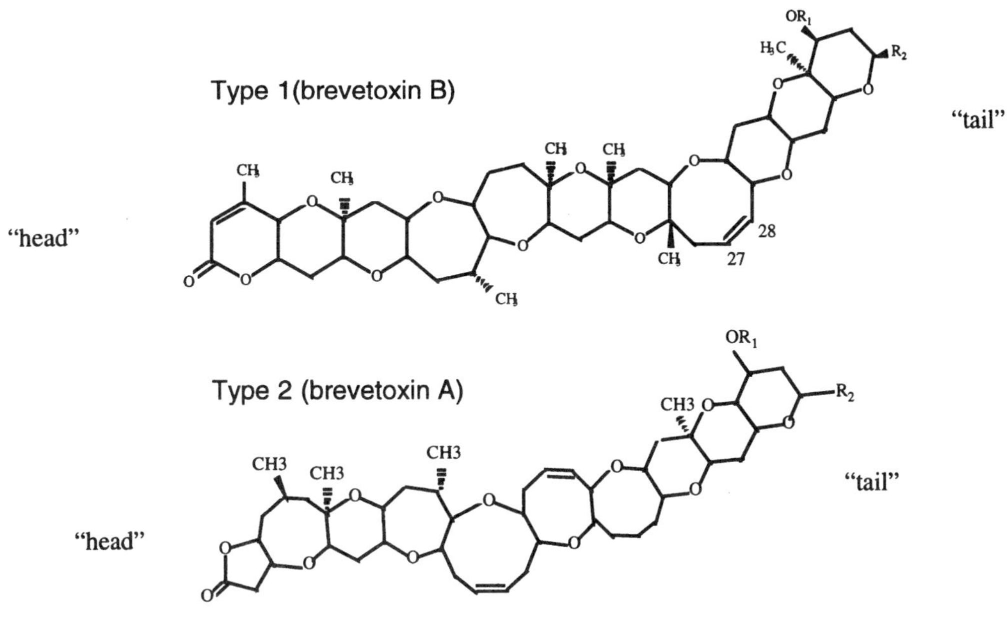

Toxin	Type	R1	R2
PbTx-1	2	H	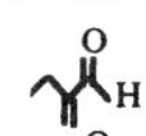
PbTx-2	1	H	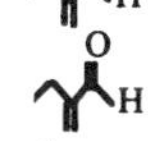
PbTx-3	1	H	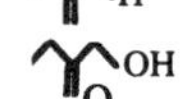
PbTx-5	1	COCH3	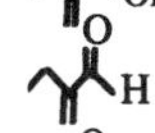
PbTx-6	1	H	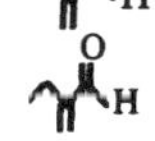 27,28 epoxide
PbTx-7	2	H	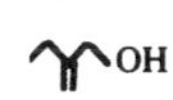
PbTx-8	1	H	
PbTx-9	1	H	
PbTx-10	2	H	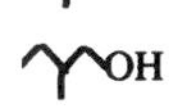

Figure 4 Structure of brevetoxins.

IV. CIGUATERA TOXINS

Another seafood intoxication caused by ladder-like polyether toxins is ciguatera fish poisoning. Ciguatera occurs circumglobally in tropical coral reef regions (Figure 5) and results from the consumption of fish which have accumulated toxins through the food web. It is estimated to affect over 50,000 people annually, and it is no longer a disease limited to the tropics owing

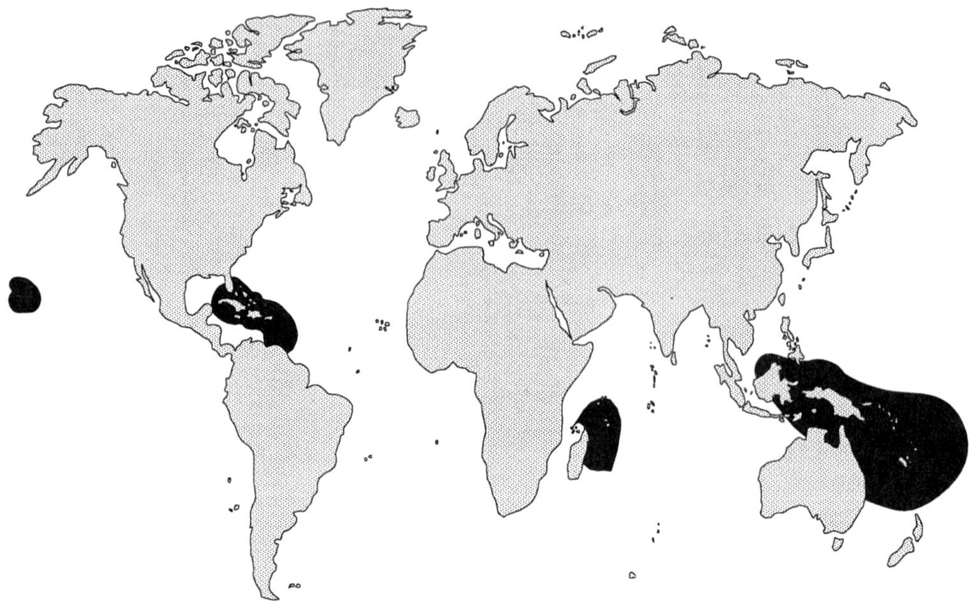

Figure 5 Worldwide distribution of ciguatera fish poisoning.

both to travel to the tropics and to shipping of tropical fish species to markets elsewhere in the world (34). Large carnivorous fishes associated with coral reefs are the most frequent source of ciguatera. Baracuda, snapper, grouper, jacks, and moray eel are particularly notorious for their potential to carry high toxin loads. However, smaller herbivorous fishes may also be ciguatoxic, particularly when viscera are consumed. The symptoms of ciguatera vary somewhat geographically, as well as between individuals and incidents, and may also vary temporally within an area, but generally include early-onset (2–6 h) gastrointestinal disturbance, including nausea, vomitting, and diarrhea, and may be followed by a variety of later onset (18 h) neurological sequelae, including numbness of the perioral area and extremities, reversal of temperature sensation, muscle and joint aches, headache, itching, tachycardia, hypertension, blurred vision, and paralysis. Ciguatera on rare occasions can be fatal. Ciguatera symptoms in the Caribbean differ somewhat from those in the Pacific in that gastrointestinal symptoms dominate, whereas in the Pacific neurological symptoms tend to dominate. This may reflect geographical differences in the toxins involved (35).

The origin of ciguatera toxins has been identified as the benthic coral reef–associated dinoflagellate *Gambierdiscus toxicus* (36), which grows as an epiphyte on filamentous macroalgae associated with coral reefs and reef lagoons. Its toxins enter the food web when these algae are grazed upon by herbivorous fishes and probably also invertebrates. *G. toxicus* produces two classes of polyether toxins, the ciguatoxins (CTXs) and maitotoxins (MTXs). The CTXs are lipophilic and are accumulated in fish through food web transfer. More than 20 CTX congeners have been isolated (37); however, only a few have been fully characterized structurally. Three classes of CTXs are currently recognized based on polyether backbone structure (Figure 6). The first CTX to be purified (38) and structurally elucidated (39) is now known as CTX-1, CTX1B, or P-CTX-1 (for Pacific CTX-1). CTX-1 is the parent compound for the type 1 CTX group, which possess 60 carbons in 13 fused ether rings. CTX-1 is believed to be responsible for 90% of the toxicity associated with most of the Pacific intoxications. Pacific CTX-2 and CTX-3 are also type 1 toxins found in fish flesh (40). The toxins found in fish flesh are more highly oxygen-

Type 1

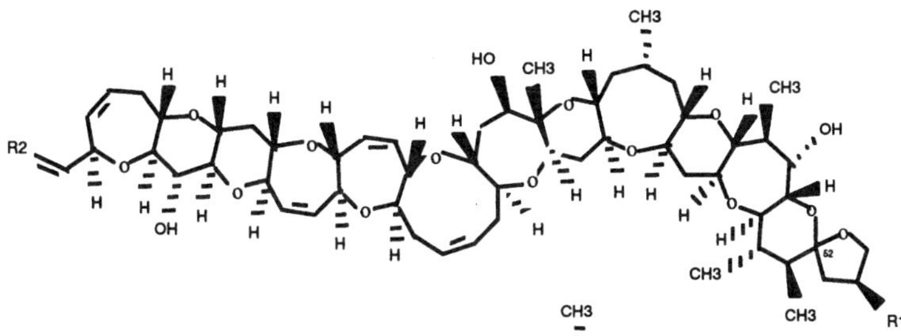

Type 2

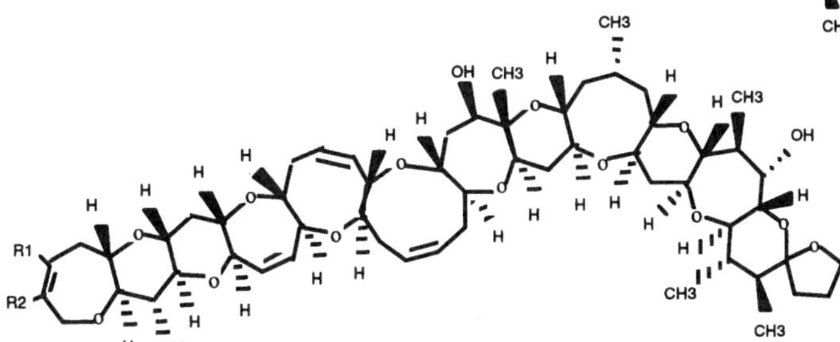

Type 3

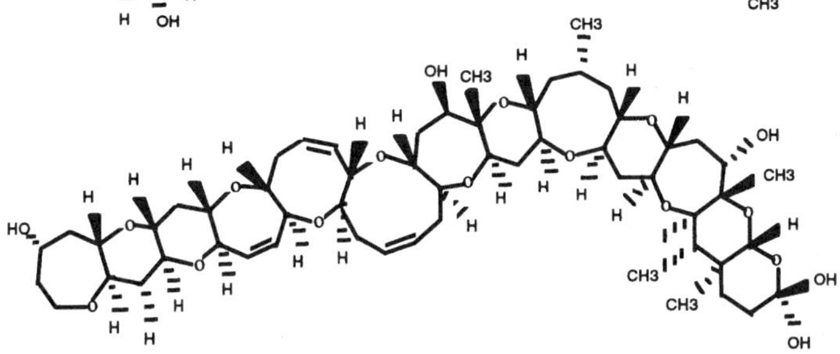

	Source	Type	R1	R2
CTX-1	fish	1	$HOCH_2CHOH$-	OH
CTX-2	fish	1	$HOCH_2CHOH$-	H
CTX-3	fish	1	$HOCH_2CHOH$-	H
CTX-4A	dino.	1	$CH_2 = CH$-	H
CTX-4B	fish,dino.	1	$CH_2 = CH$-	H
CTX-2A1	fish	2	OH	OH
CTX-3C	dino.	2	H	H
C-CTX-1	fish	3	-	-

Figure 6 Structures of the ciguatera toxins in fish flesh and their putative dinoflagellate precursors. CTX-2 and -3 and CTX-4A and -4B are epimeric pairs around the C52 spiroketal.

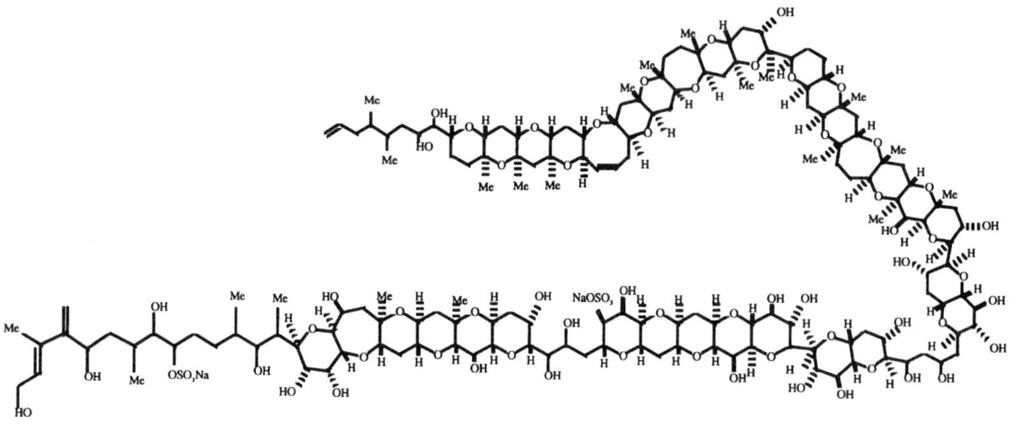

Figure 7 Structure of MTX-1.

ated than the congeners isolated from *G. toxicus*, suggesting that they are metabolites of the dinoflagellate toxins of the same backbone structure. CTX-1, CTX-2, and CTX-3 may all be derived from dinoflagllate precursors, CTX-4A and CTX4B (41). Type 2 CTX congeners possess 57 carbons in 13 fused ether rings, lack the C1-C4 side chain present in type 1, and have an eight- rather than a seven-membered ring E. Like the type 1 CTX toxins, there is evidence that CTX2A1, a congener present in fish, represents an oxygenated metabolite of a dinoflagellate precursor, CTX3C (42). Several Caribbean CTX (C-CTX) congeners have been isolated chromatographically, of which C-CTX-1, the most abundant, is the first to be fully structurally characterized (43). C-CTX-1 represents a third class of ciguatoxin that lacks the spiroketal at C52, which is replaced by an additional fused six-membered cyclic ether.

The CTXs are structurally related to the brevetoxins and compete with brevetoxin for binding bind to site 5 on the voltage-dependent sodium channel with a high affinity ($K_d \sim 0.04$–4 nM) (44). The LD50 (i.p.) in mice for CTX-1 is 0.25 μg/kg (37), whereas the potency of the Caribbean toxin C-CTX-1 is 10- fold lower (LD50 3.6 μg/g) (35). Various estimates of human toxic potency have been made. A concentration of >0.1 ppb of P-CTX-1 is estimated to be toxic to humans compared to >1.0 ppb C-CTX-1. In a study of fish implicated in ciguatera cases in French Polynesia, a minimum toxicity level to humans was estimated at 0.5 ng/g (37). Among the CTX congeners, binding affinity correlates well with toxic potency (i.p.) in mice. However, the toxic potency of CTX in mice is several orders of magnitude greater than that of the brevetoxins relative to their binding affinities at the sodium channel (e.g., for CTX1 and PbTx3, LD50 = 0.25 μg/kg vs >200 μg/kg, whereas K_d = 0.04 nM vs 2 nM, respectively). This may be related to differences in the bioavailability of the toxins or to undefined toxic effects of ciguatoxin.

The maitotoxins (Figure 7), like CTX or PbTx, are transfused ladder-like polyether toxins, but they are somewhat more polar owing to the presence of multiple sulfate groups. MTX was originally identified as a water-soluble toxin in the viscera of surgeonfishes (45), and it was later found to be the principal toxin produced by *Gambierdiscus toxicus*. The structure of MTX was first elucidated by Murata et al. (46), with complete stereochemistry resolved by Zheng et al. (47). Three MTX congeners have been identified in Pacific isolates of *G. toxicus*, MTX-1 and MTX-2 (3422 and 3298 D [dalton], respectively) and a smaller compound, MTX-3 (1060 D). MTX-1 from *G. toxicus* was found to be identical to the original MTX isolated from surgeonfish. The structure of MTX-2 has not been fully determined, but it appears to possess only one sulfate ester compared to two in MTX-1. MTX-3 possesses two sulfate esters (48). MTX isolated from Caribbean *G. toxicus* clones has not been fully characterized structurally. MTXs have not been demonstrated to bioaccumulate in fish tissues, possibly owing to their more polar structure. Thus, if MTX is involved in ciguatera poisoning, it may be implicated only in ciguatera poisonings derived from herbivorous fishes. Early hypotheses that MTX may be a metabolic precursor to CTX have not proven to be true (49).

The toxic potency of MTX exceeds that of CTX (LD50 0.05 μg/kg i.p. in mice). Its mode of action has not been fully elucidated. Its biological activity is strictly calcium dependent and causes both membrane depolarization and calcium influx in many different cell types. It was originally believed to be an activator of voltage-dependent calcium channels (for review, see ref. 50). However, voltage-dependent calcium channel antagonists can block MTX-stimulated calcium influx but not MTX-induced membrane depolarization (51). Therefore, it appears that MTX-induced activation of voltage-dependent calcium channels is a secondary effect of membrane depolarization. Despite numerous studies, the primary target of MTX has not yet been fully elucidated, although nonselective cation channels (52,53), and calcium-activated chloride channels (54,55) have received recent attention. Calcium release–activated calcium (CRAC) channels, another proposed target, do not appear to be involved based on the failure of CRAC channel antagonists to inhibit MTX activity (56). Removal of the sulfate esters causes a significant drop in toxicity (57). MTX-induced calcium influx can be inhibited by PbTx and by MTX fragments, which suggests that both hydrophobic and hydrophilic domains of the molecule are necessary for target (58).

The definition of ciguatera is complicated by the fact that *G. toxicus* is, in fact, one member of an assemblage of benthic dinoflagellates, all of which produce toxins. Unlike the planktonic dinoflagellates, toxicity in the benthic coral reef dinoflagellate assemblage appears to be quite common. Among the dinoflagellates which co-occur with *G. toxicus* are *Ostreopsis* spp., *Prorocentrum* spp., *Coolia* spp., and *Amphidinium* spp. Each of these genera produces toxins which target different pharmacological receptors (Table 1). However, with the exception of toxins derived from *Ostreopsis*, the accumulation of most of these toxins in upper trophic levels of the coral reef community to concentrations which may impact human health has not been confirmed, and therefore their contribution to ciguatera remains equivocal. However, *Ostreopsis* has been proposed to be the primary dinoflagellate responsible for ciguatera in Puerto Rico based on seasonal abundance of *Ostreopsis* versus *G. toxicus* in Puerto Rican waters (59). *Ostreopsis* produces ostreocin, an analogue of palytoxin (Figure 8). Palytoxin has been confirmed as the causative agent in ciguatera-like poisonings from crab in the Pacific (60), mackerel (61), triggerfish (62), and sardines (clupeotoxism) (63). Palytoxin is a macrocyclic polyether toxin characterized by a number of novel features, including: a C115 straight chain incorporating many functionalities; a terminal primary amine that is important for bioactivity; and an α,β-unsaturated amide, two conjugated diene systems, and a hemiketal (64). The complete structure

Table 1 Toxins Produced by Benthic Coral Reef-Associated Dinoflagellates

Species	Toxin	Pharmacological target
Gambierdiscus toxicus	Ciguatoxin	Voltage-dependent sodium channel
	Maitotoxin	(?) calcium dependent
Coolia monotis	Coolia toxin	Unknown
Ostreopsis spp.	Ostreocin	Na^+/K^+ ATPase
Prorocentrum spp.	Okadaic acid	ser/thr Protein phosphatases
Amphidinium spp.	Amphidinilides	Unknown (antifungal)

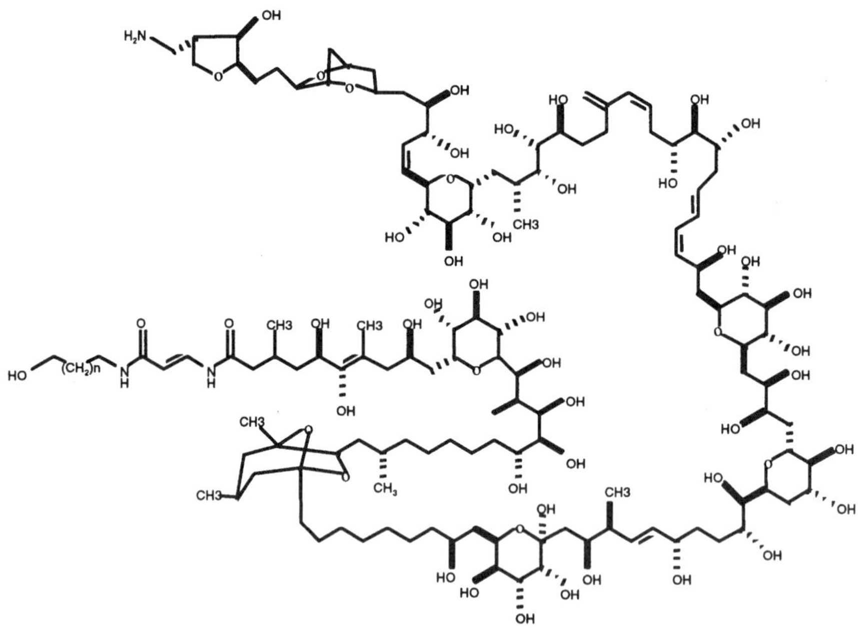

Figure 8 Structure of palytoxin.

of palytoxin was determined by Moore et al. (65). Palytoxin poisoning may be distinguishable from ciguatera by its severity (high fatality rate) and unusual taste associated with the contaminated fish. The LD50 in rodents is 0.01–0.25 μg/kg (66). The pharmacological target of palytoxin is Na^+/K^+ ATPase, which pumps Na^+ and K^+ across the cell membrane against their electrochemical gradients, such that three Na^+ ions are pumped out of the cell and two K^+ ions are pumped into the cell for each ATP hydrolyzed. In the presence of palytoxin, the pump is converted into an open channel that permits K^+ efflux and influx of monovalent cations (Na^+, $NH4^+$, Cs^+, Li^+) along their electrochemical gradients. The palytoxin-induced pore appears to reside within the protein, possibly by stabilizing a channel made up of transmembrane segments of the protein when the pump is in its open state (67–69).

V. DIARRHETIC SHELLFISH TOXINS

The diarrhetic shellfish toxins (DTXs) are a class of acidic polyether toxins produced by dinoflagellates and responsible for human illness, diarrhetic shellfish poisoning (DSP), associated with seafood consumption. This toxin class consists of at least eight congeners, including the parent compound, okadaic acid, which was first isolated from the black sponge, *Halichondria fortii* (70) (Figure 10). Okadaic acid, DTX-1, and DTX-2 are the primary congeners involved in shellfish poisoning, with the other congeners believed to be either precursors or shellfish metabolites of the active toxins. DSP is widespread in its distribution (Figure 9), with essentially seasonal occurrences in Europe and Japan. The first incidence of human shellfish-related illness identified as DSP occurred in Japan in the late 1970's, where the dinoflagellate *Dinophysis fortii* was identified as the causative organism, and the toxin responsible was termed dinophysistoxin (DTX-1) (71,72). Retrospective analysis of similar disease outbreaks in the Netherlands (73,74) and Scandanavia (75) confirmed that these were also associated with *Dinophysis*. The major

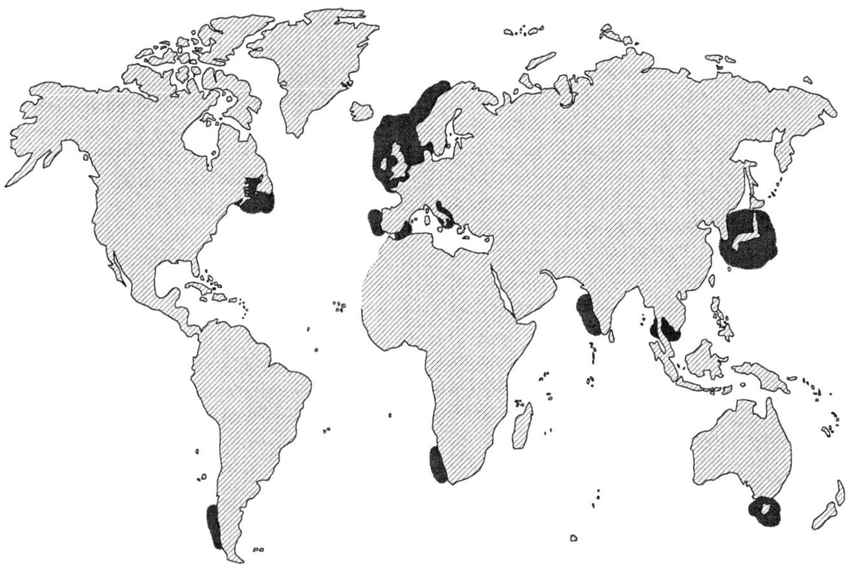

Figure 9 Distribution of diarrhetic shellfish poisoning.

toxins involved in European outbreaks are okadaic acid and DTX-1. However, incidents in Ireland and Portugal were found to include an additional toxin, DTX-2 (76). The first confirmed incident of DSP in North America occurred in 1990 in the maritime provinces of Canada (77), but was associated with the benthic dinoflagellate *Prorocentrum lima* and two toxins, DTX-1 and okadaic acid. Okadaic acid and related DTX toxins are also produced by a number of other *Prorocentrum* species, including *P. maculosum, P. concavum*, and *P. hoffmanianum*, but do not appear to be elaborated by *P. micans, P. minimum*, or *P. mexicanum* (78).

The DTXs are inhibitors of ser/thr protein phosphatases. Inhibitory activity against protein phosphatases is specific for classes PP2A (okadaic acid IC50 ~ 0.5 nM) and PP1 (okadiac acid IC50 ~ 50 nM), with PP2B being inhibited only at high concentrations (okadiac acid IC50 >10 μM) and PP2C being insensitive. The binding site for okadaic acid resides on the catalytic subunit of the protein phosphatase, at the active site of the ezyme, as determined by x-ray crystal structures (79), molecular modeling (80), and mutational analyses (81). Binding to this site requires the carboxylic acid moiety, which accounts for the inactive state of the diol esters and DTX-4.

Ser/thr protein phosphatases are critical components of signaling cascades in eukaryotic cells which regulate a diverse array of cellular processes involved in metabolism, ion balance, neurotransmission, and cell cycle regulation (82). Diarrhea associated with DSP is most likely due to the hyperphosphorylation of proteins, including ion channels, in the intestinal epithelia (83), resulting in impaired water balance and loss of fluids. In addition, okadaic acid–like polyether toxins have been identified as tumor promoters (84,85). The toxic potency of okadaic acid is much lower than that of the neurotoxin polyethers, with an LD50 of 192 μg/kg (i.p.) in mice (86).

The biosynthesis of the DTX toxins and the mechanisms by which the dinoflagellate protects itself from its toxins have received much attention. Okadaic acid is localized to the chloroplast (87). Most of the intracellular toxin in *P. lima* is present in the form of DTX-4 (88), which has been shown in metabolic labeling studies to be a biosynthetic precursor to okadaic acid (89). Thus, it is proposed that okadaic acid is released by *P. lima* as the inactive protoxin, DTX-

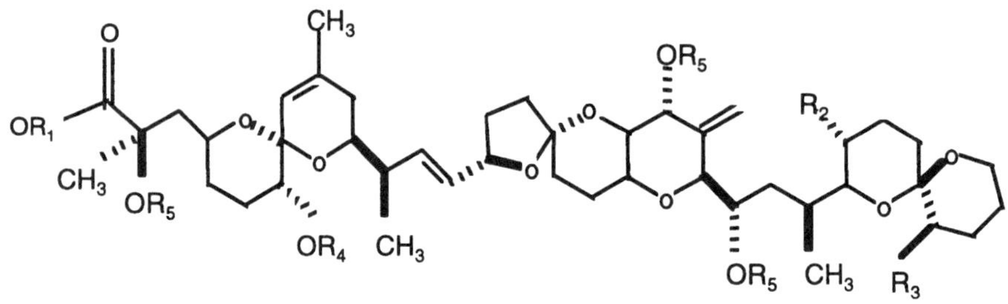

	R1	R2	R3	R4	R5
OA	H	CH3	H	H	H
DTX-1	H	CH3	CH3	H	H
DTX-2	H	H	CH3	H	H
DTX-3	H	CH3	CH3	acyl	H
DTX-4	1	H	H	H	H
DTX-5a	2	H	H	H	H
DTX-5b	3	H	H	H	H
Diol ester	4	H	H	H	H
OA methyl ester	CH3	CH3	H	H	H

1
2
3
4

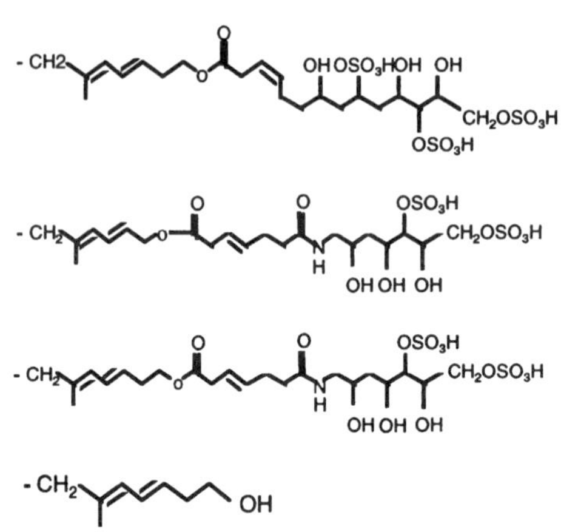

Figure 10 Structures of naturally occurring DSP toxins.

4, which is reduced extracellularly to the diol ester in the medium of *P. lima*. The diol ester may then be cleaved at the ester linkage to yield the active toxin, okadaic acid. Windust et al. (90,91) investigated the allelopathic activity of okadaic acid against other species of microalgae and found that it could inhibit the growth of diatoms at μM concentrations. However, Sugg and Van Dolah (92) tested the hypothesis that okadaic acid produced by *P. lima* confers a competitive advantage over co-occurring benthic dinoflagellates of the ciguatera assemblage, and found that the observed growth inhibition of other dinoflagellates by *P. lima* was not attributable to okadaic acid. Dinoflagellates do possess okadaic acid–sensitive protein phosphatases PP1 and PP2A (92,93,94), whereas the protein phosphatases in *P. lima* appear to be insensitive.

VI. DOMOIC ACID

Amnesic shellfish poisoning (ASP) is the only shellfish poison produced by a diatom and is currently limited in its distribution to North America (Figure 11). The first recorded occurrence of ASP was in Prince Edward Island, Canada, in 1987, when approximately 100 people became ill after consuming contaminated mussels. None of the known shellfish toxins was found to be involved in the outbreak, but rather the toxic agent was identified as domoic acid (95,96). The source of domoic acid was found to be the diatom *Pseudo-nitzschia multiseries* (formerly known as *Nitzschia pungens f. multiseries*) (97,98). Domoic acid is a water-soluble tricarboxylic amino acid of molecular weight 311, which acts an analogue of the neurotransmitter glutamate and is a potent glutamate receptor agonist. Domoic acid was previously identified in the red alga *Chondria armata* (99), but it had not previously been linked to human illness, and is related both structurally and functionally to the excitatory neurotoxin kainic acid isolated from the red macroalga *Digenea simplex* (100). Seven congeners to domoic acid have been identified (Figure 12). Of these, three geometrical isomers, isodomic acids D, E, and F and the C5′ diasteriomer are found, in addition to domoic acid, are found in small amounts in both the diatom and in shellfish tissue (101,102).

The symptoms of ASP include gastrointestinal effects (e.g., nausea, vomiting, diarrhea) and neurological effects, including dizziness, disorientation, lethargy, seizures, and permanent loss of short-term memory. Domoic acid binds with high affinity to both kainate ($K_d \sim 5$ nM)

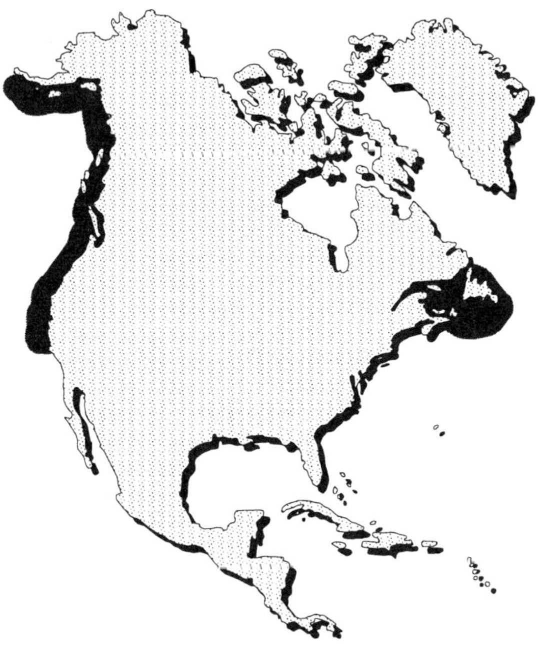

Figure 11 Distribution of ASP.

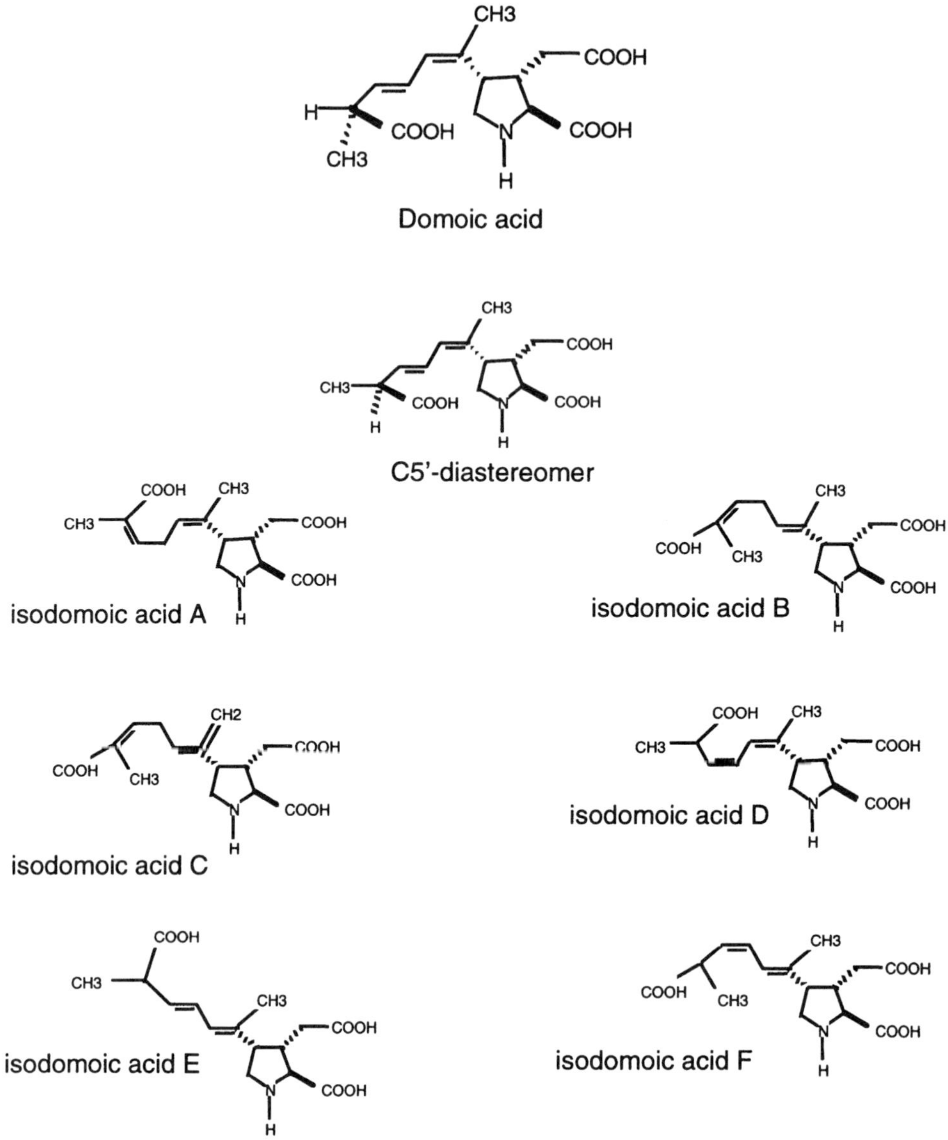

Figure 12 Structures of domoic acid and congeners.

and AMPA (K_d ~ 9 nM) subtypes of glutamate receptor (103). Persistent activation of the kainate glutamate receptor results in greatly elevated intracellular Ca^{2+} through cooperative interactions with NMDA and non-NMDA glutamate receptor subtypes and voltage-dependent calcium channels (104). Neurotoxicity due to domoic acid results in high intracellular calcium and subsequent lesions in areas of the brain where glutaminergic pathways are heavily concentrated, particularly in the CA1 and CA3 regions of the hippocampus (96), areas responsible for learning and memory processing. However, memory deficits occur at doses below those causing structural damage (105). The LD50 (i.p.) for domoic acid in rats is 4 mg/kg; however, the oral potency is substantially lower (35–70 mg/kg) (106). In the 1987 outbreak, human toxicity occurred at 1–5 mg/kg, suggesting that susceptible individuals are more sensitive than rodents to the oral toxicity of domoic acid. Individuals found most susceptible were elderly individuals and those with impaired renal function resulting in poor toxin clearance. Increased susceptibility of elderly individuals appears to be due to impaired toxin clearance, as studies in experimental animals and neonates indicate (107).

Since the 1987 outbreak, domoic acid has been identified as the causative agent in the mass mortality of pelicans and cormorants in Monterey Bay, California, in 1991 (108,109) and for the extensive die-off of California sea lions in the same region in 1998 (110). The causative organism in both the 1991 and 1998 mortality events was identified as another member of the same diatom genus, *Pseudo-nitzschia australis*. At least seven species of *Pseudo-nitzschia* are now recognized as domoic acid producers, and these toxin-producing *Pseudo-nitzschia* species have since been identified in widely diverse geographical areas around the world (for review, see ref. 111). However, none has been implicated in intoxication events.

As with the other algal toxins discussed in the chapter, the role of domoic acid in the life history of *Pseudo-nitzschia* is not clear. The production of domoic acid by *Pseudo-nitzschia* correlates with physiological stress, including silicon or phosphorus limitation or nitrogen excess. This pattern of synthesis is consistent with classic secondary metabolite biosynthesis by bacteria and other protists and differs from the constitutive pattern observed in the synthesis of polyether toxins by dinoflagellates (2) and PSP toxins (112). In culture, domoic acid is produced primarily in stationary phase, which corresponds with the depletion of Si from the medium (113). The 1987 bloom of *Pseudo-nitzschia* in Canada was associated with pulses of nitrate from riverine input or resuspended sediments (111). The biosynthesis of domoic acid involves the condensation of acetates via two intermediates, a glutamate derivative from the Krebs cycle and an isoprenoid structure likely derived from geranyl pyrophosphate (114). The precise enzymatic pathways responsible for biosynthesis have not been elucidated.

VII. CYANOBACTERIAL TOXINS

Cyanobacteria produce a wide array of secondary metabolites, a number of which are toxic to shellfish and aquaculture-reared fish when exposed to the cells or free toxins in water. Blooms of cyanobacteria result in human and wildlife intoxications due to consumption of or exposure to contaminated water, essentially on a worldwide basis. Over 40 species of freshwater cyanobacteria have been implicated in toxic blooms (115). Cyanobacterial toxins have long been considered soley a freshwater issue, but the recent identification of mycrocystin in marine waters (116) and the implication of lyngbyatoxin in a human fatality due to consumption of sea turtle (117) have added cyanobacteria to the wide array of harmful marine algae. Cyanobacterial toxins can be categorized into two general groups based on their mode of action, the neurotoxins and hepatotoxins. Hepatotoxins are the most commonly encountered toxins in cyanobacterial blooms.

Neurotoxins produced by cyanobacteria include the saxitoxins (STXs) (see Figure 2) and anatoxins (AnTx) (Figure 13). STX and neoSTX have been identified in *Aphanizomenon flos-aquae* in New Hampshire (118). An extensive bloom of *Anabaena circinalis* in southern Australia in 1991, which resulted in significant numbers of animal deaths, was found to contain STXs, most of the gonyautoxins, C1, and C2 toxins (119). Several PSP toxins were found in *Lyngbya wollei* from Alabama (120,121), including decarbamoyl STX, dcGTX2, dcGTX3, and six new STX derivatives (122). No human intoxications from cyanobacterial STXs have been reported.

AnTx-a, produced by certain strains of *Anabaena flos-aquae, Aphanizomenon flos-aquae*, and *Oscillatoria*, is a potent agonist for the nicotinic acetylcholine receptor, which causes a

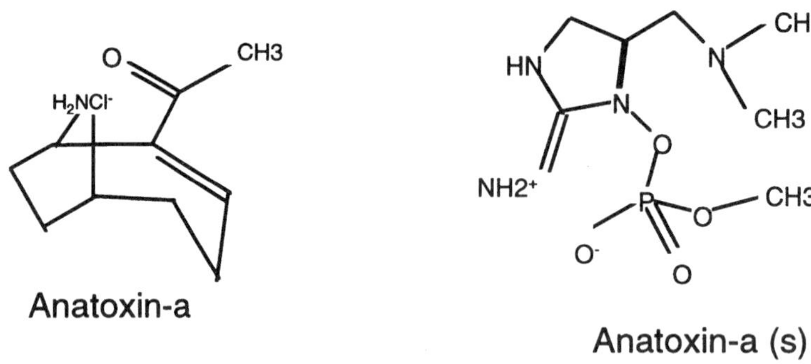

Figure 13 Structures of anatoxins.

depolarizing neuromuscular block. Toxic symptoms include staggering, gasping, convulsions, and death due to respiratory paralysis. LD50 (i.p.) in mice is 200 μg/kg. As with many of the other algal toxins discussed, no antidote is available. AnTx-a(s) (''s'' for salivation), produced by certain other strains of *Anabaena*, is an organophosphate inhibitor of acetylcholinesterase shown to be responsible for intoxication of domestic animals (123). Toxic symptoms include hypersalivation, ataxia, diarrhea, tremors, dyspnea, and cyanosis. The LD50 (i.p.) in mice is 20 μg/kg.

The hepatotoxins produced by cyanobacteria include the cyclic peptide toxins, microcystins and nodularins (Figure 14). Microcystins are produced by certain species and certain strains within these species belonging to the genera *Microcystis, Anabaena, Nodularia, Nostoc*, and *Oscillatoria*. At least 52 microcystin analogues are known (115,124). Nodularin is produced by certain strains of *Nodularia spumigena*. Cyclic peptide toxins are synthesized through a nonribosomal thio template mechanism (125).

Microcystins and nodularins are inhibitors of serine/threonoine protein phosphatases with highest potency against type 2A (PP2A; IC50 ~ 0.1 nM) and type 1 (PP1; IC50 ~ 0.1–10 nM, depending on the congener). Both the cyanobacterial cyclic peptide toxins and the dinoflagellate-derived protein phosphatase inhibitor, okadaic acid, bind to the catalytic subunit of the protein phosphatases. X-ray crystal structures (79), molecular modeling (80), and mutational analyses (81) indicate that all three toxin classes bind to the same site. However, unlike the other toxin classes, the microcystins form covalent adducts with Cys273 in the enzyme subsequent to binding (126).

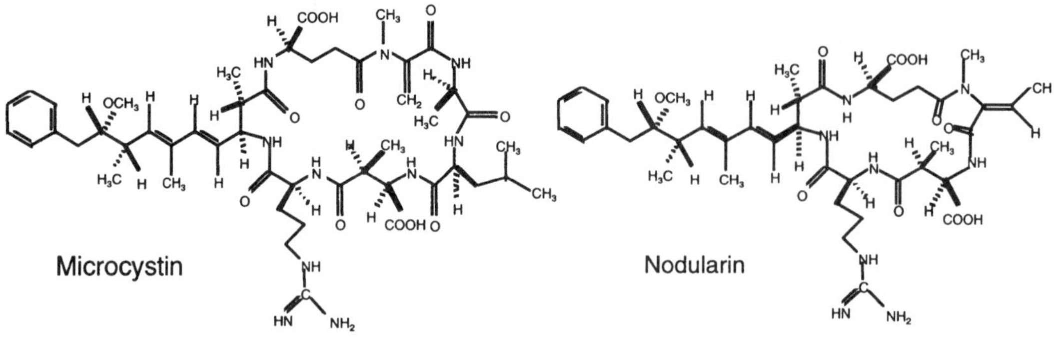

Figure 14 Cyclic peptide toxins from cyanobacteria.

Unlike okadaic acid and the DTX toxins, the microcystins and nodularins are highly selective for liver protein phosphatases. Their selective activity is conferred by their membrane impermeability to most cell types and their specific uptake in liver cells through the bile acid transport pathway. Toxicity is mediated by the depolymerization of the actin and microtubule cytoskeleton, resulting in intrahepatic hemorrhage within hours and death induced by hypovolemic shock. Microcystins were found to be the causative agent in an 1996 outbreak in which 26 patients died of liver failure at a hemodialysis clinic in Caruaru, Brazil, which drew its water form a contaminated reservoir (127). Microcystin is also implicated in net pen liver disease in pen-reared salmon (128,129).

Microcystins (130) and nodularins have been demonstrated to be liver tumor promotors in laboratory animals (131). Epidemiological studies of certain areas in China show a positive correlation between the presence of microcystin in water supplies and the incidence of human primary liver cancer (115,132). Thus, both acute and chronic exposure to cyanobacterial hepatotoxins may be significant human health risks.

VIII. SUMMARY

Algal toxins represent a major source of poisonings due to drinking water contamination and seafood consumption. Most classes of algal toxins are neurotoxins, whereas others target protein phosphatases, critical regulators of signal transduction pathways. Most algal toxins of human health significance are characterized by their high toxic potency and consist of suites of related congeners. The metabolic conversion, bioaccumulation, and food web transfer of toxins are incompletely understood in many cases. These characteristics make the analysis of algal toxins difficult and the determination of human health risk complex.

The diversity of algal toxins with impacts on human health is a reflection of the great variety of biosynthetic capabilities that have evolved in both prokaryotic and eukaryotic microalgae. Those compounds which we regard as ''toxins'' represent only a small percentage of the myriad of compounds produced by microalgae, specifically those whose selective interaction with receptors in mammalian systems results in illness. Most algal toxins may be considered secondary metabolites, since toxin expression does not appear to be required for basic cellular metabolism. Toxicity is not a phylogenetically conserved feature among microalgae, since, in most instances, species closely related to a toxigenic species (based on morphological or molecular phylogeny) may not be toxic. The selective advantage of toxin production is thus difficult to establish. Since the genes involved in toxin production have not been identified for any algal species, it is impossible currently to determine whether the capacity for toxin production is present, but silenced, in nontoxic varieties. As the molecular biology of microalgae becomes better understood, these questions will become answerable.

REFERENCES

1. JLC Wright, T Hu, JL MacLachlin, J Needham, JA Walter. Biosynthesis of DTX4: confirmation of a polyketide pathway, proof of a Baeyer-Villiger oxidation step, and evidence for an unusual carbon deletion step. J Am Chem Soc 118:8757–8758, 1996.
2. JLC Wright, AD Cembella. Ecophysiology and biosynthesis of polyether toxins. In: DM Anderson, AD Cembella, GM Hallegraeff, eds. Physiological Ecology of Harmful Algal Blooms. New York: Springer-Verlag, 1998, pp 427–452.

3. EJ Schantz, VE Ghazarossian, HKM Schnoes, FM Strong, JP Springer, JO Pezzanite, J Clardy. The structure of saxitoxin. J Am Chem Soc 97:1238–1239, 1975.
4. Y Shimizu, S Gupta, A. Prasad. Biosynthesis of dinoflagellate toxins. In: E Graneli, B Sunstrom, L Edler, DA Anderson et al., eds. Toxic Marine Phytoplankton. New York: Elsevier, 1990, pp 62–71.
5. M Kodama, T Ogata, S Sato. Bacterial production of saxitoxin. Agric Biol Chem 52:1075–1077, 1988.
6. GJ Doucette, CG Trick. Characterization of bacteria associated with different isolates of *Alexandrium tamarense*. In: P Lassus, G Arzul, E Erhard, P Gentien, C Marcaillou, eds. Harmful Marine Algal Blooms. Paris: Lavoisier, 1995, pp 33–38.
7. S Gallagher, KJ Flynn, JM Franco, EE Brueggemann, HB Hines. Evidence for the production of paralytic shellfish toxins by bacteria associated with *Alexandrium* spp. (Dinophyta) in culture. Appl Environ Microbiol 63:239–245, 1997.
8. GJ Doucette, M Kodama, S Franca, S Gallagher. Bacterial interactions with harmful algal bloom species: bloom ecology, toxigenesis and cytology. In: DM Anderson, AD Cembella, GM Hallegraeff, eds. Physiological Ecology of Harmful Algal Blooms. New York: Springer-Verlag, 1998, pp 649–652.
9. Y Shimizu. Microalgal metabolites. Chem Rev 93:1685–1698, 1993.
10. Y Oshima. Chemical and enzymatic transformation of paralytic shellfish toxins in marine organisms. In: P Lassus, G Arzul, E Erhard, P Gentien, C Marcaillou, eds. Harmful Marine Algal Blooms. Paris: Lavoisier, 1995, pp 475–480.
11. VM Bricelj, SE Shumway. Paralytic shellfish toxins in bivalve molluscs: occurrence, transfer kinetics, and biotransformation. Rev Fish Sci 6:315–383, 1998.
12. Y Oshima, SI Blackburn, GM Hallegraeff. Comparative study on paralytic shellfish toxin profiles of the dinoflagellate *Gymnodinium catenatum* from three countries. Mar Biol 116:471–476, 1993.
13. GJ Doucette, ML Logan, JS Ramsdell, FM Van Dolah. Development and preliminary validation of a microtiter plate-based receptor binding assay for paralytic shellfish poisoning toxins. Toxicon 35:625–636, 1997.
14. M Evans. Saxitoxin and related poisons: their actions on man and other animals. In: VR Locicero, eds. Proceedings of 1st International Conference on Toxic Dinoflagellates, Massachusetts Science and Technology Foundation, Wakefield, MA, 1974, pp 337–345.
15. RE Levin. Paralytic shellfish toxins: their origin, characteristics, and methods of detection: a review. J Food Biochem 15:405–417, 1992.
16. BD Gessner, P Bell, GJ Doucette, E Moczydlowski, MA Poli, F Van Dolah, S Hall. Hypertension and identification of toxin in human urine and serum following a cluster of mussel-associated paralytic shellfish poisoning outbreaks. Toxicon 35:711–722, 1997.
17. RI Huot, DL Armstrong, TC Chanh. *In vitro* and *in situ* inhibition of the sodium channel blocker saxitoxin by monoclonal antibodies. J Tox Environ Health 27:381–393, 1989.
18. TC Chanh, EB Siwak, JF Hewetson. Anti-idio–based vaccines against biological weapons. Toxicol Appl Pharmacol 108:183–193, 1991.
19. H-M Chen, C-H Lin, TM Wang. Effects of 4-aminopurine on saxitoxin intoxication. Toxicol Appl Pharmacol 141:44–48, 1996.
20. KA Steidinger, GA Vargo, PA Tester, CR Tomas. Bloom dynamics and physiology of *Gymnodinium breve* with emphasis on the Gulf of Mexico. In: DM Anderson, AD Cembella, GM Hallegraeff, eds. Physiological Ecology of Harmful Algal Blooms. Berlin: Springer-Verlag, 1998, pp 133–154.
21. PA Tester, KA Steidinger. *Gymnodinium breve* red tide blooms: initiation, transport, and consequences of surface circulation. Limnol Oceanogr 42:1039–1051, 1997.
22. A Haywood, L MacKenzie, I Garthwaite, N Towers. *Gymnodinium breve* 'look-alikes': three *Gymnodinium* isolates from New Zealand. In: T Yasumoto, Y Oshima, Y Fukuyo, eds. Harmful and Toxic Algal Blooms. Paris: International Oceanographic Committee of UNESCO, 1996, pp 227–230.
23. MD Sagir Ahmed, O Arakawa, Y Onoue. Toxicity of cultured *Chatonella marina*. In: P Lassus, G Arzul, E Erhard, P Gentien, C Marcaillou, eds. Harmful Marine Algal Blooms. Paris: Lavoisier, 1995, pp 499–504.

24. S Khan, O Arakawa, Y Onoue. Neurotoxins in a toxic red tide of *Heterosigma akashiwo* (Raphidophyceae) in Kagoshima Bay, Japan. Aquacul Res 28:9–14, 1997.
25. GM Hallegraeff, BL Munday, DG Baden, PL Whitney. *Chatonnella marina* raphidophyte bloom associated with mortality of cultured bluefin tuna (*Thunnus maccoyii*) in south Australia. In: B Reguera, J Blanco, ML Fernandez, T Wyatt, eds. Harmful Algae. Santiago del Compostela, Spain: Xunta de Galacia and IOC, 1998, pp 93–96.
26. DG Baden. Brevetoxins: unique polyether dinoflagellate toxins. FASEB J 3:1807–1819, 1989.
27. KS Rein, DG Baden, RE Gawley. Conformational analysis of the sodium channel modulator, brevetoxin A, comparison with brevetoxin B conformations, and a hypothesis about the common pharmacophore of the ''site 5'' toxins. J Org Chem 59:2101–2106, 1994.
28. MA Poli, TJ Mende, DG Baden. Brevetoxins, unique activators of voltage-sensitive sodium channels, bind to specific sites in rat brain synaprotomes. Mol Pharmacol 30:129–135, 1986.
29. KS Rein, B Lynn, RE Gawley, DG Baden. Brevetoxin B: chemical modifications, synaptosome binding, toxicity, and an unexpected conformational effect. J Org Chem. 59:2107–2113, 1994.
30. RE Gawley, KS Rein, G Jeglitsch, DJ Adams, EA Theodorakis, J Tiebes, KC Nicolau, DG Baden. The relationship of brevetoxin 'length' and a-ring functionality to binding and activity in neuronal sodium channels. Chem Biol 2:533–541, 1995.
31. K Murata, M Satake, H Naoki, HF Kaspar, T Yasumoto. Isolation and structure of a new brevetoxin analog, brevetoxin B2 from greenshell mussels from New Zealand. Tetrahedron 54:735–742, 1998.
32. GD Bossart, DG Baden, RY Ewing, B. Roberts, S Wright. Brevetoxicosis in manatees (*Trichechus manatus latirostros*) from the 1996 epizootic: gross, histologic, and immunochemical features. Tox Pathol 26:276–282, 1998.
33. Sudarsanam, S, DG Virca, CJ March, S Srinivasan. An approach to computer-based modeling: application to cathepsin. J Comput Aided Mol Design 6:223–233, 1992.
34. Ahmed, FE. Seafood Safety. Washington, DC: National Academy Press, 1991.
35. J-P Vernoux, RL Lewis. Isolation and characterization of Caribbean ciguatoxins from the horse-eye jack (*Caranx latus*). Toxicon 35:889–900, 1997.
36. T Yasumoto, I Nakajima, R Bagnis, R Adachi. Finding of a dinoflagellate as a likely culprit of ciguatera. Bull Jpn Soc Sci Fish 43:1021–1026, 1977.
37. AM Legrand. Ciguatera toxins: origin, transfer, through the food chain and toxicity to humans. In: B Reguera, J Blanco, ML Fernandez, T Wyatt, eds. Harmful Algae. Santiago del Compostela, Spain: Xunta de Galacia and IOC, 1998, pp 39–43.
38. PJ Sheuer, W Takahashi, J Tsutsumi, T Yoshida. Ciguatoxin. Isolation and chemical properties. Science 155:1267–1268, 1967.
39. M Murata, AM Legrand, Y Ishibashi, M Fukui, T Yasumoto. Structures and configurations of ciguatoxin from te moray eel *Gymnothorax javonicus* and its likely precursor from the dinoflagellate *Gambierdiscus toxicus*. J Am Chem Soc 112:4380–4386, 1990.
40. RJ Lewis, M Sellin, MA Poli, R Norton, JK MacLeod, MM Sheil. Purification and characterization of ciguatoxins from the Moray *eel (Lycodontis javonicus*, Muraenidae). Toxicon 29:1115–1127, 1991.
41. RJ Lewis, MJ Holmes. Origin and transfer of toxins involved in ciguatera. Comp Biochem Physiol 106C:615–628, 1993.
42. AM Legrand, T Teai, P Cruchet, M Satake, K Murata, T Yasumoto. Two structural types of ciguatoxin involved in ciguatera fish poisoning in French Polynesia. In: B Reguera, J Blanco, ML Fernandez, T Wyatt, eds. Harmful Algae. Santiago del Compostela, Spain: Xunta de Galacia and IOC. 1998, pp 473–475.
43. RJ Lewis, JP Vernoux, IM Brereton. Structure of Caribbean ciguatoxin isolated from *Caranx latus*. J Am Chem Soc 120:5914–5920, 1998.
44 M-Y Dechraoui, J Naar, S Pauillac, A-M Legrand. Ciguatoxins and brevetoxins, neurotoxic polyether compounds active on sodium channels. Toxicon 37:125–143, 1999.
45. T Yasumoto, R Bagnis, JP Vernoux. Toxicity of surgeonfishes II. Properties of the principle water soluble toxin. Bull Jpn Sci Fish 42:359–365, 1976.
46. M Murata, H Naoki, T Iwashita, S Matsunaga, M Sasaki, A Yokoyama, T. Yasumoto. Structure of maitotoxim. J Am Chem Soc 115:2060–2062, 1993.

47. W Zheng, JA DeMattei, J-P Wu, JJ Duan, LR Cook, H Oinuma, Y Kishi. Complete relative stereochemistry of maitotoxin. J Am Chem Soc 118:7946–7968, 1996.
48. RL Lewis, MJ Holmes, PF Alewood, A Jones. Ionspray mass spectrometry of ciguatoxin-1, maitotoxin-2, and-3 and related marine polyether toxins. Nat Toxins 2:56–63.
49. RJ Lewis, NC Gillespie, CMJ Holmes, JB Burke, AB Keys, AT Fifoot, R Street. Toxicity of lipid-soluble extracts from demersal fishes at Flinders Reef, Southern Queensland. In: JH Choat, D Barnes, MA Borowitzka, JC Coll, PJ Davies, P Flood, BG Hatcher, D Hopley, PA Hutchings, D Kinsey, GR Orme, PF Sale, PA Sammarco, CC Wallace, C Wilkinson, E Wolanski, O Billwood, eds. Proc Sixth International Coral Reef Symposium, Townsville, Australia: Executive Committee Sixth International Coral Reef Symposium 1988, Vol 3, pp. 61–65.
50. F Gusovsky, JW Daly. Maitotoxin: a unique pharmacological tool for research on calcium-dependent mechanisms. Biochem Pharmacol 39:1633–1639, 1990.
51. D Xi, FM Van Dolah, JS Ramsdell. Maitotoxin induces a voltage dependent membrane depolarization in GH4 pituitary cells via activation of type L voltage dependent calcium channels. J Biol Chem 267:25025–25031, 1992.
52. M Estacion, HB Nguyen, JJ Gargus. Calcium is permeable through maitotoxin-activated nonselective cation channel in mouse L cells. Am J Physiol Cell Physiol 39:C1145–C1152, 1996.
53. A Bielfeldackermann, C Range C Korbmacher. Maitotoxin (MTX) activates a nonselective cation channel in *Xenopus laevis* oocytes. Eur J Physiol 436:329–337, 1998.
54. RC Young, M McLaren, JS Ramsdell. Maitotoxin increases voltage independent chloride and sodium currents in GH4Cl rat pituitary cells. Nat Toxins 3:419–427, 1995.
55. M Martinez, C Salvador, JM Farias, L Vaca, LI Escobar. Modulation of a calcium-activated chloride current by maitotoxin. Toxicon 37:359–370, 1999.
56. JW Daly, J Leuders, WL Padgett, Y Shin, F Gusovsky. Maitotoxin induced calcium influx in cultured cells. Effect of calcium channel blockers. Biochem Pharmacol 50:1187–1197, 1995.
57. M Murata, F Gusovsky, M Sasaki, A Yokoyama, T Yasumoto. Effect of maitotoxin analogues on calcium influx and phosphoinositide breakdown in cultured cells. Toxicon 29:1085–1096, 1991.
58. KM Konoki, M Hashimoto, T Nonomura, M Sasaki, M Murata, K Tachibana. Inhibition of maitotoxin-induced Ca^{2+} influx in rat glioma C6 cells by brevetoxins and synthetic fragmants of maitotoxin. J Neurochem 70:409–416, 1998.
59. TR Tosteson. The diversity and origins of ciguatera fish poisoning. Puerto Rico Health Sci J 14: 117–128, 1995.
60. AC Alcala, JS Garth, D Yasumura, T Yasumoto. Human fatality due to injection of the crab *Demania reynaudii* that contained a palytoxin-like toxin. Toxicon 26:105–107, 1988.
61. A Kodama, Y Hokama, T Yasumoto, M Fukui, SJ Manea, N Sutherland. Clinical and laboraotry findings implicating palytoxin as cause of ciguatera poisoning due to *Decapterus macrosoma* (mackerel). Toxicon 27:1051–1053, 1989.
62. M Fukui, M Murata, A Inoue, M Gawel, T Yasumoto. Occurrence of palytoxin in the triggerfish *Melichthys vidua*. Toxicon 25:1121–1124, 1987.
63. Y Onuma, M Satake, T Ukena, J Roux, S Chanteau, N Rasolofonirina, M Ratsimaloto, H Naoki, T Yasumoto. Identification of putative palytoxin as a cause of clupeotoxism. Toxicon 37:55–65, 1999.
64. Y Hirata, D Uenura, Y Yizumi. Chemistry and pharmacology of playtoxin, In: AT Tu. Handbook of Natural Toxins. Vol 3. Marine Toxins and Venoms. New York: Marcel-Dekker, 1988, pp 241–258.
65. RE Moore, G Bartolini, J Barchi, AA Bothner-By, J Dodok, J Ford. Absolute stereochemistry of palytoxin. J Am Chem Soc 104:3776–3779, 1982.
66. E Haberman. Palytoxin acts through the Na+/K+ ATPase. Toxicon 27:1171–1187, 1989.
67. J Redondo, B Fiedler, G Schneider-Bobis. Palytoxin-induced Na+ influx into yeast cells expressing mammalian sodium pump is due to formation of a channel within the enzyme. Mol Pharmacol 49: 49–57, 1996.
68. G Scheinerbobis, H Schneider. Palytoxin-induced channel formation within the Na^+K^+ ATPase does not require a catalytically active enzyme. Eur J Biochem 248:717–723, 1997.
69. G Scheinerbobis. Ion-transporting ATPases as ion channels. Arch Pharmacol 357:477–482, 1998.

70. K Tachibana, PJ Scheuer, Y Tsukitani, H Kikuchi, D Van Engen, J Clardy, Y Gopichand, FJ Schmitz. Okadaic acid, a cytotoxic polyether from two marine sponges of the genus *Halichondria*. J Am Chem Soc 103:2469–2471, 1981.
71. T Yasumoto, Y Oshima, M Yamaguchi. Occurrence of a new type of shellfish poisoning in Japan and chemical properties of the toxin. In: D Taylor, HH Seliger, eds. Toxic Dinoflagellate Blooms. Amsterdam: Elsevier, 1979, pp 495–502.
72. T Yasumoto, Y Oshima, W Sugawara, Y Fukuyo, H Oguri, T Igarishi, N Fujita. Identification of *Dinophysis fortii* as the causative organism of diarrhetic shellfish posioning. Bull Jpn Soc Sci Fish 46:1405–1411, 1980.
73. M Kat. Occurrence of *Prorocentrum* species and coincidental gastrointestinal illness of mussel consumers. In: D Taylor, HH Seliger, eds. Toxic Dinoflagellate Blooms. Amsterdam: Elsevier, 1979, pp 215–220, 1979.
74. M Kat. *Dinophysis acuminata* blooms, the distinct cause of Dutch mussel poisoning. In: DM Anderson, AW White, DG Baden, eds. Toxic Dinoflagellates. Amsterdam: Elsevier, 1985, pp 73–77.
75. M Kumagai, T Yanagi, M Murata, T Yasumoto, M Kat, P Lassus, JA Rodriguez-Vazquez. Okadaic acid as the causative toxin of diarrhetic shellfish poisoning in Europe. Agric Biol Chem 50:2853–2857, 1986.
76. T Hu, J Doyle, D Jackson, J Marr, E Nixon, S Pleasance, MA Quilliam, JA Walter, JLC Wright. Isolation of a new diarrhetic shellfish toxin from Irish mussels. Chem Commun 39–41, 1992.
77. MA Quilliam, M Gilgan, S Pleasance, ASW Defrietas, D Douglas, L Fritz, T Hu, JC Marr, C Smyth, JLC Wright. Confirmation of an incident of diarrhetic shellfish poisoning in eastern Canada. In: TJ Smayda, Y Shimizu, eds. Toxic Phytoplankton Blooms in the Sea. Amsterdam: Elsevier, 1993, pp 547–552.
78. SL Morton, DR Tindall. Determination of okadaic acid content of dinoflagllate cells: a comparison of the HPLC-fluorescent method and two monoclonal antibody ELISA test kits. Toxicon 34:947–954, 1996.
79. C-M Gauss, JE Sheppock, AC Nairn, RA Chamberlin. A molecular modeling analysis of the binding interactions between okadaic acid class of natural product inhibitors and the ser-thr phospharases, PP1 and PP2A. Bioorg Medi Chem 5:1751–1773, 1997.
80. JR Bagu, BD Sykes, MM Craig, CFB Holmes. A molecular basis for different interactions of marine toxins with protein phosphatase-1. Molecular models for bound moruporin, microcystins, okadaic acid and calyculin A. J Biol Chem 272:5078–5097, 1997.
81. Z Zhang, S Zhao, F Long, L Zhang, G Bai, H Shima, M Nagao, EYC Lee. A mutant of protein phosphatase-1 that exhibits altered toxin sensitivity. J Biol Chem 269:16997–17000, 1994.
82. S Wera, B Hemmings. Ser/thr protein phosphatases. Biochem J 311:17–29, 1995.
83. P Cohen, CFB Holmes, Y Tsukitani. Okadaic acid: a new probe for the study of cellular regulation. TIBS 15:98–102, 1990.
84. K Sugunama, H Fujiki, H Sugur, S Yoshizawa, M Hirota, M Nakayasu, M Ojika, K Wakamatsu, K Yamada, T Sigimura. Proc Natl. Acad Sci 85:1768–1771, 1988.
85. H Fujiki, K Sugunama, H Suguri, S Yoshizawa, K Tagai, N Uda, K Wakamatsu, K Yamada, M Murata, T Yasumoto, T Sugimura. Diarrhetic shellfish toxin, dinophysistoxin-1, is a potent tumor promotor on mouse skin. Jpn J Cancer Res 79:1089–1093, 1988.
86. T Yasumoto, M Murata, Y Oshima, G Matsumoto, J Clardy. Diarrhetic shellfish poisoning. In: EP Regalis, ed. Seafood Toxins. Washington, DC: American Chemical Society, 1984, pp 207–214.
87. J Zhou, L Fritz. Okadaic acid localizes to the chloroplasts in DSP-toxin-producing dinoflagellates, *Prorocentrum lima* and *Prorocentrum maculosum*. Phycologia 33:455–461, 1994.
88. MA Quilliam, NW Ross. Analysis of diarrhetic shellfish poisoning toxin and metabolites in plankton and shellfish by ion-spray liquid chromatography-mass spectrometry. In: Biochemistry and Biotechnology of Electrospray Ionization Mass Spectrometry. 1996, pp 351–364.
89. J Needham, T Hu, JL MacLachlan, JA Walter, JLC Wright. Biosynthetic studies of the DSP toxin DTX-4 and an okadaic acid diol ester. J Chem Soc Chem Commun 1623–1624, 1995.
90. AJ Windust, JLC Wright, JL MacLachlin. The effects of the diarrhetic shellfish poisoning toxins, okadaic acid and dinophysistoxin-1, on the growth of microalgae. Mar Biol 126:19–25, 1996.
91. AJ Windust, MA Quilliam, JLC Wright, JL MacLachlin. Comparative toxicity of the diarrhetic

shellfish poisons, okadaic acid, okadaic acid diol-ester and dinophysistoxin-1, to the diatom *Dinophysis weissflogii*. Toxicon 35:1591–1603, 1997.
92. LM Sugg, FM Van Dolah. 1999. No evidence for an allelopathic role of okadaic among ciguatera associated dinoflagellated. J Phycol 35:93–103.
93. MP Boland, MFJR Taylor, CFB Holmes. Identification and characterization of type-1 protein phosphatases from the okadaic acid producingmarine dinoflagellate, *Prorocentrum lima*. FEBS 334:13–17, 1993.
94. MP Boland, CFB Holmes. Identification and characterization of protein phosphatase-1 and -2A in marine dinoflagellates. Adv Prot Phosphatases 8:325–342, 1994.
95. JLC Wright, RK Boyd, ASW Defrietas, M Falk, RA Foxall, WD Jamieson, MV Laycock, AW McCulloch, AG McInnes, P Odense, V Pathak, MA Quilliam, M Ragan, PG Sim, P Thibault, JA Walter, M Gilgan, DJA Richard, D Dewar. Identification of domoic acid, a neuroexcitatory amino acid, in toxic mussels from eastern P.I.E. Can J Chem 67:481–490, 1989.
96. TM Perl, L Bedard, T Kosatsky, JC Hockin, ECD Todd, RS Remic. 1990. An outbreak of toxic encephalopathy caused by eating mussels contaminated with domoic acid. N Engl J Med 322:1775–1780.
97. DV Subba-Rao, MA Quilliam, R Pocklington. 1988. Domoic acid—a neurotoxic amino acid produced from the marine daitom *Nitzschia pungens* in culture. Can J Fish Aquat Sci 45:2076–2079.
98. SS Bates, CJ Bird, ASW Defrietas, R Foxall, M Gilgan, LA Hanic, GR Johnson, AW McCulloch, P Odense, R Pocklington, MA Quilliam, PG Sim, JC Smith, DV Subba Rao, ECD Todd, JA Walter, JLC Wright. Pennate diatom *Nitzschia pungens* as the primary source of domoic acid, a toxin in shellfish from eastern Prince Edward Island, Canada. J Fish Aquat Sci 46:1203–1215, 1989.
99. T Takemoto, K Daigo. Constituents of *Chondria armata* and their pharmacological effects. Chem Pharm Bull 6:657–580, 1958.
100. S Murakami, T Takemoto, Y Shimizu. Studies on the effective principles of *Diagenea simplex* Aq. I. Separation of the effective fraction by liquid chromatography. J Pharm Soc Jpn 73:1026–1028, 1953.
101. JLC Wright, M Falk, AG McInnes, JA Walter. Identification of isodomoic acid D and two new geometrical isomers of domoic acid in toxic mussels. Can J Chem 68:22–25, 1990.
102. JA Walter, M Falk, JLC Wright. Chemistry of the shellfish toxin domoic acid: characterization of related compounds. Can J Chem 72:430–436, 1994.
103. DR Hampson, X Huang, JW Wells, JA Walter, JLC Wright. Interaction of domoic acid and several derivatives of kainic acid and AMPA binding sites in rat brain. Eur J Pharmacol 218:1–8, 1992.
104. D Xi, JS Ramsdell. Glutamate receptors and calcium entry mechanisms for domoic acid in hippocampal neurons. Neuroreport 7:1115–1120, 1996.
105. YG Peng, JS Ramsdell. Brain fos induction is a sensitive biomarker for the lowest observed neuroexcitatory effects of domoic acid. Fund Appl Toxicol 31:162–168, 1996.
106. D Xi, YG Peng, JS Ramsdell. Domoic acid is a potent neurotoxin to neonatal rats. Nat Toxins 5: 74–79, 1997.
107. F Iverson, J Truelove, E Nera, L Tryphonas, J Campbell, E Lok. Domoic acid poisoning and mussel-associated intoxication: preliminary investigations into the response of mice and rats to toxic mussel extract. Food Chem Toxicol 27:377–384, 1989.
108. L Fritz, MA Quilliam, JLC Wright, AM Beale, TM Work. 1992. An outbreak of domoic acid and poisoning attributed to the pennate diatom *Pseudonitzschia australis*. J Phycol 28:438–442.
109. TM Work, BB Barr, AM Beale, L Fritz, MA Quilliam, JLC Wright. Epidemiology of domoic acid poisoning in brown pelicans (*Pelicanus occidentalis* and Brandt's cormorants (*Phalacrocorax pencillatus*) in California. J Zool Wildlife Med 24:54–62, 1993.
110. F Gulland. Unusual mortality event—domoic acid toxicity in California sea lions (*Zalophus californianus*) stranded along the central California coast, May–October 1998. Report to U.S. National Marine Fisheries Service Working Group on Unusual Marine Mammal Mortality Events, 1999.
111. SS Bates. Bloom dynamics and physiology of domoic acid producing *Pseudo-nitzschia* species. In: DM Anderson, AD Cembella, GM Hallegraeff, eds. Physiological Ecology of Harmful Algal Blooms. New York: Springer-Verlag, 1998, pp 267–292.
112. AD Cembella. Ecophysiology and mechanism of paralytic shellfish toxins in microalgae. In: DM

Anderson, AD Cembella, GM Hallegraeff, eds. Physiological Ecology of Harmful Algal Blooms. New York: Springer-Verlag, 1998, pp 381–404.
113. Y Pan, SS Bates, AD Cembella. Environmental stress and domoic acid production by *Pseudo-nitzschia*: a physiological perspective. Nat Toxins 6:1–9, 1998.
114. MV Laycock, ASW deFrietas, JLC Wright. Glutamate agonists from marine algae. J. Appl Phycol 1:113–122, 1989.
115. WW Carmichael. The cyanotoxins. Adv Bot Res 27:1997.
116. DM Anderson. Red tides. Sci Am 271:52–58, 1993.
117. T Yasumoto. Fish poisoning due to toxins of microalgal origins in the Pacific. Toxicon 36:1515–1518, 1998.
118. JJ Ikawa. Comparison of the toxins of the blue-green alga *Aphanizomenon flos-aquae* and the *Gonyaulax* toxins. Toxicon 20:747–752, 1982.
119. AR Humpage, J Rositano, AH Bretag, R Brown, PD Baker, BC Nicholson, DA Steffensen. Paralytic shellfish poisons from Australian cyanobacterial blooms. Aust J Mar Freshwat Res 45:31–41, 1994.
120. WW Carmichael, WR Evans, QQ Yin, P Bell, E Moczydlowski. Evidence for paralytic shellfish poisons in the freshwater cyanobacterium *Lyngbia wollei* (Farlow ex Gomont) comb. Nov. Appl Env Microbiol 63:3104–3110, 1997.
121. Q Yin, WW Carmichael, WR Evans. Factors influencing growth and toxin production by cultures of the freshwater cyanobacterium *Lyngbia wollei* Farlow ex Gomont. J Appl Phycol 9:55–63, 1997.
122. H Onodera, M Satake, Y Oshima, T Yasumoto, WW Carmichael. New saxitoxin analogues from the freshwater filamentous cyanobacterium *Lyngbia wollei* Nat Toxins 5:146–151, 1997.
123. NA Mahmood, WW Carmichael. Anatoxin-a(s), an anticholinesterase from the cyanobacterium *Anabaena flos-aquae* NRC-525-17. Toxicon 25:1221–1227, 1987.
124. KL Rinehart, M Namikoshi, BW Choi. Structure and biosynthesis of toxins from blue-green algae. J Appl Phycol 6:159–176, 1994.
125. AR Arment, WW Carmichael. Evidence that microcystin is a thiotemplate product. J Phycol 32: 591–597, 1996.
126. M Craig, HA Luu, TL McCready, D Williams, RJ Andersen, CFB Holmes. Molecular mechanisms underlying the interaction of motuporin and microcystins with type-1 and type 2A protein phosphatases. Biochem Cell Biol 74:569–578, 1996.
127. EM Jochimson, WW Carmichael, JS An, DM Cardo, ST Cookson, CEM Holmes, MBD Autines, DA Demelo, TM Lyra, VST Barreto, SMFO Azevedo, WR Jarvis. Liver failure and death after exposure to microcystins at a hemodialysis center in Brazil. N Engl J Med 338:873–878, 1998.
128. DE Williams, M Craig, SC Dawe, ML Kent, RJ Andersen, CFB Holmes. 14C-labeled microcystin-LR administered to Atlantic salmon via intraperitoneal injection provides evidence for covalent binding of microcylstin-LR in salmon livers. Toxicon 35:985–989, 1997.
129. C Stephen, ML Kebt, SC Dawe. Hepatic megalocytosis in wild and farmed chinook salmon Oncorhynchus tshawtscha in British Columbia, Canada. Dis Aquat Org 16:35–39, 1993.
130. R Nishiwaki-Matsushima, T Ohta, S Nishiwaki, M Suganuma, K Kohyama, T Ishikawa, WW Carmichael, H Fujiki. Liver tumor promotion in the cyanobacterial cyclic peptide toxin microcystin-LR. J Cancer Res Clin Oncol 118:420–424, 1992.
131. T Ohta, E Sueoka, S-J Kim, WW Carmichael, H Fujiki. Nodularin, a potent inhibitor of protein phosphatases 1 and 2A, is a new environmental carcinogen in male F344 rat liver. Cancer Res 54: 6402–6406, 1994.
132. S-H Yu. Drinking water and primary liver cancer. In: ZY Tang, MC Wu, SS Xia, eds. Primary Liver Cancer. Berlin: Springer-Verlag, 1989, pp 30–37.

3
Nonneurotoxic Toxins

Juan Jesús Gestal-Otero
University of Santiago de Compostela and University Hospital of Santiago de Compostela, Santiago de Compostela, Spain

I. INTRODUCTION

Since ancient times there have been references to the awareness of existing toxins related to fish and shellfish consumption.

The first Egyptian plague (''all the water that was in the river turned to blood and the fish that were in the river died; and the river stank and the Egyptians could not drink of the water of the river; and there was blood throughout all the land of Egypt'' [1]) could very well have been due to a toxic red tide.

The first Chinese pharmacopoeia (2800 BC) warns against and makes recommendations on balloonfish consumption (2). Nevertheless, we can assume that humankind knew of the red tide dangers before the written word, given the discovery of the 26 million-year-old fossil *Gonyaulax poliedra.*

Even in pre-Colombian America, the hazards of eating shellfish extracted from the sea when it presented a red color during the day or a glow during the night were already known. To avert this danger, watchmen were placed in affected spots who would alert travellers of the such hazards. According to Halstead (3), it is the first known health quarantine in the North America.

The first scientific reference to human shellfish poisoning is probably the one in 'Ephémérides des curieux de la nature' (1689), quoted in 1851 by Chevalier et al. (4–5). However, the first written report of an outbreak in British Columbia, occurred in 1793, was reported by Vancouver in 1801 (6).

In Europe, there are scientific descriptions of paralytic shellfish poisoning (PSP) outbreaks dating back to 1689 (4–5). It is worth pointing out the Wilhelmshavem case (1885) which sparked off the scientific solving of mussel poisoning, although the precise connection of the poisoning to mussels would not be established until 1927 when an outbreak took place in the central California coast. The ensuing studies led Meyer and Sommer (1937) (7–8) to the discovery that the cause was dinoflagellates and their toxins and to the beginning of in-depth studies. Halstead (3) compiled all the cases published world wide up to 1965.

All these episodes refer to poisoning by paralyzing toxins. Awareness of the actual toxins that cause the diarrheic events is more recent. The first known diarrheic poisoning event associated with toxic-mussel consumption (described at the time as ''mussels that had ingested dinoflagellates'') took place in the Easterscheldt area in the Netherlands in 1961 (9). That same year there were other cases recorded in Waddensea. The next outbreak in Easterscheldt occurred in

1971 and affected 100 people. There are also references to the ingestion of blue mussels in Scandinavia (Norway) in 1968 (10). In the following years, several cases were reported in the Oslofjord area. They were classified as "unidentified mussel poisoning."

In October 1976, in the Netherlands, 25 people were taken ill after eating mussels from Waddensea (11). That year Yasumoto et al. described diarrheic shellfish poisoning (DSP) for the first time in a food poisoning outbreak due to the ingestion of mussels and scallops that occurred in northeastern Japan (12).

In 1976, we treated the first cases of paralytic shellfish poisoning detected in Galicia (Spain) (13), and in the summer of 1978, there were a series of diarrheic events in the Ría de Ares area. In the town of Lorbé (Oleiros, close to the city of La Coruña), we studied episodes related to mussel consumption. We ruled out microbiological causes and attributed those episodes to an unknown toxin caught by the mussels in their growing zone, as had happened with PSP in previous years.

In the following summers, we studied similar epidemiological outbreaks that, after an incubation period of a few hours presented as diarrhea, nausea, vomiting, and abdominal pain without a high temperature. This was associated with the consumption of steamed mussels. Patients recovered in 2 or 3 days. We contacted Dr. M. Kat and implemented the test she was developing in rats. The test gave inadequate results. This was probably due to methodological problems.

In 1981, the greatest diarrheic poisoning associated with mussel consumption occurred, affecting around 5000 people all over Spain, mostly in Madrid. Once again, the epidemiological analysis associated diarrhea with mussels. By then Kat's (11) and Yasumoto's (12,14–15) work was already known. These poisoning cases by the DSP toxin in Galicia as well as in the rest of Spain run parallel to those that occurred in Japan in 1976 and 1977 (12,14,16) and before and after in many other European, Asian, and American countries.

Yasumoto et al. (12,14) reported events in Japan and related them to eating mussels contaminated with the DSP toxin, pointing out that it was a lipophilic toxin and referred to it as diarrheic shellfish poison (DSP) in later works (15).

II. EPIDEMIOLOGY OF DSP

Epidemiological studies the frequency and distribution of health problems (descriptive epidemiology) as well as causes and factors influencing its onset (analytical epidemiology). It is also a necessary tool for identifying health problems and the need for health assessment and interventions (health planning).

Epidemiology is used by health professionals as a means for epidemiological surveying, causal research, and outbreak investigation, as well as for assessing interventions; all subject to the main aim of epidemiology, which is prevention. It can be stated that epidemiology is like the eyes that allow us to see ahead problems at their early stage and thus prevent them.

Epidemiological studies can be descriptive or analytical. The former can establish the frequency and distribution of health problems in the community and produce ethiological hypotheses, whereas analytical and experimental studies can check such hypotheses.

Depending on the variables of range, people, time, and place, the description of a health problem in the community allows us to assess its economic relevance as well as its harm to human health (morbidity and mortality), and to know the features of people at risk if exposed (exposed population), as well as the features of those running a higher risk of contracting the disease when exposed (vulnerable or threatened population), or the features of those at risk of suffering serious disease or even death (high-risk population).

Sometimes descriptive studies aim at updating our knowledge of a known health problem and determining whether there have been changes in relation to previous studies. In the face of health problems where the causes, the risk factors, or favorable factors are not fully known, these studies aim at creating hypotheses that later can be confirmed or ruled out by analytical or experimental studies.

There are several kinds of descriptive studies: clinical case series; descriptive studies of morbidity and mortality; time correlation; ecological correlation, proportional mortality studies; and cross-sectional or prevalence studies.

Ecological correlations are not applicable to the subject of our concern, since they use group-based observations (e.g., census units such as municipality, province, country) rather than individual-based ones, and proportional mortality studies, which are very useful in workplace epidemiology.

At times, clinical case series are the initial approach to a previously unknown problem, a new disease, or a new clinical syndrome. When the appearance of various similar cases is observed and we become aware that we face a new health problem, its etiology, diagnosis, and treatment must be established. Etiological research always begins with a descriptive study of the cases (individual, time, and place features) and observation of certain features which present a higher rate in the case series than in the general population from which the initial etiological hypotheses came.

Morbidity and mortality studies are the more traditional descriptive studies. They are carried out from data gathered systematically from several official sources, which means that their cost and the time used are low, but, on the other hand, the quality of the information is also low and hard to contrast. These data usually do not contain all the information that might be of interest. They study the frequency and distribution of health problems in the population according to individual variables (demographic, family, endogenous, and/or hereditary features, ontogenetic defects and habits); place variables, the kind of administration divisions they are (country, province, municipality): geographical variables (e.g., town/country, sea/inland, valley/mountain); and periodical and nonperiodical time variables (circadian, weekly, seasonal, yearly, and secular).

These studies allow us to know the differences in the health rate of various geographical areas and social groups and the disease trends. They are useful for assessing the effectiveness of intervention programs (time series studies). Time correlation studies are also very useful for generating etiological hypotheses.

Finally, we have prevalence or cross-sectional studies which, unlike all the others, use data gathered systematically. Data are specifically gathered for the study, which means that they are more costly and they require longer periods. But the relevant benefit is that the information generated is of a higher quality and allows all the information that is relevant to be gathered.

Analytical studies can be experimental or observational. For obvious reasons, in epidemiology, experimental studies are only possible on rare occasions. In these kinds of studies, the investigator can decide what individuals to include or exclude in the intervention group, although this allocation can be random in order to control the confusion variables. When they are actually carried out, they will always be therapy interventions (clinical essays) or prevention interventions (community studies whether randomized or not).

Observation analytical studies are preferred in epidemiology. They can be case-control or cohort studies. Case-control studies are retrospective; they go from the onset of the effects to the search for the exposure antecedent and to the causal factors. On the other hand, cohort studies are retrospective, going from the exposure to the causal effects and to the onset of the effect.

The epidemiology of human disease caused by harmful marine phytoplankton is still at an early stage. This lack of progress in phycotoxin disease epidemiology is attributed to a lack of disease biomarkers and exposure in humans. Epidemiological studies are limited to the mere description of clinically identified cases and little else. More recently, they have included laboratory testing of ingested food.

The lack of biomarkers is a hindrance to discovery of the real incidence, since it is only possible to confirm clinical cases in their acute stage, and as long as there are some food remains available or their origin is known and a sample can be obtained. Asymptomatic cases and those where this possible cause is not considered go undiagnosed.

Biomarkers that measure exposure and effect may be qualitative or quantitative. In order to be useful, they should be detected early in human biological fluids accessible and acceptable. Ideally, biomarkers should also allow for the identification of subclinical cases. Other considerations to be borne in mind are the speed in testing, precocity in its application, and price.

In order to have the exposure markers available, it is necessary to develop the toxicological analysis of toxin levels and their metabolites in body fluids. At present, effect markers are based on the clinical picture. It is important to develop markers for subclinical physiological changes.

A. Relevance

DSP is widely spread in the world, affecting particularly Japan and northwestern Europe. It is a serious economic problem for the shellfish industry and public health.

Socioeconomic Relevance

The problem of toxic bivalve shellfish (mussels, clams, cockles, scallops) affects not only public health in affected areas but also the tourist industry (exports, markets, advertising, negative publicity) and therefore can cause economic upheaval.

This problem has no easy solution, since the procedures used in the purifying plants is excellent for eliminating potential microbiological contamination, but it has no effect on the biotoxin content.

DSP has some relevant economic repercussions which are the object of study in Chapter 34. Galicia (northwestern region of Spain, comprising the land between the north of Portugal and the Bay of Biscay) produces 97% of mussels in Spain. In 1997, it produced 220,000 tons; worth around 15,000 million pesetas (US $110 million). It is the first producer in Europe, before France and Italy, and the second world producer after China, (400,000 tons), but unlike China, which barely covers its home market, 80% of Galicia mussel production is exported to countries outside the European Union, and 25% is exported to third countries. It also represents 21% of the fish production (fresh fish) first sale.

The presence of toxic red tides means having to close down the shellfish fisheries in the affected areas and having to endure during long periods a situation of economic hazard for a great number of families who directly or indirectly depend on these fishing trades. Moreover, if control failures occur and there are cases of human poisoning, the discredit and the mistrust created can lead to the loss of markets that become hard to recover.

From the beginning of the 1980s, and particularly in the second half of the decade, a progressive increase of poisoning episodes of phytoplankton origin in bivalve shellfish was observed in Galicia. Mussel extraction was prohibited for up to 200 days/year in some areas.

Health Relevance

DSP is relevant from a health viewpoint not only because of its acute effects but also because of its potential chronic effects, which are not yet fully understood.

Regarding acute effects, the gastroenteritis caused by okadaic acid and dinophysistoxins has a favorable evolution toward total recovery in 1 to 3 days, and no fatalities have been described.

Regarding chronic effects, okadaic acid (OA) and dinophysistoxin-1 have been shown to be potent tumor promoters, and given that the stomach, small intestine, and colon have binding sites of OA, this could be implicated in the growth of gastrointestinal tumors (16–17). Mutagenetic (18) and immunotoxic effects due to a marked suppression of interleukin-1 (IL-1) production have also been described (19).

It has been shown in experimental animals that pectenotoxin (PTX1) is hepatotoxic and induces rapid necrosis of hepatocytes, with a pathological action similar to that of phalloidin. In rats, intraperitonially injected with PTXs, the liver finally appears granulated and the hepatocytes contain many vacuoles.

Yessotoxins have been shown to cause heart damage. Almost all cardiac muscle cells of mice inoculated with these toxins were swollen. On the other hand, yessotoxins do not cause damage in the liver, pancreas, lungs, kidneys, and adrenergic glands.

All these chronic effects need to be studied in depth, and they underline the relevance and dimension of the problem and the need to avoid ingestion of these toxins.

B. Frequency and Distribution of DSP

The real incidence of human DSP is hard to assess, since its clinical symptoms can be mistaken for diarrhea from other causes, and it can go unrecorded owing to its benign evolution. Isolated cases usually go undetected, and they only become known in countries where outbreaks have to be reported by law.

At present, cases of human DSP should not occur. The fact that they do reflects a considerable failure in the processes of vigilence and prevention that should be operating, and to which we will refer later.

Research on outbreaks of DSP in humans has only contributed to prevention, determination of their origin, and the potential exposure of other groups of people to shellfish of the same origin. The main interest regarding prevention is watch early detection of toxic episodes in the sea that would allow the adoption of measures (forbidding shellfish extraction in the affected areas and informing people against its consumption). In countries where this surveillance network based on early detection is not possible, the surveillance based on early detection of human intoxication can be useful in detecting the problem and taking preventive measures that can avoid its spread.

Incidence

As pointed out before, the lack of biomarkers does not allow us to know the real incidence of DSP in humans, since our awareness of it is limited to notified outbreaks.

In the last 20 years, the incidence of DSP events in humans has decreased and practically disappeared in developed countries where, as an answer to the problem, watch networks have been developed to detect the presence of toxic plankton species and poisoning in shellfish. This has not been the case in countries where a watch network is not available. Nevertheless, poisoning episodes in the sea have increased, and this is partly due to a higher awareness of the disease and to setting watching schemes but also to its spreading to new, and sometimes far apart, geographical zones, aided by international trade of seafood and the chance that cysts of exotic plankton species, producers of toxins, travel with them and settle in these new zones where they were previously unknown. Thus, the phycotoxic hazard becomes a public health problem worldwide.

Although shellfish poisoning is widely spread all over the world, affecting warm and tropical zones, Europe and Japan are the most affected areas. Some reports show that DSP events have also taken place in other parts of the world (Australia, New Zealand, Indonesia) (20).

Outbreaks in Europe

Various DSP outbreaks have been reported in France during the 1980s, and they affected large numbers of people. In the Loire-Atlantique district and in Normandy, 3300 and 150 cases were detected, respectively, in 1983, and 70 and 2000, respectively, cases in 1984. Other DSP events were reported in 1985 (a few cases) and 1987 (2000 cases) (21). In 1990, there was an outbreak affecting 415 people due to mussels imported from the north Danish coast. It had a toxic load of OA: 170 μg per 100 g of meat (22). Since the watch network was set in 1984, an increase has been observed in the toxic tide frequency as well as its spreading to other previously uncontaminated areas.

The first outbreak reported in Norway due to a *Dinophysis* spp. toxin (23) took place during a long contamination period (October 1984 to April 1985) during which around 400 cases were detected among people living on the southwest coast of Norway. Coinciding with this outbreak, another one took place in October 1984 on the west coast of Sweden, where DSP events had already been taking place since 1983, affecting about 100 people (24). In 1986–1987, a monitoring program for DSP toxins was established in Norway.

Since 1990, when the watch network was set up in Denmark, mussels infected with DSP have been detected during many summers.

In Spain, the first cases were detected, as we said before, in 1978 in the Ares Estuary (25). New events also took place in the following years, the main one occurring in 1981 with 5000 cases. Since then DSP has been regularly detected in the sea water.

The first DSP event in Italy was in 1989, where affected cases were found on the north and northwest Adriatic coasts.

In Germany, in September 1978, single cases of DSP intoxication were reported in the Husum area. In November 1986, at least eight people were affected. Since 1986 DSP has regularly been detected on the coast of German Bight, but not large outbreaks of DSP have been described (21), as is the case in Portugal and Ireland.

In Portugal, since 1987, when they were spotted for the first time, DSP toxins have been detected regularly in bivalve shellfish from the northern coast including the Aveiro Estuary and the Mondego Estuary.

In Ireland, since the watch program was set up, DSP has been detected in shellfish samples almost every year: The strictness and the duration of closure of shell fisheries varies according to years. The events are recorded in the summer and autumn months (June to December).

Outbreaks in Asia

The first episodes of DSP in Asia due to the ingestion of toxic blue mussels and scallops were those mentioned above. They took place in northeastern Japan in 1976 and 1977 (12,14) and affected 164 people.

New outbreaks were recorded later. Between 1976 and 1984, Kawabata reports that there were 34 outbreaks affecting 1257 people (26). The Japanese and European shores are the most affected by toxic blooms.

Toxic red tides have also been described on the Russian east coast (*Dinophysis acuminata*, *D. acuta*, *D. fortii* and *D. norvergica*) (27) and the presence of DSP toxins in shellfish from India (28), although human cases have not been reported.

Outbreaks in America

The first reported DSP event in North America was in 1990 in Nova Scotia off the eastern coast of Canada and was due to DTX-1. It affected 16 people (29). Other on events were reported later in the same region (30). In 1989, a red tide was detected on Long Island New York (with a high number of *D. acuminata*). It was of low toxicity in shellfish (0.5 MU) and no human cases were reported (31).

In January 1991, an outbreak affecting 120 people was detected in Chile. *D. acuta* was identified as the responsible toxin (32). In January of 1992, DSP toxins were also detected on the coast of Uruguay (33).

C. Epidemiology Chain

In order to systematize an epidemiological study oriented toward the prevention of DSP, we will follow the epidemiological chain pattern developed in 1931 by Stallybarss, the great scholar of epidemiology.

The reservoir are the dinoflagellates causing the DSP; the poisoning source and transmission mechanism are the bivalve shellfish with toxins—mainly mussels, although sometimes scallops are also involved. The subjects at risk are people eating them, who usually live on the coast, have a low education, and are consumers of shellfish; moreover, they pick the shellfish themselves directly from the shore, thus bypassing any health control, which increases the risk.

Causal agent

The DSP group comprised three toxin groups: okadaic acid and dinophysotoxins (DTX); pectenotoxins (polyether lactone), and yessotoxins (toxins with sulfate groups).

The first DSP group toxin was isolated from mussel digestive glands and was called dinophysistoxin-1 (DTX-1) (34). Observation by spectral comparison showed that it was 35-R-methyl okadaic acid. Later other okadaic acid derivatives were identified (okadaic acid had been isolated for the first time from the *Halichondria okadai* sponge in 1981) (35); later it was also found in the *Prorocentrum lima* and *Dinophysis* spp. dinoflagellates, the DTX-3 (7-O-acyl-35-[R]-methylokadaic acid) in an intoxication by scallops in north-eastern Japan (36), and the DTX-2 (31, demethyl-35-methylokadaic acid) in Irish mussels (37).

Pectenotoxin-1 (PTX-1), a polyether lactone, was isolated from the digestive gland of the scallop *Patinopecten yessoensis* in northeastern Japan (36); later three homologous ones were described (PTX-2, PTX-3, and PTX6) (38).

The yessotoxin was isolated from scallops, and it turned out to be a brevetoxin-type polyether (39).

Of all these toxins, only okadaic acid and its derivatives (DTX) cause acute gastrointestinal toxicity. OA and DTX have been shown to be powerful inhibitors of phosphatase PP1 and PP2A activity; they are two of the main cytosol phosphatases in mammal cells, with the subsequent increase in phosphorylated proteins (40). Okadaic acid probably causes diarrhea by stimulating phosphorylation of proteins that control sodium secretion in intestinal cells (41–43).

Pectenotoxins and yessotoxins, although they do not present diarrheic symptoms, owing to their common lipid soluble properties, are included in the DSP toxins complex, and are subjected to the same regulations as OA and its derivatives.

Okadaic acid is the predominant toxin in most European countries, although DTX-2 has been reported in Ireland (37,44), Spain (45), and Portugal.

Reservoir and Intoxication Source

Dinophysis dinoflagellates in 1989 and later *Prorocentrum* dinoflagellates were identified as the organisms responsible for producing the DSP toxin. In western Europe, the predominant types are usually *Dinophysis* spp., whereas *Prorocentrum* spp. are more often found in Japan.

The *D. acuminata* and *D. acuta* species are the most widely spread in European waters (46–48). *D. acuminata* is the main component in the greatest algal blooms on the northwestern shores in France. In Italy, *D. fortii* has also been identified as toxic agents (49), as have *D. norvergica* (50) together with *D. acuminata*, *D. acuta*, and *P. micans* in Norway.

The presence of these dinoflagellates, even in small concentrations (hundreds of cells per liter), can lead to poisoning of shellfish. The ingestion of them, or of fish that have previously fed on small herbivorous fish that have themselves fed on toxic algae, causes poisoning in humans.

In the Galician estuaries of Spain, episodes of DSP appear to be associated mainly with a proliferation of *D. acuminata* in Ría de Ares and *D. acuminata* and *D. acuta* in Rías Bajas, although, at times, other species such as *D. caudata, D. tripos*, and *D. rotundata* can contribute significantly to the level of diarrheic toxins detected.

The source of infection as well as the transmission mechanism are bivalve shellfish, mainly mussels and scallops.

Transmission Mechanism

Bivalve shellfish, mainly mussels and less frequently scallops, are the DSP toxin vectors. They acquire the toxin when there are some species of toxic dinoflagellates in the plankton on which they feed.

Ninety-five percent of the toxin accumulates in the hepatopancreas of the mussel without it suffering any chemical changes, and apparently the toxin does not alter the mussel's physiological functioning or organoleptic properties either. The amount of toxin retained depends not only on the number of dinoflagellates present in the medium and its toxic load, but also on the amount of water filtered by the shellfish.

The ordinary cooking process, either in the home or industrial setting, does not destroy the DSP toxin. Given that cooking is the first stage in the industrial processing of bivalve shellfish, the application of detoxification is focused on it.

Vieites and Leira (51) have attempted to lower the toxicity of contaminated shellfish by cooking it 2–5 mins at 97°C in a slightly alkaline medium (pH 8.22) and adding bicarbonate salts at 2% as a technological coadjuvant. The obtained detoxification percentages (of okadaic acid) that range between 24 and 79% with residual levels of okadaic acid in all samples. They tried increasing the cooking time, but with no better results when the increase was moderate—10–15 mins. These investigators were successful in some samples with very long cooking times—over 60 mins, but this was incompatible with maintaining the optimal market quality of the seafood. Results were no better when the sodium bicarbonate concentration was increased in mid cooking. Vieites and Leira suggested the possibility of treatment by autoclave sterilization (110–120°C), as used in canning, for better results.

We have attempted various manipulations to reduce toxicity—cooking, freezing, canning (sousing)—of PSP-contaminated shellfish, but we have been only partly successful. We have reduced toxicity to a half, a third or a fourth with processes that alter the organoleptic properties and renders shellfish not apt for commercialization. These processes are a combination of a pressure of 1.5 to 2 atmospheres and a temperature of 113°C or higher (13). Many other researchers have attempted cooking treatments to destroying these toxins but without success (52,53).

To date there is no effective method for eliminating toxins from mussels and other bivalves in an economical and fast way that can be commercially profitable and free of hazards. The only solution is self-purifying or natural detoxification through metabolizing the toxins. Whereas toxin accumulation in shellfish may only require a few days, its elimination requires several weeks and on occasions months. The detoxifying process depends on various factors sometimes related to the shellfish and sometimes to its environment.

Eliminating toxins follows exactly the reverse physiological process. The toxins accumulated, mainly in the hepatopancreas and other parts of the shellfish, have to be catabolized or expelled. The more active the metabolism is and the better the physiological condition of the shellfish is, the faster elimination to proceed.

The temperature and food available can be considered the two main external factors which determine the detoxification speed. In warm or tepid waters, mussel filtering is more active, and therefore the ingestion of new food and the elimination of old remains will proceed much faster. The availability of abundant nontoxic food is a sine qua non condition for the shellfish to return to the preintoxication condition.

Very often toxic episodes take place during the autumn, but winter may arrive without the toxin having been eliminated fully from the shellfish. Water temperature is at its lowest in winter, and then phytoplankton concentrations are very low owing to a combination of excessive turbulence and low light intensity. In such circumstances, mussels may maintain low toxicity levels (possibly below the limit restricting consumption) during these months. It is even possible that the toxicity will never disappear completely before a new toxic episode takes place the following summer or autumn. This circumstance has been observed in Scandinavian mussels which have a high DSP toxin level during the autumn and have not been able to eliminate these toxins before exposure to the freezing northern waters during the following winter and spring.

Vulnerability Factors

Personal features: People most at risk are those living on the seashore of countries with underdeveloped monitoring systems, with no watch network for sea toxicity events, and a low level of health education. Traditionally, they are consumers of shellfish, which they usually pick themselves from the sea without it undergoing any health control. There are no differences in sex and age.

Time distribution: Toxic episodes generally appear in summer and autumn months, although occasionally they appear earlier—end of winter or early spring.

In The Netherlands in the years 1981, 1986, 1987, and 1989 events were recorded during September and October and one year even in December. In the Galician Estuaries (Spain), *D. acuminata* is present practically all year around. It proliferates usually in April, although some years this happens in late February or March, presenting maximum or minimum levels until mid or late autumn. *D. acuta* usually appears associated with southern winds from September to November due to the advection from towns in the surrounding coastal area, and *D. caudata* is usually found isolated in the plankton.

Space distribution: DSP is widely spread around the world. Europe's western shores and Japan's shores are affected the most.

D. Factors Associated with DSP

Studies aimed at establishing epidemiological associations between various factors and an intoxication can be carried out at two levels. The first comprises the study of the conditions that

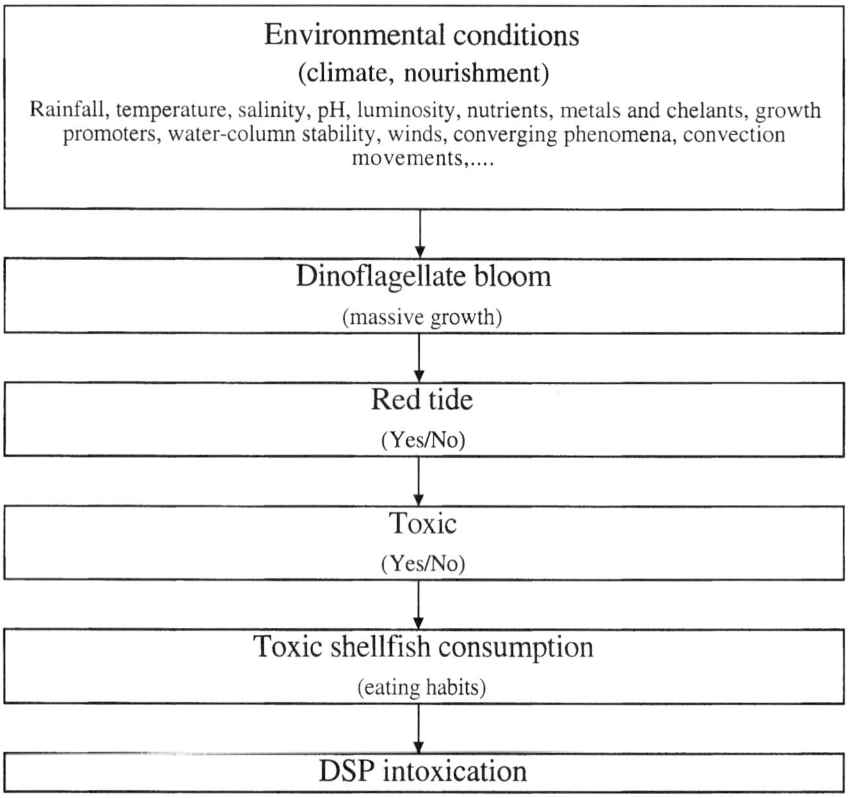

Figure 1 Factors associated with DSP intoxication.

cause the explosive growth of dinoflagellates, resulting in the production of toxins and their concentration in shellfish. The second level studies factors causing the DSP event in humans (Figure 1).

Knowing the environmental and feeding factors associated with the space distribution and occurrence of poisoning in shellfish allows us to establish risk zones of varying degrees, which together with knowing the seasonal events or cyclical phenomena will contribute to DSP prevention in humans.

The retrospective study of previous outbreaks and researching the possibility of future ones allows us to define risk groups and event-associated factors useful for monitoring and prevention toxic events.

Factors Influencing the Explosive Growth of Dinoflagellates

Red tide is a natural occurrence that consists of the massive proliferation of unicellular organisms present in phytoplankton, which presents natural growing and decreasing cycles regulated by the chemical and physical conditions of water, such as a mild temperature, a drop in the water salt content, still waters, light (long days), and concentrations of some organic and inorganic substances (nutrients), as well as biological interactions. Sometimes, and under favorable environmental conditions, some of the organisms (dinoflagellates) causing red tides multiply suddenly and cause an explosive growth that gives the well-known red coloring to water, and when they are toxic species, they cause shellfish poisoning.

Dinoflagellates reproduce themselves by partition simple or multiple, and after an intense breeding activity, they encyst and settle for long periods in the sea bottom.

The beginning and development of red tides, as well as their later disappearance, depends on the interaction of multiple biological, biochemical, hydrographic, and weather factors which are not yet fully known.

In some regions, red tides are frequent and they have become yearly events, whereas in others, they appear irregularly or occasionally.

The appearance of red tides is due to two sets of factors. Those that favor the growth of the microalgal population and those that favor their concentration.

1. *Factors influencing the increase of the microalgal population*:
   —Water that is enriched with nutrients owing to the sea-bottom waters rising to the surface, to land drainage by rainfall, and to urban and industrial waste dumping.
   —Changes in water temperature influences the encysting (temperature decrease) or excysting (temperature rise) of dinoflagellates.
   —Sunlight: essential for conditioning photosynthetic vegetative processes.
   —Salinity: its decrease, caused by freshwater flowing into the sea from rivers or by heavy rainfall, favors the presence of red tides.
   —Organic substance: comes from land drainage or marine plants, fish, and other decomposing dead matter.
   —Metals and chelants: metals decrease growth while chelants increase it.
   —Substances promoting growth (vitamin B_{12}).
   —Calm seas: favor growth and the accumulation of phytoplankton.
2. *Factors influencing concentration*: Dinoflagellates gather in clusters due to hydrological concentration phenomena. These accumulation processes are influenced by mild winds that blow the surface waters toward the shore; converging phenomena cause the dinoflagellate concentration along the front line of two masses of water of different density. The convection movements caused by the wind facilitate the dinoflagellate concentration on the converging lines.

When environmental conditions are not favorable, the haploid vegetative cells of very many dinoflagellate species form cysts that can resist very adverse conditions and remain viable in the sediment for long periods of over 15 years.

The fact that cysts are so resistant means that they can be carried viable from one zone to another where they can develop their mobile stage, going through an adverse medium in their journey.

The cyst germination is conditioned by internal factors and by external, or triggering, factors (temperature, light, and oxygenation).

Dinoflagellates have maturing and latency periods. When cysts germinate and mobile cells emerge, their survival depends on their capacity to free themselves from the sediment and get into the water column (this is favored by turbulence, which is then harmful to the dinoflagellate population development that needs still waters to avoid dispersion) and the chance of finding a favorable medium in the water column.

The relevance of the cyst population depends on their abundance in the sediment, capacity to germinate, and survival of the emerging mobile cells.

Bivalve shellfish filter water and retain the phytoplankton, which is not toxic to them. Mussels concentrate their DSP toxins in the hepatopancreas, whereas clams do it in their siphons. The degree of poisoning acquired by shellfish will depend on the toxicity per cell of the toxic phytoplankton organisms.

Toxins are secondary metabolites produced by toxic phytoplankton whose physiological and ecological functions are unknown. The amount and rate relative of toxins produced depends on intrinsic factors of the cells (genotype, age, size, cell cycle moment, general physiological state) and on environmental factors (temperature, salinity, pH, light, available nutrients, and relative rate between the various nutrients). Therefore, the toxin content per cell varies and, hence, the difficulty in answering the question: from how many cells per liter can mussels become toxic?

III. PREVENTION

In order to systematize the prevention study, we first analyze primary prevention measures based on the sea watch and the market and aimed at effectively foreseeing the toxicity phenomena before it reaches the human intoxication stage. For this a greater knowledge of the potential dinoflagellate life cycle and influencing environmental factors is necessary. We also include here the watch for human poisoning cases to be carried out in countries where there is the possibility of setting up watching networks on the sea or in the market.

Second, we study the secondary prevention measures or measures to be adopted in a human poisoning outbreak.

Both sets of measures are fundamentally based on epidemiological watching. The weakest links in the epidemiological chain have to be identified in order to focus our attention on them.

It is not possible to act on the reservoir, thus avoiding the proliferation of dinoflagellates by modifying factors favorable to them. We are equally powerless in avoiding the accumulation of toxins in shellfish or even in accelerating their detoxification. We can only act on the third chain link—the subjects at risk, by informing them of the hazards and providing all the means at our disposal to avoid toxic shellfish consumption. As we stated before, dinoflagellates do not undergo organoleptic changes that would alert the consumer, and ordinary cooking does not destroy toxins either.

In the epidemiological watch program, it is essential to have tracers for intoxication prevention. In the case of DSP, the most widely known tracer is visualizing a red tide, which, albeit its limitations, prevents the disease outbreak. Nevertheless, it is of very limited efficacy, because events can take place without previous appearance of a red tide. The most efficient watch consists of determining changes in seawater warning of a potential proliferation of toxic plankton species, watching out for these species as well as the presence of toxins in shellfish.

In short, epidemiological watching is based on the early detection of a problem (the presence of the DSP toxins in shellfish). Each watch program or scheme must be adapted to the area, region, or country where it is going to be applied, bearing in mind the following:

Incidence of the problem and its effects on the country population and economy
Legal infrastructure
Available resources (human and material)
Existing means of communication

A. Marine Biotoxins Monitoring Program

The potential monitoring of blooms requires knowledge of their ecological features. If they show signs that their increase might be due to modifiable factors (such as organic material contribution), the measures to be adopted would not be simple and would have to consider the whole fishery area as a whole.

With the data available nowadays, it seems that most blooms are controlled according to the hydrographical features of the shores, and therefore they cannot be modified, although a better understanding of them and the way they relate to phytoplankton would imply a higher capacity to predict potential events.

Given that at present predicting the exact appearance of blooms is not possible, prevention is based on setting a red tide warning network and a watch scheme in the purifying plant areas and the market in order to determine the absence or presence of DSP, at levels below those established by law, in purified shellfish for consumption.

Self-detection programs are now at an advanced stage through satellite monitoring of temperature changes in seawater. They are the object of detailed study in Chapter 35.

Marine Watching: Red Tide Warning Network

Sea-watching programs must be of several intensity levels, depending on the existence or not of toxic plankton species or of conditions favoring their proliferation. There must be a perfect coordination and a fast and fluent communication network between the authorities responsible for sea watching (Fishing Administration) and for the markets (health, agriculture, food) and health authorities. There are two subprograms:

Studies on plankton and conditions favoring its proliferation with a view to predict when toxic marine blooms are going to take place and to detect as early as possible their existence (red tides).

A program for monitoring of shellfish beds for DSP. Monitoring shellfish to check the presence of toxins in fisheries and shellfish farms before gathering is authorized as well as in purifying plants before shellfish is released on the market.

In order to achieve an efficient management causing the minimum disturbance to producers and allowing a safety warranty to consumers of the seafood, it is important to set zones and subzones in the shellfish farms, as well as fixed primary points, those which experience has shown to be most rapidly affected in the case of a toxic event; and fixed secondary points supplementing the former and allowing a more detailed knowledge of the affectation degree in the zones.

For sampling programs follow-up, and monitoring of toxic phytoplankton and DSP to work better action schemes have to be set according to the species of phytoplankton causing toxicity and to the shellfish and areas affected. Those schemes have to be set prior to assessing the information gathered in the monitoring program on the plankton and oceanographic conditions and on the shellfish biotoxins. In Galicia (Spain), there are four action schemes set on those bases (54).

Scheme A (normal situation). Oceanic conditions are not favorable for the development of toxic phytoplankton species nor are these found in significant concentrations, and there is no toxicity in bivalve shellfish.

Scheme B (alert situation), divided into three subschemes:

B1: When in spite of favorable oceanic conditions, no potentially toxic phytoplankton species are observed in significant concentrations, and there is no toxicity in bivalve shellfish either.

B2: There are favorable oceanic conditions and the presence of potentially toxic phytoplankton species but no toxicity in bivalve shellfish.

B3: There are favorable oceanic conditions, a significant increase in toxic population, and toxicity is detected in bivalve shellfish but in levels below the limits established by law.

Scheme C (extraction is forbidden). This is applied when toxic levels are above the law-established limits. It is subdivided into three subschemes:

C1: Oceanic conditions are favorable and there is a significant increase in the toxic population and in bivalve shellfish toxicity levels.

C2: Oceanic conditions are not favorable to the growth of the toxic plankton species which population is stable or decreasing. Also, toxicity levels in the bivalve shellfish is stable or decreasing.

C3: Oceanic conditions are not favorable and there is a significant decreasing and/or disappearing toxic population and toxicity levels are close to legal limits.

Scheme D: Oceanic conditions are not favorable to the development of toxic plankton species; potentially toxic phytoplankton species are in nonsignificant concentrations or absent, and toxicity stays below legal limits as a consequence of a previous event.

Bivalve shellfish samples must be gathered at different depths (1, 5, and 10 ms), since toxicity may vary with water depth. Analysis must be carried out separately or integrated, depending on the uniformity of the degree of mussels and oceanic conditions, and accumulation of toxicity at a given depth. In natural fisheries and fish farms, samples should be as significant as possible, according to the species subjected to monitoring and the area features.

Sampling frequency varies with every action scheme:

Scheme A (all year around): Weekly sampling of oceanic and phytoplanktonic conditions, as well as of biotoxins in mussels in fixed primary points, and fortnightly sampling in rock mussels.

Scheme B (alert situation):

B1: Selective sampling of phytoplankton.

B2: Increasing to twice per week the sampling for biotoxins in mussels in fixed primary points.

B3: Increase sampling to three times per week, and when biotest results advise it, sampling of other species in fixed secondary points, susceptible of being more affected.

Scheme C: Gathering is forbidden.

C1: The frequency and species to be sampled are dictated by the toxic event intensity and the existing provisions. Subzones (included within the same zone) bordering on a closed down subzone are subjected to daily sampling.

C2: Gathering samples in fixed primary or secondary points to assess the degree of affectation in the zone or subzone.

C3: Sample gathering for biotoxins at least three times per week.

Scheme D: Lifting of extraction prohibition. In fixed points where toxicity remains high, though below legal limits, samples are to be gathered twice a week.

Market Watching

Routine controls are carried out throughout the year by veterinary inspectors responsible for monitoring the salubrity and hygienic conditions of food. Surveillance should increase during periods when these problems tend to be present.

Surveillance of the Disease

Surveillance of the disease may be particularly relevant as an alternative method to primary prevention in countries that are not able to afford the costs of sustained surveillance programs in shellfish farming areas. In such circumstances, detecting the first cases of the disease has to be used to adopt preventive measures and avoid further cases. At any rate, a minimum public health infrastructure should be available, such as staff with adequate training, and laboratories with staff trained in standard techniques.

The education of medical and public health staff regarding diagnosis, treatment (symptomatic), and notification of suspect cases is very important for the success of a watch program.

Education of populations at risk about preventive measures, such as nonconsumption of shellfish when there are toxic red tides and never to eat mussels picked from cliff rocks or any shellfish picked directly on beaches, is essential and never enough.

People must be well informed and up-to-date through the most suitable means about the presence of toxic red tides, their blooming, and their disappearance.

Finally, education and cooperation with the seafood industry in everything related to poisoning risks by marine toxins as well as in primary and secondary prevention programs is necessary for the effective success of these programes.

European Regulations

In Europe, the Community Directive 91/492 (55) and 91/493 (56) from July 15th and 22nd, 1991, respectively, set health regulations applicable to producing and marketing live bivalve shellfish and seafood, respectively. They lay down requirements for marine biotoxins, and regarding DSP toxins, they establish: The usual methods of biological analysis must not give a positive reaction in the presence of DSP in the edible parts of the shellfish (the whole body or any edible part thereof). Regarding PSP toxins, there is a common agreement on the analysis method used, that is, bioassay in AOAC mice. Regarding tolerance limits, 80 μg equiv. STX/100 g of flesh, in the case of DSP toxins, there is no international consensus or toxicity criteria in the most fitting analysis process for its control.

The directive 91/492 CEE is vague, since it does not set a DSP toxin–detecting method, pointing out only that this must be biological, and it does not set either the tolerance limit, which in each case depends on the biological detection method used. Subsequently, there are important differences in the analysis methods and the toxicity criteria being applied to the health control of marine biotoxins at present in different countries of the European Union.

With a view to eliminating these discrepancies between the member states and harmonizing the European market, the European Commission named a reference national laboratory in each member country (LNRS) and a sole community reference laboratory (LCR), in order to set up and coordinate a network for exchanging information, knowledge, and experiences, and create a forum for method and toxicology agreements. The Exterior Health Laboratory in Vigo (Galicia, Spain), under the Ministry of Health and Consumption, was appointed Community Laboratory of Reference (57).

The tolerance levels have been established on the basis of the few epidemiological and toxicological studies carried out, the detection limitations of the analytical methods available, and, on occasion, they are simply based on the regulations of other countries.

The few acute toxicity studies carried out in rats and mice show that OA affects the intestinal mucosa, and that the main symptom of poisoning is diarrhea. The OA lethal dose in mice, by the intraperitoneal route, is 200 μg/kg of body weight and 160 μg/kg for DTX-1.

The minimum dose required to cause toxic symptoms in humans was initially fixed by Yasumoto at 10–12 MU of DSP, which is equivalent to 40–48 μg of OA. Severe symptoms are caused by 76–80 μg of OA. In 1981, when DSP toxins became better-known, Hamano et al. (58) pointed out that the minimum OA and DTX-1 doses that can cause diarrhea in adults are 40 and 36 μg, respectively. The first country to set a tolerance level was Japan, accepting 5 UR/100 g of meat, or 20 μg of AO/100 of meat as the top amount of toxin allowed in shellfish, which would be equivalent to 0.5 UR (19) of hepatopancreas (hp) (2 μg of AO/g of hp) assuming that the hepatopancreas is 10% of the total body.

The tolerance level should not be established only on the basis of an acute diarrhea picture caused by DSP, since the literature shows that there is the possibility of other severe effects such as promotion of tumor growth and hepatotoxic and cardiotoxic effects.

B. Epidemiological Investigation of an Acute Outbreak of DSP Disease

Human poisoning outbreaks by diarrheic toxins seem to be explosive, localized, and short-lived (holomiantic outbreaks). given the very short incubation period (between 30 min and a few hours), which depends on the amount of toxin ingested (shellfish toxic load and the amount of shellfish eaten) and the exposure to a common source.

Investigations usually start from the communication of index cases to the health authorities. That reporting, which is mandatory in many countries when there is an outbreak, must be done when clinical suspicion exists, but since this is subjective, the first step will have to be to confirm the case diagnosis. Confirming the diagnosis is easier when there are toxic red tides and when physicians and populations are alerted. Isolated cases of DSP will go undetected if DSP is not considered, given its unspecific symptomatology: diarrhea, nausea, vomiting, abdominal pain, and its mildness. And etiology identification of clusters is even easier (episodes in which two or more cases of the same disease are interrelated).

During investigation of a DSP outbreak, the four following stages are clearly marked: making or verifying the diagnosis of recorded cases and confirming the existence of an outbreak.

Identifying the intoxication source and transmission mode, other people who might have been or are exposed, as well as cases that might have appeared previously.

Description of cases according to the person, place, and time variables.

Urgent notification of the health authorities, from the mere suspicion, so that they can adopt the pertinent administrative preventive measures.

Outbreak Confirmation

First of all, In order to establish environmental exposure, existence of a toxic red tide, and then confirm the diagnosis in the laboratory (seafood testing), it has to be decided whether the signs and symptoms (gastroenteritis and no temperature) and their evolution (benign, with complete recovery in 3 days) correspond to DSP; whether the incubation period is short (30 min to a few hours); and whether there is an antecedent of shellfish consumption (appropriate seafood ingestion) and its origin.

Other causes, such as toxic infection by Bacillus cereus or by Vibrio parahaemolitic have to be excluded; the latter is also carried by shellfish. Those affected usually do not present with fever, but their incubation period is longer (12 hs), and the microorganism can be identified in feces or food remains.

Gastroenteritis associated with a clinical picture of no fever, mussel ingestion, and a short incubation period points toward the diagnosis of DSP; even more so if there is a DSP toxic red tide at the time.

Based on the study of the Japanese outbreaks, the World Health Organization (WHO) (1984) established the following symptom frequency in DSP disease: diarrhea (92%), nausea (80%), vomiting (79%), abdominal pain (53%), and chill (10%).

In order to obtain laboratory confirmation, food remains should be available (uneaten mussels), or mussel samples should be taken in the same area the eaten mussels came from (e.g., food market, shellfish purifying plant, fish farm, rocks).

The usual technique used in Belgium, Denmark, Spain is the bioassay in mice developed by Yasumoto. In Italy, United Kingdom it is used with modifications to eliminate PSP contamination. These modifications can wash off the yessotoxins present in the samples. In Portugal it is used with modifications to eliminate potential interference by fatty acids which also carry the risk of washing off weakly polar DSP group toxins such as acyl derivatives. Positive criteria range from the death of two or three mice before 24 h in France, Belgium, Denmark, and Spain

(12 h in Galicia) to under 5 h in Italy, the United Kingdom, France, and Portugal. Some countries, such as Ireland, Holland and Germany use the bioassay in rats developed by M. Kat. (59).

The bioassay in mice is the most widely used because it is more sensitive (OA 4 μg) than exposure by oral ingestion in rats (OA 10 μg), but the bioassay in mice is subject to false positives by interference of nonphycotoxic components, and it is more expensive, since mice die or are disposed of, whereas rats can be used repeatedly, although reading the test in them depends on the subjective examination of feces.

There are also chemical methods such as high-power liquid chromatography (HPLC), which is the most widely used method after bioassays; the enzyme-inhibition assays such as the fluorescent protein phosphatase inhibition assays (60), a promising new and inexpensive technique (61); the immunoassays (monoclonal antibodies to okadaic acid and DTX-1 and enzyme-linked immunosorbent assay [ELISA]), there are several commercial kits available; the cytotoxicity assays that obviates the morphological changes caused by the DSP toxin activity in various cell lines: e.g., human KB cells, salmon and rat hepatocytes, cultured neurons.

These techniques, and more so bioassays, HPLC and phosphatase inhibition, are generally used in health surveillance. Ethical and technical considerations are being focused on the development and use of health-control tests that do not require the use of animals.

Identifying Source, Transmission Mechanism, and Subjects at Risk

It is essential to know the kind of shellfish responsible for DSP intoxication and their origin, whether they were bought in the market or picked directly from fish farms or rocks, and whether there are some not yet consumed. It is also necessary to know whether they were eaten at home or in a restaurant or other public place.

It is also important to know whether other people have also eaten the product, searching for other cases, or whether they have more shellfish from the same source and have not yet consumed it. Moreover, it has to be investigated whether there have been previous cases in the area or surrounding areas.

Describing Cases According to Person, Place, and Time Variables

Regarding the person variable, information must be gathered on age, sex, and occupation; as far as the place variable is concerned, geographical distribution of the cases concerned, noting their addresses, and regarding time, it is interesting to have information on the date, time of exposure (shellfish ingestion), as well as the onset of the first symptoms and sequence of presentation. It is also advisable to collect information leading to an initial idea of the toxic load.

Administration Monitoring Measures

In the event of an outbreak, administration measures will have to be adopted forbidding shellfish gathering and setting up an area monitoring scheme. Continuous vigilance must be exercised for the occurrence of DSP poisoning, and physicians must be warned of its existence so that they can be alert for the clinical signs of gastroenteritis. At the same time, the population will have to be made aware: industrialists, health professionals, and people in general through the mass media, warning about the hazards and advising that they must not consume affected shellfish.

REFERENCES

1. Exodus 7:v. 20–21 The Bible. King James Version.
2. CY Kao. Tetrodotoxin, saxitoxin and their significance in the study of excitation phenomena. Pharmacol Rev 18:997–1049, 1966.

3. BN Halstead. Poisonnous and venomous marine animals of the world. Vol I. Invertebrates. Washington: Government Printing Office, 1965.
4. A Chevalier, EA Duchesne. Memoire sur les empoisonnements par les huîtres, les moules, les crabes, et par certains poissons de mer et de riviere. Ann Hyg Publ (Paris) 45(1):108–147, 1851.
5. A Chevalier, EA Duchesne. Memoire sur les empoisonnements par les huîtres, les moules, les crabes, et par certains poissons de mer et de riviere. Ann Hyg Publ (Paris) 45(2):387–437, 1851.
6. G Vancouver. A voyage of discovery to the North Pacific Ocean and round the world. London: John Stockdale 4:44–47, 1801.
7. KE Meyer, KF Sommer, P Schoenholz. Mussel poisoning. J Prev Med 2:365–394, 1928.
8. KF Sommer, KE Meyer. Paralytic shellfish poisoning. Arch Pathol 24(5):560–598, 1937.
9. P Korringa, RT Roskam. An unusual case of mussel poisoning. International Council of the Exploration of the Sea (ICES). Council Meeting 49, 1961.
10. L Rossebe, B Thorson, R Asse. Etiologisk uklar matforgifnig etter konsum av blaskjell. Norsk Vet Tidsskr 82:639–642, 1970.
11. M Kat. The ocurrence of Prorocentrum species and coincidental gastrointestinal illness of mussels consumers. In: D Taylor, HH Seliger, eds. Toxic Dinoflagellate Blooms. Amsterdam: Elsevier, 1979, pp 215–220.
12. T Yasumoto, Y Oshima, M Yamaguchi. Ocurrence of a new type of shellfish poisoning in Japan and chemical properties of the toxin. In: D Taylor, HH Seliger, eds. Toxic Dinoflagellate Blooms. Amsterdam: Elsevier, 1979, pp 495–502.
13. JJ Gestal Otero, JM Hernández Cochón, O Bao Fernández, L Martínez-Risco López. Brote de Mitilotoxismo en la Provincia de La Coruña. Bol Inst Espa Oceano IV (258):5–29, 1978.
14. T Yasumoto, Y Oshima, M Yamaguchi. Occurrence of a new type shellfish poisoning in the Tohoku distric. Bull Jpn Soc Sci Fish 44:1249–1255, 1978.
15. T Yasumoto, Y Oshima, W Sugawara, Y Fukuyo, H Oguri, T Igarishi, N Fujita. Identification of Dinophysis fortii as the causative organism of diarrhetic shellfish poisoning. Bull Jpn Soc Sci Fish 46:1405–1411, 1980.
16. M Suganuma, H Fujiki, H Suguri, S Yoshizawa, M Hirota, M Nakayasu, M Okika, K Wakamatsu, K Yamada, T Sugimura. Okadaic acid: an additional non phorbol-12-tetradecanoate-13-acetate-type promotor tumoral. Proc Natl Acad Sci USA 85(6):1768–1771, 1988.
17. H Fujiki, M Suganuma, H Suguri, S Yoshizawa, K Takagi, T Sassa, N Uda, K Wakamatsu, K Yamada, T Yasumoto, Y Kato, N Fusetani, K Hashimoto, T Sugimura. Diarrhetic shellfish toxin, dinophysistoxin-1, is a potent tumor promoter on mouse skin. Jpn J Cancer Res 79:1089–1093, 1988.
18. S Aonuma, T Ushijima, M Nakayasu, H Shima, T Sugimura, M Nagao. Mutation induction by okadaic acid, a protein phosphatase inhibitor in CHL cells, but not in S. typhimurium. Mutat Res 250: 375–381, 1991.
19. Y Hokama, PJ Scheuer, T Yasumoto Effect of marine toxin on human peripheral blood monocytes. J Clin Lab Anal 3:215–221, 1989.
20. B Sundstrom, L Edler, E Granéli. The global distribution of harmful effects of phytoplankton. In: E Granéli, B Sundstrom, L Edler, DM Anderson, eds. Toxic Marine Phytoplankton. Amsterdam: Elsevier, 1990, pp 537–541.
21. HP Van Egmond, T Aune, P Lassus, GJA Speijers, M Waldock. Paralytic and diarrhoeic shellfish poisons: occurrence in Europe, toxicity, analysis and regulation. J Nat Toxins 2(1):41–79, 1993.
22. B Hald, T Bjergskov, H Emsholm. Monitoring and analytical programmes on phycotoxins in Denmark. In: JM Fremy, ed. Proceedings of Symposium on Marine Biotoxins. Centre National d'Etudes Vétérinaires et Alimentaires de France, 1991, pp 181–187.
23. E Dahl, M Yndestad. Diarrheic Shellfish Poisoning (DSP) in Norway in the autumn 1984 related to the occurrence of Dinophysis spp. In: DM Anderson, AW White, and DG Baden, eds. Toxic Dinoflagellates. Amsterdam: Elsevier, 1985, pp 495–500.
24. P Krogh, L Edler, E Granéli. Outbreaks of diarrheic shellfish poisoning on the west coast of Sweden. In: DM Anderson, AW White, DG Baden, eds. Toxic Dinoflagellates. Amsterdam: Elsevier, 1985, pp 501–504.
25. MJ Campos, S Fraga, J Mariño, FJ Sánchez. Red tide monitoring program in NW Spain. Report of

1977–1981. International Council for the Exploration of the Sea (ICES), Council Meeting L:27, 1982.
26. T Kawabata. Regulatory aspectos of marine biotoxins in Japan. In: S Natori, K Hashimoto and Y Ueno, eds. Micotoxins and Phycotoxins '88. Amsterdam: Elsevier, 1989, pp 469–476.
27. GV Konovalova. Toxic and potentially toxic dinoflagellates from the far east coastal waters of the USSR. In: TJ Smayda, Y Shimizu, eds. Toxic Phytoplankton Blooms in the Sea. Proceedings of Fifth International Conference on Toxic Marine Phytoplankton. Newport (USA) 1991. Amsterdam: Elsevier, 1993, pp 275–279.
28. I Karunasagar, K Segar, I Karunasagar. Incidence of PSP and DSP in shellfish along the coast of Kamataka state (India). In: T Okachi, DM Anderson, AW White, DG Baden, eds. Red Tides: Biology, Environmental Science and Toxicology. Amsterdam: Elsevier, 1989, pp 61–64.
29. MA Quilliam, MW Gilgan, S Pleasance, ASW Defreitas, D Douglas, L Fritz, T Hu, JC Marr, C Smyth, JLC Wright. Confirmation of an incident of diarrhetic shellfish poisoning in Earstern Canada. In: TJ Smayda, Y Shimizu, eds. Toxic Phytoplankton Blooms in the Sea. Proceedings of Fifth International Conference on Toxic Marine Phytoplankton. Newport (USA), 1991. Amsterdam: Elsevier, 1993, pp 547–552.
30. MW Gilgan, C Powell, J Van de Riet, BG Burns, MA Quilliam, K Kennedy, CH McKenzie. The occurrence of a serious diarrhetic shellfish poisoning episode in mussels from Newfoundland during the late autumn of 1993. Fourth Canadian Workshop on Harmful Marine Algae. Sidney, BC, 1995, pp 3–5.
31. AR Freudenthal, J Jacobs. Observations on Dinophysis acuminata and Dinophysis norvergica in Long Island waters: Toxicity, occurrence following diatomdiscolored water and co-occurrence with Ceratium. In: TJ Smayda, Y Shimizu, eds. Toxic Phytoplankton Blooms in the Sea (abst). Proceedings of Fifth International Conference on Toxic Marine Phytoplankton. Newport (USA). Amsterdam: Elsevier, 1995.
32. G Lembeye, T Yasumoto, J Zhao, R Fernández. DSP outbreak in Chilean fjords. In: TJ Smayda and Y Shimizu, eds. Toxic Phytoplankton Blooms in the Sea. Proceedings of Fifth International Conference on Toxic Marine Phytoplankton. Newport (USA), 1991. Amsterdam: Elsevier, 1993, pp 525–529.
33. S Méndez. Update from Uruguay. Harmful Algae News. An Intergovernmental Oceanographic Commission (IOC). Newsletter on Toxic Algae and Algal Blooms 63:5, 1992.
34. M Murata, M Shimatani, H Sugitani, Y Oshima, T Yasumoto. Isolation and structure elucidation of the causative toxin of the diarrhetic shellfish poisoning. Bull Jpn Soc Sci Fish 43:549–552, 1982.
35. K Tachibana, PJ Scheuer; Y Tsukitani, H Kikuchi, D van Engen, J Clardy, Y Gopichand, FJ Schmitz. Okadaic acid, a cytotoxic polyether from two marine sponges of the genus Halichondria, J Am Chem Soc 103:2469–2471, 1981.
36. T Yasumoto, M Murata, Y Oshima, M Sano, GK Matsumoto, J Clardy. Diarrhetic shellfish toxins. Tetrahedron 41:1019–1025, 1985.
37. T Hu, ASW Defreitas, J Doyle, D Jackson, J Marr, E Nixon, S Pleasance, MA Quilliam, JA Walter, JLC Wright. Isolation of a new diarrhetic shellfish poison from Irish mussels. Chem Commun 30: 39–41, 1992.
38. M Murata, M Sano, T Iwashita, H Naoki, T Yasumoto. The structure of pectenotoxin-3, a new constituent of diarrhetic shellfish toxins. Agric Biol Chem 50:2693, 1986.
39. M Murata, M Kumagai, JS Lee, T Yasumoto. Isolation a structure of yessotoxin, a novel polyether compound implicated in diarrhetic shellfish poisoning. Tetrahedron Lett 28:5869–5872, 1987.
40. C Bialojan, A Takai. Inhibitory effect of a marine-sponge toxin, okadaic acid, on protein phosphatases. Specificity and kinetics. Biochem J 256:283–290, 1988.
41. Y Hamano, Y Hinoshita, T Yasumoto. Enteropathogenicity of diarrhetic shellfish toxins in intestinal models. J Food Hyg Soc Jpn 27:375–379, 1986.
42. K Terao, E Ito, T Yanagi, T Yasumoto. Histopathological studies on experimental marine toxin poisoning. I. Ultraestructural changes in the small intestine and liver of suckling mice induced by dinophysistoxin 1 and pectenotoxin 1. Toxicon 24:1141–1151, 1986.
43. P Cohen, CFB Holmes, Y Tsukitani. Okadaic acid: a new probe for the study of cellular regulation. Trends Biochem Sci 15:98–102, 1990.

44. EP Carmody, K James, S Kellt, K Thomas. Complex diarrhetic shellfish toxin profiles in Irish Mussels. In: P Lassus, G Arzul, LE Erard, E Denn, P Gentien, C Marcaillou Lebaut, eds. Harmful Marine Algal Blooms. Paris: Lavoisier, 1995, pp 273–278.
45. J Blanco, ML Fernández, J Mariño, B Reguera, A Miguez, J Maneiro, E Cacho, A Martínez. From Dinophysis spp toxicity to DSP outbreaks. A preliminary model of toxin accumulation in mussels. In: P Lassus, G Arzul, LE Erard, E Denn, P Gentien and C Marcaillou Lebaut, eds. Harmful Marine Algal Blooms. Paris: Lavoisier, 1995, pp 777–782.
46. M Kat. Dinophysis acuminata blooms, the distinct cause Dutch mussel poisoning. In: DM Anderson, AW White, DG Baden, eds. Toxic Dinoflagellates. Amsterdam: Elsevier, 1985, pp 73–77.
47. C Le Baut, D Lucas, L Le Dean. Dinophysis acuminata toxin status of toxicity bioassays in French. In: DM Anderson, AW White, DG Baden, eds. Toxic Dinoflagellates. Amsterdam: Elsevier, 1985, pp 485–488.
48. MA Sampayo, P Alvito, S Franca, I Sousa. Dinophysis spp. toxicity and relation to accompanying species. In: E Granéli, B Sundstrom, L Edler, DM Anderson, eds. Toxic Marine Phytoplankton. Amsterdam: Elsevier, 1990, pp 215–220.
49. A Tubaro, S Sosa, D Bussani, L Sidari, G Honsell, R Della Loggia. Diarrhoeic toxicity induction in mussels of the Gulf of Trieste. In: P Lassus, G Arzul, LE Erard, E Denn, P Gentien P, C Marcaillou Lebaut, eds. Harmful Marine Algal Blooms. Paris: Lavoisier, 1995, pp 249–254.
50. B Underdal, M Yndestad, T Aune. DSP intoxications in Norway and Sweden, Autumn 1984-Spring 1985. In: DM Anderson, AW White, DG Baden, eds. Toxic Dinoflagellates. Amsterdam: Elsevier, 1985, pp 489–494.
51. JM Vieites, FJ Leira Sanmartín. Resultados preliminares del estudio del proceso de cocción como vía de reducción de la toxicidad DSP en moluscos bivalvos. Alimentaria (October):35–38, 1995.
52. JC Medcorf, AH Leim, AB Needler, AWH Needler. Paralytic Shellfish poisoning on the Canadian Atlantic coast. Bull Fish Res Bd Can LXXV:32, 1947.
53. BW Halstead, EJ Schantz. Intoxication paralítica por moluscos. Geneva: WHO. Publ en offset n° 79, 1984.
54. Consellería de Pesca, Marisqueo y Acuicultura. Orden de 14 de noviembre de 1995 por la que se regula el programa de actuacciones para el control de biotoxinas marinas en moluscos bivalvos y otros organismos procedentes de la pesca, el marisqueo y la acuicultura. Diario Oficial de Galicia n° 221, de 17 de noviembre, pp 8454–8467.
55. Directiva 91/492/CEE, de 15 de julio de 1991, por la que se fijan las normas sanitarias aplicables a la producción y puesta en el mercado de moluscos bivalvos vivos. Diario Oficial de las Comunidades Europeas de 24-09-91, n° L 268:1–24.
56. Directiva 91/493/CEE. Pescado. Control veterinario. Fija las normas sanitarias aplicables a la producción y a la puesta en el mercado de los productos pesqueros. Diario Oficial de las Comunidades Europeas del 22-07-91, n° L 268:3849–3864.
57. Decisión 93/383/CEE de 14 de junio de 1993 relativa a los laboratorios de referencia para el control de biotoxinas marinas. Diario Oficial de las Comunidades Europeas de 08-07-93, n° L 166:31–33.
58. Y Hamano, Y Kinoshita, T Yasumoto. Suckling mice assay for diarrhetic shellfish toxins. In: DM Anderson, AW White, DG Baden, eds. Toxic Dinoflagelates. Amsterdam: Elsevier, 1985, pp 383–388.
59. M Kat. Diarrhoetic mussel poisoning in the Netherlands related to the dinoflagelle Dinophysis acuminata. Antonie van Leeuwenhoek 49:417–427, 1983.
60. OI Fontal, F Leira, JC González, MR Vieytes, JM Vieites and LM Botana. Determination of okadaic acid levels by HPLC and fluorescent microplate assay: a comparison of both methods. In: Harmful Algae. Vigo, Xunta de Galicia-Intergovernmental Oceanographic Commission of UNESCO, 1998, pp 547–548.
61. MR Vieytes, OI Fontal, F Leira, J De Sousa and LM Botana. A fluorescent microplate assay for diarrheic shellfish toxins. Anal Biochem 248:258–264, 1997.

4
Neurotoxic Toxins

Bradford David Gessner
Alaska Department of Health and Social Services, Division of Public Health
Anchorage, Alaska

I. LIMITATIONS OF EPIDEMIOLOGICAL DATA

A. Bias and Reports of Clinical Illness

Bias refers to those factors related to study design and analysis that may effect the results and conclusions of a study. Descriptions of clinical manifestations may have several sources of bias including selection bias, recall bias, diagnostic bias, and information bias. Selection bias occurs at the outset of the study. For example, persons with mild illness usually have a lower likelihood of reporting to a health care facility and thus a lower likelihood of being included in a hospital-based study or evaluation of surveillance data. Consequently, surveillance and hospital-based studies of marine neurotoxins may overstate the importance of severe sequelae such as neurological symptoms and understate the importance of mild sequelae such as gastrointestinal symptoms. One of the important contributions of outbreak investigations is actively to identify persons with mild illness or persons exposed to toxin but who did not develop illness. This in turn greatly contributes to the understanding of the spectrum of disease and illness risks associated with a particular toxin.

Recall bias refers to the extent to which patient recall of information affects the accuracy of the data collected. One can imagine that this is particularly a problem for amnesic shellfish poisoning, but it will affect all studies to a greater or lesser degree. Recall bias may also occur if ill persons recall exposures to a greater degree than nonill persons do; one result of this may be to associate falsely exposures with outcomes. The effect of recall bias may be minimized by prospective, systematic studies that use standardized interview forms to collect information from ill persons.

Recording bias is similar to recall bias and refers to the extent to which clinicians accurately and completely record the symptom history of their patients. This is particularly a problem for studies based on retrospective medical chart reviews. For example, a physician who sees a patient with paralytic shellfish poisoning may record paresthesias and vomiting as clinical symptoms; the researcher who reviews this chart at some later period will have no way to determine if other symptoms, such as dysarthria or dysphagia, were also present or even assessed. Another way that recording bias may effect the results of a study is through the linguistic and cultural barriers that may exist between the patient and clinician. For example, the floating feeling in paralytic shellfish poisoning and the temperature reversal of ciguatera represent attempts to define a health state in terms which are mutually comprehensible to the patient and clinician. These terms, however, may not have a biological basis and may not correspond with the way individual patients would describe a particular experiential state if left unprompted.

Diagnostic bias may occur when factors unrelated to the disease influence the diagnosis assigned to a specific constellation of symptoms. For marine toxin ingestions, this problem may be acute because of the lack of specific and sensitive diagnostic tests. For example, a patient who has ingested ciguatoxin and presents with gastrointestinal but not neurological symptoms may not receive a diagnosis of ciguatera unless his or her illness occurred in the context of a more extensive outbreak investigation. Conversely, a patient may have diarrhea during an outbreak investigation and receive a diagnosis of ciguatera even if the illness resulted from a different cause.

B. Incidence Rates

Incidence is a fundamental component of epidemiological inference, yet it is notoriously difficult to obtain. The source of data may be of high, low, or, most often, unknown quality. For example, surveillance systems based on passive reporting of illness from medical providers to government authorities (such as exists for individual states in the United States) will identify an unknown but incomplete proportion of all cases. Breakdowns in reporting may occur at several levels: affected individuals may fail to present to a medical provider because they have mild illness or because they fail to recognize the potentially serious nature of their illness; medical providers may fail to make an appropriate diagnosis because of lack of interest, knowledge, or diagnostic capability; or the provider may neglect to report the illness to the appropriate authorities.

Other factors may influence surveillance data as well. During an outbreak, active case finding will identify additional ill persons, particularly those with mild symptoms. Agencies or institutions which have a researcher or public health official who is particularly interested in a specific disease may pursue case finding more rigorously than other groups. Finally, public awareness and appreciation of the disease as a significant health problem will influence reporting. The most accurate surveillance data derive from systems which are prospective, systematic, active, and which have a high degree of support from the local medical community and the public.

Other methods of estimating incidence exist besides routine surveillance. Retrospective review of hospital or clinical medical charts may provide an estimate of the incidence of relatively severe illness. Unfortunately, these studies are subject to many of the types of bias discussed above. If case definitions with a high degree of specificity are used and clinical acumen in an area is high, incidence data from medical chart reviews may be regarded as lower estimates.

Investigators may design special studies, such as telephone and door-to-door surveys. If a population is systematically sampled, survey studies may provide population-based incidence data and may be the best source of data on less severe cases. The primary limitation of survey data is recall bias, which increases as the retrospective length of the study increases.

Finally, the public health relevance of incidence data may be limited by the investigator's choice of data for the denominator. For example, a study may report the incidence of ciguatera as 100 cases per 100,000 population per year. Such a figure implies that the entire population is at risk for illness. The true population at risk, however, is that which consumes marine organisms. This population may differ from other populations by proximity to coastal areas, race or culture, income, and other factors. An incidence calculated using as the denominator the number of persons consuming potentially toxic fish would allow more directed and local implementation of public health control programs. It might also allow public health officials to predict the impact on disease incidence of changes in demographic variables and dietary practices. Unfortunately, this information is rarely available.

C. Risk Factors

Risk factors refer to those factors that increase an individual's risk of a particular outcome. In marine toxin investigations, outcomes have included illness, severity of illness, and particular symptoms. For example, a study of paralytic shellfish poisoning examined risk factors for the development of paralytic shellfish poisoning among persons who ate shellfish (1). During an outbreak of domoic acid intoxication, investigators examined risk factors for severe illness among ill persons (2). During an analysis of ciguatera, investigators identified risk factors for the development of gastrointestinal versus neurological illness (3).

Risk factors for these various outcomes may be placed in the following categories: differences in toxin composition and genetic and nongenetic differences in the host response. For example, toxin composition may differ between outbreaks of ciguatera (e.g., ciguatoxin, maitotoxin, palytoxin, gambiertoxin) or paralytic shellfish poisoning (e.g., saxitoxin, neosaxitoxin, gonyautoxin); although the toxins may be structurally related, their effects are not identical, and thus the risk of illness may differ depending on the specific toxin or toxins ingested. Genetic differences in the host response may potentially influence the development or progression of illness through slight alterations in binding sites and altered immunological response. Finally nongenetic host differences—including diet, dose of toxin ingested, method of food preparation, and ingestion of particular organs—may also influence the development or progression of illness.

D. Toxic Doses

The estimation of toxic dose is limited in several ways. In particular, samples of the actual ingested animal are rarely available for testing except for occasional autopsy or vomitus samples. Many investigators estimate the toxin concentration in marine animals implicated in human illness by examining toxin from animals served at the same meal or collected later from a similar location. However, toxin may not be uniformly distributed in an animal and toxin levels may differ between animals of the same species collected at different times or places. Thus, the reported toxin concentration usually represents an estimate (of unknown accuracy) of the toxin concentration in the actual ingested animal. Additionally, most studies which attempt to calculate a toxic dose do not report dose on a per unit weight basis. Finally, most studies estimate toxic dose based on bioassays that do not differentiate the toxic dose of the different toxin components.

E. Geographical Distribution

Geographical distribution of reported illness may differ by the distribution of toxin-producing and toxin-concentrating organisms, but may also differ by factors unrelated to the true distribution of disease. Disease may be underreported in some geographical regions because the region has other health priorities, a paucity of diagnostic facilities, poorly trained medical providers, or a lack of researchers who publish data in scientific journals. For example, Guatemala appears in the medical literature on paralytic shellfish poisoning because of the occurrence of a large outbreak coupled with the involvement of the U.S. Centers for Disease Control and Prevention and the Food and Drug Administration (4).

F. Temporal Distribution

Seasonal and longer term temporal variations in neurotoxin poisonings have been reported in some instances but not in others. Temporal variations in human illness are the combined product of variations in dinoflagellate blooms, the proportion of dinoflagellate blooms which produce toxin, fish and shellfish depuration patterns, dietary preferences, fish and shellfish availability, public health measures, and surveillance system characteristics. In essentially all instances, the relative contribution of these various factors to the presence or absence of temporal variation is unknown.

G. Literature Sources

The results of any review of existing data naturally depend on the available literature sources. Scientific journals represent the primary and, in general, the most scientifically sound source of widely available epidemiological information. For many marine toxin-related illnesses, however, a large body of knowledge exists in regional and governmental publications, the proceedings of scientific meetings, and textbooks. These sources are usually less widely accessible than journals, and no attempt has been made to compile a comprehensive list for the current review. Consequently, it is entirely possible that important pieces of information have been excluded. Where possible, an attempt has been made to use only primary data sources. Occasionally, however, information is referenced from a secondary source.

II. PARALYTIC SHELLFISH POISONING

A. Incidence

Incidence data for paralytic shellfish poisoning (PSP) are rare and limited by a lack of surveillance systems, lack of active case finding in areas which have surveillance, and lack of data on appropriate denominators for public health interventions. One investigator estimated that 1600 cases of paralytic shellfish poisoning had been identified worldwide between 1689 and 1971 (5), a number which had increased by another 900 by 1984 (6). Unfortunately, such global information is of little use for public health officials in communities where paralytic shellfish poisoning occurs. In two of the largest series of cases of PSP from a specific area, Prakesh (5) reported that 80 cases of illness were identified from the Bay of Fundy region in Eastern Canada between 1889 and 1961 and 107 cases were identified from the St. Lawrence region between 1880 and 1970. Unfortunately, no attempt was made to provide population data from which incidence rates might be calculated.

Gessner et al. (1,7) have performed two investigations in Alaska from which the incidence of PSP could be calculated. The first was based on a retrospective review of surveillance data collected from 1973 to 1992 by the Alaska Division of Public Health. Based on overall population data from Alaska and 117 reported cases during this period, the estimated incidence was 1.2 per 100,000 persons per year.

The second study attempted to identify more clearly the incidence of PSP among high-risk populations by using a randomized telephone survey among two coastal populations. This study found an incidence of 150 and 1500 per 100,000 persons per year in Kodiak and Old Harbor, respectively, and 560 and 1570 per 100,000 persons per year among persons who reported consuming shellfish collected from unregulated beaches. The incidence calculated from surveillance data from the Department of Health for the same period was 6 and 170 per 100,000 persons per year for Kodiak and Old Harbor, respectively. The large difference in estimated incidence between survey and surveillance data indicates that even in a state with a high awareness of PSP, surveillance data grossly underestimate the true incidence.

Recently, investigators have postulated that PSP incidence has increased worldwide owing to increased oceanic eutrophication and possibly increased ocean commerce with subsequent dispersal of toxic dinoflagellate cysts (8,9). Although compelling evidence exists that observed red tides have increased recently, particularly in Southeast Asia, much less evidence exists for human illness.

Anderson has summarized data showing that more than twice as many areas worldwide reported outbreaks of PSP during 1970 as 1990. Summary data from Prakesh in the St. Lawrence region of Canada show that an average of nine cases of illness per decade occurred during 1900 to 1950, whereas 47 cases occurred during the 1960s alone. One investigator suggests that this must represent a true increase, since the symptoms of PSP are sufficiently unique that they would immediately be recognized (10).

Despite this faith in surveillance systems and the clinical ability of medical providers, the data of Anderson and Prakesh have several possible explanations other than a true increase in the incidence of PSP. These include better surveillance systems, an increase in the number of researchers interested in PSP, lower rates of other diseases such that PSP has assumed a greater relative priority, changes in shellfish consumption patterns (which in turn may be related to changes in social conditions and population movements, including tourism), and increased awareness of PSP symptoms through worldwide dissemination of medical knowledge.

Alaska has had a consistent PSP surveillance system in place for the past 20 years. During the 11 years from 1973 to 1983, there were 70 cases of illness and 33 outbreaks, whereas during the 11 years from 1984 to 1994, there were 73 cases of illness and 34 outbreaks (Alaska Division of Public Health, unpublished data). This despite a steady increase in coastal populations and maritime tourist activity. It is possible that pollution levels in Alaskan waters have not reached levels sufficient to cause large fluctuations in dinoflagellate blooms. What seems clear, however, is that more effort should be made to establish surveillance systems and estimate baseline incidence rates before conclusions on increases in incidence rate can be made.

B. Clinical Features

PSP toxins exert a local effect on the oral mucosa, leading to the rapid onset of perioral paresthesias following exposure. For nine studies where information was available, onset time ranged from a minimum of 5 to a maximum of 660 min. The median or mean time to illness onset ranged from 8 to 120 min (4,7,11–17).

Unlike ciguatera, the clinical presentation of PSP is reasonably consistent across populations, possibly because individual PSP toxins differ in their quantitative rather than qualitative effects. Three recent published reports include detailed descriptions of the clinical symptoms of PSP and report on at least 50 ill persons (4,7,18) (Table 1). Data from Guatemala and England come from single-outbreak investigations, whereas those for Alaska come from 20 years of surveillance. Despite this, and the wide geographical distribution of the three areas, clinical presentation remains consistent.

The most common symptom of PSP, occurring in almost all affected individuals, is perioral paresthesias, generally described as either numbness or tingling. Other than in the context of an outbreak, health care providers should be cautious about diagnosing PSP in a person who does not have this symptom. It is likely that most people have no other symptom and do not report to health care facilities. Among persons who progress to more severe illness, a minority report gastrointestinal symptoms, including nausea, vomiting, abdominal pain, and diarrhea.

Table 1 Symptoms of Paralytic Shellfish Poisoning in Guatemala, England, and Alaska

Symptom	Guatemala (4) N = 187 (%)	England (8) N = 78 (%)	Alaska (7) N = 117 (%)
Gastrointestinal			
Nausea	52		38
Vomiting		36	29
Abdominal pain	38		
Diarrhea	27		9
Neurological			
Paresthesia			97
Oral paresthesia	93	88	
Acral paresthesia	86	83	
Weakness	81	71	28
Ataxia	37	57	27
Dysarthria	66	23	14
Diplopia	39		16
Vertigo	86		24
Transient blindness	53		
Floating sensation		66	21
Headache	80	41	
Dyspnea	74	24	25
Paralysis			3
Outcome			
Death	14	0	1
Hospitalized	70		26

More severe illness leads to a variety of neurological symptoms culminating, in some instances, in respiratory arrest or death. Neurological symptoms may include weakness, dysarthria, diplopia, ataxia, and vertigo or dizziness. One of the more interesting symptoms is a dissociative feeling, which has been described as floating in various reports. It is unclear why persons from Guatemala did not describe this symptom, although it is possible investigators did not specifically inquire. Limb paralysis is an uncommon event but may occur in severe cases.

Respiratory arrest and collapse represent terminal symptoms in persons who do not have access to medical care. These symptoms may occur within minutes in a person who otherwise exhibits no evidence of respiratory difficulty (19), emphasizing the need for symptomatic individuals to seek medical attention immediately even though they have seemingly mild symptoms. Few data exist regarding how long after toxin ingestion onset of respiratory arrest may occur: For four patients in Alaska, respiratory arrest occurred from 75 to 240 min after toxin ingestion (16). It in unclear whether PSP exerts significant effects directly on the myocardium. In one reported case, where the victim consumed mussels containing a PSP toxin concentration of 19,418 μg/100 g tissue, cardiac arrest and ventricular fibrillation occurred despite prompt initiation of bag and mask ventilation at the onset of respiratory failure (16).

Signs of PSP are nonspecific and may include sluggishly reactive or dilated and fixed pupils; absent or diminished reflexes, including deep tendon reflexes; and muscle weakness or paralysis. In severe cases, PSP may clinically resemble brain death. It is possible, however, for patients to retain consciousness despite complete muscular paralysis. The author has interviewed patients who remembered the course of events from the onset of respiratory arrest through intubation. Among other things, this suggests that patients with PSP should receive sedation prior to intubation. Unlike ciguatoxin, PSP does not cause hypotension or bradycardia. Where blood pressure and heart rate have been measured, patients have instead shown a normal rhythm and hypertension (16,18). The hypertensive effects reported may result from a direct action of PSP (although by an unknown mechanism), stress, or the presence of other unidentified toxins (e.g., those with calcium channel agonist activity) (20).

Death results from respiratory arrest and occurs in a variable number of cases, ranging from zero (18) to 14% (4). Differences in the proportion of cases who die may reflect different toxin composition, different doses of toxin ingested, or, most critically, differences in access to emergency medical services.

Six studies for which information was available reported a maximum duration of symptoms of from 1 to 14 days (4,7,11,16,17,21). Prolonged illness, however, included such nonspecific symptoms as weakness, headaches, memory loss, and fatigue. In general, recovery from neurological symptoms is rapid and complete. Rodrigue et al., for example, reported that neurological symptoms resolved within 24–72 h (4). Gessner et al. described a person who progressed from an appearance of clinical brain death to almost complete recovery within 28 h (19). No long-term clinical effects of PSP have been reported, although it is worth mentioning that no reports exist in the literature of rigorous long-term neurological assessment of patients who have had PSP.

C. Toxic Dose

The minimum toxic dose of shellfish toxin represents one of the most critical pieces of information for regulatory agencies. Six studies have information available on toxic doses in humans (4,5,7,13,16,18) (Table 2). From these, the values of interest are the minimum toxic and lethal doses, the minimum toxic and lethal doses per kilogram body weight, the mean toxic dose, and the maximum asymptomatic and nonlethal doses. The estimated minimum toxic dose varies greatly from 13 to 2250 μg. Minimum dose estimates from surveillance data, such as the 13 μg presented for Alaska, should be viewed with skepticism, since they may represent cases that were not actually PSP. This is particularly true if they occurred as isolated cases and with mild symptoms. If this report is excluded, the minimum toxic dose may be estimated as something greater than 100 μg, the equivalent of eating 31 4-g mussels or 4 37-g butter clams at the U.S. regulatory limit of 80 μg per 100 g tissue.

Table 2 Toxic doses (in μg) of Paralytic Shellfish Poison

	Alaska (7)	Alaska (16)	Guatemala (4)	England (18)	British Columbia (13)	Eastern Canada (5)
Category	N = 54	N = 10	N = 5	N = 71	N = 2	N = 37
Minimum toxic dose	13			558		160
Minimum lethal dose	5863		2046		2600	
Minimum toxic dose per kg		21				
Minimum lethal dose per kg		230	89		36	
Mean toxic dose	5452	9176				
Maximum nonlethal dose	123457			5580		8272
Maximum asymptomatic dose	36580			5580		3000

Three studies present the minimum lethal dose of shellfish toxin and all three are in general agreement that somewhere in excess of 2000 μg represents a potentially lethal dose. There is more variation in the estimated minimum lethal dose per kilogram (see Table 2). The highest reported shellfish toxin levels are approximately 20,000 μg per 100 g tissue (11,16). At this level, a lethal dose may be ingested from as little as two 4-gram mussels. If it is assumed that a lethal dose per kilogram is 36 μg, then a child may consume a lethal dose from eating a fraction of a mussel.

Three studies provide information on the maximum amount of shellfish toxin that an individual may ingest without illness. This value ranges from 3000 to over 36,000 μg, placing these values considerably above the level needed to induce potentially lethal illness. The reason for this is unclear, but what is clear is the consistency of data which suggests that illness and illness severity are not related to estimated ingested toxin dose in any simple way (5,7,18).

One study attempted to determine the toxin concentration in shellfish eaten by healthy people (1). Following random selection, participants were contacted by telephone and asked if they had shellfish at their home that they had collected from unmonitored beaches. This study found that 29 people had eaten shellfish containing over 80 μg per 100 g tissue on multiple occasions, including some who had eaten shellfish that contained over 200 μg per 100 g tissue. Of these 29 persons, one may have experienced mild paresthesias.

D. Geographical Distribution

PSP has a wide distribution and cases have been reported from as far south as Chile (21,22) and as far north as Alaska (7). Other cases have occurred in Taiwan (14), South Africa (12), England (18), Guatemala (4), Costa Rico (17), Singaporc (15), Canada (5,13), Spain (23), Mexico (24), Japan (25), and other areas of the United States (11,26). Similar to incidence, Anderson has raised the prospect that the distribution of PSP is increasing. He supports this argument by identifying more than twice as many areas that reported PSP during 1990 as 1970. As mentioned above, it is problematic trying to determine if geographical and temporal trends represent changes in the occurrence of PSP or changes in the occurrence of confounding factors.

E. Temporal Distribution

Most reports of clinical illness come from isolated outbreak investigations making conclusions regarding seasonal and longer term distribution difficult. A study of 20 years of surveillance data in Alaska found that although the majority of cases of PSP occurred during late spring and early summer, cases occurred during all seasons of the year and during every month except November and December (7). Similarly, a large series in eastern Canada reported outbreaks between March and November, with the great majority occurring between June and September (5). Outbreaks in Costa Rica and Mexico, by contrast, have occurred during October and December (17,24). Some species, such as mussels, have rapid depuration of toxin, consistent with seasonal variation in outbreaks. Other species, however, such as the butter clam, *Saxidomus giganteus*, may retain toxin for up to 2 years following a single exposure to toxic dinoflagellates (5,27).

F. Risk Factors

No definitive risk factors for PSP have been identified. Particularly surprising is the consistency of reports that have failed to identify a correlation between estimated ingested toxin dose and illness. Although no studies have been conducted in humans, animal studies suggest that antibodies or binding proteins may be produced against PSP components (28–32). Genetic differences (33), differences in toxin components, and poor estimation of toxic dose may also help explain this finding.

Two studies have examined the role of alcohol consumption in PSP, with one study finding alcohol to have no relation to illness (12) and the other study finding a protective role of alcohol (7). Another study found that Alaska Native race was associated with illness (1). Finally, one study suggests that age increases the likelihood of death following exposure (4). Although PSP is a heat-stable toxin, boiling the shellfish appears to decrease the risk of illness because of elution of toxin into the water.

G. Public Health Issues

The threat of PSP resulting from the consumption of commercial shellfish has led to the implementation of shellfish monitoring programs in many areas of the world with commercial shellfish harvests. For example, in Alaska, the cost to the farmer of sampling for PSP represents approximately 5% of the total crop value. The cost to society is higher because of the added costs associated with the mouse bioassay performed by the State Health Department. Despite these costs, there seems little disagreement among producers that rigorous monitoring must continue to safeguard commercial interests as well as public health. Jensen reports on the economic consequences of PSP outbreaks (34). Another case, unrelated to PSP intoxication, provides a warning to the shellfish industry: A single death in Belgium associated with canned salmon produced in Alaska led to a decrease in market demand for all types of Alaskan salmon and the loss of an estimated $300,000,000 (35). With potential economic impacts of this magnitude, despite relatively small human health impacts, PSP monitoring programs for commercial shellfish beds will continue to receive support.

More problematic is monitoring of subsistence and recreational shellfish harvesting. Washington state and British Columbia have established a program to monitor a few recreational beaches and to close the remainder during the summer months. The implicit message from this program is that shellfish are safe to eat when a particular beach is not closed. In Alaska, however, outbreaks of illness have occurred during all seasons (7). Moreover, shellfish toxin levels may vary widely from one section of a particular beach to another (B.D. Gessner, unpublished data) and, as demonstrated above, one mussel may have sufficient toxin to cause death. Thus, in Alaska, the Health Department has taken the position that shellfish harvested from noncommercial beaches are unsafe to eat regardless of the season. Unfortunately, this position may conflict with the cultural and nutritional benefits which derive from subsistence food harvesting (36) and, by its very inflexibility, lead people to ignore it (1). It may also lead to conflict between different governmental agencies that are charged with different tasks, such as protecting the public health versus promoting tourism. If recreational shellfish monitoring programs are implemented, they should consider the following issues (27,35):

- The geographical and temporal sampling frame, taking into account issues of safety, cost, and fairness to a multitude of local communities.
- The representativeness of toxin levels in tested shellfish for an entire beach.
- The impact of a monitoring program on the public's belief in shellfish safety; an unrealistic expectation of shellfish safety may lead to an increase in PSP cases.
- Given the expense and ethical issues associated with the current mouse bioassay, the sources of funding and support for the program.
- The criteria that will be used to reopen a beach that has been closed.

Education programs also have the potential to impact the occurrence of PSP cases. A study in Alaska found that residents held erroneous beliefs regarding PSP, including that it could only occur during the winter or following a red tide and that cooking or eviscerating the shellfish rendered it safe (7).

III. CIGUATERA

A. Incidence

Several studies have attempted to define the incidence of ciguatera and some have reported incidence at multiple locations within the same geographical area and over time (Table 3). A study from Hawaii reported that the incidence varied among different islands from 3 to 34 per 100,000 persons per year during 1988 (37). Within islands, the incidence stayed relatively constant during 1984–1988, but on Kauai varied from a low of 17 to a high of 105 per 100,000 per year. The South Pacific showed variation between nations and territories, ranging from a low of 1 to a high of 653 per 100,000 per year (38). As with Hawaii, the incidence remained relatively constant within areas during 1973–1983. Lewis (38) provides an excellent visual image of the distribution of ciguatera among different areas during 1979–1983 (Figure 1). In the eastern Caribbean, disease incidence varied from 3.3 per 100,000 persons per year in Cuba

Table 3 Ciguatera Incidence by Area

Location	Incidence (per 100,000 per year)
Hawaii, 1988 (37)	
Oahu	3
Kauai	13
Maui	11
Hawaii island	34
Hawaii state	8
South Pacific, 1973–1983 (38)	
American Samoa	87
Cook Islands	1
Fiji	16
French Polynesia	545
Guam	8
Kiribati	324
Nauru	7
New Caledonia	200
Niue	130
Papua New Guinea	>1
Solomon Islands	2
Tokelau	653
Tonga	21
TIPI	173
Tuvalu	439
Vanuatu	25
Wallis and Futuna	9
Western Samoa	54
Virgin Islands (39)	365–730
Australia, 1965–1985 (41)	30

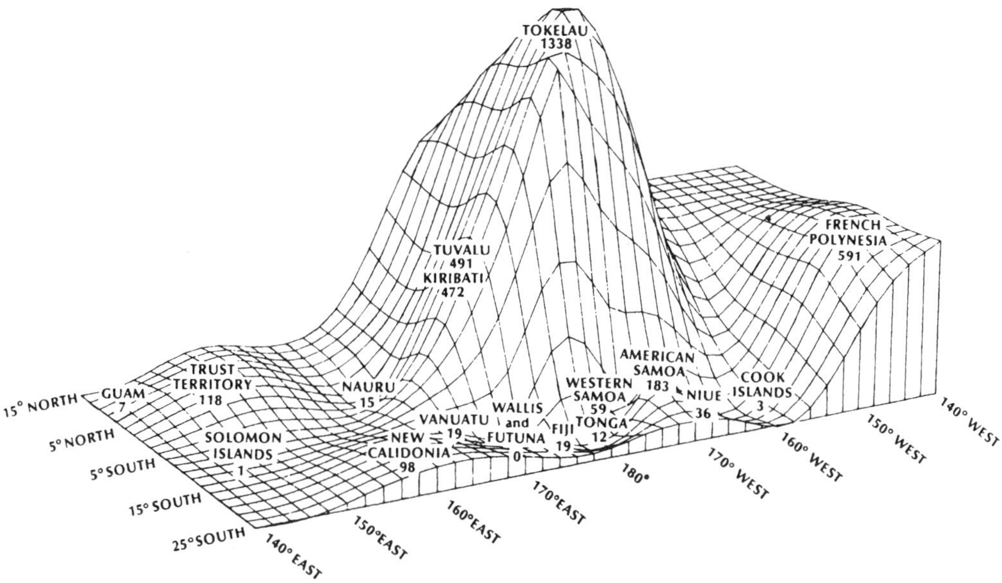

Figure 1 Mean annual incidence of ciguatera 1979–1983 (cases per 100,000). (From Ref. 38.)

to 730 per 100,000 persons per year in St. Thomas, Virgin Islands (39,40). Incidence data are also available from Australia (41), with a reported value of 30 per 100,000 persons per year.

Certain islands or island groups in the South Pacific have extraordinarily high ciguatera incidence rates. For example, on the island of Atiu in the Cook Islands, 19% of persons reported having experienced ciguatera at some point in the past and 12% of the population had developed ciguatera during the 2-year period before the study (estimated incidence 6243 per 100,000 persons per year) (42). Even more dramatically, the population of the Gambier archipelago in French Polynesia had an incidence of 22,700 cases per 100,000 per year during 1960–1984 (43). The author of this study emphasized that the incidence in the South Pacific varies dramatically among different islands and increased among certain islands during 1960–1984 (43).

These latter two observations may be related to "decreasing diversity of marine and terrestrial fauna," which in turn may be related to nuclear test explosions (43) and other environmental disruptions of marine ecosystems (44,45). As has been pointed out, however, significant disruptions to marine ecosystems have occurred without an increase in ciguatera (46). Additionally, increases in ciguatera poisoning may reflect changes in diagnostic capability and surveillance systems. Nevertheless Ruff's hypothesis (43) remains intriguing and agrees in its broad principle with the arguments of Anderson regarding PSP (8).

During outbreaks of ciguatera, some investigators have attempted to determine attack rates. Where these data are presented, the attack rates following consumption of the same fish have been high. Among five reviewed studies, the attack rates varied from 63 to 100%, emphasizing the consistent, although not necessarily uniform, distribution of ciguatoxins among the edible flesh of toxic fish (47–51).

B. Clinical Features

A plethora of studies exist documenting the clinical presentation and course of ciguatera. Additionally, at least 175 symptoms have been reported to occur during ciguatera (52), likely reflecting the variety of toxins that may cause illness (53,54). The present analysis does not attempt a comprehensive review of all studies and all symptoms. Instead, a representative sample of large case series from different areas has been gathered. Ciguatera presents with neurological, gastrointestinal, and cardiac symptoms. The median or mean time from ingestion to onset of illness is longer than for PSP and in different studies has varied from 5 to 7 h (48–50,55). The minimum time reported (except for cases where palytoxin is implicated) is 30 min, whereas the maximum time is 48 h.

Five large series (41,55–58)—two from the Atlantic and three from the Pacific Ocean areas—were reviewed (Table 4). Gastrointestinal effects include nausea, vomiting, diarrhea, and abdominal pain. The predominant neurological effects include acral and circumoral paresthesias, vertigo or dizziness, ataxia, and diaphoresis. Temperature reversal (also known as paradoxical dysasthesia) is a well-known feature of ciguatera, but one investigator points out that although

Table 4 Ciguatera Symptoms by Area

Symptom	South Pacific (56) N = 3009 (%)	Australia (41) N = 527 (%)	Hawaii (57) N = 203 (%)	Miami (55) N = 129 (%)	Virgin Islands (58) N = 33 (%)
Gastrointestinal					
Diarrhea	71	64	65	76	91
Nausea	43	55	38		
Vomiting	38	35	37	68	70
Abdominal pain	47	52	28		39
Neurological					
Acral paresthesias	89	71	63	71	33
Circumoral paresthesia	89	66	61	54	36
Temperature reversal	88	76	48		36
Vertigo or dizziness	42	45	21		21
Ataxia	38	54			
Diaphoresis	37	43	15	24	18
Tremor	27	31			9
Salivation	19	10			
Dyspnea	16	28	2		
Paresis	11	27			
Cardiovascular					
Bradycardia	13				
Hypotension	12		2		
Other					
Myalgia	81	83	64	86	
Arthralgia	86	79	4		52
Weakness	60		69	30	58
Chills	59	49	24		24
Headache	59	62	12	47	
Pruritus	45	76	21	48	58
Dental pain	25	37			
Neck stiffness	24	27			24
Watery eyes	22	41	4		21
Rash	21	26	3		9
Dysuria	19	22			
Death	0.1	0.2			

cold objects may feel hot, no reports exist of hot objects feeling cold (59); this investigator suggests that the symptom may more closely resemble hyperasthesia than dysasthesia. Other common symptoms include myalgia, arthralgia, weakness, chills, headache, and pruritus. Cardiovascular symptoms include hypotension (60) and bradycardia, as well as less commonly arrhythmias. Death is rare, with only three occurrences out of 3009 cases documented by Bagnis (56) and one occurrence out of 527 cases documented by Gillespie (41).

Unlike PSP, there is a firm body of evidence that documents the occurrence of hypotension and bradycardia with severe ciguatera. Katz, for example, reported bradycardia for 66% and hypotension for 27% of cases in Hawaii (47), whereas Bagnis reported bradycardia for 14% and hypotension for 12% of cases in the South Pacific (56). One report suggests that hypotension in ciguatera results from parasympathetic excess and sympathetic failure (60). Occasionally, tachycardia has occurred as well.

Others have compared the symptoms reported in different regions and have noted that affected persons in some areas of the South Pacific (e.g., French Polynesia) have more severe symptoms than those in Hawaii, including more temperature reversal and cardiovascular manifestations (57). To take the studies compared in Table 4 at face value, it appears that the South Pacific and Australia have a higher proportion of persons with neurological symptoms (other than paresthesias) than persons in Hawaii or the Atlantic region and a lower proportion of persons with gastrointestinal symptoms. A geographical difference in the distribution of symptoms is certainly consistent with the variety of toxins that may cause a clinical presentation consistent with ciguatera. One should exercise caution, however, in making a direct comparison. The studies of Bagnis and Lawrence were designed prospectively, whereas the others were retrospective. Moreover, it is not clear from the reports whether symptoms that were not reported did not occur or were not assessed.

Some reports have suggested that symptoms may vary by the type of fish consumed. Bagnis associated consumption of herbivores such as surgeonfish with gastrointestinal illness and consumption of carnivores such as grouper with neurological illness (3). Kodama (61) and Glaziou (62) found that, compared with consumption of herbivores, consumption of carnivores resulted in more severe disease, including increased cardiovascular symptoms; additionally, symptoms may differ following consumption of different carnivorous species. Presumably, these findings represent the accumulation of different toxins or the same toxin at different concentrations within different species.

In contrast to PSP, prolonged duration of symptoms in ciguatera is well documented. For two studies where this information was available, the mean or median length of illness was 24 and 72 days (47,63), but both of these studies included a small number of cases. Gastrointestinal symptoms resolve relatively quickly over several hours to a week (48–50,58). Neurological symptoms, including weakness, paresthesias, and temperature reversal, have been reported to last up to 180 days (49,55,58,64). In one extreme example, polymyositis was documented 11 years after a case of ciguatera (65), although the causal link in this case is questionable.

C. Geographical Distribution

Dinoflagellates that produce one or more of the toxins associated with ciguatera are found worldwide within tropical coastal waters. As has been pointed out, however, ciguatera occurs more commonly on islands rather than continental coasts, the most notable exceptions being Florida and the Great Barrier Reef of Australia (46). It is hypothesized that the implicated dinoflagellate thrives in areas most exposed to oceanic flows and does not thrive near continents or other major land masses with land runoff. A review published during 1994 documents the particular countries that have reported cases of ciguatera (66). In the United States, cases have now been reported from fish caught as far north as North Carolina (50), and a recent report documents the first outbreak in California, which was traced to fish caught off the coast of Baja, California, Mexico (67).

With increases in interstate fish transport, more outbreaks have occurred in areas without risk of indigenous ciguatera such as Canada (63), Rhode Island (68) , California (69), and Vermont (70). Additionally, clinicians in any part of the world may see patients who present after acquiring illness during travel (71,72). Finally, one report identified a case of ciguatera that resulted from the consumption of farm-raised salmon, raising the possibility of ciguatera occurring in novel locations (73).

D. Temporal Distribution

Temporal differences may occur in the rate of ciguatera fish poisoning but, as with PSP, the occurrence of seasonal variations and the seasons of high risk vary by location. In Hawaii, the greatest number of cases occurred during July, but no overall seasonal distribution was identified (57). Similarly, no seasonal variation was found in the Virgin Islands (39). In Miami, Florida, the majority of cases occurred during May, with a clear increase in cases during the spring and summer (55), whereas in Puerto Rico, the majority of cases occurred during January, March, and April (64). In Puerto Rico, investigators have found consistent increases in the number of ciguatoxic barracuda during January through April, but they also have found less consistent increases during the summer and fall (74). As one review points out, fish may remain toxic for years following exposure to ciguatoxin, an observation which likely explains the reported increased risk of illness following consumption of older and larger fish (53). This finding may also help to explain the lack of seasonal variation in human illness despite variations in dinoflagellate blooms.

E. Risk Factors

Numerous studies have attempted to identify risk factors for ciguatera, although the outcomes and risk factors measured have differed widely between these studies. One study reports an association between illness and age (47), but three other studies report no association (48,55,57); additionally, cases have been reported among persons from less than 1 to 83 years of age (39).

Similarly, Bagnis (56) and Glaziou (62) found 50–60% more males than females with ciguatera, but numerous other studies have found no association with gender (39,47,48,55,58); it is possible the described association represents gender-specific differences in fish consumption practices in the South Pacific.

Perhaps the strongest documented association is between previous exposure to ciguatera and either severity of illness relative to presumed dose (56,62) or illness (39,56). Glaziou has explained his findings by suggesting that humans may accumulate toxin. It should be kept in mind that two smaller studies report no association with previous exposure (47,48). Fish evisceration does not appear to be protective (47,58) despite concentration of toxin in specific organs, and, because the toxin is heat stable, cooking also provides no protection.

Two studies suggest that alcohol increases the severity or chronicity of symptoms (50,72), whereas a third study found no association with severity of illness (48). Two studies report no association with race (55,58). Other reported risk factors include exertion and eating large fish. All of the studies which examined risk factors suffer from one or more methodological problems, including lack of a control group, retrospective design, lack of clear case definitions, failure to use appropriate denominators (i.e., the population of persons who eat fish rather than the overall population), and lack of a systematic approach to measurement of risk factors and outcomes. The discrepancies between studies may be attributable to these methodological flaws, to differences in population characteristics, or to differences in toxin components among different areas.

F. Public Health Issues

Recently, an immunobead assay for the detection of ciguatoxin in fish has been developed (Cigua-Check, Oceanit Laboratories, Hawaii). However, because of the difficulty with testing fish for the presence of the variety of toxins that might cause ciguatera, no regulatory limits exist. Instead, governments have reacted in a variety of other ways, including banning the sale of fish suspected to cause ciguatera such as grouper, snapper, barracuda, and surgeonfish in French Polynesia (75) and barracuda in Miami (55); provision of education regarding ciguatera case identification to medical providers (55); and, despite its limitations, using an assay for fish sampling with subsequent withdrawal of toxic fish from the market (70,76–78). Additionally, where governments have not acted, the combination of legal decisions and insurance industry pressure has prompted interventions such as the placement of warnings on restaurant menus in endemic areas (79). Research is currently ongoing in the United States to develop a screening method for fish that will meet the requirements of the new Hazard Analysis and Critical Control Point approach adopted by the U.S. Food and Drug Administration. This approach specifies that preventive efforts, in this case toxin detection, be directed at the first point of encounter with the fish after the fisherman.

Ciguatera undoubtedly limits subsistence seafood harvesting in endemic regions, but the extent of its impact is unclear and likely varies by location. Factors which affect whether the presence of ciguatera limits seafood harvesting may include the economic scale of the fisheries industry relative to other industries, the availability of other protein sources, and the perceived alteration of a state of well-being from ciguatera poisoning relative to other health events. In the Pacific region, Lewis (38) and Bourdy (80) report a number of strategies which islanders have adopted to avoid ciguatera, including avoidance of high-risk species, discarding the internal organs of fish, and feeding fish to a pet and observing the reaction (38). Additional strategies have been employed which are less effective or ineffective, such as cooking the fish with plant materials or feeling the texture of the fish. Another report documents similar practices among the residents of the Dominican Republic (81). It is unclear, however, how many people avoid fish because of the presence of ciguatoxin. In Puerto Rico, an area with relatively high levels of health and a diversified economy, the threat of ciguatera has been shown to lead people to avoid eating fish entirely (64).

Similar to the case with PSP, Bagnis and Lewis have suggested that ciguatera results in the loss of hundreds of thousands of dollars in commercial fish harvesting in French Polynesia (38,75,82). The threat of ciguatera may also adversely effect the tourist industry, particularly hotels and restaurants. A single well-publicized outbreak may adversely affect income not only at one or several businesses but also for an entire circumscribed location that becomes associated with the outbreak in the public's viewpoint (82).

Eventually, inexpensive, sensitive, and specific tests may be developed for identification of ciguatoxins in fish. If this occurs, more specific regulatory intervention will not be far behind. Public health and regulatory measures to control ciguatera should consider the relative value of a fish diet compared to the risk of ciguatera, the economic and social importance of fish harvesting to a community, the anticipated intervention when toxic fish are identified (particularly if toxic fish represent a considerable portion of the total harvest), and guidelines for relaxation of specific restrictions (e.g., import or export restrictions) once implemented.

IV. AMNESTIC SHELLFISH POISONING

A. Incidence

The incidence of amnestic shellfish poisoning (ASP) is unknown but appears to be low, as only one confirmed outbreak of human illness has been reported. Following the detection of domoic acid in razor clams in the State of Washington, an epidemiological investigation was conducted to identify possible cases of human illness (83). Among 127 persons who had recently eaten razor clams, no illness was identified.

B. Clinical Features

Almost all of the information regarding clinical features of ASP derives from the originally described outbreak in Canada involving 107 people (2,84). Illness onset during this outbreak varied from 15 min to 38 h (mean 5.5 hs). The most common symptom was nausea (77%) followed by vomiting (76%), abdominal cramps (51%), diarrhea (42%), headache (43%), and memory loss (usually but not exclusively anterograde) (25%). Among hospitalized patients, symptoms included confusion, disorientation, coma, mutism, grimacing, seizures, hiccups, and emotional lability. Physical findings included no response to painful stimuli, piloerection with miosis or mydriasis, paresis, opthalmoplegia, unstable blood pressure, and arrhythmias (2,84). Three patients died. During examination several months after exposure, prolonged symptoms included memory deficits, atrophy and mild weakness of the extremities, and hyporeflexia (84).

Quick used a case-control study to identify symptoms resulting from exposure to lower doses of domoic acid (83). Initial analysis suggested that persons with mild gastrointestinal and neurological symptoms had eaten razor clams with a higher concentration of domoic acid than persons without symptoms. Unfortunately, this study was not completed and further studies have not been conducted. Consequently, the effect of low-dose exposure to domoic acid remains unknown. The hypothesis exists that exposure to environmental chemicals, such as domoic acid, underlies some human neurodegenerative disorders, including Parkinson's disease and dementia of the Alzheimer type. A recent article, however, suggests that it is unlikely that progressive neurodegenerative disorders are linked to environmental toxins (85).

C. Toxic Dose

During the Canadian outbreak of ASP, the implicated mussels contained from 31 to 128 mg of domoic acid per 100 g of tissue and total ingested dose ranged from 60 to 290 mg (2). Although not reported in the original article by Perl (2), a subsequent article reported that a dose per kilogram could be calculated for seven persons with mild symptoms and ranged from 0.9 to 2.0 mg/kg (86). Primate studies also suggest that 1 mg/kg represents a toxic dose (87). Quick found that cases and controls had eaten razor clams with 3.7 and 2.6 mg of domoic acid per 100 g of tissue, respectively; the doses for these two groups were 12.5 (range 4.2–29.0) and 6.5 mg (range 0–24.4) (83). The interpretation of this latter data is uncertain. It seems clear that ingesting 60 mg will lead to illness in some patients, but the lowest toxic dose remains to be determined.

D. Geographical Distribution

The geographical distribution of confirmed ASP remains limited to eastern Canada. Nevertheless, domoic acid has been recovered from the entire west coast of North America, including Mexico (88), California, Oregon, Washington, and Alaska and recently also in Europe (89). Additionally, Japanese researchers have identified new isomers of domoic acid (90).

E. Temporal Distribution

No temporal trend data for human ASP exists. Similar to the case with other marine biotoxins, some investigators have argued that toxic blooms are increasing (88).

F. Risk Factors

Among ill persons with ASP, males and the elderly had an increased risk of memory loss and hospitalization (2). Perl (2) has suggested that the association with age was due to increased renal disease in the elderly and thus that domoic acid is excreted through the kidneys. By contrast, Auer has suggested that increased susceptibility with age is related to the dendritic location of excitatory receptors and the increased branching of neuronal dendritic trees among the elderly (91). Because specific parts of different species may concentrate toxin—for example, the viscera of dungeness crabs or the foot of razor clams (92)—selective consumption of these parts may increase the risk of illness. In the Canadian outbreak, cooking was not protective (2); another study, however, suggests that boiling dungeness crabs significantly reduces the visceral toxin level (93). Some ethnic groups may have an increased risk of toxin exposure because of different patterns of seafood consumption; for example, the practice of eating the viscera of crabs among persons of Chinese descent in Washington state (86).

G. Public Health Issues

Based on current knowledge, domoic acid is primarily of public health concern because of its potential for widespread illness via commercial shellfish. During the original outbreak of ASP, 68% of persons became ill in 1 of 45 different restaurants (2). Based on this fear, regulatory limits for commercial shellfish have been established. The current level of 2 mg per 100 g of tissue (20 ppm) (94) was established based on animal studies. A recent study suggests that, based on consumption patterns and toxic dose estimates, a tolerable regulatory level would equal 20 ppm for razor clams and 32 ppm for dungeness crabs (86). It remains unclear whether the current regulatory limits are excessive or too low for preventing human illness. It is clear, however, that these regulatory limits will result in periodic closures of fisheries, as toxin levels in excess of 20 ppm have been found on numerous occasions (95).

V. TETRODOTOXIN

A. Incidence

No reliable incidence data exist for tetrodotoxin poisoning. The most extensive data on tetrodotoxin poisoning comes from Japan where 6386 cases of pufferfish poisoning were reported during the 78-year period 1886–1963 (59.4% were fatal) (96,97). If the average population during this time is assumed to have been approximately 60,000,000, this implies a minimum incidence of 0.14 cases per 100,000 population per year. Another report from Japan identified 2,688 *deaths* due to pufferfish ingestion during 1927–1949 (98). Using the same denominator, this implies that 0.2 deaths per 100,000 population per year occurred owing to tetrodotoxin poisoning during this time period. Interestingly, with the exception of the period during World War II, the number of reported pufferfish poisoning episodes in Japan during 1886–1963 remained relatively constant at 100–300 per year. Additionally, during the same time period, no systematic decrease in the case fatality rate occurred. More recently, 495 persons became ill from pufferfish ingestion during 1977–1986 (25).

Other Southeast Asian countries have also reported cases of tetrodotoxin poisoning. A report from the Poison Control Center in Taiwan, with a 1989 population of approximately 20,000,000, identified 20 outbreaks involving 52 patients during 1988–1995 (99). This suggests a minimum incidence of 0.03 cases per 100,000 population per year. Similarly, in Thailand, with a 1989 population estimate of approximately 55,000,000, 71 persons developed tetrodotoxin poisoning from horseshoe crab ingestion during January 1994 through May 1995 (100). This suggests a minimum incidence related to ingestion of this animal of 0.09 cases per 100,000 population per year.

B. Clinical Symptoms

Symptom onset of tetrodotoxin poisoning occurs within minutes and only rarely more than 6 hs after eating a toxic animal (99,101,102). Perioral paresthesia is the most immediate symptom and acral paresthesia is the most common symptom (Table 5). Nausea and vomiting may or may not occur. Disease may progress to dizziness or vertigo, weakness, ataxia, dyspnea, diaphoresis, and death from respiratory failure. Similar to PSP, affected persons may report a floating sensation (102). Clinical findings may include mydriasis, motor paralysis, respiratory paralysis, tachycardia, and bradycardia (103). Additionally, although hypotension has been a classic finding, at least three reports from Taiwan have documented the occurrence of hypertension (99,104,105); in one case, the blood pressure rose to 300/140 with death occurring 2 h after consumption of the implicated fish (105).

The mortality rate in tetrodotoxin poisoning is dependent on, among other things, timely access to intensive care facilities. In some series, it has approached 60% (96,106). When death results, it usually occurs within 6 h, and sometimes as rapidly as in 17 min, following toxin ingestion. Persons who have not died within 24 h generally recover completely. Similar to PSP,

Table 5 Tetrodotoxin Symptoms in Two Different Areas

Symptoms	Taiwan (mainly pufferfish) (99) N = 52 (%)	Thailand (horseshoe crab) (100) N = 71 (%)
Acral paresthesias	54	87
Perioral paresthesias	48	94
Vomiting	40	30
Dizziness	37	
Weakness	29	44
Headache	25	
Dyspnea	17	
Vertigo	12	42
Diaphoresis	9.6	
Respiratory paralysis	33	27
Ataxia	27	
Hypertension	25	
Mydriasis	15	13
Hypotension	13	
Cyanosis or tachycardia	9.6	
Death	13	2.8

symptoms of tetrodotoxin poisoning usually resolve within 1–2 days, and residual impairment has not been reported.

Fukuda and Hani, as reported by Halstead (96), have divided tetrodotoxin intoxication into four stages of progression. Stage 1 includes oral paresthesias with or without gastrointestinal symptoms. Stage 2 includes paresthesias of other areas and motor paralysis. Stage 3 includes muscular incoordination, aphonia, dysphagia, respiratory distress, precordial chest pain, cyanosis, and hypotension. Stage 4 includes depressed mental status, respiratory paralysis, and severe hypotension. As measurement of tetrodotoxin levels in implicated seafood has not usually been performed, it remains unknown whether this disease classification corresponds to the toxin dose or other biological parameters.

C. Toxic Dose

The toxic and lethal doses of tetrodotoxin are not known. In Taiwan, 30 persons became ill following consumption of the ovaries of an unknown species of fish. Subsequent testing of uneaten ovaries revealed toxin levels of 54 and 287 mouse units (MU) per gram of tissue with an estimated intake of no more than 74,000 MU (105). A second study in Taiwan measured toxin levels in the implicated marine organisms for six outbreaks with toxin varying between 13 (for gastropod mollusks) to 1200 MU (for pufferfish roe) per g of tissue; unfortunately the amount of fish or mollusk eaten was not reported (99). The lethal dose for humans has been estimated as 200,000 MU (105,107).

D. Geographical Distribution

Human intoxication from tetrodotoxin has occurred in a variety of species that live in diverse ecosystems (108), including pufferfish and other tetraodontiforme fish, the blue-ringed octopus (109,110), mollusks (99), horseshoe crabs (100), and the Oregon newt (111). Moreover, tetrodotoxin-containing fish exist in tropical waters throughout the world (102). For most populations, however, species that contain tetrodotoxin do not constitute a significant part of the diet. Consequently, illness is generally restricted to areas where potentially toxic animals such as pufferfish, gastropod mollusks, and horseshoe and other crabs are eaten: Southeast Asia and, more specifically, Japan. In addition to Japan, Taiwan, and Thailand, intoxication has also been reported from the South Pacific (112), Malaysia (113,114), Hong Kong (115,116), Singapore (117), and Australia (109,118). It is possible that fatalities from eating some species of crab on Negros Island, Philippines, also resulted from tetrodotoxin poisoning (119,120).

E. Temporal Distribution

Cases of tetrodotoxin poisoning occur during all months of the year. It is not known whether the proportion of tetrodotoxin fish has increased. Simple incidence data would not necessarily answer this question, since public health measures such as education and regulation of fugu chefs in Japan may effect incidence estimates regardless of changes in the proportion of animals containing toxin. Additionally, it is possible that improved medical care has lowered the number of lethal cases.

F. Risk Factors

No risk factors for tetrodotoxin poisoning are known. It is likely that intoxication and its severity are dose dependent (99). Age has not been shown to increase the risk of illness; in Taiwan, illness occurred in persons from 9 months to 71 years of age (99). The toxin is heat stable, so that cooking is not protective. Tetrodotoxin concentrates in the viscera and roe of some animals, such as pufferfish (108). Presumably, removal of the viscera will provide some measure of protection, although cases have been reported where only the flesh of the fish was eaten (99). Previous exposure does not provide protection (102).

G. Public Health Issues

No regulatory limits for tetrodotoxin have been established in the United States, as personal importation of pufferfish is prohibited. An agreement between the U.S. Food and Drug Administration and the Japanese Ministry of Health and Welfare has been adopted which allows importation of fugu for special occasions provided the fish is certified safe by the Japanese government before export (101). Japan and Taiwan have attempted to control tetrodotoxin poisoning through licensing of restaurants and chefs or by establishing regulatory limits for the sale of pufferfish (25). As people may ingest tetrodotoxin from fish not served at restaurants, this approach will prevent only a portion of cases. Some countries, including Japan, have enacted laws restricting the sale of certain species known to cause tetrodotoxin poisoning (102).

VI. NEUROTOXIC SHELLFISH POISONING

A. Incidence

No data on neurotoxic shellfish poisoning (NSP) incidence exist in the literature. Less than 100 cases have been reported in the United States, approximately half of which came from a single outbreak in North Carolina (121).

B. Clinical Symptoms

The most rigorous analysis of data on NSP comes from an outbreak in North Carolina involving 48 persons (121). The median latent period between ingestion and onset of illness was 3 h (range 15 min to 18 h) with a similar onset for both gastrointestinal and neurological symptoms. The most common symptoms were paresthesias (81%), vertigo (60%), malaise (50%), abdominal pain (48%), nausea (44%), diarrhea (33%), weakness (31%), ataxia (27%), chills (21%), headache (15%), myalgia (13%), and vomiting (10%). Illness lasted from 30 min to 3 days (median 17 h) and no long-term symptoms have been reported. The symptoms reported from the North Carolina outbreak in general agree with other investigations of illness from Florida (122,123), although cases in Florida tended to have a shorter incubation period and less associated nausea and vertigo. Among other causes, reported differences in symptoms may result from qualitative or quantitative differences in toxin consumption or more rigorous identification of milder cases. Death, if it occurs, is exceedingly uncommon.

The above documents the consequences of neurotoxin ingestion. A few reports suggest that inhalation of aerosolized toxin may cause conjunctival irritation, rhinorrhea, respiratory irritation and possibly exacerbate or cause symptoms similar to reactive airways disease (124,125).

C. Toxic Dose

During the outbreak in North Carolina, implicated oysters had a NSP toxin level of 35–60 MU per 100 g of tissue. Two persons became ill after consuming less than 12 oysters, but in this group the attack rate was only 13%. At 12 oysters and above, the attack rate equaled 65%. If 12 oysters are used in the calculation, and we assume a weight of 10 g per oyster, a low (but not minimum) toxic dose estimate equals 42–72 MU. No lethal toxic dose estimates based on human intoxication episodes exist.

D. Geographical and Temporal Distribution

As summarized by Fleming, the causative agents of NSP have been found in Florida, North Carolina, the Gulf of Mexico, Brazil, Spain, Japan, New Zealand, and the Solomon Islands (126). Most reports of illness and dinoflagellate blooms come from Florida, although this finding may represent differences in surveillance rather than true differences in occurrence. No information on temporal distribution exists, as published reports have relied on outbreaks rather than systematically collected surveillance data.

E. Risk Factors

No risk factors for NSP have been identified other than estimated ingested dose (121). The investigation in North Carolina examined age, gender, the presence of chronic illness, medication use, and alcohol consumption during the implicated meal and found no association with illness. As with other marine neurotoxins, brevetoxin is heat stable, and thus cooking contaminated seafood will not alter the risk of intoxication. Furthermore, the toxin is lipid rather than water soluble (127), and thus boiling or steaming contaminated food is similarly unlikely to alter the risk of intoxication.

F. Public Health Issues

As with ASP, NSP is primarily of public health concern because of its potential for large outbreaks via distribution in commercial seafood products. Because of this concern, public health agencies in Florida have routinely monitored coastal waters for the presence of *Ptychodiscus brevis* since the mid 1970s. Shellfish beds are closed when *P. brevis* concentrations exceed 5000 cells/mL (124) and reopened when shellfish toxin levels drop below 20 MU (approximately 80 μg) per 100 g tissue (9). North Carolina implemented a similar program following the occurrence of human brevetoxin intoxication (121).

REFERENCES

1. BD Gessner, M Schloss. A population-based study of paralytic shellfish poisoning in Alaska. Alaska Med 38:54–58, 68, 1996.
2. TM Perl, L Bedard, T Kosatsky, JC Hockin, ECD Todd, RS Remis. An outbreak of toxic encephalopathy caused by eating mussels contaminated with domoic acid. N Engl J Med 322:1775–1780, 1990.
3. R Bagnis. Clinical aspects of ciguatera (fish poisoning) in French Polynesia. Hawaii Med J 28:25–28, 1968.
4. DC Rodrigue, RA Etzel, S Hall, E de Porras, OH Velasquez, RV Tauxe, EM Kilbourne, PA Blake. Lethal paralytic shellfish poisoning in Gautemala. Am J Trop Med Hyg 42:267–271, 1990.
5. A Prakesh, JC Medcof, AD Tennant. Paralytic shellfish poisoning in eastern Canada. Ottawa: Fisheries Research Board of Canada, Bull 177, 1971.
6. WHO Aquatic (Marine and Freshwater) Biotoxins. Environmental Health Criteria 37. International Programme on Chemical Safety. Geneva: World Health Organization, 1984.

7. BD Gessner, JP Middaugh. Paralytic shellfish poisoning in Alaska: a 20-year retrospective analysis. Am J Epidemiol 141:766–770, 1995.
8. DM Anderson. Red tides. Sci Am 271:62–68, 1994.
9. R Viviani. Eutrophication, marine biotoxins, human health. Sci Total Environ (suppl):631–662, 1992.
10. CY Kao. Paralytic shellfish poisoning. In: I Robert, ed. Algal Toxins in Seafood and Drinking Water. Orlando, FL, 1993, pp 75–86.
11. Centers for Disease Control and Prevention. Paralytic shellfish poisoning—Massachusetts and Alaska, 1990. MMWR Morb Mort Wkly Rep 40:157–161, 1991.
12. MEE Popkiss, DA Horstman, D Harpur. Paralytic shellfish poisoning. A report of 17 cases in Cape Town. S Afr Med J 55:1017–1023, 1979.
13. E Todd, G Avery, GA Grant. An outbreak of severe paralytic shellfish poisoning in British Columbia. Can Commun Dis Rep 19:99–102, 1993.
14. HS Cheng, SO Chua, JS Jung, KK Yip. Creatine kinase MB elevation in paralytic shellfish poisoning. Chest 99:1032–1033, 1991.
15. CTT Tan, EJD Lee. Paralytic shellfish poisoning in Singapore. Ann Acad Med Sing 15:77–79, 1986.
16. BD Gessner, P Bell, GJ Doucette, E Moczydlowski, MA Poli, F Van Dolah, S Hall. Hypertension and identification of toxin in human urine and serum following a cluster of mussel-associated paralytic shellfish poisoning outbreaks. Toxicon 35:711–722, 1997.
17. L Mata, G Abarca, L Marranghello, R Viquez. Intoxicacion paralitica por mariscos (IPM) por Spondylus calcifer contaminado con Pyrodinium bahamense, Costa Rica, 1989–1990. Rev Biol Trop 38:129–136, 1990.
18. JPK McCollum, RCM Pearson, HR Ingham, PC Wood, HA Dewar. An epidemic of mussel poisoning in northeast England. Lancet 2:767–770, 1968.
19. BD Gessner, J Middaugh, GJ Doucette. Paralytic shellfish poisoning in Kodiak, Alaska. West J Med 166:351–353, 1997.
20. T Yasumoto, M Murata. Polyether toxins involved in seafood poisoning. In: S Hall, GR Strichartz, eds. Marine Toxins: Origin, Structure, and Molecular Pharmacology. American Chemical Society Symposium Series No 418. Washington, DC: American Chemical Society, 1990, pp 120–132.
21. D Montebruno. Paralytic shellfish poisoning in Chile. Med Sci Law 33:243–246, 1993.
22. LRV Bravo. Republic of Chile, Ministry of Health, National Program for the Prevention of Poisoning Caused by Red Tide. In: Proceedings of the Workshop Conference on Seafood Intoxications: Pan American Implications of Natural Toxins in Seafood, Miami FL: University of Miami, 1996, INPAAZ Country Report Section, pp 31–38.
23. DM Anderson, JJ Sullivan, B Reguera. Paralytic shellfish poisoning in northwest Spain: the toxicity of the dinoflagellate Gymnodinium catenatum. Toxicon 27:665–674, 1989.
24. OS Castaneda, JLV Castellanos, J Galvan, AS Anguiano, A Nazar. Intoxicaciones por toxina paralizante de molusco en Oaxaca. Salud Publica Mex 33:240–247, 1991.
25. T Kawabata. Regulatory aspects of marine biotoxins in Japan. In: S Natori, K Hashimoto, Y Ueno, eds. Mycotoxins and Phycotoxins 88: A Collection of Invited Papers Presented at the Seventh International IUPAC Symposium on Mycotoxins and Phycotoxins. Amsterdam: Elsevier, 1989, pp 469–76.
26. RR Long, JC Sargent, K Hammer. Paralytic shellfish poisoning: a case report and serial electrophysiologic observations. Neurology 40:1310–1312, 1990.
27. BW Halstead, EJ Schantz. Paralytic shellfish poisoning. WHO Offset Publication No. 79. Geneva: WHO, 1984.
28. B Kaufman, DC Wright, WR Ballou, et al. Protection against tetrodotoxin and saxitoxin intoxication by a cross-protective rabbit anti-tetrodotoxin antiserum. Toxicon 29:581–587, 1991.
29. TC Chanh, JF Hewetson. Polyclonal anti-idiotypes induce specific anti-saxitoxin antibody responses. Immunopharmacology 26:225–233, 1993.
30. BJ Benton, VR Rivera, JF Heweston, et al. Reversal of saxitoxin-induced cardiorespiratory failure by a burro-raised alpha-STX antibody and oxygen therapy. Toxicol Appl Pharmacol 124:39–51, 1994.

31. Y Li, E Moczydlowski. Purification and partial sequencing of saxiphilin, a saxitoxin-binding protein from the bullfrog, reveals homology to transferrin. J Biol Chem 266:15481–15487, 1991.
32. K Daigo, P Noguchi, A Miwa, N Kawai, K Hashimoto. Resistance of nerves from certain toxic crabs to paralytic shellfish poison and tetrodotoxin. Toxicon 26:485–490, 1988.
33. KJ Kontis, AL Goldin. Site-directed mutagenesis of the putative pore region of the rat IIA sodium channel. Mol Pharmacol 43:635–644, 1993.
34. AC Jensen. The economic halo of a red tide. In: VR LoCicero, ed. Proceedings of the First International Conference on Toxic Dinoflagellate Blooms. Boston, MA: Massachusetts Science and Technology Foundation, 1975, pp 507–516.
35. R Ralonde, R Painter. Living with paralytic shellfish poisoning. A Conference to Develop PSP Research and Management Strategies for Safe Utilization of Shellfish in Alaska. Alaska Department of Commerce and Economic Development. June 1995.
36. G Egeland, J Middaugh. Balancing fish consumption benefits with mercury exposure. Science 278: 1904–1905, 1997.
37. JH Gollop, EW Pon. Ciguatera: a review. Hawaii Med J 51:91–99, 1992.
38. ND Lewis. Disease and development: ciguatera fish poisoning. Soc Sci Med 23:983–993, 1986.
39. JG Morris, P Lewin, CW Smith, PA Blake, R Schneider. Ciguatera fish poisoning: epidemiology of the disease on St. Thomas, U.S. Virgin Islands. Am J Trop Med Hyg 31:574–578, 1982.
40. TR Tosteson. Ciguatera in Puerto Rico. In: Proceedings of the Workshop Conference on Seafood Intoxications: Pan American Implications of Natural Toxins in Seafood. Miami: University of Miami, 1996, Introduction and Scientific Reports Section, pp 33–48.
41. NC Gillespie, RJ Lewis, JH Pearn, ATC Bourke, MJ Holmes, JB Bourke, WJ Shields. Ciguatera in Australia: occurrence, clinical features, pathophysiology and management. Med J Aust 145:584–590, 1986.
42. W Losacker. Ciguatera fish poisoning in the Cook Islands. Bull Soc Pathol Exot 85:447–448, 1992.
43. TA Ruff. Ciguatera in the Pacific: a link with military activities. Lancet 1:201–205, 1989.
44. R Bagnis. Naissance et development d'une flambee de ciguatera un atoll des Tuamotu. Rev Corps Sante Armees 10:783–787, 1969.
45. MJ Cooper. Ciguatera and other marine poisoning in the Gilbert Islands. Pac Sci 18:411–440, 1964.
46. DM Anderson, PS Lobel. The continuing enigma of ciguatera. Biol Bull 172:89–107, 1987.
47. AR Katz, S Terrell-Percia, DM Sasaki. Ciguatera on Kauai: investigation of factors associated with severity of illness. Am J Trop Med Hyg 49:448–454, 1993.
48. NC Engleberg, JG Morris, J Lewis, JP McMillan, RA Pollard, PA Blake. Ciguatera fish poisoning: a major common-source outbreak in the U.S. Virgin Islands. Ann Intern Med 98:336–337, 1983.
49. C Frenette, JD MacLean, TW Gyorkos. A large common-source outbreak of ciguatera fish poisoning. J Infect Dis 158:1128–1131, 1988.
50. PD Morris, DS Campbell, JI Freeman. Ciguatera fish poisoning: an outbreak associated with fish caught from North Carolina coastal waters. South Med J 83:379–382, 1990.
51. Centers for Disease Control and Prevention. Ciguatera fish poisoning—Bahamas, Miami. MMWR Morb Mort Wkly Rep 31:391–392, 1982.
52. JK Sims. Theoretical discourse on the pharmacology of toxic marine ingestions. Ann Emer Med 16:1006–1015, 1987.
53. AEB Swift, TR Swift. Ciguatera. Clin Toxicol 31:1–29, 1993.
54. RJ Lewis, MJ Holmes. Origin and transfer of toxins involved in ciguatera. Comp Biochem Physiol 106c:615–628, 1993.
55. DN Lawrence, MB Enriquez, RM Lumish, A Maceo. Ciguatera fish poisoning in Miami. JAMA 244:254–258, 1980.
56. R Bagnis, T Kuberski, S Laugier. Clinical observations of 3,009 cases of ciguatera (fish poisoning) in the South Pacific. Am J Trop Med Hyg 28:1067–1073, 1979.
57. BS Anderson, JK Sims, NH Wiebenga, M Sugi. The epidemiology of ciguatera fish poisoning in Hawaii, 1975–1981. Hawaii Med J 42:326–334, 1983.
58. JG Morris, P Lewin, NT Hargrett, W Smith, PA Blake, R Schneider. Clinical features of ciguatera fish poisoning. A study of the disease in the US Virgin Islands. Arch Intern Med 142:1090–1092, 1982.

59. NA Palafox. Ciguatera and human health lessons from the Pacific. In: Proceedings of the Workshop Conference on Seafood Intoxications: Pan American Implications of Natural Toxins in Seafood. Miami, FL: University of Miami, 1996, Introduction and Scientific Reports Section, pp 14–19.
60. RJ Geller, NL Benowitz. Orthostatic hypotension in ciguatera fish poisoning. Arch Intern Med 152: 2131–2133, 1992.
61. AM Kodama, Y Hokama. Variations in symptomatology of ciguatera poisoning. Toxicon 27:593–595, 1989.
62. P Glaziou, PMV Martin. Study of factors that influence the clinical response to ciguatera fish poisoning. Toxicon 31:1151–1154, 1993.
63. AMH Ho, IM Fraser, ECD Todd. Ciguatera poisoning: a report of three cases. Ann Emerg Med 15:1225–1228, 1986.
64. RJ Holt, G Miro, A Del Valle. An analysis of poison control center reports of ciguatera toxicity in Puerto Rico for one year. Clin Toxicol 22:177–185, 1984.
65. EW Stommel, J Parsonnet, LR Jenkyn. Polymyositis after ciguatera toxin exposure. Arch Neurol 48:874–877, 1991.
66. P Glaziou, A LeGrand. The epidemiology of ciguatera fish poisoning. Toxicon 32:863–873, 1994.
67. ED Barton, P Tanner, SG Turchen, CL Tunget, A Manoguerra, RF Clark. Ciguatera fish poisoning. A Southern California epidemic. West J Med 163:31–35, 1995.
68. DJ DeFusco, P O'Dowd, Y Hokama, BR Ott. Coma due to ciguatera poisoning in Rhode Island. Am J Med 95:240–243, 1993.
69. RJ Geller, KR Olson, PE Senecal. Ciguatera fish poisoning in San Francisco, California, caused by imported barracuda. West J Med 155:639–642, 1991.
70. Centers for Disease Control and Prevention. Ciguatera fish poisoning—Vermont. MMWR Morb Mort Wkly Rep 35:263–264, 1986.
71. R Johnson, E Jong. Ciguatera: Caribbean and Indo-Pacific fish poisoning. West J Med 138:872–874, 1983.
72. WR Lange, FR Snyder, PJ Fudala. Travel and ciguatera fish poisoning. Arch Intern Med 152:2049–2053, 1992.
73. JS Ebesu, H Nagai, Y Hokama. The first reported case of human ciguatera possibly due to a farm-cultured salmon. Toxicon 32:1282–1286, 1994.
74. TR Tosteson, DL Ballantine, HD Durst. Seasonal frequency of ciguatoxic barracuda in Southwest Puerto Rico. Toxicon 26; 795–801, 1988.
75. R Bagnis, A Speigel, L N'Guyen, R Plichart. Public health, epidemiological and socioeconomic patterns of ciguatera in Tahiti. In: Tosteson TR, ed. Proceedings of the 3rd International Conference on Ciguatera Fish Poisoning. Quebec: Polyscience, 1990, pp 157–168.
76. LY Kimura, MA Abad, Y Hokama. Evaluation of the radioimmunoassay (RIA) for detection of ciguatoxin (CTX) in fish tissues. J Fish Biol 21:671–680, 1982.
77. Hokama Y. A rapid, simplified enzyme immunoassay stick test for the detection of ciguatoxin and related polyethers from fish tissues. Toxicon 23:939–946, 1985.
78. Y Hokama, AY Asahina, ES Shang, TW Hong, JL Shirai. Evaluation of the Hawaiian reef fishes with the solid phase immunobead assay. J Clin Lab Anal 7:26–30, 1993.
79. DW Nellis, GW Barnard. Ciguatera: a legal and social overview. Mar Fish Rev 48:2–5, 1986.
80. G Bourdy, P Cabalion, P Amade, D Laurent. Traditional remedies used in the Western Pacific for the treatment of ciguatera poisoning. J Ethnopharmacol 36:163–174, 1992.
81. MDAC De Blanchard. Report on outbreaks of ciguatera in the Dominican Republic. Epidemiological aspects related to public health and the laboratory. In: Proceedings of the Workshop Conference on Seafood Intoxications: Pan American Implications of Natural Toxins in Seafood. Miami FL: University of Miami, 1996, INPAAZ Country Reports Section, pp 59–74.
82. ND Lewis. Epidemiology and impact of ciguatera in the Pacific: a review. Mar Fish Rev 48:6–13, 1986.
83. R Quick. The epidemiology of domoic acid poisoning. Domoic Acid Workshop, San Pedro, CA, 1992.
84. JS Teitelbaum, RJ Zatorre, S Carpenter, D Gendron, AC Evans, A Gjedde, NR Cashman. Neurologic

sequelae of domoic acid intoxication due to the ingestion of contaminated mussels. N Engl J Med 322:1781–1787, 1990.
85. PS Spencer, AC Ludolph, GE Kisby. Are human neurodegenerative disorders linked to environmental chemicals with excitotoxic properties? Ann NY Acad Sci 648:154–160, 1992.
86. K Marien. Establishing tolerable dungeness crab (Cancer magister) and razor clam (Siliqua patula) domoic acid contaminant levels. Environ Health Perspect 104:1230–1236, 1996.
87. F Iverson, J Truelove. Toxicology and seafood toxins: domoic acid. Nat Toxins 2:334–339, 1994.
88. AS Belltran, M Palafox-uribe, J Grajales-Montiel, A Cruz-Villacorta, JL Ochoa. Sea bird mortality at Cabo San Lucas, Mexico: evidence that toxic diatom blooms are spreading. Toxicon 35:447–453, 1997.
89. S Loscutoff. West Coast experience—overview. Domoic Acid Workshop, San Pedro, CA, 1992.
90. L Zaman, O Arakawa, A Shimosu, Y Onoue, S Nishio, Y Shida, T Noguchi. Two new isomers of domoic acid from a red alga, Chondria armata. Toxicon 35:205–212, 1997.
91. RN Auer. Excitotoxic mechanisms, and age-related susceptibility to brain damage in ischemia, hypoglycemia and toxic mussel poisoning. Neurotoxicolagy 12:541–546, 1991.
92. JC Wekell, EJ Jr Gaugitz, HJ Barnett, CL Hatfield, M Eklund. The occurrence of domoic acid in razor clams (Siliqua patula), dungeness crab (Cancer magister), and anchovies (Engraulis mordax). J Shellfish Res 13:587–593, 1994.
93. CL Hatfield, EJ Jr Gauglitz, HJ Barnett, JK Lund, JC Wekell, M Eklund. The fate of domoic acid in dungeness crab (Cancer magister) as a function of processing. J Shellfish Res 14:1–5, 1995.
94. International Oceanographic Commission. Amnesic Shellfish Poisoning, ASP. HAB Publication series. Vol. 1, Manual and Guides No. 31. UNESCO, 1995.
95. JC Wekell, EJ Jr Gaugitz, HJ Barnett, CL Hatfield, D Simons, D Ayres. Occurrence of domoic acid in Washington State razor clams (Siliqua patula) during 1991–1993. Nat Toxins 2:197–205, 1994.
96. BW Halstead. Poisonous and Venomous Marine Animals of the World. Princeton, NJ: Darwin, 1978.
97. JK Sims, DC Ostman. Puffer fish poisoning: emergency diagnosis and management of mild human tetrodotoxication. Ann Emerg Med 15:1094–1098, 1986.
98. J Nowadnick. Puffers: a taste of death. Sea Frontiers 22:350–359, 1976.
99. CC Yang, SC Liao, JF Deng. Tetrodotoxin poisoning in Taiwan: an analysis of poison center data. Vet Hum Toxicol 38:282–286, 1996.
100. J Kanchanapongkul, P Krittayapoositpot. An epidemic of tetrodotoxin poisoning following ingestion of the horseshoe crab Carcinoscorpius rotundicauda. Southeast Asian J Trop Med Public Health 26:364 367, 1995.
101. Centers for Disease Control and Prevention. Tetrodotoxin poisoning associated with eating puffer fish transported from Japan—California, 1996. MMWR Morb Mort Wkly Rep 45:389–391, 1996.
102. BW Halstead. Poisonous and Venomous Marine Animals of the World, 2nd ed. Princeton, NJ: Darwin, 1988.
103. M Sorokin. Ciguatera poisoning in North-west Viti Levu, Fiji Islands. Hawaii Med J 34:207–210, 1975.
104. CC Yang, JF Deng, WJ Tsai, TJ Lin, KC Han. An outbreak of tetrodotoxin poisoning following gastropod mollusk consumption. Hum Exp Toxicol 14:446–450, 1995.
105. JF Deng, WJ Tsai, HM Chung, RL Tominack. Hypertension as an unusual feature in an outbreak of tetrodotoxin poisoning. J Toxicol Clin Toxicol 29:71–79, 1991.
106. WR Lange. Puffer fish poisoning. Am Fam Physician 42:1029–1033, 1990.
107. S Tsunenari, Y Uchimura, M Kanda. Puffer poisoning in Japan—a case report. J Forensic Sci 25: 240–245, 1980.
108. CY Kao. Tetrodotoxin, saxitoxin and their significance in the study of excitation phenomena. Pharmacol Rev 18:997–1049, 1966.
109. DG Walker. Survival after severe envenomation by the blue-ringed octopus (Hapalochlaena maculosa). Med J Austral 2:663–665, 1983.
110. DD Sheumack, MEH Howden, I Spence, RJ Quinn. Tetrodotoxin in the blue-ringed octopus. Med J Aust 1:160–161, 1978.

111. SG Bradley, LF Kilka. A fatal poisoning from the Oregon roughskinned newt (Taricha granulosa). JAMA 246:247, 1981.
112. AF Bartsch, EF McFarren. Fish poisoning: a problem in food toxication. Pacific Science 16:42–56, 1962.
113. PC Lyn. Puffer fish poisoning: four case reports. Med J Malaysia 40:31–34, 1985.
114. SK Kan, P David, MK Chan. Nine fatal cases of puffer fish poisoning in Sabah, Malaysia. Med J Malaysia 42:199–200, 1987
115. FL Lau, CK Wong, SH Yip. Puffer fish poisoning. J Accid Emerg Med 12:214–215, 1995.
116. K Sun, J Wat, P So. Puffer fish poisoning. Anaesth Intens Care 22:307–308, 1994.
117. SK Chew, CH Goh, KW Wang, PK Mah, BY Tan. Puffer fish (tetrodotoxin) poisoning: clinical report and role of anti-cholinesterase drugs in therapy. Singapore Med J 24:168–171, 1983.
118. J Tibballs. Severe tetrodotoxic fish poisoning. Anaesth Intens Care 16:215–217, 1988.
119. D Yasumura, Y Oshima, T Yasumoto, AC Alcala, LC Alcala. Tetrodotoxin and paralytic shellfish toxins in Philippine crabs. Agric Biol Chem 50:593–598, 1986.
120. RB Gonzales, AC Alcala. Fatalities from crab poisoning on Negros Island, Philippines. Toxicon 15:169–170, 1977.
121. PD Morris, DS Campbell, TJ Taylor, JI Freeman. Clinical and epidemiological features of neurotoxic shellfish poisoning in North Carolina. Am J Public Health 81:471–474, 1991.
122. EF McFarren, H Tanabe, FJ Silva, WB Wilson, JE Campbell, KH Lewis. The occurrence of a ciguatera-like poison in oysters, clams, and Gymnodinium breve cultures. Toxicon 3:111–123, 1965.
123. WH Hemmert. The public health implications of Gymnodinium breve red tides, a review of the literature and recent events. In: Locicero VR, ed. Proceedings of the First International Conference on Toxic Dinoflagellate Blooms. Boston, MA: Massachusetts Science and Technology Foundation, 1975, pp 489–497.
124. DG Baden, LE Fleming, JA Bean. Marine toxins. In: FA De Wolff, ed. Handbook of Clinical Neurology, Vol. 21 (65): Intoxications of the Nervous System, Part II. Oxford, UK: Elsevier, S, 1995, pp 141–175.
125. DG Baden. Marine food-borne dinoflagellate toxins. Int Rev Cytol 82:99–150, 1983.
126. LE Fleming, DG Baden, JA Bean, R Weisman, DG Blythe. Marine seafood toxin diseases: issues in epidemiology and community outreach. In: B Ruergo, ed. Eighth International Conference on Toxic Marine Phytoplankton. Vego, Spain: UNESCO of Intergovernmental Oceanographic Commission. Vego, Spain, 1998.
127. DG Baden. Brevetoxins: unique polyether dinoflagellate toxins. FASEB J 3:1807–1817, 1989.

5

Calcium Channels for Exocytosis: Functional Modulation with Toxins

Antonio G. García
Instituto de Farmacología Teófilo Hernando
Universidad Autónoma de Madrid
Hospital Universitario de la Princesa
Madrid, Spain

Luis Gandía, Manuela G. López and Carmen Montiel
Instituto de Farmacología Teófilo Hernando
Universidad Autónoma de Madrid
Madrid, Spain

I. INTRODUCTION

The combination of patch-clamp techniques, ω-toxins, and molecular strategies has revealed a great heterogeneity of voltage-dependent Ca^{2+} channels in neurons. Peptide toxins derived from the venoms of marine snails *Conus geographus* (ω-conotoxin GVIA), *Conus magus* (ω-conotoxins MVIIA, MVIIC, and MVIID), as well as from *Agelenopsis aperta* spider venom (FTX, ω-agatoxin IVA) are powerful diagnostic pharmacological tools to discriminate between different subtypes of neuronal Ca^{2+} channels. Thus, so-called high-threshold–activated (HVA) Ca^{2+} channels are selectively recognized by ω-conotoxin GVIA and MVIIA (N-type), by low concentrations (nanomolar) of ω-agatoxin IVA (P-type), or by high concentrations of ω-agatoxin IVA (micromolar) or the ω-conotoxins MVIIC and MVIID (Q-type). L-type HVA Ca^{2+} channels present in neurons, cardiovascular tissues, skeletal and smooth muscle, and in endocrine cells are targetted by so-called organic Ca^{2+} antagonists such as the 1,4-dihydropyridines (DHP) nifedipine or Bay K 8644, the benzylalkylamine verapamil, or the benzothiazepine diltiazem; they are also specifically blocked by snake toxins calciseptine and calcicludine. Wide-spectrum ω-toxins (ω-conotoxin MVIIC, ω-agatoxin IA, IIA, and IIIA) and organic compounds (flunarizine, dotarizine, cinnarizine, fluspirilene, R56865, lubeluzole) can block several classes of HVA Ca^{2+} channels, including the L-type. A neuronal R-type HVA channel seems to be resistant to all known toxins. Low-voltage–activated (LVA) channels (T-type) are blocked by l-octanol, amiloride, and mibefradil, and are more sensitive to Ni^{2+} than to Cd^{2+}; no toxins are known that recognize these channels.

It is interesting that a single cell can express different subtypes of HVA Ca^{2+} channels and that the quantitative expression of each channel subtype differs with the animal species. The example of adrenal medulla chromaffin cells is illustrative. In the bovine, P/Q-type (45%)

and N-type (35%) are predominant; the L-type Ca^{2+} channel carries a minor component of the whole-cell current (20%). In the rat and the mouse, the L-type predominates (50%), together with the N-type (35%), whereas the P/Q family accounts for a minor component (15%). In cat chromaffin cells, L-type Ca^{2+} channels carry 50% of the current and N-type channels 45%; P/Q account by only 5%. In human chromaffin cells, P/Q-type Ca^{2+} channels dominate (60%), whereas in pig chromaffin cells, N-type channels are predominant (80%). The functional significance if this variety of Ca^{2+} channels remains to be elucidated.

Ca^{2+} channels consist basically of a multiple subunit protein complex with a central pore-forming α_1 subunit and several regulatory and/or auxiliary subunits, which include a β subunit and the disulfide-linked α_2/δ subunit. Additionally, a fifth subunit has been reported in some tissues, such as the skeletal muscle γ subunit or the neural P95 subunit. The α_1 subunit contains the Ca^{2+} conductance pore, the essential gating machinery, and the receptor sites for the most prominent pharmacological agents. The mammalian family of Ca^{2+} channel α_1 genes may be grouped into two subfamilies. The L-subfamily consists of three genes designed as α_{1S} (cloned from skeletal muscle), α_{1C} (cardiac, smooth muscle, neuronal), and α_{1D} (neuronal, endocrine). The DHP-resistant non-L subfamily have another three genes that are expressed almost exclusively in neuronal tissues. The α_{1B} gene encodes an ω-conotoxin GVIA-sensitive N-type Ca^{2+} channel, the α_{1A} subunit is likely related to the P/Q-family of Ca^{2+} channels, and the α_{1E} subunit seems to be related to the R channel expressed in rat cerebellar granule neurons. Recently, an α_{1G} subunit that expresses T-type channels has been cloned.

Marine toxins have been invaluable tools to recognize the role of each channel subtype in controlling the Ca^{2+}-dependent exocytotic release of a given neurotransmitter. Thus, N-type Ca^{2+} channels are highly involved in the control of norepinephrine (noradrenaline) release from sympathetic neurons, as well as acetylcholine release from the electric fish muscle endplate, the myenteric plexus, and detrusor muscle. Also, N channels partially control the nonadrenergic noncholinergic neurotransmission in smooth muscle, gamma-aminobutyric acid (GABA) release in cerebellar neurons, glycine release in dorsal horn neurons of the spinal cord, epinephrine release from the dog adrenal, dynorphin release in dentate gyrus, and the synaptic neurotransmission in retinal ganglion neurons and the hippocampus. P channels dominate the release of GABA from deep cerebellar neurons, glycine from dorsal horn neurons of the spinal cord, and acetylcholine from the mammalian neuromuscular junction. They also seem to participate partially in the control of the release of other neurotransmitters. Up to now, Q channels have been implicated in the control of neurotransmission in the hippocampus and in the release of catecholamines from bovine chromaffin cells. L-type Ca^{2+} channels dominate the release of catecholamines in rat and cat chromaffin cells, and they partially control the secretory process in bovine chromaffin cells.

A critical question is why a neurosecretory cell expresses several Ca^{2+} channel subtypes. In bovine adrenal chromaffin cells, L, N, P, and Q channels have been found, yet only the L- and Q-types seem to be involved in the control of catecholamine release induced by depolarizing stimuli. Therefore, it seems that those ''exocytotic channels'' (L and Q) must be located nearby the secretory surface of the cell; however, it is uncertain what might be the functions of Ca^{2+} channels (i.e., N and P) that do not participate in the immediate control of the secretory process. Many other questions remain unanswered. For instance, the nature of the R-type channel and whether a toxin can be found to recognize this channel. A third question relates to the number of Ca^{2+} channels yet unrecognized. The functions of the Ca^{2+} channels not related to exocytosis (i.e., the neuronal L-type channels) are a mystery. Finally, it is important to stress the need of finding nonpeptide molecules to target specifically different channel subtypes; these compounds should cross the blood-brain barrier and thus serve as therapeutic drugs to treat different brain diseases.

II. ω-TOXINS AS DIAGNOSTIC PHARMACOLOGICAL TOOLS

Some static or slow animals, both terrestrial (snakes, spiders) and marine (snails), have developed venoms containing potent neurotoxins in order to capture their prey with high efficiency and speed. The efficiency of the method used to capture the prey will influence the venom content evolved by a predator.

One of the most representative examples of venomous animals is exemplified by the *Conus* marine snails (1). Of the approximately 500 *Conus* species, about 40–100 prey primarily on fish (fish-hunting species), and these snail species use two parallel physiological mechanisms requiring multiple neurotoxins to immobilize fish rapidly (2): neuromuscular block and excytotoxic shock. Fish-hunting *Conus* snails use an harpoon-like device to inject their venom into their preys. The venom contains a cocktail of neurotoxins that will cause a double-phase paralytic process (Table 1), with an initial phase characterized by a fast paralysis with tetanus, and a second phase characterized by a flaccid paralysis. Finally, the fish will be engulfed by the snail.

The fast paralysis of the phase I is mediated by two groups of neurotoxins, the δ-conotoxins, which suppress the inactivation of the voltage-dependent Na^+ channels, thus causing an increase in Na^+ influx; and the κ-conotoxins, which block K^+ channels, not allowing the cells to repolarize. This combination of toxins leads to hyperactivity of the fish, followed by a continuous contraction and extension of major fins without death. The second phase consists of a flaccid state and is caused by a different cocktail of neurotoxins (see Table 1): The α-conotoxins, which block nicotinic acetylcholine receptors; the μ-conotoxins, which block voltage-dependent Na^+ channels; the Ψ-conotoxins, which also block nicotinic acetylcholine receptors; the κ-conotoxins, which cause the blockade of K^+ channels; the δ-conotoxins, which suppress the inactivation of the voltage-dependent Na^+ channels, and the ω-conotoxins, which block voltage-dependent Ca^{2+} channels, and are the subject of this chapter.

Another example of venomous animals is the funnel-web spider, *Agelenopsis aperta*, which has a potent venom with paralytic properties. As in the *Conus*, the venom of this spider possesses a mixture of toxins with different targets, with the polyamines and the polypeptides being the main components of such venom. The polyamines group is composed of the funnel-web toxin (FTX), which targets voltage-dependent Ca^{2+} channels (3), and the acylpolyamines (α-agatoxins), most of which are blockers of glutamate receptors (4). The other group, the polypeptide toxins, is composed of the ω-agatoxins, which selectively block different subtypes of voltage-dependent Ca^{2+} channels, and the μ-agatoxins, which are potent activators of voltage-

Table 1 Paralytic Process Induced by the Venom of *Conus* Marine Snails and Neurotoxins Implied and with Their Mechanisms of Action

Phase I. Fast paralysis with tetanus (rapid immobilization)	
δ-Conotoxins	Suppression of Na^+ channel inactivation (increases Na^+ influx)
κ-Conotoxins	Blockade of K^+ channels
Phase II. Flaccid paralysis	
α-Conotoxins	Blockade of nicotinic acetylcholine receptors
μ-Conotoxins	Blockade of voltage-dependent Na^+ channels
ψ-Conotoxins	Blockade of nicotinic acetylcholine receptors
κ-Conotoxins	Blockade of K^+ channels
δ-Conotoxins	Suppression of Na^+ channel inactivation
ω-Conotoxins	Blockade of voltage-dependent Ca^{2+} channels

dependent Na^+ channels. These combinations of toxins secure a fast and reversible paralytic effect (induced by the α- and μ-agatoxins) with a slower but irreversible paralysis of the prey induced by the ω-agatoxins.

Finally, venoms from different snakes from the *Elapidae* and *Hydrophidae* families also contain a cocktail of different paralytic toxins, some of which are selective for voltage-dependent Ca^{2+} channels. For instance, the venom of the black mamba, *Dendroaspis polylepis polylepis*, contains a toxin termed calciseptine, which selectively blocks L-type Ca^{2+} channels (5), and the venom fom the green mamba, *Dendroaspis agusticeps*, contains calcicludine, a toxin that acts as a potent blocker of most of the high-voltage–activated Ca^{2+} channels (6).

III. ω-TOXINS FOR THE CHARACTERIZATION OF VOLTAGE-DEPENDENT Ca^{2+} CHANNELS

As indicated, ω-conotoxins are found in almost all fish-hunting *Conus* species examined up to now. The most thoroughly studied are the venoms from *C. geographus, C. magus, and C. striatus*. Several of these peptides have been purified, sequenced, and synthetized (with similar potency as the natural toxins), and they have become important tools for the identification and characterization of the different subtypes of voltage-dependent Ca^{2+} channels found in neuronal tissues.

A. ω-Conotoxins

ω-Conotoxins are small peptides containing 24–29 amino acid residues (Figure 1); they share several features that are common to all ω-conotoxins. The more characteristic is the presence of six Cys residues, with three intramolecular disulfide bridges, forming an structure known as a ''four-loop framework'' (7,8). This arrangement of Cys residues is similar to that observed in δ-conotoxins, which target voltage-gated Na^+ channels (9).

Although the sequence of different ω-conotoxins has great interspecies variations, they can compete for the same calcium binding site and show similar physiological effects. For

ω-Conotoxin GVIA	CKSOGSSCSOTSYNCCR.SCNOYTRKCY
ω-Conotoxin MVIIA	CKGKGAKCSRLMYDCCTGSC..RSGKC
ω-Conotoxin MVIIC	CKGKGAPCRKTMYDCCSGSC.GRRGKC
ω-Conotoxin MVIID	CQGRGASCRKTMYNCCSGSC..NRGRC
ω-Conotoxin SVIA	CRSSGSOCGVTSI.CC.GRC..YRGKCT
Disulfide linkages	C......C......CC...C.....C

Figure 1 Upper panel shows the sequence of ω-conotoxins isolated from *Conus geographus* (GVIA), *Conus magus* (MVIIA, MVIIC, and MVIID), and *Conus striatus* (SVIA). Lower panel shows the arrangement of the Cys residues that constitute de ''four-loop'' structure.

instance, ω-conotoxin GVIA (10) and ω-conotoxin MVIIA (11) have a homology lower than 30% in the non-Cys residues, but both target N-type Ca^{2+} channels (as described below) and elicit similar biological effects; the major differences are that ω-conotoxin GVIA blocks N-type Ca^{2+} channels in an irreversible manner (1,12), whereas ω-conotoxin MVIIA does it in a reversible manner (13,14).

Other ω-conotoxins have broader Ca^{2+} channel blocking properties than ω-conotoxin GVIA and ω-conotoxin MVIIA. cDNA clones encoding a previously unknown ω-conotoxin were identified from a cDNA library made from the venom duct of *Conus magus* (15). The predicted peptides ω-conotoxin MVIIC and ω-conotoxin MVIID were chemically synthetized and characterized. Both peptides inhibit N-type Ca^{2+} channels and P-type Ca^{2+} channels, but they also inhibit other Ca^{2+} channels resistant to 1,4-dihydropyridines (DHP), ω-conotoxin GVIA, and ω-agatoxin IVA (15), and thus they constitute actually an important tool for the characterization of P/Q-types of Ca^{2+} channels, as described below. Some differences between the ω-conotoxins relate to the reversibility of their blocking effects, and thus N-type Ca^{2+} channels can be blocked in an irreversible manner by ω-conotoxin MVIIC but in a reversible manner by ω-conotoxin MVIID (16).

B. ω-Agatoxins

ω-Agatoxins, derived from the venom of *Agelenopsis aperta*, are also a heterogeneous group of polypeptides (5–100 kD) that specifically target voltage-dependent Ca^{2+} channels. Four subtypes of ω-agatoxins with different blocking properties have been identified up to now (4,17,18). Type I ω-agatoxins (ω-Aga-IA, ω-Aga-IB, and ω-Aga-IC) are potent blockers of neuromuscular transmission in insects. Of these, the most studied is ω-agatoxin IA, which seems to block both L- and N-type Ca^{2+} channels (19). Type II ω-agatoxins have a spectrum of action on neuronal Ca^{2+} channels in vertebrates similar to that of ω-agatoxin IA, although they may block Ca^{2+} channels by a different mechanism (20). ω-Agatoxin IIA has been shown as a potent blocker of both L- and N-type Ca^{2+} channels (21).

Type III ω-agatoxins (ω-Aga IIIA, ω-Aga IIIB, ω-Aga IIIC, and ω-Aga IIID) have a broader spectrum of blockade than other agatoxins and block several subtypes of voltage-dependent Ca^{2+} channels. Of these, ω-agatoxin IIIA has been shown to be a potent inhibitor or L-, N-, and P/Q-type Ca^{2+} channels in neurons of rats and frogs (22–24); it shows a very high potency (IC50 <1 nM) for both inhibiting L and N channels, being more potent than ω-conotoxin GVIA for blocking N-type channels (25). Efficacy of blockade induced by ω-agatoxin IIIA is higher for L-type channels and decreases for N- and P/Q-type Ca^{2+} channels (23). In these latter channel subtypes, ω-agatoxin IIIA seems to act as a high-affinity partial antagonists, blocking less than 50% of calcium conductance (23).

Type IV ω-agatoxins (ω-Aga IVA and ω-Aga-IVB) show a different pharmacological effect to that described for other ω-agatoxins, and, in addition to N-type Ca^{2+} channels, they also block P/Q-type Ca^{2+} channels with a K_d of 2–3 nM (26–28).

C. FTX

The toxin fraction (FTX) of *Agelenopsis aperta* spider venom can be also used as a P-type Ca^{2+} channel blocker. In fact, this toxin was initially used to describe and characterize P-type Ca^{2+} channels in Purkinje cells (3,29). Although initially FTX was considered to be selective for P-type Ca^{2+} channels, later on it was shown to block other ionic channels (30).

IV. DIVERSITY OF VOLTAGE-DEPENDENT Ca^{2+} CHANNELS

Two approaches are mainly responsible for the discovery of the rich diversity of voltage-dependent Ca^{2+} channels. On the one hand, the characterization of the biophysical properties of Ca^{2+} channels (kinetics of activation, inactivation, and deactivation, voltage-range for activation, conductance) both at the single-channel and at the whole-cell level has been possible thanks to the improvement of the patch-clamp techniques (31). On the other hand, the isolation, purification, and synthesis of different neurotoxins have provided ligands with remarkable discrimination for different subtypes of high-threshold dihydropyridine (DHP)–resistant Ca^{2+} channels (1).

With the combination of the patch-clamp techniques and these pharmacological probes, at least six subtypes of voltage-dependent Ca^{2+} channels have been described up to now: T, L, N, P, Q, and R (Table 2). These channels can be classified according to their range of activation in two main groups: one with a low threshold for activation (low-voltage–activated: LVA) and other with a high threshold for activation (high-voltage–activated: HVA).

A. Low-Voltage–Activated Channels: T-Type Ca^{2+} Channels

The first attempt to identify different subtypes of voltage-dependent Ca^{2+} channels was carried out by Carbone and Lux (32), who identified two types of channels: those that open with small depolarizations from a hyperpolarized holding potential, so-called low-voltage–activated (LVA) channels and those that require higher depolarizations to open, so-called high-voltage–activated (HVA) channels.

In addition to their low threshold for activation, LVA Ca^{2+} channels (32) are characterized by a similar permeability for Ca^{2+} and Ba^{2+} (33,34). A single subtype of Ca^{2+} channel has been identified in this group (32) and has been termed T (for transient or tiny). The main characteristics of this channel are its fast inactivation, which generates a transient current, and its inactivation when the holding potential is fixed between −60 and −50 mV. The single-channel conductance has been estimated to be around 8 pS.

Table 2 Drugs and Neurotoxins Used for Characterization of Voltage-Activated Ca^{2+} Channels

Channel subtype	Blocker	Supramaximal concentration	Reversibility
L	Dihydropyridines (i.e., nifedipine, nimodipine)	3 μM	Reversible
N	ω-Conotoxin GVIA	1 μM	Irreversible
	ω-Conotoxin MVIIA	1 μM	Reversible (slowly)
	ω-Conotoxin MVIIC	3 μM	Reversible (slowly)
	ω-Conotoxin MVIID	3 μM	Reversible (fast)
P	ω-Agatoxin IVA	20 nM	Irreversible
	ω-Conotoxin MVIIC	3 μM	Irreversible
	ω-Conotoxin MVIID	3 μM	Irreversible
Q	ω-Agatoxin IVA	2 μM	Irreversible
	ω-Conotoxin MVIIC	3 μM	Irreversible
	ω-Conotoxin MVIID	3 μM	Irreversible
T	Mibefradil	1 μM	Reversible

See text for references.

Pharmacologically, T-type channels can be distinguished from other subtypes, because they are more sensitive to blockade by the inorganic Ca^{2+} channel blocker Ni^{2+} than to Cd^{2+} (33,34). It has also been reported that T-type channels can be blocked by l-octanol, amiloride, and the antihypertensive drug mifebradil (35).

B. High-Voltage–Activated Channels

HVA channels are characterized by their activation by strong depolarizing steps (33,34), a higher permeability to Ba^{2+} than to Ca^{2+}, and a higher sensitivity to Cd^{2+} than to Ni^{2+} in contrast to LVA channels. Up to now, five major subtypes (L, N, P, Q, and R) of HVA channels have been identified. The major differences between them are related to their inactivation kinetics and their pharmacological properties.

L-Type Ca^{2+} Channels

L-type (for long-lasting) Ca^{2+} channels are kinetically characterized by showing little inactivation during depolarizing steps (τ_{inact} >500 ms) and their lower sensitivity to depolarized holding potentials. Single-channel conductance was estimated to be around 18–25 pS. This subtype of Ca^{2+} channel seems to be present in all excitable cells and in many nonexcitable cells, and they constitute the main pathway for Ca^{2+} entry in heart and smooth muscle, serving also to control hormone and transmitter release from endocrine cells and some neuronal preparations. Pharmacologically, L-type Ca^{2+} channels are highly sensitive to DHPs (Table 2) and both agonists (i.e., Bay K 8644) and antagonists (i.e., nifedipine, nimodipine, furnidipine). DHP agonist effects are characterized by the prolongation of the mean open time for channel (36), typically observed in whole-cell electrophysiological recordings as a prolongation of tail currents (37).

Other organic compounds have been described to block effectively L-type Ca^{2+} channels (38,39): The arylalkylamines (i.e., verapamil) and benzothiazepines (i.e., diltiazem) are particularly useful in cardiac and smooth muscle cells, where they exert negative inotropic effects. Some piperazine derivatives (cinnarizine, flunnarizine, dotarizine, R56865) also block L-type Ca^{2+} channels, but they also block other subtypes of Ca^{2+} channels, and thus have been proposed as ''wide-spectrum'' Ca^{2+} channel blockers (40–42). As described before, some toxins have also been shown to block L-type Ca^{2+} channels either selectively (calciseptine and calcicludine) or in a nonselective manner (ω-agatoxin IA, ω-agatoxin IIA, and ω-agatoxin IIIA).

N-Type Ca^{2+} Channels

N-type Ca^{2+} channels display a faster inactivation kinetics (τ_{inact} 50–80 ms) than that of L-type channels. This relative fast inactivation usually leads to their inactivation when maintaining a depolarizing holding potential, although in some preparations, N-type Ca^{2+} channels can contain a noninactivating component even at the end of long depolarizations, for instance, in bovine chromaffin cells in which N-type channels have been described as ''nonclassical N-type'' (43). Single-channel conductance of N-type channels has been estimated to be around 13 pS.

Pharmacologically, N-type Ca^{2+} channels are characterized by the irreversible blockade induced by the *Conus geographus* toxin ω-conotoxin GVIA (1,36,44) and the reversible blockade induced by the *Conus magus* toxin ω-conotoxin MVIIA (see Table 2) (13,14). Other wide-spectrum toxins such as ω-conotoxin MVIIC and ω-conotoxin MVIID (15,45) can also block N-type Ca^{2+} channels in a nonselective manner. This is also the case for ω-agatoxin IIA, ω-agatoxin IIIA, and ω-grammotoxin SIA (isolated from the venom of the tarantula *Grammostola spatulata*).

P-Type Ca^{2+} Channels

P-type Ca^{2+} channels were first described by Llinás et al. (29) in cerebellar Purkinje cells in which Ca^{2+} currents were resistant to blockade by DHPs and ω-conotoxin GVIA. The toxin fraction from the venom of the funnel-web spider *Agelenopsis aperta* (FTX) was found effectively to block this resistant current, and these results led these Llinás et al. to suggest the existence of a new subtype of HVA Ca^{2+} channel, which was termed P (for Purkinje).

P-type Ca^{2+} channels are characterized by their relative insensitivity to changes in the holding potential, and they do not inactivate during depolarizing steps (27,46,47); multiple single-channel conductances have been described for P-type Ca^{2+} channels (48,49).

Pharmacologically, P-type Ca^{2+} channels can be blocked by FTX and its synthetic analogue sFTX and by ω-agatoxin IVA at concentrations in the nanomolar range (<30–100 nM). This toxin is actually accepted to be the selective probe to identify the presence of P-type Ca^{2+} channels (see Table 2). P-type Ca^{2+} channels can also be blocked in a nonselective manner by ω-conotoxin MVIIC (15,45), ω-conotoxin MVIID, and ω-grammotoxin SVIA (50–53).

Q-Type Ca^{2+} Channels

In many neuronal preparations, a significant component of the whole-cell current through Ca^{2+} channels is resistant to blockade with DHPs, ω-conotoxin GVIA, and ω-agatoxin IVA (<100 nM), suggesting the presence of a subtype of Ca^{2+} channel different from L-, N-, and P-types. The isolation, purification, and synthesis of the toxin from the marine snail *Conus magus* ω-conotoxin MVIIC (15,45) led to the identification and characterization of a new subtype of HVA channel termed Q (54–56).

Characterization of Q type Ca^{2+} channels is mostly based on pharmacological criteria. As described, Q-type channels are resistant to blockade by DHPs, ω-conotoxin GVIA, and low doses (<100 nM) of ω-agatoxin IVA, but they are sensitive to ω-conotoxin MVIIC (1–3 μM). Increasing concentrations of ω-agatoxin IVA (up to 2 μM) can also block Q-type Ca^{2+} channels (56). It should be noted that these toxins used to identify Q-type channels are not selective for this subtype of channel, and they also block in a nonselective manner N- and P-types. Other toxins that can also block this subtype of Ca^{2+} channel include the *Conus magus* snail toxin ω-conotoxin MVIID (45) and the *Grammostola spatulata* tarantula toxin ω-grammotoxin SIA (50–53).

R-Type Ca^{2+} Channels

In neuronal tissues, a residual Ca^{2+} current, characterized by its insensitivity to blockade by DHPs, ω-conotoxin GVIA, ω-agatoxin IVA, and ω-conotoxin MVIIC has also been described and termed R-type (for resistant (54,55). This new subtype of Ca^{2+} channel belongs to the HVA group, is rapidly inactivating (τ = 22 ms), and is more sensitive to blockade by Ni^{2+} (IC50 = 66 μM) than to Cd^{2+}. No pharmacological tools are actually available to block this subtype of channel.

V. SOME CURIOUS DIFFERENCES AMONG SPECIES

Drastic species differences in the subtypes of Ca^{2+} channels expressed by different cell types have been found. For instance, the K^+-evoked Ca^{2+} entry in brain cortex synaptosomes is controlled by N channels in the chick and by P channels in the rat (57). On the other hand, neurotransmitter release at the muscle endplate is controlled by N channels in fish (58–60) and amphibians (61) and by P channels in mammals (62).

Detailed comparative electrophysiological studies among six mammalian species have been performed only in adrenal medullary chromaffin cells (Fig. 2). L-type Ca^{2+} channels account for nearly half of the whole-cell Ca^{2+} channel current in the cat (63), rat (64), and mouse chromaffin cells (65). In pig (66), bovine (67,68), and human species (69), L channels carry only 15–20% of the whole-cell Ca^{2+} current.

The N channels also shows a high interspecies variability. In the pig, it carries as much as 80% of the whole-cell Ca^{2+} channel current (66), and in the cat, 45% (63); in bovine (70), rat (64), mouse (65), and human chromaffin cells (69), the N-type fraction accounts for 30% of the whole-cell Ca^{2+} channel current.

P channels have proven difficult to characterize in chromaffin cells. Through the use of the synthetic funnel-web toxin sFTX (67), 1 μM ω-agatoxin IVA (68), or 100 nM ω-agatoxin IVA (71) as much as 40–55% of the whole-cell Ca^{2+} channel current was attributed to P channels. Later on we learned that concentrations of ω-agatoxin IVA higher than 10–20 nM in addition of P channels (27,47) also block Q channels (56). Thus, nanomolar concentrations of ω-agatoxin IVA known to block fully and selectively P channels (27,47) cause only 5–10% blockade of Ca^{2+} channel current in bovine chromaffin cells (72). In cat chromaffin cells, combined ω-conotoxin GVIA plus nisoldipine blocked 90% of the current, leaving little room for

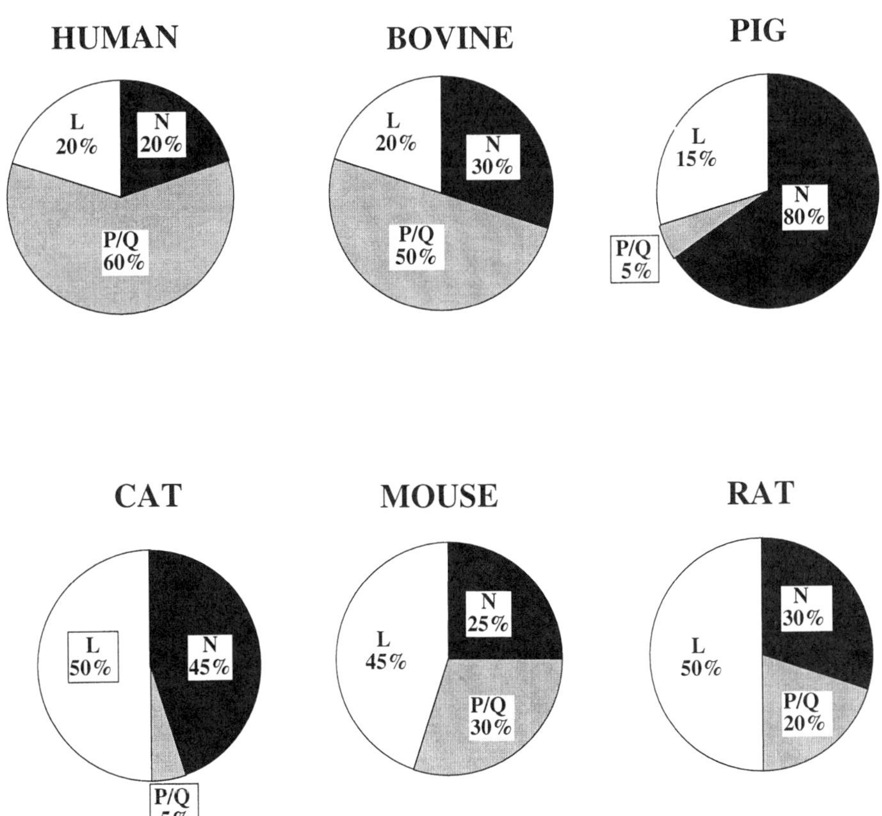

Figure 2 Relative proportions of different neuronal Ca^{2+} channel subtypes in chromaffin cells isolated from bovine, rat, mouse, cat, pig, and human adrenal medullary tissues.

P channels (63). In rat (64) and mouse (65), the ω-agatoxin IVA–sensitive current fraction was only 10–15%. Thus, in all species studied, it seems that P channels are barely expressed, if at all, in their chromaffin cells. This, together with the difficulty of separating the α_{1A} subunit into P and Q channels (73), suggest the convenience of speaking of P/Q channels rather than of two separate Ca^{2+} channel subtypes.

The P/Q channel component is pharmacologically isolated by 2 μM ω-conotoxin MVIIC or ω-conotoxin MVIID or by 2 μM ω-agatoxin IVA. In bovine chromaffin cells, ω-conotoxin MVIID blocks the N current reversibly, whereas ω-conotoxin MVIIC does so irreversibly (16); Thus, the use of ω-conotoxin MVIID followed by its washout can be a convenient tool to isolate the P/Q channel. The blocking effects of ω-conotoxin MVIIC are extraordinarily slowed down and decreased in the presence of excessive concentrations (i.e., more than 2 mM) of Ba^{2+} (72,74) or Ca^{2+} (14). Taking into consideration these methodological problems, we believe that the fraction of current carried out by P/Q channel amounts to 50% (72). This fraction is even higher (60%) in human chromaffin cells (69). The opposite occurs in pig (66) and cat chromaffin cells (63) where P/Q channels carry only 5% of the current. Finally, in rat chromaffin cells, P/Q channels contribute 20% to the current (64) and in the mouse 30% (65).

We do not know yet what is the physiological relevance of these drastic species differences. But surely that it has clear consequences for the fine control of the differential exocytotic release of epinephrine and norepinephrine in response to different stressors. Different autocrine/paracrine regulation by catecholamines and other coexocytosed vesicular components of L- and non–L-types of Ca^{2+} channels might be a reason. Other regulatory mechanisms (i.e., voltage-dependent or Ca_i^{2+}-dependent inactivation of Ca^{2+} channels could also explain the preferential expression of one or another channel type in a given species. Also, the selective segregation of a given channel type to exocytotic microdomains, and the uneven geographical distribution of other channel types, might also decide a given neurosecretory cell to express preferentially one or another channel type.

VI. MOLECULAR STRUCTURE OF CALCIUM CHANNELS

Voltage-gated Ca^{2+} channels are basically a multiple subunit protein complex consisting of a pore-forming α_1 subunit and several auxiliary subunits, which include the intracellular β subunit and a disulfide linked α_2/δ subunit. In some tissues, a fifth subunit may also form part of the channel complex, such as the transmembrane γ subunit found in skeletal muscle or the neuronal p95 subunit. Figure 3A represents the hypothetical subunit arrangement for the skeletal muscle high-voltage–activated Ca^{2+} channels. The functional diversity found among different subtypes of Ca^{2+} channels can be explained attending to molecular differences of the channels due to (1) the existence of multiple genes encoding different classes of α_1 and β subunits as well as diverse variants from a single gene generated by alternative splicing; and (2) multiple possible combinations among the subunits which make up the channel complex.

A. Structural Diversity of Ca^{2+} Channel α_1 Subunits

Expression studies have shown that voltage-activated channel function, which is typical for the L-, N-, P-, Q-, R, and T-type Ca^{2+} channels, is carried by the corresponding α_1 subunit. This subunit confers the characteristic pharmacological and functional properties of Ca^{2+} channels, although their function and properties are modulated by association with the other auxiliary subunits.

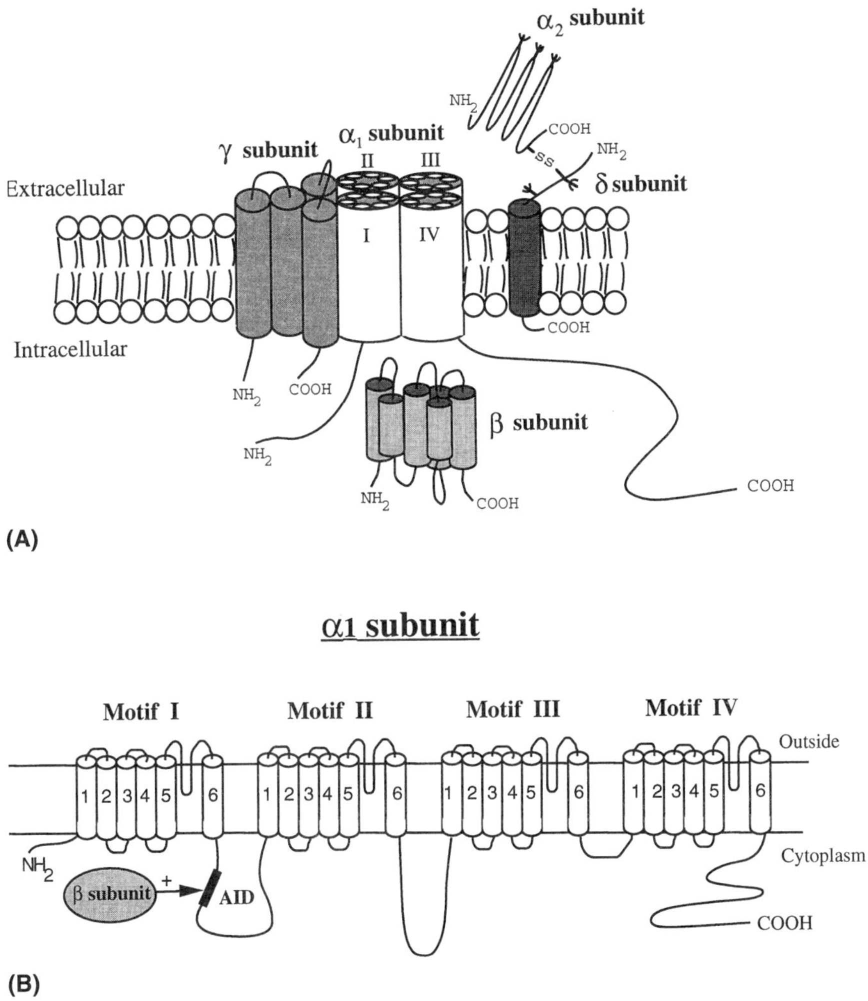

Figure 3 (A) Subunit structure arrangement of the L-type Ca^{2+} channel from skeletal muscle. (B) Proposed transmembrane folding model for the Ca^{2+} channel α_1 subunit which contains the α_1 subunit interaction domain (AID) for the β subunit binding.

The Ca^{2+} channel α_1 subunits isolated, sequenced, and expressed so far can be grouped in seven different classes (α_{1S}, α_{1C}, α_{1D}, α_{1B}, α_{1A}, α_{1E}, and α_{1G}). It is generally agreed that some of these subunits encode functional L-type (α_{1S}, α_{1C}, α_{1D}), N-type (α_{1B}), and P/Q-type (α_{1A}) Ca^{2+} channels. However, the correspondence of α_{1E} to a functional channel remains controversial, since it has been suggested that it would encode either one R-subtype Ca^{2+} channel and/or a T-subtype. At the beginning of 1998, Pérez-Reyes and colleagues cloned a new α_1 subunit referred as α_{1G}; this being the first member of the low-voltage–activated Ca^{2+} channel family, which clearly appears to support the classic T-type Ca^{2+} channel activity (75). The finding of this subunit does not mean that only one subtype of T-type Ca^{2+} channel exists, since the same group has reported a close relative gene highly expressed in heart, referred to as α_{1H}, which also supports a typical T-type channel activity. Table 3 summarizes the nomenclature for the Ca^{2+} channel α_1 genes known to date as well as the homology among their members. The Table 3 also relates each gene with its functional correlate Ca^{2+} channel type as well as the major

Table 3 Voltage-Activated Ca^{2+} Channel α_1 Subunit Genes

Homology percentage	Gene product	Functional channel	Sites of expression
20 40 60 80 100 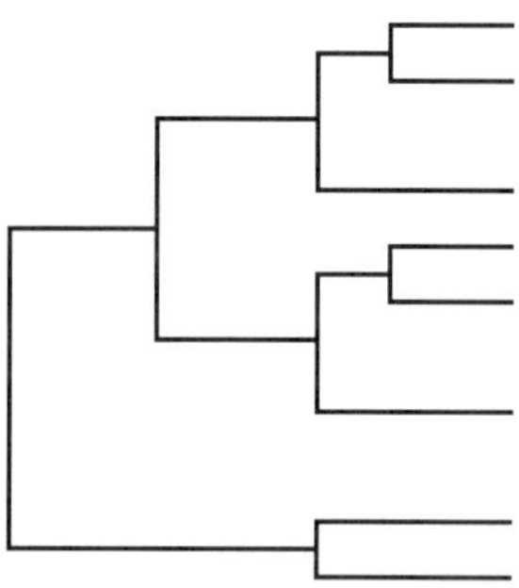		*HVA*	
	α_{1D}	L-type	Brain, pancreas, PC12 and GH3 cells
	α_{1C}	L-type	Heart, brain, aorta, lung, fibroblast, kidney, PC12 and GH3 cells
	α_{1S}	L-type	Skeletal muscle
	α_{1B}	N-type	Brain, peripheral neurons, and PC12 cells
	α_{1A}	P/Q-family	Brain, cerebellum, Purkinje and granular cells, kidney, PC12 cells.
	α_{1E}	R-type ?	Brain, heart
		LVA	
	α_{1G}	T-type	Brain, heart
	α_{1H}	T-type	Heart

HVA, high voltage-activated Ca^{2+} channel; LVA, low voltage-activated Ca^{2+} channel.

sites of expression. The diversity of α_1 genes found, together with the alternative splicing from each single gene, adds a large structural diversity to the multitude of Ca^{2+} channel α_1 gene subproducts.

In spite of the divergence among different α_1 genes, the deduced amino acid sequence of all Ca^{2+} channel α_1 subunits show a basic generalized secondary structure (see Fig. 3B) that consists of four internal repeated motifs (I–IV). Each motif being comprised of six α-helical transmembrane segments (S1–S6), including one (S4) that is the positively charged voltage sensor. Also, a specific region exists in all the cloned α_1 subunits, with the exception of α_{1G}, which is a highly conserved sequence of the cytoplasmic linker loop between motif I and II. This region is known as the α_1 subunit interaction domain (AID) for the binding of β subunits in the Ca^{2+} channel.

B. Other Ca^{2+} Channel Subunit Genes

The intracellular β subunit of voltage-activated Ca^{2+} channel is also represented by a multigene family and, up to now, a minimum of four genes have been established with alternative splicing (76–81). Therefore, potential combinations of 12 mature messages can coexist in various tissues. Table 4 summarizes the β subunit genes described so far according to the more recent consensus nomenclature (82). The Table 4 also shows the known splice variants from each gene and some, but not all, of the expression sites of mature β messenger RNAs.

The α_2/δ protein is a heavily glycosylated protein derived from a proteolytic cleavage of a single gene with alternative splicing that gives rise to the existence of five messenger RNA species ($\alpha_{2a\text{-}e}$) (83). Although some tissues, such as skeletal muscle and brain, seem to express single isoforms of this protein (α_{2a} and α_{2b}, respectively), the cardiovascular system expresses all five splice variants (84). The α_2 and δ polypeptides are connected by a disulfide bond, and recent data (85) suggest that the α_2 was entirely located outside the cell, whereas the δ polypeptide anchors the whole protein to the membrane through a single transmembrane domain (see Fig. 3A). The α_2 extracellular location restricts its interactions with the Ca^{2+} channel to the extracellular domains of the α_1 subunit. Also, one gene has been described for the transmembrane γ subunit apparently without the existence of any splice variants (80,86).

Table 4 Voltage-Activated Ca^{2+} Channel β Subunit Genes

Gene product	Splice variant	Sites of expression	Component of
β_1	β_{1a}	Skeletal muscle	DHP receptor
	β_{1b}	Brain, heart	ω-Conotoxin GVIA and MVIIC receptors
	β_{1c}	Brain, heart	?
	β_{1d}	ND	?
β_2	β_{2a}	Brain, heart	ω-Conotoxin MVIIC receptor
	β_{2b}	Brain, heart	?
	β_{2c}	Brain, heart	?
	β_{2d}	ND	?
β_3	β_{3a}	Brain, heart	ω-Conotoxin GVIA and MVIIC receptors
	β_{3b}	ND	?
β_4	β_{4a}	Brain	ω-Conotoxin GVIA and MVIIC receptors
	β_{4b}	ND	?

ND, not determined.

Extensive studies on the α_1, β, and α_2/δ subunits of voltage-dependent Ca^{2+} channels have been performed; however, little is known about the structure of the 95-kD protein associated with neuronal Ca^{2+} channels and whether it forms part of any of the neuronal Ca^{2+} channel complex; preliminary data suggested that it could be the fifth subunit of the N-type Ca^{2+} channel. However, Campbell's group reported that this protein, surprisingly, is associated with the P/Q-type Ca^{2+} channel (87) and not with the N-type channel. Furthermore, the 95-kD protein is a short form of the α_{1A} subunit which contains approximately the first half (motif I and II) of the full-length α_{1A} subunit, including the binding region (AID) for the β subunit. This finding suggests the possibility, among others, that the 95-kD protein would compete with the full-length α_{1A} subunit for the β auxiliary subunit binding, affecting the functional activity and expression of the channel.

C. Modulation of the α_1 Activity by the Other Ca^{2+} Channel Auxiliary Subunits

Although all classes of α_1 subunits direct the expression of functional Ca^{2+} channels in different expression systems (i.e., *Xenopus* oocytes, mouse fibroblast L cells, or HEK293 cells), the contribution of the Ca^{2+} channel auxiliary subunits to the native properties of the channel complex was unknown until the first α_2, β, and γ cDNAs, obtained from rabbit skeletal muscle, became available (76,86,88). In recent years, cloning of new subunit genes and splice variants has allowed a great increase in the number of potential combinations among α_1 and other subunits.

The results obtained until now show that coexpression of several classes of β-subunits with different α_1 subunits alters the voltage dependence, kinetics, and magnitude of the Ca^{2+} channel current, with different effects depending on the particular β and α_1 subunits used (89–94).

The most complete studies about the role of auxiliary subunits within the Ca^{2+} channel complex have been done with the neuronal α_{1A}, the skeletal muscle α_{1S}, and the cardiac α_{1C} subunits. Therefore, the coexpresion of four different β subunit genes (β_{1b}, β_{2a}, β_3, and β_4) with the α_{1A} subunit dramatically increases the current amplitude expressed in oocytes. All β subunits trigger significant changes in the voltage-dependence of both activation and inactivation. The α_2/δ_b subunit does not modify the properties of the α_{1A} subunit in the absence of β subunits.

However, it increases the β-induced stimulation in current amplitude and also regulates the β-induced change in inactivation kinetics. Both α_2/δ_b and β_{1b} slightly modified the sensitivity of the α_{1A} subunit to ω-conotoxin MVIIC (94). The mechanisms responsible of the α_2/δ effects are not known, although it has been suggested that α_2/δ subunits may stabilize incorporation of the channel complexes in the membrane and also increase the binding affinity of channels for ω-conotoxin.

It has been also demonstrated that β subunits increase the peak current induced by the expression of cardiac α_{1C} and this effect can be correlated with an increase in the number of DHP binding sites (77,79). Since DHPs bind on the α_{1C} subunit, the increase both in the current and in the number of dihydropyridine binding sites would suggests that β subunits act by increasing the number of channels present at the plasma membrane. However, gating charge measurements (95) and immunoblot analysis (96) demonstrate that the number of α_{1C} subunits expressed at the plasma membrane was not altered by the β subunits. In order to reconcile these apparently contradictory results, it was hypothesized that β subunits could induce important conformational changes on the α_{1C} subunit which in turn would increase not only the opening probability of the channel but also the accessibility of the drug to its binding site.

VII. CALCIUM CHANNELS AND NEUROSECRETION

It has been long demonstrated that Ca^{2+} is essential for neurotransmitter release. The existence of multiple types of Ca^{2+} channels and the fact that several of them can coexist in the same cell type has raised questions about which channel (or channels) contribute to the control of the delivery of the Ca^{2+} necessary to trigger a secretory signal in a particular synapse. We will therefore review throughout this section how the different Ca^{2+} channel subtypes (defined by ω-toxin blockade of neurosecretion) control the release of neurotransmitters depending on the synapse, the neurotransmitter, and the animal species. Tables 5–8 summarize how Ca^{2+} entry through different Ca^{2+} channel subtypes controls neurotransmitter release at different sites of the central and peripheral nervous system, motor nerve terminals, and chromaffin cells.

A. Brain Synaptosomes

Transmitter release from brain synaptosomes is controlled by different Ca^{2+} channels and is greatly dependent on the animal species studied. In chick brain synaptosomes, inositol phosphate production together with norepinephrine release is highly sensitive to ω-conotoxin GVIA (97). Ca^{2+} transients measured in chick brain synaptosomes loaded with the Ca^{2+} sensitive fluorescent dye fura-2 demonstrated that increases in the $[Ca^{2+}]_i$ induced by high K^+ was almost completely supressed by ω-conotoxin GVIA (58). On the other hand, in rat brain synaptosomes, the production of inositol phosphate and secretion of norepinephrine is insensitive to ω-conotoxin GVIA, but sensitive to ω-agatoxin IVA (98). Glutamate release from rat brain synaptosomes is blocked 56% by ω-agatoxin IVA and 23% by ω-agatoxin IIIA, an L-N-P–type Ca^{2+} channel blocker (28). These results indicate that, in chick brain synaptosomes, Ca^{2+} entry and therefore neurotransmitter release is predominately controlled via an N-type Ca^{2+} channel. In rat brain synaptosomes, L- and N-type Ca^{2+} channel blockers do not modify $[Ca^{2+}]_i$ levels or transmitter release; therefore another Ca^{2+} entry pathway seems to be involved in the control of neurotransmitter release.

Turner and Dunlap (99) measured $[^3H]$-glutamate release from rat cortical synaptosomes as an assay for presynaptic calcium channel activity. In this system, they observed that the efficacies of ω-agatoxin IVA and ω-conotoxin GVIA and MVIIC were increased when calcium influx was decreased by decreasing the KCl concentration to diminish the extent of depolarization, by decreasing the external concentration of Ca^{2+}, or by partially blocking Ca^{2+} influx with one of the other toxins. Using these ω-toxins, they found at least three types of pharmacologically distinct Ca^{2+} channels that participate in exocytosis. The largest fraction of glutamate release was blocked by ω-agatoxin IVA with an IC_{50} of 12.2 nM and ω-conotoxin MVIIC with an IC_{50} of 35 nM, which are consistent with the pharmacology of a P-type Ca^{2+} channel. The N-type Ca^{2+} channel blocker, ω-conotoxin GVIA, inhibited a significant portion of the release (IC_{50} less than 1 nM) but only under conditions of reduced Ca^{2+} concentrations. These results suggest that the N-type channel in nerve terminals is different from that found in hippocampal somata, since it appears to be resistant to ω-conotoxin MVIIC. The combination of ω-conotoxin GVIA (100 nM) and either ω-agatoxin IVA or ω-conotoxin MVIIC (1μM) blocked approximately 90% of release when the Ca^{2+} concentration was reduced (0.46 mM or less), but 30–40% of release remained when the concentration of Ca^{2+} in the stimulus buffer was 1 mM or greater, indicating that a resistant channel also participates in exocytosis.

B. Striatum

In the striatum, neurotransmitter release is controlled by different Ca^{2+} channels. Dopamine release induced by K^+ is blocked approximately 30% by ω-conotoxin GVIA (98,100), although dopamine release evoked by electrical stimulation is almost completely inhibited by ω-conotoxin GVIA (100). Turner and cowokers (98), using subsecond measurements of glutamate and dopamine release from rat striatal synaptosomes, showed that P-type Ca^{2+} channels, which are sensitive to ω-agatoxin IVA, trigger the release of both neurotransmitters, although dopamine (but not glutamate) was also partially blocked by ω-conotoxin GVIA–sensitive Ca^{2+} channels. Another interesting observation by these investigators is that the blockade of neurotransmitter release is voltage dependent. With strong depolarizations (60 mM K^+), neither ω-agatoxin IVA nor ω-conotoxin GVIA was effective alone, although a combination of both produced a synergistic inhibition of 60–80% of Ca^{2+}-dependent dopamine release. With milder depolarizations (30 mM K^+), ω-agatoxin IVA (200 nM) blocked over 80% dopamine and glutamate release, whereas ω-conotoxin GVIA (1 μM) blocked dopamine release by 25% and left unaffected glutamate release. The results suggest that multiple Ca^{2+} channel subtypes coexist to regulate neurosecretion under normal physiological conditions in the majority of nerve terminals, whereas P-type and ω-conotoxin GVIA– and ω-agatoxin IVA–resistant channels coexist in glutamatergic terminals. Such an arrangement could lend a high degree of flexibility in the regulation of transmitter release under diverse conditions of stimulation and modulation.

C. Hippocampus

In the hippocampi of rabbits, Doodley at al. (101) demonstrated that electrically induced release of dopamine, 5-hydroxytryptamine, and acetylcholine was similarly blocked (about 40%) by nanomolar concentrations of ω-conotoxin GVIA. Under the same experimental conditions, dopamine release from the corpus striatum and norepinephrine release from the neocortex were also blocked 40% by ω-conotoxin GVIA (5 nM). Using a superfusion system with subsecond temporal resolution, Luebke et al. (102) studied the effects of ω-conotoxin GVIA and ω-agatoxin IVA on glutamate release from rat hippocampal synaptosomes. K^+-induced release of glutamate was inhibited 16% by ω-conotoxin GVIA and 40% by ω-agatoxin IVA; such blockade was increased when lower concentrations of K^+ were employed to induce secretion. The amplitude of excitatory postsynaptic potentials in CA1 pyramidal neurons was reduced by ω-conotoxin GVIA and ω-agatoxin IVA, although ω-agatoxin IVA was more rapid and more efficacious (102).

Table 5 Control of Neurotransmitter Release by N-Type Ca^{2+} Channels

Neurotransmitter	Preparation	ω-Conotoxin GVIA [μM]	% Inhibition	Reference
Acetylcholine	Electric fish (*Torpedo marmorata*)	5	30	59
	Electric fish (*Gymnotus carajo*)	2.5	>95	60
	Myenteric plexus (guinea pig)	0.01	92	112
	Myenteric plexus (rat)	0.1	70	62
	Detrusor (guinea pig)	0.1	88	116
	Urinary bladder (guinea pig)	1	71	113
	Urinary bladder (guinea pig)	1	71	113
	Urinary bladder (rat)	1	25	113
	Urinary bladder (rat)	1	25	114
	Atria (guinea pig)	IC50 = 0.42 μM		120
	Phrenic nerve (Rat)	0.1	47	62
	Brain slices (rat)	1	60	153
ATP	Vas deferens (rat)	IC50 = 20 nM		119
Catecholamines	Urethra (rabbit)	0.1	77	116
	Chromaffin cells (dog)	0.4 μg/min	33	142
	Chromaffin cells (bovine)	5	17	71
	Chromaffin cells (bovine)	1	10	124
	Chromaffin cells (bovine)	3	30	139
	Chromaffin cells (cat)	1	20	70
	Chromaffin cells (rat)	1	42	141
Dopamine	Striatum (rabbit)	0.005	39	101
	Striatum (rat)	1	38	100
	Brain slices (rat)	1	30	153
Dynorphin	Dentate gyrus dendrite (guinea pig)	1	79	104
	Dentate gyrus axon (guinea pig)	1	50	104
EPSCs	Retinal ganglion neurons (rat)	5	67	109
EPSPs	Hippocampal synaptic transmission (rat)	1	46	56

GABA	Deep cerebellar neurons (rat)	0.1	50	103
5-HT	Brain slices (rat)	1	30	153
Glutamate	Hippocampal CA1 pyramidal cells (rat)	3	80	103
	Hippocampal synaptosomes (rat)	1	16	102
Glycine	Dorsal horn neurons of the spinal cord (rat)	3	50	103
Norepinephrine	Brain synaptosomes (rat)	1	>90	97
	Neocortex (rabbit)	0.005	46	57
	Neocortex (rabbit)	0.005	46	57
	Sympathetic neurons (rat)	0.1	92	107
	Vas deferens (rat)	1	100	113
	Anococcygeus (guinea pig)	0.01	98	113
	Vas deferens (guinea pig)	0.01	97	112
	Atria (guinea pig)	IC50 = 0.200 μM		14
	Atria (guinea pig)	0.1	80	154
	Atria (guinea pig)	10	49	120
	Chromaffin cells (dog)	0.4 μg/min	32	142
	Mesenteric artery (rat)	0.01	92	117
	Right atria (mouse)	0.1	100	117
	Right atria (rat)	0.1	100	117
NANC	Detrusor (rabbit)	0.1	85	116
	Anococcygeus (guinea pig)	0.05	64	113
	Urinary bladder (guinea pig)	0.1	58	113
	Urinary bladder (guinea pig)	0.1	58	113
	Urethra (rabbit)	0.1	47	116
	Jejunum (guinea pig)	0.1	33	112
	Taenia caecum (guinea pig)	0.05	20	112
Oxytocin	Neurohypophysial terminals (bovine)	0.8	32	105
Vasopressin	Neurohypophysial terminals (bovine)	0.8	32	105

NANC, nonadrenergic-noncholinergic neurotransmission; EPSCs, excitatory postsynaptic currents; EPSPs, excitatory postsynaptic potentials.

Table 6 Control of Neurotransmitter Release by P-Type Ca^{2+} Channels

Neurotransmitter	Preparation	ω-Agatoxin IVA [μM]	% Inhibition	Reference
Acetylcholine	Phrenic nerve (guinea pig)	0.02	>95	155
	Phrenic nerve hemidiaphragm (mouse)	0.1	92	117
	Phrenic nerve hemidiaphragm (rat)	0.1	0	117
Catecholamines	Chromaffin cells (bovine)	0.1	35	71
GABA	Deep cerebellar neurons (rat)	0.2	98	103
Glutamate	Brain synaptosomes (rat)	0.2	56	28
	Cortex synaptosomes (rat)	IC50 = 12.2 nM		53
	Hippocampal synaptosomes (rat)	0.2	40	102
	Hippocampal CA1 pyramidal cells (rat)	0.2	25	103
Glycine	Dorsal horn neurons of the spinal cord (rat)	0.2	98	103

GABA, gamma-aminobutyric acid.

Thus, at least two Ca^{2+} channels seem to control glutamate release from hippocampal neurons, but P-type channels seem to play a major role.

Synaptic transmission between hippocampal CA3 and CA1 neurons is mediated by N-type Ca^{2+} channels together with Ca^{2+} channels whose pharmacology differs from L- and P-type channels but resembles that of Q-type Ca^{2+} channels encoded by the α_{1A} subunit gene. Using rat hippocampal slices, Wheeler et al. (56) showed that ω-conotoxin GVIA blocked excitatory postsynaptic potentials (EPSPs) by 46%, whereas P- and L-type Ca^{2+} channel antagonists had no effect. In contrast ω-conotoxin MVIIC (N-P-Q Ca^{2+} channel blocker) inhibited 100% of the EPSPs. This suggests that hippocampal synaptic transmitter release is regulated by the N- and Q-subtypes of Ca^{2+} channels. Measuring excitatory postsynaptic currents (EPSCs) from hippocampal CA1 pyramidal neurons,Takahashi and Momiyama (103) demonstrated that synap-

Table 7 Control of Neurotransmitter Release by Q-Type Ca^{2+} Channels

Neurotransmitter	Preparation	ω-conotoxin MVIIC [μM]	% Inhibition	Reference
Acetylcholine	Urinary bladder (rat)	3	54	114
	Atria (guinea pig)	IC50 = 0.28 μM		120
	Phrenic nerve hemidiaphragm (mouse)	1	80	117
	Phrenic nerve hemidiaphragm (rat)	1	57	117
	Chromaffin cells (bovine)	IC50 = 218 nM		138
ATP	Vas deferens (rat)	IC50 = 200 nM		119
Catecholamines	Chromaffin cells (bovine)	3	50	70
Glutamate	Cortex synaptosomes (rat)	IC50 = 35 nM		53
EPSPs	Hippocampal synaptic transmission (rat)	5	100	56
Norepinephrine	Atria (guinea pig)	0.5	100	120
	Atria (guinea pig)	IC50 = 0.19 μM		14
	Atria (mouse)	1	100	117
	Atria (rat)	1	100	117
	Chromaffin cells (bovine)	IC50 = 182 nM		138
Vasopressin	Neurohypophysial terminals (bovine)	0.3	25	105

ATP, adenosine triphosphate; EPSPs, excitatory postsynaptic potentials.

Table 8 Control of Neurotransmitter Release by P/Q-Type Ca^{2+} Channels (Using High Concentrations of ω-Agatoxin IVA)

Neurotransmitter	Preparation	ω-Agatoxin IVA [μM]	% Inhibition	Reference
Acetylcholine	Urinary bladder (rat)	3	46	114
	Atria (guinea pig)	3	37	120
	Brain Slices (rat)	1	50	153
GABA	Brain Slices (rat)	1	100	153
Glutamate	Brain Slices (rat)	1	100	153
Dopamine	Brain Slices (rat)	1	70	153
5-HT	Brain Slices (rat)	1	50	153
Norepinephrine	Atria (guinea pig)	3	21	120

GABA, gamma-aminobutyric acid; 5-HT, 5-hydroxytryptamine (serotonin).

tic transmission at this level is predominantly controlled by N-type Ca^{2+} channels (80% block of EPSPs by ω-conotoxin GVIA) and to a lesser extent by P-type Ca^{2+} channels (25% inhibition by ω-agatoxin IVA).

The release of the neuropeptide dynorphin is controlled by different Ca^{2+} channels depending on the release site (dendrite or axon). L-type Ca^{2+} channels mediate dynorphin release from dendrites and N-type Ca^{2+} channels mediate dynorphin release from the axons of hippocampal granule cells (104).

D. Cerebellum

Inhibitory postsynaptic currents (IPSCs) evoked in neurons of the deep cerebellar nuclei by stimulating presumptive Purkinje cell axons were reversibly abolished by bicuculline, indicating that the responses were mediated by GABA. The application of ω-agatoxin IVA (200 nM) blocked IPSCs' amplitude by 50%, whereas the L-type Ca^{2+} channel blocker nicardipine had no effect (103), indicating that GABA release from Purkinje cell axons is mediated via Ca^{2+} entry through P-type Ca^{2+} channels.

E. Neurohypophysis

Neurohypophysial terminals exhibit besides L- and N-type currents; another component of the Ca^{2+} current that is blocked by low concentrations of ω-conotoxin MVIIC or by high concentrations of ω-agatoxin IVA indicating the presence of a Q channel. In the study performed by Wang and coworkers (105), they demonstrated that secretion of vasopressin is controlled by N, L, and Q channels, whereas that of oxytocin is regulated mainly by N and L channels with no participation of Q channels.

F. Sympathetic Neurons

In rat sympathetic neurons, whole-cell recordings have provided evidence for the presence of two subtypes of Ca^{2+} channels, the N- and the L-type (36), although norepinephrine release is predominantly blocked by ω-conotoxin GVIA (106,107). In contrast to sympathetic neurons, release of substance P from peripheral sensory neurons is highly dependent on Ca^{2+} entry through L-type Ca^{2+} channels. In the sympathetic nerve endings of the iris, ω-conotoxin GVIA

(1 μM) blocked over 80% of norepinephrine synthesis induced by high K^+, whereas nicardipine had no effect, indicating that Ca^{2+} entry through N-type Ca^{2+} channels play a major role in norepinephrine synthesis (108).

G. Retinal Ganglion Neurons

Glutamatergic synaptic responses in rat retinal ganglion neurons is partially sensitive to ω-conotoxin GVIA (30% block) and insentive to ω-agatoxin IVA (109). These results indicate that the major part of synaptic glutamate release in retinal ganglion neurons is governed by a novel toxin-resistant Ca^{2+} channel that could possibly be of the Q or R-type.

H. Spinal Cord and Dorsal Root Ganglion Neurons

In cocultures of fetal neurons from the ventral half of the spinal cord (VH neurons) and from the dorsal root ganglion (DRG neurons), the synaptic transmission between pairs of spinal cord neurons from the ventral half of the spinal cord (VH-VH connections) or between dorsal root ganglion neurons and VH neurons (DGR-VH connections) were studied with two-cell recording and stimulation techniques. In 70% of the VH-VH connections and in 50% of the DGR-VH connections, Bay K 8644 failed to affect transmitter release. ω-Conotoxin GVIA produced no consistent effect on EPSPs or IPSPs elicited by VH neurons by stimulation of the nearby neurons. VH EPSPs elicited by stimulation of the nearby DGR neurons were reduced 50% by ω-conotoxin GVIA. Therefore neither sustained nor inactivating HVA Ca^{2+} channels sensitive to Bay K 8644 or ω-conotoxin GVIA, such as those measured in the neuronal cell body, are responsible for action potential–evoked transmitter release from the majority of the VH neurons; these channels may be involved in transmitter release in approximately 30% of these neurons (110).

In rat dorsal horn neurons of the spinal cord, release of glycine induces IPSCs. The IPSCs were almost completely blocked by ω-conotoxin GVIA (>95%) and partially inhibited by ω-agatoxin IVA (50%) whereas nicardipine had no effect.

I. Intestinal Tract

Electrically evoked release of acetylcholine is predominantly controlled through N-type Ca^{2+} channels at the myenteric plexus (62,111,112). ω-Conotoxin GVIA markedly reduced (70%) the evoked release of [^{3}H]-acetylcholine from the myenteric plexus of the small intestine, with an IC50 of 0.7 nmol/L; the potency was similar at 3- and 10-Hz stimulation. An increase in the extracellular Ca^{2+} concentration attenuated the inhibitory effect of ω-conotoxin GVIA (62). No species difference was observed as to the channel controlling Ca^{2+} entry for transmitter release.

In the guinea pig jejunum, ω-conotoxin GVIA blocked only partially (33%) the inhibitory nonadrenergic noncholinergic (NANC) transmission on electrical stimulation. This was also the case at the taenia caecum (20% inhibition) (112). In the proximal duodenum, the NANC transmission was insensitive to ω-conotoxin GVIA (113).

Therefore, cholinergic transmission at this level seems to be regulated by Ca^{2+} entering through N-type Ca^{2+} channels, whereas NANC transmission is regulated by another Ca^{2+} entry pathway besides N channels.

J. Lower Urinary Tract

In the rat or guinea pig isolated bladder, ω-conotoxin GVIA produced a concentration and time-dependent inhibition of twitch responses to field stimulation without affecting the response to exogenous acetylcholine. In the rat bladder, the maximal effect did not exceed 25% inhibition, whereas a much larger fraction of the response (70%) was inhibited in the guinea pig bladder. In the rat bladder, the effects of ω-conotoxin GVIA were frequency dependent; maximal effects of ω-conotoxin GVIA were observed at 2–5 Hz. Frew and Lundy (114) have demonstrated that neurotransmission in the rat urinary bladder is supported by both N- and Q-type Ca^{2+} channels. In their experiments, the resistant portion (non-N non-P) was sensitive to ω-conotoxin MVIIC, which in addition to N and P also blocks Q channels. Further experiments carried out by Waterman (115) in mouse bladder using calcium channel toxins demonstrates that acetylcholine release in these parasympathetic neurons depends primarily on N-type channels and to a lesser extent on P- and Q-type channels, whereas adenosine triphosphate (ATP) release involves predominantly P- and Q-type channels.

In the rabbit urethra and detrusor, Zygmunt et al. (116) have studied the effects of ω-conotoxin GVIA on adrenergic, cholinergic, and NANC responses induced by electrical stimulation. The adrenergic contraction (25 Hz) and NANC relaxation (10 Hz) in the urethra and the cholinergic and NANC contractions (10 Hz) in the detrusor, were inhibited in a concentration-dependent manner by ω-conotoxin GVIA. The adrenergic contraction of the urethra was 10 times and the cholinergic contraction in the detrusor was 3 times more sensitive to ω-conotoxin GVIA than the NANC responses. These results suggest that NANC transmission is less sensitive to ω-conotoxin GVIA than transmission mediated by adrenergic and cholinergic nerves in the rabbit lower urinary tract.

K. Vas Deferens

In rat or guinea pig isolated vas deferens, ω-conotoxin GVIA (1 nM to 1 μM) produced concentration and time-dependent inhibition of the response to electrical field stimulation, whereas the response to K^+, norepinephrine, or ATP was unaffected. A concentration as low as 1 nM produced almost complete inhibition of twitches, but this effect took about 1 h to be completed. With higher concentrations, the time course of the inhibition was much faster (113). In a study performed by Wright and Angus (117) in rat and mouse vas deferens, they observe that ω-conotoxin GVIA (10 nM) and ω-conotoxin MVIIC (1 μM) block completely the twitch responses when they are induced at low frequencies (0.05 Hz); but when higher frequencies are used (20 Hz), there is a ω-conotoxin GVIA–resistant component that can be blocked by 1 μM ω-agatoxin IVA or ω-conotoxin MVIIC. These results indicate that sympathetic transmission in the vas deferens is mainly controlled by Ca^{2+} entering N channels, although when high-frequency stimulation is employed (20 Hz), P/Q-type channels are also implicated (117).

As to the purinergic transmission in the vas deferens, Hata et al. (118) showed that the ATP-mediated component of the biphasic contraction was found to be more susceptible to ω-conotoxin GVIA than the adrenergic component. In a study performed 5 years later by Hirata and coworkers (119), electrically induced twitch responses of the prostatic segment of the rat vas deferens, which depends mainly on ATP release, was fully blocked by nanomolar concentrations of ω-conotoxin GVIA, MVIIA, and MVIIC; most likely by inhibiting Ca^{2+} entry through presynaptic N-type Ca^{2+} channels that control ATP release. The main conclusion we can draw from these studies is that sympathetic and purinergic transmission in the vas deferens is predominantly controlled by N-type Ca^{2+} channels.

L. Heart

The innervation in mammalian atria is both sympathetic and parasympathetic, which regulates the heart rate and the contractile strength. The subtypes of Ca^{2+} channels involved in neurotransmitter release have been studied by various investigators. Vega et al. (14) have shown that electrically stimulated guinea pig left atria are sensitive to N-type Ca^{2+} channel blockers. Thus, ω-conotoxin GVIA and ω-conotoxin MVIIA blocked the inotropic response in a concentration-dependent fashion with IC50 values of 0.20 and 0.044 μM, respectively. The N-P-Q channel blocker ω-conotoxin MVIIC showed an IC50 of 0.19 μM; ω-agatoxin IVA had no effect. These results were confirmed later by Wright and Angus (117) in the right atria of mouse and rat, where they saw full inhibition of contraction with 100 nM ω-conotoxin GVIA.

Hong and Chang (120) have studied the Ca^{2+} channel subtypes mediating the cholinergic and adrenergic neurotransmission in the guinea pig atria. In left atria paced at 2–4 Hz, the negative inotropic effect induced by electrical field stimulation on parasympathetic nerves (in the presence of propranolol) was abolished by ω-conotoxin MVIIC. On the other hand, the inotropic response resulting from electrical field stimulation of the sympathetic nerves (in the presence of atropine) was abolished by ω-conotoxin GVIA and ω-conotoxin MVIIC. None of the peptide toxins affected the chronotropic and the inotropic responses evoked by carbachol, isoprenaline, or norepinephrine (14,120).

These results suggest that, under physiological conditions, the release of acetylcholine from parasympathetic nerves to the heart is dominated by a P/Q subfamily of Ca^{2+} channels, whereas that of norepinephrine from sympathetic nerves is controlled by an N-type Ca^{2+} channel.

M. Motor Nerve Terminals

Neurotransmitter release at this level is controlled by different Ca^{2+} channels depending on the species. The electroplax of marine electric fish is highly rich in motor nerve endings; this is the reason why it has been so widely used as a model to study transmitter release from motor nerve endings. ω-Conotoxin GVIA blocks the release of acetylcholine and Ca^{2+} uptake induced by depolarization in electric organ nerve terminals of the ray; the IC50 values were 3 μM for blocking transmitter release and 2 μM for blocking Ca^{2+} entry (58). Sierra et al. (60) have also shown that N-type Ca^{2+} channels mediate transmitter release at the electromotoneuron-electrocyte synapses of the weakly electric fish *Gymnotus carapo*; ω-conotoxin GVIA (2.5 μM) blocked over 95% of the endplate potential (EPP), whereas ω-agatoxin IVA and nifedipine had no effect. In contrast to these data, in *Torpedo* synaptosomes, Fariñas et al. (59) showed that ω-conotoxin GVIA (10^{-8} to 5×10^{-5} M) had a differential effect on acetylcholine and ATP release: Nucleotide release was inhibited 90% at the highest concentration tested, whereas acetylcholine release was only moderately decreased (30%). In the frog neuromuscular junction, Jahromi et al. (61) have demonstrated that synaptic transmission is also governed by Ca^{2+} entry through an N-type Ca^{2+} channel.

In contrast, in mammalian motor nerve terminals, Ca^{2+} entry serving to discharge acetylcholine release seems to be ruled by a P-type Ca^{2+} channel rather than an N-type Ca^{2+} channel, as in fish and amphibians. So in the rat phrenic nerve, [^{3}H]-acetylcholine release was only partially inhibited by ω-conotoxin GVIA (63). In the mouse, endplate potentials (EPPs) were almost completely abolished (>95%) with 200 nM ω-agatoxin IVA. The twitch responses of the phrenic nerve hemidiaphragm were blocked in a different manner depending on the animal species. In the mouse, ω-agatoxin IVA at 100 nM blocked 92% of the twitches whereas in the rat, ω-agatoxin IVA (≤100 nM) and ω-conotoxin GVIA (≤1 μM) had little effect, although, ω-conotoxin MVIIC caused 57% blockade (117). In normal human muscles, Protti et al. (121)

have shown that transmitter release at the motor nerve terminals is mediated by a P-type calcium channel.

N. Chromaffin Cells

Chromaffin cells contain different subtypes of Ca^{2+} channels at their plasmalemmal membrane. The presence and proportion of the various subtypes of Ca^{2+} channels depends on the animal species (see Fig. 2). Therefore, secretion of catecholamines from these cells will presumably be controlled in a different manner according to the Ca^{2+} channels it contains. In this section, we will review how catecholamine secretion is controlled in four different animal species (cat, cow, rat, and dog) and how some subtypes of Ca^{2+} channels are more directly implicated in the control of exocytosis.

Cat Chromaffin Cells

Secretion of catecholamines is effectively blocked in a concentration-dependent manner by DHPs and by other drugs acting on L-type Ca^{2+} channels like verapamil and diltiazem (122). Measuring differential secretion of epinephrine and norepinephrine, Cárdenas et al. (123) demonstrated that secretion of both amines is completely blocked when secretion is induced either by high K^+ or the nicotinic agonist dimethylphenilpiperazinium (DMPP). All these data indicated that secretion in these cells was controlled by an L-type channel. But Albillos et al. (63) showed that cat chromaffin cells also contained ω-conotoxin GVIA–sensitive channels besides L type. It was then demonstrated that HVA Ca^{2+} channels in cat chromaffin cells were present in an approximate proportion of 50–50%, and that the increase in $[Ca^{2+}]_i$ induced by short (10 s) depolarizing pulses (70 mM K^+) was also reduced by 44% by furnidipine and 43% by ω-conotoxin GVIA. In the perfused adrenal gland or in isolated cat chromaffin cells, catecholamine release induced by 10 s pulses with 70 mM K^+ was blocked over 95% by furnidipine and about 25% by ω-conotoxin GVIA. These results show that, although Ca^{2+} entry through both channels (N and L type) lead to similar increments of the average $[Ca^{2+}]_i$, the control of catecholamine release is dominated by Ca^{2+} entering through L-type Ca^{2+} channels (124).

Bovine Chromaffin Cells

Catecholamine secretion from bovine chromaffin cells is greatly potentiated in the presence of the dihydropyridine agonist Bay K 8644; this increase in secretion went in parallel with an increase in ^{45}Ca uptake (125). Ceña et al. (126) showed complete blockade of catecholamine release ([^{3}H]-norepinephrine) by nitrendipine in bovine chromaffin cells stimulated with high K^+. These results disagreed with those obtained by other investigators who found that blockade of secretion in bovine chromaffin cells by DHPs did not exceed further than 40–50% (127–129).

When toxins were available to block selectively other subtypes of Ca^{2+} channels, it was demonstrated that these cells contained other Ca^{2+} channel subtypes besides L; that is N, P, and Q type (67,68,130–133). ω-Conotoxin GVIA is ineffective or very little effective in blocking catecholamine secretion (70,71,127,129,134–136).

As to the contribution of P-type Ca^{2+} channels to secretion of catecholamines, we find different results in the literature. Granja et al. (137) found that catecholamine secretion induced by high K^+ remains unaffected by ω-agatoxin IVA (100 nM); nevertheless, when secretion was activated by nicotine, catecholamine release was significantly blocked by 50%. Thus, Granja et al. (137) concluded that ω-agatoxin IVA could also be acting on the nicotinic receptor. Duarte et al. (135) showed that FTX decreased the K^+-evoked norepinephrine release to 25% and epinephrine release to 39%; the combination of FTX plus nitrendipine further decreased norepinephrine and epinephrine release to 12 and 24%, respectively, of the control. The latest results obtained by Baltazar and coworkers (138) show that the bovine chromaffin cell contains two types of ω-agatoxin IVA–sensitive Ca^{2+} channels and that the contribution of P-type channels to secretion is larger at low levels of depolarization.

The L-N-P–insensitive portion of catecholamine release seems to be ω-conotoxin MVIIC sensitive. López et al. (70) have shown that catecholamine release from superfused bovine chromaffin cells (stimul: 70 mM K^+ for 10 s) was inhibited 50% by furnidipine (3 μM). ω-Conotoxin MVIIC (3 μM) also reduced the secretory response by 50%. The combination of furnidipine and ω-conotoxin MVIIC completely abolishes secretion. On the other hand, these investigators also demonstrate that ω-conotoxin GVIA and ω-agatoxin IVA have no effect on secretion. These results strongly suggest that in bovine chromaffin cells secretion is predominantly controlled by Ca^{2+} entering through L- and Q-type Ca^{2+} channels.

Further studies performed by Lara et al. (139) suggest that Q-type channels are coupled more tightly to exocytotic active sites as compared to L-type channels. This hypothesis is founded on the fact that external Ca^{2+} entering the cell through a Ca^{2+} channel located near to chromaffin, vesicles will saturate the K^+ secretory response at both $[Ca^{2+}]_o$; that is, 0.5 mM. In contrast, Ca^{2+} ions entering through more distant channels will be sequestered by intracellular buffers and thus, will not saturate the secretory machinery at lower $[Ca^{2+}]_o$.

Rat Chromaffin Cells

DHPs block secretion in rat adrenal glands in a concentration-dependent manner. The magnitude of such blockade is related to the type of stimuli employed to induce secretion. The DHP isradipine can fully block secretion when the stimuli used is K^+ or nicotine. In contrast, when electrical field stimulation is used, only partial blockade is obtained with DHPs, such inhibition being frequency dependent (140). Measuring Ca^{2+} currents and capacitance, Kim et al. (141) have shown that ω-conotoxin GVIA (1 μM) blocks 40% and nicardipine approximately 60% of the total capacitance increase in rat chromaffin cells. Therefore, in these cells, secretion would be controlled by L- as well as by N-type Ca^{2+} channels.

Dog Chromaffin Cells

Kimura et al. (142) have studied the effects of ω-conotoxin GVIA and L-type blockers (nifedipine and verapamil) on catecholamine release in anesthetized dogs. Catecholamine release into the blood stream was induced either by electrical field stimulation of the splanchnic nerve or by intra-arterial injection of acetylcholine. Administration of 0.4 μg/mL of ω-conotoxin GVIA reduced catecholamine secretion by 30% when electrical stimulation was used; under these experimental conditions, nifedipine (1 μg/mL) or verapamil (10 μg/mL) had no effect. When catecholamine release was induced by acetylcholine, ω-conotoxin GVIA blocked secretion approximately 50% and nifedipine also reduced it by 50%. These results suggest that N- and L-type Ca^{2+} channels located in the adrenal medullary cells may contribute to the release of adrenal catecholamines in the dog adrenal.

VIII. DIFFERENT EFFICACIES OF THE ω-TOXINS DEPENDING ON THE CONCENTRATION OF CALCIUM THAT ENTERS THE CELL

In addition to this review, it has been shown that, in the majority of the synapses, several types of Ca^{2+} channels coexist together and, on the other hand, that the efficacy of the different channels to control exocytosis varies with the degree of depolarization and the concentration of external Ca^{2+} used to perform the experiments. There are different examples in the literature that demonstrate this fact. Turner and coworkers (99) observed that the efficacies of ω-agatoxin IVA and ω-conotoxin GVIA to block glutamate release from rat cortical synaptosomes increased when Ca^{2+} influx was reduced by decreasing the external concentration of KCl, to diminish the extent of depolarization, by decreasing the external concentration of Ca^{2+}, or by partially blocking Ca^{2+} influx with one of the other antagonist. For example, glutamate release was inhibited by ω-conotoxin MVIIC with an IC50 of 200 nM when stimulation of secretion was induced with 30 mM of KCl, but the same toxin had no effect when synaptosomes were stimulated with 60 mM KCl. The same investigators also found that dopamine release from rat striatal synaptosomes (98) could be blocked by ω-agatoxin IVA and ω-conotoxin GVIA when they used mild depolarizations with KCl, but with strong depolarization neither toxin was effective alone, although a combination of both toxins produced a synergistic inhibition of 60–80% of Ca^{2+}-dependent dopamine release.

Transmitter release in parasympathetic neurons in the mouse bladder shows a similar pattern when contraction of bladder strips was stimulated by single pulses or trains of 20 pulses at 1–50 Hz. Waterman (115) observed that ω-conotoxin GVIA and MVIIC inhibited contractions in a concentration-dependent manner with IC50 values of approximately 30 and 200 nM at low frequencies of stimulation; the same toxins had little effect at high frequencies of stimulation.

In a review by Dunlap et al. (143), they tried to give an explanation for why with strong depolarizations exocytosis of the neurotransmitters is unaffected when a single toxin, what is, a single Ca^{2+} entry pathway, is blocked, but a synergic inhibitory effect is observed when a combination of the toxins or two Ca^{2+} entry pathways are blocked, or why in a synapses with several calcium channel subtypes, when one tries to sum up their individual inhibitory effects, the values obtained are greater than 100%. These findings are explained by the presence of ''spare channels,'' Under conditions in which the intracellular concentration of Ca^{2+} is saturating for the acceptor, participation of multiple Ca^{2+} channels might increase the reliability of excitation-secretion coupling, since activation of a single channel becomes sufficient to maximize release probability. This spare channel model might describe excitation-secretion coupling under conditions of relatively strong stimulus, such as high-frequency trains of action potentials or with prolonged depolarizations using increasing concentrations of K^+. Biochemical modifications (such as phosphorylation) that increase the sensitivity of the Ca^{2+} acceptor would also predict an increase in the probability of release elicited by entry of Ca^{2+} through a single channel. Under such conditions, the binding affinity of Ca^{2+} channel antagonists would be underestimated by their effect on synaptic transmission, since blockade of one of several channels at the active zone would produce little or no effect on release. This would produce a rightward shift in the concentration-response relationship relative to the binding curve.

Calcium dependency of the blockade of different Ca^{2+} channel subtypes was also shown in the study performed by Lara et al. (139) in bovine chromaffin cells, where it was demonstrated that L and Q channels predominantely control catecholamine release. These investigators observed that blockade of secretion mediated by L-type channels is not dependent on the extracellular concentration of Ca^{2+}, whereas blockade of Q-type channels is. The explanation that these investigators give for these findings is that that Q-type channels could be coupled more tightly to exocytotic active sites in comparison to the L type. The physiological meaning given to this channel distribution might be found in considering the need for a regular secretory rate during normal activity of the body (L-type secretion) and explosive catecholamine secretion that occurs under stressful conditions (Q-type secretion).

IX. PERSPECTIVES

In 15 years of ω-toxins use, at least six subtypes of HVA Ca^{2+} channels have been identified and characterized. New toxins are needed to target selectively the Q-type Ca^{2+} channel without affecting the N or P. The R- or T-type channels also need new toxins to characterize their functions. Whether the P and Q channels are the same or separate entities in various cell types remains to be clarified. The question of how many Ca^{2+} channel subtypes remain to be discovered is also relevant. In addition, differences among tissues and cell types for a given Ca^{2+} channel are emerging; L-type Ca^{2+} channels differ from skeletal, to cardiac, to smooth muscles and the brain. Are the Q channels from hippocampal and chromaffin cells identical? What about the N, P, and R channels? Why are different Ca^{2+} channels required to control exocytosis of the same transmitter (i.e., acetylcholine, catecholamines) in the same cell type and in different animal species? Another important question relates to the expression of various channel subtypes in the same cell. Why does exocytosis require Ca^{2+} from different pathways? Is it a safety valve to secure the efficiency of the process? If the N channel is a part of the secretory machinery, what about the L, P, and Q channels? How close are they from exocytotic active sites? And most interesting, are the channels of a paraneuronal cell such as the chromaffin cell equally organized as those of brain synapes? Why is the release of norepinephrine controlled by N channels in sympathetic neurons and by L or Q channels in chromaffin cells? Do action potentials recruit different Ca^{2+} channel subtypes than those two catecholaminergic cell types? Will a K^+-depolarizing stimulus recruit Ca^{2+} channels different from those recruited by action potentials in neurons or by acetylcholine receptors in chromaffin cells? Does the electrical pattern of different excitable cells cause different secretion patterns by simply recruiting specific Ca^{2+} channels with particular gating and kinetic properties?

Another critical question relates to the development of a pharmacology for neuronal Ca^{2+} channels. Although L-type Ca^{2+} channels have a rich pharmacology that have provided novel therapeutic approaches to treat cardiovascular diseases, nonpeptide molecules which block or inactivate the N, P, Q, T, and R channels are lacking. Thus, a major goal for research in this field is the search for selective blockers of specific Ca^{2+} channel subtypes. The recent introduction of mibefradil as a T-type Ca^{2+} channel blocker opens new possibilities to study the functions of these channels. The knowledge of the three-dimensional structure in solution of the different toxins is very important for studying the specificity of their interactions with Ca^{2+} channel subtypes and to define active sites that can serve as models to design and synthetize nonpeptide blockers. The ω-conotoxins are small peptides containing 24–29 amino acid residues. It is interesting that the amino acid sequence of ω-conotoxin MVIIA is much more similar to that of ω-conotoxin MVIIC than to ω-conotoxin GVIA; yet the pharmacology of ω-conotoxin MVIIA is much more close to that of ω-conotoxin GVIA (blockade of N-type channels). Thus, it will be very important to define structural differences determining the toxin selectivity for N- or Q-type Ca^{2+} channels. The three-dimensional structures of ω-conotoxin GVIA (144–147), ω-conotoxin MVIIA (148), and ω-conotoxin MVIIC (149,150) have been recently elucidated. Elucidation of the structures of ω-conotoxin MVIID, ω-agatoxin IVA, and other new toxins will facilitate their comparisons and the definition of structural determinants for specific binding to Ca^{2+} channel subtypes.

Nonpeptide blockers for neuronal Ca^{2+} channels are emerging, but they lack selectivity. For instance, the piperazine derivatives flunarizine, R56865, lubeluzole, and dotarizine are "wide-spectrum" Ca^{2+} channel blockers (40,41,151). Fluspirilene, a member of the diphenylbutylpiperidine class of neuroleptic drugs (which also includes pimozide, clopimozide, and penfluridol), has antischizophrenic actions and blocks N-type Ca^{2+} channels in PC12 cells (152). It may be that its neuroleptic properties are due, at least in part, to an inhibition of neuronal

N-type Ca^{2+} channels. Thus, inhibition (or facilitation) of specific neurotransmitter release by selective blockers (or activators) of Ca^{2+} channels may have functional and therapeutic consequences. For instance, synthetic ω-conotoxin MVIIA protected hippocampal CA1 pyramidal neurons from damage caused by transient, global forebrain ischemia in the rat (13). Selective blockade of N-type Ca^{2+} channels may also be beneficial in the treatment of specific pain syndromes (1). Thus, intrathecal administration of as little as 0.3 μg of ω-conotoxin MVIIA completely suppressed the nociceptive responses in the rat hindpaw formalin test. Tactile allodynia was also selectively abolished in a rat neuropathic pain model by intrathecal administration of ω-conotoxin MVIIA at doses that did not impair motor function; the toxin was found to be 100 times more potent than morphine. It seems clear that several other neurological or psychyatric diseases will benefit from the development of drugs that interfere selectively with different Ca^{2+} channel subtypes.

Many questions have been answered during a decade of intense research in the field of neuronal voltage-dependent Ca^{2+} channels. Most of the studies performed were possible only because, particularly the group of professor Baldomero Olivera, made available to many groups of researchers potent ω-toxins to type and characterize those channels. The same is true for research on the patch-clamp techniques, which exploded at the same time as that on ω-toxins thanks to the efforts of the group of professor Erwin Neher. We are sure that some references on this topic escaped our review, we apologize to those investigators inadvertently not cited here.

ACKNOWLEDGMENTS

Supported by grants from DGICYT (PB94-0150), Fundación Teófilo Hernando and CAM (08.9/0001/1997), Spain. We thank M.C. Molinos for the typing of this manuscript.

REFERENCES

1. BM Olivera, G Miljanich, J Ramachandran, ME Adams. Calcium channel diversity and neurotransmitter release: the ω-conotoxins and ω-agatoxins. Annu Rev Biochem 63:823–867, 1994.
2. H Terlau, K-J Shon, M Grilley, M Stocker, W Stühmer, BM Olivera. Strategy for rapid immobilization of prey by a fish-hunting marine snail. Nature 381:148–151, 1996.
3. R Llinás, M Sugimori, D Hillman, B Cherksey. Distribution and functional significance of the P-type voltage-dependent Ca^{++} channels in the mammalian central nervous system. Trends Neurosci 15:351–355, 1992.
4. ME Adams, VP Bindokas, L Hasegawa, VJ Venema. ω-Agatoxins: novel calcium channel antagonists of two subtypes from Funnel web spider (Agelenopsis aperta) venom. J Biol Chem 265:861–867, 1990.
5. JR de Weille, H Schweitz, P Maes, A Tartar, M Lazdunski. Calciseptine, a peptide isolated from black mamba venom, is a specific blocker of the L-type calcium channel. Proc Natl Acad Sci USA 88:2437–2440, 1991.
6. H Schweitz, C Heurteaux, P Bois, D Moinier, G Romey, M Lazdunski. Calcicludine, a venom peptide of the Kunitz-type protease inhibitor family, is a potent blocker of high-threshold Ca^{2+} channels with a high affinity for L-type channels in cerebellar granule neurons. Proc Natl Acad Sci USA 91:878–882, 1994.
7. RA Myers, LJ Cruz, JE Rivier, BM Olivera. *Conus* peptides as chemical probes for receptors and ion channels. Chem Rev 93:1923–1936, 1993.

8. OD Uchitel. Toxins affecting calcium channels in neurons. Toxicon 35:1161–1191, 1997.
9. K-J Shon, A Hasson, ME Spira, LJ Cruz, WR Gray, BM Olivera. δ-Conotoxin GmVIA, a novel peptide from the venom of *Conus Gloriamaris*. Biochemistry 33:11420–11425, 1994.
10. BM Olivera, JM McIntosh, LJ Cruz, FA Luque, WR Gray. Purification and sequence of a presynaptic toxin from *Conus geographus* venom. Biochemistry 23:5087–5090, 1984.
11. JM McIntosh, LJ Cruz, MW Hunkapiller, WR Gray, BM Olivera. Isolation and structure of a peptide toxin from the marine snail *Conus magus*. Arch Biochem Biophys 218:329–334, 1982.
12. LM Kerr, D Yoshikami. A venom peptide with a novel presynaptic blocking action. Nature 308: 282–284, 1984.
13. K Valentino, R Newcomb, T Gadbois, T Singh, S Bowersox, S Bitner, A Justice, D Yamashiro, BB Hoffmann, R Ciaranello, G Miljanich, J Ramachandran. A selective N-type calcium channel antagonist protects against neuronal loss after global cerebral ischemia. Proc Natl Acad Sci USA 90:7894–7897, 1993.
14. T Vega, R de Pascual, O Bulbena, AG García. Effects of ω-conotoxins on noradrenergic neurotransmission in beating guinea-pig atria. Eur J Pharmacol 276:231–238, 1995.
15. DR Hillyard, VD Monje, IM Mintz, BP Bean, L Nadasdi, J Ramachandran, G Miljanich, A Azimi-Zoonooz, JM McIntosh, LJ Cruz, JS Imperial, BM Olivera. A new *Conus* peptide ligand for mammalian presynaptic Ca^{2+} channels. Neuron 9:69–77, 1992.
16. L Gandía, B Lara, JS Imperial, M Villarroya, A Albillos, R Maroto, AG García, BM Olivera. Analogies and differences between ω-conotoxins MVIIC and MVIID: binding sites and functions in bovine chromaffin cells. Pflügers Arch Eur J Physiol 435:55–64, 1997.
17. ME Adams, EE Herold, VJ Venema. Two classes of channel specific toxins from funnel web spider venom. J Comp Physiol A164:333–342, 1989.
18. JM Pocock, VJ Venema, ME Adams. Omega-agatoxins differentially block calcium channel in locust, chick and rat synaptosomes. Neurochem Int 20:263–270, 1992.
19. RH Scott, AC Dolphin, VP Bindokas, ME Adams. Inhibition of neuronal Ca^{2+} channel currents by the funnel web spider toxin omega-Aga-IA. Mol Pharmacol 38:711–718, 1990.
20. VP Bindokas, ME Adams. ω-Aga IA: a presynaptic calcium channel antagonist from venom of the funnel web spider *Agelenopsis aperta*. J Neurobiol 20:171–188, 1989.
21. VJ Venema, KM Swiderek, TD Lee, GM Hathaway, ME Adams. Antagonism of synaptosomal calcium channels by subtypes of ω-agatoxins. J Biol Chem 267:2610–2615, 1992.
22. IM Mintz, VJ Venema, ME Adams, BP Bean. Inhibition of N- and L-type Ca^{2+} channels by the spider venom toxin ω-agatoxin-IIIA. Proc Natl Acad Sci USA 88:6628–6631, 1991.
23. IM Mintz. Block of Ca channels in rat central neurons by the spider toxin omega-Aga-IIIA. J Neurosci 14:2844–2853, 1994.
24. CJ Cohen, EA Ertel, MM Smith, VJ Venema, ME Adams, MD Lebowitz. High affinity block of myocardial L-type calcium channels by the spider toxin ω-agatoxin IIIA: advantages over 1,4-dihydropyridines. Mol Pharmacol 42:947–951, 1992.
25. IM Mintz, BL Sabatini, WG Regehr. Calcium control of transmitter release at a cerebellar synapse. Neuron 15:675–688, 1995.
26. IM Mintz, BP Bean. Block of calcium channels in rat neurons by synthetic ω-Aga-IVA. Neuropharmacology 32:1161–1169, 1993.
27. IM Mintz, VJ Venema, K Swiderek, T Lee, BP Bean, ME Adams. P-type calcium channels blocked by the spider toxin ω-Aga-IVA. Nature 355:827–829, 1992.
28. TJ Turner, ME Adams, K Dunlap. Calcium channels coupled to glutamate release identified by ω-aga-IVA. Science 258:310–313, 1992.
29. R Llinás, M Sugimori, J-W Lin, B Cherksey. Blocking and isolation of a calcium channel from neurons in mammals and cephalopods utilizing a toxin fraction (FTX) from funnel-web spider poison. Proc Natl Acad Sci USA 86:1689–1693, 1989.
30. RH Scott, MI Sweeney, EM Kobrinsky, HA Pearson, GH Timms, IA Pullar, S Wedley, AC Dolphin. Actions of arginine polyamine on voltage and ligand-activated whole cell currents recorded from cultured neurones. Br J Pharmacol 106:199–207, 1992.
31. OP Hamill, A Marty, E Neher, B Sakmann, FJ Sigworth. Improved patch-clamp techniques for

high-resolution current recording from cells and cell-free membrane patches. Pflügers Arch Eur J Physiol 391:85–100, 1981.
32. E Carbone, HD Lux. A low voltage-activated, fully inactivating Ca channel in vertebrate sensory neurones. Nature 310:501–502, 1984.
33. AP Fox, MC Nowycky, RW Tsien. Kinetic and pharmacological properties distinguishing three types of calcium currents in chick sensory neurones. J Physiol 394:149–172, 1987.
34. AP Fox, MC Nowycky, RW Tsien. Single-channel recordings of three types of calcium channels in chick sensory neurones. J Physiol 394:173–200, 1987.
35. SK Mishra, K Hermsmeyer. Selective inhibition of T-type Ca2+ channels by Ro 40-5967. Circ Res 75:144–148, 1994.
36. MC Nowycky, AP Fox, RW Tsien. Three types of neuronal calcium channels with different calcium agonist sensitivity. Nature 316:440–443, 1985.
37. MR Plummer, DE Logothetis, P Hess. Elementary properties and pharmacological sensitivities of calcium channels in mammalian peripheral neurons. Neuron 2:1453–1463, 1989.
38. A Fleckenstein. History of calcium antagonists. Circ Res 52 (suppl I):3–16, 1983.
39. M Spedding. Calcium antagonist subgroups. Trends Pharmacol Sci 6:109–114, 1985.
40. L Gárcez-do-Carmo, A Albillos, AR Artalejo, MT de la Fuente, MG López, L Gandía, P Michelena, AG García. R56865 inhibits catecholamine release from bovine chromaffin cells by blocking calcium channels. Br J Pharmacol 110:1149–1155, 1993.
41. M Villarroya, L Gandía, B Lara, A Albillos, MG López, AG García. Dotarizine versus flunarizine as calcium antagonists in chromaffin cells. Br J Pharmacol 114:369–376, 1995.
42. B Lara, L Gandía, A Torres, R Olivares, R Martínez-Sierra, AG García, MG López. "Wide spectrum Ca^{2+} channel antagonists": lipophilicity, inhibition, and recovery of secretion in chromaffin cells. Eur J Pharmacol 325:109–119, 1997.
43. CR Artalejo, RL Perlman, AP Fox. ω-Conotoxin GVIA blocks a Ca^{2+} current in chromaffin cells that is not of the "classic" N type. Neuron 8:85–95, 1992.
44. H Kasai, T Aosaki, J Fukuda. Presynaptic Ca-antagonist ω-conotoxin irreversibly blocks N-type Ca channels in chick sensory neurons. Neurosci Res 4:228–235, 1987.
45. VD Monje, JA Haack, SR Naisbitt, G Miljanich, J Ramachandran, L Nadasdi, BM Olivera, DR Hillyard, WR Gray. A new *Conus* peptide ligand for Ca channel subtypes. Neuropharmacology 32: 1141–1149, 1993.
46. LJ Regan. Voltage-dependent calcium currents in Purkinje cells from rat cerebellar vermis. J Neurosci 11:2259–2269, 1991.
47. IM Mintz, ME Adams, BP Bean. P-type calcium channels in central and peripheral neurons. Neuron 9:1–20, 1992.
48. MM Usowicz, M Sugimori, B Cherksey, R Llinás. P-type calcium channels in the somata and dendrites of adult cerebellar Purkinje cells. Neuron 9:1185–1199, 1992.
49. M Umemiya, AJ Berger. Single-channel properties of four calcium channel types in rat motoneurones. J Neurosci 15:2218–2224, 1995.
50. RA Lampe, PA Defeo, MD Davison, J Young, JL Herman, RC Spreen, MB Horn, TJ Mangano, RA Keith. Isolation and pharmacological characterization of omega-grammotoxin SIA, a novel peptide inhibitor of neuronal voltage-sensitive calcium channel responses. Mol Pharmacol 44:451–460, 1993.
51. TM Piser, RA Lampe, RA Keith, SA Thayer. Omega-Grammotoxin blocks action-potential-induced Ca^{2+} influx and whole-cell Ca^{2+} current in rat dorsal-root ganglion neurons. Pflügers Arch Eur J Physiol 426:214–220, 1994.
52. TM Piser, RA Lampe, RA Keith, SA Thayer, S.A. ω-Grammotoxin SIA blocks multiple, voltage-gated, Ca^{2+} channel subtypes in cultured rat hippocampal neurons. Mol Pharmacol 48:131–139, 1995.
53. TJ Turner, RA Lampe, K Dunlap. Characterization of presynaptic calcium channels with omega-conotoxin MVIIC and omega-grammotoxin SIA: role for a resistant calcium channel type in neurosecretion. Mol Pharmacol 47:348–353, 1995.
54. AD Randall, B Wendland, F Schweizer, G Miljanich, ME Adams, RW Tsien. Five pharmacologi-

cally distinct high voltage activated Ca^{2+} channels in cerebellar granule cells Soc Neurosci Abst 19:1478, 1993.

55. A Randall, RW Tsien. Pharmacological dissection of multiple types of Ca^{2+} channel currents in rat cerebellar neurons. J Neurosci 15:2995–3012, 1995.
56. DB Wheeler, A Randall, RW Tsien. Roles of N-type and Q-type Ca^{2+} channels in supporting hippocampal synaptic transmission. Science 264:107–111, 1994.
57. D Bowman, S Alexander, D Lodge. Pharmacological characterisation of the calcium channels coupled to the plateau phase of KCl-induced intracellular free Ca^{2+} elevation in chicken and rat synaptosomes. Neuropharmacology 32:1195–1202, 1993.
58. SN Ahmad, GP Miljanich. The calcium antagonist, ω-conotoxin, and electric organ nerve terminals: binding and inhibition of transmitter release and calcium influx. Brain Res 453:247–256, 1988.
59. I Fariñas, C Solsona, J Marsal. Omega-conotoxin differentially block acetylcholine and adenosine triphosphate releases from torpedo synaptosomes. Neuroscience 47:641–648, 1992.
60. F Sierra, F Lorenzo, O Macadar, W Buño. N-type Ca^{2+} channels mediate transmitter release at the electromotoneuron-electrocyte synapses of the weakly electric fish Gymnotus carapo. Brain Res 683:215–220, 1995.
61. BS Jahromi, R Robitaille, MP Charlton. Transmitter release increases intracellular calcium in perisynaptic schwann cells in situ. Neuron 8:1069–1077.
62. I Wessler, DJ Dooley, J Werhand, F Schlemmer. Differential effects of calcium channel antagonists (ω-conotoxin GVIA, nifedipine, verapamil) on the electrically-evoked release of [^{3}H]-acetylcholine from the myenteric plexus, phrenic nerve and neocortex of rats. Naunyn Schmiedebergs Arch Pharmacol 341:288–294, 1990.
63. A Albillos, AR Artalejo, MG López, L Gandía, AG García, E Carbone. Ca^{2+} channel subtypes in cat chromaffin cells. J Physiol 477:197–213, 1994.
64. L Gandía, R Borges, A Albillos, AG García. Multiple types of calcium channels are present in rat chromaffin cells. Pflügers Arch Eur J Physiol 430:55–63, 1995.
65. JM Hernández-Guijo, R de Pascual, AG García, L Gandía. Separation of calcium channel current components in mouse adrenal chromaffin cells superfused with low- and high-barium solutions. Pflügers Arch Eur J Physiol 436:75–82, 1998.
66. N Kitamura, T Ohta, S Ito, Y Nakazato. Calcium channel subtypes in porcine adrenal chromaffin cells. Pflugers Arch Eur J Physiol 434:179–187, 1997.
67. L Gandía, A Albillos, AG García. Bovine chromaffin cells possess FTX-sensitive calcium channels. Biochem Biophys Res Commun 194:671–676, 1993.
68. A Albillos, AG García, L Gandía. ω-Agatoxin-IVA-sensitive calcium channels in bovine chromaffin cells. FEBS Lett 336:259–262, 1993.
69. L Gandía, I Mayorgas, P Michelena, I Cuchillo, R de Pascual, F Abad, JM Novalbos, E Larrañaga, AG García. Human adrenal chromaffin cell calcium channels: drastic current facilitation in cell clusters, but not in isolated cells. Pflugers Arch Eur J Physiol 436:696–704, 1998.
70. MG López, M Villarroya, B Lara, R Martínez-Sierra, AG García, L Gandía. Q- and L-type Ca^{2+} channels dominate the control of secretion in bovine chromaffin cells. FEBS Lett 349:331–337, 1994.
71. CR Artalejo, ME Adams, AP Fox. Three types of Ca^{2+} channels trigger secretion with different efficacies in chromaffin cells. Nature 367:72–76, 1994.
72. A Albillos, AG García, BM Olivera, L Gandía. Re-evaluation of the P/Q Ca^{2+} channel components of Ba^{2+} currents in bovine chromaffin cells superfused with low and high Ba^{2+} solutions. Pflügers Arch Eur J Physiol 432:1030–1038, 1996.
73. WA Sather, T Tanabe, J-F Zhang, Y Mori, ME Adams, RW Tsien. Distinctive biophysical and pharmacological properties of class A (BI) calcium channel α_1 subunits. Neuron 11:291–303, 1993.
74. SI McDonough, KJ Swartz, IM Mintz, LM Boland, BP Bean. Inhibition of calcium channels in rat central and peripheral neurons by ω-conotoxin MVIIC. J Neurosci 16:2612–2623, 1996.
75. E Perez-Reyes, LL Cribbs, A Daud, AE Lacerda, J Barclay, MP Williamson, M Fox, M Rees, J-H Lee. Molecular characterization of a neuronal low-voltage-activated T-type calcium channel. Nature 391:896–900, 1998.

76. P Ruth, A Röhrkasten, M Biel, E Bosse, S Regulla, HE Meyer, V Flockerzi, F Hofmann. Primary structure of the β subunit of the DHP-sensitive calcium channel from skeletal muscle. Science 245: 1115–1118, 1989.
77. E Pérez-Reyes, A Castellano, HS Kim, P Bertrand, E Baggstrom, AE Lacerda, X Wei, L Birnbaumer. Cloning and expression of a cardiac/brain β subunit of the L-type calcium channel. J Biol Chem 267:1792–1797, 1992.
78. A Castellano, X Wei, L Birnbaumer, E Perez-Reyes. Cloning and expression of a third calcium channel β subunit. J Biol Chem 268:3450–3455, 1993.
79. A Castellano, X Wei, L Birnbaumer, E Perez-Reyes. Cloning and expression of a neuronal calcium channel β subunit. J Biol Chem 268:12359–12366, 1993.
80. F Hofmann, M Biel, V Flockerzi. Molecular basis of Ca^{2+} channel diversity. Annu Rev Neurosci 17:399–418, 1994.
81. LL Isom, KS De Jongh, WA Catterall. Auxiliary subunits of voltage gated ion channels. Neuron 12:1183–1194, 1994.
82. L Birnbaumer, KP Campbell, WA Catterall, MM Harpold, F Hofmann, WA Horne, Y Mori, A Schwartz, TP Snutch, T Tanabe, RW Tsien. The naming of voltage-gated calcium channels. Neuron 13:505–506, 1994.
83. PF Brust, S Simerson, AF McCue, CR Deal, S Schoonmaker, ME Williams, C Veliçelebi, EC Johnson, MM Harpold, SB Ellis. Human neuronal voltage-dependent calcium channels: studies on subunit structure and role in channel assembly. Neuropharmacology 32:1089–1102, 1993.
84. T Angelotti, F Hofmann. Tisue-specific expression of splice variants of the mouse voltage-gated calcium channel α_2/δ subunit. FEBS Lett 397:331–337, 1996.
85. O Wiser, M Trus, D Tobi, S Halevi, E Giladi, D Atlas. The α_2/δ subunit of voltage sensitive Ca^{2+} channels is a single transmembrane extracellular protein which is involved in regulated secretion. FEBS Lett 379:15–20, 1996.
86. SD Jay, SB Ellis, AF McCue, ME Williams, TS Vedvick, MM Harpold, KP Campell. Primary structure of the γ subunit of the DHP-sensitive calcium channel from skeletal muscle. Science 248: 490–492, 1990.
87. VES Scott, R Felix, J Arikkath, KP Campbell. Evidence for a 95 kDa short form of the α_{1A} subunit associated with the ω-conotoxin MVIIC receptor of the P/Q-type Ca^{2+} channels. J Neurosci 18: 641–647, 1998.
88. SB Ellis, ME Williams, NR Way, R Brenner, AH Sharp, AT Leung, KP Campell, E McKenna, WJ Koch, A Hui, A Schwartz, MM Harpold. Sequence and expression of mRNAs encoding the α_1 and α_2 subunits of a DHP-sensitive calcium channel. Science 241:1661–1664, 1988.
89. Y Mori, T Friedrich, M-S Kim, A Mikami, J Nakai, P Ruth, E Bosse, F Hofmann, V Flockerzi, T Furuichi, K Mikoshiba, K Imoto, T Tanabe, S Numa. Primary structure and functional expression from complementary DNA of a brain calcium channel. Nature 350:398–402, 1991.
90. D Singer, M Biel, I Lotan, V Flockerzi, F Hofmann, N Dascal. The roles of the subunits in the function of the calcium channel. Science 253:1553–1557, 1991.
91. AE Lacerda, HS Kim, P Ruth, E Perez-Reyes, V Flockerzi, F Hofmann, L Birnbaumer, AM Brown. Normalization of current kinetics by interaction between the α_1 and β subunits of the skeletal muscle dihydropyridine-sensitive. Nature 352:527–530, 1991.
92. G Varadi, P Lory, D Schultz, M Varadi, A Schwartz. Acceleration of activation and inactivation by the β subunit of the skeletal muscle calcium channel. Nature 352:159–162, 1991.
93. PT Ellinor, J-F Zhang, AD Randall, M, Zhou, TL Schwarz, RW Tsien, WA Horne. Functional expression of a rapidly inactivating neuronal calcium channel. Nature 363:455–458, 1993.
94. M de Waard, KP Campbell. Subunit regulation of neuronal α_{1A} Ca^{2+} channel expressed in *Xenopus* oocytes. J Physiol 485:619–634, 1995.
95. A Neely, X Wei, R Olcese, L Birnbaumer, E Stefani. Potentiation by the β subunit of the ratio of the ionic current to the charge movement in the cardiac calcium channel. Science 262:575–578, 1993.
96. S Nishimura, H Takeshima, F Hofmann, V Flockerzi, K Imoto. Requirement of the calcium channel β subunit for functional conformation. FEBS Lett 324:283–286, 1993.
97. F Hofmann, E Habermann. Role of ω-conotoxin-sensitive calcium channels in inositolphosphate

production and noradrenaline release due to potassium depolarisation or stimulation with carbachol. Naunyn Schmiedeberg's Arch Pharmacol 341:200–205, 1990.

98. TJ Turner, ME Adams, K Dunlap. Multiple Ca^{2+} channel types coexist to regulate synaptosomal neurotransmitter release. Proc Natl Acad Sci USA 90:9518–9522, 1993.
99. TJ Turner, K Dunlap. Pharmacological characterization of presynaptic calcium channels using subsecond biochemical measurements of synaptosomal neurosecretion. Neuropharmacology 34:1469–1478, 1995.
100. H Herdon, SR Nahorski. Investigations of the role of dihydropyridine and ω-conotoxin-sensitive calcium channels in mediating depolarization-evoked endogenous dopamine release from striatal slices. Naunyn Schmiedeberg's Arch Pharmacol 340:36–40, 1989.
101. DJ Dooley, A Lupp, G Herting. Inhibition of central neurotransmitter release by ω-conotoxin GVIA, a peptide modulator of the N-type voltage-sensitive calcium channel. Naunyn-Schmiedeberg's Arch Pharmacol 336:467–470, 1987.
102. JI Luebke, K Dunlap, TJ Turner. Multiple calcium channel types control glutamatergic synaptic transmission in the hippocampus. Neuron 11:895–908, 1993.
103. T Takahashi, A Momiyama. Different types of calcium channels mediate central synaptic transmission. Nature 366:156–158, 1993.
104. ML Simmons, G Terman, SM Gibbs, C Chaukin. L-type calcium channels mediate dynorphin neuropeptide release from dendrites but not axons of hippocampal granule cells. Neuron 14:1265–1272, 1995.
105. G Wang, G Dayanithi, S Kim, D Hom, L Nadasdi, R Kristipati, J Ramachandran, EL Stuenkel, JJ Nordmann, R Newcomb, J Lemos. Role of Q-type Ca^{2+} channels in vasopressin secretion from neurophypopophysial terminals of the rat. J Physiol 502:351–363, 1997.
106. LD Hirning, AP Fox, EW McClesky, BM Olivera, SA Thayer, RJ Miller, RW Tsien. Dominant role of N-type Ca^{2+} channels in evoked release of norepinephrine from sympathetic neurons. Science 239:57–61, 1988.
107. DA Przywara, SV Bhave, PS Chowdhury, TD Wakade, AR Wakade. Sites of transmitter release and relation to intracellular Ca^{2+} in cultured sympathetic neurons. Neuroscience 52:973–986, 1993.
108. A Rittenhouse, RE Zigmond. ω-Conotoxin inhibits the acute activation of tyrosine hydroxylase and the stimulation of norepinephrine release by potassium depolarisation and sympathetic nerve endings. J Neurochem 56:615–622, 1991.
109. H Tashenberger, R Grantyn. Several types of Ca^{2+} channels mediate glutamatergic synaptic responses to activation of single Thy-1-inmunolabeled rat retinal ganglion neurons. J Neurosci 15: 2240–2254, 1995.
110. C Yu, PX Lin, S Fitzgerald, P Nelson. Heterogeneous calcium currents and transmitter release in cultured mouse spinal cord and dorsal root ganglion neurons. J Neurophysiol 67:561–575, 1992.
111. PM Lundy, R Frew. Evidence of ω-conotoxin GVIA-sensitive Ca^{2+} channels in mammalian peripheral nerve terminals. Eur J Pharmacol 156:325–330, 1988.
112. PM Lundy, R Frew. Effect of ω-agatoxin-IVA on autonomic neurotransmission. Eur J Pharmacol 261:79–84, 1994.
113. CA Maggi, R Patachini, P Santiciol, IT Lippe, S Guiliano, P Geppetti, E del Blanco, S Selleri, A Meli. The effect of omega-conotoxin GVIA, a peptide modulator of the N-type voltage sensitive calcium channels on motor responses produced by activation of efferent sensory nerves in mammalian smooth muscle. Naunyn Schmiedeberg's Arch Pharmacol 338:107–113, 1988.
114. R Frew, PM Lundy. A role for Q-type Ca^{2+} channels in neurotransmission in the rat urinary bladder. Br J Pharmacol 116:1595–1598, 1995.
115. SA Waterman. Multiple subtypes of voltage-gated calcium channels mediate transmitter release from parasympathetic neurons in mouse bladder. J Neurosci 16:4155–4161, 1996.
116. PM Zygmunt, PKE Zygmunt, ED Högestätt, KE Andersson. Effects of ω-conotoxin on adrenergic, cholinergic and NANC neurotransmission in the rabbit. Br J Pharmacol 110:1285–1290, 1993.
117. CE Wright, JA Augus. The effects of N-, P- and Q-type neuronal calcium channel antagonists on mammalian peripheral neurotransmission. Br J Pharmacol 119:49–56, 1996.
118. F Hata, A Fujita, K Saeki, I Kishi, T Takeuchi, O Yagasaki. Selective inhibitory effects of calcium

channel antagonists on the two components of the neurogenic responses of guinea-pig vas deferens. J Pharmacol Exp Ther 263:214–220, 1992.

119. H Hirata, A Albillos, F Fernández, J Medrano, A Jurkiewicz, AG García. ω-Conotoxin block neurotransmission in the rat vas deferens by binding to different presynaptic sites on the N-type Ca^{2+} channel. Eur J Pharmacol 321:217–223, 1997.

120. SJ Hong, CC Chang. Inhibition of acetylcholine release from mouse motor nerve by P-type calcium channel blocker, ω-agatoxin IVA. J Physiol Lond 482:283–290, 1995.

121. DA Protti, R Reisin, TA Mackinley, OD Uchhitel. Calcium channel blockers and transmitter release at the normal human neuromuscular junction. Neurology 46:1391–1396, 1996.

122. L Gandía, RI Fonteríz, MG López, CR Artalejo, AG García. Relative sensitivities of chromaffin cell calcium channels to organic and inorganic calcium antagonists. Neurosci Lett 77:333–338, 1987.

123. AM Cárdenas, C Montiel, C Esteban, R Borges, AG García. Secretion from adrenaline- and noradrenaline-storing adrenomedullary cells is regulated by a common dihydropyridine-sensitive calcium channel. Brain Res 456:364–366, 1988.

124. MG López, A Albillos, MT de la Fuente, R Borges, L Gandía, E Carbone, AG García, AR Artalejo. Localized L-type calcium channels control exocytosis in cat chromaffin cells. Pflügers Arch Eur J Physiol 427:348–354, 1994.

125. AG García, F Sala, JA Reig, S Viniegra, J Frías, RI Fonteríz, L Gandía. Dihydropyridine Bay K 8644 activates chromaffin cell calcium channels. Nature 309:69–71, 1984.

126. V Ceña, GP Nicolás, P Sánchez-García, SM Kirpekar, AG García, A.G. Pharmacological dissection of receptor-associated and voltage-sensitive ionic channels involved in catecholamine release. Neuroscience 10:1455–1462, 1983.

127. PJ Owen, DB Marriot, MR Boarder. Evidence for a dihydropyridine and conotoxin-insensitive release of noradrenaline and uptake of calcium in adrenal chromaffin cells. Br J Pharmacol 97:133–138, 1989.

128. L Gandía, P Michelena, R de Pascual, MG López, AG García. Different sensitivities to dihydropyridines of catecholamine release from cat and ox adrenals. NeuroReport 1:119–122, 1990.

129. RR Jiménez, MG López, C Sancho, R Maroto, AG García. A component of the catecholamine secretory response in the bovine adrenal gland is resistant to dihydropyridines and ω-conotoxin. Biochem Biophys Res Commun 191:1278–1283, 1993.

130. JJ Ballesta, M Palmero, MJ Hidalgo, LM Gutiérrez, JA Reig, S Viniegra, AG García. Separate binding and functional sites for ω-conotoxin and nitrendipine suggest two types of calcium channels in bovine chromaffin cells. J Neurochem 53:1050–1056, 1989.

131. LM Rosario, B Soria, G Feuerstein, HB Pollard. Voltage-sensitive calcium flux into bovine chromaffin cells occurs through dihydropyridine-sensitive and dihydropyridine- and ω conotoxin insensitive pathways. Neuroscience 29:735–747, 1989.

132. JL Bossu, M de Waard, A Feltz. Inactivation characteristics reveal two calcium currents in adult bovine chromaffin cells. J Physiol Lond 437:603–620, 1991.

133. JL Bossu, M de Waard, A Feltz. Two types of calcium channels are expressed in adult bovine chromaffin cells. J Physiol Lond 437:621–634, 1991.

134. CR Artalejo, MK Dahmer, RL Perlman, AP Fox. Two types of Ca^{2+} currents are found in bovine chromaffin cells: facilitation is due to the recruitment of one type. J Physiol Lond 432:681–707, 1991.

135. CB Duarte, LM Rosario, CM Sena, AP Carvalho. A toxin fraction FTX from the funnel-web spider poison inhibits dihydropyridine-insensitive Ca^{2+} channels coupled to catecholamine release in bovine adrenal chromaffin cells. J Neurochem 60:908–913, 1993.

136. JM Fernández, R Granja, V Izaguirre, C González-García, V Ceña. ω-Conotoxin GVIA blocks nicotine-induced catecholamine secretion by blocking the nicotine receptor-activated inward currents in bovine chromaffin cells. Neurosci Lett 191:1–4, 1995.

137. R Granja, JM Fernández-Fernández, V Izaguirre, C González-García, V Ceña. ω-Agatoxin IVA blocks nicotinic receptor channels in bovine chromaffin cells. FEBS Lett 362:15–18, 1995.

138. G Baltazar, I Ladeira, AP Carvalho, EP Duarte. Two types of ω-agatoxin IVA-sensitive Ca^{2+} channels are coupled to adrenaline and noradrenaline release in bovine chromaffin cells. Pflügers Arch Eur J Physiol 434:592–598, 1997.

139. B Lara, L Gandía, R Martínez-Sierra, A Torres, AG García. Q-type Ca^{2+} channels are located closer to the secretory sites than L-type channels: functional evidence in chromaffin cells. Pflügers Arch Eur J Physiol 435:472–478, 1997.
140. MG López, R Shukla, AG García, AR Wakade. A dihydropyridine-resistant component in the rat adrenal secretory response to splanchnic nerve stimulation. J Neurochem 58:2139–2144, 1992.
141. SJ Kim, W Lim, J Kim. Contribution of L- and N-type calcium currents to exocytosis in rat adrenal medullary cells. Brain Res 675:289–296, 1995.
142. T Kimura, A Takeuchi, S Satoh. Inhibition by ω-conotoxin GVIA of adrenal catecholamine release in response to endogenous and exogenous acetylcholine in anesthetized dogs. Eur J Pharmacol 264: 169–175, 1994.
143. K Dunlap, JI Luebke, TJ Turner. Exocytotic Ca^{2+} channels in mammalian central neurons. Trends Neurosci 18:89–98, 1995.
144. P Sevilla, M Bruix, J Santoro, F Gago, AG García, M Rico. Three-dimensional structure of ω-conotoxin GVIA determined by ^{1}H NMR. Biochem Biophys Res Commun 192:1238–1244, 1993.
145. JH Davis, EK Bradley, GP Miljanich, L Nadasdi, J Ramachandran, VJ Basus. Solution structure of omega-conotoxin GVIA using 2-D NMR spectroscopy and relaxation matrix analysis. Biochemistry 32:7396–7405, 1993.
146. PK Pallaghy, BM Duggan, MW Pennington, RS Norton. Three-dimensional structure in solution of the calcium channel blocker omega-conotoxin. J Mol Biol 234:405–420, 1993.
147. JJ Skalicky, WJ Metzler, DJ Ciesla, A Galdes, A Pardi. Solution structure of the calcium channel antagonist omega-conotoxin GVIA. Protein Sci 2:1591–1603, 1993.
148. T Kohno, JJ Kim, K Kobayashi, Y Kodera, T Maeda, K Sato. Three-dimensional structure in solution of the calcium channel blocker ω-conotoxin MVIIA. Biochemistry 34:10256–10265, 1995.
149. N Nemoto, S Kubo, T Yoshida, N Chino, T Kimura, S Sakakibara, Y Kyogoku, Y Kobayashi. Solution structure of omega-conotoxin MVIIC determined by NMR. Biochem Biophys Res Commun 207:695–700, 1995.
150. S Farr-Jones, GP Miljanich, L Nadasdi, J Ramachandran, VJ Basus. Solution structure of omega-conotoxin MVIIC, a high affinity ligand of P-type calcium channels, using ^{1}H NMR spectroscopy and complete relaxation matrix analysis. J Memb Biol 248:106–124, 1995.
151. JM Hernández-Guijo, L Gandía, R de Pascual, AG García. Calcium channel blocking properties of lubeluzole in bovine and mouse chromaffin cells. Br J Pharmacol 122:275–285, 1997.
152. CJ Grantham, MJ Main, MB Cannell. Fluspirilene block of N-type calcium current in NGF-differenciated PC12 cells. Br J Pharmacol 111:483–488, 1994.
153. J Harvey, S Wedley, JD Findlay, MR Sidell, IA Pullar. Omega-agatoxin IVA identifies a single calcium channel subtype which contributes to the potassium-induced release of acetylcholine, 5-hydroxytryptamine, dopamine, gamma-aminobutyric acid and glutamate from rat brain slices. Neuropharmacology 35:385–392, 1996.
154. M Haass, G Richardt, T Brenn, E Schömig, A Schömig. Nicotine induced release of noradrenaline and neuropeptide Y in guinea-pig heart: role of calcium channels and protein kinase C. Naunyn Schmiedeberg's Arch Pharmacol 344:527–531, 1991.
155. SJ Hong, CC Chang. Calcium channel subtypes for the sympathetic and parasympathetic nerves of guinea pig atria. Br J Pharmacol 116:1577–1582, 1995.

6
Ecobiology, Classification, and Origin

Masaaki Kodama
School of Fisheries Sciences, Kitasato University, Sanriku, Iwate, Japan

I. INTRODUCTION

Saxitoxin and its analogues (STXs) are potent neurotoxins that block voltage-gated sodium channels on excitable cells (1). These toxins are produced by several species of dinoflagellates such as *Alexandrium tamarense* and *A. catenella* (2). During a bloom of these species, filter-feeding bivalves become toxic by ingesting them. Human consumption of toxic shellfish results in severe food poisoning known as paralytic shellfish poisoning (PSP) owing to its main symptom, paralysis. Shellfish farming areas infested by these species run costly monitoring programs to check for the causative dinoflagellates in the water. When these are present, regular tests for toxins in associated seafood products such as bivalves are carried out and often result in the prohibition of harvesting of these products. Thus, the contamination of bivalves with STXs has posed severe problems for the shellfish culture industry as well as for public health.

Many studies have been carried out on the toxin-producing dinoflagellates and the toxins they produce. There is much known about the taxonomy and ecology of the toxin-producing dinoflagellates (3). The chemical structures and pharmacological properties of the toxins have been well defined (2,4). However, little is known about the biosynthetic pathway for STXs in the STX producing organisms. The only information available concerning the biosynthesis of STXs is presented by Shimizu et al. (5). They proved that the skeleton of STXs is synthesized from arginine and acetate (Fig. 1). However, the process of synthesis, the enzymes involved, and the biochemical function of the toxins in toxin-producing dinoflagellates have not been defined. In this chapter, current knowledge of the ecobiology of STXs, the toxins responsible for PSP, is reviewed, with special emphasis on the origin of STXs and the mechanism by which the organisms accumulate STXs.

II. TOXIN COMPONENTS IN ALGAL SPECIES AND SHELLFISH

STXs are produced by several species of dinoflagellates. The first dinoflagellate to be linked with PSP was *Alexandrium* (formerly *Gonyaulax*, *Protogonvaulax*) *catenella* (Whedon et Kofoid) Balech, which killed six people and made 102 people ill in 1927 near San Francisco (6). Since then, members of three genera of dinoflagellates have been reported to be involved in PSP (7): several species of *Alexandrium*, *Gymnodinium catenatum*, and *Pyrodinium bahamense* var. *compressum*. More than 20 derivatives of saxitoxin are found in these dinoflagellates and shellfish (Fig. 2).

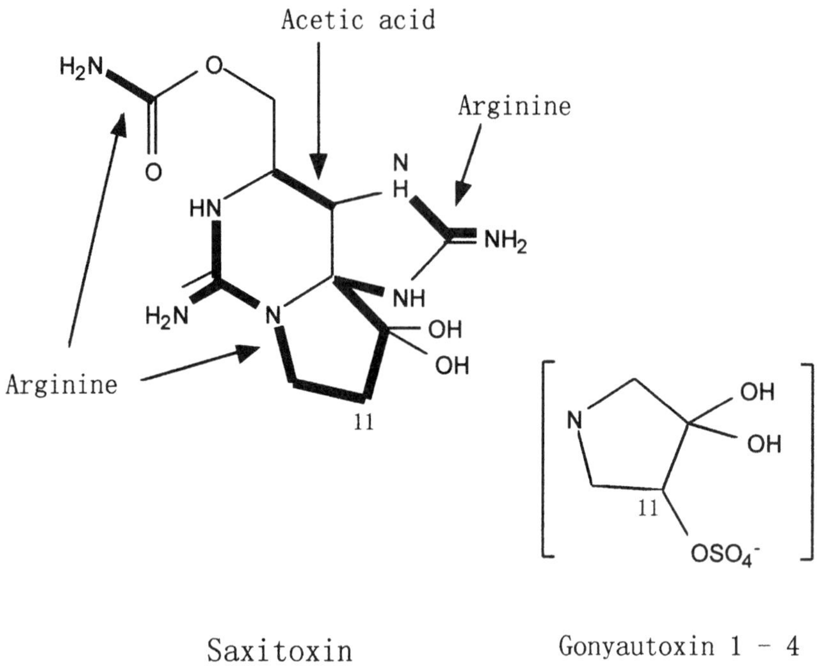

Figure 1 Biosynthesis of saxitoxin from arginine and acetate proposed by Shimizu et al.(5)

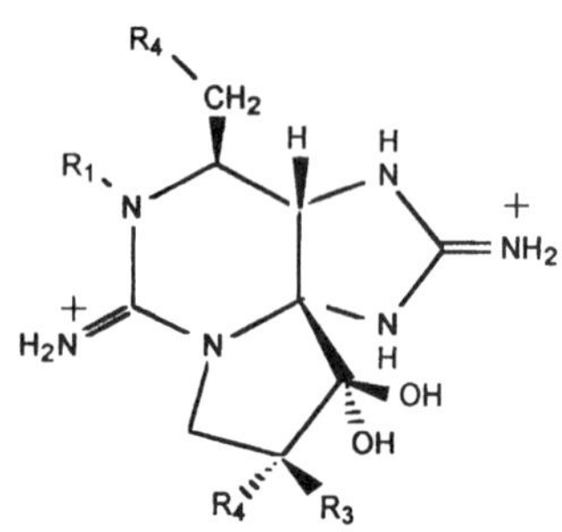

R_1	R_2	R_3	R_4: —O–C(=O)–NH_2	R_4: —O–C(=O)–NH–SO_3^-	R_4: –OH	R_4: –H
H	H	H	STX	GTX5 (B1)	dcSTX	doSTX
OH	H	H	neoSTX	GTX6 (B2)	dcneoSTX	-
OH	OSO_3^-	H	GTX1	C3	dcGTX1	-
H	OSO_3^-	H	GTX2	C1	dcGTX2	doGTX2
H	H	OSO_3^-	GTX3	C2	dcGTX3	doGTX3
OH	H	OSO_3^-	GTX4	C4	dcGTX4	-

Figure 2 Chemical structure of saxitoxin analogues.

The recent development of a sensitive method of analyzing these toxins by high-performance liquid chromatography (HPLC) has enabled us to demonstrate the complex toxin profiles in toxic dinoflagellates and also contaminated shellfish (8,9). These studies revealed that a specific group of toxin components is often found in each species. For example, the major toxin components in *G. catenatum* are those belonging to the N-sulfocarbamoyl toxin group such as C toxins and gonyautoxins (GTX) 5 and 6. No significant amounts of carbamate toxins such as GTX1–4 are detected in this species (10). *P. bahamense* var. *compressum* also shows a characteristic toxin profile which mainly consists of GTX5 and 6 and their carbamate counterparts STX and neoSTX (10). Decarbamoylsaxitoxin (dcSTX) was first identified in this species and its associated bivalves (11).

Differences in toxin profiles have been observed among the species belonging to the genus *Alexandrium*. C toxins are major components in *A. tamarense* and *A. catenella* (10), whereas these are minor components in *A. tamiyavanich* (formerly *A. cohorticula*) (12). GTX1–4 are found in high proportion in *A. tamiyavanich* (12). In contrast, *A. ostenfeldii* has GTX5 or GTX6 and their carbamate counterparts as major components, showing that its toxin profile is different from that of other *Alexandrium* spp. (13).

Differences in toxin profiles have been observed among strains of the same species collected from different geographical locations. Based on these findings, some researchers have attempted to use the toxin profile as a biochemical marker to distinguish strains of the same species (10,14,15). Toxin profiles are reported to be influenced by physiological conditions such as nutritional conditions, the stage of the cell cycle, and other factors (16,17). However, it has been confirmed by many investigators that the toxin profile of a clonal culture is constant even at different stages of growth (10,14,15). Therefore, the toxin composition is considered to be a genetically stable characteristic of a clone.

Further results supporting this idea were obtained in a study focusing on the toxin profiles of *G. catenatum* Graham populations from different geographical locations (18). All of the isolates were found to produce N-sulfocarbamoyl toxins such as C1–4 , GTX5, and GTX6. However, the strains from different geographical locations could be distinguished from each other on the basis of the relative abundance of these components. In addition, the Australian population could be easily distinguished from other populations because of the presence of characteristic toxin components, 13-deoxydecarbamoyl toxins, which were not detected in other populations (18). On the other hand, *A. minutum* isolates from different geographical locations (Taiwan, Australia, Portugal, and Spain) showed similar toxin profiles, consisting mainly of GTX1–4 (19–22). As an exceptional case, a French strain and New Zealand strains of *A. minutum* are reported to show different profiles; these strains contain C toxins (French strain) and STX and neoSTX (New Zealand strains) (21).

It has been proposed that Japanese strains of *A. tamarense* can be categorized into three groups on the basis of their toxin profiles (23). The first group is distinguished from the other two on the basis of the presence of STX, a characteristic component of this group. The other two populations could be distinguished from one another on the basis of the relative proportion of nSTX.

As described in another chapter, STXs have been detected also in three species of blue-green algae (cyanobacteria): *Aphanizomenon flos-aquae*, *Anabaena circinalis*, and *Lyngbya wollei*. The toxin profiles of these species are completely different from each other. *A. flos-aquae* was the first species for which toxin production was confirmed, although only one strain is reported to produce the toxins (24). This species shows a characteristic toxin profile with STX and neoSTX as major components and dcGTX2, dcGTX3, dcSTX, and GTX5 as minor components (25). *A. circinalis* was the second toxic species of cyanobacteria identified, and it was found to be responsible for the death of domestic animals in Australia in 1991 (26). This species

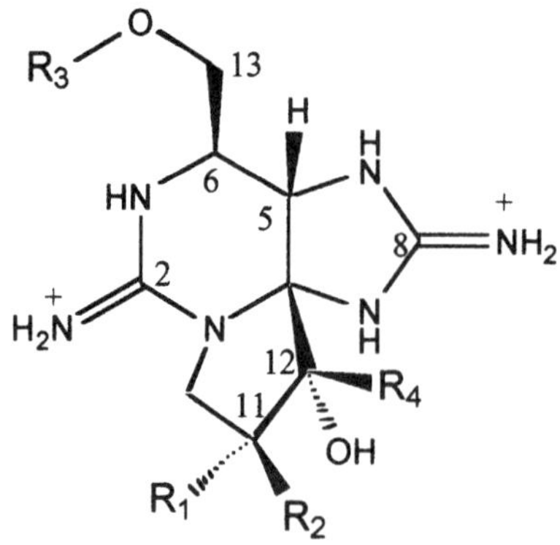

	R1	R2	R3	R4
1	H	OSO_3^-	$COCH_3$	H
2	H	OSO_3^-	$COCH_3$	OH
3	OSO_3^-	H	$COCH_3$	OH
4	H	H	H	H
5	H	H	$COCH_3$	OH
6	H	H	$COCH_3$	H

Figure 3 Newly found saxitoxin derivatives from cyanobacterium *Lyngbia wollei* (30).

shows a characteristic profile with large proportions of C1, C2, and GTX2 and GTX3 (27,28) that is quite different from that of *A. flos-aquae*. The shellfish that consume this species are reported to accumulate toxins (29). Recently, *L. wollei*, a third toxic species of cyanobacteria, has been shown to produce STXs (30). It is noteworthy that six new saxitoxin analogues were isolated from this organism together with dcSTX, dcGTX2, and dcGTX3 (Fig. 3). The presence of new toxins in *L. wollei* shows that the metabolism of STXs in this species is different from that in other cyanobacteria and dinoflagellates.

III. TRANSFORMATION OF TOXIN COMPONENTS IN SHELLFISH AND DINOFLAGELLATES

Shellfish accumulate dinoflagellate toxins via food webs, and thus the toxin profiles of bivalves seem to reflect those of dinoflagellates. However, the pattern of toxins in shellfish is not always similar to that of the causative dinoflagellates, indicating that toxin components are transformed in shellfish (10). One of the transformation is that due to chemical properties of toxins. Toxins having a hydroxysulfate moiety at position 11 undergo epimerization through keto-enol equilibration (31). Oshima (32) showed by incubation experiments that the equilibration is accelerated at higher pH and higher temperature and the N-sulfocarbamoyl toxin group equilibrated much faster than the carbamate group. The biosynthetic product of the dinoflagellates was thought to be β-epimer (GTX3, GTX4, C2, C4), since only this epimer group is detected in actively growing cells (33). In toxins transmitted to shellfish, epimerization proceeds gradually until it reaches equilibrium at a β/α ratio close to 1:3. Thus, the relative ratio of epimers provides information on how long toxins have been retained by shellfish (32).

Another chemical transformation is due to natural reductants. The scallop *Patinopecten yessoensis* retained toxins for several months, and a significant change of toxin profile was observed during the depuration period (32). Marked changes in toxin profiles are evident in scallops during the toxin depuration period; the levels of components of N-OH toxins (GTX1, GTX4, and neoSTX) decrease, whereas the levels of N-H toxins (GTX2, GTX3, and STX) increase (32). It has been confirmed by in vitro experiments that the reductive removal of 11-O-sulfate occurs in the shellfish. This reaction was found to occur in a protein-free extract of the scallop and it was accelerated by heating, indicating that the reaction is not enzymatic (32). A similar conversion occurs on treatment of N-OH toxins or 11-O-sulfate toxins with cysteine or glutathione (32,34,35). The same reductive processes have been reported to occur in a homogenate of *Patinopecten magellanicus* (36). Kotaki et al. (37) observed that a similar toxin conversion occurs when these toxins are incubated with bacterial cultures.

Most of shellfish contains less C toxins and higher carbamate toxins than the causative dinoflagellates (10). For example, the toxin profile of toxic scallops during a bloom of *A. catenella* was found to be different from that of *A. catenella* as the proportion of C toxins, the major toxin components of *A. catenella*, substantially decreased in the scallops soon after their ingestion of *A. catenella* (10). These suggest that enzymatic hydrolysis of N-sulfate occurs. However, the corresponding carbamate toxins do not always increase in parallel with the decrease in C toxins, indicating that C toxins ingested by shellfish are not converted to the corresponding carbamate toxins in shellfish. This is supported by the fact that the enzyme which converts C toxins to corresponding carbamate toxins could not be detected in 18 species of bivalves in Japan (32). On the other hand, C toxins are easily hydrolyzed at neutral pH to yield decarbamoyl derivatives (32). The reaction proceeds quickly when heated, but it also occurs very slowly at low temperature (incubation at 4°C for 1 month). These suggest that the same hydrolysis might be occurring in live shellfish, because decarbamoyl toxins were always detected in shellfish containing a large proportion of C toxins (10). It is noteworthy that the enzyme which catalyzes the same hydrolysis is reported to occur in two species of clams, *Mactra chinensis* and *Peronidia venulaso* (32).

Enzymatic conversion of toxin components is reported to occur also in dinoflagellates. *A. tamarense* possesses an oxidase that converts GTX2 and GTX3 to GTX1 and GTX4 (32). An N-sulfotransferase that converts carbamate toxins to the corresponding C toxins has been detected in both toxic and nontoxic isolates of *G. catenatum* from Tasmania in which C toxins occur in high proportion (32). In contrast, *A. tamarense* isolates that produce C2 in high abundance relative to other components do not have this enzyme (32).

These findings show that conversion of toxin components can be attributed to enzymatic reactions as well as nonenzymatic processes. Various toxin components found in dinoflagellates and shellfish seem to be products of these reactions. Ishida et al. (38) suggested that the enzymes involved in the conversion of toxin components may be those involved in biosynthesis of the toxins. However, conversion of toxin components occurs in shellfish as well as in dinoflagellates. In addition, enzymes involved in toxin conversion, such as N-sulfotransferase, for example, occur in both toxic and nontoxic strains of dinoflagellates. These observations indicate that conversion of toxin components differs from biosynthesis of STXs. However, it should be noted that at least one of the toxin components found in toxin-producing dinoflagellates could be a de novo product of biosynthesis of STXs. Thus, the enzyme which produces this component is the one involved in biosynthesis of STXs.

IV. ACCUMULATION OF STXs IN SHELLFISH; IS IT VIA FOOD WEBS?

The accumulation of STXs in shellfish via food webs is well documented. During a bloom of toxic dinoflagellates, shellfish become toxic as a result of their ingestion of dinoflagellates. Available evidence suggests that STXs in dinoflagellates are transported to shellfish via food webs, absorbed, and accumulated in the digestive gland (39). Some of the toxin absorbed is released from the shellfish. Thus, the amount of toxin in the shellfish is the difference between the amount of toxin supplied and the amount released (Fig. 4). On the other hand, there are some phenomena concerning shellfish toxicity that cannot be explained by this mechanism. In a field survey on the toxicity of the scallop *Patinopecten yessoensis* in association with an abundance of *A. tamarense*, we have observed that scallop toxicity does not increase significantly when *A. tamarense* blooms in high density. In contrast, it increases after the peak in abundance of *A. tamarense* when most of the cells have already disappeared (40). There is a 1-week time lag between the peak of scallop toxicity and the peak in abundance of *A. tamarense*. Furthermore, after the levels of the toxin in the scallop had begun to decrease, a further rise in toxicity was noted even though the causative dinoflagellate was no longer detected. This may be due to the spatial patchiness of the algal population, which makes it difficult to estimate correctly the abundance of the toxic dinoflagellates. However, this phenomenon has been observed repeatedly

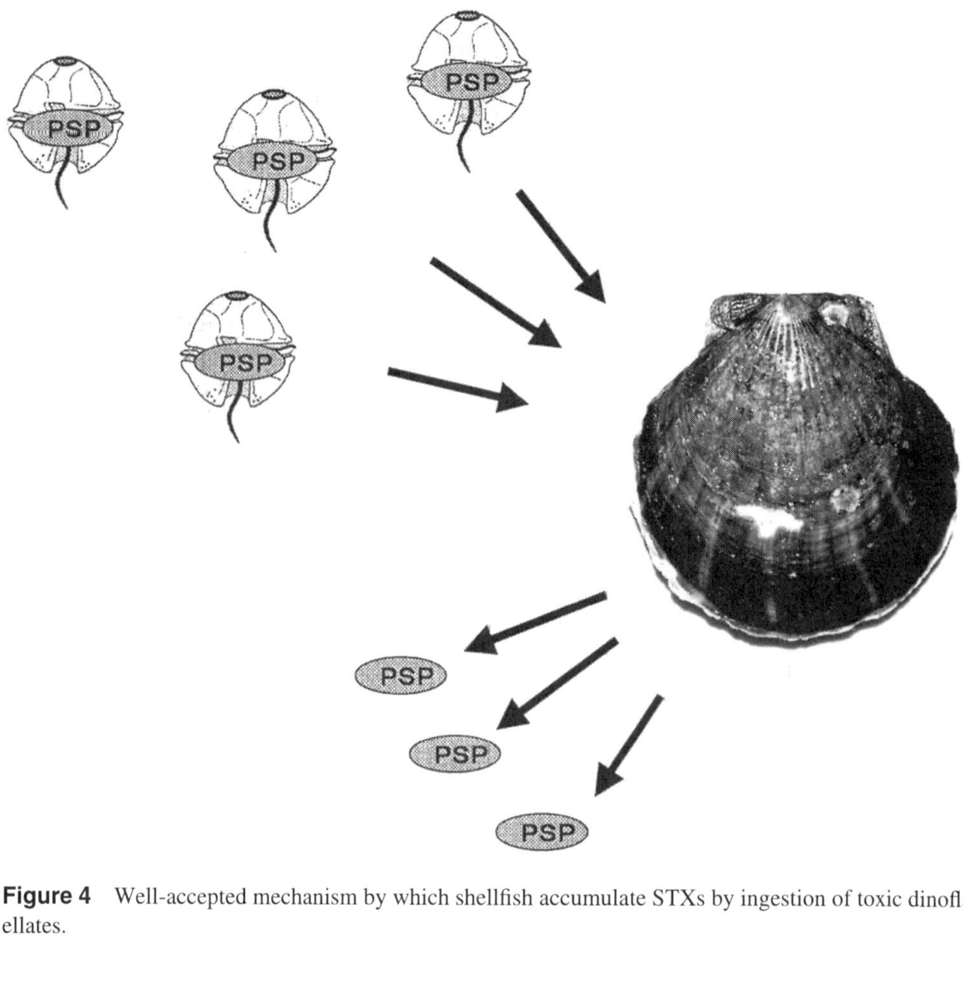

Figure 4 Well-accepted mechanism by which shellfish accumulate STXs by ingestion of toxic dinoflagellates.

in field monitoring surveys of shellfish toxicity in association with the abundance of toxic dinoflagellates (41,42). A peak in abundance of dinoflagellates always precedes the peak of shellfish toxicity. Thus, spatial patchiness of the algal population is not the only factor responsible for this phenomenon. It may be due to biotransformation of sulfocarbamate derivatives of STXs in the shellfish to their corresponding carbamate forms which are far more potent. However, no significant bioconversion of toxin components that might explain these phenomena was detected in shellfish during the period when shellfish is releasing the accumulated toxins (unpublished data).

In a monitoring survey on shellfish toxicity, we analyzed STXs in particles suspended in seawater after fractionation of the particles according to size by passing them through sieves of different pore sizes. Figure 5 shows data obtained for samples from Tanabe Bay where shellfish became toxic during a bloom of *A. catenella* (43). Surprisingly, no significant amount of toxin was detected in the particulate fraction containing natural *A. catenella* cells at a time when shellfish toxicity increased considerably (indicated by arrow in Figure 5). In contrast, significant toxicity was detected in association with particles smaller than *A. catenella* such as those in the size ranges of 5–20 and 0.45–.05 μm by HPLC analysis. The 5- to 20-μm fraction showed the highest level of toxicity when the cell density of *A. catenella* was very low. The toxins in this fraction were highly purified and identified as STXs by mass spectrometry. These findings clearly indicate that an unknown STX-producing organism(s), smaller than *A. catenella*, occurs together with *A. catenella* in the seawater. Similar results were obtained for samples from Ofunato Bay during a bloom of *A. tamarense* (44,45). A slight increase in shellfish toxicity was observed when fractions containing particles smaller than dinoflagellates showed considerable toxicity. As no significant toxicity was observed in other particle fractions, the shellfish toxicity during the period seemed to be due to the toxins associated with these particles. However, shellfish toxicity did not increase much, although high levels of STXs were detected in the fractions containing particles with a size smaller than dinoflagellates. It seems quite strange that shellfish toxicity increased significantly as *A. catenella* grew to high density; yet no significant amount of toxin was detected in *A. catenella* (43). These observations suggest that some unknown mechanism other than food webs which is associated with the dinoflagellates or other unidentified causative organism(s), plays a role in the accumulation of STXs in shellfish. It seems possible that this may involve de novo synthesis of the toxin by the shellfish, for example, through digestion of algal cells or other ingested materials together with dinoflagellates.

V. BACTERIAL PRODUCTION OF STXs

STXs have been detected in fractions of particles in the size ranges of 0.45–5.0 μm and 5–20 μm. No significant amount of phytoplankton was observed in the water sample from which the 5- to 20-μm fraction containing a high level of STXs was prepared (43). Instead, small detritus particles were observed in this sample. These observations suggest that the unknown causative organism is a bacterium. The possibility that STXs may be of bacterial origin was first suggested by Silva (46,47), who observed bacteria-like structures inside dinoflagellate cells. She found that the bacteria isolated from the algae exhibited toxicity in mice. However, these findings remain to be confirmed by other investigators. While studying toxin production in the dinoflagellate *A. tamarense*, we observed that there was a 20-fold difference in toxin yields between subclones prepared from a clonal culture grown under identical conditions, suggesting that toxin production in this species is not a hereditary characteristic of the alga (48). As this finding appeared to be consistent with Silva's idea of a bacterial source of STXs, attempts were made to isolate bacteria from this dinoflagellate. A single bacterium referred to as "*Moraxella* sp."

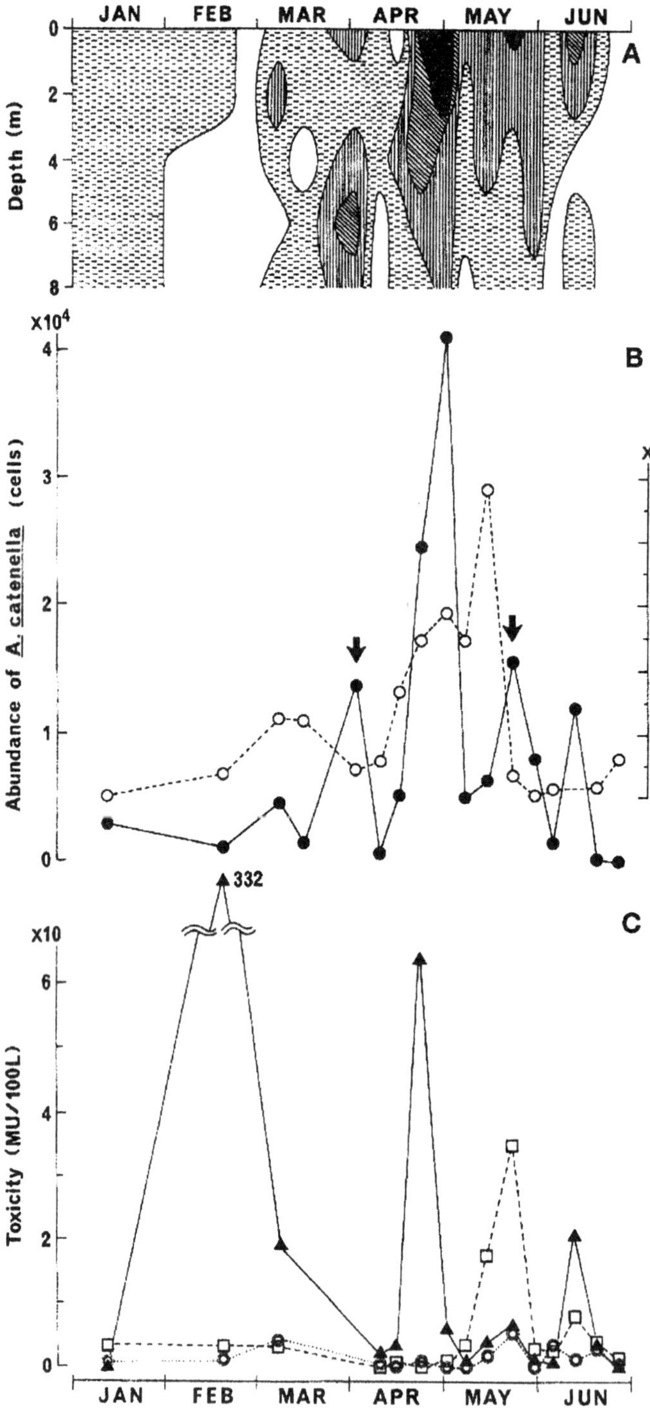

Figure 5 Seasonal variation of *Alexandrium catenella* abundance, scallop toxicity, and the toxicity of particle fractions from January to June 1990 in Tanabe Bay. (A) Vertical distribution of *A. catenella*. Cell density (cells l^{-1}): (□) 0 to 99; (■) 100 to 999; (■) 1000 to 4999; (■) 5000 to 9999; (■) >10000. (B) Abundance of *A. catenella* (●, no. of cells) and scallop toxicity (○, MU g^{-1}). *A. catenella* were collected from 11 samples taken every 2 m from the surface to 8 m depth. Arrows: peaks of *A. catenella* abundance where shellfish toxicity is decreasing. (C) Toxicity of particle fractions: (□) >20 μm; (▲) 5–20 μ; (■) 0.45–5.0 μm. (From Ref. 43.)

(later classified as a *Pseudomonas/Alteromonas* species (49) and most recently as a new genus within the α-proteobacteria (50), was obtained. An extract of this bacterium, prepared from cells grown in Marine Broth (DIFCO Laboratories, Detroit, MI), killed mice and the symptoms were typical of those elicited by STXs, although the toxicity was very low. The toxin was then purified from cell extracts prepared from a 50-L culture by the same method as that used for the purification of STXs (51). The purified toxin showed behavior identical to that of STX on HPLC-fluorometric analysis (8) as well as on analysis by thin-layer chromatography (TLC) (52) and electrophoresis (53). The dose-death time relationship in a mouse bioassay and the relationship between dose and peak area of the toxin in the HPLC analysis was found to coincide well with those of STX, indicating that the toxin is STX (51). These findings suggest that *A. tamarense* possesses intracellular bacteria which produce STXs, providing support to Silva's idea that intracellular bacteria are associated with dinoflagellate toxins. The toxicity of *Moraxella* sp. tended to increase under starvation conditions (54) and when the cells were exposed to P-limiting conditions in chemically defined medium (49). One possible interpretation of the above findings is that, since the toxin accumulates when anabolic processes in bacteria are depressed, STXs may represent catabolites of some unknown bacterial substance(s). Alternatively, the apparently close relationship between bacterial toxin yields and P availability may imply that toxin gene expression is regulated by phosphorus, as has been demonstrated for several secondary metabolites (55), or simply that excess nitrogen is available for toxin synthesis (STXs contain six to seven N atoms per molecule).

VI. INTRACELLULAR BACTERIA IN TOXIC DINOFLAGELLATES

Several studies indicate that axenic, or bacteria-free, dinoflagellate cultures retain the ability to produce these toxins at levels similar to those found in nonaxenic cultures (e.g., see Ref. 56). If bacteria are involved in toxin production in dinoflagellates, they should be present within the dinoflagellates. Several investigators have reported the direct observation of intracellular bacteria in toxic species of dinoflagellates by electron microscopy (46,57,58), although the bacterial numbers were generally few and such bacteria were not always discernible even within cells of the same algal culture. Although toxigenic bacteria could be isolated from toxic dinoflagellates, it was not clearly proven whether the isolated bacterial strains and the corresponding intracellular bacteria were the same. We have recently used an antibody against a toxin-producing bacterium to probe Western blots of an extract from the "host," an *A. tamarense* isolate (axenic culture), and increased cross reactivity was observed after the apparent disruption of residual (=accumulation) bodies present within the algal cells (59). These results, and the fact that bacteria could be isolated from an axenic culture of this dinoflagellate only after disruption of the residual bodies, led us to contend that intracellular bacteria exist within the residual bodies and these retain the ability to reproduce within the algal cells. Consistent with this idea are electron microscopic observations of partially digested bacteria in the residual bodies of several toxic dinoflagellates (58), as well as the apparently stable association of certain bacteria with an algal isolate over many years. Generally, bacteria responsible for infection are killed by defense mechanisms of the host organism. However, some specific infectious bacteria are known to possess escape mechanisms allowing them to survive within cells of the host (60). It is hypothesized that the growth of such bacteria inside residual bodies may be controlled by digestive processes, thereby harmonizing bacterial growth with that of the dinoflagellate and allowing extended maintenance of the intracellular-bacterial flora.

Recently, evidence of mixotrophy has been demonstrated in the case of thecate photosynthetic dinoflagellates, including toxic *Alexandrium* species. *A. tamarense* has been shown to survive and produce toxins, utilizing organic substances from the environment or bacteria instead of nitrate (61). As shown in Figure 6, a culture of *A. tamarense* remained in the exponential phase of growth constantly under nitrogen-limited conditions (2% of the level of nitrate in normal medium) when acclimatized to such conditions for 30 days. The toxin content of the cells decreased to about one tenth of that in the initial culture maintained in normal T1 medium (Figure 6). When nitrate (1 mM) or yeast extract (DIFCO; 20 mg/L) was added to the culture (indicated by the arrow in Figure 6), the cells grew well. The maximum yield of cells in the culture increased markedly in both cases. At the same time, toxin production by the cells was enhanced almost fivefold by these treatments. Levels of nitrate and ammonia in the yeast extract used here were measured and found to be 0.055 and 1.095 μmol/g, respectively, which is too

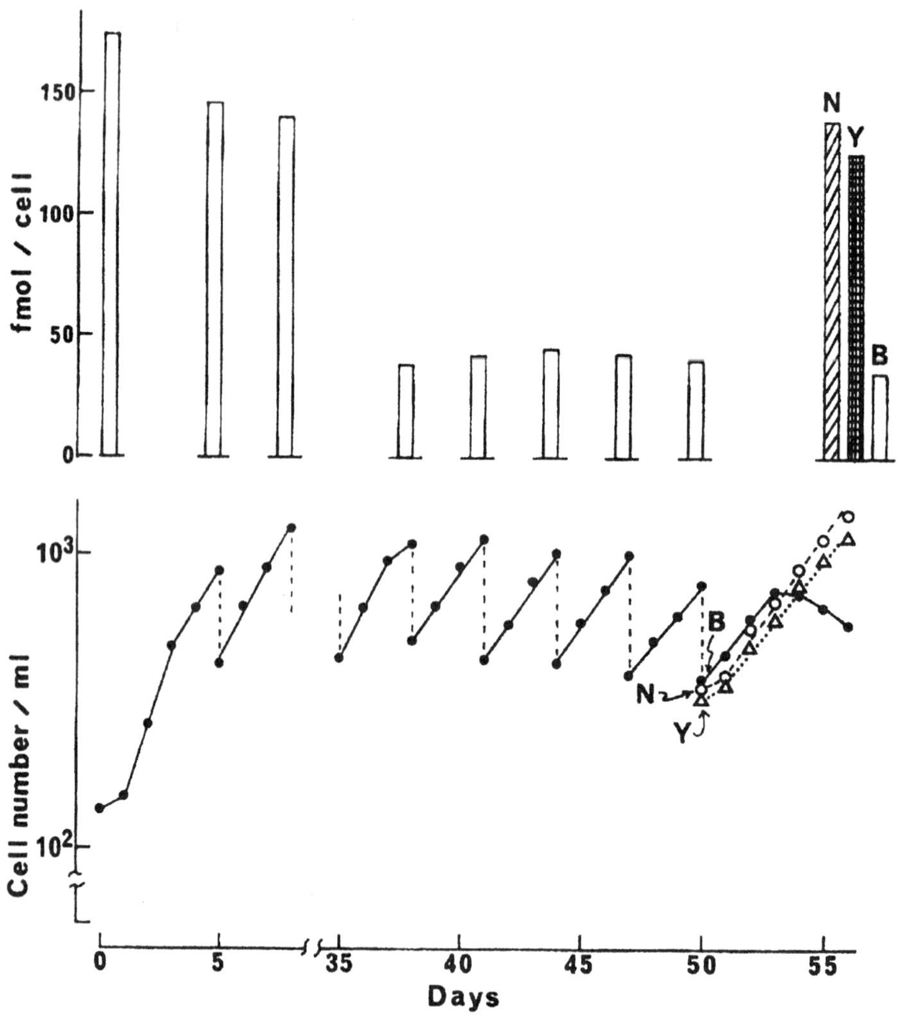

Figure 6 Change of growth (lower) and toxin production (upper) of *A. tamarense* after addition of nitrate (1 mM) or yeast extract (20 mg l^{-1}) under N-limited condition. Nitrate and yeast extract were added at a time indicated by arrow. N, Culture supplemented with nitrate; Y, culture supplemented with yeast extract; B, control culture. (From Ref. 61.)

low to be responsible for the growth observed after the addition of yeast extract. Thus, *A. tamarense* seems to utilize nitrogen-containing organic substances in the yeast extract as a nitrogen source instead of nitrate. Similar results were obtained when autoclaved or live cells of a STX-producing bacterial strain were added to the nitrogen limited culture (Figure 7). Toxin production by *A. tamarense* was induced by this treatment. These results clearly show that *A. tamarense* can grow and produce STXs by utilizing bacteria as a nitrogen source under nitrogen limited conditions. On the other hand, Jacobson and Anderson (62) presented electron micrographs clearly showing that *A. ostenfeldii* ingests other plankton as food, which is evidence of the heterotrophy of *A. ostenfeldii*, a photosynthetic toxic dinoflagellate. Bacteria coexisting inside the dinoflagellate cells may represent a self-sustaining source of organic nutrition to the algae cooperating in a mutualistic fashion.

The issue of taxonomic specificity is important, and the question remains as to whether a single taxon or multiple bacterial taxa are able to reside and reproduce within the cells of a given dinoflagellate species. Very recently, we have found that a 16s rRNA fragment prepared from *A. tamarense* is identical to 16s rRNA prepared from a STX-producing bacterium isolated from the alga (manuscript in preparation). These results show that an STX-producing bacterium is living within the *A. tamarense* cells, and that biochemical interactions between, the bacteria and the dinoflagellate, for example, digestion of the bacteria by the dinoflagellate,occur in the dinoflagellate.

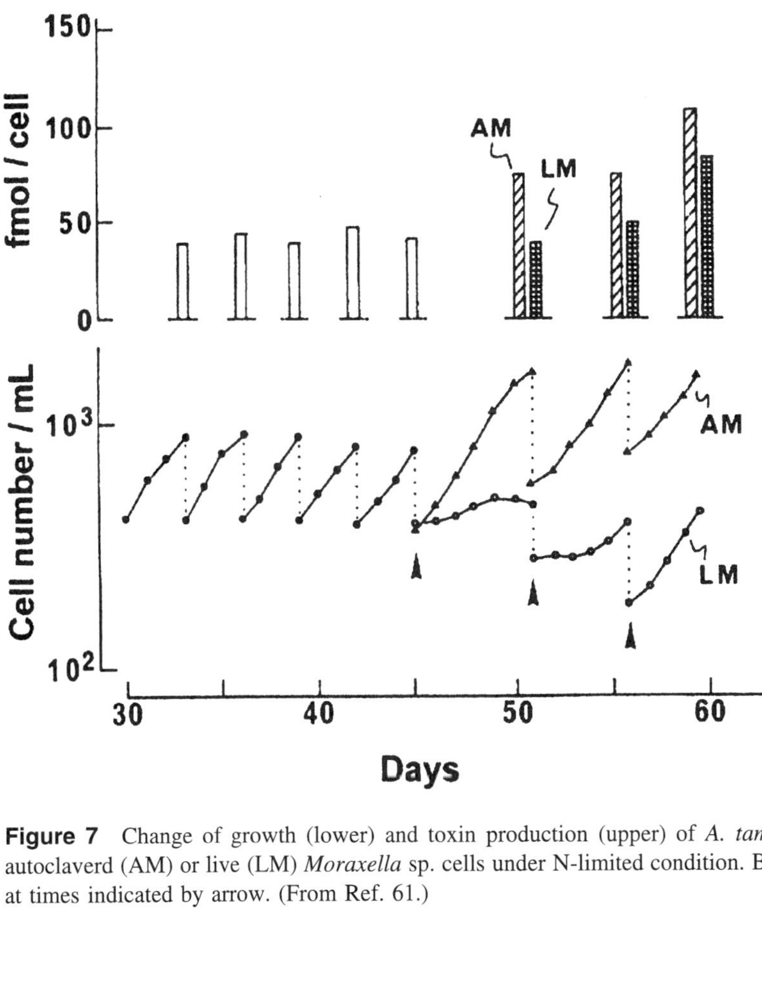

Figure 7 Change of growth (lower) and toxin production (upper) of *A. tamarense* after addition of autoclaverd (AM) or live (LM) *Moraxella* sp. cells under N-limited condition. Bacterial cells were added at times indicated by arrow. (From Ref. 61.)

VII. TOXIN-PRODUCING BACTERIA IN NATURAL ECOSYSTEMS

Since the first finding that a bacterium isolated from *A. tamarense* produces STX (49,51), several strains of STX-producing bacteria have been isolated from dinoflagellates (63,64). Also, we have detected STXs in fractions of particles smaller in size than dinoflagellates (43). Although definitive proof of a bacterial source of the toxin is lacking, STXs detected in particles smaller in size than dinoflagellates are thought to originate from bacteria, STX producing bacteria can grow in seawater (54), and thus may be living on the surface of detritus which is favorable for their growth. Probably, because of the large number of cells present, the amount of toxin in the detritus is often greater than the amount associated with free-living bacteria. However, even though the detritus fraction (5- to 20-μm fraction) showed high toxicity, shellfish toxicity did not increase much. These findings may indicate that bacterial toxins are not directly associated with shellfish toxicity. It is also noteworthy that, although shellfish became toxic during a bloom of *A. catenella*, no significant toxicity was detected in the *A. catenella* cells from this location (see Figure 5). These findings support the idea that the toxicity of shellfish cannot be attributed solely to transport of dinoflagellate toxins via the food chain. De novo synthesis of toxins in shellfish, in association with toxic dinoflagellates or bacteria may also be involved.

The distribution of STX-producing bacteria in the marine environment appears to be quite widespread. Investigators in increasing numbers are now reporting the isolation of such bacteria from locations that are spatially and temporally distinct from the locations of the dinoflagellates. Recently, Levasseur et al. (65) examined the potential for bacterial production of STXs in the St. Lawrence estuary, Canada. They screened bacterial strains isolated from the St. Lawrence by HPLC and found four strains positive for STXs, indicating that putatively toxigenic bacteria are present in that region. Interestingly, these four bacterial strains isolated from the St. Lawrence estuary by Levasseur et al. (65) originated from sites exhibiting no concurrent PSP activity, and three of these locations had no prior history of bloom formation.

They also measured ambient toxin levels in various particle size fractions during and after an *Alexandrium* bloom and the production of toxins by the heterotrophic component of the plankton community (i.e., bacteria) during dark incubation of size-fractionated material. Although ambient STXs were detected by HPLC only in the two largest size fractions (>21 and 5–21 μm) obtained during the bloom, de novo toxin synthesis was detected in most size fractions following dark incubation, including the 0.22- to 5.0-μm range regardless of the sampling time relative to the bloom. Although free-living and particle associated bacteria capable of producing STXs appear to exist in the St. Lawrence estuary, only the toxins associated with larger size fractions accounted for shellfish toxicity during this *Alexandrium* bloom. These results indicate that STX-producing bacteria ingested by shellfish could not be directly responsible for the STXs in the shellfish, although a contribution of toxin-producing bacteria to shellfish toxicity could not be ruled out.

Information on the distribution of STX-producing bacteria in marine ecosystems is important to solve the problem of the association between these bacteria and the STXs of shellfish and other organisms. In order to prove the toxigenicity of bacteria, more detailed chemical evidence such as analytical data obtained by mass spectrometry or nuclear magnetic resonance (NMR) are required. However, because of the low toxin productivity of bacteria, various investigators have applied indirect methods for toxin identification such as HPLC and bioassay of crude extracts. When crude extracts are analyzed by fluorometric-HPLC, ghost peaks with retention times identical to those of standard toxins often appear (28). However, these peaks can be distinguished from toxin peaks by changing the reaction reagent for postcolumn derivatization to water; the ghost peaks do not disappear as a result of this treatment, but the toxin peaks disappear. In addition, these impurities can be mostly removed by pretreatment of the samples, such as by partial purification of the toxins by Bio-Gel P-2 (Bio-Rad Laboratories, Richmond,

CA) or activated charcoal treatment. HPLC analysis of bacterial toxins after such treatment will provide more detailed information concerning the toxigenicity of the bacteria.

VIII. STXs IN ORGANISMS OTHER THAN DINOFLAGELLATES AND THEIR ASSOCIATED ORGANISMS

It is well known that STXs are found not only in dinoflagellates and their associated shellfish but also in phylogenetically diverse species of animals. Noguchi et al. (66) showed for the first time that STX can be detected in the xanthid crab *Atergatis floridus*, which is not associated with dinoflagellates. After this discovery, high levels of STXs were detected in various animals, such as the horseshoe crab, *Carcinoscorpius rotundicauda* (67) several species of marine snails (68) and pufferfish, *Takifugu pardalis* (69) *Tetraodonfangi* (70). These animals are not plankton feeders. Thus, STXs in these animals are unlikely to originate directly from dinoflagellates. Yasumoto (71) concluded that the STXs in the xanthid crab are not derived from dinoflagellates, as no STXs were detected in bivalves living in the same area as the toxic xanthid crab. Instead, they detected a low level of STXs in *Jania* sp., a carcacious alga which is often found in the stomach of the xanthid crab, and suggested that the origin of STXs in the xanthid crab is this alga (72). It is possible that the xanthid crab accumulates a high level of STXs by continuous ingestion of *Jania* sp. over a long period. On the other hand, the xanthid crab is known to release STXs (73). Arakawa et al. (74) have recently shown that a large proportion of the STXs injected into the xanthid crab disappears within a short period. This finding implies that the rate of release of STXs from the xanthid crab is significantly high. Therefore, it is unlikely that the xanthid crab accumulates a high level of STXs by continuous ingestion of *Jania* sp., even if over a long period.

Interestingly, these STX-bearing animals possess tetrodotoxin (TTX), a potent neurotoxin which binds to the same biological receptor as STXs (1). This suggests that organisms known to possess either STXs or TTX may possess both of these toxins. This speculation is supported by the fact that a significant amount of TTX has been detected in *A. tamarense*, a representative dinoflagellate capable of producing STXs (75). During a bloom of *A. tamarense*, considerable amounts of TTX accumulate in shellfish together with STXs (76). The levels of TTX in the shellfish was shown to be parallel with the levels of STXs (76), indicating that the origin of TTX in the shellfish is *A. tamarense*.

It is well known that TTX is produced by bacteria, although the productivity is very low (77,78). It is noteworthy that TTX-producing bacteria and STX producing bacteria display similar characteristics; the toxin productivity of TTX producing bacteria is very low and the TTX levels increase significantly under starvation conditions (77,78). Although the evidence is circumstantial, various phenomena suggest that these bacteria are widely distributed in natural ecosystems. The origin of the STXs in toxic animals not associated with dinoflagellates seems to be bacteria, as in the case of TTX, and the mechanism by which the organisms become toxic owing to the presence of TTX and STXs derived from bacteria seems to be similar in each instance.

IX. MECHANISM BY WHICH PUFFERFISH BECOME TOXIC: A MODEL BASED ON THE MECHANISM BY WHICH ACCUMULATION OF STXs OCCURS IN DINOFLAGELLATES

It has been a mystery why pufferfish possess the potent toxin TTX. Although the reason is still not clear, it is known by fish culturists that pufferfish which are artificially hatched and fed an artificial diet are nontoxic. Matsui et al. (79) demonstrated that nontoxic cultured pufferfish *T. rubripes rubripes* become toxic when fed a diet containing TTX, indicating that TTX in the pufferfish is an exogenous substance which originates from the diet. On the other hand, TTX has been shown to be produced by bacteria (77,78) and thus the source of TTX in TTX bearing animals such as pufferfish is considered to be bacteria. Hence, TTX of bacterial origin is considered to accumulate in the pufferfish via food webs (80). The screening of various marine organisms for TTX revealed that this toxin is widely distributed in marine organisms, some species of which could be components of the diet of pufferfish (81,82). These studies also showed that the trumpet shell *Charonia sauliae* accumulates TTX by ingesting the TTX bearing starfish *Astropecten polyacanthus* (83). However, such a case clearly supporting the food chain theory of TTX accumulation in toxic animals is rare. The distribution of toxic components of the diet of the pufferfish rarely overlaps with that of toxic pufferfish, indicating that the puffer toxin cannot be accumulated via a simple food web mechanism.

On the contrary, Noguchi et al. (84) proposed that pufferfish become toxic by absorbing a low level of TTX produced by intestinal bacteria over a long period. This idea seems plausible, because TTX-producing *Vibrio alginolyticus* is a dominant species in the intestinal bacterial flora of toxic pufferfish (84). However, Sugita et al. (85) reported that *V. alginolyticus* is also a dominant member of the intestinal bacterial flora of nontoxic cultured pufferfish *T. rubripes rubripes*. We analyzed TTX levels in the liver, the main toxic organ of the pufferfish, and in the intestine of cultured puffer specimens 60 days and 6 months old, applying a sensitive method of analysis of TTX (86). No TTX was detected in the liver of either juvenile or adult pufferfish specimens, whereas a low level of TTX, 0.2 MU/g (where one mouse unit [MU]) is the dose required to kill a 20-g male mouse within 30 min), apparently due to intestinal bacteria, was detected in both samples. These observations show that the level of TTX in the intestine is too low for accumulation by the pufferfish.

TTX- and STX-bearing animals are reported to be resistant to these toxins (87). However, these animals die when a substantial amount of the toxin (about 1000 times the lethal dose for common animals) is injected (88). These observations show that the toxins are also toxic to these animals. Although the mechanism is not known, pufferfish sometimes accumulate TTX at levels much higher than their lethal dose (89). On the other hand, toxic species of pufferfish are secreting this dangerous toxin into environmental water continuously (90). As shown in Figure 8, pufferfish possess developed TTX-secreting glands in the skin which consist of specific TTX-secreting cells (TTX cells) (91). When stimulated, the pufferfish secretes a considerably high amount of TTX into the environment; that is, 4 mg of TTX on a single stimulation (90), showing that the rate of TTX secretion is significantly high. A similar phenomenon is also observed in the case of the TTX bearing newt (92). These findings indicate that TTX-bearing animals actively release TTX from a specifically developed organ and the rate of release of the toxin is significantly high. Given that the TTX present in the pufferfish is accumulated via food webs, the rate of uptake of TTX should be greater than the high rate of release; that is, a considerably large amount of TTX must be supplied to the pufferfish continuously. This seems consistent with our results indicating that the cultured pufferfish *T. rubripes rubripes* did not accumulate TTX, although a low level was detected in the intestine. The supply of TTX from the intestine was too low for accumulation by the pufferfish.

Watabe et al. (93) reported that tritium-labeled TTX injected intraperitoneally into toxic pufferfish is accumulated in the liver for a short period and then transported to the skin and gallbladder to be released. About 70% of the injected TTX disappeared within 6 days. We also observed that TTX injected into the nontoxic cultured puffer, *T. rubripes rubripes*, was accumulated in the liver first, moved to the skin, and then was released into the environment (manuscript in preparation). About 50% of the injected TTX disappeared within a few weeks.

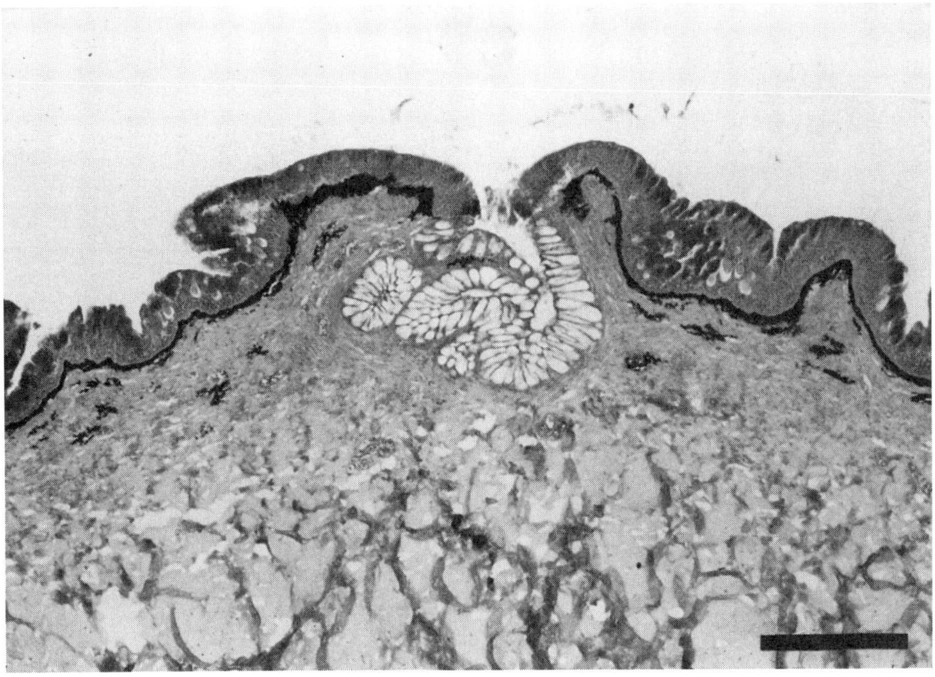

Figure 8 Tetrodotoxin secreting gland in the skin of a puffer *Takifugu pardalis*. Bar = 100 μm.

These data show that TTX taken up by the pufferfish is not retained for a long period because of the high rate of secretion of TTX. Although no significant amount of TTX has been detected in the diet of the pufferfish, some specimens of wild pufferfish accumulate substantial amounts of TTX (89). These observations lead us to contend that pufferfish accumulate TTX not via food webs but through some other mechanism which remains unknown.

Interestingly, a considerable amount of TTX was detected in the muscle of *T. rubripes rubripes* specimens 2 weeks after injection of TTX (manuscript in preparation). Watabe et al. (93) also observed that injected tritium labeled TTX is detected in significant concentrations in the muscle of the pufferfish *T. rubripes rubripes*. In contrast, no TTX was detected in the muscle of wild specimens of *T. rubripes rubripes*. Tani (94) reported that the muscle of *T. rubripes rubripes* is nontoxic, although considerable toxicity is often detected in association with the liver and other organs. Based on analyses of the toxicity of various specimens, the muscle of this species has been designated as a safe food after careful removal of the organs prone to contamination with the toxin by licensed chefs in Japan. The results of these toxicity analyses indicate that natural specimens of *T. Rubripes rubripes* never take in such an amount of TTX at once as that administered in the experiments. The results also tend to support the view that TTX in the pufferfish is not accumulated via food webs.

As described above, we have suggested that symbiotic bacteria in dinoflagellates are involved in the production of substantial amounts of STXs. Most of the toxic species of pufferfish possess STXs as well as TTX (69). In specimens of the freshwater puffer, *Tetraodon fangi*, which causes food poisoning in Thailand (95), only STX was detected as the toxic agent (70). Therefore, the approach taken in investigation of toxins from dinoflagellates was applied to pufferfish toxin; that is, attempts were made to isolate bacteria from the liver tissue of toxic and nontoxic specimens of pufferfish. Interestingly, bacteria were isolated only from toxic speci-

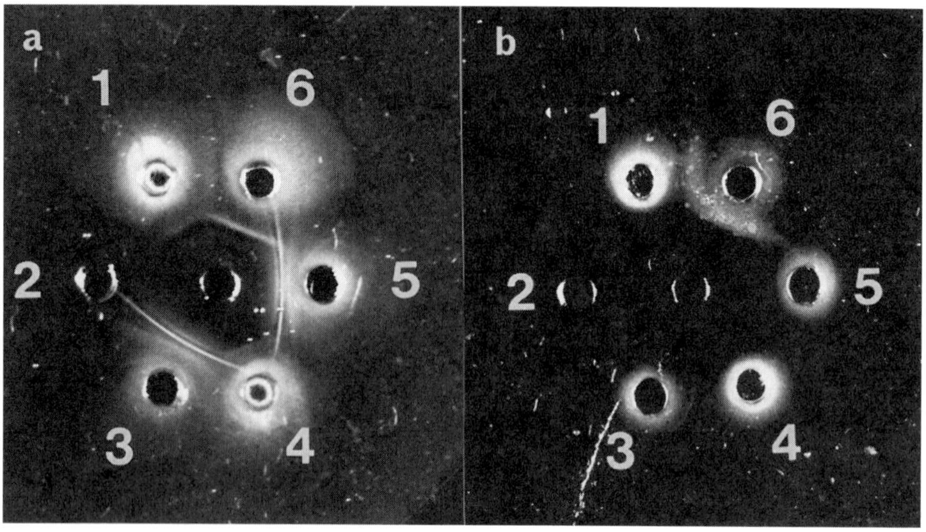

Figure 9 Double immunodiffusion of the extracts of the liver and isolated bacteria against antibody P and puffer liver. 1 and 4: the liver extract of a toxic specimen of the pufferfish *T. porphyreus*; 2 and 6: the extract of bacteria isolated from the liver, the concentration of the extracts is different; 3 and 5: the extract of antigenic bacteria to prepare antibody P. (a) Center: antibody P, (b): Center the serum of puffer specimen *T. porphyreus* from which the liver (1 and 4) and bacteria (2 and 6) were prepared. Antibody P was prepared by immunizing rabbit with homogenate of bacteria isolated from the liver of another toxic specimen of *T. porphyreus*. (From Ref. 96.)

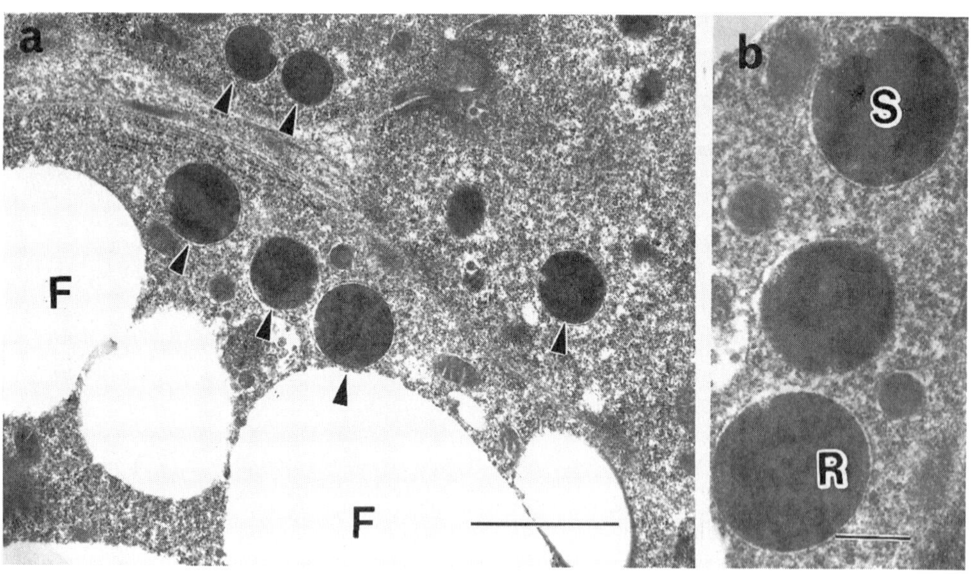

Figure 10 Electron micrograph of hepatocytes from *T. poecilonotus*. (a) There are three hepatocytes in the electron micrographs. Each cell contain many microorganism-like components (arrows). No abnormal bioreactions are seen in every hepatocyte. F, fat droplet. Bar = 2 μm. (b) Enlarged part of (a). Three round microorganisms are seen. These are surrounded by double membranes from the cytoplasm of the hepatocytes. Numerous ribosomes (R) and the thick septum (S) are demonstrable. Bar = 0.5 μm. (From Ref. 96.)

mens, whereas no bacteria were obtained from nontoxic specimens of the pufferfish (96). Polyclonal antibody against the bacterium isolated from the liver reacted with an extract of the liver from which the bacterium was isolated (Figure 9). Electron microscopic observations also supported the view that bacteria were present in the liver cells (Figure 10). An extract of the isolated bacteria was found to display sodium channel blocking activity and gave a TTX-like peak on HPLC analysis. These findings imply that liver cells in the toxic pufferfish are infected by bacteria which produce TTX or a TTX-related substance. Apparently, there was no difference in behavior between toxic pufferfish specimens with bacteria present in the liver and nontoxic specimens free of bacterial infection, suggesting that the bacteria in the liver are symbiotic rather than infectious. These findings suggest that bacterial symbiosis in liver cells is involved in the mechanism of production of pufferfish toxin as in the case of toxic dinoflagellates.

Most of the bacteria isolated from the liver of the pufferfish were gram, positive bacteria, which are usually not found in seawater but rather in the sediments (96). It is well known that

Table 1 Occurrence of TTX and STXs in Common Aquatic Animals

			Toxin composition (mg mL^{-1} solution[a])			
Species	Part	TTX	GTX1 + 4	GTX2 + 3	neoSTX	STX
Shark						
Lemna ditropis	Int	2.3	2.3	2.3		0.1
	Liv	1.0	0.3	0.5		0.1
Salmon						
Oncorhynchus keta	Int	1.1			4.2	1.9
	Liv	0.7			0.3	0.1
Conger eel						
Conger myriaster	Vis	1.3				
Saury						
Cololabis saira	Vis	1.0				
Cod						
Gadus macrocephalus	Int	0.6	0.6	0.2	3.1	
	Liv	0.2	2.1			
Parrotfish						
Ypsiscarus ovifrons	Int	1.9	0.9	0.7	0.2	0.1
	Liv	1.6	0.1	0.4		
Skipjack						
Katuwonus pelamis	Liv	1.7		0.9		
Filefish						
Nevodon modestus	Int	0.8	0.3		0.3	0.1
	Liv	2.5	1.8			0.1
Crayfish						
Procambarus clarkii	Hep	0.4	1.0			0.1
Rock crab						
Hemigrapsus sanguineus	Hep					0.1
Rock crab						
Pachygrapsus crassipes	Hep					0.1
Marine snail						
Neptunea arthritica	Dig	0.2	1.0	0.9	0.3	0.1

Int, intestine; Liv, liver; Vis, viscera; Hep, hepatopancreas; Dig, digesive gland.
[a] 1 mL solution equivalent to 100 g of organ.
Source: Ref. 99.

pufferfish tend to hide themselves in the bottom sediment (97). It seems plausible that bacterial infection may occur in this habitat of the pufferfish. This may explain why pufferfish cultured in a cage, having no contact with bottom sediments, are nontoxic.

As described above, bacteria which produce STXs or STX-like substances are widely distributed in natural ecosystems. This is also true for bacteria which produce TTX or TTX-related compounds (84,98). *Shewanella alga*, the species found to produce TTX, has been recently shown to produce STXs as well (99). This species shows wide phylogenetic and spatial distribution in natural ecosystems. These observations imply that most aquatic animals are exposed to these bacteria. Therefore, we screened the viscera of various common aquatic animals for the presence of STXs and TTX (100). HPLC analyses and bioassays of the toxins partially purified from the viscera showed that all species examined showed the presence of both TTX and STXs, although the levels were very low (Table 1). For further confirmation, an extract of the chum salmon, *Onchorhyncus keta*, in which low levels of TTX, STX, and neoSTX were detected by HPLC analysis, was prepared from a large amount of the viscera (62 kg) of specimens collected at a river mouth during the homing period. A toxin with a retention time identical to that of TTX was detected in the HPLC analysis, but was lost during the purification process; however, toxins corresponding to STX and neoSTX were obtained in a highly pure state. Mass spectrometry and NMR analyses clearly showed that these toxins were STX and neoSTX (S. Sato, T. Ogata, M. Kodama. Trace amount of saxitoxins in the viscera of chum salmon *Oncorhynchus keta*. Mar. Ecol. Prog. Ser. 175:295–298, 1998). These findings show that TTX and STXs are not substances found only in specific organisms but are widely distributed in common aquatic animals at low levels. As shown in Table 1, no significant difference was observed in the levels of these toxins in the viscera of various common animals, indicating that these toxins are not concentrated via the food web but rather they originate from intestinal bacteria. More than 90% of the homing chum salmon returning close to their home river do not consume any diet, and thus their stomachs are empty (101). All of the stomachs in the viscera used in the above experiments were empty, thus the STXs found in the viscera did not originate from the diet. These findings support the view that bacteria which produce TTX and STXs are widely distributed in natural ecosystems.

X. DE NOVO SYNTHESIS OF TOXINS IN TOXIC ANIMALS

The finding that TTX and STXs are widely distributed at low levels in common aquatic animals in natural ecosystems raises again the question as to whether the food web theory of toxin accumulation is correct, as these common animals are in various trophic levels of food webs in the ecosystem. Nevertheless, high levels of toxins are not found in common animals but are limited to specific animals. This may be due to the unusual ability of specific animals to accumulate toxins. However, it seems unlikely that this occurs considering the continuous toxin excretion by these specific toxic animals. On the other hand, it has been observed in an aquarium setting that pufferfish reared for several years maintain a high level of toxins. We have also observed that natural pufferfish specimens maintain a high level of toxicity after 1 year of rearing (unpublished data). These findings may indicate that de novo synthesis of toxins occurs in the pufferfish.

A similar phenomenon is observed in the case of STXs of shellfish. As described above, a 1-week time lag was observed between the peak of shellfish toxicity and the peak in abundance of *A. tamarense* in a field survey (40). In order to confirm further this phenomenon, the toxicity of scallops fed cultured *A. tamarense* cells was analyzed. Surprisingly, the toxins were accumulated in the scallops at levels greater than those in *A. tamarense* fed to the scallops (manuscript in preparation). The epithelial cells of the diverticula of the scallops which were fed *A. tamarense* were stained by a probe specific for the 16s rRNA of the bacteria as well as by an antibody against the STX-producing bacterium isolated from *A. tamarense* even 1 week after feeding was stopped (manuscript in preparation). In contrast, such staining was not observed in the case of nontoxic scallops prior to feeding. These observations indicate that the intracellular bacteria from *A. tamarense* are taken up by the epithelial cells of the diverticula together with *A. tamarense* and survive for a while in the cells. These bacteria move into amebacytes in the connective tissue of the scallop and survive for at least 1 month. During this period the toxicity of the scallop fluctuates, indicating that a part of the total amount of the toxins is eliminated from the tissue, but the toxins are synthesized at the same time in the scallop. STX-producing bacteria carried to the cells of the diverticula of shellfish by *A. tamarense* are thought to play a role in de novo toxin synthesis in the shellfish.

Although there is little evidence available, it appears that pufferfish do not become nontoxic when they are reared with a nontoxic artificial diet. Many laboratories have confirmed that toxic dinoflagellates maintain their toxicity even if the cultures are maintained for a long period. As described above, these organisms possess symbiotic bacteria. On the contrary, in shellfish which accumulate toxins by ingestion of toxic dinoflagellates, the toxins are known to deteriorate when the toxic dinoflagellates disappear from their environment. Different rates of deterioration of the toxins are observed among the shellfish species. The scallop *Patinopecten yessoensis* often maintains considerable toxicity for more than 6 months, whereas the toxicity of the mussel *Mytilis edulus* disappears within 1 month (102). As described above, staining indicative of the presence of STX-producing bacteria was not observed in the case of nontoxic scallops. These findings suggest that toxin-producing bacteria ingested along with toxic dinoflagellates are trapped by phagocytes of the shellfish and survive for awhile in the cells. However, these bacteria would be finally digested by phagocytes such as the epithelial cells of the diverticula and amebacytes. The different rates of deterioration of the toxin in shellfish may be due to the differing activity of these cells in the shellfish.

XI. HYPOTHESIS ON TOXIN ACCUMULATION IN TOXIC ORGANISMS

As described above, bacteria which produce low levels of toxins live in cells of the pufferfish and in dinoflagellates which display high levels of TTX and/or STXs. Toxic shellfish also possess these bacteria in their cells, carried by the vector dinoflagellates. These bacteria grow in cells of the host organisms. The host organisms kill and digest some of the bacteria that have proliferated within the host cells and thereby maintain a suitable number of bacterial cells for their survival. That is, direct biochemical interactions between the bacteria and the host occur within the cells of the host organisms. The finding that the number of bacteria found in the host organisms is few tends to support this idea. Based on these findings, the hypothesis shown in Figure 11 is proposed, suggesting that TTX and STXs are not the final products of metabolism but are intermediate products generated in the course of synthesis of some bacterial substance which seems to be important for survival of the bacteria. Previously, we reported that TTX and STXs were released from a nontoxic high molecular weight fraction prepared from toxic scallop digestive glands or toxic pufferfish livers on digestion with RNase (103,104). These findings suggest that the bacterial substance involved in production of these toxins is RNA. This idea is consistent with the observation that the toxin content of the bacteria is very low (54), and that toxin production by the bacteria increases significantly under P-limiting conditions (49) as well as under starvation conditions (54). When some of the bacterial cells that have proliferated within the host cells are killed in the host organisms, they are digested by host cells and the substance associated with the toxins (probably RNA) is also digested by host enzymes resulting in release of the toxins as catabolites. In other words, toxic animals and dinoflagellates are hosts which possess the enzyme(s) required to digest the substance and release the toxins. If the

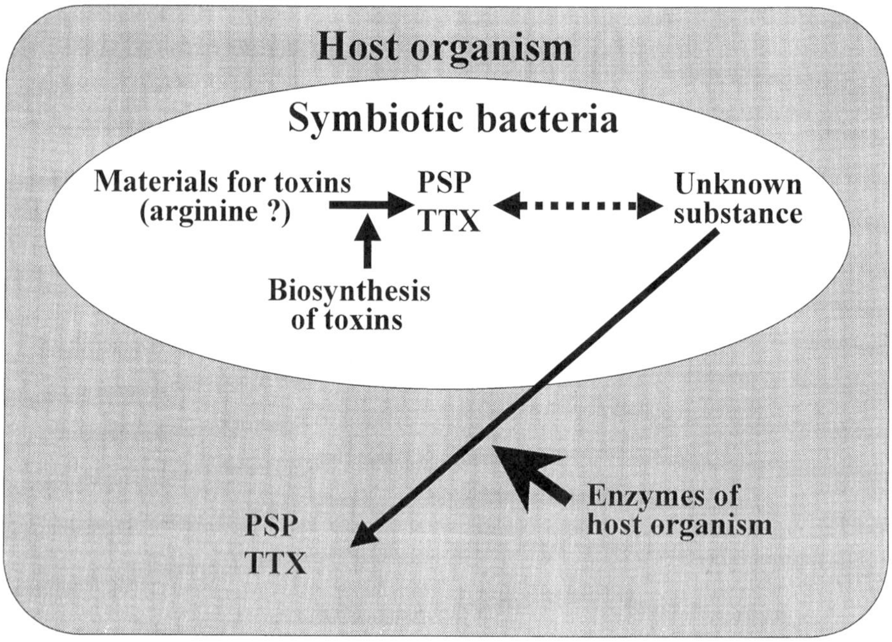

Figure 11 Hypothesis on the mechanism by which toxic organisms become toxic in associated with infectious bacteria.

substance in question is RNA, the enzyme required to digest this substance would be an RNase, an enzyme which occurs widely in living organisms. Toxic organisms might possess a specific type of RNase capable of digesting foreign RNA which contains unusual components, STXs, and/or TTX. Some strains of toxic dinoflagellates are known to be nontoxic. The RNase in these strains might lack the ability to digest foreign RNA containing unusual components which are not essential for their growth.

REFERENCES

1. CY Kao. Tetrodotoxin, saxitoxin and their significance in the study of excitation phenomena. Pharmacol Rev 18:997–1049, 1966.
2. EJ Schantz. Chemistry and biology of saxitoxin and related toxins. Ann NY Aca Sci 479:15–23, 1986.
3. KA Steidinger. Some taxonomic and biologic aspects of toxic dinoflagellates. In: IR Falconer, ed. Algal Toxins in Seafood and Drinking Water. London: Academic Press, 1993, pp 1–28.
4. CY Kao. Structure-activity relation of tetrodotoxin, saxitoxin, and analogues. Ann NY Acad Sci 479:52–67, 1986.
5. Y Shimizu, M Norte, A Hori, A Genenah, M Kobayashi. Biosynthesis of saxitoxin analogues: The unexpected pathway. J Am Chem Soc 106:6433–6434, 1984.
6. H Sommer, KF Meyer. Paralytic shellfish poisoning. Arch Pathol 24:560–598, 1937.
7. SE Shumway. A review of the effects of algal blooms on shellfish and aquaculture. J World Aquac Soc 21:65–104, 1990.
8. Y Oshima, M Machida, K Sasaki, Y Tamaoki, T Yasumoto. Liquid chromatographic-fluorometric analysis of paralytic shellfish toxins. Agric Biol Chem 48:1707–1711, 1984.

9. JJ Sullivan, MM Wekell, LL Kentala. Application of HPLC for the determination of PSP toxins in shellfish. J Food Sci 50:26–29, 1985.
10. Y Oshima, K Sugino, H Itakura, M Hirota, T Yasumoto. Comparative studies on paralytic shellfish toxin profile of dinoflagellates and bivalves. In: E Graneli, B Sundstrom, L Edler, DM Anderson, eds. Toxic Marine Phytoplankton. New York: Elsevier, 1990, pp 391–396.
11. T Harada, Y Oshima, T Yasumoto. Natural occurrence of decarbamoylsaxitoxin in tropical dinoflagellate and bivalves. Agric Biol Chem 47:191–193, 1983.
12. M Kodama, T Ogata, Y Fukuyo, T Ishimaru, S Wisessang, K Saitanu, V Panichyakarn, T Piyakarnchana. *Protogonvaulax cohorticula*, a toxic dinoflagellate found in the gulf of Thailand. Toxicon 26:707–712, 1988.
13. PJ Hansen, AD Cembella, O Moestrup. The marine dinoflagellate *Alexandrium ostenfeldii*: paralytic shellfish toxin concetration, composition, and toxicity to a tintinnid ciliate. J Phycol 28:597–603, 1992.
14. GL Boyer, JJ Sullivan, RJ Andersen, FJR Taylor, PJ Harrison, AD Cembella. Use of high-performance liquid chromatography to investigate the production of paralytic shellfish toxins by *Protogonyaulax* spp. in culture. Mar Biol 93:361–369, 1986.
15. AD Cembella, JJ Sullivan, GL Boyer, FJR Taylor, RJ Andersen. Variation in paralytic shellfish toxin composition within the *Protogonyaulax tamarensis/catenella* species complex; red tide dinofalgellates. Biochem Syst Ecol 15:171–186, 1987.
16. BA Boczar, MK Beitler, J Liston, JJ Sullivan, RA Cattolico. Paralytic shellfish toxins in *Protogonyaulax tamarensis* and *Protogonyaulax catenella* in axenic culture. Plant Physiol 88:1285–1290, 1988.
17. DM Anderson, DM Kulis, JJ Sullivan, S Hall, C Lee. Dynamics and physiology of saxitoxin production by the *Alexandrium* spp. Mar Biol 104:511–524, 1990.
18. Y Oshima, SI Blackburn, GM Hallegraeff. Comparative study on paralytic shellfish toxin profiles of the dinoflagellate *Gymnodinium catenatum* from three different countries. Mar Biol 116:471–476, 1993.
19. Y Oshima, M Hirota, T Yasumoto, GM Hallegraeff, SI Blackburn, DA Steffensen. Production of paralytic shellfish toxins by the dinoflagellate *Alexandrium minutum* Halim from Australia. Nippon Suisan Gakkaishi 55:925, 1989.
20. A Sournia, C Belin, B Berland, EE-1 Denn, P Gentien, D Grzebyk, CM-1 Baut, P Lassus, F Partensky. *Alexandrium minutum* (Dinophyceae). In: Le Phytoplancton Nuisible des Cote de France, de la Biologie a la Prevention. Centre National de La Recherche Scientifique, Service de la Documentation et des Publications, IFREMERS, 1991, pp 83–90.
21. FH Chang, DM Anderson, DM Kulis, DG Till. Toxin production of *Alexandrium minutum* (Dinophyceae) from the Bay of Plenty, New Zealand. Toxicon 35:393–409, 1997.
22. JM Franco, P Fernandez, B Reguera. Toxin profiles of natural population and cultures of *Alexandrium minutum* Halim from Galician (Spain) coastal waters. J Appl Phycol 6:275–279, 1994.
23. Y Oshima, T Hayakawa, M Hashimoto, Y Kotaki, T Yasumoto. Classification of *Protogonyaulax tamarensis* from northern Japan into three strains by toxic composition. Bull Jpn Soc Sci Fish 48: 851–854, 1982.
24. M Alam, Y Shimizu, M Ikawa, JJ Sasner. Reinvestigation of the toxins from the blue-green alga, *Aphanizomenon flos-aquae*, by high performance chromatographic method. J Environ Sci Health A13:493–499, 1978.
25. NA Mahmood, WW Carmichael. Paralytic shellfish poisons produced by the freshwater cyanobacterium *Aphanizomenon flos-aquae* NH-5. Toxicon 24:175–186, 1986.
26. L Bowling. The cyanobacterial (blue-green algae) bloom in the Darling/Barwon River system, November-December 1991. In: , ed. Department of Water Resources Technical Services Report. ISBN: 07305 78955, 1992, p 49.
27. AR Humpage, J Rositano, AH Brown, PD Baker, BC Nicholson, DA Steffensen. Paralytic shellfish poisons from Australian blue-green algal (cyanobacterial) blooms. Aust J Mar Freshwat Res 45: 761–771, 1994.
28. H Onodera, Y Oshima, MF Watanabe, M Watanabe, CJ Bolch, S Blackburn, T Yasumoto. Screening

of paralytic shellfish toxins in freshwater cyanobacteria and chemical confirmation of the toxins in cultured *Anabana circinalis* from Australia. In: T Yasumoto, Y Oshima, Y Fukuyo, eds. Harmful and Toxic Algal Blooms. Paris: Intergovernmental Oceanographic Commision of UNESCO, 1996, pp 563–566.
29. AP Negri, GJ Jones. Bioaccumulation of paralytic shellfish poisoning (PSP) toxins from the cyanobacterium *Anabaena circinalis* by the freshwater mussel *Alathyria condola*. Toxicon 33:667–678, 1995.
30. H Onodera, M Satake, Y Oshima, T Yasumoto, WW Carmichael. New saxitoxin analogues from the freshwater filamentous cyanobacterium *Lyngbya wollei*. Nat Toxins 5:146–151, 1997.
31. Y Shimizu. Paralytic shellfish poisons. In: W Herz, H Griesebach, GW Kirby, eds. Progress in the Chemistry of Organic Natural Products. Vienna: Springer-Verlag, 1984, pp 235–265.
32. Y Oshima. Chemical and enzymatic transformation of paralytic shellfish toxins in marine organisms. In: P Lassus, G Arzul, EE-L Denn, P Gentien, CM-L Baut, eds. Harmful Marine Algal Blooms. Paris: Londres, 1995, pp 475–480.
33. Y Oshima, CJ Bolch, GM Hallegraeff. Toxin composition of resting cysts of *Alexandrium tamarense* (Dinophyceae). Toxicon 30:1539–1544, 1992.
34. N Ichihara, N Kobayashi, M Kobayashi, CP Hsu, Y Shimizu. Bio-conversion of paralytic shellfish toxins. In: Abstracts of Joint Meeting of the American Society of Pharmacognosy and the Society for Economic Botany. Abstract no. 81. Boston: 1981.
35. M Asakawa, M Takagi, A Iida, M Oishi. Studies on the conversion of paralytic shellfish poison (PSP) components by biochemical reducing agents. Eiseikagaku 33:50–55, 1984.
36. Y Shimizu, M Yoshioka. Transformation of paralytic shellfish toxins as demonstrated in scallop homogenates. Science 212:547–549, 1981.
37. Y Kotaki, Y Oshima, T Yasumoto. Bacterial transformation of paralytic shellfish toxins in coral reef crabs and a marine snail. Bull Jpn Soc Sci Fish 51:1009–1013, 1985.
38. Y Ishida, C-H Kim, Y Sako, N Hirooka, A Uchida. PSP toxin production is chromosome dependent in *Alexandrium* spp. In: TJ Smayda, Y Shimizu, eds. Toxic Phytoplankton Blooms in the Sea. New York: Elsevier, 1993, pp 881–887.
39. J Maruyama, T Noguchi, Y Onoue, Y Ueda, K Hashimoto, S Kamimura. Anatomical distribution and profiles of the toxins in highly PSP-infested scallops from Ofunato Bay during 1980–1981. Bull Jpn Soc Sci Fish 49:233–235, 1983.
40. T Ogata, M Kodama, Y Fukuyo, T Inoue, H Kamiya, F Matsuura, K Sekiguchi, S Watanabe. The occurrence of *Protogonyaulax* spp. in Ofunato Bay, in association with the toxification of the scallop *Patinopecten yessoensis*. Bull Jpn Sci Soc Fish 48:563–566, 1982.
41. K Sekiguchi, N Inoguchi, M Shimizu, S Saito, S Watanabe, T Ogata, M Kodama, Y Fukuyo. Occurrence of *Protogonyaulax tamarensis* and shellfish toxicity in Ofunato Bay from 1980–1986. In: T Okaichi, DM Anderson, T Nemoto, eds. Red Tides: Biology, Environmental Science, and Toxicology. New York: Elsevier, 1989, pp 399–402.
42. S Sato, S Sakamoto, T Ogata, Y Ueda, M Kodama. What is going on the monitoring survey of shellfish toxins. Bull Coast Oceanogr 32:69–79, 1994.
43. S Sakamoto, T Ogata, S Sato, M Kodama, T Takeuchi. Causative Organism of paralytic shellfish toxins other than toxic dinoflagellates. Mar Ecol Prog Ser 89:229–235, 1992.
44. M Kodama, T Ogata, S Sato, S Sakamoto. Possible association of marine bacteria with paralytic shellfish toxicity of bivalves. Mar Ecol Prog Ser 61:203–206, 1990.
45. S Sakamoto, S Sato, T Ogata, M Kodama. Field survey of bivalve toxicity associated with toxicity of different particle sizes in seawater. In: TJ Smayda, Y Shimizu, eds. Toxic Phytoplankton Blooms in the Sea. New York: Elsevier, 1993, pp 913–917.
46. ES Silva. Intracellular bacteria, the origin of the dinoflagellates toxicity. In: ed. International IUPAC Symposium on Mycotoxins and Phycotoxins. Lausane: Pahotox, 1979, pp. 8.
47. ES Silva. Relationship between dinoflagellates and intracellular bacteria. Mar Algae Pharm Sci 2: 269–288, 1982.
48. T Ogata, M Kodama, T Ishimaru. Toxin production in the dinoflagellate *Protogonvaulax tamarensis*. Toxicon 25:923–928, 1987.
49. GJ Doucette, CG Trick. Characterization of bacteria associated with different isolates of *Alexan-*

drium tamarense. In: P Lassus, G Arzul, EE-L Denn, P Gentien, CM-L Baut, eds. Harmful Marine Algal Blooms. Paris: Londres, 1995, pp 33–38.
50. M Kopp, GJ Deucette, M Kodama, G Gerdts, C Schutt, L Medlin. Phylogenetic analysis of selected toxic and non-toxic bacterial strains isolated from the toxic dinoflagellate *Alexandrium tamarense*. FEMS Microbiol Ecol 24:251–257, 1997.
51. M Kodama, T Ogata, S Sato. Bacterial production of saxitoxin. Agric Biol Chem 52:1075–1077, 1988.
52. LJ Buckley, M Ikawa, JJS Jr. Isolation of *Gonyaulax tamarensis* toxins from soft shell clams (*Mya arenaria*) and a thin-layer chromatographic-fluorometric method for their detection. J Agric Food Chem 24:107–111, 1976.
53. WE Fallon, Y Shimizu. Electrophoresis analysis of paralytic shellfish toxins. J Environ Sci Health A12:455–464, 1977.
54. M Kodama, T Ogata, S Sakamoto, S Sato, T Honda, T Miwatani. Production of paralytic shellfish toxins by a bacterium *Moraxella* sp. isolated from *Protogonyaulax tamarensis*. Toxicon 28:707–714, 1990.
55. LC Vining. Functions of secondary metabolites. Ann Rev Microbiol 44:395–427, 1990.
56. C-H Kim, Y Sako, Y Ishida. Variation of toxin production and composition in axenic cultures of *Alexandrium catenella* and *A. tamarense*. Nippon Suisan Gakkaishi 59:633–639, 1993.
57. M Kodama. Possible links between bacteria and toxin production in algal blooms. In: E Graneli, B Sundstrom, I Edler, DM Anderson, eds. Toxic Marine Phytoplankton. New York: Elsevier, 1990, pp 52–61.
58. S Franca, S Viegas, V Mascarenhas, L Pinto, GJ Doucette. Prokaryotes in association with a toxic *Alexandrium lusitanicum* in culture. In: P Lassus, G Arzul, E Erard, P Gentien, C Marcaillou, eds. Harmful Marine Algal Blooms. Paris: Lavoisier, 1995, pp 45–51.
59. M Kodama, S Sakamoto, K Koike. Symbiosis of bacteria in *Alexandrium tamarense*. In: T Yasumoto, Y Oshima, Y Fukuyo, eds. Harmful and Toxic Alagal Blooms. Paris: Intergovernmental Oceanographic Commission of UNESCO, 1996, pp 351–354.
60. EA Groisman. Resistance to host antimicrobial peptides is necessary for *Salmonella virulence*. Proc Natl Acad Sci USA 89:11939–11943, 1992.
61. T Ogata, K Koike, S Nomura, M Kodama. Utilization of organic substances for growth and toxin production by *Alexandrium tamarense*. In: T Yasumoto, Y Oshima, Y Fukuyo, eds. Harmful and Toxic Alagal Blooms. Paris: Intergovernmental Oceanographic Commission of UNESCO, 1996, pp 343–346.
62. DM Jacobson, DM Anderson. Widespread phagocytosis of ciliates and other protists by marine mixotrophic and heterotrophic thecate dinofagellates. J Phycol 32:279–285, 1996.
63. S Franca, L Pinto, P Alvito, I Sousa, V Vasconcelos, GJ Doucette. Studies on prokaryotes associated with PSP producing dinoflagellates. In: T Yasumoto, Y Oshima, Y Fukuyo, eds. Harmful and Toxic Alagal Blooms. Paris: Intergovernmental Oceanographic Commission of UNESCO, 1996, pp 247–350.
64. S Gallacher, KJ Flynn, JM Franco, EE Brueggemann, HB Hines. Evidence for production of paralytic shellfish toxins by bacteria associated with *Alexandrium* spp. (Dinophyta) in culture. Appl Environ Microbiol 63:239–245, 1997.
65. M Levasseur, P Monfort, GJ Doucette, S Michaud. Preliminary study of bacteria as PSP producers in the Gulf of St. Lawrence, Canada. In: T Yasumoto, Y Oshima, Y Fukuyo, eds. Harmful and Toxic Algal Blooms. Paris: Intergovernmental Oceanographic Commission of UNESCO, 1996, pp. 363–366.
66. T Noguchi, S Konosu, Y Hashimoto. Identity of the crab toxin with saxitoxin. Toxicon 7:325–326, 1969.
67. N Fusetani, H Endo, K Hashimoto, M Kodama. Occurrence and properties of toxins in the horseshoe crab *Carcinoscorpius rotundicauda*. Toxicon Suppl. 3:165–168, 1983.
68. Y Kotaki, Y Oshima, T Yasumoto. Analysis of paralytic shellfish toxins of marine snails. Bull Jpn Soc Sci Fish 47:943–946, 1981.
69. M Kodama, T Ogata, T Noguchi, J Maruyama, K Hashimoto. Occurrence of saxitoxin and other toxins in the liver of the pufferfish *Takifugu pardalis*. Toxicon 21:897–900, 1983.

70. S Sato, M Kodama, T Ogata, K Saitanu, M Furuya, K Hirayama, K Kakinuma. Saxitoxin as a toxic principle of a freshwater puffer *Tetraodon fangi*, in Thailand. Toxicon 35:137–140, 1997.
71. T Yasumoto, Y Oshima, T Konta. Analysis of paralytic shellfish toxins of xanthid crabs in Okinawa. Bull Jpn Soc Sci Fish 47:957–959, 1981.
72. Y Kotaki, M Tajiri, Y Oshima, T Yasumoto. Identification of a calcareous red alga as the primary source of paralytic shellfish toxins in coral reefs crabs and gastropods. Bull Jpn Soc Sci Fish 49: 283–286, 1983.
73. T Noguchi, K Daigo, O Arakawa, K Hashimoto. Release of paralytic shellfish poison from the exoskeleton of a xanthid crab *Zosimus aneus*. In: DM Anderson, AW White DG Baden, eds. Toxic Dinoflagellates. New York: Elsevier, 1985, pp 293–298.
74. O Arakawa, T Noguchi, Y Onoue. Transformation of gonyautoxins in the xanthid crab *Atergatis floridus*. Fish Sci 64:334–337, 1998.
75. M Kodama, S Sato, S Sakamoto, T Ogata. Occurrence of tetrodotoxin in *Alexandrium tamarense*, a causative dinoflagellate of paralytic shellfish poisoning. Toxicon 34:1101–1105, 1996.
76. M Kodama, S Sato, T Ogata. *Alexandrium tamarense* as a source of tetrodotoxin in the scallop *Patinopecten yessoensis*. In: TJ Smayda, Y Shimizu, eds. Toxic Phytoplankton Blooms in the Sea. New York: Elsevier, 1993, pp 401–406.
77. T Yasumoto, D Yasumura, M Yotsu, T Michishita, A Endo, Y Kotaki. Bacterial production of tetrodotoxin and anhydrotetrodotoxin. Agric Biol Chem 50:793–795, 1986.
78. T Noguchi, J Jeon, O Arakawa, H Sugita, Y Deguchi, Y Shida, K Hashimoto. Occurrence of tetrodotoxin and anhydrotetrodotoxin in *Vibrio* sp. isolated from the intestines of a Xanthid crab, *Atergatis floridus*. J Biochem 99:311–314, 1986.
79. T Matsui, H Sato, S Hamada, C Shimizu. Comparison of toxicity of the cultured and wild puffer fish *Fugu niphobles*. Bull Jpn Soc Sci Fish 48:253, 1982.
80. T Yasumoto, M Yotsu. Biogenetic origin and natural analogs of tetrodotoxin. In: RF Keeler, NB Mandava, AT Tu, eds. Natural Toxins: Toxicology, Chemistry and Safety. Fort Collins, CO: Alaken, 1992, pp 226–233.
81. T Noguchi, J Maruyama, H Narita, K Hashimoto. Occurrence of tetrodotoxin in the gastropod mollusk *Tutufa lissostoma* (frog shell). Toxicon 22:219–226, 1984.
82. K Miyazawa, JK Jeon, J Maruyama, T Noguchi, K Ito, K Hashimoto. Occurrence of tetrodotoxin in the flatworm *Planocera multitentaculata*. Toxicon 24:645–650, 1986.
83. H Narita, M Nara, K Baba, H Ohgami, TK Ai, T Noguchi, K Hashimoto. Effect of feeding a trumpet shell, *Charonia sauliae* with toxic starfish. Shokuhin Eisei Gakkaishi 25:251–255, 1984.
84. T Noguchi, DF Hwang, O Arakawa, H Sugita, Y Deguchi, Y Shida, K Hashimoto. *Vibrio alginolyticus*, a tetrodotoxin-producing bacterium, in the intestines of the fish *Fugu vermicularis vermicularis*. Mar Biol 94:625–630, 1987.
85. H Sugita, T Noguchi, M Furuta, T Harada, O Murata, K Hashimoto, Y Deguchi. The intestinal microflora of cultured specimens of a puffer *Fugu rubripes rubripes*. Nippon Suisan Gakkaishi 54: 733, 1988.
86. S Sato, K Komaru, T Ogata, M Kodama. Occurrence of tetrodotoxin in cultured puffer. Nippon Suisan Gakkaishi 56:1129–1131, 1990.
87. K Koyama, T Noguchi, A Uzu, K Hashimoto. Resistibility of toxic and nontoxic crabs against paralytic shellfish poison and tetrodotoxin. Bull Jpn Soc Sci Fish 49:481–484, 1983.
88. T Saito, T Noguchi, T Harada, O Murata, T Abe, K Hashimoto. Resistibility of toxic and non-toxic pufferfish against tetrodotoxin. Bull Jpn Soc Sci Fish 48:1371, 1985.
89. M Kodama, T Ogata, K Kawamukai, Y Oshima, T Yasumoto. Toxicities of muscle and other organs of five species of puffer collected from the Pacific coast of Tohoku area of Japan. Bull Jpn Soc Sci Fish 50:703–706, 1984.
90. M Kodama, T Ogata, S Sato. External secretion of tetrodotoxin from puffer fishes stimulated by electric shock. Mar Biol 87:199–202, 1985.
91. M Kodama, S Sato, T Ogata, Y Suzuki, T Kaneko, K Aida. Tetrodotoxin secreting glands in the skin of puffer fishes. Toxicon 24:819–829, 1986.
92. Y Shimizu, M Kobayashi. Apparent lack of tetrodotoxin biosynthesis in captured *Taricha torosa* and *Taricha granulosa*. Chem Pharm Bull 31:3625–3631, 1983.

93. S Watabe, Y Sato, M Nakaya, N Nogawa, K Oohashi, T Noguchi, N Morikawa, K Hashimoto. Distribution of tritiated tetrodotoxin administered introperitoneally to pufferfish. Toxicon 25:1283–1289, 1987.
94. I Tani: 1945. Nihonsan fugu no chudokugakuteki kenkyu (Toxicological Studies on Japanese Puffers) (in Japanse). Tokyo: Teikoku Tosho, 1944, pp. 103.
95. S Laobhripart, K Limpakarnjanarat, K Saitanu, O Sangwonloy, B Anuchatvorakul, S Leelasitorn, K Saitanu. Food poisoning due to consumption of the freshwater puffer *Tetraodon fangi* in Thailand. Toxicon 28:1372–1375, 1990.
96. M Kodama, H Shimizu, S Sato, T Ogata, K Terao. The infection of bacteria in the liver cells of toxic puffer—a possible cause for organisms to be made toxic by tetrodotoxin in association with bacteria. In: P Lassus, G Arzul, EE-L Denn, P Gentien, and CM-L Baut, eds. Harmful Marine Algal Blooms. Paris: Londres, 1995, pp 457–462.
97. M Katayama, S Fujita. Ecological studies on the puffer, *Fugu niphobles* (Jordan et Snyder) II. Bulletin of the Faculty of Education, Yamaguchi University 15:77–84, 1965.
98. U Simidu, K Kita Tsukamoto, T Yasumoto, M Yotsu. Taxonomy of four marine bacterial strains that produce tetrodotoxin. Int J Syst Bacteriol 40:331–336, 1990.
99. GJ Doucette, M Kodama, S Franca, S Gallacher. Bacterial interactions with harmful algal bloom species: Bloom ecology, toxigenesis, and cytology. In: DM Anderson, AD Cembella, GM Hallegraeff, eds. Physiological Ecology of Harmful Algal Blooms. Berlin: Springer, 1998, pp 619–647.
100. S Sato, T Ogata, M Kodama. Wide distribution of toxins with sodium channel blocking activity similar to tetrodotoxin and paralytic shellfish toxins in marine animals. In: TJ Smayda, Y Shimizu, eds. Toxic Phytoplankton Blooms in the Sea. New York: Elsevier, 1993, pp 429–434.
101. Y Ueno. Studies on the ecology and resource of maturing chum salmon off the Pacific coast of northern Honshu. Bull Natl Inst Far Seas Fish 30:80–206, 1993.
102. Y Oshima, T Yasumoto, M Kodama, T Ogata, Y Fukuyo, F Matsuura. Features of paralytic shellfish poison occurring in Tohoku district. Bull Jpn Soc Sci Fish 48:525–530, 1982.
103. M Kodama, T Ogata, Y Takahashi, T Niwa, F Matsuura, Gonyautoxin associated with RNA-containing fraction in the toxic scallop digestive gland. J Biochem 92:105–109, 1982.
104. M Kodama, T Noguchi, J Maruyama, T Ogata, K Hashimoto. Release of tetrodotoxin and paralytic shellfish poison from puffer liver by RNase. J Biochem 93:243–247, 1983.

7
Chemistry and Mechanism of Action

Yuzuru Shimizu
University of Rhode Island, Kingston, Rhode Island

I. INTRODUCTION

This group of toxins was first recognized as being paralytic toxins found in shellfish (1). However, these toxins are known to occur in many other organisms of which some are not related to PSP. In this chapter, all chemically related naturally occurring compounds with the characteristic tricyclic skeleton will be covered and discussed.

The PSP toxins are found in different phyla of both prokaryotic and eukaryotic organisms (2). The etiology and biology of PSP and the causative organisms are discussed by other authors in this book (Chapters 6–10). Therefore, this chapter is devoted only to the chemistry, biochemistry, and chemical aspects of structure-activity relationship of the toxins.

The first toxin obtained in pure form was saxitoxin (STX), which was isolated from the Alaska butter clam, *Saxidomus giganteus*. (3). In retrospect, the pattern of saxitoxin occurrence in *S. giganteus* is exceptional rather than general. The shellfish stores a large amount of toxins in its siphons (4). Saxitoxin constitutes the major toxic component in the siphons together with only a small amount of neosaxitoxin (neoSTX) (5). The toxins seems to exist bound to the siphon tissues and remain there for a length of time. The binding mechanism of the toxin is not well understood. In other shellfish, however, the toxins are normally concentrated in the hepatopancreas through the food chain and depurated fairly fast once the supply route is removed. In most cases, STX is not the sole or major toxin, but gonyautoxins (GTXs), neosaxitoxin, and other sulfated toxins are dominant toxins (5).

II. ISOLATION AND PURIFICATION OF PSP TOXINS

Saxitoxin can be purified fairly easily from Alaska butter clams. The isolation procedure is to take advantage of the strongly basic nature of guanidinium groups in the molecule, which bind to a weakly acidic carboxylate ion-exchange resin and recovered by stepwise elution with acetate buffer solution (3). The isolation scheme is summarized in Figure 1. This procedure, however, is not applicable to most other PSP-causing toxins with sulfate groups, such as gonyautoxins, B-toxins, and C-toxins. Owing to their amphoteric nature, these toxins do not bind tightly to the carboxylate resins as STX. Fortunately, PSP toxins were found to bind rather exclusively to small matrix size gel-filtration resins such as Bio-Gel P-2 (Bio-Rad, Richmond, CA) or Sepha-

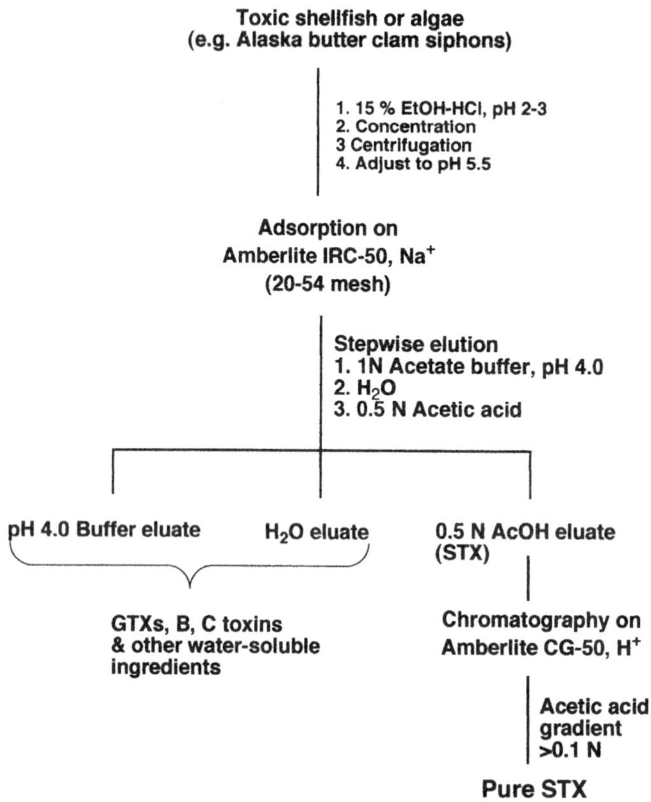

Figure 1 Isolation scheme for saxitoxin (STX) from toxic shellfish and algal cells. (Modified from ref. 3.)

dex G-10 (Amersham Pharmacia, Uppsala) at controlled pH and can be eluted with dilute acetic acid (5–9). The exact mechanism of this rather selective adsorption is not known, but this process is very effective to remove most of other water-soluble small molecules in the extract. Fractionation of the toxin mixture to individual pure toxins can be accomplished by careful and repeated chromatography on the H^+ form of carboxylate resins such as Bio-Rex 70 (Bio-Rad, Richmond, CA), using diluted acetic acid as eluant. The standard scheme for the purification of PSP toxins is shown in Figure 2 (10). The toxins with net charges of zero or negative numbers have to be separated by careful chromatography on Bio-Gel P-2 or preparative thin-layer chromatography (TLC) on silica gel plates (11). The purity of the toxins can be checked by TLC (12) and high-performance liquid chromatography (HPLC) (13). For TLC analysis, the use of high-performance plates such as Whatman HPK (Whatman, Maidstone, UK) is highly recommended, because they give fast separation and undiffused spots. The visualization of the TLC is done by spraying the plate with 1% H_2O_2 and heating. The PSP toxins appear as whitish blue spots or yellowish green spots for N1-hydroxylated toxins (neoSTX type) under ultraviolet (UV) light (254 nm). Caution has to be taken because some of the non-PSP–type compounds also give similar fluorescent spots. Aphanorphine, which coexists with PSP toxins in the blue-green alga *Aphanizomenon flos-aquae* gives an R_f value similar to those of PSP toxins and a fluorescent spot on H_2O_2 spraying (14). Similarly the HPLC analysis, which is also known to give false fluorescent compounds in the postcolumn reaction system (15,16).

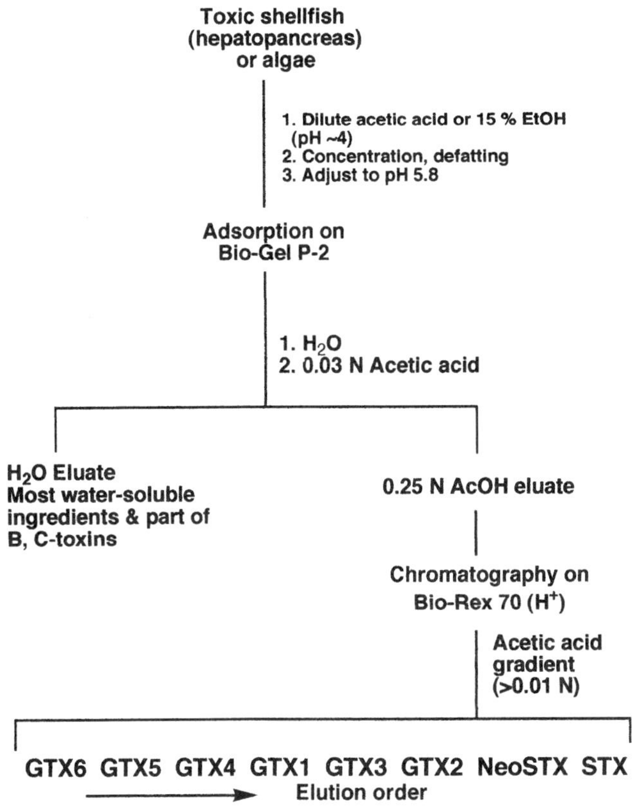

Figure 2 General isolation procedure for paralytic shellfish toxins (5,10).

Pure toxins in solid form can be obtained by concentration and final lyophilization of the dilute acetic acid solution. In this process, acetate usually becomes the major counter ion of the guanidinium groups, but other inorganic ions are also mixed. In the case of STX, the uniform HCl salt can be obtained by addition of a small amount of hydrochloric acid and lyophilization. However, neoSTX and sulfated toxins are not so stable to hydrochloric acid and tend to decompose on concentration. Another aspect of the toxin instability is the epimerization at C11 chiral center of 11-O-sulfated toxins. The epimerization, which can take place at pH $\sim$ 6 or higher, is greatly accelerated by a trace amount of base, as discussed later. Thus, it is rather difficult to store the solid forms of the compounds such as GTX-1–4 in pure form, since they usually reach equilibrium mixtures of the α- and β-epimers.

III. STRUCTURES OF PSP TOXINS

The earlier works on the structural elucidation of PSP toxins have been reviewed by a number of investigators including myself (5,12,17,18). Therefore, only a brief summary of the structural information is presented here.

The structure of STX was established first and gave the basic framework to elucidate other PSP structures. The chemistry of STX was studied extensively by Rapoport and his coworkers

(19). The highly nitrogenous structure was not easy to determine. The classic degradation studies gave fragments including aromatic purine and pyrimidine derivatives. The oxidation also gave guanidine and its derivatives as the final products. The unique tricyclic ring system was finally determined by x-ray crystallography of the *p*-bromobenzene sulfate (20). Subsequently, gonyautoxin-I–VI (GTX-1–6), neosaxitoxin (neoSTX), and B- and C-toxins were isolated from toxic shellfish and dinoflagellates and their structures determined (21–28). As often happens when several groups work independently, some of the PSP toxins were given duplicate names. For example, the B1-toxin described by Hall et al. (11) is identical with GTX-5 reported by Shimizu's group (5). The structures and names of all naturally occurring toxins including synonyms are shown in Figure 3.

The structures can be divided into two major categories: The STX series and the neoSTX series. The difference between the two groups is the presence or absence of N1-hydroxyl group, which has profound effects on the stability and ionic character of the toxins. Another structural variation come from the presence or absence of O-sulfate at C11 and N-sulfate groups on carbamoyl groups, which also greatly change the physical and chemical characters as well as pharmacological characters. Added to the combinations are the decarbamoyl compounds (29,30). In recent years, Oshima and other research groups have discovered a number of unconventional toxins which carry an acetyl group instead of a carbamoyl group and 12-deoxydecarbamoyl and 12-acetoxy compounds (31,32). Also, his group found compounds whose 12-keto (hydrated) group is reduced to alcohol (32).

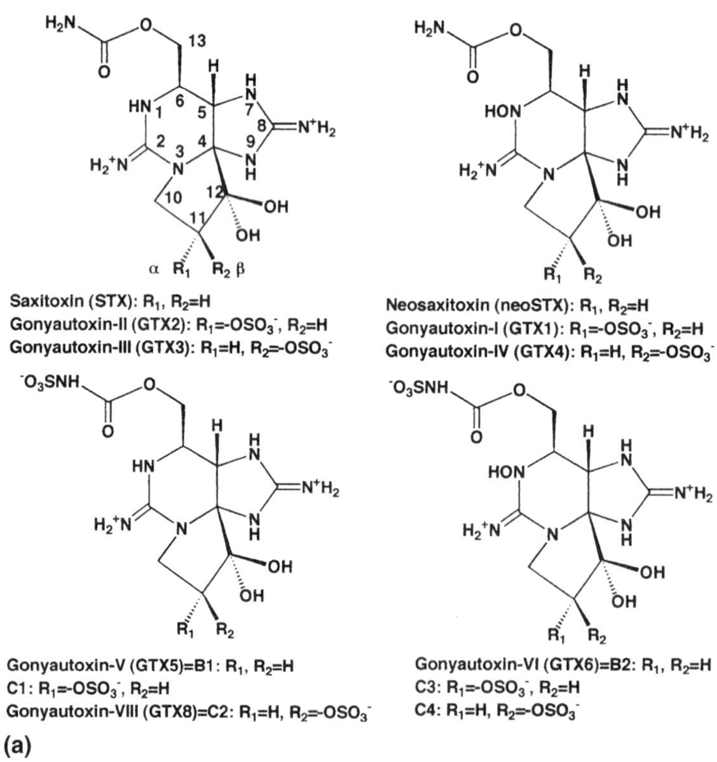

Figure 3 (a–c) Structures of naturally occurring PSP toxin derivatives: their synonyms and commonly used abbreviations.

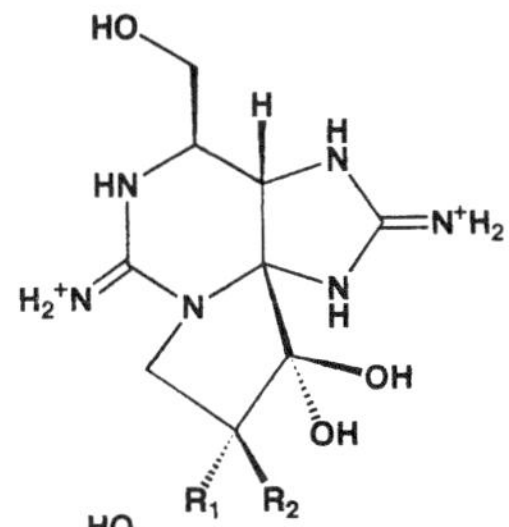

Decarbamoylsaxitoxin (dcSTX): R_1, R_2=H
Decarbamoylgonyautoxin-II (dcGTX2): R_1=$-OSO_3^-$, R_2=H
Decarbamoylgonyautoxin-III (dcGTX3): R_1=H, R_2=$-OSO_3^-$

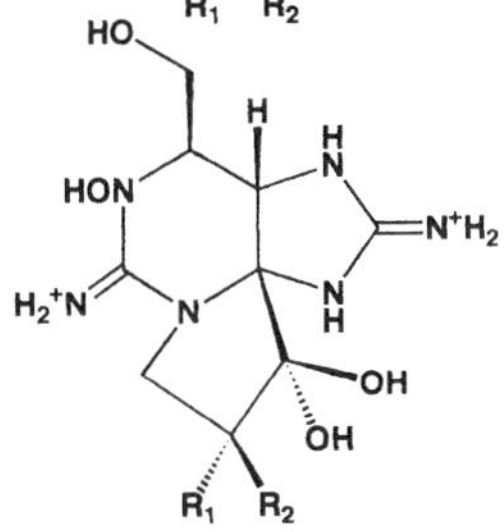

Decarbamoylneosaxitoxin (dcNeoSTX): R_1, R_2=H
Decarbamoylgonyautoxin-I (dcGTX1): R_1=$-OSO_3^-$, R_2=H
Decarbamoylgonyautoxin-IV (GTX4): R_1=H, R_2=$-OSO_3^-$

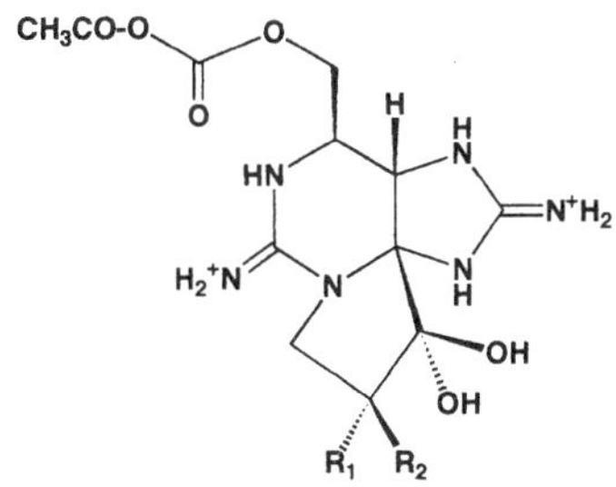

13-*O*-acetyldecarbamoylsaxitoxin (13-O-dcSTX): R_1, R_2=H
13-*O*-acetyldecarbamoylgonyautoxin-II (13-O-dcGTX2): R_1=$-OSO_3^-$, R_2=H
13-*O*-acetyldecarbamoylgonyautoxin-III (13-O-dcGTX3): R_1=H, R_2=$-OSO_3^-$

(b)

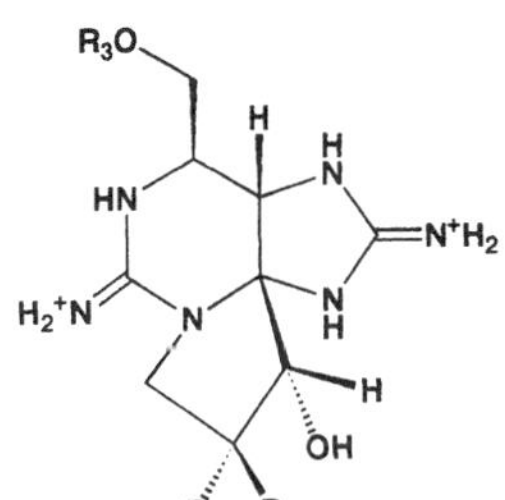

Decarbamoyl-12β-deoxysaxitoxin : R_1, R_2=H, R_3=H
(=Decarbamoyl-12α-saxitoxinol or dihydrosaxitoxin)
13-*O*-acetyldecarbamoyl-12β-deoxysaxitoxin : R_1, R_2=H
13-*O*-acetyldecarbamoyl-12β-deoxygonyautoxin-III: R_1=H, R_2=$-OSO_3^-$

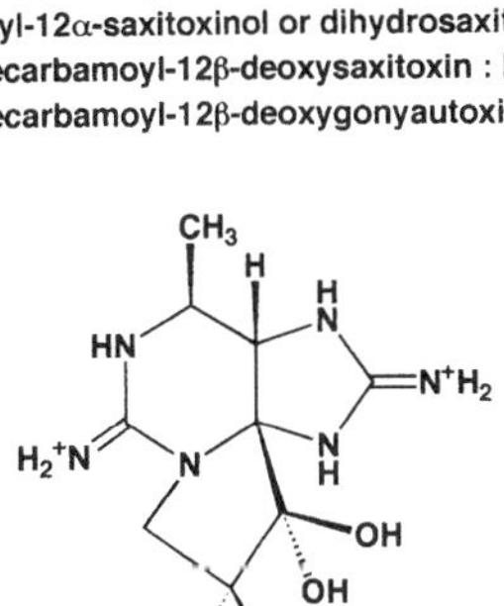

Decarbamoyloxysaxitoxin (doSTX): R_1, R_2=H
Decarbamoyloxygonyautoxin-II (doGTX2): R_1=$-OSO_3^-$, R_2=H
Decarbamoyloxygonyautoxin-III (doGTX3) (doGTX3): R_1=H, R_2=$-OSO_3^-$

(c)

IV. CHEMICAL AND PHYSICAL PROPERTIES OF PSP TOXINS

Generally, PSP toxins exist only in ionized form and are highly hygroscopic. Only a few of them are known as crystalline salts. They are saxitoxin *p*-bromobenzene sulfonate salt (20), saxitoxin hemi-ethylacetal hydrochloride (33), and C4-toxin (34), all of which were used for x-ray crystallography. The heat stability of the toxins is a very important issue with respect to food safety. Generally, PSP toxins are considered to be heat stable, and simple cooking can not eliminate the toxicity. However, the stability varies greatly depending on the structures and pH. Saxitoxin is extremely stable even at a high temperature at a low pH. The hydrochloride solution can be stored without loss of its potency for decades. However, at above pH 8, it degrades quickly even at ambient temperature, which suggests STX may not survive in seawater too long unless it is stabilized by complexation with other substances. On the other hand, 11-O-sulfate toxins undergo decomposition at elevated temperature even at an acidic pH; probably due to decomposition triggered by the solvolysis of the O-sulfate group. N1-hydroxy compounds such as neoSTX are also more labile to acid and heat than STX. The carbamoyl-N-sulfated toxins such as B-toxins and C-toxins are easily hydrolyzed with dilute mineral acids to yield desulfated toxins, which are 100–1000 times more toxic than the original compounds (11,26,27).

The basic structures of PSP toxins are composed of a 3,4-propinoperhydropurine tricyclic system. The oxidation of the ring system with H_2O_2 or other oxidants cleaves the propino ring to give fluorescent aromatic aminopurine derivatives (12,34) (Figure 4). The reaction is the basis for the detection and quantitative analysis of PSP toxins. The compounds from the STX series toxins have UV maxima at 252 and 335 nm and emission maxima at ~390 nm (17). The neoSTX series toxins seem to undergo a similar transformation, but the structures of the degradation products have not been confirmed. All PSP toxins have two guanidinium groups. STX has two pKa's. The pKa of one of the guanidine groups is somewhat anomalous and its assignment was a subject of discussion. Nuclear magnetic resonance (NMR) studies (36,37) showed that the pyrimidine guanidine has the normal pKa of ~11, but the imidazol guanidine has a very low pKa (Figure 5), suggesting that the three nitrogen atoms in the five-membered ring are not in full hybridization owing to a ring strain and possibly suppression by the other ionized guanidinium group. In fact, the x-ray crystallographic data indicate that the N8–C9 bond length is longer than the other two bonds (20), which makes the group a substituted amidine rather than a normal guanidine. Another important structural feature of PSP toxins is the presence of a hydrated ketone at C12. The hydrated form is preferential because of the presence of

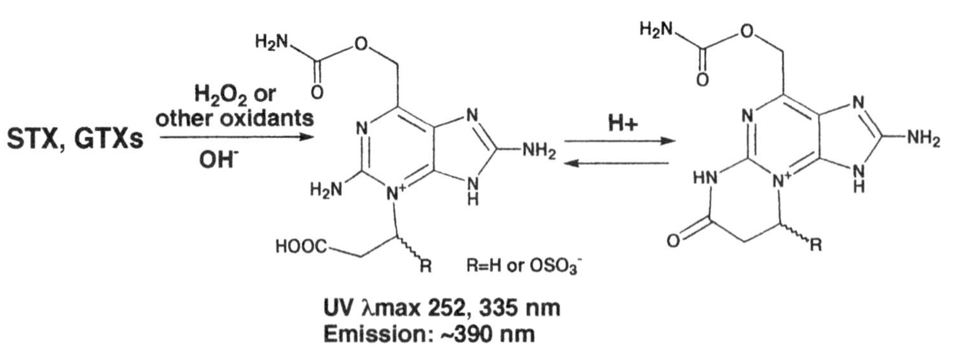

Figure 4 Oxidative degradation of PSP toxins to fluorescent purine derivatives.

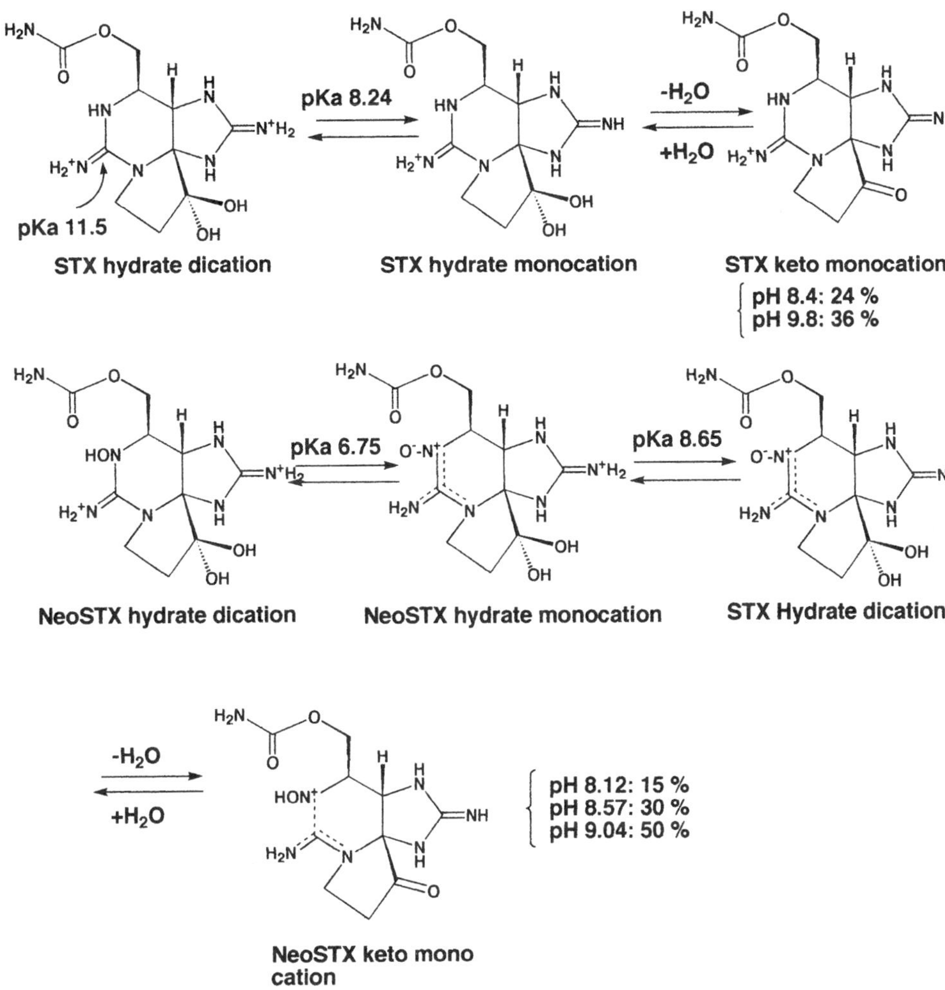

Figure 5 Molecular species of saxitoxin (STX) and neosaxitoxin (neoSTX) in equilibrium around the physiological pH and their compositions.

two electron-withdrawing guanidinium groups on the neighboring carbon, C4. The hydration alleviates the double-charge situation of the two neighboring carbons. In fact, once the imidazol guanidine is deprotonated, the keto form emerges as the dominant species (35–37) (Figure 5).

NeoSTX also has similar pKa's and an additional pKa 6.75 (17,24). This was interpreted as a result of the dissociation of the N1-OH group. Thus, after the dissociation of N1-OH, the molecule is stabilized as a 1,2-dipolar form or ylide like N-oxide. The ratios of molecular species at different pH were calculated on the basis of NMR data (36). As stated later, the dynamics of the involved molecular species is important to interpret the data on toxin binding to the receptors.

The net charge of the toxins, which is also an important factor to consider in separation strategies and the toxin binding mode, varies by pH and structures. The net charge of STX is between +2 and +1 in the pH range of 7 to 9. On the other hand, neoSTX net charge is between

+2 and 0 in the same pH range. In the sulfate-bearing toxins, the net charges are between +1 and −2 depending on the number of sulfate groups and the presence or absence of the N1-OH group (34,37). Also, the additional electron-withdrawing group at C11 is expected to increase the hydrated form of the ketone.

Although the 12-keto group of PSP toxins is always depicted as the hydrate form, it undergoes keto-enol equilibration. The keto component and enolization increase greatly at pH 9 or higher. However, it should be noted that even at pH as low as ~6, slow enolization takes place. This is the reaction used to prepare tritiated STX. It should be noted that it is a reversible process; thus the tritium-labeling can be washed out by exchange with ambient protons, especially at high pH. With the 11-O-sulfate toxins, the enolization results in the epimerization at C11, which is greatly accelerated at high pH or by the presence of bases such as sodium acetate. Because of this facile epimerization, 11-substituted toxins such as GTX-1–4 usually reach equilibrium mixtures of α- and β-isomers, where the α-isomers are dominant with a ratio of approximately 3:1 (17) (Figure 6).

The 11-O-sulfate group in GTX-2 and GTX-3 and the N-hydroxyl group in neoSTX derivatives can be removed reductively with zinc HCl or AcOH (17). The same treatment also reduces the N1-hydroxyl group in neoSTX and GTX-1 and GTX-4. Similar conversions also take place in vivo (39). Ichihara et al. looked at this in vivo conversion and discovered that sulfhydryl

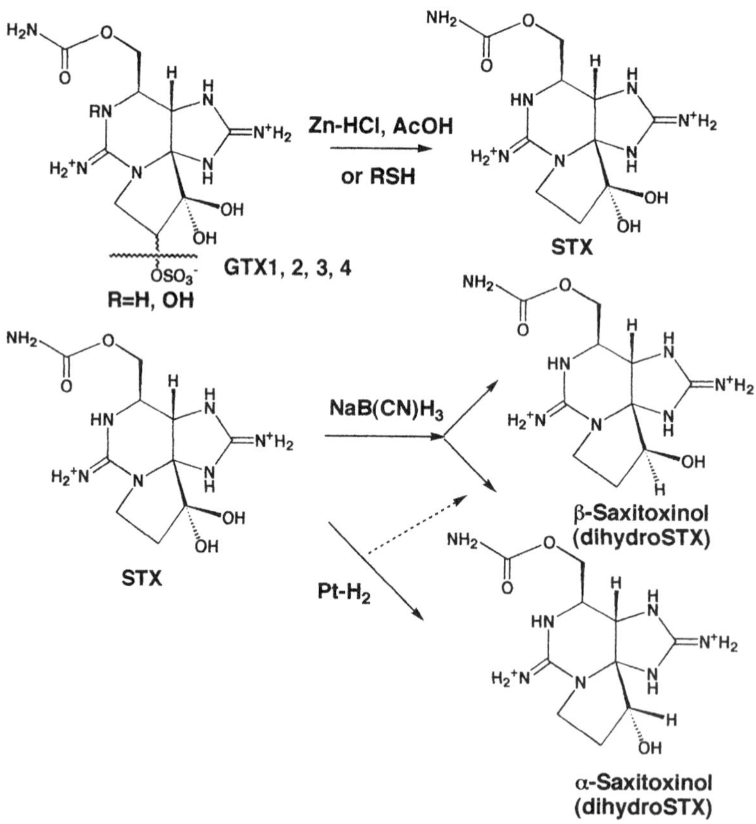

Figure 6 Facile proton exchange and epimerization at C11 by enolization of the keto-form of PSP toxins around the physiological pH.

groups can reductively remove the O-sulfate group without enzyme (40,41). The interesting reductive cleavage was further confirmed by other groups (42,43).

Reduction of STX with $NaB(CN)H_3$ gives a mixture of α- and β-isomers of 12-alcohol: saxitoxinol (or dihydrosaxitoxin, 12-deoxysaxitoxin), which can be separated by careful chromatography (36). Catalytic hydrogenation with Pt-H_2, on the other hand, gives predominantly the α-isomer (Figure 7). Interestingly, the α-ol derivatives occur naturally in the blue-green algae (32) (Figure 7).

Prolonged heating of STX in 7.5N HCl hydrolyzes the carbamoyl group to give decarbamoylsaxitoxin (dcSTX) (44). Similar drastic treatment of sulfated GTXs or neoSTX destroys the molecules. Decarbamoyl toxins have been also discovered in natural samples (28,29).

The N-sulfate groups can be hydrolyzed easily in dilute mineral acids even at ambient temperature (25,26). Interestingly, Oshima's group discovered that heating the N-sulfo toxin in near neutral solutions gives rise to decarbamoyl compounds (43). The reaction, which is clearly assisted by a neighboring group participation, is remarkable in view of the great difficulty associated with the hydrolysis of carbamoyl groups. These reactions are summarized in Figure 8.

V. BIOSYNTHESIS OF PSP TOXINS

As soon as the structure of STX was elucidated, the biosynthetic origin of the unique molecule attracted enormous interest. The conventional idea was that the tricyclic system was formed by

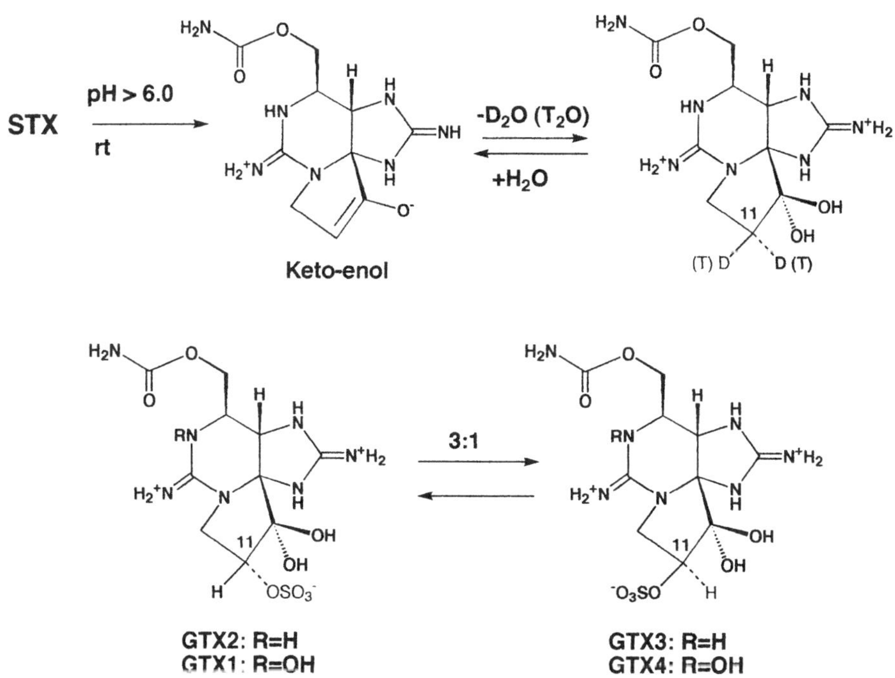

Figure 7 Reduction of PSP toxins: hydrogenolysis of 11-O-sulfate and formation of dihydrosaxitoxins (saxitoxinols).

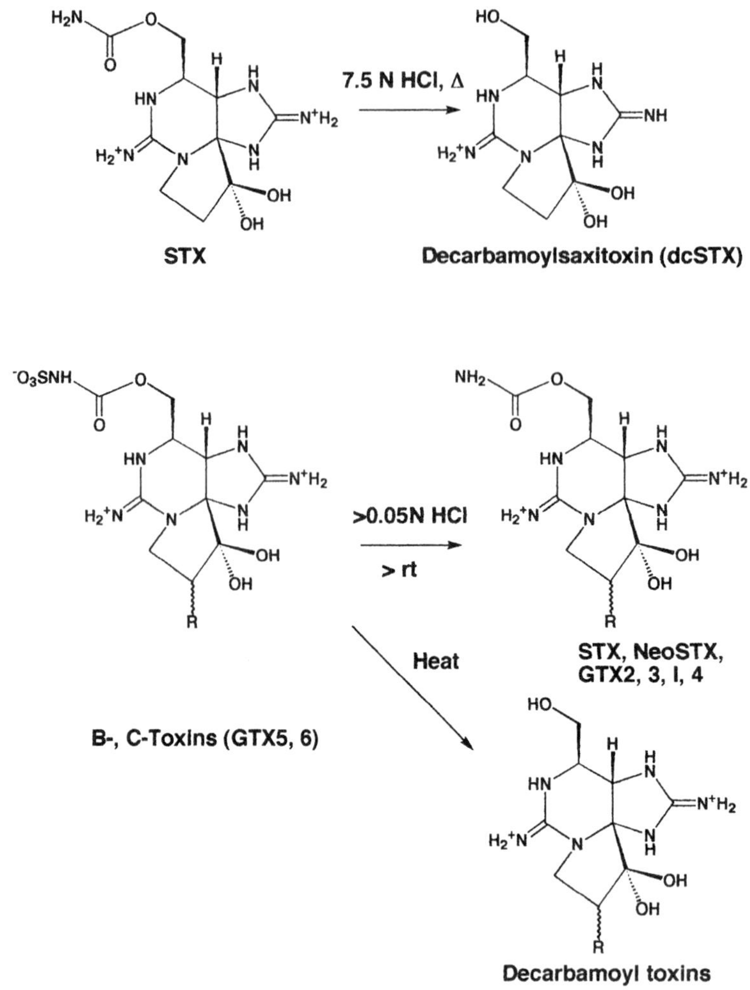

Figure 8 Acid hydrolysis of saxitoxin (STX), gonyautoxins (GTXs), and B- and C-type toxins.

addition of a C3 unit such as acrylic acid to a purine derivative (40). In fact, acrylic acid and its precursors are abundant compounds in microalgae. Feeding experiments, however, led to a totally unexpected pathway (45–48).

First, it was found out that guanido-^{14}C-arginine could be incorporated into guanidium groups and carbamate of GTX2 (40). However, 1-^{13}C-arginine was not incorporated into the molecule. The crucial information for the origin of the skeleton came from the feeding experiment using 2-^{15}N-2-^{13}C-ornithine. The immediate precursor of arginine was found to be incorporated, the connectivity between ^{15}N-^{13}C was found intact in the neoSTX molecule as proved by ^{14}N-^{13}C spin-spin coupling seen in NMR (Figure 9) (45). As said before, 1-^{13}C-arginine was not incorporated. Instead, acetate was incorporated at C5 and C6 as a unit. Therefore, it was postulated that the first step in the biosynthesis is the Claisen-type of condensation of acetyl CoA on the α-carbon of arginine followed by decarboxylation, amidine group transfer, and double cyclization. The C13 side chain carbon is derived from the C1 pool, with S-methyladenosine (SAM) as the direct precursor. The feeding of methyl-^{13}C-methionine-methyl-d_3 resulted in the

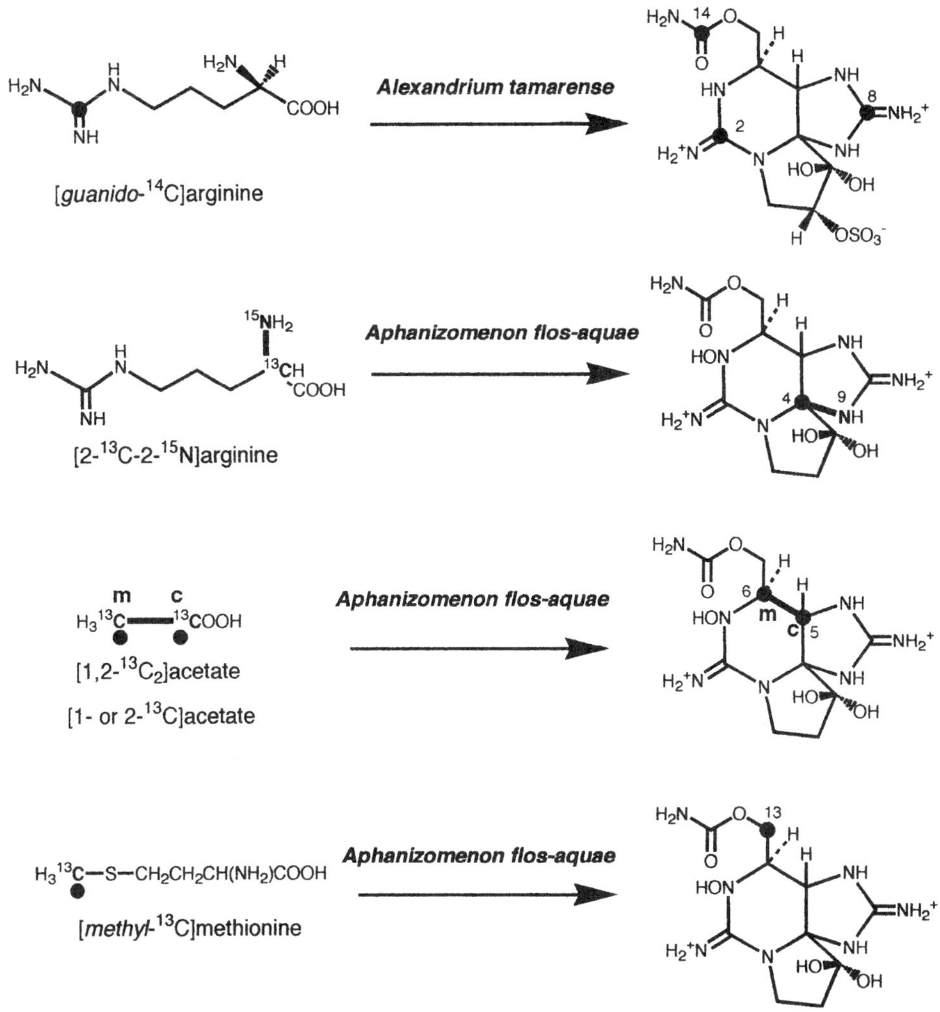

Figure 9 Incorporation patterns of some key labeled precursors into paralytic shellfish toxins.

methyl transfer with a loss of two deuterium atoms. It was also concluded by feeding $1\text{-}^{13}C$-acetate-d_3 that there is a migration of a hydrogen atom from C6 to the angular C5.

Based on these findings, the entire biosynthetic pathway was proposed as shown in Figure 10. The pathway involves several interesting features. First, the Claisen-type condensation on the α-carbon of an amino acid is rarely seen, but the key precursor in porphyrine biosynthesis, aminolevulinate, is known to be formed by the condensation of succinate on glycine. Another noticeable aspect of PSP toxin biosynthesis is a heavy involvement of arginine in the biosynthesis.

VI. BIOCONVERSION OF PSP

Although the exact biosynthetic sequence of a number of PSP toxins in the producing organism is not known, there is some information with respect to the conversion of the toxins in vivo.

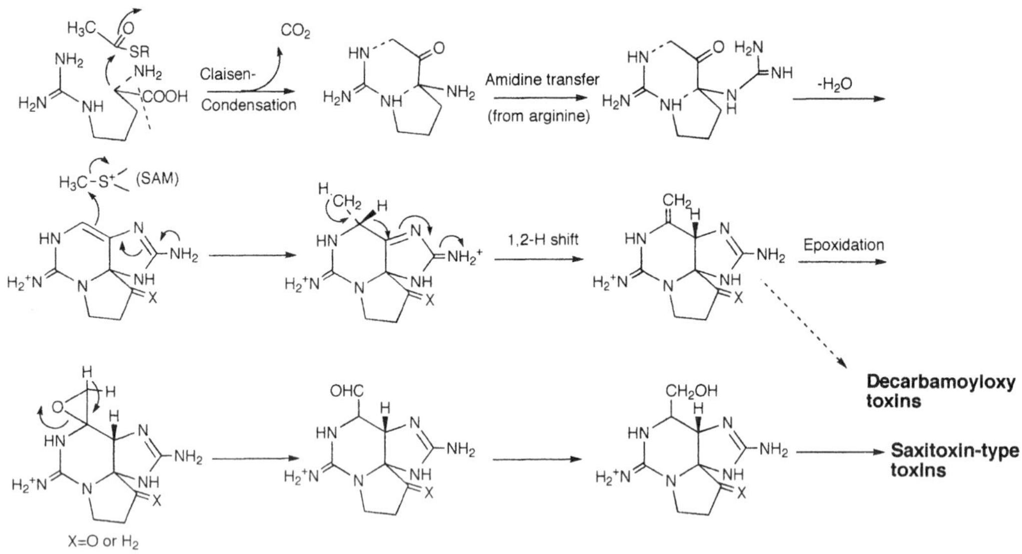

Figure 10 The biosynthetic pathway of paralytic shellfish toxins proposed by Shimizu et al (48).

Yoshioka and Shimizu first demonstrated that, in scallop tissue homogenates, GTXs and neoSTX could be converted to saxitoxin (39). In their experiment, both O-sulfate and N-hydroxyl were reduced to give saxitoxin as the final products (Figure 11). Kotaki et al. demonstrated that similar conversions could take place microbiologically (49). Sullivan et al. reported that both carbamoyl and carbamoyl-N-sulfated toxins could be converted to the decarbamoyl compounds in the littleneck clam, *Prototheca staminea* (29). Although all these reports seem to point at saxitoxin as the final product in the metabolism, Oshima et al, studying the enzymatic and nonenzymatic conversion of PSP toxins in shellfish and the causative dinoflagellates, demonstrated that the homogenate of *Alexandrium tamarense* can oxidize GTX-2 and GTX-3 to GTX-1 and GTX-4 (43). They also showed that *Gymnodinium catenatum* homogenate can introduce a sulfate group on a carbamoyl group to produce C1 and C2 toxin. These enzymatic capabilities seem to reflect the toxin profiles found in these organisms (Figure 12).

VII. TOTAL SYNTHESIS OF SAXITOXIN

At present, saxitoxin is the sole toxin that has been totally synthesized. The conversion of STX to other PSPs except decarbamoylSTX has not been achieved. The first synthesis of saxitoxin was accomplished by Kishi's group at Harvard (50). The other synthesis by Jacobi's group, which was reported some years after Kishi's accomplishment, took a quite different route, which claimed to be biomimetic (51). The key steps of both syntheses are shown in Figures 13 and 14.

As the amount in the STX depository is dwindling, the procurement of STX is becoming an important issue. However, either of the syntheses, which take more than 10 steps not counting the synthesis of the starting material, is not easily done as a routine practice. Isolation from contaminated shellfish or a cultured organism is still the easiest and the most economical method for obtaining PSP toxin.

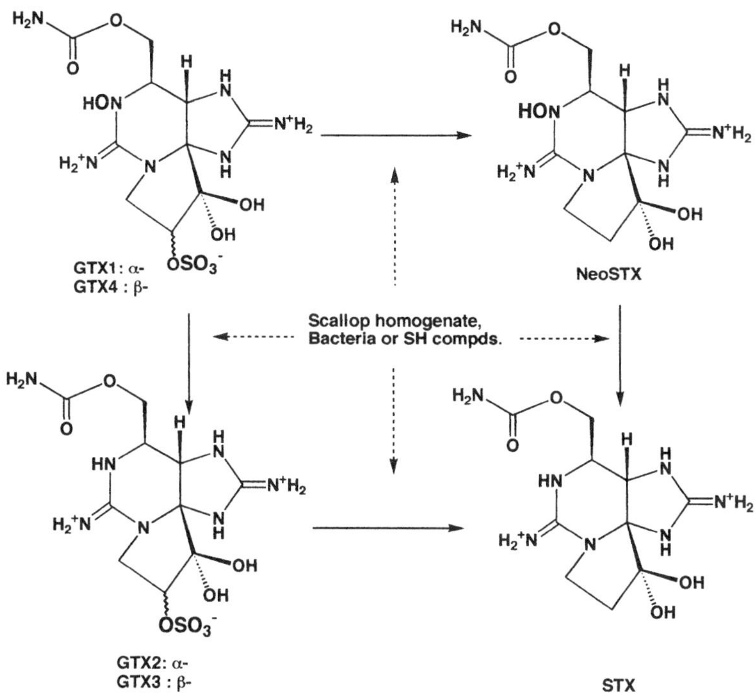

Figure 11 Reductive bioconversions of PSP toxins by shellfish and bacteria.

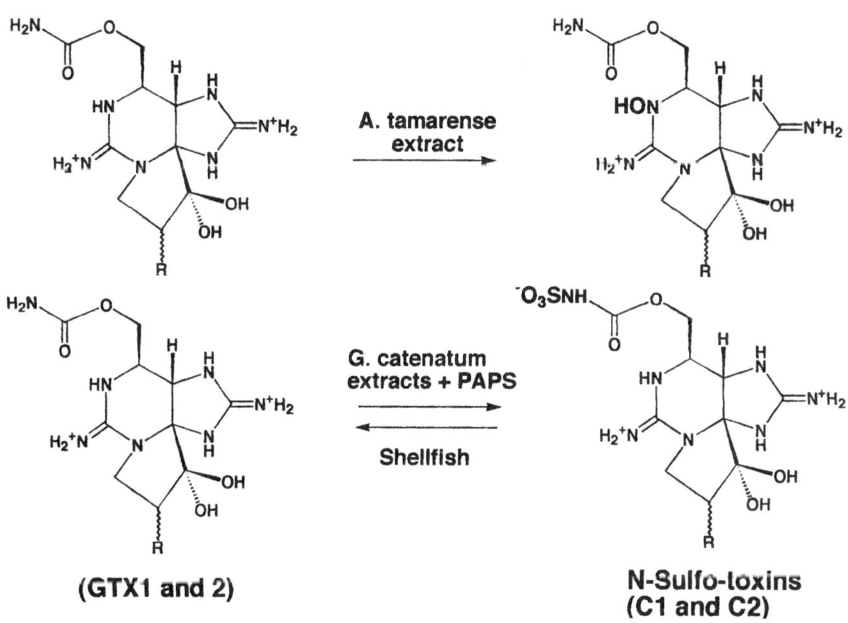

Figure 12 Enzymatic conversions of PSP toxins with dinoflagellate cell extracts.

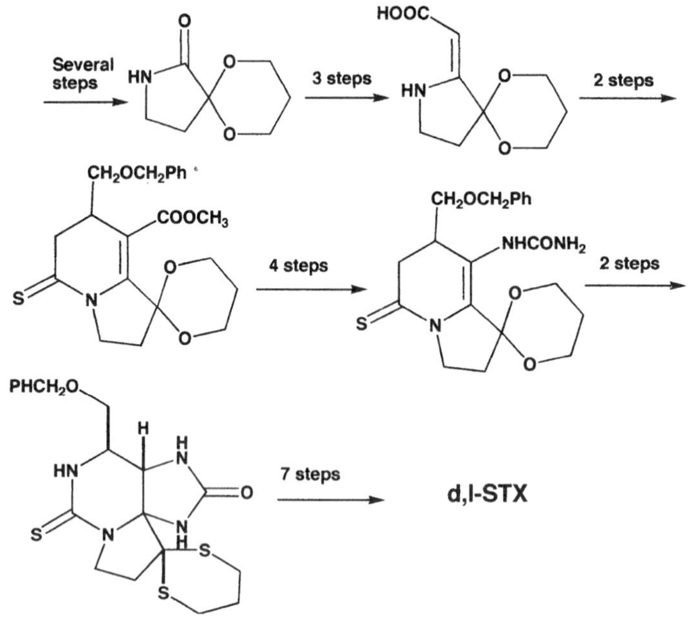

Figure 13 The major steps in Kishi's total synthesis of saxitoxin (STX).

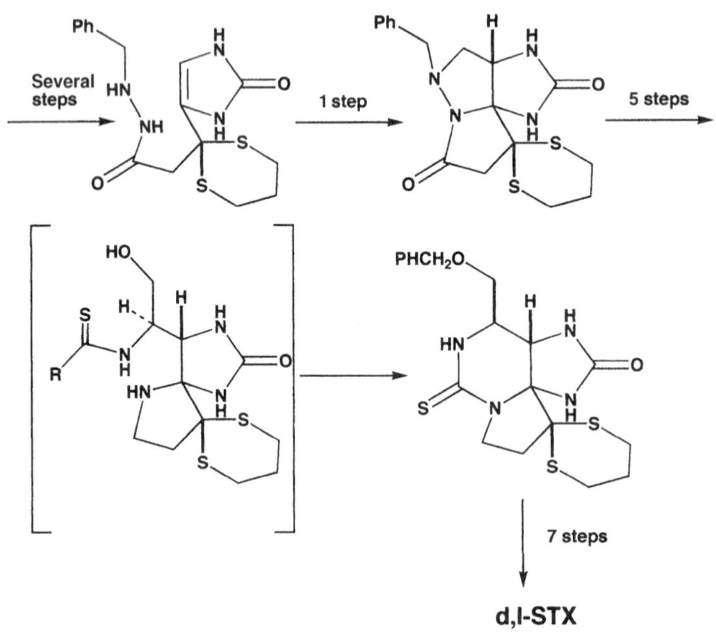

Figure 14 The major steps in Jacobi's synthesis of saxitoxin (STX).

VIII. TOXICITY AND CHEMICAL STRUCTURES

PSP toxins are neurotoxic. Manifested symptoms in paralytic shellfish poisonings include a tinkling sensation of the lips, numbness of extremities, gastrointestinal problems, and difficulty in breathing (52). The onset of the symptoms of PSP is fast, and death usually occurs within 12 hs. Lavage at the early stage of the poisoning is effective to remove the toxins in the stomach. Once absorbed, however, there is no confirmed antidote for the intoxication, and treatments are limited to symptomatic ones. Since respiratory paralysis is usually the direct cause of death, respiratory aids such as the use of an iron lung is an effective method to save lives. Once patients recover, the prognosis is good and there are no long-term ill effects sometimes seen in other neurotoxin poisonings such as intoxication by ciguatoxin.

The PSP toxins cause paralysis of neuromuscular systems. The mechanism of action is the blockage of sodium channels in excitable membranes, which results in stoppage of the propagation of neuronal impulses. The similarity of action of STX to that of another prominent marine toxin, tetrodotoxin (TTX), has been recognized from early days. Thus, the two toxins have been studied side by side and discussed together (35,53). Their high selectivity to bind to a specific site of the sodium channel made them an invaluable tool to study the function of the channel and related neuromuscular systems. Both PSP toxins and TTX are specific to the so-called site 1, one of the several known toxin-binding sites on Na^+ channels (55). In 1975, based on the newly revealed structures, Hille presented the first binding model, which depicted the toxin molecule penetrating rather deep inside of the channel and plugging the channel, making an ion pair with an anionic site which is supposed to be located near the bottom of the channel (Figure 15) (55). As the structures of gonyautoxins and other new PSP toxins were starting to be unfolded, however, it become evident that the model, as it was, could not explain the lack of anticipated steric interactions of the new toxin molecules. For example, GTX-2 and GTX-3 showed the same order of activity as STX in the voltage clamp or binding assays even though they possess O-sulfate groups on C11. In the Hille model, such bulky, charged groups in both α- and β-orientation are expected to prevent the penetration of the molecules or have at least a drastic effect on the toxin binding. All alternate models proposed the binding site on the extracellular membrane or cavity (Figure 16) (56–58). Here one controversial issue was about the nature of the bond between the channel and the toxins. Since the structure of STX was shown to have a hydrated ketone, there has been a persistent speculation that it may form an easy covalent bond with a nucleophilic group such as —OH, —NH_2, or —SH on the channel protein. The covalent bond model was an attractive one, because it could explain the tight binding of the toxins. Moreover, studies on the pH dependency of the activity showed that, in a lower pH range, in which the ketal formation is prevalent, an enhancement of activity was seen

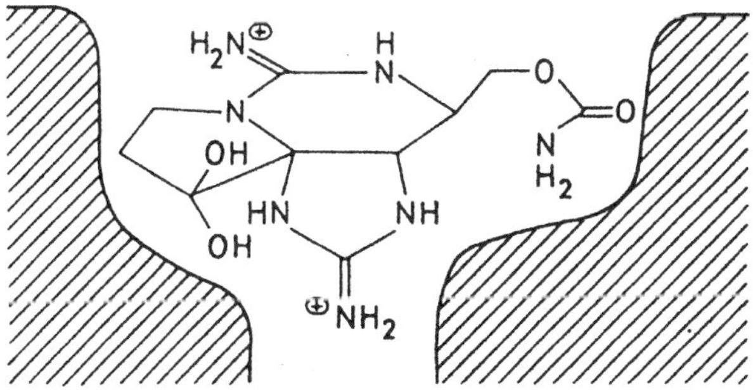

Figure 15 A plugging model of the binding of Na^+ channel with saxitoxin (55).

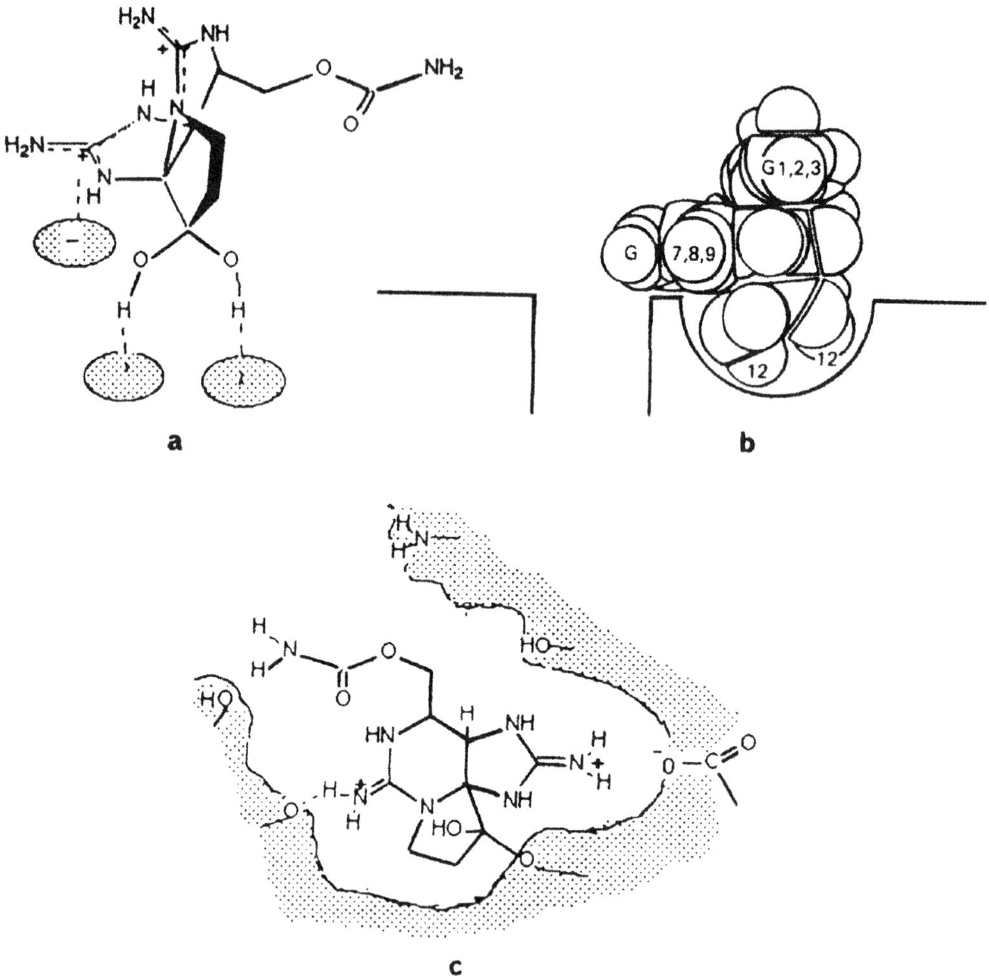

Figure 16 Alternative models proposed for saxitoxin-Na^+ channel binding: (a) by Shimizu (57), (b) Kao and Walker (56), and (c) Strichartz (58).

(56). This pH dependency, however, can be also explained by assuming that the hydrogen bond formation with either or both hydrate OH groups is crucial.

One problem with the covalent bond model is that it is difficult to find a group similar to ketal in TTX, which acts very much like STX. The chemical characteristics of the ortho-ester in TTX is quite different from those of ketal and unlikely to form a covalent bond in an easily reversible manner. On the other hand, comparison of the structure of the protonated, ortho-ester form of TTX with that of hydrated STX shows that both molecules share remarkably close arrangements of two hydroxyl groups and guanidinium moiety with STX. Thus, Kao and Walker proposed a model which placed the toxin molecules on the outside edge of the channel with the guanidinium group on the top of the channel entrance (54). Independently, Shimizu suggested a three-point binding model with two hydrogen bonds with the ketal OHs and ion pairing of the guanidinium group with an anionic site on the outside surface of the membrane (56).

There are other facts which favor the hydrogen bond models. First, dihydroSTXs (=saxitoxinols or 11-deoxySTXs), especially β-isomer, which can no longer form the covalent bond, still retain activity, although at a diminished level (58,59). In other words, the ability to form a covalent bond is not an exclusive requirement for the binding. More precise information regarding the toxin-binding mode has come from the molecular biological studies of the sodium channel.

The first cloning of the sodium channel was done by Numa's group (60) in which the channel was isolated from the electric eel. Subsequently, channels from other sources have been also cloned and sequenced. Generally, they have four repeating units (I–IV), and each repeating unit consists of six transmembrane α-helical segments (S1–S6) (61) (Figure 17). The precise arrangement of these segments and their roles in selective ion filtration and the gating mechanism have been discussed and reviewed by several investigators (54,62). In summary, the extracellular loop sections of S1–S2 (P-loops) are considered to be the PSP toxin–binding site. The functional channel can be expressed in the oocyte of the *Xenopsis* frog by introducing cRNA. Noda et al., using the point-mutation technique, replaced glutamate 387 (Glu387) with glutamine, and showed that the resulting channel was no longer inhibited by STX or TTX (63). The result shows that the dissociable glutamate carboxylic acid is essential for the effective inhibition. Since then, several other amino acids in the outside segments have been point mutated and found to be important for the toxin binding (64–67). They are, in addition to Glu387, Asp384, Glu942, and Glu945. Replacement of these acidic amino acids with the corresponding neutral amide derivatives resulted in almost complete loss of activity. These findings seem to indicate that the imidazol guanidinium group is not counterioned by a single carboxylate but probably by two or three groups. The two hydrate hydroxyl groups probably form cyclic hydrogen bonds with a carboxylic acid of Glu945 (758) in a manner similar to the dimeric form of carboxylic acid. The fifth carboxylate (Asp1717) interacts with the pyrimidine guanidinium group, which is absent in TTX. The toxin is trapped in a pocket surrounded by five carboxyl groups (Figure 18). Lipkind and Fozzard suggests that this pocket may be the Na^+ channel itself through which hydrated sodium ions penetrate (67).

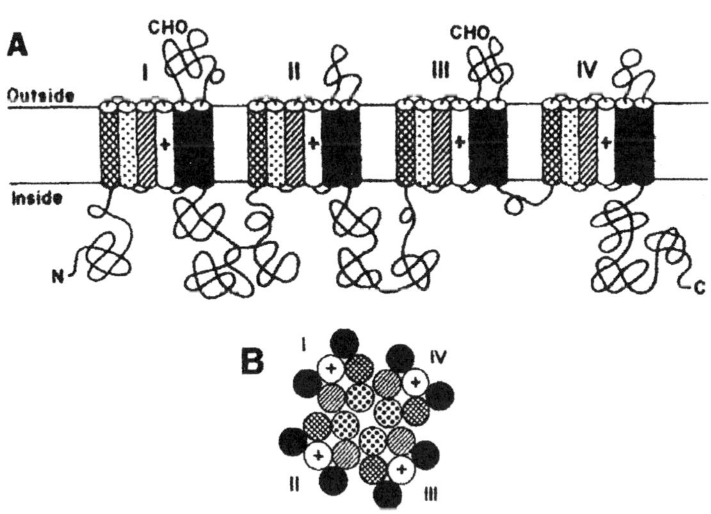

Figure 17 The topology of Na^+ channel proposed by Noda et al (61). The Roman numerals show the repeating transmembrane segments: (A) a side view and (B) an overview.

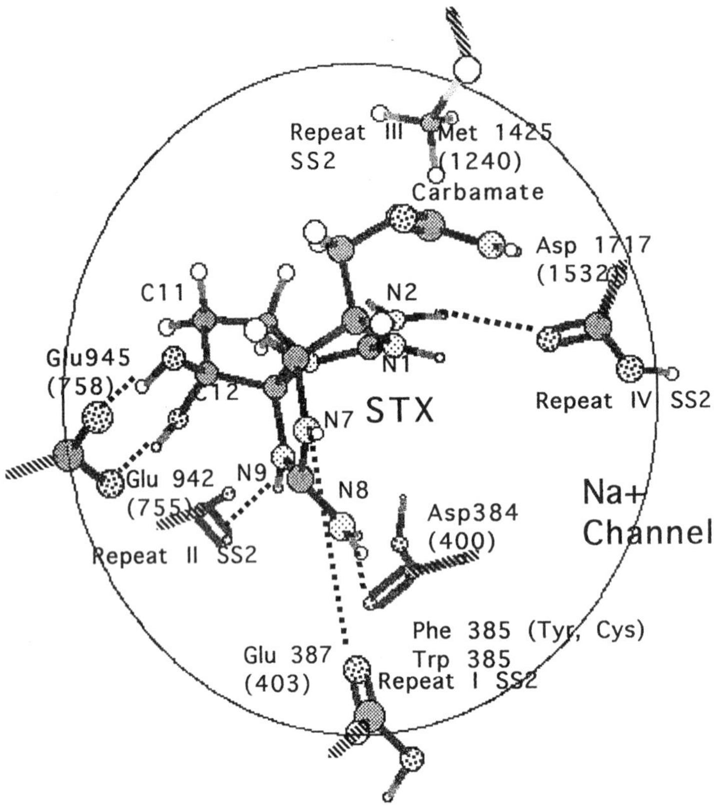

Figure 18 Possible interactions of the energy-minimized structure of saxitoxin molecule and the Na^+ channel amino acids. The positions of the amino acids are plotted according to the coordinates described by Lipkind and Fozzard (67).

IX. STRUCTURAL DIFFERENCES AND CHANNEL SUBTYPES (ISOFORMS)

An often asked question is whether all PSP toxins effect the same pharmacological action. Although all PSP- and TTX-type toxins are considered to have the same receptor site, it has been noticed that there are some significant differences in affinity depending on the structures of the toxins and also channel isoforms (35). Sodium channels (actually other ion channels too) share the majority of conserved amino acid sequences. However, there are some subtle but important differences in the channel structures in the isoforms. For example, phenylalanine 385 in the SS2 portion of the rat brain channel (type II) is replaced with tyrosine and cysteine in the sodium channels from skeletal and cardiac muscles, respectively. These differences may lead to different clinical manifestations depending on the toxin structures.

First, TTX lacks the second guanidinium group, which should void the ionic interaction with Asp1717. In fact, the point mutation of Asp1717 to Asn1717 almost wiped out the STX activity, but the activity of TTX diminished only by two orders. On the other hand, the replacement of Tyr401 (or Phe385) with Cys (in cardiac isoform) reduces TTX activity twice more than STX activity. It may indicate the importance of hydrophobic interaction in the initial approach of the toxins, and it may not be surprising that TTX, with lack of an ionic group at the side, is more affected by this. The repeat III also has a carboxylic acid moiety (Asp1426) at the corresponding location, but it does not seem to have a direct interaction with the toxin, but rather the neighboring Met1425 is recognized to have a nonbonded interaction with STX. The nature of this interaction is not known, but it is interesting that the replacement of methionine with cationic lysine was reported to have resulted in the loss of activity.

As stated earlier, the presence of both α- and β-O-sulfate groups at C11 (e.g., GTX2 and GTX3) seem to have little influence on the toxicity, suggesting C11 is completely unobstructed in α and β directions. The presented picture does not contradict this notion. One important question is the possible effect of the N1-hydroxyl group, which is found in almost half of PSP toxins.

The mouse toxicity of neoSTX is comparable to that of STX. However, it was noted that, in the frog sciatic nerve, the potency of neoSTX is almost 4.5 times that of STX, and the dissociation constant of neoSTX is about one fourth or one fifth that of STX (35,58). The interpretation of this finding may need further investigation. Interestingly, the potency of neoSTX is more dependent on pH than that of STX and is sharply enhanced at a low pH range, where the N-OH is expected to be in protonated form (58,68).

Another puzzling finding which needs further investigation is the apparent lack of mouse toxicity with 13-deoxysaxitoxin (13-doSTX) (69,70). Since the toxicity of 13-hydroxyl compounds is comparable to that of STX (e.g., dcSTX's mouse toxicity: 70% of STX's [44]), it may suggest that some kind of interaction (e.g., hydrogen bond on 13-O) may be needed in the transition state.

The carbamoyl-N-sulfated toxins (e.g., GTX5, GTX6, B- and C-toxins) have drastically diminished toxicity (26,27). It was suggested earlier that there must be a repulsive interaction with an anionic site on the membrane which prevents the settling of the toxin molecule (68). Asp1717 fits one such anionic site in the model.

REFERENCES

1. H Sommer, KF Meyer. Paralytic shellfish poisoning. Arch Pathol 24:560–598, 1937.
2. Y Shimizu. Microalgal metabolites: a new perspective. Annu Rev Microbiol 50:431–465, 1996.
3. EJ Schantz, JD Mold, DW Stanger, J Shavel, FJ Riel, JP Bowden, JM Lynch, RS Wylcr, BR Riegel, H Sommer. Paralytic shellfish poison VI. A procedure for the isolation and purification of the poison from toxic clams and mussel tissues. J Am Chem Soc 79:5230–5235, 1957.
4. EJ Schantz, HW Magnusson. Observations on the origin of the paralytic poison in Alaska butter clam. J Protozool 11:239–242, 1964.
5. Y Oshima, LJ Buckley, M Alam, Y Shimizu. Heterogeneity of paralytic shellfish poisons. Three new toxins from cultured *Gonyaulax tamarensis* cells, *Mya arenaria*, and *Saxidomus giganteus*. Comp Biochem Physiol 57c:31–34, 1977.
6. LJ Buckley, M Ikawa, JJ Sasner, Jr. Isolation *of Gonyaulax tamarensis* toxins from softshell clams (Mya arenaria) and a thin-layer chromatographic-fluorometric method for their detection. J Agric Food Chem 24:107–111, 1976.
7. Y Shimizu, M Alam, Y Oshima, WE Fallon. Presence of four toxins in red tide infested clams and cultured *Gonyaulax tamarensis* cells. Biochem Biophys Res Commun 66:731–737, 1975.
8. Y Shimizu, WE Fallon, JC Wekell, K Gerber, Jr, EJ Gauglitz Jr. Analysis of toxic mussels (*Mytilus* sp.) from the Alaskan Inside Passage. J Agric Food Chem 26:878–881, 1978.
9. CP Hsu, A Marchand, Y Shimizu, GG Sims. Paralytic shellfish toxins in sea scallops, *Placopecten magellanicus* in the Bay of Fundy. J Fish Res Board Can 36:32–36, 1979.
10. Y Shimizu. Bioactive marine natural products, with emphasis on handling of water-soluble compounds. J Nat Prod 48:223–235, 1985.

11. S Hall, PB Reichardt, RA Neve. Toxins extracted from an Alaskan isolate of *Protogonyaulax* sp. Biochem Biophys Res Commun 97:649–653, 1980.
12. Y Shimizu. Dinoflagellate toxins. In: PJ Scheuer, ed. Marine Natural Products: Chemical and Biological Perspectives. Vol 1. New York: Academic Press, 1979, pp 1–42.
13. Y Oshima. Post-column derivatization HPLC methods for paralytic shellfish poisons. In: GM Hallegraeff, DM Anderson, AD Cambella, HO Enevoldsen, eds. Manual on Harmful Marine Microalgae. Paris: UNESCO, 1995, pp 81–94.
14. N Gulavita, A Hori, Y Shimizu. Aphanorphine, a novel tricyclic alkaloid from the blue-green alga *Aphanizomenon flos-aqua*. Tetrahedron Lett 29:4381–4384, 1988.
15. H Onodera, Y Oshima, MF Watanabe, M Watanabe, CJ Block, S Blackburn, Y Yasumoto. Screening of paralytic shellfish toxins in freshwater cyanobacteria and chemical confirmation of toxins in cultured *Anabaena circinalis* from Australia. In: T Yasumoto, Y Oshima, Y Fukuyo, eds. Harmful & Toxic Algal Blooms. Paris: UNESCO, 1996, pp 563–566.
16. S Sato, Y Shimizu. Purification of a fluorescent product from the bacterium, *Moraxella*: a neosaxitoxin impostor. In: B Reguera, J Blanco, ML Fernandez, T Wyatt, eds. Paris: UNESCO, 1988, pp 465–467.
17. Y Shimizu. Paralytic shellfish poisons. In: W Herz, H Griesbach, GW Kirby, eds. Progress in the Chemistry of Organic Natural Products. New York: Springer-Verlag, 1984, pp 235–264.
18. Y Shimizu. The chemistry of paralytic shellfish toxins. In: AT Tu, ed. Handbook of Natural Toxins. Vol 3. New York: Marcel Dekker, 1988, pp 63–85.
19. JL Wong, R Oestelin, H Rapoport. The structure of saxitoxin. J Am Chem Soc 93:7344–7355, 1971.
20. EJ Schantz, VE Ghazarossian, HK Schnoes, FM Strong, JP Springer, JO Pezzanite, J Clardy. The structure of saxitoxin. J Am Chem Soc 97:1238–1239, 1975.
21. Y Shimizu, LJ Buckley, M Alam, Y Oshima, WE Fallon, H Kasai, I Miura, VP Gullo, K Nakanishi. Structure of gonyautoxin-II from the East Coast toxic dinoflagellate, *Gonyaulax tamarensis*. J Am Chem Soc 98:5414–5416, 1976.
22. CF Wichmann, GL Boyer, CL Divan, EJ Schantz, HK Schnoes. Neurotoxins of *Gonyaulax excavata* and Bay of Fundy scallps. Tetrahedron Lett 22:1941–1944, 1981.
23. CF Wichmann, WP Niemczura, HK Schnoes, S Hall, PB Reichardt, SD Darling. Structures of two novel toxins from *Protogonyaulax*. J Am Chem Soc 103:6977–6978, 1981.
24. Y Shimizu, CP Hsu, WE Fallon, Y Oshima, I Miura, K Nakanishi. The structure of neosaxitoxin. J Am Chem Soc 100:6791–6793, 1978.
25. Y Shimizu, CP Hsu. Confirmation of the structures of gonyautoxins-II–V by correlation with saxitoxin. J Chem Chem Commun 314–315, 1981.
26. FE Koehn, S Hall, CF Wichmann, HK Schnoes, PB Reichardt. Dinoflagellate neurotoxins related to saxitoxin: structure and latent activity of toxins B1 and B2. Tetrahedron Lett 23:2247–2248, 1982.
27. M Kobayashi, Y Shimizu. Gonyautoxin-VIII, a cryptic precursor of paralytic shellfish poisons. J Chem Soc Chem Commun 827–828, 1981.
28. T Harada, Y Oshima, T Yasumoto. Structures of two paralytic shellfish toxins, gonyautoxins-V and -VI, isolated from a tropical dinoflagellate, *Pyrodinium bahamense* var. *compressa*. Agric Biol Chem 46:1861–1864, 1982.
29. JJ Sullivan, WT Iwaoka, J Liston. Enzymatic transformation of PSP toxins in the littleneck (*Protothaca staminea*). Biochem Biophys Res Commun 114:465–472, 1983.
30. T Harada, Y Oshima, T Yasumoto. Natural occurrence of decarbamoylsaxitoxin in tropical dinoflagellate and bivalves. Agric Biol Chem 47:191–193, 1983.
31. Y Oshima, H Itakura, KC Lee, T Yasumoto, S Blackburn, G Hallegraeff. Toxin production by the dinoflagellate *Gymnodinium catenatum*. In: TJ Smayda, Y Shimizu, eds. Toxic Phytoplankton Blooms in the Sea. New York: Elsevier, 1993, pp 907–912.
32. H Onodera, Y Oshima, WW Carmichael. New saxitoxin analogs from the freshwater filamentous cyanobacterlum *Lyngbya wollei*. Nat Toxins 5:146–151, 1997.
33. J Bordner, WE Thiessen, HA Bates, H Rapoport. The structure of a crystalline derivative of saxitoxin. The structure of saxitoxin. J Am Chem Soc 97:6008–6012, 1975.
34. HA Bates, H Rapoport. A chemical assay for saxitoxin, the paralytic shellfish poison. J Agric Food Chem 23:237–239, 1975.

35. S Hall, G Strichartz, E Moczydlowski, A Ravindran, PB Reicharct. The saxitoxins. sources, chemistry and molecular pharmacology. In: S Hall, G Strichartz, eds. Marine Toxins. Washington, DC: American Chemical Society, 1990, pp 29–77.
36. Y Shimizu, CP Hsu, A Genenah. Structure of saxitoxin in solution and stereochemistry of dihydroxaxitoxin. J Am Chem Soc 103:605–609, 1981.
37. RS Rogers, H Rapoport. The pKa's of saxitoxin. J Am Chem Soc 102:7355–7399, 1980.
38. Y Shimizu, NK Gulavita. Recent progress in the chemistry and toxicology of paralytic shellfish poisoning. In: AE Pohland, VR Dowell Jr, JL Richard, eds. Microbial Toxins in Foods & Feeds, Cellular and Molecular Modes of Action. New York: Plenum Press, 1990, pp 605–610.
39. Y Shimizu, M Yoshioka. Transformation of paralytic shellfish toxins as demonstrated in scalop homogenates. Science 212:546–549, 1981.
40. Y Shimizu, M Kobayashi, A Genenah, N Ichihara. Biosynthesis of paralytic shellfish poisoning. In: EP Ragelis, ed. Seafood Toxins. Washington, DC: American Chemical Society, 1984, pp 151–160.
41. N Ichihara, M Kobayashi, CP Hsu, Y Shimizu, Bio-conversion of Paralytic shellfish toxins. Abstracts of Joint Meeting of the American Society of Pharmacognosy and the Society for Economic Botany, Boston, 1981, p 81.
42. M Asakawa, M Takagi, A Iida, M Oishi. Studies on the conversion of Paralytic shellfish poison (PSP) components by biochemical reducing agents. Eiseikagaaku 33:50–55.
43. Y Oshima, Chemical and enzymatic transformation of paralytic shellfish toxins in marine organisms. In: P Lassus, G Arzul, E Erard, P Gentien, C Marcaillou, eds. Harmful Marine Algal Blooms. Paris: Intercept, 1995, pp 475–480.
44. VE Ghazarossian, EJ Schantz, HK Schnoes, FM Strong. A biologically active acid hydrolysis product of saxitoxin. Biochem Biophys Res Commun 68:776–780, 1976.
45. Y Shimizu, M Norte, A Hori, A Gnenah, M Kobayashi. Biosynthesis of saxitoxin analogues: The unexpected pathway. J Am Chem Soc 106:6433–6434, 1984.
46. Y Shimizu, S Gupta, M Norte, A Hori, A Genenah, M Kobayashi. Biosynthesis of paralytic shellfish toxins. In: DM Anderson, AW White, DG Baden, eds. Toxic Dinoflagellates. New York: Elsevier, 1985, pp 271–274.
47. Y Shimizu, S Gupta, HN Chou. Biosynthesis of red tide toxins. In: S Hall, G Strichartz, eds. Marine Toxins: Origin, Structure, and Molecular Pharmacology. Washington, DC: American Chemical Society, 1990, pp 21–28.
48. Y Shimizu, S Gupta, AVK Prasad. Biosynthesis of dinoflagellate toxins. In: E Graneli, B Sundstrom, L Edler, DM Anderson, eds. Toxic Marine Phytoplankton. New York: Elsevier, 1990, pp 62–73.
49. Y Kotaki, Y Oshima, T Yasumoto. Analysis of paralytic shellfish toxins in marine snails. Bull Jpn Soc Sci Fish 47:943–946, 1981.
50. H Tanino, T Nakata, T Kaneko, Y Kishi. A stereospecific total synthesis of d,l-saxitoxin. J Am Chem Soc 99:2818–2819, 1977.
51. PA Jacobi, MJ Martinelli, S Polanc. Total synthesis of (±) saxitoxin. J Am Chem Soc 106:5594–5598, 1984.
52. BW Halsetead. Poisonous and Venomous Marine Animals of the World. Princeton: Darwin, 1978.
53. CY Kao. New perspectives on the interaction of tetrodotoxin and saxitoxin with excitable membranes. Toxicon 21(Suppl. 3):211–219, 1983.
54. WA Catterall. Cellular and molecular biology of voltage-gated sodium channels. Physiol Rev 72: S15–S48, 1992.
55. B Hille. The receptor for tetrodotoxin and saxitoxin: a structural hypothesis. Biophys J 15:615–619, 1975.
56. CY Kao, SE Walker. Active groups of saxitoxin and tetrodotoxin as deduced from action of saxitoxin analogs on frog muscle and squid axon. J Physiol (Lond.) 323:619–637, 1982.
57. Y Shimizu. Recent progress in marine toxin research. Pure Appl Chem 54:1973–1980.
58. G Strichartz. Structural determinants of the affinity of saxitoxin for neuronal sodium channels. J Gen Physiol 84:281–305, 1984.
59. CY Kao, PN Kao, MR James-Kracke, FE Koehn, CF Wichmann, HK Schnoes. Actions of epimers of 12-(OH)-reduced saxitoxin and 11-(OSO_3)-saxitoxin on squid axon. Toxicon 23:647–655, 1985.

60. M Noda, S Shimizu, T Tanabe, T Takai, T kayano, T Ikeda, M Takahashi, H Nakayama, Y Kanaoka, N Minamino, K Kangawa, J Matsuo, MA Raftery, T Hirose, S Inayama, H Hayashida, T Miyata, S Numa. Primary structure of *Electrophorus electiricus* sodium channel deduced from cDNA sequence. Nature 312:121–127, 1984.
61. S Numa, M Noda. Molecular structure of sodium channels. Ann NY Acad Sci 479:338–355, 1986.
62. HR Guy, F Conti. Pursuing the structure and function of voltage-gated channels. Trends Neurol Sci 13:201–206, 1990.
63. M Noda, S Suzuki, S Numa, W Stuhmer. A single point mutation confers tetrodotoxin and saxitoxin insensitivity on the sodium channel-II. FEBS Lett 259:213–216, 1989.
64. HS Terlau, H Heinemann, W. Stuhmer, M Pusch, F Conti, K Imoto, S Numa. Mapping the site of block by tetrodotoxin and saxitoxin on sodium channel-II. FEBS Lett 293:93–96, 1991.
65. JJ Satin, W Kyle, M Chen, P Bell, LL Cribbs, HA Fozzard, RB Rogart. A mutant of TTX-resistant cardiac sodium channels with TTX-sensitive properties. Science 256:1202–1205, 1992.
66. KJ Kontis, AL Goldin. Site-directed mutagenesis of the putative pore region of the rat IIA sodium channel. Mol Pharmacol 43:635–644, 1933.
67. GM Lipkind, HA Fozzard. A structural model of the tetrodotoxin and saxitoxin binding site of the Na^+ channel. Biophys J 66:1–13, 1994.
68. EA Moczylowski, S Hall SS Garber, GS Strichartz, C Miller. Voltage-dependent blockade of muscle Na^+ channels by guanidinium toxins. Effects of toxin charge. J Gen Physiol 84:687–704, 1984.
69. Y Oshima, K Sugino, H Itakura, M Hirota, T Yasumoto. Comparative studies on paralytic shellfish toxin profile of dinoflagellates and bivalves. In: E Graneli, B Sundstrom, L Edler, DM Anderson, eds. Toxic Marine Phytoplankton. New York: Elsevier, 1990, pp 391–396.
70. LC Yang, Y Kao, Y Oshima. Actions of decarbamoyloxysaxitoxin and decarbamoylneosaxitoxin on the frog skeletal muscle fiber. Toxicon 30:645–652, 1992.
71. Y Shimizu, M Kobayashi, A Genenah, Y Oshima. Isolation of side-chain sulfated saxitoxin analogs-their significance in interpretation of the mechanism of action. Tetrahedron 40:539–544, 1984.

8
Chemical Analysis of PSP Toxins

Bernd Luckas
University of Jena
Jena, Germany

I. INTRODUCTION

Paralytic shellfish poisoning (PSP) toxins are potential neurotoxins which specifically block the excitation current in nerve and muscle cells and this finally results in signs of paralysis. Consequently, the development of analytical methods for the determination of poisoning caused by PSP toxins was an important task. The mouse bioassay unambiguously gives evidence of the toxic potential of a sample, since the application of higher toxin concentrations yields a shortage of the time of challenge to death of the laboratory animals. However, these biological tests only reveal the total PSP toxicity of a sample expressed in MU (mouse units)/kg or PSP/kg. A specification of the toxins is impossible and the mouse bioassay is not suitable for studies on the formation and effect of individual PSP toxins (e.g., variations in the toxin profiles cannot be monitored by mouse bioassay). There is also an ethical problem associated with mouse bioassay because of the killing of large numbers of animals (1).

II. SAMPLE PREPARATION

Independent of the type of the determination method, PSP toxins have to be quantitatively extracted from the sample materials. The standard method using 0.1N hydrochloric acid as extraction solvent was suggested by the Association of Official Analytical Chemists (AOAC). The sample and the extraction solvent are heated for 5 min at 100°C. Hereby the N-sulfocarbamoyl toxins are converted into the respective carbamate toxins and B1-, B2-, and C1–C4–toxins are no longer present in the extracts (2). However, these extracts are suited both for the mouse bioassay and other methods for PSP determination (Figure 1). The extraction with 0.1N hydrochloric acid is useful in a monitoring program, because the aim is the detection of toxicity in food, but it may be an inconvenience if the objective is to study the toxin profile in the food sample. However, the natural toxin pattern could be observed by extracting samples with acetic acid (3).

Fresh sample material (1 g) was mixed with acetic acid (0.03N, 3 mL) using an Ultra-Turrax. The mixture was centrifuged for 10 min (2980 g) and passed through a 0.45-μm nylon filter. A second sample (1 g) was extracted with hydrochloric acid (0.2N, 3 mL); mixing, centrifugation, and filtration were executed as described above.

100 g shellfish meat

(1) Homogenization with a blender
(2) Add 100 mL 0.1 N HCl, and adjust pH to 2-4 (preferably 3)
(3) Boil for 5 min.
(4) Let cool and adjust pH to 2-4 (preferably 3)
(5) Dilute to 200 mL
(6) Filtrate or centrifuge to obtain a clean solution

Bioassay (1 mL of the solution intraperitoneally)

(7) Ultrafiltration (0.45 μm or molecular weight cut off filter)

HPLC injection

Figure 1 Sample preparation according to AOAC.

The extracts obtained from both procedures were injected separately into the high-performance liquid chromatography (HPLC) system. After separate injection of the different extracts into the HPLC system, determination of N-sulfocarbamoyl toxins is possible by calculating the peak height increases for the carbamoyl toxins formed (4).

III. SPECTROPHOTOMETRIC ASSAY

In 1975, a fluorometric method was recommended for PSP determination in samples in addition to the mouse bioassay (5). PSP toxins which exhibit neither ultraviolet (UV) absorption nor fluorescence were oxidized in alkaline solution to fluorescent pyrimidino purins (6). After acidification, the intensity of the fluorescence of the oxidation products was measured in the solution (Figure 2). Individual PSP toxins differ both in toxicity and fluorescence intensity after oxidation (7). As a group, the carbamate toxins (STX [saxitoxin]), NEO, and GTX1–4 [gonyautoxins]) are the most toxic, the N-sulfocarbamoyl toxin (B1, B2, and C1–4) are the least toxic, and the decarbamoyl toxins have intermediate toxicity (3). However, no correlation exists between toxicity and fluorescence intensity (Table 1). Therefore, a chromatographic separation of the PSP toxins was suggested prior to the determination of the fluorescence. Total toxicity is calculated from the individual toxin concentrations and the absolute toxicities for each of the PSP toxins using the equation:

$$T = \sum [C(i)\, T(i)\, D\, (CF) : 10]$$

where T = total toxicity (in micrograms of STX/100 g); C(i) = concentration of individual toxins (in micromolars), T(i) = toxicity of individual toxins (in mouse units per micromoles); D = dilution of extract (in milliliters per gram of meat); and CF = conversion factor (in micrograms of STX per mouse unit) (8).

IV. CHROMATOGRAPHY

Since the methods based on the oxidation/fluorescence assay does not give any information about individual toxins, it is imperative to know what PSP components are in the sample under

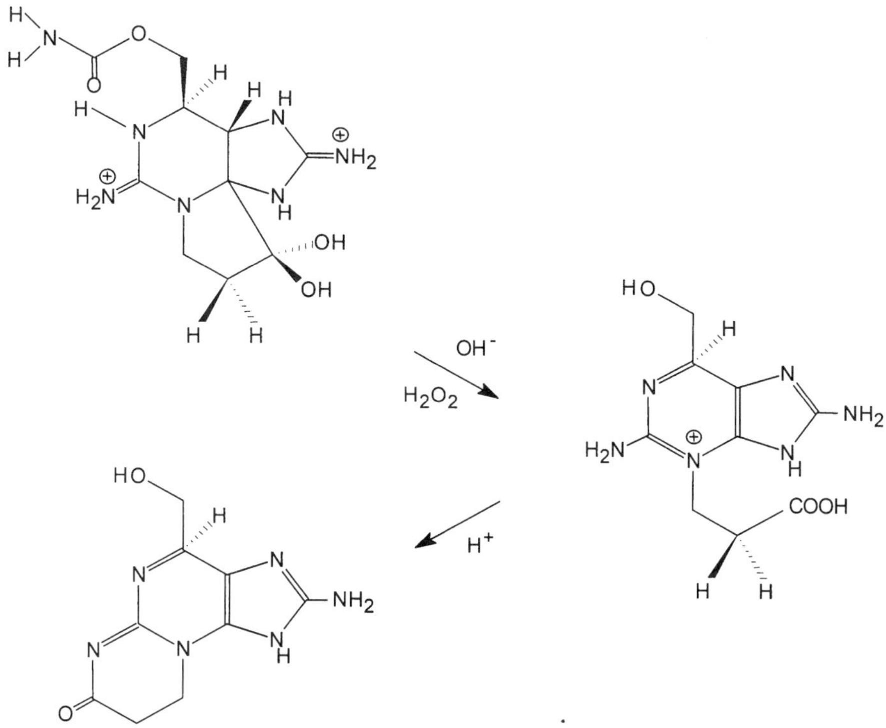

Figure 2 Alkaline oxidation of saxitoxin (5).

Table 1 Toxicity of PSP Toxins (MU/μmol) and Relative Fluorescence Intensities of PSP Toxins After Derivatization

Toxin	Relative fluorescence[a]	Toxicity[b]
STX	1.0	2100
NEO	0.04–0.3	2300
dcSTX	0.71–0.42	900
GTX-1	0.05	1900
GTX-2	1.8	100
GTX-3	1.8	1600
GTX-4	0.05	1900
B1 (GTX-5)	0.41	150
B2 (GTX-6)	0.05	150
C1 (epiGTX-8)	0.48	17
C2 (GTX-8)	0.48	258

[a] From ref. 3.
[b] From ref. 8.

study. Therefore, more sophisticated analytical methods have been developed. HPLC is the most widely used technique that allows the necessary sensitivity and separation to study each toxin (3).

There are several HPLC methods for PSP determination described in the literature. All are based on fluorescence detection of the oxidized PSP toxins. Yet the formation of a purin that becomes fluorescent in acidic solution can be done before or after the chromatographic separation.

A. HPLC with Precolumn Derivatization

Lawrence et al. (9) proposed liquid chromatographic methods with prechromatographic oxidation using both hydrogen peroxide and periodic acid. The N1-hydroxylated toxins NEO, B2, GTX-1, and C3 formed fluorescent products after periodate oxidation at ca pH 8.7, but did not form fluorescent derivatives with peroxide oxidation. The non–N1-hydroxylated toxins STX, B1, GTX-2, GTX-3, C1, and C2 formed highly fluorescent derivatives with both peroxide and periodate oxidation. The addition of ammonium formate to the periodate oxidation reaction greatly improved the yield for the N1-hydroxylated toxins (10).

Because the oxidation products of NEO and B2 could not be separated, parent compounds were separated before oxidation by using an ion-exchange cartridge.

Both periodate and hydrogen peroxide oxidations produce two fluorescent products for GTXs, dcGTXs, dcSTX, and dcNEO. However, the determination of NEO failed by application of hydrogen peroxide oxidation. Additionally, it is important to know how decarbamoyl toxins behave and whether they interfere with the quantification of other PSP toxins (Figure 3). The importance of a complete chromatographic separation of all PSP toxins which contribute to the

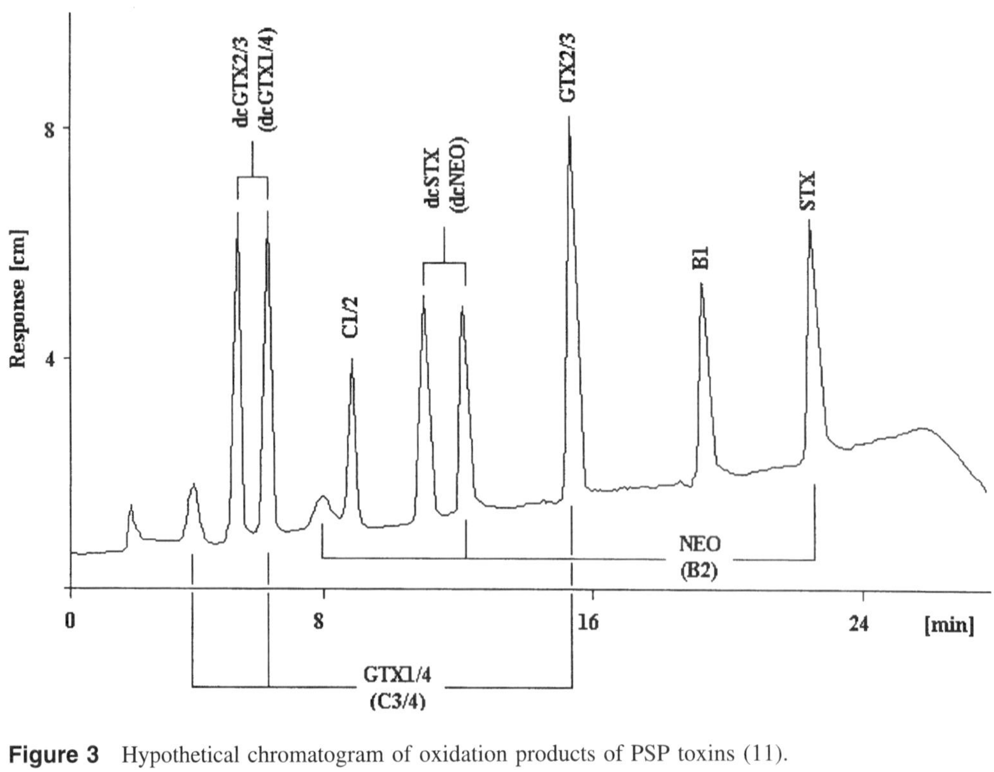

Figure 3 Hypothetical chromatogram of oxidation products of PSP toxins (11).

total PSP toxicity prior to the quantification with a fluorescence detector was recently confirmed in an intercalibration exercise. The results clearly demonstrated that HPLC methods for the determination of PSP toxins are superior to enzyme-linked immunosorbent assays (ELISAs) especially developed for the determination of STX (12).

Additionally, the intercalibration exercise demonstrated that the alkaline oxidation of the PSP toxins has to be performed after the chromatographic separation of the underivatized PSP toxins (postcolumn derivatization). On the other hand, varying results obtained with different chromatographic methods suggested for the determination of PSP toxins require a careful evaluation of the results (13). Consequently, in the following sections methods are only described on the basis of chromatographic separation of the underivatized PSP toxins and postcolumn oxidation prior to fluorescence detection.

B. HPLC with Postcolumn Derivatization

The breakthrough in the use of HPLC methods for PSP determination took place in 1984 when Sullivan et al. (14) succeeded in the separation of underivatized PSP toxins by ion-pair chromatography with alkylsulfonic acids in an HPLC device with the postcolumn derivatization system (Figure 4). The stationary phase consisted of polystyrene-divinylbenzene. A solvent gradient was used with a phosphate buffer containing n-hexane sulfonic acid and n-heptane sulfonic acid as ion-pair formers. The postcolumn derivatization used periodic acid as the oxidizing reagent. This method allows a good separation of the carbamate toxins (Figure 5).

A disadvantage of the method was coelution of STX and dcSTX resulting in a wrong total PSP toxicity, since STX is twice as toxic as dcSTX. Therefore, a separate determination of STX/dcSTX is definitely necessary for a comparison of the HPLC results with the results of the mouse bioassay (15).

Oshima et al. (16) described the entire HPLC separation of PSP toxins. In dependence on the acidity, three groups of PSP toxins were determined after isocratic elution on three HPLC systems (group 1: C1–C4; group 2: GTX-1–4, dcGTX-1–4, B1, and B2; group 3: NEO, dcSTX, and STX). An RP-C_8 column was applied as the stationary phase. Tetrabutylammonium phos-

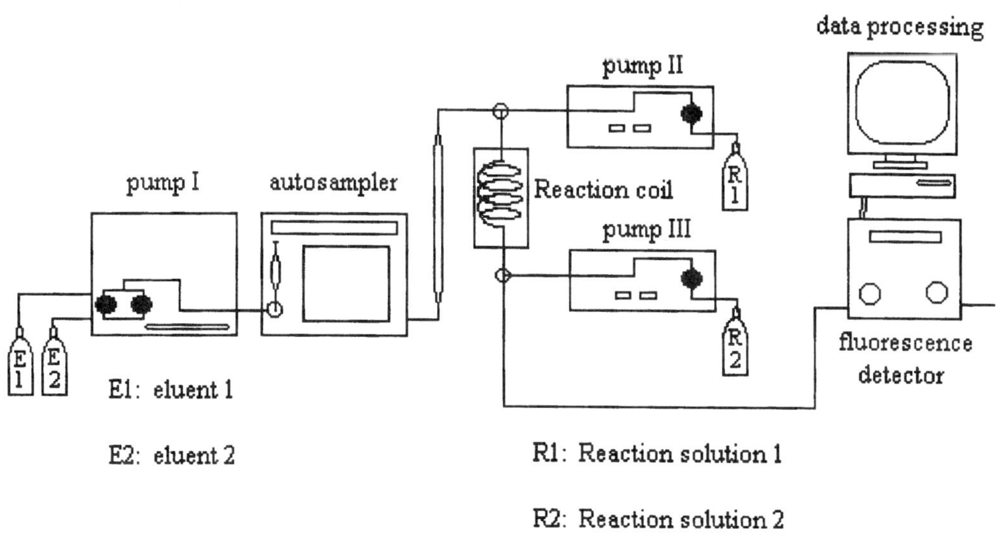

Figure 4 HPLC system with postcolumn derivatization for determining PSP toxins (4).

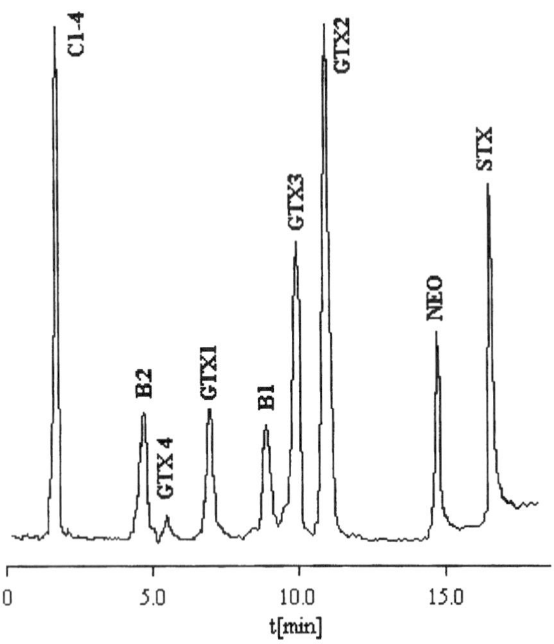

Figure 5 HPLC separation of PSP toxins according to Sullivan et al. (14).

phate (eluent A) was used for the separation of the N-sulfocarbamoyl-11-hydroxysulfate toxins C1–C4, and eluent B and eluent C contained n-heptanesulfonic acid as ion-pair formers for the separation of the carbamate and decarbamoyl toxins. This method suffers from its high expenditure; that is, three chromatographic runs for the quantitation of all PSP toxins in a sample. Figure 6 shows the HPLC separation of the carbamate toxins with eluent B and the decarbamoyl toxins with eluent C as suggested by Oshima et al.

Thielert et al. (17) introduced ion-pair chromatography on an RP-C_{18} phase using n-octanesulfonic acid and ammonium phosphate in the eluent. Isocratic elution enabled separation of STX and dcSTX but problems arose for the separation of GTX toxins.

A gradient with two buffers was introduced to overcome this problem. Both eluents contained n-octanesulfonic acid, phosphate, and tetrahydrofuran. The second eluent additionally contained acetonitrile. This HPLC method allows an extensive separation of the carbamate, decarbamoyl, and N-sulfocarbamoyl toxins. Unfortunately, GTX-1 and GTX-4 coeluted with this method (Figure 7).

To exclude inaccuracies during the PSP determination with HPLC (mainly caused by interferences), it is recommended that two different HPLC methods be applied for the determination of PSP toxins in seafood. The joint evaluation of the chromatograms obtained with the different HPLC methods results in an exact quantification of all PSP toxins relevant for the calculation of the total PSP toxicity in a sample.

Application of the HPLC methods introduced by Thielert et al. (17) and Oshima et al. (16) is a suitable solution for this problem. It is advantageous that the elution order of GTX-2 and GTX-3 is reversed by using n-octanesulfonic acid (Thielert et al.) instead of n-heptanesulfonic acid (Oshima et al.).

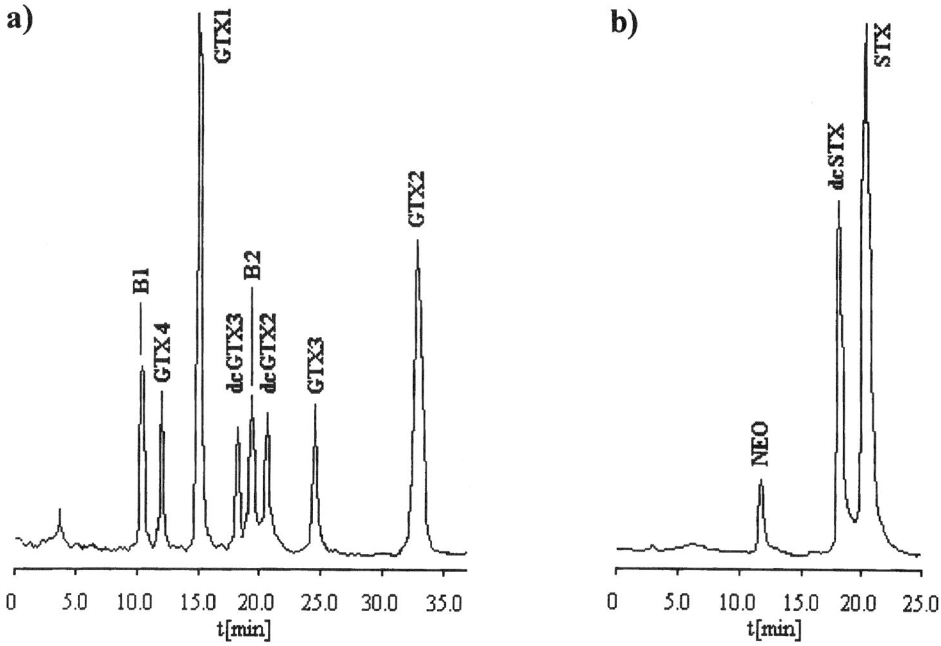

Figure 6 HPLC separation of PSP toxins according to Oshima et al. (16). (a) Eluent B. (b) Eluent C.

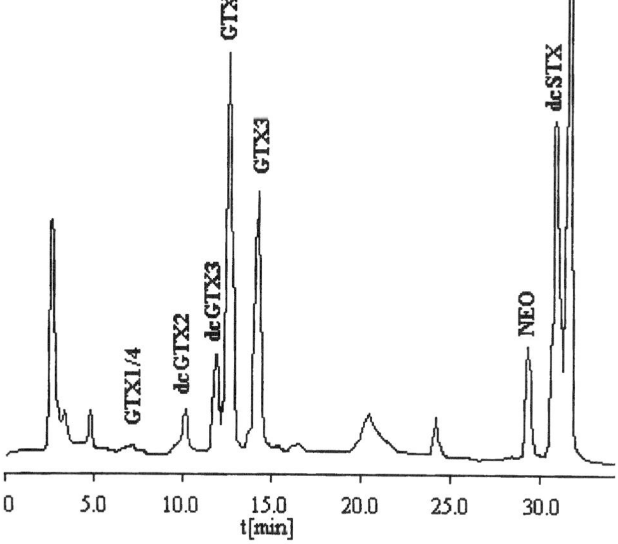

Figure 7 HPLC separation of PSP toxins according to Thielert et al. (17). Sample: mussel extract (*Mytilus edulis*), Spain 1990.

The HPLC method of Oshima et al. (16) can be modified with a two-step elution using eluent B and C. In contrast to the HPLC method of Sullivan et al., no complex gradient is necessary and the HPLC analysis of PSP toxins is performed by simply switching from the first eluent to the second, which is in agreement with the HPLC method of Thielert et al. (17).

High concentrations of phosphate and ion-pair reagent in the eluent stabilize the chromatographic system, particularly after injection of extracts containing many matrix components into the HPLC system. Consequently, neither baseline drift nor variations in the retention times are observed by application of HPLC methods with higher concentrations of ammonium phosphate in the eluents (4). Therefore, the maintenance of a concentration of 40–50 mM ammonium phosphate in the buffers used as eluents by the HPLC method of Thielert et al. (17) was considered as a matter of course by experiments directed at overcoming the shortcomings of this method; that is, the coelution of GTX-1 and GTX-4.

Hummert et. al. (18) developed a very efficient HPLC method for PSP determination based on eluents also used by the HPLC method according to Thielert et al. (17). A new HPLC column was applied and an eluent, C, was introduced besides eluents A and B. Additionally, the pH of the eluents were enhanced to the optimal value 6.9, and a four-step elution was proposed (Table 2).

Table 2 Determination of PSP Toxins by HPLC and Fluorescence Detection According to Hummert et al. (18)

HPLC						
Column:	Luna C18 (Phenomenex), 250 × 4.6 mm ID, 5 μ					
Eluent A:	98.5% 40 mM ammonium phosphate buffer (pH 6.9) with 11 mM octanesulfonic acid (sodium salt); 1.5% tetrahydrofuran					
Eluent B:	83.5% 50 mM ammonium phosphate buffer (pH 6.9) with; 15.0% acetonitrile; 1.5% tetrahydrofuran					
Eluent C:	98.5% 40 mM ammonium phosphate buffer (pH 6.9) with; 1.5% tetrahydrofuran					
Gradient:	Step	time (min)	eluent A (%)	eluent B (%)	eluent C (%)	flow (mL/min)
	1	0.0	50	0	50	1
	2	13.5	50	0	50	1
	3	15.5	0	100	0	1
	4	30.0	0	100	0	1
	5	31.0	100	0	0	1
	6	45.0	100	0	0	1
	7	46.0	50	0	50	1
	8	57	50	0	0	1
Postcolumn oxidation						
Oxidation solution:	40 mL ammonia solution (25%) and 1.14 g periodic acid in 500 mL water					
Acidifying solution:	30 mL acetic acid (100%) in 500 mL water					
Reaction conditions:	50°C, 1 min (10-m coil), flow 0.3 mL/min for both solutions					
Detection:	fluorescence, Ex.: 330 nm Em.: 390 nm					

The optimized HPLC method was successfully applied for separate determination of PSP toxins in algae and mussels from China (19). The presented chromatograms show clearly that the separation, especially for GTXs, was improved dramatically in comparison to the Thielert method (17) (Fig. 8).

It is recommended that positive findings of PSP toxins be confirmed by application of mass spectrometry. However, phosphate, ion-pair formers as well as periodic acid from the postcolumn derivatization unit prevent an efficient application of the HPLC/MS coupling.

Recently, an HPLC method was developed for the PSP determination which allows a direct coupling of the HPLC with the mass spectrometer. Separation of the PSP toxins was obtained on a weak cation-exchange resin (PRP X-200, Hamilton, no. 25041470) using an aqueous eluent with ammonium acetate as the only additive. In case of a parallel application of HPLC/MS and fluorometric detection, electrochemical postcolumn derivatization is suggested in order to avoid contamination of the ion source with chemical oxidation substances (20). This HPLC method does not only enable a determination of PSP toxins by HPLC/MS, but it is also well suited for the production of PSP standards from contaminated algae and mussels (Figures 9 and 10).

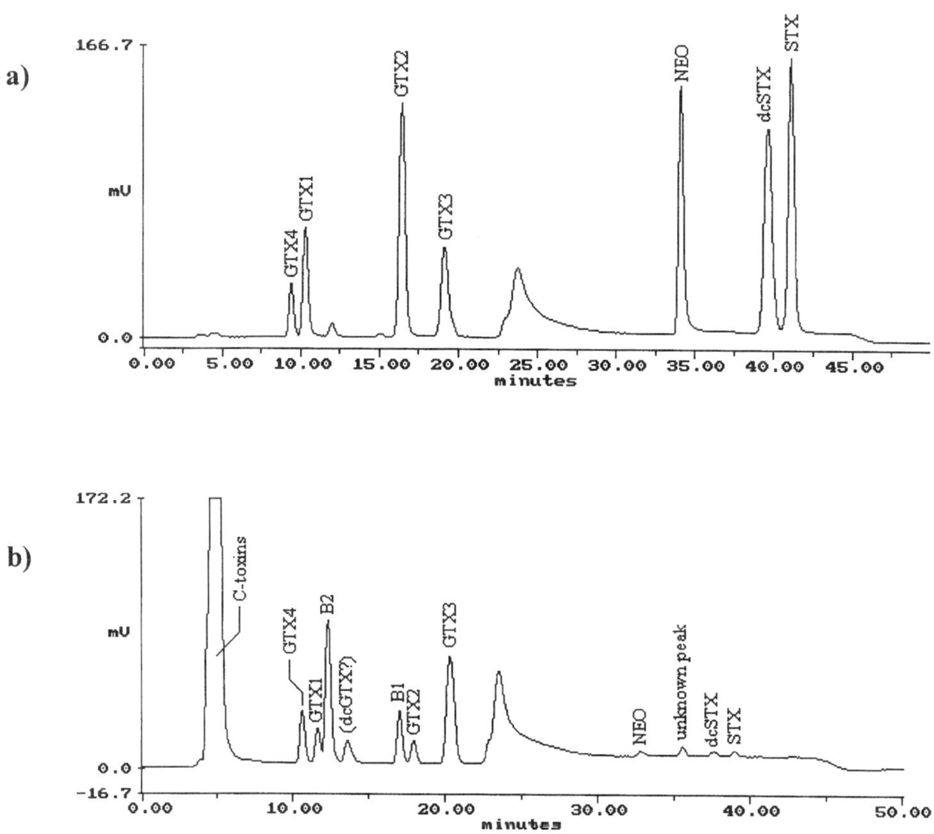

Figure 8 HPLC separation of PSP toxins according to Hummert et al. (18). (a) PSP standard mixture. (b) *Alexandrium tamarense* (extracted with 0.03N acetic acid).

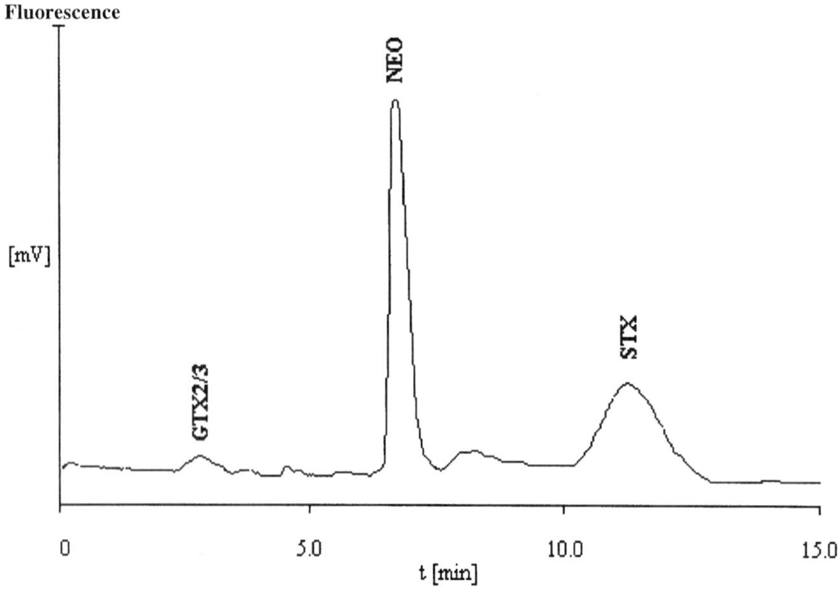

Figure 9 Separation of PSP toxins on a PRP-X200 column (20). Extract of cultivated *Aphanizomenon flos-aquae*.

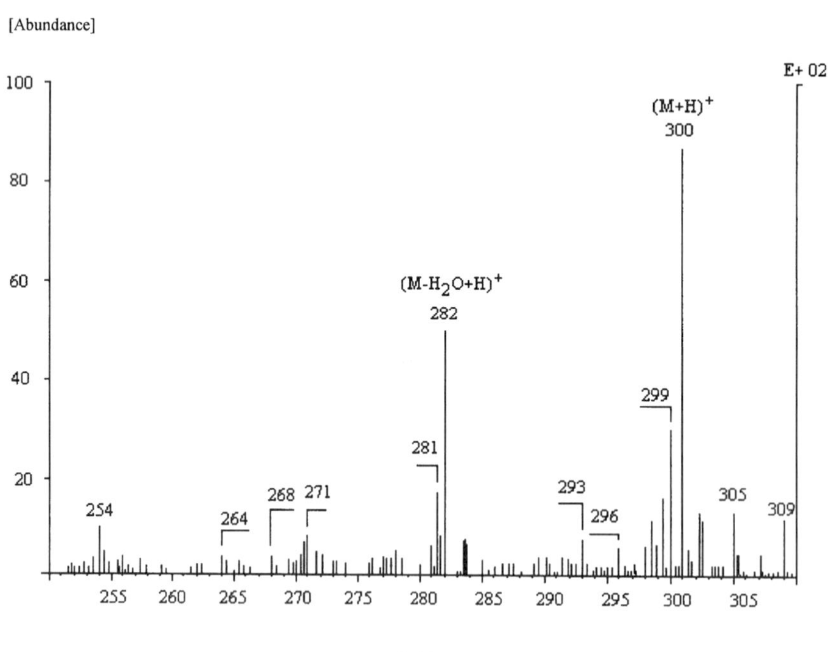

Figure 10 HPLC-atmospheric pressure ion source/electrospray ionization (API/ESI) mass spectrum of STX (20).

V. CAPILLARY ELECTROPHORESIS

There are several options for analyzing PSP toxins other than HPLC. The highly polar nature of these compounds and the lack of UV chromophores absorbing above 220 nm necessitates reversed-phase HPLC using ion-pairing reagents and postcolumn derivatization permitting fluorescence detection.

Although the HPLC methods with fluorescence detection offer good sensitivity and dynamic range for the separation and detection of the different PSP toxins, the sensitivity is dependent on parameters such as reagent concentrations, reaction time, pH, and temperature of the oxidation reaction. Therefore, a capillary electrophoresis (CE) method with UV detection was developed for the separation and determination of the underivatized PSP toxins (21).

Separation by CE is obtained by differential migration of solutes in an electric field. In CE, electrophoresis is performed in narrow-bore capillaries, which are usually filled only with buffer. The use of the capillary has numerous advantages, particularly with respect to the detrimental effects of Joule heating. The use of high electric fields results in short analysis times and high efficiency and resolution. Peak efficiency is due in part to the plog profile of the electroosmotic flow (EOF), an electrophoretic phenomenon that generates the bulk flow of solution within the capillary. The EOF also enables the simultaneous analysis of all solutes, regardless of charge.

The versatility of CE is partially derived from its numerous modes of operation. The separation mechanisms of each mode are different and thus can offer complementary information. Capillary zone electrophoresis (CZE) is fundamentally the simplest form of CE; mainly the capillary is only filled with buffer. Separation occurs because solutes migrate in discrete zones and at different velocities. In CZE, selectivity can most readily be altered through changes in running buffer pH or by the use of buffer additives.

Capillary isotachophoresis (CITP) is a ''moving boundary'' electrophoretic technique. In CITP, a combination of two buffer systems is used to create a state in which the separated zones all move with the same velocity. The zones remain sandwiched between so-called leading and terminating electrolytes.

In a single CITP experiment, either cations or anions can be analyzed (22). The application of on-column sample reconcentration with CITP and discontinuous buffer systems prior to CZE has been investigated for the analysis of PSP toxins (23).

A judicious choice of leading and terminating electrolytes for the preconcentration step has provided an improvement of the concentration detection limit of at least two orders of magnitude over that obtainable using the conventional CZE format. CITP/CZE separation of a mixture of eight PSP standards, performed using a 107 cm × 75 μm ID column with 35 mM morpholine as the leading electrolyte, is presented in Figure 11.

Buzy et al. (24) demonstrated the application of such a stacking procedure prior to the CZE for the analysis of decarbamoyl toxins. The separation conditions developed were found to be entirely compatible with electrospray mass spectrometry, which permitted the analysis of PSP toxins and their decarbamoyl derivatives in crude enzyme digests. The products released during the enzymatic digestion were identified using CE combined with tandem mass spectrometry.

However, CE separations demand highly purified extracts to obtain reproducible separation. Additionally, the handling of the extremely small volumes for injection results in about one magnitude higher limits of detection of the method compared to the HPLC with fluorescence detection.

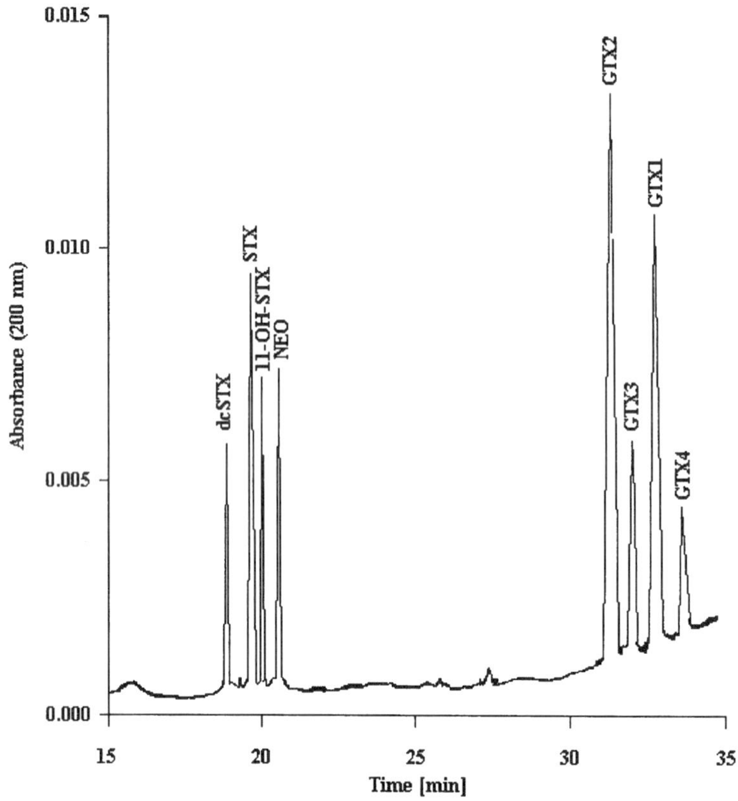

Figure 11 CITP/CEZ/UV analysis of a mixture of eight PSP standards (23). Injection of 2.2 µL of a solution containing 500 ng/mL GTX-1 and GTX-2 and 200 ng/mL for all other PSP toxins.

VI. CONCLUSIONS

The more perfect the determination of the concentrations of the respective PSP toxins, the more exactly will be the calculation of the total PSP toxicity. This follows from the fact that the total PSP toxicity after HPLC or CE separation is obtained by summing up the individual PSP toxicities. Therefore, the unambiguous assignment of the peaks in the chromatograms to definite PSP toxins is necessary for a proper quantification of PSP toxins. Then, the value of the total PSP toxicity obtained by a chemical method agrees well with the results of the mouse bioassay (1). Although the mouse bioassay is still obligate for the export of seafood, the chemical methods for the determination of PSP toxins already has reduced the number of animal experiments.

REFERENCES

1. B Luckas. Phycotoxins in seafood—toxicological and Chromatographical Aspects J Chromatogr 624: 439–456, 1992.
2. T Hollingworth, MM Wekell. Fish and Other Marine Products—959.08. Paralytic shellfish poison.

Biological method. In: K Hellrich, ed. Official Methods of Analysis of the Association of Official Analytical Chemists. 15th ed., Arlington, VA: 1990, pp 881–882.
3. LM Botana, M Rodriguez-Vieytes, A Alfonso, MC Louzao. Phycotoxins: paralytic shellfish poisoning and diarrhetic shellfish poisoning. In: LML Nollet, ed. Handbook of Food Analysis. Vol. 2. New York: Marcel Dekker, 1994, pp 1147–1169.
4. C Hummert, M Ritscher, K Reinhardt, B Luckas. Analysis of the characteristic PSP profiles of *Pyrodinium bahamense* and several strains of *Alexandrium* by HPLC based on ion-pair chromatographic separation, post-column oxidation, and fluorescence detection. Chromatographia 45:312–316, 1997.
5. HA Bates, H Rapoport. A chemical assay for saxitoxin, the paralytic shellfish poison. J Agric Food Chem 23:237–239, 1975.
6. E Hellwig, F Petuely. Bestimmung von Saxitoxin in Muschelkonserven Z Lebensm Unters Forsch 171:165–169, 1980.
7. JM Franco, P Fernández. Separation of paralytic shellfish toxins by reversed phase high performance liquid chromatography with postcolumn reaction and fluorimetric detection. Chromatographia 35: 613–619, 1993.
8. JJ Sullivan, MM Wekell, LL Kentala. Application of HPLC for the determination of PSP toxins in shellfish. J Food Sci 50:26–29, 1985.
9. JF Lawrence, C Ménard, CF Charbonneau, S. Hall. A study of ten toxins associated with paralytic shellfish poison using prechromatographic oxidation and liquid chromatography with fluorescence detection. J Assoc Off Anal Chem 74:404–409, 1991.
10. JF Lawrence, C Ménard. Liquid chromatographic deterimation of paralytic shellfish poisons in shellfish after prechromatographic oxidation. J Assoc Off Anal Chem 74:1006–1013, 1991.
11. JF Lawrence, B Wong, C Ménard. Determination of decarbamoyl saxitoxin and its analogues in shellfish by prechromatographic oxidation and liquid chromatography with fluorescence detection. J Assoc Off Anal Chem Intern 79:1111–1115, 1996.
12. HP van Egmont, HJ van den Top, WE Paulsch, X Goenaga, R Vieytes. Paralytic shellfish poison reference materials: an intercomparison of methods for the determination of saxitoxin. Food Add Contam 11:39–54, 1994.
13. MA Quilliam. Seafood toxins. J Assoc Off Anal Chem Intern 78:209–214, 1995.
14. JJ Sullivan, MM Wekell. Determination of paralytic shellfish poisoning toxins by high pressure liquid chromatography. In: EP Ragelis, ed. Seafood Toxins. ACS Symposium Series 262, American Chemical Society, Washington, DC, 1984, pp 197–205.
15. B Luckas, G Thielert, K Peters. Zur Problematik der selektiven Erfassung von PSP-Toxinen aus Muscheln. Z Lebensm Unters Forsch 190:491–495, 1990.
16. Y Oshima, K Sugino, T Yasumoto. Latest advances in HPLC analysis of paralytic shellfish toxins. In: S Natori, K Hashimoto, Y Ueno, eds. Mycotoxins and Phycotoxins '88. Amsterdam: Elsevier, 1989, pp 319–328.
17. G Thielert, I Kaiser, B Luckas. HPLC determination of PSP toxins. In: JM Fremy ed. Proceedings of Symposium on Marine Biotoxins, Paris, 30–31 January 1991, Maisons-Alfort, Centre National d'Etudes Veterinaires et Alimentaires, 1991, pp 121–125.
18. C Hummert, M Ritscher, K Reinhardt, B Luckas. A New Method for the Determination of paralytic shellfish poisoning (PSP). Proceedings of the 22nd International Symposium on Chromatography (ISC). Rome, September 13–18, 1998.
19. RC Yu, C Hummert, J Li, PY Qian, MJ Zhou, B Luckas. Analysis of PSP toxins in algae and shellfish samples from China using a modified HPLC method. Chromatographia, 1998.
20. J Kirschbaum, C Hummert, B Luckas. Determination of paralytic shellfish poisoning (PSP) toxins by application of ion-exchange HPLC, electrochemical oxidation, and mass detection. In: P Lassus, G Arzul, E Erard-LeDenn, P Gentien, C Marcaillou-LeBaut, eds. Harmful Marine Algal Blooms Paris: Lavoisier, 1995, pp 311–314.
21. P Thibault, S Pleasance, MV Laycock. Analysis of paralytic shellfish poisons by capillary electrophoresis. J Chromatogr 542:483–501, 1991.
22. DN Heiger. High performance capillary electrophoresis—an introduction. 2nd ed. Hewlett-Packard Company, Publication No. 12-5091-6199E, 1992.

23. SJ Locke, P Thibault. Improvement in detection limits for the determination of paralytic shellfish poisoning toxins in shellfish tissues using capillary electrophoresis/electrospray mass spectrometry and discontinuous buffer systems. Anal Chem 66:3436–3446, 1994.
24. A Buzy, P Thibault, MV Laycock. Development of a capillary electrophoresis method for the characterisation of enzymatic products arising from the carbamoylase digestion of paralytic shellfish poisoning toxins. J Chromatogr A688:301–316, 1994.

9
Biological Detection Methods

Benjamín A. Suárez-Isla
Institute of Biomedical Sciences
University of Chile
Santiago, Chile

Patricio Vélez
University of Valparaíso
Valparaíso, Chile

I. INTRODUCTION

Paralytic shellfish poisoning (PSP) is the accepted epidemiological description of the toxic and potentially fatal syndrome caused in humans by the ingestion of a complex of organic compounds known as saxitoxins (STXs) and the related compound tetrodotoxin (TTX) and their analogues (1,2). Saxitoxins and tetrodotoxins are guanidinium-containing highly toxic compounds (1,2) (see Chap. 7). By far the most frequent pathway of human intoxication is the consumption of shellfish that have fed on toxic phytoplankton and have accumulated these toxins above a threshold level (currently 80 μg STX equivalent per 100 g of tissue [3]). Within minutes to a few hours, PSP develops as a loss of muscle coordination, ascending paralysis, deep coma, and death by cardiorespiratory failure. Symptoms described as PSP (4) (see Chap. 10) are caused by the high-affinity binding of STXs to voltage-dependent sodium channels (VDSCs) in neuronal and muscle cells (5–11). As a result of toxin-receptor interaction, the flux of sodium ions into the cells through the channels is blocked, impairing and/or abolishing the generation and conduction of electrical impulses or action potentials.

In this chapter, we review first the molecular and biophysical basis of the action of saxitoxins and tetrodotoxins, toxins that reversibly block sodium channels at nanomolar concentrations in a dose- and time-dependent mechanism. The chemistry and analytical techniques that measure the mass of a saxitoxins/tetrodotoxins are reviewed in Chapters 7 and 8. For the purpose of this chapter, we have narrowly defined as ''biological methods'' those assays where the endpoint is related toxin-receptor interaction; that is, binding of the saxitoxin molecule(s) to site 1 in the α-subunit of the VDSC. With these criteria we discuss here bioassays (e.g., mouse or fly tests) and in vitro assays for PSP, such as citotoxicity tests, cell receptor-based assays, and electrophysiological assays.

II. ETIOLOGY

Harmful algal blooms (HABs) are natural phenomena triggered by complex environmental signals that, in several instances, have been associated with nutrient inputs from human activities (12,13). About 300 of the estimated 3400–4100 phytoplankton species produce "red tides" but only 60–70 species are actually harmful (14). Dinoflagellates are the most noxious, and it is notable that only 10–12 species are primarily responsible for the current expansion of HABs worldwide (15,16). Toxic dinoflagellates produce potent nonpeptidic neurotoxins (1,17). Among them, saxitoxins (1,18), brevetoxins (19), and ciguatoxins (20) have the sodium channel protein as their sole molecular target and bind with high affinity to specific sites of the α-subunit (21–24). Fatal PSP intoxications represent the most serious threat of marine origin to public health. PSP is caused by the ingestion of a variable mixture of saxitoxins. PSP toxins include the parent compound saxitoxin (STX) and a number of derivatives. Saxitoxins and tetrodotoxins are examples of the few highly lethal nonpeptidic toxins (1,2). Saxitoxins are synthesized by a limited number of species of marine toxic dinoflagellates (2,18), freshwater cyanobacteria (25), and marine bacteria (26,27). The related guanidinium-containing compound tetrodotoxin and its analogues are also produced by freshwater cyanobacteria (25). These toxins are reversible blockers of VDSCs and have been conventionally classified in three groups: (1) highly potent carbamates such as STX, neosaxitoxins, and gonyautoxins (GTX-1–GTX-4), (2) weakly toxic N-sulfocarbamoyl toxins (B1, B2, C1–C4), and (3) decarbamoyl analogues of intermediate toxicities (see Chap. 10). In addition to its potential lethality (PSP) has presently a significant economic impact on commercial shellfish areas worldwide (28), including the United States and Canada (29), Europe (28), and Chile (30).

III. FUNCTION OF SODIUM CHANNELS

Ion channels are integral membrane proteins that allow the passage of a few physiologically significant cations (Na^+, K^+, Ca^{2+}) and anions (Cl^-) in and out of cells. Cell membranes are highly impermeable to charged species and, in this sense, ion channels function as catalysts of ion transfer through cell membranes (31). They do this by opening a water-lined pore which is part of the ion channel molecule in response to specific stimuli, such as a change in transmembrane potential, the sudden elevation of agonist concentration in the extracellular milieu (neurotransmitters) or of internal second messengers (such as Ca^{2+}) or membrane stretch. The direction of ion flow depends on the direction of the concentration gradient across the cell membrane and on the size and sign of the transmembrane potential. The ion current through a single channel (i) will depend on the number of ions per unit time that the open channel can transfer (its single channel conductance, g) and the fraction of time that the channel spends in its open, conductive state (P_O. In general, for N channels, the total current can be expressed as:

$$I = N \times P_O \times i$$

Saxitoxin and its derivatives produce a decrease of the fractional open time (P_O) of the sodium channel, increasing the length and frequency of closed periods. This decrement depends on the type of saxitoxin derivative and its concentration (1). For STX derivatives of low toxicity, this decrement of the time the channel spends in the open state is lower than that caused by the more toxic saxitoxins.

A. Molecular Aspects

Sodium channels seemed to have appeared first phylogenetically in the jellyfish, allowing efficient signal transmission among neurons distributed in separate ganglia. In invertebrates, sodium channels are expressed mainly in neuronal cells, whereas they can be found in muscle cells in chordates. Sodium channels display the highest densities of distribution in axons and muscles. A mammalian heart muscle cell may express more than 10^5 sodium channels present in a mean cell surface of approximately 50 μm^2. In contrast, unicellular organisms express principally potassium and calcium channels (9).

Sodium channels are composed of several subunits held together by weak noncovalent interactions (Figure 1). Structure-function studies indicate that most of the essential functions of the channel molecule such as ion permeation, selectivity, gating, and voltage sensitivity reside in the α-subunit. The α-subunit has a modular architecture of four internally homologous domains (I–IV) (Figure 1) composed of six transmembrane segments, each resembling a single α-subunit of a tetrameric voltage-dependent potassium channel. The four domains are believed to fold together, delineating a central pore whose amino acidic residues may determine the selectivity and conductance properties of the sodium channel.

Several far-reaching findings have emerged from structure-function studies in recent years. One of them is the observation that the amino acid sequences that link segments 5 and 6 in each domain (called S5-S6 linkers or P segments), may form the pore itself (33) (Figure 2). It is not clear yet how these particular residues interact with the ions in the bulk solution to allow selective sodium flux over other cations by factors of 100:1 given the characteristically high throughput rate of each channel (more than 10^7 ions per second).

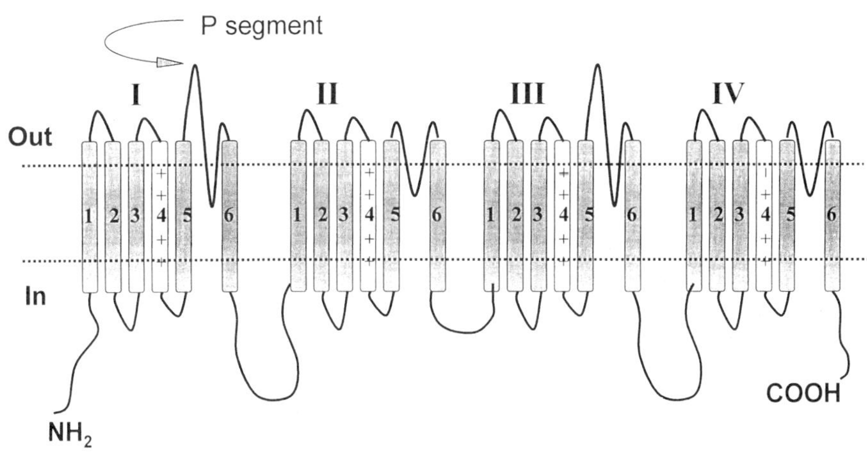

Figure 1 Schematic representation of the primary structure of the α-subunit of the sodium channel depicted as transmembrane folding diagrams. Polypeptide chains linking different segments (1–6) and domains (I–IV) are not drawn to scale. Cylinders represent putative α-helices. P segments are not equivalent in each domain. Folding of this hypothetical structure into a pseudo-tetramer (cf. Figure 2) would build an internal space lined in part by the P-loops. A subset of specific amino acids located deeper into the channel outer vestibule would form the selectivity ring. The binding site for saxitoxins is partially defined by a few crucial amino acids tentatively located in the extracellular loops linking segments I and II in each domain.

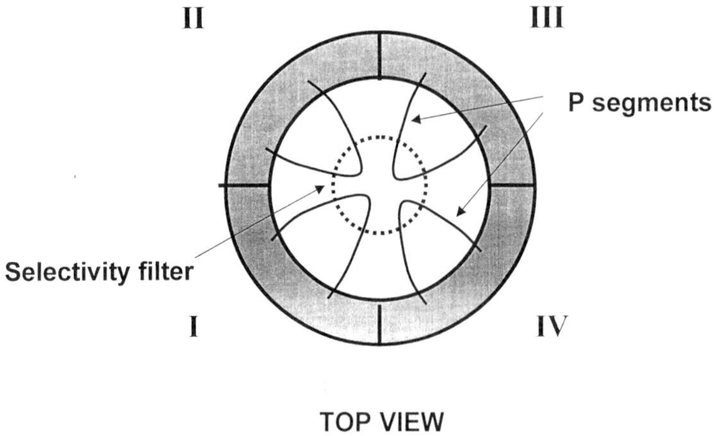

Figure 2 Schematic top view representation of the assembled α-subunit of the sodium channel to display the proposed arrangement of P segments hanging over the extracellular channel vestibule. Dotted circle indicates general location of the selectivity filter.

B. Molecular Basis of Toxin Interaction

The α-subunit contains the binding sites for several neurotoxins. These putative sites have been defined in pharmacological competition studies and systematic amino acid mutagenesis. It is intriguing that a large number of the most toxic marine toxins (saxitoxins, tetrodotoxins, brevetoxins, and ciguatoxins) have as their sole molecular target the VDSC. However, most of the understanding of sodium channel structure and function has been contributed from studies where tetrodotoxin or saxitoxins were used (7–10,32). Purification of sodium channels and their reconstitution in artificial bilayer systems was possible by the use of these compounds that bind with very high affinity to the α-subunit (23).

Previous studies from electrophysiological experiments (34,35) and from binding studies using radiolabeled toxins (22,36,37) have indicated that STX and TTX bind to an acidic site that also binds other metal and organic cations. STX and TTX mutually exclude from this site, which is an observation that has led to the proposal that STX and TTX bind to the channel molecule sharing a common mechanism.

Pharmacological studies have demonstrated that STX and TTX bind competitively to a site located on the external surface of the channel, named toxin site 1 (5,32). This assertion was based on several observations. Both toxins completely block current of the sensitive sodium channel isoforms, competitively displace each other in binding assays, and share functional groups crucial for binding; namely, one or two guanidinium groups and a diol group (24). Molecular studies located this site in the polypeptides (initially called SS1-SS2 segments) that were thought to link segments 1 and 2 in all four domains (32). This was supported after the finding that mutations of the amino acids present in the outer pore vestibule had similar effects on the affinities of both toxins (33,38,39). In 1988, Noda et al. (38) found that a single point mutation of the α-subunit of the rat sodium channel could drastically reduce its sensitivity to tetrodotoxin and saxitoxin by more than three orders of magnitude. The specific mutation that replaced glutamic acid 387 with a glutamine had minimal effects on the sodium current.

The hypothesis of a shared binding for STX and TTX was also supported by experiments that suggest that a single site in the domain I pore-forming region may be responsible for the isoform differences in STX and TTX affinities (40–43). In fact, when applied externally to cells, TTX (and STX) blocks sodium channels of neural and muscular origin with high potency. Only nanomolar concentrations of TTX are needed to block passage completely of sodium ions (sodium currents) through neuronal and skeletal muscle channel isoforms. However, two orders of magnitude higher STX or TTX concentrations need to be added to cardiac sodium channel isoforms to attain a comparable level of sodium current block. The lower affinity of the cardiac isoform is thought to be caused by a cysteine for phenylalanine or tyrosine substitution in domain I of the outer vestibule. This may give the cardiac sodium channel its known reduced sensitivity for STX and TTX and its increased affinity to calcium and group llb metals (cadmium).

Several amino acids located in the external aspect of the pore region are essential to maintain high-affinity binding. Terlau et al. (1991) (33) demonstrated that several mutations of the SS1-SS2 segments of the rat brain type II sodium channel may alter binding affinities for STX and TTX. In the rat skeletal muscle isoform, μ1, the identity of one amino acid in the P region of domain I (position 401 in the rat) accounts for the isoform-dependent TTX sensitivity. An aromatic residue in position 401 confers high affinity, whereas lack of it renders the channel TTX resistant (40,43). Other amino acid residues located in the outer aspect of the pore are known to contribute to the high specific binding of TTX and saxitoxins. These results suggests that the guanidinium-containing toxins bind to the external surface of the channel through an extensive set of noncovalent interactions (33,42,44).

Only recently (1994) has the hypothesis of a shared binding mechanism of STX and TTX been challenged by Kirsch et al. (45). In 1998, Penzotti et al. (46) compared the effects of mutations in the μ1 isoform–specific domain I Phe/Tyr/Cys location on toxin binding. They have proposed a revised binding model that considers that both TTX and STX have similar interactions with amino acid residues of the selectivity filter (P-loops) located in the outer vestibule of the channel. However, TTX has a stronger interaction with Tyr401 and STX interacts more strongly with more extracellular residues probably through the second guanidinium group that is absent in TTX.

IV. BIOLOGICAL METHODS

We have included here as "biological methods" those assays where the endpoint is related to the toxin-receptor interaction; that is, binding of the saxitoxin molecule(s) to site 1 in the α-subunit of the VDSC. With these criteria, we discuss here bioassays (e.g., mouse or fly tests) and in vitro assays for PSP, such as citotoxicity tests, cell receptor–based assays, and electrophysiological assays. Chemical methods such as HPLC with fluorescence or mass spectrometry (MS) detection are reviewed in Chapter 8.

A large effort has been directed to immunomethods to recognize saxitoxins in acid shellfish extracts. However, limited cross reactivity of available antibodies to saxitoxin mixtures of varying composition and low predictability of human oral potency have precluded their use for regulatory purposes and are not discussed here. (For detailed descriptions of these and other methods, see ref. 47.)

A. General Considerations

Detection of saxitoxins is a challenge for analytical chemists. This is due to large differences in the specific toxicity of STX derivatives (two orders of magnitude), the small total amounts present in natural samples, the variability in composition and content of STX derivatives in samples, their propensity to undergo biotransformation in the shellfish digestive glands and during sample extraction and preparation, and matrix interference. Analytical methods provide an instrumental response proportional to the concentration of each derivative in a complex sample. This proportionality is obtained running concurrent analyses with pure standards of known concentration.

In contrast, assays provide a single response proportional to the toxicity of a complex toxic mixture. This toxicity is related to the average effect of a number of chemical derivatives acting on the same functional site. The assay response (endpoint) may be a colorimetric or fluorescence change, a displacement of a radiolabeled reference compound from its binding site, or the impairment of a physiological response (cell growth or survival, ion current inactivation). In the case of PSP toxins, the interaction with their receptor produces block of ion channel function; that is, a decrement or complete blockade of ion flow. Thus, single Na channel current is decreased or totally inhibited.

In analyses and assays, it is necessary to obtain a suitable *response factor* to convert an instrumental signal (analyses) or a biological response (assays) into a useful toxicity value (1). In the case of an analysis (such as HPLC), there will be individual response factors to convert each response for each STX derivative into a concentration. The problem that still remains is to convert this concentration into a toxicity value that is related to human oral potency. In this case, a specific toxicity has to be obtained for each derivative. If this is accomplished, the estimate of total toxicity of a complex PSP mixture will still rely on the assumption that the final result is a linear combination of specific toxicities and molar percentages.

A strategy to overcome these problems has been to establish whether a statistically significant correlation exists between toxicity values determined by different methods and the standard accepted mouse bioassay. Few studies have obtained specific toxicities of pure saxitoxin and its derivatives comparing two or more methods (see refs. 1 and 48). In 1990, Hall et al. (1) found a general correspondence among relative potencies of STX and nine derivatives as determined by mouse bioassay, displacement of radiolabeled saxitoxin, and single channel recording. The general agreement between the biological mouse assay and other functional assay that report effects at the molecular level suggests that intraperitoneal (i.p.) toxicity in mice is significantly determined by the relative potencies of saxitoxins. Other factors such as bioavailability and pharmacodynamics of the toxins may have a minor role in determining final toxicity.

B. Mouse Bioassay

To protect human consumers from PSP, many countries have adopted shellfish monitoring programs. The principal regulatory method is the mouse mortality bioassay initially developed by Sommer and Meyer in 1937 (49) that was later accepted in the United States (3) as an official method to detect saxitoxins in acidic shellfish extracts. In recent years, the mouse bioassay has been sanctioned by other countries that signed agreements with the United States concerning the import/export of potentially toxic shellfish. Toxicity is expressed as microgram STX equivalents (eq)/100 g shellfish tissue standardized against a reference STX solution, and harvesting or consumption is not permitted in many countries when toxicity reaches or exceeds 80 μg/100 g. The mouse bioassay for PSP is a reliable indicator of human oral toxicity, but it is costly and time consuming, especially when a large number of samples has to be analyzed. Its main analytical limitations are its high variability (±20%), low sensitivity very near to the detection limit (about 35 μg/100 g) is too near to the accepted toxicity threshold (80 μg/100 g), and interferences. The major limitation of the mouse bioassay, however, is of ethical and political nature (i.e., the controversial use of live animals). The assay measures the time to death after intraperitoneal injection of the toxic extract, a procedure that has received increasing ethical criticisms and that cannot be carried out in some European countries (47).

C. Sources of Interferences in the Mouse Bioassay

A source of indetermination in biological assays is the low specificity of the response, a factor that, in contrast, is well appreciated in regulatory work. However, a critical weakness of bioassays is the certain possibility of interferences unrelated to the analytes being tested. This is very significant in the case of the PSP mouse test, and several sources of interferences have been reported.

High salts present in the acidic extracts can decrease apparent toxicity (47,50). In contrast, high levels of zinc can produce deaths with neurological symptoms (51,52) that can be interpreted as being caused by PSP. These interferences can be explained in part if we consider the results of early electrophysiological (34,35) and binding studies using radiolabeled toxins (22,36,37) discussed in the previous paragraphs. Those results indicated that STX and TTX could bind to an acidic site (toxin site 1) that was also the binding site for other metal and organic cations. Indeed, high concentrations of divalent cations modify STX binding (22,34–37,69) reducing its affinity; a fact that could help explain the loss of potency of acidic extracts when high salts are present (47,50).

The effects of zinc can be tentatively explained by its direct competition with STX for a restricted area of the binding site. Both cations block the Na channel and decrease or abolish ion current. It has been established that not all Na channel isoforms display the same sensitivity toward these metals. Mammalian heart Na^+ channels exhibit approximately 100-fold higher affinity for block by external Zn^{2+} than other Na^+ channel subtypes. In this case, application of external Zn^{2+} can relieve the block elicited by STX in a strictly competitive fashion suggesting direct binding competition between Zn^{2+} and STX at a single site with intrinsic equilibrium dissociation constants of 30 nM for STX and 30 μM for Zn^{2+} (75).

Another important variable to consider is the difficulty in the mouse test to assure optimal pH to generate an extract that displays maximal toxicity. It is known that complete hydrolysis of N-sulfocarbamoyl saxitoxin derivatives (less potent analogues) is pH dependent and that the AOAC recommended procedure does not result in extracts that display maximal potency (1). In addition, binding of saxitoxins to the receptor site is also pH dependent (69). Thus, the interferences of high salts, Zn, and pH might be explained by the specific modification of STX binding to its receptor.

Simultaneous occurrence of diarrhetic shellfish poisoning (DSP) and PSP toxins has been reported (52). In these cases, the mouse bioassay for diarrhetic toxins may give unusually short death times with neurological symptoms especially when the lipophilic extract is not washed (53). This can also be the case when low PSP toxin levels not detectable by the mouse bioassay are present. The mouse test for DSP toxins uses the hepatopancreas where normally toxins are more concentrated than in the rest of the shellfish tissues.

D. Alternatives to the Mouse Bioassay

The analytical shortcomings of the mouse bioassay as well as mounting public pressure to eliminate or reduce live animal bioassays have stimulated the development of several alternative methods to be used in shellfish monitoring programs. In the search of acceptable alternative living organisms, bioassays that make use of houseflies, chick embryos, brine shrimp, and bacteria have been proposed (54). The housefly assay is more sensitive than the mouse bioassay but requires considerable operator skills to perform microinjections, whereas the other bioassays do not have the appropriate sensitivity for regulatory work. More recently, a bioassay that employs the desert locust (*Schistocerca gregaria*) for the detection of saxitoxin and related compounds in cyanobacteria and shellfish has been reported (55). Results suggest that this assay may be useful for the routine screening of PSP toxins.

Extensive efforts have been done to design alternative in vitro methods that do not use live organisms to be used in regulatory work. These include cell culture bioassays, immunoassays, receptor-based assays, and voltage-clamp methods that use isolated axons from squid and frogs or cultured cells. It would be reasonable to visualize the sodium channel as the natural biosensor for PSP toxins to measure how the electrical current carried by sodium ions is impaired by the blocking toxins. Although techniques that accurately measure sodium currents in several biological preparations were developed in 1950 by Hodgkin and Huxley (57) (For voltage-clamp and patch-clamp methods, see refs. 31 and 56.), and have been used extensively since then to unravel the structural basis of sodium channel function, it is still an open problem how to reduce into practice a simple design that uses measurement of sodium currents as the endpoint signal that reports PSP effects. We describe here several approaches to this objective.

E. In Vitro Tissue Culture Bioassays

Cytotoxicity assays for PSP performed with cultured cells have been presented as potentially promising tools to be applied in screening and/or monitoring programs. Recently, a large intercalibration effort has been undertaken to certificate a commercial version (MIST, Jellet Biotek Ltd., Darmouth, Canada) (58,59). The principle of all later versions was developed by Kogure et al. (1988) (60) to assay for toxins that blocked the sodium channel. This method used an established neuroblastoma cell line (Neuro 2A) and measured cell morphology and survival after exposure to veratridine, which in the presence of ouabain enhances sodium permeability and concomitant water flow into the cells. Swelling of the cells and their eventual disruption and death was inhibited by saxitoxins and tetrodotoxin. Cell counting and assessment of morphological changes was correlated with the amount of blocking toxins and later used to assay for equivalent ''cell-protection'' potency in shellfish extracts.

Ouabain blocks the Na/K ATPase and thus inhibits the ability of the cell to maintain the physiological concentration difference for sodium and potassium that exists between the intracellular and the extracellular compartments (56) (in mammals, internal [Na] approximately 12 mM; external [Na] approximately 120 mM; internal [K] approximately 100 mM; external [K] approximately 5 mM). This leads to gradient dissipation, depolarization of the cell, Na and water influx, and cell swelling. The latter process is enhanced by the application of veratridine, an alkaloid that increases sodium permeability by shifting the inactivation curve of sodium channels to more depolarized potentials. At the single channel level, veratridine is known to cause persistent activation of Na^+ channels. Channels activated with veratridine open and close on a time scale of 1–10 s. Increasing the veratridine concentration enhances the probability of channel opening, primarily by increasing the rate constant of opening (60a).

In more recent versions, the endpoint is visualized with vital stainings that diffuse into dead cells only (crystal violet [58,59] or neutral red [61]). At present, these versions are suitable to be applied in research laboratories well equipped with aseptic tissue culture facilities. Reproducibility of results critically depend on batch-standardized serum and high-quality veratridine and ouabain. The MIST Cell Bioassay kits represent a significant step toward simplification and provide preplated cells and reagents, but plates have a limited shelf life. The semiquantitative kit version is relatively expensive and the quantitative version is very expensive. This simplified system still demands limited aseptic conditions to deliver good results.

F. Receptor-Based Assays

Under equilibrium conditions, displacement of a radiolabeled parent compound from its binding site provides a relative measure of the capacity of a pure compound or a group of analogues to compete for the same receptor. Percentage of reduction in radiolabeled binding of the reference compound is proportional to the concentration of unlabeled related species present in the sample. For PSP toxins, displacement of radioactive STX from the common binding site on the channel

(21,22,62,65) reflects the capacity of a STX mixture to share that site. Detection of saxitoxins by this procedure reflects functional activity rather than recognition of a structural component, as is the case with immunomethods. Several lines of evidence suggest that the binding of PSP toxins present in different ratios and amounts elicits biotoxicity in mouse bioassays proportional to the capacity of the toxic mixture to displace [^{3}H]STX from its binding site.

Receptor-based assays have been essential to learn about toxin-binding kinetics and to estimate sodium channel densities in excitable membranes. The first application to detect saxitoxins was done by Davio and Fontelo in 1984 (63) with samples of human samples. In 1993, Vieytes et al. (64) used microtiter plates to detect STXs in shellfish extracts, and later efforts by Doucette et al. (62,65) have simplified the technique and validated its use for a variety of matrices. We have extensively applied a radioassay (66,73) to evaluate the toxicity of various shellfish extracts in regulatory programs in Chile. As shown in Figure 3, the relationship between STX displacement and in vivo toxicity showed a good correlation (R = 0.972; N = 41). This assay was applied routinely to monitor low PSP levels in aquaculture sites in northern Chile and has served as an early warning monitoring, as subtoxic variations of PSP toxicity can be detected in advance. Nevertheless, displacement of a radiolabeled ligand does not directly assay how saxitoxins impair Na channel function. These direct observations can only be accomplished with electrophysiological methods that measure in real time the block of sodium currents by saxitoxins and/or tetrodotoxins.

G. Electrophysiological Methods

The electrical excitability of cells in all chordates and many invertebrate species is based on the voltage-dependent entry of Na^+ ions into cells (31,56). The entry of positively charged ions

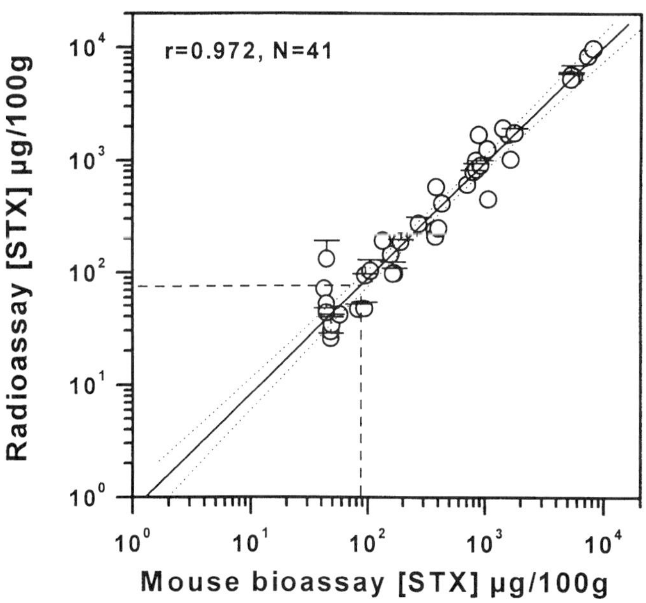

Figure 3 Correlation between PSP toxicity levels determined by mouse test and the STX receptor-based radioassay. The displacement of [^{3}H]STX by natural PSP extracts was performed with toxic samples provided by the Health Service Magallanes (40–10000 µg STX/100g). A significant correlation (R = 0.972 for 41 samples) was found between the toxicity levels determined by mouse bioassay and radioassay. The dotted lines indicate the safety limit of 80 µg STX/100 g.

rapidly shifts the membrane potential from negative values (−70 to −90 mV) to positive values (> +20 mV) within milliseconds. The delayed exit of K^+ ions brings back the membrane potential to resting values. This wave of depolarization/repolarization is called the action potential. Blockade on Na^+ entry abolishes cell excitability and conduction of the nerve impulse. Studies carried out with electrophysiological techniques constitute the main body of experimental evidence that supports current knowledge on the properties of ion channels and ion currents through them (11).

The properties of macroscopic sodium currents were elucidated first in the early 1950s by Hodgkin and Huxley (57) with a procedure that allowed the recording of sodium currents under conditions of controlled membrane voltage (the voltage-clamp method). Those studies provided the first evidence to understand how excitation regulates the entry of Na^+ ions. These investigators provided far-reaching insights into the components and functions of excitable units of the cell membrane that displayed voltage- and time-dependent conductances and that conducted sodium ions selectively and with independence of other ionic pathways. Since then a substantial body of evidence and more than 600 papers have addressed the issue of the molecular basis of channel toxin interaction. However, efforts to apply well-known techniques and knowledge on sodium channels to develop a simple functional assay for PSP toxins, have not been successful.

A very promising approach was developed by Cheun et al. in 1998 (66a). These investigators developed a tissue biosensor for STX and TTX that consisted of a Na^+ electrode covered with a frog bladder membrane integrated within a flow cell. Active Na transport that takes place from the internal to the external face was found to be TTX and STX sensitive. This procedure allowed the detection of PSP toxins well below the detection limit of the mouse bioassay.

However, the development of a simple functional assay that uses the natural target molecule, that is, the STX/TTX–sensitive sodium channel, has not been completed. Ideally, it would be desirable to measure directly the biotoxicity of PSP extracts as a direct block of sodium currents.

To this end we have developed a functional assay (73) that uses patch-clamp current recordings (74) in HEK 293 cells stably expressing STX-sensitive rat skeletal muscle Na channels (μ1) (67). These cells display robust Na currents (Figure 4), and owing to their small size (diameter 14 ± 5 μm), it is possible to avoid voltage-clamp inhomogeneities during experiments

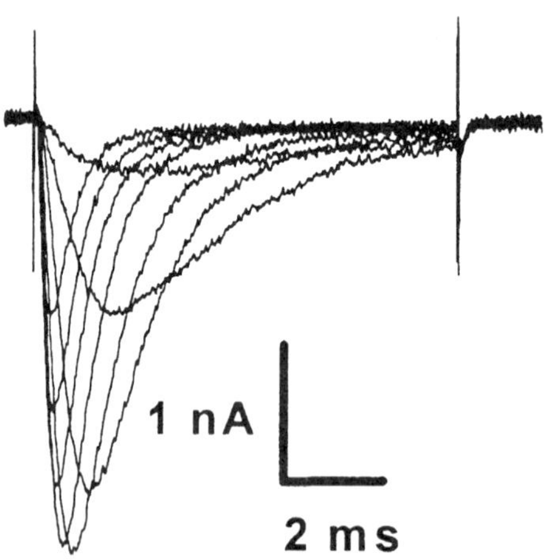

Figure 4 A family of skeletal muscle sodium currents in HEK 293 cells obtained in response to voltage pulses from −40 to +30 mV from a holding potential of −100 mV. The peak I_{Na} was elicited at −10 mV.

that may last on average more than 60 min of useful recording time. Bath-applied STX blocks peak sodium currents (I_{Na}) in a concentration- and time-dependent manner (68). As shown in Figure 5a, after perfusing the cells with a solution containing STX, I_{Na} decreased with a time constant that depended on the frequency of stimulation (68) and reached a new steady-state level determined by the STX concentration. Figure 5b displays the concentration-dependent decrease of I_{Na} amplitude in the presence of 0.3, 1.0, and 3.0 nM standard STX. Figure 5b shows a full dose-response curve with an IC50 of 1.10 ± 0.12 nM (n = 7).

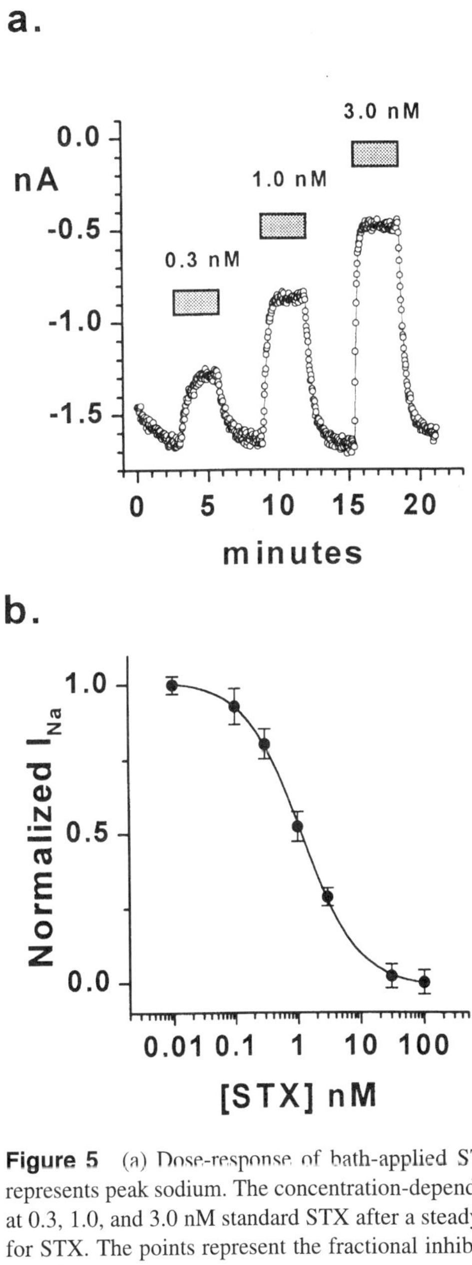

Figure 5 (a) Dose-response of bath-applied STX on sodium currents in HEK 293 cells. Each point represents peak sodium. The concentration-dependent decrease of sodium current amplitude was measured at 0.3, 1.0, and 3.0 nM standard STX after a steady-state level was reached. (b) Reference inhibition curve for STX. The points represent the fractional inhibition of peak I_{Na} and correspond to the average of 5–14 experiments.

We next used the electrophysiological assay to evaluate PSP toxicity in naturally contaminated samples. Figure 6a demonstrates the effect of application and washout of 1 nM STX after stabilization of the electrical recording. Diluted PSP samples were then applied sequentially interspersed with purified STX (1 nM). The regulatory level of 80 μg of STX equivalent per 100 g tissue is comparable to a solution of 1910 nM STX diacetate. Thus, it was necessary to dilute toxic samples to approximately 1 nM final concentration to obtain an I_{Na} blockage near 50%. This dilution factor of three orders of magnitude eliminates variations in pH and/or high concentrations of divalent cations, factors that are known to modify STX binding (22,34–37,69). In addition, matrix effects that could interfere with the assay are minimized.

To compare toxicity evaluations by mouse bioassay and the electrophysiological assay, a set of 26 samples of PSP extracts ranging from 35 to 800 μg STX eq/100 g was used. Figure 6b demonstrates a good correlation between the two assays (R = 0.958). The practical detection limit was determined by addition of decreasing STX concentrations. Perfusion of 0.1 nM STX produced a small but detectable current inhibition (not shown). Average values of peak I_{Na} before and after STX were significantly different and the observed deflection was equivalent to a signal to noise ratio > 5:1. Smaller inhibitions could still be detected, but 0.1 nM STX represents the practical detection limit. This is equivalent to 0.042 μg STX/100 g or 19100 times below the regulatory limit and 8550 times below the mouse test detection limit (based on STX diacetate). This assay is three orders of magnitude more sensitive than the mouse bioassay, inexpensive, and avoids the use of experimental animals and radioactive STX or STX analogues as internal standards. These results are of immediate utility, and they also provide the basis for simplified biosensors based on recombinant Na channels.

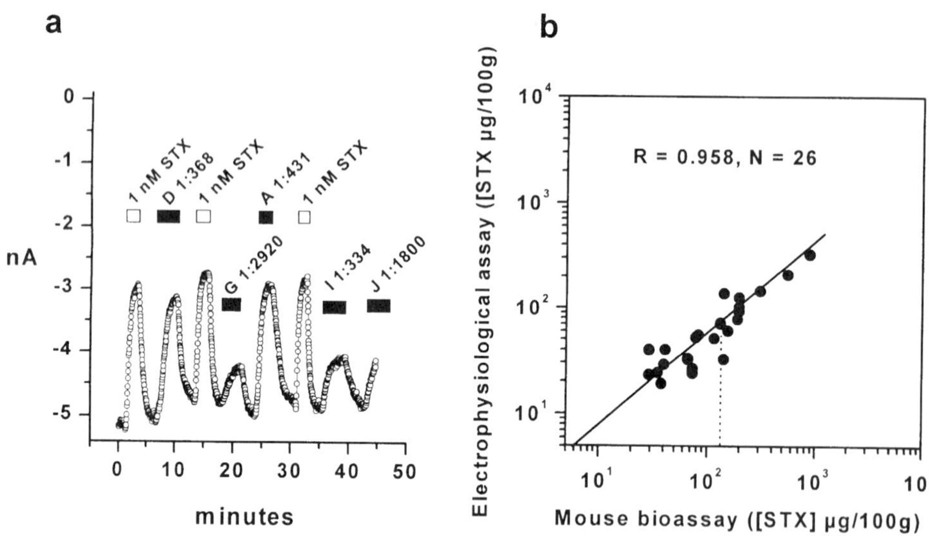

Figure 6 (a) Electrophysiological measurement of PSP toxicity of extracts from contaminated shellfish. Each point represents peak I_{Na} at −10 mV. The open bars represent bath application of standard STX; other bars indicate application of toxic PSP extracts. Extracts were first tested at 1000-fold dilution and then diluted to produce fractional inhibition of I_{Na} near 0.5. Dilution factors are shown in the figure. In this example, five samples were tested within 50 min. (b) Correlation between PSP toxicity levels determined by bioassay and the electrophysiological assay. Toxicity performed by the electrophysiological assay was determined in dilutions of shellfish extracts (range 1:800–1:8000). A significant correlation (R = 0.958 for 26 samples) was found between the toxicity levels determined by mouse bioassay and the electrophysiological assay. The dotted lines indicate the safety limit of 80μg STX eq/100 g.

V. CONCLUSIONS: THE NEED FOR ALTERNATIVE AND COMPLEMENTARY BIOLOGICAL ASSAYS

Fatal PSP intoxications represent the most serious threat of marine origin worldwide (28) with notable public health and economic impacts in Asia, Europe, North America, and South America. As a consequence, most seafood-exporting countries have established mandatory PSP toxin screening programs. The standard biological method used worldwide to monitor levels of STX and other marine toxins in seafood is the semiquantitative mouse mortality bioassay (3). This procedure is reliable for regulatory purposes, but is costly and time consuming, especially when a large number of samples has to be analyzed. Its major limitation, however, is the controversial use of live animals. The assay measures the time to death (''to the last gasp'') (3) after intraperitoneal injection of seafood extract, a procedure that has received such ethical scrutiny that it can no longer be carried out in some European countries (47).

Analytical HPLC methods require oxidation of STX analogues for fluorescent detection (48,70) and the availability of scarce sets of analogues as internal standards. Acidic extracts from naturally contaminated PSP shellfish samples may contain over 20 different analogues of STX in variable proportions (70,71). Analytical methods provide the molar composition of toxic extracts, a quantity that has to be transformed into intraperitoneal toxicity values. This calculation still relies on scales of relative toxicities obtained by mouse bioassay carried out with pure STX analogues (48). Recently, the routine application of radioassays has been further hampered by strict restrictions concerning the international transfer of STX and radiolabeled derivatives owing to biological warfare conventions (72).

It is hoped that the large body of evidence on sodium channel properties based on structure-function studies should make it feasible for the development of a biosensor for saxitoxins and other toxins directed to the sodium channel that is based on their specific molecular target, the sodium channel protein.

ACKNOWLEDGMENTS

Work done by the authors and reported in this chapter was funded by Conicyt-FONDEF Grant 2-37 to B.A. Suárez-Isla.

REFERENCES

1. S Hall, G Stricharzt, E Moczydlowski, A Ravindran, PB Reichardt. The saxitoxins: sources, chemistry and pharmacology. In: Marine Toxins: Origin, Structure and Molecular Pharmacology. American Chemical Society Symposium Ser. 418 1990, pp 29–65.
2. CY Kao. Structure activity relations of tetrodotoxin, saxitoxin and analogues. In: CY Kao, SR Levinson, eds. Tetrodotoxin, Saxitoxins and the Molecular Biology of the Sodium Channel, New York, The New York Academy of Sciences, 1986, pp. 52–67.
3. W Horwitz, ed. Paralytic Shellfish Poison. Biological Method. Final Action. In Official Methods of Analysis. AOAC, Arlington, VA: 1990, pp 881–882.
4. D Andrinolo, LF Michea, N Lagos. Toxic effects, pharmacokinetics and clearance of saxitoxin, a component of paralytic shellfish poison (PSP), in cats. Toxicon 37:447–464, 1999.
5. WA Catterall. Neurotoxins that act on voltage-sensitive sodium channels in excitable membranes. Annu Rev Pharmacol Toxicol 20:15–43, 1980.
6. WA Catterall. Cellular and molecular biology of voltage-gated sodium channels. Physiol Rev 72: S15–S48, 1992.

7. B Hille. The receptor for tetrodotoxin and saxitoxin. A structural hypothesis. Biophys J 15:615–619, 1971.
8. HA Fozzard, DA Hanck. Structure and function of voltage-dependent sodium channels: comparison of brain II and cardiac isoforms. Physiol Rev 76:887–926, 1996.
9. E Marban, T Yamagishi, GF Tomaselli. Structure and function of voltage-gated sodium channels. J Physiol (Lond.) 508.3:647–657, 1998.
10. S Cukierman. Regulation of voltage-dependent sodium channels. J Membr Biol 151:203–214, 1996.
11. CM Armstrong, B Hille. Voltage-gated ion channels and electrical excitability. Neuron 20:371–380, 1998.
12. TJ Smayda. What is a bloom: a commentary. Limnol Oceanogr 42:1132–1136, 1997.
13. TJ Smayda. Harmful algal blooms: their ecophysiology and general relevance to phytoplankton blooms in the sea. Limnol Oceanogr 42:1137–1153, 1997.
14. A Sournia. Red-tide and toxic marine phytoplankton of the world ocean: an inquiry into biodiversity. In: P Lassus, G Arzul, E Erard, P Gentian, C Marcaillou (eds.) Harmful Marine Algal Blooms. Proceedings of the 6th Internation Conference on Toxic Marine Phytoplankton. Paris: Lavoisier, 1995, pp 103–112.
15. DM Anderson, DM Kulis, GJ Doucette, JC Gallagher, E Balech. Biogeography of toxic dinoflagellates in the genus *Alexandrium* from the northeastern United States and Canada. Marine Biol 120: 467–478, 1989.
16. GM Hallegraef. A review of harmful algal blooms and their apparent global increase. Phycologia 32:79–99, 1993.
17. T Yasumoto, M Murata. Marine toxins. Chem Revi 93:1897–1909, 1993.
18. Y Shimizu. Microalgal metabolites: a new perspective. Annu Rev Microbiol 50:431–435, 1996.
19. DG Baden. Brevetoxins: unique polyether dinoflagellate toxins. FASEB J 3:1807–1817, 1989.
20. R Lewis. Detection of Ciguatoxins and related benthic dinoflagellate toxins: *in vivo* and *in vitro* methods. In: Manual of Harmful Marine Microalgae. CM Hallegraef, DM Anderson, AD Cembella, eds. IOC Manual and Guides No. 33, 1995, pp 135–161.
21. JM Ritchie, RB Rogart. The binding of saxitoxin and tetrodotoxin to excitable tissue. Rev Physiol Biochem Pharmacol 79:1–50, 1977.
22. RL Barchi, JB Weigele. Characteristics of saxitoxin binding to the sodium channel of sarcolemma isolated from rat skeletal muscle. J Physiol (Lond) 295:383–396, 1979.
23. RL Barchi, SA Cohen, LE Murphy. Purification from rat sarcolemma of the saxitoxin-binding component of the excitable membrane sodium channel. Proc Natl Acad Sci USA 77:1306–1310, 1980.
24. GR Stricharz. Structural determinants of the affinity of saxitoxin for neuronal sodium channels. J Gen Physiol 84:281–305, 1980.
25. W Carmichael. The cyanotoxins. Adv Botanical Res 27:211–256, 1997.
26. T Ogata, M Kodama, K Kodaru, S Sakamoto, S Sato, U Simidu. Production of paralytic shellfish toxins by bacteria isolated from toxic dinoflagellates. In: E Granéli, B Sundstroem, L Edler, DM Anderson, eds. Toxic Marine Phytoplankton, New York: Elsevier. 1990, pp 311–315.
27. M Kodama, T Ogata, S Sakamoto, S Sato, T Honda, T Miwatami. Production of paralytic shellfish toxins by a bacterium *Moraxella* sp. Isolated from *Protogonyaulax tamarensis*. Toxicon 28:707–714, 1990.
28. CM Hallegraeff. Harmful algal blooms: a global overview. In: Manual of Harmful Marine Microalgae. CM Hallegraef, DM Anderson, AD Cembella Eds. IOC Manual and Guides No. 33, 1995, pp 1–22.
29. DM Anderson. Bloom dynamics of toxic Alexandrium species in the northeastern U.S. Limnol Oceanogr 42:1009–1022, 1997.
30. P Uribe C, BA Suárez-Isla, RT Espejo. Ribosomal RNA heterogeneity and indentification of toxic dinoflagellate cultures by heteroduplex mobility assay. J Phycol 35:884–888, 1999.
31. B Hille, Ion Channels of Excitable Membranes. 2nd ed. Sunderland, MA: Sinauer, 1992.
32. WA Catterall. Structure and function of voltage-sensitive ion channels. Annu Rev Biochem 64:493–531, 1995.
33. H Terlau, SH Heinemann, W Stuhmer, M Push, F Conti, K Imoto, S Numa. Mapping the site of block by tetrodotoxin and saxitoxin of sodium channel-II. FEBS Lett 293:93–96, 1991.

34. W Ulbricht, HH Wagner. The influence of pH on equilibrium effects of tetrodotoxin on myelinated nerve fibers of Rana esculenta. J Physiol (Lond.) 252:159–184, 1975.
35. W Ulbricht, HH Wagner. The influence of pH on the rate of tetrodotoxin action on myelinated nerve fibers. J Physiol (Lond.) 252:185–202, 1975.
36. D Colquhoun, R Henderson, JM Ritchie. The binding of labelled tetrodotoxin to non-myelinated nerve fibers. J Physiol (Lond.) 227:95–126, 1972.
37. R Henderson, JM Ritchie, GR Strichartz. Evidence that tetrodotoxin and saxitoxin act at a metal cation binding site in the sodium channels of nerve membrane. Proc Natl Acad Sci USA. 71:3936–3940, 1974.
38. MH Noda, H Susuki, S Numa, W Stuhmer. A single point mutation confers tetrodotoxin and saxitoxin insensitivity on the sodium channel-II. FEBS Lett 259:213–216, 1989.
39. MM Stephan, JF Potts, WS Agnew. The μ1 skeletal muscle sodium channel: mutation E403Q eliminates sensitivity to tetrodotoxin but not to μ-conotoxins GIIIA and GIIIB. J Membr Biol 137:1–8, 1994.
40. P Backx, D Yue, J Lawrence, E Marban, GF Tomaselli. Molecular localization of an ion-binding site within the pore of mammalian sodium channels. Science 257:248–251, 1992.
41. I Favre, E Moczydlowski, L Schild. Specificity for block by saxitoxin and divalent cations at a residue which determines sensitivity of sodium channel subtypes to guanidinium toxins. J Gen Physiol 106: 203–229, 1995.
42. MT Pérez-García, M Chiamvimonvat, E Marban, GF Tomaselli. Structure of the sodium channel revealed by serial cysteine mutagenesis. Proc Natl Acad Sci USA 93:300–304, 1996.
43. J Satin, JW Kyle, M Chen, P Bell, LL Cribbs, RB Rogart. A mutant of TTX-resistant cardiac sodium channels with TTX-sensitive properties. Science 256:1202–1205, 1992.
44. GM Lipkind, HA Fozzard. A structural model of the tetrodotoxin and saxitoxin binding site of the Na^+ channel. Biophys J 66:1–13, 1994.
45. GE Kirsh, M Alam, HA Hartmann. Differential effects of sulfhydryl reagents on saxitoxin and tetrodotoxin block of voltage dependent Na channels. Biophys J 67:2305–2315, 1994.
46. JL Penzotti, HA Fozzard, GM Lipkind, SC Dudley Jr. Differences in saxitoxin and tetrodotoxin binding revealed by mutagenesis of the Na^+ channel outer vestibule. Biophys J 75:2647–2657, 1998.
47. AD Cembella, L Milenkovic, G Doucette, M Fernández. In vitro biochemical methods and mammalian bioassays for phycotoxins. In: Manual of Harmful Marine Microalgae. CM Hallegraef, DM Anderson, AD Cembella, eds. IOC Manual and Guides No. 33, 1995, pp 177–228.
48. Y Oshima. Post-column derivatization HPLC methods for paralytic shellfish poisons. In: Manual of Harmful Marine Microalgae. CM Hallegraef, DM Anderson, AD Cembella, eds. IOC Manual and Guides No. 33, pp 81–94, 1995.
49. H Sommer, KF Meyer. Paralytic shellfish poisoning. Arch Pathol 24:560–598, 1937.
50. EJ Shantz, EF McFarren, ML Schaffer, KH Lewie. Purified poison for bioassay standardization. J Assoc Off Anal Chem 41:160–168, 1958.
51. A Mc Culloch, R Boyd, A De Freitas, R Foxall, W Jamieson, M Laycock, M Quilliam, J Wright. Zinc from oyster tissue as causative factor in mouse deaths in official bioassay for paralytic shellfish poison. J Assoc Off Anal Chem 72:384–386, 1989.
52. A Gago-Martinez, JA Rodriguez-Vazquez, P Thibault, MA Quilliam. Simultaneous occurrence of diarrhetic and paralytic shellfish poisoning toxins in Spanish mussels in 1993. Nat Toxins 4:72–79, 1996.
53. T Yasumoto, M Murata, Y Oshima, K Matsumoto, J Clardy. Diarrhetic shellfish poisoning. In: EP Ragelis, ed. Seafood Toxins. ACS Symposium Series 262:207–214. Washington, DC: American Chemical Society, 1984.
54. MR Ross, A Siger, BC Abbott. The house fly: an acceptable subject for paralytic shellfish toxxin bioassay. In: Toxic Dinoflagellates. DM Anderson, AW White, DG Baden, eds. New York: Elsevier, 1985, pp 433–438.
55. J McElhiney, LA Lawton, C Edwards, S Gallacher. Development of a bioassay employing the desert locust (*Schistocerca gregaria*) for the detection of saxitoxin and related compounds in cyanobacteria and shellfish. Toxicon 36:417–420, 1998.
56. LJ DeFelice. Electrical Properties of Cells (Patch-Clamp for Biologists). New York: Plenum Press, 1997.

57. AL Hodgkin and AF Huxley. A quantitative description of membrane current and its application to conduction and excitation in nerve. J Physiol (Lond) 117:500–544, 1952.
58. JF Jellet, LJ Marks, JE Stewart, ML Dorey, W Watson-Wright, JF Lawrence. Paralytic shellfish poison (saxitoxin family) bioassays; automated endpoint determination and standardisation of the *in vitro* tissue culture bioassay, and comparison with the standard mouse bioassay. Toxicon 30:1143–1156, 1992.
59. JF Jellet, JE Stewart, MV Laycock. Toxicological evaluation of saxitoxin, neosaxitoxin, gonyautoxin II, gonyautoxin II plus III and decarbamoylsaxitoxin with the mouse neuroblastoma cell bioassay. Toxic In Vitro 9:57–65, 1995.
60. K Kogure, ML Tamplin, U Simidu, RR Colwell. A tissue culture assay for tetrodotoxin, saxitoxin and related toxins. Toxicon 26:191–197, 1988.
60a. SS Garber, C Miller. Single Na^+ channels activated by veratridine and batrachotoxin. J Gen Physiol 89:459–480, 1988.
61. S Gallacher, TH Birbeck. A tissue culture assay for direct detection of sodium channel blocking toxins in bacterial culture supernatants. FEMS Microbiol Lett 92:101–108, 1992.
62. FM Van Dolah, EL Finley, BL Haynes, GJ Doucette, PD Moeller, JS Ramsdell. Development of rapid and sensitive high throughput pharmacologic assays for marine phycotoxins. Nat Toxins 2: 189–196, 1994.
63. SR Davio, PA Fontelo. A competitive displacement assay to detect saxitoxin and tetrodotoxin. Anal Biochem 141:199–203, 1984.
64. MR Vieytes, AG Cabado, A Alfonzo, MC Louzao, AM Botana, LM Botana. Solid-phase radioreceptor assay for paralytic shellfish toxins. Anal Biochem 211:87–93, 1993.
65. GJ Doucette, MM Logan, JS Ramsdell, FM Van Dolah. Development and preliminary validation of a microtiter plate-based receptor binding assay for paralytic shellfish poisoning toxins. Toxicon 35: 625–636, 1997.
66. J Sierralta, M Fill, BA Suárez-Isla. Functionally heterogenous ryanodine receptors in avian cerebellum. J Biol Chem 271:17028–17034, 1996.
66a. BS Cheun, M Loughran, T Hayashi, Y Nagashima, E Watanabe. Use of a channel biosensor for the assay of paralytic shellfish toxins. Toxicon 36:1371–1381, 1998.
67. T Yamagishi, M Janecki, E Marban, GF Tomaselli. Topology of the P segments in the sodium channel pore revealed by cysteine mutagenesis. Biophys J 73:195–204, 1997.
68. F Conti, A Gheri, M Pusch, O Moran. Use dependence of tetrodotoxin block of sodium channels: a revival of the trapped-ion mechanism. Biophys J 71:1295–1312, 1996.
69. DD Doyle, Y Guo, SL Lustig, J Satin, RB Rogart, HA Fozzard. Divalent cation competition with [^{3}H]saxitoxin binding to tetrodotoxin-resistant and -sensitive sodium channels: a two site structural model of ion/toxin interaction. J Gen Physiol 101:153–182, 1993.
70. Y Oshima, CJ Bolch, GM Hallegraef. Comparative study on paralytic shellfish toxin profiles of the dinoflagellate *Gymnodinium catenatum* from three different countries. Marine Biol 116:471–476, 1993.
71. N Lagos, D Compagnon, M Seguel, Y Oshima. Paralytic shellfish toxin composition: A quantitative analysis in Chilean mussels and dinoflagellates. In: Harmful and Toxic Algal Blooms. T Yasumoto, Y Oshima, Y Fukuyo, eds. IOC UNESCO, 1996, pp. 121–124.
72. OPCW: Organization for the Prohibition of Chemical Weapons. STX was included in the list of toxic compounds qualified as potential chemical weapons. As of August 1, 1997, STX cannot be re-exported even among treaty-signing countries.
73. P Vélez, BA Suárez-Isla, J Sierralta, M Fonseca, H Loyola, DC Johns, GF Tomaselli and E Marbán. Electrophysiological assay to quantify saxitoxins in contaminated shellfish. Biophys J 76:A82, 1999.
74. OP Hamill, A Marty, E Neher, B. Sakmann, FJ Sigworth. Improved patch-clamp techniques for high-resolution current recording from cells and cell-free membrane patches. Pfluegers Arch Eur J Physiol 391:85–100, 1981.
75. L Schild, E Moczydlowski. Competitive binding interaction between Zn^{2+} and saxitoxin in cardiac Na^+ channels. Evidence for a sulfhydryl group in the Zn^{2+}/saxitoxin binding site. Biophys J 59:523–537, 1991.

10
Paralytic Shellfish Poisoning (PSP): Toxicology and Kinetics

Néstor W. Lagos and Darío Andrinolo
University of Chile
Santiago, Chile

I. INTRODUCTION

Paralytic shellfish poisoning (PSP) has been recognized for centuries as a clinical entity, and the prevention of human intoxication due to the ingestion of toxic shellfish has been a problem of mutual interest to public health and fishery authorities all over the world for the last 60 years (1–4). From the standpoint of the reported number of deaths each year, PSP cannot be classified as a major public health problem. Nevertheless, it is of considerable concern, because a fatal dose of the poison for humans, which may be obtained from a single serving of highly toxic shellfish, is only around a milligram, and there is no known antidote. Widespread distribution of poisonous shellfish products would present a hazard to every shellfish consumer, because the PSP toxins are very stable.

A. PSP Toxins

PSP is a mixture of 26 different toxins (5,6) that have a skeleton denominated by 3,4,6-trialquil tetrahidropurine (7–9).

According to the net charge that these compounds, at pH 7, they can be classified in three groups: (1) saxitoxins (STXs) with a net charge of +2, (2) gonyaulatoxins (GTXs) group with net charge of +1, and (3) N-sulfocarbamoyl-ll-hydroxysulfate toxins (C) with a net charge of 0.

STX was the first PSP toxin isolated and also the most studied one. The structure of STX was established by x-ray analysis of crystalline derivatives carried out by Schantz (7). The two guanidinic groups that STX possesses in positions 2 and 8 are responsible for its positive net charge. Titulation of this molecule reveals two protonic dissociations at pKa 8.22 and 11.28. In concordant with these data, the toxicity of STX is relatively constant between pH 6.5 and 8.2, and it diminishes markedly to higher pH (1,10).

STX is a neurotoxin that blocks the sodium voltage-gated channel of excitable cells (3,10,12–18). All PSP toxins act by reversibly binding to a receptor on the voltage-gated sodium channel, blocking neuronal transmission. These toxins act only from the exterior surface of the channel, since its binding site is located on the outer opening of the sodium channel (13,19).

Table 1 Clinical Symptoms Caused by PSP

Low intoxication	Moderate intoxication	High intoxication
Nausea and vomiting	Respiratory insufficiency	Pupils dilated and unreactive
Muscular weakness	Jaw and facial muscles paralyzed	Deep dystonia
Lips paresthesia	Gag reflex absent	Respiratory paralysis
Upper extremity paresthesia	Tongue immobile	Hypotension
		Cardiac arrest

B. Paralytic Shellfish Poisoning

Accidental PSP in humans is primarily caused by the ingestion of PSP toxin–contaminated mollusks. Symptoms associated with mild to moderate levels of intoxication consist of a state of generalized malaise, facial paresthesia, asthenia, dystonia, ataxia, dyspnea, hypotension, tachycardia, vomiting, disphagia, headache, and gastrointestinal disturbance. In the case of serious intoxication, respiratory arrest and/or cardiovascular shock also occur (Table 1). Nerve conduction studies in patients reveal marked prolongation of distal motor and distal and sensory latencies, a less dramatic but significant decrease in conduction velocities, and moderately diminished motor and sensory amplitudes. All parameters become normal within 72 h, paralleling the patients' recovery (3,20).

II. PHARMACOLOGY AND TOXICOLOGY

Although STX's mechanism of action is well established at the molecular level, very little is known about its distribution in the total body, where and how these toxins are absorbed, how these toxins are excreted by mammalians that eat shellfish contaminated with PSP toxins, or if these toxins can go through the blood-brain barrier.

There are several reasons why the research on the pharmacokinetics and toxic effects of PSP toxins has been hindered. In the first place, the only available PSP toxin was STX; therefore, the most of the investigations have been done using only STX. In spite of the fact that no other PSP toxins have been studied, it is possible that all PSP toxins have similar pharmacokinetic behavior because of their similar chemical structure and molecular action.

On the other hand, the research on the pharmacokinetics of PSP toxins has been complicated because of the relative low sensitivity and nonspecificity of the methods used to detect STX in biological fluids and tissues.

The effects caused by intoxication with PSP can be reproduced by means of the administration of STX to laboratory animals (12,21–23,25,26). These effects consist of cardiovascular alterations and breathing problems, as have been reported in clinical cases.

A. Cardiovascular Effects

STXs are among the most potent hypotensive agents known. In anesthetized cats and dogs, as little as 2–3 μg/kg injected intravenously produces a fall in arterial pressure to about 50% from the normal level (21,23,25). At these doses, STX does not affect the heart significantly, and during the hypotensive event, a pressor effect can be elicited by intravenous injection of α-adrenergic drugs. The last two observations suggest that, although the vascular smooth muscle is not affected by these toxins, the neurally controlled vasomotor tone is reduced. Since STX is an extremely potent axonal blocking agent, it may block sympathetic nerves peripherally and lead to a loss of vasomotor tone. Another possibility is that STX has some specific depressant actions on the medullary vasomotor center.

When intravenous doses of STX are smaller than 1 μg/kg, STX has a hypertensive effect, as a compensatory response to the vasodilatory action of STX. This response disappears completely when doses higher than 1.5 μg/kg are administrated (12,21).

With an intravenous dose of 2.7 μg/kg of STX, the mean arterial pressure falls in the first 10 min to about 50% of the control and recovers to a normal level after 50 min (Figure 1a). Because STX causes a peripheral vasodilatation by blocking vasomotor nerves, inhibiting

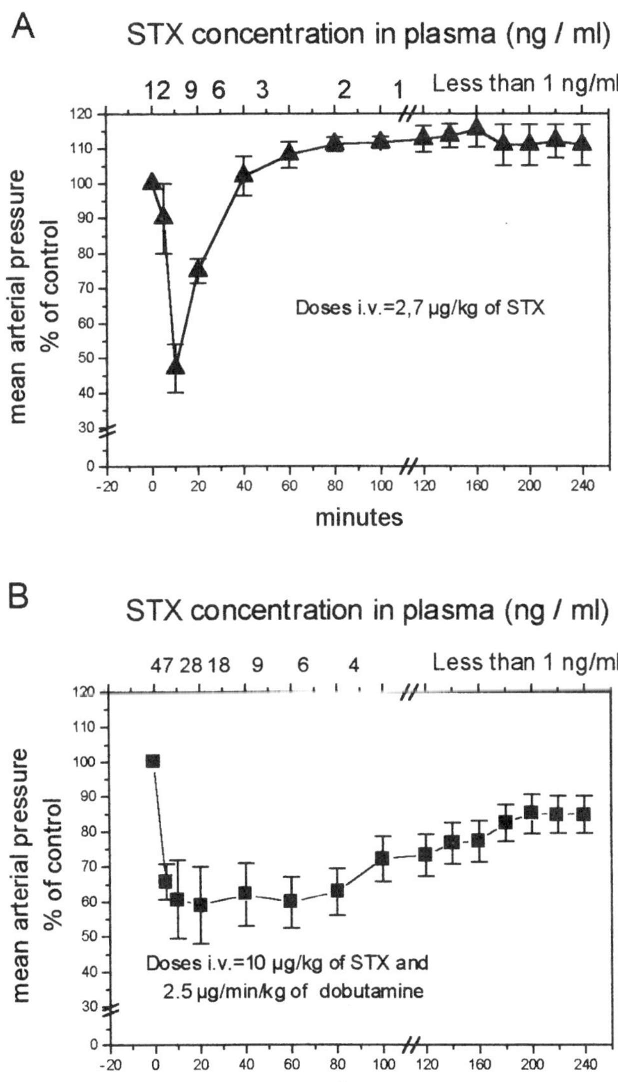

Figure 1 Effects of STX on cardiovascular performance. Arterial pressure (% of control) of cats intoxicated with low doses (2.7 μg STX/kg) (triangle) (A) and high doses (10 μg STX/kg) (filled square) (B). Means of results from four cats were graphed. Bars indicate SEM.

the conduction of impulses in nerve fibers, it is possible to have a pressor effect after the administration of dopamine (25).

The shock generated by the administration of high doses of STX (10 μg/kg) is a result of a combination of the vascular hypotensive effect of STX on the vascular net and the decrease of the cardiac output, which is produced as a final consequence of the lack of venous return and hypoxemia. Therefore, cardiovascular shock takes place (12,21,22,25). It is possible to overcome these effects in highly intoxicated cats using the adrenergic agonist dobutamine. This can be administrated intravenously as an initial bolus of 500 μg immediately after STX and followed by constant intravenous infusion of 2.5 μg/min/kg (25). Using this therapeutic method, cats can recover the mean arterial pressure to a steady state of 75% of the control mean arterial pressure (Figure 1b). The severe arrhythmia is controlled and the sinusoidal rhythms return in less than 30 min (Figure 2). Although the regulation of the arterial pressure could be made by means of the dopamine administration owing to its pressor effect on the vasculature. Dobutamine administration and infusion of physiological serum allowed better and more stable recovery of intoxicated cats (25).

Approximately one third of the guinea pigs used in studies of STX cardiorespiratory effects died of myocardial failure prior to apnea. Similar to the cat electrocardiogram (EKG) shown in Figure 2, guinea pigs also developed ventricular tachycardia and other anomalies such as right/left bundle block, t-wave inversion, and atrioventricular (AV) dissociation. Moreover, with

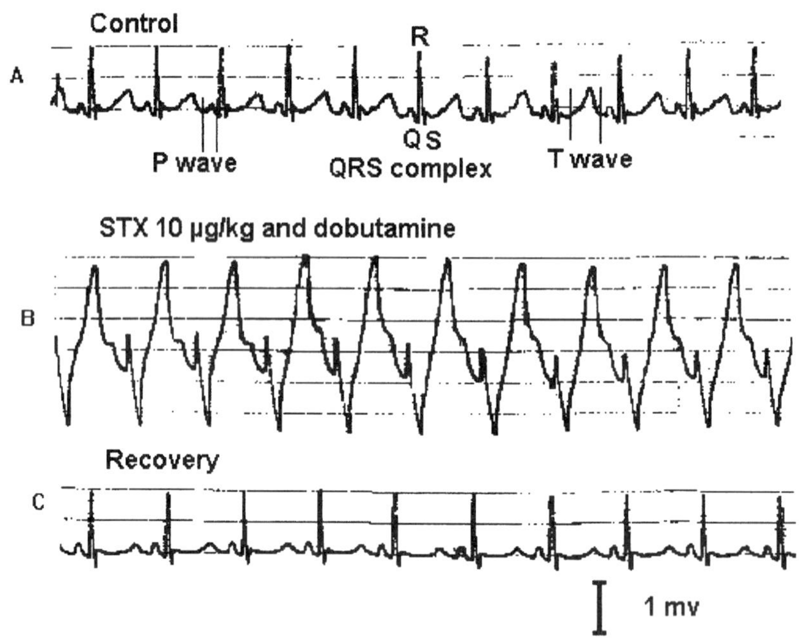

Figure 2 Sequences of an electrocardiogram corresponding to an animal intoxicated with a high dose of STX. (A) Electrocardiogram control taken before toxin administration, which shows rhythm of 136 beats/min. (B) Five minutes after administration of the toxin. These beats have a ventricular origin. (C) Electrocardiogram taken for 30 min. Normalization of the EKG is observed: P waves (sinusal node) and T waves of ventricular repolarization reappear and the QRS complex (ventricular contraction) is normal and is preceded by a P wave. Although a slight falling of the pulse is appreciated, it is within the normal values.

a dose of 10 μg of STX/kg, cats and guinea pigs died within a few minutes owing to cardiovascular shock. Ventilatory insufficiency was not the cause of death, because cats were initially anesthetized and coupled to mechanical ventilatory support (25).

With STX doses ranging from 3 to 10 μg/kg administrated intravenously, is common the findings of arrhythmia, showing that the pacemaker and/or the conduction system of the myocardium also appeared to show a high degree of sensitivity to STX (Figure 3).

The fall in the heart rate (bradycardia) observed in the sequences shown in Figure 3A (control) and 3B (in the presence of STX) can be explained by blockade of conduction in sympathetic fibers carrying impulses to the pacemaker region and not to a direct effect on cells of the sinoatrial node, as occurs with tetrodotoxin (TTX) administration in cats (27). The 2:1 AV conduction ratio shown in Figure 3D shows cases a normal sinusoidal beats alternate with ventricular beats, this means that a second-degree AV block occurs. This arrhythmia can be followed by ventricular pulses and the regular R-R tracing suggests the development of complete AV block. In this condition, it is also possible to observe an inverted QRS complex displaying an intraventricular conduction defect. The term *intraventricular conduction defect* is frequently used when the QRS complex is abnormal, but within normal limits of duration, like the one shown in Figure 3B.

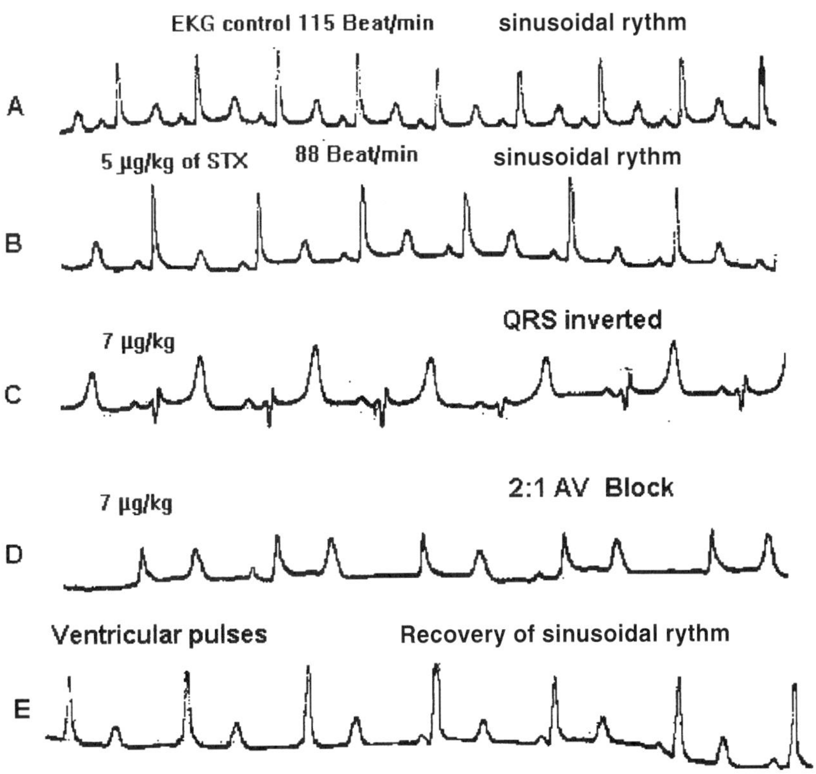

Figure 3 Arrhythmias found in cats intoxicated with doses between 3 and 7 μg of STX/kg. (A) EKG control with a normal sinusoidal rhythm. (B) Bradycardia developed after administration of 5 μg/kg. (C) EKG shows an inverted QRS complex. (D) 2:1 AV block. (E) EKG displaying the recovery of a normal sinusoidal rhythm from a total AV dissociation as is suggested by the occurrence of regular ventricular pulses.

B. Respiratory Effects of STX

Progressive failure of respiratory function has always been recognized as the most life-threatening element of PSP intoxication (3,20,26).

In studies with cats, it was found that intravenous STX doses of 3.3 μg/kg caused motor paralysis of respiration (26). This failure is shown by the loss of activity in the diaphragmatic electromyogram (EMG) and makes artificial respiration necessary. Rhythmic phrenic discharges persisted at a slower rate than the control. STX administrated intravenously at a low dose (3 μg/kg) does not affect the regular cardiac beat. The diaphragm blockade clearly represents a direct peripheral effect of STX.

When high doses of STX are administrated intravenously, in addition to a variety of peripheral cardiorespiratory effects, STX also appears to have a profound influence on the central respiratory rhythmogenic network (22,26). The mechanism by which the rate and rhythm of central discharge can be modified by peripheral paralytic chemical action is not known. Hence, the inescapable conclusion is that the slowing of phrenic respiratory discharge frequency could be due to the direct effects of STX on the central nervous system.

In fact, in 1993, Chang et al. (22) found that STX intravenously induces a prolongation in the Botzinger discharge duration. The Botzinger neurons interact synaptically with all known medullary and bulbospinal respiratory areas related to neural substrates. It is possible that STX can produce site-specific alterations at the Botzinger complex (BOT). STX effects may produce a significant increase in the amount of inhibitory amino acids (gamma-aminobutyric acid [GABA] and glycine) and diminish the release of excitatory amino acids such as glutamate. Physiologically, BOT neurons have been shown to exert an inhibitory influence over inspiratory neurons of the dorsal respiratory group, the ventral respiratory group, and the phrenic motor nucleus. Functionally, in view of its inhibitory influence on bulbospinal inspiratory neurons, the time-dependent BOT discharge could impose an increasing delay in inspiratory activation and reduce the respiratory frequency and ventilatory efficiency.

C. Central Cardiorespiratory Effects of STX

Minute amounts of STX introduced directly into the cerebrospinal fluid (CSF) surrounding the brainstem produce the following sequences of events in the respiration: slowing, apneustic activity, loss of spontaneous function, and finally a depression of responsiveness to electrical stimulation of the respiratory center (23). Likewise, STX administrated directly into the CSF produces dissociation of respiratory control components, since it causes slowing of the breathing frequency which is under central control independently of the changes in the peripheral controlled inspiratory volume.

The significant amount of STX found in total brain (1.81 ng/g of wet tissue) and medulla oblongata (2.50 ng/g of wet tissue) by Andrinolo et al. (25) showed that STX is capable of penetrating the blood-brain barrier when high levels of STX in the intravascular system are reached, as was the case of cats intoxicated with 10 μg of STX/kg (25). These findings support the data published on the central cardiorespiratory effects of STX when applied topically in the lateral cerebral ventricle and brainstem (23,28), by intravenous infusion (12,26), and by intraperitoneal administration (22). Our data reinforced the notion that, at high doses of toxin, the STX-induced lethality involves both central and peripheral cardiorespiratory system components. We agree with the suggestion of Evans et al. that the blood-brain barrier permeability can be drastically altered by such cardiovascular burdens and hypercapnia (29) and in this way allows STX to gain access to the central neuronal environment. Moreover, in 1993, Chang et al. found that uncompensated acidosis (hypercapnia), which results in a persistent decrease in the blood pH, occurs after intravenous administration of STX in guinea pigs.

III. PHARMACODYNAMIA OF STX

A. Elimination of STX

Clinicians have observed that patients intoxicated with PSP toxins who survive for 24 h have an excellent chance for a rapid and full recovery (3,20). Such observations suggest that these toxins undergo rapid excretion and/or are metabolized.

The presence of PSP toxin in the urine has been shown using the mouse bioassay for more than 50 years. The first evidence of this was shown by studies using dogs as the experimental model (1). In agreement with these studies, more recently the presence of PSP toxins has been detected in urine samples of patients intoxicated by consumption of PSP toxin–contaminated shellfish (20,30).

Nevertheless, studies to address this issue using live animal as the experimental model have been complicated by the lack of a reproducible method with sufficient sensitivity to detect picogram quantities of toxin and the high amount of toxin necessary to perform these experiments.

In 1993, Hines et al. (31) used a tritiated reduced form of STX, dihydrosaxitoxin or saxitoxinol ([H^3]STXOL), to study toxin metabolism and elimination in mammals. Saxitoxinol retains most of STX's structural features, but it has less than one hundredth of the STX activity on excitable membranes (8). When [H^3]STXOL is administrated into the rat dorsal penile vein, 60% of the injected radioactivity was excreted in the urine after 4 h and postinjection radioactivity was not detectable in feces (31).

In 1995, Stafford and Hines (24) for the first time quantified the amount of STX in urine samples. They reported that approximately 19% of the STX injected intravenously in rats was excreted during the first 4 h.

Using cats as the experimental animal model, Andrinolo et al. (25) showed STX as being the only PSP toxin excreted by urine during the 4 h of the experiment. Under doses of 2.7 μg of STX/kg, the cats excreted 25% of the total toxin injected. On the other hand, at higher doses such as 10 μg of STX/kg, during the same period of time, only 10% of the total STX injected was excreted (Figure 4).

B. Renal Clearance

The rate of urinary excretion of a substance divided by its plasma concentration is a measure of the minimal volume of plasma required to supply the amount of the substance excreted in the urine in a given period of time.

Inuline is a substance that gains access to the urine only by glomerular filtration and is neither reabsorbed nor secreted by renal tubules. Thus, inuline clearance is used as a reference to compare with the clearance of any substance, because its clearance is equal to the rate of glomerular filtration.

With doses of 10 μg of STX/kg, the measured STX renal clearance in cats was 0.81 mL/min $\times$ kg^{-1}. This value corresponds to 25% of the reported inuline renal clearance in cats (3.24 mL/min $\times$ kg^{-1}) (32). Instead, when lower doses (2.7 μg of STX/kg) were used, the STX renal clearance calculated was 3.99 mL/min $\times$ kg^{-1}. This value correlated very well with inuline clearance.

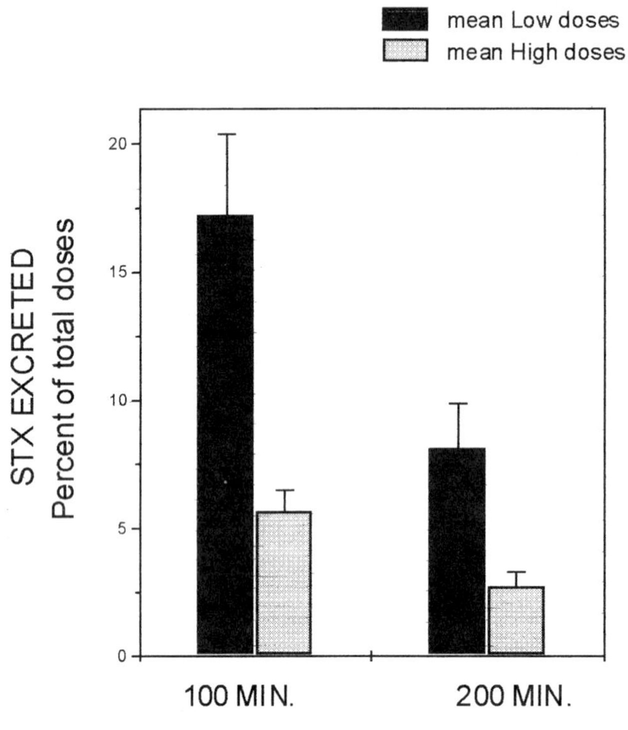

Figure 4 Percentage of STX excreted by urine during the first 200 min after intravenous toxin administration shows the toxic effect of the drugs on renal efficiency. With doses of 2.7 μg of STX/kg (low doses), the cats excreted 25% of the total toxin injected. With doses of 10 μg of STX/kg, only 10% of the total STX injected was excreted.

These data suggest that, in cats with normal cardiovascular parameters and diuresis, the STX excretion mainly involves glomerular filtration. The changes in the renal clearance of the toxin at high doses can be explained owing to cardiovascular complications and the initial hypovolemic shock produced by one bolus intravenous injection of STX. Therefore, the difference obtained between low and high intoxication doses used may be explained by STX's toxic effects. Besides, one of the most important effects of STX intoxication in vivo is vascular hypotension, so the glomerular filtration rate should be reduced. It is important to remember that there is no reported evidence of STX renal tubular reabsorption or secretion.

C. Distribution

To study STX distribution in body compartments, Andrinolo et al. (25) used male cats anesthetized and mechanically ventilated to prevent respiratory failure. The cats were intoxicated with a single-dose intravenous injection of 10 μg of STX/kg. The heart rate, arterial pressure, and central venous pressure were continuously monitored to ascertain the physiological status throughout the course of the experiment.

Andrinolo et al. collected blood samples every 10 min from the cannulated femoral vein and then the samples were analyzed by high-performance liquid chromatography (HPLC) to measure the amount of STX. The first sample was collected 5 min after the dose was administered, which showed the highest STX level (43.27 ± 0.7 ng/mL; N = 4) followed by an exponential decline to 3.57 ng/mL after 80 min (25).

Because the disappearance of STX of the serum occur in an exponential form, the general equation that represents the variation of the quantity of toxin in the organism after a unique dose in the intravascular space; it is given for

$$C = C_0 \cdot e^{-Kt} \tag{1}$$

where C_0 = initial concentration of toxin, K = first order elimination constant, expressed in this case in min^{-1}, and t = time lapsed from the administration of the toxin.

The initial concentration reached was 47 ng/mL. This concentration in blood is sufficient to kill cats in a few minutes by respiratory arrest and cardiovascular shock. When cats were intoxicated with doses of 2.7 μg/mL, the C_0 calculated was 12 μg/mL. This concentration was sufficient to cause respiratory arrest, but no significant cardiovascular problems were detected.

The toxin disappeared from the intravascular compartment with an elimination constant, $K = 0.032\ min^{-1}$, showing an elimination half-life of 21.65 min (25).

To understand the distribution of the toxin in the body, it was necessary to calculate the volume of apparent distribution (VAD). The VAD is the relationship between the serum concentration of the drug and the quantity of drug in the body. Drugs that bind to plasmatic proteins with high affinity present a smaller VAD than real distribution volume. On the other hand, drugs that bind to extravascular binding sites located in tissues show higher values of distribution volume.

In accordance with the one-compartment model, the VAD of STX can be calculated according to Eq. 2.

$$\text{VAD} = \frac{\text{Quantity of toxin in the body}}{\text{Toxin concentration in the serum}} \tag{2}$$

At initial times, the VAD for STX is 212 mL/kg, which is characteristic of acidic drugs like sulfonamides, penicillin, and salicylic that are widely distributed in the intravascular volume. This result means that STX is instantaneously distributed in the intravascular compartment. After 10 min, the VAD reaches a level of 600 mL/kg, showing a completed distribution of STX in the extravascular fluids. Finally, after 50 min, the VAD reaches higher values, showing that the STX was sequestrated by binding to receptors located on cell membranes in different tissues (Figure 5).

The one-compartment model of STX distribution is shown in Figure 6. According to the experimental data, STX goes into the intravascular and extravascular compartments, binds to membrane tissues, and is only excreted by glomerular filtration from the intravascular space. This model shows that distribution of STX in the mammalian body is extremely expeditious, being able to exchange easily from one compartment to another.

The one-compartment model implies free distribution between the vascular and extravascular spaces. With one specific binding site of high affinity, which should be the voltage-gated sodium channel, it is widely distributed in different tissues associated with the extravascular space.

Only free STX in the intravascular compartment is accessible to be filtered in renal glomerulus to be excreted. Therefore, STX and its analogues' (PSP toxins) distribution and excretion should be limited by their affinity to the binding sites, the voltage-gated Na^+ channels in the muscle and nervous tissues. According to this, less toxic PSP components, the ones that have less affinity for the sodium channel (33), should be more as free toxin, so they should be eliminated faster than the more toxic ones like STX, neoSTX, and GTX-1/4 epimers.

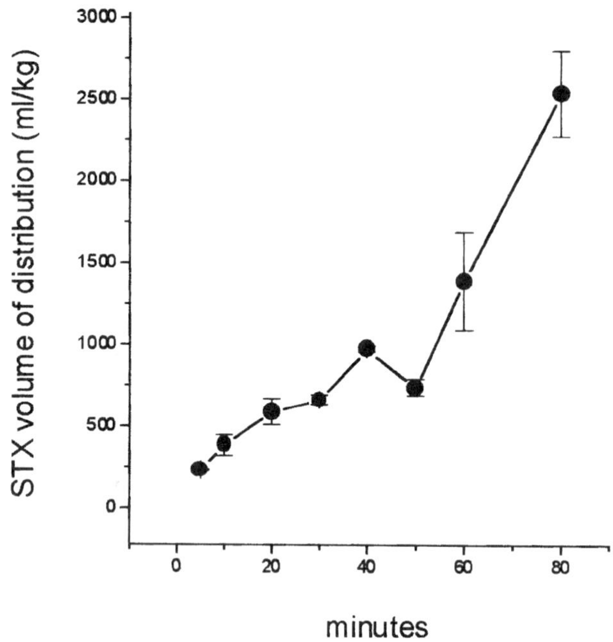

Figure 5 STX volume of distribution. The relationship between the serum concentration of the drug and the quantity of drug in the body suggesting a very fast distribution in the vascular and extravascular spaces when the toxin is administrated into the vascular space followed by the STX bound to an extravascular binding site (mean ± SEM, N = 4).

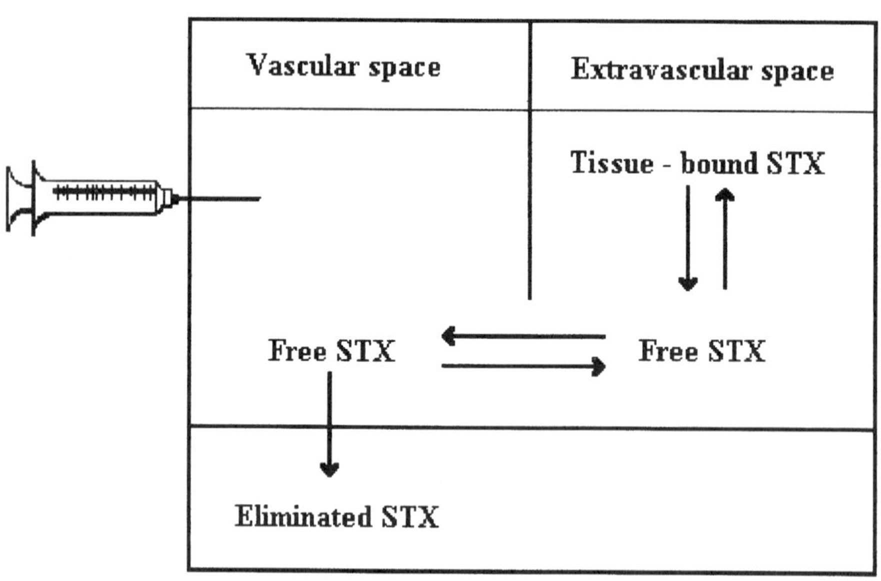

Figure 6 Model of STX distribution in the mammalian body. This is a one-compartment pharmacokinetics model in that the drug is distributed between two physiological spaces. This model has one possible binding site in the extravascular space. Only free STX in the vascular space is available to be excreted by the renal system.

The STX distribution into the extravascular fluid explains the presence of STX in different tissues such as spleen, liver, brain, and medulla oblongata. Even then, the highest STX concentration was found in spleen and liver (blood-rich tissues). Significant concentrations of STX also were found in total brain and medulla oblongata samples (25).

IV. METABOLIZATION OF STX BY MAMMALIANS

The first evidence that STX and the other PSP toxins were eliminated without loss of their toxic effects and probably not metabolized was obtained from the pioneer experiments done by Sommers et al. (34). More recently, Hines et al., using [H^3]STXOL injected into rats, revealed only one radioactive peak in the HPLC analysis of urine samples, suggesting that the [H^3]STXOL excreted was not metabolized by rats (31).

Andrinolo et al. (25), using a postcolumn derivatization HPLC method (35) to quantify the mass amount of STX and its analogues together with a cleansing procedure with a cartridge column and microcentrifuge filters, which allowed measurements as low as 1 pmol of STX in body fluids and tissues samples, were able to quantify and identify all components of PSP toxins.

In their study (25), besides STX, no other PSP toxin product of possible metabolic or chemical change was detected in any of the samples analyzed within the 4 h of the experiment. Moreover, incubation of STX for 24 h at 25°C in the presence of cat liver 100,000 $\times$ g supernatant did not show any chemical or enzymatic transformation, since 100% of the STX used in the incubation was recovered (25).

The data only showed STX and not other analogs in urine, blood, liver, spleen, bile, medulla oblongata, and brain, suggesting that mammals do not display the metabolic mechanisms necessary to transform STX.

Recently, Gessner et al. (30) reported differences between toxin composition in mussels and human serum samples of intoxicated persons on Kodiak Island, Alaska, during 1994. These cases need more investigation to be explained satisfactorily, because the difference in toxin profiles could be due to the compartmentalization of the PSP toxins by binding with different affinity to their receptor location and that in the tissue membranes. Also, the clean-up and extraction procedure of the samples before being analyzed would explain the change in composition.

V. SUMMARY

The data suggest that, in cats with normal cardiovascular parameters and diuresis, the STX excretion mainly involves glomerular filtration. The changes in the renal clearance of the toxin at high doses can be explained as being due to cardiovascular complications and the initial hypovolemic shock produced by one bolus intravenous injection of STX. Therefore, the difference obtained between low and high intoxication doses used may be explained by STX toxic effects. There is no reported evidence of STX renal tubular reabsorption or secretion.

The significant amount of STX found in total brain and medulla oblongata showed that STX is capable of passing through the blood-brain barrier when high levels of STX intoxication are reached. These findings support the data published on the central cardiorespiratory effects of STX when applied topically in the lateral cerebral ventricle and brainstem throughout intravenous infusions. These data enforced the notion that, at high doses of toxin, the STX-induced lethality involves both central and peripheral cardiorespiratory system components. STX distribution includes the intravascular and extravascular compartments and it also binds to membrane tissues. The excretion of STX is only by glomerular filtration from the intravascular space. STX distribution in mammalians is extremely expeditious, being able to exchange easily from one compartment to another.

No evidence of chemical or enzymatic transformation in mammals has been reported, suggesting that mammals do not display the metabolic mechanisms described for microalgae, bacteria, cianobacteria, and mollusks.

ACKNOWLEDGMENTS

Supported by Instituto de Ciencias Biomédicas Fellowship, Facultad de Medicina, Universidad de Chile; FONDECYT 1961122 and Fundación ANDES.

REFERENCES

1. EF McFarren, ML Schafer, JE Campbell, KH Lewis, ET Jensen, EJ Schantz. Public health significance of paralytic shellfish poison. Adv Food Res 10:135–179, 1960.
2. JM Hughes, MA Horwitz, MH Merson, WH Barker, EJ Gangarosa. Foodborne disease outbreaks of chemical etiology in the United States, 1970–1974. Am J Epidemiol 105:233–244, 1977.
3. RR Long, JC Sargent, K Hammer. Paralytic shellfish poisoning: a case report and serial electrophysiologic observations. Neurology 40:1310–1311, 1990.
4. GM Hallegraeff. A review of harmful algal blooms and their apparent global increase. Phycologia 32:79–99, 1993.
5. Y Oshima. Chemical and enzimatic transformation of paralytical shellfish toxins in marine organism. In: Harmful Marine Algal Blooms. Paris: Lavoisier, Intercept, 1995, pp 475–480.
6. H Onodera, M Satake, Y Oshima, T Yasumoto, WW Carmichael. Detection of PSP toxins and six saxitoxin analog in the fresh water filamentous cianobacterium *Lingbya wallei*. Nat Toxins 5:146–151, 1997.
7. EJ Schantz, VE Ghazarossian, HK Schones, FM Strong. The structure of saxitoxin. J Am Chem Soc 97:1238–1239, 1975.
8. Y Shimizu, H Chien-ping, A Genenah. Structure of saxitoxin in solution and stereochemistry of dyhidrosaxitoxins. J Am Chem Soc 103:605–609, 1981.
9. T Yasumoto, M Murata. Marine toxins. Chem Rev 93:1897–1909, 1993.
10. G Strichartz. Structural determinants of the affinity of saxitoxin sodium channel. J Gen Physiol 84: 281–305, 1984.
11. SL Hu, CY Kao. Interactions of neosaxitoxin with the sodium channel of the frog skeletal muscle fiber. J Gen Physiol 97:561–578, 1991.
12. CY Kao, T Suzuki, AL Kleinahus, MJ Siegman. Vasomotor and respiratory depressant actions of tetrodotoxin and saxitoxin. Arch Int Pharmacodyn 165:438–450, 1967.
13. R Henderson, JM Ritchie, GR Strichartz. The binding of labeled saxitoxin to the sodium channel in nerve membrane. J Physiol 235:783–804, 1973.
14. WA Catterall, CS Morrow, RP Hartshorne. Neurotoxin binding to receptor sites associated with voltage sensitive sodium channel in intact, lysed, and detergent solubilized brain membranes. J Bil Chem 254:11379–11387, 1979.
15. E Moczydlowski, S Hall, SS Garber, GS Strichartz, C Miller. Voltage-dependent blockade of muscle Na^+ channels by guanidinium toxins: effect of toxin charge. J Gen Physiol 84:687–704, 1984.
16. E Moczydlowski, A Uehara, S Hall. Blocking pharmacology of batrachotoxin-activated Na-channels from rat skeletal muscle. In: Ion Channel Reconstitution, C Miller, ed. New York: Plenum Press, 1986, pp 75–114.
17. X Guo, A Uehara, A Ravindran, SH Bryant, S Hall, E Moczydlowski. Kinetic basis for insensitivity to tetrodotoxin and saxitoxin in sodium channels of canine heart and denervated rat skeletal muscle. Biochemistry 26:7546–7556, 1987.

18. S Hall, G Strichartz, E Moczydlowski, A Ravindran, PB Reichardt. The saxitoxins. Sources, chemistry, and pharmacology. In: Marine Toxins. Origin, Structure and Molecular Pharmacology, S Hall, G Strichartz, eds. American Chemical Society Symposium Series 418. Washington, DC, 1990, pp 29–65.
19. CY Kao, SR Levinson. Tetrodotoxin, saxitoxin and the molecular biology of the sodium channel. Ann NY Acad Sci 479:1986.
20. D Montebruno. Intoxicación por consumo de mariscos contaminados con veneno paralizante en la XII región, Chile. Estudio anatomopatológico. Rev Med Chile 121:94–97, 1993.
21. J Nagasawa, MY Spiegelstein, CY Kao. Cardiovascular actions of saxitoxin. J Pharm Exp Ther 176: 103–120, 1971.
22. FT Chang, BJ Benton, RA Lenz, R Benedict, BR Capacio. Central and periferal cardiorespiratory effects of saxitoxin (STX) in urethane-anesthetized guinea-pigs. Toxicon 31:645–664, 1993.
23. HL Borison, LE McCarthy. Respiratory and circulatory effects of saxitoxin in the cerebrospinal fluid. Br J Pharmacol 61:679–689, 1977.
24. RG Stafford, HB Hines. Urinary elimination of saxitoxin after intravenous injection. Toxicon 33: 1501–1510, 1995.
25. D Andrinolo, LF Michea, N Lagos. Toxic effects, pharmacokinetics and clearance of saxitoxin, a component of paralytic shellfish poison (PSP), in cats. Toxicon 37:447–464, 1999.
26. HL Borison, WJ Culp, SF Gonsalves, LE McCarthy. Central respiratory and circulatory depression caused by intravascular saxitoxin. Br J Pharmacol 68:301–309, 1980.
27. MB Feinstein, BS Marve Paimre. Mechanism of cardiovascular action of tetrodotoxin in the cat. Circ Res 23:553–565, 1968.
28. PJ Jaggard, MH Evans. Administration of tetrodotoxin and saxitoxin into the lateral cerebral ventricle of the rabbit. Neuropharmacology 14:345–349, 1975.
29. CAN Evans, JM Reynolds, ML Reynolds, NR Saunders. The effect of hypercapnia on blood-brain barrier mechanism in foetal and new born sheep. J Physiol 255:701–714, 1976.
30. BD Gessner, P Bell, GJ Doucette, E Moczydlowski, MA Poli, FV Dolah, S Hall. Hypertension and identification of toxin in human urine and serum following a cluster of mussel-associated paralytic shellfish poisoning outbreaks. Toxicon 35:711–722, 1997.
31. HB Hines, SM Naseem, RW Wannemacher Jr. ^{3}H-saxitoxinol metabolism and elimination in the rat. Toxicon 31:905–908, 1993.
32. MJ Fettman, TA Allen, WL Wilke, MC Radin, MC Eubank. Single injection method for evaluation of renal function with ^{14}C-inulin and ^{3}H-tetraethilamonium bromide in dogs and cats. Am Vet Res 46:482–485, 1985.
33. GR Strichartz, S Hall, B Magnani, CY Hong, Y Kishi, JA Debin. The potencies of synthetic analogues of saxitoxin and the absolute stereoselectivity of decarbamoilsaxitoxin. Toxicon 33:723–737, 1995.
34. EF McFarren, ML Schafer, JE Campbell, KH Lewis, ET Jensen, EJ Schantz. Public health significance of paralytic shellfish poison. Adv Food Res 10:135–179, 1960.
35. Y Oshima. Postcolumn derivatization liquid chromatographic method for paralytic shellfish toxins. J AOAC Int 78:528–532, 1995.

11
Detection Methods for Okadaic Acid and Analogues

Kevin J. James, Alan G. Bishop, Eoin P. Carmody, and Seán S. Kelly
Cork Institute of Technology
Bishopstown, Cork, Ireland

I. INTRODUCTION

The diarrhetic shellfish poisoning (DSP) toxins belong to three different structural classes: (1) okadaic acid (OA) and its analogues, (2) pectenotoxins, and (3) yessotoxins. Major toxic incidents resulting from the ingestion of shellfish in which the predominant human symptom was diarrhea occurred in Japan in 1976 when 42 people suffered food poisoning after eating mussels from the Sanriku coast. Similar incidents occurred in the Tohoku District in 1977 but with a larger number of intoxications (1) from the ingestion of clams, scallops, and mussels. As no pathogenic microorganisms were found in leftover foods, it was presumed that an unknown toxin was responsible. In Europe, the first report of similar toxic incidents from the ingestion of mussels occurred in the Netherlands, although no toxin was identified (2). The main toxic symptoms were diarrhea (92%), nausea (80%), vomiting (79%), and abdominal pain (53%) and the syndrome was termed diarrhetic shellfish poisoning (DSP). Symptoms occurred between 30 mins to a few hours after shellfish consumption and lasted for a maximum of 3 days. The dinoflagellate *Dinophysis fortii* was abundant in the marine shellfish cultivation areas in Japan around the time of toxicity, and this was subsequently confirmed to be the toxin-producing organism. The toxin responsible was isolated and identified as 35-methylokadaic acid (3,4) and was named dinophysistoxin-1 (DTX-1).

The acidic group of DSP toxins comprises mainly OA and its isomers, DTX-1 and the 7-O-acyl derivatives of these compounds, collectively termed DTX-3 (Figure 1). The isolation and structural elucidation of the polyether acid, OA, from the sponge *Halicondria okadaii* had previously been achieved by Tachibana et al. in 1981 (5), and DTX-2 was first isolated from Irish mussels in 1992 (6). DTX-3 analogues have not been detected in phytoplankton, suggesting that they are produced in the hepatopancreas of the shellfish (7). The OA analogues exert their acute toxic effects through their inhibition of protein phosphatases, PP1 and PP2A (8). However, the discovery that they are also potent tumor promoters (9) will significantly influence the acceptable limits for these toxins in seafood.

Various methods have been developed to detect DSP toxins in shellfish. Live animal bioassays were the first methods to be developed for the detection of DSP toxins, but problems due to the lack of sensitivity, false positives, lack of method validation, and the prohibition of such

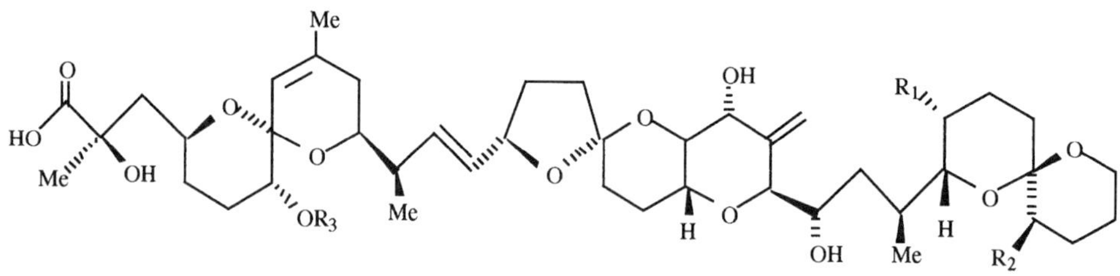

Figure 1 Structures of the DSP toxins, okadaic acid and analogues.

okadaic acid (OA)	$R_1 = CH_3$,	$R_2 = H$	$R_3 = H$
dinophysistoxin-1 (DTX-1)	$R_1 = CH_3$	$R_2 = CH_3$	$R_3 = H$
dinophysistoxin-2 (DTX-2)	$R_1 = H$	$R_2 = CH_3$	$R_3 = H$
dinophysistoxin-3 (DTX-3)	R_1, R_2 = H or CH_3		R_3 = acyl

testing in many countries owing to ethical considerations has led to an examination of a variety of alternative biochemical and chemical methods.

II. BIOASSAYS (ANIMAL)

A. Adult Mouse Bioassay

The mouse bioassay for the detection of DSP was developed by Yasumoto following outbreaks of DSP in Japan in 1976 and 1977. The acute toxicity of DSP toxins in mice led to a routine assay of DSP toxins in shellfish (1). Digestive glands were homogenized three times with acetone and the combined extracts were evaporated. The residue was made up to 2 mL with 1% Tween 60 solution and 0.5–1.0 mL aliquots of this solution, or serially diluted solutions, were given intraperitoneally to female mice (15–20 g). Initially, 1 mouse unit (MU) was defined as the minimum dose of toxin required to kill a mouse within 48 h. However, this mouse bioassay is compromised by a lack of specificity owing to the coextraction of other substances, especially free fatty acids which may give false positives (10–12). Low levels of other shellfish toxins, especially paralytic shellfish poisoning (PSP) toxins, can also lead to problems. The DSP bioassay protocol concentrates the PSP toxins yielding an extract that is highly toxic to mice even when the level of these toxins present may not produce a positive response in the official PSP bioassay (13). The high temperatures employed to remove water can also lead to low toxin recoveries (14). Another practical problem is that the final mixture is difficult to suspend to a fixed volume as it is an aqueous mixture with lipids.

To overcome the problems of PSP interferences, the above procedure has been modified by the inclusion of a diethyl ether extraction of the aqueous residue after acetone evaporation. The combined ether extracts were back washed twice with water to remove PSP contaminants and salts (15). Dichloromethane has been suggested as an alternative, because all the DSP toxins, including pectenotoxins and yessotoxins, will be extracted, but if a petroleum ether wash is not used, free fatty acids may still interfere (16). The mouse unit has been redefined by the regulatory agency in Japan as the amount of toxin required to kill two of three mice (20 g) in 24 h. One mouse unit corresponds to 4 μg OA, 3.2 μg DTX-1, and 5 μg of DTX-3 (17).

A modified mouse bioassay protocol has been employed in France. If all three mice survive for more than 5 h, the shellfish are considered safe for human consumption. However, as a precautionary measure, the mice are kept under observation for at least 24 h (18). A variation of this procedure introduced a redissolution of the residue from the extraction of shellfish tissue using 80% methanol. This is followed by an extraction with hexane to remove lipids (19). This procedure successfully reduces fatty acid levels but DTX-3 compounds are also probably removed (16) and interferences from PSP toxins and salts remain using this procedure.

A modified procedure was used in Norway (20) in which the residue from the acetone extraction was dissolved in water and extracted with petroleum ether followed by chloroform. Evaporation of the chloroform phase gave a dry residue which was then dissolved in 1% Tween 60 and injected intraperitoneally into white female mice. Two mice were used for each test and their behavior and survival times were recorded for 24 h. Toxicity was given a relative number of 0–4 according to survival times. The toxic limit corresponded to 5 MU or 20 μg OA/100 g shellfish meat, but losses of low-polarity DSP toxins (DTX-3) can occur in the petroleum ether washing step.

B. Suckling Mouse Bioassay

Hamano et al. reported a suckling mouse bioassay that provided a direct measure of diarrhetic effects. The method involves the introduction of 0.1 mL of shellfish extract in 1% Tween 60 via a Teflon tubing to 4- to 5-day-old mice. After 4 h, the mice are sacrificed and their whole intestines are removed. The fluid accumulation ratio is calculated as the ratio of intestine to that of the remaining body. Fluid accumulation was observed for OA, DTX-1, and DTX-3. A fluid accumulation ratio of 0.09 corresponds to a value of 0.1 MU and free fatty acids, which give false positives in intraperitoneal mouse bioassays, do not interfere with this assay (21). This assay also has a lower threshold limit and is more easily related to human symptomology. However, the test is difficult to use routinely and wounding of the mice is common on administration of the sample. Also, pectenotoxins and yessotoxins are not detected by this method (18).

C. Rat Intestinal Ligated Loop Bioassay

Edebo et al. determined OA and DTX-1 from toxic mussels in ligated loops of rats. The closed loop of an anesthetized rat was injected with homogenized mussel tissue suspended in phosphate-buffered saline. After 2 h, the loop was dissected out, weighed (a), and its length measured (L_a). Also, for reference, a more distal piece of intestine was similarly weighed (b) and its length measured (L_b). The secretion (mg/cm) was determined as a function of the calculation, $a/L_a - b/L_b$. The detection limit of the test corresponded to 0.05 μg OA and a linear range of 0.5–5.0 μg OA was reported. This assay was reported to be a sensitive and quantitative method for the determination of OA and DTX-1 (22).

D. Rat Fecal Bioassay

A nonlethal DSP bioassay using rats was developed in the Netherlands (23). In the published protocol, the meat of 10 mussels is mixed with normal ground rat food (6 g) and offered to a rat (100–120 g) that has been starved for 24 h. After 16 h, the consistency of feces and/or food refusal are noted and an estimate of DSP toxicity is made on these data. This test probably does not detect pectenotoxins or yessotoxins. Despite the fact that mussel meat weights can vary considerably, this assay has been used by regulatory agencies in several European countries for many years. The dose level was changed to 10 g hepatopancreas, but the lack of a consistent relationship between this assay and DSP toxin levels led to the abandonment of this test in

Ireland. This assay remains as the official method in some European countries where the lethal mouse bioassays for DSP are prohibited.

E. *Daphnia magna* Bioassay

OA is toxic to *Daphnia magno*, a species of freshwater cladoceran invertebrate (24). A bioassay was developed to determine OA concentrations in mussel extracts and a linear correlation ($r = 0.74$) with the 9-anthryldiazomethane (ADAM) fluorimetric high-performance liquid chromatograpy (HPLC) method was observed. The assay is potentially more precise than the mouse bioassay, since a large number of animals can be used and it is 10 times more sensitive than the mouse bioassay. *Daphnia magna* is also sensitive to other toxicants, and it has been proposed as a replacement for the mouse bioassay in the screening of shellfish contaminated by DSP or coextracted toxins (25).

III. CYTOTOXICITY ASSAYS

The cytotoxic effect of OA has been recognized since its isolation by Tachibana et al. (5). It was noticed that when rat hepatocytes were exposed to toxic shellfish extracts, a rapid leakage of lactate dehydrogenase occurred, whereas nontoxic extracts had no disintegrating effects on the cell membranes (26,27). A DSP toxicity test has been reported based on the morphological changes of rat hepatocytes when exposed to toxins. All three classes of DSP compounds can be determined, as they each produce different changes in cell morphology. However, the analysis is time consuming and results are confusing when mixtures of toxins are involved (28).

Amzil et al. have developed a rapid cytotoxicity assay for the detection of the OA group of toxins. The method evaluates the minimal active concentration (MAC) of shellfish extracts by the microscopic observation of morphological changes produced in human KB cells exposed to toxin. A high correlation ($r = -0.90$) was found between Log (MAC) of the tested shellfish extracts and the OA concentrations measured by the ADAM LC-FLD method. At low doses, some of the cells treated with OA showed epithelial features, whereas the remaining cells exhibited round features induced by the toxin. High doses of toxin resulted in all of the cells becoming round. The MAC of mussel extract is measured by the incubation of 50 μL of serially diluted mussel extracts with 50 μL of a 200,000 KB cells/mL suspension and the detection limit for OA was 0.125 μg (29).

A colorimetric method for quantifying the cytotoxic effect of OA on KB cells has also been developed. The method is based on the ability of metabolically active cells to reduce a tetrazolium compound during a contact period of 24 h. It was demonstrated that this assay could detect OA concentrations as low as 50 ng/g hepatopancreas and that there was also a good correlation ($r = 0.964$) between this assay and the DSP-Check ELISA (enzyme-linked immunosorbent assay) method (30). More recently, Pouchus et al. showed that the cytotoxic effect of OA and DTX-2 on KB cells were similar with MAC values, 155 nM and 115 nM, respectively, but DTX-1 was more active giving an MAC value of 19 nM (31).

OA also showed toxicity to buffalo-green-monkey kidney (BGM) cell cultures. This method is also based on the direct microscopic observation of toxin-induced morphological changes in cell cultures and showed good correlation ($r = 0.95$) with the mouse bioassay (32).

The uptake of a dye by mammalian fibroblasts in culture has been exploited in a nonspecific assay for the determination of maitotoxin and OA. More specific responses were also obtained by morphological examination. Cells exposed to OA characteristically presented a two-step morphological change in that they first became square and subsequently round. Following incubation with toxic extracts, cells were examined both for morphological changes and for neutral red uptake by measuring the absorbance at 540 nm. Concentrations of maitotoxin and OA in shellfish extracts were determined from dose-response curves (33).

IV. PROTEIN PHOSPHATASE ASSAYS

Protein kinases catalyze the addition of phosphate to intracellular proteins and protein phosphatases catalyze the dephosphorylation of phosphoproteins, thereby reversing the actions of protein kinases. Protein phosphatase-1 (PP1) and PP2A, PP2B, and PP2C are responsible for the dephosphorylation of serine and threonine residues in the cytoplasmic and nuclear compartments of eukaryotic cells. Both PP1 and PP2A are potently inhibited by OA, with PP2A exhibiting a higher sensitivity. The acute diarrhetic effects of OA and analogues are probably caused by the stimulation of the phosphorylation of the proteins that control sodium secretion by intestinal cells (34). The tumor-promoting effects of OA and its derivatives are now known to be caused by protein phosphatase inhibition (35,36). A number of PP assays with detection using radioactivity (37,38), colorimetry (39,40), bioluminescence (41), and fluorescence (42) have been developed.

A. PP Radioassay

Using a ^{32}P-labeled substrate, the protein phosphatase activity can be determined by measuring the release of inorganic ^{32}P. The reaction is stopped by the addition of trichloroacetic acid, which inactivates the phosphatase and precipitates any unused substrate. The acid fraction containing inorganic ^{32}P phosphate is counted using liquid scintillation to calculate the protein phosphatase activity (43).

Holmes reported the analysis of OA and DTX-1 in toxic shellfish by a liquid chromatography–linked protein phosphatase assay (44). The toxins were resolved by liquid chromatography and then assayed by inhibition of both PP1 and PP2A catalytic subunits in a ^{32}P phosphorylase-a. Using this method, a protein phosphatase inhibitor, in addition to OA and DTX-1, was detected in Canadian shellfish (44). Subsequent studies revealed the presence of at least six phosphatase inhibitors, distinct from known DSP toxins, in mussels from Holland and Canada. This method proved to be a sensitive bioscreen for the detection of novel PP-inhibitory compounds in shellfish (37). This approach was also used with capillary electrophoresis (CE), which improved the ultraviolet (UV) detection of OA in purified shellfish extracts (45).

B. PP Colorimetric Assay

PP2A shows a particularly high level of activity toward the substrate p-nitrophenyl phosphate (p-NPP) (46). Simon et al. exploited this to develop a colorimetric phosphatase–inhibition assay in which the ability of PP2A to dephosphorylate p-NPP to a colored product, p-nitrophenol, was used to determine OA concentrations (40). The OA detection limit was 0.4 pmol and the method was reported to be accurate, rapid, and readily performed. Good correlation data were obtained between this assay and the DSP-Check ELISA ($r = 0.933$) as well as the MTT cytotoxicity assay ($r = 0.968$). This assay was reported to be 50–100 times more sensitive than the MTT cytotoxicity and ELISA methods. A modification of this assay has been reported and, when carried out using a 96-well microplate, a large sample throughput is possible (39).

C. PP Bioluminescent Assay

Isobe et al used firefly luciferin phosphate as a substrate successfully to determine the activities of known inhibitors of PP2A, including OA. The luciferin phosphate is hydrolyzed to free luciferin by the action of the protein phosphatase. Luciferin in then immobilized onto a column where it reacts in the presence of ATP and Mg^{2+} to form oxyluciferin. During this process, light is emitted and detected by a luminescence detector. This method requires only 10 μM of substrate, which means that the sensitivity approaches that of the ^{32}P radioassay (41).

D. PP Fluorescent Assay

Fluorescent microplate PP assays have also been developed for the analysis of OA in shellfish and phytoplankton extracts. 4-Methylumbelliferyl phosphate (MUP) and fluorescein phosphate were examined as substrates for protein phosphatases and the products were quantified using fluorescence. A 100-fold increase in sensitivity was observed in the fluorescent assay using MUP when compared with p-NPP in the colorimetric assay (42). In the published protocol, this fluorescent assay had a detection limit of 13 ng of OA/g hepatopancreas and has the potential of providing a rapid and sensitive tool for shellfish monitoring. Shimizu et al have also reported MUP as being a suitable substrate for PP1 and PP2A assays, and they also found that ELF-97 was an excellent substrate for PP1 producing a high fluorescent yield (47).

V. IMMUNOSORBENT METHODS

Hokama et al. developed a monoclonal antibody during research into the development of anticiguatoxin antibodies and a stick test was produced for the detection of ciguatoxin and related polyethers (48). A monoclonal antibody was tested against OA, a fragment of OA and ciguatoxin. A modification of this stick test has led to the marketing of a test kit for the screening of ciguatera and DSP toxins called Ciguatech test kit. The presence of toxins is determined by the binding of toxins to a membrane attached to a plastic strip and exposing the toxin-laden membrane to an antibody–colored latex bead complex that has a high specificity for the toxins of interest. It was reported that 1 ng of OA could be detected on the test strip and 50 pg OA/g fish flesh could be determined after using a rapid extraction procedure (49). A recent study demonstrated the utility of this test kit in seafood safety monitoring programs by detecting the presence of OA and related DSP toxins in mussels during dinoflagellate blooms and depuration operations in France (50).

Levine et al. reported the production of antibodies and the development of a radioimmunoassay for OA. This toxin was found to be responsible for the stimulation of aracidonic acid metabolism in rabbits. Therefore, OA was conjugated to bovine serum albumin and used to immunize rabbits. The rabbits responded by producing anti-OA antibodies which neutralized the OA stimulation of arachidonic acid metabolism. Inhibition of the [^{3}H] okadaic acid binding to anti-OA by unlabeled OA occurred at concentrations as low as 0.2 ng/mL. A linear relationship was obtained between %inhibition and log (OA) in the range 0.2–9.0 ng/mL. Maitotoxin, brevetoxin B, and other marine toxins did not inhibit binding, but no data were presented for the cross reactivity with other DSP toxins (51).

Usagawa et al. prepared three monoclonal anti-OA antibodies from hybridoma clones obtained by the fusion of mouse myeloma cells with immune spleen cells sensitized to okadaic acid–ovalbumin conjugate. An ELISA was also developed and is now marketed as a kit under the name DSP-Check (Sceti, Tokyo). A constant quantity of OA conjugated to bovine serum albumin is precoated onto the wells. The test solution and horseradish-peroxidase–labeled monoclonal antibody is added simultaneously to the well. This enzyme-linked monoclonal antibody competively reacts with the fixed antigen and the free antigen from the sample. The quantity of enzyme proportional to that of fixed antibody is determined by enzyme activity which is measured colorimetrically, and DSP toxin content in the sample is inversely related to absorbance. Many samples can be rapidly analyzed using this assay, which is sensitive to OA at concentrations as low as 10 ng/mL. The antibodies show a cross reactivity of approximately 70% to DTX-1 but do not react to 7-O-acyl derivatives (DTX-3), pectenotoxins, or yessotoxins (52,53). The cross reactivity to the OA isomer, DTX-2, was approximately 40% (54). The DTX-3 group of toxins can be determined by alkaline hydrolysis prior to ELISA. The fact that OA is used in the well coating in this assay makes this kit somewhat expensive. Shestowsky et al have developed an alternative immunoassay in which two mouse monoclonal antibodies to OA were used. One of these antibodies is an anti-OA monoclonal antibody (mAb) called 6/50 (idiotype Id) and the other is an anti-anti-OA mAb called 1/59 (anti-Id). The 1/59 anti-Id mAb is an internal image of OA and can be substituted for the okadaic coating on the solid phase. 1/59 competes with free OA analyte for binding to a limited amount of anti-OA mAbs in a liquid phase. The bound 6/50 mAb is then quantified by an enzyme-conjugated antimouse antibody. The assay permits a reliable measurement of OA in the 9–81 ng/mL range and is marketed as a test kit by Rougier Bio-tech, Montreal (55). DTX-3, calyculin A, and brevetoxin-1 are not detected by this ELISA kit and, as DTX-2 and DTX-1 have a 10- to 20-fold lower affinity than OA, the general application of this kit is limited (56). The OA derivatives, 1-methyl ester, diol ester, and OA alcohol, have an activity similar to OA itself (57), suggesting that the 6/50 antibody recognizes the portion of OA that is most distant from the carboxyl acid group (56).

A specific and sensitive ELISA using immobilized mouse monoclonal antibodies to OA that are resistant to organic solvents has been developed. One antibody which binds only OA in 10% aqueous methanol was used to detect OA selectively. Another antibody which equally binds OA, DTX-1, and 7-O-palmitoyl DTX-1 in methanol was used to determine total DSP toxin content (58). In addition, an antibody which binds more strongly to 7-O-palmitoyl DTX-1 and 7-O-palmitoyl OA than to OA and DTX-1 in 50% aqueous methanol has been developed to detect DTX-3 (59).

An ELISA method with electrochemical detection has been used to determine OA. The method involves a competitive heterogeneous immunoassay and the hydrolysis of phenyl phosphate to phenol which is monitored by the electrochemical oxidation of the latter (60).

VI. LIQUID CHROMATOGRAPHY WITH FLUORESCENCE DETECTION

Sensitive and specific detection methods are needed to identify and determine trace amounts of toxins in complex matrices. Precolumn derivatization of acidic DSP toxins using various fluorescent-labeling reagents and their analysis using reversed phase HPLC with fluorescence detection (LC-FLD) has been employed for the determination of individual toxins. The fluorescent reagents that have been used for the analysis of acidic DSP toxins are shown in Figure 2 and the derivatization reaction conditions are summarized in Table 1.

A. 9-Anthryldiazomethane

The first fluorimetric HPLC method for the analysis of OA class of DSP toxins was reported in 1987 (11). The hepatopancreas of shellfish were homogenized with 80% methanol, the extract was washed with petroleum ether, and then extracted with chloroform. An aliquot of the chloro-

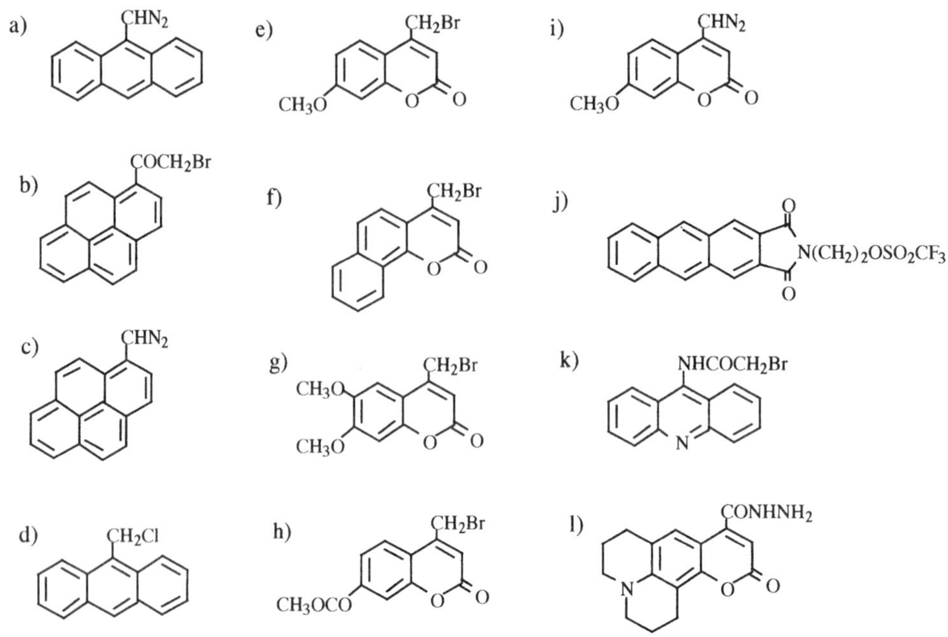

Figure 2 Structures of fluorimetric derivatizing reagents for the HPLC determination of okadaic acid and analogues; (a) 9-anthryldiazomethane (ADAM), (b) 1-bromoacetylpyrene (BAP), (c) 1-pyrenyldiazomethane (PDAM), (d) 9-chloromethylanthracene (CA), (e) 4-bromomethyl-7-methoxycoumarin (BrMmC), (f) 4-bromomethyl-7,8-benzcoumarin (BrBMC), (g) 4-bromomethyl-6,7-dimethoxy-coumarin (BrDMC), (h) 4-bromomethyl-7-acetoxycoumarin (BrMAC), (i) 4-diazomethyl-7-methoxy-coumarin (DAMMC), (j) 2,3-(anthracenedicarboximido)ethyl trifluoromethanesulphonate (AE-OTf), (k) N-(9-acridinyl)-bromoacetamide (NABA), (l) luminarine-3 (LN3).

form extract was dried under nitrogen and derivatized with 0.1% w/v 9-anthryldiazomethane (see Figure 2a). Following a clean-up using a silica solid phase extraction (SPE) cartridge, the esters of the toxins were analyzed using an octadecylsilane (ODS) column with a mobile phase of acetonitrile : methanol : water (8 : 1 : 1) with fluorimetric detection: λ_{ex} = 365 nm, λ_{em} = 412 nm (11).

There have been many attempts to improve the robustness of the 9-anthryldiazomethane (ADAM) method and changes have been proposed to the extraction, the derivatization reaction step, and especially the SPE clean-up. The ADAM reagent is a very sensitive derivatizing reagent but is notoriously unstable, which can lead to incomplete derivatization of the target analytes and the appearance of artefact peaks in chromatograms due to degradation products (61). For best results, the ADAM reagent should be stored at −70°C and maintained in complete darkness. Immediately before use, solutions may be prepared by dissolving a small quantity of ADAM in acetone (50 μL) and made up to the desired concentration using methanol (62). A modification of the ADAM protocol has been proposed by Draisci et al. which uses an exhaustive extraction of the shellfish homogenate (63). The various SPE methods proposed for ADAM are summarized in Table 1, whereas the reaction conditions are summarized in Table 2. Using a number of changes to the ADAM protocol, Stabell reported improved reproducibility when ultrasonic treatment of the samples was performed during derivatization (64). This treatment probably increased access to reactive sites on toxins owing to the disruption of micelles formed by toxins and other hydrophobic compounds. Changes to the SPE clean-up by using a smaller cartridge and different

Table 1 Fluorogenic Reaction Conditions for the Derivatization of DSP Toxins

Fluorimetric reagent	Reagent conc (% w/v)	Reagent vol (μl)	Reaction time (min)	Reaction solvent	Reaction temp (°C)	Catalyst or other reagent	Excitation wavelength λ_{ex} (nm)	Emission wavelength λ_{em} (nm)	Reference
ADAM	0.10	100	60	MeOH	25	None	365	412	11
ADAM	0.20	200	60	AcOEt	RT	None	365	415	65
ADAM	0.20	100	120	MeOH	37	None	254	412	67
ADAM	0.10	50	60	MeOH	30	None	365	412	69
ADAM	0.10	250	60	Acetone	25	None	365	412	63
ADAM	0.20	200	60	MeOH	RT	None	365	412	75
ADAM	0.10	90	60	MeOH	35	None	254	412	79
ADAM	0.10	200	30	MeOH/EtOAc	RT	None	365	400	66
BAP	0.10	400	15	MeCN	75	DIPEA	365	418	74
BAP	0.10	500	20	MeCN	75	DIPEA	365	418	62
BAP	0.20	80	20	MeCN	75	DIPEA	365	440	76
PDAM	0.10	—	30	EtOAc	75	None	340	394	78
CA	0.10	200	30	MeOH	75	TEAH	366	404	80
CA	0.02	400	60	MeCN	90	TEAH	365	412	81
BrMmC	0.15	25	120	Acetone	55	18-crown-6-ether	325	390	82
BrMmC	0.28	100	120	DMF	45	DIPEA	314	400	79
BrDMC	0.30	100	120	DMF	45	DIPEA	346	430	79
BrDMMC	0.22	100	120	DMF	45	DIPEA	346	430	79
BrBMC	0.29	100	120	DMF	45	DIPEA	346	430	79
BrMAC	0.29	100	120	DMF	45	DIPEA	346	430	79
NABA	0.03	100	15	$CHCl_3/H_2O$	90	(PT)	358	482	95
AE-OTf	0.09	100	10	MeCN	RT	TEAC	298	462	83
LN3	0.20	20	15	DMF	50	EDC	394	464	85

DIPEA, diisopropylethylamine; TEAH, tetraethyl ammonium hydroxide; RT, room temperature; TEAC, tetraethyl ammonium carbonate; EtOAc, ethyl acetate; MeCN, acetonitrile; EDC, 1-ethyl-3-(3-dimethylaminopropyl)carbodiimide; PT, phase transfer conditions.

Table 2 Solid Phase Extraction (SPE) Conditions Used with ADAM LC-FLD Method

Sorbent (quantity, mg)	Washing solvent #1 (vol, mL)	Washing solvent #2 (vol, mL)	Elution solvent (vol, mL)	Reference
Silica (1000)	1:1 $CHCl_3$/hexane (5)	$CHCl_3$ (5)	95:5 $CHCl_3$/MeOH (5)	11
Silica (100)	1:1 CH_2Cl_2/hexane (1)	9:1 CH_2Cl_2/acetone (1)	1:1 CH_2Cl_2/MeCN (2.5)	61
Silica (1000)	1:1 CH_2Cl_2/hexane (5)	9:1 CH_2Cl_2/acetone (5)	1:1 CH_2Cl_2/MeCN (8)	63
Silica (500)	1:1 $CHCl_3$/hexane (5)	$CHCl_3$ (5)[a]	95:5 $CHCl_3$/MeOH (5)	75
Silica (100)	1:1 CH_2Cl_2/hexane (7.5)	CH_2Cl_2 (2.5)	96.5:3.5 CH_2Cl_2/MeOH (2.5)	69
Silica (300)	1:1 CH_2Cl_2/hexane (4)	9:1 CH_2Cl_2/acetone (15)	1:5 CH_2Cl_2/MeCN (6)	66
Silica (500)	1:1 $CHCl_3$/hexane (5)	$CHCl_3$[a]	9:1 $CHCl_3$/MeOH[a] (5)	67
Silica (100)	1:1 CH_2Cl_2/hexane (1)	95:5 CH_2Cl_2/acetone (2)	1:1 CH_2Cl_2/MeCN (2)	65
Carbon (300)	ethyl acetate (6)	hexane (6), THF (6)	97:3 THF/toluene (20)	68

[a] Spiked with 1.15–1.2% ethanol.

eluent compositions were reported to improve the resulting chromatograms. Deoxycholic acid (DCA) was also used as an internal standard (64). However, the structure of DCA is very different from that of DSP toxins, and this can give rise to a positive bias if quantitative derivatization is not achieved for all analytes and DCA. Other attempts to improve the ADAM protocol, particularly the extraction and SPE steps, have been reported by Periera et al. (65) and by Stockemer et al. (66).

Quilliam reported a detailed examination of the ADAM method and proposed a number of modifications (67). More reliable derivatization was reported by increasing the ADAM concentration and doubling the reaction time to 2 h. The sensitivity of the procedure was improved by using $\lambda_{ex} = 254$ nm instead of 365 nm with apparently little diminution in selectivity. Also, 7-O-acetylokadaic (AcOA) was used as an internal standard, as its relative retention time was constant for different columns and temperatures. AcOA proved to be a more effective internal standard than DCA and led to a more accurate and precise method by correcting for incomplete recoveries, particularly in the extraction, derivatization, and SPE cleanup steps (67). However, AcOA is prohibitively expensive, as it must be synthesized from OA, and it has not been widely

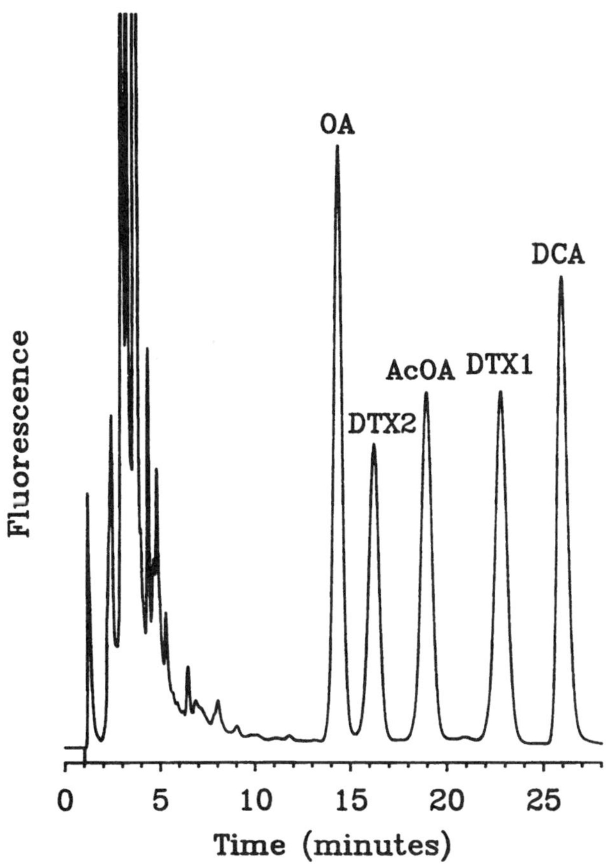

Figure 3 Fluorimetric HPLC analysis of ADAM derivatized samples; (a) calibration mixture, 1 (OA), 2 (DTX-2), 3 (DTX-1), 4 (AcOA, internal standard), (b) extract from certified mussel material reference material, MUS-2 (11 mg/g OA, 0.9 mg/g DTX-1). Conditions: Lichrospher-100 (5 μm, 250 × 4 mm) ODS column at 40°C; elution using acetonitrile/water (80:20); flow rate = 1 ml/min; fluorescence detection: $\lambda_{ex} = 254$ nm, $\lambda_{em} = 412$ nm. (From ref. 90.)

adopted. Figure 3 shows the chromatogram from the analysis of a mixture of OA, DTX-1, and DTX-2 with the internal standards, AcOA and DCA. A critical factor is the amount of ethanol stabilizer in chloroform used in the SPE steps. The optimum ethanol content of chloroform was found to be 1.15%. Thus, it was suggested that the ethanol and impurities, such as chloroform decomposition products, should be removed using an activated alumina column and then the ethanol content adjusted to the required concentration (67). However, this may more conveniently be achieved by using chloroform stabilized with amylene and then adjusting the chloroform to the optimum ethanol concentration (62). Recently, Syhre et al. proposed activated carbon as a replacement for silica for the SPE clean-up of ADAM derivatized DSP toxins, which avoided the use of halogenated solvents (68).

Aase et al. optimized the ADAM protocol by the use of an experimental design (69). The design took into account the concentration of ADAM used, the derivatization time and temperature, the amount of silica in the SPE tubes, the amount of washing solvent, the concentration of ethanol in the washing solvents, the speed of washing, and eluting solvent. Based on this design, it was suggested that derivatization should be carried out using 0.1% w/v ADAM, a 100-mg silica SPE tube, and dichloromethane instead of chloroform to minimize the effect of stabilizing alcohol. Lucas et al. described an automated HPLC method for determining DSP toxins using a column-switching device to avoid the clean-up procedures after derivatization of shellfish extracts with ADAM. However, the operation of this column-switching method is complex and there is a reduced sample throughput (70,71).

The generation of ADAM in situ has been studied in order to overcome the problems associated with reagent stability. ADAM was prepared prior to analysis by the reaction of 9-anthraldehyde hydrazone with quinuclidine and N-chlorosuccinimide in ethyl acetate (72). The quantitative results were generally in agreement with those obtained using the commercial reagent. However, a peak near to OA occasionally appeared in the chromatogram of the test solution which was probably due to an artefact formed in the derivatization reaction. Quilliam et al. have investigated this artefact peak using liquid chromatography–mass spectrometry (LC-MS) and it was found to be AcOA. The possible source of the acetyl moiety was ascribed to the reaction solvent, ethyl acetate. This is potentially a critical problem, especially if AcOA is used as an internal standard, but replacing ethyl acetate with tetrahydofuran eliminated this problem (73).

B. 1-Bromoacetylpyrene

Dickey et al. developed a method of derivatizing DSP toxins with 1-bromoacetylpyrene (BAP) (see Figure 2b). The extraction and clean-up procedures were similar to those used with ADAM. The derivatization was carried out on a shellfish extract that was reacted with BAP using diisopropylethylamine catalyst at 75°C. The HPLC analysis was performed on an ODS column with fluorescence detection; λ_{ex} = 365 nm and λ_{em} = 418 nm. BAP proved to be more stable in both the solid and solution states than ADAM or PDAM (74). Kelly et al compared the sensitivities of ADAM and BAP and the latter was found to be four times less sensitive than ADAM. However, chromatograms obtained using BAP showed less artefact peaks than with ADAM and this proved useful in the determination of the purity of DSP toxins isolated from shellfish (62). More recently, the simultaneous separation of ADAM and BAP derivatives of DSP toxins was achieved using a column designed to resolve polyaromatic hydrocarbons (Figure 4), and this was one of the methods that was used recently to confirm the presence of DTX-2 in *Dinophysis acuta* (75). An improved BAP procedure has also been reported for the determination of OA and DTX-2 in mussels that required a short reaction time (2 min) and used

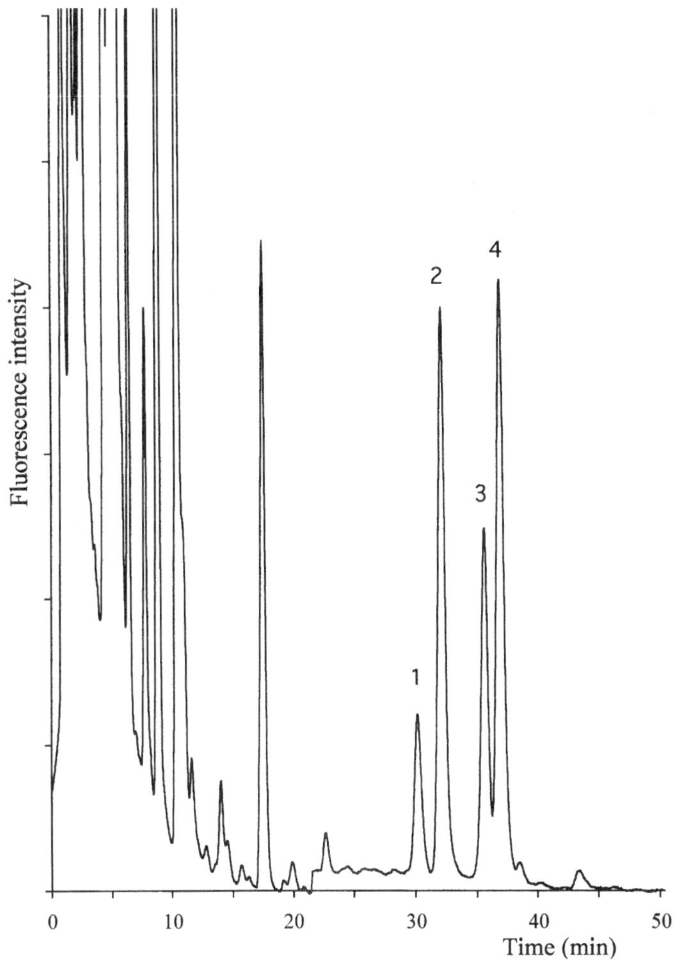

Figure 4 Fluorimetric HPLC of OA and DTX-2, separately derivatized using ADAM and BAP and mixed prior to injection. (1) ADAM-OA, (2) BAP-OA, (3) ADAM-DTX-2, (4) BAP-DTX-2. Conditions: Envirosep-PP ODS column (5 μm, 250 × 3.2 mm); linear gradient of acetonitrile/water (54–71% acetonitrile, 61.5 min); flow rate = 1.5 ml/min; fluorescence detection: λ_{ex} = 365nm, λ_{em} = 412 nm. (From ref. 75.)

solvent combinations with dichloromethane instead of chloroform in the silica SPE clean-up steps (76).

C. 1-Pyrenyldiazomethane

1-Pyrenyldiazomethane (PDAM) (see Figure 2c) has previously been used as a dervatization reagent for fatty acids and prostaglandins (77). PDAM is more stable than ADAM but elevated reaction temperatures (~75°C) are required to produce pyrenylmethyl esters with shellfish extracts. The SPE clean-up and chromatographic conditions were similar to those used in the ADAM protocol except that fluorescence detection requires λ_{ex} = 340 nm and λ_{em} = 394 nm. A similar procedure has been used for the determination of OA in *Prorocentrum* sp. (78). However, possible decomposition of labile DSP compounds under the higher temperatures required in the PDAM reaction may present a problem (79).

D. 9-Chloromethylanthracene

9-chloromethylanthracene (CA) (see Figure 2d) was used as a derivatizing reagent for the analysis of OA in mussels from Italy. The residue from a shellfish extract was treated with CA and tetramethylammonium hydroxide for 30 min at 75°C. After SPE, the derivatives were separated on a C_8 column using gradient elution. The procedure was less sensitive than the ADAM method but was sufficient to determine toxins in amounts below the regulatory limits and was also useful for the determination of free fatty acids (80). A modified procedure using CA was recently reported by Lawrence et al. for the determination of OA and DTX-1 in shellfish. The derivatization reaction was carried out for 1 h at 90°C. Clean-up was achieved using two silica SPE cartridges before LC analysis using an ODS column and isocratic elution using acetonitrile: water (75:25) mobile phase with fluorimetric detection; $\lambda_{ex} = 365$ nm and $\lambda_{em} = 412$ nm. This method was reported to give a similar sensitivity to the ADAM protocol (81).

E. Bromomethylcoumarin Derivatives

A number of bromomethylcoumarin derivatives have also been investigated as fluorimetric derivatizing reagents. 4-Bromomethyl-7-methoxycoumarin (BrMmC) (see Figure 2e) was used as a reagent for the derivatization of OA analogues (82). This method differed from the ADAM protocol in that a silica SPE clean-up was used on extracts from shellfish or phytoplankton before derivatization. Derivatization was carried out using BrMmC in the presence 18-crown-6-ether and potassium carbonate and the reaction was maintained at 55°C for 2 h. An aliquot was injected directly into the HPLC and the products were separated on an ODS column with fluorescence detection; $\lambda_{ex} = 325$ nm, $\lambda_{em} = 390$ nm. Advantages claimed for this method compared to the ADAM protocol were that lipid removal was not required and that clean-up using SPE before derivatization avoided possible decomposition of fluorescent products. However, the chromatographic times were considerably longer using the BrMmC protocol, which leads to a small sample throughput, and chromatograms also contained artefact peaks in the region where DSP toxins eluted. The BrMmC was found to be adequate for the analysis of cultured dinoflagellates or mussel tissue containing high concentrations of DSP toxins. Other bromomethylcoumarin derivatives have also been examined, including 4-bromomethyl-7,8-benzcoumarin (BrBMC) (see Figure 2f), 4-bromomethyl-7-acetoxycoumarin (BrMAC) (see Figure 2g), and bromomethyl-6,7-dimethoxycoumarin (BrDMC) (see Fig. 2h). The BrMAC was found to be weakly fluorescent. BrDMC had an 85% fluorescent yield compared with that of BrMmC but showed less chemical noise from impurities. BrDMC showed the greatest promise because of its stability, purity, and the typical ADAM SPE clean-up could be used for derivatized mussel extracts. BrDAMMC (see Figure 2i) was the least reactive of the coumarin derivatives (79).

F. 2,3-(Anthracenedicarboximide)ethyltrifluoromethanesulpfonate

Recently, 2,3-(anthracenedicarboximide)ethyltrifluoromethanesulfonate (AE-OTf) (see Figure 2j) was reported as being a very sensitive reagent for the LC determination of acidic DSP toxins. A shellfish extract, in methanol, was treated with tetraethylammonium carbonate solution and the solvent was removed under nitrogen. The samples were derivatized with AE-OTf at room

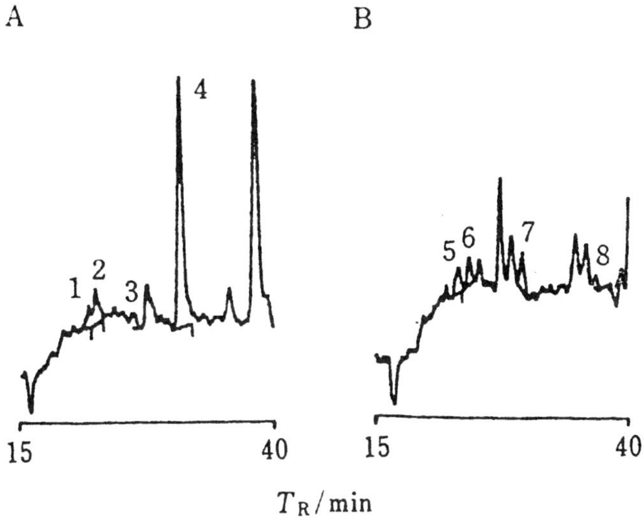

Figure 5 Fluorimetric HPLC analysis of AE-OTf derivatized DTX-3 compounds in an extract of scallop material. The acyl moieties were assigned: (1) 20:0, (2) 18:3, (3) 18:2, (4) 16:0, (5) 22:6, (6) 20:4, (7) 18:1, (8) 18:0. Conditions: First column was a Develosil ph-5 (5 μm, 150 × 4.6 mm) at 60°C; elution using methanol/water (9:1); flow rate = 1 ml/min. The target fraction 8.5–9.4 min (A) or 9.2–10.5 min (B) was loaded onto a second column, Develosil ODS K-5 (5 μm, 150 × 4.6 mm); elution using methanol/acetonitrile/water (71:24:5); fluorescence detection: λ_{ex} = 298 nm, λ_{em} = 462 nm. (From ref. 84.)

temperature and, after silica SPE, the LC separation of the derivatized toxins involved the use of a silica column and a valve-switching device to introduce a target fraction onto an ODS column (83). The reported detection limits were 0.8 pg (OA) and 1.3 pg (DTX-1), which makes this the most sensitive of the chromatographic methods for the determination of these toxins. DTX-3 compounds have also been determined by this method (84) by using a small modification to the LC mobile phase composition (Figure 5).

G. Luminarine-3

More recently, it was demonstrated that luminarine-3 (LN3) (see Figure 21), a stable hydrazine derivative, could successfully derivatize OA, DTX-2, and DTX-1 in shellfish (85). Derivatization of the DSP toxins was achieved at a moderate temperature (50°C) using 1-(3-dimethylaminopropyl)-3-ethylcarbodiimide (EDC) as a catalyst. Acceptable chromatograms were obtained in the analysis of shellfish extracts even without SPE clean-up, and the sensitivity using LN3 was comparable with that using ADAM. However, the method was particularly useful for the rapid LC determination of DSP toxin profiles in phytoplankton extracts and in chromatographic eluate fractions obtained during toxin isolation studies as good-quality chromatograms were obtained without the need for SPE (86). The derivatives were separated on an ODS column with fluorimetric detection; λ_{ex} = 394 nm and λ_{em} = 464 nm.

VII. LIQUID CHROMATOGRAPHY–MASS SPECTROMETRY

Pleasance et al. developed an LC-MS method, using an ion spray interface and atmospheric ionization, for the determination of OA and DTX-1. Using a mobile phase of acetonitrile:water

(6:4) with 0.1% trifluoroacetic acid (TFA) and an ODS column connected to the ionspray interface with a fused silica capillary, the positive ion spectra of OA and DTX-1 were simple with abundant peaks due to the protonated molecules $[M+H]^+$ occurring at m/z 805 and 819, respectively. Thus, selected ion monitoring (SIM) at these m/z values was found to give the best sensitivity and selectivity for the analysis of OA and DTX-1. Although this method was not as sensitive as the LC-FLD methods, it did not involve any troublesome derivatization and cleanup stages. The detection limit was found to be 2 ng of toxin injected and linearity was observed for up to 50 ng of injected toxin. This method was used to confirm the presence of OA in *Prorocentrum concavum* and in natural populations of dinoflagellates from Canadian waters

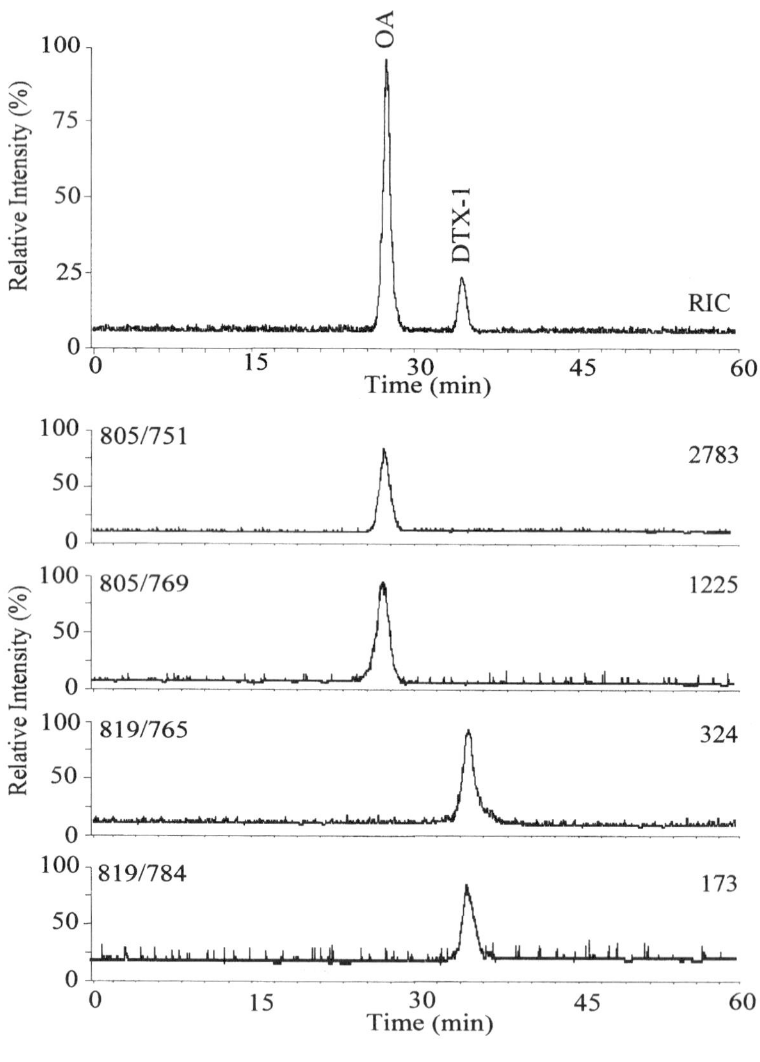

Figure 6 SRM LC-MS-MS chromatograms of an extract from Italian toxic mussel hepatopancreas; OA (1.51 μg/g), DTX-1 (0.29 μg/g). Conditions: Supelcosil LC_{18}-DB column (5 μm, 300 × 1 mm); elution using acetonitrile/water (80/20) containing 0.1% TFA; flow rate = 10 μL/min. Intensity counts are shown at the upper right of each chromatogram. (From ref. 91.)

(87). LC-MS was also used to detect DTX-2 as well as the 7-O-acyl derivatives of OA, DTX-1, and DTX-2 in Irish mussels (88).

An improvement to the LC-MS method for DSP toxins was reported (89). A microbore column was used with a linear gradient (20 min) of 40–100% acetonitrile with water containing 0.1% trichloroacetic acid and, when combined with the development of a fully articulated version of the ISP interface, this resulted in a 10-fold improvement in the detection limit. The sensitivity approaches that of the ADAM LC-FLD method. The improved detection limits also permitted the analysis of DSP toxins in whole mussel tissue (89).

A further sensitivity increase was observed when methanol:water (7:3 with 0.1% TFA) was used as the mobile phase. 7-O-acetylokadaic acid (AcOA) was successfully used as an internal standard and an aminopropyl silica phase was used for the clean-up for shellfish extracts which reduced interferences and allowed concentration of the toxins as well as helping to protect the HPLC column. OA, derivatized with ADAM, was also analyzed using LC-MS in an attempt to improve the sensitivity but without success (87). However, using the fluorescent reagent, N-(9-acridinyl)-bromoacetamide (NABA) (see Figure 2k) to derivatize OA and DTX-1, a fivefold improvement in sensitivity was observed compared to using the underivatized standards. The NABA-derivatized analytes were well resolved on an ODS column and the protonated nitrogen also improved the sensitivity, as the ion evaporation process favors preformed ions in solution (89). The LC-MS analysis of DSP toxins, derivatized with BAP, has been used to confirm the presence of OA and DTX-2 in *Dinophysis acuta* (75). DSP toxins, derivatized with 4-bromo-methyl-7-methoxycoumarin (BrMmC), have also been reported for the analysis of DSP toxins using LC-MS (82).

Collision induced dissociation (CID) before the sampling orifice of the LC-MS can provide structural information about the toxins and confirmatory fragmentation data (90,91). A sensitive and specific analytical procedure for determining four acidic DSP toxins using micro–liquid chromatography coupled with tandem mass spectrometry (micro-LC-MS-MS) has recently been reported (91). This was carried out using isocratic elution at 10 μL/min, with acetonitrile-water (80:20) containing 0.1% trifluoroacetic acid, through an ODS column. CID mass spectra of the protonated molecules were obtained in MS-MS experiments in order to identify two diagnostic fragment ions for each analyte which could be used for selected reaction monitoring (SRM) micro-LC-MS-MS analysis. This was implemented using the parent-daughter ion combinations of *m/z* 805-751; *m/z* 805-769 for OA, DTX-2, and DTX-2B; and *m/z* 819-765, *m/z* 819-784 for DTX-1 and applied to the analysis of mussels from Italy and Ireland (91). Figure 6 shows the chromatograms obtained using an extract from Italian toxic mussels.

VIII. CAPILLARY ELECTROPHORESIS

Capillary electrophoresis (CE) of DSP toxins is potentially complementary to HPLC, as it provides a different selectivity mode. Boland et al. used CE with optical detection at 200 nm for the analysis of OA in semipurified marine samples that had previously been screened using LC-protein phosphate assays (45). The surfactant, sodium dodecyl sulfate, which incorporates OA into its hydrophobic cavity, was added to the electrolyte solution used in the CE method, and this allowed OA to migrate as a charged molecule. A variety of separation principles have arisen from CE, including micellar electrokinetic chromatography (MEKC), and this was applied to the determination of DSP toxins in extracts of *Prorocentrum lima* (92), but the evidence presented for the identification of DTX-2 in *P. lima* was not convincing.

The coupling of CE to mass spectrometry (MS) has been reviewed, and it has been shown that the major advances in CE-MS have been the development of interfaces to couple electrospray and continuous fast atom bombardment (93). Pleasance et al. have used a coaxial interface for the analysis of OA by CE-MS, but significant improvements in sensitivity need to be made before this can offer an alternative to the current fluorimetric HPLC methods for the determination of DSP toxins (94).

REFERENCES

1. T Yasumoto, Y Oshima, M Yamaguchi. Occurrence of a new type of toxic shellfish poisoning in the Tohoku district. Bull Jpn Soc Sci Fish 44:1249–1255, 1978.
2. M Kat. The occurence of *Prorocentrum* species and the coincidental gastrointestinal illness of mussel consumers. In: D Taylor, HH Seliger, eds. Toxic Algae Blooms. Amsterdam: Elsevier, 1979, pp 215–220.
3. T Yasumoto, Y Oshima, W Sugawara, Y Fukuyo, H Oguri, T Igarashi, N Fujita. Identification of dinophysis fortii as the causative organism of diarrhetic shellfish poisoning. Bull Jpn Soc Sci Fish 46:1405–1411, 1980.
4. M Murata, M Shimatani, H Sugitani, Y Oshima, T Yasumoto. Isolation and structural elucidation of the causative toxin of the diarrhetic shellfish poisoning. Bull Jpn Soc Sci Fish 48:549–552, 1982.
5. K Tachibana, PJ Scheuer, Y Tsukitani, H Kikuchi, D Van Engen, J Clardy, Y Gopichand, FJ Schmidtz. Okadaic acid, a cytotoxic polyether from two marine sponges of the genus *Halichondria*. J Am Chem Soc 103:2469–2471, 1981.
6. T Hu, J Doyle, D Jackson, J Marr, E Nixon, S Pleasance, MA Quilliam, JA Water, JLC Wright. Isolation of a new diarrhetic shellfish poison from Irish mussels. J Chem Soc Chem Commun 30: 39–41, 1992.
7. JS Lee, T Igarashi, S Fraga, E Dahl, P Hovgaard, T Yasumoto. Determination of diarrhetic shellfish toxins in various dinoflagellate species. J Appl Phycol 1:147–152, 1989.
8. JJ Cohen, M Al-Rubeai. Apoptosis-targeted therapies: the ‘next big thing’ in biotechnology. Trends Biochem Tech 13:281–283, 1995.
9. Suganuma, Fujiki. Diarrhetic shellfish poisoning and okadaic acid. Jikken Igaku 8:1243–1247, 1990.
10. T Takagi, K Hayashi, Y Habashi. Toxic effect of free fatty acids in the mouse bioassay of diarrhetic shellfish toxin by intraperitoneal injection. Bull Jpn Soc Sci Fish 50:1413, 1984.
11. JS Lee, T Yanagi, R Kenma, T Yasumoto. Fluorometric determination of diarrhetic shellfish toxins by high-performance liquid chromatography. Agric Biol Chem 51:877–881, 1987.
12. T Suzuki, R Yoshizawa, T Kawamura, M Yamasaki. Interference of free fatty acids from the hepatopancreas of mussels. Lipids 31:641–645, 1996.
13. A Gago-Martinez, JA Rodriguez-Vasquez, P Thibault, MA Quilliam. Simultaneous occurrence of diarrhetic and paralytic shellfish poisoning toxins in Spanish mussels in 1993. Nat Toxins 4:72–79, 1996.
14. L Croci, L Toti, D De Medici, L Cozzi. Diarrhetic shellfish poison in mussels: comparison of methods of detection and determination of the effectiveness of depuration. Int J Food Microbiol 24:337–342, 1994.
15. T Yasumoto, M Murata, Y Oshims, GK Matsumoto, J Clardy. Diarrhetic shellfish poisoning. In: EP Ragelis, ed. Seafood Toxins. Washington, DC: AOAC International, 1984, pp 214–217.
16. ML Fernandez, A Miguez, E Cacho, A Martinez. Sanitory controls of marine biotoxins in the European Union. National references laboratory network. In: T Yasumuto, Y Oshima, Y Fukuyo eds. Harmful and Toxic Algal Blooms. Paris: International Oceanographic Commission of UNESCO, 1996, pp 11–14.
17. T Yasumoto, M Fukui, K Sasaki, K Sugiyama. Determinations of marine toxins in foods. J AOAC Int 78:574–582, 1995.
18. C Marcaillou-LeBaut, D Lucas, L Le Dean. Dinophysis acuminata toxin: status of toxicity bioassays

in France. In: DM Anderson, AW White, DG Baden, eds. Toxic Dinoflagellates. New York: Elsevier, 1985, pp 485–488.
19. C Marcaillou-LeBaut, P Masselin, M Bohec. Comparison des methodes chimiques et biologiques pour la determination des toxines diarrheques dans les moules. In: JM Fremy, ed. Proceedings of Symposium on Marine Biotoxins. Paris: CNEVA, 1991, pp 89–92.
20. E Dahl, A Rogstad, T Aune, V Hormazabal, B Underdal. Toxicity of mussels related to occurrence of *Dinophysis* species. In: P Lassus, G Arzul, E Erard, P Gentien, C Marcaillou, eds. Harmful Marine Algal Blooms. Paris: Lavosier, 1995, pp 783–788.
21. Y Hamano, Y Kinoshita, T Yasumoto. Suckling mice assay for diarrhetic shellfish toxins. In: DM Anderson, DW White, DG Baden, eds. Toxic Dinoflagellates. New York: Elsevier, 1985, pp 383–388.
22. L Edebo, S Lange, XP Li, S Allenmark, K Lindgren, R Thompson. Toxic mussels and okadaic acid induce rapid hypersecretion in the rat small intestine. APMIS (Acta Pathol Microbiol Immunol Scand) 96:1029–1035, 1988.
23. M Kat. *Dinophysis acuminata* blooms in the Dutch coastal area related to diarrhetic mussel poisoning in the Dutch Waddensa. Sarsia 68:81–84, 1983.
24. JP Vernoux, C Le Baut, P Masselin, C Marais, B Baron, R Choumiloff, F Proniewski, G Nizard, M Bohec. The use of *Daphnia magna* for detection of okadaic acid in mussel extracts. Food Addit Contam 10:603–608, 1993.
25. C Marcaillou-LeBaut, Z Amzil, J-P Vernoux, Y-F Pouchus, M Bohec, J-F Simon. Studies on the detection of okadaic acid in mussels: preliminary comparison of bioassays. Natural Toxins 2:312–317, 1994.
26. B Underdal, M Yndestad, T Aune. DSP intoxications in Norway and Sweden, autumn 1984–spring 1985. In: DM Anderson, DW White, DG Baden, eds. Toxic Dinoflagellates. New York: Elsevier, 1985, pp 457–460.
27. T Aune. Toxicity of marine and freshwater algal biotoxins towards freshly prepared hepatocytes. In: S Natori, K Hashimoto, Y Ueno, eds. Mycotoxins and Phycotoxins '88. Amsterdam: Elsevier, 1989, pp 461–468.
28. T Aune, T Yasumoto, E Engeland. Light and scanning electron microscopic studies on effects of marine algal toxins toward freshly prepared hepatocytes. J Toxicol Environ Health 34:1–9, 1991.
29. Z Amzil, YF Pouchus, J Le Boterff, C Roussakis, J-F Verbist, C Marcaillou-Lebaut, P Masselin. Short-time cytotoxicity of mussel extracts: a new bioassay for okadaic acid detection. Toxicon 30: 1419–1425, 1992.
30. A Tubaro, C Florio, E Luxich, R Vertua, R Della Loggia, T Yasumoto. Suitability of the MTT-based cytotoxicity assay to detect okadaic acid contamination of mussels. Toxicon 34:965–974, 1996
31. YF Pouchus, Z Amzil, C Marcaillou-Le Baut, KJ James, J-F Verbist. Specificity of the test based on modification of cell morphology for detection of lipophilic inhibitors of protein phosphatases. Toxicon 35:1137–1142, 1997.
32. L Croci. Algal biotoxins and their biological determinations. Rapp ISTISAN, Contaminazione Chimica e Biologica dei Prodotti della Pesca 5:56–67, 1997.
33. G Diogene, V Fessard, M Ammar, A Dubreuil, S Puiseux-Dao. Cytotoxic responses of Maitotoxin extract and okadaic acid. In: P Lassus, G Arzul, E Erard, P Gentien, C Marcaillou, eds. Harmful Marine Algal Blooms. Paris: Lavosier, 1995, pp 285–289.
34. P Cohen, CFB Holmes, Y Tsukitani. Okadaic acid: a new probe for the study of cellular regulation. Trends Biochem Sci 15:98–102, 1990.
35. H Fujiki, M Suganuma, H Suguri, S Yoshizawa, K Takagi, N Uda, K Wakamatsu, K Yamada, M Murata. Diarrhetic toxin DTX-1 is a potent tumor promotor on mouse skin. Jpn J Cancer Res 79: 1089–1093, 1988.
36. M Suganuma, H Fujiki, H Suguri, S Yoshizawa, M Hirota, M Nakayasu, M Ojika, K Wakamatsu, K Yamada. Okadaic acid: an additional non-phorbol-12-tetradecanoate-13-acetate-type tumor promoter. Proc Natl Acad Sci USA 85:1768–1771, 1988.
37. HA Luu, DZ Chen, J Magoon, J Worms, J Smith, CF Holmes. Quantification of diarrhetic shellfish toxins and identification of novel protein phosphatase inhibitors in marine phytoplankton and mussels. Toxicon 31:75–83, 1993.

38. RE Honkanen, DE Mowdy, RW Dickey. Detection of DSP-toxins, okadaic acid, and dinophysistoxin-1 in shellfish by serine/threonine protein phosphatase assay. J AOAC Int 79:1336–1343, 1996.
39. A Tubaro, C Florio, E Luxich, S Sosa, R Della Loggia, T Yasumoto. A protein phosphatase 2a inhibition assay for a fast and sensitive assessment of okadaic acid contamination in mussels. Toxicon 34:743–752, 1996.
40. JF Simon, JP Vernoux. Highly sensitive assay of okadaic acid using protein phosphatase and paranitrophenyl phosphate. Natl Toxins 2:293–301, 1994.
41. M Isobe, Y Sugiyama, T Ito, II Ohtani, Y Toya, Y Nishigohri, A Takai. New analysis method for protein phosphatase type 2a inhibitors using the firefly bioluminescence system. Biosci Biotech Biochem 12:2235–2238, 1995.
42. MR Vieytes, OI Fontal, F Leira, J Baptista de Sousa, M, V., LM Botana. A fluorescent microplate assay for diarrheic shellfish toxins. Anal Biochem 248:258–264, 1997.
43. P Cohen, JG Foulkes, CFB Holmes, G Nimmo, NK Tonks. Protein phosphatase inhibitor-1 and inhibitor-2 from rabbit skeletal muscle. Methods Enzymol 159:427–437, 1988.
44. CBF Holmes. Liquid chromatography-linked protein phosphatase bioassay, a highly sensitive marine bioscreen for okadaic acid and related diarrhetic shellfish toxins. Toxicon 29:469–477, 1991.
45. M Boland, M Smillie, D Chen, C Holmes. A unified bioscreen for the detection of diarrhetic shellfish toxins and microcystins in marine and freshwater environments. Toxicon 31:1393–1405, 1993.
46. A Takai, G Mieskes. Inhibitory effect of okadaic acid on the p-nitrophenyl phosphate phosphatase activity of protein phosphatases. Biochem J 275:233–239, 1991.
47. Y Shimizu, M Kinoshita, F Oi. Highly sensitive, non radioactive assays for protein phosphatase 1 and protein phosphatase 2A. VIIIth International Conference on Harmful Algae, Vigo, 1997 (abst).
48. Y Hokama, S Smith. Review: Immunological Assessment of Marine Toxins. Food Agric Immunol 2:99–105, 1990.
49. Y Hokama, SAA Honda, MN Kobayashi, LK Nakagawa, AY Asahina, JT Miyahara. Monoclonal antibody (MAb) in detection of ciguatoxin (CTX) and related polyethers by the stick-enzyme immunoassay (S-EIA) in fish tissues associated with ciguatera poisoning. In: S Natori, K Hashimoto, Y Ueno, eds. Mycotoxins and Phycotoxins '88. Amsterdam: Elsevier, 1989, pp 303–310.
50. DL Park, JM Fremy, C Marcaillou-LeBaut, E Gleizes, P Masselin, CH Goldsmith. Immunobead assay to monitor fish and shellfish harvesting areas for ciguatera fish and diarretic shellfish poisoning toxins. In: P Lassus, G Arzul, E Erard, P Gentien, C Marcaillou-LeBaut, eds. Harmful Marine Algal Blooms. Paris: Lavosier, 1995, pp 327–332.
51. L Levine, H Fujiki, K Yamada, M Ojika, HB Gijka, H Van Vunakis. Production of antibodies and development of a radioimmunoassay for okadaic acid. Toxicon 26:1123–1128, 1988.
52. T Uda, Y Itoh, M Nishimura, T Usagawa, M Murata, T Yasumoto. Enzyme immunoassay using monoclonal antibody specific for diarrhetic shellfish poisons. In: S Natori, K Hashimoto, Y Ueno, eds. Mycotoxins and Phycotoxins '88. Amsterdam: Elsevier, 1989, pp 335–342.
53. T Usagawa, M Nishimura, Y Itoh, T Uda, T Yasumoto. Preparation of monoclonal antibodies against okadaic acid prepared from the sponge, *Halichondria okadaii*. Toxicon 27:1323–1330, 1989.
54. EP Carmody, KJ James, SS Kelly. Diarrhetic shellfish poisoning: evaluation of enzyme-linked immunosorbent assay methods for dinophysistoxin-2 determination. J AOAC Int 78:1403–1408, 1995.
55. WS Shestowsky, MA Quilliam, HM Sikorski. An idiotypic-anti idiotypic competitive immunoassay for quantitation of okadaic acid. Toxicon 30:1441–1448, 1992.
56. JD Chin, MA Quilliam, JM Fremy, SK Mohapatra, HM Sikorska. Screening for okadaic acid by immunoassay. J AOAC Int 78:508–513, 1995.
57. WS Shestowsky, CFB Holmes, T Hu, J Marr, JLC Wright, J Chin, H Sikorski. An anti-okadaic acid-anti-idiotypic antibody bearing an internal image of okadaic acid inhibits protein phosphatase PP1 and PP2A catalytic activity. Biochem Biophys Res Commun 193:302–310, 1993.
58. S Matsuura, Y Hamano, H Kita, Y Takagaki. An ELISA for okadaic acid and its analogs among the diarrhetic shellfish toxins using mouse monoclonal anti-okadaic acid antibodies which are resistant to organic solvents. Jpn J Toxicol Environ Health 40:365–373, 1994.
59. S Matsuura, H Kita, Y Takagaki. Specificity of mouse monoclonal anti-okadaic acid antibodies to okadaic acid and its analogs among diarrhetic shellfish toxins. Biosci Biotech Biochem 58:1471–1475, 1994.

60. G Ramsay, GJ Lubrano, LL Nordyke, GG Giulbault (1990) Electrode probes for the rapid assay of seafood toxicants. Universal Sensors, Metairie, LA.
61. OB Stabell, AD Cambella. Standardising extraction and analysis techniques for marine phytoplankton toxins. In: E Graneli, B Sundstrom, L Edler, DM Anderson, eds. Toxic Marine Phytoplankton. New York: Elsevier, 1990, pp 518–521.
62. SS Kelly, AG Bishop, EP Carmody, KJ James. Isolation of dinophysistoxin-2 and the high-performance liquid chromatographic analysis of diarrhetic shellfish toxins using derivatisation with 1-bromoacetylpyrene. J Chromatogr 749:33–40, 1996.
63. R Draisci, L Lucentini, L Giannetti, A Stacchini. Diarrhetic shellfish toxins in mussels: optimization of HPLC method for okadaic acid determination. Riv Soc Ital Sci Alim 4:443–454, 1993.
64. OB Stabell, V Hormazabal, I Steffenak, K Pedersen. Diarrhetic shellfish toxins: improvement of sample clean-up for HPLC determination. Toxicon 29:21–29, 1991.
65. A Pereira, D Klein, K Sohet, G Houvenaghel, JC Braekman. Improved HPLC analysis for the determination of acid DSP toxins. In: P Lassus, G Arzul, E Erard, P Gentien, C Marcaillou-LeBaut, eds. Harmful Marine Algal Blooms. Paris: Lavosier, 1993, pp 333–338.
66. J Stockemer, M Geurke. Optimierung einer HPLC-methode zur bestimmung von DSP toxinen. Arch Lebensmittelhyg 44:146–147, 1993.
67. MA Quilliam. Analysis of diarrhetic shellfish poisoning toxins in shellfish tissue by liquid chromatography with fluorimetric and mass spectrometric detection. J AOAC Int 78:555–570, 1995.
68. M Shyre, E Unger. An alternative for the clean-up for ADAM derivatives of DSP toxins without halogenated solvents. In: T Yasumoto, Y Oshima, Y Fukuyo, eds. Harmful and toxic algal blooms. Paris: International Oceanographic Commission of UNESCO, 1996, pp 535–538.
69. B Aase, E Rogstad. Optimisation of sample cleanup procedure for the determination of diarrhetic shellfish poisoning toxins by use of experimental design. J Chromatogr 764:223–231, 1997.
70. B Luckas. Phycotoxins in seafood—toxological and chromatographic aspects. J Chromatogr 624: 439–456, 1992.
71. C Hummert, JL Shen, B Luckas. Automatic high-performance liquid chromatographic method for the determination of diarrhetic shellfish poison. J Chromatogr 729:387–392, 1996.
72. JA Rodriguez-Vazquez, A Gago-Martinez, A Ibanez-Panielo, P Burdaspal-Perez, T Legara-Gomez. High performance liquid chromatography of DSP toxins in bivalves by *in situ* formation of 9-anthryldiazomethane. In: TJ Smayda, Y Shimizu, eds. Toxic phytoplankton blooms in the sea. Amsterdam: Elsevier, 1993, pp 571–579.
73. MA Quilliam, A Gago-Martinez, A Rodriguez-Vazquez. Improved method for preparation and use of 9-anthryldiazomethane for derivatisation of hydroxy carboxylic acids: application to diarrhetic shellfish poisoning toxins. J Chromatogr 807:229–239, 1998.
74. RW Dickey, HR Granade, FA Bencsath. Improved analytical methodology for the derivatisation and HPLC-fluorometric determination of okadaic acid in phytoplankton and shellfish. In: TJ Smayda, Y Shimizu, eds. Toxic Phytoplankton Blooms in the Sea. Amsterdam: Elsevier, 1993, pp 495–499.
75. KJ James, AG Bishop, M Gillmann, SS Kelly, C Roden, R Draisci, L Lucentini, L Giannetti, P Boria. High-performance liquid chromatography with fluorimetric, mass spectrometric and tandem mass spectrometric detection to investigate the seafood toxin producing phytoplankton, *Dinophysis acuta*. J Chromatogr 777:213–221, 1997.
76. JC Gonzalez, MR Vieytes, JM Vieites, LM Botana. Improvement on sample clean up for high-performance liquid chromatographic-fluorimetric determination of diarrhoetic shellfish toxins using 1-bromoacetylpyrene. J Chromatogr 793:63–70, 1998.
77. N Nimura, T Kinoshita, T Yoshida, A Uetake, C Naki. 1-Pyrenyldiazomethane as a fluorescent labelling reagent for the liquid chromatographic determination of carboxylic acids. Anal Chem 60: 2067–2070, 1988.
78. SL Morton, JW Bomber. Maximising okadaic acid content from *Prorentrum hoffmannianum* Faust. J Appl Phycol 6:41–44, 1994.
79. JC Marr, LM McDowell, A Quilliam. Investigation of derivatization reagents for the analysis of diarrhetic shellfish poisoning toxins by liquid chromatography with fluorescence detection. Nat Toxins 2:302–311, 1994.

80. F Zonta, B Stancher, P Bogoni, P Masotti. High-performance liquid chromatography of okadaic acid and free fatty acids in mussels. J Chromatogr 594:137–144, 1992.
81. JF Lawrence, S Roussel, C Ménard. Liquid chromatographic determination of okadaic acid and dinophysistoxin-1 in shellfish after derivatization with 9-chloromethylanthracene. J Chromatogr 721:359–364, 1996.
82. C Hummert, B Luckas, J Kirschbaum. HPLC/MS determination of diarrhetic shellfish poisoning (DSP) toxins as 4-bromo-7-methoxycoumarin derivatives. In: P Lassus, G Arzul, E Erard, P Gentien, C Marcaillou-LeBaut, eds. Harmful Marine Algal Blooms. Paris: Lavosier, 1995,pp 297–302.
83. K Akasaka, H Ohrui, H Meguro, T Yasumoto. Fluorimetric determination of diarrhetic shellfish toxins in scallops and mussels by high-performance liquid chromatography. J Chromatogr 729:381–386, 1996.
84. K Akasaka, H Ohrui, H Meguro, T Yasumoto. Determination of dinophysistoxin-3 by the LC-LC method with fluorometric detection. Anal Sci 12:557–560, 1996.
85. KJ James, A Furey, B Healy, SS Kelly, M Twohig. A new fluorimetric reagent for the HPLC analysis of diarrhoeic shellfish poisoning toxins. In: B Reguera, J Blanco, ML Fernandez, T Wyatt, eds. Harmful Algae. Santiago de Compestella: Xunta de Galicia & Intergovernmental Oceanographic Commission of UNESCO, 1998, pp 519–520.
86. KJ James, AR Bishop, BM Healy, C Roden, IR Sherlock, M Twohig, R Draisci, C Giannetti, L Lucentini. Efficient isolation of the rare diarrhoeic shellfish toxin, dinophysistoxin-2, from marine phytoplankton. Toxicon 37:343–357, 1999.
87. S Pleasance, MA Quilliam, ASW de Freitas, JC Marr, AD Cembella. Ion-spray mass spectrometry of marine toxins II. Analysis of diarrhetic shellfish toxins in plankton by liquid chromatography/mass spectrometry. Rapid Commun Mass Spectrom 4:206–213, 1990.
88. JC Marr, T Hu, S Pleasance, MA Quilliam, JLC Wright. Detection of new 7-O-acyl derivatives of diarrhetic shellfish poisoning toxins by liquid chromatography-mass spectrometry. Toxicon 30:1621–1630, 1992.
89. S Pleasance, MA Quilliam, JC Marr. Ionspray mass spectrometry of marine toxins. IV. Determination of diarrhetic shellfish poisoning toxins in mussel tissue by liquid chromatography/mass spectrometry. Rapid Commun Mass Spectrom 6:121–127, 1992.
90. MA Quilliam, JLC Wright. Methods for diarrhetic shellfish poisons. In: GM Hallegraeff, DM Anderson, AD Cembella, eds. Manual on Harmful Marine Microalgae. Paris: UNESCO, 1995, pp 95–111.
91. R Draisci, L Lucentini, L Giannetti, P Boria, KJ James, A Furey, M Gillman, SS Kelly. Determination of diarrhoetic shellfish toxins in mussels by micro high-performance liquid chromatography with tandem mass spectrometric detection. J AOAC Int 81:441–447, 1998.
92. N Bouaicha, M.-C. Hennion, P Sandra. Determination of okadaic acid by micellar electrokinetic chromatography with ultraviolet detection. Toxicon 35:273–281, 1997.
93. J Cai, J Henion. Capillary electrophoresis-mass spectrometry. J Chromatogr 703:667–692, 1995.
94. S Pleasance, P Thibault, J Kelly. Comparison of liquid-junction and coaxial interfaces for capillary electrophoresis-mass spectrometry with application to compounds of concern to the acquaculture industry. J Chromatogr 591:325–329, 1992.
95. S Allenmark, M Chelminska-Bertilsson, RA Thomson. N-(9-acridinyl)-bromoacetamide—a powerful reagent for phase-transfer-catalised fluorescence labeling of carboxylic acids for liquid chromatography. Anal Biochem 185:279–285, 1990.

12
Mechanism of Action and Toxicology

Mercedes R. Vieytes, M. C. Louzao, A. Alfonso, A. G. Cabado, and Luis M. Botana
University of Santiago de Compostela, Lugo, Spain

I. OVERVIEW

Several organisms produce protein phosphatase (PP)–inhibiting toxins. Okadaic acid (OA) and calyculin A (both from marine sponges *Halichondria okadaii* or *H. melanodocia*), tautomycin (an antibiotic from *Streptomyces spiroverticillatus*), microcystin and nodularin (both cyanobacterial hepatotoxins), and cantharidin (from blister beetles) are all potent inhibitors of the catalytic subunits of protein phosphatases (see Chapters 13 and 30 in this book). There are four major protein phosphatases identified in the cytosol of eukaryotic cells, the PP1, PP2A, and PP2B group, which belong to the PPP family, and PP2C (calcineurin, a member of the PPM family [Barford, 1996]), and they are responsible for the dephosphorylation of serine and threonine residues in proteins (Cohen and Cohen, 1989). Many other protein phosphatases have been reported (for a review, see Barford, 1996; Burke and Zhang, 1998).

Under certain environmental conditions, marine and freshwater phytoplankton may produce phycotoxins which are inhibitors of serine/threonine PP 1, 2A, and 3. Marine dinoflagellates of the species *Dinophysis* and *Prorocentrum* produce fatty polyethers, particularly okadaic acid and its derivatives, the dinophysistoxins (DTXs 1–5), all of which are responsible for gastrointestinal tract symptoms (Yasumoto et al., 1985). In freshwaters, the functionally equivalent and chemically distinct toxins are microcystins and nodularin, which are seven– or five–amino acid cyclic peptides and are hepatotoxic (see Chapters 26–31). All these toxins are potent tumor promoters but belong to a new class, different from the PMA (phorbol esters) class, because they do not act on protein kinase C. Okadaic acid (OA) is a 38-carbon polyether derivative fatty acid, which accumulates in marine sponges such as *H. okadaii* and in the digestive glands (the hepatopancreas) of bivalve mollusks, where it is highly concentrated by filter feeding.

In several parts of the world, algal toxins leading to diarrhea (diarrheic shellfish poisons, DSP) are found in mussels for extended periods each year. DSP is a rapid-onset intoxication (see epidemiology of DSP toxins in Chapter 3) in humans caused by the consumption of shellfish contaminated with OA or the DTXs 1–5) from certain species of marine phytoplankton (Aune, 1997; Hu et al., 1992a; Tachibana et al., 1981; Yasumoto et al., 1985). DSP is a serious problem to aquaculture worldwide.

The predominant human symptoms of DSP are mostly gastrointestinal disturbances, such as diarrhea, nausea, vomiting, and abdominal pain, and no fatal cases have been so far reported. The symptoms start between 30 min and a few hours after eating the contaminated seafood, and the victims usually recover within 3 days (Aune, 1997; Yasumoto et al., 1978a, 1978b). The current method to monitor the toxins is the mouse bioassay (for a review, see Botana et al., 1996; Quilliam, 1999), which uses the mouse unit (MU) as the measure to quantify the toxin. One MU is defined as the amount of toxin sufficient to kill one mouse of 20 g within 24 h. It was concluded that 12 MU was sufficient to cause mild symptoms in adults (Aune and Yundestad, 1993). At the west coast of Sweden, a dinoflagellate bloom occurred in October 1984; several people who had consumed blue mussels from this area developed symptoms of DSP. The reported toxin concentration was higher than 170 MU kg-1 mussel flesh (Krogh et al., 1985). The minimum doses of OA and DTX-1 producing diarrhea in adults have been estimated to be 40 and 36 μg, respectively (Hamano et al., 1985). In another toxic episode, Danish mussels exported to France in 1990 caused intoxications to 415 people (Aune and Yundestad, 1993; Hald et al., 1991). These mussels contained 170 μg OA per 100 g of mussel meat. In August 1990, 16 people who ate cultivated mussels from eastern Nova Scotia became ill with symptoms of nausea, vomiting, and diarrhea. The analysis established that there was up to 1 μg DTX-1 in the flesh of one mussel but no OA (Quilliam et al., 1993). Two patients developed chills, nausea, abdominal pain, vomiting, and diarrhea 1–2 h after each one had consumed 10 mussels (approximately 200 g), which were found to contain 20–30 μg of OA per 100 g (Scoging and Bahl, 1998). In general, shellfish containing more than 2 μg OA and/or 1.8 μg DTX-1 per gram of hepatopancreas are considered unfit for human consumption.

The DSP toxins are produced primarily by algae belonging to the genus *Dinophysis (D. acuta, D. norvegica, D. acuminata*, and *D. fortii)* and species in the genus *Prorocentrum (P. lima, P. minimum)*. Although OA and DTX-1 production is restricted to a few species in the dinoflagellate genera *Prorocentrum* and *Dinophysis*, DSP is a global problem (Shumway et al., 1994). Since the first report by Yasumoto et al. in 1978, the occurrence of DSP has been reported in many parts of the world. DSP occurs predominantly in Europe and Japan, but affected areas now also include North and South America, Australia, Indonesia, and New Zealand (Sundstrom et al., 1990) due to increased global spreading of toxic dinoflagellates in recent years. *Dinophysis* blooms occur widely in European seawaters in the summer months, when conditions of pH, temperature, and salinity permit their proliferation. It has been reported that 200–2000 cells of *Dinophysis* spp per liter of seawater resulted in significant accumulation of cells in blue mussels (20,000–30,000 cells per digestive gland) and a high toxicity level in the mussels in Norway (Séchet et al., 1990). On the other hand, only blooms greater than 20,000 cells/L were associated with diarrheic mussel poisoning in the Dutch Wadden Sea (Kat, 1983).

Windust et al. (1996) have demonstrated that at micromolar concentrations OA is a potent microalgal growth inhibitor, effective against several different species of microalgae, and this may explain its biological role, as a secondary metabolite aimed to allow the competitive growth of the toxin-producer species. Nevertheless, DSP-producing dinoflagellates are not affected even at much higher concentrations.

II. CHEMICAL STRUCTURES

The so-called diarrheic toxins are a rather heterogeneous group, which need to be reclassified, since the only criterion to group them all is their lipophilicity, whereas their mechanism of action is completely different (i.e., whereas OA inhibits phosphatases, the action of yessotoxin shows a heavy component of cAMP modulation (unpublished results).

The toxins can be separated into three groups according to their basic skeletons (see Chapters 11 and 14):

1. Okadaic acid and dinophysistoxins
2. Yessotoxins (YTXs)
3. Pectenotoxins (PTXs)

Chemically, the structures of OA and dinophysistoxins, yessotoxins, and pectenotoxins are polyethers. The first group includes (OA), the polyether macrolides DTXs 1–5, and diol esters (Figure 1) (Hu et al., 1995b; Lawrence et al., 1998). By spectral comparison with OA, the DTXs were identified as OA derivatives (Figure 1). The toxin profile is different depending on the zone to be sampled; that is, DTX-2 is an isomer of OA isolated from Irish mussels (Hu et al., 1992a), and it is the predominant toxin on those coasts, whereas DTX-1 is the predominant one in Japan (Carmody et al., 1996) and OA in Spain. Several DTX-2 isomers have been already

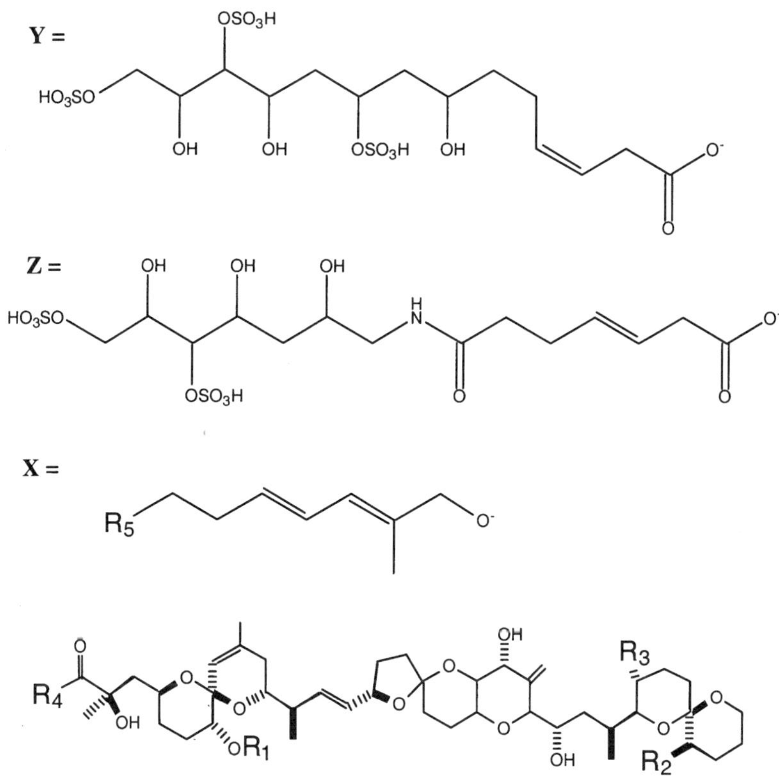

Toxin	R1	R2	R3	R4	R5
OA	CH3	H	H	OH	—
DTX-1	CH3	CH3	H	OH	—
DTX-2	H	CH3	H	OH	—
DTX-3	Acyl	H or HC3	H or CH3	OH	—
DTX-4	CH3	H	H	X	Y
DTX-5	CH3	H	H	X	Z
DIOL ESTERS	CH3	H	H	X	OH

Figure 1 Chemical structure of Okadaic acid and analogs.

reported (James et al., 1999). DTX-1 and DTX-3 structures were determined by Murata et al. (1982) and Yasumoto et al. (1985), respectively. DTX-3 is a general name for a group of 7-acyl derivatives of OA. OA and DTX-1, DTX-2, and DTX-3, or diol esters are lipophilic, whereas DTX-4 and DTX-5 are water-soluble sulfate derivatives of OA (see Figure 1) (Hu et al., 1995b). Although DTX-4 and DTX-5 are weakly toxic, they are readily hydrolyzed to form OA.

OA the main toxin of the diarrheic shellfish poisoning (DSP) group, was first isolated from the marine sponges of the genus *Halichondria (H. okadaii)* and *H. melanodocia*) and from *Prorocentrum lima*. The structure of OA was determined by Tachibana et al. (1981) after isolation of the component from sponges. OA is a polyether derivative of a C38 fatty acid with one carboxyl group and four hydroxyl groups. The major DTX toxin responsible for DSP was isolated in 1982 by Murata et al. (1982) and named dinophysistoxin 1. Spectrometric analysis identified dinophysistoxin 1 as 35 s-methyl okadaic acid (Murata et al., 1982). Since DTX-3 has not been detected in *Dinophysis* species, it is suggested that acylation of DTX-1 to DTX-3 occurs in the hepatopancreas of scallops (Murata et al., 1982). Suzuki et al. (1999) obtained the first direct evidence of the transformation of DTX-1 to 7-O-acyl-dinophysistoxin-1 (DTX-3) in the tissues of scallop (*Patinopecten yessoensis*).

III. MECHANISM OF ACTION

After OA was first identified as a specific inhibitor of serine/threonine protein phosphatases 1 and 2A (PP1 and PP2A), Fujiki and Suganuma (1993) found 30 compounds of the okadaic acid class, all potent inhibitors of PP1 and PP2A. These investigators classified these compounds into four types according to their structural differences:

1. Okadaic acid itself
2. Calyculin A
3. Microcystin-LR
4. Tautomycin

OA and tautomycin are serine/threonine phosphatase inhibitors of the polypropionate class. Although structurally unrelated to microcystin, OA and tautomycin have also shown inhibition of PP2A in a dose-dependent manner. OA inhibited PP2A with an IC50 of 0.45 nM, which was substantially lower than that for tautomycin (IC50 = 10 nM). The IC50 of OA and tautomycin for PP1 was found to be 1.24 and 0.67 nM, respectively.

Fujiki and Suganuma (1993) propose a conformational change of the OA molecule through hydrogen bond formation (between C-1 and C-24), which gives a structure closely resembling that of microcystin. This might explain the binding of the different compounds to the same cellular receptors. OA and analogues are highly active phosphatase inhibitors effective against the serine- and threonine-specific phosphatases PP1 and PP2A (Bialojan and Takai, 1988; Nishiwaki et al., 1990a). Serine/threonine phosphatases (Millward et al., 1999) comprise a unique class of enzymes consisting of four primary subclasses based on their differences in substrate specificity and environmental requirements. Phosphatases are highly conserved proteins, and OA has been found to interfere with phosphatase-kinase-based control mechanisms in a broad range of eukaryotes (MacKintosh et al., 1990). PP2A appears to play a role in the regulation of most major metabolic pathways, as well as translation, transcription, and control of transition from G2 to the M phase of the cell cycle; PP2A may function as either a tumor promoter or tumor suppressor depending on the cell type or the transforming agent (Coggeshall, 1999; Millward et al., 1999).

In the interaction of OA and protein phosphatase (Nishiwaki et al., 1990b), OA binds specifically to the catalytic subunit. The catalytic subunit of type 1 phosphatase is the natural and specific target for inhibitor 2 and, together with the catalytic subunit of type 2A phosphatase, the target of OA (Cohen and Cohen, 1989). PP2A consists of two heterogeneous regulatory subunits and one catalytic subunit. The results strongly supported the evidence that OA binds to its protein receptor, the catalytic subunit of PP2A, and inhibits the activity of their catalytic subunits (Erdodi et al., 1988; Hescheler et al., 1988; Ishihara et al., 1989). Serine/threonine PP1 and PP2A share sequence identity of both enzyme catalytic subunits (50% for residues 23–292; 43% overall). Perhaps the most interesting link between PP1 and PP2A is their shared sensitivity toward a structurally diverse family of natural products, such as calyculins, tautomycin, nodularins, motuporin, cantharidin, fostriecin, acanthifolicin, the microcystins, and OA. The tertiary structure of these toxins is similar, and explains why they all inhibit these protein phosphatases (Bagu et al., 1997). In vitro competition experiments suggest that inhibition of PP1 and PP2A by these natural products is mutually exclusive with respect to the natural substrates in addition to one another (DilipDeSilva et al., 1992; Holmes et al., 1990; Laidley et al., 1997; MacKintosh and Klumpp, 1990; Takai et al., 1995; Walsh et al., 1997). As to the nature of inhibition, OA acts as a noncompetitive or mixed inhibitor on the OA sensitive protein phosphatases (Bialojan and Takai, 1988). Several studies (Haystead et al., 1989; Takai et al., 1992) established OA as a potent and specific inhibitor of PP1 (IC50 = 3–150 nM) and PP2A (IC50 = 0.03–0.2 nM). The purified catalytic subunits of PP1 and PP2A were potently inhibited by OA with 50% inhibitory concentration values of 20 and 0.2 nM, respectively.

The IC50 values for different serine/threonine phosphatase inhibitors from previous experiments vary over a wide concentration range. In addition to variances in assay conditions (e.g., temperature and duration), differences in the choice of substrate also has an effect on the observed inhibition profiles, and it has been well reported (An and Carmichael, 1994; Fontal et al., 1999; Gupta et al., 1997; Honkanen et al., 1990, 1991; Vieytes et al., 1997) an enzyme concentration dependence of IC50 values for both PP1 and PP2A. This observation is characteristic of tight-binding enzyme-inhibitor interactions. Optimal concentration values of 1 and 0.1 unit/mL were subsequently established for PP1 and PP2A. Okadaic acid is a much more potent (200-fold) inhibitor of PP2A than PP1 (IC50 of 0.07 and 3.4 nM, respectively). In fact, OA has been previously used to differentiate between PP2A and other phosphatases owing to its high affinity for PP2A (Bialojan and Takai, 1988; Haystead et al., 1989; Hescheler et al., 1988; Takai et al., 1992).

Regarding other protein phosphatase inhibitors, like calyculin A (Ishihara et al., 1989) or microcystin-LR (MacKintosh et al., 1990) the order of potency is microcystin>calyculin A>tautomycin, and the IC50 values go from 0.1 to 0.7 nM. Initially, the inhibition IC50 for microcystin, calyculin A, and tautomycin are in the same range (0.1–0.7 nM) for both PP1 and PP2A. The inhibitory dose response on the catalytic subunit of PP1 for the OA class compounds is microcystin>calyculin A>tautomycin>okadaic acid (Suganuma et al., 1992). With regard to the inhibition of the PP2A catalytic subunit by the same drugs, a similar order of potency was obtained for all the compounds except okadaic acid: okadaic acid>microcystin>calyculin A>tautomycin.

None of the OA class compounds has been reported to inhibit protein tyrosine phosphatase 1 activity at concentrations up to 0.01 mM. In several reports, OA was also shown to lack inhibitory activity toward any protein tyrosine phosphatase (Hescheler et al., 1988; Ishihara et al., 1989; Suganuma et al., 1992). Thus, the compounds of the OA class exhibit selective inhibition of protein serine/threonine, but no tyrosine phosphatase activity.

With regard to protein phosphatase purified from bovine brain (PP3), the reported observations show that, like PP1 and PP2A, the activity of PP3 is potently inhibited by OA, microcystins, nodularin, calyculin A and tautomycin (Honkanen et al., 1994). Further characterization of the toxins employing the purified catalytic subunits of PP1, PP2A, and PP3 under identical experimental conditions indicates that:

1. Okadaic acid, microcystin-LR, and microcystin-LA inhibit PP2A and PP3 more potently than PP1 (order of potency PP2A>PP3>PP1);
2. Nodularin inhibits PP1 and PP3 at a similar concentration that is slightly higher than that which affects PP2A, and
3. Both calyculin A and tautomycin show little selectivity among the phosphatases tested. This study also shows that the chemical modification of the (C1) carboxyl group of OA can have a profound influence on the inhibitory activity of this toxin.

Although most of the actions of OA are due to inhibition of PP1 or PP2A, the effects are also mediated, at least in some models, to inhibition of PP4 or PP5, this being a field that must be studied in much more detail (Millward et al., 1999).

IV. STRUCTURE-FUNCTION RELATIONSHIPS

The study on structure-function relationships with OA derivatives revealed that the carboxyl group, as well as four hydroxyl groups on C-2, C-7, C-24, and C-27 of OA are important for activity (Nishiwaki et al., 1990a). The relationship between the chemical structure of OA and its inhibitory effect on the protein phosphatases has been studied to some extent by experiments with OA derivatives (Nishiwaki et al., 1990a). Alteration of the C-1 carboxylic acid or 24-OH greatly reduces the activity. Other structural changes (e.g., hydrogenation at C-14 = C-15 or deoxidation at C-2) which affect the pseudocyclic conformation formed by interaction between C-1 carboxylic acid and C-24 hydroxyl group also reduce the potency. The structure of OA is similar to acanthifolicin (9,10-epithio-okadaic acid) (Holmes et al., 1990). Methyl esterification of the single carboxyl group on acanthifolicin (and OA) drastically reduced the potency of these compounds as protein phosphatase inhibitors, demonstrating that the region comprising this carboxyl group is important for effective inhibition of the enzymes. Nishiwaki et al. (1990a) have examined the effect of 17-OA derivatives on the protein phosphatase (mainly PP2A) activity of a cytosolic fraction of mouse brain. They have reported that OA decreases its affinity to the phosphatase when the carboxylic acid is either removed or modified (Takai et al., 1992). It has been reported that OA tetramethylether (methylated at 2-, 7-, 24-, and 27-carbon hydroxyl groups) does not promote intracellular hyperphosphorylation in human fibroblasts and is not active as a tumor promoter (Issinger et al., 1988; Suganuma et al., 1989). Experimental evidence for the importance of the four hydroxyl groups of OA for its inhibitory effect on protein phosphatases was obtained by Sassa et al. (1989), who reported that 2,7,24,27-tetramethoxy-OA, in which all the four hydroxyl groups were methoxylated, had no inhibitory effect on the protein phosphatase (mainly PP2A) activity in a cytosolic fraction of mouse brain (see also Nishiwaki et al., 1990a). The latter observation indicates also the crucial importance of the hydroxyl groups for the interaction of OA with protein phosphatases. Takai et al. (1992) have examined the effect of some other OA derivatives, including 2-deoxy-OA and 7-deoxy-OA, on the affinity for purified preparations of PP1 and PP2A. They found that selective removal of the hydroxyl groups at the 2 and 7 positions increases the K_i values for the interaction with PP2A 30-fold and 2-fold, respectively. According to the model of the three-dimensional conformation of OA on the basis of x-ray analyses, the 27-hydroxyl group thus appears to be present in a position relatively free from intramolecular bond formation, in comparison with the other three hydroxyl groups, which are thought to form hydrogen bonds either with the carboxyl group or with the cyclic-ether oxygens to stabilize the circular conformation (Taylor et al., 1996). Therefore, of the specific modifications of the hydroxyl groups examined (Sasaki et al., 1994), oxidation of the 27-hydroxyl group causes the most marked reduction of the affinity to the phosphatases. In contrast with methoxylation of all four hydroxyl groups (see above), however, this modification does not completely abolish the inhibitory effect. Therefore, it seems that the 27-hydroxyl group serves as one of the multiple binding sites, forming a hydrogen bond with some electron-rich portion of the PP1 and PP2A molecules.

V. MOLECULAR INTERACTION WITH PHOSPHATASES

The importance of a cyclic system in modulating binding affinity of microcystin and nodularin has been highlighted, since the acyclic inhibitors appear to fold onto themselves, resulting in cyclic conformations. This observation is supported by the existence of intramolecular hydrogen bonds between N[3]-H and O(1) in calyculin A (Kato et al., 1986) and the C-1 carboxylate and C-24 hydroxy oxygen in OA that form 21- and 18-membered pseudorings, respectively. A similar analysis of OA suggests that the C-4–C-12-spiroketal unit is responsible for the effective turn of the molecule. These spirocyclic features reinforce adoption of ''circular'' conformations despite the acyclic primary structures of OA and calyculin. Gupta et al. (1997) have reported a computational study performed on the inhibitors calyculin A, nodularin, and OA to binding to PP1. These investigators show that the natural toxins, although having diverse structures, possess remarkably similar three-dimensional motifs on superimposition and Van der Waals minimization within the PP1 active site (Gupta et al., 1997).

VI. TOXIN TRANSFORMATION AND METABOLISM

The importance of PP2A in cellular homeostasis is underscored by the fact that several organisms produce PP2A-inhibiting toxins, presumably as a self-defense mechanism. It has been demonstrated that OA inhibits the growth of a variety of microalgae in a concentration-dependent manner, but not that of a DSP-producing dinoflagellate, *Prorocentrum lima* (Windust et al., 1996), suggesting an allelopathic role for the toxin (Windust et al., 1996). This activity of OA is consistent with an ability to inhibit serine- and threonine-specific phosphatases (Bialojan and Takai, 1988), thereby disrupting normal eukaryotic cell functions. Indeed, OA is extremely potent and enhances overall levels of protein phosphorylation when present at nanomolar levels within the cell (Haystead et al., 1989). Derivatives of DSP toxins have been recovered from cultures of *P. lima*. These include OA diol ester (Hu et al., 1992b) and the water-soluble DTX-4 (Hu et al., 1995b). Importantly, both OA diol-ester and DTX-4 or DTX-5 can be hydrolyzed to yield the parent toxin OA (see Figure 1), and the OA methyl ester is 4500 times less potent than OA against PP2A. In vitro enzyme assays have shown that DTX-4 is about 50 times less effective as a phosphatase inhibitor than OA (Hu et al., 1995a, 1995b) and that OA diol-ester is relatively inactive (Hu et al., 1992b). The low activity of these ester derivatives of OA is consistent with previous studies which have concluded that high activity against phosphatases is correlated with a free carboxyl group (Holmes et al., 1990; Lawrence et al., 1998; Nishiwaki et al., 1990a).

Although the fundamental importance of phosphatase to cell function make them ideal targets for allelopathic agents, producing cells must also avoid autotoxicity (MacKintosh and MacKintosh, 1994). This function in DSP toxin–producing dinoflagellates may be fulfilled by the synthesis of weakly active sulfates, such as DTX-4 or DTX-5. This contention is supported by the fact that at least 80% of intracellular DSP exists as DTX-4 in *P. lima* (Quilliam et al., 1996) and by biosynthetic evidence that implicates DTX-4 as the initial product of DSP synthesis in *P. lima* (Needham et al., 1994). The allelopathic potential of OA toward microalgae has been demonstrated when the toxin is dissolved in seawater (Windust et al., 1996), and the biosynthesis of water-soluble DSP compounds such as DTX-4 by *P. lima* has been proposed as a means of facilitating excretion (Hu et al., 1995a). DTX-4 is rapidly hydrolyzed to OA diol-ester and OA by enzymes released following lysis of *P. lima* cells (Quilliam and Roos, 1996); therefore the DSP compounds may also be effective against predators that rupture producing cells during feeding (Windust et al., 1997).

It has been proposed that the relative toxicity of the DSP compounds in vivo may be primarily explained by their widely differing solubilities (Windust et al., 1997). This is indirectly supported by several lines of evidence: the much higher concentrations of OA required to cause an effect in vivo versus in vitro (Botana and MacGlashan, 1993; Estévez et al., 1994; Vieytes et al., 1997), the propensity of the esters to hydrolyze (albeit at a slow rate), and the oxidative metabolism of OA diol-ester by *Thalassiosira weissflogii*.

The concentrations of OA required to inhibit phosphatases in extracts (~1 nM) are three orders of magnitude lower than those required to affect the biochemistry of whole cells (~1 μM). This is believed to be due to the barriers imposed by cell walls and membranes to the diffusion of OA (Cohen et al., 1990). In general, the lipid solubility of a toxin will predict its propensity to accumulate within cells and is therefore a significant factor in determining the potency. Although often described as a lipid-soluble polyether, OA still possesses a carboxyl group likely to be ionized at physiological pH; this would hinder diffusion across membranes. In contrast, OA diol-ester has no ionizable group and would be expected to move across cell membranes with relative ease. Therefore, the minor hydrolysis of OA diol-ester to OA observed (Windust et al., 1997) (~3%) is in fact very important in determining in vivo toxicity as the diol-ester could provide an effective carrier for transporting OA into a cell. OA liberated by hydrolysis within the cell would be 1000 times more potent than that outside the cell, and this could explain reported observations (Windust et al., 1997), contrary to data derived from in vitro experiments, showing a high relative toxicity of OA diol-ester to *T. weissflogii*.

The ability of OA diol-ester to enter cells readily is indirectly supported by the rapid and inducible hydroxylation of this molecule by *T. weissflogii*. Cells of *T. weissflogii* challenged with OA diol-ester rapidly metabolized most of the toxin to a more water-soluble product. From interpretation of mass spectral data obtained using ion-spray liquid chromatography–mass spectrometry (LC-MS), the metabolite was identified as an oxygenated diol-ester of OA, implying that it was the product of a monooxygenase-detoxification pathway. It is postulated that OA diol-ester, as a lipid-soluble, uncharged molecule with a propensity to hydrolyze to OA, may facilitate the transfer of OA across cell walls and membranes (Windust et al., 1997).

If OA diol-ester serves as an effective intermediate for the transport of DSP toxins across cell membranes, it could prove to be a significant factor in DSP transfer and toxicity within the food web, including human consumers. This is because DTX-4 is the major DSP toxin found within *P. lima*, and it will be hydrolyzed in seconds to OA diol-ester, but then only slowly to OA, following cell lysis (Quilliam and Roos, 1996). DTX-4 can be sequentially hydrolyzed through uncharged, lipophilic intermediates ultimately to yield the active, free acid toxin OA. Thus, as *P. lima* cells are ingested and lysed by predators, OA diol-ester will be the primary DSP toxin available for uptake and potential conversion to OA.

VII. CELLULAR TOXICITY

Okadaic acid is a potent tumor promoter (Fujiki et al., 1990), it is not an activator of protein kinase C, and it is a very specific inhibitor of protein phosphatases. OA has been the focus of major interest since the first reports that phosphorylation and dephosphorylation of proteins are among the most important regulatory processes in eukaryotic cells (Cohen, 1989). Phosphatases are present in all eukaryotic cells, and their control presents a number of challenges to medicinal chemists. The protein phosphatases are integral regulatory proteins and play significant roles in signal transduction pathways pertaining to metabolism, cell proliferation, gene expression, neurotransmission, membrane transport, contractility, morphogenesis, development, cell cycle progression, or secretion (Cohen, 1989; Wera and Hemmings, 1995), but the mechanisms by which PP2A affects such diverse cellular functions are not well understood. However, over the past 10 years, it became clear that one of the major classes of phosphatases substrates are in fact the protein kinases themselves (Millward et al., 1999). The toxic properties of OA have so far been attributed to its ability to inhibit the activity of these enzymes, which in turn affects all kind of intracellular processes. Inhibition of other serine/threonine protein phosphatases, such as PP3 (Brewis et al., 1993; Honkanen et al., 1991) may also contribute to the toxic effects of OA; however, to date, little is known about the physiological or pathological roles of these enzymes. Recently, several groups have discovered additional PP2A-like phosphatases that, although they represent only a small fraction of the total cellular serine/threonine phosphatase activity, they also show sensitivity to OA. OA inhibits PP4 in vitro with an IC50 comparable to that of PP2A, whereas PP5 is inhibited with an IC50 of 1–10 nM (for a review, see Millward et al., 1999).

The first insight into the mechanism of action of OA came from the discovery that OA enhanced the contraction of smooth muscle fibers, and it caused long-lasting contraction of smooth muscle from human arteries (Shibata et al., 1982). This appeared to be mediated by a phosphatase from smooth muscle with activity toward isolated myosin P light chains. Smooth muscle contraction is triggered by phosphorylation of a subunit of myosin, and the effect of OA was due to inhibition of myosin light chain phosphatase (Takai et al., 1987). Later, OA was shown to be a specific inhibitor of two of the four major protein phosphatases in cytosol in mammalian cells, type 1 and type 2A (Bialojan and Takai, 1988). PP2B, a Ca^{2+}-dependent protein phosphatase (a calcium/calmodulin–dependent enzyme) whose catalytic subunit is homologous to those of PP1 and PP2A, is also inhibited by OA, but with much lower potency (half-maximal effect at 5 μM) (Bialojan and Takai, 1988), whereas PP2C, a magnesium-dependent enzyme, is unaffected. OA can be applied to cells at levels of up to 1 μM without any detectable inhibitory effect on PP1 or on the other major serine/threonine–phosphatase activities.

Along with the original observations that OA caused contraction of smooth muscle, OA has been shown to modulate dramatically several metabolic parameters in intact cells. The toxin stimulates glucose output and gluconeogenesis in hepatocytes, stimulates lipolysis, and inhibits fatty acid synthesis in adipocytes (Haystead et al., 1989).

Proteins whose phosphorylation is increased in the presence of OA include (1) acetyl-CoA carboxylase and ATP-citrate lyase in adipocyte and hepatocyte cytosol, (2) pyruvate kinase and 6-phosphofructo-2-kinase/fructose-2,6-biphosphatase (PFK2/FBPase) in hepatocyte cytosol, and (3) glycogen phosphorylase and glycogen synthase in the hepatocyte-glycogen fraction. Because of this inhibitory effect, OA produces in intact cells a great increase in phosphorylation of several proteins, that is, myosin light chain, elongation factor 2, acetyl-CoA carboxylase, and tyrosine hydroxylase, and modulates a variety of cellular functions such as smooth muscle contraction, fatty acid biosynthesis, protein synthesis, and cathecholamine synthesis and secretion (Felix et al., 1990; Haystead et al., 1989).

The finding that OA converts acetyl-CoA carboxylase completely into the phosphorylated form in adipocytes shows that protein phosphatase 2A is responsible for dephosphorylation in the intact cells. Given the dramatic effects of okadaic acid on protein phosphorylation, it might be expected that all manner of toxic effects secondary to increases in protein phosphorylation would occur in long-term incubations.

Although OA antagonized the actions of insulin on fatty acid synthesis and lipolysis, it mimicked the action of insulin in dramatically enhancing the incorporation of radioactivity from [^{3}H]glucose into total lipid. OA is known to provoke insulin-like effects on GLUT-4 translocation and glucose transport, and causes a significant (three- to six-fold) increase of the rate of 2-deoxyglucose (an index of glucose transport and phosphorylation) accumulation by isolated rat adipocytes (Corvera et al., 1991; Haystead et al., 1989). Similar insulin-like results were obtained in skeletal muscles from mice that show, with 5 μM OA, a response as efficient as a maximally effective insulin concentration (Tanti et al., 1991). On the other hand, pretreatment of cells with OA acutely inhibits the effects of insulin to stimulate the rate of glucose transport, principally by blocking the insulin-stimulated redistribution of GLUT-4 transporters to the plasma membrane. This effect is similar to growth hormone (GH) since OA initially mimics the actions of insulin (Corvera et al., 1991), but subsequently, after being removed, the ability of adipocytes to accelerate glucose metabolism in response to insulin is severely reduced. OA produces such refractoriness without affecting the ligand affinity or tyrosine kinase activity of the insulin receptor. Similarly, refractoriness to the insulin-like effect of GH is also produced without a change in the affinity or abundance of GH-binding sites (Gaur et al., 1998; Grichting et al., 1983).

OA has been reported to exert potent tumor-promoting activity (Fujiki and Suganuma, 1993) and to cause apoptotic neuronal death in vitro (see Chapter 13). The effect of OA in human fibroblasts or neuroblastoma cells show clear apoptotic effects regarding cell morphology, mitochondrial membrane potential, actin or DNA organization, and free radical (Leira et al., 2000). The neurotoxic effect that OA induces in cerebellar neurons in primary culture can be observed at concentrations (0.5 nM) that do not affect nonneuronal cells (Fernández et al., 1991). Since in cultured neurons OA behaves as a potent neurotoxin, the regional selectivity and mechanisms underlying its toxicity were investigated by Runden et al. (1998) in hippocampal slice cultures. The studies revealed that OA caused a dose- and time-dependent injury to hippocampal neurons. Treatment with OA led to rapid and sustained tyrosine phosphorylation of the mitogen-activated protein kinases ERK1 and ERK2 (p44/42 [mapk]). The phosphorylation was markedly reduced after treatment of the cultures with the microbial alkaloid K-252a (a nonselective protein kinase inhibitor) or the MAP kinase (MEK1/2) inhibitor PD98059. Both K-252a and PD98059 also ameliorated the OA-induced cell death.

In order to asses its neurotoxicity in vivo, Arias et al. (1998) studied the effect of different doses of OA administered into the dorsal rat hippocampus. Hippocampal sections revealed that a notable neurodegeneration occurred as early as 3 h after injection of 300 ng OA. Neuronal death was more evident at 24 h, and at this time the extent of damage was dose dependent. The process of neuronal death was accompanied by a loss of the microtubule-associated protein MAP2 and also an increase in the expression of the inducible heat shock protein 72 in the surviving neurons.

In fact, the neurotoxic effect of OA may be related to the nervous modulation of its diarrheic effect, as suggested by the clearly observable nausea and vomiting during DSP episodes. Nevertheless, in addition to this, the induction of diarrhea by OA is a result of phosphorylation of proteins controlling the sodium secretion by intestinal cells and by enhanced phosphorylation of cytoskeletal elements which influence the permeability of cell membranes (Aune, 1997; Cohen et al., 1990). Cytoskeletal integrity in cells requires intense protein phosphatase activity, and OA produces profound morphological alterations in a variety of cells. The changes include reorganization of the microfilament and intermediate-filament networks, organelle segregation, and surface blebbing (Eriksson et al., 1990).

Initially OA was noted for its Ca^{2+}-independent activation of smooth muscle (Ozaki et al., 1987). Since then it has been shown to have a variety of effects on Ca^{2+} signaling. OA has been described to potentiate calcium currents, possibly via phosphorylation of voltage-dependent calcium channels, in experimental systems as different as the protozoan *Paramecium tetraaurelia* (Klumpp et al., 1990) or isolated guinea pig cardiac myocites (Hescheler et al., 1988). It has been shown to increase Ca^{2+} currents in isolated myocytes (Hescheler et al., 1988), to potentiate Ca^{2+} influx by I_{CRAC} (Parekh et al., 1993), to inhibit agonist-stimulated lns(1,4,5) P3 production and release of Ca^{2+} from stores in rat hepatocytes (Mattingly and Garrison, 1992), and to induce the release of Ca^{2+} from the agonist- and thapsigargin-sensitive intracellular stores in ECV304 endothelial cells (Hepworth et al., 1997). OA has complex patterns of cell modulation depending on the transduction pathways under study, that is, the combination of OA and cholera toxin (which permanently blocks Gs protein), provides the strongest stimulus for the Na^+-H^+ exchanger in comparison to any other cell modulator (Alfonso et al., 1994). Also, OA shows an exquisite selectivity to inhibit IgE-mediated responses in mast cells and basophils (Botana and MacGlashan, 1993; Estévez et al., 1994), this being more evidence of the powerful and widespread effect of these toxins and the potential they may have as lead molecules on the search for new drugs (see Chapter 33).

The effects of OA acid as tumor promoter and apoptotic inducer are thoroughly covered in Chapter 13.

VIII. TOXICITY IN ANIMALS AND EXPERIMENTAL SYSTEMS

Since OA class compounds are widespread contaminants, they can induce both acute and chronic toxicity in humans. Their acute toxicity has been well documented. The toxic effects of OA, DTX-1, and DTX-3 on experimental animals or in vitro systems have been examined biochemically, pharmacologically, and pathologically, and the three toxins seem to induce similar effects.

Traditionally, the toxicity of the DSP toxins is measured by means of intraperitoneal injections of extracts from contaminated mussels into the mice (Yasumoto et al., 1978b). The effects of intraperitoneal injections of DSP toxins in mice are inactivation and general weakness. The symptoms appear 30 min to several hours after injection, and at sufficiently high concentrations the animals die from minutes up to 2 days. When the algal toxins are given by the oral route, the lethal dose is 16–50 times higher, but the symptoms are the same. Chickens and cats are less sensitive to DSP toxins than mice.

Although nonfatal, DSP is recognized as a worldwide health problem. Diarrhea associated with shellfish poisoning is poorly understood, and very little information exists regarding how the toxin affects intestinal epithelial function. The first toxin isolated from mussels contaminated by *Dinophysis fortii* was DTX-1 (Murata et al., 1982). Edebo et al. (1988) reported that the instillation of dinophysistoxin 1 into ligated loops of rat small intestine induced a rapid accumulation of fluid. Mice receiving this toxin at 160 $\mu g\ kg^{-1}$ body weight died within 24 h while suffering from constant diarrhea. Further studies on suckling mice (7–10 g) were performed by Terao et al. (1986, 1993), who found morphological changes 1 h after the administration of DTX-1 by intraperitoneal injection. The duodenum and upper portion of the small intestine became distended and contained mucoid, but not bloody, fluid. At the ultrastructural level, three sequential stages of changes of the villi were observed: extravasation of villi vessels, degeneration of absorptive epithelium and degeneration of microvilli, and desquamation of the degenerated epithelium from the lamina propria. Interestingly, the crypts appeared to be protected from the toxic effects.

When OA, DTX-1, or DTX-3 were given orally to infant mice, they caused excessive fluid accumulation in the intestine. This effect is similar to the watery secretion caused by well-recognized bacterial enterotoxins like cholera toxin, the *Escherichia coli* enterotoxins LT and ST, and *Clostridium difficile* enterotoxin. The diarrheic activity of algal toxins has been quantified in ligated intestinal loops of the rat. Toxic tissue homogenates, liquid recovered after steaming of toxic mussels, and purified OA produced maximum fluid accumulation in the loops within 2 h. Thus, the fluid accumulation kinetic was faster than the 6 h described for cholera toxin and the heat-labile enterotoxin (LT) from *E. coli* but equaled that of the heat-stable enterotoxin (ST) (Lange, 1982; Lange and Lönnroth, 1983). In this way, toxins with lower molecular weights such as OA (molecular weight = 736) and ST (MW = 2500) have shown faster action than cholera toxin (MW = 84,000) and LT (MW = 90,000). The minimum quantity of OA which produced significant secretion in the rat intestinal ligated loop test was approximately 0.5 μg. On a body weight basis, therefore, humans are estimated to be at least four times as sensitive as the rat to enteral challenge with OA (Edebo et al., 1988).

According to Lange et al. (1990), the rat small intestine functions as the most sensitive and reproducible organ for studies of the diarrheal effects of marine toxins. OA injected in ligated loops from the middle duodenum of male rats show changes within 15 min after injection. After 60–90 min, most of the enterocytes of the villi were shed into the lumen and large parts of the flattened villi were covered by globet cells. The degree of damage was dose dependent: 3 μg OA affected only the top of the villi, whereas 5 μg led to collapse of the villous structure. Intravenous injection of OA induced similar, but less extensive, changes, indicating that the enterocytes are specific target cells for OA.

After the DSP effect, the degenerated cells and debris are adsorbed and the villi begin to regenerate within 2 h (Terao et al., 1993). Within 24 h, degenerated parts of the small intestine are completely regenerated. It is worth pointing out that the pathomorphological changes induced by intraperitoneal administration are essentially the same as those induced by oral administration, except for DTX-3, which is not as potent, although the severity of the intraperitoneal administration is about seven times greater (Terao et al., 1993). In a similar fashion to microcystin, oral and intraperitoneal administration of OA and DTX-1 in mice and rats also induces liver damage expressed as degeneration of endothelial lining cells at the sinusoid, this hepatotoxicity seems to be more intense with DTX-3 (Terao et al., 1993). Again, the acyl group at R1 of DTX-3 may play an important role. The degeneration of the hepatocytes occurred 24 h after the administration of DTX-3, whereas the injuries induced by the phycotoxin in the small intestine appeared at the same doses within 15 min.

The injuries and repair processes in the intestines of mice induced by DTX-3 were compared to those induced by OA and DTX-1 (Ito and Terao, 1994). DTX-3 impaired intestinal villi by the oral route only, whereas OA and DTX-1 caused intestinal injury with both oral and intraperitoneal exposures (oral toxicity was about 25–50 times lower than toxicity obtained by intraperitoneal injections). The character of the lesions caused by the three toxins and the recovery processes were highly similar, which was to be expected owing to their shared mechanism of action.

Regarding the effect of OA on intestinal ion transport, which might account for the accumulation of fluid, it has been reported (Tripuraneni et al., 1997) that OA increased paracellular permeability in cultured colon epithelium. OA does show effects on both Cl^- secretion and barrier function on cultured human intestinal epithelial T84 cell monolayers, although it does not directly stimulate Cl^- secretion, but attenuates Cl^- secretion in response to secretagogues in both T84 cells and in stripped rabbit colonic mucosa. Also, OA increases the paracellular permeability of intestinal epithelia, which may contribute to the diarrhea associated with shellfish poisoning. These data suggest that the diarrhea associated with DSP is not the result of secretagogue-like activity that characterizes the actions of other toxins such as cholera or the heat labile toxin of *E. coli*. Hosokawa et al. (1998) reports that OA promptly evokes mucosal damage and that the damage is self-limited, leaving no after effect within 7 days. Although over 90% of patients experience diarrhea, with an incubation period seldom exceeding 12 h, from an epidemiological perspective, DSP-intoxicated patients recovered within 3 days (Yasumoto et al., 1984).

IX. PHARMACOKINETICS DSP TOXINS

Very little is known about the toxicokinetics of the toxins in the DSP complex, both in animals and humans. Owing to their lipophilicity, they might be expected to be easily absorbed and distributed in the human body. Accumulation in numerous body compartments, most notably in intestinal tissue, stomach, liver, and lung, has been demonstrated for several toxins, although there is no specific target organ known for this toxin. Nevertheless, the fact that an acute intoxication shows clear diarrheic symptoms suggests that the main target for OA is the gastrointestinal tract. The variations in the distribution of [^{3}H]OA in organs and biological fluids of mice related to diarrheic syndrome has been studied by Matias et al. (1989) following an oral dose of 50 μg/kg OA, which could be considered as a case of mild OA intoxication. The results indicate a very low metabolization, the capability to reach any tissue, the existence of an enterohepatic cycle, and an important accumulation in intestinal tissues, which accounts for the intestinal cytotoxicity and diarrhea.

This distribution was even more pronounced in intestinal tissue when animals were given OA orally, 90 μg/kg, which induced diarrhea. The high concentrations of [^{3}H]OA in intestinal tissues and contents 24 h after administration demonstrates a slow elimination of OA. When the dose of OA was increased from 50 to 90 μg/kg, the concentrations of the toxin in the intestinal content and feces increased proportionally. A good correlation was found between an increase of OA in the intestinal tissue and the diarrhea in animals given 90 μg/kg orally. These data also revealed that in acute OA intoxication the concentration of the toxin in the intestinal tissues reaches cytotoxic concentrations in accordance with the diarrhea, which is the main symptom of OA poisoning. When toxic doses are ingested, the effects on the gastrointestinal tract are the main cause of death. In mice dosed 90 μg/kg, which is still a nonlethal dose, diarrhea takes place but does not wash out the toxin. The toxin was rather accumulated in the intestinal tissue, in intestinal content, and in kidney and liver, whereas the distribution was of the same extent in blood, urine, feces, and gallbladder. It could be that diarrhea and subsequent hypovolemia did not permit a normal blood circulation, urine elimination, and enterohepatic cycle. This finding is in accordance with the symptomatology of shellfish diarrhea, which occurs in consumers after ingestion of OA-contaminated shellfish and in animals given OA by gastric intubation (Matias et al., 1999).

REFERENCES

A Alfonso, MA Botana, MR Vieytes, LM Botana. Functional characterization of the Na+-H+ exchanger in rat mast cells: crosstalks between different kinase pathways. Eur J Pharmacol 267:289–296, 1994.

J An, WW Carmichael. Use of a colorimetric protein phosphatase inhibition assay and enzyme linked immunosorbent assay for the study of microcystins and nodularins. Toxicon 32:1495–1508, 1994.

C Arias, F Becerra-Garcia, I Arrieta, R Tapia. The protein phosphatase inhibitor okadaic acid induces heat shock protein expression and neurodegeneration in rat hippocampus in vivo. Exp Neurol 153:242–254, 1998.

T Aune. Health effects associated with algal toxins from seafood. Arch Toxicol 19(Suppl):389–397, 1997.
T Aune, M Yundestad. Diarrhetic shellfish poisoning. In: *Algal Toxins in Seafood and Drinking Water*. London: Academic Press, 1993, pp 87–104.
JR Bagu, BD Sykes, MM Craig, CFB Holmes. A molecular basis for different interactions of marine toxins with protein phosphatase-1—molecular models for bound motuporin, microcystins, okadaic acid, and calyculin A. J Biol Chem 272:5087–5097, 1997.
D Barford. Molecular mechanisms of the protein serine/threonine phosphatases. TIBS 21:407–412, 1996.
C Bialojan, A Takai. Inhibitory effect of a marine-sponge toxin, okadaic acid, on protein phosphatases. Specificity and kinetics. Biochem J 256:283–290, 1988.
LM Botana, DJ MacGlashan. Effect of okadaic acid on human basophil secretion. Biochem Pharmacol 45:2311–2315, 1993.
LM Botana, MR Vieytes, A Alfonso, MC Louzao. Phycotoxins. Paralytic shellfish poisoning. diarrhetic shellfish poisoning. In: L Nollet, ed., Handbook of Food Analysis. New York: Marcel Dekker, 1996, pp 1147–1170.
ND Brewis, AJ Street, AR Prescott, PTW Cohen. PPX, a novel protein serine/threonine phosphatase localised to centrosomes. EMBO J 12:987–996, 1993.
TR Burke Jr, ZY Zhang. Protein-tyrosine phosphatases: Structure, mechanism, and inhibitor discovery. Biopolymers 47:225–241, 1998.
EP Carmody, KJ James, SS Kelly. Dinophysistoxin-2: the predominant diarrhoetic shellfish toxin in Ireland. Toxicon 34:351–360, 1996.
KM Coggeshall. Negative signaling in health and disease. Immunol Res 19:47–64, 1999.
P Cohen. The structure and regulation of protein phosphatases. Annu Rev Biochem 58:453–508, 1989.
P Cohen, PTW Cohen. Protein phosphatases come of age. J Biol Chem 264:21435–21438, 1989.
P Cohen, CF Holmes, Y Tsukitani. Okadaic acid: a new probe for the study of cellular regulation. Trends Biochem Sci 15:98–102, 1990.
S Corvera, S Jaspers, M Pasceri. Acute inhibition of insulin-stimulated glucose transport by the phosphatase inhibitor, okadaic acid. J Biol Chem 266:9271–9275, 1991.
E DilipDeSilva, DE Williams, RJ Andersen, H Klix, CFB Holmes, TM Allen. Motuporin, a potent protein phosphatase inhibitor isolated from the Papua New Guinea sponge *Theonella swinhoei* Gray. Tetrahedron Lett. 33:1561–1564, 1992.
L Edebo, S Lange, XP Li, S Allenmark. Toxic mussels and okadaic acid induce rapid hypersecretion in the rat small intestine. Apmis 96:1029–1035, 1988.
F Erdodi, A Rokolya, SJ Di, M Barany, K Barany. Effect of okadaic acid on phosphorylation-dephosphorylation of myosin light chain in aortic smooth muscle homogenate. Biochem Biophys Res Commun 153:156–161, 1988.
JE Eriksson, D Toivola, JA Meriluoto, H Karaki, YG Han, D Hartshorne. Hepatocyte deformation induced by cyanobacterial toxins reflects inhibition of protein phosphatases. Biochem Biophys Res Commun 173:1347–1353, 1990.
MD Estévez, MR Vieytes, MC Louzao, LM Botana. Effect of okadaic acid on immunologic and nonimmunologic histamine release in rat mast cells. Biochem Pharmacol 47:591–593, 1994.
MA Felix, P Cohen, E Karsenti. Cdc2 H1 kinase is negatively regulated by a type 2A phosphatase in the Xenopus early embryonic cell cycle: evidence from the effects of okadaic acid. EMBO J 9:675–683, 1990.
MT Fernández, V Zitko, S Gascón, A Novelli. The marine toxin okadaic acid is a potent neutorotoxin for cultured cerebellar neurons. Life Sci 49:157–162, 1991.
OI Fontal, MR Vieytes, JMV De Sousa, MC Louzao, LM Botana. A fluorescent microplate assay for microcystin-LR. Anal Biochem 269:289–296, 1999.
H Fujiki, M Suganuma. Tumor promotion by inhibitors of protein phosphatases 1 and 2A: The okadaic acid class of compounds. Adv Cancer Res 61:143–194, 1993.
H Fujiki, M Suganuma, S Nishiwaki, S Yoshizawa, B Winyar, T Sugimura, FJ Schmitz. A new pathway of tumor promotion by the okadaic acid class compounds. Adv Second Messenger Phosphoprotein Res 24:340–344, 1990.
S Gaur, Y Schwartz, LR Tai, GP Frick, HM Goodman. Insulin produces a growth hormone-like increase in intracellular free calcium concentration in okadaic acid-treated adipocytes. Endocrinology 139: 4953–4961, 1998.

G Grichting, LK Levy, HM Goodman. Relationship between binding and biological effects of hGH in rat adipocytes. Endocrinology 113:1111–1120, 1983.

V Gupta, AK Ogawa, X Du, KN Houk, RW Armstrong. A model for binding of structurally diverse natural product inhibitors of protein phosphatases PP1 and PP2A. J Med Chem 40:3199–3206, 1997.

B Hald, T Bergskov, H Emsholm. Monitoring and analytical programmes on phycotoxins in Denmark. In: JM Premy, ed., Proceedings Symposium on Marine Biotoxins. Paris, France, Centre National d'Etudes Vétérinaires et Alimentaires, 1991, pp 181–187.

Y Hamano, Y Kinoshita, T Yasumoto. Suckling mice assay for diarrhetic shellfish toxins. In: Toxic Dinoflagellates. Amsterdam: Elsevier, 1985, pp 383–388.

TA Haystead, AT Sim, D Carling, RC Honnor, Y Tsukitani, P Cohen, DG Hardie. Effects of the tumour promoter okadaic acid on intracellular protein phosphorylation and metabolism. Nature 337:78–81, 1989.

TJ Hepworth, AM Lawrie, AWM Simpson. Okadaic acid induces the release of Ca^{2+} from intracellular stores in ECV304 endothelial cells. Cell Calcium 21:461–467, 1997.

J Hescheler, G Mieskes, JC Ruegg, A Takai, W Trautwein. Effects of a protein phosphatase inhibitor, okadaic acid, on membrane currents of isolated guinea-pig cardiac myocytes. Pflugers Arch 412: 248–252, 1988.

CF Holmes, HA Luu, F Carrier, FJ Schmitz. Inhibition of protein phosphatases-1 and -2A with acanthifolicin. Comparison with diarrhetic shellfish toxins and identification of a region on okadaic acid important for phosphatase inhibition. FEBS Lett 270:216–218, 1990.

RE Honkanen, BA Codispoti, K Tse, AL Boynton, RE Honkanen. Characterization of natural toxins with inhibitory activity against serine/threonine protein phosphatases. Toxicon 32:339–350, 1994.

RE Honkanen, M Dukelow, J Zwiller, RE Moore, BS Khatra, AL Boynton. Cyanobacterial nodularin is a potent inhibitor of type 1 and type 2A protein phosphatases. Mol Pharmacol 40:577–583, 1991.

RE Honkanen, J Zwiller, RE Moore, SL Daily, BS Khatra, M Dukelow and AL Boynton. Characterization of microcystin-LR, a potent inhibitor of type 1 and type 2A protein phosphatases. J Biol Chem 265: 19401–19404, 1990.

M Hosokawa, H Tsukada, T Saitou, M Kodama, M Onomura, H Nakamura, K Fukuda, Y Seino. Effects of okadaic acid on rat colon. Dig Dis Sci 43:2526–2535, 1998.

T Hu, J Doyle, D Jackson, J Marr, E Nixon, S Pleasance, MA Quilliam, JA Water, JL Wright. Isolation of a new diarrhetic shellfish poison from Irish mussels. J Chem Soc Commun 30:39–41, 1992.

T Hu, J Marr, ASW DeFreitas, MA Quilliam, JA Walter and JLC Wright. New diol esters isolated from cultures of the dinoflagellates Prorocentrum lima and Prorocentrum concavum. J Nat Products 55: 1631–1637, 1992b.

T Hu, JM Curtis, JA Walter, JL McLachlan, JLC Wright. Two new water-soluble DSP toxin derivatives from the dinoflagellate *Prorocentrum maculosum*: possible storage and excretion products. Tetrahedron Lett 36:9273–9276, 1995a.

T Hu, JM Curtis, JA Walter, JLC Wright. Identification of DTX-4, a new water-soluble DSP toxin derivatives from the dinoflagellate *Prorocentrum lima*. J Chem Soc Chem Commun 1995:597–599, 1995b.

H Ishihara, BL Martin, DL Brautigan, H Karaki, H Ozaki, Y Kato, N Fusetani, S Watabe, K Hashimoto, D Uemura, et al. Calyculin A and okadaic acid: inhibitors of protein phosphatase activity. Biochem Biophys Res Commun 159:871–877, 1989.

OG Issinger, T Martin, WW Richter, M Olson and H Fujiki. Hyperphosphorylation of N-60, a protein structurally and immunologically related to nucleolin after tumour-promoter treatment. EMBO J 7: 1621–1626, 1988.

E Ito, K Terao. Injury and recovery process of intestine caused by okadaic acid and related compounds. Nat Toxins 2:371–377, 1994.

KJ James, AG Bishop, BM Healy, C Roden, IR Sherlock, M Twohig, R Draisci, L Giannetti, L Lucentini. Efficient isolation of the rare diarrhoeic shellfish toxin, dinophysistoxin-2, from marine phytoplankton. Toxicon 37:343–357, 1999.

M Kat. Dinophysis acuminata blooms in the Dutch coastal area related to diarrhoeic shellfish poisoning the Dutch Waddensea. Sarsia 68:81–84, 1983.

Y Kato, N Fusetani, S Matsunaga, K Hashimoto, S Fujita, T Furuya. Calyculin A, a novel antitumor metabolite from the marine sponge, *Discodermia calyx*. J Am Chem Soc 108:2780–2781, 1986.

S Klumpp, P Cohen, JE Schultz. Okadaic acid, an inhibitor of protein phosphatase 1 in Paramecium, causes sustained Ca^{2+}-dependent backward swimming in response to depolarizing stimuli. EMBO J 9:685–689, 1990.

P Krogh, L Edler, L Graneli. Outbreaks of diarrheic shellfish poisoning on the west coast of Sweden. In: G Anderson, AW White, DG Baslen, eds., Toxic Dinoflagellates. Amsterdam:Elsevier, 1985, pp 501–504.

CW Laidley, E Cohen, JE Casida. Protein phosphatase in neuroblastoma cells: [3H] cantharidin binding site in relation to cytotoxicity. J Pharmacol Exp Ther 280:1152–1158, 1997.

S Lange. A rat model for an in vivo assay of enterotoxic diarrhea. FEMS Microbiol Lett 15:239–242, 1982.

S Lange, I Lönnroth. Hypersecretion induced by Escherichia coli heat-stable enterotoxin in intestinal loops in rat and mouse. J Diarrh Dis Res 2:94–98, 1983.

S Lange, GL Andersson, E Jennishe, I Lönnroth, XP Li, L Edebo. Okadaic acid produces drastic histopathologic changes of the rat intestinal mucosa and with concomitant hypersecretion. In: E Granéli, B Sundstrom, eds., Toxic Marine Phytoplankton. Amsterdam: Elsevier, 1990, pp 356–361.

JE Lawrence, AD Cembella, NW Ross, JLC Wright. Cross-reactivity of an anti-okadaic acid antibody to dinophysistoxin-4 (DTX-4), dinophysistoxin-5 (DTX-5), and an okadaic acid diol ester. Toxicon 36:1193–1196, 1998.

F Leira, JM Vieites, MR Vieytes, LM Botana. Apoptotic events induced by the phosphatase inhibitor okadaic acid in normal human lung fibroblasts. Toxicans (in press).

C MacKintosh, S Klumpp. Tautomycin from the bacterium *Streptomyces verticillatus*, another potent and specific inhibitor of protein phosphatases 1 and 2A. FEBS Lett 277:137–140, 1990.

C MacKintosh, RW MacKintosh. Inhibitors of protein kinases and phosphatases. Trends Biochem Sci 19: 444–448, 1994.

C MacKintosh, KA Beattie, S Klumpp, P Cohen, GA Codd. Cyanobacterial microcystin-LR is a potent and specific inhibitor of protein phosphatases 1 and 2A from both mammals and higher plants. FEBS Lett 264:187–192, 1990.

WG Matias, A Traore, EE Creppy. Variations in the distribution of okadaic acid in organs and biological fluids of mice related to diarrhoeic syndrome. Hum Exp Toxicol 18:345–350, 1999.

RR Mattingly, JC Garrison. Okadaic acid inhibits angiotensin II stimulation of Ins(1,4,5)P3 and calcium signalling in rat hepatocytes. FEBS Lett 296:225–230, 1992.

TA Millward, S Zolnierowicz, BA Hemmings. Regulation of protein kinase cascades by protein phosphatase 2A. Trends Biochem Sci 24:186–191, 1999.

M Murata, M Shimatani, H Sugitani, Y Oshima, T Yasumoto. Isolation and structural elucidation of the causative toxin of the diarrhetic shellfish poisoning. Bull Jpn Soc Sci Fish 48:549–552, 1982.

J Needham, JL McLachlan, JA Walter, JLC Wright. Biosynthetic origin of C-37 and C-38 in the polyether toxins okadaic acid and dinophysistoxin-1. J Chem Soc Chem Commun 1994:2599–2600, 1994.

S Nishiwaki, H Fujiki, M Suganuma, H Furuyasuguri, R Matsushima, Y Iida, M Ojika, K Yamada, D Uemura, T Yasumoto, FJ Schmitz, T Sugimura. Structure activity relationship within a series of okadaic acid derivatives. Carcinogenesis 11:1837–1841, 1990a.

S Nishiwaki, H Fujiki, M Suganuma, M Ojika, K Yamada, T Sugimura. Photoaffinity labeling of protein phosphatase-2A, the receptor for a tumor promoter okadaic acid, by ⟨27-H-3⟩methyl 7-O-(4-azidobenzoyl)okadaate. Biochem Biophys Res Commun 170:1359–1364, 1990b.

H Ozaki, H Ishihara, K Kohama, Y Nonomura, S Shibata, H Karaki. Calcium-independent phosphorylation of smooth muscle myosin light chain by okadaic acid isolated from black sponge (Halichondria okadai). J Pharmacol Exp Ther 243:1167–1173, 1987.

AB Parekh, H Terlau, W Stühmer. Depletion of InsP3 stores activates a Ca^{2+} and K^+ current by means of a phosphatase and a diffusible messenger. Nature 364:814–819, 1993.

M Quilliam, MW Gilgan, S Pleasance, ASW deFreitas, D Douglas, L Fritz, T Hu, JC Marr, C Smyth, JLC Wright. Confirmation of an incident of diarrhetic shellfish poisoning in eastern Canada. In: TJ Smayda, Y Shimizu, eds., Toxic Phytoplankton Blooms in the Sea. Amsterdan: Elsevier, 1993, pp 574–572.

M Quilliam, NW Roos. Analysis of diarrhetic shellfish poisoning toxins and metabolites in plankton and shellfish by ion-spray liquid chromatography-mass spectrometry. In: AP Snyder, ed., *Biochemical* and *Biotechnological Applications of Electrospray Ionization Mass Spectrometry*. Washington, DC: American Chemical Society, 1996, pp 351–364.

MA Quilliam, WR Hardstaff, N Ishida, JL McLachlan, AR Reeves, NW Ross, AJ Windust. Production of diarrhetic shellfish poisoning (DSP) toxins by *Prorocentrum Lima* in culture and development of analytical methods. In: T Yasumoto, Y Oshimo, Y Fukuyu, eds., Harmful and Toxic Algal Blooms. Paris: Intergovernmental Oceanographic Commission of UNESCO, 1996, pp 289–292.

MA Quilliam. Committee on natural toxins. J AOAC Int 82:773–781, 1999.

E Runden, PO Seglen, FM Haug, OP Ottersen, T Wieloch, M Shamloo, JH Laake. Hippocampal slice cultures: evidence for a MAP kinase-dependent mechanism. J Neurosci 18:7296–7305, 1998.

K Sasaki, M Murata, T Yasumoto, G Mieskes, A Takai. Affinity of okadaic acid to type-1 and type-2A protein phosphatases is markedly reduced by oxidation of its 27-hydroxyl group. Biochem J 298: 259–262, 1994.

T Sassa, WW Richter, N Uda, M Suganuma, H Suguri, S Yoshizawa, M Hirota, H Fujiki. Apparent ''activation'' of protein kinases by okadaic acid class tumor promoters. Biochem Biophys Res Commun 159:939–944, 1989.

A Scoging, M Bahl. Diarrhetic shellfish poisoning in the UK. Lancet 352:117, 1998.

V Séchet, P Safran, P Hovgaard, T Yasumoto. Causative species of diarrhetic shellfish poisoning (DSP) in Norway. Mar Biol 105:269–274, 1990.

S Shibata, Y Ishida, H Kitano, Y Ohizumi, J Habon, Y Tsukitani, H Kikuchi. Contractile effects of okadaic acid, a novel ionophore-like substance from black sponge, on isolated smooth muscles under the condition of Ca deficiency. J Pharmacol Exp Ther 223:135–143, 1982.

SE Shumway, SA Sherman, AD Cembella, R Selvin. Accumulation of paralytic shellfish toxins by surf-clams, Spisula solidissima (Dillwyn, 1897) in the Gulf of Maine: seasonal changes, distribution between tissues, and notes of feeding habitats. Nat Toxins 2:236–251, 1994.

M Suganuma, H Fujiki, S Okabe, S Nishiwaki, D Brautigan, TS Ingebritsen, MR Rosner. Structurally different members of the okadaic acid class selectively inhibit protein serine/threonine but not tyrosine phosphatase activity. Toxicon 30:873–878, 1992.

M Suganuma, M Suttajit, H Suguri, M Ojika, K Yamada, H Fujiki. Specific binding of okadaic acid, a new tumor promoter in mouse skin. FEBS Lett 250:615–618, 1989.

B Sundstrom, L Edler, E Granéli. The global distribution of harmful effects of phytoplankton. In: E Granéli, B Sundstrom, et al., eds., Toxic Marine Phytoplankton. Amsterdam: Elsevier, 1990, pp 537–541.

T Suzuki, H Ota, M Yamasaki. Direct evidence of transformation of dinophysistoxin-1 to 7-O-acyl-dinophysistoxin-1 (dinophysistoxin-3) in the scallop Patinopecten yessoensis. Toxicon 37:187–198, 1999.

K Tachibana, PJ Scheuer, H Tsukitani, D Van Engen, J Clardy, Y Gopichand, FJ Schmitz. Okadaic acid, a cytotoxic polyether from two marine sponges of the genus a *Halichondria*. J Am Chem Soc 103: 2469–2471, 1981.

A Takai, K Sasaki, H Nagai, G Mieskes, M Isobe, K Isono, T Yasumoto. Inhibition of specific binding of okadaic acid to protein phosphatase 2A by microcystin-LR, calyculin-A and tautomycin: method of analysis of interactions of tight-binding ligands with target protein. Biochem J 306:657–665, 1995.

A Takai, M Murata, K Torigoe, M Isobe, G Mieskes, T Yasumoto. Inhibitory effect of okadaic acid derivatives on protein phosphatases: a study on structure-affinity relationship. Biochem J 284:539–544, 1992.

A Takai, C Bialojan, M Troschka, JC Ruegg. Smooth muscle myosin phosphatase inhibition and force enhancement by black sponge toxin. FEBS Lett 217:81–84, 1987.

JF Tanti, T Gremeaux, OE Van, MBY Le. Effects of okadaic acid, an inhibitor of protein phosphatases-1 and -2A, on glucose transport and metabolism in skeletal muscle. J Biol Chem 266:2099–2103, 1991.

C Taylor, RJ Quinn, P Alewood. Inhibition of protein phosphatase 2A by cyclic peptides modelled on the microcystin ring. Bioorg Med Chem Lett 6:2113–2116, 1996.

K Terao, E Ito, M Ohkusu, T Yasumoto. A comparative study of the effects of DSP-toxins on mice and rats. In: TJ Smayda, Y Shimizu, eds., Toxic Phytoplankton Blooms in the Sea. New York: Elsevier, 1993, pp 581–586.

K Terao, E Ito, T Yanagi, T Yasumoto. Histopathological studies on experimental marine toxin poisoning. I. Ultrastructural changes in the small intestine and liver of suckling mice induced by dinophysistoxin-1 and pectenotoxin-1. Toxicon 24:1141–1151, 1986.

J Tripuraneni, A Koutsouris, L Pestic, P De Lanerolle, G Hecht. The toxin of diarrheic shellfish poisoning, okadaic acid, increases intestinal epithelial paracellular permeability. Gastroenterology 112:100–108, 1997.

MR Vieytes, OI Fontal, F Leira, J De Sousa, LM Botana. A fluorescent microplate assay for diarrheic shellfish toxins. Anal Biochem 248:258–264, 1997.

AH Walsh, AY Cheng, RE Honkanen. Fostriecin, an antitumor antibiotic with inhibitory activity against serine/threonine protein phosphatases types 1 (PP1) and 2A (PP2A), is highly selective for PP2A. FEBS Lett 416:230–234, 1997.

S Wera, BA Hemmings. Serine/threonine phosphatases. Biochem J 311:17–29, 1995.

AJ Windust, MA Quilliam, JLC Wright, JL McLachlan. Comparative toxicity of the diarrhetic shellfish poisons, okadaic acid, okadaic acid diol-ester and dinophysistoxin-4, to the diatom Thalassiosira weissflogii. Toxicon 35:1591–1603, 1997.

AJ Windust, JLC Wright, JL McLachlan. The effects of the diarrhetic shellfish poisoning toxins, okadaic acid and dinophysistoxin-1, on the growth of microalgae. Marine Biol 126:19–25, 1996.

T Yasumoto, M Murata, Y Oshima, M Sano. Diarrhetic shellfish toxins. Tetrahedron 41:1019–1025, 1985.

T Yasumoto, M Murata, Y Oshima, GK Matsumoto, J Glardy. Diarrhetic shellfish poisoning. In: EP Ragelis, ed., ACS Symposium Series, No 262. Seafood Toxins. Washington, DC: American Chemical Society, 1984, pp 207–214.

T Yasumoto, Y Oshima, M Yamaguchi. Occurrence of a new type of toxic shellfish in Japan and chemical properties of the toxin. Bull Jpn Soc Sci Fish 44:1249, 1978a.

T Yasumoto, Y Oshima, M Yamaguchi. Occurrence of a new type of shellfish poisoning in the Tohoku Distric Bull Jpn Soc Sci Fish 44:1249–1255, 1978b.

13

Neoplastic Activity of DSP Toxins: The Effects of Okadaic Acid and Related Compounds on Cell Proliferation: Tumor Promotion or Induction of Apoptosis?

Gian Paolo Rossini
University of Modena and Reggio Emilia
Modena, Italy

I. INTRODUCTION

The functional states of cells in multicellular organisms may be of proliferation, differentiation, or apoptosis. The decision as to whether at any instant a cell enters in and/or proceeds through one of these states is based on the genetic program and chemical signals recognized by that cell.

It is well established that recognition of extracellular signals and their transduction in living cells, to trigger appropriate responses, involves either plasma membrane (1–3) or intracellular (4) receptors. The control of cell proliferation by endogenous hormones and growth factors (GFs), in particular, may involve signal transduction through both types of receptors, and, in any case, the effects include the controlled expression of genes responsible for cell growth (5). A major difference in the mechanisms of signal transduction involving intracellular or plasma membrane receptors is that the former are ligand-regulated transcription factors (6), whereas the latter trigger intracellular cascades of events, culminating with the activation of a defined set of protein kinases, which will eventually control the functioning of transcription factors. Figure 1 is a schematic representation of major steps currently believed to participate in the mechanisms of signal transduction. For a detailed description of the phenomena and possible cross-talk among signal transduction pathways, the reader may refer to excellent reviews which have appeared on those topics (2–9).

A number of hormones and several natural components trigger responses in target cells by binding to intracellular proteins of the steroid/thyroid hormone receptor superfamily (4,6). Once the ligand (S) has bound to its cognate receptor (SR), dimerization of ligand-receptor complexes, and their direct interaction with DNA sequences located in the promoter region of controlled genes, is responsible for either positive or negative transcriptional responses (6,10,11).

Signaling through plasma membrane receptors (Rs) displaying protein tyrosine kinase (TK) activity, as in the case of most GFs (3,9), in turn, is triggered by ligand binding to receptor

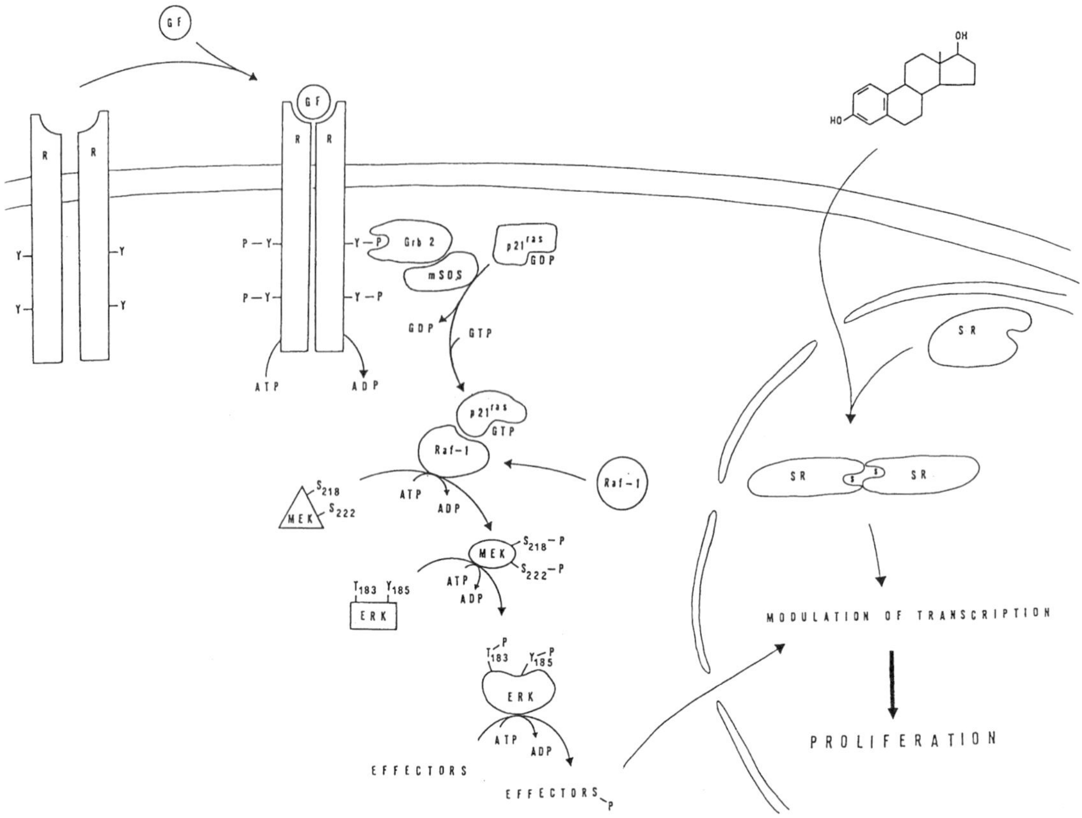

Figure 1 Schematic representation of major mechanisms of signal transduction involved in the control of cell proliferation.

proteins and receptor dimerization, with the consequent activation of TK activity. The intracellular TK domain of one receptor subunit is then believed to trans-phosphorylate selected tyrosine residues in the carboxy-terminal portion of the other subunit, giving rise to docking sites which are recognized and bound by proteins (adapters) containing SH2 domains, displaying adequate affinity for the phosphotyrosines and the surrounding amino acids (9,12). The proteins bound to docking sites then form transducing complexes (including, for instance, GRB2, SHC, SOS) capable of being able further to process the extracellular signals by binding and activating other components of the transduction machinery, such as the GTP-binding protein Ras, and the ser/thr protein kinase Raf-1 (9). This kinase is then believed to induce, either directly or by phosphorylation and activation of some intermediate ser/thr protein kinase (7,9,13,14), the phosphorylation of Ser218 and Ser222 of mitogen-activated/extracellular signal–regulated kinase-activating kinase (MEK). This kinase then becomes capable of catalyzing the phosphorylation of both Thr183 and Tyr185 residues of mitogen-activated protein kinase (MAPK)/extracellular signal–regulated kinase (ERK), leading to activation of the kinase. The active ERKs, or some other protein kinase(s) activated by ERKs (14,15), in turn, will phosphorylate transcription factors, leading to changes in their functional properties, and then bringing about the final step of extracellular signal-induced modulation of transcription at the nuclear level.

As it will be apparent from some issues discussed in Section IV, these two major modes of signal transduction could influence each other. Indeed, cross-talk among different mechanisms of signal transduction involved in the control of cell proliferation is fairly common, as exemplified by signaling via G protein-coupled receptors and activation of the ser/thr protein kinase C (PKC), which appear to converge over the Raf-1/MEK/MAPK pathway (7,8,16). Although these pathways will not be discussed here, it seems important to point out that PKC isoforms are believed individually to play some role in proliferative responses through phosphorylation of proteins which may or may not be substrates of ERKs (7,17,18).

If these concepts are put in perspective, it is clear that a key aspect of the molecular mechanisms involved in the control of cellular functions is related to defined sets of phosphorylative events. The phosphorylation of specific substrates catalyzed by individual protein kinases, in fact, changes the functional properties of target proteins, which will be the agents of responses. Removal of phosphate groups from target proteins catalyzed by phosphoprotein phosphatases (PPases), in turn, will lead to reversion of their functional properties to those existing before stimulation. Within this framework, the phosphorylation state, that is, the balance between phosphorylated and unphosphorylated forms of individual proteins, will dictate the overall trend of the function they perform (19).

As far as the control of cell proliferation is concerned, genetic changes in one or more components of the signal transduction machinery, resulting in alteration of their functioning, are believed to represent the molecular basis of neoplastic transformation, often leading to cells displaying a phenotype characterized by deregulated protein phosphorylation (19,20). Furthermore, this type of outcome can be operationally due to either altered kinase activity, phosphatase activity, or both. The events leading to cancer development and outgrowth then include both genetic and epigenetic factors.

This chapter will describe one type of these epigenetic factors, namely, the diarrhetic shellfish poisoning (DSP) toxins, which include those chemicals sharing the capacity to bind and inhibit type 1 and type 2A ser/thr PPases (21–26). The proximal molecular consequence of this inhibition is then a stabilization of phosphorylated forms of proteins being substrates of PP1 and 2A.

The chemistry and mechanism of action of okadaic acid (OA) and related compounds are dealt with in other chapters of this book, and will not be discussed here. Attention will be rather focused on the effects of OA class compounds on molecular and cellular events involved in proliferation of normal and tumor cells.

II. TUMOR PROMOTION BY OKADAIC ACID CLASS COMPOUNDS

The multistage skin carcinogenesis model is a useful system to study the molecular events underlying the development of skin cancer and, in a broader view, other types of tumors (reviewed in ref. 27).

The ontogeny of skin cancer can be operationally divided into three stages: initiation, promotion, and progression (27). The first stage is irreversible, and involves some genetic damage due to application of a carcinogen to mouse skin. The promotion stage follows initiation, and consists of repetitive treatment with an agent (the tumor promoter) that may not have the mutagenic activity of initiators. Owing to the need of multiple applications of the promoter, this second phase is initially reversible, but eventually becomes irreversible. Finally, in the last stage of progression, the benign lesion may develop into a malignant tumor and give rise to metastases.

The two-stage model of mouse skin carcinogenesis has been extensively used by Suganuma, Fuijiki, and their collaborators to discover tumor promoters and characterize their mechanisms of action (reviewed in ref. 28). The first clear indication of tumor promoter activity of a

DSP toxin was provided in 1988, when investigators showed that application of OA twice a week on mouse skin, after a single application of the carcinogen dimethylbenzanthracene (DMBA), would result in tumor development in 80% of mice population by the 30th week of treatment (29). In line with the contention that tumor development requires both initiation and promotion, tumor incidence in mice treated with either DMBA alone or OA alone was less than 10%. Furthermore, in the same paper, evidence was provided that OA functions through a molecular mechanism which is more distinct than that used by the phorbol ester–type tumor promoters, which are known to bind and activate some isoforms of PKC (30–32). No activation of this protein kinase, in fact, could be induced by OA (29).

Since then other studies have further supported the contention that an okadaic acid class of tumor promoters exists, including several toxins of algal origin, such as derivatives of this polyether fatty acid, calyculins, microcistins, and nodularin (28,33–35). The distinct feature of this heterogeneous group of chemicals, as already mentioned, is their capacity to bind to the ''OA receptors,'' represented by PP1 and 2A (21–24). It is thus surprising that one such compound, tautomycin, has not been found to behave as a tumor promoter (36). Although dose-response studies have shown that the potencies of these inhibitors vary among the different PPases (21,22), those on their effects as tumor promoters have interesting features. As reported in Table 1, OA and calyculin A have been shown to be as powerful as phorbol ester–type tumor promoters (TPA) in the two-stage mouse skin carcinogenesis model (23,24,27). Furthermore, by development of the two-stage approach, OA has been shown to be a tumor promoter in the glandular stomach in rats (37). Still, some OA class compounds may not easily penetrate plasma membrane in most cells, as in the case of microcystin and nodularin, which may require appropriate transport systems to reach intracellular compartments (38). When this is possible, as in the liver, protocols of two-stage carcinogenesis, involving diethylnitrosamine initiation followed by administration of the tumor promoter, have resulted in an increased number of foci found to be positive for a proper marker, such as the placental glutathione S-transferase, as compared to controls (35,39). Interestingly, nodularin itself has been found to be a carcinogen (35).

The effects of OA class compounds as tumor promoters described in animal models of two-stage carcinogenesis have been confirmed in in vitro cell culture systems (40,41). When BALB/3T3 cells have been exposed to the initiator 3-methylcholanthrene for 72 h, subsequent treatments for 2 weeks with low doses (10 ng/mL) of OA promoted cell transformation (40). A 4-week exposure to such doses of OA, however, was found to be toxic to BALB/3T3 cells (see Section III).

Table 1 Tumor-Promoting Activities of the Okadaic Acid Class Compared with Those of TPA-Type Tumor Promoters

Promoter	Amount of compound/application, μg(nmole)	Maximal percentage of tumor-bearing mice	Average numbers of tumors/mouse in week 30
Okadaic acid	1.0 (1.2)	86.7	7.2
Dinophysistoxin-1	1.0 (1.2)	100.0	8.5
Calyculin A	1.0 (1.0)	93.3	4.3
TPA	2.5 (4.1)	100.0	11.0
Teleocidin	2.5 (5.7)	100.0	4.0
Aplysiatoxin	2.5 (4.0)	80.0	3.4

TPA, phorbol ester–type tumor promoters.
Source: Ref. 24.

III. EFFECTS OF OKADAIC ACID CLASS COMPOUNDS ON CELL PROLIFERATION

The conclusion that OA class compounds behave as tumor promoters poses the general question on the molecular mechanisms by which they enhance clonal expansion of initiated cells in terms of a net increase in their proliferation rate as compared to normal cells (42,43). It then seems relevant to analyze available information on the effects of OA class compounds on cell proliferation. Data are summarized in Table 2, where they have been organized according to the three

Table 2 Effects of DSP Toxins on Functional States of Cells in Culture

Response/Cell line	References
Proliferation	
BALB/c3T3	40,41
Human T lymphocytes	44
ALVA-41	45
Syrian hamster embryo fibroblasts	46
Differentiation	
NIH/3T3	47
BALB/c3T3	48
U937	49
Human breast cancer cells	50
Cell death[a]	
AU-565	51
BALB/c3T3 (apoptosis)	40,41,52
BM 13674 (apoptosis)	53
Demel human melanoma	54
Primary human fibroblasts	55
Mouse fibroblasts (apoptosis)	56–59
Syrian hamster embryo fibroblasts (apoptosis)	46,60,61
GH_3 (apoptosis)	62,63
HeLa, HeLa S_3 (apoptosis)	64,94
Primary rat hepatocytes (apoptosis)	62
HL-60 (apoptosis)	65–67
K562 (apoptosis)	66–71
Primary mouse keratinocytes	72
T lymphoblasts	73
B lymphocytes	74,75
MB-231 (apoptosis)	51
MCF-7 (apoptosis)	51,62,76,77
ML-1 (apoptosis)	78
Primary rat cerebellar neurons (apoptosis)	79
NIH/3T3	55,56
NRK-52.E rat kidney epithelium (apoptosis)	80,81
PC 12	82
Saos-2 and MG-63 human osteoblasts (apoptosis)	83
SK-N-SH (apoptosis)	62
U 937 (apoptosis)	49,65

[a] If individual cell lines have been shown to undergo apoptosis in at least one report, this type of response has been specified in parentheses.

functional states of cell proliferation, differentiation, and death. It should be also pointed out that recorded responses have been primarily observed by the use of OA, but, in most cases, they have been confirmed with other OA-type compounds, such as calyculin A (50,51,63,66,83) and, in a few cases, microcystin (62).

At a first glance, it can be noted that few reports document proliferative responses of cells in culture to OA treatment, including the original articles on the two-stage transformation of BALB/c 3T3 cells (40,41), described in the previous section.

In the study by McClure et al. (44), T lymphocytes were treated with 1 nM OA during the first hour of activation with concanavalin (ConA) and 10% fetal calf serum (FCS), the medium was then replaced with fresh medium containing ConA + FCS but lacking OA, and the incubation was continued for 3 days. When cell proliferation was measured by thymidine uptake, it was found that OA treatment was associated with a 35% increase of response after 1 day. Although the increased response was still significant on the second day of treatment (18%), it became insignificant on the last day of the experiment (44).

The experimental design of the investigation described in the second report is similar, inasmuch as ALVA-41 human prostate cancer cells preloaded with sex hormone–binding globulin (SHBG) were stimulated with androgen (1–100 nM 5α-dihydrotestosterone, DHT) during treatment with 50 nM OA (45). Under those experimental conditions of SHBG/DHT stimulation, it was found that 1 day of treatment with OA markedly enhanced thymidine incorporation as compared to control cells.

A major feature which distinguishes the article by Afshari and Barrett (46) from the previous ones, is the fact that the experimental conditions employed to evaluate the effect of OA on proliferation of normal hamster and human fibroblasts did not involve any accompanying proliferative stimulus. In this case, treatment of quiscent cells with 0.16 nM OA induced a suboptimal, albeit measurable, proliferative response. Interestingly enough, the effects of OA were dose dependent, and concentrations higher than 3 nM were found to be toxic (46). To the best of our knowledge, this is the only study providing direct evidence of OA-induced enhancement of cell proliferation in cultured cells and in the absence of costimulators or without pretreatment with initiators.

As far as the efficacy of OA to induce cell differentiation, few reports are available in which the issue has been directly evaluated. The capacity of OA to behave as a differentiating agent has been originally shown by Sakai et al. (47). In that paper, the effect of OA was tested on NIH/3T3 fibroblasts transformed by human activated *raf*, *ras*, and *ret-II*. Surprisingly, it was found that OA inhibited the proliferation of transformants and induced their reversion to a more differentiated phenotype characterized by flat morphology, contact inhibition, and increased synthesis of fibronectin (47). Similar findings were later obtained by Gupta et al. when studying the effect of OA on v-*src*–transformed BALB/c 3T3 cells (48).

In the human leukemia cell line U937, it was found that OA (50 ng/mL) treatment induced increased expression of cell surface markers Leu11 and Mac-1 detectable on maturing granulocytes and monocytes, leading to the conclusion that the toxin induced differentiation toward a more mature cell type (49). In line with this report, Kharbands et al. (84) also showed differentiation of U937 cells toward monocytes.

Treatment of human breast cancer cells with OA, in turn, induced their change to more mature, milk-producing cells. In that case, nanomolar concentrations of OA (between 9 and 15 nM depending on the cell line) led to increased production of α-lactalbumin, β-casein, and lipid droplets (50). Furthermore, it was shown that the effect was also induced by calyculin A and an active OA derivative (dinophysistoxin-1, DTX) but not by an inactive metabolite (okadaic acid tetramethyl ether, OATM) (50).

Taken together, the previous findings document that OA class tumor promoters may behave as differentiating agents under defined experimental conditions and/or in some cell lines. One aspect worth being pointed out is that the differentiative responses induced by OA were accompanied by inhibition of cell proliferation (49,50). Although it can be argued that inhibition of cell proliferation is a common feature of differentiating cells, it should be noted that OA concentrations used in those studies were in the order of 10^{-8}M. This fact suggests that toxic effects of OA class compounds might have contributed to inhibition of proliferation accompanying differentiative responses.

The overwhelming body of available evidence, in fact, leads to the conclusion that OA at concentrations higher than 10^{-9}–10^{-8}M causes cell death (see Table 2). If the more potent calyculin A compound is tested, the effects essentially coincide with those induced by OA, but the concentrations needed to observe the phenomena are between 5- and 30-fold lower than those of OA (51,63,66,83).

In most studies, the antiproliferative effect of OA class compounds has been directly evaluated by measuring cell growth in the presence of the toxin, as exemplified by Figure 2,

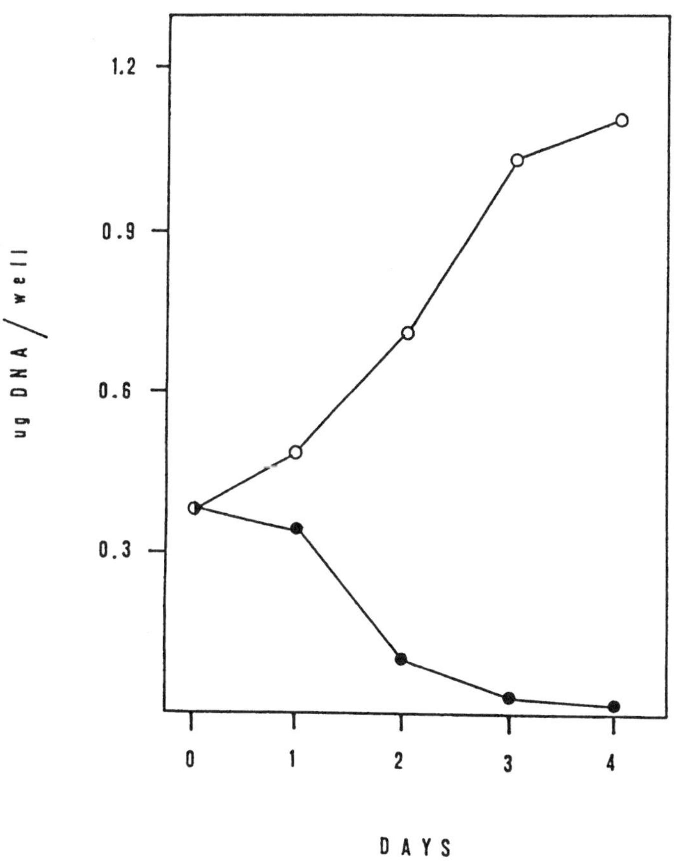

Figure 2 Effect of okadaic acid on proliferation of normal human fibroblasts. Cells in multiwell plates were incubated with (solid circles) or without (void circles) 50 nM okadaic acid and the DNA content of wells was determined at daily intervals.

and the response usually becomes detectable by the second day of treatment. By using incubation of cultured cells with tritiated thymidine instead, inhibitory responses on DNA synthesis are readily detectable after 1 day of OA treatment (54,56,58,59,68), but earlier time points may provide a clear indication of growth arrest, particularly in the case of experimental conditions employing high concentrations (10^{-8}–10^{-7}M) of OA (62,64,73).

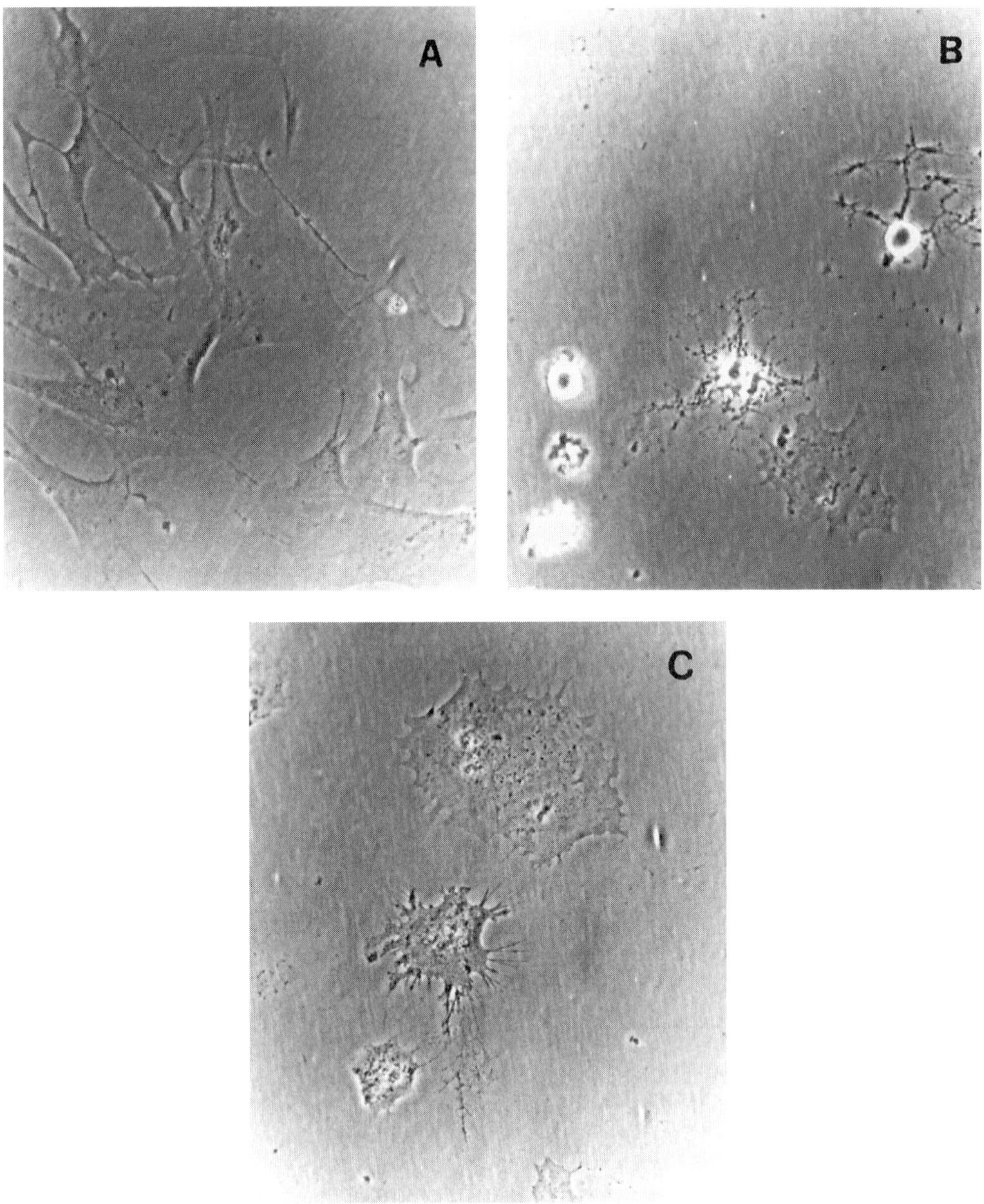

Figure 3 Effect of okadaic acid on morphology of normal human fibroblasts in culture. Phase contrast microscopy (magnification 200×) of normal human fibroblasts after a 24-h treatment with vehicle (A) or 50 nM okadaic acid (B, C).

Inspection of morphological characteristics of cells treated with OA class compounds may also be used to evaluate the appearance of toxic effects in cultured cells. Changes may become apparent within minutes or several hours, depending on the specific phenomenon and/or cell type, are dose-dependent, and involve both nuclear and cytoplasmic structures, as well as extracellular matrix components.

Gross changes in cell shape can be easily detected by inspection of culture vessels and involve the appearance of protrusions of different shapes and blebs (47,62,68,81,85–87), loss of cell-cell contacts (40,60,81,88), cell rounding (40,52,62,81,88–90), and cell detachment from substratum (40,52,62,77,81,82,88–90). Some of these changes are shown in Figure 3B and 3C, where normal human fibroblasts losing contact with the dish are rounded and highly refractile. Detection of cell rounding and detachment also can be accompanied by the appearance of cell flattening (47,60), with cells taking a fried egg–like shape (Fig. 3B and 3C), and it is not clear whether differences in altered cell shapes (rounding vs flattening) represent steps of a single process.

Nuclear structures are also affected by OA class compounds, as shown by detection of fragmented and multilobed nuclei (49,51,62,65,66,69,76,81,91) containing condensed chromatin (49,53,62,65,66,91–94).

Morphological analyses by electron microscopy reveal alterations involving different cellular structures, such as loss of microvilli at the cell surface (85), aggregates of cytoplasmic material (85), and segregation of organelles (62) without detection of gross changes in the case of mitochondria (62). Extensive alterations of cytoskeletal networks are also detected after cell treatment with OA class compounds in terms of destruction of microtubules (95,96), microfilament reorganization (85), and altered intermediate filament networks (86,88,97–100).

Most of the components and most of the events which constitute the molecular basis of these phenomena will be discussed in the next Sections IV and V where several aspects of morphological changes induced by OA class compounds will be interpreted in terms of a pseudomitotic state. Membrane blebbing, cytoplasmic and nuclear fragmentation, and cell shrinkage, however, are considered to be hallmarks of apoptosis (101,102), lending support to the notion that OA class compounds induce this type of response in treated cells.

Indeed, the conclusion that OA class compounds induce apoptosis has greatly stimulated the debate on the involvement of PPases in cell proliferation and apoptosis (20,22,103,104). Interestingly enough, however, once the apoptotic response has been already set in motion by different means, OA class compounds display antiapoptotic activity in several cell culture systems (53,54,78,93,105,106). Apparently, this antiapoptotic activity of OA class compounds could not be related to individual responses, as in the aforementioned studies apoptosis had been induced by as diverse agents as etoposide (78), γ-radiation (53,92,93), bistratene A (53), tetrandrine (53), cisplatin (53), heat (92), glucocorticoids (105), AraC (106), and VP-16 (106).

A final remark on the antiproliferative effects of OA class compounds is that available data show that these toxins apparently affect any type of mammalian cell studied from primary cultures to more aggressive tumor cells (see Table 2).

IV. PROXIMAL AND DISTAL MOLECULAR TARGETS OF OKADAIC ACID CLASS COMPOUNDS

The variety and even contradictory features of the effects triggered by OA class compounds, such as tumor promotion versus cell death or apoptotic versus antiapoptotic responses, underlie the extreme complexity of molecular events playing some role in these processes.

As the primary molecular targets of OA class compounds are PP1, PP2A, and related ser/thr PPases, it seems reasonable to approach the analyses of molecular mechanisms set in motion by DSP toxins in vivo by an evaluation of proteins whose phosphorylation state has been shown to be affected in intact cells on treatment with OA and related compounds. Table 3 includes some of these proteins, which have been arranged according to the major functional role they are believed to play in the cell. This list is certainly incomplete, as it includes only those proteins which are either directly or indirectly involved in mechanisms controlling normal and neoplastic growth, as will be discussed below. A second limitation is due to the choice to restrict the list to those proteins whose changes in the phosphorylation state induced by OA class compounds have been directly observed. Furthermore, Table 3 does not include the many enzymes generally involved in cellular metabolism whose activity is controlled by reversible phosphorylation and is affected by OA class compounds (107,108), as they are not considered to play a primary role in the control of cell proliferation.

The first group of proteins whose phosphorylation state is affected by OA class compounds consists of protein kinases, and this could be expected because of the widespread occurrence of phosphorylation cascades in mechanisms controlling cell functions. The two major forms of MAPK, that is, p44 ERK1 and p42 ERK2, which are known to be dephosphorylated by PP2A (46,75,109–115), are among them. Another protein kinase which is grouped with MAPKs, and whose phosphorylation state is affected by cell treatment with OA class compounds, is p38

Table 3 Proteins Whose Phosphorylation State Is Affected in Cells Treated with DSP Toxins

Protein kinases
ERK1
ERK2
P38MAPK
Rsk
Abl
Lck
Src
$p34^{cdc2}$
Transcription factors and cofactors
TCF/Elk-1
SRF
Egr-1
c-Jun
CREB
ATF-1
Rb
p53
Steroid/thyroid hormone receptors
Cytoskeletal components
Vimentin
Cytokeratins
Miscellaneous components
Receptor of EGF
85-kD Phospholipase A_2
Histones
cdc25

MAPK (111). Phosphorylation of this kinase is brought about by a phosphorylation cascade which closely parallels that of ERKs, although it might involve distinct kinases (13,15,118,119). p90rsk, a ser/thr protein kinase activated by phosphorylation of thr residues catalyzed by ERKs (13–15), is also stabilized in its phosphorylated forms by OA (109,111), and it represents one of the kinases believed to play a role in phosphorylation and, hence, regulation of the activity of nuclear proteins together with ERKs1 and 2.

The next three examples of protein kinases represent nonreceptor TKs regulated by both ser/thr and tyr phosphorylations (73,90,120,121). The effects of OA may be summarized in terms of stabilization of phosphorylated residues, but a careful consideration of data reveals interesting features. In the case of c-Abl, for instance, increased ser phosphorylation is observed, but this condition does not apparently affect TK activity of the protein (120). Instead the catalytic activity of c-Src and Lck kinases is affected by OA, being stimulated in the former case (90) and inhibited in the latter (73). The molecular basis of the inhibition of Lck is the stabilization of phosphorylated serines induced by OA. A more complex picture is found in the case of c-Src, as its enhanced activity was found to be due to increased dephosphorylation of a specific tyr residue (90). The effect of OA, therefore, could be mechanistically reduced in its simplest interpretation to stabilization of phosphorylated forms of some enzyme (either a TK or phosphatase) responsible for the control of tyr527 phosphorylation in c-Src (90).

An even more complex phenomenon appears to be responsible, in turn, for increased phosphorylation of $p34^{cdc2}$, the catalytic component of a ser/thr protein kinase (histone 1 kinase, H1K) and a member of the cyclin-dependent kinase (CDK) family, playing important roles in cell cycle (reviewed in refs. 122–124). Three phosphorylated residues control the activity of $p34^{cdc2}$ (123–124), including an activating phosphorylation of thr161 and two inactivating phosphorylations of thr14 and tyr15 of human cdc2 (Figure 4). Phosphorylation of thr161 is brought about by a CDK-activating kinase (CAK), whereas those of thr14 and tyr15 are catalyzed by a dual-specificity protein kinase (Weel or some homologue). Dephosphorylation of thr161 is brought about by a ser/thr PPase (possibly PP2A), whereas thr14 and tyr15 are dephosphorylated by the cdc25 dual-specificity phosphatase (125–128). A key aspect of this phenomenon is that both Wee1 and cdc25 are controlled themselves by ser/thr phosphorylation in an opposite fashion and consisting of inactive phosphorylated Wee1 and active cdc25. Hence, OA treatment of several cells results in increased phosphorylation of thr161 of $p34^{cdc2}$ due to inhibition of its dephosphorylation and decreased phosphorylation of thr14 and tyr15 as a consequence of stabilization of phosphorylated, active, cdc25 phosphatase (128–131). The functional consequence, as it will be discussed in Section V, is an enhanced activity of CD kinases.

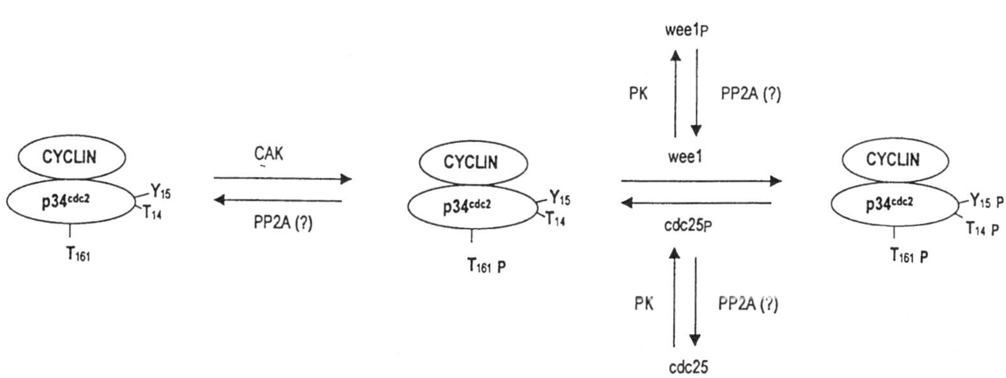

Figure 4 Molecular mechanisms involved in regulation of cyclin-dependent protein kinase activity.

If these examples of functional consequences due to OA-induced stabilization of phosphorylated residues in individual proteins are considered as a whole, they provide the two general types of responses triggered by this class of tumor promoters at the molecular level. In the first instance, one finds the direct effect caused by stabilization of either stimulatory or inhibitory phosphorylations on effector proteins. Instead the second type of responses includes the indirect effects due to components involved in the control of effectors and whose activity is regulated by OA-sensitive phosphorylations.

To give a further twist to this picture of great complexity, it should be pointed out that, as exemplified by c-Src and $p34^{cdc2}$, the functional state of proteins may be controlled by multiple phosphorylations, and hence different kinases and phosphatases may be involved in the overall control of their activity. In the case of ERKs, for instance, and in analogy to $p34^{cdc2}$, dephosphorylations are performed by both OA-sensitive and OA-insensitive PPases, whose expression and/or activity may vary among different cell types (115,133).

A second group of proteins whose phosphorylation state is affected by OA class compounds includes transcription factors (TFs), or proteins capable of interacting with TFs and controlling their functions. The ternary complex factor (TCF/Elk-1) and the serum responsive factor (SRF) are proteins whose phosphorylated forms are stabilized by OA (134), and they are the first to be considered here, as both are believed to be responsible for enhanced transcription of selected genes following mitogenic stimuli (5,134–136). The proposed mechanism is that SRF is permanently bound to a specific DNA sequence termed serum response element (SRE), but it does not stimulate the transcription of controlled genes. Cell stimulation by GFs would result in ERK activation, and this kinase then phosphorylates the third component of the complex, TCF/Elk-1, triggering transcription of selected genes (134). This type of response occurs within minutes of cell stimulation without the involvement of new protein synthesis, and it requires only conformational changes (in this case protein phosphorylation) of TF. Hence, the genes induced by this mechanism are termed immediate-early, or primary, response genes (5,137,138). When cellular responses culminate with proliferation, induced genes appear to play active roles in bringing about the response (5). Interestingly enough, SRF itself is the product of a primary response gene, as is the case of Egr-1 and c-Jun, whose phosphorylation state is also enhanced by OA class compounds (see Table 3). Primary response genes believed to be involved in cell proliferation are discussed more thoroughly in Section V.C, with reference to transcriptional responses affected by OA class compounds. c-Jun, however, requires here some attention, because it represents a key component of the AP-1 complex together with members of the Fos family (reviewed in refs. 139 and 140). Both components can be phosphorylated by members of the MAPK family, although it is believed that different sets of protein kinases are responsible for ser/thr phosphorylation of these TFs (13,136). In the case of Jun, in particular, a kinase cascade similar to that involved in ERK, and p38MAPK activation, is responsible for activation of the Jun NH_2-terminal protein kinase/stress–activated protein kinase (JNK/SAPK), catalyzing Jun phosphorylation (118,119,141). As it can be surmised by the terminology, this cascade is set in motion by noxious stimuli, and it is believed to play a role in stress responses (119,142,143). As a component of the AP-1 complex, Jun participates in the control of gene expression through binding to DNA sequences, termed TPA responsive element (TRE), which are involved in the control of expression of several genes, including Jun itself (136,144).

The activating transcription factor-1 (ATF-1) and the cAMP responsive element binding protein (CREB) represent TFs whose phosphorylation is catalyzed by several ser/thr protein kinases and primarily by the cAMP-dependent protein kinase (PKA) (145). Dephosphorylation of these proteins is known to be catalyzed by PP2A and, accordingly, OA class compounds stabilize their phosphorylated forms (146–148). Although involvement of these TFs in controlled expression of genes is well established (144), their possible role in cell proliferation is less well characterized. In any case, the DNA sequences bound by these proteins, termed cAMP responsive element (CRE), are found in the promoter region of primary response genes involved in proliferation, as in the case of members of the *fos* and *jun* families (136).

The TFs considered so far, whose phosphorylated forms are stabilized by OA class compounds, have been linked mainly to stimulation of proliferation, but other modulators exist that apparently play their major roles in inhibition of cell proliferation and induction of cell death. The components in question are Rb and p53 (see Table 3), the products of two tumor suppressor genes (149–151). As will be further discussed in Section V, covalent modification by phosphorylation has been shown to affect the activity of both proteins. Rb essentially behaves as a corepressor by forming complexes with TFs and preventing their functioning as stimulators of transcription of genes involved in cell duplication. Phosphorylation of multiple ser/thr residues then inactivates Rb and leads to gene transcription (149,152). Instead p53, is a TF itself, and multiple phosphorylations control its properties, leading to either stimulation or inhibition of *trans*-activating functions depending on specific sets of phosphorylated residues (153). Both Rb and p53 can be dephosphorylated by PP2A and, perhaps, PP1 (44,55,154,155), and hence their phosphorylation states are affected by DSP toxins (44,46,52,55,58,78,106,154–160).

The last TFs reported in Table 3 include members of the steroid and thyroid hormone receptor superfamily, such as the glucocorticoid receptor (161,162), the 1α,25-dihydroxycholecalciferol (vitamin D_3) receptor (163), and the triiodothyronine receptor (164,165). Although we are not aware of reports on OA-induced stabilization of phosphorylated forms of other members of this intracellular receptor superfamily, it has long been recognized that this toxin may potentiate transcriptional responses involving these ligand-dependent transcription factors (162,163,165–176). In some cases, it has been also shown that responses could strictly require receptor proteins but not the ligand (163,165,166,170,175,176). Hence, those results have clearly indicated the existence of cross-talk between signal transduction involving plasma membrane receptors and intracellular receptors. Although the effects observed in those experimental systems have been proposed to involve both receptor proteins and coactivators of transcription (166,167,173,175), the possibility that OA class compounds might affect proliferation of steroid hormone responsive cells by changes in receptor phosphorylation should be considered, as many functional aspects of members of the nuclear receptor superfamily are influenced by phosphorylation (reviewed in ref. 177).

More generally, phosphorylation of TFs or their cofactors is a widespread mechanism involved in the control of gene expression by extracellular signals (5,138,145). Alterations of the phosphorylation state of relevant TFs induced by OA class compounds, therefore, can be a relevant cause of perturbed cell proliferation.

Other proteins which have been found to be hyperphosphorylated after cell treatment with OA include the major components of microfilaments; that is, cytokeratins and vimentin (86,97–99,178,179). These proteins assemble in polymers to yield the fibrillar structures responsible for mechanical stability of cells (180). Under normal growth conditions, phosphorylation of these proteins is regulated during cell cycle, being maximal at mitosis, when microfilaments disassemble and form cytoplasmic aggregates (58,73,181,182). Hence, an altered mechanical stability of affected cells is the primary effect induced by OA class compounds through hyperphosphorylation of cytokeratins and vimentin. Other interpretations of the phenomenon, however, can relate that primary effect on different pathophysiological issues, such as altered cell-cell and cell-matrix interactions (see Section III) or induction by OA class compounds of pseudomitotic phenotypes (see Section V.B). With reference to this latter issue, in particular, the hyperphosphorylation of several histones induced by cell treatment with OA (91,183,184) is of interest, as maximal levels of phosphorylated H1 and H3 are observed during mitosis (185).

The last group of proteins whose phosphorylation state is affected by OA class compounds includes molecules displaying some role in mechanisms of signal transduction which may be involved in the control of cell proliferation. This is certainly the case, for instance, of the receptor for epidermal growth factor (EGF), whose functioning involves tyr phosphorylation and is affected by ser/thr phosphorylation (3). OA has been shown to stabilize the latter type of phosphorylations (156,186), leading to altered responses to EGF, such as inositol-phosphate formation, in the lack of any inhibition of TK activity (186).

The cytosolic 85-kD form of phospholipase A_2 (PLA_2) is another example of proteins whose phosphorylated forms are stabilized by OA (187). Phosphorylation of 85-kD PLA_2 is thought to be catalized by ERKs and results in activation of the enzyme (188,189), which then brings about the hydrolysis of membrane phospholipids and the release of a lysophospholipid and arachidonic acid, both of which have been shown to affect cell proliferation, mainly through activation of PKC and MAPK (190,191).

The functional roles and characteristics of the various proteins discussed above impinge on so many aspects of the molecular mechanisms involved in the control of cell proliferation that it seems appropriate to conclude that ser/thr PPases control key aspects of normal and abnormal cell growth (20,22,103).

V. MOLECULAR MECHANISMS UNDERLYING RESPONSES TO OKADAIC ACID CLASS COMPOUNDS

The experimental findings described in Section IV clearly show that OA class compounds, by inhibiting PP1 and PP2A, have a variety of molecular targets. Furthermore, the time lags occurring between the beginning of cell treatment with those toxins and the appearance of detectable changes on protein phosphorylation indicate that these agents have both proximal and distal targets, as effects may be observed within few minutes (85,90,99,107,116,134,183) or be delayed by several hours (55,58,63,75,77,83,94) after treatment. Thus, as anticipated, some of affected proteins are primary targets, whereas delayed effects may be due to secondary events.

In this section, attention will be focused on the description of the chains of events set in motion by OA class compounds in treated cells as they have been deduced by experimental data.

A. Effects of OA Class Compounds on Protein Kinases

A response frequently observed after cell treatment with DSP toxins consists of activation of protein kinases, an apparent paradox (192) due to the widespread occurrence of kinase cascades controlling cell functioning and proliferation. Indeed, the number of protein kinases whose activity has been shown to be modified in cell responses to OA class compounds (Table 4) exceeds that of kinases showing altered phosphorylation state (see Table 3). A stabilizing effect on individual phosphorylated residues, due to inhibition of PP1 and PP2A by OA and related compounds, seems to be the most reasonable mechanism underlying those effects, as discussed in Section IV.

In general, OA class compounds stimulate protein kinase activity (see Table 4), and a consideration of individual components suggests that some signal amplification might result. As outlined in Section I, a major PK cascade activated by mitogens proceeds through Raf-1, MEK, ERK, and Rsk, whose activity is stimulated by OA class compounds (13–15,75,81,110–112,114,116,117,193–195). Thus, these agents might potently enhance protein phosphorylation by this route in several systems. Surprisingly enough, stimulation of MAPK activity has not

Table 4 Effects of DSP Toxins on Activity of Selected Protein Kinases

Protein kinase	Effect
Raf-1	+
MEK	+
ERK1	+
ERK2	+
Rsk	+
JNK	+
p38MAPK	+
OSK	+
Lck	−
Src	±
$p34^{cdc2}$	+(−)

been invariably found to be correlated with cell growth (75,196–198), and it often accompanies cell death induced by OA class compounds (75,81). Thus, cell-specific responses to these toxins are clearly apparent in the case of ERK activation, and recently obtained data indicate that they might include a less tight control of ERKs in cells displaying more aggressive proliferative phenotypes, such as tumor cells (G.P. Rossini et al., unpublished observations). In any case, the lack of correlation between ERK activity and proliferative response has not been explained yet, and it may be a key aspect to understanding the opposite effects of DSP toxins on cell growth in different systems.

Stimulation of JNK (81) and p38 MAPK (81,111) may also be framed in this contradictory picture, as outlined in Section IV.

Instead the activation of a 53- to 55-kD protein kinase, has been detected after 12–24 h of cell treatment with OA (Figure 5), and, accordingly, the enzyme has been named okadaic acid–stimulated protein kinase (OSK) (77). This response also has been observed after cell treatment with calyculin A, and, based on the methodology employed to detect OSK, its activation can be explained by stabilization of phosphorylated residues due to inhibition of PPases (77). In the eight different cell lines analyzed so far, activation of OSK was always found to accompany cell death (G.P. Rossini, et al. unpublished observations). Although purification of this kinase is expected to allow unequivocal identification of the enzyme, the observation that maximal activity of OSK is detected 12–24 h after OA addition to culture dishes, when cell death becomes clearly detectable, leads to the hypothesis that OSK might play an active role in death responses (77).

Members of the Src tyrosine kinase family may also be involved in responses elicited by OA class compounds (see Table 4). The functional roles of these cytoplasmic kinases are not completely understood, but evidence obtained so far implies their participation in proliferative responses and cell adhesion (for recent reviews, see refs. 121 and 199). In the case of cell proliferation, for instance, Src and related kinases are believed to participate in the assembly of signal transducers through the interaction of their SH2 domains with phosphotyrosines of membrane receptors for GFs and the consequent synergy in stimulation of downstream components of the Raf-1, MEK, MAPK pathway (121,199). Within this framework, the inactivating effect of OA on Lck could then underlay inhibition of T-lymphoblast proliferation (73), whereas activation of Srk, overexpressed in NIH/3T3–derived cells (90), might potentially represent a mechanism responsible for stimulation of cell proliferation. As it has been already pointed out

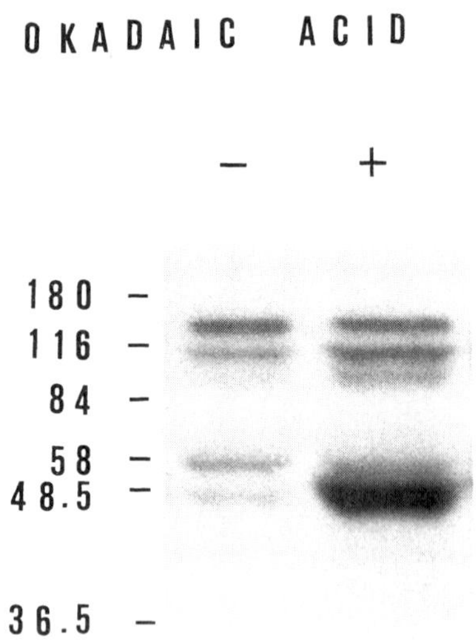

Figure 5 Activation of a 53- to 55-kD protein kinase by cell treatment with okadaic acid. MCF-7 human breast cancer cells were incubated for 24 h with (+) or without (−) 50 nM okadaic acid and protein kinase activity of total cell lysates was measured in situ by incorporation of ^{32}P-labeled orthophosphate in substrate protein, as described (77).

in the description of OA effects on ERKs, however, the responses to this agent show a remarkable cell specificity at the molecular level. Paradoxically enough, this contention might actually represent an oversimplification if one considers the effects of OA on the activity of CDKs (see Table 4). As discussed in Section IV, OA, either directly or indirectly, affects several components of the complex set of kinases and PPases involved in the control of phosphorylation states of $p34^{cdc2}$. In line with the net response to OA, consisting of enhanced activating phosphorylation of thr161 and reduced inactivating phosphorylations (thr14 and tyr15) of $p34^{cdc2}$, the functioning of this kinase has been often found to be stimulated or activated (125,126,128–132,200–203). Still, there are reports showing that, under certain experimental conditions, OA can indeed cause inhibition of cdc2 activity (55,200). Perhaps not surprisingly, owing to the role played by cdc2 in the cell cycle (122–124), different responses to OA are apparently related to the cell cycle phase of cells subjected to treatment. Thus, it seems appropriate to expand this issue and evaluate the molecular events involved in the control of the cell cycle which have been found to be altered by OA class compounds.

B. Effects of OA Class Compounds on the Cell Cycle

The control of the cell cycle in eukaryotic cells is primarily exerted at the boundary between G_1/S and G_2/M phases (122,204), and the transistion from one phase to the next is controlled by activation of CDKs following the general scheme outlined in Figure 4 (123,124). Major differences between the molecular mechanism involved in progression through cell cycle phases are believed to depend on the identity of both cyclins and the kinases of the complex (123,124). Thus, G_1/S transition would involve cyclins D and E in association with CDK4 and CDK6 and progression through S phase would require cyclin A in association with CDK2, whereas entry into mitosis would be induced by activation of CDK1 associated primarily with cyclin B, as described above. Specific roles of individual CDKs would then stem from different sets of their subunits and the functional roles they would play in the cell.

The phosphorylation state of CDKs is a key aspect in the control of their functioning and, hence, of cell cycle progression. Other components, however, participate in the overall control of the process. For instance, the different cyclins described above play a role in cell cycle progression because their expression in the cells follows a well-defined pattern of synthesis and degradation. Thus, cyclins D and E are synthesized during G_1 and are rapidly degraded during either S phase (cyclin E) or G_2/M (cyclin D), whereas accumulation of cyclin A slowly starts in late G_1, to reach its maximum in G_2, when high levels of cyclin B (whose synthesis begins in late S) are also observed (122,204).

Replication of chromosomal DNA and segregation of replicated chromosomes in daughter cells represent the processes which typify S and M phases of the cell cycle, respectively. In line with these features, the molecular components which are recognized, or putative, substrates of CDKs in vivo, and would then constitute the causal link between these enzymes and functional responses, are strikingly different during the cell cycle. The major substrate of CDKs by late G_1 and through S is constituted by the Rb protein (149,152). Thus, by phosphorylating the Rb protein, CDKs would cause release of E2F transcription factors, leading to stimulation of the transcription of genes coding for components involved in DNA duplication (205). At G_2, instead, CDK1 activation brings about phosphorylation of proteins of the nuclear lamina and cytoskeletal components involved in the disassembly of cellular ultrastructures at mitosis (122). Furthermore, other $p34^{cdc2}$ substrates are histones, whose phosphorylation is believed to be involved in chromosome condensation occurring at mitosis (122,185).

The effects of OA class compounds on the cell cycle are complex, because these toxins can directly and indirectly affect multiple components of the process and at different levels.

The most direct effects can be probably explained by an altered phosphorylation state, and hence functioning of proteins controlling the cell cycle itself. This is particularly clear in the case of $p34^{cdc2}$, whose activation by OA has long been recognized to promote a mitosis-like state. Interestingly enough, this effect, which was originally described on OA treatment of *Xenopus* oocytes (206), is observable with cultured, normal cycling cells (91) and with cells synchronized in S phase (200,201,203) but not in quiescent (G_0) cells (55). OA treatment of serum-deprived NIH3T3 cells, in fact, has been found to result in cells blocked in S phase and displaying depressed cdc2 and CDK2 activity (55).

The Rb protein (46,78,156,157), histones (91,183,184), and vimentin (86,97,178) represent other examples of components whose phosphorylation is increased following cell treatment with OA class compounds, which is in line with a stimulatory effect of these tumor promoters on cell progression through the cycle (Rb) toward mitosis (histones and vimentin).

The effects of OA on the cell cycle, however, cannot be easily reduced to a single coherent picture. The phosphorylation of the Rb protein, for instance, has been found to be decreased by OA treatment in some experimental systems (55,58) when serum-starved cells have been treated with GFs, as in the case of NIH/3T3 cells described above (55). Furthermore, OA treatment of cells in G_0 (or G_1) has led to cell cycle arrest at G_1/S (55,58,59,64,65,207).

On the basis of those results, it should then be concluded that some as yet undetermined dephosphorylative event(s) blocked by OA is mandatory for cells to progress toward S phase (55,58,59,64,65,207). Indeed, cell cycle arrest induced by OA class compounds is not confined to the G_1/S boundary. Mitotic and pseudomitotic effects induced by OA (91,94,125,126,128,129, 131,132,200–203,206), in fact, are matched by cell cycle arrest at some early or late step in M (64,65,68,94,208).

Thus, by enhancing the phosphorylation of specific amino acid residues on relevant components, such as Rb and $p34^{cdc2}$, OA class compounds would behave as stimulators of proliferation. In turn, by preventing a timely dephosphorylation of those proteins, they would eventually block the well-ordered chain of events underlying cell duplication, leading to growth arrest and cell death.

A similar conclusion may be also put forth if one considers more indirect effects of OA class compounds, as in the case of changes in the expression of genes coding for cell cycle regulatory proteins. For instance, expression of cyclin A, cyclin B, $p34^{cdc2}$, and Rb is affected by OA class compounds in several systems where the effect can be either stimulatory (64) or inhibitory (55,200,210).

Those responses, in any case, accompany cell cycle arrest rather than its progression, and the two phenomena do not show a complete correlation. It has been remarked, however, that enhanced expression of cell cycle regulatory proteins by OA occurs in tumor but not in immortalized nontransformed cells, providing evidence of differences between the molecular responses to OA in the two types of cells and suggesting that these differences may reflect fundamental alterations of cell cycle regulation in tumor cells (64). Although the implications of this conclusion remain to be determined, it seems important to point out that OA class compounds stimulate the expression of the WAF1/CIP1 gene (76,211,212) coding for a 21-kD protein which represents an endogenous inhibitor of CDKs. By inhibiting cyclin D–, cyclin E–, and cyclin A–dependent kinases, $p21^{WAF1/CIP1}$ blocks cells in G_1 and plays a fundamental regulatory role in cell cycle progression (122–124). Thus, the enhanced expression of the WAF1/CIP1 gene observed in both embryonic fibroblasts (211) and in tumor cells (212) has lent support to the contention that OA-induced apoptosis is linked to deregulated expression of $p21^{WAF1/CIP1}$ (76).

C. Effect of OA Class Compounds on Gene Expression

Modulation of gene expression is a primary response in mechanisms of signal transduction involving both plasma membrane receptors and intracellular receptors (5,137). Among the mechanisms responsible for regulation of gene expression, phosphorylation of transcription factors and co-factors is considered of paramount importance (5,138,145,205). By stabilizing phosphorylated forms of TF or associated modulators (Section IV), OA class compounds can thus influence transcriptional responses.

Table 5 includes several genes whose expression is affected by OA class compounds arranged according to the process involving their products in living cells.

The first group includes early-response genes, whose expression is enhanced in cells stimulated by mitogens, and hence are considered to be primary effectors of proliferative responses (5). The effect of OA on expression of these genes is mainly stimulation, as in the case of c-*jun* (49,50,74,82,156,213–215), c-*fos* (49,50,70,74,82,135,213,214,216), *junB* (50,82,214), *junD* (49,82,214), *fra-1* (214), *egr-1* (156,217), and fibronectin (47,48), whereas repression of transcription has been observed in the case of c-*myc* and *max* genes (49,70).

The enhanced expression of mitogen-induced genes by OA class compounds is in line with the role of tumor promotion proposed for these biotoxins. However, it should be noted, that, in several instances, the transcriptional response to OA treatment was actually accompanied by inhibition rather than stimulation of cell proliferation (47,49,70,74,82). Instead, a more coherent response, is found in the case of c-*myc* and *max*, when the inhibition of their expression was followed by the apoptotic response induced by OA class compounds (49,70).

A similar situation would apparently hold true for transcriptional responses of cell cycle–related genes (see Table 5), described previously, as OA enhanced transcription of genes coding for either cyclin A, cyclin B, $p34^{cdc2}$, or Rb but induced cell cycle arrest (46,64).

The third group of genes in Table 5 includes those coding for proteins playing active roles in the control of apoptosis. The p53 protein, for instance, is believed to represent a key modulator

Table 5 Effects of DSP Toxins on Expression of Selected Genes

Gene	Effect
Mitogen-inducible genes	
c-*jun*	+
c-*fos*	+
jun-B	+
c-*myc*	−
max	−
fra-1	+
egr-1	+
Fibronectin	+
Cell cycle–related genes	
Cyclin A	±
Cyclin B	+
$p34^{cdc2}$	±
WAF1/CIP1	+
Rb	±
Apoptosis-related genes	
bcl-2	−
bcl-x_L	−
BAX	±
p53	±

of cell proliferation and apoptosis by controlling expression of two different sets of genes (150,151). In the first case, p53 would stimulate the expression of WAF1/CIP1, leading to inhibition of CDKs and arrest of cells in G_1 (150,151). Instead induction of apoptosis by p53, would involve enhanced expression of some genes, such as that coding for the Bcl-2 associated protein X (BAX), and repression of others, as in the case of *bcl*-2 (151,218).

The molecular mechanisms underlying p53 functioning in both cell cycle arrest and apoptosis are based on enhanced levels of cellular p53 protein due to inhibition of its degradation (151) and on conformational changes induced by protein phosphorylation (153). OA class compounds have been shown to affect both processes (44,52,76,157–160,219,220), and these effects could then help to explain the death responses to DSP toxins. This straightforward interpretation, however, is questioned by the findings that cellular levels of p53 may be altered by OA class compounds both positively (76,158) and negatively (219,220). Furthermore, hyperphosphorylation of p53 due to OA has been linked to both enhanced (52,160) and depressed (158,160) transcriptional activity, whereas, in one report, the apoptotic response induced by the toxin did not even require the transcriptional activity of p53 (52). Thus, the effects of OA appear to be cell type–specific and to depend on individual promoters (*c*101), providing other examples of dual responses induced by DSP toxins.

A final remark about the effects of OA class compounds on gene expression and its role in proliferative or death responses should be focused on members of the *bcl*-2 family. *Bcl-2*, *BAX*, and *bcl-x* represent the best characterized members of the family, and code for 20- to 25-kD proteins, located mainly in mitochondrial and nuclear membranes with wide tissue distribution (218). The functional roles of these components are not fully understood, but they appear to be related to the control of apoptotic responses, with Bcl-2 displaying antiapoptotic activity, whereas BAX would rather stimulate apoptosis (218). Although the molecular events underlying the effects due to members of the Bcl-2 family remain largely undetermined, there is agreement regarding the proposal that cell fates, in terms of apoptotic or antiapoptotic processes, depend on the physical interactions between members of the family, giving rise to different dimers (218). Thus, the functional correlates of the many alternatives stemming from the combinatorial associations between members of the family can be summarized by the contention that cell susceptibility to apoptosis increases when dimers including BAX predominate, whereas high levels of Bcl-2 homodimers would protect cells from apoptosis (218). Within this framework, the inhibitory effect of OA on the expression of *bcl-2* and *bcl-*x_L (67), and its stimulation of *BAX* expression (76), would be in line with the apoptotic response induced by DSP toxins in many cells.

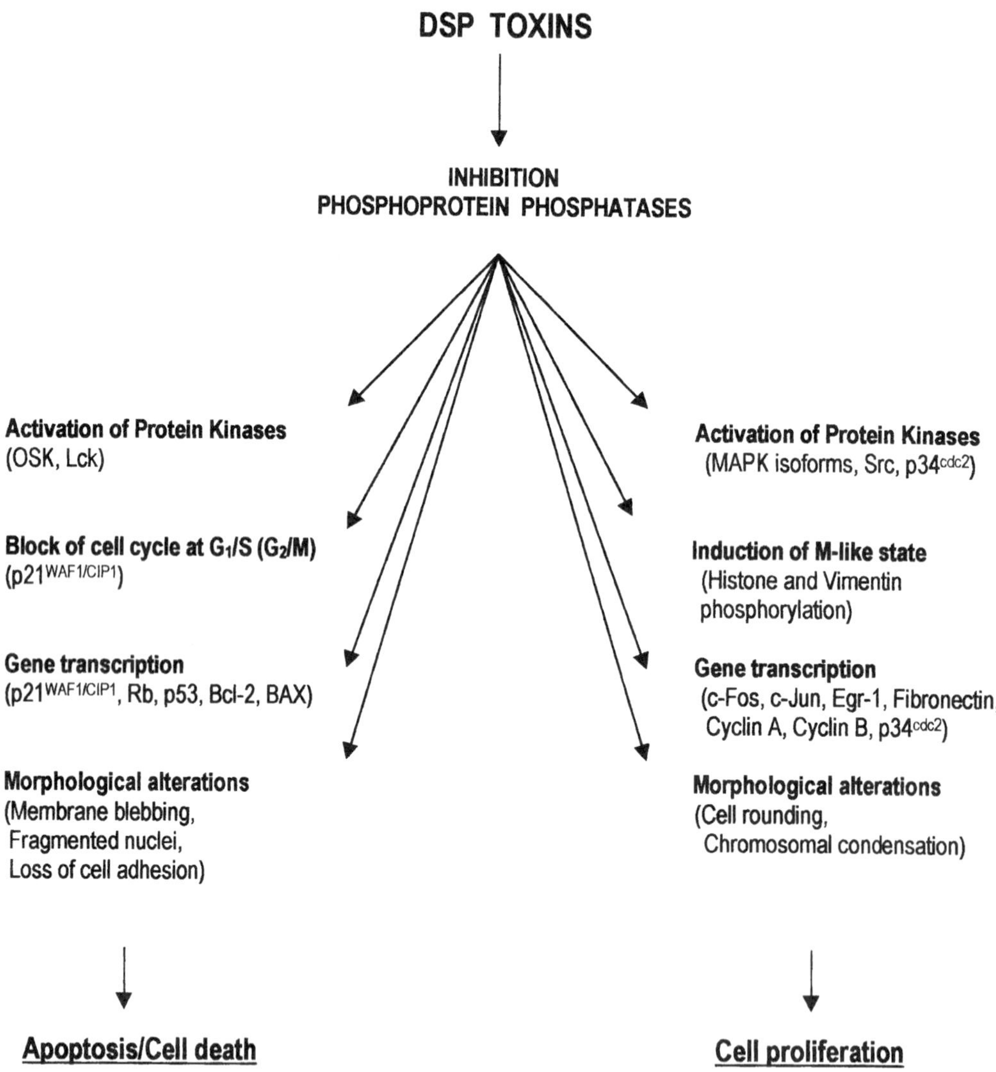

Figure 6 Simplified view of the molecular basis underlying dualistic responses triggered by DSP toxins in affected cells.

VI. CONCLUSIONS AND PERSPECTIVES

A striking feature pervading virtually every investigation performed on the effects of DSP toxins on cell proliferation is the widespread occurrence of dualistic responses. Figure 6 shows a schematic representation of the alternate fates of cells after treatment with OA class compounds, and of the processes which may underlay the two opposite responses, as it can be surmised on the basis of the experimental evidence provided in previous sections.

The sharp distinction between events favoring either cell death or proliferation should be considered as a simplified view of the molecular mechanisms which may play some causative role when DSP toxins display either apoptotic or tumor-promoting activity. Still, a model of enhanced cell proliferation due to inhibition of PPases by OA class compounds involving stimulation of MAP, Src, and cyclin-dependent protein kinases, increased expression of mitogen-induced genes, and resulting in cell division, seems in line with the accepted views on the molecular mechanisms underlying cell responses to GFs. By the same token, a response to DSP toxins involving increased levels and activity of p53, Rb, WAF1/CIP1, and BAX proteins in treated cells would be expected to induce cell cycle arrest and apoptosis.

Although it could be argued that anyone of the two opposite responses to OA class compounds might simply reflect the outcome of a stochastic process, the bulk of the data summarized in previous sections would rather support the conclusion that the effects induced by DSP toxins are cell type specific, dose related, and timing sensitive.

These three features will certainly shape and permeate investigations on the molecular mechanisms of action of DSP toxins in the next decade. A general model to explain the dualistic effects of these agents is not available yet, but information is accumulating at an ever-increasing rate, and this will eventually lead us to understand the challenging issue of why some cells die while others shift to unrestrained proliferation after being in contact with DSP toxins.

ACKNOWLEDGMENTS

The research carried out by the author's group is supported by grants from Associazione Angela Serra per la Ricerca sul Cancro, Modena, Italy, the C.N.R., and M.U.R.S.T.

REFERENCES

1. BL Stoddard, H-P Biemann, DE Koshland Jr. Receptors and transmembrane signaling. Cold Spring Harb Symp Quant Biol 57:1–15, 1992.
2. CD Strader, TM Fong. MR Tota, D Underwood. Structure and function of G protein-coupled receptors. Annu Rev Biochem 63:101–132, 1994.
3. P van der Geer, T Hunter, RA Lindberg. Receptor protein-tyrosine kinases and their signal transduction pathways. Annu Rev Cell Biol 10:251–337, 1994.
4. DJ Mangelsdorf, C Thummel, M Beato, P Herrlich, G Schütz, K Umesono, B Blumberg, P Kastner, M Mark, P Chambon, RM Evans. The nuclear receptor superfamily: the second decade. Cell 83: 835–839, 1995.
5. HR Herschman. Primary response genes induced by growth factors and tumor promoters. Annu Rev Biochem. 60:281–319, 1991.
6. M-J Tsai, BW O'Malley. Molecular mechanisms of action of steroid/thyroid receptor superfamily members. Annu Rev Biochem 63:451–486, 1994.
7. R Seger, EG Krebs. The MAPK signaling cascade. FASEB J. 9:726–735, 1995.

8. T van Biesen, LM Luttrell, BE Hawes, RJ Lefkowitz. Mitogenic signaling via G protein-coupled receptors. Endocr Rev. 17:698–714, 1996.
9. K Malarkey, CM Belham, A Paul, A Graham, A McLees, PH Scott. The regulation of tyrosine kinase signalling pathways by growth factor and G-protein–coupled receptors. Biochem J. 309: 361–375, 1995.
10. JA Katzenellenbogen, BW O'Malley, BS Katzenellenbogen. Tripartite steroid hormone receptor pharmacology: interaction with multiple effector sites as a basis for the cell- and promoter-specific action of these hormones. Mol Endocrinol. 10:119–131, 1996.
11. CK Glass, DW Rose, MG Rosenfeld. Nuclear receptor coactivators. Curr Opin Cell Biol. 9:222–232, 1997.
12. GB Cohen, R Ren, D Baltimore. Modular binding domains in signal transduction proteins. Cell 80: 237–248, 1995.
13. RJ Davis. The mitogen-activated protein kinase signal transduction pathway. J Biol Chem. 268: 14553–14556, 1993.
14. MH Cobb, EJ Goldsmith. How MAP kinases are regulated. J Biol Chem. 270:14843–14846, 1995.
15. J Blenis. Signal transduction via the MAP kinases: proceed at your own RSK. Proc Natl Acad Sci USA 90:5889–5892, 1993.
16. BE Hawes, T van Biesen, WJ Koch, LM Luttrell, RJ Lefkowitz. Distinct pathways of G_i- and G_q-mediated mitogen-activated protein kinase activation. J Biol Chem 270:17148–17153, 1995.
17. DW Meek, AJ Street. Nuclear protein phosphorylation and growth control. Biochem J. 287:1–15, 1992.
18. K Buchner. Protein kinase C in the transduction of signals toward and within the cell nucleus. Eur J Biochem. 228:211–221, 1995.
19. T Hunter. Protein kinases and phosphatases: the yin and yang of protein phosphorylation and signalling. Cell 80:225–236, 1995.
20. G Walter, M Mumby. Protein serine/threonine phosphatases and cell transformation. Biochim Biophys Acta 1155:207–226, 1993.
21. C Bialojan, A Takai. Inhibitory effect of a marine-sponge toxin, okadaic acid, on protein phosphatases. Biochem J 256:283–290, 1988.
22. P Cohen, PTW Cohen. Protein phosphatases come of age. J Biol Chem 264:21435–21438, 1989.
23. S Nishiwaki, H Fujiki, M Suganuma, H Furuya-Suguri, R Matsushima, Y Iida, M Ojika, K Yamada, D Uemura, T Yasumoto, FJ Schmitz, T Sugimura. Carcinogenesis 11:1837–1841, 1990.
24. H Fujiki, M Suganuma, S Yoshizawa, S Nishiwaki, B Winyar, T Sugimura. Mechanisms of action of okadaic acid class tumor promoters on mouse skin. Environ Health Persp 93:211–214, 1991.
25. S Wera, BA Hemmings. Serine/threonine protein phosphatases. Biochem J 311:17–29, 1995.
26. M Nagao, H Shima, M Nakayasu, T Sugimura. Protein serine/threonine phosphatases as binding proteins for okadaic acid. Mut Res 333:173–179, 1995.
27. TJ Slaga, J DiGiovanni, LD Winberg, IV Budunova. Skin carcinogenesis: characteristics, mechanisms, and prevention. Progr Clin Biol Res 391:1–20, 1995.
28. H Fujiki, M Suganuma. Tumor promotion by inhibitors of protein phosphatases 1 and 2A: the okadaic acid class of compounds. Adv Cancer Res 61:143–194, 1993.
29. M Suganuma, H Fujiki, H Suguri, S Yoshizawa, M Hirota, M Nakayasu, M Ojika, K Wakamatsu, K Yamada, T Sugimura. Okadaic acid: an additional non-phorbol-12-tetradecanoate-13-acetate-type tumor promoter. Proc Natl Acad Sci USA 85:1768–1771, 1988.
30. Y Nishizuka. Studies and perspectives of protein kinase C. Science 233:305–312, 1986.
31. H Hug, TF Sarre. Protein kinase C isoenzymes: divergence in signal transduction? Biochem J 291: 329–343, 1993.
32. AC Newton. Protein kinase C: structure, function, and regulation. J Biol Chem 270:28495–28498, 1995.
33. H Fujiki, M Suganuma, H Suguri, S Yoshizawa, K Takagi, N Uda, K Wakamatsu, K Yamada, M Murata, T Yasumoto, T Sugimura. Diarrhetic shellfish toxin, dinophysistoxin-1, is a potent tumor promoter on mouse skin. Jpn J Cancer Res 79:1089–1093, 1988.
34. M Suganuma, H Fujiki, H Furuya-Suguri, S Yoshizawa, S Yasumoto, Y Kato, N Fusetani, T Sugi-

mura. Calyculin A, an inhibitor of protein phosphatases, a potent tumor promoter on CD-1 mouse skin. Cancer Res 50:3521–3525, 1990.
35. T Ohta, E Sueoka, N Iida, A Komori, M Suganuma, R Nishiwaki, M Tatematsu, S-J Kim, WW Carmichael, H Fujiki. Nodularin, a potent inhibitor of protein phosphatases 1 and 2A, is a new environmental carcinogen in male F344 rat liver. Cancer Res 54:6402–6406, 1994.
36. M Suganuma, S Okabe, E Sueoka, R Nishiwaki, A Komori, N Uda, K Isono, H Fujiki. Tautomycin: an inhibitor of protein phosphatases 1 and 2A but not a tumor promoter on mouse skin and in rat glandular stomach. Cancer Res Clin Oncol 121:621–627, 1995.
37. M Suganuma, M Tatematsu, J Yatsunami, S Yoshizawa, S Okabe, D Uemura, H Fujiki. An alternative theory of tissue specificity by tumor promotion of okadaic acid in glandular stomach of SD rats. Carcinogenesis 13:1841–1845, 1992.
38. MT Runnegar, IR Falconer, J Silver. Deformation of isolated rat hepatocytes by a peptide hepatotoxin from the blue-green alga *Microcystis aeruginosa*. Naunyn-Schmiedeberg's Arch Pharmacol 317:268–272, 1981.
39. R Nishiwaki-Matsushima, T Ohta, S Nishiwaki, M Suganuma, K Kohyama, T Ishikawa, WW Carmichael, H Fujiki. Liver tumor promotion by the cyanobacterial cyclic peptide toxin microcystin-LR. J Cancer Res Clin Oncol 118:420–424, 1992.
40. F Katoh, DJ Fitzgerald, L Giroldi, H Fujiki, T Sugimura, H Yamasaki. Okadaic acid and phorbol esters: comparative effects of these tumor promoters on cell transformation, intercellular communication and differentiation *in vitro*. Jpn J Cancer Res 81:590–597, 1990.
41. A Sakai, H Fujiki. Promotion of BALB/3T3 cell transformation by the okadaic acid class of tumor promoters, okadaic acid and dinophysistoxin-1. Jpn J Cancer Res 82:518–523, 1991.
42. SH Yuspa, MC Poirier. Chemical carcinogenesis: from animal models to molecular models in one decade. Adv Cancer Res 50:25–70, 1988.
43. H Hennings, VA Robinson, DM Michael, GR Pettit, R Jung, SH Yuspa. Development of an *in vitro* analogue of initiated mouse epidermis to study tumor promoters and antipromoters. Cancer Res 50:4794–4800, 1990.
44. JE McClure, CA Noonan, WT Shearer. Okadaic acid inhibits dephosphorylation of cytoplasmic p53 during lymphocyte activation. Biochem Biophys Res Commun 197:1578–1584, 1993.
45. AM Nakhla, W Rosner. Stimulation of prostate cancer growth by androgens and estrogens through the intermediacy of sex hormone-binding globulin. Endocrinology 137:4126–4129, 1996.
46. CA Afshari, JC Barrett. Disruption of G_0-G_1 arrest in quiescent and senescent cells treated with phosphatase inhibitors. Cancer Res 54:2317–2321, 1994.
47. R Sakai, I Ikeda, H Kitani, H Fujiki, F Takaku, U Rapp, T Sugimura, M Nagao. Flat reversion by okadaic acid of *raf* and *ret*-II transformants. Proc Natl Acad Sci USA 86:9946–9950, 1989.
48. RW Gupta, CK Joseph, DA Foster. v-*src*-induced transformation is inhibited by okadaic acid. Biochem Biophys Res Commun 196:320–327, 1993.
49. SE Adunyah, TM Unlap, CC Franklin, AS Kraft. Induction of differentiation and c-*jun* expression in human leukemic cells by okadaic acid, an inhibitor of protein phosphatases. J Cell Physiol 151: 415–426, 1992.
50. K Kiguchi, C Giometti, CH Chubb, H Fujiki, E Huberman. Differentiation induction in human breast tumor cells by okadaic acid and related inhibitors of protein phosphatases 1 and 2A. Biochem Biophys Res Commun 189:1261–1267, 1992.
51. K Kiguchi, D Glesne, CH Chubb, H Fujiki, E Hubermann. Differential induction of apoptosis in human breast tumor cells by okadaic acid and related inhibitors of protein phosphatases 1 and 2A. Cell Growth Differ 5:995–1004, 1994.
52. Y Yan, JW Shay, WE Wright, MC Mumby. Inhibition of protein phosphatase activity induces p53-dependent apoptosis in the absence of p53 transactivation. J Biol Chem 272:15220–15226, 1997.
53. Q Song, GD Baxter, EM Kovacs, D Findik, MF Lavin. Inhibition of apoptosis in human tumour cells by okadaic acid. J Cell Physiol 153:550–556, 1992.
54. DL Coppock, JB Tansey, L Nathanson. 12-*O*-tetradecanoylphorbol-13-acetate induces transient cell cycle arrest in G_1 and G_2 in metastatic melanoma cells: inhibition of phosphorylation of $p34^{cdc2}$. Cell Growth Differ 3:485–494, 1992.

55. A Schönthal, JR Feramisco. Inhibition of histone H1 kinase expression, retinoblastoma protein phosphorylation, and cell proliferation by the phosphatase inhibitor okadaic acid. Oncogene 8:433–441, 1993.
56. HR Herschman, RW Lim, DW Brankow, H Fujiki. The tumor promoters 12-*O*-tetradecanoylphorbol-13-acetate and okadaic acid differ in toxicity, mitogenic activity and induction of gene expression. Carcinogenesis 10:1495–1498, 1989.
57. NM Dean, LJ Mordan, K Ise, SI Mooberry, AL Boynton. Okadaic acid inhibits PDGF-induced proliferation and decreases PDGF receptor number in C3H/10T1/2 mouse fibroblasts. Carcinogenesis 12:665–670, 1991.
58. T-A Kim, BR Velasquez, CE Wenner. Okadaic acid regulation of the retinoblastoma gene product is correlated with the inhibition of growth factor-induced cell proliferation in mouse fibroblasts. Proc Natl Acad Sci USA 90:5460–5463, 1993.
59. A Simm, T-A Karenberg, V Hoppe, J Hoppe. Okadaic acid bloks PDGF-induced proliferation of AKR-2B fibroblasts at the transition from G_1- to S-phase. Exp Cell Res 210:160–165, 1994.
60. E Rivedal, S-O Mikalsen, T Sanner. The non-phorbol ester tumor promoter okadaic acid does not promote morphological transformation or inhibit junctional communication in hamster embryo cells. Biochem Biophys Res Commun 167:1302–1308, 1990.
61. GA Preston, JC Barrett, JA Biermann, E Murphy. Effects of alterations in calcium homeostasis on apoptosis during neoplastic progression. Cancer Res 57:537–542, 1997.
62. R Bøe, BT Gjertsen, OK Vintermyr, G Houge, M Lanotte, SO Døskeland. The protein phosphatase inhibitor okadaic acid induces morphological changes typical of apoptosis in mammalian cells. Exp Cell Res 195:237–246, 1991.
63. F Tergau, J Weichert, I Quentin, R Optiz, C von Zezschwitz, J Marwitz, V Ritz, HJ Steinfelder. Inhibitors of ser/thr phosphatases 1 and 2A induce apoptosis in pituitary GH_3 cells. Naunyn-Schmiedeberg's Arch Pharmacol 356:8–16, 1997.
64. J You, RC Bird. Selective induction of cell cycle regulatory genes cdk1 ($p34^{cdc2}$), cyclins A/B, and the tumor suppressor gene Rb in transformed cells by okadaic acid. J Cell Physiol 164:424–433, 1995.
65. Y Ishida, Y Furukawa, JA Decaprio, M Saito, JD Griffin. Treatment of myeloid leukemic cells with the phosphatase inhibitor okadaic acid induces cell cycle arrest at either G1/S or G2/M depending on dose. J Cell Physiol 150:484–492, 1992.
66. K Sakurada, B Zheng, JF Kuo. Comparative effects of protein phosphatase inhibitors (okadaic acid and calyculin A) on human leukemia HL60, HL60/ADR and K562 cells. Biochem Biophys Res Commun 187:488–492, 1992.
67. A Benito, A Lerga, M Silva, J Leon, JL Fernandez-Luna. Apoptosis of human myeloid leukemia cells induced by an inhibitor of protein phosphatases (okadaic acid) is prevented by Bcl-2 and Bcl-X_L. Leukemia 11:940–944, 1997.
68. B Zheng, CF Woo, JF Kuo. Mitotic arrest and enhanced nuclear protein phosphorylation in human leukemia K562 cells by okadaic acid, a potent protein phosphatase inhibitor and tumor promoter. J Biol Chem 266:10031–10034, 1991.
69. B Zheng, TC Chambers, RL Raynor, PN Markham, HM Gebel, WR Vogler, JF Kuo. Human leukemia K562 cell mutant (K562/OA200) selected for resistance to okadaic acid (protein phosphatase inhibitor) lacks protein kinase C-ε, exhibits multidrug resistance phenotype, and expresses drug pump P-glycoprotein. J Biol Chem 269:12332–12338, 1994.
70. A Lerga, B Belandia, MD Delgado, MA Cuadrado, C Richard, JM Ortiz, J Martín-Perez, J León. Down-regulation of c-*myc* and *max* genes is associated to inhibition of protein phosphatase 2A in K562 human leukemia cells. Biochem Biophys Res Commun 215:889–895, 1995.
71. N Canelles, MD Delgado, KM Hyland, A Lerga, C Richard, CV Dang, J Leon. Max and inhibitory c-Myc mutants induce erythroid differentiation and resistance to apoptosis in human myeloid leukemia cells. Oncogene 14:1315–1327, 1997.
72. H Hennings, DT Lowry, VA Robinson, DL Morgan, H Fujiki, SH Yuspa. Activity of diverse tumor promoters in a keratinocyte co-culture model of initiated epidermis. Carcinogenesis 13:2145–2151, 1992.
73. Y Churcher, SE Moss. Inhibition of interleukin-2 mediated DNA synthesis in activated human T-

lymphoblasts by okadaic acid is accompanied by hyperphosphorylation of *lck*. Biochem. J 308: 777–783, 1995.
74. P Rieckmann, C Thévenin, JH Kehrl. Okadaic acid is a potent inducer of AP-1, NF-*k*B, and tumor necrosis factor-α in human B lymphocytes. Biochem Biophys Res Commun 187:51–57, 1992.
75. AM Casillas, K Amaral, S Chegini-Farahani, AE Nel. Okadaic acid activates p42 mitogen-activated protein kinase (MAP kinase; ERK-2) in B-lymphocytes but inhibits rather than augments cellular proliferation: contrast with phorbol 12-myristate 13-acetate. Biochem. J 290:545–550, 1993.
76. MS Sheikh, M Garcia, Q Zhan, Y Liu, AJ Fornace Jr. Cell-cycle independent regulation of $p21^{Waf1/Cip1}$ and retinoblastoma protein during okadaic acid–induced apoptosis is coupled with induction of Bax protein in human breast carcinoma cells. Cell Growth Differ. 7:1599–1607, 1996.
77. GP Rossini, C Pinna, R Viviani. Inhibitors of phosphoprotein phosphatases 1 and 2A cause activation of a 53 kDa protein kinase accompanying the apoptotic response of breast cancer cells. FEBS Lett. 410:347–350, 1997.
78. SJ Morana, CM Wolf, J Li, JE Reynolds, MK Brown, A Eastman. The involvement of protein phosphatases in the activation of ICE/CED-3 protease, intracellular acidification, DNA digestion, and apoptosis. J Biol Chem 271:18263–18271, 1996.
79. CM Cagnili, E Kharlamov, C Atabay, T Uz, H Manev. Apoptosis induced in neuronal cultures by either the phosphatase inhibitor okadaic acid or the kinase inhibitor staurosporine is attenuated by isoquinolinesulfonamides H-7, H-8, and H-9. J Mol Neurosci 7:65–76, 1996.
80. MA Davis, SH Chang, MW Smith, BF Trump. Morphologic and biochemical characterization of okadaic acid induced apoptosis in NRK-52E cells. Toxicol. Pathol 22:595–605, 1994.
81. MA Davis, SH Chang, BF Trump. Differential sensitivity of normal and H-*ras* oncogene-transformed rat kidney epithelial cells to okadaic acid-induced apoptosis. Toxicol Appl Pharmacol 141: 93–101, 1996.
82. C Haby, D Aunis, J Zwiller. Okadaic acid induces activator protein 1 activity and immediate early gene transcription in rat pheochromocytoma cells. Biochem Pharmacol 48:819–825, 1994.
83. Y Morimoto, T Ohba, S Kobayashi, T Haneji. The protein phosphatase inhibitors okadaic acid and calyculin A induce apoptosis in human osteoblastic cells. Exp Cell Res 230:181–186, 1997.
84. S Kharbands, R Datta, E Rubin, T Nakamura, R Hasse, D Kufe. Regulation of *c-jun* expression during induction of monocytic differentiation by okadaic acid. Cell Growth Differ. 3:391–399, 1992.
85. JE Eriksson, GIL Paatero, JAO Meriluoto, GA Codd, GEN Kass, P Nicotera, S Orrenius. Rapid microfilament reorganization induced in isolated rat hepatocytes by microcystin-LR, a cyclic peptide toxin. Exp Cell Res 185:86–100, 1989.
86. V Niggli, H Keller. Low concentrations of the phosphatase inhibitor okadaic acid stop tumor cell locomotion. Eur J Pharmacol. 324:99–108, 1997.
87. M Macias-Silva, JA Garcia-Sainz. Inhibition of hormone-stimulated inositol phosphate production and disruption of cytoskeletal structure. Effects of okadaic acid, microcystin, chloropromazine, W7 and nystatin. Toxicon 32:105–112, 1994.
88. M Serres, C Grangeasse, M Haftek, Y Durocher, B Duclos, D Schmitt. Hyperphosphorylation of β-catenin on serine-threonine residues and loss of cell-cell contacts induced by calyculin A and okadaic acid in human epidermal cells. Exp Cell Res 231:163–172, 1997.
89. GM Hatch, Y Tsukitani, DE Vance. The protein phosphatase inhibitor, okadaic acid, inhibits phosphatidylcholine biosynthesis in isolated rat hepatocytes. Biochim Biophys Acta 1081:25–32, 1991.
90. I Chackalaparampil, S Bagrodia, D Shalloway. Tyrosine dephosphorylation of $pp60^{c\text{-}src}$ is stimulated by a serine/threonine phosphatase inhibitor. Oncogene 9:1947–1955, 1994.
91. K Ajiro, K Yoda, K Utsumi, Y Nishikawa. Alteration of cell cycle-dependent histone phosphorylations by okadaic acid. J Biol Chem 271:13197–13201, 1996.
92. GD Baxter, MF Lavin. Specific protein dephosphorylation in apoptosis induced by ionizing radiation and heat shock in human lymphoid tumor lines. J Immunol 148:1949–1954, 1992.
93. Q Song, MF Lavin. Calyculin A, a potent inhibitor of phosphatases-1 and -2A, prevents apoptosis. Biochem Biophys Res Commun 190:47–55, 1993.
94. JR Paulson, WA Ciesielski, BR Schram and PW Mesner. Okadaic acid induces dephosphorylation of histone H1 in metaphase-arrested HeLa cells. J Cell Sci 107:267–273, 1994.

95. G Gurland, GG Gundersen. Protein phosphatase inhibitors induce the selective breakdown of stable microtubules in fibroblasts and epithelial cells. Proc Natl Acad Sci USA 90:8827–8831, 1993.
96. SE Merrick, JQ Trojanowski, VM Lee. Selective destruction of stable microtubules and axons by inhibitors of protein serine/threonine phosphatases in cultured human neurons. J Neurosci 17:5726–5737, 1997.
97. JE Eriksson, DL Brautigan, R Vallee, J Olmsted, H Fujiki, RD Goldman. Cytoskeletal integrity in interphase cells requires protein phosphatase activity. Proc Natl Acad Sci USA 89:11093–11097, 1992.
98. J Yatsunami, A Komori, T Ohta, M Suganuma, SH Yuspa, H Fujiki. Hyperphosphorylation of cytokeratins by okadaic acid class tumor promoters in primary human keratinocytes. Cancer Res 53:992–996, 1993.
99. K Kasahara, T Kartasova, X-Q Ren, T Ikuta, K Chida, T Kuroki. Hyperphosphorylation of keratins by treatment with okadaic acid of BALB/MK-2 mouse keratinocytes. J Biol Chem 268:23531–23537, 1993.
100. H Blankson, I Holen, PO Seglen. Disruption of the cytokeratin cytoskeleton and inhibition of hepatocytic autophagy by okadaic acid. Exp Cell Res 218:522–530, 1995.
101. JFR Kerr. Shrinkage necrosis: a distinct mode of cellular death. J Pathol 105:13–20, 1971.
102. JFR Kerr, AH Wyllie, AR Currie. Apoptosis: a basic biological phenomenon with wide ranging implications in tissue kinetics. Br J Cancer 26:1790–1794, 1972.
103. BT Gjertsen, SO Døskeland. Protein phosphorylation in apoptosis. Biochim Biophys Acta 1269: 187–199, 1995.
104. MF Lavin, D Watters, Q Song. Role of protein kinase activity in apoptosis. Experientia 52:979–994, 1996.
105. Y Ohoka, Y Nakai, M Mukai, M Iwata. Okadaic acid inhibits glucocorticoid-induced apoptosis in T cell hybridomas at its late stage. Biochem Biophys Res Commun 197:916–921, 1993.
106. QP Dou, B An, PL Will. Induction of a retinoblastoma phosphatase activity by anticancer drugs accompanies p53-independent G_1 arrest and apoptosis. Proc Natl Acad Sci USA 92:9019–9023, 1995.
107. TAJ Haystead, ATR Sim, D Carling, RC Honnor, Y Tsukitani, P Cohen, DG Hardie. Effects of the tumor promoter okadaic acid on intracellular protein phosphorylation and metabolism. Nature 337:78–81, 1989.
108. L Agius, M Peak. Interactions of okadaic acid with insulin action in hepatocytes: role of protein phosphatases in insulin action. Biochim Biophys Acta 1095:243–248, 1991.
109. TW Sturgill, LB Ray, E Erikson, JL Maller. Insulin-stimulated MAP-2 kinase phosphorylates and activates ribosomal protein S6 kinase II. Nature 334:715–718, 1988.
110. T Miyasaka, J Miyasaka, AR Saltiel. Okadaic acid stimulates the activity of microtubule associated protein kinase in PC-12 pheochromocytoma cells. Biochem Biophys Res Commun 168:1237–1243, 1990.
111. TAJ Haystead, JE Weiel, DW Litchfield, Y Tsukitani, EH Fisher, EG Krebs. Okadaic acid mimics the action of insulin in stimulating protein kinase activity in isolated adipocytes. J Biol Chem 265: 16571–16580, 1990.
112. MC Amaral, AM Casillas, AE Nel. Contrasting effects of two tumour promoters, phorbol myristate acetate and okadaic acid, on T-cell responses and activation of p42 MAP-kinase/ERK-2. Immunology 79:24–31, 1993.
113. E Sontag, S Fedorov, C Kamibayashi, D Robbins, M Cobb, M Mumby. The interaction of SV40 small tumor antigen with protein phosphatase 2A stimulates the MAP kinase pathway and induces cell proliferation. Cell 75:887–897, 1993.
114. Z-H Qiu, CC Leslie. Protein kinase C–dependent and–independent pathways of mitogen-activated protein kinase activation in macrophages by stimuli that activate phospholipase A_2. J Biol Chem 269:19480–19487, 1994.
115. DR Alessi, N Gomez, G Moorhead, T Lewis, SM Keyse, P Cohen. Inactivation of p42 MAP kinase by protein phosphatase 2A and a protein tyrosine phosphatase, but not CL100, in various cell lines. Curr Biol 5:283–295, 1995.
116. SJ Persaud, CPD Wheeler-Jones, PM Jones. The mitogen-activated protein kinase pathway in rat

islets of Langerhans: studies on the regulation of insulin secretion. Biochem J 313:119–124, 1996.
117. Y Sonoda, T Kasahara, Y Yamaguchi, K Kuno, K Matsushima, N Mukaida. Stimulation of interleukin-8 production by okadaic acid and vanadate in a human promyelocyte cell line, an HL-60 subline. Possible role of mitogen-activated protein kinase on the okadaic acid-induced NF-kB activation. J Biol Chem 272:15366–15372, 1997.
118. Z Xia, M Dickens, J Raingeaud, RJ Davis, ME Greenberg. Opposing effects of ERK and JNK-p38 MAP kinases on apoptosis. Science 270:1326–1331, 1995.
119. JM Kyriakis, J Avruch. Sounding the alarm: protein kinase cascades activated by stress and inflammation. J Biol Chem 271:24313–24316, 1996.
120. ET Kipreos, JYJ Wang. Differential phosphorylation of c-Abl in cell cycle determined by *cdc2* kinase and phosphatase activity. Science 248:217–220, 1990.
121. T Erpel, SA Courtneidge. Src family protein tyrosine kinases and cellular signal transduction pathways. Curr Opin Cell Biol 7:176–182, 1995.
122. C Norbury, P Nurse. Animal cell cycles and their control. Annu Rev Biochem 61:441–470, 1992.
123. M Dorée, S Galas. The cyclin-dependent protein kinases and the control of cell division. FASEB J 8:1114–1121, 1994.
124. DO Morgan. Principles of CDK regulation. Nature 374:131–134, 1995.
125. M-A Félix, P Cohen, E Karsenti. Cdc2 H1 kinase is negatively regulated by a type 2A phosphatase in the *Xenopus* early embryonic cell cycle: evidence from the effects of okadaic acid. EMBO J 9: 675–683, 1990.
126. MJ Solomon, M Glotzer, TH Lee, M Philippe, MW Kirschner. Cyclin activation of $p34^{cdc2}$. Cell 63:1013–1024, 1990.
127. TH Lee, MJ Solomon, MC Mumby, MW Kirschner. INH, a negative regulator of MPF, is a form of protein phosphatase 2A. Cell 64:415–423, 1991.
128. J Gautier, MJ Solomon, RN Booher, JF Bazan, MW Kirschner. Cdc25 is a specific tyrosine phosphatase that directly activates $p34^{cdc2}$. Cell 67:197–211, 1991.
129. C Jessus, H Rime, O Haccard, J Van Lint, J Goris, W Merlevede, R Ozon. Tyrosine phosphorylation of $p34^{cdc2}$ and p42 during meiotic maturation of *Xenopus* oocyte. Antagonistic action of okadaic acid and 6-DMAP. Development 111:813–820, 1991.
130. A Kumagai, WG Dunphy. Regulation of the cdc25 protein during the cell cycle in *Xenopus* extracts. Cell 70:139–151, 1992.
131. A Devault, D Fesquet, J-C Cavadore, A-M Garrigues, J-C Labbé, T Lorca, A Picard, M Philippe, M Dorée. Cyclin A potentiates maturation-promoting factor activation in the early *Xenopus* embryo via inhibition of the tyrosine kinase that phosphorylates CDC2. J Cell Biol 118:1109–1120, 1992.
132. H Rime, C Jessus, R Ozon. Tyrosine phosphorylation of $p34^{cdc2}$ is regulated by protein phosphatase 2a in growing immature *Xenopus* oocytes. Exp Cell Res 219:29–38, 1995.
133. SM Keyse. An emerging family of dual specificity MAP kinase phosphatases. Biochim Biophys Acta 1265:152–160, 1995.
134. RA Hipskind, M Baccarini, A Nordheim. Transient activation of RAF-1, MEK, and ERK2 coincides kinetically with ternary complex factor phosphorylation and immediate-early gene promoter activity in vivo. Mol Cell Biol 14:6219–6231, 1994.
135. R Zinck, RA Hipskind, V Pingoud, A Nordheim. *c-fos* transcriptional activation and repression correlate temporally with the phosphorylation status of TCF. EMBO J 12:2377–2387, 1993.
136. M Karin. The regulation of AP-1 activity by mitogen-activated protein kinases. J Biol Chem 270: 16483–16486, 1995.
137. KR Yamamoto, BM Alberts. Steroid receptors: elements for modulation of eukaryotic transcription. Annu Rev Biochem 45:721–746, 1976.
138. T Hunter, M Karin. The regulation of transcription by phosphorylation. Cell 70:375–387, 1992.
139. PK Vogt, TJ Bos. *jun*: Oncogene and transcription factor. Adv. Cancer Res 55:1–35, 1990.
140. RJ Distel, BM Spiegelman. Protooncogene *c-fos* as a transcription factor. Adv Cancer Res 55:37–55, 1990.
141. A Minden, A Lin, M McMahon, C Lange-Carter, B Dérijard, RJ Davis, GL Johnson, M Karin.

Differential activation of ERK and JNK mitogen-activated protein kinases by Raf-1 and MEKK. Science 266:1719–1723, 1994.
142. I Sánchez, RT Hughes, BJ Mayer, K Yee, JR Woodgett, J Avruch, JM Kyriakis, LI Zon. Role of SAPK/ERK kinase-1 in the stress-activated pathway regulating transcription factor c-Jun. Nature 372:794–798, 1994.
143. M Yan, T Dai, JC Deak, JM Kyriakis, LI Zon, JR Woodgett, DJ Templeton. Activation of stress-activated protein kinase by MEKK1 phosphorylation of its activator SEK1. Nature 372:798–800, 1994.
144. P Angel, K Hattori, T Smeal, M Karin. The *jun* proto-oncogene is positively autoregulated by its product, Jun/AP-1. Cell 55:875–885, 1988.
145. P Sassone-Corsi. Transcription factors responsive to cAMP. Annu Rev Cell Dev Biol 11:355–377, 1995.
146. M Hagiwara, A Alberts, P Brindle, J Meinkoth, J Feramisco, T Deng, M Karin, S Shenolikar, M Montminy. Transcriptional attenuation following cAMP induction requires PP-1–mediated dephosphorylation of CREB. Cell 70:105–113, 1992.
147. BE Wadzinski, WH Wheat, S Jaspers, LF Peruski Jr, RL Lickteig, GL Johnson, DJ Klemm. Nuclear protein phosphatase 2A dephosphorylates protein kinase A–phosphorylated CREB and regulates CREB transcriptional stimulation. Mol Cell Biol 13:2822–2834, 1993.
148. M Yoshizumi, H Wang, C-M Hsieh, NES Sibinga, MA Perrella, M-E Lee. Down-regulation of the cyclin A promoter by transforming growth factor-β1 is associated with reduction in phosphorylated activating transcription factor-1 and cyclic AMP–responsive element–binding protein. J Biol Chem 272:22259–22264, 1997.
149. RA Weinberg. The retinoblastoma gene and cell growth control. Trends Biochem Sci 15:199–202, 1990.
150. D Sidransky, M Hollstein. Clinical implications of the p53 gene. Annu Rev Med 47:285–301, 1996.
151. E Yonish-Rouach. The p53 tumor suppressor gene: a mediator of a G1 growth arrest and of apoptosis. Experientia 52:1001–1007, 1996.
152. Y Taya. RB kinases and RB-binding proteins: new points of view. Trends Biochem Sci 22:14–17, 1997.
153. SJ Ullrich, CW Anderson, WE Mercer, E Appella. The p53 tumor suppressor protein, a modulator of cell proliferation. J Biol Chem 267:15259–15262, 1992.
154. KH Scheidtmann, MC Mumby, K Rundell, G Walter. Dephosphorylation of simian virus 40 large-T antigen and p53 protein by protein phosphatase 2A: inhibition by small-t antigen. Mol Cell Biol 11:1996–2003, 1991.
155. I Takenaka, F Morin, BR Seizinger, N Kley. Regulation of the sequence-specific DNA binding function of p53 by protein kinase C and protein phosphatases. J Biol Chem 270:5405–5411, 1995.
156. GR Guy, X Cao, SP Chua, YH Tan. Okadaic acid mimics multiple changes in early protein phosphorylation and gene expression induced by tumor necrosis factor or interleukin-1. J Biol Chem 267:1846–1852, 1992.
157. J Yatsunami, A Komori, T Ohta, M Suganuma, H Fujiki. Hyperphosphorylation of retinoblastoma protein and p53 by okadaic acid, a tumor promoter. Cancer Res 53:239–241, 1993.
158. W Zhang, C McClain, J-P Gau, X-YD Guo, AB Deisseroth. Hyperphosphorylation of p53 induced by okadaic acid attenuates its transcriptional activation function. Cancer Res 54:4448–4453, 1994.
159. B Fuchs, D Hecker, KH Scheidtmann. Phosphorylation studies on rat p53 using the baculovirus expression system. Eur J Biochem 228:625–639, 1995.
160. M Lohrum, KH Scheidtmann. Differential effects of phosphorylation of rat p53 on transactivation of promoters derived from different p53 responsive genes. Oncogene 13:2527–2539, 1996.
161. DB DeFranco, M Qi, KC Borror, MJ Garabedian, DL Brautigan. Protein phosphatase types 1 and/or 2A regulate nucleocytoplasmic shuttling of glucocorticoid receptors. Mol Endocrinol 5:1215–1228, 1991.
162. ML Moyer, KC Borror, BJ Bona, DB DeFranco, SK Nordeen. Modulation of cell signaling pathways can enhance or impair glucocorticoid-induced gene expression without altering the state of receptor phosphorylation. J Biol Chem 268:22933–22940, 1993.

163. HM Darwish, JK Burmester, VE Moss, HF DeLuca. Phosphorylation is involved in transcriptional activation by the 1,25-dihydroxyvitamin D_3 receptor. Biochim Biophys Acta 1167:29–36, 1993.
164. K-H Lin, K Ashizawa, S-Y Cheng. Phosphorylation stimulates the transcriptional activity of the human β1 thyroid hormone nuclear receptor. Proc Natl Acad Sci USA 89:7737–7741, 1992.
165. MK Bhat, K Ashizawa, S-H Cheng. Phosphorylation enhances the target gene sequence-dependent dimerization of thyroid hormone receptor with retinoid X receptor. Proc Natl Acad Sci USA 91: 7927–7931, 1994.
166. LA Denner, NL Weigel, BL Maxwell, WT Schrader, BW O'Malley. Regulation of progesterone receptor-mediated transcription by phosphorylation. Science 250:1740–1743, 1990.
167. JP Somers, DB DeFranco. Effects of okadaic acid, a protein phosphatase inhibitor, on glucocorticoid receptor-mediated enhancement. Mol Endocrinol 6:26–34, 1992.
168. CA Beck, NL Weigel, DP Edwards. Effects of hormone and cellular modulators of protein phosphorylation on transcriptional activity, DNA binding, and phosphorylation of human progesterone receptors. Mol Endocrinol 6:607–620, 1992.
169. CA Beck, NL Weigel, ML Moyer, SK Nordeen, DP Edwards. The progesterone antagonist RU486 acquires agonist activity upon stimulation of cAMP signaling pathways. Proc Natl Acad Sci USA 90:4441–4445, 1993.
170. SMI Kazmi, V Visconti, RK Plante, A Ishaque, C Lau. Differential regulation of human progesterone receptor A and B form–mediated *trans*-activation by phosphorylation. Endocrinology 133: 1230–1238, 1993.
171. Y Zhang, W Bai, VE Allgood, NL Weigel. Multiple signaling pathways activate the chicken progesterone receptor. Mol Endocrinol 8:577–584, 1994.
172. KE Jones, JH Brubaker, WW Chin. Evidence that phosphorylation events participate in thyroid hormone action. Endocrinology 134:543–548, 1994.
173. SK Nordeen, ML Moyer, BJ Bona. The coupling of multiple signal transduction pathways with steroid response mechanisms. Endocrinology 134:1723–1732, 1994.
174. T Ikonen, JJ Palvimo, PJ Kallio, P Reinikainen, OA Jänne. Stimulation of androgen-regulated transactivation by modulators of protein phosphorylation. Endocrinology 135:1359–1366, 1994.
175. T Matkovits, S Christakos. Ligand occupancy is not required for vitamin D receptor and retinoid receptor-mediated transcriptional activation. Mol Endocrinol 9:232–242, 1995.
176. P Lefebvre, M-P Gaub, A Tahayato, C Rochette-Egly, P Formstecher. Protein phosphatases 1 and 2A regulate the transcriptional and DNA binding activities of retinoic acid receptors. J Biol Chem 270:10806–10816, 1995.
177. NL Weigel. Steroid hormone receptors and their regulation by phosphorylation. Biochem J 319: 657–667, 1996.
178. J Yatsunami, H Fujiki, M Suganuma, S Yoshizawa, JE Eriksson, MOJ Olson, RD Goldman. Vimentin is hyperphosphorylated in primary human fibroblasts treated with okadaic acid. Biochem Biophys Res Commun 177:1165–1170, 1991.
179. B Favre, P Turowski, BA Hemmings. Differential inhibition and posttranslational modification of protein phosphatase 1 and 2A in MCF7 cells treated with calyculin-A, okadaic acid, and tautomycin. J Biol Chem 272:13856–13863, 1997.
180. E Fuchs, K Weber. Intermediate filaments: structure, dynamics, function, and disease. Annu Rev Biochem 63:345–382, 1994.
181. A Komori, J Yatsunami, M Suganuma, S Okabe, S Abe, A Sakai, K Sasaki, H Fujiki. Tumor necrosis factor acts as a tumor promoter in BALB/3T3 cell transformation. Cancer Res 53:1982–1985, 1993.
182. DT Ho, M Roberge. The antitumor drug fostriecin induces vimentin hyperphosphorylation and intermediate filament reorganization. Carcinogenesis 17:967–972, 1996.
183. LC Mahadevan, AC Willis, MJ Barratt. Rapid histone H3 phosphorylation in response to growth factors, phorbol esters, okadaic acid, and protein synthesis inhibitors. Cell 65:775–783, 1991.
184. JR Paulson, JS Patzlaff, AJ Vallis. Evidence that the endogenous histone H1 phosphatase in HeLa mitotic chromosomes is protein phosphatase 1, not protein phosphatase 2A. J Cell Sci 109:1437–1447, 1996.

185. SY Roth, CD Allis. Chromatin condensation: does histone H1 dephosphorylation play a role? Trends Biochem. Sci 17:93–98, 1992.
186. SM Hernandez-Sotomayor, M Mumby, G Carpenter. Okadaic acid–induced hyperphosphorylation of the epidermal growth factor receptor. Comparison with receptor phosphorylation and functions affected by another tumor promoter, 12-*O*-tetradecanoylphorbol-13-acetate. J Biol Chem 266: 21281–21286, 1991.
187. Z-H Qiu, MS de Carvalho, CC Leslie. Regulation of phospholipase A_2 activation by phosphorylation in mouse peritoneal macrophages. J Biol Chem 268:24506–24513, 1993.
188. X-X Xu, CO Rock, Z-H Qiu, CC Leslie, S Jackowski. Regulation of cytosolic phospholipase A_2 phosphorylation and eicosanoid production by colony-stimulating factor 1. J Biol Chem 269:31693–31700, 1994.
189. G Sa, G Murugesan, M Jaye, Y Ivashchenko, PL Fox. Activation of cytosolic phospholipase A_2 by basic fibroblast growth factor via a p42 mitogen-activated protein kinase-dependent phosphorylation pathway in endothelial cells. J Biol Chem 270:2360–2366, 1995.
190. Y Nishizuka. Intracellular signaling by hydrolysis of phospholipids and activation of protein kinase C. Science 258:607–614, 1992.
191. WH Moolenaar. Lysophosphatidic acid, a multifunctional phospholipid messenger. J Biol Chem 270:12949–12952, 1995.
192. T Sassa, WW Richter, N Uda, M Suganuma, H Suguri, S Yoshizawa, M Hirota, H Fujiki. Apparent ''activation'' of protein kinases by okadaic acid class tumor promoters. Biochem Biophys Res Commun 159:939–944, 1989.
193. Y Gotoh, E Nishida, H Sakai. Okadaic acid activates microtubule-associated protein kinase in quiescent fibroblastic cells. Eur J Biochem 193:671–674, 1990.
194. KE Meier, KA Licciardi, TAJ Haystead, EG Krebs. Activation of messenger-independent protein kinases in wild-type and phorbol ester-resistant EL4 thymoma cells. J Biol Chem 266:1914–1920, 1991.
195. KC Gause, MK Homma, KA Licciardi, R Seger, NG Ahn, MJ Peterson, EG Krebs, KE Meier. Effects of phorbol ester on mitogen-activated protein kinase kinase activity in wild-type and phorbol ester–resistant EL4 thymoma cells. J Biol Chem 268:16124–16129, 1993.
196. DJ Withers, SR Bloom, E Rozengurt. Dissociation of cAMP-stimulated mitogenesis from activation of the mitogen-activated protein kinase cascade in Swiss 3T3 cells. J Biol Chem 270:21411–21419, 1995.
197. Y Chatani, S Tanimura, N Miyoshi, A Hattori, M Sato, M Kohno. Cell type-specific modulation of cell growth by transforming growth factor $\beta 1$ does not correlate with mitogen-activated protein kinase activation. J Biol Chem 270:30686–30692, 1995.
198. N Chajry, P-M Martin, C Cochet, Y Berthois. Regulation of p42 mitogen-activated–protein kinase activity by protein phosphatase 2A under conditions of growth inhibition by epidermal growth factor in A431 cells. Eur J Biochem 235:97–102, 1996.
199. JT Parsons, SJ Parsons. Src family protein tyrosine kinases: cooperating with growth factor and adhesion signaling pathways. Curr Opin Cell Biol 9:187–192, 1997.
200. K Yamashita, H Yasuda, J Pines, K Yasumoto, H Nishitani, M Ohtsubo, T Hunter, T Sugimura, T Nishimoto. Okadaic acid, a potent inhibitor of type 1 and type 2A protein phosphatases, activates *cdc2*/H1 kinase and transiently induces a premature mitosis-like state in BHK21 cells. EMBO J 9: 4331–4338, 1990.
201. KE Steinmann, GS Belinsky, D Lee, R Schlegel. Chemically induced premature mitosis: differential response in rodent and human cells and the relationship to cyclin B synthesis and $p34^{cdc2}$/cyclin B complex formation. Proc Natl Acad Sci USA 88:6843–6847, 1991.
202. A Picard, J-C Labbé, H Barakat, J-C Cavadore, M Dorée. Okadaic acid mimics a nuclear component required for cyclin B–cdc2 kinase microinjection to drive starfish oocytes into M phase. J Cell Biol 115:337–344, 1991.
203. W Meikrantz, S Gisselbrecht, SW Tam, R Schlegel. Activation of cyclin A-dependent protein kinases during apoptosis. Proc Natl Acad Sci USA 91:3754–3758, 1994.
204. CJ Sherr. Cancer cell cycles. Science 274:1672–1677, 1996.

205. JR Nevins. E2F: a link between the Rb tumor suppressor protein and viral oncoproteins. Science 258:424–429, 1992.
206. J Goris, J Hermann, P Hendrix, R Ozon, W Merlevede. Okadaic acid, a specific protein phosphatase inhibitor, induces maturation and MPF formation in *Xenopus laevis* oocytes. FEBS Lett 245:91–94, 1989.
207. S Metcalfe, J Milner. Phosphatases PP1 and PP2A act after the G0/G1 interface in lymphocyte activation. Immunol Lett 26:177–183, 1990.
208. JE Thomas, M Smith. JL Tonkinson, B Rubinfeld, P Polakis. Induction of phosphorylation on BRCA1 during the cell cycle and after DNA damage. Cell Growth Differ 8:801–809, 1997.
209. VL Jaramillo-Babb, JL Sugarman, R Scavetta, S-J Wang, N Berndt, TL Born, CK Glass, AH Schönthal. Positive regulation of *cdc2* gene activity by protein phosphatase type 2A. J Biol Chem 271: 5988–5992, 1996.
210. DD Vandre, VL Wills. Inhibition of mitosis by okadaic acid: possible involvement of a protein phosphatase 2A in the transition from metaphase to anaphase. J Cell Sci 101:79–91, 1992.
211. P Michieli, M Chedid, D Lin, JH Pierce, WE Mercer, D Givol. Induction of *WAF1/CIP1* by a p53-independent pathway. Cancer Res 54:3391–3395, 1994.
212. JR Biggs, JE Kudlow, AS Kraft. The role of the transcription factor Sp1 in regulating the expression of the *WAF1/CIP1* gene in U937 leukemic cells. J Biol Chem 271:901–906, 1996.
213. S-J Kim, R Lafyatis, KY Kim, P Angel, H Fujiki, M Karin, MB Sporn, AB Roberts. Regulation of collagenase gene expression by okadaic acid, an inhibitor of protein phosphatases. Cell Reg 1: 269–278, 1990.
214. C Thévenin, S-J Kim, JH Kehrl. Inhibition of protein phosphatases by okadaic acid induces AP1 in human T cells. J Biol Chem 266:9363–9366, 1991.
215. JS Lee, B Favre, BA Hemmings, B Kiefer, Y Nagamine. Okadaic acid-dependent induction of the urokinase-type plasminogen activator gene associated with stabilization and autoregulation of c-Jun. J Biol Chem 269:2887–2894, 1994.
216. A Schönthal, Y Tsukitani, JR Feramisco. Transcriptional and post-transcriptional regulation of c-*fos* expression by the tumor promoter okadaic acid. Oncogene 6:423–430, 1991.
217. X Cao, R Mahendran, GR Guy, YH Tan. Detection and characterization of cellular EGR-1 binding to its recognition site. J Biol Chem 268:16949–16957, 1993.
218. TJ McDonnell, A Beham, M Sarkiss, MM Andersen, P Lo. Importance of the Bcl-2 family in cell death regulation. Experientia 52:1008–1017, 1996.
219. FM Richards, J Milner, S Metcalfe. Inhibition of the serine/threonine protein phosphatases PP1 and PP2A in lymphocytes: effect on mRNA levels for interleukin-2, IL-2Rα, krox-24, p53, hsc70 and cyclophilin. Immunology 76:642–647, 1992.
220. J Skouv, P Østrup Jensen, J Forchhammer, JK Larsen, LR Lund. Tumor-promoting phorbol ester transiently down-modulates the p53 level and blocks the cell cycle. Cell Growth Differ 5:329–340, 1994.

14

Pectenotoxins and Yessotoxins: Chemistry, Toxicology, Pharmacology, and Analysis

Rosa Draisci, Luca Lucentini, and Alessandro Mascioni
Italian National Institute of Health
Rome, Italy

I. INTRODUCTION

Among the natural toxicants indirectly acquired from seafood, yessotoxins (YTXs) and pectenotoxins (PTXs) are currently grouped together with toxins responsible for diarrheic shellfish poisoning (DSP) (i.e., okadaic acid [OA] and dinophysistoxins [DTXs]).

Over the last decade YTXs and PTXs have caused growing concern in public health authorities and in the shellfish industry and represent a controversial topic among scientists for a number of reasons. As the phycotoxin research progresses, the DSP contamination scenario proves to be more complex. PTXs and YTXs, restricted in the past to specific aquatic environments, have been found in several marine areas, together with new toxins. It is controversial whether this is due to an actual expansion of the DSP phenomena, to the global diffusion of shellfish culture, or to the increased statutory monitoring measures and scientific investigations. However, urgent sanitary measures are required to protect consumers' health from the ingestion of PTX- and YTX-contaminated seafood and the shellfish industry from the significant economic damages due to extensive closures.

There are a number of reason for including PTXs and YTXs in the DSP group other than because they coexist in DSP toxin–contaminated shellfish with okadaic acid (OA) and its congeners and, on account of their lipophilic nature, they are coextracted from shellfish digestive glands together with acidic toxins. They are also included in the DSP group because they exert toxic effects following intraperitoneal injection in mice, thus producing positive results in mouse bioassay, largely used as a screening method in monitoring programs. In addition, YTXs and PTXs are produced by dinoflagellates; indeed one of them is produced by the same toxinogenic algal species which also transmits OA and DTX-1. On the other hand, since PTXs and YTXs do not fit the clinical case definition of DSP toxins, because they lack diarrhogenicity in mammals, it has been questioned whether they should be maintained within the DSP toxin class together with the diarrheic toxins conforming to the case definition (i.e., OA and DTXs) or, alternatively, be considered as separate groups.

At any rate, it is important to ensure that levels of PTXs and YTXs in commercial marine species do not pose a health concern, based on the information obtained from risk assessment—the standardized process designed to determine if the toxin concentration in food exceeds safe estimates of intakes.

In order to assist the process of risk assessment, the scientific data and information concerning PTXs and YTXs are reviewed in this chapter. Chemical, pharmacological, and toxicological aspects, occurrence, identification of biogenetic origins, and conventional assays for control of PTXs and YTXs are described. Particular emphasis is given to the latest developments in the field of analytical methods for research and monitoring purposes, since they are proving to be excellent tools to determine the distribution of toxins and to investigate new toxic compounds, as well as to study biotransformation of original toxins.

II. IDENTIFICATION AND STRUCTURES

The first recognized episodes of human gastrointestinal intoxication due to the ingestion of toxic shellfish led to the identification of OA and DTXs as causative toxins of DSP. Through subsequent studies, mainly carried out with hepatopancreas of DSP toxin–contaminated Japanese scallops (*Patinopecten yessoensis*), other toxins coexisting with OA and its congeners were discovered. Since they differ from the toxins of the OA group in both chemical and toxicological properties, two additional families were included in the group of DSP toxins, that is, PTXs (1) and YTXs (2), both named from the investigated toxin-transvector species. To date 10 PTXs and 6 YTXs have been isolated from naturally contaminated marine biological material.

Studies on the identification and structural features of PTXs and YTXs are described in the following paragraphs.

A. Identification and Structures of the PTXs

Eight of the 10 isolated PTXs have been chemically defined so far. In addition, two PTX analogues have been recently reported as artefacts (3). The chemical structures of PTXs and some of their chemical properties are shown in Figure 1 and Table 1, respectively.

PTX-1 and PTX-2 were originally isolated from toxic shellfish by Yasumoto in 1984 and their structure elucidated by X-ray crystallography, nuclear magnetic resonance (NMR), and mass spectrometry (MS) (1,4) together with ultraviolet (UV) and infrared (IR) detection. In the same research, PTX-3–5 were also isolated. Structurally, PTXs are cyclic polyether macrolide compounds, differing from acidic toxins in having a longer carbon backbone (C40), a C-33 lactone ring rather than an open structure, and a dioxabicyclo ring structure. Additional studies using spectroscopic techniques showed PTX-3 to be a 43-aldehyde derivative of PTX-2 (5). Acidic PTXs, PTX-6 and PTX-7, were later isolated from toxic scallops (6). The absolute configuration of PTX-6 was determined by NMR spectroscopy using a chiral anisotropic reagent (7).

Interestingly, the main difference among PTXs concerns the nature of the side chain located at C18, showing all stages of oxidation from methyl to carboxylic acid (PTX-2, CH_3; PTX-1, CH_2OH; PTX-3, CHO; PTX-6, COOH). PTX-2 was only found in phytoplankton (*Dinophysis fortii*); thus supporting the view that the oxidation occurs in the digestive glands of shellfish (Figure 2). Biotransformation of PTX-2 in digestive glands of mollusks was confirmed by the content of PTX-6 in scallops, which was found to be significantly higher than that of PTX-2, also indicating that oxidation of PTX-2 to PTX-6 in scallops occurs rapidly (8).

Identification and characterization of PTX-4 and PTX-7 as 7-*epi*-PTX-1 and 7-*epi*-PTX-6, respectively, have recently been achieved on the basis of NMR data and acid-catalyzed chemical interconversion (3). Equilibration between PTX-6 and PTX-7 and between PTX-1 and PTX-4 resulted in the formation of two additional isomeric products, coded PTX-8 and PTX-9. It was also unambiguously demonstrated that PTX-1, PTX-4, PTX-6, and PTX-7 were naturally occurring toxins rather than artefacts of the extraction process, whereas PTX-8 and PTX-9 were

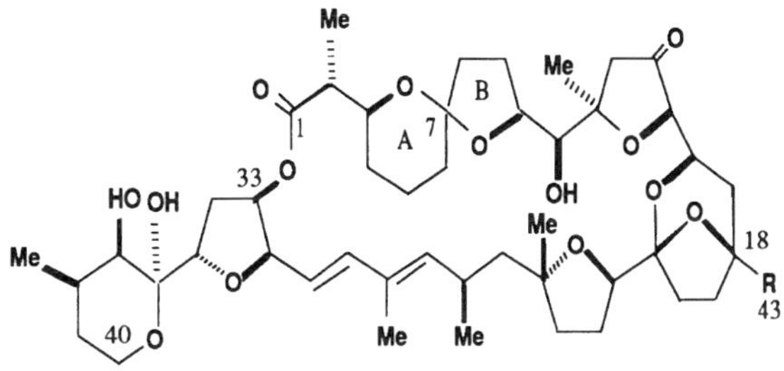

	R	C-7
PTX-1	CH_2OH	*R*
PTX-2	CH_3	*R*
PTX-3	CHO	*R*
PTX-4	CH_2OH	*S*
PTX-5	unidentified	
PTX-6	COOH	*R*
PTX-7	COOH	*S*
PTX-10	unidentified	

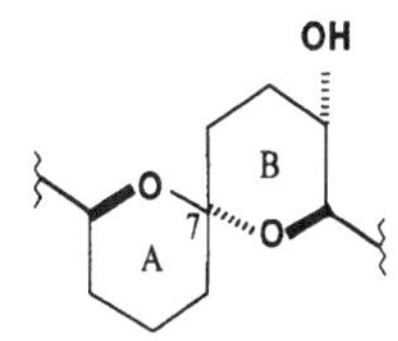

	R	C-7
PTX-8	CH_2OH	*S*
PTX-9	CO_2H	*S*

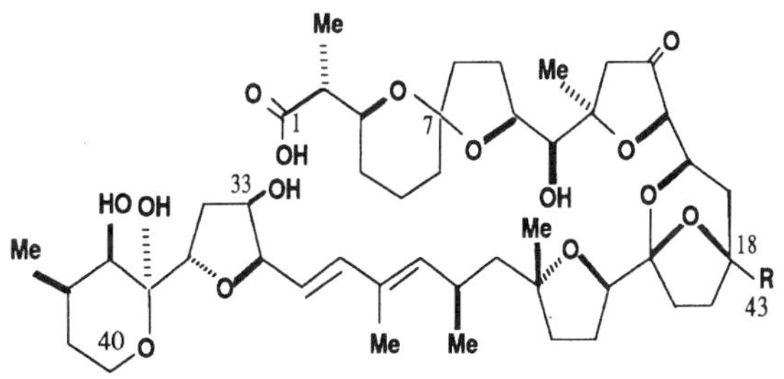

	R	C-7
PTX-2SA	CH_3	R
7-*epi* -PTX-2SA	CH_3	S

Figure 1 Structures of PTXs.

artefacts obtained by acidic interconversion, and no evidence exists of their presence in nature. Naturally occurring PTX-C-7 epimers are therefore supposed to rise from a spiroisomerization reaction within the digestive glands of the scallop, perhaps selectively mediated by a scallop enzyme.

The structures of PTX-5 and PTX-10 still remain unidentified, primarily due to the small amounts of the compounds currently available, although PTX-5 was presumed to be a dihydro derivative of PTX-1 and PTX-4 (4).

Two new acidic analogues of PTX-2, PTX-2 seco acid and 7-*epi*-PTX-2 seco acid, have been isolated and identified in dinoflagellates and shellfish (9–10). Their reactivity toward 9-anthryldiazomethane (ADAM) and 1-bromoacetylpyerene (BAP) led us to assume the presence of a carboxylic group in the molecules. Further structural elucidation by NMR, electrospray-MS, ionspray-MS, and tandem MS (MS/MS), and fast-atom bombardment (FAB) MS/MS showed that PTX-2SAs are C-7 epimers structurally differing from PTX-2 in having a carboxylic acid moiety at C-1 and a hydroxyl group at C-33 rather than the 1,33-ester bond of PTX-2.

Table 1 Chemical Properties of PTXs

Name	Molecular formula	MW	Melting point (°C)	UV λ_{max}(nm)	$[\alpha]_D^{20}$	Characteristic IR band (Kbr) (cm^{-1})	Reference
PTX-1	$C_{47}H_{70}O_{15}$	874	208–209	235	+17.1 (~0.41, MeOH)	3400, 1760, 1740, 970	1,4
PTX-2	$C_{47}H_{70}O_{14}$	858	—	235	+16.2 (~0.105, MeOH)	3400, 1760, 1740, 970	1,4
PTX-3	$C_{47}H_{68}O_{15}$	872	159–160	235	+2.22 (~0.135, MeOH)	3400, 1760, 1740, 970	1,4–5
PTX-4	$C_{47}H_{70}O_{15}$	874	—	235	+2.07 (~0.193, MeOH)	3500, 1760, 1730	1,3–4
PTX-5	$C_{47}H_{72}O_{15}$	876	—	235	—	[a]	4
PTX-6	$C_{47}H_{68}O_{16}$	888	—	235	+8.77 (~0.114, MeOH) +37.1 (~1.49, $CHCl_3$)	(3500, 1760, 1730, 1720, 1700)[b]	4,6
PTX-7	$C_{47}H_{68}O_{16}$	888	—	237[b]	+11.5[b] (~0.131, MeOH)	(3500, 1760, 1730, 1720, 1700)[b]	3
PTX-8[c]	$C_{47}H_{70}O_{15}$	874	—	237	+19.8 (~0.126, MeOH)	3500, 1760, 1730	3
PTX-9[c]	$C_{47}H_{68}O_{16}$	888	—	239[b]	+15.2[b] (~0.105, MeOH)	(3500, 1750, 1730, 1720, 1700)[b]	3
PTX-10	Undetermined	—	—	—	—	—	3
PTX-2SA	$C_{47}H_{72}O_{15}$	876	—	235	—	—	9–10
7-*epi*-PTX-2SA	$C_{47}H_{72}O_{15}$	876	—	235	—	—	9–10

[a] No absorption band at 1760 cm^{-1}.
[b] Measured on phenacyl ester derivative.
[c] No evidence exists of their presence in nature.

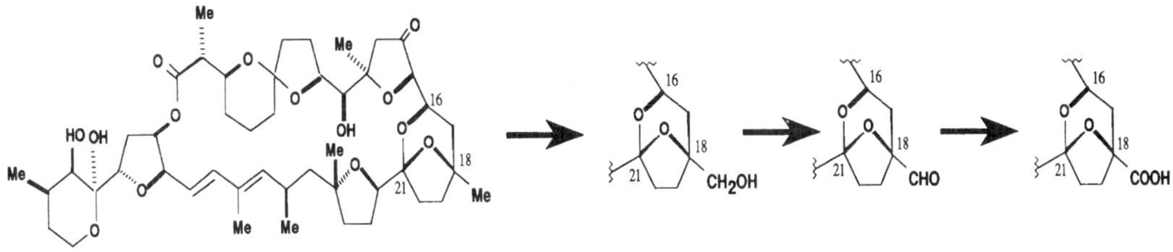

PTX-2 PTX-1 PTX-3 PTX-6

Figure 2 Oxidation of PTX-2 in shellfish digestive glands.

Both PTX-2SAs were proved to be naturally occurring toxins rather than artefacts produced during the extraction. Unsurprisingly, during the isolation from a dinoflagellate PTX-2SA slightly epimerized to 7-*epi*-PTX-2SA, as previously observed for PTX-6 and PTX-7. A third unidentified minor compound appeared during the isolation process of PTX-2SA, which was possibly related to a rearrangement of these toxins in solution (10).

B. Identification and Structures of the YTXs

Structures of YTXs are closely related to those of ciguatoxins (CTXs) and brevetoxins, since they have 11 contiguously transfused ether rings and an unsaturated side chain. Unlike brevetoxin, YTX has a longer backbone of 47 carbons, a terminal side chain of 9 carbons, and 2 sulfate esters and it lacks carbonyl groups. Thus far, six natural compounds belonging to the YTX class have been identified. Moreover, six YTX derivatives were experimentally produced by Yasumoto et al. during the set-up of a method for the structural elucidation of YTX analogues (11). The known structures and chemical properties of YTXs are reported in Figure 3 and Table 2, respectively.

YTX was first isolated in 1987 by Murata et al., and its planar structure was obtained by NMR techniques (2). Desulfated YTX (dsYTX) was produced by solvolysis of the precursor YTX in order to determine the positions of the sulfate esters by comparison between the NMR spectra.

In 1993, Naoki et al. used negative FAB MS/MS as a complementary method to NMR spectroscopy to confirm the structure of YTX (12). Satake et al. subsequently elucidated the relative stereochemistry of YTX and ring conformations by NMR experiments (13). A further study determined the absolute configuration of YTX by NMR spectroscopy using a chiral anisotropic reagent. The same research experimentally obtained five new derivatives in addition to dsYTX by applying the reagent to a monoacetate of dsYTX (see Figure 3) (11).

The first naturally occurring YTX analogue was identified in the Japanese scallop *P. yessoensis* as 45-hydroxy YTX (45-OH-YTX) on the basis of NMR and FAB MS measurements (6). The relative configuration of 45-OH-YTX was later determined by Satake et al. through electrospray MS and NMR experiments. The same study resulted in the identification of another congener of YTX, 45,46,47-trinorYTX (13).

The occurrence of toxicity in the Adriatic mussel *Mytilus galloprovincialis*, which contains only low amounts of OA and no YTX, led to the discovery of two additional YTX analogues, homoYTX and 45-hydroxyhomoYTX. Their planar structures and stereochemistry were determined by electrospray MS and NMR spectroscopy (14).

A novel natural congener of YTX, adriatoxin (ATX), has been recently identified in mussels from the Northern Adriatic Sea, which also contains YTX, homoYTX, and 45-OH-YTX.

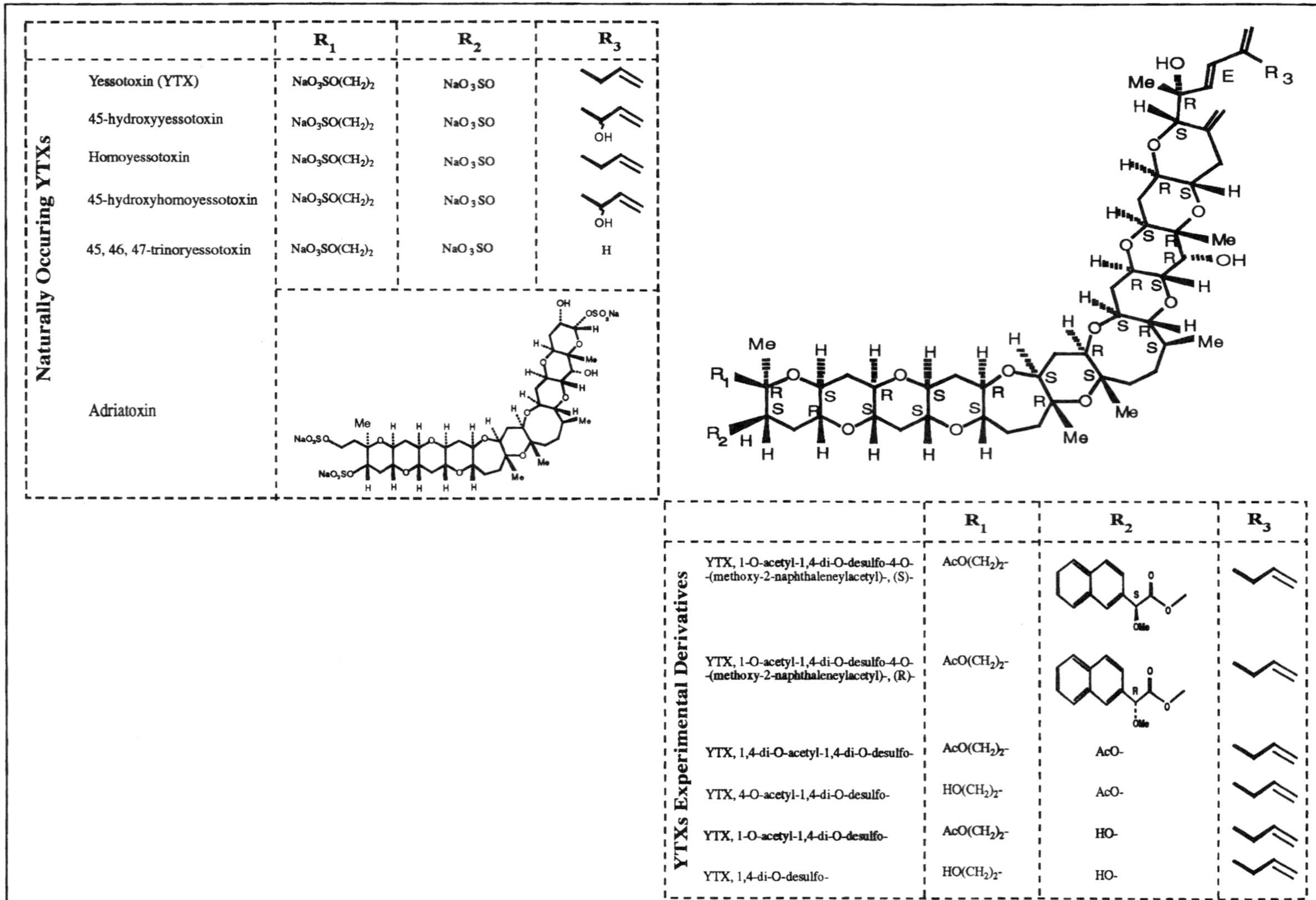

Figure 3 Structures of YTXs.

Table 2 Chemical Properties of YTXs

Name	Molecular formula	MW	UV λ_{max}(nm)	$[\alpha]^{20}{}_{D}$	Characteristic IR bands (Kbr)(cm^{-1})	Reference
YTX	$C_{55}H_{80}O_{21}S_2Na_2$	1186	230	+3.01 (~0.45, MeOH)	3400, 1240, 1220, 1070, 820	2
45-HydroxyYTX	$C_{55}H_{80}O_{22}S_2Na_2$	1202	230	—	—	6,13
45,46,47-TrinorYTX	$C_{52}H_{76}O_{21}S_2Na_2$	1146	226	—	—	13
HomoYTX	$C_{56}H_{82}O_{21}S_2Na_2$	1200	—	—	—	14
45-Hydroxy-homoYTX	$C_{56}H_{82}O_{22}S_2Na_2$	1216	—	—	—	14
ATX	$C_{42}H_{63}O_{24}S_3Na_3$	1116	—	—	—	15
Cooliatoxin	Undetermined	1062	—	—	—	16

The structural elucidation of ATX was carried out by negative FAB MS and NMR spectroscopies. Comparison of data between YTX and ATX revealed the main difference to be in the terminal part of the molecule, in which the absence of an ether ring and the presence of a third sulfate group were observed (15).

In an early study, another suspected analogue of YTX, cooliatoxin, was found in dinoflagellate through liquid chromatography (LC) coupled with MS. On the basis of molecular weight and its toxic effects, the toxin was thought to correspond to the monosulfated form of YTX (16). No other evidence concerning cooliatoxin has been subsequently reported.

III. OCCURRENCE AND BIOGENETIC ORIGIN

The development of monitoring network programs aimed at preventing human consumption of contaminated seafood is based on a number of elements. Above all, it is essential to identify the biogenetic origin of each toxic compound and to know the geographical diffusion of toxin-producing organisms, as well as to clarify the relationship between algal and shellfish contamination.

It is quite obvious, however, that we can only provisionally assume that certain toxins, including PTXs and YTXs, are absent in specific marine areas or that certain algal species are nontoxic, since few studies have been conducted on such topics owing to existing difficulties. A major hindrance concerns the unavailability of suitable analytical methods.

Since the first report on DSP toxins in Japanese marine waters, a number of advanced analytical techniques have gradually become available and investigations have been extended to different marine areas, so that the scenario of DSP toxicity is more definite. However, despite the ongoing progress made in our knowledge of toxins of the OA group, detected in algae and mollusks over the world, little information has been achieved about PTX- and YTX-producing organisms and about their spatiotemporal occurrence. Interestingly, in certain marine areas, such as in Japan, Italy, and New Zealand, detailed studies on DSP toxin–contamination events allowed us to define very complex toxin profiles—in some cases with all three groups of DSP toxins coexisting; single species of PTX- and YTX-producing organisms were also identified. It is likely that the occurrence of PTXs and YTXs will be recorded in many other marine areas as the number of specific investigations increases and the ability to detect such toxins improves.

The current state of the art concerning algal species producing PTXs and YTXs and the geographical distribution of the toxins is summarized in Table 3 and Figure 4, respectively.

Details on research concerning the occurrence and biogenetic origin of PTXs and YTXs are reported in the following paragraphs.

Table 3 Dinoflagellates Confirmed to Produce PTXs and YTXs

Species	Location	Toxins	Reference
Dinophysis fortii	Japan	PTX-2/OA/DTX-1	17,18
	Italy	PTX-2/OA	20
Dinophysis acuta	Ireland	PTX-2SA/7-*epi*-PTX-2SA/PTX-2	9,10,23
Protoceratium reticulatum	New Zealand	YTX	39
Lingulodinium polyedrum (*Gonyaulax polyedra*)	Italy	YTX/homoYTX	35–36
Coolia monotis	Australia	Cooliatoxin	16

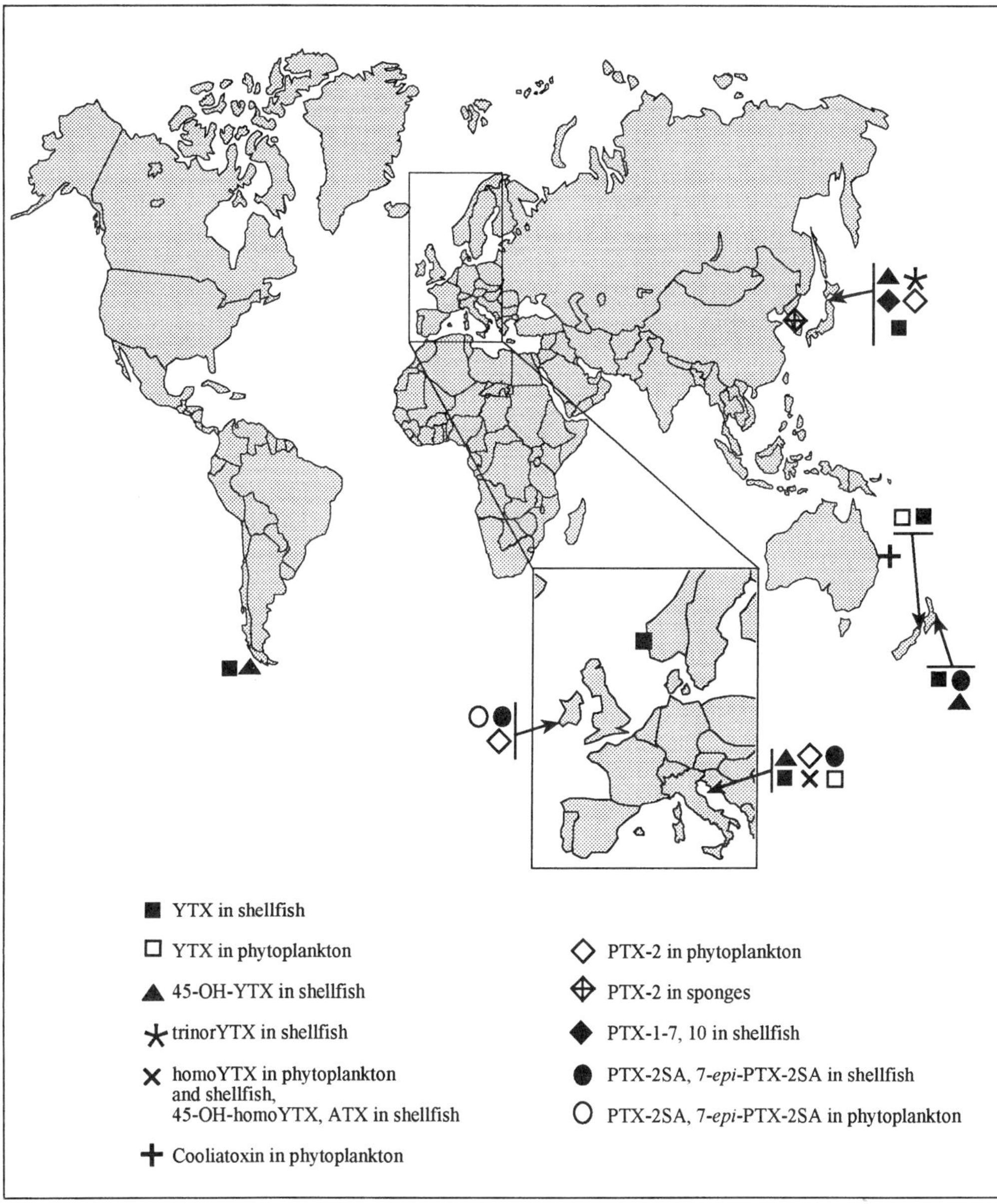

Figure 4 Geographical distribution of PTXs and YTXs in phytoplankton, shellfish, and sponges.

A. Occurrence and Biogenetic Origin of PTXs

Following the identification of PTX-1–5 in the Japanese scallop *P. yessoensis*, which also contains acidic toxins (1,4,6), the need to improve our knowledge of the biogenetic origin of toxins addressed studies to the identification of suspected toxic algal species observed during shellfish toxicity phenomena.

Toxin studies on marine phytoplankton are difficult for some species, such as those of the *Dinophysis* genus, as they cannot be cultured in laboratories. Furthermore, since algal cell concentrations could be low in marine waters even during blooms, and wild populations are always mixed with other phytoplankton species, the only practicable approach for investigation requires individual cells to be picked individually from a microscopic slide. The size of a cell of *Dinophysis* spp. can be less than 50 μm, and the laborious collection procedure makes the total amount of available unialgal samples very limited. Highly specific and sensitive analytical methods are therefore necessary for toxin profile evaluation of these samples.

By this approach, Lee et al. characterized the DSP toxin composition of unialgal samples of seven species belonging to the *Dinophysis* genus; that is, *D. acuminata, D. fortii, D. mitra, D. rotundata, D. tripos, D. acuta, and D. norvegica* (17–18). The research revealed *D. fortii* to be the only species containing PTX-2. PTX-2 alone was observed in phytoplankton, whereas other PTXs, that is, PTX-1, PTX-3, and PTX-6, were present in scallops, thus suggesting that biotransformation reactions occur in shellfish digestive glands after the ingestion of the original toxin.

Interestingly, PTX-2 and no other PTX congener has also been isolated from an association of two sponges (*Poecillastra* sp. and *Jaspis* sp.) that consume dinoflagellates (19). This supports the hypothesis that the other PTXs are oxidation products originated in shellfish hepatopancreas from the precursor toxin, PTX-2.

The first evidence of PTX-2 in Europe was reported in 1996 (20). The toxin was identified in phytoplankton populations dominated by *D. fortii* collected from the northern Adriatic Sea. The high PTX-2:OA ratio indicated a significant amount of the toxin in the examined phytoplankton samples. Interestingly, *D. fortii* strains in Italy contained only OA and PTX-2 and no DTX-1, thus differing from Japanese strains of the same algal species producing OA and/or DTX-1 and/or PTX-2. These results corroborated the view that profound differences in toxin content and composition can occur within the same algal species (17–18); maybe on account of geographical isolates, environmental conditions, composition, and the abundance of other algal species (21).

An investigation was performed on greenshell mussels (*Perna canaliculus*) cultivated at Marlborough Sound, New Zealand, following their exposure to a dinoflagellate bloom of *D. acuta*, a species known to produce OA, DTX-2, and DTX-2C in Ireland (22). PTX-2 seco acid and 7-*epi*-PTX-2 seco acid were firstly identified both in shellfish and in *D. acuta* cells (9–10). The presence of PTX-2-seco acid has also been recently discovered in mussels collected in Italy (23), coexisting in a few cases with OA and YTXs (R. Draisci, unpublished results).

B. Occurrence and Biogenetic Origin of YTXs

The first evidence of YTX in Japan (Mutsu Bay) (2) prompted investigations on its potential occurrence in other marine areas. It was shown that toxin profiles of European mussels can be as complex as in Japanese shellfish. Particularly, profound differences in toxin profiles were observed between mussels collected from two different areas of Norwegian coasts: Arendal, on the south coast, and Sogndal, in the Sognefiord, Norway (24). Although the same plankton species, *D. acuta* and *D. norvegica*, were probably responsible for mussel infestations in both areas, the Arendal samples showed a toxin profile similar to that of other European countries, with OA being the principal toxin. On the other hand, the Sogndal mussels resembled Japanese shellfish in showing multiple toxin profiles, with the simultaneous presence of DTX-1, YTX, and an unidentified PTX-like compound.

Significant research progress made in recent years on the occurrence of YTX in shellfish from different marine areas has indicated that YTX contamination of mollusks can be more significant than previously thought. In addition to YTX, a number of its analogues have also been identified during investigations.

45-OH-YTX and 45,46,47-trinorYTX were identified together with YTX in Japanese *P. yessoensis* scallops collected in 1993 in Mutsu Bay. Although they were minor components in the tested samples, their relative abundance was supposed to vary depending on locations and seasons (13).

The presence of YTX was suspected in shellfish from Chile and Italy on account of discrepancies observed in some samples between mouse bioassay and LC results. However, identification of YTX was precluded at that time because of the small amount of the available samples (25). Subsequent research confirmed the occurrence of YTX in mussels from Chile. Samples of *M. chilensis* collected in 1991 at the Chonos Archipelago, Chile, were proved to be infested by YTX and 45-OH-YTX (26).

The situation of DSP toxicity in Italy is very interesting, as toxic phenomena have occurred with alarming frequency since the end of the 1980s (27). Although complex toxin profiles had repeatedly been reported in shellfish (28–32) and phytoplankton (20,32), sometimes with the suspected presence of new unidentified lipid-soluble toxic compounds (25,29,31,33), the first report of YTX in Italy appeared in 1997 (34). The toxin was identified in mussels (*M. galloprovincialis*) collected from the northern Adriatic Sea in 1995 during a phytoplankton bloom where the simultaneous presence of the following toxic algae was determined: *D. sacculus, D. fortii, Prorocentrum micans, P. minimum, P. compressum, Alexandrium pseudogonyalax*, and *Gymnodinium* spp. The digestive glands of contaminated shellfish were shown to contain OA and YTX in relatively large quantities; thus proving that the two toxins are the major agents responsible for DSP contamination in the area in the period under investigation.

Studies were then carried out on samples collected in the same marine area the following year. Interestingly, two novel YTX congeners, homoYTX and 45-OH-homoYTX, were found in shellfish (14). The difference between the phytoplankton composition of 1995 and 1996 probably accounted for the seasonal variation of shellfish toxin profiles; also suggesting that YTX and homoYTX were produced by different plankton species.

HomoYTX was also unambiguously identified in mussels collected from different marine areas of the Adriatic Sea exposed to a bloom of *Lingulodinium polyedrum (Gonyaulax polyedra)*. The toxin was also suspected in the phytoplankton *L. polyedrum* (35), but specific investigations on the phytoplankton specimens only detected the presence of YTX-like toxins, since the adopted LC method was unable to distinguish between YTX and homoYTX. Subsequent studies by LC-MS and LC-MS-MS unambiguously showed the simultaneous production of YTX and homoYTX by the algal species *L. polyedrum (G. polyedra)* (36).

Conspicuous YTX contamination of Adriatic mussels from two adjacent marine areas also occurred in 1997 when a prolonged DSP toxicity was recorded. YTX levels ranged from 0.18 to 9.02 μg/g of hepatopancreas, with the highest values being in samples collected in midsummer, which also contained homoYTX. OA coexisted with YTXs at comparatively lower concentrations ranging from 0.13 to 2.31 μg/g (36).

45-OH-YTX was identified for the first time in Adriatic mussels collected in 1997, which also contained YTX and homo-YTX (37). Analysis of a chromatographic toxic fraction from the same samples allowed the isolation of a new congener of YTX, ATX (15).

Complex toxin profiles of Adriatic mussels were again defined in 1998 when the simultaneous presence of OA, YTXs, and PTXs was noticed in some samples (R. Draisci, unpublished results).

Interesting findings also resulted from studies on shellfish from New Zealand. YTX and 45-OH-YTX were first identified in greenshell mussels *(P. canaliculus)* collected in Colomandel in the winter of 1993 (26). In the autumn of 1996, greenshell mussels from Marlborough Sound, where a large shellfish formation industry is located, were exposed to a dinoflagellate bloom in which *D. acuta* and *P. reticulatum* were dominant species. The toxicity levels of mussels were very high, but only trace amounts of OA and DTX-1 were detected, whereas high levels of YTX were observed (38). Furthermore, YTX was detected in phytoplankton populations from the same area in which *P. reticulatum* was a dominant species. Interestingly, wild blue mussels *(M. edulis aoteanus)* collected from the same location and feeding on the same dinoflagellates exhibited very high contents of OA and DTXs, thus suggesting that the different absorption of toxins depends on the shellfish species (38).

Subsequent studies were then carried out on axenic unialgal cultures of *P. reticulatum*; thus confirming the species to be one of the biogenetic origins of YTX (39). Interestingly, none of the known analogues, 45-OH-YTX, 45,46,47-trinorYTX, or homoYTX, was detected in *P. reticulatum*. Since 45-OH-YTX was occasionally found to be more abundant than YTX in mussels (26), it was hypothesized that, as observed for the biotransformation of PTXs, oxidation of YTX in mussel digestive glands can occur, leading to the presence of 45-OH-YTX. On the other hand, the other YTXs were suspected to be produced from different dinoflagellates (39).

The involvement of *P. reticulatum* and *L. polyedrum* in the production of YTXs implies a high worldwide risk of YTX contamination of shellfish, since these species are very common in phytoplanktonic communities of coastal waters and their cysts are ubiquitous in marine sediments.

Cooliatoxin was identified from cultured microalgal species *Coolia monotis* isolated from Platypus Bay, Australia, in 1995 (16). More recently, *C. monotis* has been shown to be toxic to larvae of *Artemia salina* and *Haliotis virginea* (40).

IV. TOXICOLOGY AND PHARMACOLOGY

The evidence that PTXs and YTXs are not diarrheic (2,41–42), unlike acidic toxins of the OA class, led to several debates as to whether to still include YTXs and PTXs in the DSP group of toxins together with diarrheic toxins. Alternatively, PTXs and YTXs should be considered separately and specific acceptable levels for each of the three classes should be established.

In order to reduce the risk of hazards to as minimal a level as possible within an established risk-cost framework, risk analysis should be used. This well-established discipline significantly contributes to food safety by making the best use of available scientific data and information to implement appropriate regulatory actions. Its foundation is the process of risk assessment, a systematic scientific process of documenting potential hazards and characterizing the risk of adverse events associated with the hazards. The formal process of risk assessment includes four steps; that is, hazard identification, hazard characterization, exposure assessment and risk characterization.

For PTXs and YTXs, data that are crucial in the risk assessment process are incomplete. Owing to the small availability of pure toxins necessary for toxicological evaluations, the knowledge of the acute and chronic effects of PTXs and YTXs is very limited.

Some data concerning the exposure to PTXs and YTXs are available. A few species of PTX- and YTX-producing organisms have been identified and the diffusion of the toxins in a small number of shellfish-producing areas has been studied. Improved knowledge of contamination for other areas, including potential production sites, and detailed information on toxicogenic organisms and temporal and positional (the depth in the water column is particularly important) production of toxins within each algal species, are required.

As a result, tolerance levels presently adopted for these toxins are not deduced from robust toxicological studies, or from exposure and epidemiological data, which should determine the relationship between the magnitude of exposure and adverse effects (i.e., dose-response assessment), the not observable effect levels (NOAEL), and the lowest adverse effect level (LOAEL). Current limits are established on the basis of the scarce toxicity data available and, especially, on the sensitivity of the biological assay. Toxicity of DSP toxins was originally estimated by means of intraperitoneal administration of toxic material in mice. Although the suitability of this assessment is controversial, it is the basis of the mouse bioassay, which is the screening method extensively used to prevent human acute intoxication. It can therefore be useful to compare intraperitoneal toxicity data of the toxins; the values are reported in Table 4 together with some pharmacological/toxicological actions of the toxins.

PTXs and YTXs present a singular chemical structure and seem to be unique in their mode of action. Thus, they can be regarded as being particularly interesting tools for probing biological and pharmacological phenomena. Instances of such a research approach are the use of saxitoxin in ion channel studies (43), maitotoxin in studying calcium-dependent mechanisms on cell membrane (44), and okadaic acid as a protein phosphatase inhibitor (45). However, relatively little effort has been devoted to the understanding of the molecular basis of the mechanisms of actions or to exploration of the potential medicinal value of DSP toxins other than OA and its derivatives.

Research on the toxicology and pharmacology of PTXs and YTXs is reviewed in the following paragraphs.

A. Toxicology and Pharmacology of the PTXs

Among PTXs, PTX-2 shows the highest lethality when injected into mice; thus suggesting that the oxidation of substituent in C-18 of the PTX molecule may play a role in diminishing the

Table 4 Mouse Lethality and Main Pathological Effects of PTXs and YTXs

Toxin	Lethality to mice (μg/kg b.w. i.p.)	Pathological effect and biological activity
PTX-1	250 (6)	Hepatotoxic (42) apoptogenic (52)
PTX-2	230 (6)	Hepatotoxic[b]
PTX-2SAs	—	—
PTX-3	350 (6)	Hepatotoxic[b]
PTX-4	770 (3,6)	Hepatotoxic[b]
PTX-6	500 (6)	—
PTX-7	>5000 (3)	—
PTX-8	>5000 (3)	—
PTX-9	>5000 (3)	—
YTX	100 (2)	Cardiotoxic (54)
45-OH-YTX	500 (13)	—
45,46,47-trinorYTX	220 (13)	—
homoYTX	100 (14)	—
45-hydroxyhomoYTX	500 (14)	—
dsYTX	[a]	Liver and pancreas toxicity (54) Ichthyotoxic (53), Antifungal (48)

[a] A dose of 301 μg/kg b.w., i.p. was lethal at 48 h (54).
[b] Presumed from the toxicity of PTX-1.

toxicity of these compounds. Interestingly, intraperitoneal (i.p.) toxicity data for PTX-2 are comparable with the oral toxicity values estimated for the toxin (T. Yasumoto, personal communication, 1997). On the other hand, spiroisomerization at C-7 for PTX-4 and PTX-7 resulted in a significant decrease in lethal toxicity; the lethality value for PTX-4 was 770 μg/kg body weight (b.w.) (mouse, i.p.), which is approximately one third of that of PTX-1; toxicity of PTX-7 also diminished to more than 5000 μg/kg b.w., which is about one order of magnitude higher than PTX-6 lethality (i.e., 500 μg/kg b.w., mouse, i.p.). Lethality values similar to those of PTX-7 were recorded for PTX-8 and PTX-9. No lethality data have been estimated for PTX-5, PTX-10, and PTX-2SAs on account of the small availability of isolated toxic material.

The first toxicological evaluation of PTX-1 was carried out by Hamano et al. (41). Using the fluid accumulation ratios in suckling mice as an index, they found that the toxin does not cause fluid accumulation in rabbit and mouse intestinal loops; thus proving its lack of diarrhogenicity. Cytotoxicity was also observed for PTX-1, with minor potency compared to OA or DTXs.

Terao et al. carried out histopathological studies (42) on suckling mice administered PTX-1 and DTX-1. Although no abnormality was caused by PTX-1 in small and large intestines or in other visceral organs, the liver was the target organ. Pathomorphological changes induced by intraperitoneal injection of PTX-1 (500–1000 μg/kg b.w.) consisted of the formation of nonfatty vacuoles in the hepatocytes around the periportal triads. Electron microscopic observations demonstrated that these vacuoles originated from invaginated plasma membrane of the hepatocytes. Incubation time from the injection of the toxin to the pathological changes in the target organ were rather short, and this resembles the onset of human DSP. Only slight hepatic injures were noticed in mice administered 150–200 μg/kg b.w. of PTX-1.

Very interestingly, vacuole formation induced by PTX-1 was similar to what had been observed for intoxications caused by several cyclic oligopeptides, such as phalloidin (46) and cyclochlorotine (47), produced by the mushroom *Amanita phalloides* and the mold *Penicillium islandicum*, respectively. This similarity in ultrastructural changes led to the supposition that similar mechanisms are involved for these toxins in the intracytoplasmic nonfatty vacuolisation. The short incubation time was also characteristic of both intoxications. Nevertheless, the exact mechanism of the toxicity of PTX-1 at the molecular level remains unknown.

Further studies investigated the antimicrobial activities of PTX-1 and other 11 polyether compounds of dinoflagellate origin (48). PTX-1 was shown to be inactive against the fungi *Aspergillus niger* and *P. funiculosum*, even though it shares some structural features with the polyether macrolide goniodomin A (49), which is highly potent against some fungi other than the investigated species. No antibacterial activity was shown for PTX-1 against *Escherichia coli, Staphylococcus aureus*, or *Batillus megaterium*.

Cytotoxic effects of PTX-1 toward freshly prepared hepatocytes resulted in morphological changes: Following exposure at 5 μg/mL or higher concentrations, cells maintained their spherical shape, but the surface appeared almost smooth and devoid of most microvilli. The effects were similar to those induced by DTX-1 and YTX, if observed by light microscopy, whereas they were easily discernible using scanning electron microscopy (50).

Zhou et al. explored PTX-1–induced damage of the cytoskeleton in liver cells in culture through fluorescence microscopy (51). In vitro application of PTX-1 resulted in reduction of the cell number, loss of radial arrangements in microtubules, disruption of stress fibers, and accumulation of actin at the cellular peripheries. Cell damage was reversible following treatment with toxin concentrations lower than 0.5 μg/mL for less than 4 h. Histopathological changes in the hepatocytes appeared to be somewhat different from those induced by phalloidin.

PTX-2 was proved to be the active constituent responsible for the brine shrimp (*Artemia salina*) lethality (19). As brine shrimp lethality is correlated with cytotoxicity and antitumor activity, PTX-2 was assayed for cytotoxicity against three human tumor cell lines. The toxin displayed very potent cytotoxicity against human colon, lung, and breast carcinomas, which is much higher than that of the anthracycline antineoplastic agent lethality (19). Antitumor activity of PTX-2 against 60 human cancer cell lines was therefore investigated and a selective effect on several cell lines of ovarian, renal, lung, colon, central nervous system (CNS), melanoma, and breast cancer was discovered. Remarkable differences (100-fold or more) in LC50 values were found between sensitive and resistant cell lines. No selective cytotoxicity was observed against any leukemia or prostate cancer cell lines.

A recent investigation of PTX-1 evaluates its capability to induce apoptotis in primary rat and salmon hepatocytes in suspension cultures (52). Currently, apoptosis (regulated cell death) is an actively studied biological phenomenon. Indeed, intense research activities are aimed at identifying novel apoptogenic compounds to use as cures for different diseases, such as cancer or autoimmune diseases, as well as to elucidate the apoptotic signaling pathways within cells. Electron and light microscopic observation have proved that, in both salmon and rat hepatocytes, apoptosis occurs in less than 2 h as a result of exposure to 2 μM of PTX-1. The apoptopic phenotype induce by the toxin is marked by more extreme nuclear and cellular shrinkage and chromatin hypercondensation in both rat and salmon hepatocytes. The apoptotic nature of the death was confirmed by the observation that the apoptotic protease (caspase) inhibitor ZVAD-fmk and the Ca^{21}/calmodulin kinase II inhibitor KN-93 were able to inhibit the action of PTX-1.

Interestingly, apoptotic changes in the morphological features of rat and salmon hepatocytes in response to other toxins under investigation (i.e., mycrocistin-LR, OA, DTX-1, nodularin, calyculin A) were similar; thus indicating a phylogenetically well-conserved action mechanism for PTX-1 and the other compounds.

Toxicity evaluation of PTX-2SAs was hampered by the small availability of isolated material. Similarly to PTX-2, they do not inhibit protein phosphatase PP2A (10). Neither PTX-2SA nor 7-*epi*-PTX-2SA showed cytotoxicity against KB cells at a dose of 1.8 μg/mL, whereas PTX-2 did at 0.05 μg/mL; thus suggesting the cyclic structure exerts the potency (9).

In conclusion, it is generally accepted that, since PTXs do not conform to the clinical case definition of DSP in terms of toxicological effects, they should not be classified as DSP toxins. On the other hand, although the exact role played by each toxin of the PTX group in human pathogenicity is virtually unknown, the potent hepatoxicity recognized for these toxins requires severe vigilance of PTX contamination in seafood. PTXs should therefore still remain included in monitoring programs.

It is apparent that an improved knowledge on the acute and chronic effects of PTXs on human health as well as on their mode of action is imperative. The former necessity is particularly urgent in order to define the NOAEL and the LOAEL for PTXs and through which appropriate PTX thresholds in shellfish could be established. Moreover the elucidation of the molecular basis of PTX mechanism of action could prove PTXs to be useful as pharmacological tools, thus contributing to life science.

B. Toxicology and Pharmacology of the YTXs

In spite of its lack of diarrhogenic behavior, YTX shows high lethality when injected in mice, and compared with PTXs and with the other DSP toxins, it appears to be the most toxic compound on account of its high mouse lethality (LD99 = 80–100 μg/kg b.w., mouse, i.p.) (2,53). YTX-injected mice showed normal behavior for the first hours following administration and then died owing to sudden dyspnea (54). Symptoms following injection of YTX in mice resemble those typical of neurotoxins and are characterized by short survival time together with convulsions and jumping before death. For these reasons, an involvement of the nervous system in YTX toxicity was originally presumed. The hypothesis was also strengthened by the resemblance of the YTX structure to the brevetoxins and CTXs for which pharmacological evidence exists of induction of an increase in cell membrane permeability to sodium ion (55).

The level of intraperitoneal toxicity for homo-YTX is presumably comparable with that of YTX (14), whereas intraperitoneal toxicity values for the other YTX analogues are higher than those observed for YTX itself. Lethality data after intraperitoneal injection in mice are estimated to be 500 μg/kg b.w. and 220 μg/kg b.w. for 45-OH-YTX and 45,46,47-trinorYTX, respectively (13). Lethality for 45-OH-homo-YTX is comparable with that of 45-OH-YTX (26).

YTX toxicity is supposed to be related to the presence of the sulfate groups which give the molecule different affinity for lipidic membranes. The increase in polarity in the side chains of 45-OH-YTX and 45,46,47-trinorYTX can therefore be responsible for their reduced toxicity owing to a decreased affinity for the lipidic membrane (13). For the same reason, the presence of three sulfate groups in ATX, increasing the polarity of the molecule, causes the toxin to be less toxic than YTX (15). Unexpectedly, however, dsYTX, lacking two sulfate groups, was proved to be less toxic by intraperitoneal injection to mouse than YTX (54).

Terao et al. (54) compared the istopathological response of YTX in mice with that of dsYTX. Such comparison seems necessary, since intestinal bacteria are known to desulfate different compounds, such as steroids or bile acids.

After 3 hs from intraperitoneal administration of YTX (300 μg/kg b.w.), the target organ proved to be the heart, and marked intracytoplasmic edema in cardiac muscle cells were observed. No discernible changes were caused in liver, pancreas, lungs, adrenal glands, kidney, spleen, or timus. Interestingly, istopathological changes in the heart induced by YTX appeared similar to those provoked by CTX, which modifies cell membrane permeability (56).

Although for other marine toxins, such as PTX-2, DTX-1, DTX-3 and CTX, oral toxicity is comparable with their intraperitoneal toxicity [T. Yasumoto, personal communication, 1997], YTX seems to exert significantly less toxic effect by oral administration. Indeed, YTX did not cause death in mouse by oral intake even at 1 mg/kg, which is 10 times the lethal intraperitoneal dose (53).

Unlike the effects following YTX intraperitoneal administration, oral treatment with 500 μg/kg b.w. provoked no discernible changes in the heart or in any other organ (54). On the other hand, intraperitoneal administration of dsYTX (300 μg/kg b.w.) caused severe fatty degeneration and intracellular necrosis in liver and pancreas but not in the heart, and similar injures to these organs were also observed following oral administration of dsYTX even if they were less than those observed after intraperitoneal injection.

Further investigations showed that some other properties of YTXs were markedly affected by slight modifications in their structure similarly to those observed for other polyether toxins (48). Because of the resemblance between polyether dinoflagellate toxins and some polyether antibiotics of *Actynomyces*, Nagai et al. tested growth-inhibiting activities against fungi, yeasts, and bacteria (48). In addition to YTX, dsYTX was also tested for comparison. Toxicity of YTX against fungi and yeasts proved to be weak, as noted for other tested polyether toxins; that is, OA esters, prorocentrolide, PTX-1, and maitotoxin and its desulfated derivative. Nevertheless, desulfation of YTX markedly enhanced the antifungal activity and reduced mouse lethality. Interestingly, CTX, with structural features resembling those of dsYTX, showed the most potent antifungal activity among the tested polyether toxins. Ogino et al. (53) confirmed the inability of YTX and dsYTX to inhibit bacterial growth, whereas growth-inhibiting activities were observed against fungi and yeasts, with YTX being less potent than its analogue.

The above findings give rise to significant expectations regarding the possibility to induced specific modifications to molecules of YTX and other polyether compounds in order to improve desirable activities while reducing their toxicity. And they seem to confirm dinoflagellates as a promising source of novel and potent antifungals (48).

The two sulfate groups at one terminus of the hydrophobic polyether skeleton of YTX were thought to give the molecule an amphiphilic property. Thus, a hemolytic activity of YTX was expected. Moreover, an ichthyotoxic property was hypothesized because of the close structural resemblance between YTX and brevetoxin B, a well-known ichthyotoxin (57). Hemolytic tests were performed on mouse blood cells using the plant saponin as a positive control. YTX was tested at the concentration of 50 ppm, which caused no hemolytic effect under conditions in which the plant saponin causes complete hemolysis at only 10 ppm. No ichthyotoxic property was observed following exposure of killyfish (*Oryzias latipes*) to up to 1 ppm of YTX (53). A weak inhibition of PP2A was observed for YTX. However, the concentration required to reduce the enzymatic activity by 50% (IC50) was lower than that of OA by four orders of magnitude; hence no interference in the enzyme inhibition assay occurs between the two toxins (53).

Early observations showed morphological effects of YTX toward freshly prepared rat hepatocytes starting at about 10 μg/mL, which is quite similar to those observed for OA and DTX-1 at 1 μg/mL (53). Cytotoxicity of YTXs was also tested by Ogino et al. (53). They measured the effect of YTX on cell viability using a method based on the reduction of MTT (3-[4,5-dimethylthiazol-2-yl]-2,5-diphenyltetrazolium) to a blue formazan product by metabolically active cells. Although the use of MTT was complicated by the detachment of cells exposed to YTX, even at concentrations as low as 10 ng/mL, the test showed YTX to be less toxic than OA by three orders of magnitude.

In conclusion, although there are reasons to believe that YTX imports are much less hazardous than OA or DTX-1, other toxicological data, such as the cell detachment, the lethal effect of oral administration of YTX on infant mice (53), and the severe heart injures following intraperitoneal administration should be considered. In addition, histopathological changes in liver and pancreas due to the oral administration of dsYTX (54) have to be taken into account owing to the likely biotransformation of ingested YTX within the gastro-intestinal tract.

Thus, even if YTX could be not classified as a DSP toxin, the existing surveillance measures for YTX should be maintained until a complete toxicological evaluation is performed and acceptable levels of YTX in seafood by risk assessment methods are established.

V. METHODS FOR IDENTIFICATION AND DETERMINATION OF PTXs AND YTXs

The mouse bioassay currently represents the basic tool for most PTX and YTX monitoring programs. Nevertheless, there are several issues against this method: ethical considerations, lack of accuracy and specificity, occasional occurrence of false-positive and false-negative results, the difficulties of comparing results due both to the variety of toxin extraction procedures used, and of the mouse survival times chosen as the toxicity criterion.

On the other hand, a number of factors combine to present serious hurdles to the development, validation, and diffusion of analytical methodologies alternative to the biological assays. Primarily, the lack of satisfactory natural sources of analytical standards has hampered the production of reference material for several toxins. Among the species known to produce YTXs and PTXs, only YTX-producing phytoplankton can be maintained in culture, and analytical calibrants of YTX have only recently been made commercially accessible. Conversely, the inability to maintain *Dinophysis* species in culture prevents the large-scale production of PTX standards. Moreover, only limited progress has been achieved in the synthesis of these compounds, and although the total synthesis of the well-known OA has been obtained, only little effort has been devoted to the synthesis of PTXs, and only the partial synthesis of the common C-31–C-40 fragment has been reported (58).

Additional difficulties which inhibit the development of alternative methods are the frequent identifications of new PTX and YTX analogues, the changes in the original structures of toxins due to biotransformation by the vector mollusks, the complexity and variability of marine biological material, the availability, in many cases, of only small amounts of samples for the analytical determination, and the frequent coexistence of multiple toxins in samples.

Nevertheless, a number of biochemical and chemical methods have been proposed as specific and sensitive analytical tools for the determination of PTXs and YTXs in both shellfish and phytoplankton in order to support basic science research and monitoring programs. These are reviewed in the following paragraphs.

A. Biological Methods

Most countries use the biological method to test overall DSP toxicity in shellfish for human consumption for regulatory purposes. However, significant differences in the testing procedures and in the toxicity criteria exist. These mainly involve the type of assay and the extraction/purification procedure, resulting in different performances of the methods. Moreover, various toxicity criteria are adopted, so that interpretations of positive results (and therefore tolerance limits) are not equivalent among different countries.

The original biological mouse bioassay (59) involves the extraction of all DSP toxins with acetone, removal of the solvent, suspension of the residue in Tween 60, and intraperitoneal injection of the suspension in mice. Although the method is capable of detecting PTXs and YTXs, the possible presence of PSP toxins and high concentrations of free fatty acids, which are harmless to humans by oral intake but are lethal to mice by intraperitoneal injection, can interfere and cause false-positive results (41).

Since the usual procedure involves intraperitoneal injection of 1 mL of extract (corresponding to 5 g of digestive glands) in a ~20 g mouse, the following theoretical detection limit, based on LD99 of each toxin, could be estimated for the mouse bioassay (losses of toxins during the extraction procedure are not considered): ~1.0 μg/g for PTX-1, ~0.9 μg/g for PTX-2, ~1.4 μg/g for PTX-3, ~3.1 μg/g for PTX-4, ~2.0 for PTX-6, more than 20 μg/g for PTX-7,8,9, ~0.4 μg/g for YTX and homo YTX, ~2.0 μg/g for 45-OH-YTX and 45-OH-homoYTX and ~0.9 μg/g 45,46,47-trinorYTX.

A modification of the assay was subsequently proposed by the same investigator (1). It consists of the dissolution of the residue of acetone extract with water followed by its extraction with diethyl ether. The water layer containing PSP toxins is then discarded. The ether solution is finally evaporated and the residue dissolved in Tween 60 and injected in mice. However, YTX losses due to its poor solubility in diethyl ether can occur; thus leading to false-negative results (36,60). The replacement of diethyl ether with dichloromethane, which is suitable to solubilize all DSP toxins, including YTX, has therefore been proposed.

A further modification of the Yasumoto method implements redissolution in methanol 80% of the residue obtained after the removal of acetone. In order to remove nonpolar lipids, the methanol solution is partitioned with hexane. The methanol phase is then recovered, evaporated, and the residue redissolved in Tween 60 and injected into mice. The elimination of free fatty acids by hexane prevents false-positive results. Nevertheless, a major drawback of this method is that some hydrophobic toxins (acyl-OA and acyl-DTX-2) can be partially solubilized in the hexane phase. Moreover, PSP interferences are not removed (61).

Mice survival time is used as the toxicity criterion or regulatory limit for all DSP toxins, but the observation times considered to be appropriate to assess the toxicity of samples are not harmonized. The results are usually regarded to be positive if two of three mice die before 5 h or, in some countries, in less than 24 h.

Other proposed biological methods, such as the suckling mouse bioassay (41) or the rat bioassay (62), based on diarrhea induction, are not able to detect nondiarrhogenic toxins (i.e., PTXs and YTXs).

B. Chemical Methods

Extraction and Purification of PTXs and YTXs

Owing to their lipophilic nature, all DSP toxins are usually isolated from marine biological material by acetone extraction, although aqueous methanol or chloroform are also used. Individual toxins are then isolated through a complex fractionation and purification procedure of raw extracts based on a combination of chromatographic techniques.

The first method used to isolate PTXs and YTXs was reported in 1987 by Murata et al. (2). The basic scheme for the isolation of PTX-1–10, YTX, and 45-OH-YTX is shown in Figure 5. PTX-1–5 are obtained from neutral fraction and PTX-6, PTX-7 and PTX-10, together

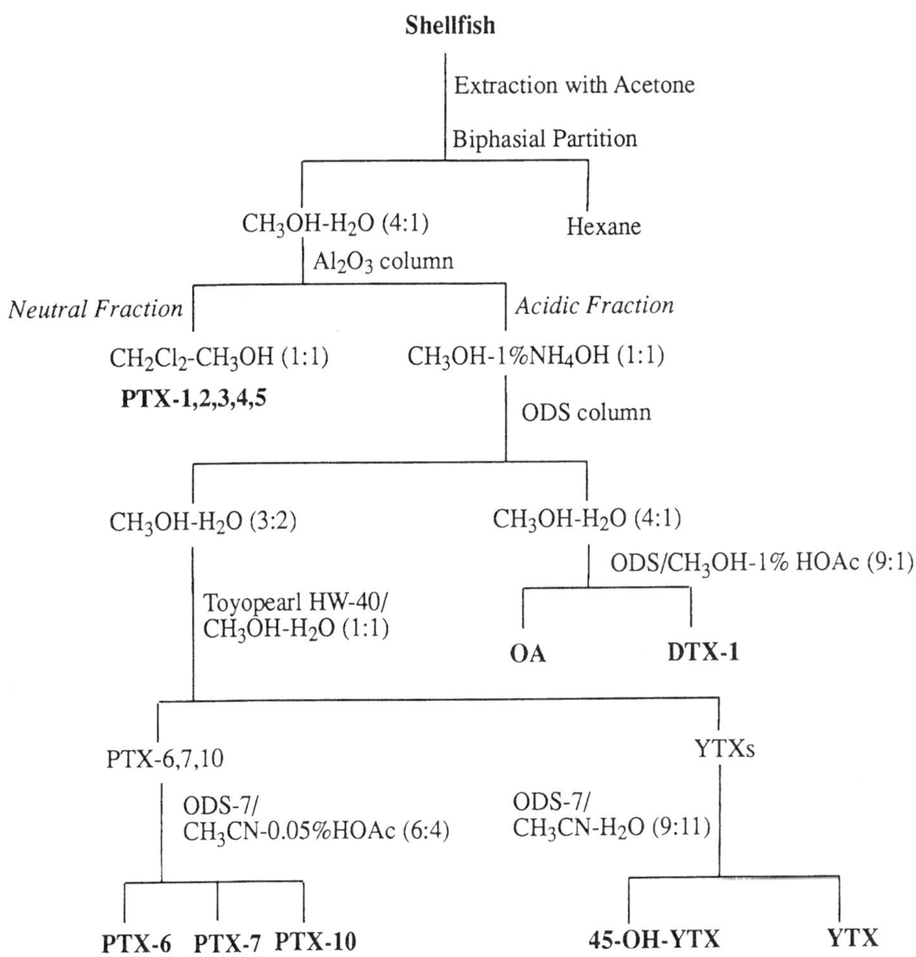

Figure 5 Fractionation and purification procedure for DSP toxins.

with YTX and 45-OH-YTX, from the acidic fraction of a chromatographic separation of the shellfish extract on alumina (6).

Satake et al. reported a modification of the final step of the purification procedure using a chromatography on ODS-7 to isolate YTX ($CH_3OH:CH_3CN:H_2O$, 4:2:7), 45-OH-YTX and 45,46,47-trinorYTX ($CH_3OH:CH_3CN:H_2O$, 4:2:9) (13).

PTX-2SAs were extracted from shellfish hepatopancreas with $CH_3OH:H_2O$ (4:1) and the extract partitioned between EtOAc and H_2O. The toxins in the organic layer were chromatographed on alumina with $CHCl_3:CH_3OH$ (1:1), CH_3OH, and then 1%$NH_4OH:CH_3OH$ (1:1). Successive chromatographic separations were performed: (1) sylica gel column with $CHCl_3$ and $CHCl_3:CH_3OH$ (7:3, 1:1); (2) reversed phase ODS column $CH_3OH:H_2O$ (1:1, 6:4, 8:2) and CH_3OH; (3) toyopearl column $CH_3OH:H_2O$ (1:1). Final purification of PTX-2SA was performed by LC using a ODS-7 column with $CH_3OH:CH_3CN:H_2O$ (2:1:2) (9).

A simplified procedure for the isolation of PTX-2SAs from marine phytoplankton using chloroform as extractant has been recently reported (10).

Homo YTX and its hydroxylated analogue were isolated by the following procedure (15): (1) extraction of toxins from digestive glands with acetone; (2) partitioning of the extract between $CH_3OH:H_2O$ (4:1) and hexane; (3) partitioning of the residue obtained in the hydromethanolic layer between butanol and H_2O; (4) evaporation of the butanolic fraction followed by dissolution in $CHCl_3:CH_3OH$ (1:1); (5) clean up on alumina column washing with $CHCl_3$: CH_3OH (1:1) and CH_3OH and eluting with 1% NH_4OH/CH_3OH (1:1); (6) chromatography on silica gel column with $CHCl_3:CH_3OH$ (9:1, 1:1); the toxins were obtained in the second eluate; (7) evaporation of the solvent and dissolution of the residue in $CH_3OH:H_2O$ (3:7); (8) clean up on silica gel ODS-Q3 column washing with $CH_3OH:H_2O$ (3:7, 1:1) and eluting with $CH_3OH:H_2O$ (8:2); (9) chromatography on toyopearl column with $CH_3OH:H_2O$ (1:1); (10) final LC purification on ODS-7 column with $CH_3OH:CH_3CN:H_2O$ (2:1:2).

Extraction of YTXs from phytoplankton using methanol as extractant was achieved with a simplified procedure described by Satake et al. (39).

Biochemical Assay

Protein Phosphatase 2A Inhibition Assay Unlike OA and its derivatives, PTXs do not inhibit protein phosphatase 2A (PP2A). YTX showed a weak inhibiting effect against PP2A, which is lower than that of OA by four orders of magnitude. Therefore, interference of YTX on PP2A assay for OA and DTX-1 has been disregarded (53) and methods based on such inhibition are not effective to detect DSP toxins other than OA and its analogues in DSP-contaminated material.

Immunoassay In order to develop a method for rapid testing of PTXs, a monoclonal antibody using PTX-6 as hapten was developed and its cross reactivity against OA, DTX-1, PTX-1, PTX-6, and YTX evaluated (63). However, the suitability of the immunological method to measure PTXs in contaminated biological material still has to be demonstrated.

Cytotoxicity In vitro biological methods can provide useful toxicity information and represent convenient tools for screening and research on toxins. Using light and scanning electron microscopy, Aune et al. (50) observed the morphological effects on freshly prepared hepatocytes from rats caused by their exposure to four purified algal toxins: OA, DTX-1, PTX-1, and YTX. The exposure to PTX-1 at concentrations of up to 5 μg/mL led to evident formation of dose-dependent vacuolization. The cells kept their spherical appearance and no blebs were observed on the cell surface. Effects of YTX following exposure to concentrations of up to 10 μg/mL consisted of the appearance of tiny round blebs on the cell surface. Effects were dose dependent and, as the YTX concentration increased to 30 and 50 μg/mL, the cells maintained their spherical shape, but their surfaces were studded with semilarge round blebs. We proposed hepatocytes as being a possible means to discriminate diarrhogenic toxins from YTXs and PTXs as well as to indicate the presence of non-DSP toxins. However, several disadvantages related to the hepatocyte assay, such as time consumption and confusing results in case of mixtures of algal toxins, suggest its use may be combined with chemical methods.

In a different approach, PTX-1 and other toxins (i.e., microcystin-LR, nodularin, OA, DTX-1, and calyculin A) were tested for their ability to induce apoptotic regulated cell death in animal hepatocytes (52). Since sea animals are more directly exposed to marine toxins than terrestrial animals, sensitivity of salmon and rat hepatocytes were compared. Electron and light microscopic observation proved all the toxins under investigation to induce apoptosis in both hepatocytes in less than 2 h. Effects of PTX-1 on cells were evident following exposure to 2 μM of the compound. Salmon hepatocytes showed slightly higher overall sensitivity toward algal toxins than rat hepatocytes. The apoptotic phenotype induced by PTX-1 was distinct from that induced by other toxins and consisted of an extreme nuclear and cellular shrinkage and chromatin hypercondensation in both rat and salmon cells. An assay was then suggested using hepatocytic apoptosis as parameter for the detection of algal toxins. Since hepatocytes can easily be cultured at volumes of 50 μL, the detection limit of the assay for PTX-1 was estimated to be ~10–20 pmol. Moreover, since subapoptotic concentrations of the toxins inhibited aggregation of cells, the latter might be eventually used as screening criterion enhancing the sensitivity of the method.

Instrumental Methods

Instrumental methods are the most promising alternative to biological assays for control of all groups of aquatic toxins. They are sensitive and amenable to full automation, and they allow precise quantitation of known toxins for screening and confirmatory purposes. More importantly, methods using chromatographic and spectroscopic techniques, such as LC-MS and LC-MS-MS, are unsurpassed tools for unambiguous confirmation of known toxins on a structural basis and identification of novel compounds. They are particularly useful in PTX and YTX analysis, since they can identify the analytes even when reference standards are unavailable.

Conventional Chromatographic Methods LC techniques using conventional detection allow the separation and the specific and sensitive detection of individual toxins in complex biological materials. Thus, they represent a great promise for routine measurement of DSP toxins in phytoplankton and shellfish both for monitoring and research purposes. A major drawback of conventional LC techniques is the lack of analytical standards for several DSP toxins, especially for PTX and YTX groups.

Like OA and its derivatives, PTX and YTX molecules lack a strong chromophore, useful for sensitive UV absorption, and they are not naturally fluorescent. Chemical modification of the original toxin molecule is therefore required to improve the detectability of PTXs and YTXs by conventional techniques. They are usually derivatized with fluorescent labeling reagents before LC analysis and highly selectively and sensitively detected by measuring the fluorescence intensities of the toxin adducts. Depending on the functional groups present in the molecules, different fluorescent tags are used. Figure 6 shows the derivatization reactions adopted to produce fluorescence adducts for PTXs and YTXs.

Toxins having carboxylic groups, that is, PTX-6, PTX-7, and PTX-2SAs, are derivatized by esterification with ADAM (6,9–10,63) or BAP (10). The primary hydroxil group of PTX-1 and PTX-4 is derivatized with 1-anthroylnitrile (AN) (6,63). Alternatively, the conjugated diene present in the molecules of PTXs and of almost all the YTXs is derivatized with the dienophyle reagent 4-[2-(6,7-dimethoxy-4-methyl-3-oxo-3,4-dihydroquinoxalinyl)ethyl]-1,2,4-triazoline-3,5-dione (DMEQ-TAD) (26,64).

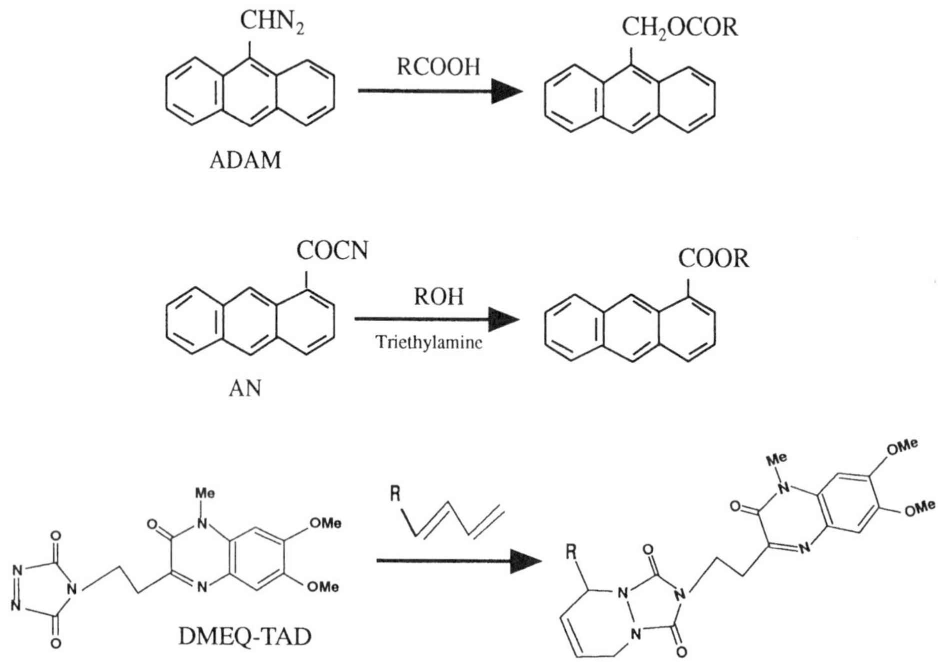

Figure 6 Derivatization reactions for PTXs and YTXs.

Conventional Chromatographic Methods for the PTXs. Determination of PTX-6 and PTX-7 in shellfish and PTX-2SAs in shellfish and phytoplankton was achieved by precolumn derivatization with ADAM (6,9–10,63) or BAP (10) reagent as for the OA class of DSP toxins (65).

Toxins were extracted from shellfish hepatopancreas and treated with ADAM, which reacts with carboxylic acid to give a fluorescent ester. The reaction was carried out without catalyst at room temperature in 60 min. Esterified toxins were purified on a silica cartridge column and finally injected in a HPLC-FLD equipped with a C_{18} column using $CH_3CN/CH_3OH/H_2O$ (8:1:1) as mobile phase. The fluorescence intensities of the analytes were monitored by setting excitation and emission wavelength at 365 and 412 nm, respectively. PTX-6 and PTX-7 eluted before OA in the chromatogram of a DSP toxin mixture. The method can be affected by interference owing to the instability of the reagent and the presence of free fatty acids in the digestive glands, which can produce fluorescent adducts with ADAM, thus resulting in complex chromatographic profiles. The selective extraction of toxins and clean-up procedures are therefore crucial to obtain specific and accurate results.

PTX-2SAs have been detected in phytoplankton and shellfish following derivatization with either ADAM (9–10) or BAP (10). In both cases, LC separation of the esterified analytes from OA and DTX-2 was achieved isocratically on an ODS column with a mobile phase of $CH_3CN/CH_3OH/H_2O$ (80:5:15) at a flow rate of 0.5 mL/min. Both ADAM and BAP derivatives of OA, PTX-2SA, and 7-*epi*-PTX-2SA eluted before the corresponding DTX-2 esters in the chromatograms.

PTXs containing hydroxylic groups, such as PTX-1 and PTX-4, have been derivatized using AN (6,63). AN reacts with primary hydroxyl groups under very mild conditions, producing highly fluorescent esters. It had been previously used for LC analysis of bile acids in blood (66). The extraction procedure and pretreatment of the derivatives were nearly the same as those adopted for acidic PTXs. The mixture was heated at 60°C for 30 min in acetonitrile in the presence of triethylamine. The derivatized toxins eluted on a C_{18} column using CH_3CN/H_2O (8:2) as mobile phase. Fluorescence detection was performed at 365 and 465 nm for excitation and emission, respectively.

Owing to the lack of carboxylic acid and hydroxylic group, PTX-2 and PTX-3 cannot be derivatized using the labeling reagent mentioned above. First, these toxins were analyzed using a low-sensitivity LC-UV method. Samples were eluted on C_{18} column using CH_3CN/H_2O (6.5:3.5) mixture as mobile phase and monitoring elution at 235 nm (6).

Because of the presence of a conjugated diene at the 28–31 position of PTXs, also the dienophile reagent, DMEQ-TAD, successfully used for YTX analysis, has been recently proposed as a possible alternative labeling reagent for PTXs (64).

Conventional Chromatographic Methods for the YTXs. As YTXs possess a conjugated diene in the side chain, formation of fluorescent adducts with a dienophile reagent has been hypothesized. Yasumoto and Takizawa successfully proposed DMEQ-TAD, previously employed in assaying vitamin D metabolites (67), for the fluorimetric measurement of YTXs in shellfish (26).

YTXs were extracted from homogenated scallops (1 g) with 9 mL of CH_3OH/H_2O (8:2). An aliquot (0.5 mL) of the centrifuged extract was added with 1.5 mL of 20 mM phosphate buffer (pH 5.8) and purified on Sep-pack C_{18} column using 6 mL of CH_3OH/H_2O (2:8) and 6 mL of CH_3OH/H_2O (7:3) for washing and elution, respectively. After solvent evaporation, the residue was added with a 50 μL of DMEQ-TAD solution (0.1% solution in CH_2Cl_2) and the reaction mixture was kept at room temperature in the dark for 2 h. Excess of reagent was quenched with methanol (50 μL). After solvent removal, the reaction residue was loaded onto Sep-pack C_{18} column using 0.5 mL of CH_3OH/H_2O (3:7). Purification was performed using 8 mL of CH_3OH/H_2O (3:7) and 8 mL CH_3OH/H_2O (7:3) for washing and elution, respectively. After solvent removal, the residue was suspended in CH_3OH (50 μL) and injected (10 μL) into the LC system.

Chromatographic conditions were the following: C_{18} column; temperature: 35°C; mobile phase of 40 mM phosphate buffer (pH 5.8) and CH_3OH (3:7) used for LC separation of YTX and 45-OH-YTX, and of 35 mM phosphate buffer (pH 5.8) and CH_3OH (4:6) for separation of 45,46,47-trinorYTX; flow rate of 1 mL/min; excitation and emission wavelengths: 370 and 440 nm, respectively.

DMEQ-TAD-YTX adducts gave two peaks due to the formation of two C-42 epimers (Figure 7) in a ratio approximately equal to 3:1 (Figure 8). The method proved linear and free from interference. It is also able to detect and distinguish YTX among its congeners 45-OH-YTX and 45,46,47-trinorYTX but does not allow separation between YTX and homoYTX (35). The detection limit was estimated to be 1 ng of injected YTX, which results in a 2000-fold greater sensitivity than the mouse bioassay. The method was also applied for analysis of YTX in phytoplankton (35).

LC-MS and LC-MS-MS Methods. Improvements in analytical methods and instruments play a critical role in expanding the opportunities of knowledge in any scientific field. The success of MS as an analytical technique is due to its potential to supply definitive qualitative and quantitative information on molecules based on their structural composition. Despite the extensive modifications and developments in MS, its original basis remains unchanged, the technique allowing the separation, detection, and recording of the ions formed by ionization of molecules. Depending on the nature of the ionization and on the molecular structure, intact ionized molecules and/or fragments of original molecules are produced. Significant advances in MS, which are having a tremendous impact on the investigation of biomolecules, mainly concern initial sample introduction, including prior separation via chromatography, soft ioniza-

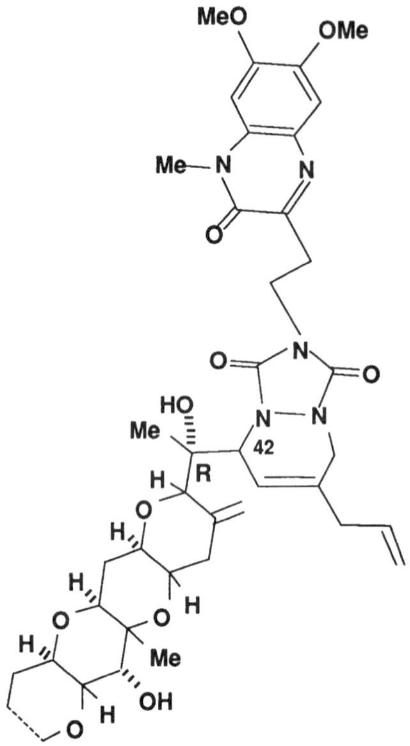

Figure 7 Structure of the fluorescent adduct DMEQ-TAD-YTX.

tion techniques compatible with biomolecule lability, and computerized techniques for processing data.

In the last two decades the on-line coupling of the two powerful analytical techniques, LC and MS, has become a robust, routinely applicable and widespread analytical tool. This can be mainly attributed to the introduction of soft ionization techniques, such as FAB and atmospheric pressure ionization (API), which elude the difficulties of ionization of nonvolatile and/or thermally labile analytes and give abundant molecular related ions with negligible fragmentation. Additionally, the development of API LC-MS interfaces, such as electrospray and pneumatically assisted electrospray (i.e., ionspray [68]), allowed us to overcome the difficulties due to the introduction of the high flow rate of LC effluent into the high vacuum of MS. API interfaces are based on ionization of analytes in solution and transferring of ionic species by ion evaporation for MS analyses.

Soft ionization LC-MS can provide many benefits, such as the prompt measurement of molecular mass and/or peak purity assessment. Moreover, the on-line coupling of API LC-MS with conventional detectors, such as UV-DAD, supplies the researcher with a fast integration of different analytical data for the compounds under investigation.

Although the lack of fragmentation typical of mild ionization techniques represents a significant advantage in many circumstances, it can hamper the achievement of other useful structural information. The latter can, however, be produced with relative simplicity through CID, which is usually performed by tandem mass spectrometry, employing two stages of mass analysis in order to examine the fragmentation of particular ions in ion mixtures.

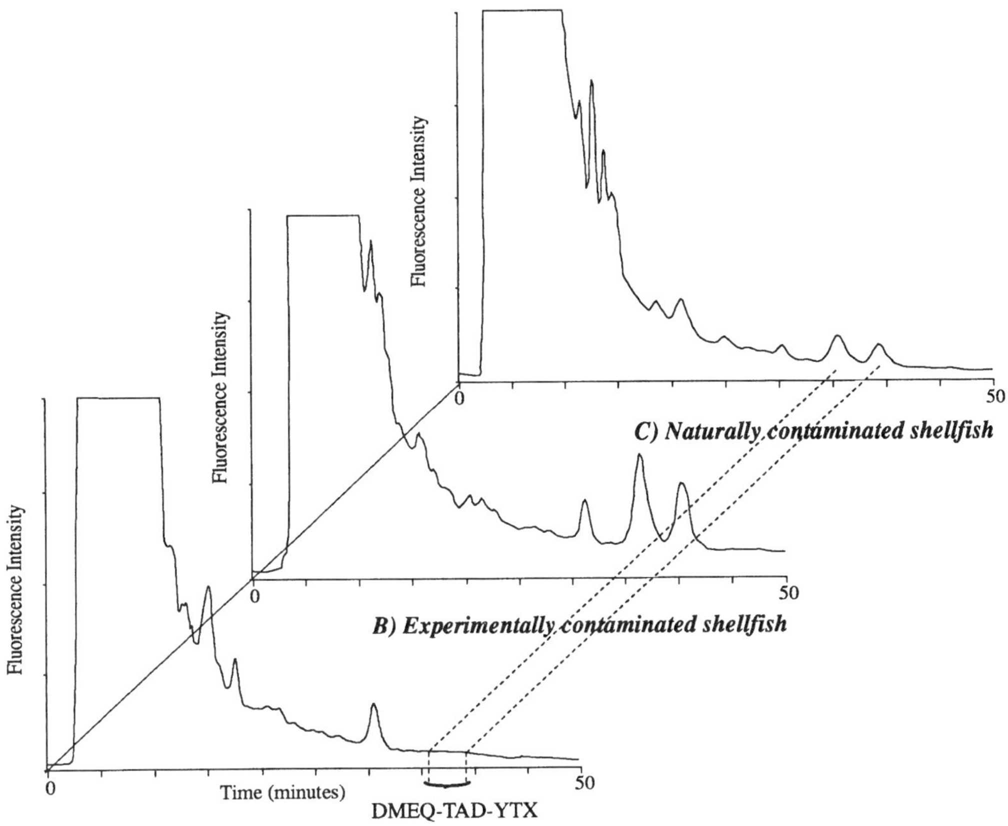

Figure 8 Fluorometric LC chromatogram of blank mussels, *M. galloprovincialis* (A), of the same mussels experimentally contaminated with YTX (B) and of naturally contaminated toxin mussels from the Adriatic Sea (C).

Triple quadrupole is a widely used tandem mass spectrometer in which only the first and third quadropoles are mass analyzers operating with a combination of both r.f. and d.c. potentials necessary for mass separation and detection. The second quadrupole is used as a collision cell with ion-focusing properties. Since it only operates with a fixed r.f. voltage, ions of any *m/z* can pass this quadrupole and undergo collision induced dissociation (CID) by ion-molecule interactions with a collision neutral gas. Among the different scan modes available using MS-MS, the most common is the product ion scan. Ions of a given *m/z* value, the so-called precursors, are selected by the first mass spectrometer and focused into the collision cell where CID is obtained. The population of fragment ions thus produced is then passed into the second mass analyzer, which is set to scan over an appropriate mass range, thus producing full-scan product ion mass spectra.

An alternative to MS-MS, CID information can also be gained from a single MS analyzer in various ways, including accelerating the voltage in the free-jet expansion region, which is the so-called *up-front* CID.

Several reports on API LC-MS and MS-MS showed the technique fit to meet a number of complex analytical needs, such as identifying known and unknown products unambiguously, profiling impurities in isolated material, investigating biotransformation of bioactive molecules, characterizing metabolites, and performing highly sensitive and specific qualitative and quantitative analyses with minimal sample preparation and short analysis time even when only a limited quantity of sample is available.

The attention of scientists from many different research fields is therefore increasing being focused on the API LC-MS and LC-MS-MS approach. Furthermore, the production of many manufacturers of robust and user-friendly automated instruments, requiring only minor operator skills, and the reduction of prices have combined to broaden API LC-MS techniques from the esoteric field of MS science to the routine activity of many analytical chemistry laboratories.

These developments grant important benefits in the field of DSP toxin control in which several analytical problems exist. Toxins significantly differing in chemical structures are frequently present in trace amounts in complex and variable biological matrices in which biotransformation of the original compounds can also occur. In addition, new toxins are frequently discovered and analytical standards are available only for a few known toxins. In many cases, particularly for phytoplankton, only small amounts of samples are available for the analytical determination; thus precluding the employment of other spectrometric techniques such as NMR. It should also be considered that methods for official control of DSP toxins are required to provide analytical performances suitable for legal actions. Finally, the necessity to monitor high numbers of samples requires versatile, rapid, and highly productive instrumental tools, preferably automated, requiring only minimal sample preparation.

The LC-MS and LC-MS-MS approaches can offer many advantages for the identification and quantification of DSP toxins, particularly for PTXs and YTXs. These, however, are polar, intractable compounds considered to be unlikely candidates for direct MS analysis using classic techniques, such as electron or chemical ionization, requiring analyte volatility. On the other hand, PTXs and YTXs are amenable to the acquisition of charge in solution and therefore well suited to soft ionization LC-MS techniques. FAB MS was originally used for structural confirmation of PTX-2 by Lee et al. (17). In this study, intact PTX-2 was collected from the outlet of a UV detector during LC analyses of toxic phytoplankton. Confirmation of PTX-2 was obtained by negative ion FAB MS and using glycerol as matrix, through the observation of the signal due to the deprotonated molecule, $[M\text{-}H]^-$, of the analyte. API LC-MS and MS-MS techniques were first used by Draisci et al. for the identification of PTX-2 in phytoplankton without a reference standard (20). The presence of PTX-2 was initially suspected following the detection of a chromatographic peak with strong UV adsorption at 235 nm in the LC-UV-DAD analyses of crude methanol extracts of toxic phytoplankton. Studies were carried out by ionspray-positive ion LC-MS on toxic phytoplankton under mild ionization conditions. Positive ionspray mass spectra of the suspected PTX-2 were dominated by the protonated molecule $[M+H]^+$ of the analyte; thus showing the same molecular weight as PTX-2 for the investigated compound. Confirmation of the toxin was finally performed through identification of structurally informative fragments by low-energy full-scan LC-MS-MS experiments.

Similar studies based on a combined approach involving MS and MS-MS are particularly useful for the development of selected reaction monitoring (SRM) LC-MS-MS analytical methods. In this approach, the first mass spectrometer is set to pass precursor ions with a particular, selected *m/z* value. After CID of precursor ions, which is performed into the collision cell, only ions with a given *m/z*, the so-called diagnostic, or confirmatory, ions, are selected by the second mass analyzer and transmitted to the detector. The SRM LC-MS-MS allows a high degree of specificity, since analyte identification is based on retention time, molecular weight (if a molecular related ion is chosen as precursor), and structural information, such as the presence of diagnostic fragments for each analyte and ion ratios. Interestingly, CID in the collision cell of MS-MS instruments usually produces the same fragment ions obtained by CID between the ion sampling orifice and the mass analysis region; thus allowing very high specificity in confirmatory analyses also using the less expensive single MS instruments. In some instances, however, owing to the lack of precursor ion selection before CID, a chromatographic separation of the analyte is required to remove sample interference.

The suitability of the SRM LC-MS-MS method was proven by the analysis of DSP toxins in crude methanol algal extracts of *D. fortii* and of the neutral and acidic fractions obtained by the same extract fractionated on alumina column (Figure 9). The ability to identify DSP toxins of both acidic and neutral classes with high specificity, minimal sample preparation, and no reference standard is easily inferred from similar data. Quantitative analyses are, however, precluded without reference standards. LC-UV on-line with electrospray MS was also used for determination of PTX-2 from toxic phytoplankton and shellfish (8). In addition to the protonated molecule of PTX-2 which was generated exclusively by the ionspray LC-MS technique, other related molecular ions, such as ammonium and sodiated ions, were observed in positive electrospray ion mass spectra; maybe on account of differences in the droplet desolvation process during ionization (8).

Investigations on new DSP toxins are typically based on a combined approach involving single MS and MS-MS. After the isolation of interesting compounds, the first simple stage consists of the introduction of analyte solutions into MS under mild ionization conditions. This is usually performed without chromatographic separation by direct infusion or flow injection analysis (FIA). These techniques allow fast optimization of source/interface parameters and mobile phase composition, as well as accumulation of several spectra, which can reduce background interference and facilitate data interpretation. Once the precursor ion of interest has been determined, the same experiments may be repeated under CID conditions, and the corresponding product ion scan mass spectra obtained. Interpretation of spectra can be performed by reference to model compounds and to related mass spectral behavior from known structures. The creation and updating of MS and MS-MS spectra databases of marine toxins and related compounds by laboratories is therefore particularly helpful in this context.

The above analytical strategy was successfully used for the identification of two new PTX-2 analogues in phytoplankton and shellfish (9–10). Molecular weight for the compounds under investigation was determined through electrospray, ionspray, and FAB MS and resulted in being 18 mass units larger than PTX-2 (molecular weight 858) for both compounds. CID from each precursor ion, as performed both in the positive and negative ion mode, produced mass spectra that were very similar to those obtained for authentic PTX-2. MS and MS-MS data were interpreted together with the observation that the compounds gave fluorescent derivatives with ADAM and BAP; thus showing the presence of carboxylic acid in the molecule. It was therefore supposed that the two analytes were stereoisomers and that they possessed an open chain structure rather than the large lactone ring typical of PTXs. The hypothesis was also confirmed by the MS-MS fragmentation pattern proposed for the analytes (9–10) and by the complete structural elucidation on the compounds finally performed by NMR (9).

The first MS studies on YTX were performed by Murata et al. (2). The negative FAB MS spectra for purified YTX exhibited two molecular related ions, including the desodiated ion.

A number of toxic fractions isolated from Japanese shellfish were identified on a structural basis through negative FAB LC-MS using a reversed phase column and methanol with 1% glycerol as mobile phase (24). Confirmation of YTX was obtained by the observation of the desodiated ion of the analyte in the higher field region of the negative ion mass spectrum obtained from the FAB LC-MS analysis of the suspected fraction.

Similar negative FAB MS data for YTX were also produced by Yasumoto et al. (6). Negative FAB MS was also used in the same research to achieve the molecular weight of 45-OH-YTX, although no detailed spectral data for the toxins were provided.

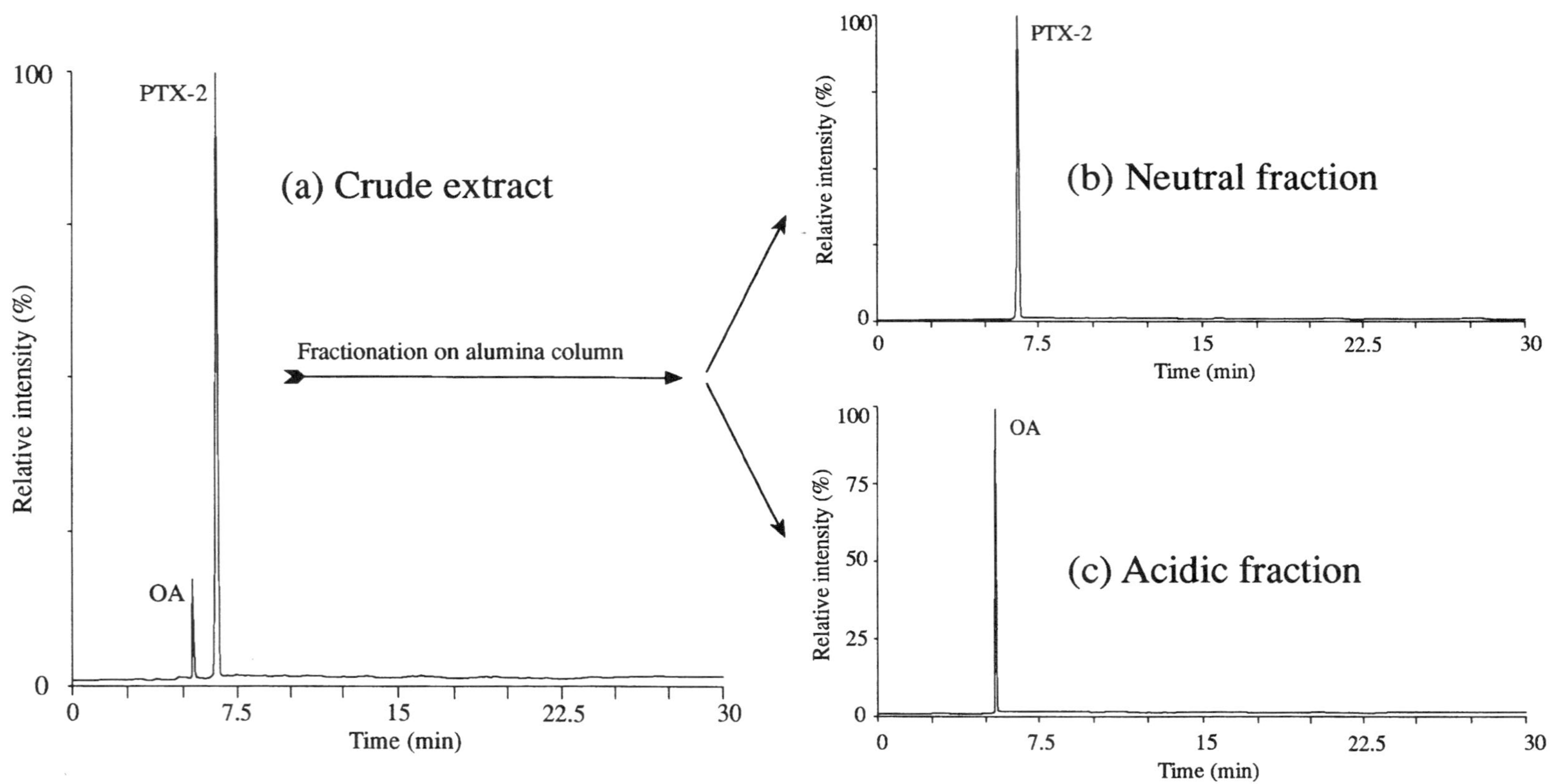

Figure 9 SRM LC-MS-MS chromatograms of the analysis of a crude methanolic *D. fortii* extract (a), and of the neutral (b) and acidic (c) fraction obtained by fractionation of the extract on basic alumina column (20).

The applicability of tandem MS to the structural elucidation of polyether compounds was well demonstrated by Naoki et al. for YTX (12) as well as for several polyhydroxy-polyether compounds including OA, DTX-1, DTX-2, and brevetoxins (69). FAB MS-MS represents a useful complementary method to NMR for structural elucidation of toxins having large molecular size, repetitive structural units, and/or poor crystallinity. NMR methodology alone could be insufficient for structural elucidation of these compounds on account of the heavy overlapping of NMR signals. Extensive fragmentation of YTX produced FAB MS-MS spectra which were particularly informative, especially regarding repetitive portions of the toxin structure giving a regular fragmentation pattern. Thus, on the basis of MS-MS data alone, the authors succeeded in assigning the number and location of 11 rings of YTX, confirming previous NMR data (12).

Negative electrospray MS was used for molecular weight determination of a number of YTX analogues, 45-OH-YTX, 45,46,47-trinorYTX (13), homoYTX, and 45-OH-homoYTX (14), isolated from infested shellfish. For each toxin, the desodiated molecular related ion was observed in the MS spectra. Confirmation of YTX by electrospray MS was also performed on toxins isolated from phytoplankton organisms (39).

The first LC-MS and MS-MS method for direct identification of YTX in shellfish samples has been recently reported by Draisci et al. (60). The method, using an API source and an ionspray interface, required preliminary studies by FIA-MS and MS-MS on a pure standard of YTX. Ionspray FIA MS spectra exhibited a weak signal corresponding to the deprotonated molecule of the analyte, which had not been previously observed through other MS ionization techniques. Other analytically useful molecular-related ions were observed. Among these, the $[M\text{-}2Na+H]^-$ ion was the most abundant and was therefore selected as precursor for CID MS-MS experiments. Structurally informative MS-MS spectra were then produced showing fragment ions similar to those produced by FAB MS-MS.

On the basis of MS and MS-MS characterization of YTX, the SRM LC-MS-MS method was finally developed. It is based on a simple acetone extraction of YTX from shellfish followed by reversed-phase LC-MS-MS with isocratic elution using aqueous acetonitrile containing ammonium acetate 4 mM as mobile phase. An instrumental detection limit of ~50 pg is estimated; it is ~400 times lower than the limit of the mouse bioassay, and proves the sensitivity of the SRM LC-MS-MS method similar to that of the fluorimetric LC assay. The method was demonstrated to be specific, sensitive, and rapid for the determination of YTX in shellfish and phytoplankton (Figure 10), and it also allowed for unambiguous showing of false-negative results of mouse bioassay in a number of mussel samples (36,60).

In our further research on the use of LC-MS methods for the analysis of DSP toxins, we have recently developed a new LC-MS approach using an API source and an ionspray interface for the direct determination of known and new DSP toxins in mussels and phytoplankton (23). The proposed approach involves a first reversed-phase negative ion mode SIM LC-MS analysis in order to detect eleven DSP toxins (i.e., OA, DTX-1, DTX-2, DTX-2B, DTX-2C, PTX-2, PTX-2SA, 7-*epi*-PTX-2SA, YTX, homoYTX, 45-OH-YTX) in the same chromatographic run with a total time of 20 min. Chromatographic separation is performed isocratically with a mobile phase of acetonitrile-water (80:20 v/v) containing 2 mM ammonium acetate. As shown in Figure 11a, under the adopted conditions, a significant overlapping is observed for YTXs as well as for OA and its isomers, which eluted around the same retention time as the unresolved PTX-2SAs. Nevertheless, the specificity of SIM LC-MS allows the unambiguous identification of toxins having different molecular masses even if they are chromatographically unresolved. Thus, the above SIM LC-MS method is appropriate as a first-stage analysis for the unambiguous identification of DTX-1, PTX-2, and YTXs and for the screening of OA isomers and PTX-2 analogues. Complete chromatographic separations of the latter toxins are achieved by reversed-phase, positive ion mode LC-MS (Figure 11b) using the same LC column and mobile phase

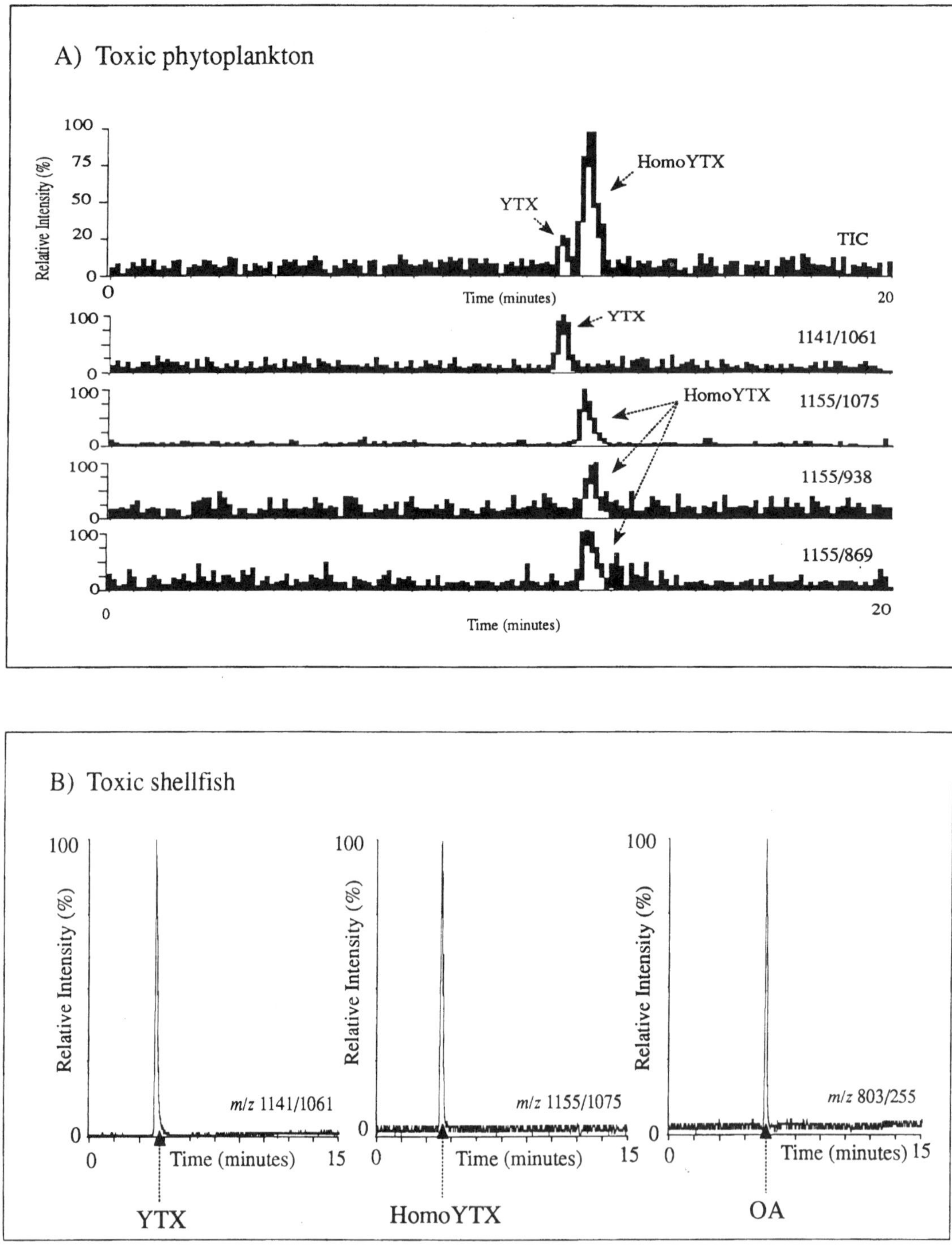

Figure 10 SRM LC-MS-MS chromatograms of the analysis of extracts of toxic phytoplankton, *L. polyedrum* (A) and mussels, *M. galloprovincialis* (B), from the Adriatic Sea.

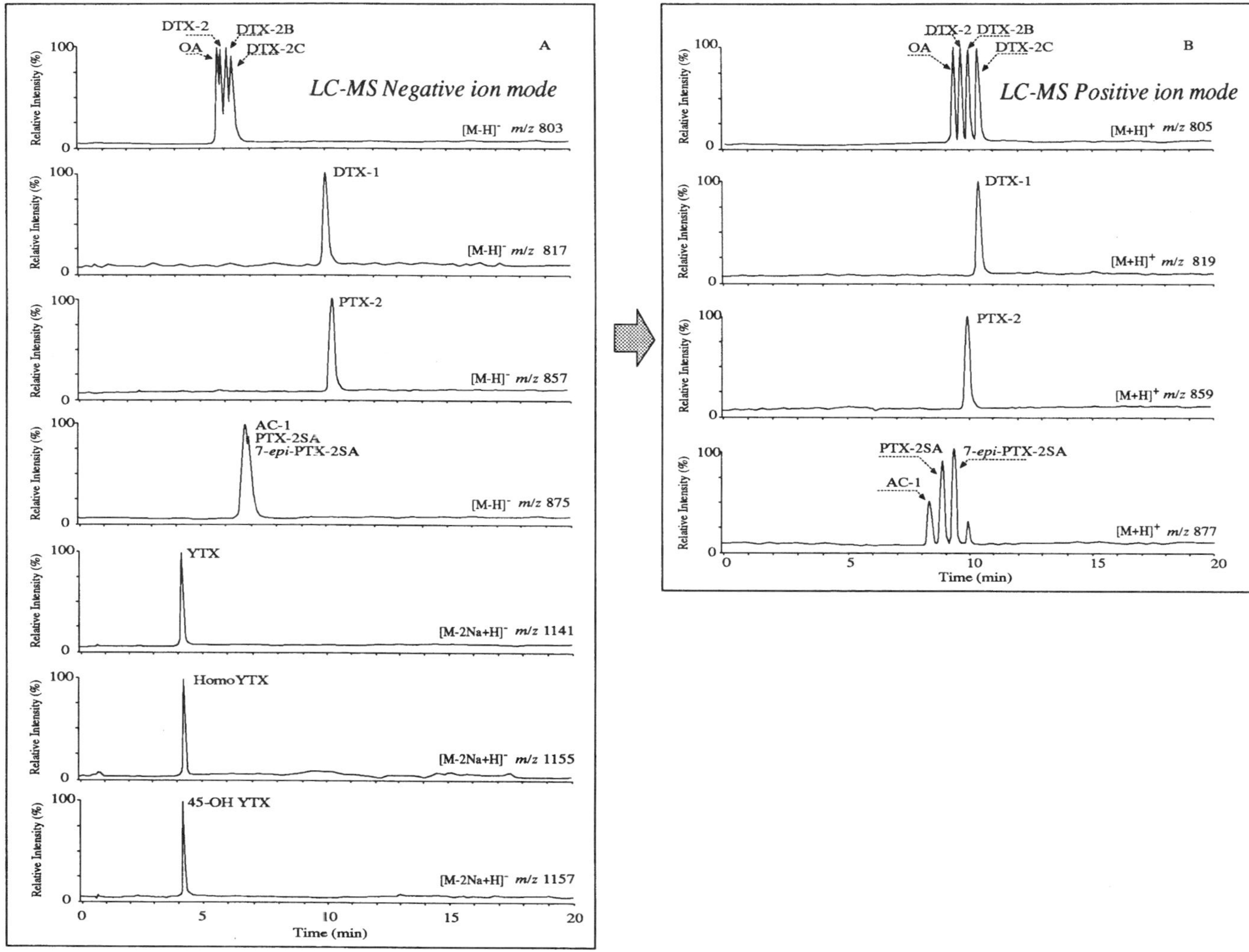

Figure 11 SIM LC-MS chromatograms of the analysis of a mixture of DSP toxins isolated from toxic marine material in (A) negative and (B) positive ion mode.

composition except that TFA is added instead of ammonium acetate. Detection limits for the investigated toxins range from 100 to 400 pg of injected material; thus showing high sensitivity of the method for all the analytes. Although a full validation of the LC-MS methods is still hampered by the limited availability of calibrants, the above analytical approach represents a very promising and effective alternative to the existing biological assays for the analyses of known and new DSP toxins in marine biological material both in monitoring and research work.

In conclusion, the need for replacing animal testing requires the development, validation, and diffusion of sensitive and specific instrumental methods. The production of reference materials is a crucial support for this task, and pilot studies already conducted by Japanese regulatory agencies (70) should be replicated in other countries.

API LC-MS proved to be a powerful approach for direct determination not only of PTXs (8–10,20) and YTXs (60), but also for DSP toxins of the OA class (22,31–32,71–76). Moreover, recent developments in LC-MS methodology allow the simultaneous detection of new and known toxins of all classes with high sensitivity and specificity, minimal sample preparation, and high throughput (23). Technological improvement involving both refined design and lower costs of LC-MS instruments as well as highly automated systems are making this technique a viable analytical tool in many laboratories (77). The LC-MS approach has thus been recently proposed as a universal method for marine toxins (78), and its increasing use in the field is apparent in the periodic AOAC reports on phycotoxin research (79–80).

REFERENCES

1. T Yasumoto, M Murata, Y Oshima, GK Matsumoto, J Clardy. Diarrhetic shellfish poisoning. In: EP Ragelis, ed. Seafood toxins, ACS Symposium Series no. 262. Washington, DC: American Chemical Society, 1984, pp 207–214.
2. M Murata, M Kumagai, JS Lee, T Yasumoto. Isolation and structure of yessotoxin, a novel polyether compound implicated in diarrhetic shellfish poisoning. Tetrahedron Lett 28:5869–5872, 1987.
3. K Sasaki, JLC Wright, T Yasumoto. The identification and characterization of pectenotoxin (PTX) 4 and PTX7 as spiroketal stereoisomers of two previously reported pectenotoxins. J Org Chem 63: 2475–2480, 1998.
4. T Yasumoto, M Murata, Y Oshima, M Sano, GK Matsumoto, J Clardy. Diarrhetic shellfish toxins. Tetrahedron 41:1019–1025, 1985.
5. M Murata, M Sano, T Iwashita, H Naoki, T Yasumoto. The structure of pectenotoxin-3, a new constituent of diarrhetic shellfish toxins. Agric Biol Chem 50:2693–2695, 1986.
6. T Yasumoto, M Murata, JS Lee, K Torigoe. Polyether toxins produced by dinoflagellates. In: S Natori, K Hashimoto, T Ueno, eds. Mycotoxins and Phycotoxins '88. Amsterdam: Elsevier, 1989, pp 375–382.
7. K Sasaki, M Satake, T Yasumoto. Identification of the absolute configuration of pectenotoxin-6, a polyether macrolide compound, by NMR spectroscopic method using a chiral anisotropic reagent, phenylglycine methyl ester. Biosci Biotech Biochem 61:1783–1785, 1997.
8. T Suzuki, T Mitsuya, H Matsubara, M Yamasaki. Determination of pectenotoxin-2 after solid-phase extraction from seawater and from the dinoflagellate *Dinophysis fortii* by liquid chromatography with electrospray mass spectrometry and ultraviolet detection. Evidence of oxidation of pectenotoxin-2 to pectenotoxin-6 in scallops. J Chromatogr A 815:155–160, 1998.
9. M Daiguji, M Satake, KJ James, A Bishop, L Mackenzie, H Naoki, T Yasumoto. Structures of new pectenotoxin analogs, pectenotoxin-2 seco acid and 7-*epi*-pectenotoxin-2 seco acid, isolated from a dinoflagellate and greenshell mussels. Chem Lett 7:653–654, 1998.
10. KJ James, AG Bishop, R Draisci, L Palleschi, C Marchiafava, E Ferretti, M Satake, T Yasumoto. Liquid chromatographic methods for the isolation and identification of new pectenotoxin-2 analogues from marine phytoplankton and shellfish. J Chromatogr A 844:53–65, 1999.

11. H Takahashi, T Kusumi, Y Kan, M Satake, T Yasumoto. Determination of the absolute configuration of yessotoxin, a polyether compound implicated in diarrhetic shellfish poisoning, by NMR spectroscopic method using a chiral anisotropic reagent, methoxy-(2-naphtyl) acetic acid. Tetrahedron Lett 37:7087–7090, 1996.
12. H Naoki, M Murata, T Yasumoto. Negative-ion fast-atom bombardment tandem mass spectrometry for the structural study of polyether compounds: structural verification of yessotoxin. Rapid Commun Mass Spectrom 7:179–182, 1993.
13. M Satake, K Terasawa, Y Kadowaki, T Yasumoto. Relative configuration of yessotoxin and isolation of two new analogs from toxic scallops. Tetrahedron Lett 37:5955–5958, 1996.
14. M Satake, A Tubaro, JS Lee, T Yasumoto. Two new analogs of yessotoxin, homoyessotoxin and 45-hydroxyhomoyessotoxin, isolated from mussels of the Adriatic Sea. Nat Toxins 5:107–110, 1997.
15. P Ciminiello, E Fattorusso, M Forino, S Magno, R Poletti, R Viviani. Isolation of Adriatoxin, a new analog of yessotoxin from mussels of the Adriatic Sea. Tetrahedron Lett 39:8897–8900, 1998.
16. MJ Holmes, RJ Lewis, AWW Hoy. Cooliatoxin, the first toxin from *Coolia monotis* (Dinophyceae). Nat Toxin 3:355–362, 1995.
17. JS Lee, T Igarashi, S Fraga, E Dahl, P Hovgaard, T Yasumoto. Determination of diarrhetic shellfish toxins in various dinoflagellate species. J Appl Phycol 1:147–152, 1989.
18. JS Lee, M Murata, T Yasumoto. Analytical methods for determination of diarrhetic shellfish toxins. In: S Natori, K Hashimoto, T Ueno, eds. Mycotoxins and Phycotoxins '88. Amsterdam: Elsevier, 1989, pp 327–334.
19. JH Jung, CJ Sim, CO Lee. Cytotoxic compounds from a two-sponge association. J Nat Prod 58: 1722–1726, 1995.
20. R Draisci, L Lucentini, L Giannetti, P Boria, R Poletti, First report of pectenotoxin-2 (PTX-2) in algae (*Dinophysis fortii*) related to seafood poisoning in Europe. Toxicon 34:923–935, 1996.
21. KA Steidinger. Some taxonomic and biologic aspects of toxic dinoflagellates. In: IR Falconer, ed. Algal Toxins in Seafood and Drinking Water. New York: Academic Press, 1993, pp 1–28.
22. R Draisci, L Giannetti, L Lucentini, C Marchiafava, KJ James, AG Bishop, BM Healy, SS Kelly. Isolation of a new okadaic acid analogue from phytoplankton implicated in diarrhetic shellfish poisoning. J Chromatogr. A 798:137–145, 1998.
23. R Draisci, L Palleschi, L Giannetti, L Lucentini, KJ James, AG Bishop, M Satake, T Yasumoto. New approach to the direct detection of known and new diarrhoeic shellfish toxins in mussels and phytoplankton by liquid chromatography-mass spectrometry. J Chromatogr. A 847:213–221, 1999.
24. JS Lee, K Tangen, E Dahl, P Hovgaard, T Yasumoto. Diarrhetic shellfish toxins in Norwegian mussels. Nippon Suisan Gakk 54:1953–1957, 1988.
25. J Zhao, G Lembeye, G Cenci, B Wall, T Yasumoto. Determination of okadaic acid and dinophysistoxin-1 in mussel from Chile, Italy and Ireland. In: TJ Smayda, Y Shimizu, eds. Toxic Phytopl Blooms in the Sea. Amsterdam: Elsevier, 1993, pp 587–592.
26. T Yasumoto, A Takizawa. Fluorimetric measurement of yessotoxins in shellfish by high-pressure liquid chromatography. Biosci Biotech Biochem 61:1775–1777, 1997.
27. L Boni, L Mancini, A Milandri, R Poletti, M Pompei, R Viviani. First cases of diarrhoetic shellfish poisoning in the northern Adriatic Sea. Sci Total Environ (suppl):419–426, 1992.
28. E Fattorusso, P Ciminiello, V Costantino, S Magno, A Mangoni, A Milandri, R Poletti, M Pompei, R Viviani. Okadaic acid in mussels of Adriatic Sea. Mar Poll Bull 24:234–237, 1992.
29. R Draisci, L Lucentini, L Giannetti, A Stacchini. Biotossine algali nei molluschi bivalvi: ottimizzazione di un metodo HPLC per la ricerca dell'acido okadaico. Riv Sci Aliment 4:443–454, 1993.
30. R Draisci, L Croci, L Giannetti, L Cozzi, L Lucentini, D De Medici, A Stacchini. Comparison of mouse bioassay, HPLC, and enzyme immunoassay methods for determining diarrhetic shellfish poisoning toxins in mussels. Toxicon 32:1379–1384, 1994.
31. R Draisci, L Lucentini, L Giannetti, P Boria, A Stacchini. Detection of diarrhoetic shellfish toxins in mussels from Italy by ionspray liquid chromatography-mass spectrometry. Toxicon 33:1591–1603, 1995.
32. R Draisci, L Lucentini, L Giannetti, P Boria, A Stacchini, R Poletti. Biotossine algali DSP in molluschi bivalvi e fitoplancton del mare Adriatico. Riv Soc Ital Sci Alim 25:7–16, 1996.
33. R Draisci, L Giannetti, L Lucentini, P Boria, R Poletti, L Boni, E Funari. Diarrhoeic shellfish poison-

ing (DSP) toxins in *Dinophysis fortii* samples from the Adriatic Sea. In: M Miraglia, H van Egmond, C Brera, J Gilbert, eds. Mycotoxins and Phycotoxins—Developments in Chemistry, Toxicology and Food Safety. Fort Collins, CO: Alaken, 1998, pp 455–462.
34. P Ciminiello, E Fattorusso, M Forino, S Magno, R Poletti, M Satake, R Viviani, T Yasumoto. Yessotoxin in mussels of the northern Adriatic Sea. Toxicon 35:177–183, 1997.
35. A Tubaro, L Sidari, R Della Loggia, T Yasumoto. Occurrence of Yessotoxin-like toxins in phytoplankton and mussels from northern Adriatic sea. Proceedings of the VIII International Conference of Harmful Algae, Vigo, Spain, 1997, pp 470–472.
36. R Draisci, E Ferretti, L Palleschi, C Marchiafava, R Poletti, A Milandri, A Ceredi, M Pompei. High levels of yessotoxin in mussels and presence of yessotoxin and homoyessotoxin in dinoflagellates of the Adriatic Sea. Toxicon 37:1187–1193, 1999.
37. P Ciminiello, E Fattorusso, M Forino, S Magno, R Poletti, R Viviani. Isolation of 45-hydroxyyessotoxin from mussels of the Adriatic Sea. Toxicon 37:689–693, 1999.
38. L MacKenzie, P Truman, M Satake, T Yasumoto, J Adamson, D Mountfort, D White. Dinoflagellate blooms and associated DSP-toxicity in shellfish in New Zealand. Proceedings of the VIII International Conference of Harmful Algae, Vigo, Spain, 1997, pp 74–77.
39. M Satake, L MacKenzie, T Yasumoto. Identification of *Protoceratium reticulatum* as the biogenetic origin of yessotoxin. Nat Toxins 5:164–167, 1997.
40. LL Rhodes, AE Thomas. *Coolia monotis (Dinophyceae)*: a toxic epiphytic microalgal species found in New Zealand. NZ J Mar Fresh Res 31:139–141, 1997.
41. Y Hamano, Y Kinoshita, T Yasumoto. Suckling mice assay for diarrhoetic shellfish toxins. In: DM Anderson, AW White, DG Baden, eds. Toxic Dinoflagellates. New York: Elsevier, 1985, pp 383–388.
42. K Terao, E Ito, T Yanagi, T Yasumoto. Histopathological studies on experimental marine toxin poisoning. I. Ultrastructural changes in the small intestine and liver of suckling mice induced by dinophysistoxin-1 and pectenotoxin-1. Toxicon 24:1141–1151, 1986.
43. CY Kao. Tetrodotoxin, saxitoxin and their significance in the study of excitation phenomenon. Pharm Rev 18:997–1049, 1966.
44. F Gusovsky, JW Daly. Maitotoxin: a unique pharmacological tool for research on calcium-dependent mechanisms. Biochem Pharmacol 39:1633–1639, 1990.
45. TAJ Haystead, ATR Sim, D Carling, RC Honnor, Y Tsukitani, P Cohen, DG Hardie. Effects of the tumor promoter okadaic acid on intracellular protein phosphorylation and metabolism. Nature 337: 78–81, 1989.
46. T Wieland. Poisonous principles of mushrooms of the genus *Amanita*. Science 159:946–952, 1968.
47. M Saito. Liver cirrhosis induced by metabolites of *Penicillium islandicum* Sopp. Acta Pathol Jpn 9: 785–790, 1959.
48. H Nagai, M Satake, T Yasumoto. Antimicrobial activities of polyether compounds of dinoflagellate origins. J Appl Phycol 2:305–308, 1990.
49. M Murakami, K Makabe, K Yamaguchi, S Konosu. Goniodomin A, A novel poliether macrolide from the dinoflagellate *Goniodoma pseudogoniaulax*. Tetrahedron Lett 29:1149–1152, 1988.
50. T Aune, T Yasumoto, E Engeland. Light and scanning electron microscopic studies on effects of marine algal toxins toward freshly prepared hepatocytes. J Toxicol Environ Health 34:1–9, 1991.
51. ZH Zhou, M Komiyama, K Terao, Y Shimada. Effects of pectenotoxin-1 on liver cells *in vitro*. Nat Toxins 2:132–135, 1994.
52. KE Fladmark, MH Serres, NL Larsen, T Yasumoto, T Aune, SO Doskeland. Sensitive detection of apoptogenic toxins in suspension cultures of rat and salmon hepatocytes. Toxicon 36:1101–1114, 1998.
53. H Ogino, M Kumagai, T Yasumoto. Toxicologic evaluation of yessotoxin. Nat Toxins 5:255–259, 1997.
54. K Terao, E Ito, M Oarada, M Murata, T Yasumoto. Histopathological studies on experimental marine toxin poisoning-5. The effects in mice of yessotoxin isolated from *Patinopecten yessoensis* and of a desulfated derivative. Toxicon 28:1095–1104, 1990.
55. CH Wu, T Narahashi. Mechanism of action of novel marine neurotoxins on ion channels. Ann Rev Pharmacol Toxicol 28:41–61, 1988.

56. K Terao, E Ito, T Yasumoto, K Yamaguchi. Enterotoxic, hepatotoxic and immunotoxic effects of dinoflagellates toxins on mice. In: E Graneli, B Sundstroem, L Edler, DM Andreson, eds. Toxic Marine Phytoplankton. New York: Elsevier, 1990, pp 418–423.
57. YY Lin, M Risk, SM Ray, D VanEngen, J Clardy, JC James, K Nakanishi. Isolation and structure of brevetoxin B from the ''red tide'' dinoflagellate *Ptychodiscus brevis* (*Gymnodinium breve*). J Am Chem Soc 103:6773–6775, 1981.
58. S Amano, K Fujiwara, A Murai. The synthesis of the common C31-C40 fragment of pectenotoxins. Synlett 11:1300–1302, 1997.
59. T Yasumoto, Y Oshima, Yamaguchi. Occurrence of a new type of shellfish poisoning in the Tohoku district. Bull Jpn Soc Sci Fish 44: 1249–1255, 1978.
60. R Draisci, L Giannetti, L Lucentini, E Ferretti, L Palleschi, C Marchiafava. Direct identification of yessotoxin in shellfish by liquid chromatography coupled with mass spectrometry and tandem mass spectrometry. Rapid Commun Mass Spectrom 12:1291–1296, 1998.
61. C Marcaillou-Le Baut. Toxins involved in diarrhetic shellfish poisoning D.S.P. Oceanis 16:359–374, 1990.
62. M Kat. Diarrhetic mussel poisoning in the Netherlands related to the dinoflagellate *Dinophysis acuminata*. Antonie v Leeuwenhoek 49:417–427, 1983.
63. T Yasumoto, M Fukui, K Sasaki, and K Sugiyama. Determinations of marine toxins in foods. J AOAC Int. 78:574–582, 1995.
64. T Yasumoto, M Satake. New toxins and their toxicological evaluations. Proceedings of the VIII International Conference of Harmful Algae, Vigo, Spain, 1997, pp 461–464.
65. JS Lee, T Yanagi, R Kenma, T Yasumoto. Fluorometric determination of diarrhetic shellfish toxins by high performance liquid chromatography. Agric Biol Chem 51:877–881, 1987.
66. J Goto, M Saito, T Chikai, N Goto, T Nambara, Studies on Steroids. CLXXXVII. Determination of serum bile acids by high performance liquid chromatography with fluorescence labelling. J Chromatogr 276:289–300, 1983.
67. M Shimizu, S Kamachi, Y Nishii, S Yamada. Synthesis of a reagent for fluorescence-labeling of vitamin D and its use in assaying vitamin D metabolites, Anal Biochem 194:77–81, 1991.
68. AP Bruins, TR Covey and D Henion. Ionspray interface for combined liquid chromatography/athmospheric pressure ionization mass spectrometry. Analyt Chem 59:2642–2646, 1987.
69. H Naoki, T Fujita, M Satake, T Yasumoto. Identification of marine polyether compounds by tandem mass spectrometry. In: M Miraglia, H van Egmond, C Brera, J Gilbert, eds. Mycotoxins and Phycotoxins—Developments in Chemistry, Toxicology and Food Safety. Fort Collins, CO: Alaken, 1998, pp 469–474.
70. H Goto, T Igarashi, R Sekiguchi, K Tanno, M Satake, Y Oshima, T Yasumoto. A Japanese project for production and distribution of shellfish toxins as calibrants for HPLC analysis. Proceedings of the VIII International Conference of Harmful Algae, Vigo, Spain, 1997, pp 216–219.
71. R Draisci, L Lucentini, L Giannetti, P Boria, KJ James, A Furey, M Gillman, SS Kelly. Determination of diarrheic shellfish toxins in mussels by micro liquid chromatography-tandem mass spectrometry. J AOAC Int 81: 1–7, 1998.
72. S Pleasance, MA Quilliam, SW DeFreitas, GC Marr, AD Ciambella. Ion-spray mass spectrometry of marine toxins II: Analysis of diarrhoetic shellfish toxins in plankton by LC-MS. Rapid Commun Mass Spectrom 4:206–213, 1990.
73. S Pleasance, MA Quilliam, JC Marr, Determination of diarrhetic shellfish poisoning toxins in mussel tissue by liquid chromatography ionspray mass spectrometry. Rapid Commun Mass Spectrom. 6: 121–127, 1992.
74. JC Marr, T Hu, S Pleasance, MA Quilliam, LC Wright. Detection of 7-O-acyl derivatives of diarrhoetic shellfish poisoning toxins by liquid chromatography-mass spectrometry. Toxicon 30:1621–1630, 1992.
75. MA Quilliam, Analysis of diarrhoetic shellfish poisoning toxins in shellfish tissue by liquid chromatography with fluorimetric and mass spectrometric detection. J AOAC Int 78:555–570, 1995.
76. KJ James, AG Bishop, M Gillman, SS Kelly, C. Roden, R Draisci, L Lucentini, L Giannetti, P Boria. Liquid chromatography with fluorimetric, mass spectrometric and tandem mass spectrometric detection for the investigation of the seafood-toxin-producing phytoplankton, *Dinophysis acuta*. J Chromatogr 777:213–221, 1997.

77. WMA Niessen Advances in instrumentation in liquid chromatography-mass spectrometry and related liquid-introduction techniques. J Chromatogr. A 794:407–435, 1998.
78. MA Quilliam Liquid chromatography-mass spectrometry: a universal method for analysis of toxins? Proceedings of the VIII International Conference of Harmful Algae, Vigo, Spain, 1997, pp 509–514.
79. MA Quilliam. General referee reports—Phycotoxins. J AOAC Int 81:142–151, 1998.
80. MA Quilliam. General referee reports—Phycotoxins. J AOAC Int 82:773–781, 1999.

15

Ecobiology, Clinical Symptoms, and Mode of Action of Domoic Acid, an Amnesic Shellfish Toxin

Mohinder S. Nijjar
Atlantic Veterinary College
University of Prince Edward Island
Charlottetown, Prince Edward Island, Canada

Satnam S. Nijjar
University of Toronto
Toronto, Ontario, Canada

I. ECOBIOLOGY

A. Introduction

In late 1987, 250 individuals with gastrointestinal (GI) distress and central nervous system (CNS) disorders appeared at different clinics in Montreal, Canada. The medical history revealed that these patients had one thing in common. They all consumed steamed blue mussels (*Mytilus edulis*) at dinner the previous night. The tainted mussels originated from Prince Edward Island (PEI), and had been cultivated in three river estuaries on the east coast of PEI (1). The intoxicated patients exhibited GI disorders, including nausea, vomiting, abdominal cramps, and diarrhea, about 24 h after the consumption of mussels. This was followed by the neurological symptoms of headache, confusion, disorientation, loss of short-term memory, seizures, and coma within 48–72 h (2). Some patients were treated and allowed to go home, whereas other were admitted into hospitals. In fact, some patients were so seriously ill that they required intensive care. Four patients died as a result of amnesic shellfish poisoning. An autopsy of dead patients revealed marked neuronal lesions and loss, which was predominantly localized in the hippocampus and amygdala (3). The toxin in mussels causing poisoning was identified as domoic acid (DA) which is a tricarboxylic amino acid (4). It was later realized that DA toxin was produced by a pennate diatom, *Nitzschia pungens* (5), which was filtered from water by mussels and accumulated in their stomach (6). The human poisoning associated with cultivated mussels was reported for the first time ever in North America. DA was previously reported to be present in a seaweed, *Chondria armata*, and its extract was used to expel intestinal worms in young children in Japan (7). It was also the first time DA was ever found in shellfish cultivated in Canadian waters. Equally surprising were the findings that (1) *Nitzschia pungens* has never been associated with DA production, (2) nor do we know the circumstances which produced *Nitzschia pungens* bloom, and (3) the conditions favorable for DA synthesis by the phytoplankton. We also do not know how this diatom migrated to waters in this part of the world.

Table 1 Diversity of Phytoplankton and Their Toxigenocity

Location	Diatom spp.	DA level	Time of year	Reference
Cardigan Bay, PEI	*P. multiseries*	790 μg/g mussel	Nov–Dec 1987	5
Bay of Fundy, NB	*P. delicatissima*	74 μg/g	Sept 1990	15
Hood Canal, WA	*P. pungens, P. multiseries* *P. australis*	10 μg/g	Oct–Nov 1990	10
Monterey Bay, CA	*P. australis*	2300 μg/g	Sept 1991	12,13
Sechelt Inlet, BC	*P. pungens/multiseries* *P. australis*		Sept 1992	11

It is now known that different diatom species produce variable amounts of DA (Table 1). These diatom species are distributed worldwide, and they show variable environmental and nutritional requirements for optimal growth and DA production. This topic has been recently reviewed and the reader is referred to these articles for detailed information (8,9). For this reason, only a brief discussion of the ecobiology of phytoplankton producing DA is included in this chapter. Also, there is a limited information about the clinical symptoms in human patients intoxicated with this mussel toxin, because the poisoning episode in 1987 was the first ever described in the literature. The behavioral changes in experimental animals in response to the administration of DA are described in this chapter. In addition, different biochemical and pharmacological mechanisms(s) by which DA may produce GI disorders and CNS dysfunction are described.

B. Source of Domoic Acid Toxin

Mussels cultivated in the Cardigan, Brudnell, and Murray Rivers were found to be toxic and to be responsible for human poisoning with DA in Montreal in late 1987. The phytoplankton count was high in these rivers during the same time. The phytoplankton collected from these rivers was characterized morphologically and classified as the diatom *Nitzschia pungens*, which was found to be the predominant species in affected rivers. In view of the positive correlation between the phytoplankton counts and mussel toxicity, *Nitzschia pungens* was considered to be the likely source of toxin responsible for human poisoning (L. Hanic, personal communication, 1987). The toxin was later identified chemically as DA which is an unusual tricarboxylic amino acid (4). Since the human poisoning in 1987, DA has been found in other parts of the world probably because of our awareness of DA being present in the North American waters and its potential to produce toxicity in humans.

C. Phytoplankton Blooms and Domoic Acid Production

Similar blooms in the phytoplankton and DA toxicity have been reported in different parts of the world (8). The infested areas include Washington state (10), British Columbia, Canada (11), Monterey Bay, California (12,13), and Europe (14). However, there were striking differences in the ecobiology of mussels at different locations, which generated much interest in mechanism(s) underlying the phytoplankton blooms and optimal environmental conditions for diatoms to produce DA (8). Equally interesting was the finding that Boughton River in PEI contained high *Nitzschia pungens* counts, but mussels from these rivers lacked DA. Another intriguing observation was that the rise in DA concentration in mussels occurred 10 days after the increase in the phytoplankton counts. The reason(s) for this lag in DA production by the phytoplankton is not known, but it may be related to the genetic, sex, and environmental conditions (9). The source of DA in PEI mussels appears to be the pennate diatom *Nitzschia pungens* (8,9). Accordingly, *Nitzschia pungens* species producing DA have been further characterized and classified (16). In 1987, *Pseudonitzschia pungens* were present at a cell density of more than 10^6 cells/L in PEI waters and produced DA to cause human intoxication. Similar density of *Pseudonitzschia pungens* in waters off the west coast of British Columbia did not produce DA (11). Interestingly, other diatom species, *Pseudonitzschia seriata* and *P. delicatissima*, were found to produce DA. In another incidence, some fatalities of pelicans and Brandt's cormorants were reported in Monterey Bay, California. This was apparently due to the consumption of northern anchovies that had fed on the diatom *Pseudonitzschia australis*, producing DA (12). Domoic acid was also found in razor clams and Dungeness crabs in Washington state and north Oregon, but mussels from the same area were relatively free from DA. It may be that DA was cleared by mussels faster than razor clams, resulting in variable level of DA in these species living in the same area of water under similar environmental conditions. The mortality in razor clams and Dungeness crabs was accompanied by a bloom in *P. pungens*, which was followed by another bloom in *P. australis*. It is not clear which one of these diatom species was responsible for DA production and toxicity (11).

Domoic acid has also been found in other parts of the world and produced by the pennate diatom *P. pungens* (8). However, the source of DA in mussels cultured in Newfoundland during 1994, in sea scallops on Georges-Browns Bank, Nova Scotia, in 1995, razor clams and Dungeness crabs in Oregon, Washington, and British Columbia since 1991 has not been clearly established. In fact, *P. pungens* was nontoxigenic in other locations, including Atlantic Canada (17), Monterey Bay, California, (12,13), and Europe (14). However, episodes of DA poisoning had occurred in these areas. These observations suggest that other organism(s) may be responsible for DA production either independently or synergistically with the phytoplankton. The isolates of *P. pungens* from different geographical locations produced negligible amounts of DA which were inadequate to threaten human health. It is also not clear whether *P. seriata* produces DA under all conditions, since the phytoplankton produced variable toxicity depending on the geographical location and environmental conditions of water. It would appear that toxicity of the phytoplankton is influenced by genetic differences among strains of the same *Pseudonitzschia* and *Amphora* species from different geographical locations (8). The environmental factors such as nutrients, bacteria, and growth characteristics of organisms may influence DA production (9). The variable production of DA may arise from our difficulty in clearly identifying diatoms morphologically and/or to measure DA reliably. DA analysis by high-performance liquid chromatography (HPLC) seems to be adequate and reliable for DA estimation in mussel extracts and does not appear to be responsible for variable toxogenecity of different phytoplankton species. However, DA analysis in aqueous oyster extract by HPLC revealed the presence of additional component(s) which coeluted with DA and produced false-positive results for DA (M.S. Nijjar, unpublished data). The eluting buffer was modified to resolve the unknown component from DA in oyster extract. Similarly, we need further development of morphological criteria that would clearly distinguish between different *Pseudonitzschia* species. Thus, investigation into the environmental, nutritional, and growth factors that influence DA production by *Pseudonitzschia* species are warranted. The development of new techniques such as lectin binding (13), immunofluorescence (18), and DNA sequencing (19) may allow differentiation between the toxogenic and nontoxic strains of phytoplankton.

D. Environmental Conditions Affecting Phytoplankton Blooms and Domoic Acid Production

P. australis counts were low in the spring and autumn in Monterey Bay, California, but increased in the summer to quite high numbers ($>10^6$ cells/L) (13). Similarly, the DA level was also high

(>10 μg/L). However, the rise in DA was delayed by about 10 days after the rise in *P. australis* counts. The cell counts of other species such as *P. multiseries, P. pseudodelicattisima*, and *P. pungens* increased concomittantly, but DA was predominently produced by *P. australis* (13). The bloom in *P. australis* in southern California usually occurs in late spring and is associated with the intrusion of cool water with high nutrients, but was not accompanied by any poisoning episode. Although *P. pungen* has been shown to produce DA in culture, later studies revealed that it was, in fact, *P. pungens f. multiseries* diatom which produced DA and coincided with DA concentration in mussels (8). These findings explain earlier observations that the rise in DA concentration in mussels lagged 10 days behind the increase in *P. pungens* counts. It is now clear that *P. pungens* counts were high in PEI rivers during the summer of 1987 but declined as the temperature of water dropped during that autumn. Instead, *P. pungen f. multiseries* was the predominant diatom species present in late autumn and their cell density correlated with the increase in DA concentration in mussels (8,17). In fact, there was a reciprocal relationship between *P. pungens* and *P. multiseries* cells. The increase in *P. multiseries* coincided with the rise of DA in mussels (5). Similar relationship between *P. pungens, P. australis*, and *P. multiseries* was noted in the western coastal water (11). On the basis of these observations, it was suggested that *P. pungens* may transform into *P. multiseries* as the temperature of water drops. In fact, *P. pungens* counts in some PEI estuaries increased significantly in 1987, but the diatom and mussels lacked DA and toxicity (8,17).

The criteria of demonstrating the presence of DA in mussels or anchovies may not be reliable in view of the findings that DA is depurated from mussels with time in their natural river habitat (6) or in laboratory conditions (20). In addition, mussels may not be an appropriate vector for measurement of environmental contamination, since they stop filtering water which contains undesirable substance(s) in the environment. Dichlorvos was applied to salmonids to control sealice and its concentration measured in mussels placed at increasing distance from the site of application to assess the impact on ecobiology of marine life. Most interestingly, the DA concentration in mussels at different sites did not decrease as the distance from the site of dichlorvos application had increased (21). Similarly, DA at a concentration of 2300 μg/g wet weight was present in offshore anchovies which also contained *P. australis*. However, bivalves from nearshore were totally devoid of DA, indicating that inshore bivalves are not good indicators of toxicity in offshore anchovies or other aquatic species (22). It is also known that diatoms undergo bloom in offshore deep water, but strong tidal waves and vigorous mixing of water with strong winds may bring toxic phytoplankton to inshore waters. This may contaminate marine life and present a potential for human and aquatic toxicity (8). Also, DA accumulated in bivalves declines with time and totally disappears in about 4 weeks under their natural habitat in river estuaries or in laboratory conditions (6). Apparently toxins in bivalves are depurated variably with time from the peak DA concentration (8). The source of DA in razor clams (*Siliqua patula*) and Dungeness crabs (*Cancer magister*) in Oregon and Washington coast was not established (10), since the phytoplankton was not collected from the offshore waters. However, low counts of toxic phytoplankton species were found in beach water. This poisoning episode with DA occurred after record high temperatures and a dry period for 45 days followed by heavy rainfall and windy conditions. These environmental conditions were similar to those found in eastern PEI in 1987. Domoic acid levels varied considerably among clams from the same water. *P. pungens, P. multiseries, P. pseudodelicatisma*, and *P. australis* were present in low numbers ($1–5 \times 10^3$ cells/L) most of the time of year and transiently increased after warm dry periods without incidence of poisoning. The seed of *Pseudonitzschia* spp. in coastal water comes from deep offshore waters where the *Pseudonitzschia* population seems to originate. It appears that *Pseudonitizschia* blooms and DA-producing diatom species usually grow at colder temperatures in deep offshore waters. In the autumn when heavy rainfalls and high winds are commonly encountered, these phytoplankton spread to inshore waters and river estuaries, contaminating bivalves and other aquatic species (11). *P. multiseries* seems to be able to adapt to colder temperatures and is able to survive, whereas all other phytoplankton species are eliminated (8).

During *P. multiseries* bloom in PEI rivers, silicate and phosphate levels were very low. As a result, the DA production might have increased. At the same time, nitrogen was high, which is known to inhibit cell division. In culture, DA production by *P. multiseries* was maximum when either silicate or phosphate was low and nitrogen was high (23). The bloom in *P. multiseries* and enhanced production of DA were preceded by a bloom in another diatom, *Skeletonema costatum*. This might have depleted the water of silicate and phosphate that would trigger DA production by *P. multiseries* (24). The nitrate level in river water may have also risen as a result of heavy rainfall collecting runoff water from fields or have been due to resuspension of sediment with vigorous mixing of water by high winds (9). It is interesting to note that DA production was maximum when *P. pungens* reached the stationary phase of growth, which is brought about by limiting substrates, including low phosphate and silicate (25–27). During the stationary phase of growth, diatom cells will be in close proximity to each other, increasing the probability of cell pairing. The blooms of *P. multiseries* in river waters may reflect similar close proximity for cell pairing to occur, influencing the cell division and toxin production. It may take years for cells to decrease to a suitable size for reproduction, resulting in a long-term periodicity in bloom intensity and toxin production (28).

Different species of *Pseudonitzschia* produce variable amounts of DA toxin in different parts of the world under different environmental conditions. For example, *P. multiseries* produced DA in PEI, whereas it was nontoxigenic in other parts of the North American continent (9). The growth of *P. multiseries* and DA production was optimal during the stationary phase of the growth cycle and was triggered by low levels of silicate and phosphate and by high level of nitrates, which reduces cell division. The differential production of DA by the same diatom species under identical culture conditions may be related to genetic differences, bacterial contamination, and/or cell morphology. During the stationary phase, cells undergo deformation and a reduction in their cell size prior to sexual reproduction. *P. pungens* produces DA 10 days after it has entered the stationary phase. Ammonium ions slow down diatom cell division in culture but induce DA production prior to the stationary phase. These results indicate that reduced cell division in the stationary phase may somehow initiate DA production by *P. multiseries*. These varying environmental and nutritional conditions in waters of different parts of the world may be responsible for variable amounts of DA production.

Results of culture studies revealed that DA production began 10 days after the diatoms had reached the stationary phase and are consistent with the observations of a field study where peak *Nitzschia pungens* cell counts preceded 10–15 days before the maximum rise in DA concentration in mussels or phytoplankton (8,9). Apparently, the stationary phase of growth of *N. punges* was most likely brought about by the limiting amounts of phosphorus, producing a deficiency in phospholipids and resulting in poorly formed membranes and reduced cell integrity (29). The diatom integrity appeared to be compromised during low silicates, since DA leaked out easily from the phytoplankton during tow as compared with phytoplankton which retained DA after its filteration over membranes (9). The lack of silicates prevents diatoms from their normal progression of the cell cycle, prolonging the G_1 and G_2 phases. As a result, the rate of cell division slows down and DA synthesis is enhanced (30). Similarly, *P. seriata* and *P. multiseries* in culture produced a negligible amount of DA during the exponential phase, but most toxin was produced in the stationary phase (14), which was essentially similar to the growth and DA production by *N. pungens* in culture. Under similar culture conditions, *P. australis* began producing DA during the exponential phase, which remained either stable or declined during the stationary phase (31).

The amount of light also has considerable influence on DA synthesis and production by the phytoplankton in culture. During the light period, uptake of silicate by phytoplankton and cell DNA were high. When the light was turned off, there was a rapid increase in cell division which was accompanied by reduced DNA/cell, enhanced uptake of silicate, and increased production of DA (32). It would appear that the physiological stress caused by reduced silicate triggers DA production by the phytoplankton. During the exponential phase, photosynthesis is high because of the increased demand for adenosine triphosphate (ATP) for different metabolic processes. However, as the phytoplankton growth is decreased during the stationary phase, ATP becomes available for the synthesis of DA (27). Domoic acid synthesis by *P. multiseries* was increased by low phosphate in culture medium in addition to stimulation by low silicates. DA production by *P. multiseries* continues as the temperature is dropped from 15 to 0°C. Under similar culture conditions, *P. seriata* produced higher levels of DA at 4 than at 15°C. The differential effects of temperature on DA production by the two diatom species may be related to their genetic differences and tolerance to the variable temperatures of surrounding water (33).

The presence of bacteria in the phytoplankton culture increased DA production considerably. In fact, the number of bacteria in phytoplankton culture was positively correlated with the increased DA production (34). The bacteria may produce precursors of DA synthesis or release some factors which stimulate DA production by *P. multiseries*. There appears to be a dynamic interaction between the diatoms and live bacteria, since extracts of bacteria failed to enhance DA production by the phytoplankton as in the presence of live bacteria (34).

II. CLINICAL SYMPTOMS

A. Human Patients

About 250 patients were seen in different clinics in the Montreal area between November 19 and December 5, 1987 (Table 2). The medical history revealed that all patients had dinner the previous night which included steamed blue mussels (*Mytilus edulis*). The toxic mussels originated from three river estuaries in eastern PEI. About 76% of patients developed vomiting, 50%

Table 2 Bloom in *Nitzschia pungens* versus Domoic Acid Concentration November to December, 1987

Date	*Nitzschia pungens* (Cells/L $\times 10^{-3}$)	Domoic acid (mg/L)
November, 1987		
01	550	10
05	400	12
15	1200	50
20	900	100
25	625	287
December, 1987		
01	600	220
15	200	50
30	<100	20

Table 3 Clinical Symptoms and Their Incidence in Patients Intoxicated with Domoic Acid

Symptom	No. of patients	%
Nausea	(99 patients)	77
Vomiting		76
Abdominal cramps		50
Diarrhea		42
Headache		43
Memory loss		25
Confusion, disorientation	(of 12 ICU patients)	100
Grimacing		50
Mute		92
Seizures		67
Coma		75

abdominal cramps, and 42% had diarrhea within 24 h of eating the mussels (Tables 3 and 4). The neurological disorders developed within 48 h of mussel consumption and included symptoms such as headache in 43% of the patients, and short-term memory loss was noted in 25% of the patients (2). Other patients showed symptoms of vertigo, disorientation, confusion with signs of abnormal gait, extraocular and jaw movements, grimacing, seizures, and coma. In addition, symptoms like sweating, urinary retention, fluctuations in blood pressure, and tachycardia were noted in some patients (2). About 56% of the patients were males with a median age of 65 years (Table 4). Males in increasing age correlated positively with hospitalization and memory loss. Twelve of 19 patients who were admitted to the hospital required intensive care for seizures, coma, profuse respiratory secretions, and/or unstable blood pressure. Four patients died as a result of mussel intoxication (2).

In the acute phase of mussel-induced intoxication, the patients presented with unsteadiness, generalized weakness, headache, symmetrical transient hyperreflexia, fasciculations, seizures, hemiparesis, ophthalmoplegia, and abnormalities of arousal ranging from agitation to coma. The acute response to mussel toxin reflects a neural hyperexcitation syndrome resulting from the stimulation of central and possibly peripheral neurons and leading to seizures, myoclonus, cardiovascular instability, pupillar abnormalities, and piloerection (3). This was followed

Table 4 Epidemiology (Age, Sex, and Geographical Distribution) of Intoxicated Patients

Criteria	No. of patients	% Patients
	(107)	
Age (40–59 years)	49	56
(>60 years)	38	44
Male	47	44
Female	60	56
Quebec	71	66
PEI	13	12
Ontario	15	14
British Columbia	7	7
New Brunswick	1	1

by chronic loss of function in neural systems susceptible to excitotoxic degeneration of hippocampal and anterior horn cells of the spinal cord. After 4–6 months, patients showed distal atrophy and mild weakness of extremities and hyporeflexia. Twelve months later, the patients had severe anterograde memory deficits with relative preservation of other cognitive functions (3). Some of the hospitalized patients showed generalized, psychomotor, and focal motor types of seizures. The seizures were resistant to phenytoin therapy but responded to intravenous administration of diazepam and phenobarbital, became less frequent with time, and ceased entirely in about 4 months after intoxication (3). It would appear that the mussel intoxication induced an acute nonprogressive neuronopathy involving anterior horn cells or a diffuse axonopathy predominantly affecting the motor axons. Most interestingly, the rate of glucose metabolism in the amygdala and hippocampus was reduced in severely affected patients and correlated with the patients' memory scores (3).

Three patients died in hospital after consuming the tainted mussels. A fourth patient was discharged from the hospital but died of myocardial infarction within 3 months of mussel consumption. Histological examination of their brains showed neuronal necrosis and astrocytosis which was most prominent in the hippocampus and amygdaloid nucleus (3).

The loss of memory in patients intoxicated with mussel toxin appeared to be similar to patients with Alzheimer's disease (35). However, the loss of memory in mussel-intoxicated patients was not affected by the age of patients, whereas symptoms of Alzheimer's disease intensify with advancing age and are generally noted in older people (3). Further, the findings that intellect and higher cortical functions were not influenced by DA intoxication distinguishes the mussel-induced intoxication from Alzheimer's disease (35). This difference between DA-induced intoxication and Alzheimer's disease, which is related to the age and involves activation of glutaminergic receptors, was further supported by the cytopathological findings of complete lack of neurofibrillary tangles, granulovacuolar degeneration, and senile plaques in DA-induced damage in the hippocampus (3). Unlike amyotrophic lateral sclerosis in Alzheimer's disease, neuronal death of anterior horn cells and anterior horn axonal spheroids were not detected in mussel-induced intoxication (3). It would appear that DA produces a widespread neurological dysfunction followed by chronic residual memory deficits and motor neuropathy or axonpathy (3).

B. Experimental Animals

The contaminated blue mussels (*Mytilus edulis*) from PEI were extracted in water and an intraperitoneal injection of 1 mL extract into mice produced nausea, vomiting, diarrhea, dizziness, confusion, weakness, lethargy, somnolence, loss of short-term memory, seizures, and coma (36). Intraperitoneal administration of the toxic mussel extract into mice produced (1) scratching behind the ears with hind limbs, (2) uncontrolled rolling, twisting motions culminating in the loss of righting reflex, and (3) tremors, spontaneous seizures, and paddling motions followed by (4) cyanosis and death. The toxin in contaminated blue mussels that causes toxicity in humans as well as in rats was characterized chemically to be a heat-stable amino acid which is structurally similar to kainic acid (KA) and glutamic acid (GLU) (4) (Figure 1). Pure DA and DA in mussel extracts binds to KA receptors in rat hippocampus, and neuropathology observed in patients intoxicated with mussel poisoning was consistent with those produced in rats by the parenteral administration of KA, further support the concept that DA is a neuroexcitotoxin (3). The systemic injection of KA in rats produced signs of neural excitation such as wet-dog shakes, masticatory movements, rearing, and limbic seizures followed by neuronal degeneration in the olfactory cortex, amygdaloid complex, hippocampus, and related parts of the thalamus and cortex (37,38). The damage was primarily localized to the CA_3 and CA_1 regions of hippocampus, which are homologous of fields H_3 and H_1 in humans (3). Kainic acid is a potent neuroexcitotoxin which produces brain damage by activating non-NMDA (N-methyl-D-aspartate) (KA/AMPA) α-amino-3-hydroxy-5-methyl-4-isoxazol propionic acid (AMPA) receptors (39,40). Domoic acid structurally resembles KA and produces identical behavioral and neurochemical effects which are mediated by KA receptors (39,40).

Intraperitoneal administration of pure DA produced identical symptoms in mice and rats. The onset times for scratching, seizures, and death were reciprocally related to DA concentration in the mussel extracts (6,41). A scale of behavioral scores has been developed on the basis of onset times for different behaviors after the administration of DA in different formulations and is used to measure DA concentration in the mussel extracts (40). Extracts of contaminated mussels produced identical changes in behavior but elicited higher in vivo toxicity than pure DA. These results suggested the presence of additional toxic component(s) in mussel extract that was responsible for additional toxicity in mice. The toxic components may include DA isomers, other excitatory amino acids (EAAs) such as GLU, aspartate (ASP), and zinc (40). Administration of KA by a similar route in mice or rats produced identical changes in the behavior of animals as DA (39,40). On the basis of these observations, DA toxicity is believed to be mediated by activation of KA receptors in the CNS (39,40). The damage to hippocampus and amygdala observed in patients intoxicated with the mussel toxin were consistent with those produced in rats and mice by the parenteral administration of KA and DA (3). However, the current evidence indicates some differential effects of DA and KA on the brain in experimental animals (40–43).

Intraperitoneal administration of extracts of toxic mussels from PEI into mice produced a neurovisceral response which includes symptoms of nausea, vomiting, neurogenic gastric distress, gastric bleeding, diarrhea, dizziness, confusion, weakness, lethargy, somnolence, loss of short-term memory, seizures, and coma (31). Although there has been considerable discussion about neurological sequelae of DA-induced intoxication, very little is known about the gastroenteric pathology induced by DA. The excitatory amino acids such as GLU, ASP, and KA are considered to be potent emetic compounds which produce vomiting via activation of chemoreceptors in the area postrema (44). Domoic acid–induced lesions in the nucleus of the solitary tract may be responsible for the GI symptoms and possibly for cardiovascular disorders associated with DA toxicity (43). Oral administration of low doses of DA to monkeys caused vomiting only. However, DA at higher doses produced GI as well as neurotoxic effects. Apparently higher elevation in serum DA is required for neurotoxic effects, whereas at low levels, DA produced GI disorders only. The variable toxicity in humans following the ingestion of mussels contaminated with DA (3) and in animals after the administration of DA in different formulations (39–42) may be related to variable DA pharmacokinetics which reflects absorption, distribution, and its elimination via kidneys (63–65).

When extracts of contaminated mussels were given orally to mice, lesions in the antral portion of the stomach and duodenum were noted, along with scarring and hyperemia as well as massive proteinaceous peritoneal ascites (48). The severity of these lesions was dependent on the DA concentration in the extracts. Administration of 0.5 mg of DA/kg in infant mice produced GI bleeding and duodenal and gastric ulcers. In addition, DA-treated mice showed subpleural petechial hemorrhages on the lung surface (36). Treatment of mice with kynurenic acid prior to DA administration reduced gastropathy in a dose-dependent manner. A high dose of kynurenic acid also blocked DA-induced lung damage and inhibited acid secretion in the stomach. Kynurenic acid may achieve its GI-protective effects by blocking NMDA receptors which have been reported to be present in the guinea pig ileum (49).

Domoic acid was localized in the stomachs of contaminated mussels harvested from PEI (6). These mussels had engorged digestive glands which contained a large amounts of algae

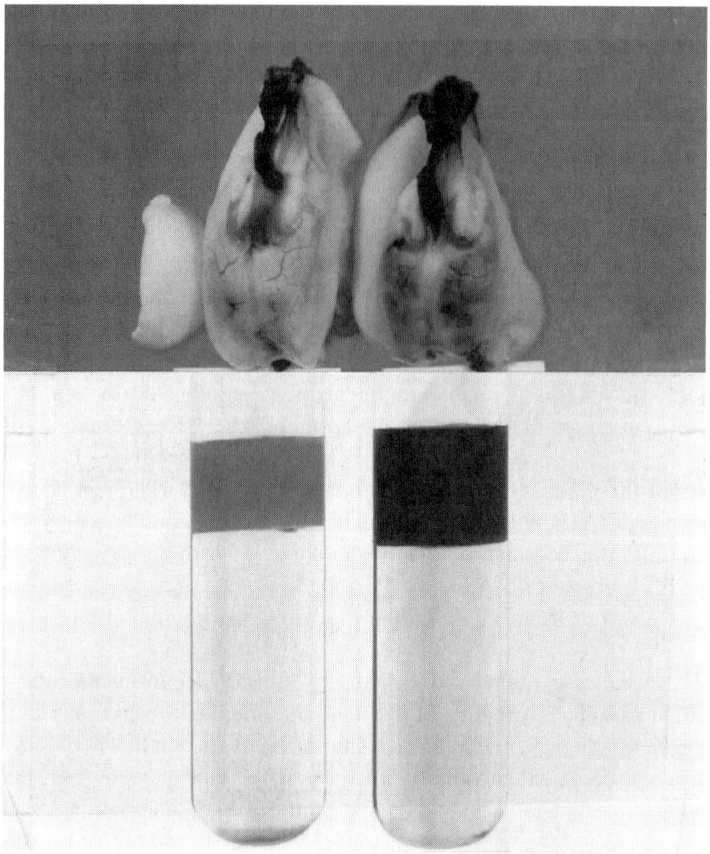

Figure 1 (Top panel) The control (left) and toxic (right) mussels are shown after their removal from shells. The grayish black coloration reflects the amount of algae present in their stomachs. The stomach of a toxic mussel is clearly engorged with algae. (Bottom panel) Tissue from the control (left) and toxic (right) mussels was homogenized in a mixture of chloroform and methanol and centrifuged. Denatured proteins form a distinct band between the aqueous and organic layers. The toxic mussel contains more proteinaceous material than the control mussel.

(Figure 1) that were later morphologically characterized as the pennate phytoplankton diatom *P. pungens f. multiseries* (8,23,27). The morphological changes were confirmed by the biochemical data (Table 5). While the mussel weight and protein were not altered, the weight of hepatopancreas (stomach) of toxic mussels was twice as much as that of the control mussels, and chlorophyll concentration, which reflects algal biomass, was elevated five fold in toxic mussels as compared with the control mussels (Table 5). Intraperitoneal administration of either whole extracts of phytoplankton collected from the Cardigan River or their subfractions after HPLC into mice produced behavioral changes that were identical to those seen after the administration of toxic mussel extracts or purified DA (50). In vivo toxicity of the phytoplankton extracts or its purified fractions in mice was directly related to DA concentration (6). The phytoplankton isolated from toxic mussels were grown in culture and DA production by different isolates of *P. pungens* was measured. *P. pungens f. multiseries* isolates produced DA, whereas other isolates produced variable amounts of DA (8,23).

The behavioral responses to systemically injected DA and KA in mice produced identical

Table 5 Biochemical Characteristics of the Control and Toxic Mussels from Cardigan River in PEI

Parameters	Control	Toxic
Mussel weight (g)	3.03 ± 1.01	2.99 ± 0.26
Hepatopancreas weight (g)	0.57 ± 0.17	1.06 ± 0.20*
Chlorophyll A (ug/g)	4.29 ± 0.80	24.80 ± 3.10*
Protein (mg/g)	7.65 ± 3.05	11.20 ± 5.40

* Significantly different from control ($p < 0.01$).
** The toxic mussel contains more proteinaceous material than the control mussels. The data show that the toxic mussels were heavier and contained more chlorophyll, an indicator of the algal biomass, than the control mussels. Algae collected from the stomachs of toxic mussels contained domoic acid, whereas algae obtained from the control mussels was relatively free from domoic acid.

behavioral changes such as scratching, circling, or head weaving, loss of postural control, tremors, and convulsions followed by death (40). Domoic acid was 8–11 times more potent than KA. Mice injected with NMDA or quisqualic acid (QA) showed sufficiently different patterns of behavior, indicating the involvement of different receptors. Interestingly, the scratching behavior was specifically noted after DA administration, in addition to other behavioral changes that were noted in response to other excitotoxins (40). Other investigators have also reported DA being three times more potent than KA in toxicity, and they suggested that the relative potency of DA and KA may reflect variable pharmacokinetics and different receptor dynamics (40).

Administration of mussel extracts containing DA produced higher toxicity in rats than the administration of purified DA, suggesting the presence of other excitotoxins such as GLU and ASP in addition to DA isomers and zinc in the mussel extract (40). There seems to be a parallel between human mussel poisoning and rat poisoning induced by mussel extracts containing DA, pure DA, and KA. In fact, administration of DA or KA in mice produced identical behavioral changes in animals (40,42) and produced similar lesions in hippocampus (39–41). On the basis of these findings, it was proposed that DA produces neurotoxicity via activation of KA receptors (39–42). However, current evidence indicates that there are some differences between DA- and KA-induced lesions in hippocampus (41) and in DA- and KA-induced release of GLU and ASP from rat brain synaptosomes (51). Apparently further studies are warranted to define the molecular mechanism(s) underlying DA- and KA-induced neurotoxicity and the cellular basis of CNS dysfunction.

Human intoxication with tainted mussels showed a great deal of variation among the patients in terms of severity of GI and CNS pathophysiology (2,3). Similarly, a considerable variation in DA toxicity has been reported among different animal species (41,45,46) or within the same animal species depending on the formulation of DA (40,51). The variable toxicity in humans following the ingestion of toxic blue mussels and in animals in response to the administration of mussel extract and/or pure DA appears to be related to variable absorption and its elimination via kidney (40,41). Domoic acid was not detected in the serum samples obtained from patients intoxicated with blue mussels after 2 days (2). Most of the seriously affected patients were older patients, and some had preexisting medical complications such as renal disease, cardiac dysfunction, and hypertension. It is possible that the reduced excretion of DA in such patients resulted in the elevated levels of DA which caused neurotoxicity (2). Recent work of Preston and Hynie confirmed this notion in that the DA level in blood samples from the nephrectomized rats was much elevated compared to the sham-operated controls, demonstrating the significance of renal function in the circulating levels of DA and its toxicity (53). Iverson et al. have shown that monkeys are more sensitive to DA intoxication after orally admin-

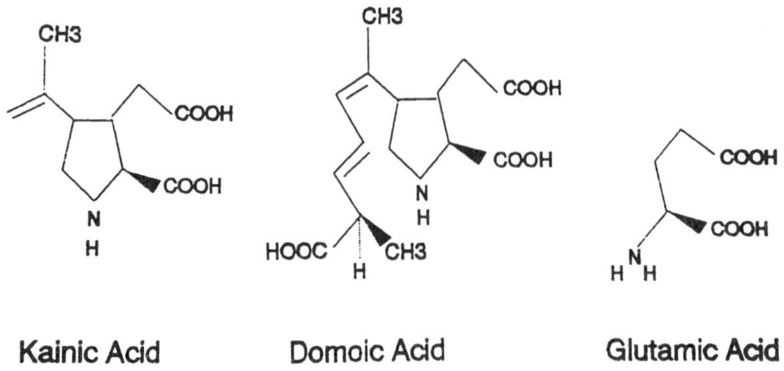

Figure 2 Chemical structure of domoic, glutamic, and kainic acids, showing the resemblance between three excitotoxins.

istered DA as compared with rats (54). Rats did not exhibit overt symptoms, whereas monkeys showed overt symptoms to the same dose of DA, indicating rapid absorption and clearance of DA in rats as compared with monkeys. As a result monkeys showed a higher sensitivity to DA toxicity than rats. The fact that vomiting center in the monkey lies outside the blood-brain barrier as compared with hippocampus, which lies within the blood-brain barrier, may account for different behavioral responses to DA in these two species (54).

Glutamate, ASP, NMDA, and KA are potent excitotoxins and produce brain lesions and CNS dysfunction. The cellular basis of brain lesions and CNS dysfunction induced by these agonists has been investigated extensively and is well defined (see ref. 44 for review). Since DA structurally resembles KA and GLU (Figure 2) and produces excitotoxicity in brain, it would be useful to investigate the effects of DA on different brain preparations and compare them with those induced by KA or GLU. This approach would assist in deliniating mechanism(s) underlying DA action on the GI and CNS systems.

The behavioral presentation of patients, and the hippocampal damage noted in patients who died as a result of DA intoxication, appeared to parallel those seen in animals treated with KA (3). Administration of KA into the amygdala of conscious freely moving rats via chronically implanted fused silica cannula produced limbic seizures (mastication, forelimb clonus, and rising on hind limbs) of short duration. In three cases, recurrent seizures culminated in behavioral status epilepticus 60–90 min after injection, which were blocked by diazepam. After 30 min of status epilepticus, there was a marginal neuronal damage with slight gliosis in hippocampal tissue. The limbic motor seizures do not lead to histopathological changes in the brain provided they do not culminate in a state of permanent seizure activity (55).

KA injected systematically or locally into limbic brain areas produces potent convulsive behavior which is accompanied by the destruction of perikarya and dendrites, while sparing glial cells and axons traversing through the treated areas of the brain (56). KA-induced distant lesions in the hippocampus, the pyriform cortex, and the amygdala appear to be related to KA-induced epileptiform activity which is similar to that in temporal lobe epilepsy of humans and can be prevented by pretreatment of animals with anticonvulsants (56). Apparently, KA causes a permanent decrease in the seizure threshold of animals. Animals treated with KA become susceptible to spontaneous seizures occurring even months after treatment (56). Less than 1% of systemically injected KA passes through the blood-brain barrier, and due to its low bioavailability, KA may produce variable responses in different animals. In addition, differences in age, sex, strain, and weight of rats may be responsible for the variable response (56). Older rats respond at lower concentrations of subcutaneously applied KA and exhibit more severe brain damage than younger rats at the same KA dose (57).

KA-induced epileptiform activity may arise in the entorhinal cortex, and is then immediately propagated to the CA_3 region of the hippocampus, and finally to the amygdaloid complex. The epileptiform activity then progresses to other limib structures, including the median thalamic complex, the CA_1 region, and the medial frontal cortex before diffusely spreading to the cortical region. Administration of KA systemically is accompanied by an increase in the volume of the hemispheres; the temporal lobes especially appear to be swollen (57). Histopathological changes in the acute stages (90 min after injection) cause dendritic swelling and shrinkage and pyknosis of perikarya along with signs of general brain edema and swelling of astrocytes. Acute seizures are due to dendritic or glial swelling and disappear from the neocortex, striatum, or cerebellum in 12–24 h after KA injection. KA may induce direct axon-sparing lesions either by exerting a strong excitatory action on certain neuronal pathways, resulting in seizures and brain damage in certain target neurons. Alternatively, KA-induced brain damage may be secondary to seizure activity and be related to neuropathological events such as hypoxia, hypoglycemia, and edema (57).

Excessive damage to the hippocampus and amygdala by KA may be due to the presence of excess KA receptors presynaptically on mossy fibers to induce GLU release, which causes neuronal damage (57). The transection of mossy fibers prior to KA administration prevents KA-induced neurotoxicity, and there appears to be a cooperative interaction between KA and mossy fiber terminals to produce damage to CA_3 pyramidal cells (57). However, the available evidence strongley suggests a direct relationship between KA-induced seizures, lesions in the brain, and CNS dysfunction.

At low doses, intracereberoventricular administration of DA in rats and mice induced preconvulsive behaviors such as wet-dog shakes, hypermotility, and mild facial clonus. At higher doses, DA caused clonic convulsions followed by the status epilepticus syndrome, which includes wet-dog shakes, forelimb clonus, rearing, and salivation. KA at low concentrations produced almost identical symptoms as DA. However, KA at higher doses induced convulsive behavior similar to DA but did not produce status epilepticus. Almost all excitatory agonists produce generalized seizures in mice, but animals treated with DA or KA produced a prolonged sequence of seizures with forelimb clonus, implying that DA and KA act at the same receptor subtypes (42). Agonists acting on NMDA receptors produce a limited incidence of preconvulsive behavior and totally lack status epilepticus, which is different from non-NMDA receptor agonists such as DA and KA. Administration of KA and DA produces maximal damage in the CA_3 and CA_1 regions of the hippocampus of animals, and treatment with anticonvulsants prevents DA- and KA-induced seizure activity and damage to brain tissue (56). It would appear that the seizure activity is a prerequisite for KA- and DA-induced neuroexcitotoxicity.

The seizure-induced brain damage involves massive edema, hemorrhages, and necrosis of neurons and nonneuronal elements in amygdaloid nuclei, pyriform and entorhinal cortices, and olfactory areas of the forebrain (56). Perivascular hemorrhages are accompanied by capillary sprouting, the occurrence of small cystic lesions in the amygdala and pyriform cortex, reactive gliosis, loss of oligodendrocytes, and demyelination. Early cytotoxic edema caused by severe seizures may result in compression of drainage vessels and a disturbed blood supply of affected brain areas. This is indicated by decreased accumulation of 2-deoxyglucose (56). Mannitol-induced diuresis prevents development of cytotoxic edema, seizure activity, and nonselective brain damage induced by KA (58). Intraperitoneal administration of KA resulted in an increase in the blood-brain barrier permeability throughout the forebrain, septum, thalamus, and amygdala. It has been suggested that the augmented cerebrovascular blood flow together with an elevation in blood pressure may be related to the seizure-induced alterations in the permeability of blood-brain barrier. There was also a good correlation of the time course and regional distribution of local glucose consumption, seizure activity, and the pattern of neuronal damage following the administration of KA (59).

Intraperitoneal administration of KA in mice produced four- to fivefold accumulation of

2-deoxyglucose in the CA_3 sector of the hippocampus (60). This was accompanied by a concomittant decrease in glucose consumption in nonlimbic cortical structures as well as most other regions of the forebrain, mesencephalon, and brainstem. The time course and pattern of expression of c-*fos* in response to intraperitoneal injection of KA followed the same profile as did the accumulation of 2-deoxyglucose and the propagation of electrographic seizure activity (60). At the onset time of limbic motor seizures, the granular cell layer and interneurons of the fascia dentata exhibited the highest concentration of c-*fos* gene product. (61). Glutamate toxicity in brain slices has been shown to be potentiated by glucose deprivation (62). Glucose deprivation–potentiated GLU toxicity appeared to be mediated by an elevation in the intracellular level of Ca^{2+} in brain tissue (80).

KA-induced seizures were accompanied by the enhanced release of GLU, taurine, and phosphoethanolamine in the hippocampus and only GLU in the pyriform cortex (64). Subcutaneous injection of KA markedly reduced GLU uptake in the hippocampus, amygdala, and pyriform cortex (65). Thus, KA-induced brain damage may be mediated by elevated levels of GLU, resulting from either reduced uptake or enhanced GLU release from the presynaptic elements (65). If this was true, one would expect GLU to produce identical behavioral changes in mice and similar distribution of lesions in the brain as in response to KA. This does not appear to be the case, suggesting different modes of actions for GLU and KA in brain. KA may induce neuronal damage by activating either non-NMDA metabotropic receptors coupled to G protein and phospholipase C (67) or ionotropic receptors which stimulate ion channels postsynaptically (68). Glutamate is known to stimulate NMDA receptors to induce alterations in ion fluxes across the postsynaptic membranes, whereas KA activates non-NMDA (KA/AMPA receptors) presynaptically and postsynaptically to enhance cation transport across membranes (Figure 3). Kainic acid activates the metabotropic ($mGluR_1$ and $mGluR_5$) receptors, resulting in the stimulation of phospholipase C, which catalyzes the hydrolysis of phosphatidylinositol-4,5-bisphosphate (PIP_2) to inositol-1,4,5-triphosphate (IP_3) and diacylglycerol (DAG). IP_3 is known to mobilize Ca^{2+} from the intracellular stores to stimulate numerous important intracellular processes (69–71).

Recent reports indicate differences between KA and DA in terms of the distribution of lesions in brain (41) and their ability to stimulate GLU and ASP release from synaptosomes (51). Although KA produced severe damage to the CA_3 region plus a diffuse and milder damage to $CA_4 > CA_1 > CA_2 >$ dentate granular cells in the hippocampus, DA produced severe damage to the CA_3 region but essentially had no detrimental effects on other parts of hippocampus in mice (41). The percentage of necrotic cells (82.1%) exhibiting damage by the administration of DA was greater than 58.8% cell necrosis produced by the administration of KA (41). Domoic acid stimulated GLU release from the isolated brain synaptosomes and had no effect on ASP release, whereas KA enhanced GLU and ASP release from the synaptosomes under identical conditions (51). It would appear that DA and KA interact with different receptor subtypes on the presynaptic and postsynaptic nerve terminals to activate voltage-dependent Ca^{2+} channels, resulting in the enhanced influx of Ca^{2+} targeted for different pools of neurotransmitter release (51).

Glutamate mediates neuropathology induced by cerebral ischemia, hypoxia, hypoglycemia, mechanical trauma, epilepsy, and other chronic neurodegenerative disorders such as Alzheimer's, Huntington's, and Parkinson's diseases (71). NMDA receptor antagonists prevent brain damage from these disorders (71). The loss of memory in patients intoxicated with mussels contaminated with DA was similar to patients with Alzheimer's disease. However, the loss of memory in mussel-intoxicated patients was not affected by the age, since young and older patients were equally affected (3). In addition, relative preservation of intellect and higher cortical functions distinguished the mussel-induced intoxication syndrome from Alzheimer's disease and other neurodegenerative diseases which involve GLU activation of NMDA receptors (35). This conclusion was further supported by the cytopathological observations of the complete lack of neurofibrillary tangles, granulovacuolar degeneration, and senile plaques in DA-induced damage in the hippocampus (3). Unlike amyotrophic lateral sclerosis in Alzheimer's disease, neuronal death of anterior horn cells and anterior horn axonal spheroids were not detected in mussel-induced intoxication (3). Using autoradiographic radioligand binding technique, different distribution of NMDA, AMPA, and KA receptors has been described in the brain (72). High-affinity KA binding sites are preferentially located in the stratum lucidum of the CA_3 region, the commissural layer of the dentate gyrus, and the pyramidal layer of the CA_1 and CA_3 regions of the hippocampus. It would appear that DA produces a widespread neurological dysfunction followed by chronic residual memory deficits and motor neuropathy by a mechanism(s) which is different from those involved in neuropathy induced by GLU (3).

KA produced a marked decrease in glutamate decarboxylase (GAD) activity which catalyzes the formation of gamma-aminobutyric acid (GABA) from GLU in rat hippocampal slices (73). A KA-induced reduction in GAD was found in the amygdala, pyriform cortex, lateral septum, and hippocampus (56). Evidently there is a loss of GABAergic inhibitory function in the hippocampus (73), which is responsible for KA-induced seizure activity. The distribution of damage to GABA neurons follows the distribution of general neuronal damage induced by KA (73).

Intracerebroventricular administration of DA in rats inhibited GAD activity, resulting in the accumulation of GLU in the brain with a concomitant decrease in the GABA level and producing seizure activity (74,75). Since DA is structurally similar to KA (4), DA has been reported to activate non-NMDA (KA/AMPA) receptors to elecit behavioral responses, including wet-dog shakes and seizure activity which is followed by damage to the CA_3 region of the hippocampus, amygdala, and cerebral cortex. Interestingly, the seizure activity following either the consumption of contaminated mussels or intracerebroventricular injection of DA or KA is necessary for brain damage to ensue (2,3). Thus, treatment of mussel-intoxicated patients with anticonvulsant drugs would prevent seizure activity and protects the brain from EAA-induced damage and CNS dysfunction (3,56). Indeed, anticonvulsants such as diazepam, phenobarbital, and trimethadione are quite effective in preventing DA-induced seizures and brain damage. Treatment with KA receptor antagonists prevent local brain injury due to DA (48).

Administration of KA in rats results in a marked decrease in GAD activity in the amydgala, pyriform cortex, lateral septum, and hippocampus (56). Similarly, GABAergic pathways are reduced in temporal lobe epilepsy in humans (76). Apparently the GABA level in the brain controls seizure activity (epilepsy). KA-induced inhibition of GAD activity is totally recovered in 1 month after termination of treatment with KA, representing increased synthesis of the enzyme and increased activity of GABAergic neurons (56). Increased levels of mRNA encoding for GAD or the neuropeptides neurokinin B and neuropeptide Y have been found in GABA neurons after KA-induced seizures (77).

Microinjection of DA into the hippocampal CA_3 region produced bilateral electrical seizure discharge, and local administration of GABA resulted in neuronal recovery from DA-induced seizures (74). Apparently DA-induced inhibition of GAD and the resulting diminished GABA level seems to be responsible for seizure activity induced by DA in rats (74). Thus, GABA mimetic drugs such as aminooxyacetic acid, muscimol, and progabide antagonize KA-induced seizure activity. Baclofen, a $GABA_b$ agonist, has potent anticonvulsive action and prevented KA-induced seizures and brain damage (56). Excitatory receptor antagonist are not as effective as GABA mimetic drugs, indicating the essential role of reduced GABAergic function during early epileptic and pathological events. Recently developed competitive (AP5, CPG 39551, D-AP7) and noncompetitive (ketamine, phencyclidine, MK-801) inhibitors of NMDA

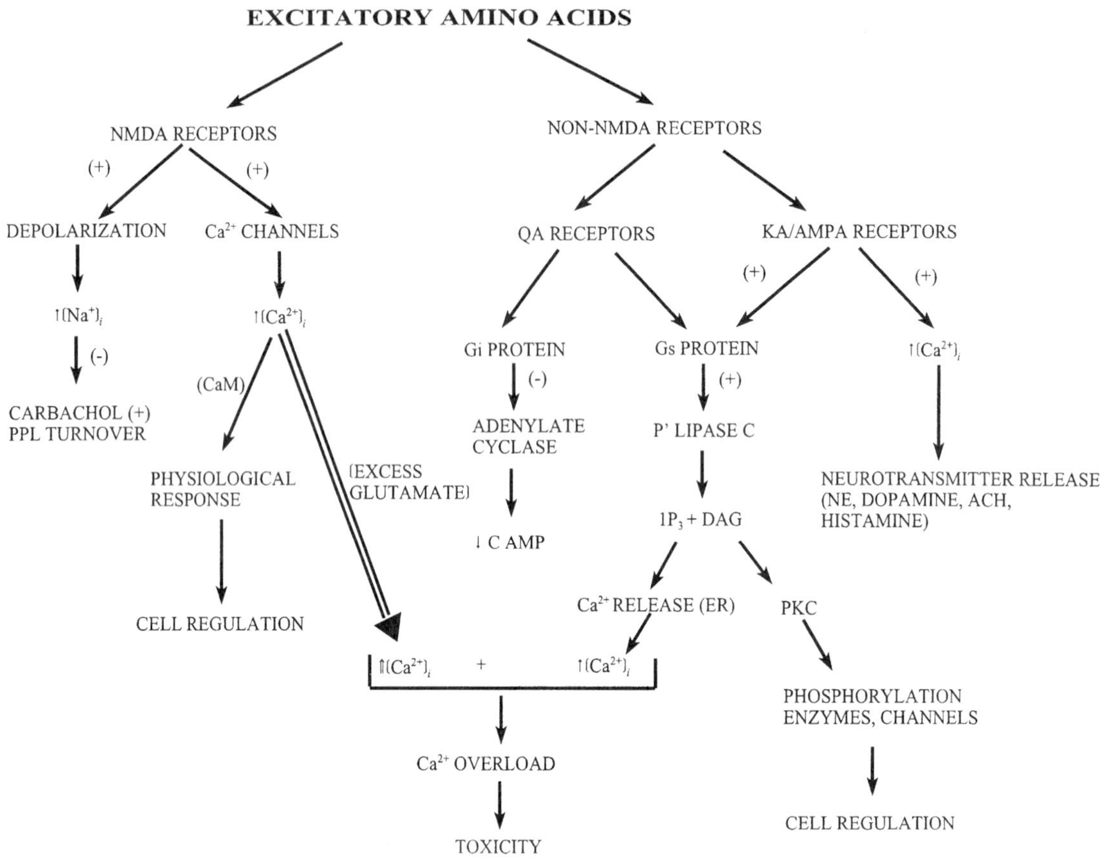

Figure 3 A schematic diagram illustrating different receptor-mediated intracellular responses to excitatory amino acids. Apparently Ca^{2+} mediates different intracellular processes. Accumulation of glutamate produces Ca^{2+} overload which disrupts ion homeostasis and water movements across membranes, causing cell swelling, tissue edema, and necrosis followed by CNS dysfunction.

receptors showed variable potency to block seizure activity and damage in different areas of brain. The non-NMDA receptor antagonists (gamma-GAMS) acting preferentially on KA receptors exhibited potent anticonvulsant action on seizures induced by KA, quisqualic (QA), and quinolinic acids in mice, whereas NMDA receptor antagonists were not as effective, indicating that KA/AMPA receptors mediate KA-induced seizure activity and brain damage. The stimulation of NMDA receptors also produced seizure activity, but damage was localized to the CA_1 region instead of the CA_3 region, which is specifically affected after KA or DA administration (56).

Treatment of experimental animals with prazosin or L-propranolol, blockers of alpha- and beta-adrenergic receptors, respectively, inhibited norepinephrine (NE) release from the presynaptic nerve terminals and prevented KA-induced seizure activity (56). An inhibitor of dopaminergic receptors, haloperidol produced a weak anticonvulsant effect. However, extensive destruction of catecholaminergic neurons drastically increased vulnerability to KA-induced seizures and brain damage. These effects appear to be mediated by intracellular Ca^{2+}, since a blocker of Ca^{2+} channels, verapamil prevented seizure activity in kindled rats, whereas nifedipine reduced the duration of seizures and total seizure time after intracerebroventricular injection of

KA (56). It would appear that KA and DA activate Ca^{2+} channels in the presynaptic terminal membranes to enhance Ca^{2+} transfer intracellularly (Figure 3). The rise in intracellular Ca^{2+} stimulates the release of catecholamines, which are partly responsible for the seizure activity and brain damage that follows KA administration (56). Beta-oxalylamino-2-alanine (L-BOAA), a specific inhibitor of AMPA receptors, inhibits binding of AMPA and KA in hippocampal subfields in the micromolar range, whereas it also inhibits binding of DA at nanomolar concentrations. Further, microinjection of DA into the CA_3 region of the hippocampus produced generalized seizure activity, and the seizure latency was inversely related to the dose of DA (74,75). It is clear from the above discussion that the effects of DA and KA are mediated by non-NMDA (KA/AMPA) receptors, whereas GLU produces its effects by activation of NMDA and non-NMDA receptors (56).

III. MODE OF DOMOIC ACID ACTION

A. Receptor Mechanisms

Depolarization of presynaptic membranes is known to activate voltage-dependent Ca^{2+} channels elevating $[Ca^{2+}]_i$, which promotes GLU release into neuronal synapses (78). Glutamate then interacts with NMDA receptors on the postsynaptic membranes to elicit an excitatory postsynaptic potential, which involves a rapid Na^+ influx and K^+ efflux and slow influx of Ca^{2+}, resulting in the excitation of postsynaptic neurons (79). The rapid influx of Na^+ causes membrane depolarization and dissociation of Mg^{2+} from NMDA receptors, resulting in activation of latent NMDA receptors and a further increase in Ca^{2+} influx (79). The resulting rise in the intracellular concentration of Ca^{2+} and Na^+ activates calcium and sodium pumps which utilize ATP for their function to reestablish the ion homeostasis. Glutamate may accumulate in the synapse either due to its excessive release or inhibition of its uptake by presynaptic elements and astrocytes. Accumulation of GLU in the synapse produces continuing discharge of the postsynaptic elements, markedly disturbing the ion homeostasis, and causing activation of Na^+ and Ca^{2+} pumps, and ultimately resulting in an irreversible depletion of ATP. As a result of an imbalance in Na^+ concentration, water movement across cell membranes is impaired, resulting in the cytoplamic vacuolation, cell swelling, and rupture (80) followed by tissue necrosis and CNS dysfunction (79). Similar lesions in brain tissue are produced by high levels of GLU (79) and KA (81) and, during anoxia, ischemia, hypoglycaemia, and stroke (70).

The NMDA receptor is activated by GLU, ASP, NMDA, L-homocysteic acid, and quinolinic acid (82) and is modulated by the endogenous agonists by competitively binding to the glycine modulatory site, a voltage-dependent Mg^{2+} binding site in the Na^+/Ca^{2+}–permeable channel, a voltage-insensitive channel site with an affinity for phenylcyclidine-like agents, and a putative polyamine site (83). Magnesium blocks the transfer of Na^+ and Ca^{2+} through the channel, but this block is removed by membrane depolarization. The putative endogenous agonist is L-GLU, but glycine is obligatory for channel activation by GLU and NMDA (82,84). The NMDA receptor is blocked by 7-chloro-kynurenic acid (7-Cl-KYN) and 6-cyano-7- nitroquinoxaline-2,3-dione (CNQX) (84).

The alpha-amino-3-hydroxy-5-methylisoxazole-4-priopionic acid (AMPA) receptor mediates fast excitatory neurotransmission via ligand-gated channels. It has no modulatory sites, but the permeability of the channel to Na^+ and K^+ is voltage independent (84). AMPA and QA are potent agonists of AMPA receptors, and these ligands do not interact with NMDA receptors. Kynurenic acid, CNQX, and DNQX are nonselective antagonists, since they also bind to KA and glycine receptors. The kainate receptor is very similar to the AMPA receptor, and it is stimulated by KA, DA, QA, and AMPA. The KA receptor is primarily located on axons (85) and is distinguishable pharmacologically from AMPA receptors (84). The AMPA- and GLU-stimulated channels are voltage independent, whereas the KA- and DA-stimulated ion channel may be stimulated electrically (86).

The metabotropic EAA receptor (mGluR) is not coupled to an ion channel, but it activates phospholipase C catalyzing the hydrolysis of PIP_2 to form IP_3, which mobilizes Ca^{2+} from the intracellular stores, and DAG, which stimulates protein kinase C, resulting in the phosphorylation of channels and other membrane proteins (87). GLU-stimulated PIP_2 breakdown is antagonized by AP3 (87). There are six subtypes of metabotropic GLU receptors (mGluR) (88). mGluR1 and mGluR5 are positively coupled to phospholipase C, resulting in the production of IP_3. Activation of mGluR2, mGluR3, mGluR4, and mGluR6 inhibits forskolin-stimulated adenylate cyclase activity and accumulation of intracellular cyclic AMP (88). Apparently activation of mGluR1 enhances Ca^{2+} influx from the extracellular compartment and stimulates Ca^{2+} release from the endoplasmic reticulum (ER) in response to IP_3 (89). 1-aminocyclopentane-15,3r-dicarboxylic acid (1S-3R-ACPD) stimulates mGluR, inducing opening of dihydropyridine (DHP)–sensitive Ca^{2+} channels (L-types) (90). Pretreatment with pertussis toxin (PTX) did not block ACPD-induced Ca^{2+} channel activation and conductance. In fact, it enhanced Ca^{2+} response to ACPD, indicating that the ACPD effect on intracellular Ca^{2+} was mediated by the PTX-sensitive G protein (91). Apparently activation of mGluR1 is positively coupled to Ca^{2+} channels in cerebellar granular cell membranes. These findings implicate a positive coupling between mGluR1 and a complex family of DHP-sensitive Ca^{2+} channels, and a diffusible second messenger was seemingly involved in the activation of Ca^{2+} channels by metabotropic receptors (91). Quisqualate is the most active agonist on mGluR1 and mGluR5 receptors which are positively coupled to phospholipase C and almost inactive on other mGluRs receptors (mGluR2, mGluR3, and mGluR4) which are negatively coupled to the adenylate cyclase system (90). Results of our studies show that DA inhibits adenylate cyclase activity in rat brain membranes, and that DA-induced inhibition of adenylate cyclase was mediated by a G protein (92).

B. Role of Calcium in Excitotoxicity

It is generally believed that EAA-induced excitotoxicity is mediated by alterations in intracellular Ca^{2+} which are brought about by different mechanisms (70,81). These include Ca^{2+} influx via Ca^{2+} channels, Ca^{2+} efflux by the pump, Ca^{2+} uptake and release by ER, and reversible Ca^{2+} binding to the cytosolic and membrane proteins (70,81). Depolarization of presynaptic membranes activates Ca^{2+} channels, inducing the influx of Ca^{2+} which triggers the release of GLU into the synapse. GLU then interacts with NMDA receptors on the postsynaptic membrane to induce rapid fluxes of Na^+ and K^+ which cause excitation of the postsynaptic neurons. GLU also activates Ca^{2+} channels in the postsynaptic membranes, enhancing Ca^{2+} transfer intracellularly. The resulting rise in intracellular Ca^{2+} removes the Mg^{2+} block of NMDA receptors and thereby further potentiates the excitatory effects of GLU on the postsynaptic neurons. Accumulation of GLU in the synapse may occur as a result of either its excessive release and/or reduced uptake by the presynaptic elements and astrocytes. Continuous over stimulation of the postsynaptic neurons by excess GLU would deplete cells of energy, ATP which is essential for the function of pumps to restore ion homeostasis. An irreversible depletion of ATP results in the permanent failure of neurons to reestablish ion gradients across membranes. Consequently, water movement across membranes is impaired, resulting in the cytoplasmic vacuolation, cell swelling and rupture, brain tissue necrosis, and CNS dysfunction (79,81).

Glutamate has been demonstrated to enhance Ca^{2+} uptake by slices from cortices of the guinea pig (93), rat (94), and mouse striatum (95). Neither KA nor NMDA stimulated Ca^{2+} uptake by mouse striatal slices (96). Glutamate-induced Ca^{2+} uptake was inhibited by Mg^{2+}

ion, but insensitive to tetrodotoxin (TTX) (96). Apparently GLU, KA, and NMDA interact with distinct receptor subtypes to produce their effects on Ca^{2+} uptake in different regions of the brain (95). Glutamate interaction with NMDA receptors directly activates Ca^{2+} channels, resulting in ionic and energy imbalances and thereby interfers with the ability of neurons to maintain their osmolar integrity (97). Prolonged exposure of striatal slices to KA produces ionic and energy imbalances similar to those observed in response to GLU in the synapse (96), but these effects of KA on Ca^{2+} channels may be secondary to produce neurotoxicity (95,96).

Kainate-induced brain damage occurs in animals anaesthetized with ketamine, a drug which blocks the NMDA-operated ion channels. Similarly, another blocker of NMDA receptors, MK-801, did not modify KA-induced brain damage. Apparently KA impaired dopamine release without involvement of NMDA receptors (98). DNQX specifically blocks AMPA/KA receptors (99), also blocked the ability of KA to inhibit electrically stimulated dopamine release from nigrostriatal boutons. It is now clear that AMPA and KA receptors are distinct entities (100) located presynaptically on striatal axon terminals (101). Prior destruction of the glutamatergic cortical projections to the striatum abolished reduction of the electrically evoked [^{3}H] dopamine release by the intrastriatal injection of KA. Evidently KA activates KA receptors in striatum, including those situated on the dopaminergic presynaptic terminals. In striatum, KA is capable of stimulating GLU release in vivo as well as in vitro (102). KA receptors are present in striatum, and DNQX confers protection primarily by blocking KA receptors located on the glutamatergic nerve terminals. Apparently, KA-induced toxicity requires intact afferent glutamatergic pathways. This conclusion is supported by findings that KA-induced reduction in the activity of GAD and choline acetyltransferase, markers for intrinsic striatal GABAergic and cholinergic neurons, is completely prevented by cortical ablation (103), although the postsynaptic excitatory response to KA is maintained (104). The time course of KA-induced dopamine release seems to parallel the decrease in GAD and choline acetyltransferase activities after intrastriatal KA administration (105). The capability of KCl or veratridine to release similar amounts of dopamine from lesioned and unlesioned striatum reveals that dopamine presynaptic terminals remain susceptible to the depolarizing agents (106). The effectiveness of veratridine to depolarize dopaminergic terminals tends to exclude impairment of the voltage-dependent Na^+ channels. The impairment may be localized at the axon-terminal junction, a site which is likely to play a crucial role in allowing the traffic of electrical impulses from the axon to the terminal boutons (98).

Domoic acid was applied by microinontophoresis to rat hippocampus and unitary extracellular recordings from pyramidal neurons in the CA_1 and CA_3 regions were obtained and compared with those produced by KA, QA, and NMDA (39). Domoic acid was three times more effective in the CA_1 and CA_3 regions of hippocampus as compared with KA (39,40). Most interestingly, DA was 20 times more effective in the CA_3 than in the CA_1 region of the hippocampus, whereas QA and NMDA were equipotent in both regions and did not produce differential effects in these regions of brain (39). The differential regional response of CA_1 and CA_3 pyramidal neurons to KA and DA is due to variable density of KA receptors in the CA_3 region. Compared with the CA_1, the CA_3 region presents a lower threshold for KA-induced seizures, has a higher density of [^{3}H]KA binding sites (107), shows a higher increase in local cerebral glucose utilization after systemic administration of KA (108), and is more sensitive to neurotoxic effects of KA (109). These findings confirm the results of in vivo electrophysiological studies that KA was 20 times more toxic in the CA_3 than in the CA_1 region of the hippocampus (39). Domoic acid was also more potent in the CA_3 region than in the CA_1, region, whereas GLU, ASP, IBO, QA, and NMDA produced similar toxic response in both the CA_1 and CA_3 regions of the hippocampus. These observations provide evidence in support of the view that DA-induced neurotoxicity is mediated by activation of KA receptors (39,40). However, results of the histological studies with rat hippocampus (41) and biochemical studies with synaptoneurosomes (51) revealed some differences between KA and DA. Further studies are warranted to establish whether or not KA and DA produce excitotoxicity by activation of the same receptor subtype.

The damage to the CA_1 region in the hippocampus is also induced by transient forebrain ischemia, and appears to be mediated by EAAs (80). Elevated levels of EAAs induce fatal Ca^{2+} influx via EAA-operated Ca^{2+} channels (109). Exposure of hippocampal slices to hypoxic-hypoglycemic conditions resulted in nonselective release of GLU followed by selective elevation of the intracellular Ca^{2+} content in the CA_1 region of the hippocampus (110). Interestingly, neuronal cells were relatively resistant to hypoxic-hypoglycemic conditions but sensitive to surrounding EAAs which are released from either astrocytes and/or glial cells. Although the endogenous content of ASP is one fifth of GLU, ASP was released as much as GLU in response to hypoxia-hypoglycemia (110). The ischemia-induced neuronal damage is mediated by EAAs, which causes membrane depolarization leading to the disruption of ion homeostasis. Eventually, water movement across membranes is impaired causing cell swelling and shrinkage of extracellular space (80).

The stimulus-secretion coupling by Ca^{2+}, as discussed above, is considered to be analogous to the stimulus-contraction coupling in the muscle (111). DA produces muscle contraction in the hindgut of the American cockroach, which is mediated by Ca^{2+} (112). DA activates Ca^{2+} channels, enhancing Ca^{2+} influx into the nerve terminal. The resulting rise in intracellular Ca^{2+} promotes GLU release into the neuromuscular junction to initiate muscle contraction (112). Verapamil, a Ca^{2+} channel blocker, blocked the effect of DA in a medium containing Ca^{2+} but had no effect on GLU-induced muscle contraction. In fact, DA treatment sensitizes GLU-induced muscle contraction (112). An inhibitor of a cysteine proteinase, strepin P-1, partially blocks DA-induced hindgut muscle contraction (112). DA-enhanced Ca^{2+} influx into postsynaptic elements stimulates calpain, which may unmask latent GLU receptors, increasing sensitivity to GLU, and further potentiating muscle contraction and toxicity (113). Domoic acid has also been shown to enhance Ca^{2+} influx in the postsynaptic elements of rat pyriform cortex slices. Calcium then activates a neutral cysteine proteinase which produces toxic factors that cause death of neuronal cells (114).

C. Inositol-1,4,5-Triphosphate

Excitatory amino acids stimulate the hydrolysis of inositol phospholipids in rat cerebellum (91), hippocampal slices (115), mouse striatum (116), cultured neurons (117) rat cortical astrocytes (118), and glia (119). The receptor-mediated activation of PI metabolism may involve activation of G protein which subsequently activates phospholipase C, catalyzing the hydrolysis of PIP_2 to IP_3 and DAG. Inositol-1,4,5-triphosphate then mobilizes Ca^{2+} from the ER, whereas DAG activates protein kinase C, which catalyzes the phosphorylation of membrane proteins, including Ca^{2+} channels, and modulates intracellular processes (120). Apparently, this membrane signal transduction system is important for the regulation of cell metabolism and function (121). Receptors mediating EAA-enhanced phospholipase C activity and PIP_2 hydrolysis appears to be of the QA type, since 2-amino-4-phosphonovaleric acid (APV) inhibited NMDA receptor–mediated phosphoinositide turnover in cultured striatal neurons (116). Neither tetrodotoxin (TTX), which blocks voltage-dependent Na^+ channels and depolarization-induced acetylcholine release, nor atropine sulfate, which is an inhibitor of muscarinic cholinergic receptors, blocked EAA-induced phosphoinositides turnover in different brain preparations, suggesting the insignificance of cholinergic mechanism(s) in EAA-induced IP formation (122). Blockers of histaminergic (H_1), alpha-1-adrenergic, serotininergic ($5HT_2$) and $GABA_a$ receptors did not block EAA-enhanced

IP formation in rat hippocampal slices (123). The role for cyclic AMP and cyclic guanosine monophosphate (GMP) in EAA-induced PI turnover is not clear, but alterations in $[Ca^{2+}]_i$ appear to play a central role in PI turnover by EAAs (115,124).

In the rat hippocampus, ibotenate (IBO) and to lesser extent GLU and ASP induced PI turnover and IP_3 formation, whereas KA and NMDA were ineffective (115). APV did not block IBO-induced IP formation, but 2-amino-4-phosphonobutyric acid (APB) inhibited responses to IBO, suggesting that IBO-induced PI turnover was not mediated by the NMDA receptor, but other novel types of receptors may be involved (115,124). Ibotenate activates QA receptors which are sensitive to APB (125). Interestingly, APB did not affect KA-, QA-, and GLU-induced PI turnover in cerebral granular cells (123). This was in marked contrast to receptor types in hippocampal slices. These results indicate either the involvement of different receptors subtypes or their differential development in different regions of brain (123–125). EAA also enhanced IP formation in rat cortical astrocytes (118). It would appear that either EAA stimulates receptors which are coupled to the G protein to activate phospholipase C, resulting in the production of IP_3, or their interaction with NMDA receptors results in the activation of Ca^{2+} channels to enhance the influx of Ca^{2+} which stimulates PI turnover (69).

Interestingly, Mg^{2+} or phenylcyclidine inhibited PI turnover stimulated by GLU, NMDA, ASP, and KA in cultured cerebellar granular cells (126). Phenylcyclidine is known to block NMDA-associated ion channels, whereas Mg^{2+} induces a voltage-dependent block of NMDA-evoked membrane current by acting as a channel blocker in cultured CNS neurons (117). The Mg^{2+} block of the NMDA receptor is relieved by membrane depolarization by KC1 or veratridine (121,123). PIP_2 may be associated with the ion channels, and its hydrolysis profoundly influences channel activity, thereby regulating the intracellular level of Ca^{2+} (112,127). At low levels, Ca^{2+} stimulates many intracellular processes which are critical for normal cell function, but at high concentrations, Ca^{2+} inhibits critical enzymes and produces toxicity (70).

Domoic acid may interact with KA receptors on the cell surface to activate Ca^{2+} channels, allowing the enhanced transfer of Ca^{2+} intracellularly (125). An interaction of DA with KA receptors may be coupled to the Gi protein which inhibits adenylate cyclase activity and production of second-messenger cyclic AMP (92). Alternatively, DA may interact with the metabotropic receptors, resulting in activation of the Gs protein and stimulation of phospholipase C which catalyzes the hydrolysis of PIP_2 to IP_3 and DAG (69). We have investigated the membrane signal transduction pathways in rat brain that may be involved in DA-induced toxicity. Results of our studies revealed that (1) the intracellular Ca^{2+} level was elevated by DA and glucose deprivation (63), probably as a result of failure of pumps which extrude intracellular Ca^{2+}; (2) DA partially inhibited adenylate cyclase activity in brain membranes, and the reduced enzyme activity represented inhibition of Ca-CaM–stimulated adenylate cyclase activity (92); and (3) DA reduced Ca^{2+}-stimulated IP_3 formation in brain tissue (128). It would appear that DA interferes with Ca^{2+}-mediated intracellular processes in brain. DA may achieve these effects either by its interaction with specific receptors on the cell surface, or via modulation of G protein activity, and/or by chelation of intracellular Ca^{2+}.

The removal of extracellular Ca^{2+} completely inhibited the rise in intracellular Ca^{2+} induced by GLU (129). Inhibitors of NMDA receptors (CNQX, MK800), and blockers of Ca^{2+} channels (verapamil and nifedipine) prevent GLU-induced excitotoxicity in the brain (84,130). PI turnover in cortical neurons enhanced by membrane depolarization induced by KCl and vertaridine was mediated by dihyropyridine-sensitive Ca^{2+} channels (122,129). Apparently intracellular Ca^{2+} mediates EAA-induced excitotoxicity in brain tissue (81). Phosphoinositides may be associated with channels and show high affinity to bind Ca^{2+}. The phosphoinositide-Ca^{2+} complex closes the channels and prevents cation transport across membranes (131). However, phosphoinositides hydrolysis by phospholipase C or by phosphoinositide phosphomonoesterase would dissociate Ca^{2+} from the channel (131,132), facilitating the transfer of cations across membranes (127).

Recent evidence shows that EAAs inhibit IP_3 production stimulated by carbachol, NE, histamine, and KCl-induced depolarization in brain tissue (133,134). The enhanced PI metabolism has an absolute requirement for Ca^{2+}, but Ca^{2+} fluxes across membranes are not significant (134). However, the removal of Na^+ from the incubation medium reduced carbachol-induced IP_3 accumulation and virtually eliminated the inhibitory effects of EAAs on phosphoinositides turnover (134). Veratridine, which opens Na^+ channels, also reduced stimulation of PI turnover by carbachol. EAAs also inhibited NE-stimulated hydrolysis of PIP_2 in rat brain slices (133). EAAs are known to produce their cellular effects by interacting with NMDA receptors, inducing alterations in cation fluxes across membranes, but this interaction may influence the function of other receptors for hormones and neurotransmitters. EAAs suppressed PI turnover stimulated by other neurotransmitters, suggesting a possible cross-talk between different receptor types on cell membranes or between different second messengers. This cross-talk may be mediated by Na^+ and Ca^{2+} fluxes (133). The cross-talk between receptors and/or second messengers may be also mediated by transducer proteins such as the guanine nucleotide binding protein (G protein), including the Na-Ca exchanger and calcium pump in membranes (69).

The membrane signal transduction resulting in the formation of IP_3 in response to agonists involves its interaction with receptors on the cell surface, leading to the dissociation of the G protein complex into an alpha subunit from beta and gamma subunits. The alpha subunit of G protein undergoes ADP ribosylation, interacts with GTP, and then the Gp alpha-GTP complex activates phospholipase C, which catalyzes the hydrolysis of PIP_2 into IP_3 and DAG (69). The transduction system involved in response to EAAs appears to be distinct from that evoked by adrenergic and muscarinic agoninsts (121). Potassium-induced IP formation is not additive with GLU or QA but is additive with NE and more than additive with that triggered by carbachol (135). Other depolarizing agents such as veratridine, batrachotoxin, and monensin lack an additive effect with EAAs. In fact, EAAs appear to inhibit IP formation enhanced by carbachol, and by NE in some systems, whereas in other systems, EAAs had no effect on NE-induced IP formation (121). There are several possibilities to explain the variable effects of EAAs on IP formation, which include (1) GLU and other depolarizing agents stimulate the release of neuroactive substance(s) which stimulates IP formation, (2) depolarizing agents may stimulate the release of GLU which stimulates IP formation, (3) EAAs may stimulate a specific G protein and phospholipase C enzyme distinct from those activated by other neurotransmitters, and (4) depolarization and Na^+ influx represent a specific common step in the multistep phosphoinositide transduction pathway activated by EAAs and other depolarizing agents to stimulate IP formation (121).

Excitatory agonists are known to stimulate the release of neurotransmitters (98–101). Tetrodotoxin, which blocks voltage-dependent Na^+ channels and action potential-induced release of neurotransmitters, does not inhibit IP formation elicited by QA, IBO, and NMDA in different brain preparations, suggesting the insignificance of other neurotransmitters. In the absence of Ca^{2+}, there was a decrease in the basal or QA-, IBO-, or GLU-induced IP accumulation, indicating that Ca^{2+} was required for optimal stimulation of phospholipase C (136). It would appear that EAAs produce their stimulatory effects on IP formation by acting directly on Ca^{2+} channels in membranes rather than indirectly via release of other neurotransmitter substance(s).

A variety of excitatory agonists are known to stimulate the release of GLU which binds to EAA metabotropic receptors to enhance IP formation (137). However, NMDA stimulates GLU release but has no influence on IP formation. In the absence of Mg^{2+}, IBO enhances the release of endogenous GLU from guinea pig cerebrocortical synaptosomes, which is mediated by the ionotropic NMDA receptor subtype. In the presence of Mg^{2+}, IBO stimulates GLU release and IP formation in hippocampal slices and was inhibited by L-AP4, indicating mediation by IBO metabotropic receptors (138). Quisqualate is the most potent agonist in stimulating IP formation in rabbit retinal slices (139), but it does not release GLU from the chick retina (140). KA triggers GLU release in chick retina (198), in guinea pig cerebrocortical synaptosomes (141) or ASP release from cerebellar neurons (137), and is a weak stimulator of IP formation in different brain preparations (115,123,140). Apparently GLU release is not involved in the enhanced IP formation in response to excitatory agonists.

The exposure of cerebellar granular cells in primary culture to PTX reduces GLU- and QA-induced PI turnover but not carbachol-induced IP formation (142). Similarly, treatment of striatal and hippocampal neurons with PTX markedly inhibited IP formation in response to QA but had no effect on IP formation stimulated by carbachol (143). These data show that EAA metabotropic receptors of QA types are coupled with PTX-sensitive G protein which mediates activation of phospholipase C and IP_3 formation in the hippocampus and striatal neurons.

Membrane depolarization induced by K^+ involves enhanced sodium influx and causes enhanced IP formation in rat and guinea pig forebrain synaptoneurosomes (135). Membrane depolarization also promotes release of GLU which interacts with appropriate receptors subtypes linked to phosphoinositide turnover. IP formation in the presence of GLU and K^+ is significantly lower than GLU alone. In contrast, IP formation in response to K^+ and carbachol is additive (135). K^+-induced IP formation increases with the age of the rats, whereas GLU-elicited IP accumulation decreases (119), indicating GLU release is not entirely responsible for K^+-induced IP formation, and that there may exist a common step or mechanism between depolarizing agents and GLU; that is, G protein(s) and/or second messengers.

Glutamate and other excitatory agonists induce membrane depolarization by enhancing Na^+ influx in the striatal (145) and hippocampal slices (146) with the following order of potency, NMDA $>$ KA $>$ QA $\gg$ GLU, which does not correlate with the potency of IP responses elicited by these agents in striatal (147) or hippocampal slices (115,124) and in other tissues (116,123,148). EAA-induced Na^+ fluxes are probably associated with the activation of ionotropic EAA receptor subtypes (147) and were sensitive to Cd^{2+} but resistant to TTX, excluding the possible involvement of exchange mechanisms (149). Na^+ may stimulate G protein which activates phospholipase C (149). EAA-induced membrane depolarization is mediated by ionotropic and metabotropic EAA receptors and is accompanied by enhanced influx of Ca^{2+} (117) which activates phospholipase C activity in membranes (136). Classic Ca^{2+} channel antagonists, that is, verapamil, nifedipine, and diltiazem, did not block GLU-enhanced IP formation in different brain preparations, whereas Cd^{2+} blocked IP formation in response to GLU (121), suggesting the involvement of another type of voltage-dependent Ca^{2+} channels in synaptosomes, which is sensitive to Cd^{2+} (121,150).

It is evident from the above discussion that EAA-induced phosphoinositide breakdown is mediated by distinct receptor subtypes mGluR1 and mGluR5 metabotropic receptors, whereas QA and IBO stimulate phosphoinositide turnover via activation of distinct cation channels. Receptor interaction induces alterations in the fluxes of Na^+ and Ca^{2+} across neuronal membranes. The Na^+ influx neither originates from channel opening associated with EAA ionotropic receptor activation, nor from stimulation of the TTX-sensitive voltage-dependent Na^+ channels, nor from the increased functioning of the Na^+/Ca^{2+} or the Na^+/H^+ exchangers. The transduction mechanism linked to the QA metabotropic receptor subtype is sensitive to Cd^{2+}. This specific second-messenger system associated with a specific EAA receptor subtype stimulates intracellular Ca^{2+} mobilization via protein kinase C, resulting in the control of receptors and channels activities, and may constitute an important feedback mechanism for the rapid adaptation of neuronal cells to permanent variations in electrical activity. This adaptive capability may change with age as does the QA-induced IP response (123) and is probably important for the development and the survival of some neuronal cells (121).

D. Cyclic GMP

Cyclic GMP (cGMP) level is elevated manyfold in response to EAAs in cerebral granular cells and is suggested to be involved in EAA-induced neurotoxicity (151). Both KA and GLU increased cGMP levels in cerebellar neurons. The effect of KA was inhibited by piperidine dicarboxylic acid (PDA), whereas GLU-induced stimulation of cGMP formation was blocked by a NMDA antagonist, 2-amino-5-phosphonovaleric acid (AP5). Activation of NMDA receptors enhances influx of Ca^{2+} via activated Ca^{2+} channels, resulting in an elevation of intracellular Ca^{2+} which stimulates guanylate cyclase and cGMP production (152,153). In the absence of Mg^{2+}, NMDA, ASP, and QA increased cGMP levels in cerebellar granular cells (151). The removal of Mg^{2+} was probably facilitated by the enhanced Ca^{2+} influx either via Ca^{2+} channels or NMDA-operated cation channels (117). The role of cAMP in EAA-induced neurotoxicity is not clear, since cAMP levels were unaltered in neural tissue in one study and increased in another study in the presence of EAA (154).

E. Glycine

Transmitters mediating fast synaptic processes in the veterbrate central nervous system exhibit no interaction with the receptor. The inhibitory transmitters such as glycine and GABA are considered to act only on receptors mediating an increase in chloride conductance, whereas excitatory transmitters such as GLU and ASP are considered to activate receptors mediating an increase in cationic conductance. The response activated by NMDA agonists is unique in that it exhibits a voltage-dependent Mg^{2+} block which is potentiated by glycine. This potentiation is not mediated by the inhibitory strychnine-sensitive glycine receptors, and it can be observed in outside-out patches as an increase in the frequency of opening of the channels activated by NMDA agonists. It would appear that in addition to a role as an inhibitory transmitter, glycine may facilitate excitatory transmission in the brain through an allosteric activation of the NMDA receptors (155).

IV. PROTECTION OF BRAIN DAMAGE FROM DOMOIC ACID—INDUCED TOXICITY

A. Anticonvulsants

Seizure activity following the administration of DA or KA is a prerequisit for brain damage to ensue. Accordingly, control of seizure activity following exposure to excitotoxins prevents brain damage and CNS dysfunction (3). The control of seizure activity can be achieved by intravenous administration of anticonvulsants such as phenobarbital and phenylhydantion, but diazepam was ineffective in controlling EAA-induced seizures (3). The endogenous levels of GABA (74), kynurenic acid (48,156), and picolininic acid (157) are important for the control of EAA-induced seizure activity which results in brain damage and CNS dysfunction. It is clear that excitotoxicity involves GLU release from the presynaptic elements in response to enhanced Ca^{2+} influx via voltage-dependent Ca^{2+} channels. Glutamate then interacts with NMDA and non-NMDA receptors, inducing rapid fluxes of Na^+, K^+, and Ca^{2+} and producing an ionic imbalance across membranes. Failure of pumps due to lack of ATP results in the failure of cation homeostasis, allowing water movement intracellularly, producing tissue edema, necrosis, and CNS dysfunction (70,79,81). A similar sequence of cellular events is apparently involved in brain damage caused by the cerebral ischemia, anoxia, hypoglycemia (stroke), epilepsy, and Alzheimer's disease. However, the damage to the brain can be prevented by blocking NMDA receptors and Ca^{2+} overload resulting from the failure of pumps due to lack of ATP. Verapamil and nifedipine block Ca^{2+} channels and prevent Ca^{2+} overload in response to EAAs. As a result, brain cells are protected from EAA-induced excitotoxicity (78–81).

B. Calcium-Binding Proteins

The intracellular Ca^{2+} buffering capacity of cells seems to be highly significant in the regulation of cell metabolism and functions and is achieved by reversible binding of Ca^{2+} to the cytoplasmic and membrane proteins (158) and acidic phospholipids (131). Neurons containing Ca^{2+}-binding protein calbindin-D28K are more resistant to EAA-induced damage and neurodegeneration than those cells lacking CaBP (159). CaBP is a high-affinity Ca^{2+} binding protein which is expressed only in Purkinje cells in the adult cerebellum, but they are also transiently found in neurons of the cerebellar nuclei and provides Ca^{2+} buffering in cell cytoplasm (159). In response to KA application by ionophoresis in the cerebellar cortex, CaBP immunoreactivity appeared in Purkinje cells everywhere except at the site of application where Purkinje cells had undergone degeneration. Immunoreactivity of CaBP in Purkinje cells increased as a function of KA concentration, and the increase in immunoreactivity occurred both in the soma and dendrites (159). CNQX blocked KA-induced CaBP immunoreactivity and Ca^{2+} influx, indicating that the effects of KA are receptor mediated to promote entry of Ca^{2+}, which stimulates increased production of CaBP in Purkinje cells (160). It has been shown in cultured hippocampal tissue that the KA-induced rise in intracellular Ca^{2+} declines more rapidly in neurons with CaBP than in those which lack CaBP, but the initial rise in Ca^{2+} was the same in both types (161). When KA was removed, CaBP levels decreased, implying the return to normal buffering of Ca^{2+}. Apparently EAA-induced neurotoxicity is influenced by the amount of CaBP and buffering capacity of neurons for Ca^{2+} (159–163). Relative lack of CaBP in CA1-3 pyramidal cells and its presence in dentate granular cells reflects variable protection from DA-induced injury (163). Similarly, the heat shock protein 70 is induced in the hippocampus following exposure to KA but is limited to those regions that are protected from KA-induced degeneration (164). The intracellular Ca^{2+} buffering capacity of neuronal cells, which is represented by CaBP in the cytoplasm and membranes of neuronal cells, provides protection from excitotoxicity induced by KA and DA (128,159,163).

Calmodulin (CaM) is a ubiquitous cytoplasmic protein which has a high affinity to bind intracellular Ca^{2+}. Ca^{2+} either alone or in combination with CaM modulates activities of adenylate cyclase and cyclic AMP phosphodiesterase in a variety of tissues (165). At low levels, intracellular Ca^{2+} stimulates adenylate cyclase activity which catalyzes the formation of cyclic AMP, whereas at high levels, Ca^{2+} in combination with CaM activates cyclic AMP phosphodiesterase which degrades cyclic AMP (166). This dual control of enzymes by intracellular Ca^{2+} is believed to be responsible for oscillations in cyclic AMP which modulate numerous intracellular processes and is an important regulator of cell metabolism and functions (165,166). Results from our laboratory have demonstrated that DA abolished Ca^{2+}-stimulated IP formation by brain tissue preparations (128) and Ca-CaM–stimulated adenylate cyclase activity in brain membranes (92). These observations support our premise that DA-induced toxicity is mediated by intracellular Ca^{2+} (63,92,128). An association between Ca^{2+} + CaM-stimulated adenylate cyclase in sensory neurons and learning in *Aplysia* has been reported (170). Other investigators have also noted that mRNA encoding for CaM-sensitive adenylate cyclase is expressed in areas of rat brain that are implicated in learning and memory (171). All these findings support our concept that DNA induces neurodegeneration and CNS dysfunction (a loss of short-term memory) by inhibiting Ca^{2+}-stimulated intracellular processes, including Ca^{2+} + CaM-stimulated adenylate cyclase activity in neuronal cell membranes (63,92,128,168),

C. Glucose and Cell Energy

Glutamate toxicity in brain tissue is potentiated by the lack of glucose (78). These in vitro findings may be comparable to detrimental in vivo effects of hypoglycemia in the brain (167). We have demonstrated that glucose deprivation elevated the intracellular level of Ca^{2+} in brain tissue slices (63), and most likely glucose deprivation-potentiated GLU toxicity is mediated by an elevation in intracellular Ca^{2+} (63). Further studies have revealed that inhibition of glucose metabolism and resulting diminished ATP production are important in the regulation of Ca^{2+} homeostasis, DA-induced toxicity, and CNS dysfunction (168). Others have shown that ischemia and hypoglycemia result in a lack of substrate(s) for cellular respiration and ATP production and failure of Ca^{2+} homeostasis. This sequence of events leads to brain damage and CNS dysfunction (128,168,169).

ACKNOWLEDGMENTS

This work was supported in part by grants from the Natural Science and Engineering Research Council, National Research Council, Atlantic Canada Opportunity Agency, and the University of Prince Edward Island. I am grateful to Dr. Stephen Bates, Department of Fisheries and Oceans, Moncton, New Brunswick, for providing me with his in press publications (8–9). I also appreciate the assistance of Donna Smith, and Simrit Nijjar with the preparation of this manuscript.

REFERENCES

1. T Perl, L Bedard, R Remis, T Kosatsky, J Hoey, R Masse. Intoxication following mussel ingestion in Montreal. Can Dis Weekly Rep. 13–49:224–225, 1987.
2. TM Perl, L Bedard, T Kosatsky, JC Hockin, ECD Todd, RS Remis. An outbreak of toxic encephalopathy caused by eating mussels contaminated with domoic acid. N Engl J Med 322:1775–1780, 1990.
3. JS Teitelbaum, RJ Zatorre, S Carpenter, D Gendron, AC Evans, A Gjedde, NR Cashman. Neurologic sequelae of domoic acid intoxication due to the ingestion of contaminated mussels. N Engl J Med 322:1781–1787, 1990.
4. JLC Wright, RK Boyd, ASW DeFreitas, M Falk, RA Foxall, WD Jamieson, MW Laycock, AW McCuloch, AG, McInnis, P. Odense, VP Pathak, MA Quilliam, MA Regan, PG Sim, P. Therault, JA Walter, M Gilgan, DAJ Richard, D Dewar. Identification of domoic acid, a neurotexcitatory amino acid, in mussels from eastern Prince Edward Island. Can J Chem 67:481–490, 1989.
5. SS Bates, CJ Bird, ASW deFreitas, R Foxall, M Gilgan, LA Hanic, GR Johnson, AW McCuloch, P Odense, R Pocklington, MA Quilliam, PG Sim, JC Smith, DV Subba Rao, ECD Todd, JA Walter, JLC Wright. Pennate diatom *Nitzschia pungens* as the primary source of domoic acid, a toxin in shellfish from eastern Prince Edward Island, Canada. Can J Fish Aquat Sci 46:1203–1215, 1989.

6. B Grimmelt, MS Nijjar, J Brown, N MacNair, S Wagner, GR Johnson, JF Amend. Relationship between domoic acid levels in the blue mussel (*Mytilus edulis*) and toxicity in mice. Toxicon 28: 501–508, 1990.
7. T Takemoto, K Daigo. Constituents of *Chondria armata*. Chem Pharmaceut Bull 6:578–580, 1958.
8. SS Bates. Ecophysiology and metabolism of ASP toxin production. In: DM Anderson, AD Cembella, GM Halegraeff, eds. The Physiological Ecology of Harmful Algal Blooms. Heidelberg: Springer-Verlag, 1998, pp 405–426.
9. SS Bates, DL Garrison, RA Horner. Bloom dynamics and physiology of domoic acid producing *Pseudo-nitzschia* species. In: DM Anderson, AD Cembella, GM Hallegraeff, eds. The Physiological Ecology of Harmful Algal Blooms. Heidlberg: Springer-Verlag, 1998, pp 267–292.
10. RA Horner, JR Postel. Toxic diatoms in Western Washington waters (US west coast). Hydrobiology 269:197–205, 1993.
11. FJR Taylor, RA Haigh. Spatial and temporal distributions of microplankton during the summers of 1992–1993 in Barkley Sound, BC, with emphasis on harmful species. Can J Fish Aquat Sci 53: 2310–2322, 1996.
12. TM Work, B Barr, AM Beale, L Fritz, MA Quilliam, JLC Wright, Epidemiology of domoic acid poisoning in brown pelican (*Pelecanus occidentalis*) and Brant's cormorants (*Pharlacrocorax penicillatus*) in California. J Zoo Wildlife Med 24:54–62, 1993.
13. L Fritz, MA Quilliam, JLC Wright, A. Beale, TM Work. An outbreak of domoic acid poisoning attributed to the pennate diatom *Pseudonitzschia australis*. J Phycol 28:439–442, 1992.
14. N Lundholm, J Skov, R Pocklington, O Moestrup. Domoic acid, the toxic amino acid responsible for amnesic shellfish poisoning, now in Pseudonitzschia seriata (Bacillariophyceae) in Europe. Phycologia 33:475–478, 1994.
15. JL Martin, K Haya, LE Burridge, DJ Wildish. *Nitzschia pseudodelicatissima*—a source of domoic acid in the Bay of Fundy, eastern Canada. Mar Ecol Prog Ser 67:177–182, 1992.
16. GR Hasle. Pseudo-nitzschia pungens and P. multiseries (Bacillariophyceae): nomenclatural history, morphology, and distribution. J Phycol 31:425–435, 1995.
17. JC Smith, P Odense, R Angus, SS Bates, CJ Bird, P Cormier, ASW de Freitas, C Leger, D O'Neil, P Pauley, J Worms. Variation in domoic acid levels in Nitzschia species: implication for monitoring programs. Bull Aquacult Assoc Can 90-4:27–31, 1990.
18. SS Bates, C Leger, BA Keafer, DM Anderson. Discrimination between domoic acid producing and nontoxic forms of the diatom *Pseudonitzschia pungens* using immunofluorescence. Mar Ecol Prog Ser 100:185–195, 1993.
19. CA Scholin, MC Villac, KR Buck, KM Krupp, DA Powers, GA Fryxell, FP Chavez. Ribosomal DNA sequences discriminate among toxic and non-toxic *Pseudonitzschia species*. Nat Toxins 2: 152–165, 1994.
20. I Novaczek, MS Madhyastha, RF Ablett, G. Johnson, MS Nijjar, DE Sims. Uptake, disposition and depuration of domoic acid by blue mussels (*Mytilus edulis*) Aquat Toxicol 21:103–118, 1991.
21. R. Cusak, G Johnson. A study of dichlorvos (Nuvan; 2,2-dichloroethenyldime thylphosphate), a therapeutic agent for the treatment of salmonids infected with sea lice (Lepeophtheirus salmonis). Aquaculture 90:101–112, 1990.
22. GW Langolois, KW Kizer, Hansgen, KH, R Howell, Loscutoff, SM. A note on domoic acid in California coastal molluscs and crabs. J Shellfish Res 12:467–468, 1993.
23. SS Bates, ASW deFreitas, JE Milley, R. Pocklington, MA Quilliam, JC Smith, J Worms. Controls on domoic acid production by the diatom *Nitzschia pungens f. multiseries* in culture: nutrients and irradiance. Can J Fish Aquat Sci 48:1136–1144, 1991.
24. JC Smith, R Cormier, J Worms, CJ Bird, MA Quilliam, R Pocklington, R Angus, L Hanic. Toxic blooms of the domoic acid containing diatom *Nitzschia pungens* in the Cardigan River, Prince Edward Island. In E. Graneli, B. Sundstrom, L Edler, DM Anderson, eds. Toxic Marine Phytoplankton. New York: Elsevier, 1990, pp 227–232.
25. DV Subba Rao, MA Quilliam, R Pocklington. Domoic acid—a neurotoxic amino acid produced by the marine diatom, *Nitzschia pungens* in culture. Can J Fish Aquat Sci 45:2076–2079, 1988.
26. SS Bates, J Worms, JC Smith. Effects of ammonium and nitrate on domoic acid production by Nitzschia pungens in batch culture. Can J Fish Aquat Sci 50:1248–1254, 1993.

27. Y Pan, DV Subba Rao, KH Mann, RG Brown, R. Pocklington. Effects of silicate limitation on production of domoic acid, a neurotoxin, by the diatom *Pseudonitzschia multiseries*. 1. Batch culture studies. Mar Ecol Prog Ser 131:225–233, 1996.
28. JR Rosowski, LM Johnson, DG Mann. On the report of gametogenesis, oogamy, and uniflagellated sperm in the pennate diatom *Nitzschia pungens*. J Phycol 28:570–574, 1992.
29. CC Parrish, ASW de Freitas, G Bodennec, EJ MacPherson, RG Ackman. Lipid composition of the toxic marine diatom, *Nitzschia pungens*. Phytochemistry 30:113–116, 1991.
30. MA Brzezinski, RJ Olson, SW Chisholm. Silicon availability and cell cycle progression in marine diatoms. Mar Ecol Prog Ser 67:83–96, 1990.
31. DL Garrison, SM Conrad, PP Eilers, EM Waldron. Confirmation of domoic acid production by *Pseudonitzschia australis* (Bacillariophyceae) cultures. J Phycol 28:604–607, 1992.
32. SS Bates, J. Richard. Domoic acid production and cell division by *P. multiseries* in relation to a light: dark cycle in silicate-limited chemostat culture. In: R Penney, ed. Proceeding's of the Fifth Canadian Workshop on Harmful Marine Algae. Can Tech Rep Fish Aquat Sci 2138, pp 140–143.
33. NI Lewis, SS Bates, JL McLachlan, JC Smith. Temperature effects on growth, domoic acid production, and morphology of the diatom *Nitzschia pungens f. multiseries*. In: TJ Smayda, Y. Shimizu, eds. Toxic Phytoplankton Blooms in the Sea. Amsterdam: Elsevier, 1993, pp 601–606.
34. M Osada, JE Stewart. Gluconic acid/gluconolactone: physiological influences on domoic acid production by bacteria associated with *Pseudonitzschia multiseris*. Aquat Microbial Ecol 12:203–209, 1997.
35. G McKhann, D. Drachman, N Folstein, R Katzman, D Price, EM Stadlan. Clinical diagnosis of Alzheimer's disease: report of the NINCDS-ADRDA work group under the suspices of Department of Health and Humn Services Task Force on Alzheimer's disease. Neurology 34:939–944, 1984.
36. GB Glavin, C Pinsky, R Bose. Domoic acid–induced neurovisceral toxi syndrome: Characterization of an animal model and putative antidotes. Brain Res Bull. 24:701–703, 1990.
37. JE Schwob, T Fuller, JL Price, JW Olney. Widespread patterns of neuronal damage following systemic or intracerebral injections of kainic acid: a histological study. Neuroscience 5:991–1014, 1980.
38. De Montigny, D Tardif. Differential excitatory effects of kainic acid on CA_3 and CA_1 hippocampal pyramidal neurons: further evidence for the excitotoxic hypothesis and for a receptor-mediated action. Life Sci 29:2103–2111, 1981.
39. G Debonnel, L Beauchesne, C DeMontigny. Domoic acid, the alleged 'mussel toxin', might produce its neurotoxic effect through kainate receptor activation: an electrophysiological study in the rat dorsal hippocampus. Can J Physiol Pharmacol 67:29–33, 1989.
40. RAR Tasker, BJ Connel, SM Strain. Pharmacology of systemically administered domoic acid in mice. Can J Physiol Pharmacol 69:378–382, 1991.
41. SM Strain, RAR Tasker. Hippocampal damage produced by systemic injections of domoic acid in mice. Neuroscience 44:343–352, 1991.
42. C Chiamulera, S Costa, E Valerio, A Reggiani. Domoic acid toxicity in rats and mice after intracerebroventricular administration: comparison with excitatory amino acid agonists. Pharmacol Toxicol 70:115–120, 1992.
43. YG Peng, TB Taylor, RE Finch, RC Switzer, JS Ramsdell. Neuroexcitatory and neurotoxic actions of the amnesic shellfish poison, domoic acid. NeuroReport 5:981–985, 1994.
44. JW Olney. Excitoxicity: an overview. Can Dis Weekly Rep 16SIE(Suppl, Sept):47–57, 1990.
45. L Tryphonas, J Truelove, E Todd, E Nera, F Iverson. Experimental oral toxicity of domoic acid in cynomolgus monkeys (*Macac Fascicularis*) and rats. Food Chem Toxicol 28:707–715, 1990.
46. J Truelove, F Iverson. Serum domoic acid clearance and clinical observation in the Cynomologus monkey and Sprague-Dawley rat following a single iv dose. Bull Environ Contam Toxicol 52:479–486, 1994.
47. CA Suzuki, SL Hierlihy, SL. Renal clearance of domoic acid in the rat. Food Chem Toxicol 31: 701–706, 1993.
48. GB Glavin, R Bose, C Pinsky. Kynurenic acid protects against gastroduodenal ulceration in mice injected with extracts from poisonous atlantic shellfish. Prog Neuropsychopharmacol and Biol Psychiatry 13:569–572, 1989.

49. B Sawyer, H Shannon. N-methyl-D-aspartate receptors in the guinea pig ileium myenteric plexus. Pharmacologist 30:A184, 1988.
50. MS Nijjar, B Grimmelt, J Brown. Purification of domoic acid from toxic blue mussels (*Mytilus edulis*) and phytoplankton. J Chromatogr 568:393–406, 1991.
51. JA Brown, MS Nijjar. The release of glutamate and aspartate from rat brain synaptosomes in response to domoic acid (amnesic shellfish toxin) and kainic acid. Molec Cell Biochem 151:49–54, 1995.
52. MS Nijjar, MS Madhyastha. Effect of pH on domoic acid toxicity in mice. Molec Cell Biochem 167:179–185, 1997.
53. E Preston, I Hynie. Transfer constants for blood-brain barrier permeation of the neuroexcitatory shellfish toxin, domoic acid. Can J Neurol Sci 18:39–44, 1991.
54. F Iverson, J Truelove, L Tryphonas, E Nera. The toxicology of domoic acid administered systemically to rodents and primates. Can Dis Weekly Rep 16S1E:15–19, 1990.
55. ML Berger, H Lassmann, O Hornykiewicz. Limbic seizures without brain damage after injection of low doses of kainic acid into the amygdala of freely moving rats. Brain Res 489:261–271, 1989.
56. G Sperk. Kainic acid seizures in the rat. Prog Neurobiol 42:1–32, 1994.
57. DF Wozniak, GR Stewart, JP Miller, JW Olney. Age-related sensitivity to kainate neurotoxicity. Exp Neurol 114:250–253, 1991.
58. H Baran, H Lassmann, G Sperk, F Seitelberger, O Hornykiewicz. Effect of mannitol treatment on brain neurotransmitter markers in kainic acid induced epilepsy. Neuroscience 21:679–684, 1987.
59. Y Ben-Ari, E Tremblay, D Richie, G Ghilini, R Naquet. Electrographic, clinical and pathological alterations following systemic administration of kainic acid, bicuculline and pentetrazole metabolic mapping using the deoxyglucose method with special reference to the pathology of epilepsy. Neuroscience 6:1361–1391, 1981.
60. EW Lothman, RC Colins. Kainic acid–induced limbic motor seizures: metabolic, behavioral, electroencephalographic and neuropathological correlates. Brain Res 218:299–318, 1981.
61. T Popovici, A Repressa, V Crepel, G Barbin, M Beaudoin, Y Ben-Ari. Effects of kainic acid–induced seizures ad ischemia on c-fos like proteins in rat brain. Brain Res 536:183–194, 1990.
62. G Garthwaite, GD Williams, J Garthwaite. Glutamate toxicity: an experimental and theoretical analysis. Eur J Neurosci 4:353–360, 1992.
63. MS Nijjar. Effects of domoate, glutamate, and glucose deprivation on calcium uptake by rat brain tissue *in vitro*. Biochem Pharmacol 46:131–138, 1993.
64. A Lehmann, H Isacsson, A Hamberger. Effect of *in vitro* administration of kainic acid on the extracellular amino acid pool in the rabbit hippocampus. J Neurochem 40:1314–1320, 1983.
65. DE Heggli, D Malthe-Sorenssen. Systemic injection of kainic acid: effect on neurotransmitters markers in piriform cortex, amygdaloid comples and hippocampus and protection by cortical lesioning an anticonvulsants. Neuroscience 7:1257–1264, 1982.
66. JW Ferkany, R Zaczek, JT Coyle. Kainic acid stimulates excitatory amino acid neurotransmitter release at presynaptic receptors. Nature 298:757–759, 1982.
67. AI Sacaan, DD Schoepp. Activation of hippocampal metabotropic excitatory amino acid receptors leads to seizures and neuronal damage. Neurosci Lett 139:77–82, 1992.
68. B Meldrum, J Garthwaite. Excitatory amino acid neurotoxicity and neurodegenerative disease. Trend Pharmacol Sci 11:379–387, 1990.
69. TG Smart. Excitatory amino acids: The involvement of second messengers in the signal transduction process. Cell Molec Neurobiol 9:193–206, 1989.
70. BK Siesjo, F Bengtsson, W Grampp, S Theander. Calcium, excitotoxins, and neuronal death in the brain. Ann NY Acad Sci 568:234–251, 1990.
71. SA Lipton, PA Rosenberg. Excitatory Amino Acids as a final common pathway for neurologic disorders. N Engl J Med 330:613–622, 1994.
72. AB Young, GE Fagg. Excitatory amino acid receptors in the brain: membrane binding and receptor autoradiographic approaches. Trends Pharmacol Sci 11:126–133, 1990.
73. RS Fisher, BE Alger. Electrophysiological mechanisms of kainic acid–induced epileptoform activity in the rat hippocampal slice. J Neurochem 4:1312–1323, 1984.

74. K Dakshinamurti, SK Sharma, M Sundaram. Domoic acid induced seizure activity in rats. Neurosci Lett 127:193–197, 1991.
75. K Dakshinamurti, SK Sharma, M Sundaram, T Watanabe. Hippocampal changes in developing postnatal mice following intrauterine exposure to domoic acid. J Neurosci 13:4486–4495, 1993.
76. CR Houser, AB Harris, JE Vaughn. Time course of the reduction of GABA terminals in a model of focal epilepsy: a glutamic acid decarboxylase immunocytochemical study. Brain Res 383:129–145, 1986.
77. S Feldblum, RF Ackermann, AJ Tobin. Long-term increase of glutamate decaroxylase mRNA in a rat model of temporal lobe epilepsy. Neuron 5:361–371, 1990.
78. J Garthwaite, G Garthwaite, F Hajos. Amino acid neurotoxicity: relationship to neuronal depolarization in rat cerebellar slices. Neuroscience 18:449–460, 1986.
79. JW Olney. Neurotoxicity of excitatory amino acid neurotransmitters, In: EG McGeer, JW Olney, PL McGeer, eds. Kainic Acid as a Tool in Neurobiology. New York, Raven Press, 1978, pp 95–122.
80. Y Katayama, T Tamura, DP Becker, T Tsubokawa. Early cellular swelling during cerebral ischemia in vivo is mediated by excitatory amino acides released from nerve terminals. Brain Res 577:121–126, 1992.
81. MA Verity. Ca^{2+} dependent processes as mediators of neurotoxicity. Neurotoxicology 13:139–147, 1992.
82. PL Herrling. Synaptic physiology of excitatory amino acids. Arzheim Forsch/Drug Res 42:202–208, 1992.
83. K Williams, C Romano, MA Dichter, PB Molinoff. Modulation of the NMDA receptor by polyamines. Life Sci 48:469–498, 1991.
84. AB Young, SY Sakurai, RL Albin, R Makowiec, JB Putney Jr. Excitatory amino acid receptor distribution: quantitative autoradiographic studies. In: HB Wheal, AM Thomson, eds., Excitatory amino acids and synaptic transmission, London: Academic Press, 1991, pp 19–54.
85. RH Evans, SJ Evans, PC Pook, DC Sunter. A comparison of excitatory amino acid antagonists acting at primary afferent C fibres and motoneurones of the isolated spinal cord of the rat. Br J Pharmacol 91:531, 1987.
86. Watkins JC, Krogsgaard-Larsen P, Honore T. Structure activity relationship in the development of excitatory amino acid receptor agonists and competitive antagonists. Trends Pharmacol Sci 11:25–33, 1990.
87. Schoepp D, Bockaert J, Sladeczek F., Pharmacological and functional characteristics of metabotropic excitatory amino acid receptors. Trends Pharmacol Sci 11:508–515, 1990.
88. Y Tanabe, M Masu, T Ishi, R Shigemoto, S Nakanishi. A family of metabotropic glutamate receptors. Neuron 8:169–179, 1992.
89. H Sugiyama, I Ito, C Hirono. A new type of glutamate receptor linked to inositol phospholipid metabolism. Nature 325:531–533, 1987.
90. O Manozoni, L Fagni, JP Pin, F Rassendren, F Poulat, F Sladeczek, J Bockaert. Trans-1-amino-cyclopentyl-1,3, dicarboxylate are agonists of metabotropic glutamate receptors coupled to phospholipase C. Mol Pharmacol 42:322–327, 1990.
91. F Nicoletti, JT Wroblewski, A Novelli, H Alho, A Guidotti, E Costa. The activation of inositol phospholipid metabolism as a signal transducing system for excitatory amino acids in primary cultures of cerebellar granule cells. J Neurosci 6:1905–1911, 1986.
92. MS Nijjar, B Grimmelt. Domoic acid inhibits adenylate cyclase activity in rat brain membranes. Molec Cell Biochem 136:105–111, 1994.
93. RL Ramsey, H McIlwain. Calcium content and exchange in neocortical tissue during the cation movements induced by glutmates. J Neurochem 17:781–787, 1970.
94. WJ Cooke, JD Robinson. Factors influencing calcium movements in rat brain slices. Am J Physiol 221:218–225, 1971.
95. KC Retz, JT Coyle. The differential effects of excitatory amino acids on the uptake of $^{45}CaCl_2$ by slices from mouse striatum. Neurophrmacology 23:89–94, 1984.
96. KC Retz, JT Coyle. Effects of kainic acid on high-energy metabolites in the mouse striatum. J Neurochem 28:196–203, 1982.

97. JA Harvey, H McIlwain. Excitatory acidic amino acids and the cation content and sodium ion flux of isolated tissues from the brain. Biochem J 108:269–274, 1968.
98. E Fedele, P Versace, M Raiteri. Evaluation of the mechanisms underlying the kainate-induced impairment of [^{3}H]dopamine release in the rat striatum. Eur J Pharmacol 249:71–77, 1993.
99. J Drejer, T Honore. New quinoxalinediones show potent antagonis of quisqualate responses in cultured mouse cortical neurons. Neurosci Lett 87:104–108, 1988.
100. R Dingledine. New wave of non-NMDA excitatory amino acid receptors. Trend Pharmacol Sci 12: 360–367, 1991.
101. JK Wang. Presynaptic glutamate receptors modulate dompamine release from striatal synaptosomes. J Neurochem 57:819–822, 1991.
102. JW Ferkany, JT Coyle. Kainic acid selectively stimulates the release of endogenous excitaotry acidic amino acid. J Pharmacol Exp Ther 225:399, 1983.
103. EG McGeer, PL McGeer, K Singh. Kainate-induced degeneration of neostriatal neurons: dependency upon cortico-striatal tract. Brain Res 139:381–383, 1978.
104. DH McLennan. The effect of decortication on excitatory amino acid sensitivity of striatal neurons. Neurosci Lett 18:313, 1980.
105. T Hattori, T, E McGeer. Fine structural changes in the rat striatum after local injections of kainic acid. Brain Res 129:174–180, 1977.
106. G Bonanno, E Fedele, P Versace, M Raiteri. Functional damage of dopamine nerve terminals following intrstriatal kainic acid injection. Brain Res 480:242–248, 1989.
107. AC Foster, EE Mena, DT Monaghan, CW Cotman. Synaptic localization of kainic acid binding sites. Nature 289:73–75, 1981.
108. JE Chastain, F Samson, SR Nelson, TL Pazdernik. Attenuation of cerebral glucose use in kainic acid treated rats by diazepam. Eur J Pharmacol 142:215–224, 1987.
109. DW Choi. Ionic dependence of glutamate neurotoxicity. J Neurosci 7:369–379, 1987.
110. T Ogata, Y Nakamura, T Shibata, K Kataoka. Release of exciatory amino acids from cultured hippocampal astrocytes induced by a hypoxic-hypoglycemic stimulation. J Neurochem 58:1957–1959, 1992.
111. MS Nijjar, NS Dhalla. Biochemical basis of calcium handling in developing myocardium. In: B Ostadal, M Nagano, N Takeda, NS Dhalla, eds. The Developing Heart. Philadelphia, Lippincott-Raven, 1997, pp 189–217.
112. M Maeda, T Kodam, M Saito, T Tanaka, H Yoshizuma, K Nomoto, T Fujita. Neuromuscular action of insecticidal domoic acid on the American cockroach. Pest Biochem Physiol 28:85–92, 1987.
113. M Baudry, R Siman, EK Smith, G Lynch. Regulation by calcium ions of glutamate receptor binding in hippocampal slices. Eur J Pharmacol 90:161–168, 1983.
114. N Hori, JMH French-Mullen, DO Carpenter. Kainic acid responses and toxicity show pronounced Ca^{2+} dependance. Brain Res 358:380, 1985.
115. F Nicoletti, JL Meek, MJ Iadorola, DM Chuang, BL Roth, E Costa. Coupling of inositol phospholipid metabolism with excitatory amino acid recognition sites in rat hippocampus. J Neurochemistry 46:40–46, 1986.
116. F Sladeczek, J-P Pin, M Recasens, J Bockaert, S Weiss. Glutamate stimulates inositol phosphate formation in striatal neurones. Nature 317:717–719, 1985.
117. L Nowak, P Bregestovski, P Ascher, A Herbet, A Prochiantz. Magnesium gates glutamate-activated channels in mouse central neurones. Nature 307:462–465, 1984.
118. B Pearce, J Albrecht, C Morrow, S Murphy. Astrocytic glutamate receptor activation promotes inositol phospholipid turnover and calcium flux. Neurosci Lett 72:335–340, 1986.
119. J Guiramand, I Sassetti, M Recasens. Developmental changes in the chemosensitivity of rat brain synaptoneurosomes to excitatory amino acids, estimated by inositol phosphate formation. Neuroscience 7:257–266, 1989.
120. Y Nishizuka. The role of protein kinase C in cell surface signal transduction and tumor promotion. Nature 308:693–698, 1984.
121. M Recasens, J Guiramand, M Vignes. The putative molecular mechanism(s) responsible for the enhanced inositol phosphate synthesis by excitatory amino acids: an overview. Neurochem Res 16: 659–668, 1991.

122. SR Nahorski, DA Kendall, I Batty. Receptors and phosphoinositide metabolism in the central nervous system. Biochem Pharmacol 35:2447–2453, 1986.
123. M Recasens, I Sassetti, A Nourgat, F Sladeczek, J Bockaert. Characterization of subtypes of excitatory amino acid receptors involved in the stimulation of inositol phosphate synthesis in rat brain synaptoneurosomes. Eur J Pharmacol 141:87–93, 1987.
124. Shoepp, DD, and Johnson, BG. Excitatory amino acid agonist antagonist interaction at 2-amino-4-phosphonobutyric acid-sensitive quisqualate receptors coupled to phosphoinoistide hydrolysis in slices of rat hippocampus. J Neurochem 50:1605–1613, 1988.
125. P Krogsgaard-Larsen, T Honore, JJ Hansen, DR Curtis, D Lodge. New class of glutamate agonist structurally related to ibotenic acid. Nature 284:64–66, 1980.
126. Nicoletti F, Wroblewski JT, Novelli A, Alho H, Guidotti A, and Costa E. The activation of inositol phospholipid metabolism as a signal transducing system for excitatory amino acids in primary cultures of cerebellar granule cells. J Neurosci 6:1905–1911, 1986.
127. JT Buckley, JN Hawthorne. Erythrocyte membrane polyphosphinositide metabolism and the regulation of calcium binding. J Biol Chem 247:7218–7223, 1972.
128. MS Nijjar. Cellular basis of CNS dysfunction induced by domoic acid (Amnesic shellfish toxin). In: BK Sharma, PK Singal, eds. Adaptation Biol Med. Vol 1. New Delhi: Narosa Publishing, 1997, pp 172–186.
129. DA Kendall, SR Nahorski. Dihydropyridine calcium channel activators and antagonists influence depolarization-evoked inositol phospholipid hydrolysis in brain. Eur J Pharmacol 115:31–36, 1985.
130. MS Nijjar, LL Hart, GN Pierce, NS Dhalla. Characterization of ATP-induced elevation in intracellular Ca^{2+} in freshly isolated rat cardio-myocytes. Cardiovasc Pathobiol 1:152–159, 1996.
131. M Kai, JN Hawthorne. Physiological significance of polyphosphoinositides in brain. Ann NY Acad Sci 165:761–773, 1969.
132. MS Nijjar, JN Hawthorne: Purification and properties of polyphosophonositide phosphomonoesterase from rat brain. Biochim Biophys Acta 480:490–502, 1979.
133. RS Jope, S Li. Inhibition of inositol phospholipid synthesis and norepinephrine-stimulated hydrolysis in rat brain slices by excitatory amino acid. Biochem Pharmacol 38:589–596, 1989.
134. M Baudry, J Evans, G Lynch. Excitatory amino acid inhibit stimulation of phosphatidylinositol metabolism by aminergic agonists in hippocampus. Nature 319:329–331, 1986.
135. J Guiramand, A Nourigat, I Sasetti, M Recasens, M. K^+ differentially affects the excitatory amino acids- and carbachol-elecited inositol phosphate formation in rat brains synaptoneurosmes. Neurosci Lett 98:222–228, 1989.
136. CF Bennet, ST Crooke. Purification and characterization of a phosphoinositide-specific phospholipase C from guinea pig uterus. J Biol Chem 262:13789–13797, 1987.
137. V Gallo, R Suergiu, C Giovannini, G Levi. Glutamate receptor subtypes in cultured cerebellar neurons: modulation of glutamate and aminobutyric acid release. J Neurochem 49:1801–1809,1987.
138. PG Jones, PJ Roberts. Ibotenate stimulates glutamate release from guinea pig cerebrocortical synaptosomes: inhibition of L-2-amino-4-phosphobutyrate (L-AP4). Neurosci Lett 111:228–232, 1990.
139. NN Osborne. Stimulatory and inhibitory actions of excitatory amino acids on inositol phospholipid metabolism in rabbit retina. Evidence for a specific quisqualate receptor subtype associated with neurones. Exp Eye Res 50:397–405, 1990.
140. P Campochiaro, JW Ferkany, JT Coyle. Excitatory amino acid analogs evoke release of endogenous amino acids and acetylcholine from chick retina in vitro. Vision Res 25:1375–1386, 1985.
141. JM Pockock, HM Murphie, DG Nicholls. Kainic acid inhibits the synaptosomal plasma membrane glutamate carrier and allows glutamate leakage from the cytoplasm but does not affect glutamate exocytosis. J Neurochem 50:745–751, 1988.
142. F Nicoletti, JT Wroblewski, E Fadda, E Costa. Pertussis toxin inhibits signal transduction at a specific metabotropic glutamate receptor in primary cultures of cerebellar granule cells. Neuropharmacology 27:551–556, 1988.
143. A Ambrosini, J Meldolesi. Muscarinic and quisqualate receptor-induced phosphoinositide hydrolysis in primary cultures of striatal and hippocampal neurons. Evidence for differential mechanism of activation. J Neurochem 53:825–833, 1989.

144. G Levi, F Aloisis, MT Ciooti, V Gallo. Autoradiographic localization and depolarization-induced release of amino acids in differentiating cerebellar granule cell cutures. Brain Res 290:77–86, 1984.
145. A Luini, O Goldberg, VI Teichberg. Distinct pharmacological properties of exciatory amino acid receptors in the rat striatum: study by Na^+ efflux assay. Proc Natl Acad Sci USA 78:3250–3254, 1981.
146. H Sugiyama, I Ito, C Hirono. A new type of glutamate receptor linked to inositol phospholipid metabolism. Nature 325:531–533, 1988.
147. P Ascher, L Nowak. Quisqualate- and KA-activated channels in mouse central neurones in culture. J Physiol (London) 399:227–245, 1988.
148. F Gusovsky, ET McNeal, JW Daly. Stimulation of phosphoinositide breakdown in brain synaptoneurosomes by agents that activate sodium influx: anatagonism by teterodotoxin, saxitoxin, and cadmium. Molec Pharmacol 32:479–487, 1987.
149. JF Renaud, T Kazazoglou, A Lombet, R Chicheportiche, E Jaimovitch, G Romey, M Lazdunski. The Na^+ channels in mammalian cardiac cells. Two kinds of tetrodotoxin receptors in rat heart membranes. J Biol Chem 258:8799–8805, 1983.
150. JB Suszkiw, MM Murawsky, M Shi. Further characterization of phasic calcium influx in rat cerebrocortical synaptosomes: inferences regarding calcium channel type(s) in nerve endings. J Neurochem 52:1260–1269, 1989.
151. A Novelli, F Nicoletti, JT Wroblewski, H Alho, E Costa, A Guidotti. Excitatory amino acid receptors coupled with guanylate cyclase in primary cultures of cerebellar granule cells. J Neurosci 7:40–47, 1987.
152. JC Watkins, HJ Olverman. Agonists and antagonists for excitatory amino acid receptors. Trends Neurosci 10:265–272, 1987.
153. EW Troyer, IA Hall, JA Ferrendelli. Guanylate cyclases in CNS: enzymatic characteristics of soluble and particulate enzymes from mouse cerebellum and retina. J Neurochem 31:825–833, 1978.
154. J Garthwaite. Excitatory amino acid receptors and guanosine 3′,5′-cyclic monophosphate in incubated slices of immature and adult rat cerebellum. Neuroscience 7:2491–2497, 1982.
155. JW Johnson, P Ascher. Glycine potentiates the NMDA response in cultured mouse brain neurons. Nature 325:529–531, 1987.
156. BA Wirsching, RJ Beninger, K Jhamandas, RJ Boegman, M Bialik. Kynurenic acid protects against neurochemical and behavioral effects of unilateral quinolinic acid injections into nucleus basalis of rats. Behav Neurosci 103:90–97, 1989.
157. J Cockhill, K Jhamandas, RJ Boegman, RJ Beninger. Action of Picolininic acid and structurally related pyridine carboxylic acid on quinolinic acid-induced cholinergic damage. Brain Res 599:57–63, 1992.
158. E Carafoli. The signalling function of calcium and its regulation. J Hypertension 12(suppl):S47–S56, 1994.
159. C Batini, M Palestini, M Thomasset, R Vigot. Cytoplasmic calcium buffer, Calbindin-D28K, is regulated by excitatory amino acids. NeuroReport 4:927–930, 1993.
160. ML Mayer, RJ Miller. Excitatory amino acid receptors, second messengers and regulation of intracellular Ca^{2+} in mammalian neurons. Trend Pharmacol Res 11:254–260, 1990.
161. MP Mattson, B Rychlik, C Chu, S Christakds. Evidence for calcium-reducing and excitoprotective roles for the calcium-binding protein Calbindin-D28K in cultured hippocampal neurons. Neuron 6: 41–51, 1991.
162. YG Peng, TB Taylor, RE Finch, RC Switzer, JS Ramsdell. Neuroexcitatory and neurotoxic actions of the amnesic shellfish poison, domoic acid. NeuroReport 5:981–985, 1994.
163. A Iacopino, S Christakos, D German, PK Sonsalla, CA Altar. Calbindin-D28K-containing neurons in animal models of neuro degeneration. Possible protection from excitotoxicity. Brain Res 13:251–261, 1992.
164. K Vass, ML Berger, TS Nowak, WJ Welsh, H Lassman. Induction of stress protein HSP70 in nerve cells after status epilepticus in the rat. Neurosci Lett 100:259–264, 1989.
165. WY Cheung. Calmodulin plays a pivotal role in cellular regulation. Science 207:19–27, 1980.
166. CO Brostrom, YC Huang, BM Breckenridge, DJ Wolff. Identification of a calcium-binding protein

as a calcium-dependent regulator of brain adenylate cyclase. Proc Natl Acad Sci USA 72:64–68, 1975.

167. WO Whetsell, JR, NA Shapira. Biology of Disease. Neuroexcitation, excitotoxicity and human neurological disease. Lab Invest 68:372–387, 1993.

168. MS Nijjar, RL Belgrave. Regulation of Ca^{2+} homeostasis by glucose metabolism in rat brain. Molec Cell Biochem 176:317–326, 1997.

169. K Hashimoto, H Kikuchi, M Ishikawa, S Kobayasmi. Changes in cerebral energy metabolism and calcium levels in relation to delayed neuronal death after ischemia. Neurosci Lett 137:165–168, 1992.

170. TW Abrams, KA Karl, ER Kandel. Biochemical studies of stimulus convergence during classical conditioning in *Aplysia*: Dual regulation of adenylate cyclase by Ca^{2+}/calmodulin and transmitter. J Neurosci 11:2655–2665, 1991.

171. Z Xia, CD Refsdal, KM Merchant, DM Dorsa, DR Storm. Distribution of mRNA for the calmodulin-sensitive adenylate cyclase in rat brain: expression in areas associated with learning and memory. Neuron 6:431–443, 1991.

16
Pharmacology of Domoic Acid

Adam Doble
Rhône-Poulenc S.A.
Antony, France

I. INTRODUCTION

Domoic acid is an excitatory amino acid that has been implicated as the toxin responsible for the outbreak of amnesic shellfish poisoning in Canada in 1987. Domoic acid is a highly potent and selective agonist at a subpopulation of the AMPA/kainate subtype of excitatory amino acid receptor. These receptors are found in high densities in the mammalian hippocampal formation. Activation of these receptors leads to seizures and degeneration of neurons due to excitotoxicity. Hippocampal neurons are particularly sensitive to this toxin. Long-term consequences of domoic acid administration include disturbances of learning and memory, which can be explained by damage to the hippocampus. The clinical syndrome in individuals afflicted by amnesic shellfish poisoning can be understood in terms of the neuropharmacology of domoic acid, and appropriate antidotes proposed.

II. DOMOIC ACID AS A DIETARY EXCITOTOXIN

Domoic acid (Figure 1) is a tricarboxylic acid produced by certain marine organisms, of which perhaps the best characterised are the red alga *Chondria armata* and the planktonic diatom *Nitzschia pungens*. Extracts of *C. armata* containing domoic acid have been used traditionally as antihelminthic agents (Takemoto and Daigo, 1960), and the compound is also a potent insecticide (Maeda et al., 1984). Like a number of related amino acids, including kainic and quisqualic acids, domoic acid appears to owe its acaricidal and insecticidal properties to an interaction with glutamatergic neurotransmission, this being the principal excitatory neuromuscular neurotransmitter system in invertebrates (Maeda et al., 1987). Electrophysiological studies in the 1970s showed that domoic acid also has potent excitatory properties on mammalian central neurons (Biscoe et al., 1975, 1976). It has since been shown (Slevin et al., 1983; Stewart et al., 1990) to activate preferentially excitatory amino acid receptors responding to its close analogue, kainic acid (Figure 1). The pharmacological properties of domoic acid have generally been assimilated to those of kainic acid, which have been studied in considerably more detail.

Interest in domoic acid as a potential dietary toxin was sparked by an outbreak of food poisoning in Canada in 1987 that affected over 100 individuals, which has since been named amnesic shellfish poisoning. This was traced to the consumption of contaminated mussels farmed

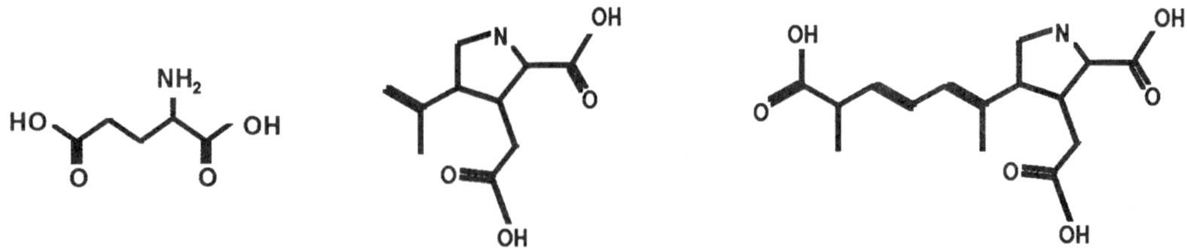

Glutamic Acid **Kainic Acid** **Domoic Acid**

Figure 1 Structure of excitatory amino acids: glutamic, kainic, and domoic acids.

off Prince Edward Island (Perl et al., 1990; Teitelbaum et al., 1990). Sufferers presented with severe gastrointestinal disturbances (diarrhea and vomiting), and cardiovascular, respiratory and neurological symptoms were also present. It soon became clear that the intoxication was due to domoic acid produced by *Nitzschia pungens*, which had been concentrated by the mussels (Wright et al., 1989). Injection of an extract of the contaminated mussels into mice produced seizures and a hyperexcitability syndrome reminiscent of that seen following administration of kainic acid. The total amounts of domoic acid consumed were in the range of 60–290 mg. The neurological syndrome produced appeared to reach its acme and to persist well after the domoic acid had been cleared from the body.

This chapter will discuss the mechanism of action of domoic acid on the mammalian central nervous system, its principal neuropharmacological and psychopharmacological properties in experimental animals, and compare the latter with the neurological syndrome produced by accidental ingestion of domoic acid in humans.

III. INTERACTION OF DOMOIC ACID WITH EXCITATORY AMINO ACID RECEPTORS

Domoic acid appears to produce its neuroexcitatory effects by activation of certain subtypes of excitatory amino acid receptors. These receptors are ligand-gated ion channels, activated by glutamic acid, that mediate fast excitatory synaptic transmission in the mammalian central nervous system (Collingridge and Lester, 1989).

Domoic acid was first demonstrated to be an excitant on mammalian central neurons by Biscoe and coworkers in 1975, who compared the potencies of a series of excitatory amino acids on frog and rat spinal neurons in vivo (Biscoe et al., 1975, 1976). Excitations produced by domoic acid were seen at much lower concentrations than those produced by glutamic acid, and, in contrast to the latter, were sustained. In this respect, domoic acid was one of the most potent excitatory amino acids to have been described.

The later discovery of selective antagonists allowed excitatory amino acid receptors to be differentiated into subtypes. The use of ω-phosphonocarboxylic acid derivatives, in particular L-2-amino-5-phosphonovaleric acid, could differentiate between responses to N-methyl-D-aspartic acid (NMDA), which were blocked, and those to quisqualic and kainic acids, which were not (Davies et al., 1981). The receptors responsible were classified as NMDA and non-NMDA receptors, respectively (Watkins and Evans, 1981). Certain neuronal populations in the spinal cord differed in their responses to quisqualic acid and kainic acid (Davies et al., 1979; McLennan and Lodge, 1979), leading to the subdivision of non-NMDA receptors into quisqualate and kainate receptors. Quisqualate receptors have since been renamed AMPA receptors, since AMPA (α-amino-3-hydroxy-5-methyl-4-isoxazolepropionic acid) is much more selective than quisqualic acid for this receptor subtype. According to this classification, domoic acid was clearly an agonist at non-NMDA receptors of the kainate type.

The study of kainate receptors has been hampered by the fact that kainic acid itself also activates AMPA receptors in a persistent fashion. Responses to kainic acid mediated by kainate receptors have lower unitary conductance and desensitize rapidly. This means that most of the response to kainic acid on neurons that contain both receptor populations is, in fact, mediated by AMPA receptors. Fortunately, certain neuronal populations, such as dorsal root ganglion cells (Huettner, 1990; Wong et al., 1994), appear to express predominantly kainate receptors rather than AMPA receptors, and they can thus be used to characterize the former. Domoic acid can excite dorsal root ganglion cells with particularly high potency (Huettner et al., 1990). Owing to its higher selectivity for kainate receptors, domoic acid has also been used to reveal functional receptors in cultured cerebellar granular cells (Sciancalepore et al., 1990).

Radioligand binding studies using [^{3}H]AMPA and [^{3}H]kainic acid can also be used to explore structure-activity relations at the two receptor subtypes (Foster and Fagg, 1984). Domoic acid has the highest affinity of any excitatory amino acid for [^{3}H]kainic acid binding sites in the rat brain (Slevin et al., 1983). The high affinity of domoic acid for the kainic acid receptor has been exploited in attempts to purify the kainate receptor from brain extracts using affinity chromatography (Hampson and Wenthold, 1988).

The subsequent cloning of the various subunits that make up the different excitatory amino acid receptors has put the pharmacological specificity of domoic acid on a firm footing, although

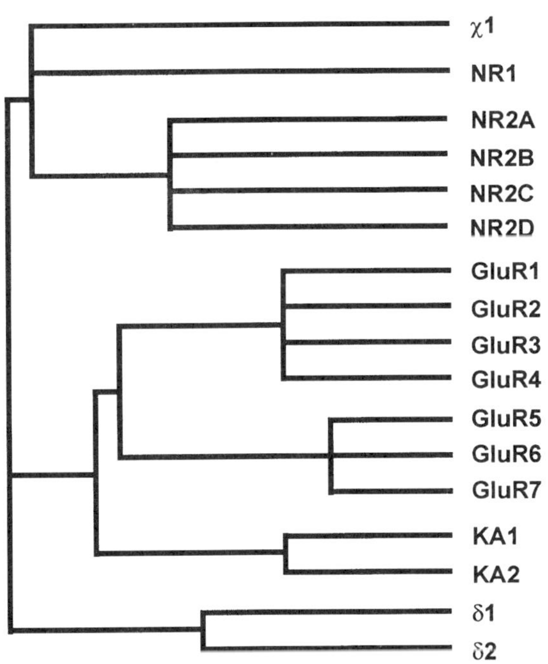

Figure 2 Classification of excitatory amino acid receptors. This dendrogram shows the amino acid sequence homology between the different subunits of excitatory amino acid receptors that have been cloned in the rat. (Sequence homology data from Hollmann and Heinemann, 1994.)

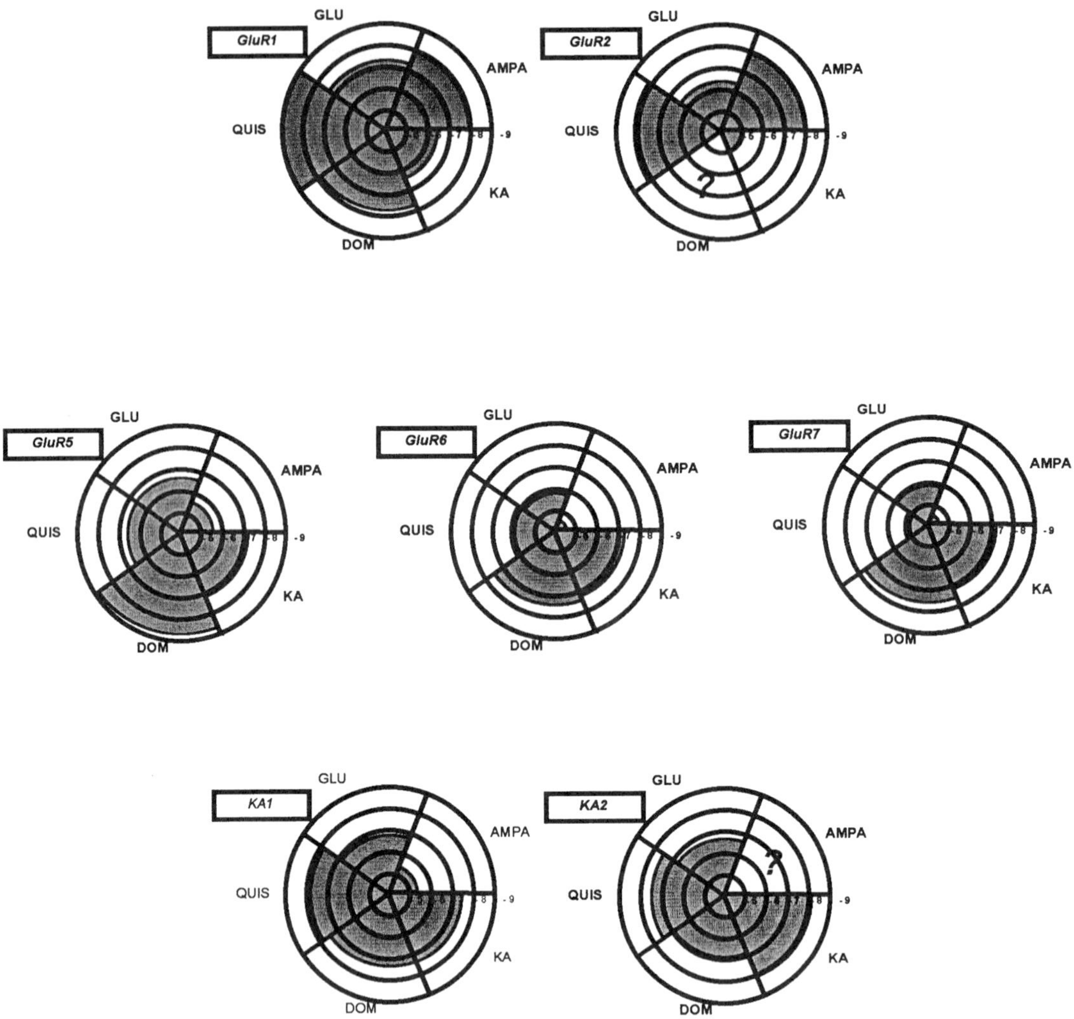

Figure 3 Affinity and selectivity of excitatory amino acids. Affinity of domoic acid and other excitatory amino acids for various recombinant excitatory amino acid receptors. Data are presented as target plots, where each segment represents an amino acid whose affinity (M) is represented by the radius of the segment. The affinities are IC50 values from binding experiments except for the GluR1 receptor, where the values represent EC50 values for activation of cation currents in oocytes. Each target plot represents a different recombinant excitatory amino acid receptor homomer. (Data from Hollmann and Heinemann, 1994.)

it has done little to simplify receptor classification (Hollmann and Heinmann, 1994). Fourteen different subunits that between them make up mammalian NMDA, AMPA and kainate receptors and three "orphan" subunits of unknown function and pharmacology have been identified (Figure 2). Most, if not all, of the subunits appear to exist as a variety of splice variants. The NMDA receptor is made up of combinations of the NR1 subunit and one of the four NR2 subunits. No recombinant NMDA receptor, regardless of subunit composition, can be activated by domoic acid.

AMPA receptors are composed of the GluR1, GluR2, GluR3, and GluR4 subunits either assembled as homomers or as heteromers. In order to account for the biophysical properties of native AMPA receptors, it would appear that the majority of these are in fact heteromers containing the GluR2 subunit. Domoic acid and kainic acid both activate recombinant homomeric AMPA receptors to produce large nondesensitizing responses, with domoic acid being the more potent agent (Hollman et al., 1989; Nakanishi et al., 1990; Sakimura et al., 1990) (see Figure 3).

Five subunits whose expression appears to lead to the expression of kainate receptors have been cloned. These are GluR5, GluR6, GluR7, KA1, and KA2. Homomeric expression of GluR5–7 yields receptors that respond to kainic acid, domoic acid, and glutamic acid but only weakly, if at all, to AMPA. However, GluR7 has only very low affinity for any excitatory amino acid and is insensitive to domoic acid (Schiffer et al, 1997). Responses to kainic acid and domoic acid desensitize rapidly. Radioligand binding studies show these recombinant receptors to correspond to the low-affinity kainic acid binding sites observed in membrane preparations from rat brain. Homomeric expression of KA1 or KA2 produces high-affinity binding sites for kainic acid but does not lead to the formation of functional ion channels. However, it appears that heteromers of KA2 and GluR5/6 are functional receptors (Herb et al., 1992), and these combinations may correspond to the composition of native kainate receptors in the brain (Wenthold et al., 1994). These heteromeric receptors have lower affinity for kainic acid and domoic acid than homomers of either KA2 or GluR5 (Herb et al., 1992).

IV. DISTRIBUTION OF KAINATE RECEPTORS IN THE MAMMALIAN CENTRAL NERVOUS SYSTEM

The distribution of kainate receptors in the central nervous system has been evaluated by autoradiography, which can yield quantitative determinations of receptor density (Foster et al., 1981; Monaghan and Cotman, 1982; Unnerstal and Walmsley, 1983). Kainate receptors are concentrated in forebrain areas, with high densities being found in the cortex, hippocampus, and amygdala. Receptor distribution shows a laminar pattern in the cortex and the hippocampus. In the cortex, the highest levels of binding sites are observed in cortical layers V and VI. The particularly high density of kainate receptors in the CA_3 region of the hippocampus is associated both with pyramidal cells and with the terminal fields of the mossy fiber input in the stratum lucidum. In the midbrain, receptor density is high in the hypothalamus, the mammilary bodies, the caudate-putamen, and certain thalamic nuclei. In the hindbrain and spinal cord, the density of kainate binding sites is generally speaking low. Kainate receptors in the cerebellum are most prevalent in the granular cell layer.

These autoradiographic studies of the distribution of kainate receptors identified by radioligand binding studies has been complemented by in situ hybridization studies of messenger RNAs encoding the different receptor subunits (Wisden and Seeburg, 1993a,b). See Table 1. In general, the distribution of the kainate binding sites seems to mirror most closely the expression of the KA2 receptor subunit, and it may be that the presence of the latter is necessary to confer high affinity on the binding site. Immunohistochemical studies with antibodies directed against the GluR6 receptor subunit show that this protein is expressed in axon terminals of the mossy fiber input to the CA_3 field of the hippocampus (Petralia et al., 1994).

In the human brain, data from the limited number of studies available suggests the distribution of kainate receptors to be similar (Ball et al., 1994; Tremblay et al., 1985; Shaw and Ince, 1994; Kunig et al., 1995), and once again, high densities of domoic acid–sensitive kainic binding sites are observed in the hippocampus.

In spite of the high density of kainate binding sites observed in many regions of the central nervous system, unequivocal electrophysiological demonstration of the presence of discrete functional kainate receptors is meagre. Kainate receptors have been difficult to study, since they

Table 1 Distribution of Kainic Acid Binding Sites and Kainate Receptor mRNA in the Rodent Central Nervous System

	KA binding	GluR5	GluR6	KA1	KA2
Olfactory bulb		High	High	NA	NA
Cortex	High	High	Low	Low	High
Hippocampus	High	Low	High	Low	High
CA_1 pyr	High	Low	High	Absent	High
CA_2 pyr	High		High	Low	High
CA_3 pyr	Very high		High	High	High
Dentate gyrus	Moderate	Low	High	Low	High
Subiculum	Moderate	High		Low	
Pyr cortex	High	High	High		High
Caudate-putamen	Moderate	Absent	Intermediate	Absent	High
Globus pallidus	Moderate				Absent
Amygdala	Moderate	High		Absent	
Thalamus	Low	Low	Low	Absent	Low
Hypothalamus	Low	High	Low	Absent	
Brainstem	Low	Low	Low		
Pons	Low	High	Low		
Cerebellum	High	High	High	Low	High
Cb Purk	Moderate	High	Low	High	Low
Cb Gran	High	Low	High	Low	High
Cb Mol	Moderate	Low	Low	Low	Low
Substantia gelatinosa	Moderate	Very low	Very low	Very low	Low
Anterior horn cells	Low	Very low	Absent	Low	Absent
PAG	Low	Absent	Very low	Low	Low
Reticular foramen	Low	Absent	Absent	Very low	Very low

Source: Data are taken from Hollmann and Heinemann, 1994 (and original work cited therein); Tolle et al., 1993; Shaw and Ince 1994; Unnerstall and Walmsley, 1982; and Wisden and Seeburg, 1993b.

have lower unitary conductances than AMPA receptors and desensitize rapidly to kainic acid. This means that in cells where both AMPA and kainate receptors coexist, the dominant response to kainic acid is mediated by AMPA receptors.

The most extensive characterisation of functional kainate receptors has been performed on dorsal root ganglion cells, where this subtype appears to be the dominant receptor expressed. These receptors are exquisitely sensitive to domoic acid (Huettner et al., 1990). Functional kainate receptors also appear to be present on nerve endings of these sensory neurons in the primary afferents of the spinal cord (Agrawal and Evans, 1986; Davies et al., 1979).

In the hippocampus, which is the brain region containing the highest density of kainate receptors, rapidly desensitising responses to kainic and domoic acid have been observed (Lerma et al., 1993), with an appropriate pharmacological specificity for kainate, as opposed to AMPA receptors (Paternain et al., 1995; Wilding and Huettner, 1997). Paradoxically, activation of kainate receptors on glutamatergic nerve terminals in the hippocampus apparently leads to a *depression* of glutamic acid release (Chittajallu et al., 1996), and a similar inhibition of synaptic transmission has subsequently been observed at hippocampal GABA(γ-aminobutyric acid)ergic terminals (Clarke et al., 1997). Overall, activation of kainate receptors probably contributes to the excitatory synaptic potential generated by high-frequency stimulation of the mossy fiber input to CA_3 neurons (Castillo et al., 1996; Vignes and Collingridge, 1996).

The presence of functional kainate receptors on cerebellar granular cells has also been attested to by electrophysiological experiments using domoic acid (Sciancalepore et al., 1990).

V. NEUROPHARMACOLOGICAL EFFECTS OF DOMOIC ACID

A. Convulsant Effects

Subcutaneous administration (2.5–3.0 mg/kg) of domoic acid to rats provoked clonic seizures, particularly of the head and forelimbs (Stewart et al., 1990) which eventually generalized into status epilepticus. Similar findings were described following intracerebroventricular administration to rats and and mice (Dakshinamurti et al., 1991; Chiamulera et al., 1992). Electroencephalographic recordings have implicated the hippocampus as being a potential source of the epileptiform activity (Nakajima and Potvin, 1992).

Exposure in utero to domoic acid can also modify seizure thresholds. In a study by Dakshinamurti et al. (1993), pregnant rats were given a single subconvulsant dose of domoic acid, and their pups were evaluated postnatally. Although apparently neurologically normal, these animals showed spike and burst activity on the electroencephalogram (EEG) and had lower seizure thresholds. Histopathological examination of the hippocampus revealed cell necrosis in the CA_3 and dentate gyrus fields.

As is the case for other convulsant agents, subconvulsant doses of domoic acid lead to an increase in expression of the early-onset gene c-*fos* (Peng et al., 1994; Peng and Ramsdell, 1996). Although an extensive anatomical study was not reported, early c-*fos* expression is particularly prominent in the hippocampal formation. Expression was also observed in certain brainstem nuclei, notably the area postrema and the nucleus tractus solitarius, which are not substrates of the convulsant or excitotoxic effects of domoic acid.

B. Excitotoxic Effects

The concept of excitotoxicity was developed by Olney (1978) who proposed that the known neurotoxicity of glutamic acid was a direct consequence of its interaction with the receptors that mediated its excitatory effect on neurons. If receptor activation was for whatever reason excessive or prolonged, this hyperexcitation would provoke irrversible damage to the neuron. Evidence for the excitotoxic hypothesis came largely from the observation that there was a close correlation between those amino acids known to provoke neuronal lesions and those known to activate excitatory amino acid receptors. These amino acids included both NMDA receptor agonists and AMPA/kainate receptor agonists, among them domoic acid.

Direct injection of domoic acid into the central nervous system of rodents produces necrotic lesions (Zaczek and Coyle, 1982). After subcutaneous administration to rats (2.5–3.0 mg/kg), domoic acid causes widespread damage, particularly to the cortex, hippocampus, amygdala, substantia nigra, and to certain thalamic nuclei (Stewart et al., 1990). This pattern of damage corresponded to what had been observed previously with high doses of kainic acid. Nonetheless, domoic acid is several times more potent than kainic acid, and is indeed the most potent of all excitotoxic amino acids known. A confounding problem in this study was the fact that all the rats had severe convulsions following the administration of the domoic acid, given that seizure activity alone can cause brain lesions. The investigators in this study concluded that it was more than likely that the lesions produced by domoic acid were caused indirectly by local release of glutamic acid during seizure activity. Nonetheless, if seizures are prevented with a benzodiazepine, neuronal damage is still manifest in the hippocampus. For example, Sutherland et al.

(1990) showed that local injection of domoic acid into the hippocamus in rats treated with a benzodiazepine to prevent indirect neuronal damage due to seizures produced extensive damage to all regions of the hippocampus and the subiculum.

Pyramidal cells in the CA_3 layer of the hippocampus, which, as we have seen, express very high levels of kainic acid binding sites, are particularly sensitive to the excitotoxic effects of kainic (Nadler et al., 1978) and domoic acids (Debonnel et al., 1989a; Tryphonas et al., 1990a). It has been demonstrated that the pyramidal cells in this region are some 20 times more sensitive to domoic acid than their counterparts in the CA_1 field (Debonnel et al., 1989a). A similarly high sensitivity of the CA_3 region of the hippocampus to the excitotoxic effects of domoic acid has also been described in mice (Strain and Tasker, 1991) and in monkeys (Tryphonas et al., 1990b; Skallet et al., 1993; Schmued et al., 1995). Excitotoxic damage to the hippocampus can also be prevented by a range of AMPA/kainate receptor antagonists, including NBQX and NS-102, but not by NMDA receptor antagonists (Stewart et al., 1990; Tasker et al., 1992).

Although the hippocampus has received the most attention, neuronal degeneration has been reported in many other brain regions, including the neocortex, amygdala, thalamus, and septum (Stewart et al., 1990; Friedberg and Ross, 1993; Peng et al., 1994; Schmued et al., 1995).

It has been known for 20 years that the excitotoxic effects of kainic acid in vivo depend on an intact glutamatergic innervation, suggesting that this is mediated by the release of endogenous glutamate (Bizière and Coyle, 1978; Campochiaro and Coyle, 1978; McGeer et al., 1978), and this is likely to be the case for domoic acid as well (Debonnel et al., 1989b). It has indeed been demonstrated that both kainic acid (Ferkany et al., 1982) and domoic acid (Dakshinamurti et al., 1991; Terrian et al., 1991; Malva et al., 1996) can facilitate the release of glutamic acid from the hippocampus and from synaptosomes prepared from whole rat brain (Brown and Nijjar, 1995). Further evidence for a role for the release of endogenous glutamic acid in the excitotoxicity of domoic acid has come from in vitro studies with cultured cerebellar granule cells. In this system, the neurotoxicity of domoic acid can be attenuated by NMDA receptor antagonists, implying that excitotoxicity is due to release of glutamic acid from nerve endings and subsequent activation of NMDA receptors rather than direct AMPA/kainate receptor–mediated neurotoxicity (Berman and Murray, 1997).

VI. BEHAVIORAL EFFECTS OF DOMOIC ACID IN ANIMALS

The behavioral syndrome in rats produced by subcutaneous administration of domoic acid was described by Stewart et al. (1990). At low doses (0.6 mg/kg), this was limited to hypoactivity and staring, whereas higher doses (2.5–3.0 mg/kg) generated a more complex pattern. This consisted of an initial period of hypoactivity followed by behavioral excitation characterized by wet-dog shakes, repetitive head scratching, salivation, and clonic seizures (see above). To a large extent, this behavioral syndrome can be considered to be prodromal to the convulsant state provoked by domoic acid. Although this syndrome bore certain similarities to the behavior evoked by kainic acid administration, the bouts of head scratching were unique to domoic acid. Such a syndrome has also been observed in mice administered domoic acid intraperitoneally (Strain and Tasker, 1991; Tasker et al., 1991). Similar findings were oberved after intracerebroventricular administration in both species (Chiamulera et al., 1992), although the head scratching was not described. These investigators compared the effects of domoic acid, kainic acid, AMPA, and NMDA and concluded that the first three excitatory amino acids produced a qualitatively identical syndrome, whereas NMDA was different. Of all the excitatory amino acids tested, domoic acid was much the most potent.

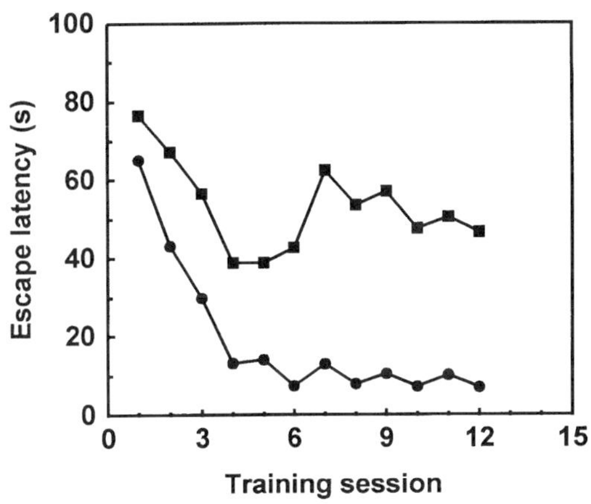

Figure 4 Performance deficit in a Morris maze following domoic acid. These data show the performance deficit in a Morris water maze of rats injected 10 days previously with domoic acid (25 ng) bilaterally into the dorsal hippocampus. The vehicle-treated rats (●) rapidly learn to escape from the maze, as shown by the decreased escape latencies in consecutive training sessions, whereas the rats treated with domoic acid (■) fail to learn and their performance remains mediocre. (Data from Sutherland et al., 1990.)

The behavioral syndrome, as well as mortality, produced by domoic acid can be prevented by a several AMPA/kainate receptor antagonists including NBQX and NS-102 (Stewart et al., 1990; Tasker et al., 1992). Generally speaking, NMDA receptor antagonists have not been found to be efficaceous, although some protection has been described with both kynurenic acid (Pinsky et al., 1989) and CPP (Tasker et al., 1992). More surprisingly, it has also been observed that morphine will attenuate the behavioral effects of domoic acid in mice (Tasker and Strain, 1992).

Although the acute behavioral effects of domoic acid are related to the convulsant properties of this amino acid, chronic behavioral signs are also seen, which are likely to be a consequence of excitotoxic damage to the hippocampus. This was first explored by Sutherland et al. (1990), who injected domoic acid locally into the hippocampus of rats, treated the animals with a benzodiazepine to prevent convulsions, and measured performance in a spatial memory test (Morris water maze) 10 days later. They observed a severe impairment of learning in this test (Figure 4), which they attributed to neuronal degeneration in the hippocampal formation, a structure known to be vital for correct performance in this sort of memory paradigm. A similar deficit is observed after intraperitoneal administration of domoic acid (Petrie et al., 1991). Spatial learning deficits have also been described using a radial maze (Nakajima and Potvin, 1992).

VII. EFFECTS OF DOMOIC ACID IN HUMANS

Insight into the effects of domoic acid on the human central nervous system can be gained from the neurological syndrome observed in the victims of the Canadian mussel poisoning epidemic (Teitelbaum et al., 1990). These individuals, who were, generally speaking, elderly, presented with an acute confusional state, disorientation, and perturbations of short-term memory. Although, in most patients, decreased arousal and somnolence were seen, in some cases leading to coma, in less severely effected individuals, agitation was observed. In some cases, the state of the patients deteriorated with the development of motor symptoms, such as buccofacial dyskinesias, myoclonus, or even seizures.

Cardiovascular symptoms (tachycardia, hypotension, and cardiac arrhythmias) may have been a consequence of dysfunction of the central autonomic control centers, since there was no evidence of primary cardiovascular impairment. Other peripheral signs of autonomic dysfunction were also observed.

In most cases, symptoms resolved over a few weeks, but in some individuals a residual memory impairment persisted. One sufferer who had presented with a severe acute syndrome with generalized convulsions, which regressed after 3 weeks, went on to develop an unremitting temporal lobe epilepsy 1 year after intoxication (Cendes et al., 1995). Following the death of the patient, autopsy revealed advanced bilateral degeneration of the hippocampus similar to that often observed postmortem in individuals afflicted by status epilepticus.

Motor impairment and electromyographic anomalies in many individuals suggested some degree of permanent damage to motoneurons, although these neurons were, in fact, preserved in those patients on whom autopsies were later performed. In these patients who died, lesions were seen in the hippocampus and amygdala. Neuropathologically, these resembled experimental excitotoxic neurodegeneration in animals, and the regions affected were indeed those containing the highest densities of kainate receptors. In particular, cortical damage was restricted to the laminae V and VI in which the expression of kainate binding sites is the most developed.

VIII. SUMMARY

Domoic acid first became a subject worthy of neuropharmacological investigation following the outbreak of amnesic shellfish poisoning in Canada in 1987 (Perl et al., 1990; Teitelbaum et al., 1990). Although this substance was known at the time to be an excitatory amino acid with some specificity for kainate receptors, advances in the molecular biology of excitatory amino acid receptors have since demonstrated that domoic acid is a selective, and highly potent, activator of excitatory amino acid receptors containing GLuR5/6 and KA1/2 subunits. In common with other AMPA/kainate receptor agonists, domoic acid is a powerful convulsant agent in experimental animals, and causes excitotoxic degeneration of neurons in several brain regions, in particular the hippocampus. The extreme sensitivity of certain subfields of the hippocampus to domoic acid is probably related to the extremely high density of kainate binding sites in this brain region (Foster et al., 1981).

Deficits in spatial memory have been demonstrated in experimental animals administered domoic acid, and this is compatible with the principal clinical symptom described in humans afflicted with amnesic shellfish poisoning. These memory deficits are probably attributable to excitotoxic damage to the hippocampal formation. Other aspects of the human syndrome can also be understood in terms of the described neuropharmacology of domoic acid in animals. For example, the attention deficit could be correlated with excitotoxic damage to the thalamic reticular nucleus (Friedberg and Ross, 1993) and the gastrointestinal manifestations to activation of the area postrema and nucleus tractus solitarius (Peng et al., 1994).

The idea that the clinical tableau of amnesic shellfish poisoning can be explained by our knowledge of the interaction of domoic acid with a subtype of excitatory amino acid receptors of the AMPA/kainate type opens the door to the identification of appropriate treatment strategies were *N. pungens* to bloom again in Canada or elsewhere and contaminate the human food chain.

REFERENCES

SG Agrawal, RH Evans. The primary afferent depolarizing action of kainate in the rat. Br J Pharmacol 87:345–355, 1986.

EF Ball, PJ Shaw, PG Ince, M Johnson. The distribution of excitatory amino acid receptors in the normal human midbrain and basal ganglia with implications for Parkinson's disease: a quantitative autoradiographic study using [^{3}H]MK-801, [^{3}H]glycine, [^{3}H]CNQX and [^{3}H]kainate. Brain Res 658:209–218, 1994.

FW Berman, TF Murray. Domoic acid neurotoxicity in cultured cerebellar granule neurons is mediated predominantly by NMDA receptors that are activated as a consequence of excitatory amino acid release. J Neurochem 69:693–703, 1997.

TJ Biscoe, RH Evans, PM Headley, M Martin, JC Watkins. Domoic and quisqualic acids as potent amino acid excitants of frog and rat spinal neurones. Nature 255:166–167, 1975.

TJ Biscoe, RH Evans, PM Headley, MR Martin, JC Watkins. Structure-activity relations of excitatory amino acids on frog and rat spinal neurones. Br J Pharmacol 58:373–382, 1976.

K Biziere, JT Coyle. Effects of cortical ablation on the neurotoxicity and receptor binding of kainic acid in the striatum. J Neurosci Res 4:383–398, 1978.

JA Brown, MS Nijjar. The release of glutamate and aspartate from rat brain synaptosomes in response to domoic acid (amnesic shellfish toxin) and kainic acid. Molec Cell Biochem 151:49–54, 1995.

P Campochiaro, JT Coyle. Ontogenic development of kainate neurotoxicity: correlates with glutamatergic innervation. Proc Natl Acad Sci USA 75:2025–2029, 1978.

PE Castillo, RC Malenka, RA Nicoll. Kainate receptors mediate a slow postsynaptic current in hippocampal CA3 neurons. Nature 388:182–186, 1996.

F Cendes, F Andermann, S Carpenter, RJ Zatorre, NR Cashman. Temporal lobe epilepsy caused by domoic acid intoxication: evidence for glutamate receptor-mediated excitotoxicity in humans. Ann Neurol 37:123–126, 1995.

C Chiamulera, S Costa, E Valerio, A Reggiani. Domoic acid toxicity in rats and mice after intracerebroventricular administration: comparison with excitatory amino acid agonists. Pharmacol Toxicol 70:115–120, 1992.

R Chittajallu, M Vignes, KK Dev, JM Barnes, GL Collingridge, JM Henley. Regulation of glutamate release by presynaptic kainate receptors in the hippocampus. Nature 379:78–81, 1996.

VRJ Clarke, BA Ballyk, KH Hoo, A Mandelzys, A Pellizzari, CP Bath, J Thomas, EF Sharpe, CH Davies, PL Ornstein, DD Schoepp, RK Kambo, GL Collingridge, D Lodge, D Bleakman. A hippocampal GluR5 kaiante receptor regulating inhibitory synaptic transmission. Nature 4:599–603, 1989.

GL Collingridge, RAJ Lester. Excitatory amino acid receptors in the vertebrate central nervous system. Pharmacol Rev 40:143–207, 1989.

K Dakshinamurti, SK Sharma, M Sundaram. Domoic acid induced seizure activity in rats. Neurosci Lett 127:193–197, 1991.

K Dakshinamurti, SK Sharma, M Sundaram, T Watanabe. Hippocampal changes in developing postnatal mice following intrauterine exposure to domoic acid. J Neurosci 13:4486–4495, 1993.

J Davies, RH Evans, AA Francis, JC Watkins. Excitatory amino acid receptors and synaptic excitation in the mammalian central nervous system. J Physiol (Lond.) 75:641–654, 1979.

J Davies, AA Francis, AW Jones, JC Watkins. 2-Amino-5-phosphonovalerate (2APV), a potent and selective antagonist of amino-acid-induced and synaptic excitation. Neurosci Lett 21:77–81, 1981.

G Debonnel, L Beauchesne, C De Montigny. Domoic acid, the alleged "mussel toxin", might produce its neurotoxic effect through kainate receptor activation: an electrophysiological study in the rat dorsal hippocampus. Can J Physiol Pharmacol 67:29–33, 1989a.

G Debonnel, M Weiss, C De Montigny. Reduced neuroexcitatory effect of domoic acid following mossy fiber denervation of the rat dorsal hippocampus: further evidence that toxicity of domoic acid involves kainate receptor activation. Can J Physiol Pharmacol 67:904–908, 1989b.

AC Foster, GE Fagg. Acidic amino acid binding sites in mammalian neuronal membranes: their characteristics and relationship to synaptic receptors. Brain Res Rev 7:103–164, 1984.

AC Foster, EE Mena, DT Monaghan, CW Cotman. Synaptic localisation of kainic acid binding sites. Nature 281:73–75, 1981.

EB Friedberg, DT Ross. Degeneration of rat thalamic reticular neurons following intrathalamic domoic acid injection. Neurosci Lett 151:115–119, 1993.

DR Hampson, RJ Wenthold. A kainic acid receptor from frog brain purified using domoic acid affinity chromatography. J Biol Chem 263:2500–2505, 1988.

A Herb, N Burnashev, P Werner, B Sakmann, W Wisden, PH Seeburg. The KA-2 subunit of excitatory amino acid receptors shows widespread expression in brain and forms ion channels with distantly related subunits. Neuron 8:775–785, 1992.

M Hollmann, S Heinemann. Cloned glutamate receptors. Annu Rev Neurosci 17:31–108, 1994.

M Hollmann, A O'Shea-Greenfield, SW Rogers, S Heinemann. Cloning by functional expression of a member of the glutamate receptor family. Nature 342:643–648, 1989.

JE Huettner. Glutamate receptor channels in rat DRG neurons: activation by kainate and quisqualate and blockade of desensitization by Con A. Neuron 5:255–266, 1990.

K Kainanen, W Wisden, B Sommer, P Werner, A Herb, TA Verdoon, B Sakmann, PH Seeburg. A family of AMPA-selective glutamate receptors. Science 249:556–560, 1990.

G Kunig, J Hartmann, F Krause, J Deckert, H Heinsen, G Ransmayr, H Beckmann, P Riederer. Regional differences in the interaction of the excitotoxins domoate and 1-β-oxalyl-amino-alanine with [^{3}H]kainate binding sites in human hippocampus. Neurosci Lett 187:107–110, 1995.

J Lerma, AV Paternain, JR Naranjo, B Mellstrom. Functional kainate-selective glutamate receptors in cultured hippocampal neurons. Proc Natl Acad Sci USA 90:11688–11692, 1993.

M Maeda, T Kodama, T Tanaka, K Ohfune, K Nomoto, K Nishimura, T Fujita. Insecticidal and neuromuscular activities of domoic acid and its related compounds. J Pestic Sci 9:27, 1984.

M Maeda, T Kodama, M Saito, T Tanaka, H Yoshizumi, K Nomoto, T Fujita. Neuromuscular action of insecticidal domoic acid on the american cockroach. Pestic Biochem Physiol 28:85–92, 1987.

JO Malva, AP Carvalho, CM Carvalho. Domoic acid induces the release of glutamate in the rat hippocampal CA3 sub-region. NeuroReport 7:1330–1344, 1996.

EG McGeer, PL McGeer, K Singh. Kainate-induced degeneration of neostriatal neurons: dependency upon corticostriatal tract. Brain Res 139:381–383, 1978.

H McLennan, D Lodge. The antagonism of amino acid–induced excitation of spinal neurones in the cat. Brain Res 169:83–90, 1979.

DT Monaghan, CW Cotman. The distribution of [^{3}H]kainic acid binding sites as determined by autoradiography. Brain Res 191:387–403, 1982.

JV Nadler, BW Perry, CW Coleman. Intraventricular kainic acid preferentially destroys hippocampal pyramidal cells. Nature 271:676–677, 1978.

S Nakajima, JL Potvin. Neural and behavioural effects of domoic acid, an amnesic shellfish toxin, in the rat. Can J Psychol 46:569–581, 1992.

N Nakanishi, NA Schneider, R Axel. A family of glutamate receptor genes–evidence for the formation of heteromultimeric receptors with distinct channel properties. Neuron 5:569–581, 1990.

JW Olney. Neurotoxicity of excitatory amino acids. In: EG McGeer, JW Olney, PL McGeer, eds. Kainic Acid as a tool in Neurobiology. New York: Raven Press, 1978, pp 95–112.

AV Paternain, M Morales, J Lerma. Selective antagonism of AMPA receptors unmasks kainate-receptor mediated responses in hippocampal neurons. Neuron 14:185–189, 1995.

YG Peng, TB Taylor, RE Finch, RC Switzer, JS Ramsdell. Neuroexcitatory and neurotoxic actions of the amnesic shellfish poison domoic acid. NeuroReport 5:981–985, 1994.

YG Peng, JS Ramsdell. Brain Fos induction is a sensitive biomarker for the lowest observed neuroexcitatory effects of domoic acid. Fund Appl Toxicol 31:162–168, 1996.

TM Perl, L Bedard, T Kosatsky, JC Hockin, ECD Todd, RS Remis. An outbreak of toxic encephalopathy caused by eating mussels contaminated with domoic acid. N Engl J Med 322:1775–1780, 1990.

RS Petralia, YX Wang, RJ Wenthold. Histological and ultrastructural localization of the kainate receptor subunits, KA2 and GluR6/7, in the rat nervous system using selective antipeptide antibodies. J Comp Neurol 349:85–110, 1994.

BF Petrie, C Pinsky, NM Standish, R Bose, GB Glavin. Parenteral domoic acid impairs spatial learning in mice. Pharmacol Biochem Behav 41:211–214, 1991.

C Pinsky, GB Glavin, R Bose. Kynurenic acid protects against neurotoxicity and lethality of toxic extracts from contaminated Atlantic coast mussels. Prog Neuropsychopharmacol Biol Psychiatry 13:595–598, 1989.

K Sakimura, H Bujo, E Kushiya, et al. Functional expression from cloned cDNAs of glutamate receptor species responsive to kainate and quisqualate. FEBs Lett 272:73–80, 1990.

AC Scallet, Z Binienda, FA Caputo, S Hall, MG Paule, RL Rountree, L Schmued, T Sobotka, W Slikker.

Domoic acid treated cynomolgus monkeys (M. fascicularis): effects of dose on hippocampal neuronal and terminal degeneration. Brain Res 627:307–313, 1993.

HH Schiffer, GT Swanson, SF Heinemann. Rat GluR7 and a carboxy-terminal splice variant, GluR7b, are functional kainate receptor subunit with a low sensitivity to glutamate. Neuron 19:1141–1146, 1997.

LC Schmued, AC Scallet, W Slikker. Domoic acid-induced neuronal degeneration in the primate forebrain revealed by degeneration specific histochemistry. Brain Res 695:64–70, 1995.

M Sciancalepore, Z Galdzicki, X Zheng, O Moran. Kainate activated single channel currents as revealed by domoic acid. Eur Biophys J 19:63–68, 1990.

PJ Shaw, PG Ince. A quantitative autoradiographic study of [^{3}H]kainate binding sites in the normal human spinal cord, brainstem and motor cortex. Brain Res 641:39–45, 1994.

JT Slevin, JF Collins, JT Coyle. Analogue interactions with the brain receptor labeled by [^{3}H]-kainic acid. Brain Res 265:169–172, 1983.

GR Stewart, CF Zorumski, MT Price, JO Olney. Domoic acid: a dementia-inducing excitotoxic food poison with kainic acid receptor specificity. Exp Neurol 110:127–138, 1990.

SM Strain, RAR Tasker. Hippocampal damage produced by systemic injections of domoic acid in mice. Neuroscience 44:343–352, 1991.

J Sutherland, JM Hoesing, IQ Whishaw. Domoic acid, an environmental toxin, produces hippocampal damage and severe memory impairment. Neurosci Lett 120:221–223, 1990.

T Takemoto, K Daigo. Constituents of Chondria armata and their pharmacological effects. Arch Pharmacol 293:627–633, 1960.

RAR Tasker, SM Strain. Morphine differentially affects domoic acid and kainic acid toxicity in vivo. NeuroReport 3:789–792, 1992.

RAR Tasker, BJ Connel, SM Strain. Pharmacology of systemically administered domoic acid in mice. Can J Physiol Pharmacol 69:378–382, 1991.

RA Tasker, SM Strain, J Drejer. Selective reduction in domoic acid toxicity in vivo by a novel non-N-methyl-D-aspartate receptor antagonist. Can J Physiol Pharmacol 74:1047–1054, 1996.

JS Teitelbaum, RJ Zatorre, S Carpenter, D Gendron, AC Evans, A Gjedde, NR Cashman. Neurotoxic sequelae of domoic acid intoxication due to the ingestion of contaminated mussels. N Engl J Med 322:1781–1787, 1990.

DM Terrian, TA Conner-Kerr, TH Privette, RL Gannon. Domoic acid enhances the $K+$-evoked release of endogenous glutamate from guinea pig hippocampal mossy fibre synaptosomes. Brain Res 551: 303–307, 1991.

E Tremblay, A Represa, Y Ben-Ari. Autoradiographic localization of kainic acid binding sites in the human hippocampus. Brain Res 343:378–82, 1985.

L Tryphonas, J Truelove, E Nera, F Iverson. Acute neurotoxicity of domoic acid in the rat. Toxicol Pathol 18:1–9, 1990a.

L Tryphonas, J Truelove, F Iverson. Acute parenteral neurotoxicity of domoic acid in Cynomolgus monkeys (M. fascicularis). Toxicol Pathol 18:297–303, 1990b.

JR Unnerstall, KJ Walmsley. Autoradiographic localization of high affinity [^{3}H]kainic acid binding sites in the rat forebrain. Eur J Pharmacol 86:361–371, 1982.

M Vignes, GL Collingridge. The synaptic activation of kainate receptors. Nature 388:179–182, 1996.

JC Watkins, RH Evans. Excitatory amino acid transmitters. Ann Rev Pharmacol Toxicol 21:165–204, 1981.

RJ Wenthold, VA Trumpy, WS Zhu, RS Petralia. Biochemical and assembly properties of GluR6 and KA2, two members of the kainate receptor family, determined with subunit-specific antibodies. J Biol Chem 269:1332–1339, 1994.

TJ Wilding, JE Huettner. Activation and desensitisation of hippocampal kainate receptors. J Neurosci 17: 2713–2721, 1997.

W Wisden, PH Seeburg. Mammalian ionotropic glutamate receptors. Curr Opin Neurobiol 3:291–298, 1993a.

W Wisden, PH Seeburg. A complex mosaic of high-affinity kainate receptors in rat brain. J Neurosci 13: 3582–3598, 1993b.

LA Wong, ML Mayer, DE Jane, JC Watkins. Willardines differentiate agonist binding sites for kainate-versus AMPA-preferring glutamate receptors in DRG and hippocampal neurons. J Neurosci 14: 3881–3897, 1994.

JLC Wright, RK Boyd, ASW DE Freitas, M Falk, RA Foxall, WD Jamieson, MV Laycock, AW McCulloch, AG McInnes, P Odense, VP Pathak, MA Quilliam, MAA Ragan, PG Sim, P Thibault, JA Walter. Identification of domoic acid, a neuroexcitatory amino acid in toxic mussels from eastern Prince Edward Island. Can J Chem 67:481–490, 1989.

R Zaczek, JT Coyle. Excitatory amino acid analogues: neurotoxicity and seizures. Neuropharmacology 21:15–26, 1982.

17

Molecular Biology of Kainate Receptors: Targets of Domoic Acid Toxicity

Donald A. Skifter, Mark P. Thomas and Daniel T. Monaghan
University of Nebraska Medical Center
Omaha, Nebraska

I. INTRODUCTION

Domoic acid produced by diatoms is thought to be responsible for an outbreak of mussel-induced toxicity in humans (1). The potent neurotoxic actions of domoic acid are primarily due to the activation of the kainate class of glutamate neurotransmitter receptors. L-glutamate is now recognized as the major excitatory neurotransmitter in the vertebrate central nervous system (CNS) and its actions are mediated by at least three classes of receptors that gate ion channels, the so-called NMDA (N-methyl-D-aspartate), AMPA (α-amino-3-hydroxy-5-methylisoxazole-4-propionate), and kainate receptors (2). In addition, glutamate activates at least eight distinct types of G protein–linked transmitter receptors (3). When kainic acid, or the closely related domoic acid, activates the kainate class of glutamate receptor, brain cells are potently activated (depolarized) and, with excessive activation, kainate and domoate lead to seizures and neuronal cell death. In this chapter, we will provide an overview of the current understanding of the kainate receptor and how its physiological properties and anatomical/cellular distribution in the CNS appear largely to account for the unique neurotoxic actions of kainate receptor agonists such as domoic acid.

The potent excitotoxic actions of kainic acid have been known for many years. Kainate has been thought of as being a useful tool to kill neurons without damaging axons passing through the lesioned area (4). However, the kainate-induced brain lesions display a highly specific anatomical distribution, with the CA_3 region of the hippocampus being the most sensitive region of the brain (5). This finding was consistent with the observations that low concentrations of kainate selectively activate CA_3 neurons (6) and that the stratum lucidum of CA_3 has the highest concentrations of kainate receptors in the entire brain (7).

Despite this early progress, the existence of a distinct class of kainate receptors has been controversial owing to the extensive overlap in pharmacology that kainate receptors display with AMPA receptors. Thus, many studies during the late 1980s and early 1990s refer to kainate-induced responses as being mediated by non-NMDA receptors, because NMDA receptors are readily distinguished from AMPA and kainate receptor responses. With the cloning of distinct subunit families corresponding to AMPA and kainate receptors, it is now clear that kainate receptors are a distinct receptor class. Kainate receptors are multimeric complexes composed of a family of homologous proteins that are specific to kainate receptors. More recently, it has been possible to develop antagonists that are selective for kainate receptors, and thus directly demonstrate that kainate receptors mediate synaptic transmission within the CNS. With kainate receptor subunit probes and kainate-selective pharmacological agents, a picture is now emerging of how kainate receptor activation mediates the well-known seizures and excitotoxicity initiated by kainate and domoate.

II. MOLECULAR COMPOSITION OF KAINATE RECEPTORS

The initial cloning of a glutamate receptor (8) lead to the identification of a family of four AMPA receptor subunits termed GluR1 to GluR4 or GluR-A to GluR-D (9,10) which can be activated by kainic acid. This caused some confusion about the separate identities of kainate and AMPA receptors until the discovery of five kainate receptor subunits (GluR5–7 and KA1 and KA2) (11–16) which display significantly higher affinities for kainate and lower affinities for AMPA than the GluR1–4 subunits.

Kainate receptors consist of a complex containing four, or perhaps five, receptor subunits of unknown stoichiometry (Figure 1). Together, these subunits form a ligand-gated ion channel that is permeable to sodium and sometimes calcium ions. These subunits are divided into two families by sequence identity. The kainate receptor GluR5–7 subunits display around 75% intra-subunit identity and around 45% identity with the AMPA receptor subunits, GluR1–4 (11–13). The other kainate receptor subunits, KA1 and KA2, display 68% identity in DNA sequence with each other but only around 45% identity with GluR5–7 and 30–35% identity with AMPA receptor subunits (GluR1–4) (14–16). It is clear that some native kainate receptors contain subunits of both kainate receptor subunit families. Thus, there is a large number of possible subunit combinations producing tetrameric or pentameric structures.

The heterogeneity of kainate receptor subunits is further enhanced by the mechanisms of alternative splicing and RNA editing. In the former mechanism, specific stretches of RNA are either inserted into, or deleted out of, the mRNA. Thus, specific domains of polypeptides can be inserted or deleted out of the resulting receptor subunit protein. In RNA editing, enzymes modify sequence-specific RNA bases such that the RNA now codes for an alternative amino acid. Of relevance to domoic acid toxicity is that kainate receptor subunits with differing RNA editing can display differing calcium permeability and channel conductance. This is significant because calcium permeability and channel flux are key factors in excitotoxicity. Presently, the impact of alternative splicing on kainate receptor function is less clear.

Splicing of kainate receptor subunits results in two splice variants of GluR5 (11) differing by the presence (GluR5-1) or absence (GluR5-2) of a 15–amino acid insert located in the N-terminal extracellular domain. A second region of splicing occurs in the carboxyl-terminal domain resulting in three splice variants (GluR5-2a, GluR5-2b, and GluR5-2c) (16). These three splice forms differ by the movement of a stop codon resulting in increased length of coded protein from 2a to 2c. Two splice variants of GluR6 (GluR6-1 and GluR6-2) (17) and two splice variants of GluR7 (GluR7a and GluR7b) (18) are also characterized by splice sites in the carboxyl-terminal domain.

Editing of pre-mRNA by mammalian RNA-dependent adenosine deaminases has been well characterized for the glutamate receptors (see ref. 19 for review) and has been shown to significantly affect calcium permeability and toxicity of neurons. Adenosine deaminases acting on double-stranded pre-mRNA result in the change of a gene-specified glutamine (Q) codon (CAG) to an arginine (R) codon (CIG) in the M2 channel-forming region of GluR2, GluR5, and GluR6 subunits in a developmentally regulated fashion. In the GluR2 subunit, Q to R editing

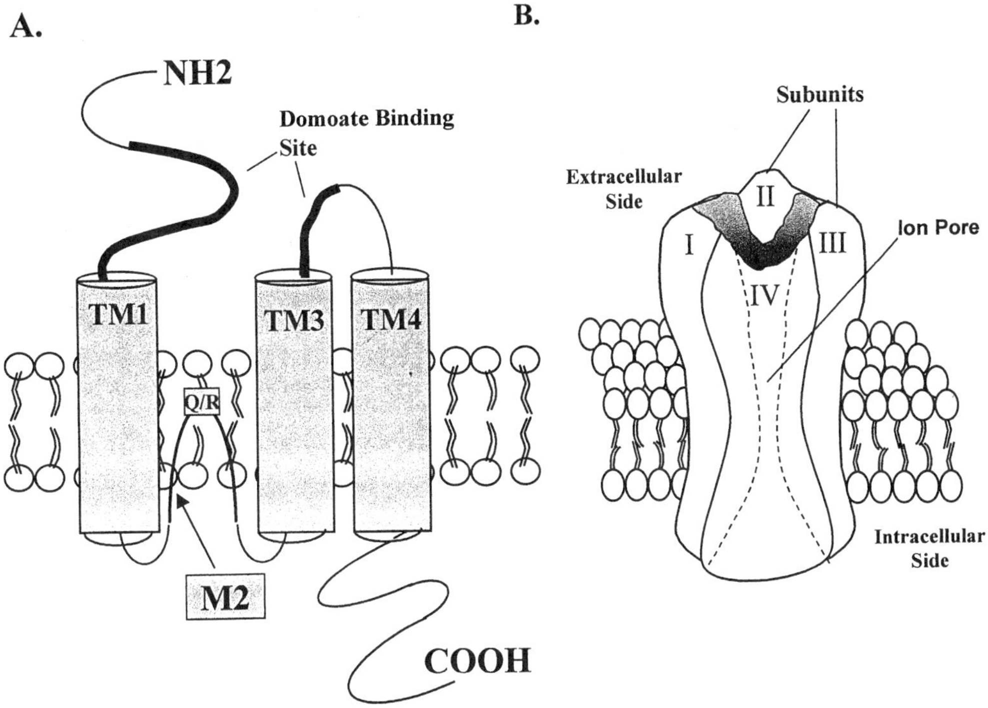

Figure 1 (A) Kainate receptor subunit structure. Results from a variety of experimental approaches supports the model wherein the kainate receptor subunit's N-terminal (NH2) is located extracellularly. The protein then passes through the membrane through a lipophilic domain–transmembrane domain 1 (TM1). The peptide then reenters the membrane as a loop (M2). After two more transmembrane domains (TM3 and TM4), the C-terminal of the subunit (COOH) is located in the intracellular compartment where the protein interacts specifically with a variety of scaffolding and signaling proteins. The critical RNA edited Q/R site is located in the M2 region. Domoate (and kainate) is thought to bind to two extracellular peptide domains (noted by the heavy black line) that are adjacent to TM1 and TM3. This dual binding action presumably opens the ion channel. (B) The kainate receptor complex has been proposed to be a tetrameric complex (subunits I–IV) which forms an ion pore. The M2 region of each subunit is thought to line the pore region of the receptor complex. In addition, portions of TM1 and TM3 also appear to contribute to channel pore formation.

at the Q/R site results in a subunit that is relatively impermeable to calcium in relation to the unedited version (20). Most AMPA receptors in native tissues contain the GluR2 subunit in its edited form, resulting in relatively calcium impermeable receptor complexes (19), but this phenomenon is region and cell type specific. In an elegant set of experiments, Seeburg et al. (21) generated mice that lacked the ability to have the GluR2 pre-mRNA edited from a glutamine codon to an arginine codon. These mice developed seizure activity and died within 3 weeks. This study demonstrates the importance of RNA editing of glutamate receptor subunits for normal brain function.

In contrast to the AMPA receptor GluR2 subunit, calcium permeability of the kainate receptor GluR6 subunit is enhanced by RNA editing at the Q/R site, especially if RNA editing has occurred at additional sites in TMI (22,23). Different subunits are edited at different frequencies (19,22), with the majority (about two thirds) of GluR6 subunits being in the fully edited form and about 10% being in the fully unedited form. Thus, there appear to be cell-specific and developmentally regulated mechanisms which control the extent of kainate receptor RNA editing. Thus, by extension, these mechanisms also modulate the calcium permeability and excitotoxicity mediated by kainate receptors.

The specific subunit compositions and stoichiometries of native receptor complexes are unknown. In recombinant expression systems, some complexes composed of a single subtype of subunit are functional. Expression of homomeric GluR5(R) (edited version) (24), KA1 (25), and KA2 (14) receptors does not result in the formation of functional ion channels. However, GluR5(Q) (unedited version) (16) and GluR6 (12) homomeric channels do form functional ion channels. GluR7 was found to be nonfunctional in many early studies (13,26), but it has recently been reported to form functional homomeric receptors (18).

Coassembly of differing kainate receptor subunits into heteromeric receptor complexes has been demonstrated in recombinant systems and by coimmunoprecipitation of multiple kainate receptor subunits in native receptor complexes. Functional ion channels are produced when KA2 is coexpressed with GluR5 or GluR6 and the functional properties are different than when any of these subunits are expressed alone (14–16). Thus, subunit coassembly alters the properties of receptors in unpredictable ways.

Coimmunoprecipitation studies generally indicate that kainate receptor subunits do not coassemble with either AMPA or NMDA receptor subunits (27); however, GluR kainate receptor subunits can be coimmunoprecipitated with KA family kainate receptor subunits (28). Furthermore, kainate receptor antibodies will immunoprecipitate [^{3}H]kainate binding activity but not [^{3}H]AMPA binding from rat brain membranes.

Ligand binding affinities vary widely between the two families of kainate subunits. KA1 and KA2 show high-affinity binding of [^{3}H]kainate (K_d = 5–15 nM), whereas GluR5–7 have lower affinity (K_d = 50–100 nM). Among GluR5–7, domoic acid displays significantly lower affinity at GluR7 receptors with an affinity of 37 nM (26), whereas GluR5 and GluR6 have higher domoate affinities (2 and 9 nM, respectively). KA1 displays a domoate affinity of 40 nM (25), whereas KA2 is lower affinity at 275 nM (14). Since subunit composition has complex effects on agonist affinity, it is difficult to predict domoate affinity in different native kainate receptors of varied heteromeric composition. Nevertheless, domoate displays very high affinity for each of the kainate receptor subunits, and modest variations in affinities among native receptors can be accounted for by subunit variations in domoate affinity.

III. DISTRIBUTION OF KAINATE RECEPTORS

Autoradiographic studies with radioligands demonstrate that KA and AMPA receptors display widespread, but distinct, anatomical distributions (Figure 2). KA binding sites are enriched in the hippocampal CA_3 stratum lucidum, striatum, deep cortical layers, reticular nucleus of the thalamus, and the cerebellar granular cell layer (7,29). This pattern is reproduced by the combined patterns of the five kainate receptor subunits (30). The KA1 subunit displays the most restricted distribution, being predominately expressed in the CA_3 region of the hippocampus and dentate gyrus. In contrast, the KA2 subunit has widespread distribution, being expressed throughout the cerebral cortex, hippocampus, striatum, and cerebellar granular cell layer. GluR5 subunits are especially enriched in cerebellar Purkinje cells and scattered cells throughout the CNS and peripheral nervous system (PNS) (including the sensory ganglia). GluR6 and GluR7 subunits appear to contribute significantly to the pattern of [^{3}H]kainate binding sites with high expression of GluR6 in the striatum, CA_3 region of the hippocampus, dentate gyrus, and cerebel-

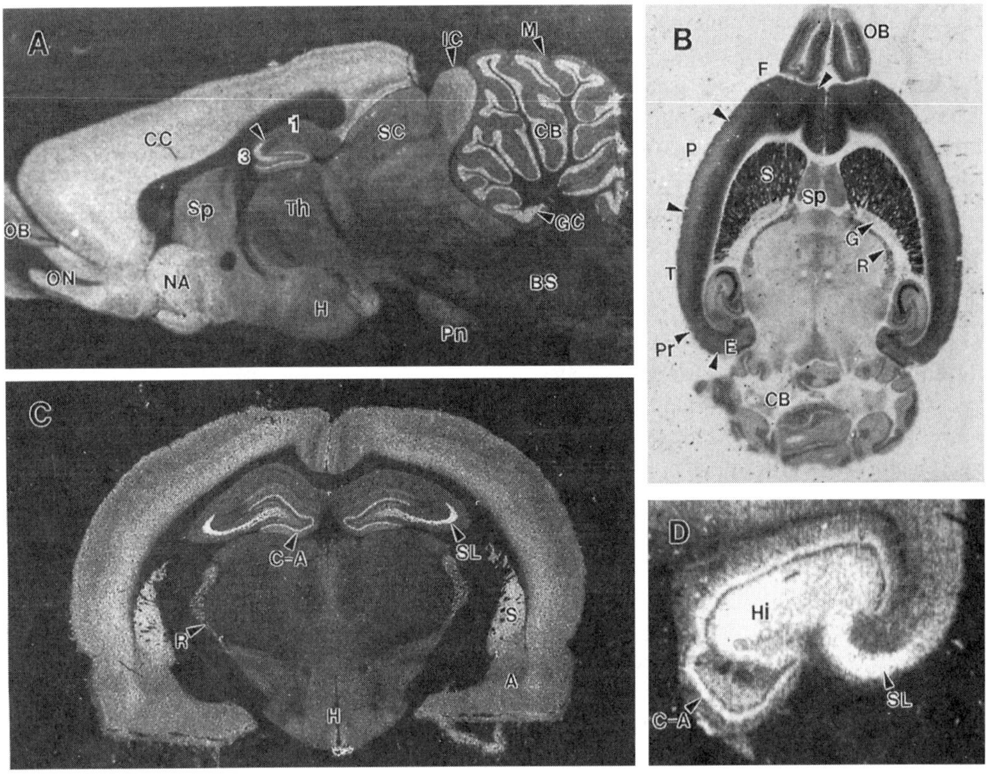

Figure 2 Distribution of kainate receptors in rat brain (A, sagittal section; B, horizontal section; and C, coronal section) and human hippocampus (D). In the darkfield autoradiographs of $[^3H]$kainate binding sites (A, C, and D), bright regions correspond to high binding site densities; in the bright field photograph in (B), high densities are dark. Highest densities of kainate receptors are found in the terminal field (stratum lucidum, SL) of the mossy fiber pathway of the hippocampus. High densities are also found in layers V and VI of the cerebral cortex (CC), the striatum (S), and cerebellar granular cell layer (GC). Anatomical abbreviations: 1, CA_1 hippocampus; 3, CA_3 hippocampus; A, amygdala; BS, brainstem; C-A, commissural-associational layer of the dentate gyrus; CB, cerebellum; CC, cerebral cortex; E, entorhinal cortex; F, frontal cortex; G, globus pallidus; GC, granular cell layer of cerebellum; H, hypothalamus; Hi, hilus of dentate gyrus; I, inferior colliculus; M, molecular layer of cerebellum; NA, nucleus accumbens; OB, olfactory bulb; ON, anterior olfactory nuclei; P, parietal cortex; Pn, Pons; Pr, perirhinal cortex; R, reticular nucleus of thalamus; S, stratum; SC, superior colliculus; SL, stratum lucidum of hippocampus; Sp, septum; T, temporal cortex; Th, thalamus.

lar granular cell layer, whereas GluR7 is enriched in the deep cortical layers, dentate gyrus, thalamic reticular nucleus, and pons.

IV. KAINATE RECEPTOR PHYSIOLOGY

A. Identification of Kainate Receptor-Mediated Responses

Prior to the mid 1990s, many studies had attempted to examine the physiological properties of kainate receptors. However, with only a few exceptions (most notably studies of the dorsal root ganglia (31,32) and CA_3 region of the hippocampus (33)), these studies were often confounded by the ability of kainate to interact with AMPA receptors (see discussion in ref. 2). The characterization of functional kainate receptors in the brain has been greatly facilitated by the development of kainate and AMPA receptor-selective antagonists. Currents attributed unequivocally to activation of kainate receptors in cultured postnatal hippocampal cells were described in a study using the selective AMPA receptor antagonists GYKI52466 and GYKI53655 (34). Potentially relevant to understanding domoate toxicity is the observation that AMPA receptor responses can appear even larger in response to kainate and domoate than to AMPA (e.g., see ref. 9). This is due to the much faster desensitization rate of AMPA receptors in response to AMPA as compared to kainate or domoate. In terms of excitotoxicity, this could potentially become significant at calcium-permeable AMPA receptors (those receptors missing edited GluR2 subunits). However, AMPA receptor activation by domoate requires higher concentrations, and it is not clear if such concentrations are attained during domoate toxicity.

The demonstration of functional kainate receptors in the brain thus raised the question as to whether they play a role in mediating or modifying synaptic transmission. Recently, two studies using the hippocampal slice preparation have provided strong evidence that postsynaptic kainate receptors located on the proximal dendrites of CA_3 pyramidal neurons are activated by stimulation of mossy fibers that project from granular cells in the dentate gyrus (35,36). In both studies, evoked currents recorded in the presence of a selective AMPA receptor antagonist (GYKI53655) and an NMDA receptor antagonist (APV) were only observed following high-frequency stimulation of the mossy fibers. Further, responses to localized application of kainate to CA_3 neurons were only observed in the stratum lucidum layer where mossy fibers terminate (35). Pharmacological evidence suggests that GluR5 subunits contribute to the postsynaptic kainate receptors mediating the mossy fiber synaptic response (36). Interestingly, work in Lerma's laboratory failed to demonstrate any role for kainate receptors in mediating excitatory transmission at CA_1 synapses (37). Taken together, the results from these studies correlate well with the distribution of high-affinity kainate binding in the hippocampus (7,38) and with the CA_3-selective neuropathology in human domoate poisoning. In rodents and humans, significant toxicity is also seen in the CA_1 region of the hippocampus, a region with relatively sparse densities of kainate receptors. In rodents, higher concentrations of kainate are necessary to cause damage in the CA_1 than in the CA_3 region. Toxicity in the CA_1 region may be related to the ability of kainate receptor activation to shut down inhibitory neurotransmission.

B. Kainate Receptor Modulation of Inhibitory Synaptic Transmission

Neuronal activity in central neurons is regulated by the balance between excitation and inhibition. An excessive or abnormal increase in excitation or a decrease in inhibition can lead to pathological hyperexcitability, with the consequence being neurotoxicity and, ultimately, neuronal death. With regard to fast synaptic transmission, excitation is mediated primarily through glutamatergic synapses, whereas inhibition is mediated by GABA(gamma-aminobutyric acid)ergic interneurons. Several reports have been published suggesting that activation of kainate receptors can increase excitability in the hippocampus by decreasing inhibitory transmission (33,39). Several recent reports have begun to reveal mechanisms of kainate receptor modulation of GABAergic inhibitory transmission in the hippocampus. Kainate acting on presynaptic kainate receptors inhibits GABA-mediated inhibitory synaptic responses of CA_1 pyramidal neurons (40). This presynaptic effect may be mediated by a GluR5 subtype of kainate receptor (41). In vivo, kainate inhibition of GABAergic inhibitory activity was associated with an epileptic electroencephalographic (EEG) pattern (40). These studies provide strong support for the idea that the neurotoxic effects of kainate may be mediated, at least in part, by an inhibitory action at presynaptic sites on GABAergic neurons which ultimately leads to a pathological increase in neuronal excitability.

V. SUMMARY

Domoic acid is one of the most potent activators of kainic acid receptors. Recent results demonstrate that kainate receptors normally mediate glutamate neurotransmission in selective CNS pathways. Unfortunately, excessive activation of these synaptic glutamate receptors can lead to seizures and neurotoxicity, especially in the hippocampus. The excitotoxicity mediated by domoic acid represents both a direct activation of kainate receptors with a consequent depolarization of the cell and also represents a kainate receptor–mediated reduction of inhibitory activity. In turn, loss of inhibition causes seizure activity and excitotoxicity via multiple types of glutamate receptors (in addition to kainate receptors). The well-known amnesic effects of domoic acid are readily explained by the significant damage to the CA_1 and CA_3 regions of the hippocampus. With the high density of kainate receptors found in the CA_3 region, a direct kainate action may be contributing to toxicity in this region (along with disinhibition), whereas kainate toxicity in the CA_1 region of the hippocampus is probably largely mediated by kainate receptor–mediated disinhibition and seizure activity. With the identification of receptors and mechanisms by which domoate toxicity occurs, it should now be possible to predict antidotes to domoate toxicity and rapidly to identify other potential toxins acting at the kainate receptor.

REFERENCES

1. TM Perl, L Bedard, T Kosatsky, JC Hockin, EC Todd, RS Remis. An outbreak of toxic encephalopathy caused by eating mussels contaminated with domoic acid [see comments]. N Engl J Med 322: 1775–1780, 1990.
2. DT Monaghan, RJ Bridges, CW Cotman. The excitatory amino acid receptors: their classes, pharmacology, and distinct properties in the function of the central nervous system. Annu Rev Pharmacol Toxicol 29:365–402, 1989.
3. PJ Conn, J-P Pin. Pharmacology and functions of metabotropic glutamate receptors. Annu Rev Pharmacol Toxicol 37:205–237, 1997.
4. JT Coyle, ME Molliver, MJ Kuhar. In situ injection of kainic acid: a new method for selectively lesioning neural cell bodies while sparing axons of passage. J Comp Neurol 180:301–323, 1978.
5. JV Nadler, BW Perry, CW Cotman. Intraventricular kainic acid preferentially destroys hippocampal pyramidal cells. Nature 271:676–677, 1978.
6. JH Robinson, SA Deadwyler. Kainic acid produces depolarization of CA3 pyramidal cells in the vitro hippocampal slice. Brain Res 221:117–127, 1981.
7. DT Monaghan, CW Cotman. The distribution of [^{3}H]kainic acid binding sites in rat CNS as determined by autoradiography. Brain Res 252:91–100, 1982.
8. M Hollmann, A O'Shea-Greenfield, SW Rogers, S Heinemann. Cloning by functional expression of a member of the glutamate receptor family. Nature 342:643–648, 1989.
9. J Boulter, M Hollmann, A O'Shea-Greenfield, M Hartley, E Deneris, C Maron, S Heinemann. Molecular cloning and functional expression of glutamate receptor subunit genes. Science 249:1033–1037, 1990.
10. K Keinanen, W Wisden, B Sommer, P Werner, A Herb, TA Verdoorn, B Sakmann, et al. A family of AMPA-selective glutamate receptors. Science 249:556–560, 1990.
11. B Bettler, J Boulter, I Hermans-Borgmeyer, A O'Shea-Greenfield, ES Deneris, C Moll, U Borgmeyer, et al. Cloning of a novel glutamate receptor subunit, GluR5: expression in the nervous system during development. Neuron 5:583–595, 1990.

12. J Egebjerg, B Bettler, I Hermans Borgmeyer, S Heinemann. Cloning of a cDNA for a glutamate receptor subunit activated by kainate but not AMPA. Nature 351:745–748, 1991.
13. B Bettler, J Egebjerg, G Sharma, G Pecht, I Hermans Borgmeyer, C Moll, CF Stevens, et al. Cloning of a putative glutamate receptor: a low affinity kainate-binding subunit. Neuron 8:257–265, 1992.
14. A Herb, N Burnashev, P Werner, B Sakmann, W Wisden, PH Seeburg. The KA-2 subunit of excitatory amino acid receptors shows widespread expression in brain and forms ion channels with distantly related subunits. Neuron 8:775–785, 1992.
15. K Sakimura, T Morita, E Kushiya, M Mishina. Primary structure and expression of the gamma 2 subunit of the glutamate receptor channel selective for kainate. Neuron 8:267–274, 1992.
16. B Sommer, N Burnashev, TA Verdoorn, K Keinanen, B Sakmann, PH Seeburg. A glutamate receptor channel with high affinity for domoate and kainate. EMBO J 11:1651–1656, 1992.
17. P Gregor, BF O'Hara, X Yang, GR Uhl. Expression and novel subunit isoforms of glutamate receptor genes GluR5 and GluR6. Neuroreport 4:1343–1346, 1993.
18. HH Schiffer, GT Swanson, SF Heinemann. Rat GluR7 and a carboxy-terminal splice variant, GluR7b, are functional kainate receptor subunits with a low sensitivity to glutamate. Neuron 19:1141–1144, 1997.
19. PH Seeburg, M Higuchi, R Sprengel. RNA editing of brain glutamate receptor channels: mechanism and physiology. Brain Res Brain Res Rev 26:217–229, 1998.
20. M Hollmann, M Hartley, S Heinemann. Ca^{2+} permeability of KA-AMPA–gated glutamate receptor channels depends on subunit composition. Science 252:851–853, 1991.
21. R Brusa, F Zimmermann, DS Koh, D Feldmeyer, P Gass, PH Seeburg, R Sprengel. Early-onset epilepsy and postnatal lethality associated with an editing-deficient GluR-B allele in mice. Science 270:1677–1680, 1995.
22. J Egebjerg, SF Heinemann. Ca^{2+} permeability of unedited and edited versions of the kainate selective glutamate receptor GluR6. Proc Natl Acad Sci USA 90:755–759, 1993.
23. M Kohler, N Burnashev, B Sakmann, PH Seeburg. Determinants of Ca^{2+} permeability in both TM1 and TM2 of high affinity kainate receptor channels: diversity by RNA editing. Neuron 10:491–500, 1993.
24. GT Swanson, D Feldmeyer, M Kaneda, SG Cull-Candy. Effect of RNA editing and subunit co-assembly single-channel properties of recombinant kainate receptors. J Physiol (Lond) 492:129–142, 1996.
25. P Werner, M Voigt, K Keinanen, W Wisden, PH Seeburg. Cloning of a putative high-affinity kainate receptor expressed predominantly in hippocampal CA3 cells. Nature 351:742–743, 1991.
26. H Lomeli, W Wisden, M Kohler, K Keinanen, B Sommer, PH Seeburg. High-affinity kainate and domoate receptors in rat brain. FEBS Lett 307:139–143, 1992.
27. N Brose, GW Huntley, Y Stern Bach, G Sharma, JH Morrison, SF Heinemann. Differential assembly of coexpressed glutamate receptor subunits in neurons of rat cerebral cortex. J Biol Chem 269:16780–16784, 1994.
28. RJ Wenthold, VA Trumpy, WS Zhu, RS Petralia. Biochemical and assembly properties of GluR6 and KA2, two members of the kainate receptor family, determined with subunit-specific antibodies. J Biol Chem 269:1332–1339, 1994.
29. JR Unnerstall, JK Wamsley. Autoradiographic localization of high-affinity [3H]kainic acid binding sites in the rat forebrain. Eur J Pharmacol 86:361–371, 1983.
30. W Wisden, PH Seeburg. A complex mosaic of high-affinity kainate receptors in rat brain. J Neurosci 13:3582–3598, 1993.
31. JC Watkins, RH Evans. Excitatory amino acid transmitters. Annu Rev Pharmacol Toxicol 21:165–204, 1981.
32. J Huettner. Glutamate receptor channels in rat DRG neurons: Activation by kainate and quisqualate and blockade of desensitization by Con A. Neuron 5:255–266, 1990.
33. RS Fisher, BE Alger. Electrophysiological mechanisms of kainic acid-induced epileptiform activity in the rat hippocampal slice. J Neurosci 4:1312–1323, 1984.
34. AV Paternain, M Morales, J Lerma. Selective antagonism of AMPA receptors unmasks kainate receptor-mediated responses in hippocampal neurons. Neuron 14:185–189, 1995.

35. PE Castillo, RC Malenka, RA Nicoll. Kainate receptors mediate a slow postsynaptic current in hippocampal CA3 neurons. Nature 388:182–186, 1997.
36. M Vignes, D Bleakman, D Lodge, GL Collingridge. The synaptic activation of the GluR5 subtype of kainate receptor in area CA3 of the rat hippocampus. Neuropharmacology 36:1477–1481, 1997.
37. J Lerma, M Morales, MA Vicente, O Herreras. Glutamate receptors of the kainate type and synaptic transmission. Trends Neurosci 20:9–12, 1997.
38. AC Foster, EE Mena, DT Monaghan, CW Cotman. Synaptic localization of kainic acid binding sites. Nature 289:73–75, 1981.
39. GL Westbrook, EW Lothman. Cellular and synaptic basis of kainic acid-induced hippocampal epileptiform activity. Brain Res 273:97–109, 1983.
40. A Rodriguez-Moreno, O Herreras, J Lerma. Kainate receptors presynaptically downregulate GABAergic inhibition in the rat hippocampus. Neuron 19:893–901, 1997.
41. VR Clarke, BA Ballyk, KH Hoo, A Mandelzys, A Pellizzari, CP Bath, J Thomas, et al. A hippocampal GluR5 kainate receptor regulating inhibitory synaptic transmission [see comments]. Nature 389:599–603, 1997.

18

Chemical and Biological Detection Methods

Antonello Novelli and M.T. Fernandez-Sanchez
University of Oviedo
Oviedo, Spain

T. A. Doucette and R. A. R. Tasker
Atlantic Veterinary College
University of Prince Edward Island
Charlottetown, Prince Edward Island, Canada

I. CHEMICAL ANALYSIS TECHNIQUES

A. Chemical properties

Domoic acid is a naturally occurring secondary amino acid. It belongs to the kainoid class of compounds and is structurally very similar to the parent compound, kainic acid, as well as to better-known amino acids such as glutamate and proline. The structure of domoic acid is shown in Figure 1. Chemically, domoic acid is (2S,3S,4S)-2-carboxy-4-l-methyl-5(R)-carboxyl-l(Z), 3(E)-hexadienyl pyrrolidine-3-acetic acid ($C_{15}H_{21}NO_6$) and has an anhydrous molecular weight of 311.14. Ultraviolet (UV), infrared (IR), and nuclear magnetic resonance (NMR) data for domoic acid are pH-dependent and have been described in detail elsewhere (1–3).

Original studies of the chemical (4) and insecticidal (5) properties of domoic acid used material extracted from the red alga *Chondria armata*. In addition to the parent compound, seaweed samples also contain minor amounts of three isomers, isodomoic acid A, B, and C (6) and domoilactones A and B (7). In contrast, none of these isomers has been reported in either plankton or shellfish contaminated with domoic acid (8), although plankton and shellfish samples have been shown to contain small amounts of three geometrical photoisomers (isodomoic acids D, E, and F) following exposure to UV light (254 nm) (9,10). These isomers can be separated from the parent compound using high-performance liquid chromatography (HPLC) (see below).

B. Sample Preparation

Domoic acid can be isolated from contaminated shellfish by boiling shellfish tissue in an equal volume of 0.1N HCl and filtering the supernatant. This technique is also the standard extraction procedure for the mouse bioassay of paralytic shellfish poisoning (PSP) toxins (11) and has been used for routine regulatory analysis of both PSP and domoic acid by either bioassay (see below) or by HPLC (12–14). Although simple and convenient, however, this method yields a crude extract that is not well suited to detailed quantitative analysis.

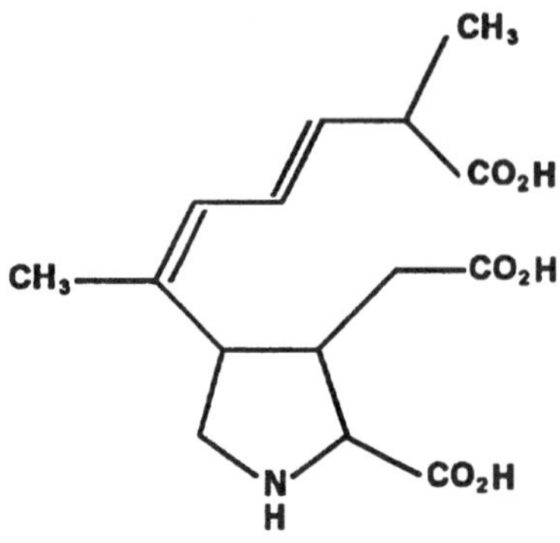

Figure 1 Structure of domoic acid.

At the time of the 1987 amnesic shellfish poisoning (ASP) toxicity outbreak, domoic acid was isolated from mussel tissue using a water/methanol (1 : 1 v/v) extraction (15). This technique was subsequently modified to use aqueous methanol and boiling water either alone or in combination with a solid-phase extraction (SPE) clean-up (16). These procedures yield reproducible, high recoveries and prolonged HPLC column lives relative to the original AOAC method (11). Subsequently, Quilliam et al. (17,18) demonstrated high recoveries of even trace amounts of domoic acid using an aqueous methanol (1:1 v/v) extraction followed by passage of the supernatant through a strong anion exchange (SAX) column preconditioned with methanol, water, and extraction solvent. This procedure eliminates many of the interferences that confound HPLC analysis.

An alternative method of sample preparation involves a modification of the common lipid extraction method of Bligh and Dyer (19), which was adapted to isolate domoic acid from shellfish and plankton (20). This method involves homogenization of the sample in dilute hydrochloric acid prior to partitioning between two volumes of methanol and one volume of chloroform. Domoic acid resides almost exclusively in the water/methanol fraction. This procedure has also been applied successfully to the preparation of guinea pig and rat serum samples for analysis of domoic acid concentrations by HPLC (21). This technique can be applied to serum samples as small as 50 μL and yields recoveries between 87 and 89% (21).

C. Analysis Techniques

There are a number of quantitative and semiquantitative techniques that have been used to detect domoic acid in a variety of biological matrices. These are listed in Table 1 and briefly described below.

Thin-Layer Chromatography

Following sample clean-up as described above, domoic acid can be analyzed semiquantitatively using thin-layer chromatography (TLC) (8,22). Direct or concentrated samples or standards (50–100 μL) can be applied to one-dimensional TLC plates (250-μm thick silica gel 60 with F254 indicator) and eluted with a butanol/acetic acid/water (3 : 1 : 1 v/v) solvent. After drying, the plate can be examined under 254-nm light for the presence of a UV-quenching spot. Spraying the plate with a solution of 1% ninhydrin will cause domoic acid to appear as a yellow spot. If for some reason it is not possible to clean up the sample prior to analysis, the TLC technique can still be used if the gel is run in two dimensions. Although TLC is only semiquantitative,

Table 1 Summary of Analytical Techniques for the Detection of Domoic Acid (ASP) in Seafood

Technique	Limit of detection	Features	References
Thin-layer chromatography	10 μg/g	Semiquantitative Applicable to a variety of biological matrices Can be used with crude extracts Inexpensive	8,12
High-Performance liquid chromatography	20–30 ng/g (UV) 15 pg/g (fluorescence)	Quantitative Extensive sample clean-up required for maximum sensitivity Derivatization required for fluorescence Isomers can be quantified Expensive equipment required	8,12,16,21,23
Capillary electrophoresis	3 pg/injection 150 ng/g	Quantitative Minimal clean-up required High resolution Very small injection volumes Expensive equipment required	24
Mass spectrometry	1 μg/g	Quantitative in some modes Requires prior separation by chromatography or capillary electrophoresis May require derivatization of sample Very expensive equipment required	15,25,26

it has some distinct advantages over other methods when applied to routine screening procedures of contaminated shellfish; namely, the ability to analyze multiple samples in a single run at relatively low cost. The detection limit by this technique is approximately 0.5 μg domoic acid (10 μg/g in shellfish).

High-Performance Liquid Chromatography

HPLC is the preferred method of quantifying domoic acid for both regulatory and research applications. Detection can be accomplished using either UV absorbance or fluorescence, with the latter being more sensitive but also more complex.

Column Conditions. Following sample extraction and clean-up using any of the procedures outlined in Section B (above), the sample can be injected (usually 20 μL volume) onto a standard (25 cm × 4.1 mm ID) column packed with 5–10 μm C18 bonded silica gel. The use of a guard column is recommended. The analytical column will function adequately at ambient temperature, although maintaining a column temperature of 40°C is recommended by some workers (8). If the HPLC is equipped to accommodate millibore columns a 25.0 cm × 2.1 mm (ID) column and a 5-μL injection volume can be used.

Effective separation of domoic acid can be accomplished using isocratic conditions with mobile phase being delivered at 1.0–1.5 mL/min. Acidic mobile phases are preferred, and most workers have used a mixture of acetonitrile and water acidified with either trichloroacetic (TCA) or phosphoric acid (12,16,21). For analysis of domoic acid in shellfish, a retention time of 7–10 min can be achieved with an acetonitrile/water mixture of 1:4 (v/v) (16), whereas serum samples require that domoate elute at 10–20 min to avoid interfering peaks, so an acetonitrile/water mixture of 1:10 (v/v) is preferred (21).

Detection. Detection is most commonly accomplished using UV absorbance at 242 nm. Domoic acid absorbs strongly at this wavelength, and aqueous standards containing as little as 10 ng/mL can be detected using UV absorbance (8). The limit of detection in biological matrices varies with the sample and the method of sample clean-up, but Wright and Quilliam (8) have reported limits of detection of 20–30 ng/g in shellfish following SAX-SPE clean-up, and Blanchard and Tasker (21) reported a detection limit of 200 ng/mL in rat serum using the modified Bligh and Dyer extraction procedure (see above). Although good sample clean-up considerably reduces the concentrations of compounds that could potentially interfere with domoic acid quantitation by UV absorbance, other compounds in shellfish and serum, notably other acidic amino acids, can also absorb light at 242 nm and may elute close to domoic acid. The use of a photodiode array detector will reliably account for such problems if the equipment is available.

In those cases where extreme sensitivity is required, UV absorbance is replaced by fluorescence detection of derivatized domoic acid. Formation of a fluorenylmethoxycarbonyl (FMOC) derivative of domoic acid prior to injection on the HPLC column results in detection limits of about 0.5 ng/mL in seawater and plankton using isocratic HPLC and as low as 15 pg/mL using gradient elution (23). In theory, this technique can also be used for sensitive detection of domoic acid in biological matrices, although complex sample clean-up procedures are required.

Capillary Electrophoresis

Capillary electrophoresis (CE) is a relatively new technique that is particularly useful for quantifying components of complex mixtures, especially when only small samples are available. A recent publication by Zhao et al. (24) described the use of CE for analysis of domoic acid and domoate isomers in seafood. Briefly, crude homogenates were prepared in methanol and applied to a preconditioned strong anion exchange (SAX) column. Domoate and isomers were eluted with formic acid and the eluent was then applied to a strong cation exchange (SCX) column and eluted with a sodium tetraborate/acetonitrile solvent. The clean extract was then applied to a CE system interfaced to a variable wavelength UV detector operating at 242 nm. Effective resolution of domoic acid and five domoate isomers was achieved using a liquid phase composed of 22.5 mM sodium tetraborate containing 20 mM β-cyclodextrin and a separation potential of 30 kV. Mass detection limits for domoate are very low using CE (about 3 pg), but the small injection volumes used makes the technique less sensitive than HPLC for shellfish samples (see Table 1). Recent advances in technology, however, are leading to unprecedented detection limits. For example, the use of a helium/cadmium laser in combination with CE to detect DOM that has been derivatized with dansyl chloride has resulted in detection limits in the atomole (10^{-18}) range (M.A. Quilliam, personal communication). The major advantage of the technique in routine applications, however, remains the high degree of resolution for both domoate and its isomers in a variety of shellfish homogenates.

Mass Spectrometry

Confirmation that the toxin responsible for ASP was domoic acid was accomplished using fast-atom bombardment (FAB) mass spectrometry and direct probe insertion (15,25). For routine regulatory and research applications, however, it is necessary to interface mass spectrometry with an analytical separation technique such as gas chromatography (GC), HPLC, or CE (26). Historically, mass spectrometry has been an expensive and complicated procedure, so most laboratories have not explored the use of this technique for the routine detection of shellfish toxins such as domoic acid. In recent years, however, LC-MS and LC-MS-MS have become increasing common tools for many contract analytical laboratories, so it is quite conceivable that these powerful analytical techniques will become much more commonplace in the near future. Moreover, mass spectrometry remains the technique that most jurisdictions consider the legally accepted method of confirmation for domoic acid (DA) and most other seafood toxins. Also it must be noted that of all the techniques described, LC-MS or LC-MS-MS, hold the greatest promise for development of a ''universal toxin monitor'' that could detect any known marine toxin in a single sample (M.A. Quilliam, personal communication).

II. IN VITRO BIOLOGICAL TECHNIQUES

A. General Characteristics

The development of in vitro techniques, including receptor binding assays, immunoassays, enzyme assays, and cytotoxicity tests, offers one of the most promising approaches to the detection of phycotoxins in complex organic matrices, such as extracts of shellfish tissues and toxic microalgae. In principle, such methods have several inherent advantages over sophisticated chemical analytical alternatives for routine phycotoxin screening in that they can be configured to yield extremely high sensitivity and specificity for the target toxin. In some cases, a relatively crude sample extract may be used for toxin detection by in vitro techniques, thereby facilitating their handling and the number of samples that can be screened in a working day.

One interesting characteristic of in vitro techniques is that they can be developed in the form of diagnostic test kits for rapid and easy manipulation by any technician and competitive unitary cost with respect to other methods. Nevertheless, such kits can be costly to develop and to patent, and they must be sold in large quantities in order to produce a net benefit to the manufacturer. Among potential users, not only regulatory personnel but also mariculturists, fishermen, and field technicians may find advantages in the use of diagnostic test kits for the rapid detection of phycotoxins in shellfish, finfish, and phytoplankton. Such kits may lend themselves to routine application in different types of facilities, such as on shipboard, or at the dockside, or in local laboratories. Occasionally, seafood restaurants may also be interested in the use of diagnostic test kits to guarantee the quality of their product to customers. Furthermore, in vitro techniques may represent an interesting alternative to in vivo bioassays, particularly in those countries with important legislative restrictions on animal use. Thus, in vitro techniques may be useful in monitoring programs and in reducing the number of samples that will have to be further tested by chemical alternatives or in vivo bioassays.

Several types of in vitro bioassays exist. They fall in two categories depending on the functional or nonfunctional nature of the measured response. The latter category includes (1) receptor binding assays and (2) immunoassays; whereas the former includes (3) receptor-activation/inhibition assays, (4) enzymatic assays, and (5) cytotoxicity assays. It is worth noting that functional assays facilitate more valuable information about the potential risk of phycotoxins for human health, because in nonfunctional assays toxin analogues devoided of functional properties are not detected, whereas toxin analogues that are not recognized by antibodies or have a too low receptor affinity to be detected by receptor-binding methods may still be evaluated for their functional activity. This makes the information obtainable by in vitro functional bioassay similar to that of in vivo bioassays. Despite their usefulness, the main limitation of in vitro bioassays is the absence of drug metabolism. Lack of metabolism reduces the toxic potential of precursor molecules and can permit the expression of toxicity of usually harmless molecules. However, such pharmacokinetic problems do not affect in vitro bioassays only, since in general, in vivo bioassays may also provide different toxicological information, depending on the animal species that is used or the route of drug administration.

B. Specific Tests

With respect to the possibility of monitoring for amnesic shellfish toxins in food and other biological samples by in vitro assays, we should mention several methods that may reveal the presence of domoic acid and its structural analogues, which are, as far as we know, the only causative agents of amnesic shellfish poisoning (27). Some of these methods have already been proposed for commercial use, whereas others are still under development. We are presenting them starting with those that may fall into the category of nonfunctional assays according to the classification proposed earlier.

Nonfunctional Assays

Receptor Binding Assays. The first method is based on domoic acid displacement of [^{3}H]kainic acid from glutamate receptors. Kainic acid binding to rat brain membranes can be described by two dissociation constants (K_d), one of high affinity with a K_d ~3–5 nM, and the other of low affinity with a K_d ~25–50 nM (28). Membranes from the cerebellum are an exception, as [^{3}H]kainic acid binding is only described by one K_d of ~25 nM (28). In one study, domoic acid was shown to displace [^{3}H]kainic acid from rat cerebellar membranes with an inhibition constant (K_i) ~13 nM, although this value was considered as an approximation due to a Hill coefficient <1 (29). Both kainic acid and domoic acid had been classified as excitatory amino acids (EAAs) based on early studies (30). Data obtained in mammals (31–35), cultured neurons (32,36–38), as well as from molecular biology studies (39–42) have further confirmed that both kainic acid and domoic acid are agonists at the ionotropic non-NMDA type of EAA receptors (43,44). Based on functional, pharmacological, and molecular biologic studies, EAA receptors may be divided in two main groups termed metabotropic and ionotropic receptors

(44). Metabotropic receptors are coupled to G proteins linked to phospholipases C and D and adenylate cyclase activities (45), whereas ionotropic receptors contain a cation-specific ion channel. Ionotropic EAA receptors are further divided in two types, NMDA and non-NMDA receptors, according to their affinity for the synthetic glutamic acid analogue N-methyl-D-aspartic acid. The NMDA receptor–associated channel is blocked by extracellular Mg^{2+} at resting membrane potentials, possesses multiple regulation sites, and it allows the passage of Ca^{2+} (43,46). Several subunits of the NMDA receptor have been cloned, and their combination may lead to functional NMDA receptors (44). Non-NMDA receptors are also generated by the assembly of subunits that can be classified into three groups (44). Table 2 reports the relative affinity for the agonists of non-NMDA receptors formed by each group of subunits according to the literature data (38–40,42,47). The first group includes the subunits GluR1 through GluR4 and generates functional non-NMDA receptor channels with high affinity for α-amino-3-hydroxy-5-methylisoxazole-4-propionic acid (AMPA) and lower affinity for domoic acid and kainic acid. The second group, formed by GluR5 through GluR7, generates functional non-NMDA receptor channels with high affinity for domoic acid and little or no affinity for AMPA, whereas the third group includes only two subunits, KA1 and KA2, with high-affinity binding for kainic acid but the need to combine with GluR5 or GluR6 in order to form functional non-NMDA receptor channels, which can also be activated by domoic acid (see refs. 48 and 49 for review).

Non-NMDA receptor distribution in rat brain has shown that the cerebellum is highly enriched in GluR6 and GluR7 subunits, thus explaining earlier binding data suggesting only one population of low-affinity binding sites for kainic acid (42).

Based on this knowledge, Van Dolah and collaborators developed an elegant receptor assay to detect domoic acid by utilizing a cloned glutamate receptor (50). They took advantage of the large experience of the group in the isolation and cloning of kainic acid and domoic acid receptors (51–55) to prepare membranes from SF9 insect cells containing the GluR6 receptor and use them to determine the presence of domoic acid in shellfish based on domoic acid displacement of [^{3}H]kainic acid from this receptor. The GluR6-expressing SF9 cell membrane preparation was shown to possess a single-affinity kainic acid binding site (K_d ~12.5 nM), and competition assays yielded a binding affinity (K_i) for purified domoic acid of 4.3 nM, with a Hill coefficient very close to 1, thereby confirming previous results from the same group (54) as well as from other investigators using a different expression system (40). When used with domoic acid–contaminated mussels, domoic acid determination was reliably achieved after glutamate interference was minimized by adding glutamate decarboxylase, thus transforming glutamate to GABA. The amount of domoic acid detected by this method showed a high correlation with values obtained by HPLC, although some caution is necessary owing to the limited number of mussel samples tested and the fact that that they did not contain other domoic acid isomers,

Table 2 Subunit Composition of Non-NMDA Receptors. Relationship to Distinct Relative Affinity for Selected Agonists

Subunits	Non-NMDA receptors Relative affinity for the agonist
GluR1, GluR2, GluR3, GluR4	QUIS > AMPA > DA > GLU > KA
GluR5, GluR6, GluR7[a]	DA > KA > QUIS > GLU > AMPA
KA-1, KA-2	KA > DA

QUIS, quisqualic acid; AMPA, alpha amino-3-hydroxy-5-methylisoxazole-4-propionic acid; DA, domoic acid; GLU, glutamic acid; KA, kainic acid.
[a] For receptors formed by the subunit GluR7, GLU > QUIS.

making impossible a proper correlation between the domoic acid content and domoic acid equivalents as measured by the binding assay. The investigators did not estimate the sensitivity of their method, although it should be lower than 100 ng of domoic acid per gram of fresh tissue, according to the data reported.

Immunoassays. Within the category of nonfunctional bioassays, several immunoassays for the determination of domoic acid in body fluids, water samples, and shellfish should be mentioned (56–60). The assay of Garthwaite et al. (60) appears to be particularly promising because of its high sensitivity. The investigators developed two different methods in order to produce domoic acid either carboxy-linked to bovine serum albumin (A) or N-linked to ovalbumin, thyroglobulin, or fetuin (B). Then they raised antibodies against B in sheep and developed an indirect competitive enzyme-linked immunoadsorbent assay (ELISA) using A as the immobile antigen. The full characterization of the properties of this immunoassay is progressing and Garthwaite et al. claim that the assay has the potential to be used for routine screening of domoic acid in a variety of samples. Although the reported detection limit of 38 ng domoic acid per gram shellfish flesh (~0.5 nM domoic acid) appears to be slightly optimistic for routine use in shellfish screening, and no information is available about the possibility of detecting other domoic acid isomers, this may be regarded as the most successful attempt so far to raise antibodies against domoic acid, and it may be of relevance for basic neuroscience research as well as for the development of bioassays.

Functional Assays

Displacement of kainic acid binding to excitatory amino acid receptors and immunological assessment of domoic acid can be considered as complementary methods for establishing the presence of this toxin in the examined samples. Failure to obtain a positive result in both methods may reveal either nonfunctional isomers of domoic acid or the presence of structurally novel toxins with binding affinity for excitatory amino acid receptors, respectively. It should be noted that the latter case does not necessarily imply a health risk, since receptor binding is not always associated with either receptor activation (i.e., inverse agonists) or toxic effect (i.e., desensitizing vs nondesensitizing response). Thus, an in vitro assay for domoic acid and functionally similar molecules should be based on functional assays that rely on the measurement of the response elicited following receptor binding.

Electrophysiological Techniques. One example of a functional assay is the essential and elegant method developed in 1991 by Minami and colleagues (61). The method takes advantage of the receptor-activated response amplification that is an intrinsic property of ionotropic receptors when they are present either in a cell membrane or in a lipid bilayer. The investigators purified a ion channel containing glutamate binding glycoprotein that was incorporated into a planar bilayer membrane and measured the electrical signal generated by the ion flux through the channel following exposure to glutamate. They claimed a sensitivity to L-glutamate in the order of 10 nM, but they did not perform any experiments with other glutamate analogues. Thus, this method may have good potential for the development of domoic acid biosensors using a purified GluR6 protein inserted in lipid bilayer membranes. The same bioassay principle may be applied to GluR6 proteins expressed in cells transfected with GluR6 cDNA, thus converting the previously presented binding method by Van Dolah et al. (50) into a functional bioassay.

It should be noted that this electrophysiological approach is technically well established and is commonly used in basic research. Besides its application to either purified or cloned receptors in artificial membranes or cell lines, respectively, this method can be useful for the detection of domoic acid (and other toxins) using slices from several brain areas. Fountain and colleagues have proposed such an application, focusing on a hippocampal slice preparation as the experimental system, for detection of toxins such as trimethyltin, triethyltin, methylpyridines, and aspartame (62). However, this experimental system has been widely used for studying the electrophysiological response to all types of excitatory amino acids, including domoic acid (63–65) because of its importance in cognitive processes. The utilization of this assay for the detection of domoic acid as well as other marine toxins either of algal or chemical origin in seafood should be straightforward after solving the initial problems related to sample preparation. Because of their electrical excitability, neurons are the natural target of domoic acid and other seafood toxins acting on either ionotropic receptors or ion channels, and therefore the electrophysiological approach should be regarded as one of the most sensitive and promising methods for toxin screening, particularly when neuronal cultures are used, thus coupling the power of the technique to the flexibility of the biological system (38,66).

Neuronal Cultures. Neuronal cultures are possibly the ideal in vitro biological system to test for the presence of seafood toxins. Neurons can be maintained in primary culture for several weeks, allowing any type of biochemical and physiological process activated following exposure to domoic acid (or other toxins) to be monitored starting from steady-state conditions, assuming that the proper detection method is provided. The electrophysiological response is, of course, the earliest event that follows domoic acid exposure, but if more time is available, the simplest response that can be measured is the toxicological one. Neurons in culture are exquisitely sensitive to any compound that alters their physiological equilibrium, whether this may occur through the induction of excessive ion fluxes (36,37) or by altering their biochemical steady state, as with toxins that are phosphatase 2A inhibitors (67,68). Neuronal culture exposure to excitatory amino acids such as domoic acid results in an excitatory amino acid concentration–dependent neuronal degeneration (37). The degenerative process begins within a few minutes after drug administration (69,70) and is complete within 12–18 h. Neuronal survival can be easily quantified by using fluorescent dyes specifically labeling live cells and counting them (37,69). This in vitro toxicological approach has been widely used in neuroscience in the last 10 years both for basic research and for the screening of drugs that may provide neuroprotection against either acute or chronic degeneration of the central nervous system mediated by excitatory amino acids (see ref. 71 for review).

Thus, a simple mathematical description of concentration-dependent neurotoxicity by excitatory amino acids has been proposed to allow for normalization of toxicological data and to facilitate quantitative measurements and data comparisons (72). The specific application of neuronal cultures to the assessment of shellfish toxicity by domoic acid was first presented at the Domoic Acid Symposium held in Ottawa, Canada, in 1989 (36), and a detailed description of the results obtained was published later (37).

More recently, based on the results obtained using neuronal cultures for characterizing the toxicity of the diarrheic shellfish toxin okadaic acid (67,70), and the possibility of inducing neurodegeneration by activating voltage-sensitive sodium channels, neuronal cultures have been proposed for the assessment of three types of shellfish toxins: amnesic, diarrheic, and paralytic (73). For each of the toxins, the assessment of their presence in seafood is based on the quantification of neuronal death following 18–24 h of toxin exposure. For amnesic and diarrheic toxins, a direct toxicological assay is used, whereas for paralytic toxins, a reverse toxicological assay is employed, because protection from neurodegeneration induced by voltage-sensitive sodium channel activators is measured. In various shellfish samples, the domoic acid content (as measured by HPLC) and domoic acid equivalents (as measured by the neuronal bioassay) were well correlated. The only exceptions were some toxic samples containing a low concentration of domoic acid. However, the neurotoxicity of these samples could not be prevented by treatment with specific domoic acid receptor antagonists, suggesting that in addition to a low concentration of domoic acid, shellfish samples did contain other undetected toxins, possibly contributing to overall toxicity. Thus, it is worth noting that one advantage inherent to the use of both cell cultures or tissue slices is the possibility of easily obtaining a pharmacological identification of the toxin when specific antagonists are available. This is an important attribute, since it is not uncommon to observe the simultaneous presence of two or more toxins in the same sample, and only alive biological systems may allow for an accurate detection of all of toxins present. This is particularly true when new toxins start to contaminate seafood, as was the case for domoic acid in 1987. Among live biological systems, neuronal culture and brain slices offer the advantage of easy and rapid drug access to the target without the problems of blood-brain barrier permeability that severely affect pharmacological testing in animals, particularly when the toxin and its antagonists are structurally similar to domoic acid and excitatory amino acids. Another advantage offered by neuronal cultures is the unique possibility of unifying the bioassay for three different toxins in one experimental system, and the possibility that such a system could be easily adapted to the detection of ciguatera and other neuronal toxins.

C. Future Perspectives

All the in vitro biological methods mentioned herein can be very useful for the detection of domoic acid; however, caution should be used when generalization is made to amnesic toxins, because substances other than domoic acid may eventually produce the amnesic clinical syndrome. Such substances may either act at excitatory amino acid receptors other than those activated by domoic acid, or directly activate one of the several biochemical pathways that control proper neuronal activity, leading to memory formation (74,75). The progressive understanding of the molecular mechanisms involved in learning and memory will offer the possibility to develop new methods for the detection of amnesic toxins.

III. IN VIVO BIOLOGICAL TECHNIQUES

Exposure to domoic acid (DA) produces a toxic syndrome with clinical signs that are remarkably consistent among different mammalian species (76–79). Overt behavioral symptoms of DA toxicity include sluggishness, stereotypies, vomiting, profuse salivation, tremors, seizures, and death. Postmortem examination of brain tissue reveals widespread neuronal necrosis, being most severe in the hippocampus and associated regions. In adult rodents, the occurrence of spontaneous recurrent seizures and impaired learning and memory performance, particularly on spatial tasks, are among the persistent consequences of DA intoxication. To a large extent, these effects parallel the sequelae observed in the human population following the episode of mussel poisoning that affected approximately 100 eastern Canadians in November 1987 (79). DA, an algae-derived contaminant, was later identified as the causative toxin responsible for the outbreak. Since then numerous experimental methods have been developed or adapted for assessing DA toxicity in vivo. Subsequent sections will discuss the evaluation techniques which use the biobehavioral response of an animal for toxicity determinations.

A. Mammalian Bioassays

Conventional whole-animal bioassays provide an easy, rapid, robust, and sensitive screen for phycotoxin analysis, offering distinct advantages over sophisticated physiochemical procedures and many in vitro methods. In standard mammalian bioassays, evaluations of toxicity are based on the biological response of the animal to the toxin, with behavioral and physiological effects that are immediately relevant in determining human toxic effects. Mammalian bioassays also inadvertently screen for the presence of unknown toxic components in the extract matrix. For instance, the first indications of the amnesic shellfish poisoning (ASP) syndrome, later attributed to DA toxicity, were revealed in the course of a routine mouse bioassay for paralytic shellfish toxin (PSP) toxicity. Test animals displayed atypical behaviors that were distinct from those associated with the classic signs of PSP intoxication and consequent death. Although it is recognized that a number of variables can affect the results of mammalian bioassays, such as the pH of the injected extracts or specific animal characteristics, researchers ensure that procedures are carefully standardized, thus offering reliable and reproducible assessments of toxicity (80).

As with all whole-animal models, both ethical considerations and legislative restrictions in some jurisdictions often limit the use of mammalian bioassays for evaluating toxicity. Additional deterrents include the high capital investment and maintenance costs that are associated with the installation and operation of live animal facilities, and the fact that mammalian bioassays are labor intensive to perform. As previously mentioned, many variables can bias the validity of the results obtained from conventional bioassays. For instance, substances coextracted during the sample preparation can result in false reactions, as can the inappropriate choice of extraction solvents. In general, mammalian bioassays have a poor dynamic range. However, this usually becomes problematic only with phycotoxins showing low acute toxicity (e.g., those which yield a high LD50 and long latencies until death). A final concern with the use of whole-animal bioassays for toxicity determination is that they tend to be less sensitive and precise when compared to instrumental analytical procedures and provide no qualitative information concerning the nature of the toxin components. However, despite the technical problems and ethical considerations associated with the use of mammalian bioassays, such assays are widely and successfully used for phycotoxin monitoring in commercial shellfish (80).

AOAC Mouse Bioassay for ASP toxicity

Procedures for quantitating DA concentrations are similar to those used in the Association of Official Analytical Chemists (AOAC) mouse bioassay for PSP toxicity and are described in detail elsewhere (11,14). In brief, DA is isolated from contaminated shellfish by boiling the tissue in an equal volume of 0.1N HCl for 5 min. Once the mixture has cooled to room temperature, the pH is adjusted to 2–4. The mixture is then diluted, stirred to homogeneity, and permitted to settle until a portion of the supernatant is translucent and can be filtered. The general principle of the standard ASP mouse bioassay depends on intraperitoneal (i.p.) inoculation of mice (19–21 g) with varying dilutions of the shellfish extract. Following injection of the extract, time- and dose-dependent appearance of typical symptoms are recorded. The standard observation period for assessing DA toxicity is 4 h. As with other phycotoxins, the presence of DA in a sample extract elicits a series of abnormal behaviors which cumulate in convulsions in mice followed by recovery or death (depending on the concentration of DA in the sample preparation). The distinguishing feature of DA toxicity, however, is that seizures are preceded by a hindlimb scratching syndrome directed at the shoulder; an automatism unique to DA intoxication. Toxicity values are generally interpolated from standard curves, LD50 determinations over a fixed time interval, or are derived from standard toxicity tables relating dosage to time until death (80). Global toxicity, measured as the survival time of mice, is expressed in mouse units (MUs) which are then converted into toxin-specific units based on the toxicity response calibrated with reference to toxin standards. The AOAC mouse bioassay can reliably detect DA in shellfish tissue at concentrations of 40 ppm. However, the tolerance level, established in Canada and subsequently adopted by other countries, is 20 ppm. Therefore, the relative insensitivity of this assay precludes its use for regulatory purposes to quantify DA (80).

Behavioral Rating Scale for DA Toxicity in Mice

In a standard mouse bioassay for evaluating DA toxicity, crude mussel extracts are injected intraperitoneally in mice and DA concentrations are predicted based on the mean time until death or time of onset of seizures (14). Although this procedure is convenient and rapid, it is clearly inadequate for identifying low levels of contamination, since these particular behaviors are observed only at relatively high doses of DA. As such, the lower level of sensitivity of this analytical method (~100 μg/mouse) approaches the unofficial LD50 for DA in mice.

In mice, systemic injections of DA elicit a graded series of behavioral change prior to seizure onset (76,81,82). The sequence of behavioral events are both highly reproducible and predictably dose related. Hypoactivity is the first aberrant behavior observed followed by sedation-akinesia, rigidity, stereotypy, loss of postural control, convulsions, and death (depending on the administered dose). Based on these behaviors, a seven-point rating scale was developed for assessing toxicity, using the time of onset of the various atypical behaviors and a cumulative toxicity score calculated over a 60-min interval (76). This rating scale can reliably quantitate DA concentrations as low as 20 μg/mL in mussel extracts, thus representing a fivefold increase in sensitivity over the more conventional mouse bioassay procedure for screening commercial shellfish (76).

B. Acute Toxicity of DA in Rats

As with mice, DA elicits a graded series of behavioral changes in rats. The specific type of behavior and the number of signs seen are dose-dependent, with a time of onset that is inversely proportional to the magnitude of the administered dose. In the adult rat, doses of 2 mg/kg (i.p.) produce transient clinical signs of nervous system disturbances. Soon after injection, these animals exhibit hypoactivity, which is then followed by periods of increased purposeless exploratory behavior. At this concentration, some rats also showed intermittent unilateral scratching of the scapular region by the ipsilateral hind paw. At this concentration, however, rats generally recover within 24 h. Rats administered higher concentrations of DA (4.0–7.5 mg/kg) progress through a characteristic series of behavioral changes. Following the initial period of hypoactivity, these animals develop signs of minor seizure activity, including a hunched posture, piloerection of the vibrissae, wet-dog shakes, and bilateral scratching of the scapular region by the ipsilateral hind paw, chewing and salivation, head nodding, rearing, generalized fine tremors, rigidity, loss of postural control, and circling. Subsequently, rats develop intermittent status epilepticus and, most often, death ensues in severely affected animals (77).

C. Acute Toxicity of DA in Cynomolgus Monkeys

The clinical course of DA toxicity in monkeys can be characterized as four phases; presymptomatic, prodromal, symptomatic, and recovery (78,83). With increasing dose, monkeys tend to show a longer symptomatic period, an increase in the number of behavioral toxicity signs, and a longer duration of clinical symptoms. Oral administrations (5–10 mg/kg), intravenous injections (0.05–0.5 mg/kg), and intraperitoneal dosing (4 mg/kg) of DA produce a myriad of behavioral signs, including anorexia, licking and smacking of the lips, empty mastications, salivation, neck spasms and shaking, retching, vomiting, hypothermia, diarrhea, loss of balance, an unsteady gait, fatigue, fine generalized tremors, and rigidity of movement. Depending on the route of administration and the dose received, the first appearance of symptoms may occur within 2 min following dosing and, in some cases, can last intermittently, for as long as 70 h (78,83,84).

D. Neurohistopathology

Whether administered experimentally to laboratory animals or involuntarily consumed by humans in concentrations which are sufficient to produce seizure activity, domoic acid produces very similar and widespread neuronal damage in rodents (77,85,86) and nonhuman primates (78,87,88). In general, lesions first occur in circumventricular brain regions (area postrema, subfornical organ, arcuate nucleus, subcomissural organ, and organum vaculosum of the lamina terminalis). Subsequently, neuronal degeneration occurs in other structures, including the amygdala, basal ganglia, thalamus, olfactory cortex, cortical mantle (including the neocortex, insular, perirhinal, piriform, and entorhinal cortices), and retina. Hippocampal pyramidal cells are a characteristic in vivo target for DA, with the CA_3 region being most severely affected and lesser degrees of damage being observed in other hippocampal subfields.

In mice, neuronal damage in the hippocampus is evident as early as 4 h following systemic injections of DA (4 mg/kg). Typical early tissue changes revealed with light microscopy include (1) edema in the vicinity of the CA_3 cells, resulting in a spongy vacuolated appearance of the tissue; and (2) the appearance of densely staining pyknotic cells. In areas most affected by the excitotoxin, cells are surrounded by one of two types of dilated processes associated with either the dendritic component of the synapse or astroglial processes (85).

Postmortum evaluation of brain tissues from four patients who died after consuming mussels contaminated with DA showed similar patterns of neuronal damage. Extensive tissue abnormalities were observed in the hippocampal region consisting of severe neuronal necrosis or loss of pyramidal neurons and astrocytosis. Considerable damage was also present in the amygdaliod nucleus in all examined tissues, and necrosis or loss of dentate cells was observed in three of the four patients. Other abnormalities included lesions in the claustrum, secondary olfactory areas, the septal area, and the nucleus accumbens septi. Neuronal damage was also observed in the dorsal medial nucleus of the thalamus, as well as the insular and subfrontal cortex (79).

The similarities in the patterns of neuronal damage observed in humans and other species is not only remarkable, but the availability of human data also serves to validate the use of animal models for studies of both the mechanisms of DA-induced toxicity and the physiological and pathological properties of non-NMDA receptors.

E. Cognition and Behavior

When DA is present in concentrations sufficient to induce seizures, damage to neurons within the CNS is unavoidable. The hippocampal subfields and the entorhinal cortex are interconnected and are known to participate in memory formation. Human patients who presented with abnormal EEG patterns following consumption of DA showed a predominantly anterograde memory disorder, with particular deficits in visuospatial tasks (79). Similar deficits are also observed in rodents following administration of doses of DA sufficient to produce mild seizures. In experimental animals, this type of memory can be assessed using particular maze paradigms, such as the Morris water maze (89), in which an animal must find a submerged ''escape'' platform using only external cues. Studies of DA in both rats and mice have consistently shown that doses capable of producing mild seizures also produce deficits on both acquisition and retention of the water maze task (90,91).

F. Summary

Whole-animal models can be used for both regulatory and investigative studies of both DA and contaminated seafood suspected of containing ASP. In vivo biological detection methods have some unique advantages over chemical or in vitro biological methods and are essential for proper scientific investigation of the pharmacological and toxicological properties of DA, as well as the physiological and pathological significance of non-NMDA receptors. For regulatory purposes or routine toxicity screening, however, the use of whole-animal tests should probably be restricted to those occasions where essential scientific rigor clearly requires an intact animal model.

IV. SUMMARY

The amnesic shellfish toxin, domoic acid, can be reliably detected and quantified using a wide variety of chemical and biological techniques. Accurate detection is required for two main purposes: (1) scientific investigation of the pharmacological and toxicological properties of domoic acid, and (2) detection of contamination in seafood by both regulatory agencies and producers interested in quality assurance for their products. Each of the techniques described in this chapter has application in both of these environments, but each is characterized by both advantages and disadvantages. Selection of an appropriate chemical analysis method often involves balancing the need for sensitivity and selectivity against the inconvenience of elaborate sample preparation or costly equipment. Existing in vitro biological techniques often feature extreme sensitivity but may be characterized by false-positive reactions and are only recently becoming available in commercial test kits appropriate to routine use. In vivo methods are well suited to scientific investigations but are increasingly unacceptable as methods of regulatory evaluation. Ongoing development and refinement of ASP detection methods in all of these three domains is therefore essential to meeting the challenges of reliable detection and the assurance of a safe food supply for the consuming public.

ACKNOWLEDGMENTS

We thank Dr. V. Zitko of the St. Andrews Biological Station, N.B., Canada, for his helpful suggestions and R. Díaz-Trelles for his help. Elements of the work described herein were supported by CICYT, Grant SAF94-0394 and by the Natural Sciences and Engineering Research Council of Canada.

REFERENCES

1. M Falk, JA Walter, PW Wiseman. Ultraviolet spectrum of domoic acid. Can J Chem 67:1421–1425, 1989.
2. M Falk. The infrared spectrum of domoic acid. Can J Spectrosc 33:117–121, 1988.
3. JA Walter, DM Leek, M Falk. NMR study of the protonation of domoic acid. Can J Chem 70:1156–1161, 1992.
4. K Daigo. Studies on the constituents of *Chondria armata*. II. Isolation of an anthelmintical constituent. Yakugaku Zasshi 79:353–356, 1958.
5. M Maeda, T Kodama, T Tanaka, Y Ohfune, K Nomoto, K Ishimura, T. Fujita. Insecticidal and neuromuscular activities of domoic acid and its related compounds. J Pestic Sci 2:27–32, 1984.
6. M Maeda, T Kodama, T Tanaka, H Yoshizumi, T Takemoto, K Nomoto, T Fujita. Structures of isodomoic acids A, B and C, novel insecticidal amino acids from the red alga *Chondria armata*. Chem Pharm Bull 34:4892–4895, 1986.
7. M Maeda, T Kodama, T Tanaka, H Yoshizumi, T Takemoto, K Nomoto, T Fujita. Structures of

domoilactones A and B, novel amino acids from the red alga *Chondria armata*. Tetrahedron Lett 28:633–636, 1987.
8. JLC Wright, MA Quilliam. Methods for domoic acid, the amnesic shellfish poisons. In: GM Hallegraeff, DM Anderson, AD Cembella, eds. Manual on Harmful Marine Microalgae. IOC Manuals and Guides No. 33, UNESCO, 1995, pp 113–127.
9. JLC Wright, M Falk, AG McInnis, JA Walter. Identification of isodomoic acid D and two new geometrical isomers of domoic acid in toxic mussels. Can J Chem 68:22–25, 1990.
10. JA Walter, M Falk, JLC Wright. Chemistry of the shellfish toxin domoic acid: characterization of related compounds. Can J Chem 72:430–436, 1994.
11. AOAC. Mouse bioassay for PSP toxins. Official Methods of Analysis. Association of Official Analytical Chemists, secs 18.086–18.092, 1984.
12. JF Lawrence, CF Charbonneau, C Menard, MA Quilliam, PG Sim. Liquid chromotographic determination of domoic acid in shellfish products using the AOAC paralytic shellfish poison extraction procedure. J Chromatogr 462:349–356, 1989.
13. JF Lawrence, CF Charbonneau, C Menard. Liquid chromatographic determination of domoic acid in mussels, using AOAC paralytic shellfish poison extraction procedure: collaborative study. J Assoc Off Anal Chem 74:68–72:1991.
14. AOAC. Domoic acid in mussels, liquid chromatographic method, first action 1991. Official Methods of Analysis. Association of Official Analytical Chemists, secs 991.26, 1991.
15. JLC Wright, RK Boyd, ASW Defreitas, M Falk, RA Foxall, WD Jamieson, MV Laycock, AW McCulloch, AG McInnis, P Odense, V. Pathak, MA Quilliam, M Ragan, PG Sim, P Thibault, JA Walter, M Gilgan, DJA Richard, D Dewar. Identification of domoic acid, a neuroexcitatory animo acid, in toxic mussels from eastern P.E.I. Can J Chem 67:481–490, 1989.
16. MA Quilliam, PG Sim, AW McCulloch, AG McInnes. High performance liquid chromatography of domoic acid, a marine neurotoxin, with application to shellfish and plankton. Int J Environ Anal Chem 36:139–154, 1989.
17. MA Quilliam, M Xie, WR Hardstaff. (1991) A rapid extraction and cleanup procedure for the determination of domoic acid in tissue samples. Technical Report No. 64. Institute for Marine Biosciences, National Research Council Canada, Halifax, NS.
18. MA Quilliam, M Xie, WR Hardstaff. A rapid extraction and cleanup procedure for the liquid chromatographic determination of domoic acid in unsalted seafood. J AOAC Int 78:543–554, 1995.
19. EG Bligh, WJ Dyer. A rapid method of total lipid extraction and purification. Can J Becam Physiol 37:911–917, 1959.
20. B Grimmelt, MS Nijjar, J Brown, N MacNair, S Wagner, GR Johnson, JF Amend. Relationship between domoic acid levels in blue mussels (*Mytilus edulis*) and toxicity in mice. Toxicon 28:501–508, 1990.
21. JRT Blanchard, RAR Tasker. High-performance liquid chromatographic assay for domoic acid in serum of different species. J Chromatogr/Biomed Appl 526:546–549, 1990.
22. MA Quilliam, K Thomas, JLC Wright. Analysis of domoic acid in shellfish by thin-layer chromatography. Nat Toxins 6:1–6, 1998.
23. R Pocklington, JE Milley, SS Bates, CJ Bird, ASW Defreitas, MA Quilliam. Trace determination of domoic acid in seawater and phytoplankton by high-performance liquid chromatography of the fluorenylmethoxycarbonyl (FMOC) derivative. Int J Environ Analyt Chem 38:351–368, 1990.
24. JY Zhao, P Thibault, MA Quilliam. Analysis of domoic acid and isomers in seafood by capillary electrophoresis. Electrophoresis 18:268–276, 1997.
25. P Thibault, MA Quilliam, WD Jamieson, RK Boyd. Mass spectrometry of domoic acid, a marine neurotoxin. Biomed Env Mass Spectrom 18:373–386, 1989.
26. S Pleasance, M Xie, Y Leblanc, MA Quilliam. Analysis of domoic acid and related compounds by mass spectrometry and gas chromatography/mass spectrometry as N-trifluoroacetyl-O-silyl derivatives. Biomed Environ Mass Spectrom 19:420–427, 1990.
27. I Hynie, ECD Todd, eds. Domoic acid toxicity. Can Dis Wkly Rep 16(Suppl 1E):1–123, 1990.
28. ED London, JT Coyle. Specific binding of [^{3}H]kainic acid to receptor sites in rat brain. Molec Pharmacol 15, 492–505, 1979.
29. JT Coyle. Neurotoxic action of kainic acid. J Neurochem 41:1–11, 1983.

30. TJ Biscoe, RH Evans, PM Headley, M Martin, JC Watkins. Domoic and quisqualic acids as potent amino acid excitants of frog and rat spinal neurones. Nature 255(5504):166–7, 1975.
31. PL Wood, JW Richard, C Pilapil, NP Nair. Antagonists of excitatory amino acids and cyclic guanosine monophosphate in cerebellum. Neuropharmacology 21:1235–8, 1982.
32. GR Stewart, CF Zorumski, MT Price, JW Olney. Domoic acid: a dementia-inducing excitotoxic food poison with kainic acid receptor specificity. Exp Neurol 110(1):127–38, 1990.
33. L Tryphonas, J Truelove, F Iverson, EC Todd, EA Nera. Neuropathology of experimental domoic acid poisoning in non-human primates and rats. Can Dis Wkly Rep 16(Suppl 1E):75–81, 1990.
34. JL Wright, CJ Bir, As de Freitas, D Hampson, J McDonald, MA Quilliam. Chemistry, biology, and toxicology of domoic acid and its isomers. Can Dis Wkly Rep 16(Suppl 1E):21–26, 1990.
35. R Zaczek, JT Coyle. Excitatory amino acid analogues: neurotoxicity and seizures. Neuropharmacology 21:15–26, 1982.
36. A Novelli, J Kispert, A Reilly, V Zitko. Excitatory amino acids toxicity in cerebellar granule cells in primary culture. Can Dis Wkly Rep 16(Suppl. 1E):83–89, 1990.
37. A Novelli, J Kispert, MT Fernandez-Sanchez, A Torreblanca, V Zitko. Domoic acid-containing toxic mussels produce neurotoxicity in neuronal cultures through a synergism between excitatory amino acids. Brain Res 577:41–48, 1992.
38. J Lerma, AV Paternain, JR Naranjo, B Mellstrom. Functional kainate-selective glutamate receptors in cultured hippocampal neurons. Proc Natl Acad Sci USA 90:11688–11692, 1993.
39. A Herb, N Burnashev, P Werner, B Sakman, W Wisden, PH Seeburg. The KA-2 subunit of excitatory amino acid receptors shows widespread expression in brain and forms ion channels with distantly related subunits. Neuron 8:775–785, 1992.
40. H Lomeli, W Wisden, M Kohler, K Keinanen, B Sommer, PH Seeburg. High-affinity kainate and domoate receptors in rat brain. FEBS Lett 307(2):139–43, 1992.
41. B Sommer, N Burnashev, TA Verdoorn, K Keinanen, B Sakmann, PH Seeburg. A glutamate receptor channel with high affinity for domoate and kainate. EMBO J 11(4):1651–6, 1992.
42. B Bettler, J Egebjerg, G Sharma, G Pecht, I Hermans-Borgmeyer, C Moll, CF Stevens, S Heineman. Cloning of a putative glutamate receptor: a low affinity kainate-binding subunit. Neuron 8:257–265, 1992.
43. GL Collingridge, RAJ Lester. Excitatory amino acid receptors in the vertebrate central nervous system. Pharmacol Rev 41:143–210, 1989.
44. S Nakanishi. Molecular diversity of glutamate receptors and implications for brain function. Science 258:597–603, 1992.
45. PJ Conn, JP Pin. Pharmacology and functions of metabotropic glutamate receptors. Annu Rev Pharmacol Toxicol 37:205–237, 1997.
46. SA Lipton. Prospects for clinically tolerated NMDA antagonists: open-channel blockers and alternative redox states of nitric oxide. Trends Neurosci 16(12):527–32, 1993.
47. B Sommer, M Kohler, R Sprengel, PH Seeburg. RNA editing in brain controls a determinant of ion flow in glutamate-gated channels. Cell 67(1):11–9, 1991.
48. R Chittajallu, S Braithwaite, V Clarke, J. Henley. Kainate receptors: subunits, synaptic localization and function. Trends Pharmacol Sci 20:26–35, 1999.
49. R Dingledine, K Borges, D Bowie, SF Traynelis. The glutamate receptor ion channels. Pharmacol Rev 51:7–61, 1999.
50. FM Van Dolah, TA Leighfield, BL Haynes, DR Hampson, JS Ramsdell. A microplate receptor assay for the amnesic shellfish poisoning toxin, domoic acid, utilizing a cloned glutamate receptor. Anal Biochem 245:102–105, 1997.
51. DR Hampson, D Huie, RJ Wenthold. Solubilization of kainic acid binding sites from rat brain. J Neurochem 49:1209–1215, 1987.
52. DR Hampson, RJ Wenthold. A kainic acid receptor from frog brain purified using domoic acid affinity chromatography. J Biol Chem 263:2500–2505, 1988.
53. K Wada, CJ Dechesne, S Shimasaki, RG King, K Kusano, A Buonanno, DR Hampson, C Banner, RJ Wenthold, Y Nakatani. Sequence and expression of a frog brain complementary DNA encoding a kainate-binding protein. Nature 342:684–9, 1989.

54. FA Taverna, DR Hampson. Properties of a recombinant kainate receptor expressed in baculovirus-infected insect cells. Eur J Pharmacol 266:181–6, 1994.
55. L-Y Wang, FA Taverna, X-P Huang, JF MacDonald, DR Hampson. Phosphorylation and modulation of a kainate receptor (GluR6) by cAMP-dependent protein kinase. Science 259:1173–1175, 1993.
56. H Newsome, JF Truelove, L Hierlihy, P Collins. Determination of domoic acid in serum and urine by immunochemical analysis. Bull Environ Contam Toxicol 47:329–334, 1991.
57. DS Smith, DD Kitts. A competitive enzyme-linked immunoassay for domoic acid determination in human body fluids. Food Chem Toxicol 32:1147–1154, 1994.
58. JF Lawrence, C Cleroux, JF Truelove. Comparison of high-performance liquid chromatography with radioimmunoassay for the determination of domoic acid in biological samples. J Chromatogr A 662: 173–177, 1994.
59. M Osada, LJ Marks, JE Stewart. Determination of domoic acid by two different versions of a competitive enzyme-linked immunosorbent assay (ELISA). Bull Environ Contam Toxicol 54:797–804, 1995.
60. I Garthwaite, KM Ross, CO Miles, R Hansen, D Foster, NR Towers. An immunoassay for determination of domoic acid in shellfish and sea water. In: B Reguera, J Blanco, ML Fernandez, T Wyatt, eds., Harmful Algae. Xunta de Galicia and Intergovernmental Oceanographic Commission of UNESCO, 1998, pp 559–562.
61. H Minami, M Sugawara, K Odashima, Y Umezawa, M Uto, EK Michaelis, T Kuwana. Ion channel sensors for glutamic acid. Anal Biochem 63:2787–2795, 1991.
62. SB Fountain, Y-LT Ting, TJ Teyler. The in vitro hippocampal slice preparation as a screen for neurotoxicity. Toxic in Vitro 6:77–87, 1992.
63. I Bureau, S Bischoff, SF Heinemann, C Mulle. Kainate receptor-mediated responses in the CA1 field of wild-type and GluR6-deficient mice. J Neurosci 19:653–63, 1999.
64. H Kamiya, S Ozawa. Kainate receptor-mediated inhibition of presynaptic Ca^{2+} influx and EPSP in area CA1 of the rat hippocampus. J Physiol (Lond) 509:833–45, 1998.
65. TM Polischuk, CR Jarvis, RD Andrew. Intrinsic optical signaling denoting neuronal damage in response to acute excitotoxic insult by domoic acid in the hippocampal slice. Neurobiol Dis 4:423–37, 1998.
66. DK Patneau, L Vyklicky, ML Mayer. Hippocampal neurons exhibit cyclothiazide-sensitive rapidly desensitizing responses to kainate. J Neurosci 13:3496–3509, 1993.
67. MT Fernandez, V Zitko, S. Gascon, A. Novelli. The marine toxin okadaic acid is a potent neurotoxin for cultured cerebellar neurons. Life Sci 49:PL157–162, 1991.
68. MT Fernandez-Sanchez, A Garcia-Rodriguez, R Díaz-Trelles, A Novelli. Inhibition of protein phosphatases induces IGF-1–blocked neurotrophin-insensitive neuronal apoptosis. FEBS Lett 398:106–112, 1996.
69. A Novelli, JA Reilly, PG Lysko, RC Henneberry, Glutamate becomes neurotoxic via the N-methyl-D-aspartate receptor when intracellular energy levels are reduced. Brain Res 451:205–212, 1988.
70. MT Fernandez-Sanchez, A Novelli, Characterization of neurotoxicity by the amnesic shellfish toxin domoic acid in cultured neurons. Neurologia 11:29–39, 1996.
71. JG Green and JT Greenamyre. Bioenergetics and glutamate excitotoxicity. Prog Neurobiol 48:613–634, 1996.
72. S Finkbeiner, CF Stevens. Application of quantitative measurements for assessing glutamate neurotoxicity. Proc Natl Acad Sci USA 85:4071–4074, 1988.
73. A Garcia-Rodriguez, MT Fernandez-Sanchez, MI Reyero, JM Franco, K Haya, J Martin, V Zitko, C Salgado, F Arevalo, M Bermudez, ML Fernandez, A Miguez, A Novelli. Detection of PSP, ASP and DSP toxins by neuronal bioassay; comparison with HPLC and mouse bioassay. In: B Reguera, J Blanco, ML Fernandez, T Wyatt, eds. Harmful Algae. Xunta de Galicia and Intergovernmental Oceanographic Commission of UNESCO, 1998, pp 554–557.
74. AJ Silva, JH Kogan, PW Frankland, S Kida. CREB and memory. Annu Rev Neurosci 21:127–148, 1998.
75. NJ Wolf. A structural basis for memory storage in mammals. Prog Neurobiol 55:59–77, 1998.
76. RAR Tasker, BJ Connell, SM Strain. Pharmacology of systemically administered domoic acid in mice. Can J Physiol Pharmacol 69:378–382, 1991.

77. L Tryphonas, J Truelove, E Nera, F Iverson. Acute neurotoxicity of domoic acid in the rat. Toxicol Pathol 18:1–9, 1990.
78. L Tryphonas, J Truelove, F Iverson. Acute parenteral neurotoxicity of domoic acid in cynomolgus monkeys (*M. fascicularis*). Toxicol Pathol 18:297–303, 1990.
79. JS Teitelbaum, RJ Zatorre, S Carpenter, D Gendron, AC Evans, A Gjedde, NR Cashman. Neurologic sequelae of domoic acid intoxication due to the ingestion of contaminated mussels. N Engl J Med 322:1781–1787, 1990.
80. ML Fernandez, AD Cemebella. Mammalian bioassays. In: GM Hallegraeff, DM Anderson, AD Cembella, eds. Manual on Harmful Marine Microalgae. IOC Manuals and Guides No. 33, UNESCO, 1995, pp 213–224.
81. RAR Tasker, SM Strain. Synergism between NMDA and domoic acid in a murine model of behavioral neurotoxicity. Neurotoxicology 19:593–598, 1998.
82. A Tasker, SM Strain. Morphine differentially affects domoic acid and kainic acid toxicity in vivo. NeuroReport 3:789–792, 1992.
83. L Tryphonas, J Truelove, E Todd, E Nera, F Iverson. Experimental oral toxicity of domoic acid in cynomolgus monkeys (*Macaca fascicularis*) and rats. Preliminary investigations. Food Chem Toxicol 28:707–715, 1990.
84. L Tryphonas, F Iverson. Neuropathology of excitatory neurotoxins: The domoic acid model. Toxicol Pathol 18:165–169, 1990.
85. SM Strain, RAR Tasker. Hippocampal damage produced by systemic injections of domoic acid in mice. Neuroscience 44:343–352, 1991.
86. C Chiamulera, S Costa, E Valeria, A Reggiani. Domoic acid toxicity in rats and mice after intracerebroventricular administration: comparison with excitatory amino acid antagonists. Pharmacol Toxicol 70:115–120, 1992.
87. LC Schmued, AC Scallet, W Slikker Jr. Domoic acid-induced degeneration in the primate forebrain revealed by degeneration specific histochemistry. Brain Res 695:64–70, 1995.
88. AC Scallet, Z Binienda, FA Caputo, S Hall, MG Paule, RL Rountree, L Schmued, T Sobotka, W Slikker Jr. Domoic acid-treated cynomolgus monkeys (*M. fascicularis*): effects of dose on hippocampal neuronal and terminal degeneration. Brain Res 627:307–313, 1993.
89. RGM Morris. Spatial location does not require the presence of local cues. Learn Motiv 12:239–260, 1981.
90. BF Petrie, C Pinsky, NM Standish, R Bose, GB Glavin. Parenteral domoic acid impairs spatial learning in mice. Pharmacol Biochem Behav 41:211–214, 1991.
91. RJ Sutherland, JM Hoesing, IQ Whishaw. Domoic acid, an environmental toxin, produces hippocampal damage and severe memory impairment. Neurosci Lett 120:221–223, 1990.

19

Ciguatera Toxins: Chemistry and Detection

Sonia E. Guzmán-Pérez and Douglas L. Park
Louisiana State University
Baton Rouge, Louisiana

I. INTRODUCTION

The term *ciguatera fish poisoning* (CFP) was first used in the Caribbean to describe an intoxication induced by the ingestion of a marine snail, *Turbo pica* (called cigua by the Cuban natives), involved in food poisoning outbreaks during the Spanish Conquest (1). Now the word is used to describe the intoxication caused by consumption of certain fish, primarily reef fish from the tropical and subtropical areas of the Caribbean Sea and the Pacific Ocean, which have accumulated specific toxins via their diet.

CFP is considered a world health problem (2). An accurate assessment of the incidence of CFP is not available; however, it has been estimated that approximately 50,000 cases occur each year (3).

The toxins produce painful and unusual symptoms in humans (4). The more than 175 ciguateric symptoms that have been reported (5) can be classified primarily in four categories: gastrointestinal, neurological, cardiovascular, and general symptoms. This multiphase intoxication is thought to be due to the presence of different ciguatera-related toxins at different ratios (6).

II. CHEMISTRY

Several toxins may be responsible for CFP (7–13). Recent studies suggest that an excess of 20 toxins may be involved in CFP (14–16). Even though CFP toxins have been defined as those interacting with sodium channels (17), the involvement of other toxins cannot be excluded.

The toxin isolated from moray eels by Scheuer's group at the University of Hawaii and characterized to be a polyether compound was named Pacific ciguatoxin-1 (P-CTX-1). It is considered to be the principal toxin involved in CFP (18). The structure of P-CTX-1 was elucidated by Yasumoto's group in 1989 (19). It is a highly oxygenated, white solid lipid with a molecular weight (MW) of 1111.7 ± 0.3 Da, determined by spectrometry. Its toxicity (LD_{50}) was initially found to be 0.45 μg/kg (intraperitoneally [ip]) to mice; however, Lewis and coworkers later proposed a toxicity of 0.25 μg/kg from an extract of moray eel (20). P-CTX-1 is stable in water, pyridine, acetic acid, 1N NaOH at 100°C for 10 min, exposure to sunlight for 1 h, and in metallic solutions for 36 h. The toxicity is lost with 1N HCl for 10 min and with treatment with acetic acid in pyridine (21).

Lewis and coworkers isolated two less polar CTXs (P-CTX-2 and P-CTX-3) from moray eels. They are less potent to mice (LD_{50} 2.3 and 0.9 μg/kg, respectively, versus 0.25 μg/kg of P-CTX-1) (20). Using [^{1}H] nuclear magnetic resonance (NMR) and mass spectrometry (MS), it was found that the diastereoisomers P-CTX-2 and P-CTX-3 were less oxidized forms than P-CTX-1 (6), lacking the secondary hydroxyl group on C-54 (20). Figure 1 shows the structure of P-CTX-1 and its congeners.

Nine gambiertoxins (GTXs) have been isolated from wild *Gambierdiscus toxicus*; however, just the structure of GTX-4B and its stereoisomer, GTX-4A (52 epi-GTX-4B), are known (Figure 2). GTX-4C (unknown structure) was identified as the GTX with the highest mouse lethality (22), whereas GTX-4B has a toxicity of less than 10 times that of P-CTX-1 (20,22,23).

It has been proposed that the type B toxins P-CTX-1 and P-CTX-3 may arise from the oxidative biotransformation of GTX-4B (20,23). P-CTX-2 (52-*epi*-CTX-3), with a type A backbone, may arise from the biotransformation of GTX-4A. Both CTXs and GTXs have a spiroacetal associated with carbon 52 that can undergo spiroisomerization (24,25) under acidic conditions either during their path through the food chain (e.g., the acid stomach of certain fish) or during purification (e.g., acidic solvents such as chloroform).

A biotransformation scheme of GTXs and CTXs has been proposed by Lewis and Holmes (Figure 3) (26). The horizontal lines symbolize biotransformation of the toxins, whereas the vertical arrows represent the acid-catalyzed spiroisomerization. P-CTX-4 (52-*epi*-P-CTX-1) is a presumed toxin (26). The structural closeness between GTX-4B and P-CTX-1 indicated that the former toxin is oxidized at two terminals of the molecule while moving up to higher trophic levels. The result strongly supports that *G. toxicus* is the origin of CTXs (27). The biotransforma-

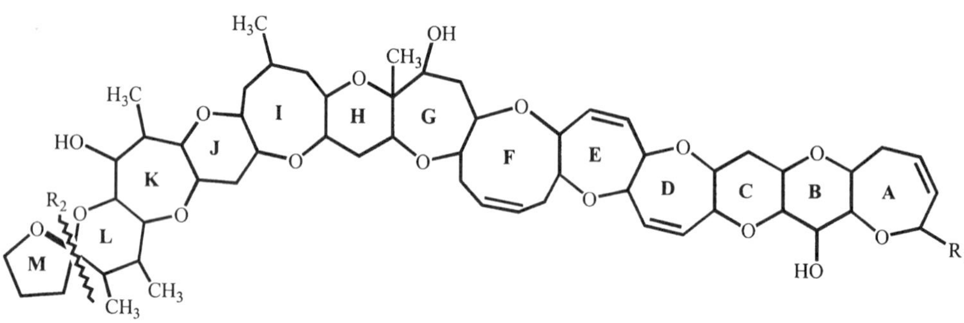

	R1	R2
P-Ciguatoxin-1 (P-CTX-1)	-CH=CH-CH(OH)-CH2-OH	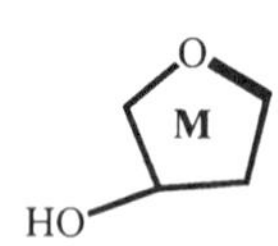
P-Ciguatoxin-2 (P-CTX-2)	-CH=CH-CH(OH)-CH2-OH	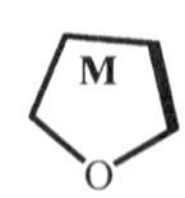
P-Ciguatoxin-3 (P-CTX-3)	-CH=CH-CH(OH)-CH2-OH	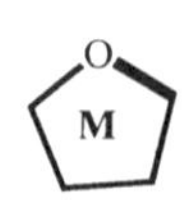

Figure 1 Structure of Pacific ciguatoxins.

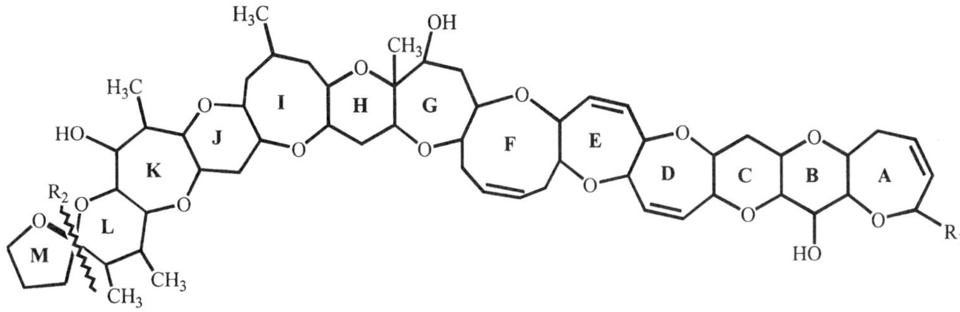

	R1	R2
Gambiertoxin-4A (GTX-4A)	-CH=CH-CH=CH2	
Gambiertoxin-4B (GTX-4B)	-CH=CH-CH=CH2	

Figure 2 Structure of gambiertoxins 4A and 4B.

tion of GTX-4B to P-CTX-3 to P-CTX-1 remains to be established, but it likely occurs in the liver of fish (20).

Two new toxic fractions, RGT-1 (1.2 mg) and RGT-2 (0.7 mg), were isolated from 1100 L of *G. toxicus* strain RGI-1 from the Rangiroa Atoll, French Polynesia (28,29). The structure of RGT-1 (gambierol, Figure 4) resembles that of P-CTX-1; both have many ether units aligned in a ladder shape. Since gambierol also resembles the symptoms in mice (diarrhea, dyspnea, and convulsions before death) and chromatographic properties of P-CTX-1 (28), it is possible that it is implicated in CFP (30). Its MW was reported to be 779.43 Da and its LD_{50} to mice (ip) was found to be 50 μg/kg (28).

RGT-2 structurally resembles GTX-4B isolated from wild *G. toxicus*, as shown by [^{1}H] NMR spectra analysis (29). This toxin was named ciguatoxin-3C (CTX-3C) (1,2,3,4-nor-E-homo-GTX-4B) (Figure 5). An original feature is the absence of the butadiene side chain typical of CTXs; however, rings D, F, and G agree well with those of P-CTX-1. Its LD_{50} to mice (ip)

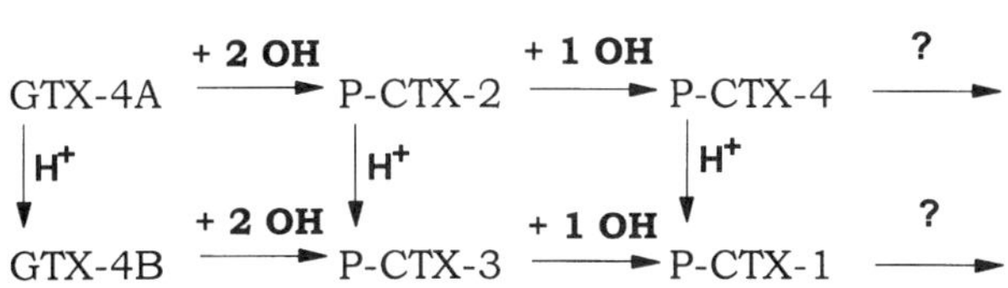

Figure 3 Biotransformation of gambiertoxins (GTXs) and Pacific ciguatoxins (P-CTXs). GTX-4A and/or GTX-4B produced by *G. toxicus* are likely to give rise to P-CTX-2, P-CTX-3, P-CTX-1, and the putative P-CTX-4 (52-*epi*-P-CTX-1) in fish. Acid-catalyzed spiroisomerization (vertical arrows) results in toxins with the thermodynamically favored type B backbone predominating (i.e, GTX-4B, P-CTX-3, or P-CTX-1). The ? indicates suspected further biotransformation to more polar and probably lower potency forms. The relative importance of the three possible acid-catalyzed spiroisomerizations indicated in this model remains to be determined. (Adapted from ref. 26.)

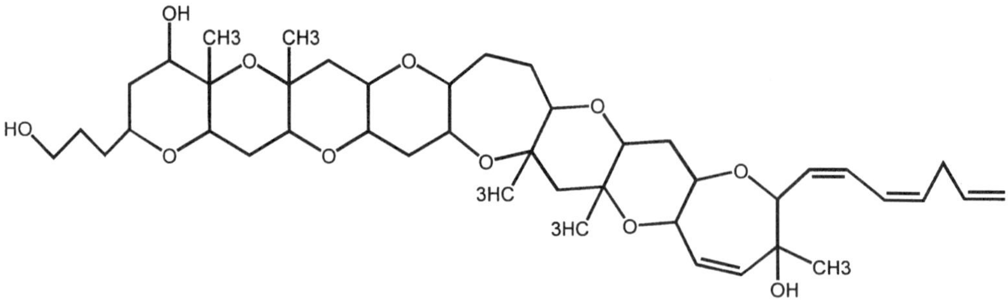

Figure 4 Structure of gambierol.

was reported to be 1.3 μg/kg, and its MW 1023.568 D (29). CTX-3C has been recently detected in herbivorous fish from Tuamotu Island (31). The structural resemblance of CTX-3C and GTX-4B and of moray eel–derived P-CTX-1 may confirm the theory that *G. toxicus* is the true source of toxins implicated in CFP.

Several new CTXs have been characterized from horse-eye jack (*Caranx latus*) caught in the French West Indies in the Caribbean. Each of these toxins was found to coelute with the P-CTXs at each chromatographic step prior to high-performance liquid chromatography (HPLC) (20), suggesting that these toxins are related but represent a new family of CTXs. In fact, Crouch's group has recently indicated that Caribbean CTX-1 (C-CTX-1) and P-CTX-1 have closely related structures (32); however, the exact differences in their structure are dependent on purification of sufficient C-CTX-1 for NMR studies (33). Toxins separated by HPLC include C-CTX-1 (LD_{50} 3.6 μg/kg [ip] mouse; MW 1141.7 D), C-CTX-2 (LD_{50} 1 μg/kg [ip] mouse; MW 1141.7 Da), a minor toxin, a sleep-inducing fraction, and a hydrophobic fast-acting toxin. Using ion-spray mass spectrometry, C-CTX-1 and C-CTX-2 were found to have the same mass, which differs from P-CTXs by 30 Da. Intraperitoneal injection of each toxin into mice induced signs typical of site 5 channel activator toxins such as P-CTXs and brevetoxins (33). C-CTX-1 is less polar than P-CTX-1, and so it may accumulate more in fish owing to greater assimilation and lower depuration rates (26). The presence of this new family of CTXs has important implications for the detection of ciguateric fish. Antibody detection based on P-CTX-1 or its fragments may not be suitable for detecting C-CTXs (33); however, if both groups of toxins are structurally related in at least some portions, cross reactivity may allow for management of risks from these naturally occurring toxicants.

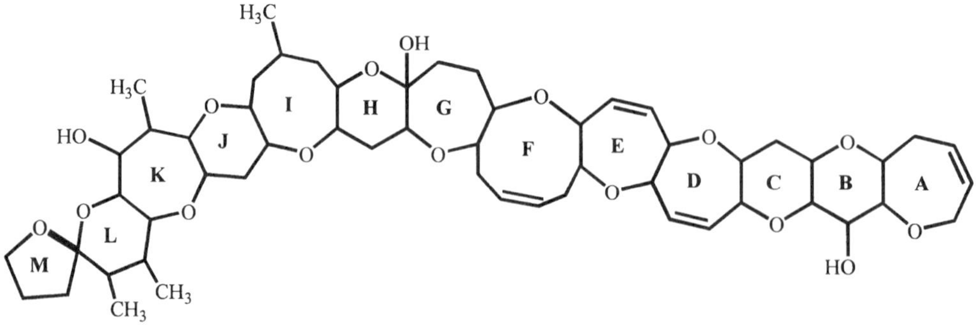

Figure 5 Structure of ciguatoxin-3C.

Marquais and coworkers have recently reported the isolation of two main pure toxins from *Caranx latus* flesh caught in the Caribbean. Both toxins have the same MW (1142 Da), raising the possibility that they are diastereoisomers. The close resemblance to the molecular weight of P-CTXs (MW 1112 Da) suggests that they may be closely related (34).

III. DETECTION

It is necessary to provide sufficient quantities of purified toxins for studies on biosynthesis, structural analysis, pharmacology, biotransformation, and detection (35). Rapid detection of CTX-like compounds has been hampered by the lack of standard material. This is due to the low amounts of CTXs present in fish. Even in the moray eel, which is the most toxic species, the CTX content is extremely low; usually only several μg/kg in whole bodies (23,36). Thus, purification of large quantities of CTXs from fish has not been successful.

Production of GTXs from mass-cultured dinoflagellates was once thought to be an alternative source of toxins. However, obtaining large quantities of the dinoflagellate CTX congeners has been limited by the ability to identify and mass culture a clone of *G. toxicus* or possibly other dinoflagellates capable of producing high levels of these toxins (35). Production of GTXs in cultured *G. toxicus* appears to be strain dependent, with most clones only producing water-soluble maitotoxins (37). So far, no isolate has produced abundant quantities of GTXs in culture, with reported yields of less than 1 mouse units (MU)/million cells (37,38). It might be possible that the size of *G. toxicus* populations does not reflect the potential for these populations to cause CFP. Only relatively low concentrations of GTXs have so far been detected from cultured compared with wild *G. toxicus* cells (39).

P-CTX-1 typically contributes approximately 90% of the toxicity of ciguateric carnivorous fish in the Pacific (6,22); therefore, it should be considered the primary target for a screening program. It is then necessary to develop a method that can rapidly and selectively screen P-CTX-1 which is present between 0.1 and 5 ppb in the flesh of fish that cause CFP (36) or focus on a group of structurally similar toxins using the toxin of highest potency as the safety factor (40). Based on a twofold risk factor to secure public health, the assay needs to be capable of reliably detecting P-CTX-1 in fish flesh at 0.05 ppb (50 ppt) and above (41). In a study developed by Lewis and Sellin, P-CTX-1 was detected only at levels higher than 0.5 ppb in fish flesh using the mouse bioassay (42).

The method used to measure CTXs should (1) be simple and readily available, (2) be easy to use and interpret, (3) have a long shelf-life, (4) have a high affinity for P-CTX-1 and related toxins, (5) have a linear response over the CTX concentration, (6) be accurate (±20%) at the level of human intoxication, (7) have an acceptable cost to the consumer, (8) be able to test a large number of samples in a short time, (9) be able to distinguish between toxic and nontoxic fish, (10) have few false negatives and positives, (11) be reproducible and repeatable, (12) function for all fish regardless of the region, harvest method, or handling procedures, and, if possible, (13) confirm the identity for the toxic products (41,43).

A. In Vivo Assays

Ciguatera toxins are odorless, tasteless, and cannot be detected by any simple chemical test. Old "tests" for toxicity testing, such as toxic fish turning a silver coin black when cooked, avoidance of toxic fish by flies or ants, or rubbing the liver on the gums to determine if it caused a tingling effect, are unreliable (44). Thus, bioassays have been used traditionally to monitor suspicious toxic fish; however, they have disadvantages such as low sensitivity, relatively large amounts of samples required for analysis, lack of specificity to distinguish toxins, and prolonged times required for analysis (45,46).

A major area of concern has been the development of simple assay methods for the detection of CTXs, especially for use in mass screening of fish from endemic areas (45). Of particular importance was the search for an animal whose route of exposure was oral so that it could reflect the mechanism of human intoxication. Of the 37 species of animals tested for use in bioassays, only cats, mongooses, mice, turtles, and crayfish appeared to respond noticeably to oral administration of toxic fish flesh (47).

Cats have been used but tend to hide when ill, regurgitate toxic foods, and also require large doses (up to 10% of body weight), thus making them impractical for the screening of numerous or small fish. Furthermore, cats are readily susceptible to viral infections that can lead to a extremely high mortality rate (48).

Mongooses (*Herpestes mangosta*) have been used to screen for toxicity by feeding them with raw fish (15% of body weight) or an equivalent amount of extract residue mixed with eggs. They were selected for their local availability in Hawaii, reluctance to regurgitate toxic flesh, and symptomatology resembling that of humans. Mongooses often carry parasites or diseases which could complicate this bioassay (49).

The brine shrimp (*Artemia* sp.) bioassay was developed with the aid of confirmed ciguateric fish from the Caribbean. Rayner demonstrated that at least one ciguatera toxin caused a marked increase in the passive transport of sodium ions across certain cell membranes (50). Based on this fact and with the known well-developed sodium pump of this animal, it was thought that it could be highly sensitive to such a toxin. This prompted an investigation of the brine shrimp as a possible test organism. The brine shrimp bioassay was carried out using fresh larvae in artificial seawater. They were exposed to extracts and were observed frequently for 20 min and then at 1, 2, 8, and 24 h. Many shrimp exhibited erratic, nonproductive swimming movement prior to death (50). Even though the small amount of suspect material required, the low cost, and speed make this an attractive option to the mouse bioassay (51), this assay has not been widely used for screening fish in the marketplace. However, several investigators report its usefulness as a screening tool in the laboratory (16,52,53).

Mosquitoes (*Aedes aegypti*) have also been tested to screen ciguateric fish. Each portion of extract of flesh sample was suspended in phosphate-buffered saline solution containing 0.5% gelatin and 5% heated calf serum. A volume of 0.5 μL of twofold serial dilutions was intrathoracically injected into eight groups of 10 mosquitoes each. A control group just received eluant. Mortality was observed 1 h after injection. This assay can be used for acute toxicity determination and is a good tool for laboratory use owing to the small amount of sample needed, the low cost of mosquitoes, and the rapidity in obtaining a reliable LD_{50} (2 h) (54).

Compared to cats and mongooses, mice offer a more reliable bioassay model. They are easily housed, can be readily dosed in several ways, and manifest diverse symptoms, some of which are similar to those of humans. The mouse bioassay is carried out with approximately 20 mg of a diethyl ether test fraction suspended in Tween-saline and using 20 $\pm$ 2 g mice of either sex. After intraperitoneal injections of the test samples, mice are observed for signs of intoxication together with measurement of rectal body temperature and time to death. The relationship between dose and time to death is used to quantify each fraction (Table 1) (55). For a mixture of CTXs, the dose–time of death relationship is $\text{Log MU} = 2.3 \text{ Log}(1 + T^{-1})$ (56). These dose versus time to death relationships should be used just as guidance, since it is necessary to obtain specific relationships using the same mice as those used for the experiment and for each toxin analyzed. This specific dose versus time to death relationship is obtained by injecting approximately 12 mice with doses between 1 and 20 MU of toxin (55).

Table 1 Effects of Ciguatoxin and Its Congeners Administered Intraperitoneally to 20-g Mice (for doses of toxin ≤20 MU)

Toxin	LD50 (ip) (μg/kg)	MU (ng)	Signs of intoxication	Dose vs time to death relationship	Min/max death time
P-CTX-1[a]	0.25	5	Hypothermia <33°C, piloerection, diarrhea, lacrimation, hypersalivation, dyspnea, wobbly upright gait, gasping, terminal convulsions with tail arching, death from respiratory failure	Log MU = 3.3 Log$(1 + T^{-1})$	37 min/ ~24 h
P-CTX-2[a]	2.3	9	As for P-CTX-1 plus progressive hind limb paralysis	Log MU = 2.4 Log$(1 + T^{-1})$	53 min/ ~100 h
P-CTX-3[a]	0.9	18	As for P-CTX-1 plus progressive hind limb paralysis	Log MU = 3.9 Log$(1 + T^{-1})$	60 min/ ~26 h
CTX-3C[b]	1.3	26	—	—	—
GTX-4B[c]	4.0	80	As for P-CTX-1 plus hind limb paralysis	—	—

1 MU = LD_{50} dose for a 20-g mouse.
[a] Refs. 20 and 23.
[b] Ref. 29.
[c] Refs. 22, 23, and 37.
Source: Adapted from ref. 55.

Laboratory mice have become the standard test animal for a number of toxins such as paralytic shellfish poisoning toxins (57), and the intraperitoneal injection assay is considered to be reliable (58). Some disadvantages of this bioassay are lack of specificity for CTXs, the time-consuming process of obtaining the lipid-soluble extracts, and the costs associated with maintaining a mouse colony (59).

In the search for a nonrodent species possibly more sensitive to CTXs, a comparative study with chickens was made. Chickens are known to have fewer nonspecific sudden lethal reactions and are less expensive to obtain and maintain. After injection into the axillary vein in one wing, the animals were autopsied to determine whether or not death was due to physical causes. In 7 of 10 comparisons of intravenous toxicity in mice and chickens, the chickens were more sensitive to CTXs (60). Vernoux and coworkers tested again the chicken assay (61). A liver mixture (10% of body weight) was force-fed by pushing it into the chicken's crop through a tube. Response to liver feeding was checked after a 48-h period. It was found that chickens did not require as large a test sample, were more sensitive, easier to handle, and did not regurgitate the sample. Symptoms expected include hypersalivation, decrease in weight, and acute motor ataxia (61). As liver is a more toxic tissue than flesh, it has been used for the screening of toxic fish at Saint Barthélémy since 1980 (62).

B. In Vitro Assays

Immunoassays

Probably the immunological detection of no other marine toxin has evolved as much as with CTXs and related polyethers (63). An immunological approach to examine low-dalton lipid marine toxins was initiated in 1977 with the development of a modified radioimmunoassay (RIA) employing sheep anti-CTX antibody (64,65). The preparation of the antibody was made possible from a highly purified CTX extract (isolated from toxic moray eels) conjugated to human serum albumin as a conjugate. This antibody was coupled to ^{125}I as a label and used for the RIA.

In 1982, Kimura and coworkers evaluated the above-mentioned RIA in its ability to distinguish toxic from nontoxic fish tissues (65). Of the total of 42 clinically documented specimens, involving 12 species, 39 (93%) gave borderline or positive results. Based on this information, the researchers developed a 2-year marketplace monitoring program for the amberjack (*Seriola dumerili*), one of the major ciguatera-implicated species in Hawaii. The RIA was used to screen a total of 5529 samples for a total weight of 40,662 kg. Eight hundred twenty-five specimens (15%) were rejected during that period, and as a result of the program, no single case of CFP was caused by RIA-tested fish (65,66). However, some false positives were reported and have subsequently been attributed to the presence of other marine toxins with similar structural properties to CTXs (64,67,68). Although proven to be effective for screening of *S. dumerili*, the RIA is economically impractical for testing fish weighing less than 9 kg (65).

An enzyme immunoassay (EIA) using the same sheep anti-CTX was developed in 1983 (69,70) to avoid the costly and tedious RIA procedure, but the antibody was coupled to horseradish peroxidase (HRP) instead of ^{125}I (69). The method included the preparation of small sample disks (3-mm thickness × 3-mm diameter) placed in a well of a 96-well polystyrene microtiter plate, fixed with 0.3% hydrogen peroxide (H_2O_2) in methanol (MeOH), labeled with a 1:1000 dilution of anti–CTX-HRP conjugate, and quantified with 4-chloro-1-naphthol as a substrate at 405 nm (69). The colorimetric determination of absorbance following the enzymatic reaction gave values that could be compared to controls. Values ≥ 1.5 were positive (toxic) and those ≤ 1.3 were negative (nontoxic) (71). The major contribution of this procedure was the determination of the cross reactivity of similar marine toxins structurally characterized as polyethers (19,70,72,73). The cross reactivity between CTXs and polyethers such as okadaic acid, brevetoxin, maitotoxin, and monensin (70) with sheep anti-CTX suggested a need for the development of monoclonal antibodies for increased specificity to purified CTX (70,74) and okadaic acid (75).

Hokama and coworkers compared the EIA with the RIA (69). The EIA appeared to be less sensitive and/or more specific than the RIA in the 78 *S. dumerili* samples examined. With the RIA, 25 samples were in the positive and borderline ranges, whereas with the EIA, only 9 samples were in that range. This may be due, in part, to the difference in molecular size between ^{125}I and HRP and to the position of the latter in the sheep–anti-CTX conjugate (69).

The EIA procedure using fish tissue proved to be tedious and time consuming, although slightly less cumbersome than the RIA; however, it clearly differentiated between clinically documented toxic and nontoxic fish tissues of several different species with a reported sensitivity of 5 pg free CTX (69).

A simplified stick enzyme immunoassay (S-EIA) was then developed (45,74–76) using correction fluid-coated skewered bamboo sticks (length 21 cm × 3 mm diameter) (45,71,74–76). This assay used the same polyclonal sheep–anti–CTX-HRP as the EIA procedure (63). Fish samples were poked with the skewer end of the stick coated with correction fluid at the dorsal, ventral, anterior, and posterior sections of a side or both sides of the fish. Then the skewer was immersed in the fixing solution (MeOH-H_2O_2) and later in the anti–CTX-HRP. The substrate (4-chloro-1-naphthol with H_2O_2) was added and the intensity of the color was measured by comparing with standardized colors as follows: 0, essentially no color; 1.0–1.5, slightly bluish purple; 1.5–2.0, lightly bluish purple; 2.0–2.5+, moderately bluish purple; 3.0–

5.0+, moderately to intense bluish purple color. The results of the colored reactions were scored as follows: 0–2, negative; 2.1–2.4, borderline; values greater than 2.5, positive (45,71).

The adaptation of the EIA to a rapid stick test appeared to be feasible (45). This method showed to be not only simple but also specific, sensitive (>1 ng pure CTX/mL methanol), and performed with no special instrumentation; however, it was found to be impractical in the field and on board ships (77).

Because of reported false-positive results experienced in the above-mentioned methods, which utilized the same polyclonal sheep anti-CTX antibody preparation, Hokama and coworkers developed monoclonal antibodies (mAbs) to CTX by the hybridoma technique using BALB/C mice and a nonimmunological secreting X63-Ag8 strain of myeloma cells for fusion in an attempt to overcome the cross reactivity phenomenon (75). They designated the monoclonal antibody from the hybridoma as mAb-CTX (5C8). However, on further testing, the 5C8 mAb-CTX was revealed to be a mixture of polyclonal and monoclonal antibodies. The S-EIA remained the same as the original format except that the HRP was now conjugated to this new anti-CTX antibody (78). This new S-EIA format was used to examine clinically implicated fish and to prescreen two fish species, *Caranx* sp. and *S. dumerili*, supplied by sports fishermen in Hawaii (79). Fish samples with S-EIA values of ≤1.2 were considered to be edible (nontoxic), whereas S-EIA values of ≥1.3–1.9 were considered to be borderline; S-EIA values ≥2.0 were considered to be positive (toxic). Nevertheless, all samples ≥1.3 were considered to be nonedible and rejected. All of the clinically implicated fish from the Department of Health gave S-EIA values ≥1.3. The *Caranx* sp. and *S. dumerili* that were considered safe (≤1.2) and consumed after testing gave no false-negative results (79). The investigators reported more specificity (less cross reactivity) and lower background color in assessing toxicity values in toxic samples. They noticed that there was still a moderate cross reactivity of the anti-CTX antibody with okadaic acid, which was not surprising, since it has been demonstrated that part of the structure of CTX is very similar to okadaic acid (20,79,80) and due to the fact that okadaic acid may accumulate in the flesh of contaminated fish (81).

This procedure has been used extensively for fish testing since 1986 for a variety of analyses of reef fish and offshore larger species such as *Caranx* sp. and *Seriola* sp. obtained by sports fishermen in Hawaii (67). The S-EIA procedure using 5C8 anti-CTX proved to be relatively specific, sensitive, and simple to use in the laboratory without any special instrumentation; unfortunately, it was not adaptable for use in the field where high temperatures and sunlight prevailed (79). The success of this procedure in part resides in the selective extraction or adsorption of the toxin in the tissue by a constituent(s) in the correction fluid (45).

In 1990, Hokama adopted the particulate solid-phase immunobead assay (S-PIA) approach, known as the "paddle test," in dealing with the detection of CTX and related polyethers. In this case, the antigen-antibody complex was evaluated through the ability of colored latex beads, to which the 5C8 anti-CTX was bound, to recognize toxins attached to correction fluid–coated bamboo paddles previously exposed to toxic fish (77). When compared to the stick method (45) in evaluating 26 documented toxic samples, a 100% agreement was found; but the S-PIA was shown to be more sensitive than the S-EIA. The results suggested the applicability of the S-PIA for field use (77).

Based on the same principle, HawaiiChemtect International, Pasadena, California, has developed a kit in which the original paddle test format was modified to a rapid S-PIA (Ciguatect) for the detection of toxins associated with ciguatera and diarrheic shellfish poisoning with application to screening programs (82,83). A plastic strip with a membrane attached on one end was used instead of a bamboo paddle device. The presence or absence of the toxins was determined by binding the toxins to the membrane and exposing the membrane-bound toxins to the anti-CTX–colored latex bead complex. The intensity of the color, compared with control strips, qualitatively represented the presence of the toxins. The level of detection was 1 ng okadaic acid equivalents (59,84). The S-PIA can be used at the harvesting, receiving docks, processing plants, distribution organizations, retail outlets, and regulatory agencies (84). The Ciguatect kit can be used for a pass/fail response depending on the level of sensitivity desired. One advantage of this method is that the kit can be used outside the laboratory and by untrained personnel (84,85). This kit has been used to map toxic and nontoxic harvesting areas in Hawaii (84). Since the antibodies used in testing for CFP in the Ciguatect kit were shown to be a mixture of polyclonal and monoclonal antibodies (5C8), studies are underway on the development and characterization of monoclonal antibodies to ciguatoxin-like compounds (S.E. Guzmán-Pérez and D.L. Park, unpublished results).

The necessity to confirm test results prompted the development of a rapid extraction method (REM, HawaiiChemtect International, Pasadena, California). The conventional extraction procedure takes approximately 48 h to perform and requires an overnight freezer precipitation step, solvent partition, and the purification on a silica gel column (86). The REM utilizes a chloroform, water, and methanol mixture to vary the polarity so that the toxins can be extracted and partitioned in the same step. This procedure takes about 30 min to perform. The level of detection for the S-PIA using this format is 50 pg okadaic acid equivalents/g fish flesh. REM extracts have been used to confirm toxicity (brine shrimp bioassay) or could be used for chemical confirmation of identity; that is, HPLC procedures (59,85). With the S-PIA, the toxins can be detected directly on fish flesh or after extraction of the toxins.

Park and coworkers submitted the S-PIA for a minicollaborative study in order to assess precision parameters through the Association of Official Analytical Chemists (AOAC) International/International Union for Pure and Applied Chemistry (IUPAC) interlaboratory validation mechanism (87). For determination of CTX and related polyethers, parrotfish, surgeonfish, and amberjack fillets and fish extracts (REM) test portions containing various concentrations of toxins were distributed to participating laboratories for analysis. The method demonstrated excellent reproducibility values; that is, the ability of different analysts to read the same color intensity on test strips. In general, it was found that the method was easy to run and to differentiate between distinct color intensities on test strips; AOAC recommended a full collaborative study (85).

The early studies all employed polyclonal antibodies to CTX. This is a disadvantage, since the animals must be immunized constantly, which is a problem especially with precious antigens such as CTXs. On the other hand, mAbs, can provide a continuous supply of selected antibodies. CTXs are small haptens; therefore, they are not able to elicit an immune response by themselves. Fortunately, they possess a relatively reactive primary hydroxyl group which can react with succinic anhydride to yield a hemisuccinate. The latter has an available carboxyl group through which CTX can now be linked to a carrier protein using water-soluble carbodiimide cross-linking reagent to produce a conjugate suitable for immunization (55). A problem with the CTX molecule though is the presence of just one reactive group in the molecule, limiting the possibilities for the development of antibodies with different specificities for the various CTX analogues (55). Even though the production of a hemisuccinate of brevetoxin-3 has been possible (88), the production of a hemisuccinate of P-CTX-1 has not been successful (R.J. Lewis, Potomski, and N.C. Gillespie, unpublished results).

The AB fragment (89) and the JKLM rings of P-CTX-1 (90) have been synthesized. They may be useful as haptens for the production of anti-CTX antibodies. This effort is currently underway by Pauillac and coworkers, who report the conjugation of the JKLM fragment to bovine serum albumin in an effort to produce mAbs which can cross react with P-CTX-1 (91).

Ideally, the chosen antibodies should have an affinity that is directly proportional to the oral potency to humans to contaminating toxins and should not cross react with compounds normally present in nontoxic fish (41). A tremendous problem with CFP is the presence of CTXs of lower potency which must be considered in the development of antibody-based screens to identify accurately ciguateric fish (26,55). The possibility of false-positive results must be weighed against the risk of human intoxication. This problem could be resolved by confirmatory analysis using assays which are based on biological activity or chemical characteristics rather than on structure.

Cytotoxicity Assay

Cell culture techniques offer the possibility of using sensitive bioassays without the use of animal testing. Kogure and coworkers developed a cell-based assay for the detection of sodium channel blocking toxins based on their ability to antagonize the effects of veratridine (sodium channel activator) and ouabain (inhibitor of sodium efflux) on murine neuroblastoma cells (Neuro-2A) (92). The original method was simplified by incorporating a colorimetric endpoint based on the ability of metabolically active cells to reduce a tetrazolium compound, MTT (3-[4,5-dimethylthiazol-2-yl]-2,5-diphenyl-tetrazolium), to a blue-colored formazan product (93). This assay was further modified to be able to detect sodium channel activating toxins such as CTXs and brevetoxins. Neuro-2A cells are plated in 96-well tissue culture–treated plates and are incubated with veratridine, ouabain, and serial dilutions of test samples. Following incubation, cells are exposed to MTT and, after color development, the plates are read using a spectrophotometer. P-CTX-1 produced ID_{50} values (dose causing 50% decrease in cell viability) of approximately 1 pg and <0.25 pg at 7 and 22 h incubation, respectively. The activity of CTX-3C was ID_{50} = 3 pg and ≤1 pg at 7 and 22 h incubation, respectively. This threefold potency difference of P-CTX-1 to CTX-3C in vitro, is consistent with their relative potency in vivo using the mouse bioassay (LD_{50} values). This cell assay is four orders of magnitude (10^{-4} MU) more sensitive than the mouse bioassay (94,95). A disadvantage of this assay is the need for specialized laboratory equipment and trained personnel. Moreover, it is not cost effective for the routine screening of individual fish (55).

Cytotoxicity Assay with a c-*fos*–Luciferase Receptor Gene

A new cytotoxicity assay is based on a reporter gene that uses luciferase-catalyzed light generation as an endpoint and a microplate luminometer for quantification. The c-*fos*–luciferase reporter gene was incorporated into murine neuroblastoma cells (Neuro-2A). The assay is based on the addition of veratridine and toxins into a 96-well luminometer plate seeded with transfectant Neuro-2A cells. After incubation, cells are lysed and then exposed to a mixture of luciferase and adenosine triphosphate (ATP). Finally, the luminiscence generated from each well is measured. This assay has been used in the detection of brevetoxins and CTXs, providing results similar to those using the traditional cytotoxicity assay. It was found that both P-CTX-1 and CTX-3C at 3 pg/mL enhanced luciferase activity. This assay is still under evaluation for its performance using fish and shellfish matrices and dinoflagellate extracts. However, these initial studies have shown that its sensitivity may well exceed the sensitivity of the existing cytotoxicity assay (96).

Receptor Binding Assay

Brevetoxins and CTXs bind to site 5 of the voltage-dependent sodium channel; however, CTXs have a higher affinity (ED_{50} 0.23–0.85 ng/mL) (20). The binding affinity of each CTX for the sodium channel is proportional to its LD_{50} in mice (ip), indicating that the effect of CTXs likely arises from their action on sodium channels (20).

Membrane preparations from rat brain are used to measure the binding affinity of toxins to the sodium channel. Competition assays are carried out with a constant amount of ^{3}[H] brevetoxin-3 and membrane preparations and serial dilutions of CTXs. It has been found that CTXs competitively inhibit the binding of ^{3}[H] brevetoxin-3, whose activity can be measured by the addition of solid scintillant to a 96-well filter mat containing bound ^{3}[H] brevetoxin-3. In contrast, maitotoxin does not cause inhibition, marking a big step to differentiate it from CTXs; these two, CTXs and MTXs, give similar results with the mouse bioassay. This assay is still under development to assess quantitatively the responses using fish and algal extracts (97).

C. Chemical Methods

High-Performance Liquid Chromatography

HPLC methodology has been reported for CTX and related toxins (20,23). Legrand and coworkers used HPLC methodology to isolate multiple ciguatera toxins from wild *G. toxicus* and toxic fish (herbivorous and carnivorous) (14,98).

Pure P-CTX-1 can be detected to levels below 5 ng by monitoring HPLC eluants with a sensitive ultraviolet (UV) detector. However, CTXs do not possess a distinctive UV chromophore, hampering the development of a method to detect them selectively in crude lipid extracts from fish (55).

The major CTXs found in fish possess a relatively reactive primary hydroxyl group through which labels could be attached to enhance detectability by HPLC. HPLC techniques linked to fluorescence detectors have demonstrated the ability to detect in relatively crude extracts ≥40 ng diarrheic shellfish poisoning toxins/g shellfish (99) and ≥13 ng aflatoxin/g peanut butter or corn (100). However, these approaches would require transformations and notable improvements in sensitivity if the CTXs in crude extracts from fish are to be detected. A simple clean-up procedure that minimizes interference from lipids at both the labeling and detection steps is critical for the development of a sensitive test for CTXs (41). During a study performed by Yasumoto and coworkers, 10 minor toxins were obtained from moray eels after being labeled with anthroylnitrile. These were characterized by retention times in HPLC and mass spectral data. There was good linearity between the dose and the fluorescence in the range from 1 to 100 ng (27). Although sensitive detection of anthroylnitrile-labeled CTX has been reported (22), these approaches have not been extended to the detection of CTXs in crude fish extracts.

Dickey and coworkers reported promising results by labeling P-CTX-1 with a novel coumarin-based fluorescent reagent. The diethylaminocoumarin-carbamate of P-CTX-1 was detected using HPLC. Even though the derivatization process was successful, the quantum yield of the derivative did not appear to provide the level of sensitivity required. To achieve this, either the reaction conditions should be changed to improve the yield of the derivative or a more highly fluorescent reagent must be found (101).

High-Performance Liquid Chromatography–Mass Spectrometry

HPLC/MS is able to identify P-CTX-1 and its congeners in highly purified extracts from ciguateric moray eel viscera. Fourteen congeners were identified, two with $[M + H]^+$ m/z of 1095.7 Da, six with $[M + H]^+$ m/z of 1111.6 Da, and six with $[M + H]^+$ m/z of 1127.7 Da. Three of these congeners were identified as P-CTX-1, P-CTX-2, and P-CTX-3; the remaining congeners were not readily detected using the mouse bioassay. The use of acetonitrile-water gradients buffered with 1 mM ammonium acetate improved the separation and detection of the minor

CTXs compared with an acetonitrile-water gradient modified with 0.1% trifluoroacetic acid (102). The use of turbo-assisted HPLC/MS further improved the sensitivity and allowed for the detection of 4 ppb P-CTX-1 in a crude extract of Australian ciguateric fish flesh (102).

Ion-spray mass spectrometry (ISMS) is a sensitive method capable of the determination of the molecular weight of polyether toxins such as CTXs, brevetoxins, and maitotoxins. In fact, HPLC coupled to ISMS has shown considerable potential for the detection of labeled diarrheic shellfish poisoning toxins (103). Initial studies with P-CTX-1 indicate that such an approach could form the basis of a confirmatory analytical assay for CTXs in fish (104). HPLC/ISMS has been employed for the analysis of P-CTX-1 isolated from moray eel viscera. It is possible to detect the parent ions in addition to ions representing Na^+ adducts and the loss of water molecules. The detection of an ion at $m/z = 1111.8$ corresponded to the $[M + H]^+$ ion of P-CTX-1, which had been previously observed by FAB MS (fast atom bombardment–mass spectrometry) (20,23). The amount of P-CTX-1 needed to give a spectrum with a signal to noise ratio of 2:1 was approximately 1 ng. ISMS was also used to try to characterize the major CTX naturally occurring in the flesh of ciguateric fish (*Scomberomorus commersoni*, *Plectropomus* spp., and *Pomadasys maculatus*) and purified by HPLC. It was not possible to detect P-CTX-1 in spiked (500 ng P-CTX-1/g extract, equivalent to 1.5 ng P-CTX-1/g fish flesh) crude extracts of fish owing to interferences from lipids which may suppress the ionization of P-CTX-1 (104). Even though this is a sensitive method, improvements are needed until this toxin can be detected at levels typically found in the flesh of ciguateric fish (0.1–5.0 ng P-CTX-1/g flesh) (36). Improvements may include modified clean-up procedures and/or derivatization to possibly diethylaminocoumarin-carbamate or a similar nitrogenous derivative of P-CTX-1 (101,104).

HPLC/MS/MS techniques have also been used to detect and quantify CFP toxins. Purified fish extracts were spiked with pure P-CTX-1 and C-CTX-1 and quantified by multiple reactant ion monitoring relative to a brevetoxin internal standard. This method was able to detect CTXs at levels below 0.1 ppb, which is the level that can cause CFP (105).

Atmospheric pressure ionization mass spectrometric detectors have been improved in the last years. Musser and coworkers have reported the simultaneous differentiation of purified preparations of C-CTX-1 and P-CTX-1 using this method. The limit of detection for C-CTX-1 was reported to be 1 ppb (106).

Nuclear Magnetic Resonance

If sufficient unknown toxin is available (>25 μg), NMR approaches can be utilized for confirmation or characterization of unknown toxins (41). One-dimensional and two-dimensional homonuclear Hartman-Hahn (HOHAHA) spectra using NMR have been obtained for P-CTX-1. They enable the identification of unknown samples by comparison with the P-CTX-1 spectrum (55). The short-range inverse-detected heteronuclear (HMQC) spectrum of P-CTX-1 has been used for confirmation of the ^{13}C assignments given by Murata and coworkers (19,23). The long-range inverse-detected heteronuclear (HMBC) spectrum confirmed the position of carbons 33 and 52 and the two quaternary carbons and the location of four of the 13 ether-linked rings (107). NMR has been used in the analysis of fish viscera (20,23) and flesh (6) and in wild and cultured *G. toxicus* extracts (23,108).

D. Method Validation

After a new method has been developed, it should undergo an interlaboratory validation process to determine method performance parameters. Method validation programs can be administered by AOAC International or IUPAC. The aim of a collaborative study is to obtain an assessment of the attributes of a method to be expected when the it is used in actual practice, focusing primarily on the possible sources of deviations (between days, between runs, and between calibration curves). The process is divided in two parts: one includes a ruggedness test, a feasibility study, and a limited interlaboratory study; the second, if appropriate, involves a full interlaboratory evaluation to determine accuracy and precision parameters (repeatability [RSD_r] and reproducibility [RSD_R]) (57,85). It should be pointed out that none of the methods described in this chapter has undergone the rigors of a full interlaboratory validation process administered by AOAC International and IUPAC.

REFERENCES

1. EW Gudger. Poisonous fishes and fish poisoning with special reference to ciguatera in the West Indies. Am J Trop Med 10:43–55, 1930.
2. FE Ahmed. Naturally occurring fish and shellfish poisons. In: FE Ahmed, ed. Seafood Safety. Washington, DC: National Academy Press, 1991, pp 87–110.
3. EP Ragelis. Ciguatera seafood poisoning. In: EP Ragelis, ed. Seafood Toxins. Washington, DC: American Chemical Society, 1984, pp 22–36.
4. BW Halstead. Fish poisonings—their diagnosis, pharmacology, and treatment. Clin Pharmacol Ther 5:615–627, 1964.
5. JK Sims. Theoretical discourse on the pharmacology of toxic marine ingestions. Ann Emer Med 16:1006–1015, 1987.
6. RJ Lewis, M Sellin. Multiple ciguatoxins in the flesh of fish. Toxicon 30:915–919, 1992.
7. T Yasumoto, PJ Scheuer. Marine toxins from the Pacific-VIII ciguatoxin from moray eel livers. Toxicon 7:273–276, 1969.
8. R Bagnis, ME Loussan, S Thevenin. Les intoxications par poisons perroquets aux Iles Gambier. Med Trop 34:523–527, 1974.
9. E Chungue, R Bagnis, T Yasumoto. Le complexe toxinique des poissons perroquets. Biochimie 59: 739–741, 1977.
10. K Tachibana. Structural studies on marine toxins. PhD dissertation, University of Hawaii, Honolulu, Hawaii, 1980.
11. DR Tindall, RW Dickey, RD Carlson, G Morey-Gaines. Ciguatoxigenic dinoflagellates from the Caribbean Sea. In: EP Ragelis, ed. Seafood Toxins. Washington, DC: American Chemical Society, 1984, pp 225–240.
12. T Yasumoto, M Murata. Polyether toxins implicated in ciguatera and seafood poisoning. Faculty of Agriculture, Tohoku University, Tsutsumidori, Sendai 980, Japan, 1988.
13. T Yasumoto, M Murata. Polyether toxins produced by dinoflagellates. Faculty of Agriculture, Tohoku University, Tsutsumidori, Sendai 980, Japan, 1988.
14. AM Legrand. Les toxines de la ciguatera. Proceedings of Symposium on Marine Biotoxins, 30–31 January, Paris, France, 1991, pp 53–60.
15. LR Juranovic, DL Park. Foodborne toxins of marine origin: ciguatera. Rev Environ Contam Toxicol 117:51–94, 1991.
16. DL Park, LR Juranovic, SM Rua, CA Nielsen, CE Ayala. Mutagenic and toxic potential of toxins produced by *Gambierdiscus toxicus* and *Prorocentrum concavum*. J Aquatic Food Prod Technol 6: 27–42, 1997.
17. RJ Lewis. Ciguatera. Harmful Algae News 16:23, 1997.
18. PJ Scheuer, W Takahashi, J Tsutsumi, T Yoshida. Ciguatoxin: Isolation and chemical nature. Science 155:1267–1268, 1967.
19. M Murata, AM Legrand, Y Ishibashi, T Yasumoto. Structures of ciguatoxin and its congener. J Am Chem Soc 111:8929–8931, 1989.
20. RJ Lewis, M Sellin, MA Poli, RS Norton, JK MacLeod, MM Sheil. Purification and characterization of ciguatoxins from moray eel (*Lycodontis javanicus, Muraenidae*). Toxicon 29:1115–1127, 1991.

21. M Nukina, LM Koyanagi, PJ Scheuer. Two interchangeable forms of ciguatoxin. Toxicon 22:169–176, 1984.
22. AM Legrand, M Fukui, P Cruchet, Y Ishibashi, T Yasumoto. Characterization of ciguatoxins from different fish species and wild *Gambierdiscus toxicus*. Proceedings Third International Conference on Ciguatera Fish Poisoning, Puerto Rico, 1992, pp 25–32.
23. M Murata, AM Legrand, Y Ishibashi, M Fukui, T Yasumoto. Structures and configurations of ciguatoxin from the moray eel *Gymnothorax javanicus* and its likely precursor from the dinoflagellate *Gambierdiscus toxicus*. J Am Chem Soc 112:4380–4386, 1990.
24. P Deslongchamps, DD Rowan, N Pothier, T Sauv, JK Saunders. 1,7-Dioxaspiro[5,5]undecanes. An excellent system for the study of stereoelectronic effects (anomeric and exo-anomeric effects) in acetals. Can J Chem 59:1105–1121, 1981.
25. F Perron, KF Albizati. Chemistry of spiroketals. Chem Rev 89:1617–1661, 1989.
26. RJ Lewis, MJ Holmes. Origin and transfer of toxins involved in ciguatera. Comp Biochem Physiol 106C:615–628, 1993.
27. T Yasumoto, M Satake, M Fukui, H Nagai, M Murata, AM Legrand. A turning point in ciguatera study. In: TJ Smayda, Y Shimizu, eds. Toxic Phytoplankton Blooms in the Sea. New York: Elsevier, 1993, pp 455–461.
28. M Satake, M Murata, T Yasumoto. Gambierol: A new toxic polyether compound isolated from the marine dinoflagellate *Gambierdiscus toxicus*. J Am Chem Soc 115:361–362, 1993.
29. M Satake, M Murata, T Yasumoto. The structure of CTX3C, a ciguatoxin congener isolated from cultured *Gambierdiscus toxicus*. Tetrah Lett 34:1979–1980, 1993.
30. T Yasumoto, M Murata. Marine toxins. Chem Rev 93:1897–1909, 1993.
31. T Teai, P Cruchet, T Bracchi, AM Legrand. Ciguatoxin profiles of herbivorous and carnivorous fish muscles from highly endemic zones of French Polynesia. Presented at the 8th International Conference on Harmful Algae, Vigo, Spain, June 25–29, 1997.
32. RC Crouch, GE Martin, SM Musser, HR Grenade, RW Dickey. Improvements in the sensitivity of inverse-detected heteronuclear correlation spectra using micro inverse probes and micro cells: HMQC and HMBC spectra of Caribbean ciguatoxin—preliminary structural inferences. Tetrah Lett 36:6827–6830, 1995.
33. JP Vernoux, RJ Lewis. Isolation and characterization of Caribbean ciguatoxins from the horse-eye jack (*Caranx latus*). Toxicon 36:889–900, 1997.
34. M Marquais, JP Vernoux, J Molgo, MP Sauviat, RJ Lewis. Isolation of new ciguatoxins from Caribbean fish. Presented at the 8th International Conference on Harmful Algae, Vigo, Spain, June 25–29, 1997.
35. JA Babinchak, PDR Moeller, FM van Dolah, PB Eyo, JS Ramsdell. Production of ciguatoxins in cultured *Gambierdiscus toxicus*. Mem Qd Mus 34.447–453, 1994.
36. RJ Lewis. Ciguatoxins are potent ichthyotoxins. Toxicon 30:207–211, 1992.
37. MJ Holmes, RJ Lewis, MA Poli, NC Gillespie. Strain dependent production of ciguatoxin precursors (gambiertoxins) by *Gambierdiscus toxicus* (*Dynophyceae*) in culture. Toxicon 29:761–775, 1991.
38. L Micouin, M Chinain, P Asin, AM Legrand. Toxicity of French Polynesian strains of *Gambierdiscus toxicus* in cultures. Bull Soc Pathol Exot 85:474–477, 1992.
39. MJ Holmes, RJ Lewis. Multiple gambiertoxins (ciguatoxin precursors) from an Australian strain of *Gambierdiscus toxicus* in culture. In: P Gopalakrishnakone, CK Tan, eds. Recent Advances in Toxinology Research. Vol 2. Singapore: National University of Singapore, 1992, pp 520–529.
40. DL Park, L Stoloff. Aflatoxin control—how a regulatory agency managed risk from an unavoidable natural toxicant in food and feed. Reg Toxicol Pharmacol 9:109–130, 1989.
41. RJ Lewis. Immunological, biochemical, and chemical features of ciguatoxins: Implications for the detection of ciguateric fish. Mem Qd Mus 34:541–548, 1994.
42. RJ Lewis, M Sellin. Recovery of ciguatoxin from fish flesh. Toxicon 31:1333–1336, 1993.
43. DL Park. Prediction of aquatic biotoxin potential in fish and shellfish harvesting areas: ciguatera and diarrheic shellfish poisoning. In: ND Croce, S Connell, R Abel, eds. Coastal Ocean Space Utilization III. London: Chapman & Hall, 1993, pp 271–282.
44. AH Banner, P Helfrich, PJ Scheuer, T Yoshida. Proceedings 16th Gulf Caribbean Fish Institute, 1963, p 84.

45. Y Hokama. A rapid, simplified enzyme immunoassay stick test for the detection of ciguatoxin and related polyethers from fish tissues. Toxicon 23:939–946, 1985.
46. JS Lee, M Murata, T Yasumoto. Analytical methods for the determination of diarrhetic shellfish toxin. In: S Natori, K Hashimoto, Y Ueno, eds. Mycotoxins and Phycotoxins '88. New York: Elsevier, 1989, pp 327–334.
47. AH Banner. Ciguatera: A disease from coral reef fish. In: OA Jones, R Endean, eds. Biology and Ecology of Coral Reefs. Vol. 3. New York: Academic Press, 1976, pp 177–213.
48. DW Hessel, BW Halstead, NH Peckham. Marine biotoxins I. Ciguatera poison: Some biological and chemical aspects. Ann NY Acad Sci 90:788–797, 1960.
49. AH Banner, S Sasaki, P Helfrich, CB Alender, PJ Scheuer. Bioassay of ciguatera toxin. Nature 189:299, 1961.
50. MD Rayner. Mode of action of ciguatoxin. Fed Proc 31:1139–1145, 1972.
51. HR Granade, PC Cheng, NJ Doorenbos. Ciguatera I: Brine shrimp (*Artemia salina* L.) larval assay for ciguatera toxins. J Pharm Sci 65:1414–1415, 1976.
52. LR Juranovic. Determination of the toxic/mutagenic potential of toxins associated with ciguatera dinoflagellates. Master of Science Thesis, University of Arizona, Tucson, Arizona, 1989.
53. MB Kosa. Biotransfer/accumulation of toxins produced by dinoflagellate *Prorocentrum concavum* to domino damsel (*Dascyllus trimaculatus*) fish. Master of Science Thesis, University of Arizona, Tucson, Arizona, 1991.
54. E Chungue, R Bagnis, F Parc. The use of mosquitoes (*Aedes aegypti*) to detect ciguatoxin in surgeon fishes (*Ctenochaetus striatus*). Toxicon 22:161–164, 1984.
55. RJ Lewis. Detection of ciguatoxins and related benthic dinoflagellate toxins: *in vivo* and *in vitro* methods. In: GM Hallegraeff, DM Anderson, AD Cembella, eds. Manual on Harmful Marine Microalgae. Paris, France: UNESCO, 1995, pp 135–162.
56. RJ Lewis, M Sellin, R Street, MJ Holmes, NC Gillespie. Excretion of ciguatoxin from moray eels (*Muraenidae*) of the central Pacific. Proceedings of 3rd International Conference on Ciguatera Fish Poisoning, Puerto Rico, 1992, pp 131–143.
57. K Hellrich. Official Methods of Analysis. Association of Official Analytical Chemists. 15th ed. Arlington, Virginia: AOAC, 1990, pp 673–684, 881–882.
58. LH Kimura, Y Hokama, MA Abad, M Oyama, JT Miyahara. Comparison of three different assays for the assessment of ciguatoxin in fish diseases: radioimmunoassay, mouse bioassay and *in vitro* guinea pig atrium assay. Toxicon 20:907, 1982.
59. DL Park. Evolution of methods for assessing ciguatera toxins in fish. Rev Environ Contam Toxicol 136:1–20, 1994.
60. TJ Kosaki, BJ Stephens, HH Anderson. Marine toxins from the Pacific: III. Comparative bioassay of ciguatoxin(s) in mouse and chicken. Proc West Pharmac Soc 11:126, 1968.
61. JP Vernoux, N Lahlou, LP Migras, JB Greaux. Chick feeding test: a simple system to detect ciguatoxin. Acta Trop 42:235–240, 1985.
62. JP Vernoux. L'ichtyosarcotoxisme de type ciguatera aux Antilles et en Polynesie Francaise: test de ciguatoxicite et chaine tropique ciguaterigene. PhD dissertation, University of Bordeaux, France, 1981.
63. Y Hokama. Recent methods for detection of seafood toxins: recent immunological methods for ciguatoxin and related polyethers. Food Addit Contam 10:71–82, 1993.
64. Y Hokama, AH Banner, DB Boyland. A radioimmunoassay for the detection of ciguatoxin. Toxicon 15:317, 1977.
65. LH Kimura, MA Abad, Y Hokama. Evaluation of the radioimmunoassay (RIA) for detection of ciguatoxin (CTX) in fish tissues. J Fish Biol 21:671, 1982.
66. PM Gamboa Pulido. The development of a diagnostic tool for ciguatera fish poisoning in human serum. PhD dissertation, University of Arizona, Tucson, 1993.
67. Y Hokama. Ciguatera fish poisoning. J Clin Lab Anal 2:44–50, 1988.
68. T Yasumoto. Marine microorganisms. Toxins-an overview. In: E Graneli, B Sundstrom, L Edler, DM Anderson, eds. Toxin Marine Phytoplankton. New York: Elsevier, 1989, pp 3–8.
69. Y Hokama, MA Abad, LH Kimura. A rapid enzyme immunoassay (EIA) for the detection of ciguatoxin in contaminated fish tissues. Toxicon 21:817–824, 1983.

70. Y Hokama, LH Kimura, MA Abad, L Yokochi, PJ Scheuer, M Nukina, T Yasumoto, BG Baden, Y Shimizu. An enzyme immunoassay for the detection of ciguatoxin and competitive inhibition by related polyether toxins. In: EP Ragelis, ed. Seafood Toxins. Washington, DC: American Chemical Society, 1984, p 307.
71. Y Hokama. Immunological studies using monoclonal antibodies for detection of low dalton marine toxins. Food Addit Contam 10:83–95, 1993.
72. K Tachibana, M Nukina, YG John, PJ Scheuer. Recent developments in the molecular structure of ciguatoxin. Biol Bull 172:122–127, 1987.
73. T Yasumoto. Recent progress in the chemistry of dinoflagellates and related toxins. In: P Gopalakrishnakone, CK Tan, eds. Progress in Venom and Toxin Research. Singapore: National University of Singapore, 1987, pp 348–365.
74. Y Hokama, SAA Honda, AY Asahina, FML Fong, CM Matsumoto, TS Gallacher. Cross-reactivity of ciguatoxin, okadaic acid, and polyethers with monoclonal antibodies. Food Agric Immunol 1: 29–35, 1989.
75. Y Hokama, AM Osugi, SAA Honda, MK Matsuo. Monoclonal antibodies in the detection of ciguatoxin and other toxic polyethers in fish tissues by a rapid poke stick test. Proceedings of the 5th International Coral Reef Congress. Antenne Museum–EPHE, Moocea, Tahiti, 4:449–456, 1985.
76. Y Hokama, LK Shirai, LM Iwamoto, MN Kobayashi, CS Goto, LW Nakagawa. Assessment of a rapid enzyme immunoassay stick test for the detection of ciguatoxin and related polyether toxins in fish tissues. Biol Bull 172:144–153, 1987.
77. Y Hokama. Simplified solid-phase immunobead assay for detection of ciguatoxin and related polyethers. J Clin Lab Anal 4:213–217, 1990.
78. Y Hokama, SAA Honda, MN Kobayashi, LK Nakagawa, AY Asahina, JT Miyahara. Monoclonal antibody (MAb) in detection of ciguatoxin (CTX) and related polyethers by the stick-enzyme immunoassay (S-EIA) in fish tissues associated with ciguatera poisoning. In: S Natori, K Hashimoto, Y Ueno, eds. Mycotoxins and Phycotoxins '88. New York: Elsevier, 1989, pp 303–310.
79. Y Hokama, AY Asahina, TWP Hong, ES Shang, JT Miyahara. Evaluation of the stick enzyme immunoassay in *Caranx* sp. and *Seriola dumerili* associated with ciguatera. J Clin Lab Anal 4:363–366, 1990.
80. RW Dickey, SC Bobzin, DJ Faulker, FA Bencsath, D Ardrzejewski. Identification of okadaic acid from a Caribbean dinoflagellate, *Prorocentrum concavum*. Toxicon 28:371–377, 1990.
81. PM Gamboa, DL Park, JM Fremy. Extraction and purification of toxic fractions from barracuda implicated in ciguatera poisoning. Proceedings of the 3rd International Conference on Ciguatera Fish Poisoning, Puerto Rico, 1992, pp 13–24.
82. DL Park, CH Goldsmith. Inter-laboratory validation solid-phase immunobead assay for the detection of toxins associated with ciguatera poisoning. Presented at the 5th International Conference on Toxic Marine Phytoplankton, Newport, RI, October 28–November 1, 1991.
83. DL Park, JM Fremy, C Marcaillou-Lebaut, PM Gamboa, E Gleizes, P Masselin, CH Goldsmith. Innovative rapid solid-phase immunobead assay for the detection of okadaic acid and related DSP toxins in shellfish. Proceedings of the 2nd International Conference on Mollusc Shellfish Depuration, 1993, pp 113–118.
84. DL Park. Reef management and seafood monitoring programs for ciguatera. Mem Qd Mus 34:587–594, 1993.
85. DL Park. Detection of ciguatera and diarrheic shellfish toxins in finfish and shellfish using Ciguatect kit. J AOAC Int 78:533–537, 1995.
86. J McMillan, H Granade, P Hoffman. Ciguatera fish poisoning in the United States Virgin Islands: preliminary studies. J Coll Virgin Isl 6:84–107, 1980.
87. DL Park, PM Gamboa, CH Goldsmith. Rapid facile solid-phase immunobead assay for screening ciguatoxic fish in the market place. Bull Soc Pathol Exot 85:504–507, 1992.
88. DG Baden, TJ Mende, J Walling, DR Schultz. Specific antibodies directed against toxins of *Ptychodiscus brevis* (Florida's red tide dinoflagellate). Toxicon 22:783–789, 1984.
89. O Sato, M Hirama. Synthesis of the AB fragment of ciguatoxin: ligand controlled regioselective addition of osmium tetroxide to double bonds. Synlett September: 705–707, 1992.

90. M Sasaki, M Inoue, K Tachibana. Synthetic studies toward ciguatoxin. Stereocontrolled construction of the KLM ring fragment. J Org Chem 59:715–717, 1994.
91. S Pauillac, M Inoue, M Sasaki, J Naar, M Murata, K Tachibana, M Chinain, AM Legrand. Production of monoclonal antibodies to ciguatoxin using a tetracyclic synthetic fragment (JKLM) conjugated to carrier proteins and cross-reactivity studies towards related polyether compounds. Presented at the 8th International Conference on Harmful Algae, Vigo, Spain, June 25–29, 1997.
92. K Kogure, ML Tamplin, U Simidu, RR Colwell. A tissue culture assay for tetrodotoxin, saxitoxin and related toxins. Toxicon 26:191–197, 1988.
93. T Mosmann. Rapid colorimetric assay for cellular growth and survival: Application to proliferation and cytotoxicity assays. J Immunol Meth 65:55–63, 1983.
94. RL Manger, LS Leja, SY Lee, JM Hungerford, MM Wekell. Tetrazolium-based cell bioassay for neurotoxins active on voltage-sensitive sodium channels: semiautomated assay for saxitoxins, brevetoxins, and ciguatoxins. Anal Biochem 214:190–194, 1993.
95. RL Manger, LS Leja, SY Lee, JM Hungerford, MM Wekell. Cell bioassay for the detection of ciguatoxins, brevetoxins, and saxitoxins. Mem Qd Mus 34:571–576, 1995.
96. ER Fairey, JSG Edmunds, JS Ramsdell. A cell-based assay for brevetoxins, saxitoxins and ciguatoxins using a stably expressed c-fos–luciferase receptor gene. Anal Biochem 251:129–132, 1997.
97. FM van Dolah, EL Finley, BL Haynes, GJ Doucette, PD Moeller, JS Ramsdell. Development of rapid and sensitive high throughput pharmacologic assays for marine phycotoxins. Nat Toxins 2: 189–196, 1994.
98. AM Legrand, M Fukui, P Cruchet, Y Ishibashi, T Yasumoto. Characterization of toxins from different fish species and wild *G. toxicus*. Proceedings 3rd International Conference on Ciguatera Fish Poisoning, Puerto Rico, 1990, pp 25–32.
99. JS Lee, T Yanagi, R Kenma, T Yasumoto. Fluorometric determination of diarrheic shellfish toxins by high-performance liquid chromatography. Agric Biol Chem 51:877–881, 1987.
100. DL Park, S Nesheim, MW Trucksess, ME Stack, RF Newell. Liquid chromatographic method for determination of aflatoxins B1, B2, G1, and G2 in corn and peanut products: Collaborative study. J AOAC 73:260–266, 1990.
101. RW Dickey, FA Bencsath, HR Granade, RJ Lewis. Liquid chromatographic-mass spectrometric methods for the determination of marine polyether toxins. Bull Soc Pathol Exot 85: 514–515, 1992.
102. RJ Lewis, A Jones. Characterization of ciguatoxins and ciguatoxin congeners present in ciguateric by gradient reverse-phase high-performance liquid chromatography/mass spectrometry. Toxicon 35:159–168, 1997.
103. S Pleasance, MA Qulliam, JC Marr. Ionspray mass spectrometry of marine toxins. IV. Determination of diarrhetic shellfish poisoning toxins in mussel tissue by liquid chromatography/mass spectrometry. Rapid Commun Mass Spectrom 6:121–127, 1992.
104. RJ Lewis, MJ Holmes, P Alewood, A Jones. Ionspray mass spectrometry of ciguatoxin-1, maitotoxin-2 and -3, and related marine polyether toxins. Nat Toxins 2:56–63, 1994.
105. RJ Lewis, A Jones, JP Vernoux. Detection of multiple ciguatoxins by HPLC/MS and HPLC/MS/MS. Presented at the 8th International Conference on Harmful Algae, Vigo, Spain, June 25–29, 1997.
106. SM Musser, HR Granade, RJ Dickey. Proceedings of the 43rd meeting of the American Society for Mass Spectrometry and Allied Topics, Atlanta, 1995.
107. RJ Lewis, IM Brereton. Inverse-detected nuclear magnetic resonance of ciguatoxin-1: Quaternary C locations confirmed in CTX-1. Mem Qd Mus 34(3):555–559, 1993.
108. M Satake, T Ishimaru, AM Legrand, T Yasumoto. Isolation of a ciguatoxin analog from cultures of *Gambierdiscus toxicus*. In: TJ Smayda, Y Shimizu, eds. Toxic Phytoplankton Blooms in the Sea. New York: Elsevier, 1993, pp 575–579.

20

Ciguatera Toxins: Pharmacology of Toxins Involved in Ciguatera and Related Fish Poisonings

Richard J. Lewis
The University of Queensland, Brisbane, Australia

Jordi Molgó
Laboratoire de Neurobiologie Cellulaire et Moléculaire, CNRS, Gif sur Yvette, France

David J. Adams
The University of Queensland, Brisbane, Australia

I. INTRODUCTION

The increased harvesting of tropical marine resources has contributed to an increased incidence of human intoxication associated with fish consumption. Of these fish poisonings, ciguatera is arguably the most significant, both in terms of the number and the severity of poisoning episodes. The disease is associated with the consumption of many species of tropical and subtropical fishes from the Indo-Pacific Oceans and Caribbean Sea that have become contaminated by ciguatoxins which arise from blooms of certain strains of the benthic dinoflagellate *Gambierdiscus toxicus*. Environmental degradation may play a role in the increased incidence of ciguatera, although the precise factors involved remain elusive (1). The role played by other marine toxins in ciguatera, including other toxins produced by benthic dinoflagellates, has not been demonstrated, although mild cases of palytoxin poisoning may be mistaken for ciguatera (1). Over the last decade there have been rapid advances in our knowledge of the precise chemical and pharmacological properties of the toxins which contribute to the different forms of fish poisoning, especially those toxins involved in ciguatera.

Ciguatera can be defined as an illness caused by the consumption of polyether sodium channel activator toxins (ciguatoxins, CTXs) that accumulate to levels that cause acute human intoxication characterized by neurological, gastrointestinal, and cardiovascular disorders. The ciguatoxins are a family of lipid-soluble, highly oxygenated, cyclic polyether molecules (2–5). Several ciguatoxins have been isolated from biodetritus containing wild *G. toxicus* (4,6–8), from toxic strains of cultured dinoflagellate isolated from different parts of the world (8–10), or from various ciguateric fish (6,11–14) (Table 1). Using nuclear magnetic resonance (NMR) techniques, the chemical structures of a number of Pacific ciguatoxins from fish and *G. toxicus* have been elucidated (4,7,9,10,15–17). Recently, the structures of C-CTX-1 and C-CTX-2 from

Table 1 Characteristics[a] of Structurally Defined Ciguatoxins Found in Fish and *G. toxicus*

Name	Alternative name	Source	$[M + H]^+$	Potency (μg/kg)	References
P-CTX-1	CTX CTX-1B	Carnivore	1111	0.25	4,5
P-CTX-2		Carnivore	1095	2.3	16
P-CTX-3		Carnivore	1095	0.9	15
P-CTX-3C		*G. toxicus*	1045	2	9
2,3-dihydroxy P-CTX-3C	CTX-2A1	Carnivore	1057	1.8	17
51-hydroxy P-CTX-3C		Carnivore	1039	0.27	17
P-CTX-4A		*G. toxicus* Herbivore	1061	2	10
P-CTX-4B	GT-4B	*G. toxicus* Herbivore	1061	4	4
C-CTX-1		Carnivore	1141	3.6	5,12
C-CTX-2		Carnivore	1141	1	5,12

[a] The protonated molecular mass ($[M + H]^+$) and potency following intraperitoneal injection are given.

Caribbean fish have been elucidated (5). The structures of known ciguatoxins are given in Figure 1.

The ciguatoxins are the most potent sodium channel toxins known (see Table 1). Despite their potency, ciguatoxins rarely accumulate in fish to levels that are lethal to humans. All the ciguatoxins isolated to date have a structural framework that is reminiscent of the brevetoxins (PbTx), another family of potent lipid-soluble polyether toxins produced by the marine dinoflagellate *Gymnodinium breve* (formerly *Ptychodiscus brevis*) that include PbTx-1 to PbTx-10 (18,19) (see Chap. 23). From our knowledge of the chemical structure of ciguatoxins found at different trophic levels, it is evident that P-CTX-1, the dominant and most potent ciguatoxin extracted from the moray eel, *Gymnothorax javanicus*, arises from the acid-catalyzed spiroisomerization and oxidative modification of P-CTX-4A produced by the dinoflagellate *G. toxicus* (1,4,17,20). The chemical structures of known ciguatoxins are compared with PbTx-2 in Figure 1.

The pharmacology of the ciguatoxins is characterized by their ability to cause the persistent activation of voltage-sensitive sodium channels (VSSCs), to increase neuronal excitability and neurotransmitter release, to impair synaptic vesicle recycling, and to modify Na^+-dependent mechanisms in numerous cell types. It is these effects that are believed to underlie the complex of symptoms associated with ciguatera. Ciguatera can be distinguished from scombroid fish poisoning, a preventable intoxication that results mainly from consumption of scombroid fish containing unusually high levels of histamine due to inappropriate storage or handling. Ciguatera also differs from tetrodotoxin intoxication, one of the most lethal seafood toxins associated with the consumption of most pufferfish species (family *Tetraodontidae*). Even though pufferfish are easily recognized, there are still cases of tetrodotoxin poisoning reported each year. Interestingly, pufferfish toxins selectively block sodium channels in excitable membranes, an action that antagonizes the action of ciguatoxin. This blockade of sodium channels prevents action potentials from propagating along axons, nerve terminals, and muscle fibers, leading to inhibition of nerve–evoked neurotransmitter release at chemical synapses.

In this chapter, we examine the pharmacology of ciguatoxins and related marine toxins that accumulate in fish to levels that cause human illness. Several related reviews have recently been published on the actions of marine neurotoxins (21–24), (see Chap. 23).

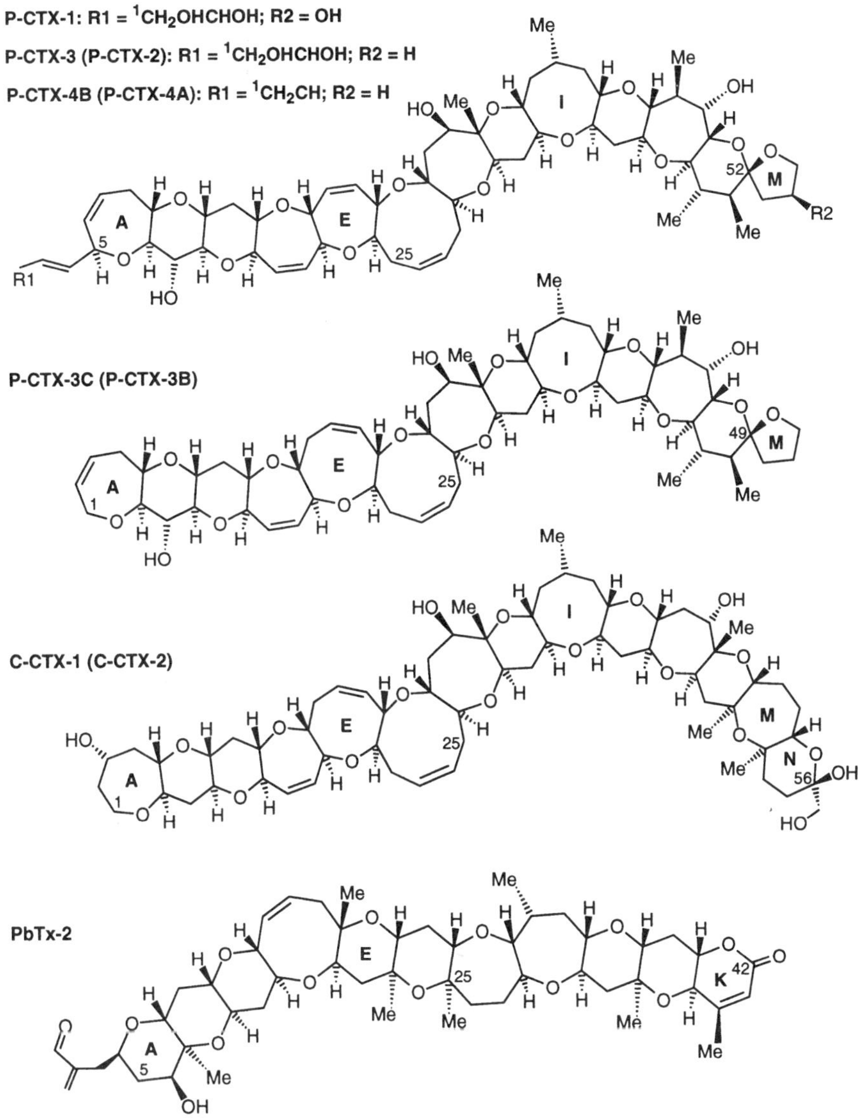

Figure 1 Structures of ciguatoxins from the Pacific Ocean and Caribbean Sea. Shown are P-CTX-1 (4), P-CTX-3 (15), GT-4B (4), P-CTX-3C (9), and C-CTX-1 (12). The less energetically favorable epimers P-CTX-2 (52-*epi* P-CTX-3) (16), P-CTX-4A = 52-*epi* P-CTX-4B) (10), P-CTX-3B (49-*epi* P-CTX-3C), and C-CTX-2 (56-*epi* C-CTX-1) (12) are indicated in parenthesis. 2,3-Dihydroxy P-CTX-3C and 51-hydroxy P-CTX-3C have also been isolated from Pacific fish (17). Brevetoxin (PbTx-2) is shown for comparison.

II. EFFECTS OF CIGUATOXINS AND RELATED TOXINS IN ANIMALS

A. Ciguatera in Humans

Although records are incomplete owing to the paucity of specialized centers for disease control in most ciguatera endemic regions, available evidence indicates that ciguatera is responsible for >25,000 cases of intoxications worldwide, and it thus represents the single largest cause of unpreventable fish poisoning. The disease is both spatially and temporally unpredictable (25).

Approximately 400 species of warm-water marine fish may at times be poisonous to humans after ingestion, with most being associated with ciguatera. Many of these ciguateric fish species, but not all, are found in coral reef waters. Usually their distribution is highly localized, being often restricted to a given island or side of an island. Most ciguateric fish are nonmigratory reef herbivores or reef carnivores. In most risk areas, ciguateric fish of each species may comprise 10–<0.01% of those captured. Certain locations harbor fish that are always toxic (e.g., moray eels from South Tarawa in the Republic of Kiribati, 1987–1990). Ciguatera in the Pacific has its greatest impact in communities inhabiting atoll islands, where fish are the primary source of dietary protein (26).

Epidemiological characterization of ciguatera has been limited by the lack of laboratory tests to confirm the presence of ciguatoxins. The pathophysiological features and symptoms of the disease in different areas of the world have been extensively examined (25,27–34) and compared with those of other seafood-related syndromes (35). The clinical picture is characterized by a variety of gastrointestinal, neurological, and, to a lesser extent, cardiovascular symptoms (Table 2). The duration, severity, and number of ciguateric symptoms depend on the quantity and type of ciguatoxin consumed. This is influenced by the type of fish consumed (herbivore, carnivore) and the location of capture. Three chemically distinct classes of ciguatoxins most likely underlie the clinical differences in ciguatera that arise from consumption of fish captured in the Caribbean Sea, Pacific Ocean, and Indian Ocean. The ciguatoxin(s) from the Indian Ocean remain to be characterized.

The neurological symptoms tend to be the most distinctive and enduring features of ciguatera, and they often include sensory disturbances such as generalized pruritus, circumoral numbness, long-lasting weakness and fatigue, and a distinctive reversal or exaggerated responses to hot and cold sensation that may have an origin in peripheral (36) and/or central (37) nerves. Patients with bradycardia and/or hypotension may require urgent care, because cardiovascular symptoms may indicate a poor prognosis (38). Significant slowing of sensory conduction velocity and prolongation of the absolute refractory, relative refractory, and supernormal periods have been recorded in the sural and common peroneal nerves of rats, supporting the contention that ciguatoxins act in vivo to open sodium channels in nerves (39).

Although the symptoms are relatively well documented, instances of ciguatera often goes unreported or misdiagnosed. Early symptom recognition has improved the identification and clinical management of ciguatera in endemic and nonendemic areas. However, there is still a need for better diagnostic, preventive, and reporting protocols to study more accurately and understand this diverse clinical syndrome. Although difficult to implement, preventive strategies remain the best form of management for ciguatera. At present, complete prevention of intoxica-

Table 2 Clinical Symptoms of Ciguatera Fish Poisoning

Neurological	Gastrointestinal	Cardiovascular
Paresthesias of mouth and extremities	Nausea	Bradycardia
Dysesthesia/hyperesthesia	Diarrhea	Hypotension
Ataxia	Vomiting	Tachycardia
Arthralgia	Abdominal pain	
Myalgia		
Headache, vertigo, dizziness		
Tremors		
Neck stiffness		
Salivation		
Perspiration		

tion depends on the avoidance of all potential vectors. Immunoassays and bioassays, as well as methods based on mass spectrometry, are being developed to detect levels of ciguatoxins less than one part per billion in suspect fish flesh prior to consumption.

B. Treatment of Ciguatera

Treatment of ciguatera is primarily supportive in nature. However, intravenous D-mannitol has evolved as a unique remedy for ciguatera, particularly those suffering acute poisoning from Pacific and Caribbean fish (40–44). A striking edema of adaxonal Schwann cells observed in severe ciguatera (45) may be reversed by D-mannitol and explain its beneficial action (41). Despite the availability of this treatment, the management of long-term chronic symptoms continues to be problematic. The pathophysiological basis of symptoms that can persist for weeks to years remains to be elucidated, but it may involve either permanent neurological damage associated with edema or perhaps slow detoxification. The reason mannitol is not consistently effective in treating ciguatera (46) is also unclear, but it may relate to the timing of administration, severity of poisoning, and individual variation in the response to ciguatoxin and/or mannitol.

A number of other treatments have also been reported to give benefit to sufferers of ciguatera in isolated instances. These include the antidepressant amitriptyline (47,48), the L-type calcium channel inhibitor nifedipine (48), and the sodium channel inhibitor tocainide (49). Symptomatic improvement of chronic fatigue associated with ciguatera has also been obtained with the antidepressant fluoxetine (50). Traditional remedies for ciguatera may also give benefit (51), but as with most other potential therapies, there is only restricted evidence of efficacy. Despite compelling evidence for the use of hyperosmotic mannitol in humans, rat and mouse models of ciguatera have failed to show the expected benefits of mannitol (52,53). The use of tocainide, an orally active lidocaine derivative, is supported by whole-animal studies (54) but not by an in vitro study (55).

C. In Vivo Effects of Ciguatoxins in Animals

Ciguatoxins have been found to be toxic in a range of animal species, especially mammals (see ref. 25 for a review). One of the most prominent effects of ciguatoxins in mice is the rapidly developing hypothermia (52). Neuroexcitatory effects in regions of the brain stem receiving vagal afferents and ascending pathways associated with visceral and thermoregulatory responses accompanied the intraperitoneal administration into mice of an uncharacterized ciguatoxin isolated from *G. toxicus* (37).

The effects of ciguatoxin have been studied in pentobarbital anesthetized cats (56). Intravenous injections of increasing doses of ciguatoxin evoke respiratory and cardiovascular disturbances comprising hyperventilation at low doses and respiratory depression leading to respiratory arrest at high doses; bradycardia and impaired atrioventricular conduction at low doses; arrhythmias and ventricular tachycardia with transient hypertension at sublethal doses; and falling arterial pressure leading to complete heart failure at high doses. The use of hexamethonium, atropine, propranolol, phentolamine, and bilateral adrenalectomy indicated that ciguatoxin has both central and peripheral effects (56).

Light and electron microscopy has revealed that intraperitoneal or oral administration of P-CTX-1 or P-CTX-4C to male mice causes effects on the heart, the adrenal medulla, and autonomic nerves (57). Continuous erection of the penis was observed in about 15% of the mice suffering from ciguatoxin. Although severe diarrhea was brought about by the administration of these ciguatoxins, no morphological alterations were seen in the mucosa and muscle layers of the small intestine except in autonomic nerve fibers and synapses. Atropine suppressed the symptoms of diarrhea but had no effect on the injury to the cardiac muscle, whereas reserpine aggravated the clinical signs and pathological findings. Guanethidine, 5-hydroxydopamine, or bilateral adrenalectomy did not alter the effect of ciguatoxin. In a subsequent study (58), repeated intraperitoneal and oral administrations of 0.1 μg/kg of P-CTX-1 or P-CTX-4C to mice over 15 days resulted in marked swelling of cardiac cells and the endothelial cells lining blood capillaries of the heart. These effects reversed within 1 month after ceasing toxin administration.

D. Animal Assays for Ciguatoxins

Since ciguatoxins are odorless and tasteless, feeding tests in animals have been traditionally used to monitor suspect fish samples. Of the 37 animal species tested by Banner et al. (59), only 5 were found to be sensitive to the oral administration of ciguateric fish. Tests based on feeding cats, mongooses, chickens, mosquitoes or Diptera larvae portions of the flesh or viscera of suspect fish appear most useful. Observations of signs of intoxication, growth, body temperature, and survival time of the animals over time are used to characterize and quantify the presence of ciguatoxins (59–64). Cats may regurgitate highly toxic fish, making quantitation problematic.

Perhaps the simplest animal assay to be developed for ciguatoxins is the fly assay (64). The meat-eating *Parasarcophaga argyrostoma* (Diptera, *Sarcophagidae*) larvae were able to detect ciguatoxin at levels above ~0.2 ng/g in fish flesh or liver. An inability to detect low-toxicity fish may limit its usefulness. A mosquito assay for ciguatoxins in crude fish extracts is also relatively simple to implement (60).

Despite the development of these nonmammalian assays, the mouse bioassay remains the most widely used assay to establish levels of ciguatoxin in extracts of fish (reviewed in ref. 63). The method quantifies lethal and sublethal doses of ciguatoxin in crude extracts administered intraperitoneally to mice. The signs of intoxication are well described and include hypothermia, hypersalivation, lacrimation, penal erection, hind-limb paralysis for certain ciguatoxins, respiratory difficulties, and asphyctic convulsions preceding death that are the result of respiratory failure (52,63,65). The establishment of the dose time to death relationship for the major toxins allows the numbers of mice required for quantitation to be reduced to as few as two per sample and avoids testing LD50 doses which are lethal at ~24 h (63). Despite these considerations, ethical concerns remain, and the assay is not sufficiently sensitive to detect low-toxicity fish in crude extracts. In addition, false-positive results may be obtained when high levels of lipids are administered (>20 mg/mouse) (66). Ciguatoxins are also potent fish toxins, but they differ from the brevetoxins, which are more potent to fish than to mammals (67).

III. RELATED FISH POISONINGS

A. Maitotoxin Poisoning

Maitotoxin, originally isolated from the gut content of the surgeonfish, *Ctenochaetus striatus*, is a potent highly polar, water-soluble polyether toxin (68,69). A family of maitotoxins of relatively high molecular mass (M_r = 1000–3500 D) has been isolated from the dinoflagellate *G. toxicus* (68–70). The maitotoxins isolated to date are all highly potent when administered intraperitoneally to mice (~0.1 μg/kg), and all have a similar in vitro pharmacology (70). Maitotoxin is transmitted through the food chain to the viscera of herbivorous fishes, and thus may contribute to ciguatera. However, symptoms distinctive of maitotoxin have not been characterized despite the fact that maitotoxin and ciguatoxin have a distinct pharmacology. In addition, if maitotoxin was a significant human health risk, it would be expected to be a relatively common syndrome given that all blooms of *G. toxicus* produce maitotoxin but few blooms produce ciguatoxin. Absence of a maitotoxin syndrome may arise because maitotoxins are ~100-fold less potent given orally than intraperitoneally and appear not to accumulate in fish flesh. In contrast, the ciguatoxins have similar potency by either of these routes.

Maitotoxin applied extracellularly induces a Ca^{2+} influx into virtually all cells studied. As a consequence of this action, maitotoxin causes the release of hormones and neurotransmitters from a range of secretory cells and nerve terminals (reviewed in ref. 71). The actions of maitotoxin are dependent on the presence of extracellular Ca^{2+}, and they are antagonized in some cells by L-type calcium channel blockers (72,73). However, the effects of maitotoxin on calcium uptake and phosphoinositide breakdown do not appear to be solely dependent on the activation of voltage-dependent L-type calcium channels, and the activation of such channels is probably an indirect consequence of membrane depolarization through maitotoxin-sensitive channels (74). Maitotoxin-induced Ca^{2+} entry is not the result of an ionophore-like action, since, unlike calcium ionophores, maitotoxin does not elicit Ca^{2+} uptake in liposomes (73), and fluorescent indicator measures of intracellular Ca^{2+} show that maitotoxin causes an increase that is different from that induced by calcium ionophores such as ionomycin. In membranes and permeabilized cells, maitotoxin has been reported to have no action on phospholipase C (75), indicating that both intact cells and extracellular Ca^{2+} are needed for maitotoxin to have a detectable effect. Although it appears that the primary target for maitotoxin is not phospholipase C, a ubiquitous maitotoxin-sensitive transmembrane protein remains to be identified. Since the compound SK&F 96365, an inhibitor of receptor-mediated calcium entry (76), was reported to inhibit maitotoxin-induced intracellular Ca^{2+} elevation, Ca^{2+} influx, phosphoiniositide breakdown, and hormone release (77), the primary action for maitotoxin may be the receptor-mediated calcium-entry pathway. In those cell types studied, the initial action of maitotoxin is the induction of a nonselective cation channel which requires the presence of extracellular Ca^{2+} (78,79). The subsequent rise in intracellular Ca^{2+} is at least in part caused by another SK&F 96365–sensitive Ca^{2+} entry pathway, which may be activated as a result of, or independently of, nonselective cation channel activation (80). Whatever its molecular target, maitotoxin has become a valuable tool for studying the modulation of calcium-dependent mechanisms in cells.

B. Palytoxin Poisoning

Palytoxin is a highly toxic polyhydroxy macromolecule isolated originally from zoanthid coelenterates of the genera *Palythoa* and *Zoanthus*, which are widespread throughout Caribbean, Pacific, and western Atlantic waters (81,82). Palytoxin is also found in sea anemones (83), certain crabs (82,84), and fish such as the triggerfish. The entry of palytoxin into the human food chain through the consumption of fish and marine invertebrates is responsible for incidences of human morbidity and mortality.

Triggerfish of the family Balistidae are common inhabitants of tropical shores. The high risk of illness associated with the consumption of these fish has resulted in a ban on their sale in French Polynesia and Hawaii. Human poisoning caused by triggerfish has long been regarded as a form of ciguatera (85). However, the acuteness and severity of the poisoning distinguish palytoxicoese from ciguatera. Clinical symptoms of palytoxin poisoning include a spinal seizure-like syndrome with tonic contractions of all muscle groups, muscle spasms associated with markedly elevated levels of serum enzymes associated with tissue damage (e.g.,creatinine phosphokinase, lactic acid dehydrogenase, and glutamic oxaloacetic transaminase), convulsions, extreme pain, myoglobinuria, respiratory distress, dyspnea, and respiratory failure. Death can occur within 2–4 days of intoxication.

The occurrence of palytoxin in the viscera of the triggerfish *Melichtys vidua* has been demonstrated by high-performance liquid chromatography and thin-layer chromatography comparisons with reference palytoxin isolated from the zoanthid coelenterate *Palythoa tuberculosa* (86). Palytoxin has also been shown to be present in the viscera of a filefish, *Alutera scripta*, of the family *Monacanthidae* which feed on *Palythoa* (87). Palytoxin has also been implicated in human poisoning after consumption of smoked mackerel (*Decapterus macrosoma*) of the family *Scombridae* (88). The presence of palytoxin can be detected in the fish extracts by chromatographic analysis by the method of Yasumoto et al. (84) and confirmed by radioimmunoassay (89). Similar food poisoning incidents due to the consumption of parrotfish containing palytoxin have been reported (90,91).

Palytoxin produces a broad range of pharmacological effects that are due to (1) binding to the Na^+/K^+-ATPase and consequent inhibition of Na^+/K^+-ATPase activity, (2) formation of nonselective channels (7–25 pS) associated with the Na^+/K^+-ATPase, which trigger the secondary activation of voltage-sensitive Ca^{2+} channels and Na^+-Ca^{2+} exchange, (3) a large Na^+ influx and cell depolarization, (4) the increase and mobilization of intracellular Ca^{2+} leading to enhanced neurotransmitter release from nerve terminals irrespective of the neurotransmitter involved, and (5) contractions of skeletal and smooth muscle cells (92–98).

C. Clupeotoxism

Clupeotoxism is a form of fish poisoning caused by consumption of sardines and herrings of the family *Clupeidae* and anchovies of the family *Engraulidae*. This poisoning is characterized by a high mortality rate and can be differentiated from ciguatera on the basis of clinical signs and symptoms. Symptoms include nausea, a sharp metallic or bitter taste, vomiting, hypersalivation, abdominal pain, severe diarrhea, feeble pulse, tachycardia, chills, cold clammy skin, hypotension, cyanosis, vertigo, tingling of the tongue and lips, numbness, nervousness, dilated pupils, violent headache, muscular cramps, dyspnea, progressive muscular paralysis, convulsions, coma, and death. The onset of symptoms is usually rapid and death may occur within a few hours. During the last 30 years, several outbreaks have been reported in the tropical Pacific and the Caribbean. Owing to its sporadic occurrence the causative toxin(s) has been difficult to identify. However, based on the fact that clupeoid fish usually contained bottom sediments in their gills and esophagus, it was inferred that the implicated fish had recently fed on the bottom, and they therefore could have obtained the toxin from a benthic organism. The benthic dinoflagellate *Ostreopsis siamensis* that produces palytoxin and its analogues was inferred as the probable progenitor of the toxins (99). In a recent study of a human intoxication due to ingestion of clupeoid fish, palytoxin has been definitively identified as the cause of clupeotoxism based on its cytotoxicity, delayed hemolysis, neutralization with an antipalytoxin antibody, chromatographic properties, and mass spectra (100).

D. Shark Poisoning

The flesh of some sharks like the Greenland shark, *Somniosus microcephalus*, are known to be toxic to humans and animals, especially when consumed fresh. That trimethylamine oxide contamination and clinical symptoms are consistent with acute trimethylamine poisoning arising from intestinal reduction of trimethylamine oxide appears to explain shark toxicity (101).

A large common-source outbreak of shark poisoning (98 deaths and a case fatality ratio of 20%) was reported after the ingestion of a single shark (*Carcharinus leuca*) in November, 1993 in Manakara, a middle-sized town on the southeast coast of Madagascar (102,103). The first clinical signs in the Madagascar outbreak appeared within 5–10 h after ingestion. Patients presented with neurological symptoms almost exclusively, the most prominent being a constant, severe ataxia, whereas gastrointestinal problems of diarrhea and vomiting were rare (104). Two lipid-soluble toxins have been isolated from this shark's liver and named carchatoxin-A and carchatoxin-B (105). The structural properties of the shark and related Indian Ocean ciguatoxins are currently under investigation.

IV. IN VITRO PHARMACOLOGY OF CIGUATOXIN

Pharmacological studies have revealed that ciguatoxin acts on voltage-sensitive sodium channels (VSSCs) present in most excitable cells and are also in some nonexcitable cells, such as glial cells. Sodium channels are large membrane-spanning proteins (Fig. 2) that mediate the rapid increase in membrane Na^+ conductance responsible for the depolarizing phase of action potentials in many excitable cells. As a consequence of ciguatoxin binding to site 5 of the sodium channel, there is a shift in the voltage dependence of activation that results in Na^+ influx through the channel at membrane potentials where the sodium channel is normally closed (Fig. 3). This ciguatoxin-induced activation of sodium channels at the resting membrane potential is responsible for numerous Na^+-dependent effects (106; reviewed in ref. 21). Table 3 summarizes our present understanding of the mode of action of ciguatoxin.

A. Effects of Ciguatoxin on Isolated Heart and Smooth Muscle

Ciguatoxin causes the rapid onset of negative inotropy (reduced strength of contraction) followed by a dominant positive inotropic effect in the left atria and positive inotropic effects in papillary muscle. These responses arise because ciguatoxin stimulates the release of excitatory and/or inhibitory neurotransmitters from the nerves innervating the heart (107–109). A second and more slowly developing phase of the P-CTX-1 effect on guinea pig atria and papillary muscle results from an additional direct action of P-CTX-1 on the myocardium (108,110). Both the direct and indirect effects of P-CTX-1 are reversed by tetrodotoxin (TTX), a selective blocker

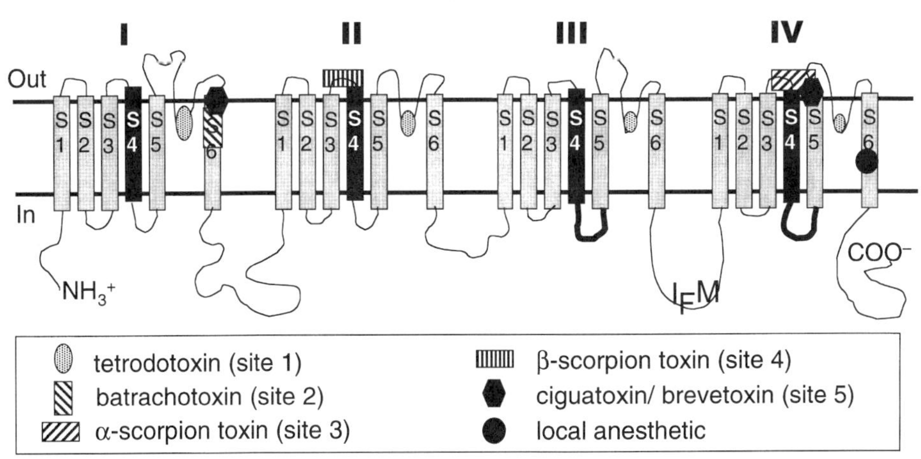

Figure 2 Toxin-binding sites 1–5 identified on the neuronal voltage-sensitive sodium channel. The sodium channel comprises the α-subunit of about 240–280 kD (~2000 amino acids) organized in four repeated homologous domains (I–IV), each containing six putative transmembrane spanning α-helical segments (S1–S6). A location of local anesthetic interaction with the VSSC is also indicated.

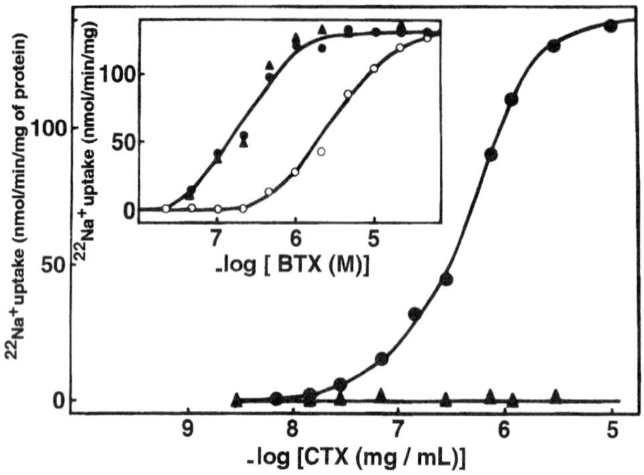

Figure 3 Ciguatoxin (CTX) enhances sodium influx into neuroblastoma cells. The ciguatoxin enhancement of $^{22}Na^+$ influx in the presence of veratridine (●) was inhibited by 10^{-7} M tetrodotoxin (closed triangles). The inset reveals that 3 ng/mL ciguatoxin (●) increases the potency of batrachotoxin (BTX) to enhance $^{22}Na^+$ influx to a similar extent as a saturating concentration of sea anemone toxin AS_{II} (closed triangles). (From ref. 136.)

of VSSCs. P-CTX-1 induced a depolarization of stimulated or quiescent atrial cells that was TTX sensitive, confirming that P-CTX-1 opens cardiac sodium channels. The link between the P-CTX–induced increase of intracellular concentration of Na^+ and the positive inotropic response can be explained as follows. The enhanced Na^+ influx reduces the effectiveness of the Na^+-Ca^{2+} exchanger to remove intracellular Ca^{2+}, leading to the enhanced Ca^{2+}-induced release of Ca^{2+} from calcium stores (a major source of Ca^{2+} for cardiac contraction) and stronger contractions (108). At high doses, P-CTX-1 additionally causes negative inotropic and arrhythmic effects in guinea pig atria and papillary muscles that are also TTX sensitive (111).

P-CTX-1 also causes a large, sustained and concentration-dependent positive inotropy in human atrial trabeculae. However, in this tissue, the inotropic effects arise through stimulation of intrinsic nerves, resulting in the release of norepinephrine (noradrenaline) (109). The addition of mannitol (50 mM) was unable to reverse the positive inotropic effects of P-CTX-1 in human atrial trabeculae. In frog atria, C-CTX-1 does not affect the membrane potential but caused a TTX-sensitive reduction in the cardiac action potential duration (112), possibly by stimulating the release of acetylcholine from intrinsic nerves.

Table 3 Modes of Action of Ciguatoxins

- Activate sodium channels of excitable and nonexcitable (Schwann cells and glial cells) cells
- Induce membrane depolarization and spontaneous and repetitive action potentials in excitable cells
- Alter the Na^+ gradient driving the Na^+-Ca^{2+} exchanger, leading to an elevation in intracellular Ca^{2+} concentration
- Induce repetitive, synchronous, and asynchronous neurotransmitter release
- Produce transient increases and decreases in the quantal content of synaptic responses
- Cause spontaneous and tetanic muscle contractions
- Impair synaptic vesicle recycling that exhausts neurotransmitters available for release
- Cause swelling of axons, nerve terminals, and perisynaptic Schwann cells

P-CTX-1 (0.001–1.0 nM) causes a sustained, dose-dependent contraction of the longitudinal smooth muscle of the guinea pig ileum (113). At high doses of P-CTX-1, P-CTX-2 and P-CTX-3, these contractions last for over 20 min (114) and are associated with bursts of contractile activity. The response to P-CTX-1 is completely blocked by atropine, TTX, and a low extracellular Na^+ bath solution; is enhanced by the acetylcholinesterase inhibitor eserine; but is unaffected by hexamethonium or mepyramine (113). Repeated doses of P-CTX-1 are tachphylactic and responses to nicotine are irreversibly reduced by prior exposure to P-CTX-1, P-CTX-2, or P-CTX-3 (113,114). P-CTX-1 does not alter ileal responses to histamine, acetylcholine, and 5-hydroxytryptamine. These results indicate that P-CTX-1 causes the release of acetylcholine from cholinergic nerve terminals via the excitation of postganglionic nerves in the ilea. The tachyphylactic nature of the action of ciguatoxin suggests that nerve stimulation is followed by nerve blockade, probably as a consequence of further nerve depolarization.

P-CTX-1 also induces a contraction of isolated guinea pig vas deferens that is inhibited or abolished by treatment with TTX, procaine, cold storage, incubation in a low Na^+ bath solution, phentolamine, or guanethidine but is not affected by atropine or mecamylamine (115). From these results, and the observation that P-CTX-1 caused a TTX-sensitive release of norepinephrine from the adrenergic nerves, it was determined that P-CTX-1–induced contractions occur via the release of norepinephrine from adrenergic nerve terminals. P-CTX-1 also markedly potentiates contraction of vas deferens induced by norepinephrine, acetylcholine, or high-K^+, but it inhibits contractile responses arising from electrically stimulated neural elements. In a subsequent study, P-CTX–induced potentiation was shown to be inhibited or abolished by TTX and a Na^+-deficient bath solution, and suggested that P-CTX-1 may potentiate vas deferens contractile responses through a direct action on smooth muscle sodium channels (116). However, subsequent studies have failed to confirm this result instead indicating that P-CTX-1 caused potentiation only through the effects of neurally released norepinephrine (R.J. Lewis and A.W. Wong Hoy, unpublished results).

The effects of P-CTX-1 have also been examined on the membrane potential of smooth muscle cells of isolated rat proximal tail arteries (117). P-CTX-1 ($\geq$10 pM) increased the frequency of spontaneous excitatory junction potentials, whereas 100–400 pM P-CTX-1 additionally caused a maintained depolarization of up to 20 mV. Although the threshold and latency of the excitatory junction potential were not affected by these concentrations of P-CTX-1, propagated impulses were blocked with $\geq$100 pM P-CTX-1. Spontaneous activity and the depolarization produced by P-CTX-1 were reduced in 0.1 mM Ca^{2+}/25 mM Mg^{2+} or 6 mM Ca^{2+} bath solution, or in the presence of 300 nM TTX, 100 nM ω-conotoxin GVIA (a selective N-type calcium channel inhibitor), or 0.1 mM Cd^{2+}. Subsequent addition of 1 μM phentolamine (an α-adrenoceptor inhibitor) repolarized the membrane, whereas 1 mM suramin (a nonselective inhibitor of purinergic receptors) selectively abolished excitatory junction potentials caused by P-CTX-1. Thus, P-CTX-1 releases norepinephrine and adenosine triphosphate (ATP) by initiating the asynchronous discharge of postganglionic perivascular nerves. In the presence of 100–400 pM P-CTX-1, the smooth muscle was depolarized to levels resembling those recorded in vivo, indicating that P-CTX-1 may be used to produce models of smooth muscle activity that are reminiscent of the in vivo situation (117).

B. Ciguatoxin Mobilizes Intracellular Ca^{2+} in Nerve Cells

Intracellular Ca^{2+} plays a key second-messenger role in numerous physiological processes in excitable cells (118). Low concentrations of P-CTX-1 (2.5–25.0 nM) increase the intracellular Ca^{2+} concentration in cultured, differentiated NG108-15 mouse neuroblastoma × rat glioma hybrid cells (119,120). The increase of intracellular Ca^{2+} concentration, assessed by fura-2–

based microfluorometric recordings, occurred either in cells bathed in a standard medium containing Ca^{2+} or after exposure to a Ca^{2+}-free medium supplemented with the Ca^{2+} chelator EGTA (ethylene glycol-bis-[β-aminoethyl ether]-N,N,N′,N′-tetraacetic acid) (Fig. 4A,B). Blockade of sodium channels by TTX completely prevented the P-CTX-1–induced increase in intracellular Ca^{2+} concentration (Fig. 4D), suggesting that the effect of the toxin on Ca^{2+} mobilization depends on Na^+ influx through sodium channels.

P-CTX-1–induced Ca^{2+} mobilization prevents the subsequent action of bradykinin, suggesting that the intracellular Ca^{2+} storage site stimulated by P-CTX-1 is the same as that activated by bradykinin (see Fig. 4C). Bradykinin has been shown in NG108-15 cells to increase the degradation of phosphatidylinositol-4,5-bisphosphate leading to the production of inositol 1,4,5-trisphosphate ($InsP_3$) and diacylglycerol (121). The mechanism of Ca^{2+} mobilization by bradykinin is considered to be predominantly through $InsP_3$-induced Ca^{2+} release from intracellular Ca^{2+} stores (118). It appears that P-CTX-1 and bradykinin may elevate intracellular Ca^{2+} through the same $InsP_3$-sensitive Ca^{2+}-stores in nerve cells. Substantial evidence indicates that enhanced influx of Na^+ can stimulate the production of $InsP_3$ in synaptosomes (122–124) and in cardiac myocytes (125), presumably by activation of phospholipase C. A Na^+-dependent mobilization

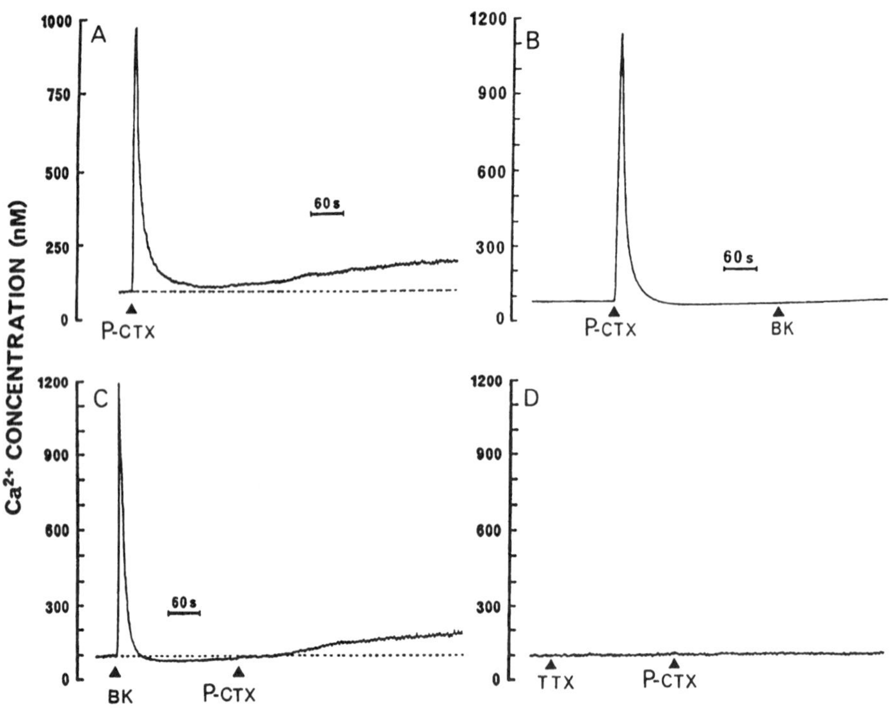

Figure 4 Changes of intracellular Ca^{2+} concentration ($[Ca^{2+}]_i$) caused by P-CTX-1 (25 nM) in neuroblastoma NG108-15 cells perfused either with a physiological solution containing Ca^{2+} (A,C) or with a Ca^{2+}-free solution supplemented with 1 mM EGTA (B,D) and the interaction between bradykinin (B,C) and tetrodotoxin (D) with P-CTX-1. Note in (A) that the transient increase followed by a slow rise of intracellular Ca^{2+} during the application of P-CTX-1 and in (B) the absence of action of 1 μM bradykinin following the transient mobilization of intracellular $[Ca^{2+}]$ by P-CTX-1. In (C) intracellular $[Ca^{2+}]$ mobilization induced by bradykinin (BK; 1 μM) prevents the transient, but not the slow, increase of intracellular $[Ca^{2+}]$ evoked by P-CTX-1, and in (D) that blockade of VSSCs with 1 μM tetrodotoxin (TTX) completely suppressed intracellular $[Ca^{2+}]$ mobilization caused by P-CTX-1. (Modified from ref. 120.)

of Ca^{2+} from $InsP_3$-sensitive stores could explain the increase in asynchronous quantal neurotransmitter release from motor nerve terminals that have been exposed to P-CTX-1 in a Ca^{2+}-free bath solution (126,127).

C. Role of Na^+-Ca^{2+} Exchange in the Action of Ciguatoxin

One transport system that plays a critical role in the control of Ca^{2+} homeostasis in many cells is the Na^+-Ca^{2+} exchanger, which harnesses the Na^+ gradient developed by the Na^+/K^+-ATPase to lower intracellular Ca^{2+} concentration. The plasma membrane Na^+-Ca^{2+} exchanger acts in concert with the plasma membrane ATP-driven Ca^{2+} pumps, Ca^{2+} channels, and intracellular Ca^{2+} sequestration systems to control cytoplasmic Ca^{2+} levels. The Na^+-Ca^{2+} exchanger is an electrogenic transporter whose operation depends on the prevailing ionic conditions, allowing it to act normally as a Ca^{2+} efflux pathway, or under certain conditions may even promote Ca^{2+} influx. Through the activation of sodium channels and consequent reduction in the Na^+ gradient, ciguatoxin reduces the ability of the Na^+-Ca^{2+} exchanger to extrude Ca^{2+} ions.

Exposure of cholinergic synaptosomes isolated from the electric organ of the fish *Torpedo marmorata* to P-CTX-1 (0.1 pM–10.0 nM) revealed that the toxin increases Na^+ influx into synaptosomes, thus enhancing acetylcholine release triggered by Ca^{2+} (128). This action does not result from P-CTX-1–induced depolarization of the synaptosomal membrane to levels above those needed to activate Ca^{2+} channels, because simultaneous blockade of Ca^{2+} channel subtypes by Gd^{3+}, ω-conotoxin GVIA, and FTX (a P-type calcium channel inhibitor from the venom of the American funnel-web spider, *Agelenopsis aperta*) did not prevent acetylcholine release caused by P-CTX-1 on addition of Ca^{2+}. In addition, little transmitter release was detected when Na^+ was replaced by Li^+, which is consistent with the fact that Li^+ cannot replace Na^+ in the Na^+-Ca^{2+} exchanger (129). Furthermore, inhibitors of the Na^+-Ca^{2+} exchanger, such as bepridil and cetiedil, completely prevented the Ca^{2+}-dependent acetylcholine release induced by P-CTX-1 (130). Therefore, P-CTX-1 causes an increase in Na^+ levels in synaptosomes that leads to an elevation of cytoplasmic Ca^{2+} via the Na^+-Ca^{2+} exchanger, and as a consequence promotes Ca^{2+}-dependent neurotransmitter release.

D. Ciguatoxin Induces Swelling of Nerve Cells

Another consequence of the persistent activation of voltage-dependent Na^+ channels induced by ciguatoxins at the resting membrane potential is the swelling of nerve cells. Swelling has been observed during the action of P-CTX-1, P-CTX-4B, or P-CTX-3C on myelinated axons, motor nerve terminals, and perisynaptic Schwann cells in situ, as determined by confocal laser scanning microscopy (131–134). These toxins induced a marked nodal swelling of single myelinated axons and spontaneous action potentials without apparent modification of the morphology of the internodal parts of nerve fibers characterized by the presence of myelin sheath layers (Fig. 5). Similarly, motor nerve terminals innervating skeletal muscle fibers, as well as perisynaptic Schwann cells in situ swelled when exposed to either P-CTX-1 or P-CTX-4B. Quantification of the swelling revealed that Pacific ciguatoxins cause an approximately twofold increase in the nodal volume of myelinated axons and in motor nerve terminal area per unit length. These effects were reversed, increasing the osmolality of the external solution by ~50% either with 100 mM D-mannitol, 50 mM tetramethylammonium chloride, or 100 mM sucrose. Although the hyperosmolar external solutions almost completely reversed the nodal swelling of myelinated axons induced by ciguatoxin, the solutions only partially decreased the nerve terminal swelling, which results from both the incorporation of synaptic vesicle membranes into the nerve terminal axolemma during stimulated neurotransmitter release and from osmotic changes.

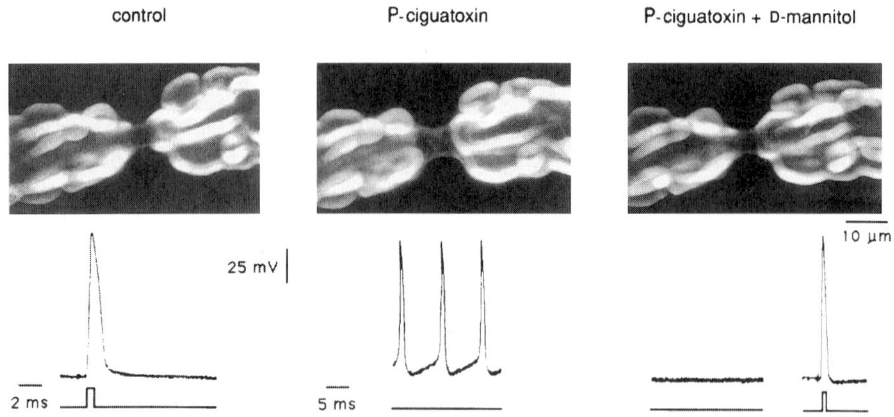

Figure 5 Effects of P-CTX-1 (10 nM) and hyperosmolar D-mannitol (110 mM) on the morphology of a single myelinated axon (A) and on action potentials recorded from a node of Ranvier (B). Note in (A) that the nodal swelling observed after 53-min exposure to P-CTX-1 is reversed by subsequent exposure for 55 min with hyperosmolar D-mannitol. The single myelinated fiber was stained with the styryl dye FM1-43 (2µM) and washout with dye-free medium for 15 min before optical sectioning and imaging with a confocal laser scanning microscope. Notice in (B) that the addition of D-mannitol to the external physiological solution containing P-CTX-1 suppressed the spontaneous action potentials recorded in the presence of P-CTX-1 but did not affect the action potential evoked by a depolarizing stimulus of 0.5-ms duration. (Modified from ref. 132.)

In the continuous presence of TTX, ciguatoxins did not cause significant changes in the nodal volume of myelinated axons or in the motor nerve terminal area per unit length. Thus, it is likely that ciguatoxins promote the entry of Na^+ through ciguatoxin-modified sodium channels that are permanently activated at the resting membrane potential and through unmodified sodium channels that open during ciguatoxin-induced spontaneous action potential discharges. The resulting increase in intracellular Na^+ concentration directly and indirectly disturbs the osmotic equilibrium between the intra- and extracellular medium and is associated with an influx of water that acts to restore both the osmotic equilibrium and the intracellular Na^+ concentration to basal levels. The cellular basis for the beneficial action of hyperosmotic D-mannitol in the treatment of ciguatera thus appears to be its ability to both shrink nerve cells swollen by ciguatoxins and to inhibit sodium channels activated by ciguatoxin (132).

E. Ciguatoxin Enhances Excitability of Nerves

As a consequence of the persistent activation of sodium channels at the resting membrane potential, ciguatoxin increases the excitability of nerve membranes. P-CTX-1 evokes a membrane depolarization which in turn causes spontaneous and/or repetitive action potential discharges at high frequencies (60–100 Hz) in myelinated axons and motor nerve terminals (21,126,132,135). In neuroblastoma cells and rat parasympathetic neurons, P-CTX-1 also induces membrane depolarization and, under appropriate conditions, induces spontaneous oscillations in the resting membrane potential and repetitive action potentials (136,137,) (Fig. 6). Research on the other ciguatoxins is less extensive, mainly due to difficulties in obtaining purified toxin. P-CTX-4B, from the dinoflagellate *G. toxicus*, has effects on myelinated nerve fibers that are similar to those of P-CTX-1, but it is approximately 50-fold less effective than P-CTX-1 (138). In contrast to P-CTX-1, P-CTX-4B decreases the amplitude and increases the duration

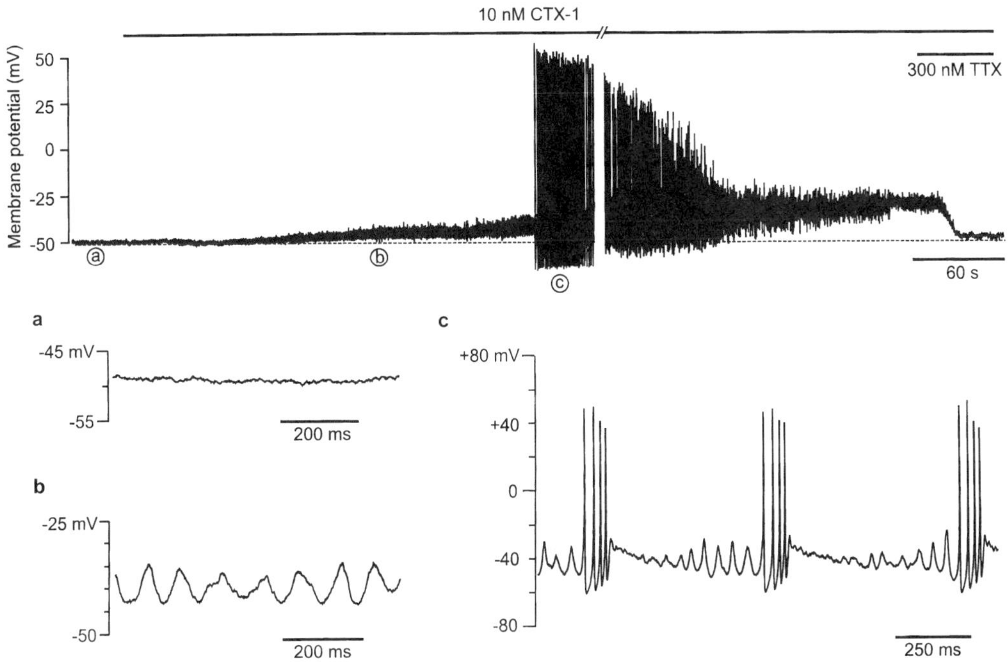

Figure 6 Effect of 10 nM P-CTX-1 on resting membrane potential and action potential firing in rat parasympathetic neurones (record is truncated at the gap by 180 s). The records shown at a faster time base reveal control membrane potential (a) and membrane oscillations (b) that develop into action potentials (c). Note the appearance of oscillations is briefly suppressed after a train of action potentials (c) and completely by tetrodotoxin (300 nM TTX). (Modified from ref. 137.)

of spontaneous action potentials, suggesting it may also block K^+ channels. The physiological consequence of any K^+ channel–blocking action of ciguatoxins remains to be elucidated. The ability of ciguatoxins to increase membrane excitability is consistent with studies on nerve conduction which show that partially purified extracts of ciguateric fish act to prolong the supernormal period of the nerve excitability (39,139).

It is noteworthy that in myelinated axons, either the local anesthetic lidocaine (50 μM), an increase in external Ca^{2+} concentration (1.8–5.4 mM), or an increase in external osmolality with D-mannitol (100 mM), tetramethylammonium chloride (50 mM), or sucrose (100 mM), can first decrease the frequency of spontaneous and repetitive action potentials induced by P-CTX-1 or P-CTX-4B, before progressively suppressing these action potentials (132,135,138). However, in the presence of these inhibitors, the action potentials evoked by depolarizing stimuli are still maintained, suggesting these agents act specifically to reverse the effects of ciguatoxin on VSSC nerves.

F. Effects of Ciguatoxin on Synaptic Transmission

Membrane depolarization and increased excitability of presynaptic nerve terminals induced by P-CTX-1 and C-CTX-1 markedly alter neurotransmission efficacy in various chemical synapses and secretory terminals. As a consequence of the changes in presynaptic excitability induced by ciguatoxins, volleys of action potentials can reach nerve terminals either spontaneously or in response to a single stimulus. Voltage-sensitive calcium channels open during each presynaptic action potential, and the ensuing rise of intracellular Ca^{2+} concentration triggers the exocytosis of synaptic vesicles, resulting in synchronous spontaneous or repetitive multiquantal neurotransmitter release (106; reviewed in ref. 21).

At vertebrate neuromuscular junctions, nanomolar concentrations of Pacific and Caribbean CTX-1 trigger spontaneous or repetitive endplate potentials at frequencies of up to 40 and 100 Hz, respectively, in response to a single nerve stimulus (112,126). Such synaptic activity at neuromuscular junctions is transient, and it is probably due to ciguatoxin-induced depolarization of motor nerve terminals that eventually has an indirect inhibitory effect. In neuromuscular junctions equilibrated with a low Ca^{2+}–high/Mg^{2+} bath solution, P-CTX-1 first increases the mean quantal content of endplate potentials, then reduces it, before finally blocking nerve-evoked transmitter release in an irreversible manner. It is likely that the increase in the averaged number of quanta released during P-CTX-1 action is not due to an enhanced Ca^{2+} influx through voltage-dependent calcium channels, but is due to the reversed-mode operation of the Na^+/Ca^{2+} exchanger, which allows Ca^{2+} entry in exchange for intracellular Na^+ as reported in isolated nerve endings (126,128,130).

Studies of guinea pig sympathetic ganglia using intracellular recording techniques showed that low concentrations of P-CTX-1 (0.2–0.8 nM) generated a marked increase in the frequency of spontaneous excitatory synaptic potentials, which often occurred in bursts (15–66 Hz) of similar amplitudes (Fig. 7) (140). However, the passive electrical properties, the amplitude and threshold of action potentials evoked by depolarizing current, and the threshold, latency, and form of the initial responses to nerve stimulation were not affected by P-CTX-1. Single stimuli to incoming nerves produced long-lasting repetitive synaptic responses arising from preganglionic

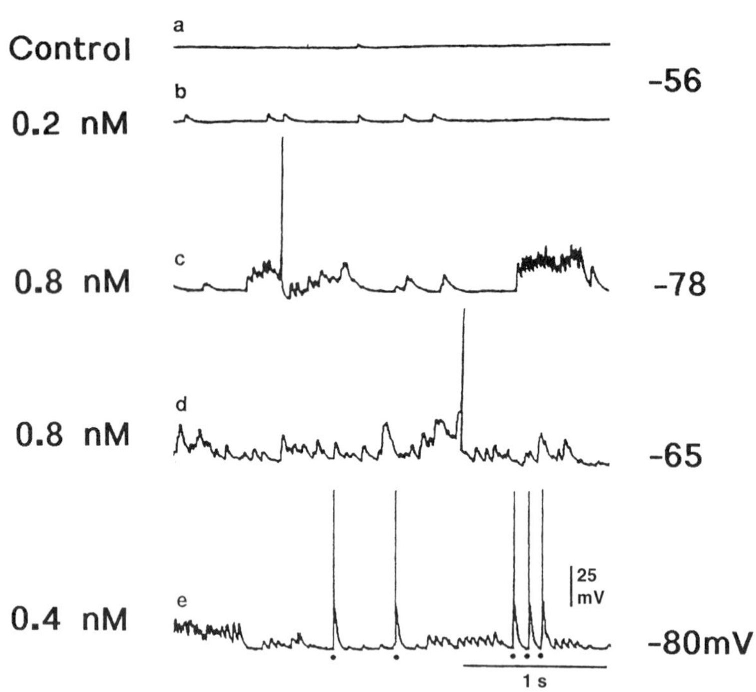

Figure 7 Concentration-dependent increase in the frequency of spontaneous excitatory potentials produced by P-CTX-1 in sympathetic ganglia. The concentration of ciguatoxin (nM) and cell resting membrane potential (mV) are given to the left and right of each trace, respectively. Several strong (suprathreshold) action potentials are indicated at the dots in (e). (Modified from ref. 140.)

axons. All activity generated by P-CTX-1 was significantly reduced or abolished by (+)-tubocurarine, hexamethonium (10 μM), TTX (≥0.1 μM), ω-conotoxin GVIA (100 nM), reduced Ca^{2+} (0.1 mM Ca^{2+})/10 mM Mg^{2+} Ringer), and elevated Ringer Ca^{2+} (6 mM) or Mg^{2+} (25 mM). The results of this study suggest that P-CTX-1 activates preganglionic axons by activating a subpopulation of VSSCs (140). These effects occurred at lower concentrations in unmyelinated than in myelinated axons, suggesting that many of the symptoms of ciguatera poisoning may be explained by activity at unmyelinated nerves.

Ciguatoxins also increase spontaneous quantal acetylcholine release from motor nerve terminals, which is detected as an increase in miniature endplate potential (MEPP) frequency. The increase in MEPP frequency occurred even in a Ca^{2+}-free medium supplemented with the calcium chelator EGTA, and leads to an exhaustion of neurotransmitter stores. Ultrastructural examination of these motor nerve terminals revealed both a marked depletion of synaptic vesicles and an increase in the nerve terminal perimeter, suggesting that the synaptic vesicle recycling process is impaired during the action of CTX-1 (127). The activation of VSSCs by ciguatoxins and the subsequent entry of Na^+ ions into nerve terminals appear to be responsible for the enhancement of asynchronous neurotransmitter release and the depletion of synaptic vesicles, since blockade of Na^+ influx into the terminals by TTX prevents these effects. These results further suggest that ciguatoxin-induced increases in intracellular Na^+ concentration in motor nerve endings can (i) mobilize intracellular stores of Ca^{2+}, (ii) directly activate the asynchronous neurotransmitter release process in motor nerve terminals, and (iii) impair the synaptic vesicle recycling process that is crucial for the maintenance of a functional population of synaptic vesicles during prolonged periods of transmitter release.

G. Mode(s) of Activation of Sodium Channels by Ciguatoxin

Nanomolar concentrations of ciguatoxin alter the electrical properties of excitable cells of various tissues causing increased Na^+ permeability and spontaneous action potential firing (reviewed in ref. 21). Following exposure to P-CTX-1 (0.4–10.0 nM), a fraction of the nodal Na^+ current of single myelinated axons failed to inactivate during a long depolarization step (132,135,141). Neither the amplitude of the peak Na^+ current nor its voltage dependence were modified by P-CTX-1 in the range of concentrations studied. In contrast, the activation threshold potential, the potential giving maximum inward current, and the reversal potential were more negative for the late Na^+ current than for the peak Na^+ current. Therefore, in the presence of P-CTX-1, a late Na^+ current is activated at the resting membrane potentials (−70 mV). These results indicate that P-CTX modifies a fraction of VSSCs which remain permanently open even at the resting membrane potential. Whether the Na^+ channels responsible for the late Na^+ current are a subset of channels distinct from those which are responsible for the peak Na^+ current remains to be determined. Although it is clear that ciguatoxin affects the biophysical properties of a fraction of the VSSCs, the pharmacological properties of toxin-modified Na^+ channels remain largely unaffected (132,135). For example, both peak Na^+ current (i.e., unmodified VSSCs) and late Na^+ current (i.e., toxin-modified VSSCs) were inhibited in a similar manner by the local anesthetic lidocaine by increasing the external Ca^{2+} concentration or by increasing the external osmolality with D-mannitol, tetramethylammonium, or sucrose.

Using whole-cell patch-clamp recording techniques, the effects of P-CTX-1 (0.2–20.0 nM) were examined on TTX-sensitive and TTX-resistant VSSCs present in acutely dissociated rat dorsal root ganglion neurons (142). P-CTX-1 had no effect on the time course of activation and inactivation; however, a concentration-dependent reduction in peak current amplitude was observed for both channel types (Fig. 8). At TTX-sensitive sodium channels, P-CTX-1 caused a 13-mV hyperpolarizing shift in the voltage dependence of activation, a 22-mV hyperpolarizing

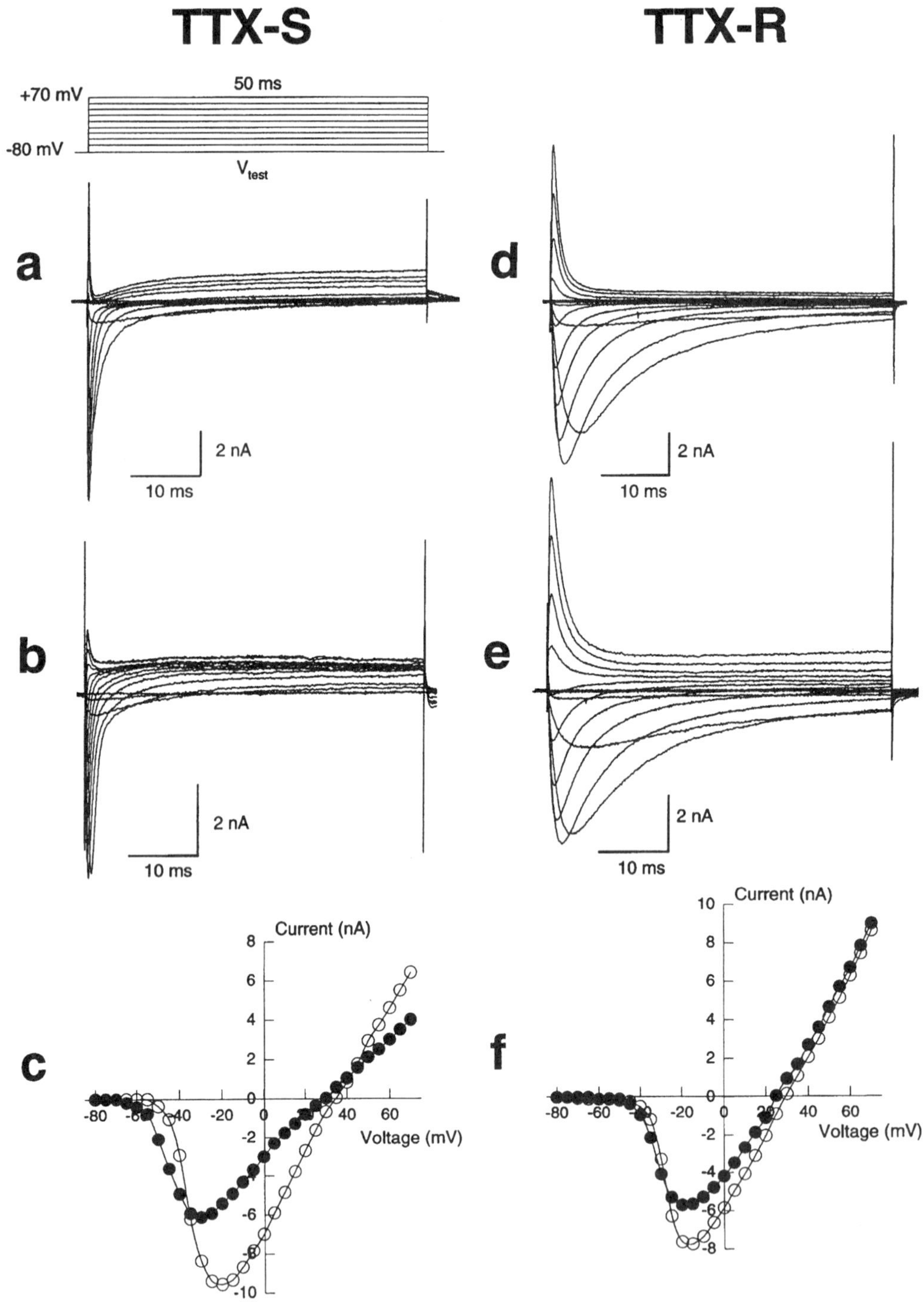

Figure 8 Typical effects of P-CTX-1 on the voltage-dependence of activation of TTX-sensitive and TTX-resistant Na^+ currents. Test pulse protocols (a) used 50-ms depolarizing test pulses from a holding potential of −80 mV to +70 mV in 5-mV increments before (a, d) and 10 min after (b, e) perfusion with 5 nM P-CTX-1. For clarity, only Na^+ currents recorded in 20-mV steps are presented. The peak Na^+ current at each voltage step was measured and plotted as a function of membrane potential (c, f) in the absence (○) and presence of 5 nM P-CTX-1 (●). (From ref. 142.)

shift in steady-state inactivation (h_∞), and a TTX-sensitive leakage current that also arose from ciguatoxin-modified TTX-sensitive sodium channels (Fig. 9). In contrast, the major effect of P-CTX-1 on TTX-resistant sodium channels was to increase the rate of recovery from sodium channel inactivation. These results indicate that P-CTX-1 modifies voltage-gated sodium channels present in peripheral sensory neurons to increase nerve excitability (142). A differential effect of ciguatoxin on TTX-sensitive and TTX-resistant sodium channels may contribute to the diversity of sensory neurological disturbances associated with ciguatera.

To gain a clearer understanding of how ciguatoxin increases neuronal excitability, the action of P-CTX-1 was examined on single VSSCs in isolated parasympathetic neurons from rat intracardiac ganglia using patch-clamp recording techniques (137). Under current clamp, bath application of 1–10 nM P-CTX-1 caused a gradual membrane depolarization and tonic action potential firing. Action potential firing ceased as the membrane became depolarized beyond ~−35 mV, and application of 300 nM TTX repolarized the cell to its control resting potential (see Fig. 6). In cell-attached membrane patches, 1–10 nM P-CTX-1 in the patch pipette markedly increased the open probability of single TTX-sensitive sodium channels in response to depolarizing voltage steps (Fig. 10B) but did not alter the unitary conductance (10 pS) or reversal potential from those of the unmodified channel. In about half the patches under steady-state conditions, P-CTX-1 caused spontaneous opening of single VSSCs which did not close even at membrane potentials hyperpolarized to −160 mV (Fig. 10C) (137). Thus, P-CTX-1 increases neuronal excitability by shifting the voltage dependence of activation of TTX-sensitive sodium channels to more negative potentials and by creating a persistent, noninactivating Na^+ current. From the single channel and whole-cell studies described above, it appears that P-CTX-1 acts on TTX-sensitive sodium channels to create two types of modified sodium channel, one where the voltage sensitivity of activation is shifted modestly in the hyperpolarizing direction (see Fig. 8c) and inactivation is impaired (Fig. 10B), and one where the channel fails to stay

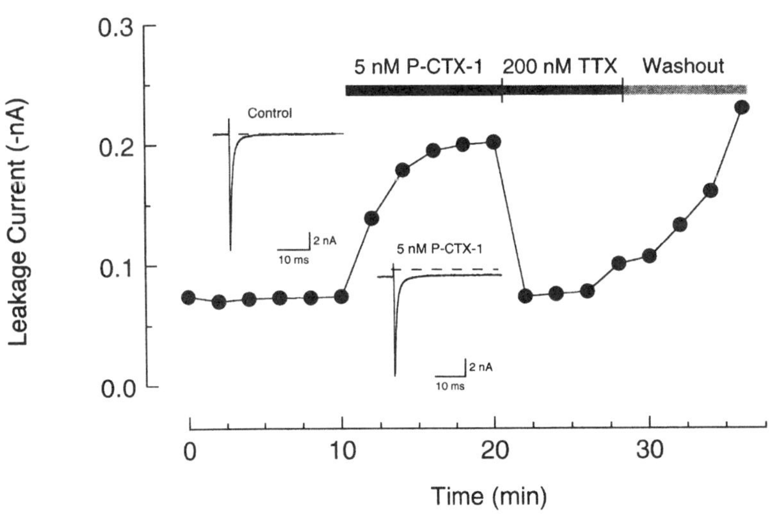

Figure 9 Leakage current produced by P-CTX-1 in dorsal root ganglion neurons. The leakage current induced by 5 nM P-CTX-1 in a cell expressing TTX-S Na^+ currents was reversed by TTX (200 nM). The effects of TTX, but not those of P-CTX-1, were reversed on perfusing the preparation with toxin-free bath solution. The insets show typical currents recorded in the absence (left) and presence (right) of 5 nM P-CTX-1. The dotted line represents zero current. Significant leakage current was not associated with the ciguatoxin-modified TTX-resistant sodium channels. (Modified from ref. 142.)

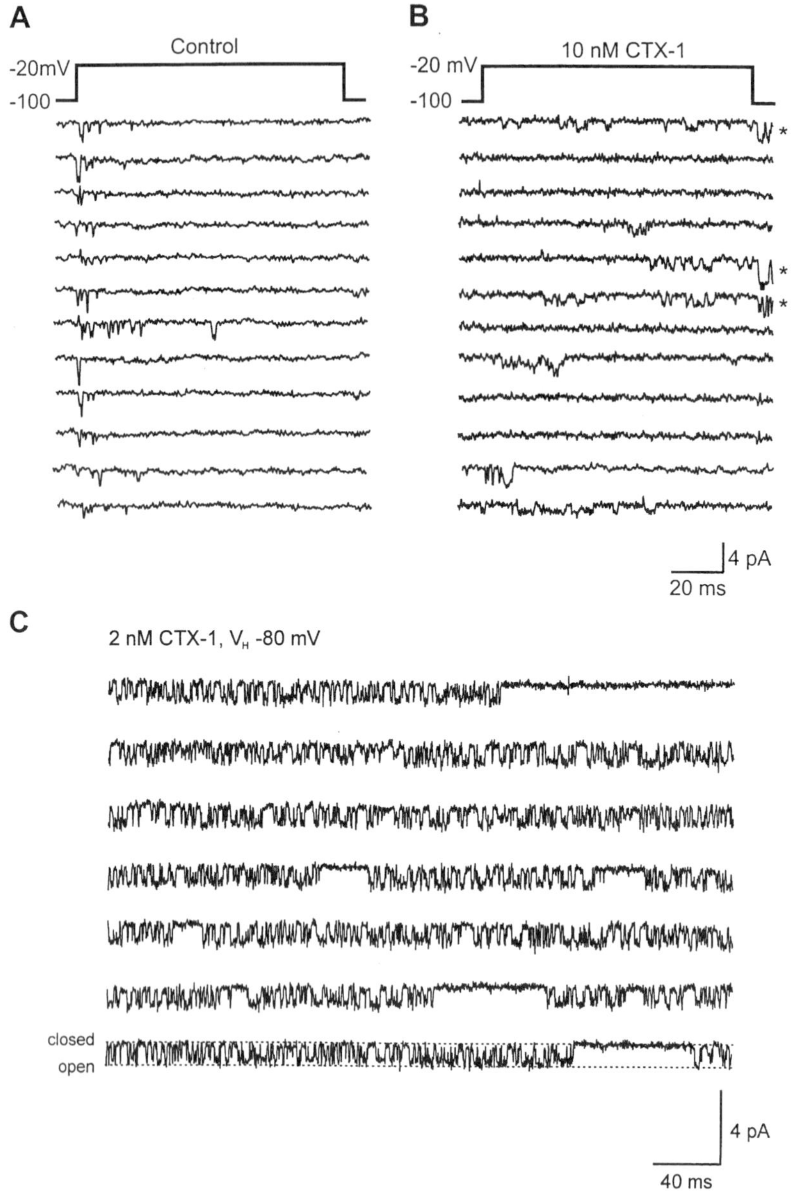

Figure 10 Single Na^+ channel currents evoked by step depolarizations and under steady-state conditions in rat parasympathetic neurons. (A) Sodium channel currents recorded from a cell-attached patch in the absence of P-CTX-1 (control). (B) Unitary Na^+ currents elicited in the presence of 10 nM P-CTX-1 in the patch pipette. Asterisks indicate channel openings observed on repolarization to −100 mV. The membrane potential of the patch in (A) and (B) was stepped from a holding potential of −100 to −20 mV for 900 ms (membrane currents filtered at 5 kHz). (C) Spontaneous sodium channel openings obtained from a cell-attached membrane patch held at −80 mV in the presence of 2 nM P-CTX-1 under steady-state conditions. These sodium channels failed to stay closed even at −160 mV. The inward currents in each of these records were ~10 pS, and are shown as downward deflections. (Modified from ref. 137.)

closed even at hyperpolarized membrane potentials. This latter effect most likely underlies the observed leakage current. Experiments examining the effects of ciguatoxin on expressed sodium channel clones are required to determine if ciguatoxin takes the sodium channel into two different gating modes or if these phenomena arise through differential effects on two different populations of sodium channel. Different populations of sodium channels could arise because of differences in the auxiliary subunits, different extents of posttranslational modification, and/or different sodium channel α-subunits (e.g., different splice variants). The origin of the TTX-sensitive membrane potential oscillations observed in several cell types also remains to be determined.

H. Ciguatoxin Binding to Sodium Channels

In addition to having similar polyether structures (see Fig. 1), ciguatoxins and brevetoxins both selectively target a common binding site on the neuronal VSSC protein (15,18,19,143–145). Ligand-binding studies have revealed more than 12 distinct classes of biologically active neurotoxins that interact with at least 6 specific receptor sites identified on the α-subunit of neuronal sodium channels (see Fig. 2). Brevetoxins bind with high affinity to neurotoxin receptor site 5 of the sodium channel, as revealed by direct binding studies using radiolabeled [^{3}H]PbTx-3 and by binding assays that show noncompetitive interactions between site 5 and a variety of toxin probes specific for sites 1–4 (136,144,146,147). Functional studies, such as increasing the potency of batrachotoxin-induced Na^+ influx into cells (see Fig. 3), further show that ciguatoxin does not act as a competitor at sites 1–4 on the VSSC (136).

Ciguatoxins are not available as radiolabeled analogues for direct investigation of their specific binding; therefore, [^{3}H]PbTx-3 has been used in homologous and heterologous displacement experiments to define site 5 toxins. Competition studies between purified ciguatoxin (148) or pure P-CTX-1, P-CTX-2, or P-CTX-3 (15,143,149) and [^{3}H]PbTx-3 indicate that the ciguatoxins are competitive inhibitors of brevetoxin binding, and thus have overlapping binding at site 5 on the sodium channel (see Fig. 2).

Using a photolabeled derivative of PbTx-3 and site-directed antibody mapping, the partial localization of the receptor site 5 of voltage-dependent Na^+ channels (from rat brain) has been suggested to be in the region of interaction of segments S6 and S5 of domains I and IV, respectively (150) (see Fig. 2). Further details of the site 5 binding site on the sodium channel is described further in Chapter 23.

I. Sodium Channel Assays for Ciguatoxin

Manger et al. (151,152) have shown that neuroblastoma cells in culture can be developed as a simple and sensitive method to detect sodium channel–specific marine toxins using a mitochondrial dehydrogenase activity endpoint. Cells are sensitized to the action of ciguatoxin by the addition of veratridine (a partial agonist at site 2) and ouabain (an inhibitor of the Na^+/K^+ ATPase). The assay responds in a dose-dependent manner to ciguatoxins, brevetoxins, and saxitoxins and delineates the toxic activity as either sodium channel–enhancing or sodium channel–blocking toxins. Within 6 h, brevetoxins can be detected at ≥250 pg and purified CTX-1 can be detected at subpicogram levels. The results obtained from cell bioassay of ciguateric fish extracts correlate with those obtained by mouse bioassay. A sensitive cell-based assay for brevetoxins, saxitoxins, and ciguatoxins has also been reported that employs a c-*fos*–luciferase reporter gene stably expressed in cells (153). [^{3}H]PbTx-3 binding to brain membrane can also be used to detect ciguatoxins in crude extracts (145) and can be developed into high-throughput assays. These sodium channel assays, and possibly antibody-based assays (154), have the potential to replace animal testing for ciguatoxins, and they warrant further validation to determine the potential of such assays to be developed into rapid screens for public health protection.

V. SUMMARY

Considerable progress has been made in determining the chemical structures of the principal toxins involved in ciguatera and in characterizing their cellular modes of action. However, only rarely are diagnoses based on the identification of the bioactive agents present in the remains of the fish implicated in the poisoning. This is a concern, since the differential diagnosis of the various types of fish poisoning depends on clinical recognition of specific signs and symptoms, but it is complicated by difficulties in distinguishing ciguatera from other forms of seafood poisoning and other illnesses. For most types of fish poisoning described here, treatment remains nonspecific, symptomatic, and supportive. Various substances and herbal remedies have been used for the treatment of ciguatera, including local anesthetics, calcium gluconate, vitamins, antidepressants, and glucose. However, the efficacy of these therapeutic agents remains uncertain, and antagonists effective at specifically reversing the pathophysiological basis of ciguatoxin action remain elusive. One exception is intravenous hyperosmolar D-mannitol, which has evolved as a unique remedy for acutely poisoned patients (40,41). The development of specific tests that detect the presence of ciguatoxin and related toxins in fish prior to consumption will significantly improve the management of ciguatera and overcome present limitations of diagnosis and existing therapies.

The action of ciguatoxin is to increase cell Na^+ permeability through VSSCs which open at normal resting membrane potentials. As a consequence, ciguatoxins affect various Na^+-dependent mechanisms to enhance membrane excitability, activate Na^+-Ca^{2+} exchange, induce mobilization of intracellular Ca^{2+}, and produce cell swelling. The effects of ciguatoxin are most prominent in nerves. These neurocellular actions of ciguatoxin are consistent with the generalized disturbance of nerve conduction, synaptic transmission, and cellular morphology observed in intoxicated patients (45,139), and they provide a basis for understanding the beneficial action of D-mannitol.

ACKNOWLEDGMENTS

Marine toxin research at CNRS was funded by grant no. 98/1051 from Direction des Systèmes des Forces et de la Prospective (to J.M.), and in Australia by a grant from the Australian Research Council (to D.J.A. and R.J.L.). We thank Drs R. Hogg and E. Benoit for helpful discussions.

REFERENCES

1. RJ Lewis, MJ Holmes. Origin and transfer of toxins involved in ciguatera. Comp Biochem Physiol 106C:615–628, 1993.
2. PJ Scheuer, W Takahashi, J Tsutsumi, T Yoshida. Ciguatoxin isolation and chemical nature. Science 155:1267–1268, 1967.
3. K Tachibana, M Nukina, Y Joh, P Scheuer. Recent developments in the molecular structure of ciguatoxin. Biol Bull 172:122–127, 1987.
4. M Murata, AM Legrand, Y Ishibashi, M Fukui, T Yasumoto. Structures and configurations of cigua-

toxin from the Moray eel *Gymnothorax-javanicus* and its likely precursor from the dinoflagellate *Gambierdiscus-toxicus*. J Am Chem Soc 112:4380–4386, 1990.
5. RJ Lewis, JP Vernoux, IM Brereton. Structure of Caribbean ciguatoxin isolated from *Caranx latus*. J Am Chem Soc 120: 5914–5920, 1998.
6. AM Legrand, M Fukui, P Cruchet, Y Ishibashi, T Yasumoto. Characterization of ciguatoxins from different fish species and wild *Gambierdiscus toxicus*. In: TR Tosteson, ed. Proceedings of the 3rd International Conference on Ciguatera Fish Poisoning: Polyscience Publications, Quebec 1992, pp 25–32.
7. M Murata, AM Legrand, Y Ishibashi, T Yasumoto. Structures and configurations of ciguatoxin and its congener. J Am Chem Soc 111:8929–8931, 1989.
8. MJ Holmes, RJ Lewis, MA Poli, NC Gillespie. Strain dependent production of ciguatoxin precursors (gambiertoxins) by *Gambierdiscus toxicus* (Dinophyceae) in culture. Toxicon 29:761–775, 1991.
9. M Satake, M Murata, T Yasumoto. The structure of CTX3c, a ciguatoxin congener isolated from cultured *Gambierdiscus toxicus*. Tetrahedron Lett 34:1975–1978, 1993.
10. M Satake, Y Ishibashi, AM Legrand, T Yasumoto. Isolation and structure of ciguatoxin-4A, a new ciguatoxin precursor, from cultures of dinoflagellate *Gambierdiscus toxicus* and parrotfish *Scarus gibbus*. Biosci Biotech Biochem 60:2103–2105, 1997.
11. AM Legrand, M Litaudon, JN Genthon, R Bagnis, T Yasumoto. Isolation and some properties of ciguatoxin. J Appl Phycol 1:183–188, 1989.
12. JP Vernoux, RJ Lewis. Isolation and characterisation of Caribbean ciguatoxins from the horse-eye jack (*Caranx latus*). Toxicon 35:889–900, 1997.
13. RJ Lewis, M Sellin. Recovery of ciguatoxin from fish flesh. Toxicon 31:1333–1336, 1993.
14. RJ Lewis, A Jones. Characterization of ciguatoxins and ciguatoxin congeners present in ciguateric fish by gradient reverse-phase high-performance liquid chromatography/mass spectrometry. Toxicon 35:159–168, 1997.
15. RJ Lewis, M Sellin, MA Poli, RS Norton, JK Macleod, MM Sheil. Purification and characterization of ciguatoxins from moray eel (*Lycodontis javanicus, Muraenidae*). Toxicon 29:1115–1127, 1991.
16. RJ Lewis, RS Norton, IM Brereton, CD Eccles. Ciguatoxin-2 is a diastereomer of ciguatoxin-3. Toxicon 31:637–643, 1993.
17. M Satake, M Fukui, AM Legrand, P Cruchet, T Yasumoto. Isolation and structures of new ciguatoxin analogs, 2,3-dihydroxyCTX3C and 51-hydroxyCTX3C, accumulated in tropical reef fish. Tetrahedron Lett 39:1197–1198, 1998.
18. DG Baden. Brevetoxins: unique polyether dinoflagellate toxins. FASEB J 3:1807–1817, 1989.
19. RE Gawley, KS Rein, M Kinoshita, DG Baden. Binding of brevetoxins and ciguatoxin to the voltage-sensitive sodium channel and conformational analysis of brevetoxin B. Toxicon 30:780–785, 1992.
20. T Yasumoto, M Murata. Marine toxins. Chem Rev 93:1897–1909, 1993.
21. J Molgó, E Benoit, JX Comella, AM Legrand. Ciguatoxin: a tool for research on sodium-dependent mechanisms. In: PM Conn, ed. Methods in Neuroscience, Neurotoxins. Vol 8. New York: Academic Press, 1992, pp 149–164.
22. DG Baden, LE Fleming, JA Bean. Marine toxins. In: FA de Wolff, ed. Handbook of Clinical Neurology. Vol 21: Intoxications of the Nervous System, Part II. New York: Elsevier, 1995, pp 141–175.
23. D Gordon. Ion channels in nerve and muscle cells. Curr Opin Cell Biol 2:695–707, 1990.
24. D Gordon. Sodium channels as targets for neurotoxins. In: Y Gutman, P Lazarowici, ed. Toxins and Signal Transduction. The Netherlands: Harwood, 1997, pp 119–149.
25. NC Gillespie, RJ Lewis, J Pearn, ATC Burke, MJ Holmes, JB Bourke, WJ Shields. Ciguatera in Australia: occurrence, clinical features, pathophysiology and management. Med J Aust 145:584–590, 1986.
26. RJ Lewis. Socioeconomic impacts and management of ciguatera in the Pacific. Bull Soc Pathol Exot 85:427–434, 1992.
27. R Bagnis, T Kuberski, S Laugier. Clinical observations on 3,009 cases of ciguatera (fish poisoning) in the South Pacific. Am J Trop Med Hyg 28:1067–1073, 1979.
28. DN Lawrence, MB Enriquez, RM Lumish, A Maceo. Ciguatera fish poisoning in Miami. JAMA 244:254–258, 1980.

29. JH Gollop, EW Pon. Ciguatera: a review. Hawaii Med J 51:91–99, 1992.
30. R Lewis, T Ruff. Ciguatera: ecological, clinical, and socioeconomic perspectives. Crit Rev Environ Sci Technol 23:137–156, 1993.
31. AEB Swift, TR Swift. Ciguatera. J Toxicol Clin Toxicol 31:1–29, 1993.
32. WR Lange. Ciguatera fish poisoning. Am Fam Physician 50:579–584, 1994.
33. P Glaziou, AM Legrand. The epidemiology of ciguatera fish poisoning. Toxicon 32:863–873, 1994.
34. JP Quod, J Turquet. Ciguatera in Reunion Island (SW Indian Ocean): epidemiology and clinical patterns. Toxicon 34:779–785, 1996.
35. DZ Levine. Ciguatera: current concepts. J Am Osteopath Assoc 95:193–198, 1995.
36. J Cameron, MF Capra. The basis of the paradoxical disturbance of temperature perception in ciguatera poisoning. J Toxicol Clin Toxicol 31:571–579, 1993.
37. YG Peng, TB Taylor, RE Finch, PD Moeller, JS Ramsdell. Neuroexcitatory actions of ciguatoxin on brain-regions associated with thermoregulation. Neuroreport 6:305–309, 1995.
38. TY Chan, AY Wang. Life-threatening bradycardia and hypotension in a patient with ciguatera fish poisoning. Trans R Soc Trop Med Hyg 87:71, 1993.
39. J Cameron, AE Flowers, MF Capra. Effects of ciguatoxin on nerve excitability in rats (Part I). J Neurol Sci 101:87–92, 1991.
40. NA Palafox, LG Jain, AZ Pinano, TM Gulick, RK Williams, IJ Schatz. Successful treatment of ciguatera fish poisoning with intravenous mannitol. JAMA 259:2740–2742, 1988.
41. JH Pearn, RJ Lewis, T Ruff, M Tait, J Quinn, W Murtha, G King, A Mallett, NC Gillespie. Ciguatera and mannitol: experience with a new treatment regimen. Med J Aust 151:77–80, 1989.
42. NA Palafox. Review of the clinical use of intravenous mannitol with ciguatera fish poisoning from 1988 to 1992. Bull Soc Pathol Exot 85:423–424, 1992.
43. DG Blythe, DP De Sylva, LE Fleming, RA Ayyar, DG Baden, K Shrank. Clinical experience with i.v. mannitol in the treatment of ciguatera. Bull Soc Pathol Exot 85:425–426, 1992.
44. PJ Fenner, RJ Lewis, JA Williamson, ML Williams. A Queensland family with ciguatera after eating coral trout. Med J Aust 66:473–475, 1997.
45. JL Allsop, L Martini, H Lebris, J Pollard, J Walsh, S Hodgkinson. Neurologic manifestations of ciguatera. 3 cases with a neurophysiologic study and examination of one nerve biopsy. Rev Neurol (Paris) 142:590–597, 1986.
46. R Bagnis, A Spiegel, JP Boutin, C Burucoa, L Nguyen, JL Cartel, P Capdevielle, P Imbert, D Prigent, C Gras, J Roux. Evaluation de l'efficacite du mannitol dans le traitment de la ciguatera en Polynesie Francaise. Med Trop 52:67–73, 1992.
47. RT Davis, LA Villar. Symptomatic improvement with amitriptyline in ciguatera fish poisoning. N Engl J Med 315:65, 1986.
48. GM Calvert, DO Hryhorczuk, JB Leikin. Treatment of ciguatera fish poisoning with amitriptyline and nifedipine. J Toxicol Clin Toxicol 25:423–428, 1987.
49. WR Lange, SD Kreider, M Hattwick, J Hobbs. Potential benefit of tocainide in the treatment of ciguatera: report of three cases. Am J Med 84:1087–1088, 1988.
50. RM Berlin, SL King, DG Blythe. Symptomatic improvement of chronic fatigue with fluoxetine in ciguatera fish poisoning. Med J Aust 157:567, 1992.
51. Bourdy G, Cabalion P, Amade P, Laurent D. Traditional remedies used in the western Pacific for the treatment of ciguatera poisoning. Ethnopharmacology 36:163–174, 1992.
52. RJ Lewis, AW Hoy, M Sellin. Ciguatera and mannitol: *in vivo* and *in vitro* assessment in mice. Toxicon 31:1039–1050, 1993.
53. CE Purcell, MF Capra, J Cameron. Action of mannitol in ciguatoxin-intoxicated rats. Toxicon 37: 67–76, 1999.
54. J Cameron, AE Flowers, MF Capra. Modification of the peripheral nerve disturbance in ciguatera poisoning in rats with lidocaine. Muscle Nerve 16:782–786, 1993.
55. AW Wong Hoy, RJ Lewis. The effect of potential therapeutics on ciguatoxin-inhibition of the rat phrenic nerve. In: P Gopalokrishnakone, CK Tan, eds. Recent Advances in Toxinology Research. Vol. 2. National University of Singapore, Venom and Toxin Research Group, 1992, pp 509–519.
56. AM Legrand, M Galonnier, R Bagnis. Studies on the mode of action of ciguateric toxins. Toxicon 20:311–315, 1982.

57. K Terao, E Ito, M Oarada, Y Ishibashi, AM Legrand, T Yasumoto. Light and electron microscopic studies of pathologic changes induced in mice by ciguatoxin poisoning. Toxicon 29:633–643, 1991.
58. K Terao, E Ito, T Yasumoto. Light and electron microscopic studies of the murine heart after repeated administrations of ciguatoxin or ciguatoxin-4c. Nat Toxins 1:19–26, 1992.
59. AH Banner, S Sasaki, P Helfrich, CB Alender, PJ Scheuer. Bioassay of ciguatera toxin. Nature 189:229–230, 1961.
60. E Chungue, R Bagnis, F Parc. The use of mosquitoes (*Aedes aegypti*) to detect ciguatoxin in surgeon fishes (*Ctenochaetus striatus*). Toxicon 22:161–164, 1984.
61. JP Vernoux, N Lahlou, LP Magras, JB Greaux. Chick feeding test: a simple system to detect ciguatoxin. Acta Trop 42:235–240, 1985.
62. H Labrousse, S Pauillac, C Jehl-Martinez, AM Legrand, S Avrameas. Techniques de détection de la ciguatoxine *in vivo* et *in vitro*. Océanis 18:189–191, 1992.
63. RJ Lewis. Detection of ciguatoxins and related benthic dinoflagellate toxins: in vivo and in vitro methods. In: GM Hallegraph, DM Anderson, AD Cembella, eds. Manual on Harmful Marine Microalgae IOC Manuals and Guides UNESCO, Paris, France. No. 33, 1995, pp 135–161.
64. H Labrousse, L Matile. Toxicological biotest on Diptera larvae to detect ciguatoxins and various other toxic substances. Toxicon 34:881–891, 1996.
65. PA Hoffman, HR Granade, JP McMillan. The mouse ciguatoxin bioassay: a dose–response curve and symptomatology analysis. Toxicon 21:363–369, 1983.
66. RJ Lewis, M Sellin. Recovery of ciguatoxin from fish flesh. Toxicon 31:1333–1336, 1993.
67. RJ Lewis. Ciguatoxins are potent ichthyotoxins. Toxicon 30:207–211, 1992.
68. M Murata, H Naoki, S Matsunaga, M Satake, T Yasumoto. Structure and partial stereochemical assignments for maitotoxin, the most toxic and largest natural non-biopolymer. J Am Chem Soc 116:7098–7107, 1994.
69. RJ Lewis, MJ Holmes, PF Alewood, A Jones. Ionspray mass spectrometry of ciguatoxin-1, maitotoxin-2 and -3, and related marine polyether toxins. Nat Toxins 2:56–63, 1994.
70. MJ Holmes, RJ Lewis. Purification and characterisation of large and small maitotoxins from cultured *Gambierdiscus toxicus*. Nat Toxins 2:64–72, 1994.
71. F Gusovsky, JW Daly. Maitotoxin: a unique pharmacological tool for research on calcium-dependent mechanisms. Biochem Pharmacol 39:1633–1639, 1990.
72. M Takahashi, Y Ohizumi, T Yasumoto. Maitotoxin, a Ca^{2+} channel activator candidate. J Biol Chem 257:10944, 1982.
73. M Takahashi, M Tatsumi, Y Ohizumi, T Yasumoto. Ca^{2+} channel activating function of maitotoxin, the most potent marine toxin known, in clonal pheochromocytoma cells. J Biol Chem 258:10944–10949, 1983.
74. M Taglialatela, LM Canzoniero, A Fatatis, G Di Renzo, T Yasumoto, L Annunziato. Effect of maitotoxin on cytosolic Ca^{2+} levels and membrane potential in purified rat brain synaptosomes. Biochim Biophys Acta 1026:126–132, 1990.
75. F Gusovsky, JA Bitran, T Yasumoto, JW Daly. Mechanism of maitotoxin-stimulated phosphoinositide breakdown in HL-60 cells. J Pharmacol Exp Ther 252:466–473, 1990.
76. JE Merritt, WP Armstrong, CD Benham, TJ Hallam, R Jacob, A Jaxa-Chamiec, BK Leigh, SA McCarthy, KE Moores, TJ Rink. SK&F 96365, a novel inhibitor of receptor-mediated calcium entry. Biochem J 271:515–522, 1990.
77. DG Soergel, T Yasumoto, J Daly, F Gusovsky. Maitotoxin effects are blocked by SK&F 96365, an inhibitor of receptor-mediated calcium entry. Mol Pharmacol 41:487–493, 1992.
78. RJ Lang, F Vogalis, MJ Holmes, RJ Lewis. Maitoxin induces muscle contraction and a non-selective cationic current in single smooth muscle cells of the guinea-pig proximal colon. Mem Qld Mus 34: 533–540.
79. LI Escobar, C Salvador, M Martínez, L Vaca. Maitotoxin, a cationic channel activator. Neurobiology 6:59–74, 1998.
80. P Dietl, H Völkl. Maitotoxin activates a nonselective cation channel and stimulates Ca^{2+} entry in MDCK renal epithelial cells. Mol Pharmacol 45:300–305, 1994.
81. RE Moore, PJ Scheuer. Palytoxin: a new marine toxin from a coelenterate. Science 172:495–498, 1971.

82. S Gleibs, D Mebs, B Werding. Studies on the origin and distribution of palytoxin in a Caribbean coral reef. Toxicon 33:1531–1537, 1995.
83. VM Mahnir, EJ Kozlovskaja. Isolation of palytoxin from the sea anemone *Radianthus macrodactylus*. Bioorg Khim 18:751–752, 1992.
84. T Yasumoto, D Yasumura, Y Ohizumi, M Takahashi, AC Acala, AL Alcala. Palytoxin in two species of Xanthid crab from the Philippines. Agric Biol Chem 50:163–167, 1986.
85. BW Halstead. Poisonous and Venomous Marine Animals of the World. Vol 2. Washington, DC: U.S. Government Printing Office, 1967, pp 679–699.
86. M Fukui, M Murata, A Inoue, M Gawel, T Yasumoto. Occurrence of palytoxin in the trigger fish *Melichtys vidua*. Toxicon 25:1121–1124, 1987.
87. Y Hashimoto, N Fusetani, S Kimura. Aluterin: a toxin of filefish, *Alutera scripta*, probably originated from a zoantharian, *Palythoa tuberculosa*. Bull Jpn Soc Scient Fish 35:1086, 1969.
88. M Kodama, Y Hokama, T Yasumoto, M Fukui, SJ Manea, N Sutherland. Clinical and laboratory findings implicating palytoxin as a cause of ciguatera poisoning due to *Decapterus macrosoma* (mackerel). Toxicon 27:1051–1053, 1989.
89. DZ Levine, H Fujiki, HB Gjika, H Van Vunakis. A radioimmunooassay for palytoxin. Toxicon 26: 1115–1121, 1988.
90. T Noguchi, DF Hwang, O Arakawa, K Daigo, S Sato, H Ozaki, N Kawai, M Ito, K Hashimoto. Palytoxin as the causative agent in the parrotfish poisoning. In: P Gopalakrishnakone, CK Tan, eds. Progress in Venom and Toxin Research. Singapore: Kent Ridge, 1987, pp 325–335.
91. H Okano, H Masuoka, S Kamei, T Seko, S Koyabu, K Tsuneoka, T Tamai, K Ueda, S Nakazawa, M Sugawa, H Suzuki, M Watanabe, R Yatani, T Nakano. Rhabdomyolysis and myocardial damage induced by palytoxin, a toxin of blue humphead parrotfish. Intern Med 37:330–333, 1998.
92. CH Wu, T Narahashi. Mechanism of action of novel marine neurotoxins on ion channels. Ann Rev Pharmacol Toxicol 28:141–161, 1988.
93. E Habermann. Palytoxin acts through Na^+,K^+-ATPase. Toxicon 27:1171–1187, 1989.
94. T Shimahara, J Molgo. Palytoxin enhances quantal acetylcholine release from motor nerve terminals and increases cytoplasmic calcium levels in a neuronal hybrid cell line. Life Sci Adv Pharmacol 9:785–792, 1990.
95. C Frelin, C Van Renterghem. Palytoxin. Recent electrophysiological and pharmacological evidence for several mechanisms of action. Gen Pharmacol 26:33–37, 1995.
96. SY Kim, KA Marx, CH Wu. Involvement of the Na,K/ATPase in the induction of ion channels by palytoxin. Naunyn-Schmiedeberg's Arch Pharmacol 351:542–554, 1995.
97. G Scheiner-Bobis, H Schneider. Palytoxin-induced channel formation within the Na+/K+-ATPase does not require a catalytically active enzyme. Eur J Biochem 248:717–723, 1997.
98. X Wang, JD Horisberger. Palytoxin effects through interaction with the Na,K-ATPase in *Xenopus oocyte*. FEBS Lett 409:391–395, 1997.
99. M Usami, M Satake, S Ishida, A Inoue, Y Kan, T Yasumoto. Palytoxin analogs from the dinoflagellate *Ostreopsis siamensis*. J Am Chem Soc 117:5389–5390, 1995.
100. Y Onuma, M Satake, T Ukena, J Roux, S Chanteau, N Rasolofonirina, M Ratsimaloto, H Naoki, T Yasumoto. Identification of putative palytoxin as the cause of clupeotoxism. Toxicon 37:55–65, 1999.
101. U Anthoni, C Christophersen, L Gram, NH Nielsen, P Nielsen. Poisonings from flesh of the Greenland shark *Somniosus microcephalus* may be due to trimethylamine. Toxicon 29:1205–1212, 1991.
102. P Boisier, G Ranaivoson, N Rasolofonirina, B Andriamahefazafy, J Roux, S Chanteau, M Satake, T Yasumoto. Fatal ichthyosarcotoxism after eating shark meat. Implications of two new marine toxins. Arch Inst Pasteur Madagascar 61:81–83, 1994.
103. GG Habermehl, HC Krebs, P Rasoanaivo, A Ramialiharisoa. Severe ciguatera poisoning in Madagascar: a case report. Toxicon 32:1539–1542, 1994.
104. A Ramialiharisoa, R Rafenoherimanana, L De Haro, J Jouglard. Collective poisoning of ciguateric type after ingestion of shark in Madagascar. Data collected by the Antananarivo medical team. Presse Med 25:1350, 1996.
105. P Boisier, G Ranaivoson, N Rasolofonirina, B Andriamahefazafy, J Roux, S Chanteau, M Satake,

T Yasumoto. Fatal mass poisoning in Madagascar following ingestion of a shark (Carcharhinus leucas): clinical and epidemiological aspects and isolation of toxins. Toxicon 33:359–364, 1995.
106. J Molgo, FA Meunier, MY Dechraoui, E Benoit, C Mattei, AM Legrand. Sodium-dependent alterations of synaptic transmission mechanisms by brevetoxins and ciguatoxins. In: B Reguera, J Blanco, ML Fernández, T Wyatt, eds. Harmful Microalgae. Santiago De Compostella: Xunta de Galicia and Intergovernmental Oceanographic Commission of UNESCO, 1998, pp 594–597.
107. AM Legrand, R Bagnis. Effects of ciguatoxin and maitotoxin on isolated rat atria and rabbit duodenum. Toxicon 22:471–475, 1984.
108. RJ Lewis, R Endean. Direct and indirect effects of ciguatoxin on guinea-pig atria and papillary muscles. Naunyn-Schmiedeberg's Arch Pharmacol 334:313–322, 1986.
109. RJ Lewis, AW Hoy, DC McGiffin. Action of ciguatoxin on human atrial trabeculae. Toxicon 30: 907–914, 1992.
110. A Seino, M Kobayashi, K Momose, T Yasumoto, Y Ohizumi. The mode of inotropic action of ciguatoxin on guinea-pig cardiac muscle. Br J Pharmacol 95:876–882, 1988.
111. RJ Lewis. Negative inotropic and arrhythmic effects of high doses of ciguatoxin on guinea-pig atria and papillary muscles. Toxicon 26:639–649, 1988.
112. M Marquais, JP Vernoux, J Molgo, MP Sauviat, RJ Lewis. Isolation and electrophysiological characterisation of a new ciguatoxin extracted from Caribbean fish. In: B Reguera, J Blanco, ML Fernández, T Wyatt, eds. Harmful Microalgae, Santiago De Compostella: Xunta de Galicia and Intergovernmental Oceanographic Commission of UNESCO, 1998, pp 476–477.
113. RJ Lewis, R Endean. Mode of action of ciguatoxin from the Spanish Mackerel, *Scomberomorus commersoni*, on the guinea-pig ileum and vas deferens. J Pharmacol Exp Ther 228:756–760, 1984.
114. RJ Lewis, AW Hoy. Comparative action of three major ciguatoxins on guinea-pig atria and ilea. Toxicon 31:437–446, 1993.
115. Y Ohizumi, S Shibata, K Tachibana. Mode of the excitatory and inhibitory actions of ciguatoxin in the guinea-pig vas deferens. J Pharmacol Exp Ther 217:475–480, 1981.
116. Y Ohizumi, Y Ishida, S Shibata. Mode of the ciguatoxin-induced supersensitivity in the guinea-pig vas deferens. J Pharmacol Exp Ther 221:748–752, 1982.
117. JA Brock, EM McLachlan, P Jobling, RJ Lewis. Electrical-activity in rat tail artery during asynchronous activation of postganglionic nerve-terminals by ciguatoxin-1. Br J Pharmacol 116:2213–2220, 1995.
118. P Kostyuk, A Verkhratsky. Calcium stores in neurons and glia. Neuroscience 63:381–404, 1994.
119. J Molgo, T Shimahara, JX Comella, Y Morot-Gaudry-Talarmain, AM Legrand. Ciguatoxin-induced changes in acetylcholine release and in cytosolic calcium levels. Bull Soc Pathol Exot 85:486–488, 1992.
120. J Molgo, T Shimahara, AM Legrand. Ciguatoxin, extracted from poisonous moray eels, causes sodium-dependent calcium mobilization in NG108-15 neuroblastoma × glioma hybrid cells. Neurosci Lett 158:147–150, 1993.
121. K Yano, H Higashida, R Inoue, Y Nozawa. Bradykinin-induced rapid breakdown of phosphatidylinositol 4,5-bisphosphate in neuroblastoma × glioma hybrid NG108-15 cells. J Biol Chem 259: 10201–10207, 1984.
122. F Gusovsky, EB Hollingsworth, JW Daly. Regulation of phosphatidylinositol turnover in brain synaptoneurosomes: stimulatory effects of agents that enhance influx of sodium ions. Proc Natl Acad Sci USA 83:3003–3007, 1986.
123. F Gusovsky, ET McNeal, JW Daly. Stimulation of phosphoinositide breakdown in brain synaptoneurosomes by agents that activate sodium influx: antagonism by tetrodotoxin, saxitoxin, and cadmium. Mol Pharmacol 32:479–487, 1987.
124. MA Carrasco, Y Morot-Gaudry, J Molgo. Ca^{2+}-dependent changes of acetylcholine release and IP_3 mass in *Torpedo* cholinergic synaptosomes. Neurochem Int 29:637–643, 1996.
125. PM McDonough, D Goldstein, JH Brown. Elevation of cytoplasmic calcium concentration stimulates hydrolysis of phosphatidylinositol bisphosphate in chick heart cells: effect of sodium channel activators. Mol Pharmacol 33:310–315, 1988.
126. J Molgó, JX Comella, A-M Legrand. Ciguatoxin enhances quantal transmitter release from frog motor nerve terminals. Br J Pharmacol 99:695–700, 1990.

127. J Molgo, JX Comella, T Shimahara, A-M Legrand. Tetrodotoxin-sensitive ciguatoxin effects on quantal release, synaptic vesicle depletion, and calcium mobilization. Ann NY Acad Sci 635:485–489, 1991.
128. J Molgo, Y Morot-Gaudry-Talarmain, AM Legrand, N Moulian. Ciguatoxin extracted from poisonous moray eels (*Gymnothorax javanicus*) triggers acetylcholine release from *Torpedo* cholinergic synaptosomes via reversed Na^+-Ca^{2+} exchange. Neurosci Lett 160:65–68, 1993.
129. M Hermoni, A Barzilai, H Rahamimoff. Modulation of the Na^+-Ca^{2+} antiport by its ionic environment: the effect of lithium. Isr J Med Sci 23:44–48, 1987.
130. Y Morot-Gaudry-Talarmain, J Molgo, FA Meunier, N Moulian, AM Legrand. Reversed mode Na^+-Ca^{2+} exchange activated by ciguatoxin (CTX-1b) enhances acetylcholine from *Torpedo* cholinergic synaptosomes. Ann NY Acad Sci 779:404–406, 1996.
131. J Molgo, P Juzans, A-M Legrand. Confocal laser scanning microscopy: a new tool for studying the effects of ciguatoxin (CTX-1b) and mannitol at motor nerve terminals of the neuromuscular junction in situ. Mem Qld Mus 34:577–585, 1994.
132. E Benoit, P Juzans, AM Legrand, J Molgo. Nodal swelling produced by ciguatoxin-induced selective activation of sodium channels in myelinated nerve fibers. Neuroscience 71:1121–1131, 1996.
133. C Mattei, E Benoit, P Juzans, AM Legrand, J Molgo. Gambiertoxin (CTX-4B), purified from wild *Gambierdiscus toxicus* dinoflagellates, induces Na^+-dependent swelling of single frog myelinated axons and motor nerve terminals *in situ*. Neurosci Lett 234:75–78, 1997.
134. C Mattei, MY Dechraoui, J Molgo, FA Meunier, AM Legrand, E Benoit. Neurotoxins targeting receptor site 5 of voltage-dependent sodium channels increase the nodal volume of myelinated axons. J Neurosci Res 55:666–673, 1999.
135. E Benoit, AM Legrand. Purified ciguatoxin-induced modifications in excitability of myelinated nerve fibre. Bull Soc Pathol Exot 85:497–499, 1992.
136. JN Bidard, HPM Vijverberg, C Frelin, E Chungue, AM Legrand, R Bagnis, M Lazdunski. Ciguatoxin is a novel type of Na^+ channel toxin. J Biol Chem 259:8353–8357, 1984.
137. RC Hogg, RJ Lewis, DJ Adams. Ciguatoxin (CTX-1) modulates single tetrodotoxin-sensitive sodium channels in rat parasympathetic neurones. Neurosci Lett 252:103–106, 1998.
138. E Benoit, AM Legrand. Gambiertoxin-induced modifications of the membrane potential of myelinated nerve fibres. Mem Qld Mus 34:461–464, 1994.
139. J Cameron, AE Flowers, MF Capra. Electrophysiological studies on ciguatera poisoning in man (Part II). J Neurol Sci 101:93–97, 1991.
140. PA Hamblin, EM McLachlan, RJ Lewis. Sub nanomolar concentrations of ciguatoxin-1 excite preganglionic terminals in guinea-pig sympathetic-ganglia. Naunyn-Schmiedeberg's Arch Pharmacol 352:236–246, 1995.
141. E Benoit, AM Legrand, JM Dubois. Effects of ciguatoxin on current and voltage clamped frog myelinated nerve fibre. Toxicon 24:357–364, 1986.
142. LC Strachan, RJ Lewis, GM Nicholson. Differential actions of Pacific ciguatoxin-1 on sodium channel subtypes in mammalian sensory neurons. J Pharmacol Exp Ther 288:379–388, 1999.
143. S Pauillac, J Bléhaut, P Cruchet, C Lotte, AM Legrand. Recent advances in detection of ciguatoxins in French Polynesia. In: P Lassus, G Arzul, E Erard, P Gentien, C Marcaillou, eds. Harmful Algal Blooms. Paris, France, Lavoisier, Intercept, 1995, pp 801–808.
144. MA Poli, TJ Mende, DG Baden. Brevetoxins, unique activators of voltage-sensitive sodium channels, bind to specific sites in rat brain synaptosomes. Mol Pharmacol 30:129–135, 1986.
145. MA Poli, RJ Lewis, RW Dickey, SM Musser, CA Buckner, LG Carpenter. Identification of Caribbean ciguatoxins as the cause of an outbreak of fish poisoning among U.S. soldiers in Haiti. Toxicon 35:733–741, 1997.
146. RG Sharkey, E Jover, F Couraud, DG Baden, W Catterall. Allosteric modulation of neurotoxin binding to voltage-sensitive sodium channels by *Ptychodiscus brevis* toxin 2. Mol Pharmacol 31: 273–278, 1987.
147. S Cestéle, F Sampieri, H Rochat, D Gordon. Tetrodotoxin reverses brevetoxin allosteric inhibition of scorpion alpha-toxin binding on rat brain sodium channels. J Biol Chem 271:18329–18832, 1996.
148. A Lombet, JN Bidard, M Lazdunski. Ciguatoxin and brevetoxins share a common receptor site on the neuronal voltage-dependent Na^+ channel. FEBS Lett 219:355–359, 1987.

149. M-Y Dechraoui, J Naar, S Pauillac, A-M Legrand. Ciguatoxins and brevetoxins, neurotoxic polyether compounds active on sodium channels. Toxicon 37:125–143, 1999.
150. VL Trainer, DG Baden, WA Catterall. Identification of peptide components of the brevetoxin receptor site of rat brain sodium channels. J Biol Chem 269:19904–19909, 1994.
151. RL Manger, LS Leja, SY Lee, JM Hungerford, MM Wekell. Tetrazolium-based cell bioassay for neurotoxins active on voltage-sensitive sodium channels: semiautomated assay for saxitoxins, brevetoxins, and ciguatoxins. Anal Biochem 214:190–194, 1993.
152. RL Manger, LS Leja, SY Lee, JM Hungerford, Y Hokama, RW Dickey, HR Granade,R Lewis, T Yasumoto, MM Wekell. Detection of sodium channel toxins: directed cytotoxicity assays of purified ciguatoxins, brevetoxins, saxitoxins, and seafood extracts. JAOAC Int 78:521–527, 1995.
153. ER Fairey, J Stewart, G Edmunds, JS Ramsdell. A cell-based assay for brevetoxins, saxitoxins, and ciguatoxins using a stably expressed c-*fos*–luciferase reporter gene. Anal Biochem 251:129–132, 1997.
154. Y Hokama, WE Takenaka, KL Nishimura, JS Ebesu, R Bourke, PK Sullivan. A simple membrane immunobead assay for detecting ciguatoxin and related polyethers from human ciguatera intoxication and natural reef fishes. JAOAC Int 81:727–735, 1998.

21
Ciguatera Toxins: Toxinology

Kiyoshi Terao
Yamawaki-Gakuen Junior College and Chiba University, Tokyo, Japan

I. INTRODUCTION

Ciguatera fish poisoning is a peculiar food poisoning which results from the consumption of any of a large variety of tropical and subtropical fish associated with coral reefs. The disease affects in excess of 25,000 persons annually. Ciguatera fish poisoning has long been reported to be of regional occurrence, being limited to definite regions of certain islands in tropical or subtropical zones of the world and, possibly, to narrow regions of those islands. The outbreaks of the disease are not seasonal but occur consistently throughout the year. Moreover, other investigators have cited the changing toxicity over the years. Further, it should be emphasized that the occurrence of ciguatera fish poisoning does not depend on the species of fish consumed. The same species in very close geographical proximity may be toxic in one area and nontoxic in another. In order to analyze the mechanisms of the outbreaks of such complex and discrepant disease, the food chain concept has been introduced. From observations of the ecological phenomena and those of food habits of fish in coral reefs, Randall (1) suggested that the causative agents of ciguatera fish poisoning may produce a benthic microorganism, most likely a blue-green alga, that is a source of the toxin. Herbivorous fish consume the benthic organism that contains the toxin. These fish may then be consumed by larger carnivorous fish. Thus the toxin passes up the food chain to the human consumer. Later, Yasumoto and his coworkers (2) indeed found that a benthic microalga, *Gambierdiscus toxicus*, produces potent causative toxins involved in ciguatera fish poisoning. Thereafter, a number of investigators have isolated several causative agents of ciguatera fish poisoning from the viscera of toxic coral reef fish or from dinoflagellates in culture. Chemical structures of these toxins are polyethers such as ciguatoxins and maitotoxin. Among them, lipid-soluble ciguatoxin and water-soluble maitotoxin are the most potent marine toxins known.

The biological activities of both toxins have been extensively studied by many investigators, and it is now widely accepted as being potent activators of Na^+ and/or Ca^{2+} influxes into the cytoplasm in various cell types. The increased contents of Na^+ or Ca^{2+} in the cytoplasm resulted in functional as well as in morphological changes of the target cells.

Using pure ciguatoxin as well as maitotoxin, we describe the pharmacological and pathological bases of experimental ciguatoxicosis and/or maitotoxicosis in mice in this chapter.

Table 1 Short History of Ciguatera Research

Year	Investigators	Location of outbreak	Suspected medium and/or remarks	References
Age of description of ciguatera fish poisoning (medical ichthyology and epidemiology)				
1511	Martyr of Anghera	Caribbean Sea	Manchineel berry	3
1601	Harmansen	Mauritius	Bonntabe	3
1606	Quiros	New Hebrides	*Lutjanidae*	3
1667	Du Tertre	Antilles	Triggerfish	3
1725	Sloane	Jamaica	Barracuda	3
1748	Boscawen	Mauritius	Vieille (*Serranus lutra*)	3
			Largest ciguatera fish poisoning outbreak in the world	3
1754	De la Motte	Mauritius	Wrasse	3
1774	Cook	New Hebrides	Red pargo	3
1787	Parra	West Indies (Cuba)	"Ciguatera fish poisoning"	3
1799	Thomas	West Indies	Cu, manchineel berry	3
1808	Chisholm	West Indies	Barracuda, snapper, etc.	3
1856	Guillon, Chevallier	Juan Fernández Island	Horse mackerel, etc., 34/51 died	3
1861	Naval Medical School of Brest, France		A manual on poisonous fish	3
1886	Savtshenko		A manual on medical ichthyology	3
1895	Steinbach	Marshall Islands	A species found in one part of the lagoon may be edible, but toxic when found elsewhere	3
1933	Costa Mandry	Puerto Rico	Pathogenic micro-organism	3
1943	Hiyama	Micronesia	Atlas of toxic and venomous fishes	3
1945	Vonfraenkel and Krick	Mariana Islands	Barracuda	3
			30 American naval personnel	
1945	Charnot	Morocco	Dolphin, *Coryphaena hippurus*	3
1946	Watanabe	Okinawa (Japan)	Red snapper: 39 persons	3
1946	Cohen	Saipan, Mariana Islands	Barracuda: 41 persons	3
1947	Ross	Fanning, Line Islands	95 in a population of 224	3
Age of analysis of ciguatera fish poisoning (ecology, chemistry and toxinology)				
1956	Hashimoto		Isolation of a poison from barracuda	3
1958	Randall	South Pacific and Caribbean	Concept of food chain	1
1962	Bartsh, McFarren	Marshall Islands	Epidemiological-clinical analysis	3
1964	Halstead		Poisonous and venomous marine animals in the world	3
1966	Scheuer et al.	South Pacific	Isolation of "ciguatoxin" from toxic fish	4
			Moray eel, *Gymnothorax javanicus*	
1971	Yasumoto et al.	French Polynesia and Okinawa	Maitotoxin	5

Table 1 Continued

Year	Investigators	Location of outbreak	Incriminated living and/or remarks	References
1977	Yasumoto et al.	French Polynesia	Dinoflagellate *Gambiediscus toxicus*	2
1979	Bagnis	South Pacific	Clinical Symptoms: 3009 cases	6
1981	Ohizumi et al.		Mode of excitor and inhibitory actions of ciguatoxin	7
1982	Legrand et al.		Mode of actions of ciguatoxin	8
1989	Palafox et al.	Marshall Islands	Mannitol treatment	9
1989	Murata et al.		Chemical structure of ciguatoxin and congener	10
1992	Gillespie et al.	Queensland (Australia)	Ecological, clinical, and socioeconomic	11
1993	Murata et al.		Chemical structure of maitotoxin	12

II. HISTORY OF CIGUATERA FISH POISONING RESEARCH

Table 1 summarizes the history of ciguatera fish poisoning, which can divide into two parts. The first four and a half centuries of the history of this type of poisoning are characterized by descriptions of toxic fish and outbreaks of ciguatera fish poisoning. Research on ciguatera fish poisoning has expanded significantly over the last three decades. This increase is due to the identification of a benthic dinoflagellate as a toxin producer, and equally important have been advances in the analytical techniques and equipment needed to characterize the toxins chemically (13).

III. CIGUATOXIN POISONING

A. Clinical Symptoms in Humans

The symptoms of human ciguatera fish poisoning appear from about 1–10 h after toxic fish are eaten. The clinical manifestations of ciguatera fish poisoning reveal the diverse spectrum of the symptoms. When ciguateric fish are consumed, more than 150 clinical manifestations are possible (14). Shortly after eating a toxic fish, at first there is nausea, which is followed by vomiting and severe stomach pains. This is often accompanied by watery diarrhea. Neurological disorders are also among the most prominent complaints of the victims. Myalgia, paresthesias of the extremities, and circumoral paresthesia are common symptoms of ciguatera fish poisoning. Comparing disorders reported in sensory nervous system to those found in motor nerves, the former is severe than the latter.

Randall (1) summarized the symptoms of ciguatera fish poisoning (Table 2). He stated that the most diagnostic symptoms of ciguatera fish poisoning are the tingling sensations in the hands and feet and the feeling of heat when cold objects are touched or cool liquids are taken into the mouth (''dry ice syndrome'').

Usually, relatively few cases of cardiovascular disorders such as bradycardia and hypotension are reported in endemic areas. Ciguatera fish poisoning is generally a self-limiting disease,

Table 2 Symptoms of Ciguatera Fish Poisoning

Gastrointestinal Disorders
Intense gastric pain
Diarrhea
Nausea
Neurological Disorders
Motor Nerve Disorders
Weakness or prostration
Difficulty in breathing
Inability to coordinate voluntary muscular movements
Sensory Nerve Disorders
Tingling or numbness of lips and hands and feet
Confusion of sensations of heat and cold
Joint and muscular pain
Itching
Burning urination

but death may occur as a result of cardiac dysrhythmias, hypotension, shock, or cerebral edema (15).

Information from 129 cases of ciguatera fish poisoning in the Caribbean revealed gastrointestinal and neurological disorders (15). Clinical symptoms of patients suffering from ciguatera fish poisoning in South Pacific islands were collected by Bagnis et al. (6). Between 1964 and 1977, clinical observations were recorded for 3009 cases of presumed ciguatera fish poisoning. The information was principally gathered from patients in French Polynesia and New Caledonia Pacific island groups in which ciguatera fish poisoning was known to be a relatively common problem. He noted that digestive and neurological symptoms are predominant in patients who became ill following the eating of surgeonfish (herbivorous fish), whereas the syndrome involves other symptoms, including cardiovascular disorders, when carnivorous fish were consumed. According to Quod and Turquet (17), 477 victims in the island of Reunion (SW Indian Ocean) showed the clinical pattern of the disease includes various neurological, gastrointestinal, and general symptoms. Neurological signs and symptoms seemed to persist longer than other symptoms.

The clinical symptoms documented from the endemic areas in the Caribbean, Indian Ocean, Australia, and the South Pacific are similar.

Symptoms usually last about 1 week, although they may persist for months or even years. Usually, the recurrence of the signs and symptoms described was experienced in the endemic areas.

B. Pathomorphological Findings

There has been only limited information about postmortem examination or histopathological changes, because only a few fatal cases of ciguatera fish poisoning were available. Studies on gross pathological changes caused by ciguatera fish poisoning have been of a superficial nature. The few reports available show evidence of acute visceral congestion. No discernible histological changes were described with the light microscope (3). In their classic paper, Bagnis and his coworkers (6) reported only one autopsy case, which was unremarkable grossly and histologically except that the liver showed eosinophilic necrotic lesions.

Using biopsy materials of sural nerve from a patient who suffered from ciguatera fish poisoning, Allsop et al. (18) observed the changes in the sural nerve with an electron microscope. There was striking edema of the adaxonal Schwann cell cytoplasm. These ultrastructural changes are very similar to those seen following the injection of scorpion and spider venoms into the peripheral nerve of experimental animals. Both these venoms and ciguatoxin increase the permeability of the membrane to sodium.

C. Experimental Pathology (Morphological Response)

Causative agents of ciguatera fish poisoning have been isolated from both the viscera of toxic fish and dinoflagellates in culture by several research groups (2,4,19,20). Among them, ciguatoxin is one of the most potent marine toxins known. Terao et al. (21) undertook to study the sequential ultrastructural changes of experimental mice after the administrations of highly purified ciguatoxin.

Single-Dose Administration of Ciguatoxin

Signs and Symptoms Defecation followed by severe diarrhea with a watery stool occurred within 15 min after the administration of 700 ng of ciguatoxin per kilogram of body weight (5 mouse units, MU), which was followed by loss of activity, lacrimation, and excessive salivation that continued for only about 90 min. Several hours after the administration of the toxin, the mice seemed to recover from the illness. However, severe dyspnea and cyanosis suddenly appeared, and about 70% died within 24 h. Incubation periods between the administration of the toxin and onset of dyspnea became shorter in a dose-dependent manner. The other approximately 30% of the surviving mice showed mostly paralysis of the paws. Often conclusive spasms was seen. About half of the surviving mice showed continuous penis erection accompanied by a markedly dilated and filled urinary bladder.

Postmortem Examination On macroscopic examination, intoxicated mice in the state of dyspnea or dead animals had a dilated heart and edema of the lungs. Marked congestion of all other organs was also observed.

Light microscopy showed that there were many single cell necroses in cardiac muscle tissues in the septum, both in the left and right ventricles. Marked congestion of right ventricle and coronary veins was seen.

Electron microscopy revealed that among the necrotic cardiocytes there were many swollen myocardial cells (Figure 1). Mitochondria became rounded and most organelles, including the myofibrils, sarcoplasmic reticulum, and T system, were separated from each other. Often firm cohesion of the successive cellular units of the myocardium were broken by exudation at the interstitial space of the myocardial cells. Space between adjacent fibers were dilated with exudation. Endothelial lining cells of the blood capillaries were more sensitive to the toxin than other tissues in the heart. Occasionally, an aggregation of platelets was seen in the lumen of many blood capillaries. Blood plasma exuded from the capillaries and cardiocytes were separated from each other. Often swollen erythrocytes were seen in the cardiac tissue. A functional disorder of the heart was reported by several investigators. Both Li (22) and Rayner (23) have described an initial biphasic cardiovascular response to intravenous injections of lethal doses of ciguatoxin: a phase of hypotension with bradycardia is followed by a phase of hypertension with tachycardia. The morphological changes may support such a functional disorder.

Mice with severe dyspnea showed marked edema of the lungs. Histologically, alveolar spaces and bronchioles were filled with plasma-like effusion and prominent congestion similar to human cardiac asthma.

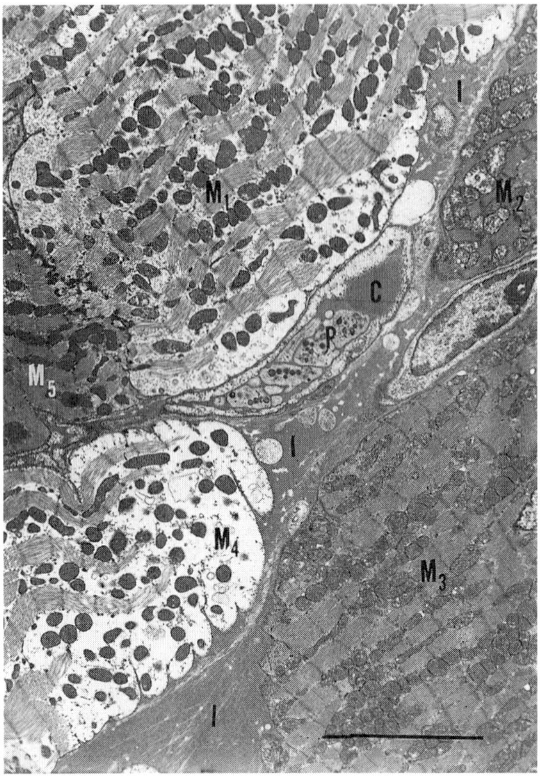

Figure 1 Electron micrograph of the left ventricle from a mouse given a single injection of ciguatoxin at a dose of 0.7 μg/kg of body weight 50 min prior to sacrifice. Five cardiac muscle cells (M_1–M_5) are seen. Two muscle cells (M_1 and M_4) are markedly swollen. Mitochondria in these cells had become round. Myofibrils in several cells (M_1 and M_4) have partially disappeared. Cells M_2 and M_3 are condensed and most mitochondria are degenerated. Aggregation of blood platelets (P) can be seen in a capillary located at the center of the micrograph. In the interstitial space (I), there is marked effusion of plasma-like material. The scale bar = 5 μm.

There were no discernible morphological changes in the small intestine such as the duodenum, jejunum, and ileum (Figure 2a) except amyelinated nerve cells in submucosal plexus of Meissner as well as those in the myenteric plexus of Auerbach (Figure 2b). Marked swelling of nerve fibers and disappearance of synaptic vesicles from synapses was prominent. Ito et al. (24) observed that ciguatoxin accelerated mucus secretion and peristalsis in the colon and stimulated defecation at the rectum, resulting in prominent diarrhea. Epithelial damage in the colon was observed in the upper portion of the large intestine but not in the lower half. The morphological changes caused by ciguatoxin in the upper portion of the large intestine were similar to those seen with cholera toxin.

Histologically, no significant changes were seen in the cortical layer of the adrenal glands, whereas degeneration of nerve cells in the medulla was prominent (Figure 3a). Electron microscopically, vacuolation in the cytoplasm and a decrease in the number of secretion granules were often seen in cells of the medulla (Figure 3b).

There were many thrombus formations in the veins of various organs such as the heart, liver, and penis.

(a)

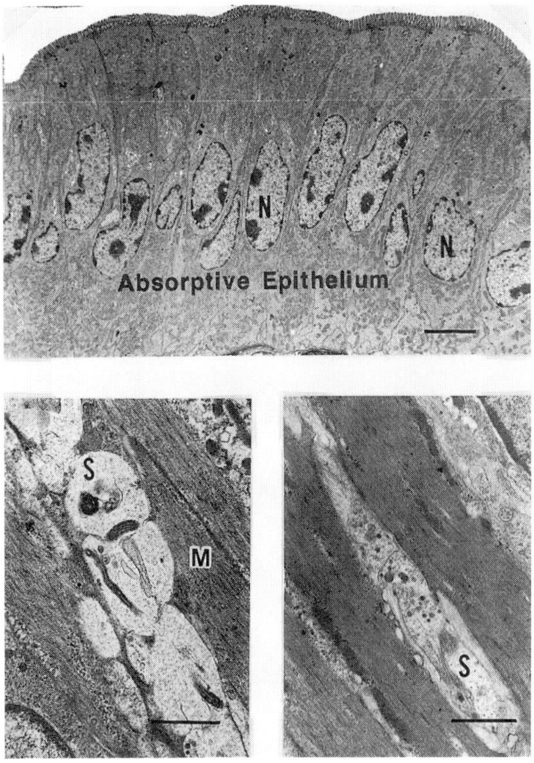

(b) (c)

Figure 2 Electron micrographs of the ileum from a mouse given i.p. injection of ciguatoxin at a dose 0.7 ng/kg 60 min prior to sacrifice (a and b) and from a control (c). (a) Absorptive epithelium of the ileum. No morphological changes are seen. N, nucleus of the absorptive epithelium. The scale mark represents 10 μm. (b) Amyelinated nerves of Auerbach's plexus from a mouse receiving ciguatoxin. Marked swelling of nerve fibers and disappearance of synaptic vesicles (S) is prominent. M, Smooth muscle. The scale mark represents 1 μm. (c) Amyelinated nerves (S) of Auerbach's plexus from a control mouse. The scale bar = 1 μm.

In addition to cardiac muscle cells, the target tissues of the phycotoxin are the endothelial lining cells of the blood capillary in the myocardium, nerve cells in intestinal ganglions, and the rectum. Swelling of endothelial lining cells of the blood capillaries results in exudation and bleeding into myocardium. Therefore, in mouse ciguatoxicosis, cardiac failure was the most prominent response, whereas neurological disorders were predominant in human cases (21).

Repeated Administrations of Ciguatoxin

Bagnis et al. (6) have reported that ciguatera fish poisoning apparently produces no immunity, and it has been suggested that second attacks may be more severe. Recurrent or multiple attacks of ciguatera fish poisoning seem to result in a clinically more severe illness compared to that in patients experiencing the disease for the first time. Results of repeated administration of ciguatoxin (25) agreed well with the clinical observation by Bagnis et al. (6). Using ICR mice, Terao et al. (25,26) injected ciguatoxin at low doses (100 ng/kg body weight) into the abdominal cavities of the mice for 15 days. A single dose of 100 ng/kg of ciguatoxin caused no discernible morphological changes in macroscopic, light microscopic, and at even ultrastructural level in the hearts of mice (Figure 4). In contrast, repeated administrations of ciguatoxin resulted in

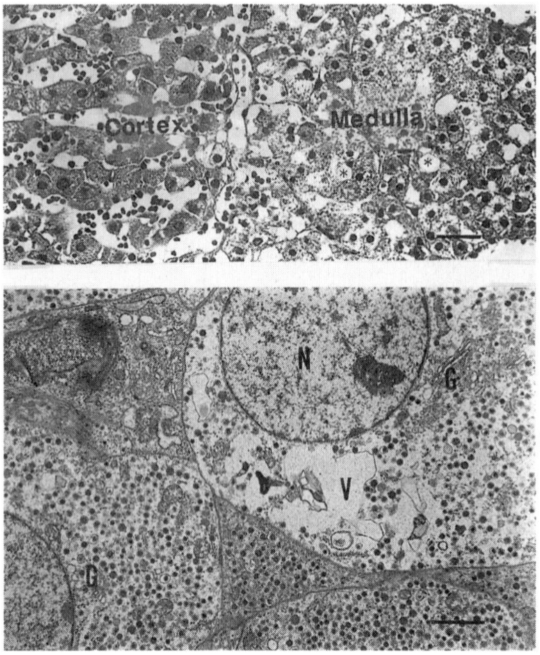

Figure 3 Light (a) and electron micrograph (b) of the adrenal gland from a mouse given i.p. 0.7 ng/kg of ciguatoxin 24 h prior to sacrifice. (a) The adrenal gland of a ciguatoxin-treated mouse. Cells in the cortex show no discernible changes, whereas cells at medulla contain many vacuoles. The scale mark represents 25 μm. (b) Electron micrograph of the adrenal gland from the same animal as in (a). Cytoplasm is occupied by a large irregular vacuole (V). N, nucleus; G, secretion granules. The scale bar = 10 μm.

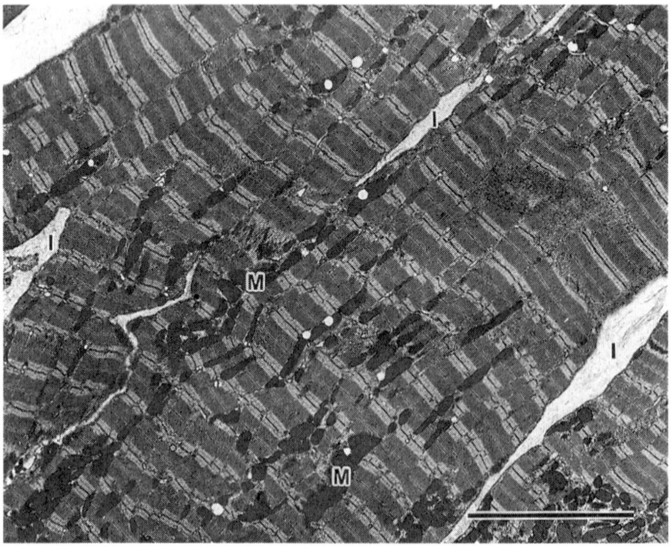

Figure 4 Electron micrograph of left ventricle from a mouse 24 h after receiving a single dose of ciguatoxin (0.1 ng/kg) No discernible changes are seen. I, interstitium; M, mitochondria. The scale bar = 8 μm.

severe morphological changes in the heart. After 15 doses, all animals receiving the injections did not show any apparent change but there were marked dilatation of both ventricles of the heart at necropsy. Changes revealed at the ultrastructural level were similar to those of the hearts of mice receiving a single dose at 700 ng/kg or more (Figure 5a,b). In the case of the repeated administrations of ciguatoxin, endothelial lining cells of the blood capillaries in the heart were attacked with ciguatoxin more severely than other components. There are no visible organelles except small vesicles at the basal portion. (Fig. 6a–c). The cardiac lumen was extremely nar-

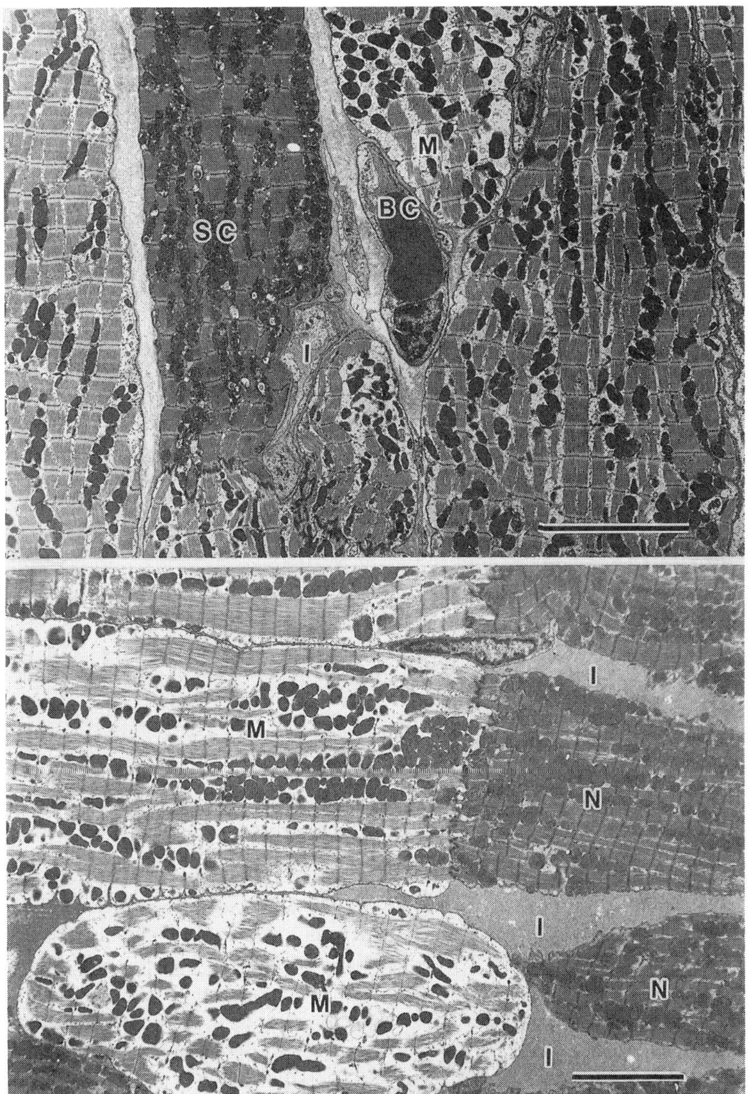

(a)

(b)

Figure 5 Electron micrographs of left ventricle from a mouse given ciguatoxin at a dose of 0.1 μg/kg 10 times. (a) Similar to single administration at a dose of 0.7 ng/kg, single-cell necrosis (SC) and swollen cardiac muscle cells (M) are seen. Amorphous exudates are present in the interstitium (I) of the myocardial tissue. BC, blood capillary. Scale bar = 10 μm. (b) Electron micrograph of the heart of a mouse given 15 repeated injections of 0.1 μg/kg of ciguatoxin. Almost all muscle cells are swollen (M). Intermuscular spaces (I) are filled with plasma-like materials. N, necrotized cardiac cells. Scale bar represents 10 μm.

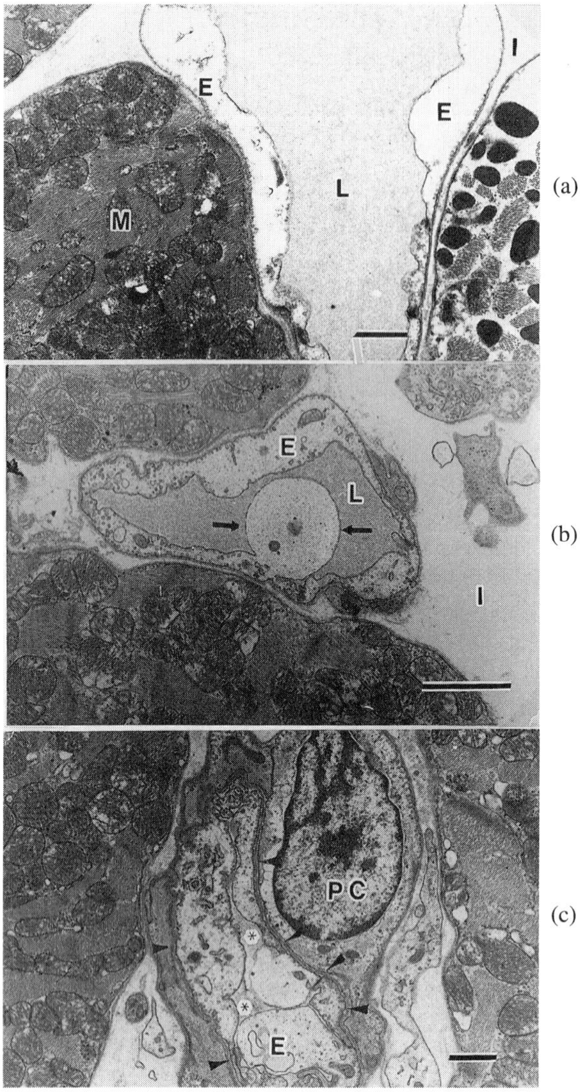

Figure 6 Electron micrographs of blood capillaries in the heart. (a) Part of a cardiac muscle cell (M) and a blood capillary. Endothelial lining cells (E) of the capillary are markedly swollen. There are no microorganelles except for vesicles. In the interstitium (I), there is plasma-like, electron-dense material as in the lumen (L) of the blood capillary. The cardiac muscle cell (M) is also swollen and mitochondria are rounded. Scale bar = 1 μm. (b) A small vessel in the interstitium of the heart. Endothelial lining cells, (E) are highly swollen. A part of the endothelial lining cell is protruded (*arrow*) into the lumen (L). Scale bar = 2 μm. (c) Degenerated endothelial lining cells (E) are surrounded by basement membrane (*arrowheads*). In contrast, a pericyte (PC) is less affected. Lumen (asterisks) is extremely narrowed. Scale bar = 1 μm.

rowed. Occasionally, marked swollen endothelial cells were surrounded by the basement membrane (Figure 6b). In contrast to the endothelial lining cells, pericytes were less affected (Figure 6c).

Recovery from Experimental Ciguatoxicosis

Within 1 week after the cessation of the administrations of ciguatoxin, the marked swelling of the cardiac muscle cells as well as that of endothelial lining cells of the blood capillaries in the heart still persisted.

About 1 month following withdrawal of the administration of ciguatoxin, morphological changes of myocardial cells as well as those in endothelial lining cells of the blood capillaries in the heart were recovered. The microorganelles in the cardiac muscle cells were completely normal with the exception of the triads of the sarcoplasmic reticulum, which remained unusually dilated. After that time, amorphous electron-dense material was replaced gradually by strands of fine and/or dense collagen fibers (Figure 7a). However, the hearts of mice given repeated injections of ciguatoxin differed in several respects from control at the histological level. (i) The effusion of serum and erythrocytes persisted in the interstitium of the heart (Figure 7b); (ii) numerous macrophages were still found in the space between the bundles of myocardial muscles; (iii) the blood components were absorbed by the macrophages and became gradually replaced by bundles of collagen fibers, and therefore groups of muscle cells were separated by these dense collagen fibers; (iv) although capillaries were in direct contact with cardiac muscle cells in control of the mouse heart, mice treated with repeated doses of ciguatoxin had wide layers of collagen fibers (Figure 7c); and (v) very often there were aggregations of blood platelets in the blood capillaries in the myocardium (Fig. 7d).

Considerable hypertrophy of cardiac muscles was also noted. The mean weight of the hearts of ciguatoxin-treated mice at 14 months (280 $\pm$ 36 mg) was significantly heavier than that of control (207 $\pm$ 15 mg) ($P < .05$). If such conditions continued for long periods, it must have resulted in "cardiac failure." Marked effusion in the alveolar spaces and bronchial lumens like asthma cardiale support this hypothesis.

The experiments with repeated administrations by Terao et al. (26) indicate that ciguatoxin has a cumulative effect on the cardiac tissue. Marked swelling of cardiac muscle cells and that of endothelial lining cells were reversible, but organized effusion was not revertible.

D. Mode of Action of Ciguatoxin

Although the earlier reports on the mechanism of activities of ciguatoxin noted an anticholinesterase in vitro, Rayner and his coworkers of Hawaii University pointed out for the first time that its pharmacological activity is a direct action on excitable membranes. Using removed pedal ganglions from the sea snail (27) or frog skin (28), they observed that ciguatoxin primarily increases the passive permeability of the plasma membrane to sodium ion. Ciguatoxin produces a positive inotropic response in rat atria (29) and guinea-pig atria (30). Ohizumi et al. (31) observed that ciguatoxin caused a marked release of norepinephrine (*noradrenaline*) from the isolated guinea-pig vas deferens, which was blocked after treatment with tetrodotoxin or incubation in a calcium ion-free medium.

Bidard et al. (32) demonstrated that ciguatoxin inhibits the net accumulation of neurotransmitters by brain synaptosomes from neuroblastoma culture cells in vitro. Electrophysiological studies indicated that ciguatoxin induced a membrane depolarization which is prevented by tetrodotoxin and which was due to an action that increases sodium permeability. The membrane depolarization and transmitter release is neither due to an inhibition by ciguatoxin of the

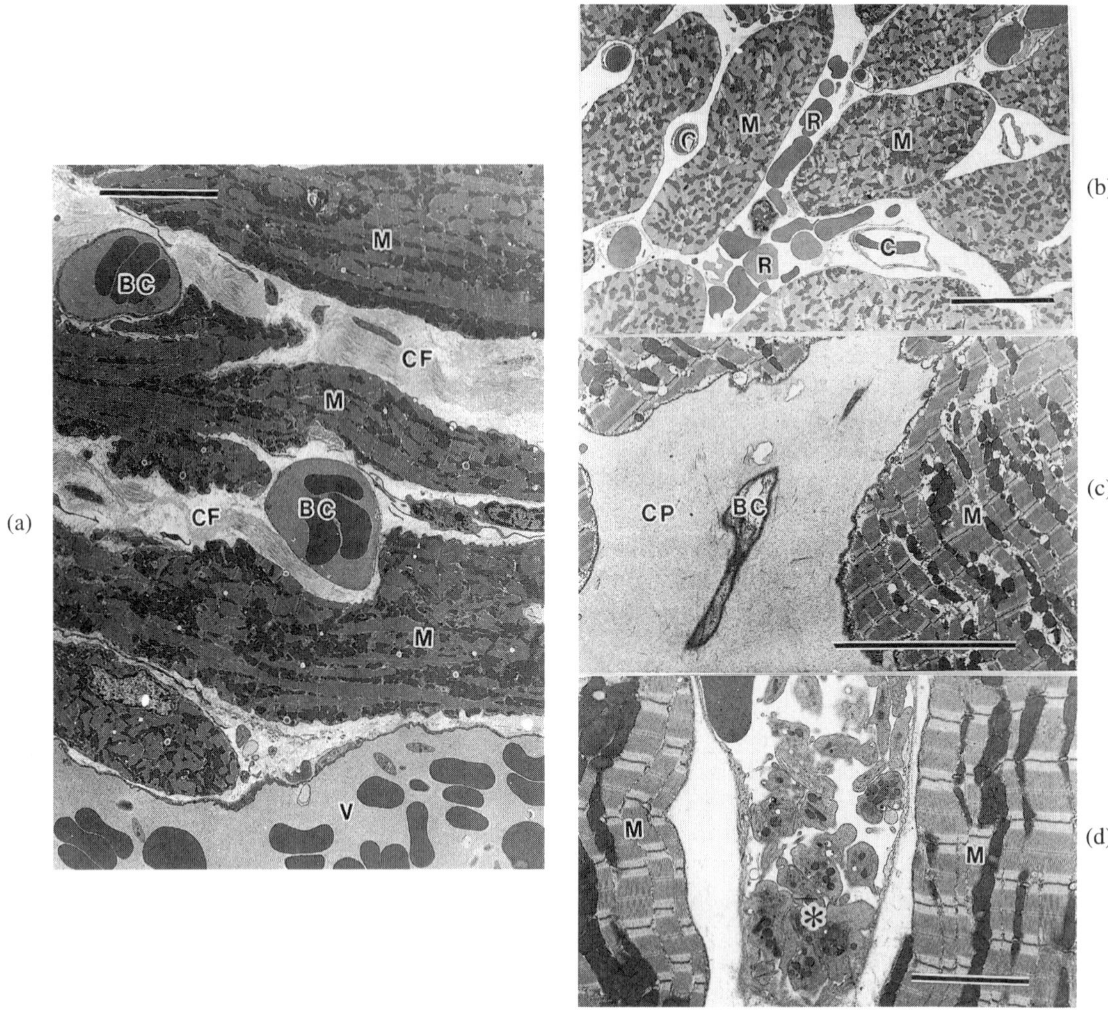

Figure 7 Electron micrographs of the hearts from mice given repeated injections of ciguatoxin at doses of 1/15 mouse unit 15 times. After cessation of injections, animals kept with ordinary feeding for 21 through 72 weeks. (a) Almost all cardiac muscle cells (M) and blood capillaries (BC) are recovered from acute changes induced by ciguatoxin. However, between myocardial cells there are thick depositions of collagen fibers (CF). Therefore, direct contact between cardiac muscle bundles and capillaries are lost. V, vein, Scale bar = 10 μm. (b) There are many erythrocytes (R) in the widened interstitium. M, muscle cell; (C), blood capillary. Scale bar = 10 μm. (c) Note, a blood capillary (BC) is embedded in thick collagen bundle (CF). M, cardiac muscle cells. Scale bar = 20 μm. (d) Very often aggregation of blood platelets (*) in blood capillaries is seen. M, cardiac muscle. Scale bar = 10 μm.

(Na^+,K^+)-ATPase, since ciguatoxin is without effect on this enzyme, nor an effect on Ca^+ channels.

Brevetoxin is produced by a dinoflagellate and is of a polyether chemical structure. The toxin is known as the causative agent of neurotoxic shellfish poisoning. Both brevetoxin and ciguatoxin are effectors of the voltage-dependent sodium channel. Lombet et al. (33) compared the mode of action of both toxins. They indicated that both toxins allosterically enhanced the influx of Na^+ ion into the cytoplasm in a very similar way to their binding to the neural sodium channel protein. Albeit, the affinity of ciguatoxin for the sodium channel is at least 20–50 times higher than that of brevetoxin. Seino et al. (34) investigated the mechanism of the inotropic action of ciguatoxin by electrophysiological and pharmacological techniques. The inotropic action of lower concentrations is primarily due to an indirect action via norepinephrine release, whereas that of higher concentrations is caused not only by an indirect action but also by the direct action of voltage-dependent sodium channels of cardiac muscle. It is also suggested that ciguatoxin activates cardiac muscle sodium channels by modifying the voltage dependence of channel activation to increase sodium inward currents.

Pathomorphological changes after the administration of ciguatoxin may occur in the following manner. Sodium ion channels of the target cells are bound by ciguatoxins. Allosterically degenerated proteins of the sodium channels cannot close against the sodium influx into the cytoplasm of the target cells. Increased concentration of sodium in the cells induces high intracytoplasmic osmolality. High osmolarity results in an influx of water into the target cells.

E. Treatment of Ciguatera Fish Poisoning

Treatment of ciguatera fish poisoning remains supportive, since most attempts at specific therapy are unsatisfactory and have no proved efficacy. Mannitol therapy is currently the treatment of choice during the acute phase of ciguatera fish poisoning. Palafox et al. (9) reported for the first time that 24 patients suffering from acute ciguatera fish poisoning were treated with intravenous mannitol, and each patient's condition improved dramatically. Neurological symptoms recurred in 5 of 23 cases after discharge. Cloyd et al. (35) demonstrated that there was a strong, positive correlation between serum osmolarity and serum mannitol concentration. Using surgically removed human atrial specimens, Lewis et al. (36) demonstrated that mannitol reverses an edema of various cells; however, other actions such as a scavenging of free radicals and/or involving a direct effect on the ability of ciguatoxin to bind sodium channels could not be discounted. From observations with the confocal laser scanning microscope, Molgo et al. (37) also confirmed such an effect of mannitol at motor nerve terminals of the neuromuscular junctions from frogs.

As shown pathologically, the most prominent change after the administration of ciguatoxin is a swelling of various cell types with excitable membranes. The swelling of these cells results in dysfunction or failure of the nervous system or the cardiovascular system. Clinical improvement observed in acute ciguatera fish poisoning after mannitol treatment would result in a shift of water from the intracellular to the extracellular compartment owing to its high osmolarity. However, mannitol cannot remove ciguatoxin from the proteins of the sodium channel. From these facts, it is suggested that mannitol treatment may be not causal but one of the most effective palliative treatments.

IV. MAITOTOXIN POISONING

It is well known that ciguatera fish poisoning shows a wide range of variability in symptoms. Therefore, in addition to ciguatoxin, there must be other causative toxins present. Among such toxins, a water-soluble toxin was judged to be the most important one. This toxin has been designated maitotoxin after the Tahitian name ''maito'' for surgeonfish (2). Bagnis et al. (38) have shown that the surgeonfish, *Ctenochaetus striatus*, is responsible for 61% of the cases of ciguatera fish poisoning in Tahiti. Yasumoto et al. (1971) (5) postulated that the reasons for the high incidence of poisoning by this fish may be because it is the most abundant fish of edible size on reefs in the Society Islands and it is highly esteemed by the Tahitian people as food. *C. striatus* in Tahiti contained both fat-soluble and water-soluble toxins. Purification of the fat-soluble toxin revealed it was ciguatoxin and purification of the water-soluble toxin revealed it was maitotoxin (5). The fat-soluble toxin (ciguatoxin) was detectable in all tissues tested of the surgeonfish, with the concentration being highest in the liver and lowest in the flesh, whereas the water-soluble toxin (maitotoxin) was found only in the liver. Ingested materials proved to be the more toxic (5). This means that ciguatoxin is absorbed from the gut and metabolized in the liver of herbivorous as well as carnivorous fish. In contrast, maitotoxin may be absorbed only in small quantities and metabolized completely in the liver. Therefore, maitotoxin may not penetrate into flesh. At present, no maitotoxin has been isolated from carnivorous fishes.

A. Experimental Pathology

Maitotoxin was first isolated by Yasumoto et al. (5) from visceral organs of the Tahitian toxic surgeonfish, *C. striatus*. This toxin is similar to ciguatoxin produced by a marine dinoflagellate, *Gambierdiscus toxicus*, and ingested by herbivorous fish. Biological activities of maitotoxin have been extensively studied by many investigators, and it is now widely accepted as being a potent activator of calcium channels in various cell types.

B. Single-Dose Administration of Maitotoxin

Terao et al. (41) examined the effects of maitotoxin on mice morphologically. Within a few minutes after injections of maitotoxin, the mobility of all animals receiving over 100 ng/kg was limited to a laborious crawling movement. The larger the doses of the toxin, the shorter were the incubation periods between the injections and the onset of the symptoms. During the ensuing 2 h, the experimental animals showed piloerection and cyanosis, huddled together in corners of the cage, and then gradually began to recover from the effects of the illness.

Postmortem Examination

Macroscopically, mice receiving 200 ng/kg or mice and rats treated with 400 ng/kg within 4 h showed distended stomachs and small intestines often accompanied by ascites in moderate quantities. There was always severe congestion in the gastrointestinal tracts. Fibrinous adhesion between the serosal surface of the stomach and the omentum was occasionally seen. The thymus always showed congestion. Scattered focal erosions were observed on the glandular mucosa of the stomach of about 70% of both mice and rats treated with a dose of at least 200 ng/kg of maitotoxin 4 h prior to examination.

Histopathological examination showed focal necroses in the mucosal layer and marked edema in the submucosal layers of the stomach. Occasionally, perforation of ulcers into the serosa and peritonitis were observed. Within 30 min, mice receiving 200 or 400 ng/kg of maitotoxin showed scattered coagulative necroses in the cardiac muscle of the left ventricle. At the ultrastructural level, a marked swelling of the endothelial lining cells was occasionally seen. The lumens of the capillaries were extremely narrowed or even closed.

At 4 h after injection with 200 or 400 ng/kg of maitotoxin, the thymus of mice as well as rats showed marked atrophy. Usually, several medium-sized lymphocytes had clumped chromatin and apoptoses were found in the cortex of the thymus. Electron microscopic observation revealed that cortical tissue in the reticular networks was filled with debris (Figure 8). Macrophages with debris of phagocytized lymphocytes were abundant. Most lymphocytes in the medulla at this time were still intact. During the next 4 h, almost all the medullary lymphocytes showed apoptosis. Epithelial reticular cells had vacuoles in the cytoplasm. Mice given 100 ng/kg of maitotoxin showed changes in the thymus essentially similar to those in the 200 or 400

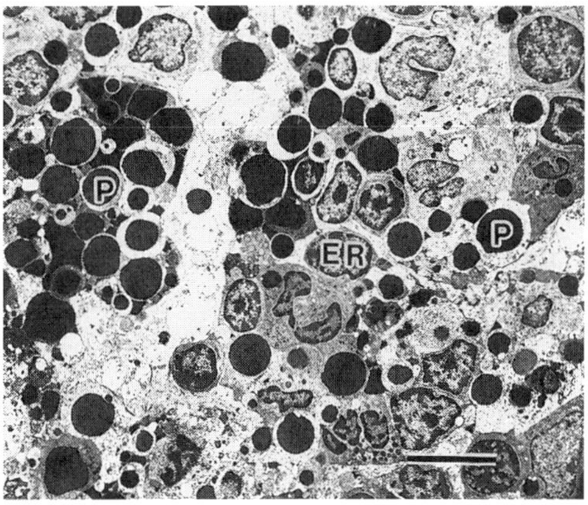

Figure 8 An electron micrograph of cortical zone of the thymus from a mouse injected four times with 45 ng/kg of maitotoxin. There are a few intact lymphocytes. Much debris of lymphocytes and apoptoses (P) is seen. Epithelial reticular cells (ER) survived. Scale bar = 10 μm.

ng/kg series but to a less severe extent. At 24 h after the injection, both the cortex and medulla were completely occupied by lymphocytic debris.

By 2 h after injection with a dose of 400 ng/kg of maitotoxin, there were numerous apoptoses of lymphocytes and debris among the hematopoietic cells in the reticular networks of the bone marrow. These patterns remained for 24 h.

As early as 4 h from the beginning of the experiments, mice injected with 100 ng/kg or more of maitotoxin had their spleens filled with masses of lymphocytic debris in the red pulp. In addition to the presence of destroyed lymphocytes in the venous sinuses, there were numerous macrophages in the white pulp as well as in the marginal zones at 24 h.

The content of calcium in the stomach, thymus, and cardiac muscle was measured after the injection of 400 ng/kg of maitotoxin into the peritoneal cavities of mice. Within 2 h after the injection, only the content of total calcium in the stomach increased and remained at high levels for 8 h. These levels were two to three times higher than that of physiological saline–injected control.

C. Repeated Administration of Maitotoxin

Terao et al. (42) observed pathomorphological changes after repeated administrations of maitotoxin. During the first 10 days, there was no significant difference between the percentage of lymphocytes in the circulating blood cells of control mice and of mice repeatedly receiving maitotoxin (45 ng/kg). After the tenth injection, however, this percentage suddenly decreased from 63 to 13%. In contrast, the number of neutrophils was only reduced slightly, and therefore the relative percentage was increased to 83% (Figure 9).

The most prominent feature of mice receiving repeated injections of maitotoxin was the marked atrophy of the organs related to immune system such as the thymus, spleen, and lymph nodes on day 13. It is generally accepted that polyvalent cations such as Co^{2+}, Mn^{2+}, Mg^{2+}, and La^{3+} selectively inhibit calcium channel function. Therefore, Terao et al. (41) examined the effect of Co^{2+} on the thymus and spleen of maitotoxin-treated animals. The decrease in the weight of these organs was encountered in mice injected with maitotoxin only. In contrast, maitotoxin-treated mice drinking water containing $CoCl_2$ throughout the duration of the experiment did not reveal any significant differences in comparison with control animals (Figure 10).

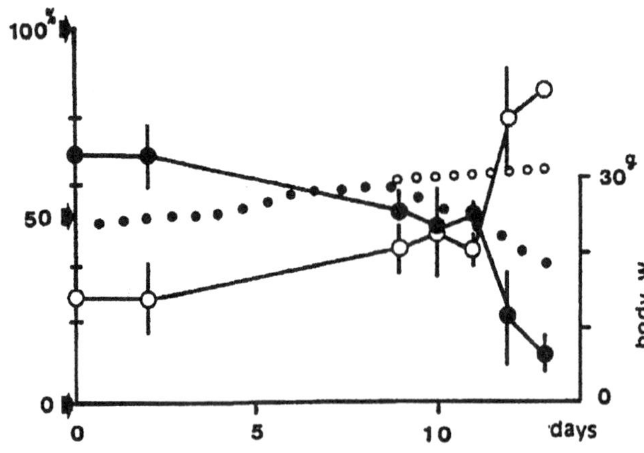

Figure 9 Effect of MTX on mouse blood lymphocytes. The time course of the percentage of lymphocytes in the mouse blood, large solid circles; percentage of lymphocytes, large open circles; percentage of polymorphic leukocytes, small circles; body weights (solid, experimental; open, control).

Histological examinations of the thymus of mice injected 13 times with maitotoxin showed a marked reduction of lymphocytes in the cortical layer. In the medulla of the thymus, however, there were several large-size lymphocytes and many dilated vessels. At the ultrastructural level, almost all of lymphocytes had disappeared, and there were many cells with spindle or bizarre nuclei accompanying masses of collagen fibers (Figure 11). Occasionally, degeneration of the endothelial lining cells of the small vessels of the cortical layer of the thymus was seen. The

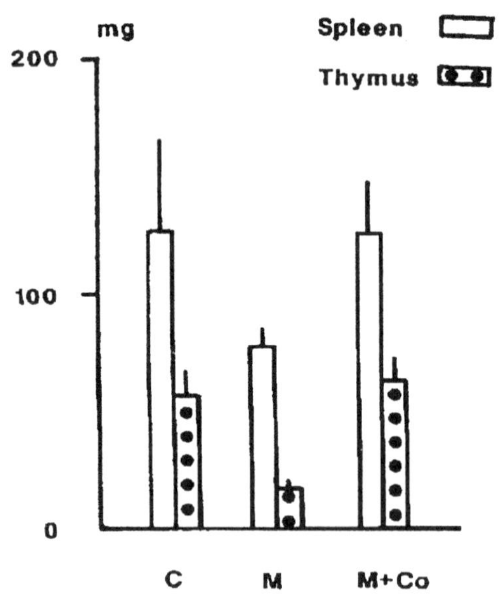

Figure 10 Weight of the thymus and spleen on day 13 of mice given various treatments. M, mice given a total of 13 repeated i.p. injections of 45 ng/kg of maitotoxin. M + Co, mice given maitotoxin injection similar to that in M. In addition to the injections, mice were drinking water containing 10 ppm of $CoCl_2$ throughout the experiment. C, control. Each column and vertical bar = the mean ± SD (n, control = 5, M and M + Co^{2+} 10). Statistical evaluation by Student's *t*-test.

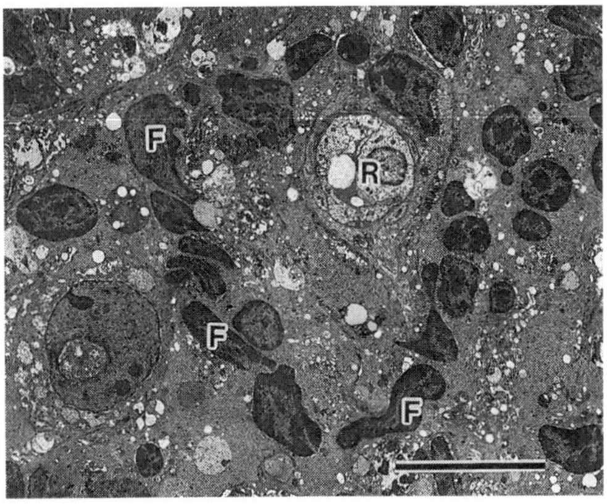

Figure 11 Electron micrograph of the thymus from a mouse injected i.p. with 45 ng/kg of maitotoxin 13 times. There are no lymphocytes in the cortex of the thymus. The parenchyma is occupied by connective tissue. F, fibroblast; R, reticular epithelial cell. The scale bar = 10 μm.

morphological appearance of the thymus from the mice treated with $CoCl_2$ together with maitotoxin as well as that of adrenalectomized animals with maitotoxin treatment were similar to that of control animals.

Light microscopic observation of the spleen of the animals treated with maitotoxin only also revealed a marked reduction of lymphocytes both in the white and red pulp. Most of the periarterial lymphoid sheaths in the white pulp were incompletely formed. Electron microscopy showed that the endothelial cells of the venous sinus in mice given maitotoxin only were markedly increased in height. T-cell–dependent areas in various lymphoid tissues, including lymph nodes and Peyer's patches in the small intestine, became atrophic.

Marked cytoplasmic vacuolation of cells in the zona fasciculata of the adrenal glands were frequently seen in the mice receiving 13 injections of maitotoxin only, whereas the mice drinking water containing $CoCl_2$ during the experimental period showed no discernible changes (Figure 12a,b). The content of the vacuoles in the cells of the zona fasciculata of the adrenal gland were ultrastructurally autophagic vacuoles up to 10 μm in diameter. The content of the vacuoles consisted mainly of degenerated mitochondria, smooth-surfaced endoplasmic reticulum, free ribosomes, membrane-limited flocculent materials, and myelin figures (Figure 12a).

D. Biochemical Changes After Administrations of Maitotoxin

Total Calcium Content of the Adrenal Glands, Thymus, and Spleen

Table 3 shows the total calcium content of the adrenal glands, thymus, and spleen of mice given an intraperitoneal injection of maitotoxin, mice receiving the toxin together with $CoCl_2$, and control animals. In mice injected with maitotoxin only, the total calcium content of the adrenal glands was increased significantly when compared with that of control ($P < .05$), whereas that of the adrenal glands of mice pretreated with $CoCl_2$ showed no significant increase after the injection of maitotoxin. The total calcium content of the thymus and spleen did not differ among the three groups.

Table 4 shows the amount of immunoglobulin M in the serum of control mice and animals receiving consecutive injections of maitotoxin for either 7 or 13 days with or without pretreatment with $CoCl_2$. There were no significant differences between immunoglobulin M values in the groups at the seventh day. At day 13, however, immunoglobulin M in the serum of mice

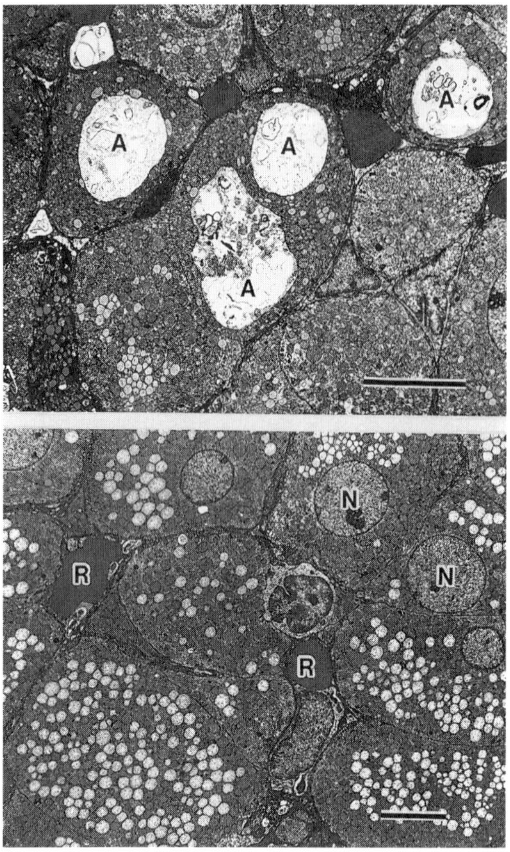

(a)

(b)

Figure 12 Electron micrographs of the cortical zone (zona fasciculata) of the adrenal gland. (a) Cells in zona fasciculata of the adrenal gland from a mouse injected 13 times with 45 ng/kg of maitotoxin. Note many autophagic vacuoles (A) in the cytoplasm of cells. N, nucleus of epithelial cell. Scale bar = 10 μm. (b) Cells from a mouse given maitotoxin injections similar to that of (a). In addition to the injections, the mice were given drinking water containing 10 ppm of $CoCl_2$ throughout the experiment. Except for small fat droplets, no autophagic vacuolation are seen in the cytoplasm of the cells. N, nucleus of cell in the zona fasciculata. Scale bar = 10 μm.

Table 3 Total Content of Calcium in the Thymus, Spleen, and Adrenal glands 1 h After Administration of Maitotoxin with or Without $CoCl_2$

	Calcium (μg/mg protein)		
Treatment	Thymus	Spleen	Adrenal glands
Control	1.71 ± 0.20	0.95 ± 0.09	2.18 ± 0.09
MTX	1.33 ± 0.25	1.00 ± 0.10	3.09 ± 0.09[a]
MTX + $CoCl_2$	1.31 ± 0.31	1.08 ± 0.12	2.29 ± 0.11

Total calcium content was measured at 1 h after i.p. injection of 200 ng/kg of maitotoxin. Mean ± SEM (n = 14).

MTX, mice given an i.p. injection 200 ng/kg of maitotoxin without $CoCl_2$-pretreatment; MTX + $CoCl_2$, mice given a 200 ng/kg of maitotoxin with $CoCl_2$ pretreatment.

[a] Treatment is significant ($P < .05$); statistical evaluation by Student's *t*-test.

Table 4 Amount of Immunoglobulin M in Serum After Treatment with Maitotoxin and $CoCl_2$

	Relative Amount of IGM	
Treatment	7th day	13th day
Control	1.00 ± 0.05	1.00 ± 0.05
MTX	0.78 ± 0.09	0.32 ± 0.01[a]
MTX + $CoCl_2$	0.94 ± 0.05	0.87 ± 9.06

The relative intensity of bands (control value: 1.00) was the average value of triplicate experiments. Mean ± SEM (n, control = 5: MTX and MTX + $CoCl_2$ = 10).
MTX, mice given a total of 13 repeated i.p. injections 45 ng/kg of maitotoxin; MTX + $CoCl_2$, mice given maitotoxin injections similar to that in MTX. In addition to the injections, the mice were drinking water containing 10 ppm of $CoCl_2$ throughout the experiment.
[a] Treatment is significant ($P < .05$): statistical evaluation by Student's *t*-test.

treated with maitotoxin decreased significantly ($P < .05$) when compared to the control, whereas there was no significant difference between control and $CoCl_2$-pretreated mice plus repeated maitotoxin.

Cortisol in Blood Plasma

Maitotoxin resulted in a sudden marked increase in the plasma cortisol concentration at 30 min after injection and then decreased gradually (Fig. 13).

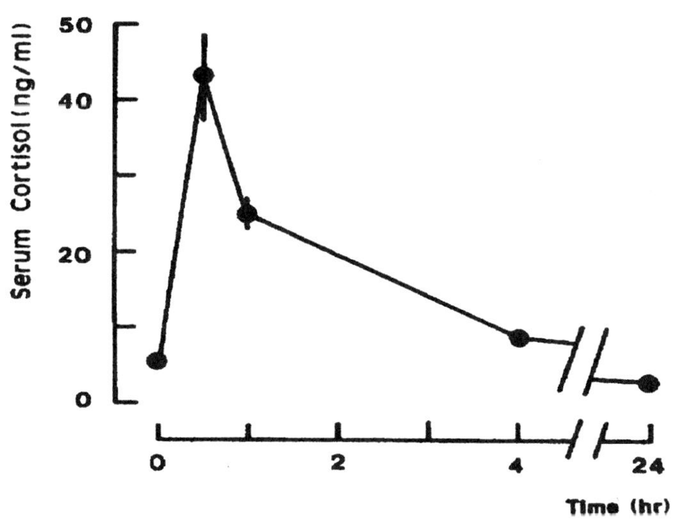

Figure 13 The time course of effects of i.p. injection of maitotoxin (200 ng/kg) on plasma cortisol levels in mice. Each point and vertical bar = the mean + SEM (n = 6). Where no standard error is given, it is less than the radius of the point.

Conclusions

Terao et al. (41) demonstrated that there were no significant increases in the total calcium content of the thymus, although that of the stomach of maitotoxin-treated mice increased about two to three times more than control. From these results, it is postulated that there may be some difference in the maitotoxin poisoning between the thymus and stomach.

Maitotoxin may first stimulate calcium influx into the cortical cells of the adrenal glands, which would in turn result in the release of cortisol from these organs into the blood. The marked reduction of lymphocytes in the thymus was probably caused by an excessive amount of cortisol in the serum after the administration of maitotoxin. The considerable reduction observed in T-cell–dependent areas in other lymphoid tissues is also not a phenomenon directly induced by the phycotoxin, because the total calcium content in these tissues constantly remained at a low level even after the high-dose administration of maitotoxin. This view was also strongly supported by the findings that pretreatment with $CoCl_2$, a potent inhibitor of calcium channels, as well as unilateral or bilateral adrenalectomy, prevented the lymphoid lesion. Gastric ulcers that occurred shortly after the administration of maitotoxin may be closely related to the excess of cortisol in the blood. Human clinical signs and symptoms induced by pure maitotoxin are not known, although surgeonfish that contain both ciguatoxin and maitotoxin are responsible for over half of the human cases of ciguatera fish poisoning in Tahiti. Peculiar clinical signs and symptoms of ciguatoxicosis may be superimposed upon those induced by maitotoxin.

E. Comparison of the Pathology of Ciguatoxicosis and Maitotoxicosis

In contrast to experimental ciguatoxicosis, maitotoxin poisoning in mice may be mainly caused by an indirect effect of this toxin on various target organs such as lymphoid tissues and the gastric tract. Table 5 summarizes the pathomorphological findings of ciguatoxin and maitotoxin poisoning in various target organs. Ciguatoxicosis apparently differs from maitotoxicosis in the pathomorphological responses of mice. Ciguatoxin stimulated an increased sodium permeability across the tetrodotoxin-sensitive sodium channels of excitable membranes such as cardiac cells

Table 5 Comparison of Ciguatoxicosis and Maitotoxicosis in Mice

Target organs	Ciguatoxicosis	Maitotoxicosis
Heart	Marked swelling of cardiac muscle cells Effusion into interstitium	Single-cell necroses Narrowing of lumen of blood capillaries
Stomach and intestine	No discernible changes in mucosa Swelling of amyelinated nerve cells in submucosal plexus and in the myentric plexus	Multiple gastric erosions and ulcers
Adrenal glands	Vacuoles in the medulla	Multiple autophagosomes in cells of zona fasciculata
Lymphoid tissue	—	Massive necroses of lymphocytes in the thymic cortex Atrophy of T-cell–dependent areas in the spleen and lymph nodes
Penis	Continuous erection Thrombosis in cavern	—
Mode of action	Increase in Na^+ influx results in swelling of target cells	Increase in Ca^{2+} influx Ca^{2+} acts as secondary messenger or excessive Ca^{2+} acts as lethal factor

as well as skeletal and smooth muscle cells, unmyelinated nerve cells in the small intestine, and vas deferens.

Results of the repeated injection experiments indicated that both ciguatoxin and maitotoxin have cumulative effects.

F. Functional Enhancement and Disorders Induced by Maitotoxin

Maitotoxin stimulates a broad spectrum of calcium-dependent physiological processes within limited doses; however, an excessive influx of Ca^{2+} into the cells is lethal.

The biological activities of maitotoxin were first examined by Miyahara et al. (43). Thereafter followed much research on the effects of maitotoxin on cells with excitable membranes and hormone-producing cells in in vitro systems.

Miyahara et al. (43) demonstrated that maitotoxin enhanced the contractility of the atria of the guinea pig in in vitro system. However, the mode of action of the phycotoxin was not clear. The first evidence on the mode of action of maitotoxin was given by Takahashi et al. in 1982 (44). Using pheochromocytoma cells, they showed that this phycotoxin induces a profound increase in Ca^{2+} influx into the cells and Ca^{2+}-dependent release of [^{3}H]norepinephrine even in the absence of external Na^+, and that these effects are nearly abolished by calcium channel blockers.

Maitotoxin caused a dose-dependent contraction of the guinea pig isolated ileum and/or tenia caeci at concentration of 100 pg to 30 ng/mL. (45). They postulated that maitotoxin activates Ca^{2+} channels in the smooth muscle membrane of both tissues to increase in Ca^{2+} influx and thus induces contractions. Ohizumi et al. (45) also found slower contraction effects of maitotoxin on isolated guiana pig vas deferens. In 1985, Kobayashi et al. (46) confirmed that the excitatory effects of maitotoxin on guinea pig heart muscle are caused by a direct action on the cardiac muscle membrane mainly due to an increase in Ca^{2+} permeability. Kim et al. (47) reported on the presynaptic and postsynaptic action of maitotoxin at the mouse neuromuscular junction studied with intracellular microelectrode recording techniques.

Kobayashi et al. (48) reported the detailed toxinological properties of maitotoxin on cardiac muscle from guinea pig and suggested that maitotoxin increases the calcium ion influx through the cardiac muscle membrane to create the calcium-overloaded state and thus caused the cardiotoxic effects.

Sladeczek et al. (49) found that Ca^{2+} can stimulate the hydrolysis of inositol phospholipids in membranes of muscle cells in culture. Furthermore, they suggested that the primary action of maitotoxin in the muscle cell line was a pore-forming or channel-forming activity of a nonclassic type.

Hormonal release by anterior pituitary cells is a complex process involving many components such as calcium ion, phospholipids, and cyclic nucleotides. Maitotoxin-stimulated hormonal release was blocked by the calcium channel blocker.

Schettini et al. (50) demonstrated that maitotoxin stimulates the release of prolactin, growth hormone, thyroid-stimulating hormone, and luteinizing hormone from anterior pituitary cells in culture in a dose-dependent manner. A rise in the cytosolic concentration of calcium ions has been widely accepted to be an intracellular messenger-coupling stimulus to insulin release from pancreatic B cells. Niki and his coworkers (51) demonstrated that maitotoxin stimulates insulin release via activating voltage-dependent calcium ion channels of pancreas B cells in rats. Soergel et al. (52) also confirmed the stimulatory effects of maitotoxin on insulin release. The theory of stimulus secretion coupling postulates that an increase in the cytosolic Ca^{2+} concentration underlies the activation of exocytosis in a variety of tissues. The parathyroid gland is unusual among secretory systems in that parathyroid hormone release is stimulated by low extracellular Ca^{2+} content ratios and inhibited by high extracellular concentrations (53). Further, Fitzpatrick et al. (54) demonstrated that maitotoxin by its action on calcium channels allows entry of extracellular calcium and inhibits parathyroid hormone release, and they suggested that calcium channels are involved in the release of the hormone. Nishiyama et al. (55) also demonstrated that maitotoxin induces calcitonin secretion from primary culture derived from fetal rat thyroid C cells.

The effects of maitotoxin on an invertebrate were studied by Miyamoto et al. (56) using skeletal muscle of the insect *Tenebrio molitor*. They confirmed electrophysiologically that maitotoxin activates the voltage-dependent calcium ion channels in insect muscle. The mussel *Mytilus edulis* and the starfish *Asterias amurensis* are also sensitive to maitotoxin (57).

Pin et al. (58) reported that maitotoxin induced a large and concentration-dependent release of endogenous gamma-aminobutyric acid (GABA). They suggested that the maitotoxin-induced GABA is only in part due to a Co^{2+}-sensitive Ca^{2+} influx. Maitotoxin also increased the release of numerous amino acids. The release of the putative amino acid neurotransmitters (taurine, hypotaurine, and GABA) was particularly augmented in the presence of maitotoxin.

REFERENCES

1. JE Randall. A review of ciguatera, tropical fish poisoning, with a tentative explanation of its cause. Bull Mar Sci Gulf Carib 8:236–267, 1958.
2. T Yasumoto, I Nakajima, R Bagnis, R Adachi. Finding a dinoflagellate as a likely culprit of ciguatera. Bull Jpn Soc Sci Fish 43:1021–1026, 1977.
3. BW Halstead. Poisonous and Venomous Marine Animals of the World. Vol 2. Washington, DC: Government Printing Office, 1964, pp 1–155.
4. PJ Scheuer, W Takahashi, J Tsutsumi, T Yoshida. Ciguatoxin: isolation and chemical nature. Science 155:1267–1268, 1967.
5. T Yasumoto, Y Hashimoto, R Bagnis, JE Randall, AH Banner. Toxicity of the surgeonfishes. Bull Jpn Soc Sci Fish 37:724–734, 1971.
6. R Bagnis, T Kuberski, S Laugier. Clinical observations on 3,009 cases of ciguatera (fish poisoning) in the South Pacific. Am J Trop Med Hyg 28:1067–1073, 1979.
7. Y Ohizumi, S Shibata, K Tachibana. Mode of excitatory and inhibitory actions of ciguatoxin in the guinea-pig vas deferens. J Pharmacol Exp Ther 217:475–480, 1981.
8. AM Legrand, M Galonnier, R Bagnis. Studies on the mode of action of ciguatic toxins. Toxicon 20: 311–315, 1982.
9. NA Palafox, LG Jain, AZ Pinao, TM Gulick, RK Williams, IJ Schatz. Successful treatment of ciguatera fish poisoning with intravenous mannitol. JAMA 259:2740–2742, 1988.
10. M Murata, AM Legrand, Y Ishibashi, T Yasumoto. Structures of ciguatoxin and its congener. J Am Chem Soc 111:8929–8931, 1989.
11. NC Gillespie, RJ Lewis, JH Pean, ATC Bourke, WJ Shields. Ciguatera in Australia. Occurrence, clinical features, pathophysiology and management. Med J Aust 145:584–590, 1986.
12. M Murata, H Naoki, T Iwashita, S Matsunaga, M Sasaki, A Yokoyama, T Yasumoto. Structure of maitotoxin. J Am Chem Soc 115:2060–2062, 1993.
13. DM Anderson, PS Lobel. The continuing enigma of ciguatera. Biol Bull 172:89–107, 1987.
14. DM Anderson, JK Sims, NH Wiebenga, M Sugi. The epidemiology of ciguatera fish poisoning in Hawaii. Hawaii Med J 42:326–334, 1983.
15. JH Gollop, EW Pon. Ciguatera: a review. Hawaii Med J 51:91–99, 1992.
16. DN Lawrence, MB Enriquez, RM Lumish, A Maceo. Ciguatera fish poisoning in Miami. JAMA 244:254–258, 1980.
17. JP Quod, J Turquet. Ciguatera in Reunion Island (SW Indian Ocean). epidemiology and clinical patterns. Toxicon 34:779–785, 1996.

18. JL Allsop, L Martini, H Lebris, J Pollard, J Walsh, S Hodgkinson. Les Manifestations neurologiques de la ciguatera. Rev Neurol (Paris) 142:590–597, 1986.
19. T Yasumoto, A Inoue, R Bagnis, M Garcon. Ecological survey on a dinoflagellate possibly responsible for the induction of ciguatera. Bull Jpn Soc Fish 45:395–399, 1979.
20. T Yasumoto, A Inoue, T Ochi, K Fujimoto, Y Oshima, Y Fukuyo, R Adachi, R Bagnis. Environmental studies on a toxic dinoflagellate responsible for ciguatera. Bull Jpn Soc Sci Fish 46:1397–1404, 1980.
21. K Terao, E Ito, M Oarada, Y Ishibashi, AM Legrand, T Yasumoto. Light and electron microscopic studies of pathologic changes induced in mice by ciguatoxin poisoning. Toxicon 29:633–643, 1991.
22. KM Li. Ciguatera fish poisoning: a cholinesterase inhibitor. Science 147:1580–1581, 1965.
23. MD Rayner. Mode of action of ciguatera. Fed Proc 31:1139–1145, 1972.
24. E Ito, T Yasumoto, K Terao. Morphological observations of diarrhea in mice caused by experimental ciguatoxicosis. Toxicon 34:111–122, 1996.
25. K Terao, E Ito, T Yasumoto. Pathomorphological studies on experimental maitotoxicosis and ciguatoxicosis in mice. In TR Tosteson, ed. 3rd International Conference on Ciguatera Fish Poisoning, Puerto Rico, 1990. Quebec: Polyscience Publishing, 1992, pp 53–70.
26. K Terao, E Ito, T Yasumoto. Light and electron microscopic studies of the murine heart after repeated administrations of ciguatoxin or ciguatoxin-4c. Nat Toxins, 1:19–26, 1992.
27. LL Boyarsky, MD Rayner. The effect of ciguatera toxin on Aplysia neurons. Proc Soc Exp Biol Med 134:332–335, 1970.
28. JA Setliff, MD Rayner, SK Hong, Effect of ciguatoxin on sodium transport across the frog skin. Toxicol Appl Pharmacol 18:676–684, 1971.
29. H Ohshika. Marine toxins from the Pacific. XI. Effects of ciguatoxin on isolated mammalian atria. Toxicon 9:337–343, 1971.
30. JT Miyahara, CK Aku, T Yasumoto. Effects of ciguatoxin and maitotoxin on the isolated guinea pig atria. Res Commun Chem Pathol Pharmacol 25:177–180, 1979.
31. Y Ohizumi, S Shibata, K Tachibana. Mode of the excitatory and inhibitory actions of ciguatoxin in the guinea-pig vas deferens. J Pharmacol Exp Ther 217:475–480, 1981.
32. J-N Bidard, HPM Vijverberg, C Frelin, E Chungue, AM Legrand, R Bagnis, M Lazdunski. Ciguatoxin is a novel type of Na^+ channel toxin. J Biol Chem 259:8353–8357, 1984.
33. A Lombert, J-N Bidard, M Lazdunski. Ciguatoxin and brevetoxins share a common receptor site on the neuronal voltage-dependent Na^+ channel. FEBS Lett 219:355–359, 1987.
34. A Seino, M Kobayashi, K Momose, T Yasumoto, Y Ohizumi. The mode of inotropic action of ciguatoxin on guinea pig cardiac muscle. Br J Pharmacol 95:876–882, 1988.
35. JC Cloyd, BD Snyder, B Cleeremans, SR Bundlie. Mannitol pharmacokinetics and serum osmolality in dogs and humans. J Pharmacol Exp Ther 236:301–306, 1986.
36. RJ Lewis, AW Wong Hoy, M Sellin. Ciguatera and mannitol: in vivo and in vitro assessment in mice. Toxicon 31:1039–1050, 1993.
37. J Molgo, P Juzans, AM Legrand. Confocal laser scanning microscopy: a new tool for studying the effects of ciguatoxin (CTX-1B) and D-mannitol at motor nerve terminals of the neuromuscular junction in situ. Mem Queensland Mus 34:577–585, 1994.
38. R Bagnis. Clinical aspects of ciguatera (fish poisoning) in French Polynesia. Hawaii Med J 28:25–28, 1968.
39. T Yasumoto, R Bagnis, JP Vernoux. Toxicity of the surgeonfishes—II. Properties of the principal water-soluble toxin. Bull Jpn Soc Sci Fish 42:359–365, 1976.
40. RJ Lewis, TA Ruff. Ciguatera: ecological, clinical, and socioeconomic perspectives. Crit Rev Environ Sci Technol 23:137–156, 1993.
41. K Terao, E Ito, Y Sakamaki, K Igarashi, A Yokoyama, T Yasumoto. Histopathological studies of experimental marine toxin poisoning. II. The acute effects of maitotoxin on the stomach, heart and lymphoid tissues in mice and rats. Toxin 26:395–402, 1988.
42. K Terao, E Ito, Y Kakinuma, K Igarashi, M Kobayashi, Y Ohizumi, T Yasumoto. Histopathological studies on experimental marine toxin poisoning. 4. Pathogenesis of experimental maitotoxin poisoning. Toxicon 27:978–988, 1989.

43. JT Miyahara, CK Akau, T Yasumoto. Effects of ciguatoxin and maitotoxin on the isolated guinea pig atria. Res Comm Chem Pathol Pharmacol 25:177–180, 1979.
44. M Takahashi, Y Ohizumi, T Yasumoto. Maitotoxin, a Ca^{2+} channel activator candidate. J Biol Chem 257:7278–7289, 1982.
45. Y Ohizumi, T Yasumoto. Contractile response of the rabbit aorta to maitotoxin, the most potent marine toxins. J Physiol (Lond.) 337:711–721, 1983.
46. M Kobayashi, Y Ohizumi, T Yasumoto. The mechanism of action of maitotoxin in relation to Ca^{2+} movements in guinea-pig and rat cardiac muscles. Br J Pharmacol 86:385–391, 1985.
47. Y Kim, I Login, T Yasumoto. Maitotoxin activates quantal transmitter release at the neuromuscular junction: evidence for elevated intraterminal Ca^{2+} in the motor nerve terminal. Brain Res 346:357–362, 1985.
48. M Kobayashi, S Kondo, T Yasumoto, Y Ohizumi. Cardiotoxic effects of maitotoxin, a principal toxin of seafood poisoning on guinea pig and rat cardiac muscle. J Pharmacol Exp Ther 238:1077–1083, 1986.
49. F Sladeczek, BH Schmidt, R Alonso, L Vian, A Tep, T Yasumoto, RN Cory, J Bockaert. New insights into maitotoxin action. Eur J Biochem 174:663–670, 1988.
50. G Schettini, K Koike, IS Login, AM Judd, MJ Cronin, T Yasumoto, RM MacLeod. Maitotoxin stimulates hormonal release and calcium flux in rat anterior pituitary cells in vitro. Am J Physiol 247:E520–E525, 1984.
51. I Niki, T Tamagawa, H Niki, A Niki, T Koide, N Sakamoto. Stimulation of insulin release by maitotoxin, an activator of voltage-dependent calcium channels. Biomed Res 7:107–112, 1986.
52. DG Soergel, F Gusovsky, T Yasumoto, JW Daly. Stimulatory effects of maitotoxin on insulin release in insulinoma HIT cells: role of calcium uptake and phosphoinositide breakdown. J Pharmacol Exp Ther 255:1360–1365, 1990.
53. DM Shoback, J Thacher, R Leomburno, EM Brown. Relationship between parathyroid hormone secretion and cytosolic calcium concentration in dispersed bovine parathyroid cells. Proc Natl Acad Sci USA 81:3113–3117, 1984.
54. LA Fitzpatrick, T Yasumoto, GD Aurbach. Inhibition of parathyroid hormone relesse by maitotoxin, a calcium channel activator. Endocrinology 124:97–103, 1989.
55. I Nishiyama, T Yasumoto, T Fujiki. Maitotoxin induces calcitonin secretion from rat thyroid C-cells, in vitro. Horm Metab Res 22:258–259, 1990.
56. T Miyamoto, Y Ohizumi, H Washio, T Yasumoto. Potent excitatory effect of maitotoxin on Ca channels in the insect skeletal muscle. Pfluegers Arch 400:439–441, 1984.
57. I Nishiyama, T Matsui, T Yasumoto, S Ohshio, M Hoshi. Maitotoxin, a presumed calcium channel activator, induced the acrosome reaction in mussel spermatozoa. Dev Growth Differ 28:443–448, 1986.
58. JP Pin, T Yasumoto, J Bockaert. Maitotoxin-evoked gamma aminobutyric acid release is due not only to the opening of calcium channels. J Neurochem 50:1227–1232, 1988.

22

Ciguatera Toxins: Mechanism of Action and Pharmacology of Maitotoxin

Mark Estacion
MetroHealth Medical Center, Cleveland, Ohio

I. INTRODUCTION

A. Source and Structure of Maitotoxin

Maitotoxin (MTX) was originally described as a compound associated with seafood poisoning in South Pacific coral reef fish. ''Maito'' is the Tahitian name for the surgeonfish in which the toxin commonly accumulates and from which the toxin was first isolated and characterized. Similar to other red tide toxic episodes, the toxin is actually biosynthesized by a dinoflagellate known as *Gamberdiscus toxicus*, which is also the source of the lipid soluble ciguatoxins.

The structure of MTX is very complex, so that even though it was able to be purified in the late 1970s, its structure was not solved until 1994 (1). As shown in Figure 1, MTX is a polycyclic molecule of 3424 D molecular weight that is kept water soluble by numerous hydroxyl groups and sulfate groups. Since the molecular weight was not known at first, most of the literature reports MTX doses on a weight basis, with active doses usually in the nanogram/milliliter range. A 1 ng/mL MTX dose corresponds to a 0.3 nM MTX concentration. For consistency, MTX doses originally reported on a weight basis will be recalculated to a molar basis in this chapter. The complicated structure of MTX has frustrated attempts to construct high-affinity radioactivity labeled forms (2,3) of MTX for use in the biochemical identification and characterization of the receptor for MTX.

Multiple forms of maitotoxin (MTX-1, MTX-2, and MTX-3) can be generated (4–6) with different molecular weights and chromatographic properties, yet they share similar bioassay properties. Some of the variation of MTX potencies reported in the same preparation (7) may be due to different lots of MTX, or possibly as loss of activity due to storage of diluted MTX in aqueous buffers (M. Estacion, unpublished observations). These different forms are likely strain dependent (4), but the form of MTX commercially available is MTX-1.

Being water soluble and having low oral potency, MTX is not thought to be a significant factor in cases of seafood toxicity compared to ciguatoxins (8). MTX, however, is extremely potent, which makes it a promising research reagent. In addition, natural toxins have repeatedly been useful probes for the identification and molecular characterization of cellular mechanisms critical for normal function. Since MTX effects are widely observed both in various tissues and across species, it is likely that the MTX receptor represents an important cellular regulator.

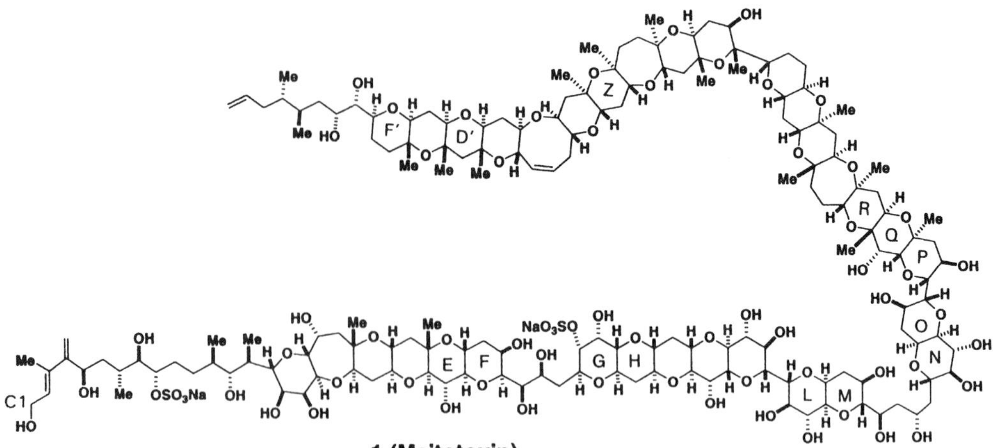

1 (Maitotoxin)

Figure 1 Structure of maitotoxin. The complete structure of maitotoxin consists of a ladder-shaped polyether skeleton. The ring structures have been labeled from A to F′. (From ref. 90.)

B. Chronological Development of Theories of MTX Action

The first assay for MTX was the LD50 of mice, which proved to be the most potent toxin identified on a weight basis: able to kill mice at less than 0.2 μg/kg (intraperitoneally) and at least 50-fold more toxic than tetrodotoxin (TTX), which was previously the most potent toxin (9). Sublethal doses of MTX were shown to have profound effects on cardiovascular function both in vivo and in isolated heart preparations. Pathological examination of MTX-treated mice identified clear abnormalities in heart cells as well as in the stomach, liver, and thymus. Ultrastructural examination of the ventricular myocytes showed disruption of the myofilament structure and swollen mitochondria indicative of calcium overload.

As will be described in detail below, these observations were followed up in numerous studies utilizing more defined cell populations and better experimental methods. It was quickly determined that MTX acts directly on cells. In the isolated heart preparation, blockers of neurotransmitter release had no effect on MTX-induced contraction and, furthermore, isolated myocyte preparations contracted on exposure to MTX, which was abolished when extracellular calcium was removed. MTX stimulates uptake of $^{45}Ca^{2+}$ in myocytes as well as other excitable cells such as neurons and neuroendocrine cells. It appeared that MTX exposure leads to a stimulated influx of calcium.

The response to MTX was shown to be reduced in the presence of the voltage-gated calcium channel blockers verapamil or nifedipine, leading to the general hypothesis that MTX acts as an activator of voltage-gated calcium channels (10). In these cells, the influx caused by the calcium ionophore A23187 was insensitive to calcium channel blockers, suggesting that MTX does not act simply as an ionophore but through voltage-gated calcium channels. This was also supported by the report of a lack of calcium ionophore activity of MTX in mitochondria and liposomes (11,12).

This hypothesis (like all hypotheses) was soon found to need revision. Directly measuring voltage-gated calcium currents with electrophysiological approaches showed a MTX-induced loss of both low-voltage–activated and high-voltage–activated calcium channels with a time course delayed from the appearance of a linear leak current and a rise in intracellular calcium (13). A revised hypothesis would be that MTX binding transformed the voltage-gated calcium channels into the nonselective current observed. However, MTX was unable to alter [^{3}H]nitrendipine binding to calcium channels in synaptosomes (14). Another difficulty was the finding that MTX caused calcium influx into cells that do not express voltage-gated calcium channels such as hepatocytes and fibroblasts. Thus, MTX must be able to activate an additional calcium influx pathway. Subsequent electrophysiological studies confirmed that MTX induces a nonselective cation current which precedes the rise of intracellular calcium. Activation of a nonselective cation current would lead to depolarization and thus indirectly activate voltage-gated calcium channels.

Meanwhile the importance of inositol phospholipid metabolism on intracellular calcium signaling was demonstrated. The minor membrane phospholipid phosphatidylinositol bisphosphate (PIP_2) can be cleaved by phospholipase C (PLC) to generate two signaling molecules, inositol trisphosphate (IP_3) and diacylglycerol (DG). IP_3 causes release of calcium from intracellular stores such as the endoplasmic reticulum; observable as an increase of intracellular calcium concentration. Methods for detecting and quantifying inositol trisphosphate IP_3 and its breakdown products inositol bisphosphate (IP_2) and inositol monophosphate (IP) demonstrated this signaling pathway is generalized to all cells (15). Examination of the effect of MTX on phosphatidylinositol metabolism demonstrated that MTX was even more potent at generating inositol phosphates (IPs) than for elevating intracellular calcium and occurred in every MTX-responsive cell type (16). A distinct plasma membrane calcium influx pathway can be activated by generation of IP_3 and depletion of intracellular stores now termed capacitative calcium entry (CCE) (17). Thus, calcium influx may be stimulated in MTX-treated cells through a CCE mechanism. The pharmacology of MTX-induced calcium influx, especially in unexcitable cells that do not express voltage-gated calcium channels, is similar to the pharmacology of the CCE pathway. Most PLC isoforms, however, require a threshold level of calcium concentration and are stimulated by elevation of intracellular calcium (18). Like the chicken and the egg paradox, it is difficult to determine if MTX first stimulates calcium influx leading to PLC activation, or if MTX stimulates calcium influx subsequent to PLC-mediated IP_3 generation and activation of the CCE pathway.

The number of cell types found to respond to MTX as well as the kinds of MTX-induced responses was steadily growing. Many cell lines representative of neural or neuroendocrine origin were stimulated to secrete by MTX. For example, MTX stimulated neurotransmitter release from PC12 cells and cultured striatal neurons. Pituitary tumor cell lines GH3 and GH4 secreted prolactin in a Ca^{2+}-dependent manner to MTX. Pancreatic islets and HIT and RIN cell lines were stimulated to secrete insulin in response to MTX. The MTX-induced responses in these cells were inhibited by voltage-gated calcium channel blockers consonant with calcium-regulated secretion. Various nonexcitable cells were also stimulated to secrete by MTX. Basophils were stimulated to release histamine, arachidonic acid, and leukotriene C4 independently of intracellular calcium increases, suggesting a more direct stimulation of secretion (19). Other responses to MTX are probably subsequent to MTX-induced intracellular calcium increases. Platelets were stimulated to clump and release adenosine triphosphate (ATP) in response to MTX (20). The calcium-dependent protease calpain is activated in MTX-treated neurons (21). Hepatocytes and lymphocytes were shown to be susceptible to MTX-induced cell necrosis (22,23).

The variety of cell types and responses elicited in response to MTX has made the elucidation of the mechanism or site of MTX action difficult. One possibility is that different moieties on the MTX molecule can directly stimulate voltage-gated calcium channels, phospholipase C, and nonselective cation channels. More satisfying, however, would be to find a common upstream regulator of these proteins which would be demonstrated as the high-affinity binding site for MTX. Understanding the pathophysiology of MTX will be one of the major questions addressed by researchers in this field in the near future. Even though the precise mechanism of MTX action is not yet known, MTX is continuing to be used by researchers as a potent and reliable stimulator of calcium influx.

II. MAITOTOXIN EFFECTS IN VARIOUS CELL TYPES

A. Visceral Organs

An in vivo histopathological study (24) of MTX poisoning identified the effects of MTX on a number of sites and cell types. At sublethal doses (50–400 ng/kg), the MTX-treated mice exhibited piloerection and cyanosis and were limited to laborious crawling for at least 2 h after which they slowly recovered. The animals showed distended stomachs and small intestines, and microscopic examination of the stomach revealed focal necrosis in the mucosal layer and marked edema in the submucosal layers. This ulceration was not due to acid secretion, since acid production was actually decreased by MTX and blockers of acid secretion did not block MTX-induced stomach ulceration. Examination of the thymus revealed an abundance of degenerating lymphocytes and an abundance of macrophages with phagocytosed lymphocytic debris. There were also numerous pyknotic cells and debris among the hematopoietic cells in the reticular networks of the bone marrow. Microscopic examination of the heart revealed focal coagulative necrosis in the left ventricle. These regions were often associated with marked swelling of the endothelial lining cells leading to narrowed or closed capillaries. The mitochondria in the neighboring muscle cells showed matrix changes suggestive of degeneration. This study suggested that lymphocytes and endothelial cells may be especially vulnerable to MTX-induced pathology. A subsequent study, however, reported that injection of 45 ng/kg MTX into mice induced a sevenfold increase of plasma cortisol which produces acute lymphocyte cell death (22).

B. Hematopoietic Cells

MTX (1.0–10.0 nM) stimulated platelet aggregation and increased intracellular calcium which was dependent on extracellular calcium (20). The dose-response curve for aggregation was shifted to the left and much steeper than the dose response for intracellular calcium. Examination of the data traces (Figure 2) suggests that aggregation is autocatalytic, so that once the aggregation reaction is started it sustains itself, probably due to release of ATP from lysed platelets. Since a rise of intracellular calcium is necessary and sufficient to aggregate platelets, this illustrates how a dose-response curve can appear shifted to the left compared to the dose response for MTX action on intracellular calcium. Barium was able to substitute for calcium, whereas Cd^{2+}, Co^{2+}, and Ni^{2+} blocked MTX-induced aggregation. The organic calcium channel blockers verapamil and nifedipine were ineffective blockers of MTX-induced platelet aggregation at 1 μM. Since voltage-gated calcium channels have not been reported in platelets, the authors conclude that MTX acts via activation of a voltage-independent calcium influx pathway.

MTX was used as a stimulator of increased intracellular calcium to study calcium-dependent release of histamine from human basophils (19). Increasing doses of MTX (0.3–3.0 nM) resulted in a nearly complete release of stored histamine, which occurred slowly over 45 min and was also temperature dependent, being completely ineffective at 4°C. There was no release in low external calcium media and strontium was able to substitute for calcium. The calcium channel blocker nifedipine was ineffective at inhibiting MTX-induced histamine release, whereas verapamil and diltiazem blocked at moderate to high (1–10 μM) concentrations.

The human leukemic cell line HL-60 was used to study MTX stimulation of both calcium influx and generation of inositol phosphates (25,26) in cells that do not express voltage-gated

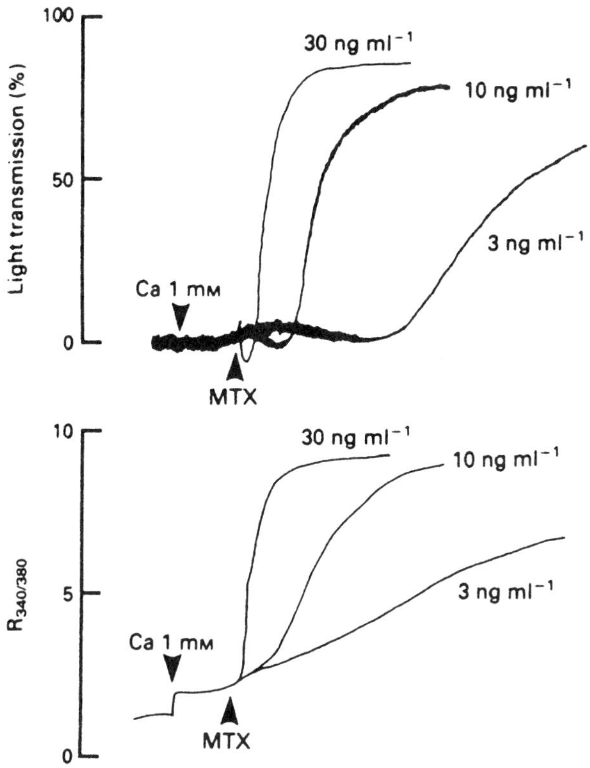

Figure 2 MTX stimulates platelets to aggregate and increase intracellular calcium. Simultaneous recordings of aggregation (upper panel) and fura-2 fluorescence (lower panel) of the washed rabbit platelets in response to 3–30 ng/mL (1–10 nM) maitotoxin. (From ref. 20.)

calcium channels. One study showed pharmacological differences between MTX-induced calcium influx versus receptor-stimulated calcium influx. The other demonstrated that MTX-induced generation of inositol phosphates has distinct properties and is additive to receptor-mediated activation of PLC. These distinctions will be described in detail later in this chapter.

C. Cardiovascular Cells

MTX was shown to have profound effects on cardiac myocytes at both the organ level and at the single-cell level. In isolated beating heart preparations, MTX showed both positive and negative ionotropic effects. At low doses (3 nM), both the amplitude and frequency of contractions were increased, whereas at higher doses (15 nM), these increases were shortly followed by a progressive decrease leading to an irreversibly contracted nonbeating preparation (27). These effects of MTX are not likely due to release of neurotransmitters, since blockers of adrenergic receptors (propranolol) and serotonin receptors (methysergide) did not block MTX effects. It is also unlikely that MTX effects are mediated via sodium channels, because tetrodotoxin was ineffective nor via the Na-K-ATPase, since oubain was also ineffective and MTX did not directly affect pumping activity (28). MTX increased the tissue calcium content and this response was inhibited by Co^{2+} or verapamil. The dose response to MTX of the atria is shifted to lower concentrations compared to ventricular myocytes when measured in the same study (28). Isolation of single ventricular myocytes by enzymatic dispersion still showed pronounced enhancement of contraction at low doses (30 pM to 3 nM) and then arrhythmogenic effects at higher doses (1.5–10 nM) leading to irreversibly contracted cells (29,30). These effects were extracellular calcium dependent, being absent in nominally calcium-free media, whereas being augmented when the extracellular calcium was raised from 1.2 to 2 mM. This study also reported that 1 μM verapamil antagonized the number of irreversibly contracted myocytes after MTX treatment. As might be expected, irreversible contracture and calcium accumulation predict myocardial cell injury. Necrotic cell death was confirmed using release of lactate dehydrogenase (LDH) in MTX-treated cultured cardiomyocytes (31). This study also demonstrated depletion of ATP as a marker of myocardial cell injury. Both of these measures were inhibited by reducing extracellular calcium concentration or with high doses of the calcium channel blocker verapamil (100 μM). They also noted that both ATP depletion and LDH release occurred later than increases of intracellular calcium and that the block of calcium influx shifted from 10 to 100 μM verapamil after a similar delay.

MTX is also active in smooth muscle preparations. Enhanced contraction was seen in aortic strips which was not reversed on washout of MTX (1 nM) (32). As in the cardiac myocytes, the contraction was extracellular calcium dependent and not mediated via neurotransmitter release, since contraction was not affected by phentolamine, methysergide, chlorpheniramine, tetrodotoxin, or indomethacin. Irreversible contraction was observed in guinea pig intestinal smooth muscle strips to MTX (1 nM) (10). Irreversible contraction was also observed in rabbit duodenum intestinal smooth muscle strips to MTX (3 nM) (33). The same dependence on extracellular calcium and the lack of a role of neurotransmitter release (tetrodotoxin and atropine) was also seen. MTX stimulated an intracellular calcium rise whose rate of rise and final value showed an EC50 of 0.3 nM (34). Interestingly, in this same preparation, the EC50 for inositol phosphate generation was shifted to the left to 60 pM. This study also reported that MTX decreased cell viability as measured by trypan blue exclusion. MTX binding occurred irreversibly in the absence of extracellular calcium, because after washing away MTX in calcium-free buffer the effect of MTX can still be observed by adding back extracellular calcium to see stimulation of IPs (34). Influx of $^{45}Ca^{2+}$ and generation of IPs was seen in the BC3H1 muscle cell line (35) Most of these preparations showed sensitivity to low concentrations of voltage-gated calcium channel blockers. The calcium channel blockers verapamil (1 μM (32)) and methoxyverapamil (D600, 0.5 μM (10)) antagonized MTX contraction.

Vascular endothelial cells were also found to respond to MTX. In bovine aortic endothelial cells, MTX (0.2 nM) stimulated generation of IPs and release of arachidonic acid (36). Similar results were found in bovine microvascular endothelial cells (37). The threshold for MTX stimulation of arachidonic acid release was at 100 pM and required extracellular calcium. MTX-induced arachidonic acid release was blocked in these cells by 50 μM SKF 96365 but not by 50 μM nifedipine. An observation that has not been subsequently pursued was that ganglioside GM1 was able to inhibit MTX-induced arachidonic acid release, IP generation, and stimulation of calcium influx. GM1 did not inhibit ionomycin-induced calcium influx nor arachidonic acid release, suggesting that GM1 may act at the level of the plasma membrane to inhibit MTX binding or activation (36).

D. Cultured Cells as Model Systems

Neuronal Cells

NG108-15 is a neuroblastoma-glioma cell line which can be differentiated to express the neuronal phenotype, including voltage-gated calcium channels. MTX stimulated $^{45}Ca^{2+}$-uptake into

NG108-15 cells at the relatively high dose of 30 nM (14). Similarly, 30 nM MTX induces $^{45}Ca^{2+}$ flux and Quin-II fluorescence calcium signals in isolated synaptosomes isolated from rat medulla oblongata (38). In this preparation, 10 μM nifedipine was ineffective, whereas dibutyryl cAMP or cholera toxin pretreatment potentiated $^{45}Ca^{2+}$ flux stimulated by MTX. It was proposed that activation of PKA by these treatments would lead to enhanced activity of voltage-gated calcium channels stimulated by MTX. Similar stimulation of calcium influx using fura-2–loaded rat brain synaptosomes was found with the much lower dose of 0.3 nM MTX, which was not seen in EGTA (ethylene glycol-bis-(β-aminoethylether)-N,N,N′,N′-tetraacetic acid) solution but was rapidly observed on readdition of calcium (39), suggesting that binding of MTX may occur in reduced calcium. Inositol polyphosphate accumulation was seen in MTX-treated rat cerebral cortical slices that appeared additive to charbachol-induced generation of IPs (40) and appeared to affect various IP isoforms differently (41). NCB-20 cells are derived from the fusion between mouse neuroblastoma and Chinese hamster embryonic brain cells and show robust stimulation of phosphoinositide breakdown in response to MTX with a peaked dose-response curve at 150 pM (42). Generation of IPs by MTX was unaffected by the organic calcium channel blockers 10 μM nifedipine, or 5 μM D600. MTX-stimulated IPs were also unaffected by the inorganic calcium channel blockers at 1 mM Co^{2+}, 0.5 mM Mn^{2+}, or 1 mM Cd^{2+}, whereas calcium-free buffer totally inhibited MTX-induced phosphoinositide breakdown (42).

Events that are probably distal to MTX-induced calcium rise have been examined in neuronal cells. MTX stimulated release of gamma-aminobutyric acid (GABA) from striatal neurons differentiated in primary culture (43). In this preparation, MTX could bind irreversibly in Ca^{2+}-free solution, washed out, and then the response measured when calcium-containing buffer was restored. In SH-SY5Y neuroblastoma cells, the MTX-induced calcium rise was followed by cell permeability to propidium iodide and activation of calpain as markers for neuronal cell damage (21,44). The MTX-induced calcium influx occurred over 0.1–1.0 nM and was not blocked by 10 μM nifedipine but was inhibited by 50 μM SKF 96365. Although the calcium influx signal was maximal by 10 min, the calpain activation signal took 60 min to reach maximum and was blocked by SKF 96365 (21).

Neuroendocrine Cells

The rat GH3 cell line is derived from a pituitary tumor yet still shows regulated secretion of prolactin (45–47) and growth hormone (45). The rat PC12 cell line is derived from an adrenal tumor and is used to study both regulated secretion (9) and growth factor–induced differentiation. HIT and RIN cells are derived from pancreatic beta cells and can be stimulated to secrete insulin (48). All of these cell lines have been shown to be sensitive to MTX. These cell lines gave essentially identical responses to MTX as acutely isolated anterior pituitary cells (47,49–51), adrenal chromaffin cells (52), and pancreatic beta cells (53–55), justifying their use as model systems. As noted previously, MTX effects required the presence of extracellular calcium and the binding is not reversible. The secretion is subsequent to the MTX-induced increases of intracellular calcium. Further description of MTX-induced secretion using these cells will be described later in this chapter.

Nonexcitable Cells

Cell lines derived from tissues considered electrically nonexcitable have also been shown to respond to MTX. A number of mouse (56,57) and human fibroblast lines show MTX-stimulated calcium influx (58,59). Glial cells (2,12,60), hepatocytes (23), and epithelial cells (61) have also been studied for MTX effects. Since these cells do not express voltage-gated calcium channels, the MTX-stimulated calcium influx pathway must be different and its unique properties can be studied in the absence of a voltage-gated calcium channel contribution. MTX-induced calcium responses were seen in 3T3 fibroblasts (56,62) and primary human skin fibroblasts (59). As might be predicted, these cells either show no sensitivity to calcium channel blockers or are blocked at high concentrations which are not specific for voltage-gated calcium channels. In addition, these cells tend to have little or no expression of other voltage-gated ion channels allowing for a direct observation of the MTX-induced nonselective cation currents. MTX induces a calcium-permeable nonselective cation current in mouse L cells (57) which preceded the increase of intracellular calcium. This same temporal sequence of MTX-activated nonselective cation current followed by a rise of intracellular calcium was also seen in MDCK cells (63). This suggests that the receptor for MTX is the channels underlying the MTX-activated nonselective cation current or an upstream regulator of these channels. The pharmacological properties of these channels and their regulation will be described more fully later in this chapter.

III. MECHANISMS AND PHARMACOLOGY OF MAITOTOXIN ACTION

A. Elevation of Intracellular Calcium

In the late 1970s and early 1980s $^{45}Ca^{2+}$ flux was commonly used to study the activity of calcium channels. This assay reveals unidirectional fluxes with high sensitivity. MTX was shown to stimulate an increase in calcium permeability in excitable cells such as cardiac and vascular muscle cells. This increase in calcium permeability was blocked by the voltage-gated calcium channel antagonists nifedipine and verapamil. This stimulated increase of intracellular calcium explained the increased contractility observed to MTX exposure. In the mid 1980s, the fluorescent calcium indicator dyes became available and the elevation of intracellular calcium could be directly observed and quantified. The increase of intracellular calcium measured by fura-2 from stimulation of receptors that generate IP_3 was found to be biphasic, showing a transient peak arising from release of calcium from intracellular stores and a sustained but smaller elevation that is now termed capacitative calcium entry (17). This sustained calcium entry has been proposed to be regulated by depletion of intracellular calcium stores as revealed by thapsigargin, an inhibitor of endoplasmic reticulum calcium pumps.

The MTX-stimulated calcium influx may have multiple components. In excitable cells, which express voltage-gated calcium channels, most of the MTX-induced calcium influx can be blocked by low and specific concentrations of calcium channel blockers. Since MTX also causes robust generation of inositol phosphate species, intracellular calcium should rise both from release of intracellular stores and activation of the CCE pathway due to store depletion. This is seen in unexcitable cells such as fibroblasts and glioma cells. The pharmacology of MTX-induced calcium influx in these cells, however, is not identical to CCE and may be due to the additional activation of calcium-permeable nonselective cation channels.

Activation of Voltage-Gated Calcium Channels

The early observations that MTX stimulated calcium influx that was blocked by voltage-gated calcium channel blockers led to the hypothesis that MTX was an activator of these calcium channels. This result was found in muscle cells derived from cardiac ventricle and atria as well as vascular smooth muscle. Similar results were also found using cell lines of neuroendocrine origin such as GH4 and PC12 cells. The hypothesis seemed even stronger when Freedman et al. reported (14) that $^{45}Ca^{2+}$ flux was not stimulated in 3T3 fibroblasts, which is an unexcitable cell line not thought to express voltage-gated calcium channels. This report, however, used very high concentrations of MTX which may have led to loss of cell integrity, as will be discussed later in this chapter. It is now known that 3T3 fibroblasts, mouse L-cell fibroblasts, C6 glioma cells, and numerous other cell types that respond to MTX with a stimulation of calcium influx do not express any voltage-gated calcium channels. Although this seems to rule out voltage-gated calcium channels as the MTX receptor, there are numerous carefully done studies in cells that express voltage-gated calcium channels in which MTX seems to stimulate the nifedipine-sensitive calcium influx pathway (64,65). Indeed, MTX only stimulated calcium influx in human neuroblastoma LAN-1 cells after differentiation by retinoic acid, a treatment that induces expression of ω-CgTx GVIA-sensitive voltage-gated calcium channels (66).

The hypothesis that MTX activates voltage-gated calcium channels was directly tested in the mouse neuroblastoma cell line N1E-115 using patch-clamp techniques (13). This study monitored both the T-type and L-type voltage-gated calcium channels using step pulse protocols and monitoring Ba^{2+} current. Contrary to expectation, MTX at 0.3 nM caused the disappearance of Ba^{2+} current with a time course that was delayed and not correlated with the appearance of a ''leak'' conductance (Figure 3). This result apparently refutes the hypothesis that MTX directly stimulates voltage-gated calcium channels. In addition, it does not give strong support to a modified hypothesis that MTX acts by transforming voltage-gated calcium channels into the observed ''leak'' conductance. Very similar results were reported in a patch-clamp study using isolated rat ventricular myocytes (67). MTX (6 nM) caused an irreversible loss of voltage-gated calcium current and the appearance of a background current which was carried predominantly by Na^+ and could be inhibited by 3 mM Co^{2+}, 1.0 mM Ni^{2+}, and 0.5 mM quinidine.

Since MTX does not stimulate voltage-gated calcium channels as measured by voltage-clamp pulse protocols and instead leads to the disappearance of calcium current, what mechanisms could account for stimulation of voltage-gated calcium channels by MTX? One possibility is that there may be a shift in voltage-dependence of gating that favors opening of voltage-gated

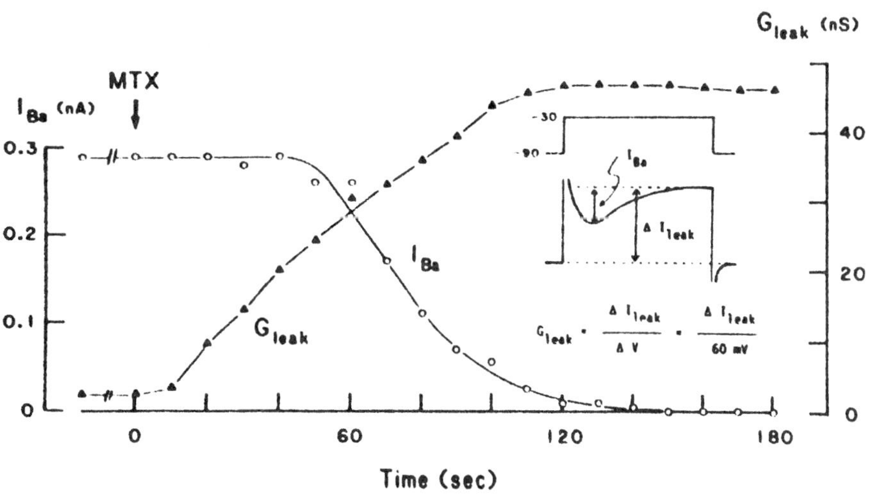

Figure 3 MTX induces ''leak'' conductance and removes calcium current in N1E-115 cells. The difference in the time courses between the amplitude of the Ba^{2+} current (I_{Ba}) and the conductance of the steady-state membrane current (G_{leak}) plotted as a function of time after application of 1 ng/mL (0.3 nM) maitotoxin. As shown in the inset, I_{Ba} was measured as a difference in the current amplitude between the peak and the end of the current trace. G_{leak} was obtained as a chord conductance by measuring steady-state currents at −90 and −30 mV. (From M Yoshi, A Tsunoo, Y Kuroda, CH Wu, T Narahashi. Maitotoxin-induced membrane current in neuroblastoma cells. Brain Res 424:119–125, 1987, with permission of Elsevier Science.)

calcium channels at resting membrane potentials. It is more likely, however, that MTX activates voltage-gated calcium channels indirectly via membrane depolarization as a consequence of activating a nonselective cation current.

Stimulation of Phospholipase C

Measurements of inositol phospholipid hydrolysis in cells exposed to MTX showed that generation of IP_3 is stimulated in every cell type assayed (16,68). In cells where both elevation of intracellular calcium and IP_3 were measured, the dose response of IP_3 generation was elicited at lower concentrations, suggesting that stimulation of PLC may be the primary site of action. In PC12 cells, for example, the threshold for arachidonic acid release was 9 pM, whereas the threshold for IP generation was 30 pM, and the threshold for calcium influx was 100 pM (69). The full dose-response curves for MTX-induced arachidonic acid release and IP generation are shown in Figure 4. Note that even though arachidonic acid was measurable at lower doses, both assays report peak responses at 1 ng/mL (300 pM). Both of these responses were insensitive to organic calcium channel blockers (10 μM nifedipine, verapamil, and diltiazem) even though these agents partially inhibited calcium influx. This suggests that MTX stimulates a calcium-influx pathway independently of voltage-gated calcium channels which is sufficient for activation of phospholipase C and phospholipase A_2. Full block of MTX-induced calcium influx by the inorganic blockers Cd^{2+} or Mn^{2+}, however, was effective at inhibiting arachidonic acid release and IP generation. Removing extracellular calcium or treating with EDTA to chelate calcium appears to inhibit fully the MTX-induced increase of intracellular calcium and generation of IPs and yet arachidonic acid release could still be measured. Quinacrine, a putative phospholipase A_2 inhibitor, was capable of inhibiting MTX-induced arachidonic acid release at 100 μM. Quinacrine, however, blocked both MTX-induced calcium influx and IP generation at similar doses. Activation of both PLC and PLA_2 could be subsequent to calcium influx, but calcium ionophore A23187 could not fully stimulate IP generation nor arachidonic acid release.

In primary cultures of rat aortic myocytes, MTX stimulates hydrolysis of inositol phospholipids generating inositol phosphates with a reported EC50 of 100 pM (68). In this study, the generation of IPs was not blocked by 1 μM nifedipine nor by 10 μM diltiazem or verapamil. Elevation of calcium using K^+ depolarization or with the calcium ionophore A23187, however, did not stimulate generation of IPs, which argues against an indirect stimulation of PLC via a rise of intracellular calcium. These observations suggest that activation of PLC is a separate MTX effect and not subsequent to activation of voltage-gated calcium channels (68).

In differentiated HL-60 human leukemic cell line, MTX stimulates phosphoinositide breakdown that does not involve a pertussis toxin–sensitive G-protein, nor is it modulated by phorbol ester pretreatment (16). PLC is known to require and is stimulated by calcium, so the intracellular calcium signals stimulated by MTX, fMLP, or ionomycin were manipulated by loading the cells with the calcium chelator BAPTA. In BAPTA-loaded HL-60 cells, none of these stimuli were found to alter IP generation (25) despite reducing the calcium responses. The authors conclude that if calcium is involved in MTX stimulation of PLC activity, then it must enter the cell close to sites of PLC hydrolysis without causing general cytoplasmic elevation. MTX may be able to activate PLC directly or indirectly through activating G-proteins. MTX, however, was not able to stimulate PLC activity in HL-60 membrane preparations either alone or in the presence of guanosine triphosphate (GTP), whereas fMLP was active when in the presence of GTP (25).

In primary cultures of rat cerebellar granular neurons, the IP generation stimulated by MTX was additive to that of norepinephrine, histamine, and serotonin, and MTX treatment inhibited the IP generation due to charbachol and glutamate, possibly due to desensitization via PKC activation subsequent to the MTX-stimulated diacylglycerol generation (70). One possible

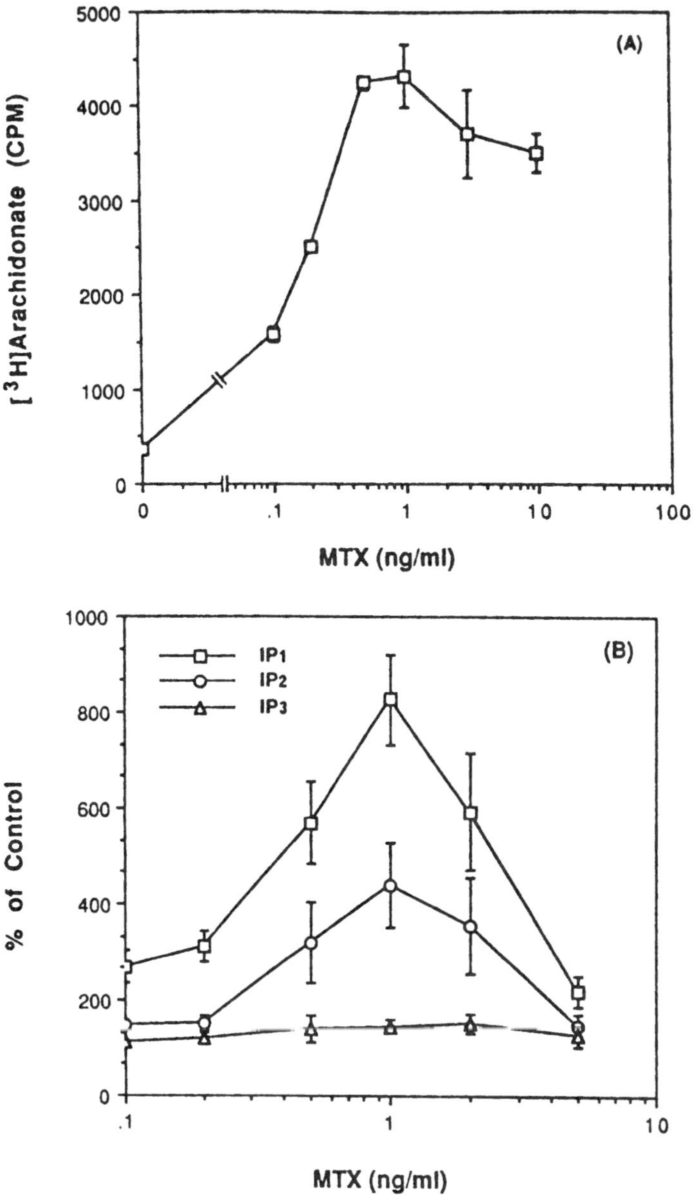

Figure 4 Dose-response of arachidonic acid and inositol phosphate generation to MTX in PC12 cells cells prelabeled with either [^{3}H]AA (A) or [^{3}H]inositol (B) were incubated with different concentrations of MTX for 30 min. MTX at 1 ng/mL corresponds to 0.3 nM concentration. Results are means ± standard errors (three experiments). (From ref. 69.)

interpretation of additive IP generation between MTX and norepinephrine is that the two receptors have access to different pools of inositol phospholipids. Alternatively, these results may support the hypothesis that MTX and norepinephrine activate two separate PLC activities.

These studies taken together suggest that MTX either stimulates PLC directly or MTX can activate a calcium influx pathway that is insensitive to organic channel blockers which is closely situated to PLC, so that MTX-induced calcium influx can activate PLC despite a lack of a generalized cytoplasmic calcium elevation. A direct effect is unlikely, because MTX had no effect on PLC activity in permeabilized cells or on purified PLC in vitro (25). However PLCγ is a type of PLC which is reportedly activated by tyrosine phosphorylation of the enzyme and is implicated in growth factor–mediated phosphoinositide turnover. MTX was found to increase tyrosine phosphorylation of PLCγ, and this may explain the lack of dependence of MTX-stimulated IP generation on the G protein–mediated PLC pathways (71).

Stimulation of Capacitative Calcium Entry

MTX-stimulated calcium influx is blocked by SKF 96365, a blocker of receptor-mediated calcium entry (60) also termed capacitative calcium entry (CCE). This agent significantly blocked MTX-stimulated calcium influx in both RIN insulinoma (Figure 5) and C6 glioma cells at 10 μM. SKF 96365 also blocked phosphoinositide breakdown and hormone release in these cells, whereas still allowing endothelin-1 stimulation, indicating that SKF 96365 effects are not nonspecifically blocking PLC directly. In these cells, 10 μM nifedipine had little inhibitory effect, indicating that the SKF 96365 pathway is not through activation of voltage-gated calcium channels. The shared pharmacology and sustained influx characteristic between CCE and MTX-induced influx suggests the possibility that MTX induces activation of the CCE pathway subsequent to generation of IP_3 and depletion of intracellular calcium stores. Similar results were found using pancreatic beta cells (55). Alternatively, it may be that MTX directly stimulates

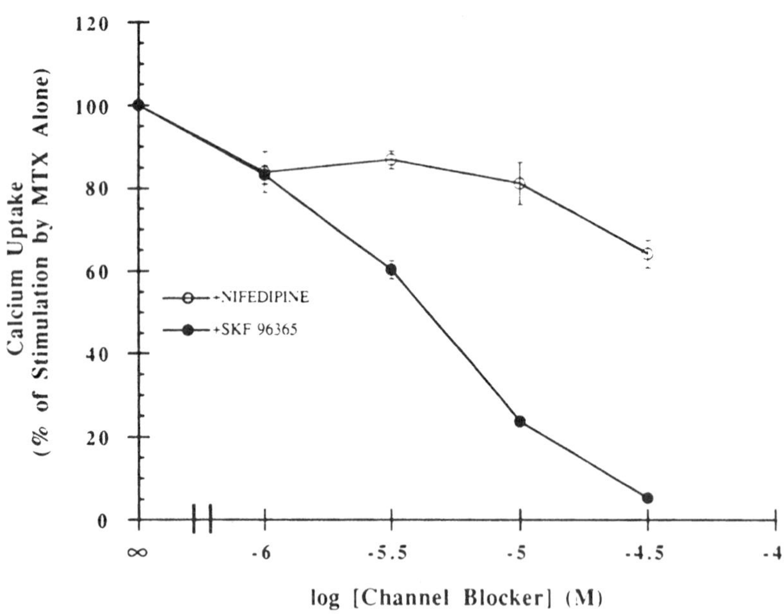

Figure 5 SKF 96365 blocks MTX-stimulated calcium uptake in RIN cells. Cells were preincubated in the presence of the indicated concentrations of SKF 96365 or nifedipine for 10–15 min before addition of 1 ng/mL (0.3 nM) MTX and $^{45}Ca^{2+}$-containing solution. After 15 min, cells were washed and calcium content was determined. MTX alone elicited an uptake of 20.89 ± 6.97 nmol of calcium used to normalize the responses to drug and the points correspond to means ± standard errors of three independent experiments. (From ref. 60.)

CCE channels as a candidate MTX receptor, since the size of the MTX-induced calcium influx is larger than that stimulated by receptor or thapsigargin.

The sensitivity of CCE and MTX-induced calcium influx to various blockers is not identical. In differentiated human leukemic HL-60 cells, the MTX-induced calcium influx pathway was compared to the one stimulated by the chemotactic peptide fMLP (26) which releases intracellular calcium from stores leading to activation of a CCE calcium influx pathway. The pharmacological properties of the calcium influx pathway stimulated by each agent suggests that each agent stimulates a separate pathway. The MTX pathway was more sensitive to SKF 96365 and less sensitive to Gd^{3+} than the fMLP pathway. For example, 10 μM Gd^{3+} nearly completely blocked CCE, whereas having no effect on MTX-induced calcium rise. Conversely, SKF 96365 at 10 μM was clearly more effective against MTX-induced calcium influx than against CCE.

A whole panel of calcium channel blocking drugs active against both voltage-gated calcium channels and receptor-mediated calcium channels was tested in the same laboratory, so that relative potencies are easily compared on various MTX-induced activities (62). $^{45}Ca^{2+}$ influx was tested in NIH 3T3 fibroblasts, and both MTX and ATP stimulated sustained calcium signals measured using fluo-3 were observed in HL-60 cells. The results are shown in Table 1 (62). Note that all compounds that inhibit $^{45}Ca^{2+}$ influx also inhibited fluo-3 calcium influx in response to MTX. There were differences, however, between compounds that block CCE and MTX-induced fluo-3 signals. For instance, verapamil or quinidine at 30 μM was found to inhibit MTX-induced calcium influx, whereas having no effect on CCE. Some of these compounds also caused changes of intracellular calcium directly, obscuring the quantitation of block on MTX- or ATP-stimulated calcium signals. Clotrimazole, for instance, increased the intracellular calcium signal alone, possibly due to a stimulation of phosphoinositide breakdown (62). Given that there were clear differences in the pharmacology between the CCE pathway and the MTX-induced calcium influx pathway, the authors conclude that MTX must activate a distinct calcium influx pathway in addition to CCE.

B. Stimulation of Secretion

MTX has been used on cell systems that are models for cell secretion. MTX causes release of insulin from acutely isolated pancreatic beta cells (53) as well as from HIT (48) and RIN cell lines (60). The dose response of MTX on these cell types seems very potent, acting at much lower doses than in most cells. The threshold for MTX-stimulated $^{45}Ca^{2+}$ influx in isolated beta cells was 3 pM (53) compared to 60 pM for ventricular myocytes (31) and 100 pM for hepatocytes (23). MTX also stimulates the neuroendocrine cell lines GH3 (45,49) and PC12 (9,11) but with MTX sensitivities similar to neuronal cells with active doses at 3–10 nM. It is likely that most of the secretion is subsequent to a MTX-induced intracellular calcium rise. In PC12 cells, for instance, 3 nM MTX potently released [^{3}H]norepinephrine in the presence of millimolar calcium, whereas even 300 nM MTX could not stimulate release in the absence of extracellular calcium (9).

Isolated pituitary cells stimulated by MTX were inhibited by reserpine (51) and dopamine (47). Dopamine fully inhibited MTX-induced prolactin release yet only inhibited MTX-stimulated $^{45}Ca^{2+}$ flux by 50% (47). It is now known that voltage-gated calcium channels can be modulated by G protein–coupled receptor activation (72). This occurs in many cases because the βγ-subunits released during activation of G protein–coupled receptors directly interact with voltage-gated calcium channels. Inhibition of secretion implies that voltage-gated calcium channels are preferentially located at sites of vesicle release, whereas the MTX-induced calcium influx is ineffective. Alternatively, since these experiments were performed on acutely isolated pituitary cells which comprise many cell types, the MTX-induced calcium influx not blocked

Table 1 Effects of Calcium Channel Blockers on MTX-Elicited Calcium Influx and CCE Influx

Agent	MTX-elicited $^{45}Ca^{2+}$ influx in 3T3 cells[a] IC50 (μM)	Fluo-3 fluorescence in HL-60 cells (Tested at 30 μM): Effect alone[b]	MTX-elicited signal[c]	CRAC-channel signal[d]
Imidazoles				
Econazole	2.7 ± 0.2	↑↑	Blocks	Inhibits
Miconazole	2.2 ± 0.3	↑	Blocks	Inhibits
SKF 96365	1.8 ± 0.2	None	Blocks	Inhibits
Clotrimazole	0.56 ± 0.12	↑	Blocks	Inhibits
Calmidazolium	2.3 ± 0.4	↑↑↑	Blocks	Indeterminate
Phenylalkylamines				
R-Verapamil	21 ± 3	ND	ND	ND
S-Verapamil	25 ± 7	ND	ND	ND
Methoxyverapamil (D-600)	16 ± 5	None	Inhibits	No effect
Dihydropyridines				
Nitrendipine	21 ± 5	None	Inhibits	No effect
Benzothiazepines				
Diltiazem	105 ± 23	None	No effect	No effect
Diphenylbutylpiperidines				
Fluspirilene	3.9 ± 0.1	↑	Blocks	No effect
Pimozide	16 ± 4	None	Blocks	No effect
Penfluridol	3.2 ± 0.7	↑↑↑	Blocks	No effect
Diphenylpropylpiperidines				
Loperamide	1.6 ± 0.2	None	Blocks	↑↑
Diphenoxylate	2.9 ± 0.9	None	Inhibits	No effect
Piperazines				
Cinnarizine	13 ± 1	None	Inhibits	No effect
Flunarizine	12 ± 2	ND	ND	ND
Other blockers				
WB4101	22 ± 5	None	Inhibits	Inhibits
W-7	7.2 ± 0.6	None	ND	No effect
Bepridil	6.7 ± 0.6	↑	Blocks	Inhibits
Proadifen (SKF 525a)	3.9 ± 0.3	None	Blocks	No effect
RMI 12330A	1.8 ± 0.2	None	Blocks	No effect
Benzamil	7.4 ± 0.5	None	Inhibits	No effect
Quinidine	12 ± 3	None	Inhibits	No effect
Trifluoperazine	4.6 ± 0.4	None	Blocks	No effect

ND, not determined.

[a] Inhibition of MTX-elicited $^{45}Ca^{2+}$ influx was measured in NIH 3T3 cells. The IC50 values are means ± SEM (N = 3).

[b] Effect of agent at 30 μM on fluo-3 fluorescence was determined in HL-60 cells. Arrows indicate an increase with the magnitude of the increase as follows ↑↑↑ > ↑↑ > ↑.

[c] Effect of agent at 30 μM on MTX-elicited increase of fluo-3 fluorescence was determined in HL-60 cells.

[d] Effect of agent at 30 μM on CRAC channel-dependent increase in fluo-3 fluorescence after ATP stimulation in HL-60 cells. Arrows indicate an increase in the CRAC channel-dependent increase in fluo-3 fluorescence.

Source: JW Daly, J Lueders, WL Padgett, Y Shin, F Gusovsky. Maitotoxin-elicited calcium influx in cultured cells, effect of calcium-channel blockers. Biochem Pharmacol 50:1187–1197, 1995, with permission of Elsevier Science.

by dopamine arises from a subpopulation of cells that may not express the dopamine receptor. The GH4 pituitary cell line was also stimulated to release prolactin in response to the relatively high dose of 30 nM MTX (65). In these cells, the MTX-stimulated prolactin release was inhibited with low and specific doses of nimodipine (IC50 of 25 nM), strongly suggesting a tight coupling of voltage-gated calcium channels with vesicle secretion. MTX also stimulated depolarization in these cells which was not fully blocked even with 2 μM nimodipine. A subsequent electrophysiological study confirmed that MTX acts by first activating a nonselective current which leads to depolarization subsequently activating L-type voltage-gated calcium channels and further depolarizing these cells (73).

A similar mode of action of MTX on vesicle-mediated secretion was seen in PC12 cells, a model for adrenal chromaffin cells. In PC12 cells, MTX (3 nM) causes rapid (within 90 s) $^{45}Ca^{2+}$ uptake which was not blocked by 1 μM tetrodotoxin (TTX) but was inhibited by high (30–100 μM) verapamil (9). When both secretion of ATP and generation of IPs was measured in the same study (74), the generation of IPs was found to occur at 150 pM, whereas secretion of ATP required 0.6 nM. Stimulation of IP generation was insensitive to the calcium channel blockers nifedipine, D600, and cobalt ions, whereas ATP release was fully inhibited by nifedipine. Again arguing that MTX-stimulated secretion is subsequent to activation of voltage-gated calcium channels which may be optimally located at release sites.

The PC12 cell line was also used to demonstrate nonvesicle-mediated effector release. MTX stimulated arachidonic acid release which was found to be temporally and pharmacologically separated from the well-characterized MTX responses such as calcium influx and generation of inositol phosphates. MTX-stimulated arachidonic acid release could be detected at 10 pM, whereas the PI breakdown threshold was at 30 pM and calcium influx responded to MTX concentrations above 100 pM (69). Curiously, at a given dose (300 pM) of MTX, calcium influx occurs most rapidly (30 s) followed closely by IP generation (1 min) and then much later (20 min) by arachidonic acid release (69). The calcium channel blockers nifedipine, diltiazem, and verapamil blocked calcium influx but did not block IP generation or arachidonic acid release. Indeed, arachidonic acid release occurred even in the presence of EDTA to remove extracellular calcium, which abolished both calcium influx and IP generation. Quinacrine, a reported inhibitor of phospholipase A2 (PLA2), was able to block all three MTX responses with nearly identical IC50 values of 10 μM (69). Activation of both PLC and PLA2 could be subsequent to MTX-induced calcium influx, but calcium ionophore A23187 could not equally stimulate IP generation or arachidonic acid release. The MTX-stimulated arachidonic acid release was reduced on removal of extracellular Na^+ (69), which would be expected to rise intracellularly subsequent to MTX stimulation of a nonselective cation current.

C. Membrane Potential Depolarization

MTX causes membrane potential depolarization. Direct microelectrode recordings and membrane potential–sensitive fluorescent dyes show a temporal correlation between membrane depolarization and MTX-induced intracellular calcium rise. As alluded to above, the neuroendocrine cell lines PC12 and GH4 demonstrated a membrane potential change induced by MTX that was substantially (but not fully) blocked by the calcium channel blockers nifedipine and nimodipine (64,65). These studies argued that the primary site of MTX action was to activate L-type calcium channels directly whereas other studies could not exclude a new type of channel or pore induced by MTX (75). Indeed, these same investigators later demonstrated that, in GH4 cells, the MTX-induced rise of intracellular calcium dose-response curve is biphasic. The calcium rise at the lowest doses was blocked by low doses of calcium channel blockers (perhaps due to depolarization-induced activation of L-type calcium channels) followed by a calcium influx pathway at higher MTX doses that is not as sensitive to calcium channel blockers (73,76). This shift in pharmacology may also reflect the reported inactivation of voltage-gated calcium channels as seen in patch-clamp recordings (13,67) leaving behind the MTX-activated nonselective cation current.

MTX-induced depolarization was also seen in studies utilizing neuronal cells. In the crayfish giant axon, 10 nM MTX reduced the resting membrane potential from −85 to −70 mV (75) resulting in an altered action potential. The human neuroblastoma cell line LAN-1 recorded MTX-induced depolarization as monitored by bisoxonol fluorescence at 1–3 nM (77). Synaptosomes prepared from rat cerebellum also showed MTX-stimulated depolarization recorded using bisoxonol (39). The threshold for depolarization was 100 pM MTX, and this depolarization was abolished in the absence of extracellular divalents. Interestingly, both Sr^{2+} and Ba^{2+} could substitute for Ca^{2+} in the MTX-induced depolarization. The investigators speculated that (at least part of) the mechanism of MTX action does not reflect activation of calcium-dependent lipases or proteases, since barium could substitute.

D. Nonselective Cation Channel Activation

The properties of the voltage-gated calcium channel blocker–insensitive MTX-activated current and/or MTX-activated calcium influx pathway are consistent with activation of a nonselective cation channel. Activation of nonselective cation channels (whose reversal potential is near 0 mV) would depolarize the cells membrane potential. TTX has been used in early studies of MTX action to argue against a role for Na^+ influx, since it is a specific blocker of voltage-gated sodium channels. It is this specificity which explains why TTX has no effect on MTX-induced nonselective cation current nor on the MTX-induced rise of intracellular calcium. Characterization of nonselective ion channels is difficult, but the MTX-induced current seems to carry the monovalent cations Na^+ and K^+ better than the divalents Ca^{2+}, Ba^{2+}, and Sr^{2+}. The MTX-induced depolarization does not occur in calcium-free buffer but is equally induced by barium or strontium (39). Large organic cations such as choline or N-methyl-D-glucamine do not seem to carry current, yet MTX-induced depolarization is still evident (39). In addition, there is evidence for anion permeability induced by MTX (73).

MTX stimulated a nonselective cation current in mouse L cells (57) which had previously been shown to express a mitogen-regulated nonselective cation channel. The MTX-activated current preceded elevation of intracellular calcium as monitored by the activity of an endogenous calcium-activated potassium current. Once the nonselective current was activated by MTX, the intracellular calcium signal was dependent on the driving force for calcium influx. Either removing extracellular calcium or changing the holding potential to very depolarized potentials was effective at reducing intracellular calcium levels in a reversible manner. This suggests that the MTX-activated nonselective cation current is permeable to calcium. An interesting speculation of this study is that MTX activates an endogenous channel involved in cell proliferation.

Candidate MTX-activated single channels have been observed utilizing isolated patch-clamp techniques. The first published reports of MTX-activated single channels were described using ventricular myocytes (78). The activity was seen reliably in cell-attached patches with 30 nM MTX included in the pipette solution. The channel activity appears as long bursts with short flickery closures separated by periods of no activity. The gating kinetics, as well as the ion selectivity observed during whole-cell recordings, suggested that this MTX-activated channel was not related to voltage-gated calcium channels. A subsequent study recorded MTX-activated single channels utilizing the outside-out patch configuration (67). Stimulation of single-channel activity in the outside-out patch strongly suggests that the receptor for MTX is located in the plasma membrane and that the steps between binding to the MTX receptor and channel activation

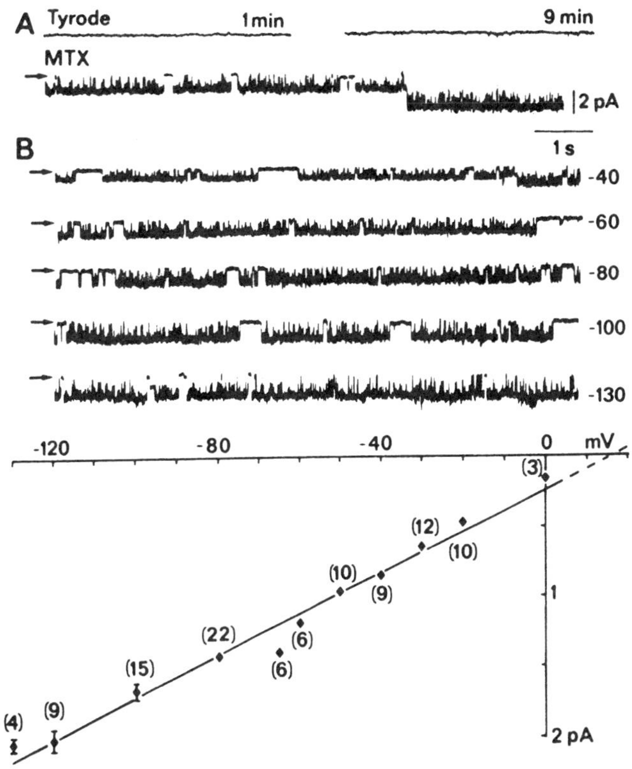

Figure 6 Single nonselective cation channels induced by MTX recorded by outside-out patch-clamp. (A) MTX-activated channel activity recorded in Tyrode solution (upper trace) and 30 s after addition of 20 ng/mL (6 nM) maitotoxin (lower trace). Holding potential is −80 mV. (B) Upper part: MTX-activated channel activity at the membrane potentials indicated at the end of each trace. Channel closed states are indicated by arrows. Lower part: current-voltage relationship of MTX-activated channels measured in three different cells. The points show means and SEM. Numbers in parenthesis indicate number of measurements. Line drown by linear regression ($r^2 = 0.989$) gives a conductance of 14.4 pS and an extrapolated reversal potential of 16.3 mV when 120 mM K-aspartate was in the pipette. (From JF Faivre, E Deroubaix, A Coulombe, AM Legrand, E Coraboeuf. Effect of maitotoxin on calcium current and background inward current in isolated ventricular rat myocytes. Toxicon 28:925–937, 1990, with permission of Elsevier Science.)

are membrane delimited and do not require cellular metabolism. As shown in Figure 6, a long flickery bursting channel activity could be induced with a conductance of 14 pS and a reversal potential of 16 mV, which is consistent with a nonselective cation channel (67). Once activated, the channel activity remained after washout of MTX. In addition, these MTX-activated channels appeared practically unchanged on removal of external calcium, but increasing external calcium resulted in a reduction of single-channel amplitude. These channels were also directly inhibited by Ni^{2+} and quinidine at similar doses that block MTX-induced calcium influx. In whole-cell recordings, the ability of MTX to induce the background nonselective current required extracellular calcium. Interestingly, a subsequent patch-clamp study of MTX-induced single channels suggests that channel activity can be induced in the absence of external calcium (79). Both of these studies found that only a few MTX-activated channels per patch were observed despite a wide range of MTX concentrations or pipette resistance (i.e., irrespective of patch membrane area), suggesting that MTX does not form pores itself, as has been reported for other toxins (80), but instead activates preexisting membrane proteins to form a channel.

The MDCK renal epithelial cell line was studied electrophysiologically in response to MTX (63). This cell line does not express voltage-gated calcium channels, yet it clearly responded to MTX with first an activation of nonselective cation current followed by elevation of intracellular calcium. The dose response for activation of nonselective cation current was steep with an EC50 of 1 nM. The nonselective cation current required external calcium for induction but, once induced, it no longer required calcium. This study also recorded MTX-stimulated single nonselective cation channels with similar flickery gating properties as seen in cardiac myocytes. The receptor site for MTX is exposed to the extracellular space, because MTX added to the intracellular face of membrane patches was without effect. Not explained, however, was the observation that SKF 96365 did not inhibit the MTX-induced nonselective cation current, whereas this compound was effective at reducing MTX-induced intracellular calcium rise. This suggests that the MTX-induced nonselective cation current was not the only calcium influx pathway stimulated by MTX.

In HIT insulinoma cells, glucagon-like peptide-1 (GLP-1) stimulates glucose-dependent insulin secretion and also activates a nonselective cation current, suggesting a role for these currents in the complex regulation of beta cell oscillations and insulin release. The properties of the GLP-1–stimulated current were compared with the MTX-induced nonselective cation channels induced in these cells (81). The nonselective nature of the MTX-induced current was confirmed by the finding that extracellular Na^+ could be replaced by Cs^+, K^+, Li^+, $choline^+$, or $NMDG^+$ ions. They also concluded that the MTX-activated current was permeable to calcium by monitoring changes of intracellular calcium in response to changing the driving force for calcium either by membrane depolarization or reducing extracellular calcium. Strong buffering with EGTA in the pipette blocked the ability of MTX to activate the nonselective cation current, suggesting that physiological levels of intracellular calcium are required for the activation and that this activation may not be related to depletion of intracellular stores. A lack of a role for intracellular stores in the regulation of the MTX-induced nonselective cation current was suggested by a lack of effect of intracellular dialysis with heparin (to block IP_3 receptor calcium release) or with ryanodine (to block caffeine-sensitive calcium release). Activation of an essentially identical nonselective current was seen in response to thapsigargin, however, and both the thapsigargin- and MTX-stimulated nonselective current was sensitive to block by SKF 96365 and to econazole (81). The investigators also reported that the tyrosine kinase inhibitor genestein (100 μM) blocked activation of nonselective cation current by MTX (81). Similar results have been seen in acutely isolated pancreatic beta cells (82,83). This channel may be upregulated in these cells to perform this physiological role of regulating secretion and thus may explain the enhanced sensitivity of insulin-secreting cells to MTX.

E. Cell Death

Numerous observations are scattered throughout these studies indicating that MTX treatment can lead to cellular damage and death. On reflection it is not surprising that an agent which can stimulate large and sustained calcium influxes would be harmful to cells. The measures of cell damage include loss of plasma membrane integrity as indicated by vital dye exclusion as well as depletion of intracellular ATP indicating a loss of metabolic homeostasis. The observations of MTX-induced cell damage occur at the higher doses and with a delayed time course compared with the calcium influx or generation of inositol phosphates.

An excellent study showed MTX-induced permeabilization to intracellular macromolecules in a temporal sequence related to molecular weight (23). In rat hepatocytes, 100 pM MTX induces a fall of intracellular ATP levels within 30 min followed shortly by the appearance of labeled adenine nucleotides into the bath. This release was not simply explained by cell lysis, because release of lactate dehydrogenase (LDH) occurred still more slowly. As was found for most MTX effects, the release of adenine nucleotides or lactate dehydrogenase (LDH) release was dependent on extracellular calcium. This study also explained why MTX may cause a peaked dose-response relation. As the MTX dose was increased to 1 nM, the time to release over 50% of LDH was shortened to 1 h and the calcium influx as measured with $^{45}Ca^{2+}$ showed a pronounced bell-shaped curve with time, as shown in Figure 7. This loss of $^{45}Ca^{2+}$ signal is due to the loss of cell integrity and rerelease of the $^{45}Ca^{2+}$ back into the media. As the MTX dose increases, the cells become permeabilized and lysed leading to loss of the cell-associated label such as $^{45}Ca^{2+}$ or [^{3}H]IP_3. This could explain how an earlier and highly influential paper (14) could report a lack of effect of MTX in 3T3 fibroblasts because high (30 nM) doses of MTX were used in a $^{45}Ca^{2+}$ flux assay. Many papers have reported peaked IP_3 dose-response curves to MTX (2,42,61,64,70). The time course of cell permeabilization suggests that cell permeabilization is a secondary effect of MTX that is subsequent to earlier actions of MTX such as massive calcium influx. A study comparing calcium influx, cellular ATP levels, and LDH release in rat ventricular myocytes showed similar results (31). High-dose (100 μM) verapamil was able markedly to inhibit these MTX-induced responses, including cell death.

Trypan blue is a vital dye that is excluded by healthy cells and its accumulation in a cell is a marker for cell damage. MTX was reported to cause trypan blue uptake in smooth muscle cells (34), whereas no change in trypan blue staining was observed in acutely isolated pituitary cells (49). Neutral red has also been used as a screening tool for testing cytotoxicity of known (MTX) and uncharacterized dinoflagellate extracts (58). Propidium iodide has become popular as a tool to monitor cytotoxicity, because it drastically increases its fluorescence on binding to DNA and is normally not able to cross the plasma membrane of healthy cells. Staining by propidium iodide is an additional marker for necrotic cell lysis, generally correlating very well with LDH release as stimulated by MTX (21,44)

Calcium-induced cell toxicity plays a role in pathophysiological conditions such as stroke and cardiac reperfusion. It is also becoming appreciated that cell death occurs along an axis with necrotic lysis at one end and apoptosis at the other end (80). MTX can clearly lead to

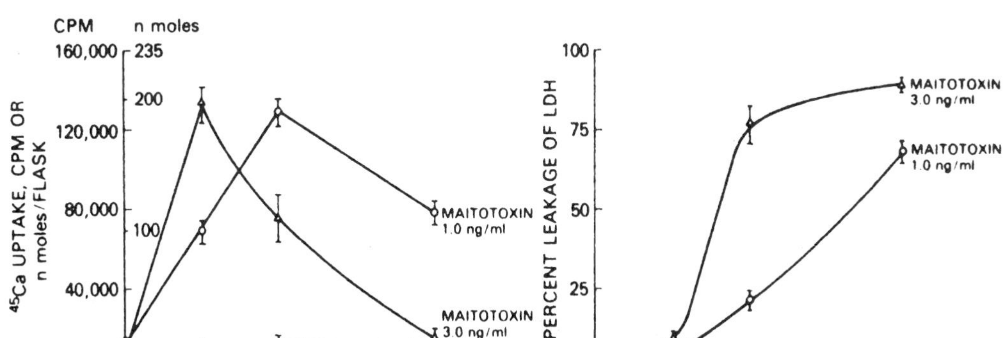

Figure 7 MTX-induced $^{45}Ca^{2+}$ uptake and LDH leakage in isolated liver cells. Hepatocytes were treated with indicated concentrations of maitotoxin and the $^{45}Ca^{2+}$ uptake and LDH leakage were measured at various times. MTX at 1 ng/mL corresponds to 0.3 nM concentration. The data are means ± standard error of four experiments. (From ref. 23.)

necrotic cell death due to unbalanced Na^+ and Ca^{2+} influx leading to depleted ATP levels as the cellular pumps are maximally activated and then overwhelmed leading to osmotic imbalance and lysis. At lower doses of MTX which do not lead to immediate cell lysis, however, MTX stimulation can be shown to activate intermediate measures of apoptosis such as protease activation (21,84). In addition, since a cell's decision to undergo apoptosis due to a variety of signals (such as DNA damage or trophic factor withdrawal) is influenced by the cells ATP/ADP ratio (80), low-dose MTX can shift the likelihood of cell death by apoptosis.

IV. HYPOTHESIS OF PRIMARY MAITOTOXIN MECHANISM

A. Common Themes of MTX Action

Does MTX interact with many targets or just one target? There are multiple reports that demonstrate that MTX stimulates membrane depolarization, increased calcium influx, PI hydrolysis, and multiple other effects that are temporally subsequent to these primary responses. Is it possible that MTX is multifunctional and able to bind to PLC, voltage-gated calcium channels, and nonselective cation channels? On the other hand, specific action in the low nanomolar range is strong evidence for a single receptor to which the toxin has evolved high-affinity binding. Normally, the highest affinity binding site would identify the receptor of an agent such as MTX. Since dose-response curves to MTX are based on assays, however, an argument could be made that the most probable site of MTX action is related to the earliest observable response temporally. Thus, even though arachidonic acid release and generation of inositol phosphates are measurable at lower doses of MTX than stimulated influx of calcium or activation of nonselective cation channels, these earlier responses are more likely directly related to the site of MTX binding.

MTX responses have been reported in many vertebrate and invertebrate cell types (75,85,86) suggesting that the receptor for maitotoxin is widely distributed and highly conserved in nature. The diversity of cell types and organisms that respond to MTX is shown in Table 2 along with the assay system used. It has become clear that MTX is able to elicit responses in cell types that are not known to express voltage-gated calcium channels, including increased calcium influx and activation of nonselective cation current. If MTX binds to a single receptor, that receptor cannot be the voltage-gated calcium channels. As suggested above, however, in cells that express voltage-gated calcium channels, a significant component of MTX-induced calcium influx will flow through these channels. An additional question is what is the normal function of the MTX receptor? Is there a natural endogenous ligand that can stimulate large and potentially toxic calcium influx? Or did MTX exploit a latent channel activity of a ubiquitous membrane protein similar to the action of palytoxin (87,88) on Na-K-ATPase?

B. Primary Role of Nonselective Cation Channels

A case could be made for a primary role of nonselective cation channels in the action of MTX. It appears to be the earliest response to MTX, and it could give rise to the subsequent calcium influx either via membrane depolarization and activation of voltage-gated calcium channels or directly owing to a partial permeability to calcium. A stimulation of calcium influx could plausibly be involved in many, if not most, of the subsequent MTX-induced responses, including generation of inositol phosphates and stimulated secretion. Activation of nonselective cation channels has been recorded in all cells responsive to MTX. The first single-channel recordings of MTX-activated ion channels was reported in guinea pig cardiac myocytes (78). Subsequent studies have described MTX-induced activation of nonselective cation currents in neuronal cells

Table 2 Cells Studied with MTX

Tissue	Type	Organism	Assay	References
	Whole animal	Mouse	LD50	4
Stomach, heart, lymphoid	Various	Mice and rats	Histopathology	24
Mitochondria	Organelle		^{45}Ca flux	12
Liposomes	Artificial		^{45}Ca flux	12
Sea urchin egg	Egg	Sea urchin	^{45}Ca flux, fertilization reaction	98
Xenopus oocytes	Oocyte	Frog	Fura-2, IP accumulation, voltage clamp	86,89
Erythrocytes and erythrocyte ghosts	Blood	Human	Fura-2	90
Platelets	Blood	Rabbit	Aggregation, ATP release, fura-2	20
Lymphocytes primary	Lymphatic	Mouse	Cell loss	22
HL-60 cell line	Lymphoma	Human	Fura-2, IP accumulation	25,26
Basophils	Monocyte	Human	Fura-2, histamine release	19
Fetal cerebral culture	Brain neurons	Rat	Protease activation	84
Synaptosomes	Brain neurons	Rat	Bisoxonol, fura-2	39
Synaptosomes	Brain neurons	Rat	^{45}Ca influx, Quin-II	38
Striatal neurons in culture	Brain neurons	Rat	GABA release	43
Cerebellar granules primary	Brain neurons	Rat	IP accumulation	70
Cerebral cortex slices	Brain neurons	Rat	IP accumulation	40,41
Neuro-2A cells	Neurons	Rat	IP accumulation, neurite outgrowth	99
Giant axon	Neurons	Crayfish	Membrane depolarization	75
NCB-20 cell line	Neuroblastoma	Rat	IP accumulation	16
LAN-1 cell line	Neuroblastoma	Human	Fura-2, bisoxonol	66,77

Table 2 Continued

Tissue	Type	Organism	Assay	References
N1E-115 cell line	Neuroblastoma	Mouse	Whole-cell voltage-clamp	13
NCB-20 cell line	Neuroblastoma	Mouse × hamster	IP accumulation	42
NG108-15 cell line	Neuroblastoma	Rat	^{45}Ca influx	14
SH-SY5Y cell line	Neuroblastoma	Human	Fura-2, calpain activation, protease activation, tTG activity	21,44,100 101
P19 cell line	Embryonal carcinoma	Mouse	ACHE activity	102
BAEC cultures	Endothelial	Bovine	IP accumulation, arachidonic release	36
Microvascular primary	Endothelial	Bovine	IP accumulation, arachidonic release	37
Alveolar epithelial	Epithelial	Bovine	Deploarization	103
Tracheal epithelium	Epithelial	Rabbit	^{45}Ca flux, fura-2, indo-1, IP accumulation	61,104
MDCK cell line	Renal epithelial	Canine	Patch clamp, fura-2	63
CHO with m5 receptor	Expression system	Hamster	^{45}Ca influx, fura-2, P-Tyr, shape change	71,105
Hepatocytes primary	Liver	Rat	^{45}Ca influx, LDH release	23
3T3 cell line	Fibroblast	Mouse	^{45}Ca influx, Indo-1	14,56,62
BHK cell line	Fibroblast	Hamster	Indo-1	56
Fibroblast cell lines	Fibroblast	Mammalian	Dye uptake	58
Skin fibroblasts primary	Fibroblast	Human	Fura-2	59
L-cell line	Fibroblast	Mouse	Patch-clamp, fura-2	57
C6 glioma cell line	Glioma	Rat	^{45}Ca influx, fura-2, IP accumulation	2,12,60 90
Skeletal muscle	Myocyte	Insect	Microelectrode spiking	85
BC3H1 cell line	Myocyte	Rat	^{45}Ca influx, IP accumulation	35
Atria	Myocyte	Guinea pig	Contraction	27
Atria	Myocyte	Rat	Contraction	33,106
Cardiac ventricle	Myocyte	Rat	Contraction, ^{45}Ca flux, ATP depletion, LDH release, patch-clamp	29,31

Cardiac ventricle	Myocyte	Rat and guinea pig	Ca content, contractions	28,30
Cardiac ventricle	Myocyte	Guinea pig	Patch-clamp	78,79
Aortic muscle	Smooth muscle	Rabbit	Contraction	32
Aortic muscle	Smooth muscle	Rats	Fura-2	34,68
Duodenum	Smooth muscle	Rabbit	Contraction	33
Ileum	Smooth muscle	Guinea pig	Ca content, contractions	10
Smooth muscle primary	Smooth muscle	Rat	IP accumulation	16
Anterior pituitary primary	Neuroendocrine	Rat	^{45}Ca influx, IP accumulation, neurotransmitter release, prolactin release, GnRH release	47,49,51 50,107,108
MMQ pituitary cell line	Neuroendocrine	Mouse	Indo-1	7
7315a pituitary tumor line	Neuroendocrine	Rat	Arachidonic metabolism, prolactin release	46
GH3 cell line	Neuroendocrine	Rat	^{45}Ca influx, fura-2, membrane depolarization, neurotransmitter release, prolactin release, cell cycle analysis	45,76,109 65
GH4 pituitary cell line	Neuroendocrine	Rat	Patch clamp	73
Pancreatic islets	Neuroendocrine	Rat	^{45}Ca flux, insulin release	53
Beta cells primary	Neuroendocrine	Rat	Fura-2	54
Beta cells primary	Neuroendocrine	Mouse	Fura-2, patch-clamp	55,83
RIN cell line	Neuroendocrine	Rat	^{45}Ca influx, fura-2, IP accumulation, insulin release	60
HIT insulinoma cell line	Neuroendocrine	Hamster	^{45}Ca influx, fura-2, IP accumulation, insulin release, patch-clamp	2,12,48 81
Adrenal chromaffin primary	Neuroendocrine	Bovine	IP accumulation	52
PC12 cell line	Neuroendocrine	Rat	^{45}Ca flux, fura-2, IP accumulation, arachidonic metabolism, cGMP generation, ATP release, neurotransmitter release, membrane depolarization	9,11,74 16,69,110 12,64,111

(13), neuroendocrine cells from pituitary (73) and pancreas (55,81), nonexcitable cells (57,63), including *Xenopus* oocytes (89).

MTX-induced depolarization can ''activate'' voltage-gated calcium channels in cells independent of specific binding to voltage-gated calcium channels. Examination of [^{3}H]PN200-110 binding to L-type calcium channels showed no competitive interaction with MTX and indeed was consistent with membrane depolarization (65). Close examination of nifedipine block of MTX-induced intracellular calcium signals shows a small nifedipine-resistant component, which is likely due to influx through the activated nonselective cation current. It may also still be true that MTX stimulation leads to a shift in voltage gating of the voltage-gated calcium channels which makes them more sensitive to small depolarizations. It is clear, however, that MTX does not lead to increased conductance and in fact seems to lead to time-dependent loss of voltage-gated calcium channel activity (13).

Still to be explained is why some cell types seem to be consistently more sensitive to MTX than others. For any given assay, such as stimulation of calcium influx, some cell types, such as vascular endothelial cells and insulin-secreting cells, seem to respond at lower doses of MTX than do pituitary cells or neuronal cells. Variation might be expected for a distal response such as cell death, because cell death will also include factors such as a cell's ability to handle elevations of intracellular calcium and the cells sensitivity to undergoing apoptosis in response to toxic stimuli. If, hypothetically, the nonselective cation channel protein were the actual MTX receptor, could the variation of MTX sensitivity be due to the expression levels of the channel protein? Another plausible idea is that most ion channel types occur in families, as has been demonstrated for voltage-gated and ligand-gated ion channels. Thus, just as different members of the voltage-gated calcium channels show differential sensitivities to specific blockers, so might different members of the nonselective cation channel family show a different affinity to MTX.

V. FUTURE DIRECTIONS OF RESEARCH WITH MAITOTOXIN

A. Molecular Characterization of MTX Receptor

A number of observations localize the MTX receptor site to the plasma membrane. First is the fact that MTX is very large and water soluble so that it cannot cross membranes. Second, MTX does not act strictly as an ionophore, since it is not able to induce calcium fluxes in mitochondria or in liposomes. MTX has been checked for direct effects on intracellular enzymes such as the Na-K-ATPase and PLC but none was found. The patch-clamp studies also strongly localize the MTX receptor to the plasma membrane, since MTX can activate channel activity in isolated membrane patches. The MTX receptor is exposed to the extracellular face of the plasma membrane, since direct application of MTX to the cytoplasmic face of membrane patches failed to induce nonselective channel activity (63). These results argue that MTX directly activates nonselective cation channels without requiring a soluble second-messenger pathway. The MTX receptor shows trypsin-mediated degradation, again suggesting a plasma membrane protein receptor (59).

The large size and chemical complexity of MTX has frustrated attempts to generate a radioactive form that maintains high specificity. Many of the effects of MTX are inhibited by ganglioside GM1 (36), which may represent a competitive inhibitor of MTX to its as yet unknown high-affinity receptor. Chemical fragments of MTX and the structurally related dinoflagellate toxin brevotoxin has been reported to inhibit MTX effects in a competitive manner (90). The nature of these fragments suggests that MTX interacts with a plasma membrane protein both in the extracellular space and within the membrane. Closer examination of the chemical structure of MTX reveals that the hydrophilic hydroxyl groups and sulfate groups are clustered at one end of the molecule, whereas the other end is nonpolar and may insert into membranes. This association with the plasma membrane could colocalize MTX to its plasma membrane receptor and explain why MTX action is difficult to wash out. Finally, the dependence of MTX effects on extracellular calcium may give clues to the nature of the MTX receptor. It appears that MTX can bind to its receptor in the absence of extracellular calcium, but many of the assay systems used do not record changes until calcium is restored, suggesting that the transduction of the MTX receptor into an active complex requires calcium.

The MTX receptor protein is ubiquitously expressed and it need not normally function as an ion channel, since palytoxin can ''transform'' the Na-K-ATPase into a sodium-permeable channel (87,88,91). However, the leading candidate MTX receptors are the nonselective cation channels. These channels are integral plasma membrane proteins whose activity is induced by MTX which precedes elevation of intracellular calcium. In addition, since these channels appear to be calcium permeable, their activation could directly explain a component of MTX-induced calcium influx. Nonselective cation channels have been reported to be regulated in many ways, including intracellular calcium (92), tyrosine kinase family growth factor receptors (93), and GTPγS (94). The molecular characterization of these nonselective cation channels and how they are regulated is currently incomplete. Thus, it is also possible that the MTX receptor binds to an upstream regulator of the nonselective cation channels. Further work is needed in this area.

B. MTX as a Tool to Stimulate Calcium-Dependent Processes

Since MTX reliably elicits a sustained elevation of intracellular calcium in nearly every cell type examined, the calcium-dependent processes in the cells can be studied. The advantages of MTX over the calcium ionophores ionomycin and A23187 is that MTX causes a sustained elevation of intracellular calcium and that there is no fluorescence interference due to MTX. Despite the generation of inositol phosphates by MTX, it also appears that MTX induces calcium influx without depletion of the intracellular stores (59), which may complicate the interpretation of an experiment. The calcium influx induced by MTX causes contraction in myocytes. MTX has already shown utility for studying regulation of secretion. Recent studies have used MTX to study the role of elevating intracellular calcium on calpain activation (84,95). The MTX-induced calcium influx has many properties in common with capacitative calcium entry, yet it has some notable differences. One hypothesis is that the nonselective cation channels activated by MTX are normally store-operated channels and that MTX is sufficient for activity rendering these channel independent of regulation by intracellular stores.

C. MTX as a Tool to Study the Necrosis/Apoptosis Cell Death Fate

It is becoming increasingly clear that as the MTX dose is increased, many of the cell types tested show signs of decreased cell viability and increased cell death. Interestingly, there seems to be different kinetics depending on the assay system used. Decreased cell viability as assayed by uptake of trypan blue occurs more quickly than cell necrosis, as measured by release of adenine nucleotides which is then followed more slowly by true necrotic cell lysis measured as the release of LDH. It is interesting to speculate that MTX might activate a pathway similar in properties to the P_2X_7 receptor, which is reported to form pores in cell membranes leading eventually to cell death (96). MTX may also activate the caspase cascade which is normally a marker of the apoptosis cell death pathway subsequent to calpain activation (97). If MTX proves to be a potent stimulator of apoptosis, then the MTX receptor is implicated in playing a role in an as yet undefined apoptosis signaling pathway. This would launch a search to identify the endogenous ligand of the MTX receptor that could stimulate cell death.

REFERENCES

1. M Murata, H Naoki, S Matsunaga, M Satake, T Yasumoto. Structure and partial stereochemical assignments for maitotoxin, the most toxic and largest natural non-biopolymer. J Am Chem Soc 116:7098–7107, 1994.
2. M Murata, F Gusovsky, M Sasaki, A Yokoyama, T Yasumoto, JW Daly. Effect of maitotoxin analogues on calcium influx and phosphoinositide breakdown in cultured cells. Toxicon 29:1085–1096, 1991.
3. M Murata, M Sasaki, A Yokoyama, T Iwashita, F Gusovsky, JW Daly, T Yasumoto. Partial structures and binding studies of maitotoxin, the most potent marine toxin. Bull Soc Pathol Exot 85: 470–473, 1992.
4. MJ Holmes, RJ Lewis, NC Gillespie. Toxicity of Australian and French polynesian strains of Gambierdiscus toxicus (Dinophyceae) grown in culture: characterization of a new type of maitotoxin. Toxicon 28:1159–1172, 1990.
5. RJ Lewis, MJ Holmes, PF Alewood, A Jones. Ionspray mass spectrometry of ciguatoxin-1, maitotoxin-2 and -3, and related marine polyether toxins. Nat Toxins 2:56–63, 1994.
6. MJ Holmes, RJ Lewis. Purification and characterisation of large and small maitotoxins from cultured Gambierdiscus toxicus. Nat Toxins 2:64–72, 1994.
7. IS Login, SI Kuan, AM Judd, RM MacLeod. Regulation of the intracellular calcium concentration in MMQ pituitary cells by dopamine and protein kinase C. Cell Calcium 11:525–530, 1990.
8. RJ Lewis, MJ Holmes. Origin and transfer of toxins involved in ciguatera. Comp Biochem Physiol C 106:615–628, 1993.
9. M Takahashi, Y Ohizumi, T Yasumoto. Maitotoxin, a Ca^{2+} channel activator candidate. J Biol Chem 257:7287–7289, 1982.
10. Y Ohizumi, T Yasumoto. Contraction and increase in tissue calcium content induced by maitotoxin, the most potent known marine toxin, in intestinal smooth muscle. Br J Pharmacol 79:3–5, 1983.
11. M Takahashi, M Tatsumi, Y Ohizumi, T Yasumoto. Ca^{2+} channel activating function of maitotoxin, the most potent marine toxin known, in clonal rat pheochromocytoma cells. J Biol Chem 258: 10944–10949, 1983.
12. M Murata, F Gusovsky, T Yasumoto, JW Daly. Selective stimulation of Ca^{2+} flux in cells by maitotoxin. Eur J Pharmacol 227:43–49, 1992.
13. M Yoshii, A Tsunoo, Y Kuroda, CH Wu, T Narahashi. Maitotoxin-induced membrane current in neuroblastoma cells. Brain Res 424:119–125, 1987.
14. SB Freedman, RJ Miller, DM Miller, DR Tindall. Interactions of maitotoxin with voltage-sensitive calcium channels in cultured neuronal cells. Proc Natl Acad Sci USA 81:4582–4585, 1984.
15. MJ Berridge. Inositol trisphosphate and calcium signalling. Nature 361:315–325, 1993.
16. F Gusovsky, T Yasumoto, JW Daly. Maitotoxin, a potent, general activator of phosphoinositide breakdown. FEBS Lett 243:307–312, 1989.
17. JW Putney Jr. Capacitative calcium entry revisited. Cell Calcium 11:611–624, 1990.
18. DW Heinz, L Essen, RL Williams. Structural and mechanistic comparison of prokaryotic and eukaryotic Phosphoinositide-specific Phospholipases C. J Mol Biol 275:635–650, 1998.
19. M Columbo, M Taglialatela, JA Warner, DWJ MacGlashan, T Yasumoto, L Annunziato, G Marone. Maitotoxin, a novel activator of mediator release from human basophils, induces large increases in cytosolic calcium resulting in histamine, but not leukotriene C4, release. J Pharmacol Exp Ther 263:979–986, 1992.
20. A Watanabe, Y Ishida, H Honda, M Kobayashi, Y Ohizumi. Ca^{2+}-dependent aggregation of rabbit platelets induced by maitotoxin, a potent marine toxin, isolated from a dinoflagellate. Br J Pharmacol 109:29–36, 1993.

21. KW Wang, R Nath, KJ Raser, I Hajimohammadreza. Maitotoxin induces calpain activation in SH-SY5Y neuroblastoma cells and cerebrocortical cultures. Arch Biochem Biophys 331:208–214, 1996.
22. K Terao, E Ito, Y Kakinuma, K Igarashi, M Kobayashi, Y Ohizumi, T Yasumoto. Histopathological studies on experimental marine toxin poisoning—4. Pathogenesis of experimental maitotoxin poisoning. Toxicon 27:979–988, 1989.
23. RK Kutty, Y Singh, G Santostasi, G Krishna. Maitotoxin-induced liver cell death involving loss of cell ATP following influx of calcium. Toxicol Appl Pharmacol 101:1–10, 1989.
24. K Terao, E Ito, Y Sakamaki, K Igarashi, A Yokoyama, T Yasumoto. Histopathological studies of experimental marine toxin poisoning. II. The acute effects of maitotoxin on the stomach, heart and lymphoid tissues in mice and rats. Toxicon 26:395–402, 1988.
25. F Gusovsky, JA Bitran, T Yasumoto, JW Daly. Mechanism of maitotoxin-stimulated phosphoinositide breakdown in HL-60 cells. J Pharmacol Exp Ther 252:466–473, 1990.
26. IF Musgrave, R Seifert, G Schultz. Maitotoxin activates cation channels distinct from the receptor-activated non-selective cation channels of HL-60 cells. Biochem J 301:437–441, 1994.
27. JT Miyahara, CK Akau, T Yasumoto. Effects of ciguatoxin and maitotoxin on the isolated guinea pig atria. Res Commun Chem Pathol Pharmacol 25:177–180, 1979.
28. M Kobayashi, Y Ohizumi, T Yasumoto. The mechanism of action of maitotoxin in relation to Ca^{2+} movements in guinea-pig and rat cardiac muscles. Br J Pharmacol 86:385–391, 1985.
29. M Kobayashi, G Miyakoda, T Nakamura, Y Ohizumi. Ca-dependent arrhythmogenic effects of maitotoxin, the most potent marine toxin known, on isolated rat cardiac muscle cells. Eur J Pharmacol 111:121–123, 1985.
30. M Kobayashi, S Kondo, T Yasumoto, Y Ohizumi. Cardiotoxic effects of maitotoxin, a principal toxin of seafood poisoning, on guinea pig and rat cardiac muscle. J Pharmacol Exp Ther 238:1077–1083, 1986.
31. G Santostasi, RK Kutty, AL Bartorelli, T Yasumoto, G Krishna. Maitotoxin-induced myocardial cell injury: calcium accumulation followed by ATP depletion precedes cell death. Toxicol Appl Pharmacol 102:164–173, 1990.
32. Y Ohizumi, T Yasumoto. Contractile response of the rabbit aorta to maitotoxin, the most potent marine toxin. J Physiol (Lond) 337:711–721, 1983.
33. AM Legrand, R Bagnis. Effects of ciguatoxin and maitotoxin on isolated rat atria and rabbit duodenum. Toxicon 22:471–475, 1984.
34. P Berta, S Phaneuf, J Derancourt, J Casanova, M Durand-Clement, C le Peuch, J Haiech, JC Cavadore. The effects of maitotoxin on phosphoinositides and calcium metabolism in a primary culture of aortic smooth muscle cells. Toxicon 26:133–141, 1988.
35. F Sladeczek, BH Schmidt, R Alonso, L Vian, A Tep, T Yasumoto, RN Cory, J Bockaert. New insights into maitotoxin action. Eur J Biochem 174:663–670, 1988.
36. JP Bressler, L Belloni-Olivi, S Forman. Effect of ganglioside GM1 on arachidonic acid release in bovine aortic endothelial cells. Life Sci 54:49–60, 1994.
37. J Bressler, S Forman, GW Goldstein. Phospholipid metabolism in neural microvascular endothelial cells after exposure to lead in vitro. Toxicol Appl Pharmacol 126:352–360, 1994.
38. H Ueda, S Tamura, H Harada, T Yasumoto, H Takagi. The maitotoxin-evoked Ca^{2+} entry into synaptosomes is enhanced by cholera toxin. Neurosci Lett 67:141–146, 1986.
39. M Taglialatela, LM Canzoniero, A Fatatis, G Di Renzo, T Yasumoto, L Annunziato. Effect of maitotoxin on cytosolic Ca^{2+} levels and membrane potential in purified rat brain synaptosomes. Biochim Biophys Acta 1026:126–132, 1990.
40. JG Baird, SR Nahorski. Increased intracellular calcium stimulates ^{3}H-inositol polyphosphate accumulation in rat cerebral cortical slices. J Neurochem 54:555–561, 1990.
41. JG Baird, SR Nahorski. Differences between muscarinic-receptor- and Ca^{2+}-induced inositol polyphosphate isomer accumulation in rat cerebral-cortex slices. Biochem J 267:835–838, 1990.
42. F Gusovsky, T Yasumoto, JW Daly. Maitotoxin stimulates phosphoinositide breakdown in neuroblastoma hybrid NCB-20 cells. Cell Mol Neurobiol 7:317–322, 1987.
43. JP Pin, T Yasumoto, J Bockaert. Maitotoxin-evoked gamma-aminobutyric acid release is due not only to the opening of calcium channels. J Neurochem 50:1227–1232, 1988.

44. R Nath, KJ Raser, D Stafford, I Hajimohammadreza, A Posner, H Allen, RV Talanian, P Yuen, RB Gilbertsen, KK Wang. Non-erythroid alpha-spectrin breakdown by calpain and interleukin 1 beta-converting-enzyme-like protease(s) in apoptotic cells: contributory roles of both protease families in neuronal apoptosis. Biochem J 319:683–690, 1996.
45. IS Login, AM Judd, MJ Cronin, K Koike, G Schettini, T Yasumoto, RM MacLeod. The effects of maitotoxin on $^{45}Ca^{2+}$ flux and hormone release in GH3 rat pituitary cells. Endocrinology 116:622–627, 1985.
46. K Koike, AM Judd, IS Login, T Yasumoto, RM MacLeod. Maitotoxin, a calcium channel activator, increases prolactin release from rat pituitary tumor 7315a cells by a mechanism that may involve leukotriene production. Neuroendocrinology 43:283–290, 1986.
47. IS Login, AM Judd, RM MacLeod. Dopamine inhibits maitotoxin-stimulated pituitary $^{45}Ca^{2+}$ efflux and prolactin release. Am J Physiol 250:E731–E735, 1986.
48. DG Soergel, F Gusovsky, T Yasumoto, JW Daly. Stimulatory effects of maitotoxin on insulin release in insulinoma HIT cells: role of calcium uptake and phosphoinositide breakdown. J Pharmacol Exp Ther 255:1360–1365, 1990.
49. G Schettini, K Koike, IS Login, AM Judd, MJ Cronin, T Yasumoto, RM MacLeod. Maitotoxin stimulates hormonal release and calcium flux in rat anterior pituitary cells in vitro. Am J Physiol 247:E520–E525, 1984.
50. MA Sortino, TM Delahunty, T Yasumoto, MJ Cronin. Maitotoxin increases inositol phosphates in rat anterior pituitary cells. J Mol Endocrinol 6:95–99, 1991.
51. IS Login, AM Judd, MJ Cronin, T Yasumoto, RM MacLeod. Reserpine is a calcium channel antagonist in normal and GH3 rat pituitary cells. Am J Physiol 248:E15–E19, 1985.
52. N Sasakawa, T Nakaki, R Kato. Stimulus-responsive and rapid formation of inositol pentakisphosphate in cultured adrenal chromaffin cells. J Biol Chem 265:17700–17705, 1990.
53. P Lebrun, M Hermann, T Yasumoto, A Herchuelz. Effects of maitotoxin on ionic and secretory events in rat pancreatic islets. Biochem Biophys Res Commun 144:172–177, 1987.
54. CA Leech, GG Holz, JF Habener. Signal transduction of PACAP and GLP-1 in pancreatic beta cells. Ann NY Acad Sci 805:81–92, 1996.
55. JF Worley, MS McIntyre, B Spencer, ID Dukes. Depletion of intracellular Ca^{2+} stores activates a maitotoxin-sensitive nonselective cationic current in beta-cells. J Biol Chem 269:32055–32058, 1994.
56. J Berreur-Bonnenfant, M Ammar, A Dubreuil, H Kiefer, G Diogene, P Metezeau, S Puiseux-Dao. Modulation of fibroblast response to maitotoxin along the cell division cycle. Cell Biol Toxicol 10: 423–427, 1994.
57. M Estacion, HB Nguyen, JJ Gargus. Calcium is permeable through a maitotoxin-activated nonselective cation channel in mouse L cells. Am J Physiol 270:C1145–C1152, 1996.
58. V Fessard, G Diogene, A Dubreuil, JP Quod, M Durand-Clement, C Legay, S Puiseux-Dao. Selection of cytotoxic responses to maitotoxin and okadaic acid and evaluation of toxicity of dinoflagellate extracts. Nat Toxins 2:322–328, 1994.
59. D Gutierrez, DL Diaz, L Vaca. Characterization of the maitotoxin-induced calcium influx pathway from human skin fibroblasts. Cell Calcium 22:31–38, 1997.
60. DG Soergel, T Yasumoto, JW Daly, F Gusovsky. Maitotoxin effects are blocked by SK&F 96365, an inhibitor of receptor-mediated calcium entry. Mol Pharmacol 41:487–493, 1992.
61. A Venant, AC Dazy, G Diogene, P Metezeau, F Marano. Effects of maitotoxin on calcium entry and phosphoinositide breakdown in the rabbit ciliated tracheal epithelium. Biol Cell 82:195–202, 1994.
62. JW Daly, J Lueders, WL Padgett, Y Shin, F Gusovsky. Maitotoxin-elicited calcium influx in cultured cells. Effect of calcium-channel blockers. Biochem Pharmacol 50:1187–1197, 1995.
63. P Dietl, H Volkl. Maitotoxin activates a nonselective cation channel and stimulates Ca^{2+} entry in MDCK renal epithelial cells. Mol Pharmacol 45:300–305, 1994.
64. O Meucci, M Grimaldi, A Scorziello, S Govoni, S Bergamaschi, T Yasumoto, G Schettini. Maitotoxin-induced intracellular calcium rise in PC12 cells: involvement of dihydropyridine-sensitive and omega-conotoxin-sensitive calcium channels and phosphoinositide breakdown. J Neurochem 59:679–688, 1992.

65. D Xi, FM Van Dolah, JS Ramsdell. Maitotoxin induces a calcium-dependent membrane depolarization in GH4C1 pituitary cells via activation of type L voltage-dependent calcium channels. J Biol Chem 267:25025–25031, 1992.
66. A Fatatis, A Bassi, E Iannotti, N Caso, GD Mita, G Di Renzo, L Annunziato. Appearance of depolarization- and maitotoxin-induced $[Ca^{2+}]_i$ elevation in single LAN-1 human neuroblastoma cells on exposure to retinoic acid. J Neurochem 63:1900–1907, 1994.
67. JF Faivre, E Deroubaix, A Coulombe, AM Legrand, E Coraboeuf. Effect of maitotoxin on calcium current and background inward current in isolated ventricular rat myocytes. Toxicon 28:925–937, 1990.
68. P Berta, F Sladeczek, J Derancourt, M Durand, P Travo, J Haiech. Maitotoxin stimulates the formation of inositol phosphates in rat aortic myocytes. FEBS Lett 197:349–352, 1986.
69. OH Choi, WL Padgett, Y Nishizawa, F Gusovsky, T Yasumoto, JW Daly. Maitotoxin: effects on calcium channels, phosphoinositide breakdown, and arachidonate release in pheochromocytoma PC12 cells. Mol Pharmacol 37:222–230, 1990.
70. WW Lin, CY Lee, T Yasumoto, DM Chuang. Maitotoxin induces phosphoinositide turnover and modulates glutamatergic and muscarinic cholinergic receptor function in cultured cerebellar neurons. J Neurochem 55:1563–1568, 1990.
71. F Gusovsky, JE Lueders, EC Kohn, CC Felder. Muscarinic receptor–mediated tyrosine phosphorylation of phospholipase C-gamma. An alternative mechanism for cholinergic-induced phosphoinositide breakdown. J Biol Chem 268:7768–7772, 1993.
72. AC Dolphin. Mechanisms of modulation of voltage-dependent calcium channels by G proteins. J Physiol 506:3–11, 1998.
73. RC Young, M McLaren, JS Ramsdell. Maitotoxin increases voltage independent chloride and sodium currents in GH4C1 rat pituitary cells. Nat Toxins 3:419–427, 1995.
74. F Gusovsky, JW Daly, T Yasumoto, E Rojas. Differential effects of maitotoxin on ATP secretion and on phosphoinositide breakdown in rat pheochromocytoma cells. FEBS Lett 233:139–142, 1988.
75. M Nishio, S Kigoshi, I Muramatsu, T Yasumoto. Ca(2+)- and Na(+)-dependent depolarization induced by maitotoxin in the crayfish giant axon. Gen Pharmacol 24:1079–1083, 1993.
76. D Xi, DT Kurtz, JS Ramsdell. Maitotoxin-elevated cytosolic free calcium in GH4C1 rat pituitary cells nimodipine-sensitive and -insensitive mechanisms. Biochem Pharmacol 51:759–769, 1996.
77. G Sorrentino, MR Monsurro, IN Singh, JN Kanfer. Membrane depolarization in LA-N-1 cells. The effect of maitotoxin is Ca^{2+}- and Na^{+}-dependent. Mol Chem Neuropathol 30:199–211, 1997.
78. M Kobayashi, R Ochi, Y Ohizumi. Maitotoxin-activated single calcium channels in guinea-pig cardiac cells. Br J Pharmacol 92:665–671, 1987.
79. M Nishio, I Muramatsu, T Yasumoto. Na^{+}-permeable channels induced by maitotoxin in guinea-pig single ventricular cells. Eur J Pharmacol 297:293–298, 1996.
80. Y Eguchi, S Shimizu, Y Tsujimoto. Intracellular ATP levels determine cell death fate by apoptosis or necrosis. Cancer Res 57:1835–1840, 1997.
81. CA Leech, JF Habener. Insulinotropic glucagon-like peptide-1-mediated activation of non-selective cation currents in insulinoma cells is mimicked by maitotoxin. J Biol Chem 272:17987–17993, 1997.
82. MW Roe, JF Worley, F Qian, N Tamarina, AA Mittal, F Dralyuk, NT Blair, RJ Mertz, LH Philipson, ID Dukes. Characterization of a Ca^{2+} release-activated nonselective cation current regulating membrane potential and $[Ca^{2+}]_i$ oscillations in transgenically derived beta-cells. J Biol Chem 273:10402–10410, 1998.
83. CA Leech, JF Habener. A role for Ca^{2+} sensitive nonselective cation channels in regulating the membrane potential of pancreatic beta-cells. Diabetes 47:1066–1073, 1998.
84. I Hajimohammadreza, KJ Raser, R Nath, R Nadimpalli, M Scott, KK Wang. Neuronal nitric oxide synthase and calmodulin-dependent protein kinase IIalpha undergo neurotoxin-induced proteolysis. J Neurochem 69:1006–1013, 1997.
85. T Miyamoto, Y Ohizumi, H Washio, Y Yasumoto. Potent excitatory effect of maitotoxin on Ca channels in the insect skeletal muscle. Pflugers Arch 400:439–441, 1984.

86. V Bernard, A Laurent, J Derancourt, M Clement-Durand, A Picard, C le Peuch, P Berta, M Doree. Maitotoxin triggers the cortical reaction and phosphatidylinositol-4,5-bisphosphate breakdown in amphibian oocytes. Eur J Biochem 174:655–662, 1988.
87. JK Hirsh, CH Wu. Palytoxin-induced single-channel currents from the sodium pump synthesized by in vitro expression. Toxicon 35:169–176, 1997.
88. C Van Renterghem, C Frelin. 3,4-Dichlorobenzamil-sensitive, monovalent cation channel induced by palytoxin in cultured aortic myocytes. Br J Pharmacol 109:859–865, 1993.
89. A Bielfeld-Ackermann, C Range, C Korbmacher. Maitotoxin (MTX) activates a nonselective cation channel in Xenopus laevis oocytes. Pflugers Arch 436:329–337, 1998.
90. K Konoki, M Hashimoto, T Nonomura, M Sasaki, M Murata, K Tachibana. Inhibition of maitotoxin-induced Ca^{2+} influx in rat glioma C6 cells by brevetoxins and synthetic fragments of maitotoxin. J Neurochem 70:409–416, 1998.
91. MS Montal, R Blewitt, JM Tomich, M Montal. Identification of an ion channel-forming motif in the primary structure of tetanus and botulinum neurotoxins. FEBS Lett 313:12–18, 1992.
92. G Yellen. Single Ca^{2+}-activated nonselective cation channels in neuroblastoma. Nature 296:357–359, 1982.
93. AM Frace, JJ Gargus. Activation of single-channel currents in mouse fibroblasts by platelet-derived growth factor. Proc Natl Acad Sci USA 86:2511–2515, 1989.
94. M Estacion. Characterization of ion channels seen in subconfluent human dermal fibroblasts. J Physiol (Lond.) 436:579–601, 1991.
95. HQ Xie, GV Johnson. Calcineurin inhibition prevents calpain-mediated proteolysis of tau in differentiated PC12 cells. J Neurosci Res 53:153–164, 1998.
96. C Virginio, D Church, RA North, A Surprenant. Effects of divalent cations, protons and calmidazolium at the rat P_2X_7 receptor. Neuropharmacology 36:1285–1294, 1997.
97. MK Squier, AC Miller, AM Malkinson, JJ Cohen. Calpain activation in apoptosis. J Cell Physiol 159:229–237, 1994.
98. D Pesando, JP Girard, M Durand-Clement, P Payan, S Puiseux-Dao. Effect of maitotoxin on sea urchin egg fertilization and on Ca^{2+} permeabilities of eggs and intracellular stores. Biol Cell 72: 269–273, 1991.
99. KK Vaswani, GS Wu, RW Ledeen. Exogenous gangliosides stimulate breakdown of neuro-2A phosphoinositides in a manner unrelated to neurite outgrowth. J Neurochem 55:492–499, 1990.
100. M Connor, A Yeo, G Henderson. Neuropeptide Y Y2 receptor and somatostatin sst2 receptor coupling to mobilization of intracellular calcium in SH-SY5Y human neuroblastoma cells. Br J Pharmacol 120:455–463, 1997.
101. J Zhang, M Lesort, RP Guttmann, GV Johnson. Modulation of the in situ activity of tissue transglutaminase by calcium and GTP. J Biol Chem 273:2288–2295, 1998.
102. G Sberna, J Saez-Valero, K Beyreuther, CL Masters, DH Small. The amyloid beta-protein of Alzheimer's disease increases acetylcholinesterase expression by increasing intracellular calcium in embryonal carcinoma P19 cells. J Neurochem 69:1177–1184, 1997.
103. P Dietl, T Haller, B Wirleitner, H Volkl, F Friedrich, J Striessnig. Activation of L-type Ca^{2+} channels after purinoceptor stimulation by ATP in an alveolar epithelial cell (L2). Am J Physiol 269:L873–L883, 1995.
104. A Venant, AC Dazy, G Diogene, F Maran. Differential effects of maitotoxin on calcium entry and ciliary beating in the rabbit ciliated tracheal epithelium. Biol Cell 85:197–205, 1995.
105. JR Spoonster, L Masiero, SA Savage, J Probst, EC Kohn. Regulation of cell spreading during differentiation in the muscarinic M5 receptor tumor-suppressor model. Int J Cancer 72:362–368, 1997.
106. AM Legrand, R Bagnis. Effects of highly purified maitotoxin extracted from dinoflagellate Gambierdiscus toxicus on action potential of isolated rat heart. J Mol Cell Cardiol 16:663–666, 1984.
107. JA Janovick, K Natarajan, F Longo, PM Conn. Caldesmon: a bifunctional (calmodulin and actin) binding protein which regulates stimulated gonadotropin release. Endocrinology 129:68–74, 1991.
108. G Schettini, T Florio, O Meucci, E Landolfi, M Grimaldi, G Lombardi, G Scala, D Leong. Interleu-

kin-1-beta modulation of prolactin secretion from rat anterior pituitary cells: involvement of adenylate cyclase activity and calcium mobilization. Endocrinology 126:1435–1441, 1990.
109. FM Van Dolah, JS Ramsdell. Maitotoxin, a calcium channel activator, inhibits cell cycle progression through the G1/S and G2/M transitions and prevents CDC2 kinase activation in GH4C1 cells. J Cell Physiol 166:49–56, 1996.
110. A Schulick, F Gusovsky, T Yasumoto, JW Daly. Effects of maitotoxin on atrial natriuretic factor-mediated accumulation of cyclic GMP in PC12 cells. Life Sci 46:671–678, 1990.
111. OH Choi, WL Padgett, JW Daly. Effects of the amphiphilic peptides melittin and mastoparan on calcium influx, phosphoinositide breakdown and arachidonic acid release in rat pheochromocytoma PC12 cells. J Pharmacol Exp Ther 260:369–375, 1992.

23

Brevetoxins: Chemistry, Mechanism of Action, and Methods of Detection

Daniel G. Baden
Center for Marine Science Research, University of North Carolina at Wilmington, Wilmington, North Carolina

David J. Adams
The University of Queensland, Brisbane, Australia

I. INTRODUCTION

Brevetoxins are produced by the unarmored marine dinoflagellate *Gymnodinium breve* (= *Ptychodiscus brevis*), an organism linked to red tide outbreaks in the Gulf of Mexico, New Zealand, and Japan. These toxins are responsible for toxicity to marine animals, neurotoxic shellfish poisoning in humans, and respiratory distress if the toxins are inhaled in sea spray. Dinoflagellates are eukaryotes and most are photosynthetic, with at least one form in the life cycle being motile. Of the total known 2000 species of dinoflagellates, only about 20 species have been shown to produce specific toxins. *G. breve* is a unicellular alga which is the causative organism of red tide resulting from blooms of this toxic dinoflagellate along the west coast of Florida and elsewhere in the Gulf of Mexico. A similar toxic species occurs in Spain, Japan, and more recently in New Zealand waters (1). Blooms of the toxic dinoflagellate have caused massive fish kills, marine mammal and sea bird deaths, and mollusk contamination which, if consumed, results in human neurotoxic shellfish poisoning (NSP). An irritating aerosol which results from contact with *G. breve* cell particles entrapped in sea spray is also a significant public health concern (2). This species blooms sporadically, reaching concentrations of 10^7–10^8 cells/L, but during nonbloom periods, the concentration is $<10^3$ cells/L. Blooms initiate exclusively offshore and are thought to occur from seed beds of resting benthic cysts (3). Various physical and nutritional factors contribute to *G. breve* proliferation. The salinity range for optimal growth of *G. breve* is from 27 to 37%. It thrives in water of temperatures ranging from 10 to 33°C. Toxicity of this organism is due to the synthesis and intracellular maintenance of a multiplicity of polyether brevetoxins (PbTx) (4). Multiple toxic compounds have been purified both from field blooms and from laboratory cultures: nine toxins are currently known, including PbTx-1 (also designated T_{16}, GB-1, or brevetoxin A), PbTx-2 (also designated T_{47}, GB-2, T34, or brevetoxin B), and PbTx-3 (also designated GB-3 or T17) (5).

II. CHEMISTRY AND STRUCTURE/FUNCTION RELATIONSHIPS

A. Lipid-Soluble Brevetoxins and Derivatives Are Unique Biochemicals

Pure toxins can be isolated from the culture liquid of *G. breve* and crystallized. The planar and relative configuration of brevetoxin-B was determined by x-ray crystallography, which disclosed an array of trans-fused rings, 1 lactone, and 10 ether rings leading to a ladder-like molecule 30 Å long, 6 Å wide, and 6 Å high (6).

The lipid-soluble toxins produced by *G. breve* are of two types, hemolytic and neurotoxic. There are 10 naturally occurring derivatives (Figure 1), all of which have the same molecular receptor site, the same overall interaction with voltage-gated sodium channels, and all can be detected by the same immunological and receptor-based assays. Numerous different derivatives of the natural toxins have also been produced (Table 1 and 2) by microsynthetic organic chemical derivatization. Within this cadre of synthetic derivatives are toxins which mimic all of the natural toxin effects, some which have no effect on sodium channels, and some which possess a specific subset of the overall effects produced by natural toxin.

There are several unique structural features that are shared by the brevetoxins and by ciguatoxin. We and others had recognized the similar ring ''tail'' regions represented by the H-I-J-K rings in PbTx-2, the G-H-I-J rings in PbTx-1, and the E-D-C-B rings of ciguatoxin. The potential importance of the A-ring lactone in all active brevetoxins was also recognized by many groups, and the facile nature of the stereochemistry surrounding the ciguatoxin A-B–ring

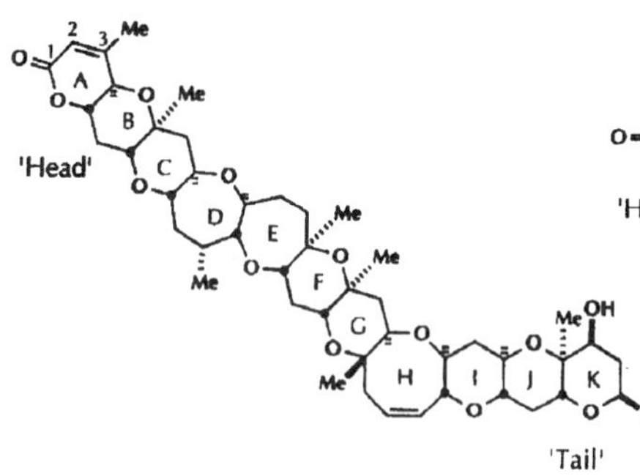

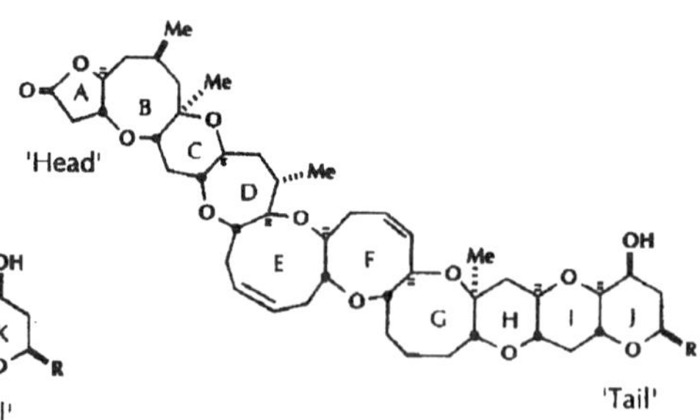

Brevetoxin B backbone:
PbTx-2, $R{=}CH_2C({=}CH_2)CHO$
PbTx-3, $R{=}CH_2C({=}CH_2)CH_2OH$
PbTx-5, [PbTx-2], C-37 O Ac
PbTx-6, [PbTx-2], C27, 28 Epoxide
PbTx-8, $R{=}CH_2COCH_2Cl$
PbTx-9, $R{=}CH_2CH(CH_3)CH_2OH$

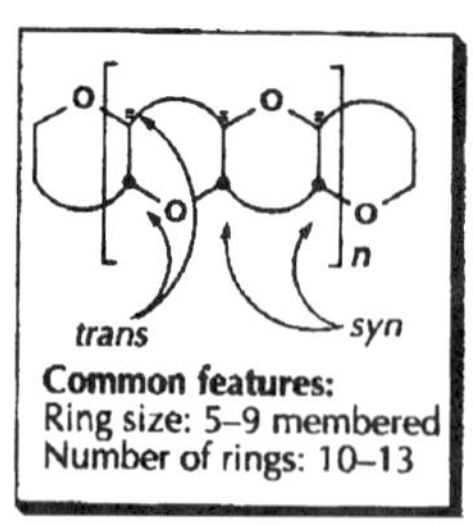

Brevetoxin A backbone:
PbTx-1, $R{=}CH_2C({=}CH_2)CHO$
PbTx-7, $R{=}CH_2C({=}CH_2)CH_2OH$
PbTx-10, $R{=}CH_2CH(CH_3)CH_2OH$

Figure 1 The brevetoxins. Florida red tide toxins are all based on two different structural backbones. All toxins contain common features including a trans-syn ordering of relatively rigid ring systems, an A-ring lactone that is required for activity, a rigid region exemplified by the H-J–ring system in brevetoxin B and the G-J–ring system in brevetoxin A, and a spacer region represented by the B-G–rings of brevetoxin B and the B-F–ring region in brevetoxin A. Derivatives include H-ring modifications in the brevetoxin B structure, K-ring esterification in brevetoxin B, and terminal ring substituent modifications in both structures (denoted as ''R''). All toxins are present during Florida red tides but proportions vary (18).

Table 1 Regions of Brevetoxin Recognized by Sodium Channel Receptors

	Toxin identity	Ring A	Ring H	C-37	Side chain	K_i (nM)
1		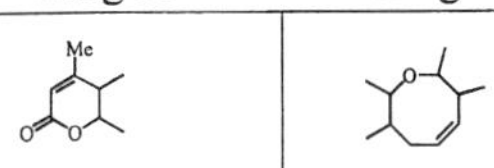		OH	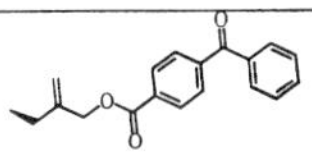	0.2
2				OH	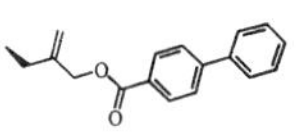	0.3
3		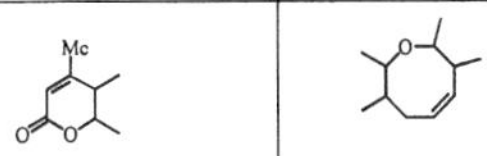		OH	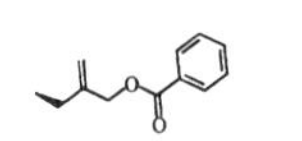	1.0
4		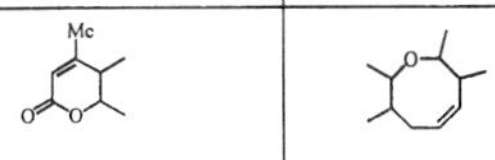		OH	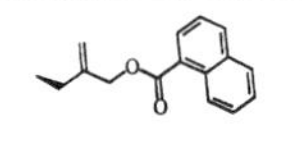	1.3
5	PbTx-3	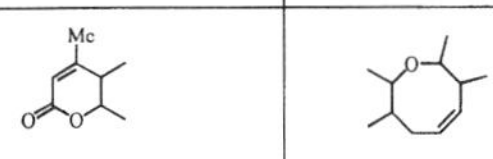		OH	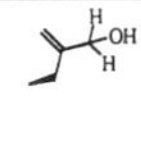	1.3
6		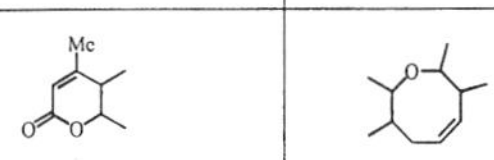		OH	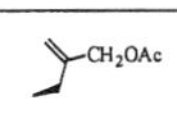	2.1
7	PbTx-2	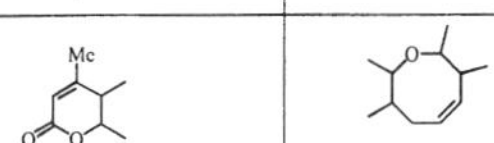		OH	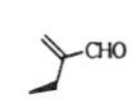	2.5
8		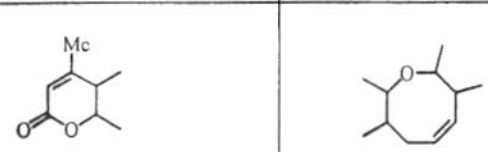		OH	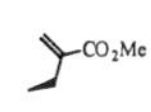	3.2
9	PbTx-9	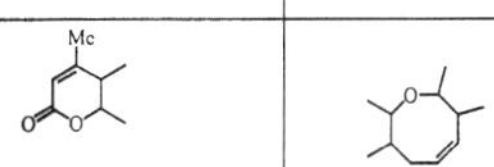		OH	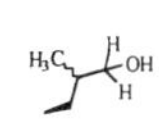	5.9
10		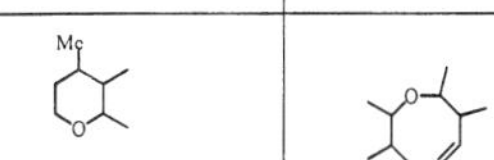		OH	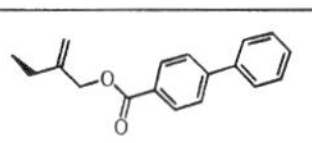	10.0
11		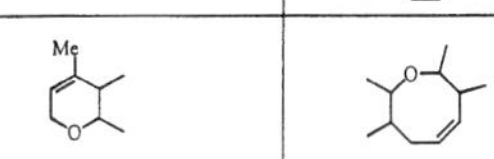		OH	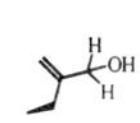	11.0
12		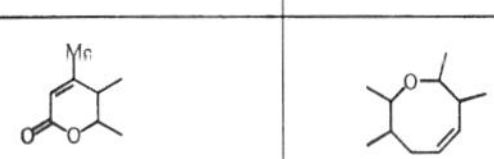		OH	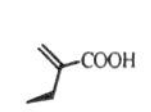	11.1
13		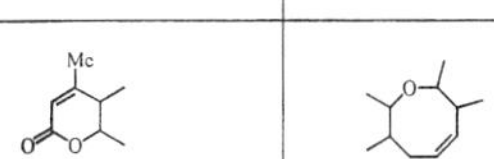		OAc	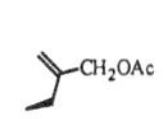	13.0
14	PbTx-6	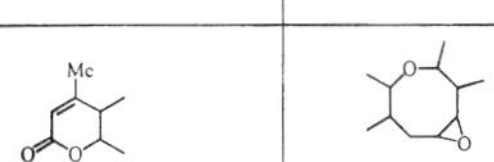		OH	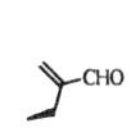	14.0
15		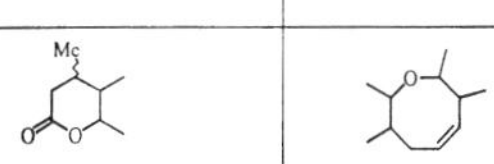		OH	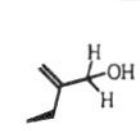	15.8
16		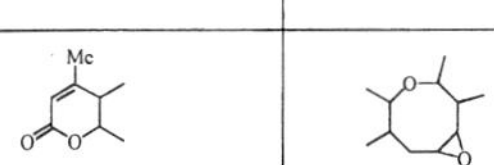		OH	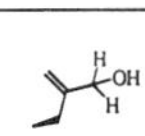	16.0

Table 1 Continued

	Toxin identity	Ring A	Ring H	C-37	Side chain	K_i (nM)
17		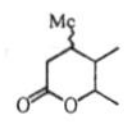	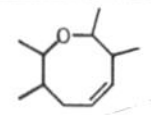	OH	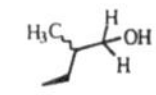	51.2
18		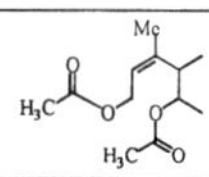	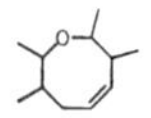	OH	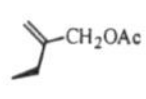	170.0
19		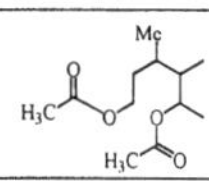	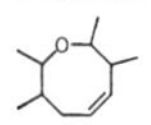	OH	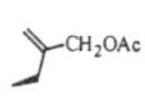	340.0
20		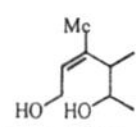	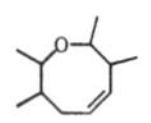	OH	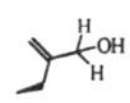	1200.0
21		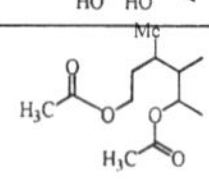	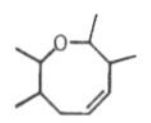	OH	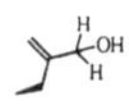	2000.0
22		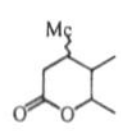	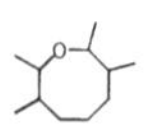	OH	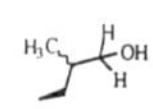	5000+

1: 2,3 dihydrobrevetoxin PbTx-3-p-benzoylbenzoate; 2: PbTx-3-p-phenylbenzoate; 3: 2,3 dihydro PbTx-3 benzoate; 4: 2,3 dihydro PbTx-3 α-naphthoate; 5: PbTx-3; 6: 2,3 dihydro PbTx-3 acetate ester; 7: PbTx-2; 8: 2,3 dihydro PbTx-3 carboxylic acid methyl ester; 9: PbTx-9; 10: 1 desoxy, 2,3 dihydro PbTx-3-p-phenylbenzoate; 11: 1 desoxy PbTx-3; 12: PbTx-3 carboxylic acid; 13: PbTx-3 C-37, C-42 diacetate; 14: PbTx-6; 15: 2,3 dihydro PbTx-3; 16: PbTx-3, 27,28 epoxide; 17: 2,3,41,43 tetrahydro PbTx-3; 18: PbTx-3, 1,5,42 triacetate; 19: 2,3 dihydro PbTx-3, 1,5,42 triacetate; 20: 1,5 dihydroxy PbTx-3; 21: 2,3 dihydro PbTx-3 1,5 diacetate; 22: 2,3,27,28,421,43 hexahydro PbTx-3.

juncture provided an indication that the A-ring opened and closed. Thus, the developing model for any "toxic" polyether ladder toxin was to provide for (1) overall relative linearity, (2) an A-ring "head" region, and (3) a rigid "tail" ring region. Conservative modification to the tail region side chain seemed to indicate that its character was probably not a stringent factor in toxicity. Likewise, the C-37 hydroxyl could be esterified (as in PbTx-5) without undue loss of toxicity. Similarly, the unsaturation in ring H of PbTx-2 could be epoxidized (as in PbTx-6), again without undue loss of toxicity. Accounting for all of these potential features in a ligand model, Figure 2 illustrates the chemical constraints surrounding any potential polyether ligand candidate.

The results provide profound evidence for "linear" rather than "bent" orientations over the entire molecule length, for the maintenance of A-ring integrity (although not necessarily of a lactone character), and implications of a requisite "spacer" region between A-ring and H-I-J-K "tail" ring region. Less clear is any correlation between hydrophobicity and toxicity. Finally, binding, toxicity, and function increase and decrease in parallel. This is in agreement with published work of Shimizu et al. (5) and Lin et al. (6).

B. Brevetoxins Bind to Orphan Receptor Site 5 (7–9)

In early molecular work with the brevetoxins, Catterall and Risk (10) observed the specific interaction of the toxins as they modulated the effects of toxins specific for other sites associated with the voltage-sensitive sodium channels (VSSC). Further work by Poli et al. (4), using tritiated brevetoxin originally synthesized for the purpose of developing radioimmunoassays for the

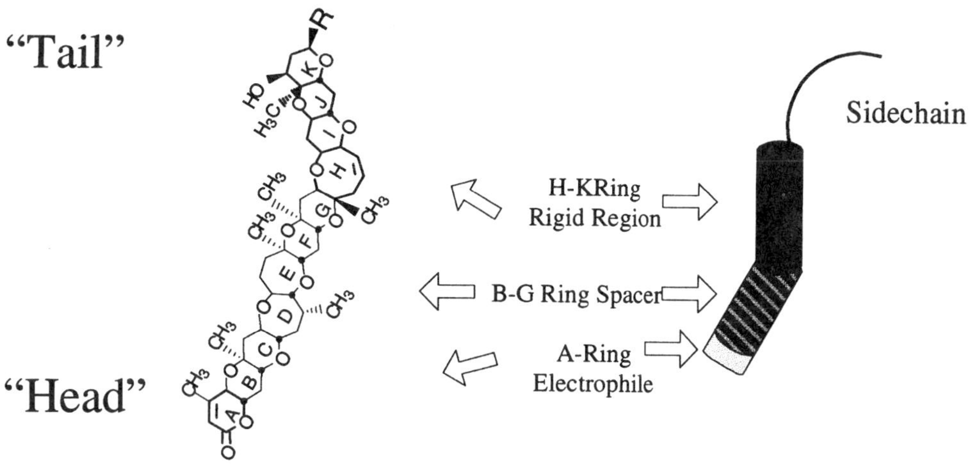

Figure 2 The brevetoxin molecule can be thought of as a long rigid cylinder with a slight bend in the middle. The molecule is composed of four regions: an A-ring electrophil thought to be requisite for activity, a B-G–ring spacer region, an H-K–ring rigid region which is required for binding to the sodium channel, and a side chain. For convenience when referring to Tables 1 and 2, each of the rings of the chemical structure is lettered and the tail and head regions identified (14).

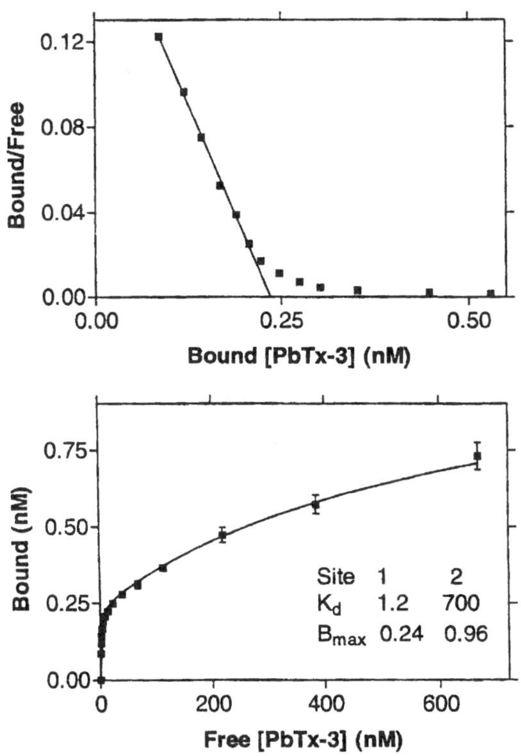

Figure 3 Binding of PbTx-3 to rat brain sodium channel. (Assays were performed according to refs. 4 and 84). Upper Panel: Scatchard analysis of brevetoxin binding up to 0.67 μM free PbTx-3 is shown with a linear least-squares line through the first six points (0.7–8.3 nM PbTx-3). Bottom Panel: Specific binding was corrected by adding the dpm (disintegrations per minute) corresponding to specific binding of labeled toxin, not displaced by unlabeled toxin in the control tubes, as calculated from the K_d and B_{max} assigned to the low- and high-affinity sites. The resulting values were fit to a two-site binding curve.

toxins in seafood, illustrated that the toxins bound to a different site than did any other sodium channel–specific neurotoxin, and the site was named site 5. Binding at this high-affinity site was characterized as a 1:1 association with the α-subunit. Characteristic high-affinity (K_d = 1.2 nM, B_{max} = 0.24–2.12 pMl/mg protein) is observed in all systems (Figure 3). The orphan receptor exhibits a Stokes radius of 55 ± 5 Å.

Studies ongoing with tritiated brevetoxin PbTx-3-(42)-p-azidobenzoate have begun to yield some detailed information concerning the recognition site for toxin in the α-subunit of the channel. Photoaffinity work has been undertaken using either rat brain synaptosomes or using solubilized sodium channel reconstituted in artificial lipid vesicles. (9).

Specific protocols have been developed in Catterall's laboratory to evaluate domain and sequence-specific binding sites for toxins on sodium channels (11). Purified sodium channel from rat brain synaptosomes is initially mixed with a radioactive photoaffinity probe followed by irradiation to induce covalent binding. The solubilized channel with bound brevetoxin photoprobe is then enzymatically cleaved to yield limit digest peptide fragments. Addition of various sodium channel antibodies, followed by protein A–sepharose specifically binds and precipitates all antibodies in solution. Those antibodies which have an affinity for the brevetoxin binding site conjugate are easily distinguished from those that do not by the specific precipitation of radioactivity in the former case and not in the latter cases. Anti–sodium channel antibodies directed against both IS6 and IVS5 were able specifically to precipitate bound brevetoxin photoaffinity probe (Figure 4) (9). More specifically, the binding domain for brevetoxin appears to reside in the interface region between domains I and IV on the S5 and S6 regions of IV. It is this hydrophobic extracellular loop which may extend into the channel and line the interior of the ion-conducting pore (7–9). Further work identified a specific 6-kD peptide from domain I, approximately corresponding to peptide residues Thr400 to Lys443 (complete limit digest) or Ala396 to Lys455 (from incomplete digests). Similarly, a second immunoprecipitated 6-kD fragment from domain IV was identified, Glu1738 to Lys1785. We take this as direct evidence for close association of transmembrane segments IS6 and IVS5 in the native conformation of the sodium channel, and further implicate this association as the topographical site of brevetoxin interaction (8). Virtually identical binding in a cardiac sodium channel system (which possess a primary structure quite distinct from rat brain sodium channel in the IS6 region but virtual primary homology in IVS5) also indicates that IVS5 is the specific binding domain (8).

C. Allostery Surrounding the Brevetoxin-Binding Site Affects Ligand Binding at Other Toxin Sites (7)

Another area of work undertaken concerns the allosteric character of the brevetoxin-binding site, especially with respect to neurotoxin binding at other specific receptor sites on the channel. These studies were undertaken using both tritiated PbTx-3 and tritiated PbTx-9, and also utilizing unlabeled toxin as a potential competitor or allosteric modulator for the other six specific neurotoxin orphan receptors on the α-subunit.

Purified and reconstituted sodium channels have previously been shown to be functional in voltage-dependent ion conductance and in high-affinity binding of tetrodotoxin and saxitoxin at neurotoxin site 1 and α-scorpion toxin at site 3. This had not been accomplished for orphan receptor sites 2, 4, and 5. The brevetoxins, by binding to receptor site 5, cause allosteric enhancement of batrachotoxin benzoate binding at site 2 in concert with the pyrethroid insecticide RU39568. The latter toxin acts at a site distinct from sites 1–5. RU39568 induces activation of sodium channels by shifting activation to more negative potentials and by inhibition of inactivation. Brevetoxin PbTx-1 (the second—more potent brevetoxin type) binds with a dissociation constant of approximately 30 nM in purified systems. Binding parameters are unaffected by membrane potential. Application of batrachotoxin and RU39568 together causes an increase in

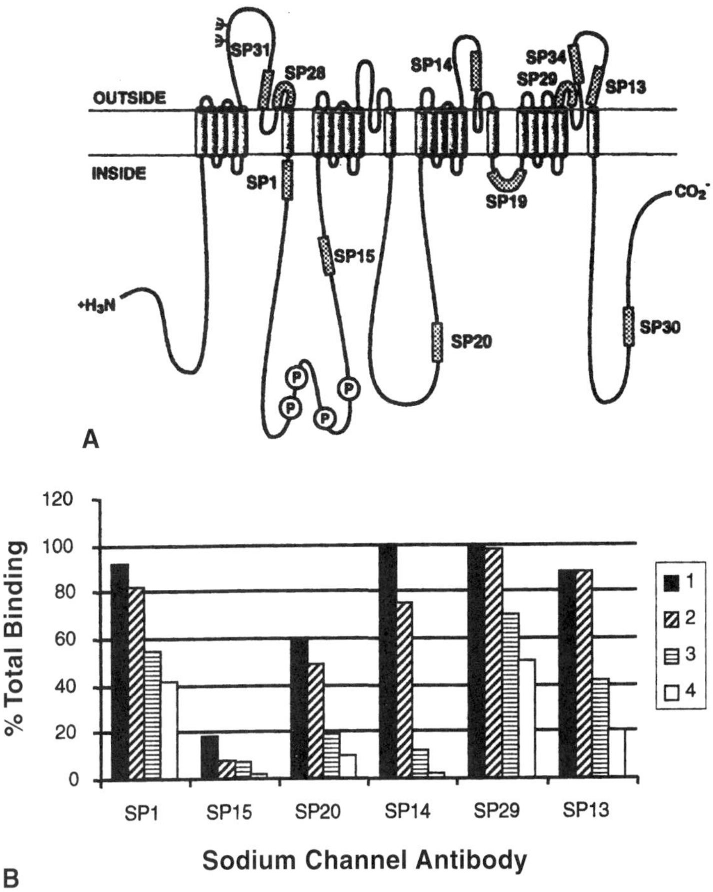

Figure 4 Immunoprecipitation of peptides covalently labeled by brevetoxin photoaffinity probe. Panel A: recognition sites of antipeptide antibodies. Antibodies directed against synthetic peptides corresponding to different sequences of the α-subunit of the type IIA sodium channel were prepared. Antibodies were directed against synthetic peptides corresponding to the amino acid sequences 355–372 (SP31), 382–400 (SP28), 427–445 (SP1), 708–722 (SP15), 1145–1164 (SP20), 1480–1498 (SP14), 1541–1561 (SP19), 1736–1753 (SP29), 1759–1777 (SP34), 1789–1798 (SP13), and 1785–2009 (SP30). Panel B: Immunoprecipitation of brevetoxin photoaffinity probe–labeled sodium channel peptide fragments from proteolytic cleavage by trypsin. Photolabeled reconstituted sodium channels were digested with increasing concentrations of TCPK-trypsin, and the resulting peptide fragments were probed with the indicated antibodies as described in Trainer et al. (8). TPCK-trypsin concentrations for each antibody treatment were as follows (left to right): 0.03, 0.1, 1.0, and 10 μg/mL. Values are expressed as the percentage of total disintegrations per minute immunoprecipitated without trypsin treatment (9).

maximum binding of about twofold, implying a strong allostery between site 2, site 5, and the pyrethroid-binding site on the sodium channel (7).

D. Brevetoxin Antibodies Detect Specific Epitopes on Toxin Derivatives (12–14)

A second type of binding phenomenon is that of brevetoxin-specific polyclonal antibodies and brevetoxin derivatives. Polyclonal antibodies are useful to develop enzyme-linked immunoassays and to produce affinity columns for purification of brevetoxin covalently bound to sodium channel peptides. Synthetic tritiated PbTx-3 was used as a probe to evaluate specific binding characteristics of goat antibodies prepared against covalent conjugates of PbTx-3 hemisuccinate, and 10 separate derivatives were examined by competitive displacement radioimmunoassay. Preliminary indications were that sera from different animals produced widely dissimilar results with respect to binding affinity and maximum. That is to say, there are multiple epitopes on the brevetoxin molecule (12,13,15). Overall antisera from different animals exhibited half-maximal binding at between 3 and 9 nM. Antibodies from two goats, for purposes of this discussion, will be designated as antiserum #1 and the other as antiserum #2. The results are presented in Table 2.

It is clear from these results that antiserum #1 is specific for ring A integrity. Progressive modification of the A-ring in PbTx-3 from saturation of the double bond 15 to ring cleavage yielding the diol 20 resulted in a five-fold reduction with each step. Likewise, reduction of the double bond in the A-ring of PbTx-9 to 17 resulted in a fivefold reduction in binding. Antiserum #2, by contrast, shows a fourfold reduction in binding affinity on reduction of the H-ring double bond 22 and more substantial reductions in binding affinity for those derivatives which are modified in the side chain of the toxin; that is, 7, 12, and 13. Antiserum #1, for our pur-

Table 2 Brevetoxin Epitope Recognition by Specific Antisera

	Toxin identity	Ring A	Ring H	C-37	Side chain	K_i (nM) **Ab1 Ab2**
5	PbTx-3	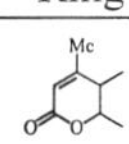	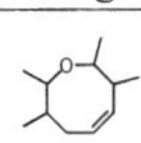	OH	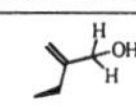	26.9 , 9.6
7	PbTx-2	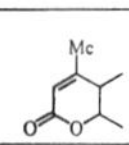	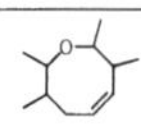	OH	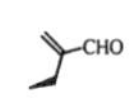	32.3 , 26.1
9	PbTx-9	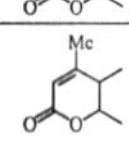	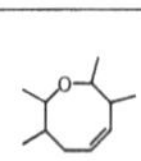	OH	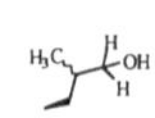	21.2 , 10.6
12		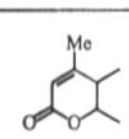	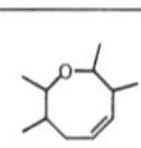	OH	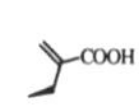	36.7 , 43.2
13		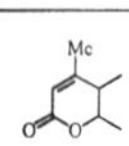	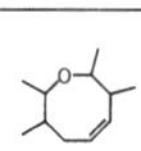	OAc	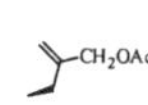	53.9 , 48.4
14	PbTx-6	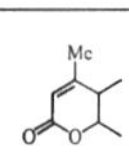	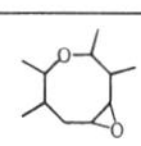	OH	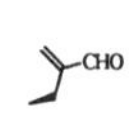	45.7 , 32.6
15		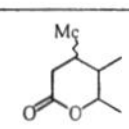	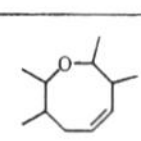	OH	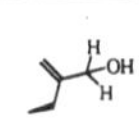	118.8 , 7.7
17		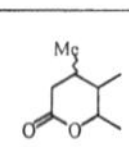	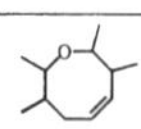	OH	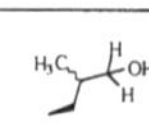	101.9 , 15.0
20		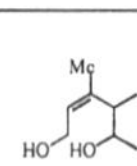	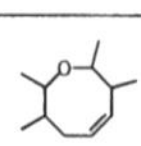	OH	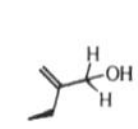	613.8 , 6.5
22		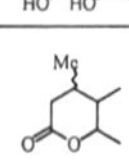	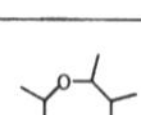	OH	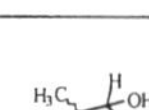	127.7 , 82.9

5: PbTx-3; 7: PbTx-2; 9: PbTx-9; 12: PbTx-3 carboxylic acid; 13: PbTx-3 C-37, C-42 diacetate; 14: PbTx-6; 15: 2,3 dihydro PbTx-3; 17: 2,3,41,43 tetrahydro PbTx-3; 20: 1,5 dihydroxy PbTx-3; 22: 2,3,27,28,421,43 hexahydro PbTx-3.

poses in initial C-42 p-azido benzoate photoaffinity experiments, is most useful for isolating toxin-conjugated peptides. Other photoaffinity probes could utilize A-ring conjugates, and so antiserum #2 will provide the most useful affinity columns. This essentially is another specific receptor-binding protocol and can provide useful tools for isolation and quantification of toxin in matrices and solution.

E. Synthetic Organic Chemistry Provides Brevetoxin Derivatives with Widely Varying Site 5 Affinities

The marine neurotoxins PbTx-1, PbTx-2, and ciguatoxin bind to the same site (site 5) on the voltage-gated sodium channel (14,16,17). This work describes efforts to identify the common pharmacophore and to develop a ligand-receptor model for the binding of these neurotoxins to site 5. Conformational analysis of PbTx-1 has been completed using an internal coordinate Monte Carlo search protocol. Within 6 kcal/mol of the global minimum (in vacuo), there are 48 conformations of PbTx-1 (14). In chloroform or water solvent, the calculated relative energies change, but no new minima appear. Like PbTx-2, PbTx-1 has both straight and bent conformers available. Elimination of several G-ring crown conformers from consideration and comparison of the two brevetoxin backbones indicates that those that match most closely in overall shape and location of functional groups are straight. We postulate that the common pharmacophore is a roughly cigar-shaped molecule (~30 Å long) bound to its receptor primarily by hydrophobic and nonpolar solvation forces, possibly aided by strategically placed hydrogen bonds near the site of the lactone carbonyl in the receptor (18).

Naturally occurring brevetoxins and synthetically modified brevetoxins were examined for their affinity for site 5 of the voltage-gated sodium channel and their toxicity for mosquitofish, *Gambusia affinis*. All but three of the toxins studied still retained some affinity for their receptor site (K_i in the range of 1–100 nM; see Table 1). Compound 22, having all three carbon-carbon double bonds reduced, is almost three orders of magnitude less strongly bound than compound 17, which has only two carbon-carbon bonds reduced. This large effect resulting from H-ring reduction was unexpected owing to the similarity of this region of the molecule to the corresponding region of PbTx-1, which has a fully saturated eight-membered G-ring and is the most strongly bound toxin of those studied. Conformational analysis revealed that the unsaturated H-ring of PbTx-2 favors the boat-chair conformation as does the saturated G-ring of PbTx-1. On reduction, the H-ring of PbTx-2 shifts to a crown conformation. This subtle change in conformational preference induces a significant change in the gross shape of the molecule, which we believe is responsible for the loss of binding affinity and toxicity. Thus, the subtle change in conformational preference induces a significant change in the gross shape of the molecule. Epoxidation of the H-ring instead of reduction does not lead to a change in conformational preference, and ring H retains the boat-chair conformation preferred by native toxin. The functional effect of such modifications is summarized below.

Brevetoxin PbTx-3 and compound 20 were investigated for their abilities to bind to the specific brevetoxin receptor site on rat brain synaptosomes and to modulate the normal function of voltage-gated sodium channels as determined by patch-clamping of cultured neurons. Compound 20 is produced from PbTx-3 by opening of the A-ring lactone to the unsaturated diol using sodium borohydride in ethanol. Natural PbTx-3 exhibited tighter binding to rat brain synaptosomes by at least three orders of magnitude as determined by competitive radioligand-binding experiments, and it was also more effective at activating voltage-gated channels. Patch-clamp studies revealed the three orders of magnitude greater potency of PbTx-3 over compound 20, although each produced delayed sodium channel opening and a pronounced delay in inactivation. Conformational modeling of the PbTx-2 backbone indicates that the two molecules are identical except for the region of the A-ring lactone. Thus, it was concluded that the brevetoxin PbTx-3 backbone requires electrophilic functionality in the region of the lactone in PbTx-3, and that opening of the ring in compound 20 is sufficient to modify binding affinities. Using PbTx-3 at a baseline binding constant of 1.3 nM, it is clear from Table 1 that addition of aromatic (hydrophobic) functions to the side chain increases binding affinity as evidenced by derivatives 1–4. Similarly, by changing the character of the side chain substituent to more hydrophilic derivatives like the acid 12 reduces affinity some 10-fold. Binding affinity can be restored by esterifying the acid as in compound 8. Detailed comparison of the PbTx-1 and PbTx-2 backbones reveal a number of similarities in gross shape. Both toxins have straight and bent conformations available to them, and both are approximately the same length from end to end (~30 Å). Most noteworthy is the fact that the G-ring of PbTx-1 and the H-ring of PbTx-2 both exhibit a preference for the boat-chair conformation, which is apparently necessary for binding. Because the G-H-I-J–rings of PbTx-1 and the H-I-J-K–rings of PbTx-2 (side chains included) have the same carbon skeleton, differing only in a double bond and the location of two angular methyl groups, and because the side chains of both toxin skeletons are structurally similar, we assume that these portions of the toxins occupy the same region of the binding site. The only other regions of the toxins which are similar are the A-ring lactones. These two regions, at opposite ends of the molecule, appear to coincide when all binding geometries are superimposed.

Since the two backbones bind competitively to the same receptor, it is presumed that the heads and tails of the two types occupy the same regions of the receptor when they are bound. Since the B-ring flexibility of PbTx-1 has virtually no effect on the spatial position of the A-ring carbonyl oxygen, and since several PbTx-1 and PbTx-2 conformers, when superimposed in the tail region, have the A-ring carbonyl oxygens of the two backbones in close proximity, the overall length is important for the observed effects and thus the spacing between the H-I-J-K ring and the A-ring may be critical for receptor recognition. As a test of this hypothesis, a synthetic toxin prepared by K.C. Nicoloau which possessed the PbTx-2 F-G-H-I-J-K–ring structure (side chain included) fused to the A-ring (in the absence of the B-E–rings) was modeled. In this synthetic derivative, all of the presumed ''business'' pieces are present, but the molecule is approximately 10 Å in length. In a similar fashion, compound 22 possesses all of the right parts and the correct length but is bent (18). Thus, we conclude that the A-ring region, the overall length, and the character of the side chain have specific characters that in concert modulate activities. To combine derivatives at each of the three regions (and conservatively at the C-37 alcohol) and trace progressive decreases in binding affinity based on sequential changes in molecular structure, see Table 1.

III. PHYSIOLOGICAL EFFECTS IN VIVO AND IN VITRO

A. In Vivo Physiological Effects of Brevetoxins

The in vivo physiological effects of *G. breve* toxins are both cardiovascular and neurological in nature. Brevetoxin induces central depression of respiratory and cardiac function. In anesthetized, artificially ventilated dogs and cats, a purified *G. breve* toxin (GBTX) produces dose-dependent acute apnea, bradycardia, and transient hypotension (19,20). At higher doses (0.1 mg/kg), GBTX produced cardiac arrhythmias and fibrillation. GBTX induces a vagal reflex (Bezold-Jarisch reflex; i.e., sinus bradycardia, hypotension, and peripheral vasodilation) on intravenous injection which is reversed by atropine, vagotomy, or ganglionic blocking drugs such as hexamethonium (19–21). These studies concluded that GBTX acts reflexly on vagally innervated receptors to evoke a Bezold-Jarisch effect but that the toxin further acts centrally to cause marked disturbances in respiratory rhythmicity and cardiovascular functions. Respiratory failure also appears to be the primary cause for lethality in guinea pigs receiving intravenous infusions of PbTx-3 (22).

Bronchoconstriction has also been shown in anesthetized, artificially ventilated guinea pigs to contribute to acute episodes of intoxication (23). Intravenous administration of T17 (PbTx-3) increased resistance to pulmonary inflation and supports the contention that death is due to asphyxiation. The bronchoconstrictor effect of brevetoxins indicates that they may be responsible for the respiratory discomfort and irritation during Florida red tides (24).

B. Effects of Brevetoxins on Nerve Conduction and Neuromuscular Transmission

The dose-dependent contractile response induced by *G. breve* toxin in isolated canine tracheal smooth muscle was inhibited by tetrodotoxin (TTX) and atropine, potentiated by neostigmine, reversed by isoproterenol but was not affected by the specific receptor antagonists chlorisondamine, phentolamine, and methysergide (25). These results indicate that airborne PbTx causes smooth muscle contraction by stimulating axonal Na channels resulting in the release of acetylcholine at postganglionic parasympathetic nerve endings. High concentrations of PbTx (up to 46 μg/mL) result in initial contraction of canine tracheal smooth muscle followed by a return to baseline tension. The repeated exposure to PbTx was ineffective even after washout of the high concentration of PbTx. However, the voltage-sensitive Na channel activators veratridine and scorpion toxin still induced contraction after administration of high concentrations of PbTx and were blocked by TTX (26). These results indicate that PbTx acts on the Na channel but at a different site from veratridine and scorpion toxin. Furthermore, given the cooperative interaction between PbTx and veratridine, which impairs inactivation (*h*-gate) of the Na channel, the PbTx receptor site(s) in parasympathetic nerve endings appears to influence Na^+ flux at the *h*-gate of the Na channel (10,26). Investigation of PbTx in isolated canine lower airway smooth muscle showed that PbTx elicited contractions at a threshold concentration of 0.15 μg/mL and a half-maximal response was obtained at 0.57 μg/mL PbTx (27). Atropine, a muscarinic receptor antagonist, blocked the response to both PbTx and acetylcholine (ACh), whereas TTX blocked the response to PbTx but did not alter the response to ACh, indicating that the action of PbTx is presynaptic rather than postsynaptic. PbTx-2 and PbTx-3 (0.01–0.07 μg/mL) were without effect in cultured airway smooth muscle cells, whereas ACh produced depolarizations which were blocked by atropine but not TTX (28). Taken together, these results confirm that PbTx-induced bronchoconstriction is due to the depolarizing effect of exogenous ACh, which is released from peripheral cholinergic nerve terminals on the airway smooth muscle. Brevetoxin-B, at concentrations of 12.5–187.0 nM, has been shown to cause positive ionotropic and arrhythmogenic (ventricular tachycardia and A-V blockade) effects on isolated rat and guinea pig preparations (29). The positive ionotropic effects of brevetoxin-B were inhibited by TTX, partially blocked by α-adrenoceptor blockade, and were not associated with inhibition of cardiac Na^+-K^+-ATPase activity. These results suggest that brevetoxin-B exerts positive ionotropic and arrhythmogenic effects by increasing cardiac sarcolemmal Na^+ permeability and by releasing catecholamines from sympathetic nerve endings.

Early in vitro investigations of the effects of brevetoxins on nerves and muscles were compromised by the purity of the toxin preparation. The studies, however, were of value, as in situ intoxications result from exposure to the entire spectrum of bioactive compounds produced by the organism. The principal action described for most toxic preparations was a depolarizing event and subsequent inexcitability of both nerve and muscle. Early studies of the effects of the brevetoxins were carried out using primarily crude fractions of the *G. breve* toxin and assayed by measurement of muscle tension development.

A major toxic fraction (fraction IVa) isolated from laboratory cultures of *G. breve* and purified by thin-layer chromatography caused inexcitability of the frog nerve–sartorius muscle preparation which was preceded by muscle fibrillations and spontaneous development of tension in the muscle (30). It was proposed, on the basis that the effect was blocked by curare, that the primary site of action was postsynaptic at the neuromuscular junction, although presynaptic effects could not be ruled out. Fraction IVa also contained some antiacetylcholinesterase activity but no hemolytic activity (30). Examination of a further purified preparation of *G. breve* toxin (T_1) indicated that bursts of spontaneous discharges occurred in the terminal region of the frog motor axon similar to that observed with perfusion of a low [Ca^{2+}] solution (31). The effects of T_1 could be overcome by increasing the external Ca^{2+} concentration tenfold over that normally present in Ringer's solution. Relatively high concentrations (>4 μg/mL) of the major toxin derived from *G. breve* was shown to partially inhibit Ca^{2+} uptake by sarcoplasmic reticulum vesicles prepared from rabbit skeletal muscle and to stimulate Ca^{2+} efflux by enhancing Ca^{2+}-Mg^{2+}–dependent ATPase activity (32).

Further studies of the effects of a crude fraction of *G. breve* toxin (GBTX) were carried out on the rat phrenic nerve diaphragm preparation using intracellular microelectrode recording techniques (33,34). GBTX depolarized the muscle fiber membrane, increased miniature endplate potential (MEPP) frequency, and blocked endplate potential (EPP) generation in a concentration-dependent manner. Indirectly stimulated muscle contractions were more sensitive to blockade than direct muscle contractions, suggesting that depolarization of the presynaptic nerve terminal by GBTX may be sufficient to inhibit neurally evoked transmitter release (i.e., EPPs). Lowering the external (Na^+) or application of TTX antagonized the GBTX-induced membrane depolarization and increased in MEPP frequency, suggesting that GBTX acts by increasing Na^+ permeability of excitable membranes (34). Although MEPP amplitude was increased slightly by GBTX, ACh-induced depolarizations were depressed more than that accounted for by depolarization of the muscle membrane, suggesting that GBTX may also act as a noncompetitive blocker of the muscle nicotinic acetylcholine receptor (34).

Purified T17 (PbTx-3) and T34 (PbTx-2) toxins, which possess the brevetoxin-B backbone, both caused a complete block of neuromuscular transmission in the rat phrenic nerve–stimulated hemidiaphragm (35). The initial contracture produced by each is prevented by TTX or curare and is enhanced by neostigmine and 4-aminopyridine. Neuromuscular block by T17 and T34 does not appear to be due to acetylcholine depletion, as electron micrographs showed no depletion of synaptic vesicles (35). Purified T17 toxin caused a large increase in spontaneous MEPP frequency when applied to the rat and frog neuromuscular junctions (36). The increase in MEPP frequency was reversed by the addition of 1 μM TTX and was not observed in a solution containing elevated Mg^{2+} and reduced Ca^{2+} concentrations, which is consistent with the idea that T17 toxin increases the probability of the sodium channel being open at the nerve terminal. The purified toxin from *G. breve*, T46 (PbTx-3), stimulated differential release of amino acid transmitters from rat cortical synaptosomes and acetylcholine release from guinea pig ileum (37). The effect was blocked by TTX, which is consistent with the proposed depolarizing action for the brevetoxin from the contraction and electrophysiological studies. Perineural waveform recordings from the motor nerve terminal of mouse skeletal muscle showed that 0.11 μM brevetoxin-B increased the components of waveforms associated with Na^+ and K^+ currents and decreased the Ca^{2+}-activated K^+ current, whereas 1.11 μM brevetoxin-B decreased all of the components of waveforms associated with Na^+, K^+, and Ca^{2+} currents (38). These effects may underlie the increase in MEPP frequency and contribute to the actions of brevetoxin on neuromuscular transmission (33). A more recent study of the actions of brevetoxin in vitro in the rat phrenic nerve–diaphragm preparation demonstrated that PbTx-3 in concentrations <60

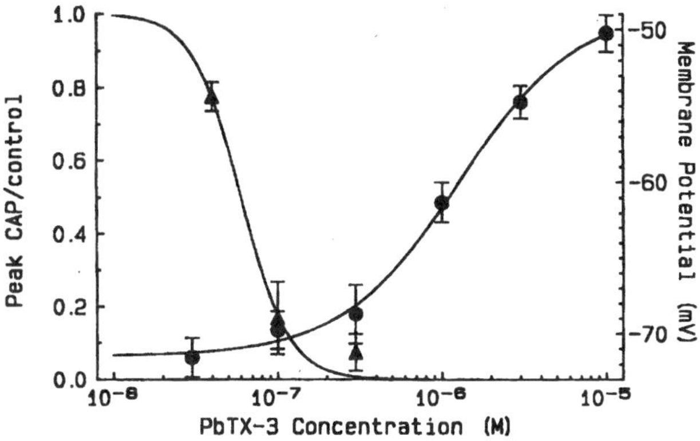

Figure 5 Effect of PbTx-3 on the compound action potential in the rat phrenic nerve and on the resting membrane potential in the diaphragm muscle. The compound action potential (CAP) value (triangles) for each concentration is expressed as a ratio of peak amplitude in toxin over control amplitudes. The IC50 for block of CAP was 0.06 μM. Note muscle membrane depolarization (circles) at PbTx-3 concentrations >0.3 μM ($P < .0001$) with respect to control values for 1.0, 3.0, and 10 μM PbTx-3. The values for peak CAP and membrane potential were obtained when maximal responses for a given concentration of PbTx-3 were achieved. Bars indicate SE of individual values recorded from five to seven muscles for each concentration (39).

nM produced a selective block of phrenic nerve compound action potentials (Figure 5). Higher concentrations (≥100 nM) were required to inhibit endplate potentials in an all or none fashion, increase MEPP frequency, and depolarize the muscle membrane (39). These results suggest that the primary cause of PbTx-3–induced failure of neuromuscular transmission in the diaphragm muscle is conduction block in the phrenic nerve.

Extracellular recordings from the CA1 region of guinea pig hippocampal slices showed that PbTx-3 produced a concentration-dependent (IC50 = 37.5 nM) depression of orthodromically evoked population spikes (40). These results suggest that depolarization of the presynaptic nerve terminals leading to failure of transmitter release could underlie the effects of brevetoxin on neurotransmission in the central nervous system.

Electrophysiological studies in squid axons showed that application of low concentrations (25–100 ng/mL) partially purified brevetoxin (GBTX) depolarized the axon and elicited a spontaneous repetitive train of action potentials which were inhibited by 100 nM TTX (41). The spikes appeared normal in shape, but the rate of recovery from the after hyperpolarization was increased. Under voltage-clamp conditions, GBTX had no discernable effect on depolarization-activated Na^+ or K^+ conductance changes, but the recovery of tail currents was accelerated. It was concluded that the most likely mechanism by which GBTX accelerates recovery after an action potential leading to repetitive firing is the induction of a small additional inward current which is blocked by TTX and presumably flows through Na channels (41). Further electrophysiological studies using the spontaneously active ventral nerve cord preparation of the crayfish showed that two (T_2 and T_4) of five toxin fractions (T_{1-5}) isolated from a crude extract of *G. breve* caused excitation to >250% of control levels at 0.05 ng/mL and blockage of firing occurred at 100-fold higher concentrations (42). Both the T_2 and T_4 fractions at concentrations of 250 ng/mL depolarized and increased the subthreshold synaptic activity and firing rate of spontaneously firing neurons of *Aplysia* abdominal ganglia (42).

Voltage-clamp of the soma membrane of *Aplysia* abdominal ganglion neurons showed that fractions of *G. breve* shifted the Na^+ conductance-voltage relationship approximately 10 mV in the hyperpolarizing direction consistent with the increased excitability observed in the presence of the toxin (43). T17 toxin (PbTx-3) also caused a concentration-dependent depolarization and transient repetitive discharges followed by depression of the action potential leading to block of excitability in crayfish giant axons (44). The T17-induced depolarization was reversed by TTX (300 nM), procaine (10 mM), or lowering the external Na^+ concentration to 1 mM and potentiated by the sea anemone toxin, anthopleurin-A. Voltage-clamp experiments in internally perfused squid axons demonstrated that T17 toxin shifted the activation voltage for the Na^+ but not K^+ current in the hyperpolarizing direction by 35 mV and greatly depressed the fast inactivation of the Na^+ current (Figure 6) (44). The results of this comprehensive study indicate that (1) T17 toxin depolarizes the membrane by selectively opening Na^+ channels at negative membrane potentials, (2) inhibits fast Na^+ channel inactivation, and (3) T17 toxin binds to a site different from TTX and sea anemone toxin but to an interior part of the Na^+ channel to which local anesthetics also bind (45).

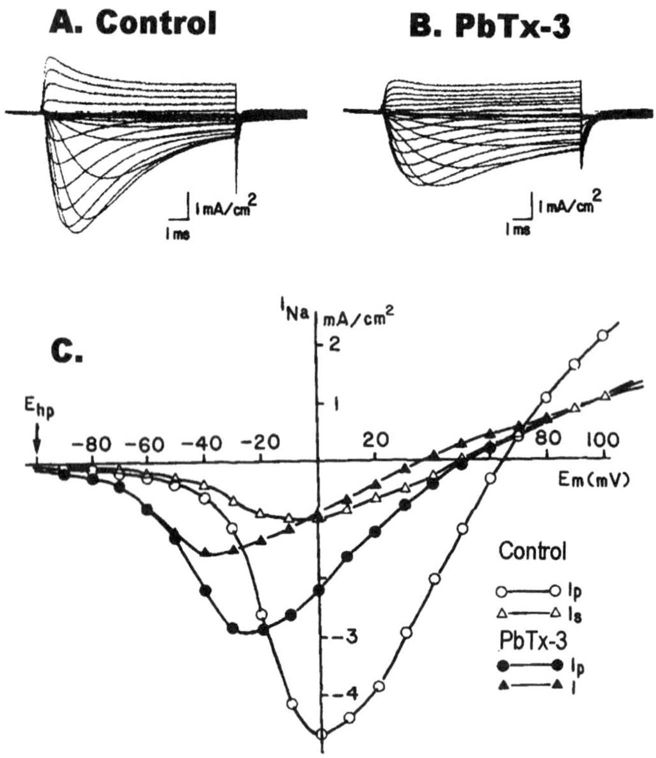

Figure 6 Families of sodium currents and the current-voltage curves before and after internal application of PbTx-3 (T17 in original ref) at 1 μg/mL at 6°C. K-free external and internal solutions were used throughout the experiment. (A) Control, depolarization ranged from −90 to +100 mV in 10-mV increments from a holding potential of −100 mV. (B) Same pulse protocol 1 h after PbTx-3 treatment. (C) I–V curves before and after PbTx-3 toxin treatment. Open circles, control peak sodium currents; filled circles, peak sodium currents after PbTx-3 toxin treatment; open triangles, control steady-state currents at the end of 10-ms pulse; filled triangles, steady-state currents after PbTx-3 toxin treatment, E_{hp}, holding potential; E_m, membrane potential (44).

C. Brevetoxin Binding and Actions on Voltage-Sensitive Sodium Channels

Direct evidence of brevetoxin action on Na^+ fluxes has been demonstrated in neuroblastoma cells (10,46) and rat brain synaptosomes (4,37). T_{47} (PbTx-2), a purified fraction from T_4, caused a triphasic dose-dependent increase in $^{22}Na^+$ uptake in a human neuroblastoma cell line (47) and T_{46} (PbTx-3), a second active fraction purified from T_4, enhanced the persistent activation of voltage-sensitive Na^+ channels by veratridine (10). The availability of radioactive toxins, either tritiated or iodinated, made binding studies of the voltage-sensitive Na^+ channel possible. The specific neurotoxin ligands and receptor-binding sites associated with the voltage-sensitive Na^+ channel are summarized in Table 3. Tetrodotoxin, saxitoxin, and μ-conotoxins bind at toxin receptor site 1 and inhibit Na^+ conductance. Grayanotoxin, batrachotoxin, and the alkaloids, veratridine, and aconitine bind to toxin receptor site 2 and cause persistent activation of Na channels. Polypeptide α-scorpion toxins (*Leiurus quinquestriatus*) and sea anemone toxins bind at site 3, slow Na channel inactivation, and enhance persistent activation of Na channels by toxins acting at site 2. β-Scorpion toxins (*Centroides suffusus suffusus*, CsTx-II) bind at toxin receptor site 4 and alter Na channel activation causing repetitive firing. Radioactive toxin–binding experiments demonstrated that PbTx-1 did not displace toxins which bind specifically to sites 1–3 of the voltage-dependent Na channel (46). Furthermore, brevetoxin was shown to enhance CsTx-II binding to receptor site 4, indicating that brevetoxins do not bind at site 4 of the Na channel (46). Catterall and Gainer (46) suggested that the brevetoxins bind at a new site, site 5, located on a region of the Na channel involved in voltage-dependent gating.

Competition experiments carried out in rat brain synaptosomes using tritiated brevetoxin ([^{3}H]PbTx-3) and natural toxin probes (saxitoxin, batrachotoxin, aconitine, sea anemone toxin II, *L. quinquestriatus* venom, *C. sculpturatus* venom) specific for sites 1–4 of the voltage-dependent Na channel confirmed that PbTx-3 binds to a unique toxin receptor site which modulates Na^+ influx in excitable membranes (4). Brevetoxins were classified on the basis of radiolabeled toxin–binding studies as allosteric modulators of Na channel binding by natural neurotoxins. PbTx-2 did not interact in any manner with either tritiated saxitoxin binding at site

Table 3 Neurotoxin Receptor Binding Sites and Brevetoxin Allosteric Interactions Associated with VSSCs

Receptor site	Ligand	Physiological effect	Allosteric modulation by brevetoxin[a]
1	Tetrodotoxin, saxitoxin, μ-conotoxins	Inhibit ion conductance	0
2	Batrachotoxin, veratridine, aconotine, grayanotoxins	Persistent activation	+
3	α-Scorpion toxins, sea anemone toxin	Delay inactivation; enhance persistent activation	−
4	β-Scorpion toxins	Shift voltage dependence of activation	+
5	Brevetoxins, ciguatoxin	Shift voltage dependence of activation; inhibit inactivation	0
6[b]	δ-Conotoxins	Delay inactivation	?
7[b]	Pyrethroids	Shift voltage dependence of activation; delay inactivation	?

[a] (+) enhances ligand binding; (−) inhibits ligand binding; (0) no allosteric effect; (?) unknown.
[b] Actual site number is equivocal.

1 or *L. quinquestriatus* toxin at site 3 in rat brain synaptosomes, but it markedly enhanced binding at sites 2 and 4 by batrachotoxin α-benzoate and CsTx-II, respectively (48). In cultured bovine adrenal medullary cells, PbTx-3 in combination with either α- or β-scorpion venom showed only additive effects on ^{22}Na influx but augmented veratridine-induced ^{22}Na influx to a greater extent than PbTx-3, α-scorpion venom, or β-scorpion venom alone (49), which is consistent with its positive allosteric interaction with site 2 of the Na channel. Given that binding at both sites 2 and 4 is use dependent, that is, toxin binding favors the open state of the channel, suggests that brevetoxins stabilize open conformations of the Na channel. These results provide evidence that brevetoxins (PbTx-1 and PbTx-2) interact with a new receptor site, site 5, on Na channels and that there are conformational interactions among sites 2, 4, and 5 mediated through the Na channel structure. More recent studies reveal that PbTx-1 (100 nM) can negatively modulate the binding of α-toxin II from the venom of the scorpion *A. australis* Hector (AaH II) in rat brain (50). Furthermore, TTX has been shown to reverse PbTx allosteric inhibition of α-scorpion toxin binding to rat brain Na channels (51). These results indicate the presence of allosteric interactions between TTX and brevetoxin binding sites that appears to be detected at the region of the α-scorpion toxin receptor site.

IV. MOLECULAR MECHANISMS OF ACTION

A. Marine Toxins Exhibit Specificity for Essential Physiological Loci

Seafood intoxications result from the interaction of toxin ligands with specific receptors associated with enzymes, neurotransmitter receptors, or voltage-sensitive ion channels (see reviews in refs. 52–55). Based on the nanomolar to picomolar potencies for these toxicants, specific toxin ligand/target receptor interactions have been postulated (56). For some intoxication types, the receptor is well described already and the natural toxicant acts as an agonist or antagonist competitive with the normal endogenous ligand. For examples, amnesic shellfish poisoning is the result of domoic acid interaction with kainate receptors, scombroid poisoning is the result of histamine interaction with histamine receptors, and diarrheic shellfish poisoning's okadaic acid interacts in an (apparently) competitive manner for the I-2 binding site on the catalytic subunit of phosphorylase phosphatase 1 and 2a, resulting in phosphatase inhibition (57).

The remaining toxins, saxitoxin, tetrodotoxin, brevetoxins, and ciguatoxin, represent toxic natural ligands for which receptors have been identified on the voltage-sensitive sodium channel α-subunit (7–9,52,58). Site 1 binds either saxitoxin and tetrodotoxin with a 1 : 1 stoichiometry at nanomolar dissociation constants, and site 5 binds either brevetoxins or ciguatoxin in a mutually competitive manner. The two sites are specific, individual, and are not allosterically linked.

B. Common Features on Polyether Toxins

For many reasons, the brevetoxins we produce occupy a unique niche in the further development of models that describe receptor site 5. First, they are routinely available in purified form. Second, they interact in a specific manner with a specific site in 1 : 1 stoichiometry and high affinity. The site is allosterically linked to sites 2, 4, 6, and 7, making toxin interaction with the site a rather pivotal event in the overall topography of the channel. Third, a variety of natural and synthetic derivatives of widely differing potencies allow for the development of structure-function hypotheses. Fourth, the toxins may be derivatized in a variety of useful ways, including photoaffinity and radioactive forms (3,5,9).

The two brevetoxin backbones and ciguatoxin represent the only three examples of a unique class of human seafood intoxicants known as the trans-fused polyether ladder toxins

(6,16,17,59,60). Ciguatoxin is the most recently characterized of the group, the structures known representing those ciguatoxins isolated from Tahiti, Australia, and Hawaii. More recently, bioactive materials have been isolated from Caribbean fish which fit ciguatoxin's general toxicological characteristics but which are somewhat different in polarity, nuclear magnetic resonance (NMR) spectra, and pharmacology. Since 1981, there has been intense interest in the pharmacology, synthesis, and biosynthesis of these compounds. The three classes of toxins have a number of structural features in common, including trans/syn stereochemistry along the polyether ladder backbone, with oxygen atoms alternating between one side and the other, and with an ether oxygen constituting a one-atom bridge between each adjacent ring; each toxin has regions of rigid and flexible character; the two brevetoxin backbones have lactone on one end with an enone, alcohol, or ketone function on the other. The structure of ciguatoxin has been published (61) and indicates the presence of a vicinal allylic diol on one end and a spirocyclic acetal on the other. The spirocyclic acetal between rings A and B of ciguatoxin, and the availability of epimers about the A-B–ring juncture indicates that terminal ring A may be opened to a dihydroxyketone, hemiacetal, or enone, and reclosed. Thus, there would appear to be extant or latent electrophilic functional groups on one or both ends of each polyether toxin. Unpublished work by Murata indicates that removal of the vicinal hydroxyl in ciguatoxin reduces toxicity 10-fold, that saturation of the H-I-J–ring system destroys activity, and that modification of the tetrahedral A-B–ring relationship decreases toxicity 10-fold (K. Murata, personal communication).

C. Molecular Mechanisms for Sodium Channel Activation

Sodium channels isolated from rat brain consist of three separate and biochemically separable protein subunits, the α-subunit, and the β-1 and β-2 subunits, which comprise the channel in a 1:1:1 stoichiometry. The α-subunit is a glycoprotein of approximately 260 kD molecular weight and is a transmembrane protein that binds neurotoxins at specific positions on its surface (seven sites are currently identified). The two β-subunits are small molecular weight peptides (~30 kD each) and are integral membrane subunit glycoproteins. There are now numerous sodium channel mutants (56) and subtypes that have been described. Little work has been completed regarding the specificity of toxins for sodium channel subtypes, although some very interesting computer modeling work has been carried out for STX and TTX on voltage-gated sodium channels (62).

Numerous models for sodium channel α-subunit structure, based on Numa's primary sequence data (68), have appeared. All are based on four homologous domains (I–IV). Each domain contains six transmembrane helices denoted S1–S6. The S5 and S6 segments of each domain are highly apolar, the S1, S2, and S3 segments are relatively apolar with just a few charged side chains, and the remaining S4 segments within each domain have the distinctive feature that every third residue is highly positively charged (mostly arginines), and these arginines protrude quite prominently when modeled. There is considerable structural homology between the three types of brain sodium channel α-subunits (I, II, and III), the μ1-sodium channel α-subunit from adult skeletal muscle, and the h1 sodium channel α-subunit from heart and denervated muscle. Particular conservation in primary structure is evident in the S1–S6 helices, but internal trans-domain connecting loops are divergent (Figure 7). Also divergent in the h1 heart primary structure is the sequence surrounding the IS5-IS6 connecting segment, which binds α-scorpion toxin in the brain and skeletal channel (56). Primary structure information such as this will be critical for all modified toxin ligand work. Investigations who couple changes in toxin structure with differences present in sodium channel subtypes or modification methods like cysteine-scanning mutagenesis may provide some unique insights into the exact loci on the channel and toxin that are required for binding and activity. For although certain parts of the

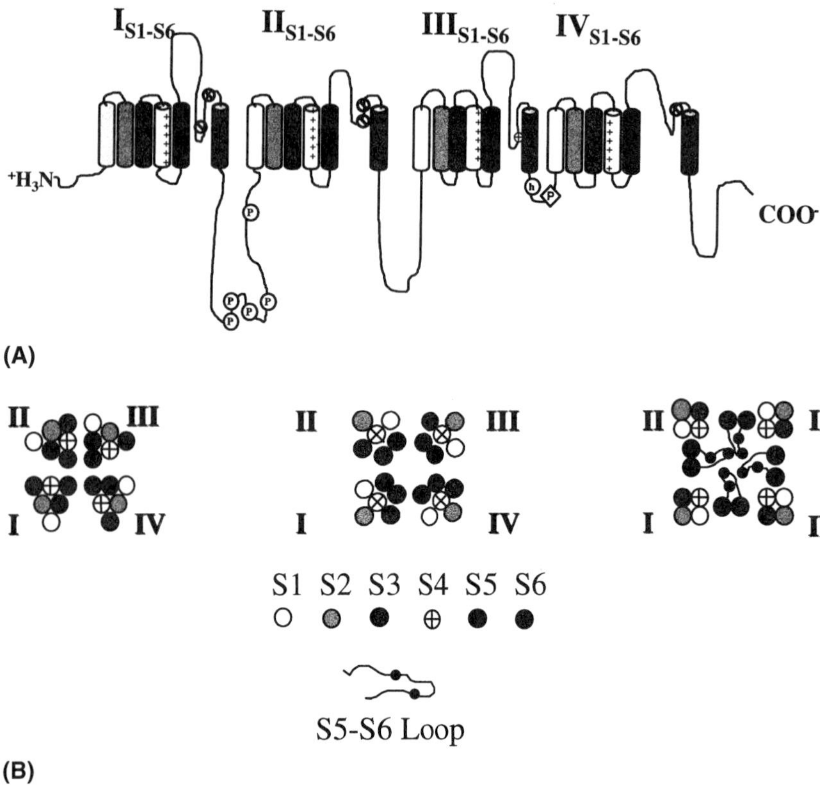

Figure 7 (A) Functional map of the voltage-sensitive sodium channel α-subunit. The VSSC is composed of three separate proteins, the α-subunit and two β-subunits. The α-subunit is a single polypeptide with fourfold internal homology-denoted domains I–IV, with each homologous domain possessing six transmembrane helices labeled S1–S6. S1–S3 are relatively neutrally charged with a few dicarboxylic amino acid residues per "S," S4 is highly "+" charged with every third amino acid usually being an arginine, and S5 and S6 are relatively hydrophobic. The S4 subunit (+++) in each domain is thought to spiral outward on depolarization by disruption of charge-charge pairing with "−" charges in S1–S3, the net result being an opening of the channel by allosteric modulation. Inactivation is accomplished by a closing of the channel pore from the inside by the IFM particle between IIIS6 and IVS1 (associated with three +++ intracellularly (h in circle) and a protein kinase C phosphorylation site by P in the box. Studies of TTX binding have been instrumental in determining the spatial arrangement between the S5–S6 intrahelix regions of domains I–IV ("−" signs in ○, site 1), the α-scorpion toxin–binding site between domains I and IV (likely the extracellular peptide between IS5 and IS6, site 3). (B) Three models of the outside views looking inside of sodium channel α-subunit helices. Several schematics of how six-transmembrane α-helices within each of the four domains might be spatially organized to form an ion-conducting pore have been postulated (56,66).

brevetoxin molecule are known to be required for specific activities, the sodium channel locus of interaction can only be speculated. This will be one of the most exciting parts of the brevetoxin story in the coming years.

Hodgkin and Huxley (63–65) first proposed that movement of the equivalence of six protein-bound charges from the inner surface of the bilayer membrane to the outer surface during activation must occur. In biochemical terms, this required that charged amino acid residues or strongly oriented dipoles within the membrane electric field of the phospholipid layer be present and able to travel in an enzymatic manner (i.e., move and recover catalytically). The S4 segments on domains I–IV of the VSSC α-subunit fulfill that role. These S4 α-helices contain a repeating primary amino acid sequence of an X-hydrophobic-hydrophobic sequence in which X = arginine or lysine (usually arginine). As reviewed by both Catterall (56) and Patlak (66), the S4 conservation across phylogeny and locale is most striking. Several investigators address aspects of the S4 helices as being regions which physically carry the charge, invoking a ''sliding helix'' model or a ''propagating helix'' model to explain how charges move across the membrane. Regardless of the model, S4 segments are proposed to be held under normal resting potential in their closed (nonconducting) configuration by a stabilizing charge complex, with each arginine residue in the helix contributing to a spiral ribbon of positive charges in pairing with negative charges carried by S1, S2, and S3 helices within each domain. The charge stabilization complex is thought to be held in place by the large relative negative potential inside on the order of −80 mV, which tends to drag the positive charges toward it. On depolarization, when the large negative potential is relieved, the S4 helices are released along a spiral sliding path (maintaining the preponderance of charge pairs but reestablishing the identity of the sets). The sliding motion is calculated to move a full positive charge across the membrane with only a 5-Å outward movement through a 60° rotation. If one includes all four domains in the calculation, a net four positive charges can be transferred.

Evidence from mutagenesis work, natural aberrant sodium channel missense mutations, and antisodium channel antibody inhibition studies indicate that it is this charge complexation in the S4 sliding helix and its ion pairs which are important for maintaining the helix in a ''cocked'' position. Domains are thought to spring open in a requisite order, the modulation of three of the four domains requiring little or no energy and sequentially resulting in three separate and electrophysiologically distinguishable closed, or C, states which differ from the closed state observed during hyperpolarization. Allosteric modification of the fourth domain, S4, causes an increase in the hydrodynamic diameter of the channel some 60 $Å^3$, and a rapid increase in Na^+ conductance followed by inactivation (66,67).

Similarly, the process of inactivation requires specific ion pairing, and indicates that the intracellular loop between domains III and IV is required for expression of normal inactivation following depolarization. Loss of the IFM (isoleucine-phenylalanine-methionine) inactivation particle (the h with a circle around it in Figure 6A), modification of the phenylalanine residue in particular, or tryptic cleavage by internal perfusion results in lost or greatly diminished inactivation kinetics (56). Natural sodium channel deletion mutations indicate that a specific methionine residue on IVS6 may be important in forming part of the ''snap'' or ''clip'' which allows the inactivation particle to fulfill its role. Substitution of the methionine for a valine totally abolishes the inactivation. Like the charge pairing complex regions of S1, S2, or S3, it appears that the nucleophilic ability of the methionyl S is important for ''directing the inactivation loop'' into the membrane channel (56,68).

Regardless of the absolute specific aspects of the charge pairing in either of these cases, it is equally apparent to all investigators working in membrane sodium channel biochemistry that charge pair complexation, relaxation, reorientation, and allosteric modulation are essential to the full functionality of voltage-sensitive sodium channels. That specific site-directed mutagenesis, voltage-depolarization, antisodium channel antibody application, natural missense mutations, and artificial and natural ligands can modulate functional kinetics of the channel is reassuring, for it provides a multitude of confirmatory and novel additive ways in which channel structure and function can be studied. That the brevetoxins can prolong the open state, inhibit onset of the inactivation state, produce multiple subconductance states, and shift the activation potential to more negative values indicate their involvement within the functional ''essence'' of channel machinery.

D. Brevetoxin (PbTx-3) and Its Derivatives Modulate Single Sodium Channel Activity

The effects of brevetoxin-3 and its derivatives have been extensively studied on TTX-sensitive sodium channels in rat sensory neurons (14,18,69). The observed effects of PbTx-3 on voltage-dependent Na^+ channels in rat nodose ganglionic neurons can be summarized as follows: (1) a shift of activation curves by $\geq$10 mV to more negative membrane potentials, (2) an inhibition of Na^+ channel inactivation at all membrane potentials whereby single Na^+ channel activity is observed during maintained depolarization (Figure 8), (3) an approximately twofold increase of the mean apparent channel open time at all membrane potentials, and (4) the appearance of subconductance states with two distinct conductance levels being detected (10.7 and 21.2 pS). The above effects were observed over the concentration range of 30–500 nM PbTx-3. All effects of PbTx-3 observed in rat sensory neurons are consistent with those observed on Na^+ channels in rat cardiac myocytes (67) but differ in detail from those observed in neuroblastoma cells (71). The lack of effect of PbTx-3 on Na^+ channel inactivation observed in neuroblastoma cells may be due, in part, to the short-duration voltage pulses used to examine the kinetics of Na^+

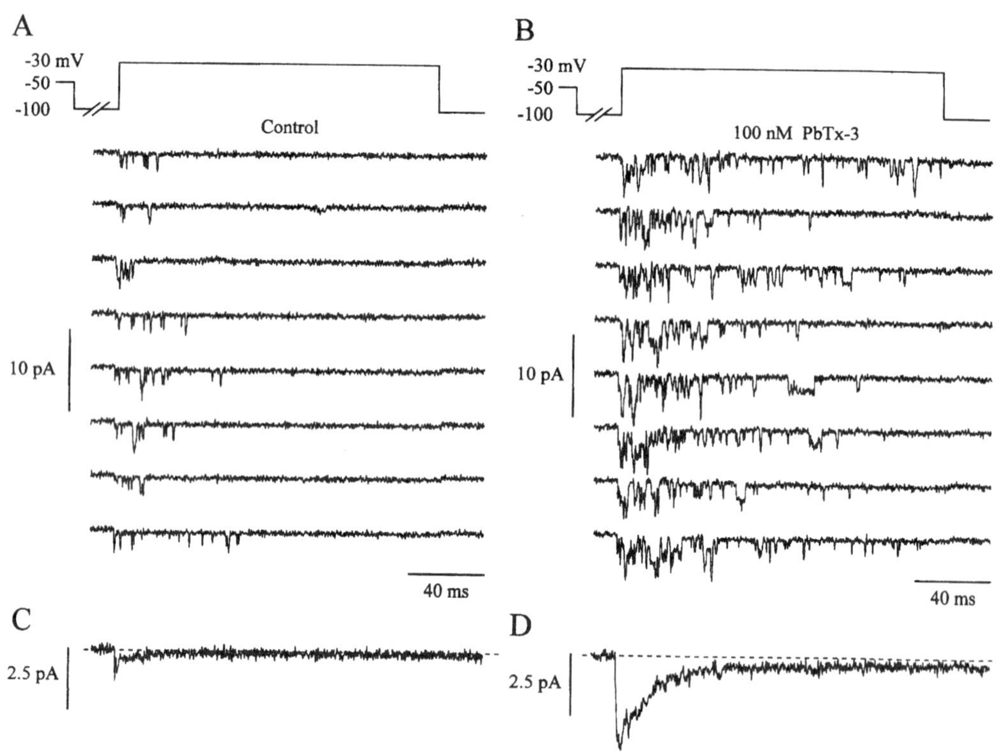

Figure 8 Depolarization-activated single sodium channel currents recorded in cell-attached membrane patches from rat sensory neurons in the absence (A) and presence (B) of 100 nM PbTx-3. (A,B) Typical consecutive traces (200-ms duration) of unitary Na^+ currents (downward deflections) obtained on depolarization to -30 from -100 mV (resting membrane potential $\cong -50$ mV) in the absence (control) and presence of 100 nM PbTx-3. The voltage jump protocol is shown above the current traces. (C,D) Ensemble average of summarized Na^+ inward current obtained in the absence (control; C) and presence of 100 nM PbTx-3 (D) by averaging 40 consecutive traces as shown in A and B, respectively. The dashed line indicates the zero-current level (69).

current inactivation in whole-cell and single-channel experiments. A more recent study of PbTx-3 on single Na channel currents in cell-attached patches from two neuronal cell lines derived from rat brain, B50 and B104, confirm the effects observed on Na channels in rat sensory neurons (72). The effects of PbTx-3 may initially cause hyperexcitability in intact brain, although chronic membrane depolarization could result in the generalized depressive effects of PbTx-3 observed in hippocampal slices (40).

All PbTx-3 derivatives (PbTx-6, 2,3,41,43-tetrahydro PbTx-3, 2,3,27,28,41,43-hexahydro PbTx-3, and 2,3-dihydro PbTx-3 A-ring diol, compounds 14 and 20, and reduction products therein) examined caused a similar shift in activation to more negative membrane potentials albeit at higher concentrations than found with PbTx-3. Similarly, inhibition of Na^+ channel inactivation was observed for all PbTx-3 derivatives but to a lesser extent than that observed with PbTx-3. The mean apparent open times were increased by 30–50% compared to control by all PbTx-3 derivatives except 1-desoxy PbTx-3, but to a lesser degree than by PbTx-3. The effect of PbTx-3 derivatives on mean channel open time appears to be related to its binding affinity for the Na^+ channel, whereby those derivatives with a low binding affinity exhibited a smaller effect (16,17). Removal of the A-ring carbonyl oxygen (1-desoxy-PbTx-3, compound 11) results in a sevenfold decrease in binding and loss of the ability to prolong the channel open time (Figure 9). Subconductance states of the Na^+ channel were also observed in the presence of PbTx-3 derivatives (18,69). In the presence of PbTx-3, only two conductance levels were observed, whereas with the reduction of the A-ring to a saturated diol (2,3-dihydro-PbTx-3 A-ring diol), a minimum of five conductance levels were observed (18).

Lipophilic polyether ladder toxins such as brevetoxins have recently been shown at micromolar concentrations to induce selective ion movements across lipid bilayers through transmembrane brevetoxin self-assemblies (73). Brevetoxin (PbTx-3) and ciguatoxin (CTX-1) applied in the absence of observed channel openings in the membrane patch failed to induce any ion channel activity, suggesting that PbTx-3 and CTX-1 molecules do not self-assemble to mediate selective ion movement through neuronal membranes (74) (S. Purkerson and D.G. Baden, unpublished observation).

The brevetoxin molecule has been shown to bind to receptor site 5 with the tail end of the molecule located near the S5–S6 extracellular loop of domain IV of the Na^+ channel α-subunit (8,9). The finding that all derivatives of brevetoxin examined to date shift the voltage dependence of activation to more negative potentials indicates that the oxygen-rich nature of the brevetoxin backbone interacts with the channel in a fashion that stabilizes the open configuration. One may speculate that the toxin may act as a "helix isostere," substituting some of its electronegative atoms for similarly arranged and spaced electronegative atoms on one of the "S" helices (56).

According to the model of the voltage-sensitive Na^+ channel by Sato and Matsumoto (74), all four extracellular loops formed by S5–S6 of all domains have to slide into the core pore formed by S2–S4 segments for full activation of the channel. In this case, the A-ring end of the PbTx-3 molecule may reach far enough into the membrane to interact with the charges of the inactivation gate, and likely leads to inhibition of Na^+ channel inactivation. In the presence of PbTx-3 and its derivatives, the open conformation is stabilized and the induction of subconductance states is observed in the modulated Na^+ channel. The induction of subconductance states may be due to the character of the A-ring and its proximity to the inactivation loop of the channel. This is consistent with results obtained with the truncated brevetoxin-B possessing all of the "essential" chemical elements except length (20.4 Å compared to ~30 Å) (75) and the five conductance states induced by the 2,3-dihydro PbTx-3 derivative (18). The multiple subconductance states may be due to the interaction of the "head" end of the toxin with the inactivation particle on the inside face of the channel associated with the loop between domains

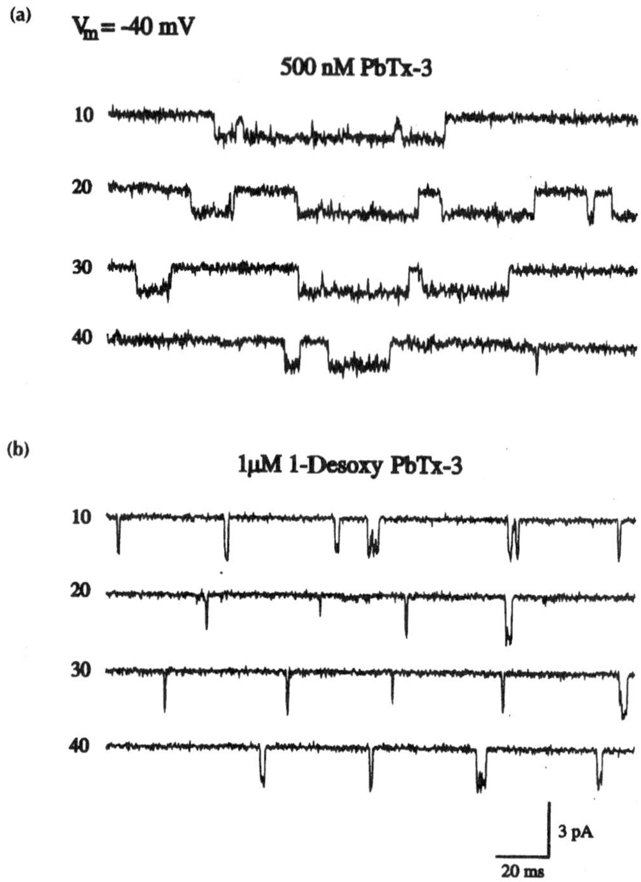

Figure 9 Steady-state recordings of single sodium channel currents in the presence of (a) 500 nM PbTx-3 or (b) 1 μM 1-desoxy PbTx-3. The membrane potential was held constant at −40 mV over a period of 1 min. Current traces (200-ms sweeps) obtained after 10, 20, 30, and 40 s are shown for each compound (18).

III and IV, resulting in partially open (or partially closed) channels by differential physical occlusion. The potential for this physical misalignment of the inactivation particle is partially supported by computer modeling which indicates a fixed angle and direction of orientation of the A-ring lactone in all closed ring toxins examined and a general free range of movement for the "arms" of any A-ring–cleaved toxin (16). However, the mechanism(s) by which brevetoxins and derivatives generate the subconductance states remains to be elucidated.

The induction of longer mean apparent open times is a characteristic of all natural brevetoxins and synthetic derivatives except the 1-desoxy PbTx-3 derivative, indicating that the carbonyl oxygen may be important for stabilizing channels in the open configuration (16). Given that voltage-dependent Na^+ channels have been proposed to activate and inactivate based on changing dipole pairs on adjacent α-helices of channel domains (56,66), the effect of 1-desoxy PbTx-3 toxin (which does not induce longer mean open times) is consistent with the importance of dipole interactions in brevetoxin-stabilizing channels in the open configuration.

The overall length of the brevetoxin molecule may be the controlling factor in inhibition of inactivation once Na^+ channels have been activated. Again, results obtained with the 1-desoxy PbTx-3 derivative indicate that an intact carbonyl moiety is not necessary for activity, whereas

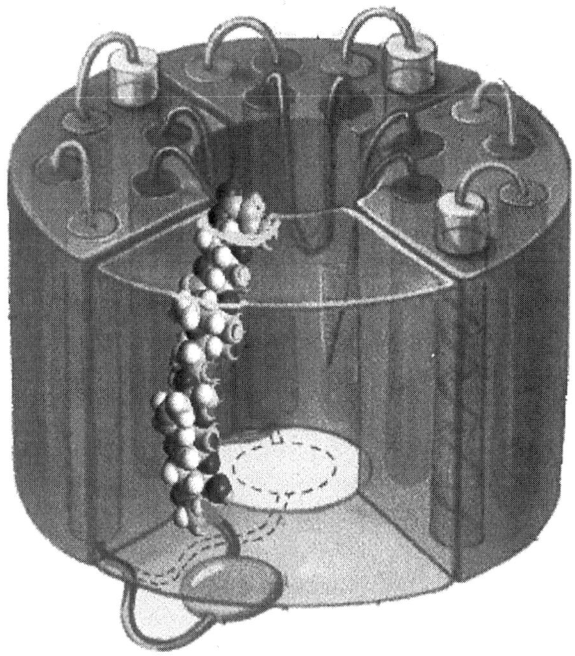

Figure 10 Hypothetical model of the binding of brevetoxin to site 5 of the voltage-sensitive sodium channel, with the tail of the model near I_{S5-S6} loop and the head near the inactivation IFM particle (18).

without the proper length (e.g., truncated brevetoxin), the presence of an intact A-ring is not sufficient to inhibit Na^+ channel inactivation (16). A model of how the brevetoxin molecule binds to the Na channel and affects conformational motion is shown in Figure 10. These results have identified some of the salient features associated with polyether toxin modification of voltage-sensitive Na^+ channel gating, and provides useful insights into the factors that modulate sodium channel function. Exactly how brevetoxins effect this modulation is unclear. Molecular modeling of brevetoxin reveals a three-dimensional topography approximately the size and shape of one of the 24 α-helices of the α-subunit of the channel. Observed end on, brevetoxins reveal exposed ring oxygen electrons in a manner similar to the ε-amino electrons exposed in the S4 helices of the four domains. It is possible that the brevetoxins act as helix mimics, intercalating into one of the domains between adjacent helices and perturbing normal ion pairing that holds the S4 in a ''trigger-ready'' conformation awaiting depolarization. The resulting perturbation reduces some of the potential energy of a closed channel, shifting the potential for activation to more negative values. Prolonging mean open time clearly resides in the carbonyl function in the A-ring, indicating a likely charge complex in this region.

V. METHODS OF DETECTION

All current detection methods for the brevetoxins have either a biological origin or an analytical chemical origin. The analytical techniques available, spectroscopy-coupled high-performance liquid chromatography (HPLC) and liquid chromatography (LC)–coupled mass spectrometry, are techniques that have not yet been completely described, and they require relatively expensive equipment. These techniques depend on specific chemical functional groups that can be detected by spectroscopy, mass determination, or functional derivatization. In general, their sensitivities are in the nanogram to microgram range. Methods of detection that have remained essential tools of experimentalists and public health officials alike all have their basis in receptor-ligand interactions. In part, the interactions are based on the pharmacological event that is essential for the onset of toxicity—specific binding of toxin to receptor. A second essential type of interaction has as its basis the principle of immune resistance through the interaction of toxin with antibody receptor. In both cases, specific tests can be developed that employ derivative ''toxin'' molecules as radioactive, fluorescent, or colorimetric markers to quantify unknown toxin in competitive displacement experiments. Direct measurement of unknowns is also possible based on knowledge of the specific receptor molecule and the availability of a derivative receptor to visualize the interaction.

Several different formats for utilization of specific antibodies against brevetoxin have been published. Radioimmunoassays (76) were developed first following synthesis of high specific activity tritium-labeled brevetoxin. Sodium channel receptor assays were developed by Poli et al. (4) soon thereafter. All biological receptor assays for the brevetoxins are based on these two reports.

The preceding discussions of SAR (structure-activity-relationships), physiological effects, and molecular mechanisms is in part to highlight the complexity of detecting a group of highly potent materials with sensitivity and specificity. For all chemical tests, it is necessary to identify a functional group that can be quantified with high precision. An example of such a functional group would be the aldehyde function of PbTx-1 and PbTx-2 (see Figure 1), which can be effectively measured using dichromate or permanganate or by Tollen's test. None, however, is specific. The ultraviolet and infrared spectra are not diagnostic of toxicity, and the NMR spectra are exceedingly complex. The mass spectral analysis of brevetoxins has some merit, but the instrumentation is prohibitive in cost, and this technique really best serves the academic research community in discovering and characterizing new toxins.

Using the receptor-based immunological tests, brevetoxin-specific serum can be induced in animals; antibodies that have specific affinities for the A-ring (12) the H-K rigid ring binding region of the toxins (12,13), and the side chain region (15). The antibodies have been used as the specific binding components for microtiter-plate assays (76,77). The antibody has been used for immunocytochemistry to confirm that brevetoxins were responsible for an epizootic disease of manatees (78). Immunofluorescent tests are forthcoming, and stable rapid techniques can be envisioned for the future. Immunosorbant techniques are especially encouraging. A comparison of the detection of antibodies and the sodium channel has been published (79).

The sodium channel receptor can be exploited and used as a receptor in simple and effective tests very much like antibody tests. The receptor assay works in membrane preparations (4,79), in rapid throughput assays in microtiter plates (80), and in solubilized form (81). Assays work on tissue extracts (80), in biological fluids, and as a surrogate assay system for the ciguatoxins (82,83). The complexities of reaching equilibrium in these assays, however, may compromise the speed with which they may yield reliable results (84). But it is clear that this system would be approachable by affinity techniques as well, and that derivatized brevetoxin in radioactive, colorimetric, and fluorescent forms would be equally useful as they are in immunological protocols. Since binding is the essential first step in the onset of toxicity, the results obtained with receptor assays would better reflect the composite toxicity of an unknown, because more potent forms of toxin bind more tightly to the receptor.

The binding affinity at site 5 must correlate directly with its potency. That is the issue that all assay development systems must verify—that the measurements taken correlate with the potential for human illness. Clearly there is much to learn.

VI. SUMMARY

The polyether toxins are becoming one of the more important types of natural marine toxins. Their identification from Florida, North Carolina, New Zealand, Japan, and the entire Gulf of Mexico indicates that their potential effects on humans will continue to increase. These toxins, being heat stable and not affected by cooking, their propensity for accumulation through food chain transvection, and their wide and varied effects by oral or inhalation exposure all tend to make their investigation more timely and compelling. Conversely, it is exactly the high potency and specificity that makes the brevetoxins important tools for neurotoxicology. Brevetoxins are being used as molecular measuring devices for quantifying the distances between specific structural components of the voltage-gated sodium channel. The toxins themselves in certain forms act as nontoxic binding antagonists, and so they may be developed as their own antidote. And finally, because of the many and varied effects of the brevetoxins, we may be able to use the toxins to sort out exactly how voltage-gated sodium channels function.

ACKNOWLEDGMENTS

This work was supported by the National Institute of Environmental Health Sciences (NIEHS) grant ES05853 (D.G.B.), NIEHS Center grant ES05705 (DGB), and Australian Research Council Grant 96/ARCL244 (D.J.A.).

REFERENCES

1. H Ishida, A Nozawa, K Totoribe, N Muramatsu, H Nukaya, K Tsuji, K Yamaguchi, T Yasumoto, H Kaspar, N Berkett, T Kosuge. Brevetoxin B1, a new polyether marine toxin from the New Zealand Shellfish *Austrovenus stuchburyi*. Tetrahedron Lett 36:725–728, 1995.
2. DG Baden. Marine food-borne dinoflagellate toxins. Int Rev Cytol 82:99–150, 1983.
3. KA Steidinger, DG Baden. Toxic marine dinoflagellates. In: DL Spector ed. Dinoflagellates. New York: Academic Press, 1984, pp 201–261.
4. MA Poli, TJ Mende, DG Baden. Brevetoxins, unique activators of voltage-sensitive sodium channels, bind to specific sites in rat brain synaptosomes. Mol Pharmacol 30:129–135, 1986.
5. Y Shimizu, HN Chou, H Bando, G VanDuyne, JC Clardy. Structure of brevetoxin-A (GB-1 toxin), the most potent toxin in the Florida red tide organism *Gymnodinium breve* (*Gymnodinium breve*). J Am Chem Soc 108:514–515, 1986.
6. Y-Y Lin, M Risk, SM Ray, D Van Engen, J Clardy, J Golik, JC James, K Nakanishi. Isolation and structure of brevetoxin B from "red tide" dinoflagellate *Gymnodinium breve* (*Ptychodiscus brevis*). J Am Chem Soc 103:6773–6776, 1981.
7. VL Trainer, E Moreau, D Guedin, DG Baden, WA Catterall. Neurotoxin binding and allosteric modulation at receptor sites 2 and 5 on purified and reconstituted rat brain sodium channels. J Biol Chem 268:17114–17119, 1993.
8. VL Trainer, DG Baden, WA Catterall. Identification of peptide components of the brevetoxin receptor site of rat brain sodium channels. J Biol Chem 269:19904–19909, 1994.
9. VL Trainer, WJ Thomsen, WA Catterall, DG Baden. Photoaffinity labeling of the brevetoxin receptor on sodium channels in rat brain synaptosomes. Mol Pharmacol 40:988–994, 1991.
10. WA Catterall, M Risk. Toxin T_{46} from *Gymnodinium breve* (formerly *Ptychodiscus brevis*) enhances activation of voltage-sensitive sodium channels by veratridine. Mol Pharmacol 19:345–348, 1981.
11. D Gordon, D Merrick, DA Wollner, WA Catterall. Biochemical properties of sodium channels in a wide variety of excitable tissues studied with site-directed antibodies. Biochemistry 27:7032–7038, 1988.

12. R Melinek, KS Rein, DR Schultz, and DG Baden. Brevetoxin PbTx-2 immunology: differential epitope recognition by antibodies from two goats. Toxicon 32:883–890, 1994.
13. MA Poli, KS Rein, DG Baden. Radioimmunoassay for PbTx-2 type brevetoxins: epitope specificity of two anti-PbTx sera. JAOAC 78:538–542, 1995.
14. DG Baden, KS Rein, RE Gawley, G Jeglitsch, DJ Adams. Is the A-ring lactone of brevetoxin PbTx-3 required for sodium channel orphan receptor binding and activity? Nat Toxins 2:212–221, 1994.
15. L Levine and Y Shimizu. Antibodies to brevetoxin B: serological differentiation of brevetoxin B and brevetoxin A. Toxicon 30:411–418, 1992.
16. KS Rein, DG Baden, RE Gawley. Conformational analysis of the sodium channel modulator, brevetoxin A, comparison with brevetoxin B conformations, and a hypothesis about the common pharmacophore of the "site 5" toxins. J Org Chem 59:2101–2106, 1994.
17. KS Rein, B Lynn, DG Baden, RE Gawley. Brevetoxin B: chemical modifications, synaptosome binding, toxicity, and an unexpected conformational effect. J Org Chem 59:2107–21013, 1994.
18. RE Gawley, KS Rein, G Jeglitsch, DJ Adams, EA Theodorakis, J Tiebes, KC Nicolaou, DG Baden. The relationship of brevetoxin 'length' and A-ring functionality to binding and activity in neuronal sodium channels. Chem Biol 2:533–541, 1995.
19. S Ellis, JJ Spikes, GL Johnson. Respiratory and cardiovascular effects of *G. breve* toxin in dogs. Proceedings of the 2nd International Conference on Toxic Dinoflagellate Blooms. New York: Elsevier North Holland, 1979, pp 431–434.
20. HL Borison, S Ellis, LE McCarthy. Central respiratory and circulatory effects of *Gymnodinium breve* toxin in anaesthetized cats. Br J Pharmacol 70:249–256, 1980.
21. HL Borison, LE McCarthy, S Ellis. Neurological analysis of respiratory, cardiovascular and neuromuscular effects of brevetoxins in cats. Toxicon 23:517–524, 1985.
22. DR Franz, RD LeClaire. Respiratory effects of brevetoxin and saxitoxin in awake guinea pigs. Toxicon 27:647–654, 1989.
23. DG Baden, TJ Mende, G Bikhazi, I Leung. Bronchoconstriction caused by Florida red tide toxins. Toxicon 20:929–932, 1982.
24. R Viviani. Eutrophication, marine biotoxins, human health. Sci Total Environ Suppl. 1:631–662, 1992.
25. S Asai, JJ Krzanowski, WH Anderson, DF Martin, JB Polson, RF Lockey, SC Bukantz, A Szentivanyi. Effects of the toxin of red tide, *Ptychodiscus brevis*, on canine tracheal smooth muscle: a possible new asthma-triggering mechanism. J Allergy Clin Immunol 69:418–428, 1982.
26. S Asai, JJ Krzanowski, RF Lockey, WH Anderson, DF Martin, JB Polson, SC Bukantz, A Szentivanyi. The site of action of *Ptychodiscus brevis* toxin within the parasympathetic axonal sodium channel *h* gate in airway smooth muscle. J Allergy Clin Immunol 73:824–828, 1984.
27. T Shimoda, J Krzanowski, R Lockey, DF Martin, M Perez-Cruet, J Polson, R Duncan. Lower airway smooth muscle contraction induced by *Ptychodiscus brevis* (red tide) toxin. J Allergy Clin Immunol 79:899–908, 1987.
28. IS Richards, AP Kulkarni, SM Brooks, R Pierce. Florida red-tide toxins (brevetoxins) produce depolarization of airway smooth muscle. Toxicon 28:1105–1111, 1990.
29. RL Rodgers, H-N Chou, K Temma, T Akera, Y Shimizu. Positive ionotropic and toxic effects of brevetoxin-B on rat and guinea pig heart. Toxicol Appl Pharmacol 76:296–305, 1984.
30. JJ Sasner, M Ikawa, F Thurberg, M Alam. Physiological and chemical studies on *Gymnodinium breve* Davis toxin. Toxicon 10:163–172, 1972.
31. BC Abbott, A Siger, M Spiegelstein. Toxins from the blooms of *Gymnodinium breve*. Proceedings of the 1st International Conference on Toxic Dinoflagellate Blooms. Boston: Massachusetts Institute of Technology, 1975, pp 355–365.
32. YS Kim, GM Padilla, DF Martin. Effect of *G. breve* toxin on calcium uptake and ATPase activity of sarcoplasmic reticulum vesicles. Toxicon 16:495–501, 1978.
33. JP Gallagher, P. Shinnick-Gallagher. Effect of *Gymnodinium breve* toxin in the rat phrenic nerve diaphragm preparation. Br J Pharmacol 69:367–372, 1980.
34. P Shinnick-Gallagher. Possible mechanisms of action of *Gymnodinium breve* toxin at the mammalian neuromuscular junction. Br J Pharmacol 69:373–378, 1980.
35. DG Baden, G Bikhazi, SJ Decker, FF Foldes, I Leung. Neuromuscular blocking action of two brevetoxins from the Florida red tide organism *Gymnodinium breve*. Toxicon 22:75–84, 1984.

36. WD Atchison, VS Luke, T Narahashi, SM Vogel. Nerve membrane sodium channels as the target site of brevetoxins at neuromuscular junctions. Br J Pharmacol 89:731–738, 1986.
37. MA Risk, JP Norris, J Coutinho-Netto, HF Bradford. Actions of *Gymnodinium breve* red tide toxin on metabolic and transmitter-releasing properties of synaptosomes. J Neurochem 39:1485–1488, 1982.
38. MC Tsai, ML Chen. Effects of brevetoxin-B on motor nerve terminals of mouse skeletal muscle. Br J Pharmacol 103:1126–1128, 1991.
39. SS Deshpande, M Adler, RE Sheridan. Differential actions of brevetoxin on phrenic nerve and diaphragm muscle in the rat. Toxicon 31:459–470, 1993.
40. JP Apland, M Adler, RE Sheridan. Brevetoxin depresses synaptic transmission in guinea pig hippocampal slices. Brain Res Bull 31:201–207, 1993.
41. M Westerfield, JW Moore, YS Kim, GM Padilla. How *Gymnodinium breve* red tide toxin(s) produces repetitive firing in squid axons. Am J Physiol 232:C23–C29, 1977.
42. JL Parmentier, T Narahashi, WA Wilson, NM Trieff, VMS Ramanujam, M Risk, SM Ray. Electrophysiological and biochemical characteristics of *Gymnodinium breve* toxins. Toxicon 16:235–244, 1978.
43. JJ Shoukimas, A Siger, BC Abbott. The action of *G breve* neurotoxin on membrane conductance. In: DL Taylor, HH Seliger, eds. Toxic Dinoflagellate Blooms. New York: Elsevier North Holland, 1979, pp 425–430.
44. JMC Huang, CH Wu, DG Baden. Depolarizing action of a red-tide dinoflagellate brevetoxin on axonal membranes. J Pharmacol Exp Ther 229:615–621, 1984.
45. B Hille. Local anesthetics: hydrophilic and hydrophobic pathways for the drug-receptor reaction. J Gen Physiol 69:497–515, 1977.
46. WA Catterall, M Gainer. Interaction of brevetoxin A with a new receptor site on the sodium channel. Toxicon 23:497–504, 1985.
47. M Risk, K Werrback-Perez, JR Perez-Polo, H Bunce, SM Ray, JL Parmentier. Mechanism of action of the major toxin from *Gymnodinium breve* Davis. In: DL Taylor, HH Seliger, eds. Toxic Dinoflagellate Blooms. New York: Elsevier North Holland, 1979, pp 367–372.
48. RG Sharkey, E Jover, F Couraud, DG Baden, WA Catterall. Allosteric modulation of neurotoxin binding to voltage-sensitive sodium channels by *Gymnodinium breve* toxin 2. Mol Pharmacol 31: 273–278, 1987.
49. A Wada, Y Uezono, M Arita, T Yuhi, H Kobayashi, N Yanagihara, F Izumi. Cooperative modulation of voltage-dependent sodium channels by brevetoxin and classical neurotoxins in cultured bovine adrenal medullary cells. J Pharmacol Exp Ther 263:1347–1351, 1992.
50. S Cestèle, RB Khalifa, M Pelhate, H Rochat, D Gordon. α-Scorpion toxins binding on rat brain and insect sodium channels reveal divergent allosteric modulations by brevetoxin and veratridine. J Biol Chem 270:15153–15161, 1995.
51. S Cestèle, F Sampieri, H Rochat, D Gordon. Tetrodotoxin reverses brevetoxin allosteric inhibition of scorpion α-toxin binding on rat brain sodium channels. J Biol Chem 271:18329–18332, 1996.
52. DG Baden. Brevetoxins: unique polyether dinoflagellate toxins. FASEB J 3:1807–1817, 1989.
53. G Strichartz, T Rando, GK Wang. An integrated view of the molecular toxinology of sodium channel gating in excitable cells. Ann Rev Neurosci 10:237–267, 1987.
54. CH Wu and T Narahashi. Mechanism of action of novel marine neurotoxins on ion channels. Annu Rev Pharmacol Toxicol 28:141–161, 1987.
55. D Gordon. Sodium channels as targets for neurotoxins. In: Y Gutman, P Lazarowici, eds. Toxins and Signal Transduction. Amsterdam, The Netherlands: Harwood, 1997, pp 119–149.
56. WA Catterall. Cellular and molecular biology of voltage-gated sodium channels. Physiol Rev 72: S15–S48, 1992.
57. DG Baden, LE Fleming, JA Bean. Marine toxins. In: FA de Wolff, ed. Handbook of Clinical Neurology. Vol 21: Intoxications of the Nervous System, Part II. New York: Elsevier, 1995, pp 141–175.
58. S Hall, G Strichartz, E Moczydlowski, A Ravindran, PB Reichardt. The saxitoxins: sources, chemistry, and pharmacology. ACS Symposium Series 418:29–65, 1990.
59. HN Chou, Y Shimizu. A new polyether toxin from *Gymnodinium breve* Davis. Tetrahedron Lett 23: 5521–5524, 1982.

60. RE Gawley, KS Rein, M Kinoshita, DG Baden. Binding of brevetoxins and ciguatoxin to the voltage-sensitive sodium channel and conformational analysis of brevetoxin B. Toxicon 30:780–785, 1992.
61. T Yasumoto, M Murata. Marine toxins. Chem Rev 93:1897–1909, 1993.
62. G Lipkind, H Fozzard. A structural model of the tetrodotoxin and saxitoxin binding site of the Na^+ channel. Biophys J 66:1–13, 1994.
63. AL Hodgkin and AF Huxley. Action potentials recorded from inside a nerve. Nature 144:710–711, 1939.
64. AL Hodgkin and AF Huxley. Currents carried by sodium and potassium ions through the membrane of the giant axon of *Loligo*. J Physiol (Lond.) 116:449–472, 1952.
65. AL Hodgkin and AF Huxley. The dual effect of membrane potential on sodium conductances in the giant axon of *Loligo*. J Physiol (Lond.) 116:497–506, 1952.
66. J Patlak. Molecular kinetics of voltage-dependent Na^+ channels. Physiol Rev 71:1047–1080, 1991.
67. N Yang, AL George, R Horn. Molecular basis of charge movement in voltage-gated sodium channels. Neuron 16:113–122, 1996.
68. M Noda, T Tanabe, T Takai, T Kayano, T Ikeda, H Takahashi, H Nakayama, Y Kanaoka, N Minamino, K Kangawa, H Matsuo, MA Rafferty, T Hirose, S Inayama, H Hayashida, T Mayata, S Numa. Primary structure of *Electrophorus eliectricus* sodium channel deduced from cDNA sequence. Nature 312:121–127, 1984.
69. G Jeglitsch, K Rein, DG Baden, DJ Adams. Brevetoxin-3 (PbTx-3) and its derivatives modulate single tetrodotoxin-sensitive sodium channels in rat sensory neurons. J Pharmacol Exp Ther 284: 516–525, 1998.
70. W Schreibmayer, G Jeglitsch. The sodium channel activator brevetoxin-3 uncovers a multiplicity of different open states of the cardiac sodium channel. Biochim Biophys Acta 1104:223–242, 1992.
71. RE Sheridan, M Adler. The actions of red tide toxin from *Gymnodinium breve* on single sodium channels in mammalian neuroblastoma cells. FEBS Lett 247:488–452, 1989.
72. SL Purkerson, DG Baden, LA Fieber. Brevetoxin modulates neuronal sodium channels in two cell lines derived from rat brain. Neurotoxicol 20:909–920, 1999.
73. S Matile, N Berova, K Nakanishi. Exciton coupled circular dichroic studies of self-assembled brevetoxin-porphyrin conjugates in lipid bilayers and polar solvents. Chem Biol 3:379–392, 1996.
74. RC Hogg, RJ Lewis, DJ Adams. Ciguatoxin (CTX-1) modulates single tetrodotoxin-sensitive sodium channels in rat parasympathetic neurones. Neurosci Lett 252:103–106, 1998.
75. KC Nicolaou, J Tiebes, EA Theodorakis, FPJT Rutjes, K Koide, M Sato, E Untersteller. Total synthesis of truncated brevetoxin B (AFGHIJK). J Am Chem Soc 116:9371–9372, 1994.
76. VL Trainer, DG Baden. An enzyme immunoassay for the detection of Florida red tide brevetoxins. Toxicon 29:1387–1394, 1991.
77. DG Baden, VL Trainer. Mode of action of toxins of seafood poisoning. In: IR Falconer, ed. Algal Toxins in Seafood and Drinking Water. San Diego: Academic Press, 1993, pp 49–74.
78. G Bossart, DG Baden, RY Ewing, B Roberts, and SD Wright. Brevetoxicosis in manatees from the 1996 epizootic: gross, histologic, and immunohistochemical features. Toxicol Pathol 26:276–282, 1998.
79. DG Baden, TJ Mende, AM Szmant, VL Trainer, RA Edwards, LE Roszell. Brevetoxin binding: molecular pharmacology versus immunoassay. Toxicon 26:97–103, 1988.
80. F Van Dolah, EL Finley, BL Haynes, GJ Doucette, PD Moeller, and JS Ramsdell. Development of rapid and sensitive high throughput assays for marine phycotoxins. Nat Toxins 2:189–196, 1994.
81. VL Trainer, DG Baden, WA Catterall. Detection of marine toxins using reconstituted sodium channel. JAOAC 78:570–573, 1995.
82. M-Y Dechraoui, J Naar, S Pauillac, A-M Legrand. Ciguatoxins and brevetoxins, neurotoxic polyether compounds active on sodium channels. Toxicon 37:125–143, 1999.
83. A Lombet, J-N Bidard, M Lazdunski. Ciguatoxins and brevetoxins share a common receptor site on the neuronal voltage-dependent Na^+ channel. FEBS Lett 219:355–359, 1987.
84. PL Whitney, DG Baden. Complex association and dissociation kinetics of brevetoxin binding to voltage-sensitive rat brain sodium channels. Nat Toxins 4:261–270, 1996.

24
Chemistry and Detection

Chee-Hong Tan and Ching-Ong Lau
National University of Singapore, Singapore

I. INTRODUCTION

Seafood is a good source of nutrients and constitutes a major part of the human diet, particularly for those living near the sea. However, many cases of human intoxication and fatalities have resulted following ingestion of seafood contaminated with marine toxins. The toxins are accumulated through food chains in the flesh and viscera of fishes, crabs, prawns, shellfish, and other marine invertebrates in concentrations high enough to cause deaths among consumers.

One of the most potent marine toxins connected with seafood poisoning is palytoxin. The elucidation of the complete structure of palytoxin and its mechanism of action proves to be most challenging in the field of toxin research. This chapter describes the chemical properties of palytoxin and its methods of detection.

II. SOURCE AND ORIGIN

Palytoxin was first discovered, isolated, and purified from zoanthids belonging to the genus *Palythoa*. The zoanthids are true corals classified under the family Zoanthidae, order Zoantharia, and phylum Coelenterata. Moore and Scheuer (1) described the purification of palytoxin from a marine coelenterate located in tidal pools around the islands of Hawaii and known to the Hawaiians as "limu-make-o-Hana" (the deadly seaweed of Hana). This toxic zoanthid was later identified as *Palythoa toxica* (2). Palytoxin was also found to be present in *P. tuberculosa* (3) located among the reefs in the tropical waters of the Pacific regions including Japan (4) and in *P. mammilosa* (5) and *P. caribaeorum* (6) from the islands of the West Indies such as Jamaica, Puerto Rico, and the Bahamas. Other toxic zoanthids include *P. vestitus* (7) and the Tahitian *Palythoa* sp. (8,9).

The first indication of the presence of palytoxin or a palytoxin-like substance in marine animals other than the zoanthids appeared in a report by Hashimoto et al. (10). They found a compound resembling palytoxin in the gut of the file fish *Alutera scripta* to be the causative agent for toxicity and death of pigs in Okinawa, Japan, after the animals had been previously fed with this fish. Fukui et al. (11) showed the presence of palytoxin or one of its congeners in the flesh and viscera of the triggerfish, *Melichthys vidua*. Another group of investigators reported that a palytoxin-like substance caused severe intoxication in consumers after they had eaten livers of the parrotfish, *Ypsiscarus oviforns* (12). Clinical and laboratory findings of

Figure 1 The coral reef crab *Lophozozymus pictor*.

ciguatera poisoning due to consumption of smoked mackerel (*Decapterus macrosoma*) had demonstrated palytoxin to be the causative agent (13). Palytoxin had also been found in the sea anemone *Radiantus macrodactylus* (14) and in the polychaete worm *Hermodice carunculata* (15).

Occasional outbreaks of palytoxin poisoning as a result of eating xanthid crabs have been reported in the Philippines (16) and in Singapore (17). Much earlier, in 1970, Teh and Gardiner (18) reported the presence of a toxin from the coral reef crab *Lophozozymus pictor* which possesses properties different from those of tetrodotoxin or saxitoxin (Figure 1). This toxin was subsequently purified and found to be a palytoxin congener (19–21). Yasumoto and coworkers also reported the isolation of palytoxin from two species of xanthid crabs (*L. pictor* and *D. alcalai*) from the Philippines (22). Subsequently there was a further report of a palytoxin-induced fatality due to ingestion of the hairy crab, *Demania reynaudii* (23).

The origin of palytoxin is still a matter of speculation. The poison appears to be produced within the body of the zoanthids rather than in their stinging nematocysts. Hirata et al. (4) reported that female polyps and mature eggs of *P. tuberculosa* exhibited higher toxicities when collected from the period May through September from Ishigaki Island in the southern part of Japan. A recent study of the origin and distribution of palytoxin in a Caribbean coral reef showed that several marine animals, including crustaceans and polychaete worms living in close association with *Palythoa* colonies, had high concentrations of palytoxin in their bodies (15). The fact that ingestion of these toxic coelenterates did not harm these animals showed that they have developed considerable resistance to the lethal effect of this toxin. There were speculations that marine microorganisms could be the source of palytoxin for these zoanthid filter feeders. Although marine bacteria have been suggested (24), experimental evidence in support of this is still lacking. Algae living symbiotically with zoanthids were considered possible candidates as secondary products similar to palytoxin could be synthesized by them (25). However, in a study on chlorophyll-*a* and palytoxin concentration in 32 *Palythoa caribaeorum* colonies which contain considerable amounts of symbiotic algae, there was no correlation between chlorophyll-*a* content (which measures the amount of algae in the zoanthid) and palytoxin concentration (15). Thus, it remains unclear if symbiotic algae really are the producers of palytoxin.

III. STRUCTURE AND PROPERTIES

Palytoxins are large organic substances. The basic palytoxin molecule consists of a long, partially unsaturated aliphatic backbone with spaced cyclic ethers, 64 chiral centers, and 40–42 hydroxyl and 2 amide groups. Molecular weights of palytoxins range from 2659 to 2680 depending on the species from which they are obtained. The toxin from the Tahitian *Palythoa* sp. has a molecular formula of $C_{129}H_{221}N_3O_{54}$ (M_r 2659), whereas that from *P. toxica* has two possible structures with a molecular formula of $C_{129}H_{223}N_3O_{54}$ (M_r 2677). The Hawaiian *P. tuberculosa* appears to have a mixture of palytoxin structures of the Tahitian *Palythoa* sp. and that of *P. toxica* (8). The toxin from *P. caribaeorum* has a molecular mass of 2680 (21). The gross structures of the palytoxins have been obtained through ordering of palytoxin fragments after sodium periodate oxidation and ozonolysis treatment followed by the use of nuclear magnetic resonance (NMR) and mass spectrometry (8). The absolute stereochemistry of palytoxin has been assigned to all 64 chiral centers through a combination of x-ray crystallography, NMR, and circular dichroic spectroscopy (8,26) (Figure 2). Moore (9) gave a detailed description of the various methods and steps used in palytoxin structure and stereochemistry determination. Chemical synthesis of palytoxin carboxylic acid (PTC) by Armstrong et al. (27) led to the successful synthesis of palytoxin from PTC without the use of protecting groups (28).

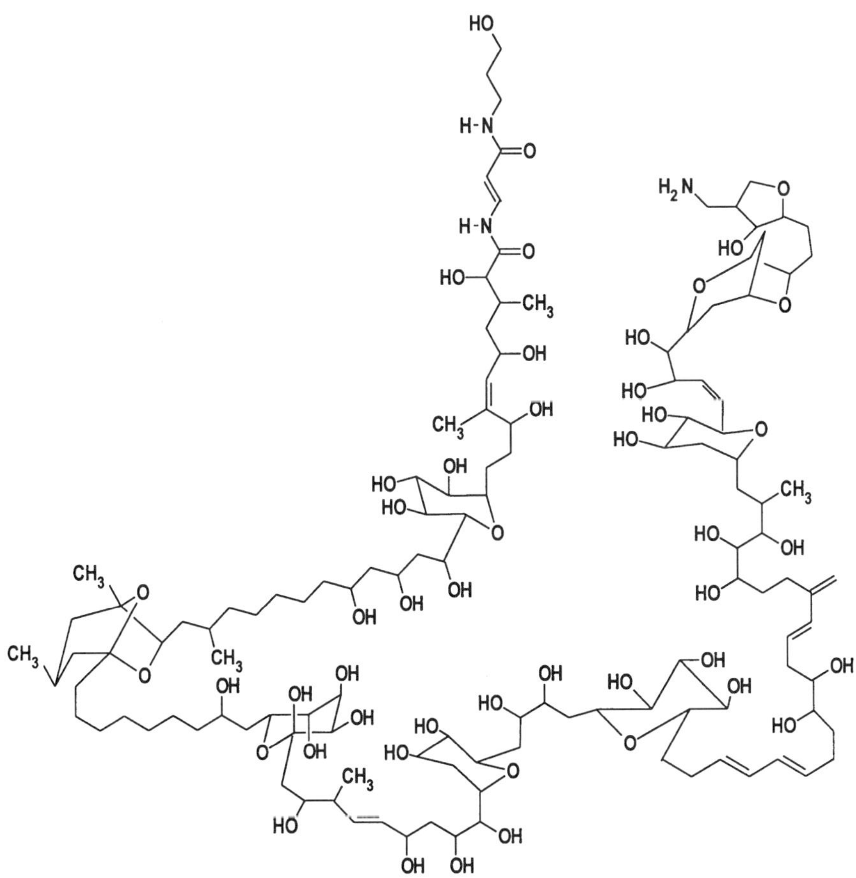

Figure 2 Structure of palytoxin. (From ref. 30.)

Moore and Scheuer (1) described their purified palytoxin as a white, amorphous hygroscopic solid. The toxin has not yet been crystallized. It is insoluble in nonpolar solvents such as chloroform, ether, and acetone but sparingly soluble in methanol and ethanol, and soluble in pyridine, dimethyl sulfoxide, and water. The toxin exhibits no definitive melting point and is resistant to heat but chars at 300°C. Because of its amphipathic nature, an aqueous solution of palytoxin foams on agitation somewhat like that of a steroidal saponin. It is optically active, having a specific rotation of +26°C ± 2° in water. The optical rotatory dispersion curve of palytoxin exhibits a positive Cotton effect with $[\alpha]_{250}$ being +700° and $[\alpha]_{215}$ being +600°. All palytoxins exhibit similar ultraviolet (UV) absorption spectra with λ_{max} at 233 (ε 47000) and 263 nm (ε 28000) (4). The first absorption maximum is due to three olefinic double bonds, whereas the second absorption peak is due to the trans-3-aminoacrylamide moiety. The ratio of their UV absorbances at 233 and 263 nm is 1.71. The λ_{263} chromophore is both acid and alkali labile and its destruction results in toxin inactivation (29,30). Although the UV data suggest similarity among palytoxins isolated from different species of *Palythoa*, subtle differences in proton magnetic resonance (PMR) and carbon magnetic resonance (CMR) spectra were observed among *P. toxica*, *P. mammilosa*, and the Tahitian *Palythoa* sp. (6,29).

IV. EXTRACTION AND PURIFICATION

Since palytoxin is quite soluble in water or water-miscible solvents, ethanol and methanol were commonly used to extract the toxin. The first report of the isolation and purification of palytoxin from the Hawaiian species of *Palythoa* employed 70% ethanol for extraction at room temperature, flash evaporation to remove the alcohol, defatting of the extract with benzene and 1-butanol, and desalting prior to further purification by ion-exchange chromatography (1). Subsequently, other investigators used aqueous methanol or ethanol and a variety of chromatographic and electrophoretic procedures to extract and purify palytoxin from zoanthids or from other marine animals (6,14,20,22). Lau and coworkers adapted the hot aqueous extraction procedure of Teh and Gardiner (18) in their attempt to isolate palytoxin-like substances from the crab *Lophozozymus pictor* (19). The hot aqueous extraction method was used because it had been documented that victims of crab poisoning ingested soup prepared from these crabs. Apparently, the toxin was rather resistant to heat under the boiling conditions employed. A subsequent modification using aqueous ethanol for extraction instead of heat resulted in higher yields and an increase in specific activities (20).

A combination of gel filtration on Sephadex G-50 and ion-exchange chromatography on QAE-Sephadex and SP-Sephadex or CM-cellulose had been used to purify palytoxin from *P. caribaeorum* (6). This purification procedure employed an initial precipitation of the toxin with acetone and a final gel-filtration step using Biogel P-6 (Figure 3). Other methods introduced for purification of palytoxin and palytoxin-like substances from marine crabs include the use of desalting columns and high-performance liquid chromatography (22) and two-dimensional thin-layer chromatography on silica gel–coated plates (20).

V. DETECTION AND ASSAYS

Sensitive and highly specific analytical methods are needed not only for the purpose of detecting, identifying, and distinguishing palytoxin or palytoxin-like substances from other marine toxins but also for monitoring the toxin during purification from its sources. Assay methods also allow

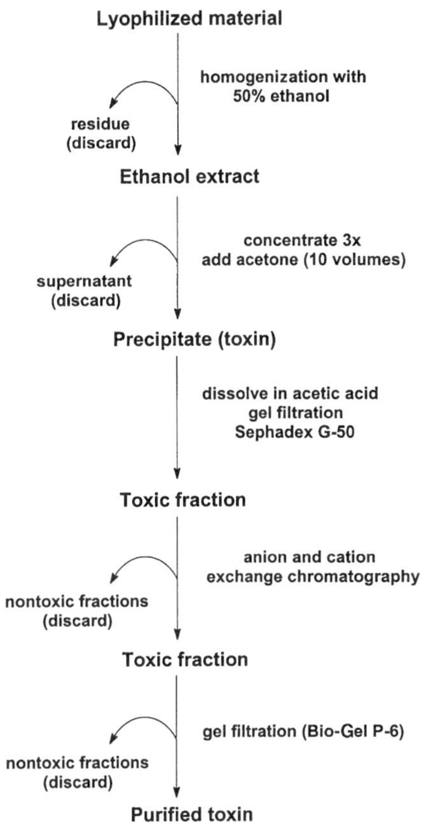

Figure 3 Simplified scheme for purification of palytoxin.

for toxin quantitation in purified samples or in crude biological extracts. Palytoxin can be detected and quantitatively measured by either chemical analytical means or by the use of biological assays. It is often necessary to use a combination of methods to confirm the presence of palytoxin.

A. Chemical Analytical Methods

Several chemical analytical methods available for the detection of palytoxin and/or palytoxin-like substances are based on chemical properties characteristic and intrinsic to the toxin. These methods include (1) infrared analysis, (2) ultraviolet absorption, (3) mass spectrometry, (4) high-performance capillary electrophoresis, (5) high-performance liquid chromatography, and (6) thin-layer chromatography.

Infrared Spectrometry

The infrared (IR) spectrum of purified palytoxin or its congener shows a characteristic band at 1670 cm^{-1} which is due to the possession of an α,β-unsaturated amide carbonyl group (1). This observation is supported by similar IR data obtained from purified palytoxin from *P. caribaeorum* (6) and is also seen in the toxin purified from *L. pictor* (Figure 4).

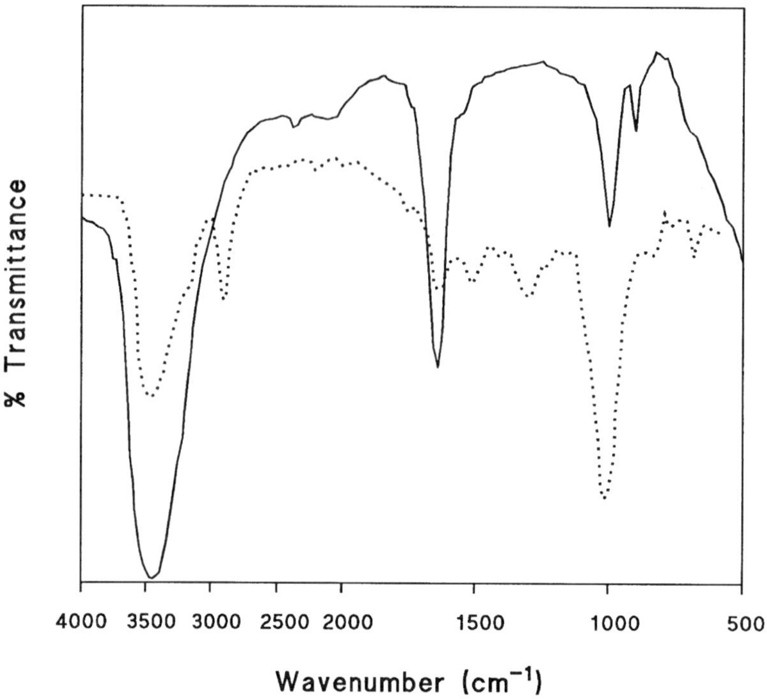

Figure 4 Infrared spectrum of *L. pictor* toxin (—) and *P. caribaeorum* (····).

Ultraviolet Spectrometry

Palytoxin's ultraviolet (UV) spectrum shows two absorption peaks contributed by the two chromophores at 233 and 263 nm, respectively (Figure 5). The ratio of their absorbance (233 vs 263 nm) is approximately 1.7. The 263-nm chromophore which contributes to the toxin's toxicity is sensitive to methanolic 0.05 M HCl or aqueous 0.05 M NaOH (1). This characteristic UV absorption profile of palytoxin is one parameter that could be used to verify the presence of the toxin.

Mass Spectrometry

Ion-spray mass spectrometry could be used to detect the presence of palytoxin as well as to analyze the structural differences in palytoxins. The ion-spray mass spectrometric profile of the Caribbean palytoxin (Figure 6) revealed a molecular mass of 2680.0 D (21). The analysis also showed that the toxin lost up to 13 molecules of water, suggesting that this toxin possessed a large number of hydroxyl and/or ether moieties. The Caribbean palytoxin gave a fragment ion at $m/z = 327.6$, which could arise from the cleavage between carbons eight and nine of the toxin molecule (31) and the additional loss of one molecule of water. Several palytoxins have been reported which differ at this end of the molecule (31). Mass spectrometry has also been useful in elucidating the difference in the chemical structure between *L. pictor* toxin (LPTX) and the Caribbean palytoxin. Using MS/MS spectrometry, the palytoxin-like LPTX was estimated to possess a molecular mass of 2681.0 D, but unlike the Caribbean palytoxin, it did not give a fragment at or near to m/z 327, indicating that the two toxins are probably structural isomers (21).

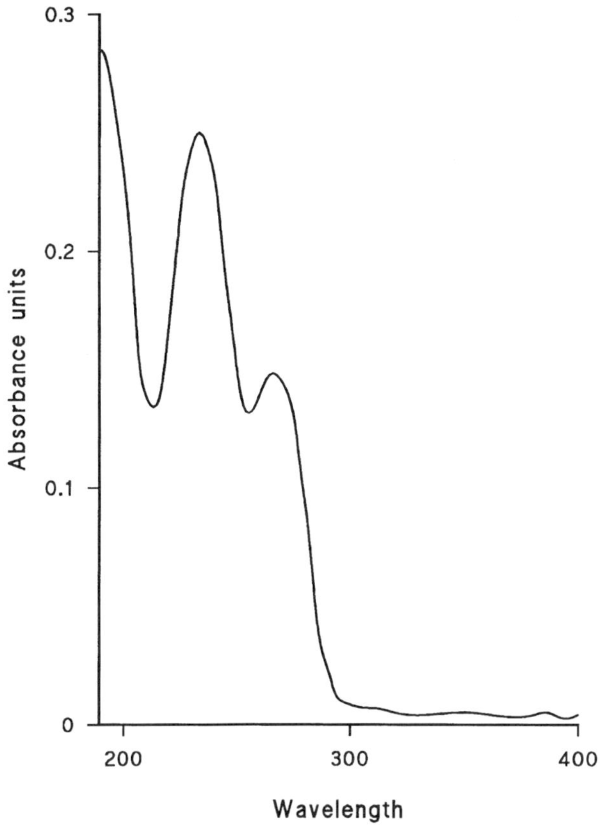

Figure 5 Ultraviolet spectrum of palytoxin from *P. caribaeorum*.

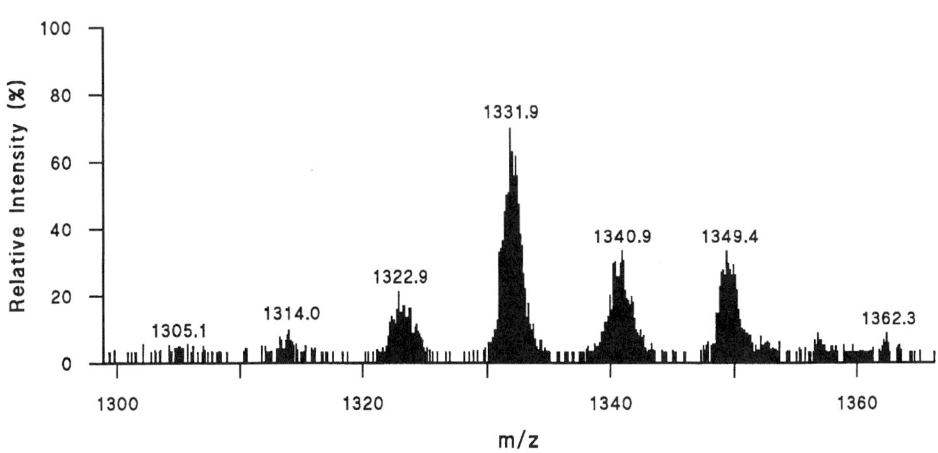

Figure 6 Ion-spray mass spectrometric profile of palytoxin from *P. caribaeorum*.

High-Performance Capillary Electrophoresis

High-performance capillary electrophoresis (HPCE) is a very sensitive tool for the identification and detection of palytoxin. In one procedure, capillary electrophoresis was performed with an untreated capillary column at 15 kV and at 25°C in 25 mM sodium borate buffer (pH 8.5), with UV detection at 230 nm, using a P/ACE System 2000 (Beckman Instruments, CA, USA). The detection limit was found to be 0.5 pg and the detection sensitivity was two-fold higher at 230 than at 263 nm (32). Similar results were obtained with the Caribbean palytoxin using the Biorad Biofocus 3000 HPCE System (Biorad Laboratories, CA, USA) (Figure 7). The ratio of the absorbances at 230 and 263 nm which was normally used for peak identification and verification of sample homogeneity (33) was calculated to be 1.6 (19).

High-Performance Liquid Chromatography

High-performance liquid chromatography (HPLC) is another analytical tool used in the detection and identification of palytoxin extracted from zoanthids and other sources. Palytoxin and/or palytoxin-like substances have been analyzed and identified by HPLC from extracts obtained from marine crabs (19,22). The crab toxin, which was chromatographed on an ERC-ODS (Erma

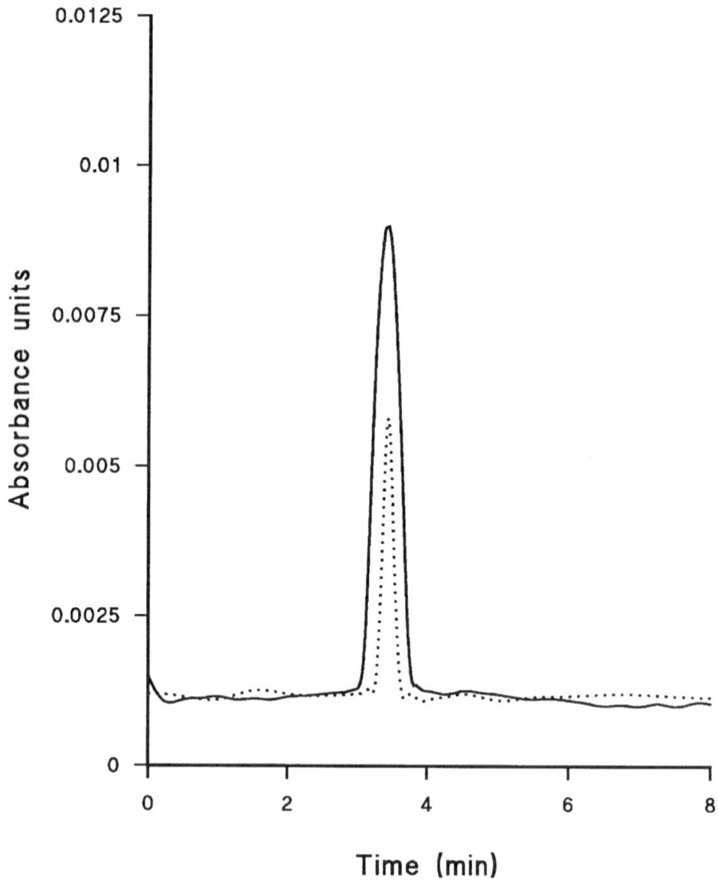

Figure 7 High-performance capillary electrophoresis of palytoxin from *P. caribaeorum* monitored at 230 nm (—) and 263 nm (····).

Optical Works Co., Japan) reversed phase column, had a retention time comparable to that of an authentic palytoxin (22). A detection limit of 125 ng for palytoxin was obtained using a Bio-Sil 5 ODS column and a mobile phase containing a mixture of acetonitrile:water (52:48) and 0.1% trifluoroacetic acid and analyzed at 230 nm (32). Lau and coworkers (19) carried out the HPLC analysis of the Caribbean palytoxin using a Novapak C18 reversed phase column, and the toxin was eluted out using a gradient system comprising two solvent systems. The eluted toxin was detected at 230 nm (Figure 8). The use of a mixed-mode HPLC method for the rapid analysis of palytoxin (34) detected several palytoxin congeners (21).

Thin-Layer Chromatography

Many researchers have used thin-layer chromatography (TLC) in the detection of palytoxin as well as in the examination of the homogeneity of extracted palytoxin (11,19,35). TLC is usually carried out on silica gel $60F_{254}$ or NH_2F_{254} plates in 1-propanol:pyridine:water (7:7:6) or 1-butanol:acetic acid:water:pyridine (15:13:12:10). The toxin can be detected by spraying the TLC plates with concentrated sulfuric acid:methanol (1:1) and charring at 150°C or by spraying with ninhydrin reagent and baking at 80°C, which stains the toxin purple. Alternatively, the

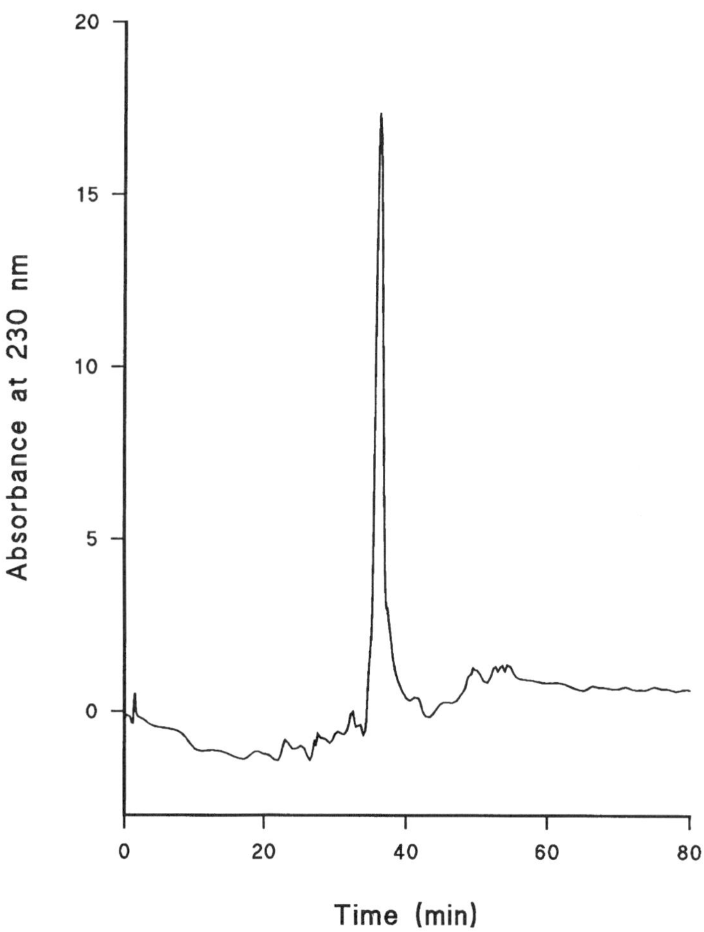

Figure 8 High-performance liquid chromatography of palytoxin from *P. caribaeorum*.

toxin could also be detected by exposing the plates to iodine vapor in a chamber. Palytoxin is stained yellow. Two-dimensional TLC has been particularly useful during the isolation of a florescent palytoxin congener from the crab *L. pictor* (19,20).

B. Biological Assays

Although physicochemical methods for toxin detection rely on the physical and chemical properties or characteristics of the toxin molecule, biological methods take advantage of palytoxin's functional properties or biological activities. The methods listed below have been used successfully, and some of them are extremely sensitive.

Animal Toxicity Assay

The simplest way to detect and quantitate the presence of palytoxin is to use the animal toxicity assay. One such method for the quantitation of palytoxin is the use of a dose-death time relationship in a mouse bioassay system developed by Teh and Gardiner (36). In their assay, a sample of the toxin (0.25 mL) was injected intraperitoneally into a mouse and the death time was measured. The amount of toxin (y) can be calculated by substituting the death time in minutes (x) into the equation:

$$y = 225.19\, x^{-0.99}$$

where y = number of mouse units and 1 MU is defined as the time taken to kill a mouse weighing 20 g in 4 h. This method has been applied successfully in the determination of palytoxin in toxic crabs (20,22). The mouse responds to the injected toxin by exhibiting several characteristic symptoms prior to death. Immediately after a lethal dose, the mouse exhibits sudden jerks and stretching of its hind limbs and lower back, weakening of its forelimbs, ataxia, decreased locomotion, convulsions, gasping for breath, and finally death.

Cell Toxicity Assay

Based on the earlier observation that the crab toxin had a characteristic effect on cells in culture (37), this cytotoxic property of palytoxin was used to monitor the purification of palytoxin-like substances from crabs (19,20). Cultured cells, including HeLa cells, incubated with either crab toxin or palytoxin showed characteristic changes in morphology such as increased granulation and rounding of the cells culminating in extensive cell damage and lysis (Figure 9). The cell

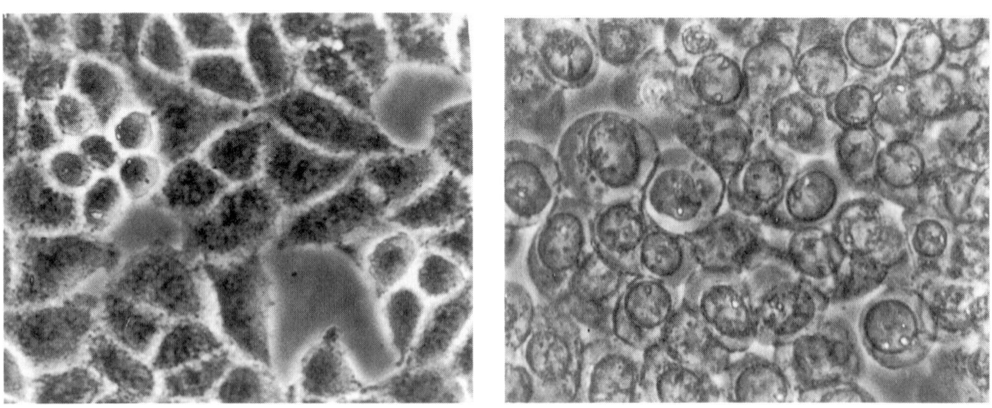

(a) (b)

Figure 9 *L. pictor*-induced cytotoxicity in HeLa cells; (a) control (b) toxin-treated. Magnification 300×.

toxicity assay serves as a convenient and rapid screening method for the detection of toxin in crude tissue samples or purified fractions. The assay is sensitive and can detect the presence of 0.006 MU of toxin in crude tissue samples (37). Quantitation of cell damage could be followed through the use of dyes taken up by intact cells (e.g., neutral red) or damaged cells (e.g., trypan blue). These dyes could then be released and the intensity measured in a spectrophotometer. The cell toxicity assay thus serves as an alternative to the use of animals in toxicity testing and is extremely useful for the monitoring of toxic fractions during toxin purification.

Hemolysis

Because palytoxin has a cytotoxic effect on cells, it is not surprising that it also functions as a powerful hemolysin (35). Unlike saponins, palytoxin does not directly destroy membrane integrity to effect hemolysis. Rather the toxin induces a prelytic release of K^+. The swelling of the erythrocytes eventually results in hemolysis. Palytoxin-induced hemolysis come long after the release of K^+ (35,38) (Figure 10). There is species specificity in relation to susceptibility to hemolysis (35). Rat and hog erythrocytes are very sensitive to palytoxin-induced hemolysis. The amount of hemoglobin released is found to be toxin concentration dependent (38,39). The quantitation of palytoxin in an unknown sample can be determined by incubating the erythrocytes with the sample and a range of palytoxin standards followed by measuring the release of hemoglobin from the cells after a fixed time period.

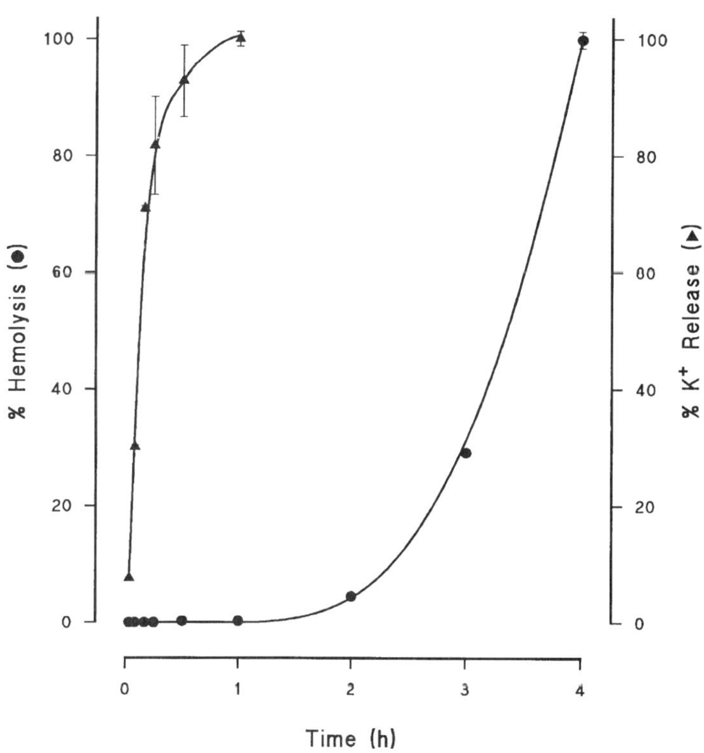

Figure 10 Time course of palytoxin-induced hemolysis and K^+ release in rat erythrocytes.

Potassium Release

Palytoxin has been shown to cause a rapid release of K^+ from living cells (38,39) as well as from nerve ending particles (40). This appears to be the primary action of the toxin in causing cytotoxicity. Palytoxin induces the release of potassium before the onset of other secondary effects such as hemolysis or the inhibition of Na^+-dependent processes (40,41). The efflux of K^+ is inhibited by ouabain, suggesting the involvement of Na^+-K^+-ATPase. However, palytoxin does not affect the activity of this enzyme at concentrations that cause K^+ release. The toxin is envisaged to function in binding to the enzyme and creating an ionic channel across the membrane of the cell (39). Measurement of K^+ release from cells is perhaps the simplest but sensitive method for the assay of palytoxin and could be carried out using a flame photometer or an atomic absorption spectrometer. The release is concentration dependent (Figure 11). The sensitivity of the method is approximately 1 pM for palytoxin-induced release of K^+ from rat and human erythrocytes(39).

Radioimmunoassay

With the availability of specific antibodies to palytoxin (42), a radioimmunoassay (RIA) method using ^{125}I-labeled palytoxin was developed for detection and quantitation of palytoxin in biological samples (43) (Figure 12). The method is very sensitive and can detect the presence of palytoxin or palytoxin-like substances in biological fluids and in tissue extracts in the picomolar range (ED50 = 0.27 pmol). It can also be used for the monitoring of toxic fractions during

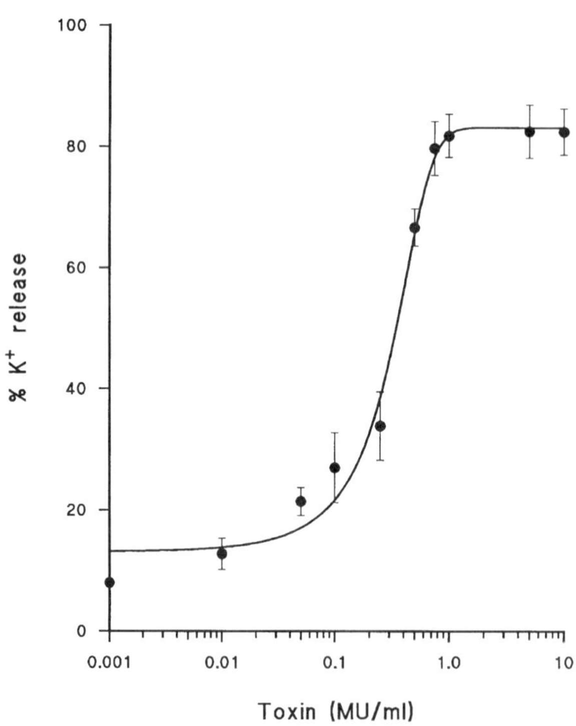

Figure 11 Release of K^+ from rat erythrocytes as a function of palytoxin concentration. MU, mouse unit.

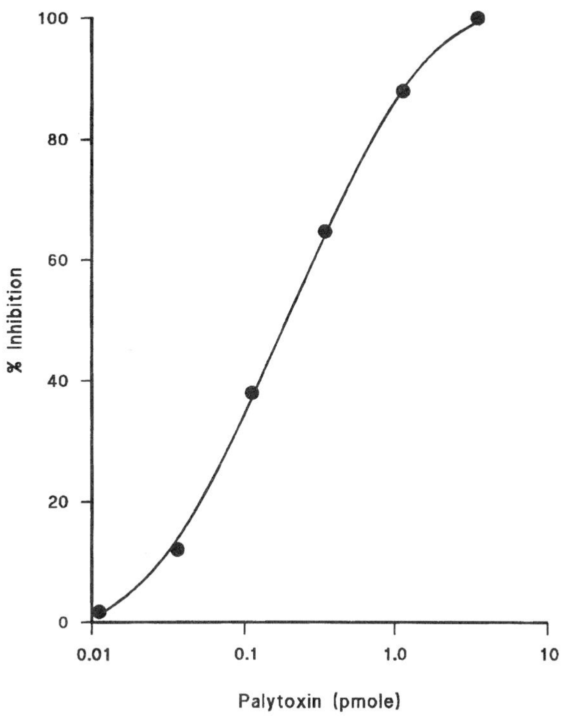

Figure 12 Inhibition of ^{125}I-palytoxin binding to rabbit antiserum by unlabeled palytoxin. (From ref. 43 with permission from Elsevier Science)

toxin purification to confirm the presence of palytoxin. According to the investigators, radioisotopic labeling of palytoxin with ^{125}I-Bolton-Hunter reagent enabled the production of a labeled ligand with high specific activity (43). The RIA method appears to be quite specific for palytoxin, since 10- to 100-fold higher concentrations of maitotoxin, telocidin, okaidic acid, debromoaplysiatoxin, and 12-tetradecanoylphorbol-13-acetate had no effect on ^{125}I-palytoxin binding to its antibody. The method, however, is not able to distinguish between biologically active and inactive palytoxins.

Monoclonal Antibody–Based Enzyme-Linked Immunoassay

Several monoclonal antibody-based enzyme-linked immunoassays (EIAs) have been developed for the measurement of palytoxin in biological samples (44). The antibodies were produced against conjugates of keyhole limpet hemocyanin and chemically defined palytoxin haptens. The investigators described five formats for their enzyme-linked immunoassays. The enzyme-linked immunoassays are specific for palytoxin and do not cross react with other marine toxins. However, they do cross react with certain nontoxic preparations of palytoxin. There is the added advantage of a shortened assay time (2–4 h) compared with the 24-h assay time required by the RIA method. The EIA methods are sensitive and can detect low levels of palytoxin in biological samples (Figure 13). An added advantage in the use of EIAs is the elimination of the use of radiolabeled palytoxin. The EIA methods have been used successfully in the quantitation of palytoxin in crude extracts of *P. tuberculosa* (44). In a sandwich immunoassay using the antipalytoxin antibody 73D3, the toxin of the marine crab *L. pictor* was demonstrated to

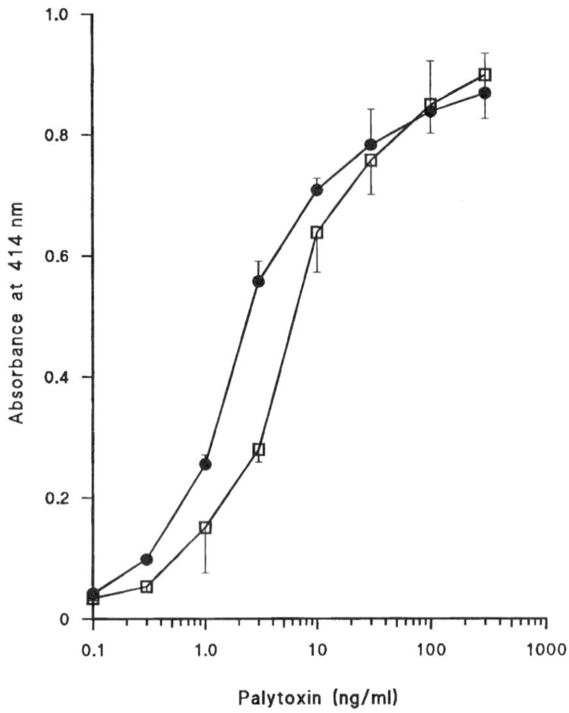

Figure 13 Representative standard curves for palytoxin sandwich ELISA systems. (From ref. 44 with permission from Elsevier Science.)

possess the same epitopes as the palytoxin from *P. tuberculosa*, suggesting that the two toxins share common structural features (21). Hence, the immunoassay technique is useful for the detection and identification of palytoxin and palytoxin-like substances in biological samples.

Antibody-Based Hemolysis Neutralization Assay

As with RIA, EIA methods do not discriminate between biologically active and inactive forms of palytoxin. Subsequently, Bignami took advantage of the hemolytic property of palytoxin and the specificity of the antipalytoxin monoclonal antibody 73D3 to develop a rapid and sensitive hemolysis neutralization assay for palytoxin which detects only pharmacologically active palytoxin (45). The neutralizing antibody was able to inhibit palytoxin-induced hemolysis using an antibody concentration of 25 µg/mL and an incubation period of 1–4 h at 37°C. The sensitivity of the assay could be increased to a detection limit of 1 pg/mL (0.37 pM) by using a longer incubation period of 24 h. This assay has been used to confirm the presence of palytoxin or palytoxin-like substances in crude tissue extracts (45) and in the purified toxin fraction from *L. pictor* (21).

ACKNOWLEDGMENTS

The authors thank F. H. Ng for excellent technical assistance and help in the preparation of illustrations and photographs for this manuscript. The project on *L. pictor* toxin was supported by a National University of Singapore grant (RP 900315) to the Principal Investigator (C.H.T.).

REFERENCES

1. RE Moore, PJ Scheuer. Palytoxin: a new marine toxin from a coelenterate. Science 172:495–498, 1971.
2. GE Walsh, RE Bowers. A review of Hawaiian zoanthids with descriptions of three new species. Zoo J Linn Soc 50:161–181, 1971.
3. S Kimura, Y Hashimoto. Purification of the toxin in a zoanthid *Palythoa tuberculosa*. Publ Seto Mar Biol Lab 20:713–718, 1973.
4. Y Hirata, D Uemura, K Ueda, S Takano. Several compounds from *Palythoa tuberculosa* (coelenterata). Pure Appl Chem 51:1875–1883, 1979.
5. DH Attaway, LS Ciereszko. Isolation and partial characterization of Caribbean palytoxin. Proceedings of 2nd International Coral Reef Symposium I, Great Barrier Reef Community, Brisbane, 1974, pp 497–504.
6. L Beress, J Zwick, HJ Kolkenbrock, PN Kaul, O Wassermann. A method for the isolation of the Caribbean palytoxin (C-PTX) from the coelenterate (zoanthid) *Palythoa caribaeorum*. Toxicon 21: 285–290, 1983.
7. RJ Quinn, M Kashiwagi, RE Moore, TR Norton. Anticancer activity of zoanthids and the associated toxin, palytoxin, against Ehrlich ascites tumor and P-388 lymphocytic leukemia in mice. J Pharm Sci 63:257–260, 1974.
8. RE Moore, G Bartolini. Structure of palytoxin. J Am Chem Soc 103:2491–2494, 1981.
9. RE Moore. The structure of palytoxin. Prog Chem Org Nat Prod 48:81–202, 1985.
10. Y Hashimoto, N Fusetani, S Kimura. Aluterin: a toxin of filefish, *Alutera scripta*, probably originating from a zoantharian, *Palythoa tuberculosa*. Bull Jpn Soc Sci Fish 35:1086–1093, 1969.
11. M Fukui, M Murata, A Inoue, M Gawel, T Yasumoto. Occurrence of palytoxin in the trigger fish *Melichtys vidua*. Toxicon 25:1121–1124, 1987.
12. N Fusetani, S Sato, K Hashimoto. Occurrence of water soluble toxin in a parrotfish *Ypsiscarus oviforns* which is probably responsible for parrotfish liver poisoning. Toxicon 23:105–112, 1985.
13. AM Kodama, Y Hokama, T Yasumoto, M Fukui, SJ Manea, N Sutherland. Clinical and laboratory findings implicating palytoxin as cause of ciguatera poisoning due to *Decapterus macrosoma* (mackerel). Toxicon 9:1051–1053, 1989.
14. VM Mahnir, EP Kozlovskaya, AI Kalinovsky. Sea anemone *Radianthus macrodactylus*—a new source of palytoxin. Toxicon 30:1449–1456, 1992.
15. S Gleibs, D Mebs, B Werding. Studies on the origin and distribution of palytoxin in a Caribbean coral reef. Toxicon 33:1531–1537, 1995.
16. RB Gonzales, AC Alcala. Fatalities from crab poisoning on Negros Island, Philippines. Toxicon 15: 169–170, 1977.
17. CTT Tan, EJD Lee. A fatal case of crab toxin (*Lophozozymus pictor*) poisoning. Asia Pac J Pharmacol 3:7–9, 1988.
18. YF Teh, JE Gardiner. Toxin from the coral reef crab, *Lophozozymus pictor*. Pharmacol Res Commun 2:251–256, 1970.
19. CO Lau, HE Khoo, R Yuen, M Wan, CH Tan. Isolation of a novel fluorescent toxin from the coral reef crab, *Lophozozymus pictor*. Toxicon 31:1341–1345, 1993.
20. CO Lau, CH Tan, HE Khoo, QT Li, R Yuen. Ethanolic extraction, purification, and partial characterization of a fluorescent toxin from the coral reef crab, *Lophozozymus pictor*. Nat Toxins 3:87–90, 1995.
21. CO Lau, CH Tan, HE Khoo, R Yuen, RJ Lewis, GP Corpuz, GS Bignami. *Lophozozymus pictor* toxin: a fluorescent structural isomer of palytoxin. Toxicon 33:1373–1377, 1995.
22. T Yasumoto, D Yasumura, Y Ohizumi, M Takahashi, AC Alcala, LC Alcala. Palytoxin in two species of xanthid crab from the Philippines. Agric Biol Chem 50:163–167, 1986.
23. AC Alcala, LC Alcala, JS Garth, D Yasumura, T Yasumoto. Human fatality due to ingestion of the crab, *Demania reynaudii* that contained a palytoxin-like toxin. Toxicon 1:105–107, 1988.
24. RE Moore, P Helfrich, GML Patterson. The deadly seaweed of Hana. Oceanus 25:54–63, 1982.
25. H Nakamura, T Asari, Y Ohizumi, J Kobayashi, T Yamasu, A Murai. Isolation of zooxanthellatoxins,

novel vasoconstrictive substances from the zooxanthella *Symbiodinium sp.* Toxicon 31:371–376, 1993.

26. JK Cha, WJ Christ, JM Finan, H Fujiko, Y Kishi, LL Klein, SS Ko, J Leder, WW McWhorter Jr, KP Pfaff, M Yonaga, D Uemura, Y Hirata. Stereochemistry of palytoxin. Complete structure. J Am Chem Soc 104:7369–7371, 1982.
27. RW Armstrong, JM Beau, SH Cheon, WJ Christ, H Fujioka, WH Ham, LD Hawkins, H Jin, SH Kang, Y Kishi, MJ Martinelli, WW McWhorter Jr, M Mizuno, M Nakata, AE Stutz, FX Talamas, M Taniguchi, JA Tino, K Ueda, J Uenishi, JB White, M Yonaga. Total synthesis of a fully protected palytoxin carboxylic acid. J Am Chem Soc 111:7525–7530, 1989.
28. EM Suh, Y Kishi. Synthesis of palytoxin from palytoxin carboxylic acid. J Am Chem Soc 116: 11205–11206, 1994.
29. RE Moore, RF Dietrich, B Hatton, T Higa, PJ Scheuer. The nature of the λ 263 chromophore in the palytoxins. J Org Chem 40:540–542, 1975.
30. AR Ibrahim, WT Shier. Palytoxin: mechanism of action of a potent marine toxin. J Toxicol Toxin Rev 6:159–187, 1987.
31. D Uemura, Y Hirata, T Iwashita, H Naoki. Studies on palytoxin. Tetrahedron 41:1007–1017, 1985.
32. KA Mereish, S Morris, G McCullers, TJ Taylor, DL Bunner. Analysis of palytoxin by liquid chromatography and capillary electrophoresis. J Liq Chromatogr 14:1025–1031, 1991.
33. Y Hirata, D Uemura, Y Ohizumi. Chemistry and pharmacology of palytoxin. In: A Tu, ed. Handbook of Natural Toxins. Vol 3. New York, Marcel Dekker, 1988, pp 241–258.
34. GP Corpuz, PG Grothaus, DF Waller, GS Bignami. An HPLC method for rapid analysis of palytoxin congeners and prodrugs. Proceedings of the International Symposium Ciguatera and Marine Natural Products, Honolulu, 1995, pp 143–151.
35. E Habermann, G Ahnert-Hilger, GS Chhatwal, L Beress. Delayed haemolytic action of palytoxin. Biochim Biophys Acta 649:481–486, 1981.
36. YF Teh, JE Gardiner. Partial purification of *Lophozozymus pictor* toxin. Toxicon 12:603–610, 1974.
37. CH Tan, YF Teh. The effect of *Lophozozymus pictor* toxin on HeLa cells. Experientia 28:46, 1972.
38. CO Lau, CH Tan, QT Li, FH Ng, R Yuen, HE Khoo. Bioactivity and mechanism of action of *Lophozozymus pictor* toxin. Toxicon 33:901–908, 1995.
39. E. Habermann. Palytoxin acts through Na^+, K^+-ATPase. Toxicon 27:1171–1187, 1989.
40. CO Lau, FH Ng, HE Khoo, R Yuen, CH Tan. Inhibition of sodium-dependent uptake processes in purified rat brain synaptosomes by *Lophozozymus pictor* toxin and palytoxin. Neurochem Int 28: 385–390, 1996.
41. CH Tan, SJ Abkowitz. *Lophozozymus pictor* toxin: a potent inhibitor of synaptosomal GABA uptake. Neurosci Lett 14:339–342, 1979.
42. L Levine, H Fujiki, HB Gjika, H Van Vunakis. Production of antibodies to palytoxin: neutralization of several biological properties of palytoxin. Toxicon 25:1273–1282, 1987.
43. L Levine, H Fujiki, HB Gjika, H Van Vunakis. A radioimmunoassay for palytoxin. Toxicon 26: 1115–1121, 1988.
44. GS Bignami, TJG Raybould, ND Sachinvala, PG Grothaus, SB Simpson, CB Lazo, JB Byrnes, RE Moore, DC Vann. Monoclonal antibody-based enzyme-linked immunoassays for the measurement of palytoxin in biological samples. Toxicon 30:687–700, 1992.
45. GS Bignami. A rapid and sensitive hemolysis neutralization assay for palytoxin. Toxicon 31:817–820, 1993.

25
Mechanism of Action, Pharmacology, and Toxicology

Magdalena T. Tosteson
Harvard Medical School, Boston, Massachusetts

I. INTRODUCTION

Limu make o Hana, or deadly seaweed of Hana, was described by Malo (1) as a poisonous moss growing close to the ocean, with the umbrella disclaimer as shown in the Preface, which has been reproduced below:

Preface by the Author

I do not suppose the following history to be free from mistakes, in that the material for it has come from oral traditions; consequently it is marred by errors of human judgment and does not approach the accuracy of the word of God.

Davida Malo

In the words of Malo: . . .''The servants of the king were known under the following designations: *malalaioa*, . . . *hamohamo*. . . .'' (p 257). In the translator's notes, on p 266, when describing the meaning of the word *Hamohamos*, he relates,

> . . . I am told (by Kapule) that in Muolea, in the district of Hana, grew a poisonous moss in a certain pool or pond close to the ocean. It was used to smear on the spear points, to make them fatal. These men were the ones who did the job, hence they were called *hamohamo*, the smearers. This moss is said to be of a reddish color and is still to be found. It grows nowhere else than at that one spot. Kapule thinks it was about the year 1857 that he was in Hana and saw this moss. It was shown him by an old man named Peelua, the father in law of S.M. Kamakau. This is a revelation and a great surprise to me. I never heard of such a thing before. Manu covered it with stones.

This poisonous "moss" was located much later by Professors Moore and Scheuer (2), and a very interesting account of their findings has been published (3). Moore and collaborators identified the organism as an animal, phylum *Coelenterata*, order *Zoantharia*, family *Zoanthidae* and assigned to the genus *Palythoa*. Since that time, the complete structure of palytoxin has been elucidated and the molecule synthesized (see ref. 4 and references therein), as reviewed in the previous chapter.

II. BIOSYNTHESIS AND DETECTION

As described above, palytoxin (PTX) was found to be produced by the genus *Palythoa* (2), and reported to be present in organisms from Okinawa (*P. tuberculosa*), Hawaii (*P. toxica, P. Vestitus*), and the Caribbean (*P. caribdea, P. mamillosa*) (5; see references in ref. 6). The chemical differences between the various palytoxin molecules were found to be minor, but they might play an important role in the affinity for the receptor (see Section IV.B). Furthermore, since the finding that there are minor toxins which are also produced by *Palythoa* (homopalytoxin, bishomopalytoxin, and deoxypalytoxin) (7,8) and that there is a fluorescent structural isomer of palytoxin (see refs. 9 and 10 and references therein), it was suggested that palytoxin might be biosynthesized by symbiotic microorganisms and that PTX might be of bacterial origin (3).

Studies to try to elucidate the biosynthetic pathway of palytoxin were further prompted by the fact that PTX was also suspected of being the toxic agent in fatal human poisoning after ingestion of fish and crabs even as early as 1969 (11) (see Section VI). Further, since the presence of palytoxin has been detected in various fish species, crabs, and sea anemones, it can be assumed that this toxin can enter into the food chain (12–17).

The development of immunoassays for detection of PTX (18,19) together with the thorough descriptions of the actions of PTX in tissues and cells (see 20 and 21 and references therein) have allowed the identification of the production of the toxin by organisms other than *Palythoa* spp. Thus, PTX has been found in the zoanthid species (*Zoanthus solanderi and Z. sociatus* (22)), competitors of *Palythoa* in the coral reefs, as well as in dinoflagellates (23,24), in extracts, and in pure cultures of *Ostreopsis lenticularis* and their associated bacterium (see ref. 6 and references therein) and in *O. siamensis* (8), which had been previously implicated in the production of ciguatera toxin. PTX was also found in red algae, as reported by Ito and collaborators (26), whose studies constitute an important step in the determination of the differences in the toxicologies of ciguatoxin (cf. Chapter 21) and palytoxin.

This chapter will focus on the possible molecular mechanism leading to the opening of the palytoxin channels through (apparently) the Na^+,K^+-ATPase (Section II). It further deals with the consequences that the opening of these channels has on cells, tissues and organisms, from being an anticancer agent to being an agent of death (sections III–IV, and V). Quite an impressive repertoire for a small cation channel!

III. PHYSIOLOGY

Studies of the toxic effects of palytoxin on isolated organs were started in 1977, when Weidmann (27) (see also ref. 28) exposed rabbit and dog muscles to ~4 nM of an acetone-extracted, partially purified PTX. Monitoring the response of the muscles to stimuli which evoked action potentials, he found that after a relatively long lag time (15 min), the threshold for excitation increased, the resting potential decreased, the "overshoot" of the action potential decreased and became less pronounced, the rise time became longer, and the total duration of the action potential shortened. The toxic effects could be reversed in a concentration- and time of exposure–dependent manner. He considered the possibility of an effect on the Na^+,K^+-ATPase but discarded it on the basis of the results of Garcia-Castineiras (29), who reported that the purification of a crude extract of PTX led to the separation of two distinct fractions, one of which had powerful Na^+,K^+-ATPase inhibition capabilities and was subsequently identified as serotonin.

Anhert-Hilger et al. (30) found that borate increases the hemolytic activity of palytoxin by a factor of ~80, and the enhancing of PTX's action by this ion has since been amply demonstrated. This effect of borate is one of the most intriguing aspects of the mode of action of palytoxin, and the molecular mechanism for it has not yet been established. Nonetheless, it is interesting to note that reports on the borate concentration in the foraminifera of the Pacific Ocean show that the concentration of boron in the seawater is quite high (~400 μM) (31) and very close to the concentration which yields a maximum effect in red cells and the Na^+,K^+-ATPase (0.5–1.0 mM) (32,33) and, further, that boron incorporates into the foraminifera (31). The implications of these findings in the biosynthetic pathway of palytoxin and/or on the effects of palytoxin in cells is not clear, but it is prudent to consider this added complexity.

Since the finding that ouabain antagonizes the effect of palytoxin in red blood cells, it was inferred that palytoxin binds to the Na^+,K^+-ATPase (34). Since then, a large body of experimental evidence has provided corroborative circumstantial evidence that the Na^+,K^+ pump protein is the receptor for the toxin, based on work with inhibitors and activators of the Na^+,K^+-ATPase (see below). In fact, the data obtained to date indicates that PTX is a very specific inhibitor of the Na^+,K^+-ATPase, since no other ATPase has been found to date to be susceptible to the toxin. In contrast, ouabain has been found to inhibit not only the Na^+,K^+-ATPase but also the H^+-, Ca^{2+}-, and H^+/K^+-transporting ATPases (35). Direct demonstration that the Na^+,K^+-ATPase is the receptor for palytoxin came from the work of Yoda et al. (36) and Anner and Moosmayer (37), who have shown that when the purified enzyme is incorporated into liposomes, there is a PTX-dependent, ouabain-sensitive increase in the Na^+ and K^+ fluxes from these vesicles; and from the work of Scheiner-Bobis (38), who showed that yeast become sensitive to palytoxin in a borate-, ouabain-, and [PTX]-dependent manner upon expression of both the α and β subunits of the Na^+,K^+-ATPase.

A. (Na^+,K^+)-ATPase

In order to focus the subsequent discussion on the effect that PTX has on the Na^+,K^+ pump, it is necessary to indicate the minimum steps in the hydrolysis of ATP which seem to be involved in the phenomenon under consideration. In this context, the overall cycle of the Na^+,K^+-ATPase which leads to the hydrolysis of ATP and the efflux of 3 Na^+ and the uptake of 2 K^+ can be represented by a simplified scheme which represents the states through which the enzyme is assumed to pass in each cycle of transport (modified from Jørgensen et al. (39)).

$$\begin{array}{l} 2Na \cdot E_1ATP \leftrightarrow (3Na)E_1 \sim P \leftrightarrow (2Na) \cdot E_2P \leftrightarrow 2Na \cdot E_2 \\ \upharpoonleft\!\downharpoonright \qquad\qquad\qquad\qquad\qquad\qquad\qquad\qquad\qquad\qquad\quad \upharpoonleft\!\downharpoonright \\ E_1ATP \leftrightarrow 2K \cdot E_1ATP \leftrightarrow (2K)E_2 \leftrightarrow 2K \cdot E_2P \leftrightarrow E_2P \end{array}$$

where E_1 indicates different allosteric forms of the enzyme and the ions in parenthesis represent the temporarily occluded ions (40), an important step in the transport of the ion. The occluded ions cannot exchange with ions in the medium, whereas the free ions (outside parenthesis) can readily interact with the ion-binding cavity in the enzyme, with E_1 through a cytoplasmic gate, and with E_2 via an extracellular gate.

From studies of the ion binding to the isolated, purified Na^+,K^+-ATPase, it is known that dephosphorylation leads to the binding (occlusion) of 2 K^+ (with a high apparent affinity (25)) to the E_2 form of the enzyme, whereas phosphorylation leads to the occlusion of 3 Na^+ ions by the E_1 form of the enzyme. Associated with the Mg^{2+}-dependent, Na^+ and K^+–activated ATPase activity of the enzyme there is a K^+-activated phosphatase activity influenced by K^+, Na^+, ATP, and ouabain and whose activity correlates with the activity of the Na^+,K^+-ATPase. These and other observations have led to the proposal that the phosphatase is an intrinsic activity of the Na^+,K^+-ATPase protein (41) and that its activity probably arises when the enzyme is in the E_2 form.

Studies with various inhibitors of the Na^+ pump have helped in the determination of some of the crucial steps governing the transitions from one of the forms of the enzyme to the others. Thus, cardiac glycosides, specific inhibitors of the Na^+,K^+-ATPase, bind and inhibit optimally when the enzyme is in the E_2P form. Vanadate, a transition-state analogue of the phosphate bound to the phosphorylation site, also inhibits the Na^+,K^+-ATPase and stabilizes the occluded K^+ form of the enzyme (40). The antibiotic oligomycin inhibits the enzyme by decreasing the rate of deocclusion of Na^+ from the (3 Na^+) $E_1 \sim P$ form (stabilizing the occlusion of Na^+) (42) and thus preventing the cycle from proceeding.

IV. MOLECULAR MECHANISMS OF ACTION

A. Studies on Palytoxin Binding

The first attempts to measure the binding of palytoxin to red blood cells, as well as to membranes from various organs (brain, liver, and kidney), were done first using an indirect assay (43) and subsequently using iodinated palytoxin (44). These first attempts yielded one binding site with an affinity of 20 pM and 200 as the number of receptors at the surface of human red blood cells (determinations of the number of ouabain-binding sites in the human erythrocyte range from 200 to 400). Subsequently, through modeling of the ion fluxes induced by PTX as a function of [PTX], as well as through determination of bound PTX using an indirect method, it was found that there are at least two binding sites for the toxin with widely different affinities (33,45):

$$K_{m1} = 25 \text{ pM} \qquad J_{m1} = 1.3 \text{ m/lc} \cdot \text{h} \qquad R_{t1} = 300$$
$$K_{m2} = 1500 \text{ pM} \qquad J_{m2} = 10 \text{ m/lc} \cdot \text{h} \qquad R_{t2} = 4000$$

Böttinger et al. (44) and Kim et al. (46) have found that the binding of palytoxin depends on the functional state of the Na^+,K^+-ATPase. In particular, their data indicate that the presence of high internal ATP (millimolar range) enhances the sensitivity as well as the binding of ^{125}I-labeled PTX to red cells and promotes the opening of single channels in inside-out patches of frog erythrocytes. The role that ATP plays is not clear at the moment, since recent results indicate that PTX can form ion channels in yeast cells containing a mutated form of the Na^+,K^+-ATPase which does not catalyze hydrolysis of ATP and which does not bind ouabain. In fact, PTX produces the same increase in cation permeability in yeast expressing either the mutated or the wild-type Na^+,K^+-ATPase, but it produces no effect in yeast cells that have not been transfected with the enzyme (47).

The effect of palytoxin on the cellular permeability occurs at concentrations of the toxin which are several orders of magnitude lower than those which inhibit the Na^+,K^+-pump *in the same cell* (33,48), suggesting that the primary effect of palytoxin is to bind to its receptor (the Na^+,K^+-ATPase) and induce a large ion permeability change (high-affinity binding site). Thus, a working hypothesis would be that as the toxin concentration is increased, binding to the low-affinity binding site would then lead to the inhibition of the Na^+,K^+ pump, and that this step involves the binding to approximately 10 times the number of ouabain-binding sites. These suggestions are currently being pursued in several laboratories.

B. Structure-Function Studies

Chemical modifications of the palytoxin molecule have been rare, since its synthesis is very complicated and it has been mainly targeted to the end groups. Early on it was shown that N-acetylation reduces the action of palytoxin to 1% of the original (49) and leads to a loss of sensitivity to borate (33). Radiolabeling with iodine also reduces the apparent affinity of palytoxin to induce potassium loss by approximately the same amount (44). The consequences of systematic changes at either or both terminal regions of PTX for both the binding sites and for the competition with [^{3}H]ouabain binding have been studied, and these studies show that all of the derivatives induce a maximum increase in the K^+ flux within an order of magnitude of that which obtains with palytoxin (50). The analogues, for the most part, seem to bind in a way such that the conductive pathway which they produce has a much lower permeability than the one induced by PTX and with only one binding site. Interestingly, if the block at the N-terminal end is a dansyl group, the flux curves are again fitted by a two-site model. All of the actions of the analogues were found to be reversed by ouabain, and they were all found to be two to three orders of magnitude less potent than PTX to compete with ouabain for a binding site (50). The results of these studies indicate that modifications at both ends of the palytoxin molecule (and probably modifications of the double bonds as well) yield significant alterations in its binding and in its pharmacological actions on cell membranes.

C. Interactions of Palytoxin and Ouabain in Red Cells

The bulk of the experimental evidence for the action of palytoxin and the interaction with ouabain (and other inhibitors and agonists) comes from work on red cells from diverse species. Since human red cells have a lower number of Na^+/H^+ and Na^+/Ca^{2+} exchangers and of Ca^{2+}-activated K^+ channels, they seem to be better suited for the study of the action of palytoxin on the cationic permeability of cells. As a consequence, the results described in this section will be primarily those obtained using human red cells.

The universality of the inhibition by ouabain of the primary (and secondary) effects of palytoxin has been amply documented (21). The data available indicate that when cells are first exposed to ouabain, subsequent exposure to PTX (in the presence of ouabain) leads to a shift of approximately two to three orders of magnitude in the apparent affinity of the cells for PTX (51,52), whereas the maximum flux seems to be unchanged (M.T. Tosteson, unpublished data), suggesting that there is a competitive inhibition of the actions of palytoxin by ouabain, as has already been described in work on the isolated enzyme (53). Further, it has also been shown that palytoxin competes for the ouabain-binding site with an affinity which is 10 times higher than that of ouabain (33,44). As indicated above, this characteristic of palytoxin is entirely dependent on the integrity of the PTX molecule in not having either (or both) of the terminal ends blocked and, probably, on the presence of intact double bonds as well (44,50).

A detailed study of the competition (displacement) of palytoxin and ouabain was done taking into account the kinetics of the loss of K^+ from human erythrocyte (D.R.L. Scriven and M.T. Tosteson, unpublished data). In one series of experiments, palytoxin was "chased" by ouabain (added during the initial phase of K^+ loss). In another set of experiments, the measurements were done with the simultaneous addition of palytoxin and a very high concentration of ouabain (1 mM) in an effort to study the displacement (or replacement) of PTX by ouabain, and in a third set of experiments, human red cells were first preincubated with ouabain (for 3 h at 37°C to achieve complete, irreversible binding) followed by washing and determination of PTX-induced K^+ efflux in order to determine the possibility that PTX could replace/displace bound ouabain. Collectively, these experiments yield the following data:

$$k_{on} \text{ for PTX} = (4 \pm 0.6)10^8 \text{ M}^{-1} \text{ min}^{-1}$$

and

$$k_{off} \text{ for PTX} = (0.04 \pm 0.02)\text{min}^{-1}$$

Both rates are within a factor of 10 of those obtained by Grell et al. using a purified enzyme from pig kidney with FITC bound (32) and those of Kim et al. for the time constant of single-channel induction by palytoxin (46).

The corresponding values for the constants for ouabain:

$$k_{on} \text{ for ouabain} = (1.4 \pm 0.7)10^7 \text{ M}^{-1} \text{ min}^{-1}$$

and

$$k_{off} \text{ for ouabain} = (0.6 \pm 0.3)\text{min}^{-1}$$

which agree with those found for the isolated enzyme by Grell et al. (54).

We determined the rate at which ouabain comes off its binding site(s) when human red cells preloaded with [^{3}H]ouabain are exposed to palytoxin. It was found that exposure to PTX does indeed accelerate the rate at which ouabain comes off its binding site (k_{off} = 0.003/min as compared to 0.6/min), but that this is due to a change in the internal ionic composition of the cells from high internal K^+ to high internal Na^+ (D.R.L. Scriven and M.T. Tosteson, unpublished data).

D. Inhibitors of Palytoxin's Actions

The study of compounds that inhibit palytoxin's action on the cationic permeability of cells has focused primarily on the inhibitors of the Na^+,K^+-ATPase, but there have been some studies conducted on the action of inhibitors of other transport systems. In general, it is clear now that the action of those inhibitors on the PTX-induced actions on cells can be explained assuming that those transport systems are activated by the primary action of PTX; namely, the increase in cation permeability of cell membranes.

Of the inhibitors of the Na^+,K^+ pump, ouabain (the compound most studied) has been found to inhibit the action of palytoxin in all cells in which it has been tested, provided that the concentration of ouabain is high enough. Habermann (21) has reported that rat red cells seem to have a portion of their palytoxin-activated K^+ loss insensitive to ouabain, but this is probably due to the activation of the Ca^{2+}-activated K^+ efflux (55) caused by the increase in Ca^{2+} influx into human red blood cells (and in other cells as well) which occurs in the presence of PTX (56) (D.R.L. Scriven and M. Tosteson, unpublished data).

Studies on the role that the sugar moiety of the cardiac glycoside has on its competition with palytoxin have shown that monoglycosides (ouabain, convallatoxin, cymarin) are more potent in their capacity to inhibit the PTX-induced K^+ release of rabbit red cells than are the triglycosides (digoxin, digitoxin) (52). They further found that the IC50 for K^+ release induced by PTX is 2.3×10^{-6} M for ouabain and $\sim 9 \times 10^{-5}$ M for digoxin and digitoxin. On the other hand, the IC50 for inhibition of the Na^+,K^+-ATPase is of the order of 10^{-7} M for all the same compounds. Molecules devoid of the glycoside moiety (ouabagenin, strophanthidin, dogoxigenin, digitoxigenin) show a substantial reduction in their capability to prevent (or reverse) the effect of PTX, but they can inhibit the Na^+,K^+-ATPase with an IC50 ~ 10^{-7} M, just like the sugar-containing cardiotonic steroids. These results once again reinforce the notion that either the binding sites for PTX and ouabain are not identical, or that the conformation that the Na^+,K^+-ATPase adopts in the presence of the cardiac glycosides lowers the affinity of the PTX site for its substrate.

Two other inhibitors of the Na^+,K^+-ATPase, vanadate and oligomycin, were tested for their effects on the palytoxin-induced actions on cells. Vanadate was found by Habermann not to interfere with PTX's actions on red blood cell ghosts (57), but it was subsequently shown that it does inhibit the K^+ efflux from human red blood cells and the single-channel currents recorded in frog red cells (33,46). On the other hand, oligomycin does not prevent palytoxin's actions in human or rabbit red blood cells (33,52).

From the data on the action of inhibitors of the Na^+,K^+ pump on the actions of palytoxin in cells, it has been suggested that the toxin binds to the E_1P form of the enzyme (93). On the other hand, Habermann (21) argues that palytoxin recognizes both the E_1 and the E_2 form of the enzyme, Wang (48) has argued for binding to the E_2 form, and Anner (37) argues that palytoxin stabilizes a leaky E_1 conformation, in contrast to ouabain, which stabilizes a tight E_2 form. The antagonism between these two compounds would then be a consequence of allosteric interactions of the toxins via their respective binding sites. However, it has been shown recently that palytoxin promotes the release of occluded K^+ in the dog kidney preparation (45), suggesting that either K^+ occlusion occurs both in the E_1 and in the E_2 forms of the enzyme, and PTX deoccludes the ''leaky'' E_1 conformation, or that palytoxin destabilizes the E_2 form of the enzyme, classically considered the K^+-occluding form. These results do not shed light on the form(s) of the enzyme to which PTX binds, but they are clearly the beginning steps toward this goal. Needless to say, more work on the kinetic steps of the isolated preparation is required in order to clarify these issues.

E. Effects of Palytoxin on the Purified Na^+,K^+-ATPase

Palytoxin inhibits the hydrolysis of ATP by the Na^+,K^+-ATPase from preparations obtained from various sources at concentrations which are orders of magnitude higher than those required to activate the cationic fluxes in the same tissues, suggesting once again a dual role for PTX's actions (33,48). Thus, it has been reported that PTX inhibits the catalytic activity of purified preparations from guinea pig heart and pig cerebral cortex (58), from pig and dog kidney (32,45), and from eel electroplax (50).

Although it has not been systematically studied, it is apparent that the affinity of the enzyme for palytoxin depends on the source of the enzyme, and it is a function of both the species as well as tissue from which it has been isolated. Thus, from the data available, it emerges that the affinity of the various enzymes for PTX follows approximately the sequence: electric eel ≈ pig brain > dog kidney > pig kidney ≥ guinea pig heart ⋙ shark. Interestingly, it was found that even though the enzyme from rectal glands of the shark is not inhibited by palytoxin at concentrations ≥30 μM, the shark erythrocytes respond to low concentrations of palytoxin (nanomolar range) with an increase in the permeability to K^+, as has been described for the erythrocytes of other species (21).

Finally, it has been found that, even though the pNPPase from the pig kidney is inhibited by palytoxin, the dog kidney pNPPase is activated by PTX (32,45)! This is not the first time that a difference has been found between the pig kidney and the dog kidney ATPase (59), and it underscores the need for caution in the analysis and interpretation of results obtained in different enzyme preparations. Further, the data presented (with differences and similarities between isoforms of the (Na,K)-ATPase suggests the need to study the effects of palytoxin on the catalytic steps of the Na^+ pump taking into account the existence of the isoforms and to compare results in different species as well.

V. EFFECTS OF PALYTOXIN ON TISSUES AND CELLS

The work that followed the identification of palytoxin as a marine toxin in the mid 1960s strongly indicates that most of the pathological actions ascribable to PTX can be explained, as Weidmann had suggested (27), by a change in the internal ionic composition of the cells through changes in the ionic permeability of the cell membrane (60). Depolarization of the cell membrane and contracture do not only occur in cardiac muscle (27) (cf. Table 2 in ref. 21) but also in smooth muscle (61), myelinated nerves (62), and skeletal muscle (63).

In 1974, Quinn et al. (64) found that extracts of eight species of Hawaiian zoanthids inhibited growth of Ehrlich ascites carcinoma in mice at doses as low as 5–80 ng/kg. These findings and those that followed have led to the notion that the activation of signal transduction pathways by palytoxin might be ascribed to palytoxin's primary action, the change in the internal cation composition of the cells, and consequently to the search for the cellular switches which become inappropriately regulated in carcinogenesis (see below).

A. Cytotoxicity

The evidence from the extensive work performed to determine the mode of action of palytoxin as a regulator of cell growth indicates that the biochemical mechanism of cytotoxicity is probably different from that of the phorbol esters. Thus, even though exposure of cells to PTX can induce signaling pathways resulting in downregulation of epidermal growth factor receptors even in the absence of Ca^{2+} (in Swiss 3T3 fibroblasts) or leading to arachidonic acid metabolism (rat liver cells), as is the case with 12-O-tetradecanoylphorbol-13-acetate (TPA), exposure to PTX does not produce HL-60 cell adhesion in vitro, nor does it bind and activate protein kinase C and is nonmitogenic in Swiss 3T3 fibroblasts (65–67). Palytoxin has been shown to be cytotoxic for normal human bronchial epithelial cells, a human lung tumor cell line, and immortalized human bronchial epithelial cells (BEAS-2B), where it produces an increase in the steady-state level of c-*myc* mRNA accompanied by an increase in thymidine uptake. The effects of palytoxin could be seen even in a ouabain-resistant cell line HUT292-DM (68).

Recent studies on the mechanism by which palytoxin produces its anticancer effects have revealed that PTX induces prolonged activation of Jun N-terminal kinase (JNK) in Swiss 3T3 cells, and that the effect is dependent on external Na^+ (69). This was further substantiated by the fact that the cytotoxic effects of palytoxin can be mimicked by gramicidin, a Na^+ ionophore. Moreover, it was shown that Ca^{2+} influx (or increase in internal Ca^{2+} concentration via the palytoxin channel) is not required for the activation of JNK by palytoxin. Thus, these results suggest that either the PTX-induced sodium influx per se or the ensuing change in the internal cation composition and/or the depolarization of the cell membrane are probably the trigger for the JNK activation. Further, since using ouabain as the activator of JNK yields results different than those seen with PTX, this provides further circumstantial evidence for the internal ion hypothesis, since ouabain inhibition of the pump does not lead to fast changes of the internal composition of the cell, which is the hallmark of PTX's action. It is of interest to mention that an amphipathic polypeptide, melittin, which produces accumulation of Na^+ and release of K^+ from human red blood cells (70) also stimulates DNA synthesis in quiescent cultures of mouse cells, which is consistent with the idea that either ion fluxes or the change in the cellular ionic composition (change in membrane potential?) signal the initiation of mitogenesis of quiescent cells (71).

Most exciting are the results of Lordanov and Magun (72) who have shown that exposure of Rat-1 cells to palytoxin leads to the activation of stress kinases (through ribosome-mediated signaling) and inhibition of protein synthesis. This phenomenon is probably a direct consequence of the loss K^+ due to PTX's action, since replacement of 75% of the external Na^+ for K^+ suppresses the effect of PTX just described. The response to PTX can be restored, however, if Na^+ is replaced by a poorly permeant cation, choline. Alternatively, the effect could also be due to the change in the membrane potential owing to the heteroexchange of cations; substantiated by the data indicating that choline replacement seems to yield larger activation of the stress kinases.

An interesting finding in the work just described is that relatively high concentrations of external K^+ (116 mM) do not prevent palytoxin's actions on Rat-1 cells, as has been suggested to be the case in red cells (21). Moreover, washout of palytoxin (100 pM) in a medium containing high K^+ fails to reverse the effects of the toxin (cf. Figure 7 in ref. 71), as has been shown to be the case in red cells (21,33). It is of interest to determine if this is due to red cells being anucleated or to the biding characteristics of palytoxin to the receptor in Rat-1 cells in particular, or to nucleated cells in general. From the data presented in this study, it is not evident that borate was present in the medium. Precedent for clear responses to PTX at relatively low concentrations (0.01–1.0 nM) exists in the work prior to 1982 (when it was found that borate increases the affinity of the receptor for PTX) and in data from excitable tissues and from whole organs as well (cf. Sections V.B and VI). The question then arises as to the role of borate, if any, on the mechanism of action of PTX in nucleated cells, since its action might be more than the trivial one of increasing the affinity of the receptor for the toxin.

B. Excitable Tissues

Effects other than membrane depolarization have been found to occur as a consequence of PTX binding to its receptor, perhaps secondary to the depolarizing effect or to the change in the cellular ionic composition. These include the stimulation of bone resorption at low PTX concentrations (0.5–10.0 pg/mL) through an increase in prostaglandin and Ca^{2+} and inhibition of the resorption at higher concentrations (>50 pg/mL) presumably due to its cytotoxicity (73). Another (perhaps secondary) action is the opening of verapamil-sensitive Ca^{2+} channels, stimulation of the release of prostaglandins, and stimulation of Ca^{2+}-independent catecholamine release from nerve terminals (74). These last results led to the proposal that the primary effect of PTX is to release endogenous transmitters, and their eventual depletion, without substantial structural damage to the muscle (75,76). However, the depolarization induced by PTX (a consequence of the exchange of K_i for Na_o presumably through the Na,K pump) even in the absence of external Ca^{2+} could lead to the release of Ca^{2+} from intracellular stores and hence stimulation of transmitter release. Clearly, more work is required in order to sort out this problem.

Other reported effects of PTX in excitable cells include a Cl^- or bicarbonate-independent acidification of chick cardiac cells observed even in the presence of ouabain (77). The cellular acidification induced by PTX at physiological pH is dependent on the presence of amiloride (78). The implication of these studies is that the Na^+/H^+ is involved in the Na^+ influx induced by PTX. However, it is possible that the flux of H^+ is a secondary effect to the influx of Na^+ or to the depolarization. Furthermore, the Na^+ pump is partially inhibited at low pH, and the steps which are H^+ sensitive have not been established as yet, but it has been shown that the action of palytoxin on the cationic permeability of human red cells is affected by pH as well (cf. Section V.D). The effects on the toxic action of palytoxin that inhibition of the pump by means other than PTX are not known at the moment, but they are important to determine in order to establish the primary effects of PTX in excitable cells (79).

An interesting aspect of the studies reported in this section is that they were conducted in the absence of borate, which immediately raises the possibility of increasing the affinity of excitable cells to palytoxin by its addition (but see discussion in Section V.A).

C. Single Channels Induced by Palytoxin

The permeability pathway induced by palytoxin was assumed to be a channel by Habermann in 1983 (57), and, as was first shown by Ikeda (80), is addition of PTX to ventricular cells clearly induces single channels permeable to cations with a slope conductance of approximately 9 pS. This was subsequently confirmed and expanded in other laboratories and this approach to the study of the action of palytoxin has led to the determination of the characteristics of the cation-permeable channels induced by the toxin in several different preparations.

The single-channel conductance of palytoxin-induced permeability pathway varies from 8–12 pS in single ventricular cells of the rat, in human erythroleukemic cells, and in oocytes and in erythrocytes of the frog (80–83) to 26 pS in mouse neuroblastoma cells (84). It has also been observed that the channels are permeable to Na^+, K^+, Cs^+, and Li^+ (see Section V.D), that the single-channel conductance is ohmic, and that the number of open channels increases with the concentration of PTX. The role of external Na^+ and of external Ca^{2+} in the opening of PTX-induced channels is dependent on the preparation used. Thus, single channels have been observed in the total absence of Na^+ in the case of rat aortic smooth muscle cells, albeit with low channel conductance (83), but are completely absent in neuroblastoma and in erythroleukemic cells (80,81), perhaps because the conductance of the channels under these conditions is below the detection limit in these preparations. The role of Ca^{2+} in the opening of the PTX-induced channels is also equivocal. Thus, total Na^+ replacement by Ca^{2+} promotes the formation of 1–2 pS channels (80,83) and estimates of the relative permeability of Ca^{2+} indicate it is very low (see Section V.D). These results then indicate that although Ca^{2+} might increase the affinity of the system for palytoxin, Ca^{2+} itself is not permeable through the palytoxin-induced permeability pathway.

Ouabain was tested as a potential inhibitor of the channels induced by palytoxin in neuroblastoma cells and in frog erythrocytes. If the cardiac glycoside was present together with palytoxin inside the patch pipette, no PTX-induced channels could be found ([PTX] = 0.1 pM, [ouabain] = 10 μM). Vanadate (another inhibitor of the Na^+,K^+-ATPase) was also effective in inhibiting the palytoxin-induced channels in a reversible way (46). On the other hand, no inhibition of the channels by ouabain was observed in the case of human erythroleukemic cells or in aortic myocytes (81,83), perhaps because the ratio of PTX to ouabain concentrations was not low enough for ouabain to overcome the action of the toxin in these preparations ([PTX] = 100 nM, [ouabain] = 200 μM).

Likewise, the role of external ions, in particular Na^+, Ca^{2+}, and H^+, is not clear at present, primarily due to the diversity of the preparations studied and the requirement of each of them for specific ions in order to maintain viability. The presence of other systems which act in parallel also precludes the identification of one of the transport systems as the primary target for palytoxin action. Thus, PTX depolarizes the neuronal cell membrane through an increase in the Na^+ (and K^+) permeability through the opening of cation-selective channels and exchange of internal potassium for external Na^+. It has been suggested that this depolarization leads to Na^+ influx and to the opening of voltage-dependent Ca^{2+} channels with the end result of an increase in internal Ca^{2+}. The Na^+ flux might also stimulate the Na/Ca exchanger or, as has also been reported, the Na/H exchange process which leads to an acidification of cells since these actions can be partially prevented by amiloride (ethylisopropylamiloride), with the rest of the PTX-activated Na^+ flux eliminated by an amiloride derivative (3,4 dichlorobenzamil) which inhibits Na^+/Ca^{2+} exchange (83).

From this brief summary, it is evident that sorting out the events which follow the binding of palytoxin to its receptor and establishing the relative importance of each of those events in excitable tissues has been and is still difficult. The basic question of whether PTX acts at one site on the cell membrane (the Na,K-ATPase) or more sites is still a matter of controversy even though there is no evidence for a receptor different from the Na,K pump. Further, the question of which of the secondary effects are due to the inhibition of the pump protein, to depolarization of the cell membrane, or to a change in the internal composition of cations has not been clarified to date. There are no data on the concentrations of PTX which produce the inhibition of the pump as opposed to the concentrations at which there is a change in the permeability of the cell membrane to cations other than in red cells and in the *Xenopus* oocyte. In human red cells, the increase in permeability induced by PTX occurs at the low end of the concentration range (picomoles to nanomoles), whereas the Na^+,K^+-ATPase is inhibited by relatively high concentrations of the toxin (micromolar) (33,85).

D. Role of Ions in Palytoxin's Action

Although there has not been any systematic study of the role of ions on the PTX effect, it has become increasingly apparent that the flux induced by the toxin requires a cation as the exchange partner. Thus, as cations in the external medium are replaced, the reversal potential changes according to the permeability of the test ion (replacing Na^+) and the increase in the permeability to Na^+ and K^+ explains the depolarizing effect of palytoxin. Replacement of Na^+ by other ions has revealed that the ion pathway opened up by palytoxin exhibits some selectivity among cations, as shown in Table 1.

The selectivity sequences obtained in the various preparations correspond to Eisenman's selectivity sequence IV, suggesting a relatively low negative electrostatic field strength of the site in the Na^+,K^+-ATPase which interacts with the ions (86).

It has also been documented that the action of palytoxin is modified by external pH, with reduction of the Na^+ permeability as the external pH is decreased, both in frog skeletal muscle (63) and in human red blood cells, in the presence of Na^+ or in the presence of Mg^{2+}, used as a Na^+ replacement (M. T. Tosteson, unpublished data). Interestingly, Forbush (87) has described that as the pH of the medium is decreased, the rate of release of the K^+ occluded by the Na,K-ATPase is decreased, which is consistent with the results in red cells if it is assumed that PTX acts on the form of the enzyme which occludes K^+, permanently opening the occlusion

Table 1 Cation Permeation Through the Palytoxin-Induced Pathway

Preparation	Selectivity sequence	Reference
Ventricular cells	$K^+ \geq Cs^+ \geq Na^+ > Li^+ \gg$ choline$^+ \approx TEA^+ \approx Ca^{2+} \approx Ba^{2+}$	80
Axon, skeletal muscle	$Na^+ > Li^+ > Cs^+ > NH^{4+}$	60,63
Neuroblastoma cells	$Na^+ \geq K^+ >$ choline$^+ >$ TMA $> Ca^{2+}$	46
Smooth muscle and human red cells	$K^+ > Rb^+ > Cs^+ > Na^+ > Li^+ \gg$ choline$^+ > TEA^+ \gg Mg^{2+}$	33,61,84

The data reported were all obtained using electrophysiological methods. The permeability ratios were calculated using the Goldman-Hodgkin-Katz equation and assuming that the contribution of anions to the palytoxin-induced current is negligible, as has been shown to be the case (33,84).

(or deocclusion) gate with the consequence of a relatively rapid release of the occluded ion (see Section IV).

VI. TOXICOLOGICAL EFFECTS

This section deals with the symptoms which arise as a consequence of intoxication due to palytoxin. The section dealing with human toxicity is a brief survey of the reported symptoms in individuals who had been allegedly poisoned with PTX even if palytoxin could not be identified as the only causative agent in an unequivocal way. Furthermore, since several isoforms of palytoxin have been identified chemically, but their individual actions have not been established, we will not make any attempt to distinguish among them, with the exception being *Lophozozymus pictor* toxin (LPTX), isolated from a xanthid crab, whose toxicity in animals has been amply documented.

In all of what follows, it is important to realize that, as far as is known, marine animals which are toxic might carry more than one toxin, so that the reported toxicologies might be the consequence of synergistic or antagonistic actions of the various toxins. Moreover, it is important to realize that different toxic species have been found to contain different proportions of toxins, making the task of establishing the potential toxic agent even more difficult.

To add to these difficulties, it has been made painfully evident that the methods of cooking the food containing the toxic agent might be a matter of life and death, since some forms of cooking increase (and others decrease) the probability of toxicity (88). The body parts of the poisonous animal ingested are also important determinants of toxicity, and thus it is of great importance that these parameters be established in each of the toxic episodes encountered.

The methodologies employed to detect toxic agents have been also quite varied, sometimes leading to conflicting results. These issues are beginning to be addressed, and with the existence of antibodies to palytoxin (18,19), the task of identification of the toxic material has been made both feasible and easier.

A. Human Toxicity

As early as 1924, there were references to poisonous fish suspected of carrying a toxin other than ciguatoxin owing to the severity of the poisoning (see refs. 12 and 89 and references therein). Since then there have been reports of toxicity of other tropical fish: mackerel (*Decapterus Macrosoma*) and blue hamphead parrotfish (*Ypsiscarus ovifrons*) as well as crabs (*Lophozozymus pictor*, *Demania alcalai*, and *Demania reynaudii*), in which the presence of palytoxin has been documented. People who were intoxicated stopped eating the poisoned animal after detecting a bitter, metallic taste. This was almost immediately followed by symptoms of toxicity such as nausea, vomiting, and diarrhea with mild to acute lethargy. Several hours after ingestion, patients showed myoblobinuria, burning sensation around the mouth and in the extremities, impairment of sensation, muscle spasms and tremor myalgia, prostration, dyspnea (shortness of breath), and dysphonia. Results of laboratory tests showed elevated CPK, LDH, and SGOT and high levels of transaminases and hyperkalemia (13,14). Autopsy results suggest that death was due to myocardial injury due to palytoxin (?) poisoning.

In cases where there were remains of the toxic food ingested, palytoxin was found mostly in the muscle, liver, ovary, and digestive tract. Interestingly, it has been found that palytoxin levels in crabs are highest in the viscera (pancreas and liver) and gills, but it was not found in the stomach. In the cases when the analysis was done using chromatography, it was found that there were other toxins present as well (12,14,16).

Very recently (90), palytoxin has been identified as the cause of clupeotoxism, or poisoning due to some species of sardines and herrings from tropical seas. A woman and a child ate two different fish, and a cat ate the remainder of what turned out to be a toxic fish. The woman stopped eating after she detected an unusual bitter taste. She also reported other symptoms including general malaise followed by uncontrollable vomiting and diarrhea during the first 2 h, plus tingling of the extremities and delirium. She died in the hospital the next day. The cat died 15 minutes after eating the cooked fish. The child did not show any sign of intoxication. The toxin was identified using the remainder of the fish (mainly the head) taking advantage of mouse bioassays, hemolysis assays, and mass spectrometry and was found to be very similar if not identical to palytoxin. The particular species of sardine investigated in this report, *Herklotsichthys quadrimaculatus*, previously identified as the vector for clupeotoxism in the Philippines, Fiji, and Madagascar, is known to be a bottomfeeder. This gives further support to the circumstantial evidence that *Ostreopsis* spp., which also grows on the bottom of tropical seas (91), might be the source of the toxin (6).

B. Animal Toxicity

There is a wide range of toxicity attributed to palytoxin, which depends on the host species, on the route of administration of the toxin, and probably on the species of *Palythoa* as well. Compounding the problem of the determination of the lethality doses (LD50) in various species is the problem of instability of the isolated material which plagued the field until relatively recently. Nonetheless, studies using terrestrial animals have shown that PTX is one of the most toxic marine compounds known, with a reported LD50 of as little as 33 ng/kg when administered intravenously (i.v.) in dogs. The interesting issue, which has not been addressed in depth to date, is that of the defense mechanism which marine organisms develop to protect themselves from the toxin's action. In this context, the report that a saxitoxin inducible protein complex has been found in crabs (92) is of interest and, as suggested by Endean (88), might indicate that fish can induce the formation of specific proteins which bind toxins to which they are exposed. The work on the actions of palytoxin on several tissues and organs has been previously reviewed by Ibrahim and Shier (20), Habermann (21), and Frelin (94).

Preliminary screening tests for toxicity of different batches of palytoxin were begun by McCreesh in 1965 (as cited in ref. 95). Vick and collaborators (96) went on to determine the degree of toxicity when PTX is administered in several animal species through various routes. Their results indicate that the LD50s 24 h after intravenous injection (route most effective in all animals tested) vary from 0.025 μg/kg in rabbits and about the same for dogs to 0.45 μg/kg in mice, with monkeys, rats, and guinea pigs around 0.9 μg/kg. Testing other routes of administration in the rat, they found that the LD50s increased following intravenous $<$ intramuscular $<$ subcutaneous $<$ intraperitoneal $<$ intrarectal $<$ intragastric from 0.09 to $>$40 μg/kg, respectively (Table 2).

Injection of sublethal doses of PTX produce elevated cortisol levels in plasma, which confers protection from subsequent lethal injections and lead to the conclusion that the probable cause of acute death caused by this toxin is due to a very high reduction in blood flow due to vasoconstriction (95). This was subsequently demonstrated by Weidmann (27) and by Ito et al. in their work on intact, anesthetized dogs (as described in ref. 97).

The pathologies of several organs have been described in studies in which mice and rats were given sublethal doses of PTX. The most salient findings are:

Lymphoid tissue: a low dose of PTX in mice (0.25 μg/kg) produced reversible damage to the lymphoid tissues, the ratio of lymphocytes to total leukocytes was reversibly

Table 2 Toxic Signs and Dose in Various Animals Injected with Palytoxin

Animal tested	Route of administration	LD50 (μg/kg)	Toxic signs
Monkeys	i.v.	0.078	Ataxia, drowsiness, weakness of limbs followed by death.
Mice	i.v.	0.45	Drowsiness, inaction, epilation followed by prostration, dyspnea, and convulsion prior to death.
Dogs	i.v.	0.033	Defecation, vomiting (sometimes bloody) followed by ataxia, weakness, collapse, and death.
	i.m.	0.08	Same as i.v. with longer onset times to toxicity and death. Local irritation and swelling at the site of injection.
Rats	i.v.	0.089	Symptoms as in mice.
	i.m.	0.24	Symptoms as in dog.
	s.c.	0.4	Symptoms as in i.m. with even longer onset.
	Percutaneous	~0.5	Blanching, swelling and necrosis of the blanched and surrounding area of the skin.
	Ocular	0.4	Tearing, irritation, swelling, edema followed by closure with exudate of pus and blood, corneal ulceration, and opacity.
	i.g.	>40	No toxic signs over 72 h.
	i.r.	>10	Same as i.g.

LD50 corresponds to the lethal concentration at 24 h; i.v., intravenous injection of a palytoxin in aqueous solutions; i.m., intramuscular; s.c., subcutaneous; i.g., intragastric; i.r., intrarectal.
Source: Data from ref. 95.

reduced, no injury to the adrenal gland, and severe damage (reversible) to thymus and spleen, which were exacerbated upon adrenalectomy. There was no recovery of lymphocytes in blood and in fibrinous changes on the serosa of liver and spleen caused by peritonitis. The investigators compared these effects with those of ciguatera poisoning, where they did not find apoptosis of lymphocytes in the thymus (98).

Intestinal tissue: a maximum nonlethal dose of PTX given intraperitoneally to mice also showed differences in the pathology when compared to the effects induced by ciguatera toxin. PTX produced diarrhea 16–24 h postinjection of a transient nature and small volume, with blood and mucus in the feces, with the site of injury being the small intestine and with very short duration and large volume, no blood in feces and the colon as the site. The insult was reversible (26,98).

These studies are important for the differential diagnosis of the type of toxin ingested after consumption of contaminated seafood, and they bring forth again the need for careful determination of the damage that each of the toxins produces in order to be able to determine the molecule responsible for intoxication. It is also necessary to realize that since, in general, the contents of toxic seafood will not be composed of only one toxin, studies such as the ones just described need to be done testing more than one compound per animal species, both one at a time and in combination. A daunting task indeed! Nonetheless, the results of this type of study will help improve human health and alleviate the economic toll that toxicity in marine animals (animals in general) takes on humankind.

REFERENCES

1. D Malo. Hawaiian antiquities. 2nd ed. Translated by Dr. NB Emerson, Honolulu: Hawaiian Gazette, 1903.
2. RE Moore, PJ Scheuer. Palytoxin: a new marine toxin from a coelenterate. Science 172:495–498, 1971.

3. RE Moore, P Helfrich, GML Patterson. The deadly seaweed of Hana. Oceanus 25:54–63, 1982.
4. RW Armstrong, J-M Beau, SH Cheon, WJ Christ, H Fuhioka, W-H Ham, LD Hawkins, H Jin, SH Kang, Y Kishi, MJ Martinelly, WW McWhorter Jr, M Mizumo, M Nakata, AE Stutz, FX Talamas, M Taniguchi, JA Tino, K Ueda, J-I Uemishi, JB White, M Yonaga. Total synthesis of a fully protected palytoxin carboxylic acid. J Am Chem Soc 111:7525–7530, 1989.
5. RE Moore and G Bartolini. Structure of palytoxin. J Am Chem Soc 103:2491–2494, 1981.
6. NM Carballeira, A Emiliano, A Sostre, JA Restituyo, IM Gonzalez, GM Colon, CG Tosteson and TR Tosteson. Fatty acid composition of bacteria associated with the toxic dinoflagellate *Ostreopsis lenticularis* and with Caribbean Palythoa species. Lipids 33:627–632, 1998.
7. D Uemura, Y Hirata, T Ewashita, H Naoke. Studies on palytoxins. Tetrahedron 41:1007–1017, 1985.
8. M Usami, M Satake, S Ishida, A Inoue, Y Kan, T Yasumoto. Palytoxin analogs from the dinoflagellates *Ostreopsis siamensis*. J Am Chem Soc 177:5389–5390, 1995.
9. CO Lau, HE Khoo, R Yuen, M Wan, CH Tan. Isolation of a novel fluorescent toxin from the coral reef crab, *Lophozozymus Pictor*. Toxicon 31:1343–1345, 1993.
10. CO Lau, CH Tan, HE Khoo, R Yuen, RJ Lewis, GP Corpuz, GS Bignami. *Lophozozymus Pictor* toxin: a fluorescent structural isomer of palytoxin. Toxicon 33:1373–1377, 1995.
11. Y Hasimoto, N Fusetani, S Kimura. Aluterin: a toxin of filefish, *Alutera scripta*, probably originating from a zoantharian *Palythoa tuberculosa*. Bull Jpn Soc Sci Fisheries 35:1086–1093, 1969.
12. M Fukui, M Murata, A Inoue, M Gawel, T Yasumoto. Occurrence of palytoxin in the trigger fish *Melichtys Vidua*. Toxicon 25:1121–1124, 1987.
13. AM Kodama, Y Hokama, T Yasumoto, M Fukui, SJ Manea, N Sutherland. Clinical and laboratory findings implicating palytoxin as cause of ciguatera poisoning due to *Decapterus macorsoma* (mackerel). Toxicon 27:1051–1053, 1989.
14. H Okano, H Masuoka, S Kimei, T Seko, S Koyabu, K Tsuneoka, T Tamai, K Ueda, S Nakazawa, M Sugawa, H Suzuki, M Watanabe, R Yatani, T Nakano. Rhabdomyolysis and myocardial damage induced by palytoxin, a toxin of blue humped parrotfish. Intern Med 37:330–333, 1998.
15. T Yasumoto, D Yasumura, Y Ohizumi, M Takahashi, AC Alcala, LC Alcala. Palytoxin in two species of xanthid crab. Philippines Agric Biol Chem 50:163–167, 1986.
16. AC Alcala, LC Alcala, JS Garth, D Yasamura, T Yasamoto. Human fatality due to ingestion of the crab *Demania Reynaudii* that contained a palytoxin-like toxin. Toxicon 26:105–107, 1988.
17. VM Mahnir, EP Kozlovskaya, AI Kalinovsky. Sea anemone *Radianthus macrodactylus*—a new source of palytoxin. Toxicon 30:1449–1456, 1992.
18. GS Bignami, TJG Raybould, ND Sachinvala, PG Grothaus, SB Simpson, CB Lazo, JB Byrnes, RE Moore, DC Vann. Monoclonal antibody-based enzyme-linked immunoassays for the measurement of palytoxin in biological samples. Toxicon 30:687–700, 1992.
19. L Levine, H Fujiki, HB Gjika, H van Vunakis. Production of antibodies to palytoxin: neutralization of several biological properties of palytoxin. Toxicon 25:1273–1282, 1987.
20. A-R Ibrahim, WT Shier. Palytoxin: mechanism of action of a potent marine toxin. J. Toxicol.-Toxin Rev 6:159–187, 1987.
21. E Habermann. Palytoxin acts through the Na^+,K^+-ATPase. Toxicon 27:1171–1187, 1989.
22. S Gleibs, D Mebs, B Werding. Studies on the origin and distribution of palytoxin in a Caribbean coral reef. Toxicon 33:1531–1537, 1995.
23. D Mebs. Occurrence and sequestration of toxins in food chains. Toxicon 36:1519–1522, 1998.
24. TR Tosteson, RF Bard, IM Gonzalez, DL Ballantine, GS Bignami. Toxin diversity in the ciguatera food chain. Proceedings of the International Symposium on Ciguatera and Marine Natural Products, Honolulu, Hawaii, 1985, pp 73–87.
25. RL Post, C Hegyvary, S Kume. Activation by adenosine triphosphate in the phosphorylation kinetics of sodium and potassium ion transport adenosine triphosphatase. J Biol Chem 247:6530–6540, 1972.
26. E Ito, M Ohskusu, T Yasumoto. Intestinal injuries caused by experimental palytoxicosis in mice. Toxicon 34:643–652, 1996.
27. S Weidmann. Effects of palytoxin on the electrical activity of dog and rabbit heart. Experientia 33: 1487–1489, 1977.
28. PF Cranefield, RS Aronson. Cardiac Arrythmias: The Role of Triggered Activity and Other Mechanisms. New York: Futura, 1988, p 201.

29. S Garcia-Castineiras, J White, E Toro-Goyco. Isolation and preliminary characterization of NaK-ATPase inhibitor from the toxin coelenterate palythoa. Fed Proc 34:505, 1975.
30. G A-Hilger, GS Chhatwal, H-J Hessler, E Habermann. Changes in erythrocyte permeability due to palytoxin as compared to amphotericin B. Biochim Biophys Acta 688:486–494, 1982.
31. A Sanyal, NG Hemming, WS Broecker, DW Lea, HJ Spero, GN Hanson. Oceanic pH control on the boron isotopic composition of foraminifera: evidence from culture experiments. Paleoceanography 11:513–517, 1996.
32. E Grell, E Lewitzki, D Uemura. Interaction between palytoxin and purified Na,K-ATPase. In: JC Skou, JG Nørby, AB Maunsbach, M Esmann, eds. Progress in Clinical and Biological Research. Vol 268B: The Na^+,K^+-Pump. Part B: Cellular Aspects. New York: Liss, 1988, pp 393–400.
33. MT Tosteson, JA Halperin, Y Kishi, DC Tosteson. Palytoxin induces an increase in the cation conductance of red cells. J Gen Physiol 98:969–985, 1991.
34. E Habermann, GS Chhatwal. Ouabain inhibits the increase due to palytoxin of cation permeability of erythrocytes. Naunyn Schmiedebergs Arch Pharmacol 319:101–107, 1982.
35. K-y Xu. Inhibition of H^+-transporting ATPase, Ca^{2+}-transporting ATPase and H^+/K^+-transporting ATPase by strophanthidin. Biochim Biophys Acta 1159:109–112, 1992.
36. A Yoda, MR Rosner, P Morrison, S Yoda. Interaction of NaK-ATPase proteoliposomes with palytoxin at picomolar concentrations. In: JH Kaplan, P De Weer, eds. The Sodium Pump: Recent Developments. Society of General Physiologists, 44th Annual Symposium. New York: Rockefeller University Press, 1991, pp 499–504.
37. BM Anner, M Moosmayer. Na,K-ATPase characterized in artificial membranes. 2. Successive measurement of ATP-driven Rb-accumulation, ouabain-blocked Rb-flux and palytoxin-induced Rb-efflux. Molec Memb Biol 11:247–254, 1994.
38. DM zu Herringdorf, E Habermann, G Scheiner-Bobis. Palytoxin induces K^+ release from yeast cells expressing Na^+,K^+-ATPase. In: E Bamberg, W Schoner, eds. Proceeding of the VIIth International Conference on the Sodium Pump. Steinkopf Verlag, Darmstadt, 385–388, 1994.
39. PL Jørgensen, JH Rasmussen, JM Nielsen, PA Pedersen. Transport-linked conformational changes in (Na,K)-ATPase. Structure-function relationships of ligand binding and E1-E2 conformational transitions. In: LA Beaugé, DC Gadsby, PJ Garrahan, eds. Na/K-ATPase and related transport ATPases. Structure, mechanism and regulation. Ann NY Acad Sci 834:161–174, 1997.
40. IM Glynn. The Na^+,K^+-transporting adenosine triphosphatase. In: A Martonosi, ed. Enzymes of Biological Membranes. New York: Plenum Press, 1985, pp 35–114.
41. JC Skou. Effect of ATP on the intermediary steps of the reaction of the (Na + K)-dependent enzyme system. III. Effect on the *p*-nitrophenylphosphatase activity of the system. Biochim Biophys Acta 339:258–273, 1974.
42. JC Skou, M Esmann. The Na^+,K^+-ATPase. J Bioenerg Biomem 24:249–261, 1992.
43. E Habermann. Action and binding of palytoxin, as studied with brain membranes. Naunyn Schmiedebergs Arch Pharmacol 323:269–275, 1983.
44. H Böttinger, L Béress, E Habermann. Involvement of (Na^+,K^+-ATPase) in binding and actions of palytoxin on human erythrocytes. Biochim Biophys Acta 861:165–176, 1986.
45. MT Tosteson, DR Scriven, A Bharadwaj, J Arnadottir, DC Tosteson. Interactions of palytoxin with the Na,K-ATPase. Where are those sites? In: LA Beaugé, DC Gadsby, PJ Garrahan, eds. Na/K-ATPase and related transport ATPases. Ann NY Acad Sci 834:424–425, 1997.
46. SY Kim, KA Marx, CH Wu. Involvement of the Na,K-ATPase in the induction of ion channels by palytoxin. Naunyn Schmiedebergs Arch Pharmacol 351:542–554, 1995.
47. G Scheiner-Bobis, H. Schneider. Palytoxin-induced channel formation within the Na^+/K^+-ATPase does not require a catalytically active enzyme. Eur J Biochem 248:717–723, 1997.
48. X Wang, J-D Horisberger. Palytoxin effects through interaction with the Na,K-ATPase in *Xenopus* oocyte. FEBS Lett 409:391–395, 1997.
49. Y Ohizumi, S Shibata. Mechanism of the excitatory action of palytoxin and *N*-acetylpalytoxin in the isolated guinea-pig vas deferens. J Pharmacol Exp Ther 214:209–212, 1980.
50. MT Tosteson, DRL Scriven, AK Bharadwaj, Y Kishi, DC Tosteson. Interaction of palytoxin with red cells: structure-function studies. Toxicon 33:799–807, 1995.

51. E Habermann, GS Chhatwal. Ouabain inhibits the increase due to palytoxin of cation permeability of erythrocytes. Naunyn Schmiedebergs Arch Pharmacol 319:101–107, 1982.
52. H Ozaki, H Nagase, N Urakawa. Interaction of palytoxin and cardiac glycosides on erythrocyte membrane and (Na^+,K^+) ATPase. Eur J Biochem 152:475–480, 1985.
53. Y Ishida, K Takagi, M Takahashi, N Satake, S Shibata. Palytoxin isolated from marine coelenterates. The inhibitory action on (Na^+,K^+)-ATPase. J Biol Chem 258:7900–7902, 1983.
54. E Grell, E Lewitzi, A Ifftner. Dynamics and mechanism of cardiac glycoside binding to Na^+,K^+-ATPase. In: IM Glynn, E Ellory, eds. The Sodium Pump. Cambridge, UK: Company of Biologists, 1995, 289–294.
55. G Gárdos. The role of calcium in the potassium permeability of human erythrocytes. Acta Physiol Acad Sci Hung 15:121–125, 1959.
56. JJ Monroe, AH Tashjian, Jr. Actions of palytoxin on Na^+ and Ca^{2+} homeostasis in human osteoblast-like Saos-2 cells. Am J Physiol 269:C582–C589, 1995.
57. GS Chhatwal, H-J Hessler, E Habermann. The action of palytoxin on erythrocytes and resealed ghosts. Formation of small, nonselective pores linked with Na^+,K^+-ATPase. Naunyn Schmiedebergs Arch Pharmacol 323:261–268, 1983.
58. Y Ishida, K Takagi, M Takahashi, N Satake, S Shibata. Palytoxin isolated from marine coelenterates. The inhibitory action on (Na,K)-ATPase. J Biol Chem 258:7900–7902, 1983.
59. M Steinberg, SJD Karlish. Studies on conformational changes in Na,K-ATPase labeled with 5-iodoacetamidofluorescein. J Biol Chem 264:2726–2734, 1989.
60. I Muramatsu, D Uemura, M Fujiwara, T Narahashi. Characteristics of palytoxin-induced depolarization in squid axons. J Pharmacol Exp Ther 231:488–494, 1984.
61. K Ito, H Karaki, N Urakawa. The mode of contractile action of palytoxin on vascular smooth muscle. Eur J Pharmacol 46:9–14, 1977.
62. JM Dubois, JB Cohen. Effect of palytoxin on membrane potential and current of frog myelinated fibers. J Pharmacol Exp Ther 201:148–155, 1977.
63. E Ecault, MP Sauviat. Characterization of the palytoxin induced sodium conductance in frog skeletal muscle. Br J Pharmacol 102:523–529, 1991.
64. RJ Quinn, M Kashiwagi, RE Moore, TR Norton. Anticancer activity of zoanthids and the associated toxin, palytoxin, against ehrlich ascites tumor and p-388 lymphocytic leukemia in mice. J Pharmaceut Sci 63:257–260, 1974.
65. L Levine, H Fujiki. Stimulation of arachidonic acid metabolism by different types of tumor promoters. Carcinogenesis 6:1631–1634, 1985.
66. H Fujiki, M Suganuma, M Nakayasu, H Hakaii, T Horuchi, S Takayamna, T Sugimura. Palytoxin is a non-12-*O*-tetradecanoylphorbol-13-acetate type tumor promoter in a two-stage mouse skin carcinogenesis. Carcinogenesis 7:707–710, 1986.
67. EV Wattenberg, H Fujiki, MR Rosner. Heterologous regulation of the epidermal growth factor receptor by palytoxin, a non–12-*o*-tetradecanoylphorbol-13-acetate-type tumor promoter. Cancer Res 47: 4618–4622, 1987.
68. C Bonnard, JF Lechner, BI Gerwin, H Fujiki, CC Harris. Effects of palytoxin or ouabain on growth and squamous differentiation of human bronchial epithelial cell *in vitro*. Carcinogenesis 9:2245–2249, 1988.
69. DW Kuroki, GS Bignami, EV Wattenberg. Activation of stress-activated protein kinase/c-Jun N-terminal kinase by the non–TPA-type tumor promoter palytoxin. Cancer Res 56:637–644, 1996.
70. MT Tosteson, SJ Holmes, M Razin, DC Tosteson. Melittin lysis of red cells. J Membrane Biol 87: 35–44, 1985.
71. E Rozengurt, TD Gelehrter, A Legg, P Pettican. Melittin stimulates Na entry, Na-K pump activity and DNA synthesis in quiescent cultures of mouse cells. Cell 23:781–788, 1981.
72. MS Lordanov, BE Magun. Loss of cellular K^+ mimics ribotoxic stress. Inhibition of protein synthesis and activation of the stress kinases SEK1/MKK4, stress activated protein kinase/c-Jun NH_2-terminal kinase 1, and p38/HOG1 by palytoxin. J Biol Chem 273:3528–3534, 1998.
73. M Lazzaro, AH Tashjian, Jr, H Fujiki, L Levine. Palytoxin: an extraordinarily potent stimulator of prostaglandin production and bone resorption in cultured mouse calvariae. Endocrinology 120:1338–1345, 1987.

74. H Nagase, H Karake. Palytoxin-induced contraction and release of prostaglandins and norepinephrine in the aorta. J Pharmacol Exp Ther 242:1120–1125, 1987.
75. J Posangi, JB Harris, MA Zar. Palytoxin-induced transmitter release in the autonomic nervous system of the rat. Toxicon 32:965–975, 1994.
76. I Amir, JB Harris, MA Zar. The effect of palytoxin on neuromuscular junctions in the anococcygeus muscle of the rat. J Neurocytol 26:367–376, 1997.
77. C Frelin, P Vigne, J-P Breittmayer. Palytoxin acidifies chick cardiac cells and activates the Na^+/H^+ antiporter. FEBS Lett 264:63–66, 1990.
78. M Yoshizumi, H Houchi, Y Ishimura, Y Masuda, K Morita, M Oka. Mechanism of palytoxin induced Na^+ influx into cultured bovine adrenal chromaffin cells: possible involvement of Na^+/H^+ exchange system. Neurosci Lett 130:103–106, 1991.
79. W Walz, EC Hinks. Extracellular hydrogen ions influence channel-mediated and carrier-mediated K^+ fluxes in cultured mouse astrocytes. Neuroscience 20:341–346, 1987.
80. M Ikeda, K Mitani, K Ito. Palytoxin induces a nonselective cation channel in single ventricular cells of rat. Naunyn Schmiedebergs Arch Pharmacol 337:591–593, 1988.
81. J Rettenger, W Schwarz. Ion-selective channels in K562 cells: a patch-clamp analysis. J Basic Clin Physiol Pharmacol 5:27–44, 1994.
82. SY Kim, KA Marx, CH Wu. Involvement of the Na,K-ATPase in the induction of ion channels by palytoxin. Naunyn Schmiedebergs Arch Pharmacol 351:542–554, 1995.
83. C Van Renterghem, C Frelin. 3,4 Dichlorobenzamil-sensitive, monovalent cation channel induced by palytoxin in cultured aortic myocytes. Br J Pharmacol 109:859–865, 1993.
84. B R-Dubois, J-M Dubois. Characterization of palytoxin-induced channels in mouse neuroblastoma cells. Toxicon 28:1147–1158, 1990.
85. K Ishii, M Ikeda, K Ito. Characteristics of palytoxin-induced cation currents and Ca^{2+} mobilization in smooth muscle of rabbit portal vein. Naunyn Schmiedebergs Arch Pharmacol 355:103–110, 1997.
86. G Eisenman, R Horn. Ionic selectivity revisited: the role of kinetic and equilibrium processes in ion permeation through channels. J Membrane Biol 76:197–225, 1983.
87. B Forbush III. Rapid release of ^{42}K and ^{86}Rb from an occluded state of the Na,K-pump in the presence of ATP or ADP. J Biol Chem 262:11104–11115, 1987.
88. R Endean, JK Griffith, JJ Robins, LE Llewellyn, SA Monks. Variation in the toxins present in ciguateric narrow-barred spanish mackerel, *Scomberomorus commersoni*. Toxicon 31:723–732, 1993.
89. R Bagnis. Concerning a fatal case of ciguatera poisoning in the Tuamotu Islands. Clin Toxicol 3: 579–583, 1970.
90. Y Onuma, M Satake, T Ukena, J Roux, S Chanteau, N Rasolofonirina, M Ratsimaloto, H Naoku, T Yasumoto. Identification of putative palytoxin as the cause of clupeotoxism. Toxicon 37:55–65, 1998.
91. MA Faust. Observations of sand-dwelling toxic dinoflagellates (Dinophyceae) from widely differing sites, including two new species. J Phycol 31:996–1103, 1995.
92. KG Barber, SS Kitts, PM Townsley, DS Smith. Appearance and partial purification of a high molecular weight protein in crabs exposed to saxitoxin. Toxicon 26:1027–1034, 1988.
93. G Scheiner-Bobis. Ion-transporting ATPases as ion channels. Naunyn-Schmiedeberg's Arch Pharmacol 357:477–482, 1998.
94. C Frelin, C Van Renterghem. Palytoxin. Recent electrophysiological and pharmacological evidence for several mechanisms of action. Gen Pharmacol 26:33–37, 1995.
95. JS Wiles, JA Vick, MK Christensen. Toxicological evaluation of palytoxin in several animal species. Toxicon 12:427–433, 1974.
96. JA Vick, JS Wiles. The mechanism of action and treatment of palytoxin poisoning. Toxicol Appl Pharmacol 34:214–223, 1975.
97. K Ito, N Urakawa, H Koike. Cardiovascular toxicity of palytoxin in anesthetized dogs. Arch Int Pharmacodyn 258: 146–154, 1982.
98. E Ito, M Ohkusu, K Terao, T Yasumoto. Effects of repeated injections of palytoxin on lymphoid tissues in mice. Toxicon 35:679–688, 1997.
99. E Ito, M Ohskusu, T Yasumoto. Intestinal injuries caused by experimental palytoxicosis in mice. Toxicon 34:643–652, 1996.

26

Freshwater Cyanobacterial Neurotoxins: Ecobiology, Chemistry, and Detection

Kaarina Sivonen
University of Helsinki, Helsinki, Finland

I. CYANOBACTERIAL NEUROTOXINS: ECOBIOLOGY

In freshwater, harmful algal blooms are mostly attributed to cyanobacteria (blue-green algae). The main types of toxins produced by cyanobacteria are hepatotoxins and neurotoxins. Three types of cyanobacterial neurotoxins are known: (homo)anatoxin-a, anatoxin-a(S) and saxitoxins (also called paralytic shellfish poisons, PSPs). Mass occurrences of neurotoxic cyanobacteria have been reported from America, Europe, and Australia where they have caused numerous animal poisonings. (Tables 1 and 2). Interestingly, in Australia, only saxitoxins are common; anatoxins have not been found. In Europe and America, the occurrence of all three types of cyanobacterial neurotoxins have been reported. Recent chemical analysis has shown anatoxin-a also to occur in freshwater in Asia (North Korea and Japan). Furthermore, it is obvious that neurotoxins are produced not only by planktic cyanobacteria but also by benthic mat-forming species (18,23).

Cyanobacterial neurotoxins block neurotransmission, but do not cause any apparent organ damage, which can be seen at autopsy. In mouse bioassays, death by respiratory arrest occurs quickly (within 2–30 min). Signs of poisonings include respiratory distress, tremors, convulsions, fasciculations and, in some toxins, salivation (34). (More details of toxicity are given in Chapter 27.) All three types of neurotoxic blooms have caused several poisonings among wild and domestic animals (Table 2). Typically, very high concentrations of toxins were detected in the samples associated with animal poisonings.

Cyanobacterial toxins also affect the water ecosystem. In several experiments, it has been shown that cyanobacterial toxins have toxic effects on fish, certain zooplankton species, and macrophytes (35). Most of these studies have been done using cyanobacterial hepatotoxins. Gentile and Maloney (36) found that fish (*Fundus heteroclitus* and *Cyprinodon variegatus*) were more sensitive to saxitoxins in *Aphanizomenon* extracts than mice by intraperitoneal injection. Also, microcrustaceans, *Notemiconus crysoleucas* and *Bosmina longirostris*, were killed by the extracts, whereas *Daphnia catawba* was much more resistant and cyclopoid copepods, ostracoids, and chydoric cladocerans were unaffected by the toxins at concentrations of 2 mg/mL. Organisms may differ in their sensitivity to the toxins depending on the strain or even a clone and the stage of the life cycle. The cyanobacterial toxins may thus change the food web's structures in the water ecosystem.

Table 1 Reported Occurrences of Neurotoxic Blooms and Specific Neurotoxins

Country	Frequency of neurotoxic blooms	Type of neurotoxins	Reference
Australia	56/231 (25%)	Saxitoxins	1,2
Australia	24/31 (77%)	Saxitoxins	3
Brazil	A sample	Saxitoxins	4
Canada	5/24 (21%) Few bloom samples and cyanobacteria isolates	Neurotoxins later identified as Anatoxin-a and Anatoxin-a(S)	5,6
Denmark	20/296 (7%)	Anatoxin-a(S), (9 samples) Saxitoxins (11 samples)	7
Finland	29/188 (15%)	Anatoxin-a (14 samples) (+ unknown neurotoxins, possibly saxitoxins)	8,9
Germany	69/329 (21%)	Anatoxin-a	10
Ireland	Detected in three lakes	Anatoxin-a	11
Italy	Few samples	Anatoxin-a	12
Japan	9/14 (64%)	Anatoxin-a (low amounts)	13
North Korea	4/26 (15%)	Anatoxin-a	14
Norway	6/64 (9%) Few samples	Neurotoxins (not analyzed) Anatoxin-a and Homoanatoxin-a	15,16
Sweden	15–25/248 (6–10%)	Neurotoxins (not analyzed)	17
UK	Few samples	Anatoxin-a	18
USA	Several samples	Neurotoxic (not analyzed) Anatoxin-a(S)	19,20, 21–24
	Several samples	Saxitoxins Anatoxin-a	

The development of accurate detection methods for cyanobacterial toxins (see Section III) has made it possible to study the toxin production of cyanobacteria, the environmental concentrations of these toxins as well as the fate of the cyanobacterial toxins in the environment and water treatment processes.

The determination of toxin concentrations in environmental samples has been mostly done from the concentrated, lyophilized bloom samples (particulate material, which contain not only cyanobacterial cells but usually also some zooplankton) and expressed as milligrams per microgram per gram of dry weight. The highest amounts of cyanobacterial toxins found in the environment are milligrams of toxins per gram of dry weight (d.w.) (37). The highest anatoxin-a concentration was reported in Finland (more than 4 mg of anatoxin-a per gram of dry weight), anatoxin-a(S) in Denmark (more than 3 mg of anatoxin-a(S) equivalents per gram of d.w.), and saxitoxins in Australia (over 3 mg of saxitoxins per gram of d.w.) (see Table 2) (37).

Anatoxin-a is produced by *Anabaena flos-aquae*, *Anabaena* spp. (*flos-aquae/lemmermannii/circinalis*-group), *Anabaena planktonica, Aphanizomenon, Cylindrospermum*, and benthic *Oscillatoria* (8,11,12,18,38). The anatoxin-a–containing blooms and isolated cyanobacterial strains have been reported mostly from North America and Europe. Anatoxin-a has been shown to be present in some Japanese *Anabaena* and *Microcystis* strains at very low concentrations (13,39). Very recently, concentrations from 0.4 up to 1.4 mg toxin per gram lyophilized bloom material were measured in bloom samples of North Korean lakes (14). Homoanatoxin-a, an anatoxin-a homologue, is produced by an *Oscillatoria formosa* strain originating from Norway (40) and also is found in Norwegian bloom samples (16). Anatoxin-a(S) was first characterized

Table 2 Animal Poisonings Caused by Cyanobacterial Neurotoxins[a]

Location	Animal affected	Toxin detected, concentrations	Methods used to detect toxin	Producer of toxin	Reference
USA, Montana, Grayling Arm of the Hebgen Lake, June–July 1977	8 dogs, 30 cattle	Anatoxin-a	Mouse bioassay, chemical comparison to anatoxin-a	*Anabaena flos-aquae*	26
Finland, L. Sääskjärvi, August 1985	2 cows	Anatoxin-a, 2.8 mg/g d.w. of bloom material	Mouse bioassay, GC-MS	*Anabaena* spp.	8,9
Finland, L. Säyhteenjärvi, August 1986	3 cows	Anatoxin-a, 3.7 mg/g d.w. of bloom material	Mouse bioassay, HPLC, GC-MS	*Anabaena* spp.	8,9
Canada, Steele Lake, Edmonton, Alberta, August 1985	Over 1000 bats, 24 mallards & American wigeons	Anatoxin-a	GC-MS	*Anabaena flos-aquae*	27
Canada, Alberta	16 cows	Anatoxin-a	GC-MS	ND (most likely *Anabaena flos-aquae*)	28
Scotland, Loch Insh, July 1990, 1991	4 dogs	Anatoxin-a	Mouse bioassay (GC-MS)	*Oscillatoria* benthic	18
Ireland, Caragh Lake, 1992, 1993, 1994	Dogs, canine	Anatoxin-a, 444 µg/L of lake water	HPLC	*Oscillatoria* benthic	11
USA, Richmond Lake, SD, August–September 1985	14 dogs, 2 calves, fish, 1 great blue heron, 1 muskrat	Anatoxin-a(S)	Mouse bioassay, HPLC, acetylcholinesterase inhibition assay	*Anabaena flos-aquae*	22
USA, Pond near Tolono, IL, July 1986	5 muscovy ducks	Anatoxin-a(S)	Mouse bioassay, acetylcholinesterase inhibition assay	*Anabaena flos-aquae*	29
USA, Pond near Griggsville, IL, September 1986	4 sows, one boar and 8 pigs	Anatoxin-a(S)	Mouse bioassay, acetylcholinesterase inhibition assay	*Anabaena flos-aquae*	29
Denmark, Lake Knud sø, 1993, 1994 (June–July)	Over 20 birds, 1 dog	Anatoxin-a(S), 0.8–3.3 mg anatoxin-a(S) equivalents/g 4.2 pure anatoxin-a(S) mg/g d.w. of bloom sample (by toxin isolation)	Mouse bioassay, acetylcholinesterase inhibition assay, HPLC, MS, NMR	*Anabaena lemmermannii*	30,31
Australia, Darling river, 1991	2000 cattle and sheep (estimate)	Saxitoxins	Mouse bioassay; subsequently identified by HPLC and MS	*Anabaena circinalis*	32
Australia	14 sheep	Saxitoxins, 1.7–2.5 mg/g d.w. bloom material	Mouse bioassay, HPLC	*Anabaena circinalis*	33

HPLC, high-performance liquid chromatography; GS, gas chromatography; MS, mass spectrometry.

[a] In the previous literature, there are numerous descriptions of poisonings caused by cyanobacterial neurotoxins but owing to missing analytic methods, or since the toxin structures were unknown, the toxins were not identified (e.g., see ref. 25). This list includes rather recent verified cases in which the presence and type of the neurotoxin was determined (and possibly quantified).

from a Canadian *Anabaena flos-aquae* strain NRC 525-17 (41), and it has been found in several bloom samples in the United States (22,29). In Denmark, it was detected in *Anabaena lemmermannii* strains and bloom samples (30,31). Saxitoxins have been found in cyanobacteria and marine dinoflagellates. They have been characterized from *Aphanizomenon flos-aquae*, *Anabaena circinalis*, *Lyngbya wollei*, and *Cylindrospermopsis raciborskii*. In Australia, saxitoxin-containing blooms are solely formed by *Anabaena circinalis* (3,42). In North America, *Aphanizomenon flos-aquae* growing in the lakes of New Hampshire (21,43,44) and *Lyngbya wollei* in reservoirs and lakes of the southeastern United States (23,45) have been reported to be saxitoxin producers. In Brazil, *Cylindrospermopsis raciborskii* was shown for the first time to produce saxitoxins (4) as well as an *Oscillatoria/Phormidium* strain from an Italian lake (46). Two species, anatoxin-a–producing *Oscillatoria* from Scotland and saxitoxin-producing *Lyngbya wollei* from the United States, are benthic, mat-forming organisms; the rest of the neurotoxin producers are planktic cyanobacteria.

The co-occurrence of toxic and nontoxic species and strains is possibly the main cause for the high variation of toxin concentrations in environmental samples. It is well documented that toxin-producing and non–toxin-producing strains exist among the same species which cannot be distinguished by microscopy (for examples in neurotoxic strains, see refs. 3 and 23). Toxin production of the strains also varies; in the laboratory, it has been found that some strains are high producers and others may produce consistantly less toxin. Cyanobacteria may also produce several toxins simultaneously. Among neurotoxic strains several saxitoxin variants are found at the same time (3,23,44). In Australia, the same main saxitoxins were found in the blooms and the cyanobacterial strains. Furthermore, simultaneous neurotoxin and hepatotoxin production has been noted; for example, the *Anabaena flos-aquae* strain NRC 525-17 originating from Canada produces anatoxin-a(S) (41) and several microcystins (47).

The effects of environmental factors on growth and toxin production of cyanobacteria have been studied in batch and continuous cultures. Most studies have dealt with cyanobacteria-producing hepatotoxins. Studies with neurotoxic strains of cyanobacteria have been done with anatoxin-a–producing *Anabaena* (48–50), *Aphanizomenon* (48), saxitoxin-producing *Aphanizomenon* (36), and *Anabaena circinalis* (3). Hepatotoxins remain mostly within cells when the conditions for the growth of the organism are favorable. In a continous culture study performed with axenic strains, anatoxin-a was found in culture medium, especially at high temperatures and in moderate light conditions (49). Similarly, Bumke-Vogt et al. (50) reported high concentrations of anatoxin-a in the growth medium of batch cultures during the exponential and stationary phases of growth. Maximal anatoxin-a production is found during the logarithmic growth phase. The amount of saxitoxins on a dry weight basis also increased with incubation time in *Anabaena circinalis*. Significant amounts of extracellular saxitoxins (5–30%) were observed throughout the growth cycle (3). The highest total saxitoxin concentrations were found in the late stationary phase of growth in New Hampshire *Aphanizomenon* and Australian *Anabaena* strains (3,36).

The few studies with neurotoxic strains (mostly anatoxin-a–producing strains have been studied) indicate that certain environmental factors have some effect on toxin production. In batch cultures, phosphorus had no effect on anatoxin-a production of *Anabaena* and *Aphanizomenon* strains, and these nitrogen-fixing species were not dependent on nitrogen in the medium for their growth or toxin production. The highest toxin production was detected at 20°C and lowest at 30°C (a threefold difference). *Anabaena* strains produced most anatoxin-a in continuous light of 26–44 $\mu mol\ m^{-2}s^{-1}$ and *Aphanizomenon* at 128 $\mu mol\ m^{-2}s^{-1}$ (three- to fourfold differences) (48). Saxitoxin production of *Aphanizomenon flos-aquae* culture was highest at 26°C (36). In continuous cultures, *Anabaena* strains produced the highest amounts of toxin at 20–25°C (four- to sevenfold differences) and under light levels higher than 40 $\mu mol\ m^{-2}s^{-1}$. In those conditions, however, a significant amount of the toxin was found in the growth medium (49).

More studies are needed to reveal how neurotoxin production is regulated in cyanobacteria. To date, the genetics and biochemistry of neurotoxin production of cyanobacteria are not understood. Several laboratories use molecular methods to characterize cyanobacteria, which produce neurotoxins. One study showed that three Finnish lakes interconnected by a river had a genetically homogeneous population of anatoxin-a–producing *Anabaena* strains (51). The new genetic tools which are applicable to cyanobacteria research make it possible to study cyanobacterial genotypes in situ and, hopefully, in the near future they will increase our understanding on how toxic blooms develop in nature.

II. CHEMISTRY OF CYANOBACTERIAL NEUROTOXINS

A. Anatoxin-a

Anatoxin-a is a low molecular weight (MW) (165) alkaloid, a secondary amine, 2-acetyl-9-azabicyclo(4-2-1)non-2-ene (52,53) (Figure 1). Homoanatoxin-a (MW 179) is an anatoxin-a homologue isolated from an *Oscillatoria formosa* strain (Figure 1). It has a propionyl group on C-2 in lieu of the acetyl group in anatoxin-a (40). The toxicities of these two compounds are similar. Interestingly, homoanatoxin-a has not been detected in anatoxin-a–containing strain and bloom samples thus far or vice versa (16,39). The mass spectrometry analysis of samples has revealed degradation products of anatoxin-a and homoanatoxin-a. They are dihydro or epoxy derivatives of the original toxins (16,28,39,54). Anatoxin-a decomposes in alkaline conditions. The breakdown of anatoxin-a is further accelerated by sunlight (55). Also, bacteria seem to be able to degrade it (56,57).

Several chemical methods used to synthesize anatoxin-a (e.g., see refs. 58–65) have been reported. More recently, homoanatoxin-a and the degradation products of both toxins also have been synthesized (54,66). Anatoxin-a and homoanatoxin-a are used as tools in neurobiological studies and they serve as standards in chemical analysis of the toxins. There are only a few studies on the biosynthesis of anatoxin-a. Gallon et al. (67,68) reported the anatoxin-a synthesis in cyanobacteria to be similar to tropane alkaloid synthesis in plants. Anatoxin-a was proposed to be formed from ornithine/arginine via putrescine, which is oxidized to pyrroline, a precursor of anatoxin-a. They also found that when a strain lost part of its plasmid, it also ceased to produce anatoxin-a. Feeding experiments with isotope-labeled substrates and subsequent nuclear magnetic resonance (NMR) studies showed that the synthesis of anatoxin a and homoanatoxin a is different from tropane alkaloids. The biosynthesis of anatoxin-a and homoanatoxin-a in cyanobacteria is a combination of the amino acid and the polyketide metabolism (69).

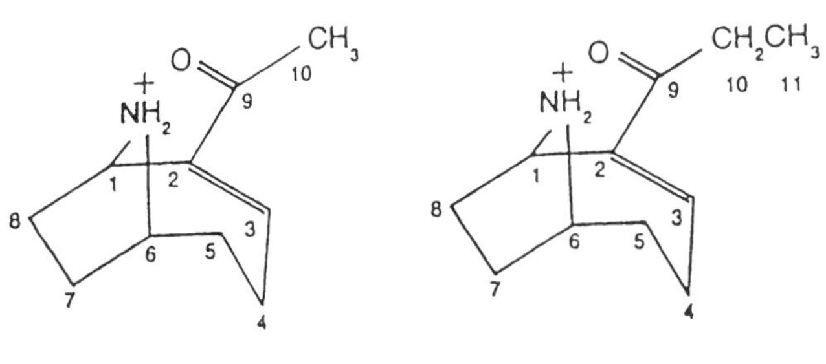

Anatoxin-a
MW 165; $C_{10}H_{15}NO$

Homoanatoxin-a
MW 179; $C_{11}H_{17}NO$

Figure 1 The chemical structures of cyanobacterial neurotoxins, anatoxin-a, homoanatoxin-a.

B. Anatoxin-a(S)

Anatoxin-a(S) is a unique phosphate ester of a cyclic N-hydroxyguanine (MW 252) (41) (Figure 2) and the only known organophosphate-type toxin found in cyanobacteria. Structural variation of anatoxin-a(S) has not been detected. During storage (at 3°C or even −20°C), and especially in basic solution, it however easily loses the monomethyl phosphate group. It is replaced by a hydroxyl group which can be derivatized further to hydrogen by catalytic hydrogenation (palladium-carbon in methanol) (31,41). The air evaporation of a methanolic solution of anatoxin-a(S) during purification was reported to result in significant hydrolysis of the compound. Anatoxin-a(S) is rather stable in neural or acidic conditions (41).

Biosynthesis of anatoxin-a(S) has been studied in the laboratory of Prof. R.E. Moore (University of Hawaii). All of the carbons of anatoxin-a(S) are derived from amino acids. Three methyl carbons arise from L-methionine or other donors to the tetrahydrofolate C-1 pool. It was shown that the carbons of the triaminopropane backbone and the guanidino function (carbons C-2, C-4, C-5, and C-6) are derived from amino acid L-arginine. (2S,4S)-4-Hydroxyarginine, also found in an anatoxin-a(S)–producing *Anabaena* strain, is an intermediate in the pathway from L-arginine to the toxin (70–72).

C. Saxitoxins

Saxitoxins (STXs) are a group of alkaloid neurotoxins. Three main classes of PSPs are known: nonsulfated (saxitoxin and neosaxitoxin), singly sulfated (gonyautoxins), and doubly sulfated (C-toxins) compounds (Table 3 and Figure 3). The potency of the toxins follows the same order, saxitoxins being the most potent and C-toxins the least potent. Toxin variants, which have a hydrogen in position R_4, are called decarbamoyl toxins.

Saxitoxins were first identified from marine dinoflagellates (red tide organisms). They are known to concentrate in seafood and have caused outbreaks of poisonings in humans (73). But saxitoxins also have been found in cyanobacteria. They have been isolated from *Aphanizomenon flos-aquae*, *Anabaena circinalis*, *Lyngbya wollei*, and *Cylindrospermopsis raciborskii*. The first reports of the occurrence of saxitoxins in cyanobacteria already appeared in the late 1960s from the lakes of New Hampshire (74,75). The North American *Aphanizomenon* strains NH-1 and NH-5 isolated from that region contained neosaxitoxin, saxitoxin, and a few unidentified neurotoxins (21,44). From the Australian bloom samples and the *Anabaena circinalis* strains, several different saxitoxins have been detected with the doubly sulfated C-1 and C-2 being the most abundant along with gonyautoxins GTX-2 and GTX-3 (3,42). Several bloom samples from Denmark contained PSPs, mostly gonyautoxins (7). In North America, the benthic, mat-forming freshwater cyanobacterium *Lyngbya wollei* produced three known saxitoxins (decarbamoysaxitoxin and decarbamoyl gonyautoxins) and six new ones (23,45). *Lyngbya* samples were found to be much less toxic in mouse bioassay than the *Aphanizomenon flos-aquae* or *Anabaena circinalis*

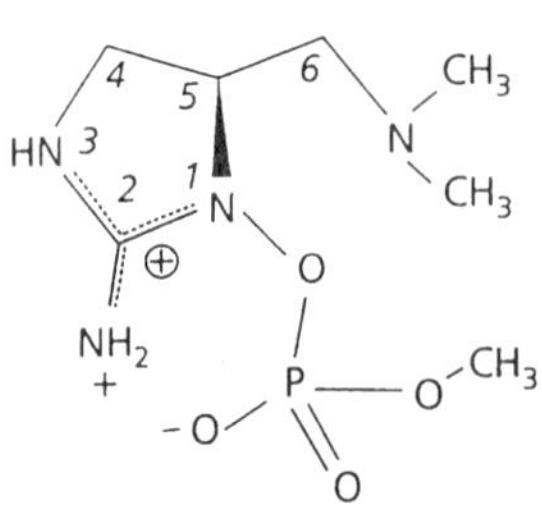

Figure 2 The chemical structure of anatoxin-a(S).

Table 3 Saxitoxins Reported to Be Found in Cyanobacteria[a]

	Variable chemical groups in toxins					Cyanobacteria			
Toxin	R_1	R_2	R_3	R_4	R_5	Aph[b]	Ana[c]	Lyn[d]	Cyl[e]
STX	H	H	H	$CONH_2$	OH	+	+		+
GTX-2	H	H	OSO_3^-	$CONH_2$	OH		+		
GTX-3	H	OSO_3^-	H	$CONH_2$	OH		+		
GTX-5	H	H	H	$CONHSO_3^-$	OH		+		
C-1	H	H	OSO_3^-	$CONHSO_3^-$	OH		+		
C-2	H	OSO_3^-	H	$CONHSO_3^-$	OH		+		
NEO	OH	H	H	$CONH_2$	OH	+			+
GTX-1	OH	H	OSO_3^-	$CONH_2$	OH		(+)		
GTX-4	OH	OSO_3^-	H	$CONH_2$	OH		(+)		
GTX-6	OH	H	H	$CONHSO_3^-$	OH		(+)		
dcSTX	H	H	H	H	OH		+	+	
dcGTX-2	H	H	OSO_3^-	H	OH		+	+	
dcGTX-3	H	OSO_3^-	H	H	OH		+	+	
LWTX-1	H	OSO_3^-	H	$COCH_3$	H			+	
LWTX-2	H	OSO_3^-	H	$COCH_3$	OH			+	
LWTX-3	H	H	OSO_3^-	$COCH_3$	OH			+	
LWTX-4	H	H	H	H	H			+	
LWTX-5	H	H	H	$COCH_3$	OH			+	
LWTX-6	H	H	H	$COCH_3$	H			+	

STX, saxitoxin; GTX, gonyautoxins; C, C-toxins; NEO, neosaxitoxins; DcSTX, decarbamoylsaxitoxin; LWTX, lyngbya-wollei toxins. +, toxin identified, (+), toxin detected based on retention time only.

[a] See Figure 3 for chemical structures of sabitoxins.

[b] Toxins found in *Aphanizomenon flos-aquae*, New Hampshire (21,44).

[c] Toxins reported in an *Anabaena circinalis* strain and bloom samples, Australia (3,33,42).

[d] Toxins detected in *Lyngbya wollei*, USA (45).

[e] Toxins thus far found in *Cylindrospermopsis raciborskii*, Brazil (4).

Source: Modified from ref. 37.

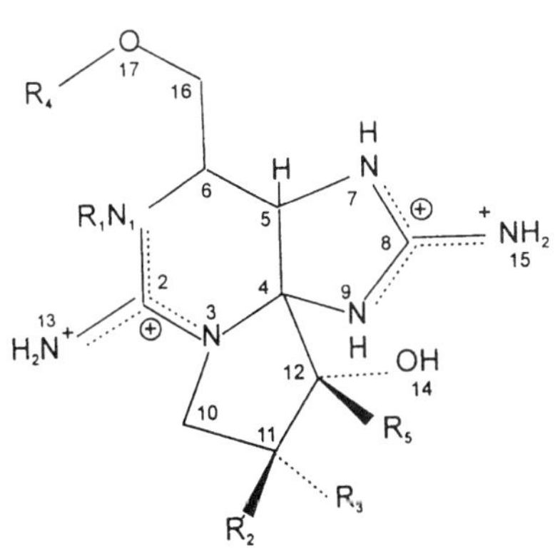

Figure 3 The general structure of saxitoxins (PSPs). Nineteen different saxitoxins have been reported in cyanobacteria (for different variants, see Table 3).

samples, which is logically explained by the detected composition of toxins in these species. *Cylindrospermopsis raciborskii* in Brazil has been found to contain mostly neoSTX and a smaller amount of STX (4). Very recently, *Oscillatoria/Phormidium* strains from an Italian lake were shown to produce saxitoxins (46).

Chemically, saxitoxins, like the other neurotoxins, seem to be unstable in alkaline conditions. It is known that in biological systems some of these toxins may transform to more poisonous variants (3,33,76). In the experiments carried out with *Anabaena circinalis* extracts and sterile water, it was shown that chemical transformation of less potent C-toxins to more potent decarbomoylgonyautoxins occurred before further degradation to nondetectable compounds. Epimerization of β-toxins to α compounds (e.g., C-2 to C-1 and GTX-3 to GTX-2) was also detected. Freeze-drying, and especially storage, of samples frozen prior to freeze-drying tends to stimulate the epimerization. In the experiments, toxins persisted as long as 90 days in freshwater. Bacteria degrading these toxins have not been isolated from freshwater thus far (3,76).

Shimizu et al. (77) studied the biosynthesis of saxitoxin analogues using an *Aphanizomenon flos-aquae* strain and proposed a new pathway for neosaxitoxin biosynthesis. The key steps of the pathway were the condensation of an acetate unit or its derivative to the amino group bearing α-carbon of arginine or an equivalent and subsequent loss of the carboxyl carbon and imidazole ring formation on the adjacent carbonylcarbon. They established the origin of all the carbons in the toxin alkaloid ring system. The side chain carbon was derived from methionine (78). Enzymes or genes involved in the process have not been studied in cyanobacteria.

III. DETECTION METHODS OF CYANOBACTERIAL NEUROTOXINS

The symptoms of poisonings (rapid death with convulsions and respiratory distress) reveal the presence of neurotoxins in the cyanobacterial scum if an animal test (mouse bioassay) is used for detection or when animal poisonings occur. In the early studies, the mouse bioassay was used to detect the toxicity of natural samples (determining the LD50 or MLD100). In low doses (near the LD50 values), salivation differentiates anatoxin-a(S) from the anatoxin-a and saxitoxins in mammals. In high doses, death occurs within a few minutes and no clear specific signs such as salivation are seen (30). Neurotoxins, unlike the hepatotoxins, do not cause any characteristic changes in the organs, which can be seen at autopsy. Alternative bioassays such as brine shrimp (*Artemia salina*) (79,80), insect larvae tests (81) for anatoxin-a, and desert locust (*Schistocerca gregaria*) for detection of saxitoxins (82) have been developed and applied to screen for the toxicity of samples. In *Artemia* tests, it has been observed that neurotoxins do not usually kill larvae but cause a typical change in their swimming behavior. The sensitivity of the *Artemia salina* test corresponds to the mouse bioassay. Both tests need rather high amounts of test material. Several immunoassay-based methods have been developed and described for the detection of saxitoxins (e.g., see refs. 83–86) and there are also commercial enzyme-linked immunosorbent assay (ELISA) tests available. Detection of anatoxin-a(S) is based on the acetylcholinesterase inhibition assay (87). A difficulty with this test is that anatoxin-a(S) standards are not commercially available and the test is not specific for anatoxin-a(S) and detects any anticholinesterase in the sample.

Several chemical isolation, separation, identification, and quantification methods have been developed and applied to cyanobacterial neurotoxins. The same methods that are used for analysis of saxitoxins from the marine environment and samples can be used to analyze saxitoxins originating from cyanobacteria. The most common methods to quantitate saxitoxins are based on high-performance liquid chromatography (HPLC), with postcolumn derivatization and fluorimetric detection (88,89). An HPLC method using prechromatographic oxidation with fluorescent detection (90) and a postcolumn electrochemical reactor for oxidation of saxitoxins have been used (91). The problems of analysis of cyanobacterial saxitoxins include: (1) the high variation of saxitoxins among species and strains of cyanobacteria, and (2) standards for all saxitoxins are not available for quantitation or identification. In scientific research, some of the problems can be overcome by using more sophisticated methods to study the toxins. Combining liquid chromatography and mass spectrometry (electrospray ionization) techniques, all types of neurotoxins (different saxitoxin variants, [homo]anatoxin-a and anatoxin-a[S]) can be identified (92).

The development of analytical methods for anatoxin-a has been rather easy, since the structure of the toxin had already been solved in the late 1970s and chemical synthesis yielded the pure compound to be used as a standard. The other advantage is that variations of the chemical structure of anatoxin-a are not found except for the homoanatoxin-a and degradation products of both these compounds (nonbioactive dihydro and epoxy anatoxins and homoanatoxins). The method of choice now depends on the samples to be analyzed (samples containing very low amounts of toxins require sensitive methods) and on the availability of equipment. Thin-layer chromatography (TLC) has been used for purification of toxins and also as an inexpensive screening method (93). Both HPLC- and gas chromatography (GC)–based methods have been used to quantitate anatoxin-a (Table 4). HPLC analyses with ultraviolet (UV) detection and

Table 4 Chemical Methods Developed to Isolate and Detect Cyanobacterial Neurotoxins Anatoxin-a and Homoanatoxin-a

Toxins analyzed	Method	Reference
Anatoxin-a	TLC (Fast Black K, in situ color reaction)	93
Anatoxin-a	Aberlite XAD-2 and liquid-liquid extraction, silica gel chromatography, HPLC with UV detection	94
Anatoxin-a	HPLC with UV detection	95
Anatoxin-a	HPLC with UV detection	96
Anatoxin-a	HPLC with UV detection, GC/MS	18
Anatoxin-a	HPLC with UV detection (aqueous samples)	97
Anatoxin-a	HPLC with fluorimetric detection	98
Anatoxin-a	LC-ESI-MS	99
Anatoxin-a, Homoanatoxin-a	HPLC and GC-MS	100
Anatoxin-a (degradation products, epoxyanatoxin-a)	TSP-LC/MS	39
Anatoxin-a, homoanatoxin-a (and degradation products)	HPLC with fluorimetric detection, ESI-MS	54
Anatoxin-a (and dihydro-anatoxin-a)	GC-MS (EI mode) (N-acetyl-anatoxin-a and N-acetyl dihydroanatoxin-a derivatives)	28
Anatoxin-a	GC-MS (SIM mode) (acetyl-anatoxin-a, derivatized with acetic acid and acetic anhydride)	101
	GC-ECD (N-trichloroacetic anatoxin-a = N-TCA-anatoxin-a)	102
Anatoxin-a (degradation product dihydroanatoxin-a)	GC-ECD (N-pentafluorobenyl-anatoxin-a derivative = N-PFB-anatoxin-a)	50
Anatoxin-a and homoana-toxin-a	GC/ECD and GC/MS (CI and NICI modes) (N-HFB-derivatives; derivation with heptafluorobutyric acid anhydride)	16

TLC, thin-layer chromatography; HPLC, high-performance liquid chromatography; GC, gas chromatography; MS, mass spectrometry, ESI, electrospray ionization; TPS, thermospray; SIM, selective ion monitoring; ECD, electron capture detection; CI, chemical ionization; NICI, negative ion chemical ionization.

with more sensitive fluorimetric detection have been used. For gas chromatographic analysis, several different derivatization procedures of anatoxin-a and homoanatoxin-a have been developed. The combination of HPLC or GC with mass spectrometry has greatly improved the identification of anatoxin-a and homoanatoxin-a and made it possible to look for the degradation products.

For solving the unknown structures of cyanobacterial neurotoxins in addition to mass spectrometry, different NMR techniques have been essential. For NMR, milligram amounts of pure compound are needed; this may sometimes limit the use of this technique.

IV. HEALTH PROBLEMS CAUSED BY NEUROTOXIC CYANOBACTERIA

Mass occurrences of cyanobacteria not only cause health risk to animals (see Table 2) but also to human beings. There are no reports of human deaths caused by cyanobacterial neurotoxins; unlike the microcystins, which have caused human deaths via hemodialysis in Garuaru, Brazil (103). Saxitoxins are known to accumulate in feeders, such as mussels and shellfish, which filter large amounts of water. The risk of ingesting toxins via freshwater mussels is low, since they are usually not used for human consumption. Cyanobacteria, even natural blooms, are sold as food supplements (so-called health food products), and if they are not tested for cyanobacterial toxins, they may pose a high risk to consumers.

The other most likely exposure routes are through the use of water for drinking and recreation. There are only a few studies where the removal of cyanobacterial neurotoxins in water-treatment systems has been studied. Activated carbon and ozonation removed anatoxin-a; the processes are also efficient for the removal of cyanobacterial hepatotoxins (104). The efficient removal of cyanobacterial cells in a treatment procedure is as important as removing the toxins, since the cyanobacterial cells may contain the highest amount of toxins. Slow sand filtration was found to be inefficient for the removal of cyanobacterial cells from water (105) but chemical flocculation was found to be efficient for this purpose.

Recreational activities in cyanobacterial blooms may pose a high risk to people. This is especially true for small children, who drink considerable amounts of water while swimming. Also, aerosols formed by water skiing or by other similar activities may also be hazardous. Experimental administration of microcystin-LR and anatoxin-a to the mouse showed that both toxins caused toxic effects with much lower doses when exposed via intranasal installation than by gastric intubation (106). Recreational waters containing cyanobacterial blooms have been reported to cause a wide variety of symptoms in water users (107). In several countries public beaches are monitored and closed during cyanobacterial bloom events.

Protecting waters from eutrophication, stopping the use of waters associated with cyanobacterial blooms for any purpose, extensive studies, and research into the problem, and high awareness of the general public and authorities are needed to prevent problems being caused by toxic cyanobacteria.

V. SUMMARY

In freshwater, the mass occurrences (blooms) toxic to animals and humans are mainly formed by cyanobacteria. Neurotoxin-producing cyanobacteria are not so widely spread as the hepatotoxin-producing ones. They are found mostly in America, Europe, and Australia where they have caused numerous animal poisonings. Cyanobacterial neurotoxins have been detected in planktic *Anabaena*, *Aphanizomenon*, *Oscillatoria/Planktothrix/Phormidium*, *Cylindrospermum*, and

Cylindrospermopsin as well as in benthic *Lyngbya* and *Oscillatoria* genera of cyanobacteria. Three types of neurotoxins are produced by cyanobacteria: (homo)anatoxin-a, anatoxin-a(S), and saxitoxins (also called paralytic shellfish poisons, PSPs). About 20 different structural PSP variants have been characterized in cyanobacteria. There are several bioassays and chemical methods that have been developed to analyze these toxins. The risks of cyanobacterial neurotoxins to human health have been briefly discussed.

REFERENCES

1. PD Baker, AR Humpage. Toxicity associated with commonly occurring cyanobacteria in surface waters of the Murray-Darling basin, Australia. Aust J Marine Freshwater Res 45:773–786, 1994.
2. DA Steffensen, AR Humpage, J Rositano, AH Bretag, R Brown, PD Barker, BC Nicholson. Neurotoxins from Australian *Anabaena*. In: GA Codd, TM Jefferies, CW Keevil, E Potter, eds. Detection Methods for Cyanobacterial Toxins. Cambridge, UK: Royal Society of Chemistry, 1994, pp 45–50.
3. AP Negri, GJ Jones, SI Blackburn, Y Oshima, H Onodera. Effect of culture and bloom development and of sample storage on paralytic shellfish poisons in the cyanobacterium *Anabaena circinalis*. J Phycol 33:26–35, 1997.
4. N Lagos, JL Liberona, D Andrinolo, PA Zagatto, RM Soares, SMFQ Azevedo. First evidence of paralytic shellfish toxins in the freshwater cyanobacterium *Cylindrospermopsis raciborskii* isolated from Brazil. In: VIII International Conference on Harmful Algae. Vigo, Spain, June 25–29, 1997, Abstract, p 115.
5. PR Gorham. Laboratory studies on the toxins produced by water blooms of blue-green algae. Am J Public Health 52:2100–2105, 1962.
6. WW Carmichael, P Gorham. Anatoxins from clones of *Anabaena flos-aquae* isolated from lakes of western Canada. Mitt Int Verein Limnol 21:285–295, 1978.
7. P Henriksen. Toxic cyanobacteria/blue-green algae in Danish fresh waters. PhD thesis. Dept of Phycology, Institute of Botany, University of Copenhagen. 1996. 43 p.
8. K Sivonen, K Himberg, R Luukkainen, SI Niemelä, GK Poon, GA Codd. Preliminary characterization of neurotoxic blooms and strains from Finland. Toxicity Assess 4:339–352, 1989.
9. K Sivonen, SI Niemelä, RM Niemi, L Lepistö, TH Luoma, LA Räsänen. Toxic cyanobacteria (blue-green algae) in Finnish fresh and coastal waters. Hydrobiologia 190:267–275, 1990.
10. C Bumke-Vogt. Cyanobakterielle Neurotoxine. Untersuchungen zum Vorkommen von Anatoxin-a in einigen deutschen Gewässern im Verlauf der Jahre 1995/1996. In: Toxische Cyanobakterien in deutschen Gewässern. Verbreitung, Kontrollfaktoren und ökologische Bedeutung. Institut für Wasser-, Boden- und Lufthygiene des Umweltbundesamtes, Berlin, 1997, pp 35–39.
11. KJ James, IR Sherlock, MA Stack. Anatoxin-a in Irish freshwater and cyanobacteria, determined using a new fluorimetric liquid chromatographic method. Toxicon 35:963–971, 1997.
12. M Bruno, DA Barbini, E Pierdominici, AP Serse, A Ioppolo. Anatoxin-a and previously unknown toxin in *Anabaena planctonica* from blooms found in Lake Mulargia (Italy). Toxicon 32:369–373, 1994.
13. P-D Park, MF Watanabe, K-I Harada, H Nagai, M Suzuki, M Watanabe, H Hayashi. Hepatotoxin (microcystin) and neurotoxin (anatoxin-a) contained in natural blooms and strains of cyanobacteria from Japanese freshwaters. Natural Toxins 1:353–360, 1993.
14. H-D Park, B Kim, E Kim, T Okino. Hepatotoxic microcystins and neurotoxic anatoxin-a in cyanobacterial blooms from Korean lakes. Environ Toxicol Water Qual 13:225–234, 1998.
15. OM Skulberg, B Underdal, H Utkilen. Toxic waterblooms with cyanophytes in Norway—current knowledge. Algol Stud 75:279–289, 1994.
16. JE Haugen, OM Skulberg, RA Andersen, J Alexander, G Lilleheil, T Gallagher, PA Borough. Rapid analysis of cyanophyte neurotoxins: an improved method for quantitative analysis of anatoxin-a and homoanatoxin-a in the sub-ppb to ppb range. Arch Hydrobiol., Algol Stud 75:111–121, 1994.

17. T Willén, R Mattsson. Water-blooming and toxin-producing cyanobacteria in Swedish fresh and brackish waters, 1981–1995. Hydrobiologia 353:181–192, 1997.
18. C Edwards, KA Beattie, CM Scrimgeour, GA Codd. Identification of anatoxin-a in benthic cyanobacteria (blue-green algae) and in associated dog poisonings at Loch Insh, Scotland. Toxicon 30: 1165–1175, 1992.
19. TA Olson. Water poisoning—a study of poisonous algae blooms in Minnesota. Am J Public Health 50:883–884, 1960.
20. WM Repavich, WC Sonzogni, JH Standridge, RE Wedepohl, LF Meisner. Cyanobacteria (blue-green algae) in Wisconsin waters: acute and chronic toxicity. Water Res 24:225–231, 1990.
21. M Ikawa, K Wegener, TL Foxall, JJ Sasner. Comparison of the toxins of the blue-green alga *Aphanizomenon flos-aquae* with the *Gonyaulax* toxins. Toxicon 20:747–752, 1982.
22. NA Mahmood, WW Carmichael, D Pfahler. Anticholinesterase poisonings in dogs from a cyanobacterial (blue-green algae) bloom dominated by *Anabaena flos-aquae*. Am J Vet Res 49:500–503, 1988.
23. WW Carmichael, WR Evans, Q-Q Yin, P Bell, E Mocauklowski. Evidence for paralytic shellfish poisons in the freshwater cyanobacterium *Lyngbya wollei* (Farlow ex Gomont) comb. nov. Appl Environ Microbiol 63:3104–3110, 1997.
24. N Kangatharalingam, JC Priscu. Isolation and verification of anatoxin-a producing clones of *Anabaena flos-aquae* (Lyngb.) de Breb. from a eutrophic lake. FEMS Microbiol Ecol 12:127–130, 1993.
25. R Ressom, FS Soong, J Fitzgerald, L Turczynowicz, O El Saadi, D Roder, T Maynard, I Falconer. Health effects of toxic cyanobacteria (blue-green algae), National Health and Medical Council. Canberra, Australia: The Australian Government Publishing Service, 1994, pp 1–108.
26. RE Juday, EJ Keller, A Horpestad, LL Bahls, S Glasser. A toxic bloom of *Anabaena flos-aquae* in Hebgen reservoir Montana in 1977. In: WW Carmichael, ed. The Water Environment: Algal Toxins and Health. Vol. 20. New York: Plenum Press, 1981, pp 103–112.
27. MJ Pybus, DP Hopson, DK Onderka. Mass mortality of bats due to probable blue-green algal toxicity. J Wildlife Dis 22:449–450, 1986.
28. RA Smith, D Lewis. A rapid analysis of water for anatoxin-a, the unstable toxic alkaloid from *Anabaena flos-aquae*, the stable non-toxic alkaloids left after bioreduction and a related amine which may be nature's precursor to anatoxin-a. Vet Hum Toxicol 29:153–154, 1987.
29. WO Cook, VR Beasley, RA Lovell, AM Dahlem, SB Hooser, NA Mahmood, WW Carmichael. Consistent inhibition of peripheral cholinesterases by neurotoxins from the freshwater cyanobacterium *Anabaena flos-aquae*: studies on ducks, swine, mice and steer. Environ Toxicol Chem 8:915–922, 1989.
30. P Henriksen, WW Carmichael, J An, Ø Moestrup. Detection of an anatoxin-a(s)–like anticholinesterase in natural blooms and cultures of cyanobacteria/blue-green algae from Danish lakes and in the stomach contents of poisoned birds. Toxicon 35:901–913, 1997.
31. H Onodera, Y Oshima, P Henriksen, T Yasumoto. Confirmation of anatoxin-a(s) in the cyanobacterium *Anabaena lemmermannii* as the cause of bird kills in Danish lakes. Toxicon 35:1645, 1997.
32. J Bartram, JC Vapnek, G Jones, L Bowling, I Falconer, GA Codd. Implementation of management plans. In: I Chorus, J Bartram, eds. Toxic Cyanobacteria in Water: A Guide to Public Health Significance, Monitoring and Management. WHO, E & FN Spon, London, UK. 1999, pp 211–234.
33. AP Negri, GJ Jones, M Hindmarsh. Sheep mortality associated with paralytic shellfish poisoning toxins from the cyanobacterium *Anabaena circinalis*. Toxicon 33:1321–1329, 1995.
34. VR Beasley, AM Dahlem, WO Cook, WL Valentine, RA Lowell, SB Hooser, K-I Harada, M Suzuki, WW Carmichael. Diagnostic and clinically important aspects of cyanobacterial (blue-green algae) toxicoses. J Vet Diagn Invest 1:359–365, 1989.
35. K Christoffersen. Ecological implications of cyanobacterial toxins in aquatic food webs. Phycologia 35(Suppl 6):42–50, 1996.
36. JH Gentile, TE Maloney. Toxicity and environmental requirements of a strain of *Aphanizomenon flos-aquae* (L.) Ralfs. Can J Microbiol 15:165–173, 1969.
37. K Sivonen, G Jones. Cyanobacterial toxins. In: I Chorus, J Bartram, eds. Toxic Cyanobacteria in Water: a Guide to Public Health Significance, Monitoring and Management. WHO, E & FN Spon, London, 1999, pp 41–111.

38. WW Carmichael, DF Biggs, MA Peterson. Pharmacology of anatoxin-a, produced by the fresh water cyanophyte *Anabaena flos-aquae* NCR-44-1. Toxicon 17:229–236, 1979.
39. K-I Harada, H Nagai, Y Kimura, M Suzuki, H-D Park, MF Watanabe, R Luukkainen, K Sivonen, WW Carmichael. Liquid chromatography/mass spectrometric detection of anatoxin-a, a neurotoxin from cyanobacteria. Tetrahedron 49:9251–9260, 1993.
40. OM Skulberg, WW Carmichael, RA Anderson, S Matsunaga, RE Moore, R Skulberg. Investigations of a neurotoxic Oscillatorialean strain (cyanophyceae) and its toxin. Isolation and characterization of homoanatoxin-a. Environmental Toxicol Chem 11:321–329, 1992.
41. S Matsunaga, RE Moore, WP Niemczura, WW Carmichael. Anatoxin-a(s), a potent anticholinesterase from *Anabaena flos-aquae*. J Am Chem Soc 111:8021–8023, 1989.
42. AR Humpage, J Rositano, AH Bretag, R Brown, PD Baler, BC Nicholson, DA Steffensen. Paralytic shellfish poisons from Australian cyanobacterial blooms. Aust J Mar Freshwater Res 45:761–771, 1994.
43. JJ Sasner Jr, M Ikawa, TL Foxall. Studies on *Aphanizomenon* and *Microcystis* toxins. In: EP Regelis, ed. Seafood Toxins. ASC Symposium Series, No. 262. Washington, DC: American Chemical Society, 1984, pp 391–406.
44. NA Mahmood, WW Carmichael. Paralytic shellfish poisons produced by the freshwater cyanobacterium *Aphanizomenon flos-aquae* NH-5. Toxicon 24:175–186, 1986.
45. H Onodera, M Satake, Y Oshima, T Yasumoto, WW Carmichael. New saxitoxin analogues from the freshwater filamentous cyanobacterium *Lyngbya wollei*. Nat Toxins 5:146–151, 1997.
46. F Pomati, S Sacchi, C Rosetti, BA Neilan. And now saxitoxin-producing cyanobacteria in Europe. In: Abstracts of the 4th International Conference on Toxic Cyanobacteria, Beufort, North Carolina, 1998, p 44.
47. K-I Harada, K Ogawa, Y Kimura, H Murata, M Suzuki, PM Thorn, WR Evans, WW Carmichael. Microcystins from *Anabaena flos-aquae* NRC 525-17. Chem Res Toxicol 4:535–540, 1991.
48. J Rapala, K Sivonen, R Luukkainen, SI Niemelä. Anatoxin–a concentration in *Anabaena* and *Aphanizomenon* at different environmental conditions and comparison of growth by toxic and non-toxic *Anabaena* strains—a laboratory study. J Appl Phycol 5:581–591, 1993.
49. J Rapala, K Sivonen. Assessment of environmental conditions that favour hepatotoxic and neurotoxic *Anabaena* spp. strains in cultured under light-limitation at different temperatures. Microbial Ecol 36:181–192, 1998.
50. C Bumke-Vogt, W Mailahn, W Rotard, I Chorus. A highly sensitive analytical method for the neurotoxin anatoxin-a, using GC-ECD, and first application to laboratory cultures. Phycologia 35(Suppl 6):51–56, 1996.
51. L Rouhiainen, K Sivonen, W Buikema, R Haselkorn. Characterization of toxin-producing cyanobacteria by using an oligonucleotide probe containing a tandemly repeated heptamer. J Bacteriol 177: 6021–6026, 1995.
52. CS Huber. The crystal structure and absolute configuration of 2,9-diacetyl-9-azabicyclo(4,2,1)non-2,3-ene. Acta Crystallogr Section B 28:2577–2582, 1972.
53. JP Devlin, OE Edwards, PR Gorham, MR Hunter, RK Pike, B Stavric. Anatoxin-a, a toxic alkaloid from *Anabaena flos-aquae* NCR-44h. Can J Chem 55:1367–1371, 1977.
54. KJ James, A Furey, IR Sherlock, MA Stack, M Twohig, FB Caudwell, OM Slulberg. Sensitive determination of anatoxin-a, homoanatoxin-a and their degradation products by liquid chromatography with fluorimetric detection. J Chromatogr A 798:147–157, 1998.
55. DK Stevens, RI Krieger. Stability studies on the cyanobacterial nicotinic alkaloid anatoxin-a. Toxicon 29:167–179, 1991.
56. J Kiviranta, K Sivonen, R Luukkainen, K Lahti, SI Niemelä. Production and biodegradation of cyanobacterial toxins—a laboratory study. Arch Hydrobiol 121:281–294, 1991.
57. J Rapala, K Lahti, K Sivonen, SI Niemelä. Biodegradability and adsorption on lake sediments of cyanobacterial hepatotoxins and anatoxin-a. Lett Appl Microbiol 19:423–428, 1994.
58. HF Campbell, OE Edwards, R Kolt. Synthesis of nor-anatoxin-a and anatoxin-a. Can J Chem 55: 1372–1379, 1977.
59. HF Campbell, OE Edwards, JW Elder, R Kolt. Total synthesis of DL–anatoxin-a and DL–isoanatoxin-a. Pol J Chem 53:27–37, 1979.

60. HA Bates, H Rapoport. Synthesis of anatoxin a via intramolecular cyclization of iminium salts. J Am Chem Soc 101:1259–1265, 1979.
61. JJ Tufarieollo, H Meckler, KPA Senaratne. Synthesis of anatoxin-a: very fast death factor. J Am Chem Soc 106:7979–7980, 1984.
62. JJ Tufarieollo, H Meckler, KPA Senaratne. The use of nitrones in the synthesis of anatoxin-a: very fast death factor. Tetrahedron 41:3447–3453, 1985.
63. JS Petersen, G Fels, H Rapoport. Chirospecific synthesis of (+)- and (−)-anatoxin-a. J Am Chem Soc 106:4539–4547, 1984.
64. RL Danheiser, JM Morin Jr, EJ Salaski. Efficient total synthesis of (±)-anatoxin-a. J Am Chem Soc 107:8066–8073, 1985.
65. NJS Huby, P Thompson, S Wonnacott, T Gallagher. Structural modification of anatoxin-a. Synthesis of model affinity ligands for the nicotinic acetylcholine receptor. J Chem Soc Chem Commun 243–245, 1991.
66. T Gallagher, PA Brough, TM Jefferies, S Wonnacott. The role of synthetic chemistry in the production of standards for toxin analysis. In: GA Codd, TM Jefferies, CW Keevil, E Potter, eds. Detection Methods for Cyanobacterial Toxins. Cambridge, UK: Royal Society of Chemistry, 1994, pp 117–121.
67. JR Gallon, KN Chit, EG Brown. Biosynthesis of the tropane-related cyanobacterial toxin anatoxin-a: role of ornithine decarboxylase. Phytochemistry 29:1107–1111, 1990.
68. JR Gallon, O Kittakoop, EG Brown. Biosynthesis of anatoxin-a by *Anabaena flos-aquae*: examination of primary enzymic steps. Phytochemistry 35:1195–1203, 1994.
69. T Hemscheidt, J Rapala, K Sivonen, OM Skulberg. Biosynthesis of anatoxin-a in *Anabaena flos-aquae* and homoanatoxin-a in *Oscillatoria formosa.* J Chem Soc Chem Commun 1361–1362, 1995.
70. BS Moore, I Ohtani, CB de Koning, RE Moore, WW Carmichael. Biosynthesis of anatoxin-a(s). Origin of the carbons. Tetrahed Lett 33:6595–6598, 1992.
71. RE Moore, I Ohtani, BS Moore, CB de Koning, WY Yoshida, MT Runnegar, WW Carmichael. Cyanobacterial toxins. Gaz Chim Ital 123:329–336, 1993.
72. T Hemscheidt, DL Burgoyne, RE Moore. Biosynthesis of anatoxin-a(s). (2*S*,4*S*)-4-hydroxyarginine as an intermediate. J Chem Soc Chem Commun 205–206, 1995.
73. DM Anderson. Red tides. Sci Am Aug 94:52–58, 1994.
74. PJ Sawyer, JH Gentile, JJ Sasner Jr. Demonstration of a toxin from *Aphanizomenon flos-aquae* (L.) Ralfs. Can J Microbiol 14:1199–1204, 1968.
75. E Jackim, J Gentile. Toxins of a blue-green alga, similarity to saxitoxin. Science 162:915–916, 1968.
76. GJ Jones, AP Negri. Persistence and degradation of cyanobacterial paralytic shellfish poisons (PSPs) in freshwaters. Water Res 31:525–533, 1997.
77. Y Shimizu, M Norte, A Hori, A Genenah, M Kobayashi. Biosynthesis of saxitoxin analogues: the unexpected pathway. J Am Chem Soc 106:6433–6434, 1984.
78. Y Shimizu. Toxigenesis and biosynthesis of saxitoxin analogues. Pure Appl Chem 58:257–262, 1986.
79. J Kiviranta, K Sivonen, SI Niemelä, K Huovinen. Detection of toxicity of cyanobacteria by *Artemia salina* bioassay. Environ Toxicol Water Qual 6:423–436, 1991b.
80. K Lahti, J Ahtiainen, J Rapala, K Sivonen, SI Niemelä. Assessment of rapid bioassays for detecting cyanobacterial toxicity. Lett Appl Microbiol 21:109–114, 1995.
81. J Kiviranta, A Abdel-Hameed, K Sivonen, SI Niemelä, G Carlberg. Toxicity of cyanobacteria to mosquito larvae—screening of active compounds. Environ Toxicol Water Qual 8:63–71, 1993.
82. J McElhiney, LA Lawton, C Edwards, S Gallacher. Development of a bioassay employing the desert locust (*Schistocerca gregaria*) for the detection of saxitoxin and related compounds in cyanobacteria and shellfish. Toxicon 36:417–420, 1998.
83. FS Chu, TSL Fan. Indirect enzyme-linked immunosorbent assay for saxitoxin in shellfish. J Assoc Off Analyt Chem 68:13–16, 1985.
84. E Usleber, E Schneider, G Terplan. Direct enzyme immunoassay in microtitration plate and test strip format for the detection of saxitoxins in shellfish. Lett Appl Microbiol 13:275–277, 1991.
85. K Dill, M Lin, C Poteras, C Fraser, DG Hafeman, JC Owicki, JD Olson. Antibody-antigen binding constants determined in solution-phase with the threshold membrane-capture system: binding con-

stants for anti-fluorescein, anti-saxitoxin, and anti-ricin antibodies. Analyt Biochem 217:128–138, 1994.
86. JA Kralovec, MV Laycock, R Richards, E Usleber. Immobilization of small molecules on solid matrices: a novel approach to enzyme-linked immunosorbent assay screening for saxitoxin and evaluation of anti-saxitoxin antibodies. Toxicon 34:1127–1140, 1996.
87. WW Carmichael, NA Mahmood, EG Hyde. Natural toxins from cyanobacteria (blue-green algae). In: S Hall, G Strichartz, eds. Marine Toxins, Origin, Structure and Molecular Pharmacology. Vol. 418. Washington, DC: American Chemical Society, 1990, pp 87–106.
88. Y Oshima, CJ Bolch, GM Hallegraeff. Toxins composition of resting cysts of *Alexandrium tamarense* (Dinophyceae). Toxicon 30:1539–1544, 1992.
89. JJ Sullivan. Methods of analysis for algal toxins. Dinoflagellate and diatom toxins. In: IR Falconer, ed. Algal Toxins in Seafood and Drinking Water. San Diego: Academic Press, 1993, pp 29–48.
90. JF Lawrence, B Wong. Evaluation of postcolumn electrochemical reactor for oxidation of paralytic shellfish poison toxins. J AOAC Int 78:698–704, 1995.
91. JF Lawrence, B Wong, C Ménard. Determination of decarbamoyl saxitoxin and its analogues in shellfish by prechromatographic oxidation and liquid chromatography with fluorescence detection. J AOAC Int 79:1111–1115, 1996.
92. H Onodera, Y Oshima. Analyses of cyanobacterial neurotoxins by LC/MS technique. In: Abstracts of the 4th International Conference on Toxic Cyanobacteria. Beufort, North Carolina, 1998, p 6.
93. I Ojanperä, E Vuori, K Himberg, M Waris, K Niinivaara. Facile detection of anatoxin-a in algal material by thin-layer chromatography with fast black K salt. Analyst 116:265–267, 1991.
94. NB Abstrachan, BG Archer. Simplified monitoring of anatoxin-a by reverse-phase high performance liquid chromatography and the sub-acute effects of anatoxin-a in rats. In: WW Carmichael, ed. The Water Environment: Algal Toxins and Health. Vol. 20. New York: Plenum Press, 1981, pp 437–446.
95. SH Wong, E Hindin. Detecting an algal toxin by high-pressure liquid chromatography. Res Technol 528–529, 1982.
96. K-I Harada, Y Kimura, K Ogawa, M Suzuki, AM Dahlem, VR Beasley, WW Carmichael. A new procedure for the analysis and purification of naturally occurring anatoxin-a from the blue-green alga *Anabaena flos-aquae*. Toxicon 27:1289–1296, 1989.
97. MW Powell. Analysis of anatoxin-a in aqueous samples. Chromatographia 45:25–28, 1997.
98. KJ James, IR Sherlock. Determination of the cyanobacterial neurotoxin, anatoxin-a, by derivatisation using 7-fluoro-4-nitro-2,1,3-benzoxadiazole (NBD-F) and HPLC analysis with fluorimetric detection. Biomed Chromatogr 10:46–47, 1996.
99. GK Poon, LJ Griggs, C Edwards, KA Beattie, GA Codd. Liquid chromatography-electrospray ionization-mass spectrometry of cyanobacterial toxins. J Chromatogr 628:215–233, 1993.
100. A Zotou, TM Jefferies, PA Brough, T Gallagher. Determination of anatoxin-a and homoanatoxin-a in blue-green algal extracts by high-performance liquid chromatography and gas chromatography-mass spectrometry. Analyst 118:753–758, 1993.
101. K Himberg. Determination of anatoxin-a, the neurotoxin of *Anabaena flos-aquae* cyanobacterium, in algae and water by gas chromatography-mass spectrometry. J Chromatogr 481:358–362, 1989.
102. DK Stevens, RI Krieger. Analysis of anatoxin-a by GC/ECD. J Analyt Toxicol 12:126–131, 1988.
103. EM Jochimsen, WW Carmichael, J An, DM Cardo, ST Cookson, CEM Holmes, MB Antunes, DA Melo Filho, TM Lyra, VST Barreto, SMFO Azevedo, WR Jarvis. Liver failure and death after exposure to microcystins at a hemodialysis center in Brazil. N Engl J Med 338:873–878, 1998.
104. A-M Keijola, K Himberg, A-L Esala, K Sivonen, L Hiisvirta. Removal of cyanobacterial toxins in water treatment processes—laboratory and pilot scale experiments. Toxicity Assess 3:643–656, 1988.
105. L Lepistö, K Lahti, J Niemi, M Färdig. Removal of cyanobacteria and other phytoplankton in four Finnish waterworks. Archiv für Hydrobiologie, Algol Stud 75:167–181, 1994.
106. RB Fitzgeorge, SA Clark, CW Keevil. Routes of intoxication. In: GA Codd, TM Jefferies, CW Keevil, E Potter, eds. Detection Methods for Cyanobacterial Toxins. Cambridge, UK: Royal Society of Chemistry, 1994, pp 69–74.
107. LS Pilotto, RM Douglas, MD Burch, S Cameron, M Beers, GR Rouch, P Robinson, M Kirk, CT Cowie, S Hardiman, C Moore, RG Attewell. Health effects of recreational exposure to cyanobacteria (blue-green algae) during recreational water activities. AZ J Public Health 21:562–566, 1997.

27

Freshwater Neurotoxins: Mechanisms of Action, Pharmacology, Toxicology, and Impacts on Aquaculture

Paul T. Smith
University of Western Sydney (Macarthur Campus)
Campbelltown, New South Wales, Australia

I. INTRODUCTION

The scientific investigation of freshwater cyanobacterial neurotoxins is a relatively new and exciting field of study which already has provided some very interesting surprises (see refs. 1–5). In particular, the first cyanobacterial neurotoxins were isolated from *Aphanizomenon flos-aquae* and were called aphantoxins (6). However, subsequent studies (7–9) have shown that the neurotoxins produced by this species and other cyanobacteria were paralytic shellfish poisons (PSPs) and include saxitoxin, neosaxitoxin, C-toxins, and gonyautoxins (10–13). The freshwater cyanobacteria that are known to produce PSPs include *Aphanizomenon flos-aquae* (7–9), *Anabaena circinalis* (11,12,14), and *Lyngbya wollei* (15). Thus, it has become clear that the PSPs found in freshwater cyanobacteria are the same as those found in the marine dinoflagellates that cause toxic red tides.

Before examining the modes of action of freshwater neurotoxins further, we should place them in perspective. The main sources of neurotoxins in freshwater habitats are planktonic (free-floating) cyanobacteria, which commonly result in blue-green algal scums (3), although benthic blooms (attached to the sediment) can also be hazardous (16). The main genera with species capable of producing neurotoxins are *Anabaena*, *Aphanizomenon*, and *Oscillatoria*. Instances of illness and death of animals from freshwater cyanobacterial blooms have been described in anecdotes from many cultures. For example, according to Elder et al. (17), a toxic scum occurred in Llangorse Lake, Wales, in the 12th century. However, the first detailed scientific report on the toxicity of cyanobacterial blooms was by Francis (18). He described a toxic bloom of *Nodularia spumigena* in Lake Alexandrina, South Australia, which caused mortalities of sheep and cattle. A history of the occurrences of toxic blooms in Lake Alexandrina has been described by Codd et al. (19).

Further, it has only been since the 1960s that blue-green algae have been appropriately classified as photosynthetic prokaryotes rather than algae. Nowadays the taxonomy of cyanobacteria continues to undergo revision as key characteristics for classification are agreed on (20–26). Also, it has only been through relatively recent studies (described in Chapter 26) that many of the types of neurotoxins produced by freshwater cyanobacteria have been clearly identified.

What we do know is that cyanobacteria are the most diverse of the photosynthetic prokaryotes in terms of structure and development. In addition to a range of neurotoxins, cyanobacteria produce hepatotoxins (see Chapter 28), volatile odor compounds (27,28), and bioactive chemicals, some of which may have potential medical value (4,29–35). Brackish water cyanobacteria and their neurotoxins are included here because most freshwater cyanobacteria tolerate different salinities (euryhaline or halotolerant) and the species in freshwater and estuarine habitats are often identical (see ref. 36, p 221 and ref. 37, p 37).

Anatoxin-a was the first neurotoxin unique to freshwater cyanobacteria to have its chemical structure elucidated (38). It was initially isolated from a species of *Anabaena flos-aquae*, but since then it has been found in species of *A. circinalis* (39,40), *Aphanizomenon* sp., *Cylindrospermum* sp., and *Oscillatoria* sp. (39) and benthic *Oscillatoria* sp. (16). Two other types of cyanobacterial neurotoxins with some similar toxicological effects to anatoxin-a also have been identified. Homoanatoxin-a has a similar mode of action to that of anatoxin-a, whereas anatoxin-a(S) has a different mode of action and causes salivation.

The largest known phenomenon of a neurotoxic cyanobacterial bloom occurred in Australia in 1991 when more than 1000 km of the Barwon and Darling Rivers were afflicted by a bloom of *Anabaena* sp. (19,41). One farmer lost 1000 ewes, 140 rams, and 30 cows during the 6 weeks in which the bloom occurred (42). Testing by researchers showed that on average, five of seven tests from the Darling River contained neurotoxin (43). *A. circinalis* was the main species collected from the blooms in both the Darling and the Barwon Rivers, and the PSP group of toxins were identified as the neurotoxins (11). In fact, a survey of the rivers over four seasons (1990–1993) found that *A. circinalis* was the only species that was neurotoxic (44,45).

Thus, the freshwater neurotoxins that already have been identified, and those that are currently unidentified, may be significant sources of seafood toxicity given (i) the increasing eutrophication of fresh and brackish waters (3) and (ii) the world's rapidly expanding demand for freshwater and brackish water fish and shellfish. This chapter reviews our current understanding of freshwater neurotoxins of cyanobacteria, with emphasis on toxicology, pharmacology, and possible impacts on aquaculture.

II. TOXICOLOGY

A. Toxicological Methods

Evidence for the toxicity of blooms is generally provided by mouse bioassay (46,47), whereas the types of toxins are determined by high-performance liquid chromatography (HPLC) techniques (48–50). The advantages of mouse bioassay are that it provides a relatively rapid result, is nonspecific for the type of toxin, and is standardized (46,47,51). In addition, the mouse bioassay is the only reference method which is accepted worldwide for PSPs (52). The test provides data for signs of poisoning, survival time, quantitative data (e.g., LD_{50}), and findings at necropsy (i.e., indications of tissue abnormalities in an autopsy of sacrificed animals). The possibility of a combination of neurotoxins and hepatotoxins can often be distinguished by the findings of necropsy; for example, increased liver weight in mice that displayed typical symptoms of neurotoxins indicates a combination of neurotoxin and hepatotoxin (53). In some countries, to prevent human poisoning from marine shellfish, shellfish are routinely monitored for PSP toxins with the mouse bioassay and a shellfish quarantine is established if levels of PSP exceed 80 μg per 100 g of tissue (54). However, the rapid expansion of the marine shellfish aquaculture industry has meant increased demand for quicker, cheaper, and more sensitive means of testing (55). As yet, freshwater fish and shellfish are not routinely monitored for PSP or other neurotoxins.

Several chromatographic methods are available for the analysis of anatoxin-a in samples from cyanobacterial blooms. These methods include HPLC with ultraviolet (UV) detection (56,57), fluorimetric HPLC analysis (58), thin-layer chromatography (TLC) (59,60), and liquid chromatography–mass spectrometry (LC-MS) using electrospray ionization (61) or thermospray (62). Other techniques which have been used to test for anatoxin-a(S) include anticholinesterase activity, high-resolution fast atom bombardment–mass spectrometry (FAS-MS), and nuclear magnetic resonance (NMR) (63).

The use of HPLC and other techniques (12,49) has enabled trace levels of PSPs to be detected and quantified. For example, toxic strains of *A. circinalis* which had PSP concentrations ranging from 50 to 3400 $\mu g\ g^{-1}$ could be detected and the bloom was found to be dominated by N-sulfocarbamoyl-1 l-hydroxysulfate C-toxins (12). The concentrations of the major toxins in 24 bloom samples averaged 44% for C-1 and 25% for C-2. The remaining PSPs included gonyautoxins GTX-2, GTX-3, and GTX-5, decarbamoylgonyautoxins DcGTX-2 and DcGTX-3, saxitoxin, and decarbamoylsaxitoxin. Negri et al. (12) suggested that samples containing low levels of PSP toxins are unlikely to be detected by conventional mouse bioassay. They point out that previous work indicated that 43% of blooms of *A. circinalis* in the Murray-Darling Basin of Australia were toxic to mice (44). However, with HPLC techniques that detect PSP toxin concentrations of 20 $\mu g\ g^{-1}$ dry weight, 77% of bloom samples were considered to be toxic (12).

Because the sensitivity of the mouse bioassay can be low and the results may be affected by test conditions (64,65), other bioassays have been used to study neurotoxicity. Cell tissue culture cytotoxicity tests have been used which gave toxicity data that agreed with those obtained by the mouse bioassay and HPLC (64,66). However, cell bioassay may be less easy to perform and may be influenced by nonspecific effects (65). Several immunoassays have been developed for saxitoxin and other PSP toxins (65,67–71). These assays have the advantages of being relatively sensitive and requiring simple apparatus. Also, they provide information on relationships between toxin structure and antibody binding (65). Saxitoxin and neosaxitoxin enzyme-linked immunosorbent assay (ELISA) kits were found to be as sensitive as the mouse bioassay and, in most cases, were able to indicate toxicity at the regulatory level for PSP toxins (72,73). DNA probes have been used to distinguish strains of *Anabaena* sp. that produce neurotoxins from those that produce hepatotoxins. This type of technique could also be used to assist in investigating toxic strain characterization.

Bioassays using animals other than mice are also useful tools for toxicology, especially when the target organism of a neurotoxin may be a crustacean or fish species used in aquaculture. In these bioassays, the test animals are usually bathed (immersed) in solutions with known concentrations of cyanobacterial extracts. The tests provide a noninvasive toxicity bioassay and can be carried out with filtered extracts, so ingestion of toxins can be eliminated from the test conditions. Cyanobacteria that have been suspected of producing neurotoxins have been tested using zooplankton such as rotifers (*Brachionus calyciflorus* and *Synchaeta pectinata*), *Artemia salina* (74,75), *Daphnia* (76), and copepods (77–79). These species of zooplankton are suitable for toxicity testing because they are relatively easy to culture and can be used without approval by an animal ethics process as required for mouse bioassay testing. Bioassay methods have also been developed which use insects such as the housefly and desert locust (*Schistocera gregaria*) for the detection of saxitoxin and related toxins (80,81).

B. Toxicological Findings

In recent times there has been a growing body of scientific evidence indicating that increases are occurring in the scale and frequency of freshwater blooms that are capable of producing neurotoxins. The enormous bloom of toxic *Anabaena* spp. that occurred in the Darling River system in Australia in 1991 has already been cited. The following examples of neurotoxic cyanobacterial blooms are not comprehensive, however they indicate the scope of the problem.

- A long-term study of plankton samples in Swedish freshwater and brackish waters revealed that 25 of 331 samples were either neurotoxic (15 samples) or both neurotoxic and hepatotoxic (10 samples) (53). Mouse bioassays showed that neurotoxic strains of *Anabaena flos-aquae* were the most frequent; however, neurotoxic strains of *Anabaena farciminiformis*, *Aphanizomenon flos-aquae*, and *Oscillatoria sancta* were also found (53).
- In Korea, progressive eutrophication in lakes used for drinking water has caused *Microcystis* and other cyanobacteria to become dominant (82). A survey revealed that anatoxin-a was detected in 15% of samples from Korean lakes, with levels of 417–1190 $\mu g\ g^{-1}$ (82), which is comparable to levels of 12–4360 $\mu g\ g^{-1}$ reported for cyanobacteria samples from lakes in Finland (39).
- Cyanobacterial scums from four Greek freshwater lakes were responsible for poisoning of wild and domestic animals (83). The samples collected in 1987 were dominated by *Microcystis*, *Oscillatoria*, *Anabaena*, and *Anabaenopsis* with lethal dose (LD50) concentrations of 40–1500 $\mu g\ g^{-1}$ by intraperitoneal (i.p.) injection of mice. Mice showed symptoms of restlessness, hunching, paralysis, muscle contractions, palpitation, coma, and death. Lanaras et al. (83) concluded from the mouse bioassays that hepatotoxins were responsible, although in reviewing the symptoms, neurotoxins could have been present as well.
- Investigation of blooms of *Anabaena lemmermannii* that were implicated in bird kills in Denmark revealed that the bloom contained a neurotoxin with anticholinesterase activity. The neurotoxin was subsequently identified as anatoxin-a(S) (63).
- A study in Finland of the effect of environmental conditions on growth rate and anatoxin-a production by neurotoxic blooms of *Anabaena* spp. revealed that a high phosphate concentration was the most significant environmental factor (84).
- Hundreds of cattle died after drinking from a lake that had a bloom of *Anabaena vigueieri* and *Oscillatoria granulata*; the symptoms of the sick cattle indicated an unidentified neurotoxin was the cause (85).
- Unidentified neurotoxins had been implicated in studies of isolates of *Anabaena* sp. (39,86). Since then Humpage et al. (11) discovered PSP toxins in many Australian isolates of *A. circinalis* and concluded PSPs may be more widespread than originally thought, and this could explain earlier reports of unidentified neurotoxins.
- Sheep mortalities in Australia have been caused by PSPs produced by toxic blooms of *A. circinalis* (87). In mouse toxicity tests (i.p. injection) symptoms included staggering (at low doses), gasping, leaping, respiratory failure, and death occurred after 4–7 min. LD100 values were 13–20 $\mu g\ mg^{-1}$.

In summary, the studies listed above are examples of work that illustrate the extent and the causes of mortalities of larger animals. On another level in the food chain, there is growing evidence suggesting that neurotoxic cyanobacteria are impacting on freshwater biodiversity. Neurotoxins may decrease the population growth rate, fecundity, and life span of zooplankton, with the degree of reduction being influenced by food availability, temperature, and other modifying environmental factors (74–77,88). *Daphnia* are sensitive to very low levels of anatoxin-a (1–5 $\mu g\ mL^{-1}$), and this may have important impacts on phytoplankton dynamics and on the numerous predators that rely on *Daphnia* as food (88). The primary impact of low levels of anatoxin-a was a decrease in fecundity of *D. pulex*, and warmer temperatures increased this effect (88). Similar studies of planktonic rotifers also found that reproduction was inhibited by low concentrations of anatoxin-a, and susceptibility increased with temperature (79,89).

LD50 values in *Artemia* bioassays were 2–14 μg mL^{-1} for filtered cyanobacterial extracts containing anatoxin-a (74). LD50 values were significantly higher in extracts from toxin-producing *Anabaena* than for purified anatoxin-a. Synergistic and/or antagonistic effects of anatoxin-a with compounds in the fractionated extracts were suggested as the likely reasons for the differences in LD50 values (74).

Further, there is evidence that unidentified toxins are also produced by cyanobacteria. Some of these compounds appear to be nontoxic to mice but are toxic to crustaceans (75,76). An EC50 value of an unidentified toxin (i.e., concentration causing 50% of animals to show atypical movement) for *Artemia* in freeze-dried extract of *Oscillatoria* was 260 μg mL^{-1}, whereas for semipurified extract, it was 60 μg mL^{-1} (76). Bioassays with *Artemia* suggest that *Oscillatoria* spp. in brackish water penaeid shrimp farms in Australia have heat-labile, water-soluble neurotoxins (75). However, in preliminary tests, PSPs, anatoxin-a, homoanatoxin-a, and microcystins could not be detected. The extracts from *Oscillatoria* had LD50 of 70 μg mL^{-1} for immersion of *Artemia* cysts and 50 μg mL^{-1} for adult *Artemia* (75).

Further, recent studies of *Lyngbya wollei* indicate that it produces unidentified toxins, because the two identified toxins, DcGTX-2 and DcGTX-3, were minor components of the total toxicity (15). Unknown peaks were identified as new saxitoxin analogues (90). Also, patch-clamping techniques have indicated that novel types of neurotoxins were present in water-soluble extracts from toxic strains of *A. flos-aquae* and *A. lemmermannii* (91).

C. Elimination and Breakdown of Neurotoxins

PSPs from blooms of *A. circinalis* were found to persist for over 3 months in dams and irrigation water, and showed a temporary increase in toxicity in the first few weeks after the bloom collapsed (13). Analysis of the composition of the PSP toxins indicated that chemical degradation of C-1 and C-2 toxins to form the more potent gonyautoxins, DcGTX-2 and DcGTX-3, occurs in test solutions and in natural water bodies (13). In a previous study of sheep mortalities, the proportion of C-toxins in the toxin profile of the cyanobacterium was 70%, whereas in the animal's small intestine the proportion was 87%, indicating that the desulfation of C-toxins occurred in the gut of sheep (87). These findings are consistent with earlier work with PSPs from marine dinoflagellates which showed that the loss of the N-sulfonate or N-sulfocarbamoyl group greatly increased the toxicity of PSPs (92). Table 1 compares the mouse toxicity of PSPs. It shows that C-toxins with double-sulfate groups are less toxic than gonyautoxins, with single-sulfate groups, and saxitoxin with no sulfate groups.

The principal route of elimination of saxitoxin from mammals has been studied using rats (95,96). Sublethal doses of saxitoxin were administered by intravenous injection, and it was found that 58% of the injected dose was eliminated in the urine, and small quantities of saxitoxin could be detected in urine up to 144 h after injection (96). The results demonstrate that the analytical method used by the investigators could be useful in diagnosing saxitoxin poisoning (96).

The decay of anatoxin-a in natural waters has been investigated in bench tests using lake sediments (97). Toxin concentrations increased dramatically in the first 4 days of the test from 50 to 100 μg L^{-1} to 400–600 μg L^{-1} and then fell to levels below 10 μg L^{-1} by day 8, whereas sterilized sediments failed to reduce levels of anatoxin-a. The initial increase was explained by incomplete lysis of cells at the start of the tests. Importantly, the results showed that degradation of anatoxin-a depends on the action of microorganisms in sediments (97).

Table 1 Paralytic Shellfish Poisoning Toxins Identified in Freshwater Cyanobacteria and Values for Mouse Toxicity

Toxin	Abbrev	Sulfate groups	Mouse toxicity (MU)	Reported for freshwater cyanobacteria
C-1		Double	15	*Anabaena circinalis* (11–13)
C-2		Double	239	*Anabaena circinalis* (11–13)
C-3		Double	33	
C-4		Double	143	
Gonyautoxin 1	GTX-1	Single	2468	*Anabaena circinalis* (11)
Gonyautoxin 2	GTX-2	Single	892	*Anabaena circinalis* (11–13)
Gonyautoxin 3	GTX-3	Single	1584	*Anabaena circinalis* (11–13)
Gonyautoxin 4	GTX-4	Single	1803	*Anabaena circinalis* (11)
Gonyautoxin 5 (B1)	GTX-5	Single	160	*Anabaena circinalis* (12)
Gonyautoxin 6 (B2)	GTX-6	Single	180	*Anabaena circinalis* (11)
Decarbamoylgonyautoxin 2	DcGTX-2	Single	1617	*Anabaena circinalis* (11–13) *Lyngbya wollei* (15)
Decarbamoylgonyautoxin 3	DcGTX-3	Single	1872	*Anabaena circinalis* (11–13) *Lyngbya wollei* (15)
Saxitoxin	STX	None	2483	*Anabaena circinalis* (11–13) *Aphanizomenon flos-aquae* (6,7)
Neosaxitoxin	NEO	None	2295	*Aphanizomenon flos-aquae* (94)
Decarbamoylsaxitoxin	DcSTX	None	1220	*Anabaena circinalis* (12)

Relative toxicity of PSPs are provided in mouse units (MU). In comparison, saxitoxin has an i.p. LD50 = 10 $\mu g\ kg^{-1}$ (92,93).

Anatoxin-a also readily decomposes in sunlight and high pH, and the degradation products have been identified as dihydroanatoxin-a and epoxyanatoxin-a, whereas the degradation products of homoanatoxin-a are dihydrohomoanatoxin-a and epoxyhomoanatoxin-a (62,98,99). The analysis of samples for these stable nontoxic degradation products has been helpful in determining the detoxification of anatoxin-a (98) and for investigating ageing of nontoxic blooms to show whether they had been toxic (99).

III. PHARMACOLOGY

Cyanobacteria have received attention as possible tools for medical research and bioscience (4,100,101) because of the range of neurotoxins, hepatotoxins, and bioactive compounds that some strains produce (102). Anatoxin-a is an extremely potent inhibitor of the nicotinic and muscarine acetylcholine receptor (AChR) (103,104), causing postsynaptic cholinergic depolarization (103). That is, anatoxin-a mimics the neurotransmitter, acetylcholine, and acts by attaching to receptor sites for acetylcholine that trigger the opening of ion channels and induce the contraction of muscle cells. Unlike acetylcholine, which is degraded by acetylcholinesterase, anatoxin-a is not broken down and so continues to open ion channels and cause depolarization and contraction of muscle cells until they are exhausted. Importantly for pharmacological studies, anatoxin-a is the most potent and stereospecific nicotinic acetylcholine receptor agonist that has been found so far (105).

Typical symptoms from anatoxin-a poisoning in mouse bioassay include loss of muscle coordination, staggering, muscle fasciculation, gasping, convulsions, and death within minutes from respiratory failure (103). In birds, opisthotonous (head bent over backwards) is a characteristic feature of poisoning by anatoxin-a (2). Anatoxin-a blocks muscle twitch at concentrations greater than 25 μM in isolated rat phrenic nerve–hemidiaphragm preparations (103), and the LD50 by intraperitoneal injection in mice is 200–250 μg kg^{-1} (Table 2) with a survival time of 4–7 min (3,38,103). Anatoxin-a is an important pharmacological tool that has been used in the study of the binding site of nAChR protein (105). It is a compound that could be used to produce a polymer-bound ligand system for affinity chromatography in the purification and characterization of nAChR protein (109). It has been considered for use as a tool to slow mental degeneration in Alzheimer's disease or disorders in which acetylcholine is deficient (4).

Homoanatoxin-a is an analogue of anatoxin-a in which there is an ethyl rather than a methyl group on the ketone carbon (C_{10}). Homoanatoxin-a has been isolated from *Oscillatoria formosa*, and it has a similar toxicity to that of anatoxin-a (108).

Anatoxin-a(S) is a powerful irreversible acetylcholinesterase inhibitor (9,106,107) exhibiting an inhibition of nicotinic acetylcholinesterase in a manner similar to that of organophosphate pesticides, such as malathion and parathion (3,4). It acts by inhibiting the breakdown of acetylcholine by acetylcholinesterase, thus causing the overstimulation of muscles, fatigue, and paralysis. In mouse bioassays, anatoxin-a(S) has an LD50 by intraperitoneal injection of 20–50 μg kg^{-1} (see Table 2) with a survival time of 10–30 min (2,3). Toxicity typically causes noticeable salivation, lacrimation (production of tears) in mice, or chromodacryorrhea (blood in tears) in rats, urinary incontinence, fasciculation, muscular weakness, convulsion, defecation, and death by respiratory failure (2,3).

PSPs are a range of hydrolated and sulfated analogues of saxitoxin; they are potent sodium channel blockers (occluders; see refs. 110 and 111). They act by reversibly blocking sodium channels in nerve and muscle membranes (112). In mouse bioassays, PSP toxins restrict signal transmission between neurons, causing muscle fasciculation, loss of muscle coordination, gasping, convulsions, and death by respiratory failure (44,103,113). These symptoms may appear to be similar to those for anatoxin-a toxicity; however, saxitoxins cause electrophysiological effects which are quite different from those of anatoxin-a. In amphibian sciatic nerve, saxitoxins produce a decrease in the peak height and rate of rise of the action potential, causing a block after 5 min (11,114).

Table 2 Freshwater Cyanobacteria that Produce Toxins Anatoxin-a, Anatoxin-a(S) and Homoanatoxin

Toxin		Mouse toxicity i.p. LD50 (μg kg^{-1}) (reference)	Reported for freshwater cyanobacteria (reference)
Anatoxin-a	Postsynaptic cholinergic depolarizing agent of nerves and muscles	200 (3,103) 250 (38)	*Anabaena flos-aquae* (38) *Anabaena circinalis* (39,90) *Aphanizomenon* sp. (39) *Cylindrospermum* sp. (39) *Oscillatoria* sp. (16,39,58)
Anatoxin-a(S)	Inhibition of acetylcholinesterase activity	20 (9,106)	*Anabaena flos-aquae* (9,53,86,107) *Anabaena lemmermannii* (63)
Homoanatoxin-a	Postsynaptic cholinergic depolarizing agent of nerves and muscles	200–250 (108)	*Oscillatoria formosa* (108)

Neurophysiological studies with various PSP compounds show that they all have the same mode of action; they only differ in potency (see Table 1). Neurotoxic species of *Aphanizomenon* have been used to resolve the biosynthetic pathway for the production of PSPs (115).

A recent report suggests that some toxic strains of *Anabaena* that were collected from lakes in Norway produce unidentified new neurotoxins, which act differently from PSPs, anatoxin-a, and anatoxin-a(S) (91). *Anabaena flos-aquae* caused death of cattle, and the clinical signs were similar to intoxication with anatoxin-a; however, depolarization of cholinergic nerve cells was caused by activation of Na^{+} channels and N-type Ca^{2+} channels (91). *Anabaena lemmermannii* caused death after 16–24 h in mouse bioassay by intraperitoneal injection, and it appeared to have a potent antagonistic effect on muscarinic and cholinergic receptors (91).

IV. POSSIBLE IMPACTS ON FOOD SOURCES AND HEALTH

A. Occurrence of Cyanobacteria in Freshwater and Brackish Water Aquaculture

Freshwater and brackish water habitats are a major source of fish, crustaceans, and shellfish for a large proportion of the world's population, with inland fisheries and farms in many Asian countries producing the main supply of protein for traditional cultures (116). Also, the overexploitation of natural fish stocks has led to the expansion of aquaculture in freshwater and brackish water habitats to meet growing demand. In 1996, world production of fish and shellfish by aquaculture accounted for nearly 23% of the combined total harvest from fishing and farming of 115.9×10^6 tons (117). The largest sector was inland freshwater farming, which produced 15.6×10^6 tons in 1996, a rise of 12.6% from the previous year. The most important species were freshwater carp, with farms producing 11.5×10^6 tons, up 12.9%. China's aquacultural production in 1996 was 17.7×10^6 tons, accounting for 67% of the world's production. The total value of foodfish in 1996 was US $41 billion (116,117). Shrimp culture is carried out in earthen ponds using brackish water, with salinities of 10–35 ppt, and in 1998 produced a total worldwide harvest of over 700,000 metric tons (118).

Evidence suggests that fish culture causes an increased occurrence of cyanobacteria in freshwater ponds. For example, cyanobacteria are the most common form of phytoplankton in Czech fish ponds and since the 1950s there has been a shift in dominance from large colonial species through to small single-celled forms (119). The progressive changes in the dominant cyanobacteria have been from *Aphanizomenon flos-aquae*, *Microcystis aeruginosa*, *Anabaena* sp., *Planktotrix agardhii*, and other small species. Analysis of data over several decades suggests that this is due to increased fish-stocking densities and resultant changes in the composition of the filtering zooplankton, particularly *Daphnia* sp. (119). Further, Pringle and Hamazaki (120) found that fish play a key role in determining the benthic algal assemblages in freshwater streams. They reported that cyanobacteria dominated algal benthos in the presence of fish, whereas in streams in which fish were excluded, benthic diatoms prevailed. Gehrke and Harris (121) contend that in Australian inland waters, the population explosion of carp since the 1970s is an important factor in the increased frequency of cyanobacterial blooms.

Further, Paerl and Tucker (122) reported that cyanobacteria are the dominant species in catfish ponds in the United States, usually accounting for 75% of the total phytoplankton biomass. Noxious, bloom-forming cyanobacteria in catfish ponds occur as filaments or multicellular colonies and mainly consist of the genera *Anabaena*, *Aphanizomenon*, *Microcystis*, and *Oscillatoria*. Factors which have been attributed to the promotion of cyanobacterial blooms and surface scums in channel catfish ponds include regular additions of nitrogen and phosphorus in supplementary feeds, high temperatures, strong sunlight, and stratification of the water column

(123,124). In aquaculture, the distinct pigment profiles of noxious cyanobacterial blooms in freshwater ponds could be incorporated into multispectral remote sensing and coupled to a geographical information system (GIS) to provide an improved method to integrate, model, and predict parameters of interest to farm managers (125).

The frequency and types of cyanobacterial blooms in fish and shrimp ponds would seem to have important implications for aquaculture. With respect to fish culture, Paerl and Tucker (122) found that there were few reports on the toxic effects of freshwater cyanobacteria. They were able to cite losses of channel catfish in South Carolina caused by *A. flos-aquae* (126) and possible links between cyanobacterial blooms and unexplained deaths of fish and crustaceans. The toxicity of *A. flos-aquae* has been tested by preparing concentrated extracts of blooms from ponds where high mortalities of channel catfish occurred (126). Tests by intraperitoneal injection showed that the toxicity was significantly increased when cells were lysed by sonication. The clinical signs of fish in affected ponds and fish used in toxicity tests were similar; activity alternated between lethargy and hyperactivity, ventilation rates increased to over 100 per minute, and fish exhibited a coughing response. English et al. (126) concluded that the clinical signs were consistent with those described for aphantoxins (since identified as saxitoxin and neosaxitoxin) produced by the blooms of *A. flos-aquae* (127).

Sevrin-Reyssac and Pletikovic (128) reviewed the literature on cyanobacteria in fish ponds and summarized cases of mortalities of fish and other aquatic organisms that were associated with cyanobacteria. There have been reports of cyanobacteria inhibiting nutrition and reproduction in rotifers (129,130) and certain cladocerans have been affected (131,132). *Microcystis* sp. have been known to cause deaths in carp and other freshwater fish (133–136), and high mortalities have occurred in carp with a bloom of *A. flos-aquae* (137). Others have also reported fast-acting lethal effects of *A. flos-aquae* (6,138,139). *A. gracile* has been associated with dwarfism in roaches; when affected fish were transported to a pond without cyanobacterium, the fish produced offspring of normal size (135). Nevertheless, Sevrin-Reyssac and Pletikovic (128) concluded that most fish mortalities in the field do not appear to be caused by cyanobacterial toxins. Depletion of oxygen and an increase in ammonia are probably the most common cause of fish mortalities in freshwater ponds containing cyanobacterial blooms (128,137). Also, gill blockage by cyanobacterial cells and high pH are additional causes of fish kills during cyanobacterial blooms (5). Therefore, although there have been cases of fish pathology caused by cyanobacterial hepatotoxins (140,141), cases of neurotoxicity do not appear to be common.

With respect to shrimp farming, Lightner (142) commented that the literature rarely refers to the toxic effects of blooms of cyanobacteria on crustaceans, and Hawser et al. (78) found the impact of *Trichodesmium* on copepods and marine food webs was largely unknown. Further, although cyanobacteria can be a major component of the phytoplankton at shrimp farms, there have been few reports of cyanobacterial toxicity. In one instance, the marine blue-green alga *Spirulina subsalsa* caused mortalities in cultured blue shrimp, *Penaeus stylirostris* (142). Affected shrimp had hemocytic enteritis characterized by necrosis of the epithelium of the midgut, dorsal cecum, and hindgut gland. It was concluded that under certain conditions, *S. subsalsa* produces an unidentified toxin that is only mildly toxic to shrimp. Moore (30) was unable to demonstrate mouse toxicity with an extract from Lightner's sample (which was reidentified as *Lyngbya* sp., after being originally identified as *S. subsalsa*). Moore (30) concluded that the levels of toxin in the sample were too low to be detected in bioassays, yet large enough to cause disease in prawns on repeated ingestion.

In Australia, Smith (75) reported the toxic effects of blooms of Oscillatoriales on *Penaeus monodon*, *P. japonicus*, and *Artemia salina* in brackish water shrimp ponds. Figure 1 shows an example of a shrimp pond similar to one in which mortalities occurred when planktonic species of Oscillatoriales were at densities of 65,000–90,000 cells mL^{-1} and the salinity was 8.2–32

Figure 1 Shrimp aquaculture is carried out in earthen ponds. Ponds are approximately 1.2–1.8 m deep and brackish water is aerated by paddlewheels (foreground). Cyanobacterial blooms are a frequent occurrence.

ppt. *Oscillatoria corakiana* was a dominant species present in each case; however, planktonic blooms and benthic mats included *Spirulina* sp., *Lyngbya* sp., *Oscillatoria* sp., and *Nodularia* sp. Two examples of commonly occurring *Oscillatoria* and *Lyngbya* are shown in Figure 2. Freeze dried extracts of laboratory cultures of cyanobacteria were toxic when injected intramuscularly into the tail muscle of shrimp, and LD50 values for *P. japonicus* and *P. monodon* were 75–180 mg extract kg^{-1} shrimp. Immersion tests with *A. salina* bathed in 50% seawater and freeze-dried cyanobacteria gave LD50 values for adult *Artemia* of 50 mg L^{-1} in supernatant solutions, whereas for cysts, the LD50 values were 70 mg L^{-1}. Smith (75) concluded that blooms of Oscillatoriales may cause disease in prawns by producing water-soluble toxins at sublethal concentrations, causing slow growth, loss of appetite, and low immunity and rendering the prawns more susceptible to secondary infection by pathogens. There was evidence to suggest that a novel neurotoxin was the primary cause of disease (75).

Cyanobacteria can also affect the health of farmed shrimp by fouling the gills and causing poor oxygen uptake. In particular, gills of shrimp often become fouled by cyanobacteria, bacteria, ciliates, detritus, and phytoplankton (Figure 3).

B. Effect on Human Health

A possible connection between toxic freshwater cyanobacteria and human illness occurred in Sweden when people ate the flesh, and particularly the liver, of fish caught in a lake that had a bloom of *Anabaena* sp. These people developed symptoms that included respiratory distress, severe muscular pain, vomiting, and brownish black urine. This condition is known as Haff disease (143). Over 1000 cases of Haff disease occurred in the 1920s and 1930s along the Baltic coast, and Yuksov-Sortlav disease may be similar (1).

Notwithstanding, it has been suggested that neurotoxins from freshwater cyanobacteria do not pose a serious threat to human health, because there are no known vectors, such as shellfish, to accumulate toxins into the human food chain (3). However, there have been recent

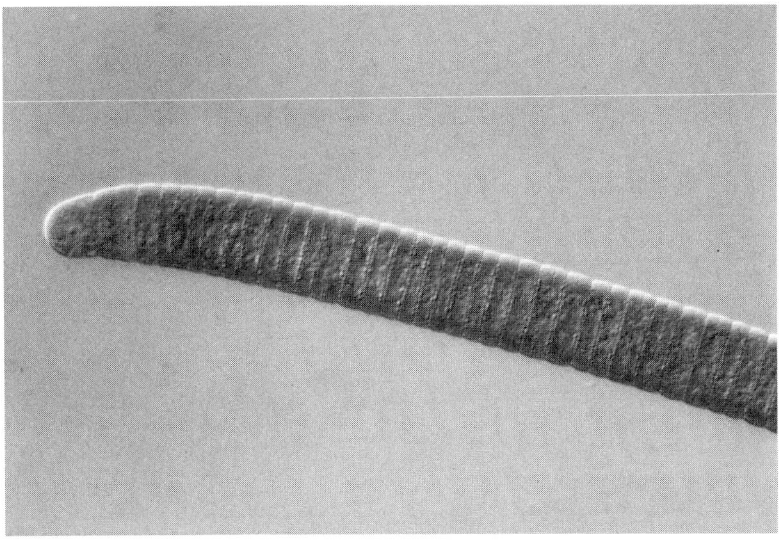

(a)

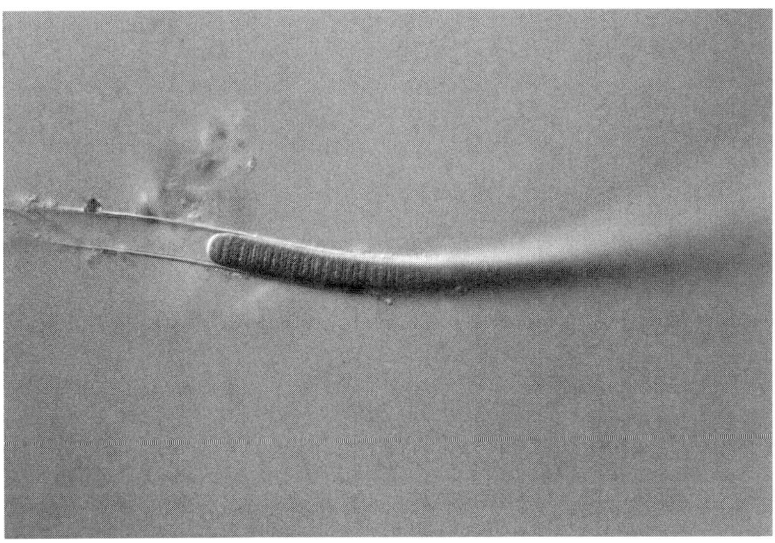

(b)

Figure 2 Examples of cyanobacteria from a shrimp farm. (a) Planktonic or benthic *Oscillatoria* are common in shrimp culture. This specimen is a planktonic form that occurred during a mass mortality of shrimp (75) at an Australian shrimp farm (400× magnification). (b) Benthic species of *Lyngbya* are common in brackish water shrimp farms. This species was not associated with shrimp mortalities (400× magnification).

reports that suggest that this position should be reviewed. Freshwater bivalves, *Elliptio complanatus* and *Corbicula fluminea*, growing in blooms of *Aphanizomenon flos-aquae* were found to have accumulated neosaxitoxin and saxitoxin at concentrations up to 40–60 μg STX eq/100 g flesh (10). Also, neosaxitoxin and gonyautoxins (GTX-1, GTX-2, GTX-3, and GTX-4) have been found in the freshwater bivalve *Corbicula fluminea* in Lake Biwa in Japan (144). The

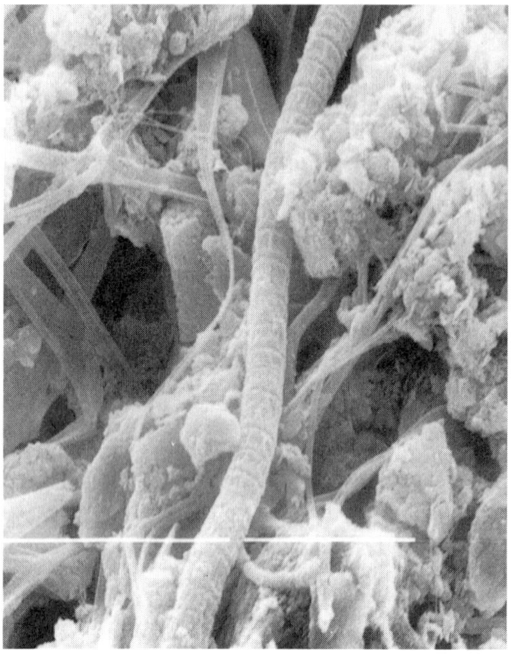

Figure 3 *Oscillatoria* sp., phytoplankton, and bacteria foul the gills of a sick shrimp. Filamentous *Oscillatoria* and many other organisms accumulate on the surface of gills of sick shrimp causing deterioration in shrimp health (right-hand bar = 5 μm).

source of the toxins was not found, although possible types of cyanobacteria were considered (144).

Further, edible mussels have been shown to accumulate hepatotoxins from blooms of *Nodularia spumigena* (145), and recent experiments have shown that, under laboratory conditions, the Australian freshwater mussel *Alathyria condola* can accumulate high levels of PSP toxins from *A. circinalis* (14). Three groups of toxins were detected in *A. circinalis*, C-toxins (C-1 and C-2), gonyautoxins (GTX-2, GTX-3, GTX-5, DcGTX-2, and DcGTX-3) and saxitoxin (STX). Humpage et al. (11) had previously identified a range of PSP toxins in *A. circinalis* and these included those found by Negri and Jones (14) plus GTX-1, GTX-4, GTX-6, and neosaxitoxin. When mussels were fed at a high concentration of *A. circinalis* (10^6 cells/mL), feeding was reduced and erratic and a maximum of 570 μg toxin/100 g flesh was accumulated (14). At a low concentration of *A. circinalis* (2×10^5 cells/mL), slightly higher levels of up to 620 μg/100 g flesh were measured (mean = 440 μg/100 g flesh). Long-term exposure (5 weeks) to lower concentrations of *A. circinalis* (at 10^4 cells/mL on every third day) resulted in only trace levels of PSP toxins being detected (14).

In marine dinoflagellates and mussels, C-toxins are converted to the more toxic gonyautoxins under acidic conditions or by enzymatic processes through the loss of the sulfocarboamyl group (146,147). Experiments with freshwater mussels showed that 96% of the PSP toxins accumulated in the visera and that there is a shift in the profile of PSP toxins that would suggest that C-toxins were converted to gonyautoxins, and it was concluded that bioaccumulation of PSP toxins in freshwater mussels may pose a health risk to humans, fish, birds, and water rats (14).

Cases of food poisoning have occurred in Thailand following the eating of freshwater pufferfish (*Tetraodon* sp.). Based on mouse bioassays and symptoms of the patients, the early reports suggested that tetrodoxin (TTX) was the causative agent (148–150). However, recent analysis using HPLC has revealed that the various tissues of freshwater pufferfish contain saxitoxin (STX), neosaxitoxin (neoSTX), and decarbamoxysaxitoxin (DcSTX) (151,152). Also, illness and fatalities have been caused by the consumption of freshwater pufferfish (*T. cutcutia* and *Chelonodon patoca*) in Bangladesh (153). The toxins were saxitoxin, DcSTX, GTX-2, GTX-3, DcGTX-2, DcGTX-3, and three unidentified components. The origin of PSP toxins in freshwater pufferfish, from both Thailand and Bangladesh, was not determined in those studies.

Finally, the previously discussed high level of cyanobacteria that commonly occurs in fish and shrimp aquaculture has important implications for the health of workers at these farms. For example, it is common for blooms of Oscillatoriales in prawn ponds to exceed greatly 20,000 cells mL^{-1} (75), which is well above the safety guideline that has been proposed for recreational waters (154). Workers at shrimp and fish farms may come into regular contact with aerosols or may be occasionally immersed in waters contaminated by toxins from cyanobacteria. At this time, the impacts on human health of chronic exposure to low levels of neurotoxins is poorly understood, and epidemiological studies are needed to demonstrate a dose-response curve between exposure to cyanobacterial toxins and human disease (155). It is suggested that the workers at freshwater and brackish water aquaculture farms would be ideal subjects for epidemiological studies to investigate this association.

V. SUMMARY

The main sources of freshwater neurotoxins are planktonic and benthic species of cyanobacteria, with the most important genera being *Anabaena*, *Aphanizomenon*, and *Oscillatoria*. Evidence suggests that eutrophication of freshwater habitats is causing increased frequencies of neurotoxic cyanobacterial blooms and that they may have significant impacts on the biodiversity of freshwater habitats and the productivity of freshwater aquaculture. The main types of freshwater neurotoxins that have been described so far are anatoxin-a, homoanatoxin-a, anatoxin-a(S), and paralytic shellfish poisons (PSPs), which include saxitoxin, neosaxitoxin, C-toxins, and gonyautoxins. However, recent studies indicate that there may be other unidentified neurotoxins produced by cyanobacteria. Toxicity tests by mouse bioassay, in combination with HPLC analysis and necropsy, are the most important means of defining the types of toxins present in samples of water, scum, or animal tissue. Anatoxin-a is unique to cyanobacteria and it mimics the neurotransmitter acetylcholine, binding tightly to nicotinic and, to a lesser extent, to muscarine acetylcholine receptor proteins, causing symptoms that include loss of muscle coordination, muscle fasciculation, gasping, convulsions, and death within minutes from respiratory failure. Anatoxin-a(S) is also unique to cyanobacteria and causes symptoms similar to those of anatoxin-a, but in addition causes excessive viscous salivation (the reason for adding the [S] label). Anatoxin-a(S) is a natural organophosphate that inhibits acetylcholinesterase, thus causing overstimulation of muscle cells by acetylcholine. PSPs that are found in freshwater cyanobacteria are better known for being produced by marine dinoflagellates that cause toxic red tides. PSPs block sodium channels, subsequently restricting signal transmission between neurons and causing death in animals by respiratory failure. Freshwater neurotoxins pose problems because of threats to (i) the health of humans and animals that depend on affected water supplies, (ii) the biodiversity of natural freshwater ecosystems, (iii) aquaculture in freshwater and brackish water, and (iv) consumers of freshwater shellfish and fish.

REFERENCES

1. GA Codd, GK Poon. Cyanobacterial toxins. Proc Phytochem Soc Eur 28:285–296, 1988.
2. WW Carmichael. Toxins of freshwater algae. In: AT Tu, ed. Handbook of Natural Toxins. Vol. 3. Marine Toxins and Venoms. New York: Marcel Dekker, 1988, pp 121–147.
3. WW Carmichael, NA Mahmood, EG Hyde. Natural toxins from cyanobacteria (blue-green algae). In: S Hall, G Strichartz, eds. Marine Toxins: Origin, Structure, and Molecular Pharmacology. Washington, DC: American Chemical Society, 1990, pp 87–106.
4. WW Carmichael. The toxins of cyanobacteria. Sci Am, Jan:64–72, 1994.
5. GA Codd. Cyanobacterial toxins: occurrence, properties and biological significance. Water Sci Technol 32(4):149–156, 1995.
6. PJ Sawyer, JH Gentile, JJ Sasner. Demonstration of a toxin from *Aphanizomenon flos-aquae* Ralfs. Can J Microbiol 14:1199–1204, 1968.
7. E Jakim, J Gentile. Toxins of blue-green algae: similarity to neosaxitoxin. Science 162:915–916, 1968.
8. M Alam, Y Shimidzu, M Ikawa, JJ Sasner. Reinvestigation of the toxins from the blue-green algae, *Aphanizomenon flos-aquae*, by a high-performance chromatographic method. J Environ Sci Health A13:493–499, 1978.
9. NA Mahmood, WW Carmichael. Paralytic shellfish toxins produced by the freshwater cyanobacterium *Aphanizomenon flos-aquae* NH-5. Toxicon 24:175–186, 1986.
10. JJ Sasner Jr, M Ikawa, TL Foxall. Studies on Aphanizomenon and Microcystis toxins. In: EP Regelis, ed. Seafood Toxins. Washington, DC: American Chemical Society, 1984, pp 391–406.
11. AR Humpage, J Rositano, AH Bretag, R Brown, PD Baker, BC Nicholson, DA Steffensen. Paralytic shellfish poisons from Australian cyanobacterial blooms. Aust J Mar Freshwater Res 45:761–771, 1994.
12. AP Negri, GJ Jones, SI Blackburn, Y Oshima, H Onodera. Effect of culture and bloom development and of sample storage on paralytic shellfish poisons in the cyanobacterium *Anabaena circinalis*. J Phycol 33:26–35, 1997.
13. GJ Jones, AP Negri. Persistence and degradation of cyanobacterial paralytic shellfish poisons (PSPs) in freshwaters. Water Res 31(3):525–533, 1997.
14. AP Negri, GJ Jones. Bioaccumulation of paralytic shellfish poisoning (PSP) toxins from the cyanobacterium *Anabaena circinalis* by the freshwater mussel *Alathyria condola*. Toxicon 33(5):667–678, 1995.
15. Q Yin, WW Carmichael, WR Evans. Factors influencing growth and toxin production by cultures of the freshwater cyanobacterium *Lyngbya wollei* Farlow ex Gomont. J Appl Phycol 9:55–63, 1997.
16. C Edwards, KA Beattie, CM Scrimgeour, GA Codd. Identification of anatoxin-a in benthic cyanobacteria (blue-green algae) and in associated dog poisonings at Loch Insh, Scotland. Toxicon 30(10): 1165–1175, 1992.
17. GH Elder, PR Hunter, GA Codd. Hazardous freshwater cyanobacteria (blue-green algae). Lancet 341:1519–1520, 1993.
18. G Francis. Poisonous Australian lake. Nature 18:11–12, 1878.
19. GA Codd, DA Steffensen, MD Burch, PD Baker. Toxic blooms of cyanobacteria in Lake Alexandrina, South Australia—Learning from history. Aust J Mar Freshwater Res 45:731–736, 1994.
20. F Drouet. Revision of the classification of the Oscillatoriaceae. Monograph 15, Acad Nat Sci Philadelphia 115:261–281, 1968.
21. R Rippka, J Deruelles, JB Materbury, M Herdman, RY Stainer. Generic assignments, strain histories and properties of pure cultures of cyanobacteria. J Gen Microbiol 111:1–61, 1979.
22. K Anagnostidis, J Komárek. Modern approach to the classification system of Cyanophytes. 1—Introduction. Arch Hydrobiol Suppl 71 (Algol Stud 38/39):291–302, 1985.
23. K Anagnostidis, J Komárek. Modern approach to the classification system of Cyanophytes. 3—Oscillatoriales. Arch Hydrobiol Suppl 80 (Algol Stud 50/53):327–472, 1988.
24. K Anagnostidis, J Komárek. Modern approach to the classification system of Cyanophytes. 5—Stigonematales. Arch Hydrobiol Suppl 86 (Algol Stud 59):1–73, 1990.

25. J Komárek, K Anagnostidis. Modern approach to the classification system of Cyanophytes. 4—Nostocales. Arch Hydrobiol Suppl 82 (Algol Stud 56):247–345, 1989.
26. J Komárek, K Anagnostidis. Modern approach to the classification system of Cyanophytes. 2—Chroococcales. Arch Hydrobiol Suppl 73 (Algol Stud 43):157–226, 1986.
27. GJ Jones, W Korth. *In situ* production of volatile odour compounds by river and reservoir phytoplankton populations in Australia. Water Sci Technol 31(11):145–151, 1995.
28. F Jüttner. Physiology and biochemistry of odorous compounds from freshwater cyanobacteria and algae. Water Sci Technol 31(11):69–78, 1995.
29. JH Cardellina II, F-J Marner, RE Moore. Seaweed dermatitis: structure of lyngbyatoxin A. Science 204:193–195, 1979.
30. RE Moore. Toxins from marine blue-green algae. In: WW Carmichael, ed. The Water Environment: Algal Toxins and Health. New York: Plenum Press, 1981, pp 15–23.
31. T Yasumoto and M Satake. Bioactive compounds from marine microalgae. Chimia 52:63–68, 1998.
32. JE Armstrong, KE Janda, B Alvarado, AE Wright. Cytotoxin production by a marine Lyngbya strain (cyanobacterium) in a large-scale laboratory bioreactor. J Appl Phycol 3:277–282, 1991.
33. H Shirahashi, T Morimoto, A Nagatsu, N Murakami, K Tatta, J Sakakibara, H Tokuda, H Nishino. Antitumor-promoting activities of various synthetic 1-*O*-acyl-3-*O*-(6′-*O*-acyl-β-D-galactopyranosyl)-*sn*-glycerols related to natural product from freshwater cyanobacterium *Anabaena flos-aquae f. flos-aquae*. Chem Pharm Bull 44(7):1404–1406, 1996.
34. MMS de Cano, MCZ de Mulé, GZ de Claire, DR de Halperin. Inhibition of *Candida albicans* and *Staphylococcus aureus* by phenolic compounds from the terrestrial cyanobacterium *Nostoc muscorum*. J Appl Phycol 2:79–81, 1990.
35. S Bloor, RR England. Antibiotic production by the cyanobacterium *Nostoc muscorum*. J Appl Phycol 1:367–372, 1989.
36. JB Waterbury, RY Stainer. Isolation and growth of cyanobacteria from marine and hypersaline environments. In: MP Starr, H Stolp, HG Trüper, A Balows, HG Schlegel, eds. The Prokaryotes: A Handbook on Habitats, Isolation and Identification of Bacteria. Berlin: Springer-Verlag, 1981, pp 221–223.
37. HJ Humm, SR Wicks. Introduction and Guide to the Marine Blue Green Algae. New York: Wiley-Interscience, 1980, p 194.
38. JP Devlin, OE Edwards, PR Gorham, NR Hunter, RK Pike, B Stravic. Anatoxin-a, a toxic alkaloid from *Anabaena flos-aquae* NRC-44h. Can J Chem 55:1367–1371, 1977.
39. K Sivonen, K Himberg, R Luukkainen, S Niemelä, GK Poon, GA Codd. Preliminary characterisation of neurotoxic cyanobacteria blooms and strains from Finland. Toxicity Assess 4:339–352, 1989.
40. DK Stevens, RI Krieger. Effects of route of exposure and repeated doses on the acute toxicity in mice of the cyanobacterial nicotinic alkaloid anatoxin-a. Toxicon 29:134–138, 1991.
41. C Creagh. What can be done about toxic algal blooms? Ecos 72:14–19, 1992.
42. S Skinner, A Catalano, C Jones. Algae spread: state waits in fear. The Sun-Herald, Dec 1, 1991, pp 12–13.
43. B Bonham. Poisoned rivers deadlier than cyanide. Sunday Telegraph, Dec. 1, 1991, pp 24–25.
44. PD Baker and AR Humpage. Toxicity associated with commonly occurring cyanobacteria in surface waters of the Murray-Darling Basin, Australia. Aust J Mar Freshwater Res 45:773–786, 1994.
45. DA Steffensen, AR Humpage, J Rositano, AH Bretag, R Brown, PD Baker, BC Nicholson. Neurotoxins from Australian Anabaena. In: GA Codd, TM Jefferies, CW Keevil, E Potter, eds. Detection Methods for Cyanobacterial Toxins. Cambridge, UK: Royal Society of Chemistry, 1994, pp 45–50.
46. IR Falconer. Measurement of toxins from blue-green algae in water and foodstuffs. In: IR Falconer, ed. Algal Toxins in Seafood and Drinking Water. London: Academic Press, 1993, pp 165–176.
47. G Premazzi, L Volterra. Microphyte toxins. EUR 14854. Commission of the European Communities, 1993, p 336.
48. Y Oshima, K Sugino, T Yasumoto. Latest advances in HPLC analysis of paralytic shellfish poisons. In: S Natori, K Hashimoto, Y Ueno, eds. Mycotoxins and Phycotoxins '88. Amsterdam: Elsevier, 1989, pp 319–326.

49. Y Oshima, H Itakura, K Lee, T Yasumoto, S Blackburn, G Hallegraeff. Toxin production by the dinoflagellate *Gymnodinium catenatum*. In: TJ Smayda, Y Shimizu, eds. Toxic Phytoplankton in the Sea. New York: Elsevier, 1993, pp 907–912.
50. MW Powell. Analysis of anatoxin-a in aqueous samples. Chromatographia 45:25–28, 1997.
51. JW Hurst, R Selvin, JJ Sullivan, CM Yentsch, RRL Guillard. Intercomparison of various assay methods for the detection of shellfish toxins. In: DM Anderson, AW White, DG Baden, eds. Toxic Dinoflagellates. Amsterdam: Elsevier Applied Science, 1985, pp 427–432.
52. JM Hungerford. AOAC official method 959.08. Paralytic shellfish poison. Official Methods of Analysis, 16th ed. Arlington, VA: AOAC International, 1995, Chapter 35.1.37.
53. T Willén, R Mattsson. Water-blooming and toxin-producing cyanobacteria in Swedish fresh and brackish waters, 1981–1995. Hydrobiologia 353:181–192, 1997.
54. GW Langlois, RE Danielson, SK Perera, MV Pickering. Toxins on the half shell: shellfish monitoring along the California coast. LC-GC 9:838–844, 1991.
55. JLC Wright. Dealing with seafood toxins: present approaches and future options. Food Res Int 28(4):347–358, 1995.
56. HS Wong, E Hindin. Detecting an algal toxin by high-pressure liquid chromatography. J Am Water Works Assoc 74:528–529, 1982.
57. K-I Harada, I Kimura, K Ogawa, M Suzuki, AM Dahlem, VR Beasley, WW Carmichael. A new procedure for the analysis and purification of naturally occurring anatoxin-a from the blue-green alga *Anabaena flos-aquae*. Toxicon 27:1289–1296, 1989.
58. KJ James, IR Sherlock, MA Stack. Anatoxin-a in Irish freshwater and cyanobacteria, determined using a new fluorimetric liquid chromatography method. Toxicon 35(6):963–971, 1997.
59. I Ojanpera, E Vuori, K Himberg, M Warris, K Niinivarra. Facile detection of anatoxin-a in algal material by thin-layer chromatography with Fast Black K salt. Analyst 116:265–267, 1991.
60. KJ Al-Layl, GK Poon, GA Codd. Isolation and purification of peptide and alkaloid toxins from Anabaena flos-aquae using high performance thin-layer chromatography. J Microbiol Methods 7: 251–258, 1988.
61. GK Poon, LJ Griggs, C Edwards, KA Beattie, GA Codd. Liquid chromatography-electrospray ionization-mass spectrometry of cyanobacterial toxins. J Chromatogr 628:215–233, 1993.
62. K-I Harada, H Nagai, Y Kimura, M Suzuki. Liquid chromatography/mass spectrometric detection anatoxin-a, a neurotoxin from cyanobacteria. Tetrahedron 49(41):9251–9260, 1993.
63. H Onodera, Y Oshima, P Hendriksen, T Yasumoto. Confirmation of anatoxin-a(S), in the cyanobacterium *Anabaena lemmermannii*, as the cause of bird kills in Danish lakes. Toxicon 35(11):1645–1648, 1997.
64. JF Jellett, LJ Marks, JE Stewart, ML Dorey, W Watson-Wright, JF Lawrence. Paralytic shellfish poison (saxitoxin family) bioassays automated endpoint determination and standardization of the in vitro tissue culture bioassay, and comparison of the standard mouse bioassay. Toxicon 30:1143–1156, 1992.
65. E Usleber, E Schneider, G Terplan, MV Laycock. Two formats of enzyme immunoassay for the detection of saxitoxin and other paralytic shellfish poisoning toxins. Food Add Contam 12(3):405–413, 1995.
66. JF Jellett, JE Stewart, MV Laycock. Toxicological evaluation of saxitoxin, neosaxitoxin, gonyautoxin II, gonyautoxin II plus III and decarbamoylsaxitoxin with mouse neuroblastoma cell bioassay. Toxic in Vitro 9(1):57–65, 1995.
67. FS Chu, TSL Fan. Indirect enzyme-linked immunosorbent assay for saxitoxin in shellfish. J Assoc Off Analyt Chem 68:13–16, 1985.
68. V Renz, G Terplan. Ein enzymimmunologischer nachweis saitoxin. Archiv für Lebensmittelhygiene 39:30–33, 1988.
69. A Cembella, Y Parent, D Jones, G Lamoureux. Specificity and cross-reactivity of an absorption inhibition enzyme-linked immunosorbent assay for the detection of paralytic shellfish toxins. In E Granéli, B Sundström, L. Edler, DM Anderson, eds. Toxic Marine Phytoplankton. New York: Elsevier Applied Science, 1990, pp 339–344.
70. E Usleber, E Schneider, G Terplan. Direct immunoassay in microtitration plate and test strip format for the detection of saxitoxin in shellfish. Lett Appl Microbiol 13:275–277, 1991.

71. FS Chu, X Huang, S Hall. Production and characterization of antibodies against neosaxitoxin. J Assoc Off Analyt Chem 75:341–345, 1992.
72. FS Chu, K-H Hsu, X Huang, R Barrett, C Allison. Screening of paralytic shellfish poisoning toxins in naturally occurring samples with three different direct competitive enzyme-linked immunosorbent assays. J Agric Food Chem 44:4043–4047, 1996.
73. E Usleber, M Donald, M Straka, E Martibauer. Comparison of enzyme immunoassay and mouse bioassay for determining paralytic shellfish poisoning toxins in shellfish. Food Add Contam 14(2): 193–198, 1997.
74. K Lahti, J Ahtiainen, J Rapala, K Sivonen, SI Niemelä. Assessment of rapid bioassays for detecting cyanobacterial toxicity. Lett Appl Microbiol 21:109–114, 1995.
75. PT Smith. Toxic effects of blooms of marine species of Oscillatoriales on farmed prawns (*Penaeus monodon* and *Penaeus japonicus*) and brine shrimp (*Artemia salina*). Toxicon 34:857–869, 1996.
76. M Reinikainen, J Kiviranta, V Ulvi, M-L Niku-Paavola. Acute toxic effects of a novel cyanobacterial toxin on the crustaceans *Artemia salina* and *Daphnia pulex*. Arch Hydrobiol 133(1):61–69, 1995.
77. PL Starkweather, PE Kellar. Utilization of cyanobacteria by *Brachionus calyciflorus*: *Anabaena flos-aquae* (NRC-44-1) as a sole or complementary food source. Hydrobiologia 104:373–377, 1983.
78. SP Hawser, JM O'Neil, MR Roman, GA Codd. Toxicity of blooms of the cyanobacterium *Trichodesmium* to zooplankton. J Appl Phycol 4:79–86, 1992.
79. JJ Gilbert. Effect of temperature on the response of planktonic rotifers to a toxic cyanobacterium. Ecology 77(4):1174–1180, 1996.
80. MR Ross, A Siger, C Abbott. The house fly: an acceptable subject for paralytic shellfish toxin bioassay. In: DM Anderson, AW White, DG Baden, eds. Toxic Dinoflagellates. Proceedings of the Third International Conference on Toxic Dinoflagellates. New York: Elsevier, 1984, pp 433–438.
81. J McElhiney, LA Lawton, C Edwards, S Gallacher. Development of a bioassay employing the desert locust (*Schistocera gregaria*) for the detection of saxitoxin and related compounds in cyanobacteria and shellfish. Toxicon 36(2):417–420, 1998.
82. H-D Park, B Kim, E Kim, T Okino. Hepatotoxic microcystins and neurotoxic anatoxin-a in cyanobacterial blooms from Korean lakes. Environ Toxicol Water Qual 13:225–234, 1998.
83. T Lanaras, S Tsitamis, C Chlichlia, CM Cook. Toxic cyanobacteria in Greek freshwaters. J Appl Phycol 1:67–73, 1989.
84. J Rapala, K Sivonen. Assessment of environmental conditions that favor hepatotoxic and neurotoxic *Anabaena* spp. Strains cultured under light limitation at different temperatures. Microb Ecol 36: 181–192, 1998.
85. G Sályi. About the toxicosis caused by blue-green algae. Magyar Állatorvosok Lapja 49:96–101, 1994.
86. WW Carmichael, PR Gorham. Anatoxins from clones of *Anabaena flos-aquae* isolated from lakes of western Canada. Mitteilungen. Internationale Vereinigung für Theoretische und Angewandte Limnologie 21:285–295, 1978.
87. AP Negri, GJ Jones, M Hindmarsh. Sheep mortality associated with paralytic shellfish poisons from the cyanobacterium *Anabaena circinalis*. Toxicon 33(10):1321–1329, 1995.
88. ME Claska, JJ Gilbert. The effect of temperature on the response of *Daphnia* to toxic cyanobacteria. Freshwater Biol 39:221–232, 1998.
89. JJ Gilbert. Susceptibility of planktonic rotifers to a toxic strain of *Anabaena flos-aquae*. Limnol Oceanogr 39(6):1286–1297, 1994.
90. H Onodera, M Satake, Y Oshima, T Yasumoto, WW Carmichael. New saxitoxin analogues from the freshwater filamentous cyanobacterium *Lyngbya wollei*. Nat Toxins 5:146–151, 1997.
91. H Ruge Holte, S Eriksen, O Skulberg, P Aas. The effect of water soluble cyanotoxin(s) produced by two species of *Anabaena* on the release of acetylcholine from the peripheral cholinergic nervous system of rat airways. Environ Toxicol Pharmacol 5:51–59, 1998.
92. Y Oshima. Latest advances in PSP studies—transformation of paralytic shellfish toxins in shellfish during accumulation and processing. In: Proceedings of the Eighth International IUPAC Symposium on Mycotoxins and Phycotoxins. Mexico City, IUPAC, 1992, pp 96–97.
93. SR Davio. Neutralization of saxitoxin by anti-saxitoxin rabbit serum. Toxicon 23:669–675.

94. M Ikawa, K Wegener, TL Foxall, JJ Sasner Jr. Comparisons of the toxins of blue-green alga *Aphanizomenon flos-aquae* with the *Gonyaulax* toxins. Toxicon 20(4):747–752, 1982.
95. HB Hines, SM Naseem, RW Wannemacher Jr. (^{3}H)-saxitoxin metabolism and elimination in the rat. Toxicon 31:905–908, 1993.
96. RG Stafford, HB Hines. Urinary elimination of saxitoxin after intravenous injection. Toxicon 33(11):1501–1510, 1995.
97. J Rapala, K Lahti, K Sivonen, SI Niemelä. Biodegradability and absorption on lake sediments of cyanobacterial hepatotoxins and anatoxin-a. Lett Appl Microbiol 19:423–428, 1994.
98. DK Stevens, RI Krieger. Stability studies on the cyanobacterial nicotinic alkaloid anatoxin-a. Toxicon 29:167–179, 1991.
99. KJ James, A Furey, IR Sherlock, MA Slack, M Twohig, FB Caudwell, OM Skulberg. Sensitive determination of anatoxin-a, homoanatoxin-a and their degradation products by liquid chromatography with fluorimetric detection. J Chromatogr A 798:147–157, 1998.
100. RE Moore, ML Patterson, WW Carmichael. New pharmaceuticals from cultured blue-green algae. In: DG Fautin, ed. Biomedical Importance of Marine Organisms. San Francisco: Academy of Sciences, 1988, pp 143–150.
101. H Fujiki, M Suganuma, H Suguri, S Yoshizawa, K Takagi, M Nakayasu, M Ojika, K Yamada, T Yasumoto, RE Moore, T Sugimura. New tumor promoters from marine natural products. In: S Hall, G Strichartz, eds. Marine Toxins. Origin, Structure, and Molecular Pharmacology. ACS Symposium Series, 418:232–240, 1990.
102. WW Carmichael. Cyanobacteria secondary metabolites—the cyanotoxins. J Applied Bacteriol 72: 445–459, 1992.
103. WW Carmichael, DF Biggs, MA Peterson. Pharmacology of anatoxin-a, produced by freshwater cyanophyte *Anabaena flos-aquae* NRC-44-1. Toxicon 17:229–236, 1979.
104. CE Spivak, B Witkop, EX Albuquerque. Anatoxin-a: a novel, potent agonist at the nicotinic receptor. Mol Pharmacol 18:384–394, 1980.
105. KL Swanson, H Rapoport, RS Aronstam, EX Albuquerque. Nicotinic acetylcholine receptor function studied with synthetic (+)-anatoxin-a and derivatives. In: S Hall, G Strichartz, eds. Marine Toxins. Origin, Structure, and Molecular Pharmacology. ACS Symposium Series 418:107–119, 1990.
106. NA Mahmood, WW Carmichael. Anatoxin-a(S), an anticholinesterase from the cyanobacterium *Anabaena flos-aquae* NRC 525-17. Toxicon 25(11):1221–1227, 1987.
107. S Matsunaga, RE Moore, WP Niemczura, WW Carmichael. Anatoxin-a(S), a potent anticholinesterase from *Anabaena flos-aquae*. J Am Chem Soc 111:8021–8023, 1989.
108. OM Skulberg, WW Carmichael, RA Anderson, S Matsunaga, RE Moore, R Skulberg. Investigations of a neurotoxic Oscillatorian strain (Cyanophyceae) and its toxins: isolation and characterisation of homoanatoxin-a. Environ Toxicol Chem 11:321–339, 1992.
109. NJS Huby, P Thompson, S Wonnacott, T Gallagher. Structural modification of anatoxin-a. Synthesis of model affinity ligands for the nicotinic acetylcholine receptor. J Chem Soc Chem Commun 1: 243–245, 1991.
110. G Strichartz, N Castle. Pharmacology of Marine Toxins. In: S Hall, G Strichartz, eds. Marine Toxins: Origin, Pharmacology, and Molecular Structure. Washington, DC: American Chemical Society, 1990, pp 12–19.
111. S Hall, G Stichartz, E Moczydlowski, A Ravindran, PB Reichardt. The saxitoxins: sources, chemistry, and pharmacology. In: S Hall, G Strichartz, eds. Marine Toxins: Origin, Pharmacology, and Molecular Structure. Washington, DC: American Chemical Society, 1990, pp 29–65.
112. MH Evans. The cause of death in experimental paralytic shellfish poisoning (PSP). Br J Exp Pathol 46:245–249, 1964.
113. WA Catterall. The voltage sensitive sodium channel: a receptor site for multiple neurotoxins. In: DM Anderson, AW White, DG Baden, eds. Toxic Dinoflagellates. New York: Elsevier, 1985, pp 329–342.
114. JJ Sasner Jr, M Ikawa, TL Foxall, WH Watson. Studies on aphantoxin from *Aphanizomenon flos-aquae* in New Hampshire. In: WW Carmichael, ed. The Water Environment: Algal Toxins and Health. New York: Plenum Press, 1981, pp 389–403.

115. Y Shimizu, M Norte, A Hori, A Genenah, M Kobayashi. Biosynthesis of saxitoxin analogues: the unexpected pathway. J Am Chem Soc 106:6433–6434, 1984.
116. MB New. Global aquaculture: current trends and challenges for the 21st century. World Aquacult 30(1):8–79, 1999.
117. FAO. Aquaculture production statistics 1987–1996. Food and Agriculture Organization Circular No. 815, Rev. 10. Rome: FAO, 1998, pp 197.
118. R Rosenberry, ed. World Shrimp Farming 1998, Number 11, San Diego: Shrimp News International, 1998, pp 328.
119. L Pechar. Long-term changes in fish pond management as 'an unplanned ecosystem experiment': Importance of zooplankton structure, nutrients and light for species composition of cyanobacterial blooms. Water Sci Technol 32(4):187–196, 1995.
120. CM Pringle, T Hamazaki. The effects of fishes on algal response to storms in a tropical stream. Ecology 78(8):2432–2442, 1997.
121. PC Gehrke, JH Harris. The role of fish in cyanobacterial blooms in Australia. Aust J Mar Freshwater Res 45:905–915, 1994.
122. HW Paerl, CS Tucker. Ecology of blue-green algae in aquaculture ponds. J World Aquacult Soc 26:109–131, 1995.
123. CS Tucker, CE Boyd. Water quality. In: CS Tucker, ed. Channel Catfish Culture. Amsterdam: Elsevier, 1986, pp 135–227.
124. CS Tucker, SW Lloyd. Phytoplankton communities in channel catfish ponds. Hydrobiologia 112: 137–141, 1984.
125. DF Millie, OM Schofield, CP Dionigi, PB Johnsen. Assessing noxious phytoplankton in aquaculture systems using bio-optical methodologies. Rev J World Aquacult Soc 26:329–345, 1995.
126. WR English, TE Schwedler, LA Dyck. *Aphanizomenon flos-aquae*, a toxic blue-green alga in commercial channel catfish, *Ictalurus punctatus*, ponds. J Appl Aquacult 3:195–209, 1994.
127. WW Carmichael, CLA Jones, NA Mahmood, WC Theiss. Algal toxins and water-based diseases. CRC Crit Rev Environ Control 15:275–313, 1985.
128. J Sevrin-Reyssac, M Pletikosic. Cyanobacteria in fish ponds. Aquaculture 88:1–20, 1990.
129. LA Erman. Alimentation et multiplication du rotifère planctonique *Brachionus calyciflorus* en cultures marines (in Russian). CR Acad Sci URSS 144:926–930, 1962.
130. R Pourriot. Recherches sur l'écologie des rotifères. Vie Milieu 21(Suppl):224, 1965.
131. JH Ryther. Inhibitory effects of phytoplankton upon feeding of *Daphnia magna* with reference to growth, reproduction and survival. Ecology 35(4):522–533, 1954.
132. D Uhlman. Uber den einfluss von planktonorganismen auf ihr milieu. Int Rev Gesamten Hydrobiol 46(1):115–129, 1961.
133. GW Prescott. Objectionable algae with reference to the killing of fish and other animals. Hydrobiology 1:1–13, 1948.
134. M Shelubsky. Observations on the properties of a toxin produced by *Microcystis*. Verh Int Verein Limnol 11:362–366, 1951.
135. M Lefèvre, H Jakob, M Nisbet. Auto et hetero-antagonisme chez les algues d'eau douce in vitro et dans les collection d'eaux naturelles. Ann Stn Cent Hydrobiol Appl 4:5–198, 1952.
136. D Barthelmes. Heavy silver carp (*Hypophthalmichthys molitrix* Val.) stocking in lakes and its influence on indigenous fish stocks. Doc Tech CECPI 42 Suppl 2 (Documents présentés au symp. sur l'amélioration des stocks dans le cadre de l'amélioration des pêcheries d'eau douce, Budapest 1982). FAO, 1984, pp 313–324.
137. EA Seymour. The effects and control of algae blooms in fish ponds. Aquaculture 19:55–74, 1980.
138. PR Gorham. Toxic algae. In: DR Jackson, ed. Algae and Man. New York: Plenum Press, 1964, pp 307–336.
139. JH Gentile, TE Maloney. Toxicity and environmental requirements of a strain of *Aphanizomenon flos-aquae* Ralfs. Can J Microbiol 15:165–173, 1969.
140. MJ Phillips, RJ Roberts, JA Stewart, GA Codd. The toxicity of the cyanobacterium *Microcystis aeruginosa* to rainbow trout, *Salmo gairdneri* Richardson. J Fish Dis 8:339–344, 1985.
141. HD Rodger, T Turnbull, C Edwards, GA Codd. Cyanobacterial (blue-green algal) bloom associated pathology in brown trout, *Salmo trutta* L., in Loch Leven, Scotland. J Fish Dis 17:177–181, 1994.

142. DV Lightner. Possible toxic effects of the marine blue-green alga, *Spirulina subsalsa*, on the blue shrimp, *Penaeus stylirostris*. J Invertebrate Pathol 32:139–150, 1978.
143. R Berlin. Haff disease in Sweden. Acta Med Scand 27:440–452, 1948.
144. T Ogata, S Sato, M Kodama. Paralytic shellfish toxins in bivalves which are not associated with dinoflagellates. Toxicon 27:1241–1244, 1989.
145. IR Falconer, A Choice, W Hosja. Toxicity of edible mussels (*Mytilus edulis*) growing naturally in an estuary during a water bloom of the blue-green alga *Nodularia spumigena*. Environ Toxicol Water Qual: An International Journal 7:119–123, 1992.
146. Y Oshima, K Sugino, H Itakura, M Hirota, T Yasumoto. Comparative studies on paralytic toxin profile of dinoflagellates and bivavles. In: E Graneli, B Sundstrom, L Elder, DM Anderson, eds. Toxic Marine Phytoplankton. New York: Elsevier, 1990, pp 391–396.
147. JJ Sullivan, WT Iwaoka, J Liston. Enzymatic transformation of PSP toxins in the Littleneck Clam (*Protothaca staminea*). Biochem Biophys Res Commun 114:465–472, 1983.
148. M Kodama, Y Ogata. Toxicity of a fresh water puffer Tetraodon leiurus. Nippon Suis Gakk 50: 1949–1951, 1984.
149. S Laobhripatr, K Limpakarnjanarat, O Sangwongloy, S Sudhasaneya, B Anuchatvotakul, S Leelasitorn, K Saitanu. Food poisoning due to consumption of the freshwater puffer *Tetraodon fangi* in Thailand. Toxicon 28:1372–1375, 1990.
150. K Saitanu, S Laobhripatr, K Limpakarnjanarat, O Sangwongloy, S Sudhasaneya, B Anuchatvorakul, S Leelasiorn. Toxicity of the freshwater puffer fish *Tetraodon fangi* and *T. palembangensis* from Thailand. Toxicon 29:895–897, 1991.
151. A Kungsuwan, O Arakawa, M Promdet, Y Onoue. Occurrence of paralytic shellfish poisons in Thai freshwater puffers. Toxicon 35(8):1341–1346, 1997.
152. S Sato, M Kodama, T Ogata, K Saitanu, M Furuya, K Hirayama, K Kakinuma. Saxitoxin as a toxic principle of a freshwater puffer, *Tetraodon fangi*, in Thailand. Toxicon 35(1):137–140, 1997.
153. L Zaman, O Arakawa, A Shimosu, Y Onoue. Occurrence of paralytic shellfish poisons in Bangladeshi freshwater puffers. Toxicon 35(3):423–431, 1997.
154. IR Falconer. Health problems from exposure to cyanobacteria and proposed safety guidelines for drinking and recreational waters. In: GA Codd, TM Jefferies, CW Keevil, E Potter, eds. Detection Methods for Cyanobacterial Toxins. Cambridge, UK: Royal Society of Chemistry, 1994, pp 3–10.
155. PR Hunter. An epidemiological critique of reports of human illness associated with cyanobacteria. In: GA Codd, TM Jefferies, CW Keevil, E Potter, eds. Detection Methods for Cyanobacterial Toxins. Cambridge, UK: Royal Society of Chemistry, 1994, pp 11–18.

28
Freshwater Hepatotoxins: Ecobiology and Classification

Akira Takai
Nagoya University, Nagoya, Japan

Ken-Ichi Harada
Meijo University, Nagoya, Japan

I. INTRODUCTION

The four orders of the division Cyanophyta (cyanobacteria) (1–5) include several freshwater or brackish water species or strains which produce potent hepatotoxins such as microcystins and nodularins (6,7), related monocyclic peptides, and cylindrospermopsin (8), a guanidine alkaloid (Table 1). Acute and chronic hazards presented by these toxins to human and animal health are now ranked among the most urgent and serious problems associated with surface water supplies all over the world (for review, see refs. 6, 9, and 10).

The aim of this chapter is to provide a brief summary of the general characteristics of these cyanobacterial hepatotoxins. For the purpose of comparison with other types of natural toxins, we will give here a short account of the mode of action of the toxins, although it is more extensively discussed in Chapter 30. The increasing awareness of the hazards associated with these toxins has stimulated the development of several specific and sensitive detection methods for the effective control of waters. This topic is dealt with in Chapter 29. Chapters 26 and 27 discuss cyanobacterial neurotoxins, which constitute another major group of cyanobacterial toxins. For the worldwide distribution of toxin-producing cyanobacteria, see Chapter 31.

II. MICROCYSTINS AND NODULARINS

Blooms (dense growths) of cyanobacteria often develop in warm and stagnant waters enriched with nutrients such as nitrogen and phosphorus that may accumulate from runoff of fertilizers or wastes from domestic animals and/or human beings. Intermittent but repeated cases of animal, bird, and fish poisonings and human illnesses associated with cyanobacterial blooms have been reported on a worldwide basis for over a century (9–11). It is now believed that a majority of such cases are attributable to the monocyclic hepta- and pentapeptide hepatotoxins, microcystins and nodularins, so called because of their original isolation from members of the genera *Microcystis* and *Nodularia*, respectively. Acute "toxic hepatitis" associated with these peptide hepatotoxins is now recognized as the most commonly encountered cyanobacterial toxicosis (6). The

Table 1 Cyanobacteria and the Hepatotoxins that They Produce[a]

Order	Genera	Hepatotoxins
Chroococcales	*Microcystis* (*M. aeruginosa, M. viridis*)	(About 40) Microcystins
Oscillatoriales	*Oscillatoria*[b] (*O. agardhii*)	Microcystins
Nostocales	*Anabaena*	Microcystins
	Nostoc	Microcystins
	Nodularia	Nodularin
	Cylindrospermopsis	Cylindrospermopsin
	Aphanizomenon	Cylindrospermopsin
Stigonematoles	*Hapalosiphon*	Microcystin-LA
	Umezakia	Cylindrospermopsin

[a] According to the currently used classification system, the division Cyanophyta (cyanobacteria or blue-green algae) is subdivided into four orders, Chroococcales, Oscillatoriales, Nostocales, and Stigonematales, mainly based on morphological differences and partly also on characteristics of reproduction and life cycle (1–5). Briefly, the morphological characteristics of each of the four orders may be summarized as follows: Chroococcales are coccal, single-celled cyanobacteria (2). They often form large colonies by secreting mucopolysaccharides. Oscillatoriales is a group of simple filamentous genera which create neither heterocytes nor akenetes (or ''spores'') (3). Nostocales are filamentous species and produce heterocytes and/or akenetes but no true branches (4). Stigonematales, which produce both heterocytes and akenetes as well as true branches, are morphologically the most differentiated and complicated cyanophytes, and they are also characterized by their complicated life and vegetation cycles (5). Note that all the four orders contain one or more species or strains which are capable of producing potent hepatotoxins. See the text for further explanations.

[b] Often called *Planktothrix.*

death of more than 50 hemodialysis patients in Caruaru, Brazil, in 1996 by accidental contamination of water by microcystins has tragically exemplified the appalling hazards of these toxins to human health (12,13).

The microcystins constitute a family of monocyclic heptapeptides of great structural diversity. The members of this family share the basic structure cyclo(-D-Ala-X-D-β-Me *iso*Asp-Y-Adda-D-*iso*-Glu-Mdha-), where X and Y represent variable L-amino acid residues, D-β-Me *iso*Asp is D-*erythro*-β-methyl-isoaspartic acid, Adda is (2*S*,3*S*,8*S*,9*S*)-3-amino-9-methoxy-2,6,8-trimethyl-10-phenyldeca-4(*E*),6(*E*)-dienoic acid, and Mdha is *N*-methyldehydroalanine (Figure 1A) (14). The unusual β-amino acid, Adda, contained in both microcystins and nodularins (see below), is known to be essential for the biological activity of these toxins (15,16). The most commonly encountered and among the most toxic (LD50 = 50 μg/kg, mice, i.p.) (7) of the family is microcystin-LR, in which the two variable amino acids are L-leucine and L-arginine (Figure 1A). To date, nearly 60 variants of microcystins, which differ primarily in the two variable L-amino acids and in the degree of demethylation of the D-β-Me *iso*Asp and/or Mdha residues, have been isolated from several bloom-forming genera belonging to different orders (6,7,17).

The nodularins are a smaller family of cyclic pentapeptides structurally related to microcystins. Nodularin, the representative form, has the structure cyclo(-D-β-Me *iso*Asp-L-Arg-Adda-D-*iso*Glu-Mdhb-), where Mdhb stands for N-methyldehydroaminobutyric acid (Figure 1B) (16). In contrast to the large number of microcystins, only several variants of nodularin have been identified. They include [desmethylAdda^3]nodularin, [6(*Z*)-Adda^3]nodularin, and [D-Asp^1]nodularin (16,18). All these three variants as well as nodularin were isolated from bloom or cultured samples of a brackish water species, *Nodularia spumigena*, which in 1878 became

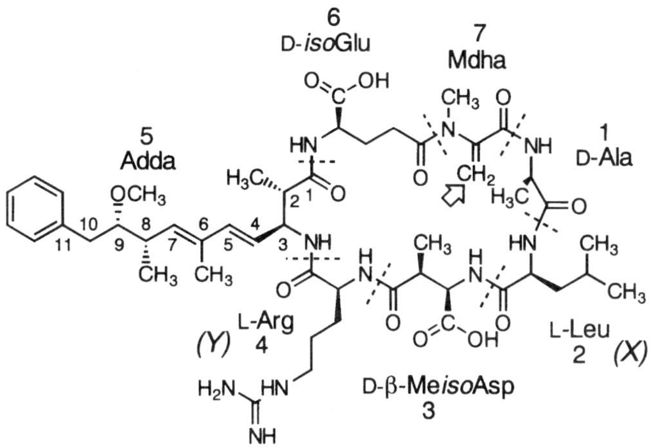

(A)

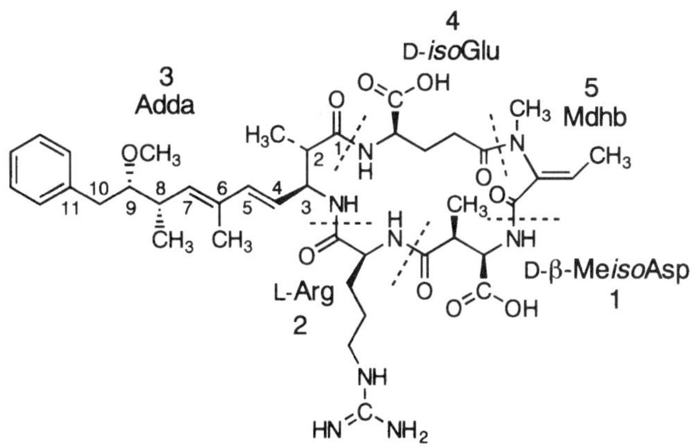

(B)

Figure 1 Chemical structure of microcystin-LR and nodularin. (A) Microcystin-LR, the most commonly encountered cyanobacterial hepatotoxin, is a monocyclic heptapeptide in which variable amino acids (*X* and *Y*) are L-leucine and L-arginine. The α,β-unsaturated carbonyl group (arrow) of the *N*-methyldehydro-alanine (Mdha) is arranged with the π-electrons polarized so that Michael addition of nucleophils (e.g., thiolates) to the β-position is favored at high pH. It is this position that is known to generate a covalent bond with Cys273 of PP1 (which corresponds to Cys266 of PP2A). (B) Nodularin is a monocyclic pentapeptide. In both microcystins and nodularins, the positions of the amino acid residues are usually indicated by numbers as shown here. See the text for further explanations.

the first cyanobacterium to be reported in the scientific literature as being toxic to livestock (19). Recently, a new member, [L-Val^2]nodularin (trivially named motuporin), which has L-valine in place of L-arginine, was isolated from the marine sponge *Theonella swinhoei* in Papua, New Guinea (20). Although isolated from the sponge, the compound itself is likely to be produced by a microbial symbiont (20).

The mechanism by which microcystins and nodularins exert their acute hepatotoxicity has been extensively studied by numerous experiments in both intact animals and isolated cells, mainly using microcystins. In experimental animals, intravenous, intraperitoneal, or oral dosing of these toxins induces symptomatically indistinguishable hepatotoxicosis with initial external signs including vomiting, diarrhea, weakness, heavy respiration, and recumbency (21–25).

Table 2 Specific Inhibitors of Protein Phosphatases 1 and 2A[a]

Inhibitor	K_i (pM) PP1	PP2A	References
Microcystin-LR	0.032	0.008	44 (38,39)
Nodularin	1.5	0.022	*, (38)
Okadaic acid	150	0.032	43,44
Calyculin-A	1	0.13	44 (45)
Tautomycin	0.4	32	44 (46,47)
Cantharidin	470	40	48,49
Fostriecin	130000	3.2	50

* Unpublished data of the authors.

[a] The dissociation constants (K_i) for the interaction of microcysin-LR and nodularin with purified catalytic subunits of protein phosphatase 1 and 2A (PP1 and PP2A) are listed in comparison with those for the other PP1/PP2A inhibitors. Since some of the inhibitors (including microcystin-LR and nodularin) interact with the enzyme at concentrations comparable with, or lower than, the enzyme concentration, the dose-inhibition curves are markedly shifted to the right as a result of the reduction of the free-inhibitor concentration in the reaction mixture resulting from binding to the inhibitors. This situation, which is generally encountered in dose-inhibition analyses with ''tight-binding inhibitors,'' must be considered in calculating the K_i values (see refs. 43 and 44). References in parentheses present ID_{50} values which are generally dependent on the enzyme concentration.

Higher doses induce massive disruption of the lobular and sinusoidal architecture of the liver, often leading to lethal intrahepatic hemorrhage (22–24,26,27). No primary toxic effects are evident in other organs even when very high doses are administered (28–32). The high degree of organ specificity is attributable to a selective uptake mechanism for the liver. Namely, these toxins are taken up by hepatocytes through one or more of the multispecific bile acid transporters (33–36). The transporters expressed specifically in this cell type are also known to be responsible for the entrance into hepatocytes of amanitin and phalloidin, the bicyclic peptide hepatotoxins from the mushroom *Amanita phalloides* (37,38).

Despite the similarity in the uptake mechanism, the cyanobacterial hepatotoxins are clearly distinct in the mode of action at the molecular level from amanitin (mRNA polymerase II inhibitor) or phalloidin (irreversible stabilizer of F-actins) (37,38). Namely, microcystins and nodularins (except for several nontoxic variants) have been established as being potent and specific inhibitors of protein phosphatases 1 and 2A (PP1 and PP2A), two of the major enzymes dephosphorylating serine and threonine residues of proteins in eukaryotic cytosol, and it is by this inhibitory action that the cyanobacterial hepatotoxins exert their toxic effects (39–41). Sequential events triggered by the inhibition of PP1 and/or PP2A include (1) hyperphosphorylation of cytoskeletal proteins, including intermediate filaments, microtubules, and actin microfilaments, (2) abnormal intracellular redistribution of these proteins, and (3) extensive cellular deformation which results in detachment of cell-to-cell connection of the hepatocytes (33,42,43). These events lead to the disruption of the hepatic architecture.

Table 2 gives the values of the dissociation constants K_i (not ID50; see the footnote) for the interaction of microcystin-LR (44–46) and nodularin (39,46) with the catalytic subunits of PP1 and PP2A in comparison with those for other potent protein phosphatase inhibitors: okadaic acid (47–50), calyculin-A (51), tautomycin (52,53), cantharidin (54,55), and fostriecin (56). As can be seen, the cyanobacterial toxins exhibit the highest enzyme affinities of the PP1/PP2A inhibitors that have been identified to date. Note that the K_i values listed are for short-time (<15

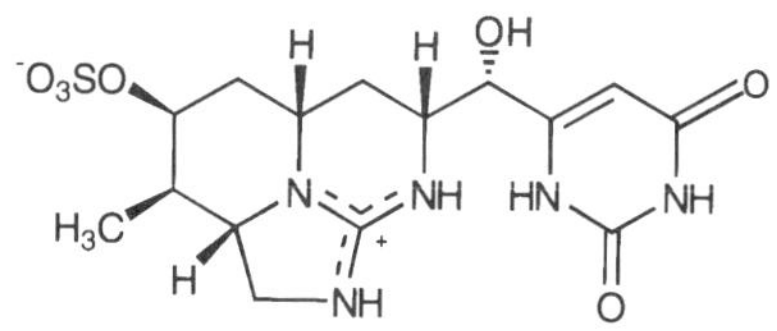

Figure 2 Chemical structure of cylindrospermopsin. Cylindrospermopsin is a cyclic alkaloid with a tricyclic guanidine moiety combined with hydroxymethyluracil.

min) interaction. It has been shown that during a prolonged reaction the unsaturated bond of the Mdha residue of microcystins (see Figure 1) can form a secondary covalent bond with PP1 at its Cys273 (which corresponds to Cys266 of PP2A) (57,58). Interestingly, results of binding-assay experiments by several groups (45,46,59,60) have strongly suggested that the phosphatase inhibitors enlisted in Table 2, except for fostriecin (56), share the same binding site on the enzyme molecules. This appears to be consistent with predictions from three-dimensional structure analysis based on x-ray crystallography (61–63), nuclear magnetic resonance (NMR) spectroscopy (64–66), and computer-assisted molecular shape modeling (67,68). Biochemical and pathophysiological consequences of the exceedingly high affinity of microcystins and nodularins to the phosphatases are further discussed in Chapter 30.

Microcystins and nodularins, together with the other PP1/PP2A inhibitors, have been shown to be potent tumor promoters (for a review, see ref. 59). This is not unexpected, because it is now widely recognized that protein phosphatases play crucial roles in cell division and proliferation (for a review, see ref. 69). The fact that the cyanobacterial hepatotoxins are tumor promoters has raised a provoking probability of the threats of long-term human exposure to low concentration of these environmental toxins. For example, epidemiological research in China over many years has shown that the incidence of primary liver cancer is significantly related to the source of drinking water (70,71). In particular, rural populations drinking surface water are more liable to suffer from hepatic cancer than those drinking well water. Recent studies have strongly suggested that this association with potable water is ascribable to contamination of water supplies by cyanobacteria (72–74).

III. CYLINDROSPERMOPSIN

A hepatotoxin completely different from microcystins and nodularins both in chemical structure and in mode of action has been isolated from three species of cyanobacteria belonging to two different orders, Nostocales and Stigonematales (see Table 1). This toxin, called cylindrospermopsin (Figure 2), is an alkaloid with a molecular weight of 415 and possessing a tricyclic guanidine moiety combined with hydroxymethyluracil (8).

In 1979, an outbreak of hepatoenteritis occurred among the people (139 children and 10 adults) drinking water from a water supply dam of Palm Island, Queensland, Australia (75). In the course of an epidemiological investigation, *Cylindrospermopsis raciborskii* (Woloszynska), a tropical cyanobacterium of the order Nostocales, was isolated from the dam. Because the extract of the organism showed strong hepatotoxicity to mice, it was suspected of being the agent responsible for the human hepatoenteritis (76). A new hepatotoxin was then isolated from the extract and named cylindrospermopsin (8,77).

In 1987, a cyanobacterium, *Umezakia natans*, was identified from net samples collected at Lake Mikata, Fukui, Japan, as a new and the only one species belonging to a new genus

Umezakia of the order Stigonematales (78). Later it was noticed that the algal extract exhibited hepatotoxicity to mice, and a toxic compound completely identical to cylindrospermopsin was isolated from a cultured material of *U. natans* (79). (Cultivation was necessary for the isolation of the toxin, because *U. natans* was not the dominating species at the above-mentioned sampling time.)

Very recently, cylindrospermopsin was also found in *Aphanizomenon ovalisporum* (Nostocales) isolated in a lake in Israel (80).

Terao et al. have examined the effects of intraperitoneal injection of purified cylindrospermopsin (0.2 mg/kg) in mice (81). They showed that the toxin initially affected the liver almost exclusively. Microscopic changes observed in the liver included fat droplet deposition and central necrosis. No special change was discernible in other organs in the early phase (until 40 h after administration). In the later stages, however, necrotic changes gradually also involved the kidneys, spleen, thymus, and also the heart in some animals. The LD50 in mice (i.p.) for purified cylindrospermopsin is reported to be 2.1 mg/kg at 24 h and 0.2 mg/kg at 5–6 days (8).

To date, there is only limited information about the mechanism by which cylindrospermopsin exerts its hepatotoxicity. Terao et al. showed that cylindrospermopsin is a potent protein synthesis inhibitor (81). Runnegar et al. (82) has observed that cylindrospermopsin lowers the intracellular level of the reduced form of glutathione in rat hepatocytes prior to any external sign of toxicity (83). They also showed the cytochrome P450 pathway is likely to be involved in the metabolism of cylindrospermopsin, and the metabolites generated may be more toxic than the parent compound, as is often the case with other metabolites of this pathway (82). Terao et al., however, reported that the total amount of cytochrome P450 was significantly decreased in the hepatic microsome of mice treated with the toxin (81). Obviously, more experiments are necessary to determine whether these seemingly complex effects of cylindrospermopsin are due to a broad spectrum of its biological effects or whether they are a consequence of an attack of some specific key regulatory point(s) by the toxin.

ACKNOWLEDGMENTS

This work was supported in part by a grant from the Shimabara Science Promotion Foundation. A.T. is a member of the Research for the Future Programme of the Japan Society for the Promotion of Science (project code number: JSPS-RFTF:96L00504).

REFERENCES

1. K Anagnostidis, J Komarek. Modern approach to the classification system of cyanophytes: 1. Introduction. Arch Hydrobiol Algol Stud 38–39(Suppl):291–302, 1986.
2. J Komarek, K Anagnostidis. Modern approach to the classification system of cyanophytes: 2. Chroococcales. Arch Hydrobiol Algological Studies 43(Suppl):157–226, 1986.
3. K Anagnostidis, J Komarek. Modern approach to the classification system of cyanophytes: 3. Oscillatoriales. Arch Hydrobiol Algological Studies 50–53(Suppl):327–472, 1988.
4. J Komarek, K Anagnostidis. Modern approach to the classification system of cyanophytes: 4. Nostocales. Arch Hydrobiol Algological Studies 56(Suppl):247–345, 1989.
5. K Anagnostidis, J Komarek. Modern approach to the classification system of cyanophytes: 5. Stigonematales. Algol Stud 59:1–73, 1990.
6. WW Carmichael. The cyanotoxins. Adv Bot Res 27:211–256, 1997.
7. KL Rinehart, M Namikoshi, BW Choi. Structure and biosynthesis of toxins from blue-green algae (cyanobacteria). J Appl Phycol 6:159–176, 1994.

8. I Ohtani, RE Moore, MTC Runnegar. Cylindrospermopsin, a potent hepatotoxin from the blue-green alga *Cylindrospermopsis racibarskii*. J Am Chem Soc 114:7941–7942, 1992.
9. WW Carmichael. Cyanobacteria secondary metabolites—the cyanotoxins. J Appl Bacteriol 72:445–459, 1992.
10. WW Carmichael. The toxins of cyanobacteria. Sci Am 270:64–72, 1994.
11. GA Codd, CJ Ward, SG Bell. Cyanobacterial toxins: occurrence, modes of action, health effects and exposure routes (review). Arch Toxicol 19(Suppl):399–410, 1997.
12. WW Carmichael, JS An, SMFO Azevedo, S Lau, KL Rinehart, EM Jochimsen, CEM Holmes, JB da Silva, Jr. Analysis for microcystins involved in an outbreak of liver failure and death of humans at a hemodialysis center in Caruaru, Pernambuco, Brazil. J Venom Anim Toxins 3:94, 1997.
13. EM Jochimsen, WW Carmichael, JS An, DM Cardo, ST Cookson, CEM Holmes, MB Antunes, DA de Melo Filho, TM Lyra, VST Barreto, SMO Azevedo, WR Jarvis. Liver failure and death after exposure to microcystins at a hemodialysis center in Brazil. N Engl J Med 338:873–878, 1998.
14. DP Botes, PL Wessels, H Kruger, MTC Runnegar, S Santikarn, RJ Smith, JCJ Barna, DH Williams. Structural studies on cyanoginosins-LR, -YR, -YA, and -YM, peptide toxins from *Microcystic aeruginosa*. J Chem Soc Perkin Trans 1:2747–2748, 1985.
15. M Namikoshi, KL Rinehart, AM Dahlem, VR Beasley, WW Carmichael. Total synthesis of Adda the unique c-20 amino acid of cyanobacterial hepatotoxins. Tetrahed Lett 30:4349–4352, 1989.
16. KL Rinehart, KI Harada, M Namikoshi, C Chen, CA Harvis, MHG Munro, JW Blunt, PE Mulligan, VR Beasley, AM Dahlem, WW Carmichael. Nodularin, microcystin, and the configuration of Adda. J Am Chem Soc 110:8557–8558, 1988.
17. M Namikoshi, M Yuan, K Sivonen, WW Carmichael, KL Rinehart, L Rouhiainen, F Sun, S Brittain, A Otsuki. Seven new microcystins possessing two L-glutamic acid units, isolated from Anabaena sp. strain 186. Chem Res Toxicol 11:143–149, 1998.
18. M Namikoshi, BW Choi, R Sakai, F Sun, KL Rinehart, WW Carmichael, WR Evans, P Cruz, MHG Munro, JW Blunt. New nodularins: a general method for structure assignment. J Org Chem 59:2349–2357, 1994.
19. G Francis. Poisonous Australian lake. Nature 18:11–12, 1878.
20. ED deSilva, DE Williams, RJ Andersen, H Klix, CFB Holmes, TM Allen. Motuporin, a potent protein phosphatase inhibitor isolated from the Papua New Guinea sponge *Teonella swinhoei* Gray. Tetrahed Lett 33:1561–1564, 1992.
21. JE Eriksson, JA Meriluoto, HP Kujari, K Osterlund, K Fagerlund, L Hallbom. Preliminary characterization of a toxin isolated from the cyanobacterium Nodularia spumigena. Toxicon 26:161–166, 1988.
22. E Ito, F Kondo, K Harada. Hepatic necrosis in aged mice by oral administration of microcystin-LR. Toxicon 35:231–239, 1997.
23. T Yoshida, Y Makita, S Nagata, T Tsutsumi, F Yoshida, M Sekijima, S Tamura, Y Ueno. Acute oral toxicity of microcystin-LR, a cyanobacterial hepatotoxin, in mice. Nat Toxins 5:91–95, 1997.
24. JA Meriluoto, A Sandstrom, JE Eriksson, G Remaud, AG Craig, J Chattopadhyaya. Structure and toxicity of a peptide hepatotoxin from the cyanobacterium *Oscillatoria agardhii*. Toxicon 27:1021–1034, 1989.
25. MT Runnegar, AR Jackson, IR Falconer. Toxicity of the cyanobacterium *Nodularia spumigena* Mertens. Toxicon 26:143–151, 1988.
26. SB Hooser, MS Kuhlenschmidt, AM Dahlem, VR Beasley, WW Carmichael, WM Haschek. Uptake and subcellular localization of tritiated dihydro-microcystin-LR in rat liver. Toxicon 29:589–601, 1991.
27. DN Slatkin, RD Stoner, WH Adams, JH Kycia, HW Siegelman. Atypical pulmonary thrombosis caused by a toxic cyanobacterial peptide. Science 220:1383–1385, 1983.
28. VR Beasley, WO Cook, AM Dahlem, SB Hooser, RA Lovell, WM Valentine. Algae intoxication in livestock and waterfowl (review). Vet Clin North Am Food Anim Pract 5:345–361, 1989.
29. SB Hooser, VR Beasley, RA Lovell, WW Carmichael, WM Haschek. Toxicity of microcystin LR, a cyclic heptapeptide hepatotoxin from *Microcystis aeruginosa*, to rats and mice (published erratum in Vet Pathol 26(6):553, 1989). Vet Pathol 26:246–252, 1989.

30. SB Hooser, VR Beasley, EJ Basgall, WW Carmichael, WM Haschek. Microcystin-LR–induced ultrastructural changes in rats. Vet Pathol 27:9–15, 1990.
31. SB Hooser, VR Beasley, LL Waite, MS Kuhlenschmidt, WW Carmichael, WM Haschek. Actin filament alterations in rat hepatocytes induced *in vivo* and *in vitro* by microcystin-LR, a hepatotoxin from the blue-green alga, *Microcystis aeruginosa*. Vet Pathol 28:259–266, 1991.
32. RA Lovell, DJ Schaeffer, SB Hooser, WM Haschek, AM Dahlem, WW Carmichael, VR Beasley. Toxicity of intraperitoneal doses of microcystin-LR in two strains of male mice. J Environ Pathol Toxicol Oncol 9:221–237, 1989.
33. JE Eriksson, GIL Paatero, JAO Meriluoto, GA Codd, GEN Kass, P Nicotera, S Orrenius. Rapid microfilament reorganization induced in isolated rat hepatocytes by microcystin-LR, a cyclic peptide toxin. Exp Cell Res 185:86–100, 1989.
34. JE Eriksson, L Gronberg, S Nygard, JP Slotte, JA Meriluoto. Hepatocellular uptake of 3H-dihydromicrocystin-LR, a cyclic peptide toxin. Biochim Biophys Acta 1025:60–66, 1990.
35. MTC Runnegar, RG Gerdes, IR Falconer. The uptake of the cyanobacterial hepatotoxin microcystin by isolated rat hepatocytes. Toxicon 29:43–51, 1991.
36. ML Wickstrom, SA Khan, WM Haschek, JF Wyman, JE Eriksson, DJ Schaeffer, VR Beasley. Alterations in microtubules, intermediate filaments, and microfilaments induced by microcystin-LR in cultured cells. Toxicol Pathol 23:326–337, 1995.
37. T Wieland, H Faulstich. Amatoxins, phallotoxins, phallolysin, and antamanide: the biologically active components of poisonous Amanita mushrooms (review). CRC Crit Rev Biochem 5:185–260, 1978.
38. T Wieland. The toxic peptides from *Amanita* mushrooms (review). Int J Pept Protein Res 22:257–276, 1983.
39. JE Eriksson, D Toivola, JA Meriluoto, H Karaki, YG Han, DJ Hartshorne. Hepatocyte deformation induced by cyanobacterial toxins reflects inhibition of protein phosphatases. Biochem Biophys Res Commun 173:1347–1353, 1990.
40. M Runnegar, N Berndt, N Kaplowitz. Microcystin uptake and inhibition of protein phosphatases: effects of chemoprotectants and self-inhibition in relation to known hepatic transporters. Toxicol Appl Pharmacol 134:264–272, 1995.
41. MT Runnegar, S Kong, N Berndt. Protein phosphatase inhibition and in vivo hepatotoxicity of microcystins. Am J Physiol 265:G224–G230, 1993.
42. S Ghosh, SA Khan, M Wickstrom, V Beasley. Effects of microcystin-LR on actin and the actin-associated proteins alpha-actinin and talin in hepatocytes. Nat Toxins 3:405–414, 1995.
43. SA Khan, ML Wickstrom, WM Haschek, DJ Schaeffer, S Ghosh, VR Beasley. Microcystin-LR and kinetics of cytoskeletal reorganization in hepatocytes, kidney cells, and fibroblasts. Nat Toxins 4:206–214, 1996.
44. RE Honkanen, J Zwiller, RE Moore, SL Daily, BS Khatra, M Dukelow, AL Boynton. Characterization of microcystin-LR, a potent inhibitor of type 1 and type 2A protein phosphatases. J Biol Chem 265:19401–19404, 1990.
45. C MacKintosh, KA Beattie, S Klumpp, P Cohen, GA Codd. Cyanobacterial microcystin-LR is a potent and specific inhibitor of protein phosphatases 1 and 2A from both mammals and higher plants. FEBS Lett 264:187–192, 1990.
46. R Matsushima, S Yoshizawa, MF Watanabe, K Harada, M Furusawa, WW Carmichael, H Fujiki. In vitro and in vivo effects of protein phosphatase inhibitors, microcystins and nodularin, on mouse skin and fibroblasts. Biochem Biophys Res Commun 171:867–874, 1990.
47. C Bialojan, A Takai. Inhibitory effect of a marine-sponge toxin, okadaic acid, on protein phosphatases: specificity and kinetics. Biochem J 256:283–290, 1988.
48. A Takai, C Bialojan, M Troschka, JC Rüegg. Smooth muscle myosin phosphatase inhibition and force enhancement by black sponge toxin. FEBS Lett 217:81–84, 1987.
49. A Takai, G Mieskes. Inhibitory effect of okadaic acid on the *p*-nitrophenyl phosphate phosphatase activity of protein phosphatases (published erratum in Biochem J 281(Pt 3):879, 1992). Biochem J 275:233–239, 1991.
50. A Takai, K Sasaki, H Nagai, G Mieskes, M Isobe, K Isono, T Yasumoto. Inhibition of specific binding of okadaic acid to protein phosphatase 2A by microcystin-LR, calyculin-A and tautomycin:

method of analysis of interactions of tight-binding ligands with target protein. Biochem J 306:657–665, 1995.
51. H Ishihara, BL Martin, DL Brautigan, H Karaki, H Ozaki, Y Kato, N Fusetani, S Watabe, K Hashimoto, D Uemura, DJ Hartshorne. Calyculin A and okadaic acid: inhibitors of protein phosphatase activity. Biochem Biophys Res Commun 159:871–877, 1989.
52. C MacKintosh, S Klumpp. Tautomycin from the bacterium *Streptomyces verticillatus*: another potent and specific inhibitor of protein phosphatases 1 and 2A. FEBS Lett 277:137–140, 1990.
53. M Hori, J Magae, YG Han, DJ Hartshorne, H Karaki. A novel protein phosphatase inhibitor, tautomycin: effect on smooth muscle. FEBS Lett 285:145–148, 1991.
54. YM Li, JE Casida. Cantharidin-binding protein: identification as protein phosphatase 2A. Proc Natl Acad Sci USA 89:11867–11870, 1992.
55. YM Li, C MacKintosh, JE Casida. Protein phosphatase 2A and its [^{3}H]cantharidin/[^{3}H]endothall thioanhydride binding site. Inhibitor specificity of cantharidin and ATP analogues. Biochem Pharmacol 46:1435–1443, 1993.
56. AH Walsh, A Cheng, RE Honkanen. Fostriecin, an antitumor antibiotic with inhibitory activity against serine/threonine protein phosphatases type 1 (PP1) and 2A (PP2A), is highly selective for PP2A. FEBS Lett 416:230–234, 1997.
57. RW MacKintosh, KN Dalby, DG Campbell, PTW Cohen, P Cohen, C MacKintosh. The cyanobacterial toxin microcystin binds covalently to cysteine-273 on protein phosphatase 1. FEBS Lett 371: 236–240, 1995.
58. G Moorhead, RW MacKintosh, N Morrice, T Gallagher, C MacKintosh. Purification of type 1 protein (serine/threonine) phosphatases by microcystin-Sepharose affinity chromatography. FEBS Lett 356: 46–50, 1994.
59. H Fujiki, M Suganuma. Tumor promotion by inhibitors of protein phosphatases 1 and 2A: the okadaic acid class of compounds. Adv Cancer Res 61:143–194, 1993.
60. M Runnegar, N Berndt, SM Kong, EY Lee, L Zhang. In vivo and in vitro binding of microcystin to protein phosphatases 1 and 2A. Biochem Biophys Res Commun 216:162–169, 1995.
61. D Barford, JC Keller. Co-crystallization of the catalytic subunit of the serine/threonine specific protein phosphatase 1 from human in complex with microcystin LR. J Mol Biol 235:763–766, 1994.
62. J Goldberg, HB Huang, YG Kwon, P Greengard, AC Nairn, J Kuriyan. Three-dimensional structure of the catalytic subunit of protein serine/threonine phosphatase-1. Nature 376:745–753, 1995.
63. HB Huang, A Horiuchi, J Goldberg, P Greengard, AC Nairn. Site-directed mutagenesis of amino acid residues of protein phosphatase 1 involved in catalysis and inhibitor binding. Proc Natl Acad Sci USA 94:3530–3535, 1997.
64. JR Bagu, FD Sonnichsen, D Williams, RJ Andersen, BD Sykes, CFB Holmes. Comparison of the solution structures of microcystin-LR and motuporin (letter). Nat Struct Biol 2:114–116, 1995.
65. JR Bagu, BD Sykes, MM Craig, CFB Holmes. A molecular basis for different interactions of marine toxins with protein phosphatase-1. Molecular models for bound motuporin, microcystins, okadaic acid, and calyculin A. J Biol Chem 272:5087–5097, 1997.
66. GB Trogen, A Annila, J Eriksson, M Kontteli, J Meriluoto, I Sethson, J Zdunek, U Edlund. Conformational studies of microcystin-LR using NMR spectroscopy and molecular dynamics calculations. Biochemistry 35:3197–3205, 1996.
67. T Lanaras, CM Cook, JE Eriksson, JA Meriluoto, M Hotokka. Computer modelling of the 3-dimensional structures of the cyanobacterial hepatotoxins microcystin-LR and nodularin. Toxicon 29:901–906, 1991.
68. RJ Quinn, C Taylor, M Suganuma, H Fujiki. The conserved acid binding domain model of inhibitors of protein phosphatases 1 and 2A: molecular modelling aspects. Bioorg Med Chem Lett 13:1029–1034, 1993.
69. S Shenolikar. Protein phosphatase regulation by endogenous inhibitors (review). Semin Cancer Biol 6:219–227, 1995.
70. S-Z Yu. Drinking water and primary liver cancer. In: Z-Y Tang, M-C Wu, S-S Xia, eds. Primary Liver Cancer, Berlin: Springer, 1989, pp 30–37.
71. SZ Yu. Primary prevention of hepatocellular carcinoma. J Gastroenterol Hepatol 10:674–682, 1995.
72. WW Carmichael, JW He, J Eschedor, ZR He, YM Juan. Partial structural determination of hepato-

toxic peptides from *Microcystis aeruginosa* (cyanobacterium) collected in ponds of central China. Toxicon 26:1213–1217, 1988.
73. K Harada, M Oshikata, H Uchida, M Suzuki, F Kondo, K Sato, Y Ueno, SZ Yu, G Chen, GC Chen. Detection and identification of microcystins in the drinking water of Haimen City, China. Nat Toxins 4:277–283, 1996.
74. Y Ueno, S Nagata, T Tsutsumi, A Hasegawa, MF Watanabe, HD Park, GC Chen, G Chen, SZ Yu. Detection of microcystins, a blue-green algal hepatotoxin, in drinking water sampled in Haimen and Fusui, endemic areas of primary liver cancer in China, by highly sensitive immunoassay. Carcinogenesis 17:1317–1321, 1996.
75. ATC Bourke, RB Hawes, A Neilson, ND Stallman. An outbreak of hepatoenteritis (the Palm Island mystery disease) possibly caused by algal intoxication. Toxicon 21(Suppl 3):40–42, 1983.
76. PR Hawkins, MT Runnegar, AR Jackson, IR Falconer. Severe hepatotoxicity caused by the tropical cyanobacterium (blue-green alga) *Cylindrospermopsis raciborskii* (Woloszynska) Seenaya and Subba Raju isolated from a domestic water supply reservoir. Appl Environ Microbiol 50:1292–1295, 1985.
77. RE Moore, I Ohtani, BS Moore, CB de Koning, WY Yoshida, MTC Runnegar, WW Carmichael. Cyanobacterial toxins. Gazz Chim Ital 123:329–336, 1993.
78. M Watanabe. Studies on the planctonic blue-green algae: 2. *Umezakia natans* gen. et sp. nov. (Stigonemataceae) from the Mikata Lakes, Fukui Prefecture. Bull Nat Sci Mus Tokyo B13:81–88, 1987.
79. KI Harada, I Ohtani, K Iwamoto, M Suzuki, MF Watanabe, M Watanabe, K Terao. Isolation of cylindrospermopsin from a cyanobacterium *Umezakia natans* and its screening method. Toxicon 32: 73–84, 1994.
80. R Banker, S Carmeli, O Hadas, B Teltsch, R Porat, A Sukenik. Identification of cylindrospermopsin in *Aphanizomenon ovalisporum* (Cyanophyceae) isolated from Lake Kinneret, Israel. J Phycol 33: 613–616, 1997.
81. K Terao, S Ohmori, K Igarashi, I Ohtani, MF Watanabe, KI Harada, E Ito, M Watanabe. Electron microscopic studies on experimental poisoning in mice induced by cylindrospermopsin isolated from blue-green alga *Umezakia natans*. Toxicon 32:833–843, 1994.
82. MT Runnegar, SM Kong, YZ Zhong, SC Lu. Inhibition of reduced glutathione synthesis by cyanobacterial alkaloid cylindrospermopsin in cultured rat hepatocytes. Biochem Pharmacol 49:219–225, 1995.
83. MT Runnegar, SM Kong, YZ Zhong, JL Ge, SC Lu. The role of glutathione in the toxicity of a novel cyanobacterial alkaloid cylindrospermopsin in cultured rat hepatocytes. Biochem Biophys Res Commun 201:235–241, 1994.

29

Freshwater Hepatotoxins: Chemistry and Detection

Fun S. Chu

Food Research Institute, University of Wisconsin, Madison, Wisconsin

I. INTRODUCTION

Although phycotoxins produced by algae and diatoms in the oceans have been a great concern for many years, recent investigations have revealed that certain freshwater cyanobacteria (blue-green algae), including *Anabaena flos-aquae, Microcystis aeruginosa, Nodularia spumigena*, and *Oscillatoria agardhii*, are toxic (1–9). These algae produce a series of structurally related hepta- and pentacyclic peptide toxins with molecular weights between 900 and 1200, which are called microcystins (MCYST) and nodularin (NODLN). The toxins cause intermittent but repeated poisoning of wild and domestic animals as well as liver damage, gastroenteritis, diarrhea, and dermatitis in humans in many parts of the world. Microcystin leucine-arginine (MCYST-LR), the major and most common naturally occurring variant in this group, has been found to be a potent inhibitor of protein phosphatases 1 and 2A both in vivo and in vitro (10–14). The inhibitory effect is due to its specific interaction with the enzyme through both covalent (15,16) and noncovalent binding in liver (10–16). MCYST-LR has also been shown to be a potent tumor promoter in rats (17–20) and is possibly associated with some liver cancer incidence in humans (21).

Two recent incidents have alerted health agencies to the potential of MCYST as a human hazard. In one case, MCYST was found to be involved in a human liver disease called Caruaru Syndrome, a previously undescribed form of acute liver disease in humans with a high lethality rate that occurred in Northern Brazil in 1996 and more than 50 persons died. The cause was identified to be contamination with MCYST in the water reservoir (22,23). The contaminated water was then introduced into the kidney dialysis clinic. Using enzyme-linked immunosorbent assay (ELISA) methods developed in our laboratory, we have confirmed the presence of MCYST in the active carbon of the filter device and in human blood as well as in the livers of the patients who died of this disease (F. S. Chu and X. Huang, unpublished observations). The other incident was the wide contamination by MCYST in blue-green algal health food products, which could be a threat to human health (24).

Other than MCYST and NODLN, *Anabaena flos-aquae* and *Aphanizomenon flos-aquae* also produce the paralytic shellfish poisons saxitoxin (STX) and neosaxitoxin (neoSTX), as well as two neurotoxins called anatoxin-a (ANTX) and anatoxin-a(s) (N-hydroxyguanidine derivative), which are potent acetylcholinesterase inhibitors (1,2). Details of the naturally occurring cyanobacterial blooms, ecological factors involved in these blooms, toxicology, as well as treatment of these blooms have appeared in a number of reviews including several chapters in this book. In this chapter, the chemistry of the freshwater blue-green algal hepatotoxins with emphasis on structural and functional relationships is briefly discussed and recent progress on the analytical technology for this group of toxins is reviewed.

II. CHEMISTRY

A. Structure of Microcystins and Nodularin

Basic Chemical Structures of MCYST and NODLN

Microcystins are cyclic heptapeptides with a basic structure shown in Figure 1. The amino acid structure is cyclo(D-Ala-X-D-*erythro*-β-methylasparatic acid; -Z-(2*S*,3*S*,8*S*,9*S*)-3-amino-9-methoxy-2,6,8-trimethyl-10-phenyldeca-4,6-dienoic acid)-D-glu-N-methyldehydroalanine-). Variants with different amino acids in the X and Z positions were identified in early studies. The toxins were originally called cyanoginosins (25,26). The unusual amino acid (2*S*,3*S*,8*S*,9*S*)-3-amino-9-methoxy-2,6,8-trimethyl-10-phenyldeca-4,6-dienoic acid is generally referred to as Adda. Subsequent investigations have led to the isolation and characterization of more than 50 variants produced by various cyanobacteria (27). The complexity of structural differences has led to a standard nomenclature approach for the different variants (28,29). Using the basic structure as the basis, the amino acid variations at the "X" and "Z" position are first identified in the name; thus, each MCYST has a name with XZ attached. For example, one of the most common MCYST variants which has L-leucine and L-arginine at the positions X and Y, respectively, is called microcystin-LR (see Figure 1). Variants with changes in amino acids in other positions are identified within the bracket before the common name followed with a subscript number to identify the position of this amino acid (27). For example, [D-Asp^3]MCYST-LR means that the D-methylaspartic acid at the third amino acid position of the LR variant is changed to D-aspartic acid; [D-Asp^3,D-Dha^7]MCYST-LR is a variant that the D-methylaspartic

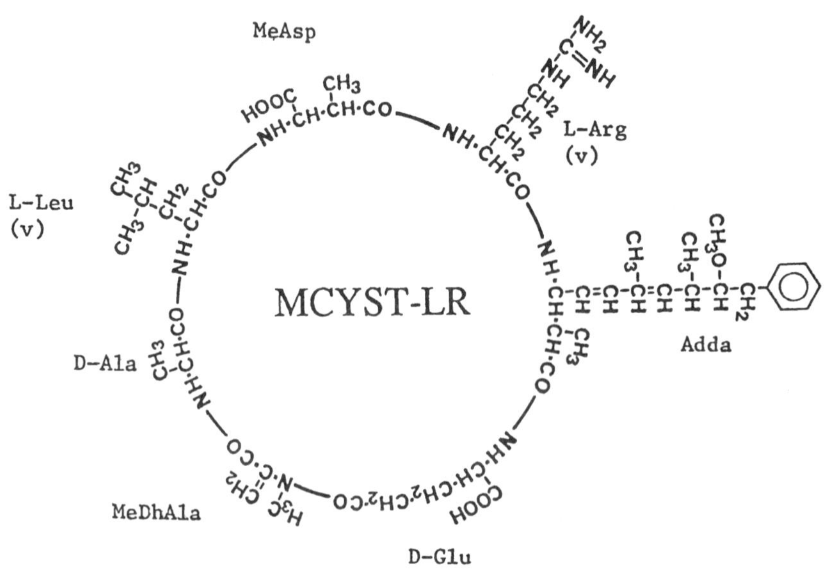

Figure 1 Chemical structure of microcystin-LR.

acid at the third amino acid position and N-methyldehydroalanine at the 7th amino acid of the LR variant are changed to D-aspartic acid and dehydroalanine, respectively. The molecular weight and general formulas of some selected variants produced by various cyanobacteria are summarized in Table 1 (30–56). MCYST-LR and MCYST-RR are the most common variants frequently found in naturally occurring algal blooms. However, some strains of cyanobacteria produce primarily other special variants. Structurally similar to MCYST, nodularin (NODLN) is a pentapeptide which also contains Adda and D-*erythro*-β-methylaspartic acid (D-MeAsp). Instead of having a dehydroalanine unit in the MCYST, NODLN contains a 2-(methylamino)-2-dehydrobutyric acid (Figure 2). It is produced by *Nodularia spumigena*, and several structurally related NODLNs have been identified. Although it is less toxic than MCYST, NODLN is also a PP1 and PP2A inhibitor (2–8,12,28,29,43,57,58).

Methods of Isolation and Characterization

Both toxic bloom and toxic cultures have been used as the starting materials for the isolation of toxins. After extracting the toxins from the culture materials/algal bloom with appropriate solvents, a combination of several chromatographic methods is generally used in the isolation of cyanobacterial toxins (37,59). For example, in the preparation of MCYST-LR and MCYST-RR, Harada et al. (37) extracted the toxins from algae with 5% acetic acid. The crude extract was subjected to an ODS-silica gel column and followed by silica gel column chromatography using a solvent system of CH_3:MeOH:water (13:5:1) to separate LR and RR. Further purification of LR was achieved by preparative high-performance liquid chromatography (HPLC). Several chromatographic steps, including silica gel (AcOEt:iso-PrOH:W, 4:3:7 or 6:4:9), ODS-silica gel (MeOH:0.05M Na_2SO_4; 6:4), and Toyoperal HW-4OF gel (MeOH), were used for further purification of RR.

For the isolation of nodularin, Namikoshi et al. (58) extracted the toxins from the dried cells with methanol. A combination of ODS-silical gel, LH-20 column, HPLC, and preparative thin-layer chromatography (TLC) was used for isolation of different toxins. For large-scale isolation of MCYST, C-18 reversed-phase flash column/cartridge chromatography is generally used. For example, Edward et al. (60,61) extracted the toxins from cyanobacterial scum with methanol. After removing the cell debris, the solution was diluted with water to a final 20% of methanol, which was then loaded to a C-18 reversed-phase column. Microcystins were eluted from the column with 70% aqueous methanol. Subsequent purification of individual compounds was achieved on an analytical column with gradient elution or isocratic elution and scaled up to a 15 × 7.5 cm ID column. Closed-loop recycling was also used to maximize yield and purity. Other than acetic acid (5–10%) or methanol, mixture of acetic acid in ethanol (0.05 M acetic acid in ethanol) (33,62), 5% 1-butanol–20% methanol (42), N-butanol–methanol-water (5:20:75 (41,51), and water have also been used for the extraction of the toxins.

Most standard chemical methods, including chemical degradation by hydrolysis, ozonolysis, gas chromatographic (GC) analysis for amino acids, and derivatization, nuclear magnetic resonance (NMR) and mass spectrometry (MS) have been used in the characterization and determination of toxin structures (28,29,63,64). Mass spectrometric methods, especially fast atom bombardment ionization (FAB) MS, FABMS/collisionally induced dissociation (CID)/MS, are used more commonly in recent studies (58,64). Mouse bioassay is used to assess the toxicity of the purified materials in most studies.

B. Structure-Function Relationship

Extensive studies on the structure-function relationship have been conducted since the discovery of MCYST. Both the toxicity of different MCYSTs in animals and the inhibitory effects on

Table 1 Structures of Selected Microcystins

Name	MW	Formula	Organisms	Reference
MCYST-LA	909	$C_{46}H_{67}N_7O_{12}$	*M. aeruginosa, Hapalosiphon hibernicus*	25,26,30,31
MCYST-LAba	923	$C_{47}H_{69}N_7O_{12}$	*M. aeruginosa*	32
MCYST-AR	952	$C_{46}H_{68}N_{10}O_{12}$	*Microcystis* spp.	30
MCYST-YA	959	$C_{49}H_{65}N_7O_{13}$	*M. aeruginosa*	26,30
[D-Asp^{3},Dha^{7}]MCYST-LR	966	$C_{47}H_{70}N_7O_{16}$	*M. aeruginosa*	33–35
[D-Asp^{3},L-Ser^{7}]MCYST-EE(OMe)	969	$C_{47}H_{63}N_{10}O_{12}$	Anabaena sp.	36
[D-Asp^{3}]MCYST-LR	980	$C_{48}H_{72}N_{10}O_{12}$	*A. flos-aquae, M. viridis, O. agardhii*	30,34,37–40
[Dha^{7}]MCYST-LR	980	$C_{48}H_{72}N_{10}O_{12}$	*M. aeruginosa, Ana, O. agardhii*	33–35,40,41
[DMAdda^{5}]MCYST-LR	980	$C_{48}H_{72}N_{10}O_{12}$	*Microcystis spp., Nostoc sp.*	30,42
[D-Asp^{3},Dha^{7}]MCYST-E(OMe)E(OMe)	983	$C_{47}H_{65}N_7O_{16}$	*Anabaena* sp.	36
[Dha^{7}]MCYST-EE(OMe)	983	$C_{47}H_{65}N_7O_{16}$	*Anabaena* sp.	36
MCYST-LR	994	$C_{49}H_{74}N_{10}O_{12}$	*M. aeruginosa, A. flos-aquae*	26,30,38,39,41,43
[D-Asp^{3},D-Glu(OCH$_3$)^{6}]MCYST-LR	995	$C_{49}H_{75}N_{10}O_{12}$	*Anabaena flos-aquae*	39
[Dha^{7}]MCYST-E(OMe)E(OMe)	997	$C_{48}H_{67}N_7O_{16}$	*Anabaena* sp.	36
[L-Ser^{7}]MCYST-LR	999	$C_{48}H_{75}N_{10}O_{13}$	*Anabaena* sp.	41
[D-Asp^{3},L-Ser^{7}]MCYST-XR	999	$C_{48}H_{75}N_{10}O_{13}$	*Anabaena sp.*	41
MCYST-LY	1001	$C_{52}H_{71}N_7O_{13}$	*M. aeruginosa*	44,45
[D-Asp^{3},L-Ser^{7}]MCYST-E(OMe)-E(OMe)$C_{47}H_{67}N_7O_{17}$	1001		*Anabaena* sp.	42
[L-Ser^{7}]MCYST-EE(OMe)	1001	$C_{47}H_{67}N_7O_{17}$	*Anabaena* sp.	36
[D-Asp^{3},ADMAddA5]MCYST-LR	1008	$C_{49}H_{72}N_{10}O_{13}$	*Nostoc* sp.	30,46
[D-Glu(OCH$_3$)^{6}]MCYST-LR	1009	$C_{50}H_{77}N_{10}O_{12}$	*Anabaena flos-aquae*	39,47
[D-Asp^{3},DhA^{7}]MCYST-RR	1009	$C_{47}H_{71}N_{13}O_{12}$	*M. aeruginosa, O. agardhii, Anabaena* sp.	30,34,40,42
[MSer^{7}]MCYST-LR	1012	$C_{49}H_{76}N_{10}O_{13}$	*Microcystis* spp.	30

[Dha^{7}]MCYST-FR	1014	$C_{51}H_{70}N_{10}O_{12}$	*M. aeruginosa*	34
[L-Ser^{7}]MCYST-E(OMe)E(OMe)	1015	$C_{48}H_{69}N_{7}O_{17}$	*Anabaena* sp.	36
[ADMAddA5]MCYST-LR	1022	$C_{50}H_{74}N_{10}O_{13}$	*Microcystis* spp.	30,46
[D-Asp^{3},ADMAddA^{5}]MCYST-LHar	1022	$C_{50}H_{74}N_{10}O_{13}$	*Nostoc* sp.	42,55
[Dha^{7}]MCYST-RR	1023	$C_{48}H_{73}N_{13}O_{12}$	*M. aeruginosa, Ana, O. agardhii*	34,40,41,48
[D-Asp^{3}]MCYST-RR	1024	$C_{48}H_{74}N_{13}O_{12}$	*M. aeruginosa, O. agardhii, Anabaena sp.*	30,34,40,49,50
MCYST-FR	1028	$C_{52}H_{72}N_{10}O_{12}$	*Microcystis* spp.	30
MCYST-M(O)R	1028	$C_{48}H_{72}N_{10}O_{13}S$	*Microcystis* spp.	30
[Dha^{7}]MCYST-HphR	1029	$C_{52}H_{73}N_{10}O_{12}$	*Anabaena* sp.	51
[Dha^{7}]MCYST-YR	1030	$C_{51}H_{70}N_{10}O_{13}$	*M. aeruginosa, Microcystis* spp.	34,35
[D-Asp^{3}]MCYST-YR	1031	$C_{51}H_{71}N_{10}O_{13}$	*M. aeruginosa*	50
[D-Asp^{3},Dha^{7}]MCYST-HtyR	1031	$C_{51}H_{71}N_{10}O_{13}$	*Anabaena* sp.	51
MCYST-YM(O)	1035	$C_{51}H_{69}N_{7}O_{14}S$	*M. aeruginosa*	26,30
[ADMAddA^{5}]MCYST-LHar	1036	$C_{51}H_{76}N_{10}O_{13}$	*Nostoc* sp.	30,46
MCYST-RR	1037	$C_{49}H_{75}N_{13}O_{12}$	*M. aeruginosa, M. viridis*	30,34,39,52–54
[D-Ser^{1},ADMAddA^{5}]MCYST-LR	1038	$C_{50}H_{74}N_{10}O_{14}$	*Nostoc* sp.	42,55
[ADMAddA^{5},MSer^{7}]MCYST-LR	1040	$C_{50}H_{76}N_{10}O_{14}$	*Nostoc* sp.	42,55
[L-Ser^{7}]MCYST-RR	1042	$C_{48}H_{76}N_{13}O_{13}$	*Anabaena* sp., *Microcystis* spp.	41
[D-Asp^{3},Mser^{7}]MCYST-RR	1041	$C_{48}H_{75}N_{10}O_{13}$	*O. agardhii*	34,40
MCYST-YR	1044	$C_{52}H_{72}N_{10}O_{13}$	*M. aeruginosa, M. viridis*	26,54
[D-Asp^{3}]MCYST-HtyR	1044	$C_{52}H_{72}N_{10}O_{13}$	*A. flos-aquae, Anabaena* sp.	51,56
[Dha^{7}]MCYST-HtyR	1045	$C_{52}H_{73}N_{10}O_{13}$	*Anabaena* sp.	51
[D-Glu(OC H^{6}]MCYST-LR	1052	$C_{52}H_{80}N_{10}O_{13}$	*Microcystis* spp.	30
MCYST-HtyR	1058	$C_{53}H_{74}N_{10}O_{13}$	*A. flos-aquae, Anabaena* sp.	56
[L-Ser]MCYST-HtyR	1063	$C_{52}H_{75}N_{10}O_{14}$	*Anabaena* sp.	51
MCYST-WR	1067	$C_{54}H_{73}N_{11}O_{12}$	*Microcystis* spp.	30

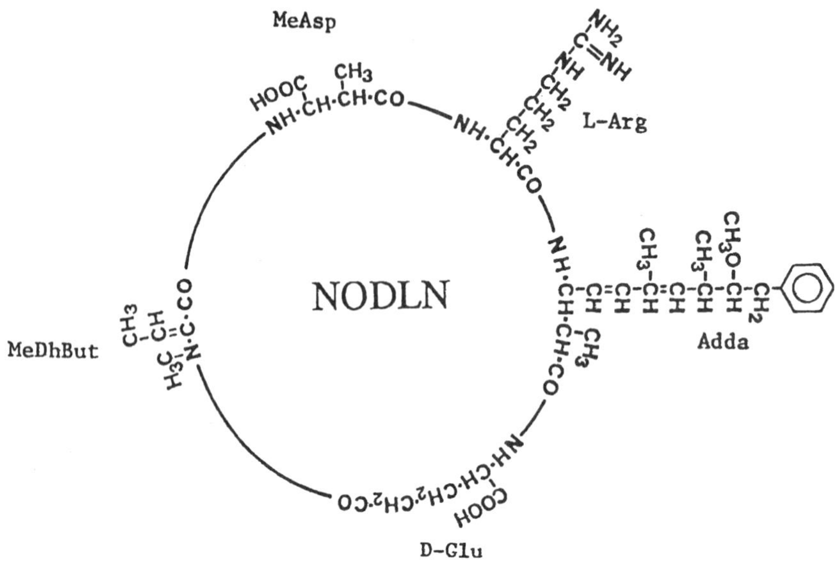

Figure 2 Chemical structure of nodularin.

PP1 or PP2A were used as the criteria for such studies. Variation of the amino acids at different positions has resulted in different effects. In general, amino acids at the 5th and 6th positions, that is, Adda and D-Glu, are more sensitive to changes than the replacement/modifications of amino acids at other positions (65). The LD50 for MCYSTs with replacement of the leu at the second position to other amino acids such as Phe, Ala, Trp, and Tyr were similar (171–249 μg/kg) (33,65–67). But, when a basic amino acid, such as Arg or a sulfoxide of methionine, was substituted at position 2, the LD50 increased to 600–650 μg/kg and 750 μg/kg, respectively (65–67). The toxicity of MCYSTs with change at the 3rd amino acid D-MeAsp such as demethylated Me-Asp LR, RR, [D-Asp,Dha]MCRR, and [D-Asp,ADMAdda]MCLR (33,46,47) also retained their toxicity. Likewise, variation in amino acid at the 4th amino acid position (L-Arg) had no effect on toxicity (2,46,66). The N-methyl group from Mdha at the 7th position was also found to play a minor role in the toxicity (33,65). Reduction of the double bond in methyldehydroalanine did not significantly reduce the toxicity of MCYST-LR (65,68,69,70). The two stereoisomers of dihydro derivatives of MCYSY-LR and NODL prepared by reduction with a sodium borohydride method were also found to have similar toxicity in mice (69). Nevertheless, recent studies showed that the site of covalent binding of MCYST with PP1 was between the methyldehydroalanine (Mdha) residues of the toxins and the Cys-273 in the PP1 molecule (15).

All the studies showed that both Adda and D-Glu play a significant role in the toxicity of MCYSTs. Although minor modifications in the Adda structure still retain the toxicity of MCYSTs and NODLN (48), removal or saturation of the Adda structure greatly reduces the toxicity (70). Modification of the carboxylic acid group of D-Glu or other changes at this position also resulted in reduced toxicity and all the toxic MCYSTs have D-Glu conserved at the 6th position. Similar data were obtained on the inhibitory effect of different MCYSTs to PP1. An and Carmichael (71) found that MCYST-LR and nodularin, with the (z) form of Adda at the C-6 double bond or having the monoester of glutamic acid, did not inhibit PP1. Because these three toxins were also nontoxic in the mouse bioassay, PP1 inhibition is related to mode of action of the toxin. The importance of these two residues in maintaining a necessary steric configuration for subsequent interaction with specific carrier proteins in the active site was postulated by Stotts et al. (65,72). The importance of the confirmation of the structure for some biologically active cyclopeptides, including MCYSTs and mushroom toxins, cannot be overemphasized. Once the structural integrity is broken, these compounds are no longer active. The linear structure of MCYST-LR has much less toxic effect than the parent molecule (73).

C. Stability and Possible Detoxification Approaches

General Stability of Microcystins

Microcystins and related toxins generally are very heat stable. As with many naturally occurring cyclopeptides, they are subject to acid hydrolysis. Several unusual amino acids in the peptides are also subject to oxidation and reduction. Ozonization of the double bond at the Adda position generally results in loss of activity. Reduction of the double bond at the DMHda position (72), however, has no effect on the toxicity.

Degradation by Chemical Agents

Although chlorination was shown to be ineffective for degradation of algal toxins in some early studies (74–76), Nicholson et al. (77) in a recent study found that chlorination at a concentration above 0.05 mg residual chlorine/liter with pH levels below 8 and a contact time of 30 min was effective for MCYST-LR removal. Inadequate chlorine doses and higher pH caused some of the previous negative results. A systematic study on the effect of chlorination on decomposition of MCYST was conducted by Tsuji et al. (78). The toxins were easily decomposed by chlorination with sodium hypochlorite, and the decomposition depended on the free chlorine dose. One of the reaction products was dihydroxymicrocystin, which was formed as a result of hydrolysis of the chloronium ion–conjugated diene of Adda [3-amino-9-methoxy-O-phenyl-2,6,8-trimethyldeca-4(E),6(E)-dienoic acid]. Stereoisomers and/or regioisomers of this compound may also be present. No noxious products were detected from the chlorination process. Although these results suggested that chlorination at an adequate chlorine dose is very effective for the removal of MCYST in raw water, preoxidation of the cells with chlorine must be avoided, because it frequently causes toxin release from algae and produces trihalomethanes during water treatment.

Degradation by Bacteria

Degradation of MCYSTs by surface water has been reported. Natural populations of microorganisms from lakes were also found to be capable of degradation of both cyanobacterial hepatotoxins and anatoxin-a (79). Implication of specific bacteria in such degradation was not demonstrated until Jones et al. (80) isolated a gram-negative rod bacterium, which was tentatively identified as a *Pseudomonas* sp. This bacterium could degrade MCYST-LR and MCYST-RR but not NODL, and it has since been classified as a new *Shingomonas* species. At least three intracellular hydrolytic enzymes were postulated in such degradation (81). The first step in the degradation involved an enzyme, designated as microcystinase, to cleave the cyclic ring at the 3-amino-9-methoxy-2,6,8-trimethyl-10-phenyldeca-4,6-dienoic acid (Adda)-Arg peptide bond to form a linearized (acyclo-) MCYST-LR with a molecular weight of 1012 D. This was followed by a second enzymatic hydrolysis to form a tetrapeptide NH_2-Adda-isoGlu-Mdha-Ala-OH (TTP) with a molecular weight of 614 D. Both of these intermediates were nontoxic to mice at doses up to 250 μg/kg. Although both were less active toward PP, the TTP had a higher inhibitory activity to PP than the linearized MCYST. The concentrations causing 50% inhibition of PP1 activity by MCYST, linear MCYST, and the TTP were 0.6, 90, and 12 nm, respectively. Further degradation of TTP by the third enzyme produced nontoxic small peptides and amino acids.

Degradation of MCYST-LR by hydrolytic enzymes from the porcine or the human gastrointestinal tract was studied by Takenaka (82,83). Among three enzymes tested, only trypsin was capable of hydrolyzing MCYST-LR to 3-amino-2,6,8-trimethyl-10-phenyldeca-4E,6E-dienoic acid (DMADDA); pepsin and chymotrypsin were not effective. DMADDA was detected in the livers of ddY male mice orally administered MCYST-LR.

Removal/Degradation of Microsystin by Physical Methods

Most detoxification processes are based on these chemical properties. Several physical treatments have been tested for the removal of MCYST from water. Whereas coagulation and filtration were ineffective in removing MCYSTs (74,76), conventional processes could be utilized for the removal of cyanobacterial cells after they were lysed by bloom collapse or chemical treatment. Adsorption with activated carbon and ozonation have been found to be effective in removing the toxin (84). Lambert et al. (85) found that conventional treatment processes combined with activated carbon removed more than 80% of MCYST-LR from raw water at concentrations of 0.27–2.9 μg/L; but a small amount of toxin was still found in the drinking water (0.1–0.5 μg/L). Removal of MCYST from water with active carbon treatment was also demonstrated by Lawton et al. (86). A low-cost activated carbon from the pan-tropical multipurpose tree *Moringa oleifera* was found to be effective in removing microcystin-LR from water in batch adsorption trials (87). Studies in our laboratory over a 3-year period indicated that most of the MCYST (when MCYST levels in the intake were less than 0.2 μg/L) were removed in the regular water treatment processes (88) (F. S. Chu and X. Huang, unpublished data).

The stability of MCYSTs under sunlight irradiation with photosynthetic pigments (89), ultraviolet (UV) irradiation (90), and the effect of pH and temperature on decomposition were studied by Harada's group (91). UV irradiation is particularly attractive in the detoxification process. Tsuji et al. (90) found that the toxins were easily decomposed by UV light at wavelengths around the absorption maxima of the toxins, and the decomposition depended on the intensity of the light. Microcystin-LR was completely degraded after exposure to high-intensity UV for 10 min. Isomerization occurred with weaker UV light and with sunlight. Several products, including three geometrical isomers of the conjugated diene of Adda, were detected. MCYST-RR showed almost the same behavior as the LR variant. Photodegradation of NODLN was also observed when the cell-free toxin or whole filaments were exposed to full sunlight or irradiated with UV light with a laboratory lamp (92). A rapid photodegradation of MCYST occurred when titanium dioxide was used as a photocatalyst (93). All the MCYST-LR in an aqueous solution at pH 4 was degraded within 40 min of exposure of UV light (330–450 nm) when high concentration of TiO_2 (200 μM) was used. Since no noxious products were formed, it is possible to use UV irradiation in combination with other methods for removing microcystins from raw water. Three major photodetoxified MCYST-LR products were recently identified by Kaya and Sano (94). Two were the geometrical isomers of Adda, that is, [4 (E),6(Z)-Adda5]- and [4-(Z)-,6(E)-Adda5]-MCYST-LR and one was a novel compound, [tricyclo-Adda5]-MCYST-LR, which is considered to be the intermediate for further breakdown of the molecule.

D. Transmission into Food Chains

Several studies have shown that the cyanobacterial toxins, like the other phycotoxins, can be transferred into the food chain (7,95–100). Studies in our laboratory and others have shown that MCYST-LR and other MCYSTs are accumulated in the hepatopancreas of clams collected during toxic cyanobacterial blooms (X. Huang and F. S. Chu, unpublished data). Accumulation of MCYST in the salt water mussel, *Mytilus edulis*, fed with the toxin-containing water has been observed by Williams et al. (96). In a later study, these investigators found the toxin was covalently bound to salmon liver and dungeness crab larvae (97). Likewise, MCYST-LR and MCYST-RR have been identified in the freshwater mussels and the clam *Anadonta woodiana*; MCYST was not found in fish, including *Cyprinus carpio*, *Carassius carassius*, and *Hypomesius transpacificus*, collected from the same lake (98). The fate of accumulation and elimination of the toxin was studied by Prepas et al. (99). Whereas feeding cyanobacteria to the signal crayfish (*Pacifastacus leniusculus*) had no negative impact on the fish, toxins were detected in the hepatopancreas of 50% of the animals tested (100). Thus, the toxins can be transferred in the food chain.

III. ANALYSIS—GENERAL CONSIDERATIONS AND CHEMICAL METHODS

A. General Considerations

To decrease the risk of human and animal exposure to the toxins, monitoring of the toxins in water, algae, and foods/feed is essential. A number of chemical and biological methods have been developed for the determination of phycotoxins both in algae and in water blooms (3–7,9,101–105). Because of their low absorption under UV (28–58), extensive sample clean-up treatment is needed for most chemical methods for the analysis of MCYSTs and NODLNs. Thus, most chemical methods to be discussed below, including HPLC, TLC, and more recently capillary electrophoresis (CE) and MS-HPLC, are very time consuming (105–109). Several biological methods are available for monitoring MCYST, but most of these methods, including mouse assay, brine shrimp assay, and neuroblastoma cell culture assay or other cell culture techniques, are nonspecific. Nevertheless, the inhibitory effect of MCYST on protein phosphatase (PP) has been used as a tool for MCYST analysis with reasonable specificity (7). With recent advances in the immunochemical methods for the detection of low molecular weight toxins, investigations in our laboratory and others have led to methods for the production of specific antibodies against MCYSTs. Subsequent studies have led to the development of several different immunochemical methods for analysis of these toxins; immunoassay kits for some phycotoxins, including MCYST, are also available commercially.

B. Sample Treatments and Cleanup

Water Samples

Two major factors should be considered regarding treatment of the sample before analysis: (1) Because the toxins can be present both intracellularly and extracellularly, release of toxins into the water phase as a result of lysis of the cell may occur. Thus, if the water is withdrawn from lakes where cyanobacterial cells are present, the cells should be removed and analyzed separately for the presence of toxins in both the cells and in the water; (2) If the sample has already been subjected to several steps of treatment by the water utilities, that is, finished water, a large volume of the sample should be taken and a concentration step made before the analysis. Thus, a solid-phase extraction protocol, which serves for the purpose of both clean-up and concentration, as needed.

Centrifugation, Filtration/Extraction. In a typical analysis, the sample is generally subjected to a centrifugation step to remove the debris, the supernatant is then subjected to a filtration step using a glass fiber filter, for example, GF/C (Whatman, Maidstone, Kent, UK). The combined solids and the filtrates are then subjected to the analysis. Thus, the data include the toxin that has already been released into the water as well as the intracellular toxin level (107). Modification of this general protocol has been made by several investigators. To ensure that the toxins are adequately released into the water after the centrifugation step, the water is subjected to a freeze and thaw process (107) or subjected to the extraction step with either methanol or bicarbonate (111) solution.

Clean-up/Concentrations. In most cases, after the above treatment, water is subjected to an ODS C-18 reversed-phase type of solid-phase extraction (SPE) cartridge. The water sample (10–200 mL) is syringed/slowly pumped into a pretreated cartridge, washed with water (20 mL) and 20–25% methanol, and then eluted with 80–100% methanol (3–5 mL) (107,111). Variations of this general protocol were made. For example, several investigators equilibrated the cartridge with 5% acetic acid and adjusted the pH of the sample in the acidic range; others had the pH of the water sample adjusted to 6 before transfer to the cartridge. Both conditions were found to be satisfactory. Although Water's (Milford, MA) C-18 reversed-phase Sep-Pak (Milford, MA) is used most frequently, Harada et al. (112) in a comparative study found that the Baker 10 cartridge is better than Analytical's BondElut (Analytichem International, Harbor City, CA, USA) or the Sep-Pak. A second clean-up step using a regular silica gel cartridge in addition to the ODS cartridge step has been used to improve the sensitivity of the liquid chromatography (LC) method, which can detect as low as 0.02 ppb of MCYST in water (113). Most recently, Tsuchiya and Watanabe (114) found that disk type's (Empore (3M Co., St. Paul, MN, USA) disks C-18 and SDB-XD) solid-phase was very effective for the extraction/clean-up of microcystin in environmental water. For example, the SDB-XD disk using 10%, H_2O-methanol as an eluent showed good recovery and reproducibility.

Algae Samples

Several protocols have been used for the extraction of MCYST from the algae, and most of these approaches are very effective. The most common and versatile one is the repeated extraction of the samples with methanol, which can be done with a tissue homogenizer (111), shaken for a period of time or by ultrasonication for 5 min, and then extraction two times with MeOH or sonication 2 h to overnight (115,116). Samples have also been subjected to drying at 90°C (110) or with a freeze and thaw process as those for the water samples. Other solvents used in the extraction included 5–10% acetic acid (0.5 g/50 mL) for 30 min (106,112) or 0.1 M ammonium bicarbonate (111). The efficacy of different solvent systems for the extraction of MCYSTs from the field samples was compared by Fastner et al. (117) most recently. They found that whereas methanol is good extraction solvent for a field sample containing water, 75% aqueous methanol should be used for the lyophilized samples. Water and ultrasonication and supercritical fluid extraction have been used to extract the toxins from samples for analysis (118). John et al. (118) extracted freeze-thawed samples with CO_2 at 400 atms and 40°C for 60 min statically and 60 min dynamically. With the exception of an immunochemical or other biological assay, almost all the algal sample extracts need a clean-up treatment before subjecting them to the subsequent chemical analysis. The methods described above for the clean-up/concentration of the water sample are widely used.

Biological Samples

Although several studies have shown the presence of MCYST in biological samples, the analytical recovery for the extraction of the toxins or their equivalent has not been critically evaluated. The major problem is the covalent interaction of MCYST with PP1 or PP2A in the samples such as livers and other organs. Nevertheless, solvent systems, including n-butanol-methanol-water (5:20:75) and methanol, have been used for the extraction of MCYST from lyophilized mussel (105) and livers (119,120). Kondo et al. (95) extracted MCYST metabolites from mouse and rat livers by homogenizing the tissues with Tris-HCl buffer (pH 7.2).

C. Chromatographic Methods

A number of chromatographic methods have been used for MCYSTs and NODLNs, and an excellent review on this subject has been made recently by Harada (121) and Meriluoto (105). However, extensive clean-up of samples is necessary before subjecting them to chromatography.

Thin-Layer Chromatography

Thin layer chromatography (TLC) is one of the most simple approaches for initial screening of MCYSTs but not for quantitative purposes. Kieselgel 60 F_{254} (0.25 mm) (E. Merck, Darmstadt, Germany) is most commonly used as absorbent in the TLC plate (121). The following solvent systems have been used for the separation of various MCYSTs and NODLNs by a number of investigators: chloroform ($CHCl_3$); methanol (MeOH): water (W) (26:15:3 or 13:7:2; ref. 33), ethyl acetate (EtOAc): 2 propanol (2-PrOH)-W: W (8:4:3 or 4:3:7; ref. 33), EtOAc-(2-PrOH)-W, (4:3:2) and 1-butanol (1-BuOH)-acetic acid (AcOH) (4:1:1) (40,55,58). For visualization, most investigators (40,55,58) sprayed the plate with phosphomolybdic acid (10% in ethanol) followed by heat and observed the spots under UV (254 mn). Iodine was used by Harada et al. (33) as the detection reagent. The Rf values of different MCYSTs in TLC in different solvent systems were reported in the respective references (33,40,55,58). In addition to the regular-phase silica gel as the coating agent, reversed-phase silica gel (C-18) has also been used (105) for separating MCYSTs in TLC, but it is not as effective as the regular-phase silica gels.

Liquid Chromatography

In general, ODS reverse-phase columns with isocratic mobile phase has been used most frequently for routine analysis of several common MCYSTs, including MCYST-LR, RR, YR, and several others (105,121). For the identification of natural blooms that contained more MCYST, gradient elution is necessary. In some cases, two different columns with varied elution conditions are used. With the development of internal surface reversed-phase (ISRP) support using glycine-L-phenylalanine-L-phenylalanine (GFF) as the partition phase that bound to internal surface of the narrow pore of the silica, this type of column has been found to be very effective for MCYST analysis. Details for the mechanism of separation and conditions for the separation of various MCYSTs were presented by Meriluoto (105). In addition to the ODS reversed-phase type of columns, ion-exchange–type columns have also been used. Various columns and solvent systems for the separations of various MCYSTs are given in Table 2.

Monitoring of Microcystins after Liquid Chromatography

Whereas UV diode-array has been shown effective in monitoring MCYSTs after they are separated from the column, both precolumn and postcolumn derivatization have also been tested (105,121). The sensitivity for UV diode-array detection for MCYSTs in water and in algae cells is generally in the range of 0.05–200 μg/L and 0.5–100 μg/g, respectively. Although the carboxylic group is a logical position for derivatization, both Meriluoto and Harada's group found difficulties for such derivatization. Reagents such as 4-bromomethyl-7-methoxycoumarin and 4-bromomethyl-7-acetoxycoumarin were tested (105). A precolumn derivatization method that can detect low pg (e.g., 15 fmol) of MCYSTs and a postcolumn reaction using peroxyoxalate chemiluminescence (PO-CL) for detection was reported by Murata et al. (129). This method involves a two-step reaction. the toxin/sample is first reacted with cysteine to form MCYST-cysteine adducts and this is followed by reaction with dansylchloride to form dansylated derivatives. After HPLC separation, these derivatives are reacted with bis[4-nitro-2-(3,6,9-trooxadesyloxycarbonyl)-phenyl] oxalate (TPPO) and hydrogen peroxide to form chemilumi-

Table 2 Conditions for Selected HPLC Systems for Microcystins Analysis

Column used	Solvent system	Detection	References
1. BioSil C_{18} (0.32 × 200 mm)	Gradient; MeCN-5mM PB (pH 2.0)	UV and MS	122
2. μBondapak C_{18} (3.9 × 300 mm)	Gradient; W: 0.05% TFA and MeCN: 0.05% TFA	UV	123
3. μBondapak C_{18}, Δ 40C	Same solvent system as above	UV	117
4. C18 reversed-phase Cosmosil 5C18-AR	MeOH: 0.05% TFA	UV	124
5. C18 reversed-phase Hypersil, 3 μm, 50 × 4.6 mm	EtOH-BtOH-$AcNH_4$	UV 238 nm	125
6. C18	Gradient; MeCN-0.008M $AcNH_4$	UV	115
7. Cosmosil 5C18-P(250 × 100 mm) Nucleosil $3C_{18}$ (4.6 × 150 mm)	MeOH:0.05% TFA (55:45) MeOH-PB (pH 3; 58:42) MeOH:05% Na_2SO_4 (6:4)	238 nm	37
8. GFF-S5-80 ISRP (4.6 × 150–250 mm)	MeCN-0.1M KH_2PO_4 (pH 6.8) (12:88 or 15:85) or Gradient at neutral and pH 2.0	238 nm	105,126,127
9. LiChrospher 100-RP-18	MeOH-0.05M KH_2PO_4 (58:42, pH 3.0)	238 nm	128
10. Spherisorb 5-μm ODS (1 × 150 mm microbore column)	Combination of gradient and isocratic Solvent A (1% HAc, W) and Solvent B (MeCN, 1% HAc).	UV at 240 nm and electrospray	106
11. Spherogel-TSK DEAE-2SW DEAE	0.04M PBS (pH 7) + 0.2M NaCl + 3% MeOH 20 mM PBS (pH 7) containing 0.3M $AcNH_4$	UV	32
12. Tosoh TSKgel ODS-80Ts column (4.6 × 150 mm)	MeCN: 0.05% TFA	PD and CLM	129
13. Ultrasphere ODS	38% aq. MeCN and 1% TFA	UV	130
14. Vydac 218TP52	Gradient, MeCN (10–60% & 0.1% TFA)	UV, LC/MS	131

$AcNH_4$, ammonium acetate; CLM, chemiluminescence; BtOH, butanol; EtOH, ethanol; HAc, acetic acid; MeCN, acetonitrile; MeOH, Methanol; PD, postcolumn derivatization; PBS, phosphate buffer; TFA, tetrafluoroacetic acid; W, water. Suppliers for columns used: (1) BioSil: BioRad, Hercules, CA, USA; (2) μBondapak C18: Waters, Milford, MA, USA; (3) Cosmosil 5C18-AR: Nacalai Tesque, Kyoto, Japan; (4) Hypersil: Life Science, Gagug, France; (5) Phase Separationsss, Ltd., Clwyd, UK; (6) Nucleosil 3C18: Chemical Scientific, Osaka, Japan; (7) GFF-S5-80 ISRP, Regis Chemical, Morton Grove, IL, USA); (8) LiChrospher 100-RP-18: E. Merck, Darmstadt, Germany; (9) Spherogel-TSK & Tosoh TSK gel: Tosoh, Tokyo; (10) Ultrasphere ODS: Beckman, Fullerton, CA; (11) Vydac 218TP52: Vydak Separation group, Hisperia, CA.

nescence in the presence of imidazole-nitrate. Although this method provides high sensitivity, it may suffer the disadvantage of low recovery through these steps of derivatization. Analytical recovery of MCYST added to the samples has not been critically evaluated. A new fluorogenic dienophil was synthesized by Shimizu et al. and also shows good reaction with MCYSTs (132). A novel amperometric method (1.20 V, Ag/AgCl) was developed by Meriluoto et al. (133) for the detection of MCYST after HPLC separation of different variants of the toxins. The electrochemical oxidation reactions were considered to originate in the arginine and tyrosine residues of the toxins.

Other than the monitoring methods discussed above, mass spectrometry has been used in conjunction with HPLC for the identification and quantification of MCYSTs in toxic cyanobacterial blooms. Erhard et al. (134) were able to identify microcystins with the matrix-assisted laser desorption/ionization time-of-flight mass spectrometry in the intact microorganisms using microgram quantities of prepared cells. Strains of various origins can easily be typed according to their cyclic peptide production, and toxic and nontoxic algal blooms can be differentiated within minutes. Harada et al. (108) used either the thermospray-liquid chromatography/mass spectrometry or electron ionization–gas chromatography/mass spectrometry at a selected ion mode to screen different MCYSTs after ozonolysis of the toxin to 3-methoxy-2-methyl-4-phenylbutyric acid. A linear relationship from 14 to 830 pmol and from 2.5 to 100 pmol of microcystin-LR was observed for signal generated from thermospray-liquid and electron ionization–gas chromatography/mass spectrometry analysis, respectively. Application of this method for monitoring microcystins in bloom and cultured samples revealed that this method could be used for both screening and accurate quantitation of microcystins in the sample.

Two novel approaches, one analyzing the PP1A and PP2A inhibitory activity (135) and the other with ELISA (119), have been used for monitoring the biological activity in the fraction after separation of the sample extracts in HPLC. These two approaches have provided good sensitivity (in the nanogram range) and specificity, but specific reagents were needed for such assays.

D. Other Chemical Methods

Although HPLC methods have been used extensively for MCYST analysis, they are very time consuming and sometimes also suffer from insensitivity, poor resolution, and inefficiency. In view of recent advancements in capillary electrophoresis (CE) for the detection of different compounds, conditions for the separation of different MCYSTs in the CE have also been studied. Onyewuenyi et al. (109) incorporated sodium dodecyl sulfate (SDS; 25 mM borate, 60 mM SDS, pH 8.85) with organic modifier solvents in the CE buffer for the separation of MCYSTs in a 50-cm long fused-silica capillary (voltage 20 kV). Separation of NODLN and MCYST-LR, MCYST-YR, and MCYST-RR in 15% MeCN-SDS was achieved in 5.8 min. In the 10% MeOH-SDS system, NODLN emerged at 5.5 min followed by MCYST-LR, MCYST-YR, and MCYST-RR at approximately 7.4 min. In a micellar electrokinetic capillary chromatography (MECC; a fused silica capillary of 50 μm ID × 70 cm), John et al. (118) were able to separate NODLN, MCYST-YR, MCYST-LR, and MCYST-RR in 6 min with retention times of 3.77, 4.42, 4.63, and 5.27 min, respectively. The borate buffer (30 mM borax; at 30 kV, pH 9.18) contained less SDS (9 mM). Detection of MCYST in both systems was made at 238 nm. Separation of MCYST in the CE was achieved by Bateman et al. (131) in bare fused-silica capillaries coated with an aqueous solution of 5% hexadimethrine bromide and 2% ethlene glycol. The column was run with 1 M formic acid; UV and mass spectral methods were used for monitoring the toxins. CE was able to resolve different desmethyl MCYST-LR.

Differential pulse polarography has been explored by Humble et al. (136) as an approach for the detection of MCYSTs with copper and zinc binding to these toxins in the aqueous phase. This electroanalytical technique measured the shift in reduction potential and decrease in height of metal polarogram peaks as a function of increasing MCYST concentration. Using MCYST-LR, MCYST-LW, and MCYST-LF as the ligands to study the complex formation, they found that all three MCYSTs are intermediate-strength metal ligands; but the substitution of arginine with either tryptophan (MCYST-LW) or phenylalanine (MCYST-LF) appeared to have no discernible effect on the overall binding capacity. Reducing the pH of the supporting electrolyte strongly influenced complexation of MCYST-LR with copper, such that at pH 5.5 only a weak labile complex was formed. Complexation between MCYST-LR and zinc, however, was not influenced as strongly by pH.

IV. BIOLOGICAL METHODS

A. Biological Methods

As with most other phycotoxins, the standard or modified version of AOAC mouse assay for the detection of paralytic shellfish poisoning toxins has been used in the early research for monitoring of toxins produced by cyanobacteria. This approach is still used in some laboratories. However, a number of other biological systems have been established for monitoring the toxins. It has well been established that MCYST induces toxic effects in hepatocytes both in vivo and in vitro. The in vitro toxic effects have been used for toxin detection. For example, Battle et al. (137) compared the toxic effects of MCYST-LR on both primary hepatocytes and on immortalized liver cells which were obtained by transfection with simian virus 40 (SV40) large T antigen–bearing plasmids. Toxicity was manifested by cytoskeletal disruption (''blebs'') using both light and scanning electron microscopy, lactate dehydrogenase release, and trypan blue dye exclusion test. At concentrations between 10^{-6} and 10^{-9} M, MCYST-LR induced bleb formation and a loss of microvilli in both primary hepatocytes and immortalized cell suspensions in comparison with controls. The immortalized cells increase the reproducibility of the assay and could reduce or curtail the use of primary cells.

Brine shrimp (*Artemia salina*) assay has been used for the detection of toxicity for many other naturally occurring toxicants and this system has also been used for the detection of MCYSTs. Kiviranta et al. (110) used this organism to detect the toxicity of cyanobacteria and found that samples from both the hepatotoxic and neurotoxic blooms were positive in this test. Feuillade et al. (138) evaluated the toxic compounds from cyanobacterial blooms and isolates of the cyanobacterium *Planktothrix* (= *Oscillatoria*) *rubescens* with brine shrimp as well as by HPLC analysis. With the exception of Nantua and Paladru strains, the toxicity of the blooms and of other strains were related to the main toxins, MCYST-RR. Similar to brine shrimp assay (139), mosquito larvae and adult mosquitoes (140) have been used as the assay organisms.

Microtox assay (MA), a commercially available toxicity test, based on the inhibition of light production by luminescent bacterium *Vibrio fisheri* or *Photobacterium phosphoreum*, was also found to be effective for the detection of toxic blooms (141,142). Lahti et al. (143) compared the efficacy of MA and shrimp assay and found that the latter is quite sensitive for the detection of hepatotoxins with LC50 (50% death) values ranging from 3 to 17 mg per liter. Although *Artemia salina* larvae were sensitive to anatoxin-a, no dose-response relationship has been demonstrated. For example, the fractionated extracts containing anatoxin-a only disrupt the mobility of the larvae with no lethality; but the filtered cyanobacterial cultures with anatoxin-a caused mortality of the larvae at a concentration of 2–4 mg/L. These investigators found that the inhibition of luminescence (Microtox test) did not correlate with the levels of hepatotoxin or anatoxin, which is in contrast with early reports by Lawton et al. (141,142) and Volterra et al. (144). For example, Lawton et al. (141) found that the LC50 values for MCYST-LR and all five MCYST-containing cyanobacterial samples in the Microtox system were between 0.02 and <0.46 mg/mL, respectively. Neither MCYSTs nor anatoxin-a inhibited the growth of *Pseudomonas putida* (143), an assay that has been used as a standardized toxicity test for the waste waters. In contrast, the growth of this bacterium was enhanced by cyanobacteria toxin fractions.

B. Enzymatic Assays—PP1/PP2A

Because cyanobacterial toxins are potent and specific inhibitors of protein phosphatases PP1 and 2A, several approaches have utilized this property in development of a sensitive method for the detection of this group of toxins. ^{32}P-labeled substrate was used in the PP1 inhibition assay in the early studies. Although these assays were sensitive and reasonably specific, because of the use of radioactive substrate and crude enzyme preparation, they are not suitable for routine MCYST assay. Intracellular phosphorylation phosphatase activity from the extracts of the bloom also interfered with the assay, thus lowering the MCYST level. A more practical approach was developed when a recombinant PP1 expressed in *Eschericia coli* was made available from Lee's group, and a colored substrate such as p-nitraphenol phosphate (pNP) was used in the assay (145). Both Carmichael's and Codd's groups have used the recombinant PP1 with the pNP substrate in a microwell system to measure the kinetics of inhibition of the enzyme by MCYSTs. Formation of a yellow color as a result of the hydrolysis of pNP to p-nitrophenol by PP1 was measured in a regular ELISA reader. Whereas An and Carmichael (71) determined the kinetics of the reaction over a period of 4 min, Ward et al. (146) took 3-min kinetics and extended to an overall time of 21 min for each sample. The concentrations of MCYST-LR, [DMAdda5] MCYST-LR, [Dha7]MCSYT-LR, and [DM Adda3] NODLN causing 50% inhibition of recombinant PP1 activity (IC50) in An and Carmichael's system were 0.3, 1.5, 5.0, and 5.0 nM, respectively. The linear concentration for MCYST assay was 1–25 ng/mL.

Ward et al. (146) further optimized the PP1 inhibition assay conditions and found that methanol at concentrations 10–80% did not affect the assay. Thus, a 70% methanolic extract of the algal bloom material could be directly used in the assay. The IC50 for MCYST-LR in this system was about 38 ng/mL of sample with a detection limit of 10–20 ng/mL. Analysis of 23 laboratory-grown strains and 25 natural bloom samples of cyanobacteria with this assay and HPLC methods showed that there was a good correlation between these two methods. To address the intracellular phosphorylation phosphatase activity problem, these investigators also analyzed the enzyme activity in the freshly prepared samples as well as the lyophilized samples; no enzyme activity was detected in the lyophilized methanolic extracts, but activity was found in extracts from the freshly harvested samples at a concentration of 25 mg of dry samples per milliliter. These investigators recommended that the colorimetric PP1 inhibition assay be used as a screening method for MCYST in cyanobacterial blooms. Several modifications of the regular protein phosphatase assay have also been suggested for MCYST detection. Rather than colored substrate, Sugiyama et al. (147) coupled the assay to the firefly bioluminescence system for the detection of protein phosphatase. Laidley et al. (148) found that the binding of cantharidin to the protein phosphatase in the neuroblastoma cells is related to the cytotoxicity; thus, this system could also be used for MCYST detection.

V. IMMUNOCHEMICAL METHODS

A. General Considerations

Immunochemical methods are based on the specific interaction between antibodies with the toxin. The key to a successful immunoassay is the availability of two important reagents, namely, the antibodies and a marker, which are used as an indicator for the binding of the antibodies. To monitor environmental exposure to MCYSTs, our laboratory and others have developed sensitive and specific immunochemical methods for the detection of the toxins in water, algae, and body fluids. Although several immunochemical methods have been developed for MCYST assay, competitive immunoassays are used most frequently. In these assays, toxin in a sample solution competes with the marker, either an enzyme conjugated with the toxin or a radioactive toxin, for binding to the same site in the antibodies. Whereas the binding is very specific, the antibodies may also cross react with analogues of the toxin or with structurally related compounds. Thus, it is important to know the cross reactivity of the antibody with toxin analogues before running the assay. In addition, it is also important to know the antibody affinity to the compound that one is interested in, because this is one of the major factors that determine the sensitivity of the assay. As with most other chemical analyses, because all the assay is based on the comparison of a set of output signals with those of standards, the purity of reference standards used in the assays is critical for quantitative determination of the toxin concentration.

B. Production and Characterization of Antibodies

Production of Antibodies

Similar to most other low molecular weight haptens, MCYSTs are not immunogenic. Thus, they must first be conjugated to a protein carrier to render them immunogenic. Since MCYST has several reactive sites, which could be directly conjugated to a protein carrier in the presence of a coupling reagent, such as water-soluble carbodiimide or N-hydroxysuccinimide (111,149), several MCYST-variants have been used in generating antibodies against this group of toxins. MCYST-LA was conjugated to polylysine (PLL) in the presence of N-acetylmuramy-L-alanyl-D-isoglutamine (NAMAG) and then used as the immunogen to generate monoclonal antibodies (mAbs) against MCYST by Kfir et al. (150). An indirect ELISA was used for MCYST determination. Subsequently, Brooks and Codd (151) generated polyclonal antibodies (pAbs) against MCYST by immunization of rabbits with MCYST-ethylenediamine–modified bovine serum albumin (EDA-BSA) (149,152). However, the sensitivity of the immunoassays for MCYST developed by both groups was low, generally in the microgram range. Polyclonal antibodies with high affinity for MCYST were later generated in our laboratory by immunizing rabbits with MCYST-LR conjugated to either to EDA-BSA or PLL, and we found that MCYST-LR-EDA-BSA was a better immunogen. Using these antibodies, sensitive immunoassays were developed (149). Rather than use rabbits or mice for antibody production, McDermott et al. (153) immunized chickens with MCYST-LR-BSA conjugate and collected the antibodies from egg yolks for subsequent assay. Within the last few years, high-affinity mAbs against MCYST were generated both in our and other laboratories (154,155).

Cross Reactivity of Antibodies

The cross reactivity of different antibodies against MCYST generated in different laboratories with various MCYST variants is summarized in Table 3. Monoclonal Antibodies generated from MCYST-LA reacted equally with six other variants: MCYST-LB, MCYST-LR, MCYST-LY, MCYST-AY, MCYST-FR, and MCYST-YR in addition to MCYST-LA (150). The pAb generated in rabbits after immunization with MCYST-LR-EDA-BSA has similar cross reactivity with a number of MCYST variants, including MCYST-LR, MCYST-RR, MCYST-HilR, [L-MeSer7]-LR, and a nontoxic [D-Glu-OCH36]-LR, and also has good cross reaction with several others (MCYST-FR, MCYST-WR, [d-asp3]-LR, MCYST-(H4)YR, MCYST-YR, and NODLN (71,149). The cross reactivities with variants MCYST-LY, MCYST-LA, MCYST-AR, MCYST-

Table 3 Cross Reactivity of Different Types of Antibodies Against Major MCYST Variants Reported as IC50 (ng/mL)

Antibodies used	LR	RR	YR	NODLN	Reference
1. pAb/r; RIA[a]	105	43 (244)[c]	112 (94)	503 (21)	149
2. pAb/r; dc-ELISA[a]	2.2	1.75 (126)	3.4 (65)	4.6 (50)	149
3. pAb/r; dc-ELISA[a]	3.1	—	10 (31)	12 (26)	71
4. mAb, idc-ELISA	0.12	0.12 (100)	0.31 (39)	0.51 (24)	154
5. mAb1, idc-ELISA	0.19	0.09 (210)	0.33 (58)	2.07 (9)	155
6. mAb1, dc-ELISA	0.29	0.38 (76)	0.83 (35)	3.97 (7)	155
7. pAb3/r, dc-ELISA[b]	0.7	24.4 (3)	19.2 (4)	24.3 (3)	157
8. pAb2/r, idc-ELISA[b]	0.37	0.66 (56)	0.72 (51)	5.55 (7)	158
9. pAb3/m, dc-ELISA[b]	0.36	0.12 (300)	1.13 (32)	152 (0.2)	158
10. mAb1/mAb2, idc-ELISA[b]	0.45	0.41 (110)	0.67 (67)	—	159

[a] The same antibodies were used in these experiments. These antibodies cross react weakly with MCYST-LA (~2%). The cross reactivities with other MCYST were detailed by An and Carmichael (71).
[b] Anti-idiotype antibodies were used in these experiments.
[c] Values in parentheses are cross-reactivities (%) relative to LR (=100%).

[DM-Asp1]-NODLN and MCYST-[Dha7]-LR, dimethyl-NODLN, and [6(z) Adda5]-LR were lower. Very little or no cross reaction was found when the ozonized MCYST-LR, MCYST-[DM-Adda3]-NODLN, [6(z) Adda3]-NODLN, and [dm-Asp3]-LR were tested. These results indicate that the arginine residue between the two unusual amino acids, that is, erytho-β-methyl-d-isoaspartate (MIA) and the β-amino acid residue of 3-amino-9-methoxy-2,6,8-trimethyl-10-phenyldeca-4,6-dienoic acid (Adda), and Adda plays an important role in immunogenicity (71,149).

The mAbs generated with the MCYST-LR conjugated to various proteins also recognize the tertiary structure around Adda. All the six mAbs obtained by Nagata et al. (154) showed good cross reactivity with MCYST-LR, MCYST-RR, MCYST-YR, MCYST-7-desmethyl-LR, and several others, but reacted poorly with 6(z)-Adda-LR and 6(z)-Adda-RR. Saito et al. (156) found that all the mAbs obtained from their laboratory reacted with MCYST-LR as well as with MCYST-RR, MCYST-3, 7-didesmethyl-LR, and 7-desmethyl-MCYST. Since these antibodies also react with some nontoxic variants, they concluded that mAbs recognize the common configuration in the molecules rather than the variant-specific structure. The mAbs obtained in our laboratory have a higher or equal affinity toward both MCYST-LR and MCYST-RR (155).

C. Alternative Approaches in Generating Immunochemical Reagents

As will be discussed later, immunochemical methods not only have served as a quick screening for toxic blooms but also have been used for quantitative analysis of the toxins. Such developments have led to a great demand for specific antibodies and related immunochemical reagents for the assays. We have developed an alternative approach for preparing immunochemical reagents for MCYSTs by generating polyclonal anti-idiotype (anti-ID) antibodies in mouse (pAb2/m) and anti–anti-idiotype in mouse (pAb3/r) for MCYSTs (157). With the availability of mAbs against MCYSTs, pAb2 were also successfully generated in rabbits (pAb2/r), and pAb3/m were then generated from pAb2/r in mice (158). Anti-ID mAbs against anti-MCYST antibody were also obtained by Tsutsumi et al. (159). Several anti-idiotype and anti–anti-idiotype antibody-based ELISAs for MCYSTs have been established in both Chu's and Ueno's laboratories (157–159). The sensitivities of the newly developed anti-ID–based ELISAs are comparable to, if not better than, the best current immunoassay systems. More recently, anti-ID antibodies against

MCYST-ID complex was obtained by Tsutsumi et al. (160), and these anti-ID antibodies have been used in a sandwich immunoassay for MCYSTs.

D. Immunochemical Assays

Radioimmunoassay (RIA)

A tritiated reduced MCYST-LR with high specific radioactivity was used by Chu et al. (149) in the RIA of MCYST. In this assay, 50 μL of radioactive marker ligand (10,000–12,000 dpm) together with 50 μL of various concentrations of different MCYST-LR or test sample were incubated with 0.15 mL of antiserum solution of various dilutions in phosphate-buffered saline (PBS) at room temperature for 30 min and then in a cold room (6°C) for 1 h or longer. Separation of bound and free ligand was achieved by an ammonium sulfate precipitation method (149). The linear response of inhibition of binding by MCYST in the RIA was found in the range of 20–500 ng/mL (1–25 ng/assay), and the minimum detection level was around 1.2 ng per assay. Although this method provides good sensitivity, specificity, and reproducibility, it requires radioactive marker. Thus, it can only be used in the laboratory and is not suitable for field use.

Indirect Competitive ELISA (idc-ELISA)

In the idc-ELISA, the immunogens are generally coated into the well of the ELISA plate, which is then incubated with antibodies together with the testing solution. In the reaction, the toxin in the test solution competes with the solid-phase immunogen for the same antibody-binding site. After removal of the unbound solution, the antibody bound to the solid-phase immunogen is determined. Early studies showed that the MCYST could be directly coated to the ELISA plate wells to perform the immunoassay. Rather than using a competitive ELISA format, Brooks and Codd (151) coated MCYST or algal extracts directly on the ELISA plate and then incubated them with the rabbit anti-MCYST antibody, which was followed by incubation with antigoat–horseradish peroxidase (HRP) to determine the amount of rabbit antibody bound to the plate. The sensitivity of this approach was poor; thus, this system was not adopted as a practical application. Nevertheless, in a recent study, McDermott et al. (153) have established an idc-ELISA by coating MCYSY-LR directly onto the wells of an ELISA plate to serve as a competitive ligand. The egg yolk anti-MCYST-LR antibodies labeled with alkaline phosphatase was used by them (153). In this case, the toxins in the samples competed with the solid-phase MCYST-LR for binding with the labeled antibodies in the solution. The standard range for this assay was between 0.1 and 3.0 ng MCYST-LR/mL and had a minimum detection level of 95 pg/mL. Analysis of 46 water samples collected from various locations in Wisconsin by these investigators with this method revealed that 87% of samples contained MCYST at levels between 0.2 and 200 ng/mL.

A more common idc-ELISA involved coating the wells of the ELISA plate with an immunogen, which is prepared by conjugating the toxin (hapten) to a large molecule. To avoid the nonspecific interaction, the immunogen to be used for coating the wells of the ELISA plate should have a carrier protein that is different from the one used for immunization of animals for antibody preparation purposes. For example, MCYST-LR-PLL or MCYST-LR–ovalbumin or MCYST–keyhole limpet hemocyanin (KLH) are generally used if MCYST-LR–EDA-BSA or MCYST-BSA was used in the immunization for antibody production. The following steps are generally involved in the idc-ELISA:

1. Coating immunogen to the wells of the plate
2. Blocking with protein to minimize the nonspecific binding

3. Incubating with anti–MCYST-LR antiserum with standard MCYST or sample solutions in the wells (approximately 1 h)
4. Incubating with goat-antirabbit IgG-HRP conjugate to determine the amount of anti–MCYST-LR antibody bound to the well (1 h)
5. Incubating with substrate (10–20 min)
6. Stopping the reaction and reading the absorbance

Thus, the idc-ELISA takes at least 2–3 h to complete. The linear range for the indirect competitive ELISA developed in our laboratory was between 0.05 and 10.0 ng of MCYST-LR/mL (2.5–500 pg/assay) and the minimum detection level was approximately 3 pg per assay. A more sensitive idc-ELISA was developed by Huang and Chu (155) when a high-affinity mAb was used, and a typical standard curve for the assay is shown in Figure 3 (top). The linear range for analysis of MCYST-LR was 0.02–1.0 ng/mL, with an IC50 of 0.2 ng MCYST-LR/mL of solution tested.

Instead of using MCYST-PLL, Nagata et al. (154,161) coated MCYST-BSA into the wells of the microtiter plate. To enhance the sensitivity, these investigators incubated MCYST-LR or test solution with mAbs against MCYST-LR at room temperature first for 1 h in a tube and then added it to a coated ELISA plate followed by an overnight incubation. Subsequent washing and incubation with second antibody-HRP (goat anti-M IgG-HRP) conjugate (2 h) and substrate (15–30 min) were similar to the above. Thus, it is possible to detect small amounts of MCYST

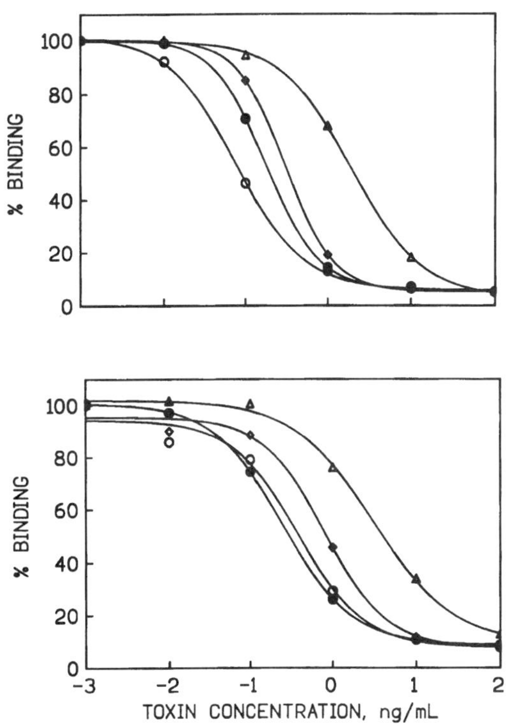

Figure 3 Cross reactivity of different MCYST variants with mAb produced by one cell line (9D74D11) in an idc-ELISA (top) using MCYST-LR-PLL as the coating immunogen and in a dc-ELISA (bottom) using MCYST-LR-HRP as the marker. MCYST-LR (●), MCYST-RR (○), MCYST-YR (◇), and NODLN (△) were used in the analysis. MCYST-LR is generally used as the standard for the preparation of standard curve in regular ELISA.

with no prior concentration. In a systematic study, the mean recovery of MCYST-LR added to tap water (20–500 pg/mL) was 101% (CV of 7.3%). The mean recovery of MCYST-LR (50–500 pg/mL) added to toxin-free cyanobacterial extracts was 93% with a CV of 12.5%. The detection limit was around 50 pg/mL. Although this system is very sensitive, it is more time consuming. Nevertheless, this system has been used in several epidemiological studies for the determination of MCYST in freshwater where high liver cancer incidence has been reported (21).

Direct Competitive ELISA (dc-ELISA)

In the direct competitive ELISA, the antibody is generally coated on the wells of the ELISA plate and the toxin and toxin-enzyme conjugate both in solution compete for binding to the same site on the antibody. Thus, this system involves the following steps:

1. Coating antibody on the ELISA plate
2. Blocking with protein to minimize the nonspecific binding
3. Incubating anti-MCYST–enzyme conjugate such as MCYST–HRP conjugate with standard MCYST or sample solutions (approximately 30 min to 1 h)
4. Incubating with substrate (10–15 min)
5. Stopping the reaction and reading the absorbance

The dc-ELISA has one less incubation step, and thus it takes less than 2 h to complete when a precoated ELISA plate is used. Using pAb developed in our laboratory, the linear range in the dc-ELISA for MCYST generally falls between 0.5 and 10.0 ng/mL (25–500 pg/assay) and the minimum detection level was approximately 0.2 ng/mL or 10 pg per assay (111,149). At the level above 0.05 ppb, a contaminated water sample could be directly used in the ELISA. The overall analytical recoveries for MCYST-LR added to water at levels of 1–20 ng/mL was 83.4%. However, a concentration step with a C-18 reversed-phase cartridge is necessary when the toxin concentration is below 0.05 ppb (111).

For analysis of cellular MCYST, the toxin can first be extracted from the algae with 0.1 M ammonium bicarbonate or methanol, diluted with PBS to less than 0.5 mg of dried algae/mL (5 mg or less of wet weight/mL), and directly used in the ELISA. The detection limit for MCYST in dried algae was approximately 0.25–0.4 μg/g (0.2–0.4 ppm) of lyophilized algae sample. The overall analytical recovery of MCYST-LR added to the algal extract by ELISA in the range of 0.25–20 ppm was found to be 83% with a CV of 11.9%. Because of the simplicity of the dc-ELISA, this method has gained wide acceptance for the analysis of MCYST both in water and in algae. We have analyzed more than 1000 water samples and 500 algae samples during the last few years. The level of MCYST in water ranged from 0.005 ppb to more than 1 ppm and in the algae ranged from 0.05 μg (after Sep-Pak concentration and clean-up) to 7 mg/g of sample.

A mAb-based dc-ELISA with a sensitivity of 0.02 ng/mL (ID50 = 0.2 ng/mL; ID80 = 0.02 ng/mL) was also established in our laboratory for the analysis of MCYST (155). The ID50 binding of mAb from 9D7 4D11 by MCYST-RR, MCYST-LR, MCYST-YR, and NODLN were found to be 0.38, 0.29, 0.83, and 3.97 ng/mL, respectively. Thus, the mAb-based dc-ELISA for MCYST is about 10 times more sensitive than the pAb-based ELISA. A typical standard curve for the mAb-based dc-ELISA of MCYST is also shown in Figure 3 (bottom). Analysis of MCYST in 22 algal samples naturally contaminated with toxic *Microcystis* showed an excellent correlation between the data obtained from the newly developed mAb-based dc-ELISA with the pAb-based dc-ELISA.

Anti-ID Antibody-Based ELISA

The efficacy of two types of anti-ID antibody-based ELISA has been tested for the analysis of MCYST. The first type involved the use of anti-ID antibodies from anti-idiotype (anti-ID) antibodies in mouse (pAb2/m) and anti–anti-idiotype in mouse (pAb3/r) as a pair for MCYSTs assay (157). The original idiotype antibodies for generating of these antibodies were from rabbits. In the pAb3/r-based ELISA, the IC50 of binding of MCYST-LR–HRP to the solid-phase pAb3 by MCYST-LR, MCYST-RR, MCYST-YR, and NODLN were 0.7, 24.4, 19.2, and 24.3 ng/mL, respectively. Analysis of 19 samples of algae, liver, and water showed that data obtained from the pAb3/r-based ELISA were consistent with those obtained from the pAb1/r-based ELISA.

The second approach involved the use of pAb2/r and pAb3/m pair, and the original idiotype antibody was a mAb (mAb1/m). This approach appeared to be more effective, because the anti-ID antibodies were generated in rabbits. In the MCYST-LR-PLL–based idc-ELISA, the concentrations causing 50% inhibition (IC50) of binding of mAb1 to the solid-phase MCYST-LR-PLL by MCYST-LR, MCYST-RR, MCYST-YR, and NODLN were found to be 0.19, 0.09, 0.33, and 2.06 ng/mL, respectively (158). In contrast, with the pAb2/r-based ELISA, the IC50 values of binding of mAb1 to pAb2/r by MCYST-LR, MCYST-RR, MCYST-YR, and NODLN were found to be 0.37, 0.66, 0.72, and 5.55 ng/mL, respectively. Analysis of MCYST in 22 water samples naturally contaminated with toxic algae showed a good correlation between the data obtained from the pAb2-based idc-ELISA with MCYST-LR-PLL–based idc-ELISA and also between the pAb2-based idc-ELISA with MCYST-pAb–based dc-ELISA. Using the affinity-purified pAb2 Fab fragment as immunogen, polyclonal anti-Ab2 antibodies (pAb3/m) were generated in BALB/c mice. The pAb3m was found to have characteristics similar to those of original mAb1. The IC50 values of binding of pAb3/m to MCYST-LR-PLL by MCYST-LR, MCYST-RR, MCYST-YR, and NODLN were found to be 0.36, 0.12, 1.13, and 152.41 ng/mL, respectively.

Sandwich Immunoassay. Although sandwich immunoassays are generally used for the analysis of large molecular weight compounds, application of this method can be done when the anti-idiotype antibodies are generated from the antibody that has been complexed with the hapten. A sandwich immunoassay, which can detect 2–100 pg MCYST/mL of test solution was recently developed by Tsutsumi et al. (160). The assay involved the use of a mAb generated from a mAb that had been complexed with MCYST (m–anti-IC). In the assay, the m–anti-IC was first coated to the wells of the microtiter plate followed by incubation with MCYST standard solution or water samples and biotinylated anti-MCYST mAb. The amount of the biotinylated mAbs reacted with the solid-phase m–anti-IC was then determined with HRP-labeled streptavidin and substrate 3,3′,5,5′-tetramethyl benzidine. Tests of 17 samples showed that 2 samples contained MCYST at a level of 4.9–13.7 pg/mL.

E. Use of Antibodies for Monitoring of Biological Samples

Antibodies against MCYST have not only been used as the analytical reagent for the analysis of toxins in algae, they have also been used as a diagnostic tool for monitoring the toxin in biological samples. ELISA protocols for the detection of MCYST in liver and serum have developed in our laboratory (119,120), and a C-18 reversed-phase clean-up of the samples was necessary for the assay. Kondo et al. (95) were able to detect the formation of MCYST-glutathione adduct formation in mouse and rat liver. Using mAb-affinity column, these investigators were able to isolate these adducts for further characterization. We have also used the ELISA as the postcolumn monitoring system for the identification of MCYST after the sample extracts were subjected to HPLC. Both pAb and mAb have used the immunochemical stains for biological samples. Shi et al. (162) were able to localize MCYST in cyanobacteria cells by first reacting with the anti-MCYST pAb raised in our laboratory and then staining with goat-antirabbit-IgG–labeled with colloid gold. Using the mAb developed in our laboratory, Solter et al. (120) were able to localize MCYST in liver sections of rats administered with MCYST. They used biotinylated goat-antimouse IgG labeled with streptoavidin-HRP as the indicator. Immunochemical stains revealed MCYST predominantly within pericanalicular regions of zone 3 hepatocytes. A discernible intensity of immunochemical in the centrilobular regions where hemorrhage and apoptosis occurred was found in the liver of mice receiving a single dose of MCYST-LR by Yoshida et al. (163). Furthermore, using Western blotting analysis, we were able to demonstrate covalent binding between PP1 and MCYST (164). Antibodies against MCYST have also been shown to have protective effects on the hepatotoxicity of MCYST-LR in vitro and in vivo, and on MCYST-LR–dependent protein phosphatase inhibition (154,164, 165).

VI. SUMMARY

The increased incidence of blue-green algal blooms in freshwater lakes and waterways has become another environmental hazard to the health of animals and humans. Research during the last decades has led to the discovery of many MCYST and NODLN analogues that are toxic to animals, and extensive studies on the chemistry of these cyanobacterial hepatotoxins have been conducted. Such studies have led to our understanding of the importance of the structural integrity on the toxic effects of these toxins. Although some detoxification methods have been developed, the effectiveness of these methods for a large scale process needs to be critically evaluated. Since effective measures for controlling algal blooms are not available, there is a need for systematic monitoring for the presence of the toxin in water and algal samples. Rapid progress in developing effective methods of toxin analysis has been made in the last few years. The scientists attempted to simplify and to improve the existing analytical procedures on the one hand; they also took some revolutionary approaches to the assay. Conditions for various HPLC methods have been optimized. New, more sensitive, and versatile instruments such as high-resolution MS, GC/tandem MS/MS, and CE are coming to the market and have been tested for MCYST analysis. Effective immunoassay techniques and protein phosphatase inhibition assays have gained more acceptance as analytical tools for this group of phycotoxins. Nevertheless, it should be pointed out that uneven distribution of naturally occurring toxins in algal samples is still a problem. Uniform standard toxins for toxin analysis are needed. Since most methods have not been critically evaluated, rigorous collaborative studies should be made. With newer instrumentation and the development of biotechnology, biosensors might be available in the foreseeable future for phycotoxin detection. This author hopes this chapter helps to generate research that develops even more simplified assay procedures to minimize errors and to increase the sensitivity and specificity of the assay procedure which may alleviate matrix interference problems.

ACKNOWLEDGMENTS

Some of the work described in this review was supported by the College of Agricultural and Life Sciences, the University of Wisconsin at Madison, and by the University of Wisconsin Sea Grant Institute under grants from the National Sea Grant College Program (grant no. R/BT-2),

National Oceanic and Atmospheric Administration, U.S. Department of Commerce, and by contract DAMD-82-C-2021 from the U.S. Army Medical Research and Development Command of the Department of Defense. The author thanks Ms. Barbara Cochrane and Ellin Doyle for their help in the preparation of the manuscript.

REFERENCES

1. DM Anderson. Red tides. Sci Am 271:62–68, 1994.
2. WW Carmichael. Toxins of freshwater algae. In: AT Tu, ed. Handbook of Natural Toxins. Vol 3. Marine Toxins and Venoms. New York: Marcel Dekker, 1988, pp 121–147.
3. WW Carmichael. Freshwater cyanobacteria (blue-green algae) toxins. In: CL Ownby, GV Odel, eds. Natural Toxins: Characterization, Pharmacology and Therapeutics. Oxford, UK: Pergamon Press, 1989, pp 3–16.
4. WW Carmichael. Cyanobacteria secondary metabolites—the cyanotoxins. J Appl Bacteriol 72:445–459, 1992.
5. WW Carmichael. A status report on planktonic cyanobacteria (blue-green algae). US Environ Protect Agency Report EPA/600R-92/079, June 1992.
6. WW Carmichael. The toxins of cyanobacteria. Sci Am 1994:78–86, 1994.
7. GA Codd, TM Jeffries, CW Keevil, E Potter. Detection Methods for Cyanobacterial Toxins. Boca Raton, FL: CRC Press, 1995.
8. MF Watanabe, K-I Harada, WW Carmichael, H Fujiki. Toxin Microcystis. Boca Raton, FL: CRC Press, 1995.
9. IR Falconer. Algal Toxins in Seafood and Drinking Water. New York: Academic Press, 1993, p 224.
10. RE Honkanen, J Zwiller, RE Moore, SL Daily, BS Khatra, M Dukelow, and AL Boynton. Characterization of microcystin-LR, a potent inhibitor of type 1 and type 2A protein phosphatases. J Biol Chem 265:19401–19404, 1990.
11. C Mackintosh, KA Beattie, S Klumpp, P Cohen, GA Codd. Cyanobacterial microcystin-LR is a potent and specific inhibitor of protein phosphatases 1 and 2A from both mammals and higher plants. FEBS Lett 264:187–192, 1990.
12. S Toshizawa, R Matsushima, MF Watanabe, KI Harada, A Ichihara, WW Carmichael, H Fujiki. Inhibition of protein phosphatases by microcystin and nodularin associated with hepatotoxicity. J Cancer Res Clin Oncol 116:609–614, 1990.
13. MT Runnegar, S Kong, N Berndt. Protein phosphatase inhibition and in vivo hepatotoxicity of microcystins. Am J Physiol 265:224–230, 1993.
14. M Runnegar, N Berndt, SM Kong, EYC Lee, LF Zhang. In vivo and in vitro binding of microcystin to protein phosphatases 1 and 2A. Biochem Biophys Res Commun 216:162–169, 1995.
15. RW Mackintosh, KN Dalby, DG Campbell, PRW Cohen, P Cohen, C Mackintosh. The cyanobacterial toxin microcystin binds covalently to cysteine-273 on protein phosphatase 1. FEBS Lett 371: 236–240, 1995.
16. S Claeyssens, A Francois, A Chedeville, A Lavoinne. Microcystin-LR induced an inhibition of protein synthesis in isolated rat hepatocytes. Biochem J 306:693–696, 1995.
17. IR Falconer. Tumor promotion and liver injury caused by oral consumption of cyanobacteria. Environ Toxicol Water Qual 6:177–184, 1991.
18. H Fujiki, M Suganuma. Tumor promotion by inhibitors of protein phosphatases 1 and 2A: the okadaic acid class of compounds. Adv Cancer Res 61:143–194, 1993.
19. IR Falconer, AR Humpage. Tumor promotion by cyanobacterial toxins. Phycologia 35:74–79, 1996.
20. R Nishiwaki-Matsushima, T Ohta, S Nishiwaki, M Suganuma, K Kohyama, T Ishikawa, WW Carmichael, H Fujiki. Liver cancer promoted by the cyanobacterial cyclic peptide toxin microcystin LR. J Cancer Res Clin Oncol 118:420–424, 1992.

21. Y Ueno, S Nagata, T Tsutsumi, A Hasegawa, MF Watanabe, HD Park, GC Chen, G Chen, SZ Yu. Detection of microcystins, a blue-green algal hepatotoxin, in drinking water sampled in Haimen and Fusui, endemic area of primary liver cancer in China, by highly sensitive immunoassay. Carcinogenesis 17:1317–1321, 1996.
22. EM Jochimsen, WW Carmichael, JS An, DM Cardo, ST Cookson, CEM Holmes, MBD Antunes, DA Demelo, TM Lyra, VST Barreto, SMFO Azevedo, WR Jarvis. Liver failure and death after exposure to microcystins at a hemodialysis center in Brazil. N Engl J Med 338:873–878, 1998.
23. S Pouria, A Deandrade, A Barbosa, RL Cavalcanti, VTS Barreto, CJ Ward, W Preiser, GK Poon, GH Neild, GA Cood. Fatal microcystin intoxication in haemodialysis unit in Caruaru, Brazil. Lancet 352:21–26, 1998.
24. J Mandelbaum-Schmid. Blue-green algae. Health Alert. March SELF, 1997. 1997, pp 144–148.
25. DP Botes, AA Tuinman, PL Wessels, CC Viljoen, H Kruger, DH Williams, S Santikam, RJ Smith, SJ Hammond. The structure of cyanoginosin-LA, a cyclic heptapeptide toxin from the cyanobacterium Microcystis. J Chem Soc Perkin Trans 1:2311–2318, 1984.
26. DP Botes, PL Wessels, H Kruger, MTC Runnegar, S Santikam, RJ Smith, JCJ Bama, DH Williams. Structural studies on cyanoginosins-LR, -YR, -YA, and -YM, peptide toxins from microcystis J Chem Soc Perkin Trans 1:2747–2748, 1985.
27. WW Carmichael, VR Beasley, DL Bunner, JN Eloff, L Falconer, P Gorham, K-I Harada, T Krishnamurthy, M-J Yu, RE Moore, KL Rinehart, M Runnegar, OM Skulberg, MF Watanabe. Naming of cyclic heptapeptide toxins of cyanobacteria (blue-green algae). Toxicon 26:971–973, 1988.
28. KL Rinehart, M Namikoshi, BW Choi. Structure and biosynthesis of toxins from blue-green algae (cyanobacteria). J Appl Phycol 6:159–176, 1994.
29. M Namikoshi, KL Rinehart. Bioactive compounds produced by cyanobacteria. J Ind Microbiol 17: 373–384, 1996.
30. M Namikoshi, KL Rinehart, R Sakai, RR Stotts, AM Dahlem, VR Beasley, WW Carmichael, WR Evans. Identification of 12 hepatotoxins from a Homer Lake bloom of the cyanobacteria *Microcystis aeruginosa*, *Microcystis viridis*, and *Microcystis wesenbergii*—nine new microcystins. J Organic Chem 57:866–872, 1992.
31. MR Prinsep, FR Caplan, RE Moore, GML Patterson, RE Honkanen, AL Boynton. Microcystin-LA from a blue-green algae belong to the Stigonematales. Phytochemistry 31:1247–1248, 1992.
32. PS Gathercole and PG Thiel Liquid chromatographic determination of the cyanoginosins, toxins produced by cyanobacterium *Microcystis aeruginosa*. J Chromatogr 408:435–440, 1987.
33. KI Harada, K Ogawa, K Matsuura, H Nagai, H Murata, M Suzuki, Y Itenono, N Nakayama, S Makoto, M Nakano. Isolation of two toxic heptapeptide microcystins from an axenic strain of *Microcystis aeruginosa*, K-139. Toxicon 29:479–489, 1991.
34. R Luukkainen, M Namikoshi, K Sivonen, KL Rinehart, SI Niemeld. Isolation and identification of 12 microcystins from four strains and two bloom samples of Microcystis spp.: structure of a new hepatotoxin. Toxicon 32:133–139, 1994.
35. K Sivonen, M Namikoshi, WR Evans, BV Gromov, WW Carmichael, KL Rinehart. Isolation and structures of five microcystins from a Russian Microcystis aeruginosa strain CALU 972. Toxicon 30:1481–1485, 1992.
36. M Namikoshi, M Yuan, S Sivonen, WW Carmichael, KL Rinehart, FR Sun, S Brittain, A Otsuki. Seven more microcystins possessing two L-glutamic acid units, isolated from Anabaena sp. strain 186. Chem Res Toxicol 11:143–149, 1998.
37. KI Harada, K Matsuura, M Suzuki, MF Watanabe, S Oishi, AM Dahlem, VR Beasley, WW Carmichael. Isolation and characterization of the minor components associated with microcystins LR and RR in the cyanobacteria (blue-green algae). Toxicon 28:55–64, 1990.
38. T Krishnamurthy, WW Carmichael, EW Sarver. Toxic peptides from fresh water cyanobacteria (blue-green algae). Isolation, purification and characterization of peptides from Microcystis aeruginosa and Anabaena flos-aquae. Toxicon 24:865–873, 1986.
39. K Sivonen, OM Skulberg, M Namikoshi, WR Evans, WW Carmichael, KL Rinehart. Two methyl ester derivatives of microcystins, cyclic heptapeptide hepatotoxin isolated from and structures of five microcystins from *Anabaena flos-aquae* strain CYA 83/1LU 972. Toxicon 30:1465–1471, 1992.

40. R Luukkainen, KM Sivonen, M Namikoshi, M Fardig, KL Rinehart, SI Niemela. Isolation and identification of 8 microcystins from 13 Oscillatoria agardhii strains and structure of a new microcystin. Appl Environ Microbiol 59:2204–2209, 1993.
41. M Namikoshi, K Sivonen, WR Evans, WW Carmichael, F Sun, L Rouhiainen, R Luukkainen, KL Rinehart. Two new L-serine variants of microcystin-LR and -RR from Anabaena ap. strain 202 A1 and 202 A2. Toxicon 30:1457–1464, 1992.
42. K Sivonen, WW Carmichael, M Namikoshi, KL Rinehart, AM Dahlem, SA Niemelii. Isolation and characterization of hepatotoxic microcystin homologs from the filamentous freshwater cyanobacterium Nostoc sp. strain 152. Appl Environ Microbiol 56:2650–2657, 1990.
43. KL Rinehart, K-I Harada, M Namikoshi, C Chen, CA Harvis, MHG Munro, JW Blunt, PE Mulligan, VR Beasley, AM Dahlem, WW Carmichael. Nodularin, microcystin and the configuration of Adda. J Am Chem Soc 110:8557–8558, 1988.
44. RD Stoner, WH Adams, DN Slatkin, HW Siegelman. The effect of single L-amino acid substitutions on the lethal potencies of the microcystins. Toxicon 27, 825–827, 1989.
45. S Rudolphbohner, JT Wu, L Moroder. Identification and characterization of microcystin-LY from Microcystis aeruginosa (strain-298). Biol Chem Hoppe-Seyler 374:635–640, 1993.
46. M Namikoshi, KL Rinehart, R Sakai, K Sivonen, WW Carmichael. Structures of three new cyclic heptapeptide hepatotoxins produced by the cyanobacterium (blue-green alga) Nostoc sp. strain 152. J Org Chem 55:6135–6139, 1990.
47. T Krishnamurthy, L Szafraniec, DF Hunt, J Shabanowitz, JR Yates, III, CR Hauer, WW Carmichael, O Skulberg, GA Codd, S Missler. Structural characterization of toxic cyclic peptides from blue-green algae by tandem mass spectrometry. Proc Natl Acad Sci USA 86:770–774, 1989.
48. J Kiviranta, M Namikoshi, K Sivonen, WR Evans, WW Carmichael, KL Rinehart. Structures determination and toxicity of a new microcystin from *Microcystis aeruginosa* strain 205. Toxicon, 30, 1093–1098, 1992.
49. JAO Meriluoto, A Sandström, JE Eriksson, G Remaud, A Grey Craig, J Chattopadhyaya. Structure and toxicity of a peptide hepatotoxin from the cyanobacterium Oscillatoria agardhii. Toxicon 27: 1021–1034, 1989.
50. M Namikoshi, K Sivonen, WR Evans, WW Carmichael, KL Rinehart. Isolation and structures of microcystins from a cyanobacterial water bloom (Finland). Toxicon 30:1473–1479, 1992.
51. M Namikoshi, K Sivonen, WR Evans, WW Carmichael, L Rouhiainen, R Luukkainen, KL Rinehart. Structures of three homotyrosine-containing new microcystins and a new homophenylalanine variant from Anabaena sp. strain 66. Chem Res Toxicol 5:661–666, 1992.
52. T Kusumi, T Ooi, MM Watanabe, H Takahashi, H Kakisawa. Cyanoviridin RR, a toxin from the cyanobacterium (blue-green alga) Microcystis viridis. Tetrahed Lett 28:4695–4698, 1987.
53. P Painuly, R Perez, T Fukai, Y Shimizu. The structure of a cyclic peptide toxin, cyanogenosin-RR, from *Microcystis aeruginosa*. Tetrahed Lett 29:11–14, 1988.
54. MF Watanabe, S Oishi, KI Harada, K Matsuura, H Kawai, M Suzuki. Toxins contained in *Microcystis* species of cyanobacteria (blue-green algae). Toxicon 26:1017–1025, 1988.
55. K Sivonen, M Namikoshi, WR Evans, M Färdig, WW Carmichael, KL Rinehart. Three new microcystins, cyclic heptapeptide hepatotoxins, from Nostoc sp. strain 152. Chem Res Toxicol 5:464–469, 1992.
56. K-I Harada, K Ogawa, Y Kimura, H Murata, M Suzuki, PM Thom, WR Evans, WW Carmichael. Microcystins from Anabaena flos-aquae NRC 525-17. Chem Res Toxicol 4:535–540, 1991.
57. WW Carmichael, JT Eschedor, GM Patterson, RE More. Toxicity and partial structure of a hepatotoxic peptide produced by the cyanobacterium Nodularia spumigena Mertens emend. L575 from New Zealand. Appl Environ Microbiol 54:2257–2263, 1988.
58. M Namikoshi, BW Choi, R Sakai, F Sun, KL Rinehart, WW Carmichael, WR Evans, P Cruz, MHG Munro, JW Blunt. New nodularins: a general method for structure assignment. J Org Chem 59: 2349–2357, 1994.
59. K Sivonen, M Namikoshi, WR Evans, WW Carmichael, F Sun, L Rouhiainen, R Luukkainen, KL Rinehart. Isolation and characterization of a variety of microcystins from seven strains of the cyanobacterial genus Anabaena. Appl Environ Microbiol 58:2495–2500, 1992.
60. C Edwards, LA Lawton, SM Coyle, P Ross. Laboratory-scale purification of microcystins using

flash chromatography and reversed-phase high-performance liquid chromatography. J Chromatogr 734:163–173, 1996.
61. C Edwards, LA Lawton, SM Coyle, P Ross. Automated purification of microcystins. J Chromatogr 734:175–182, 1996.
62. KI Harada, M Suzuki, AM Dahlem, VR Beasley, WW Carmichael, KL Rinehart, Jr. Improved method for purification of toxic peptides produced by cyanobacteria. Toxicon 26:433–439, 1988.
63. K-I Harada, K Ogawa, K Matsuura, H Murata, M Suzuki, MF Watanabe, Y Itezono, N Nakayama. Structural determination of geometrical isomers of microcystins LR and RR from cyanobacteria by two-dimensional NMR spectroscopic techniques. Chem Res Toxicol 3:473–481, 1990.
64. M Namikoshi, FR Sun, BW Choi; KL Rinehart, WW Carmichael, WR Evans, VR Beasley. Seven more microcystins from Homer Lake cells: application of the general method for structure assignment of peptides containing α, β-dehydroamino acid unit(s). J Org Chem 60:3671–3679, 1995.
65. RR Stotts, M Namikoshi, WM Haschek, KL Rinehart, WW Carmichael, AM Dahlem, VR Beasley. Structural modifications imparting reduced toxicity in microcystins from Microcystis spp. Toxicon 31:783–789, 1993.
66. DP Boths, CC Viljoen, H Kruger, PL Wessels, DH Williams. Configuration assignments of the amino acid residues and the presence of N-methyldehydroalanine in toxins from the blue green algae, Microcystis aeruginosa. Toxicon 20:1037–1042, 1982.
67. MF Watanabe, S Oishi, KI Harada, K Matsuura, H Kawai, H Nagai, M Suzuki. Toxins contained in Microcystis species of cyanobacteria (blue-green algal). Toxicon 26:1017–1025, 1988.
68. JAO Meriluoto, SE Nygard, AM Dahlem, JE Eriksson. Synthesis, organotropism, and hepatocellular uptake of two tritium-labeled epimers of dihydromicrosytins-LR, a cyanobacterial peptide toxin analog. Toxicon 28:1439–1446, 1990.
69. M Namikoshi, BW Choi, F Sun, KL Rinehart, WR Evans, WW Carmichael. Chemical characterization and toxicity of dihydro derivatives of nodularin and microcystin-LR, potentent cyanobacterial cyclic peptide hepatoxins. Chem Res Toxicol 6:151–158, 1993.
70. AM Dahlem. Structure/toxicicity relationships and fate of low molecular weight toxins from cyanobacteria. PhD dissertation, University of Illinois, Urbana, IL, 1989, pp 135–148.
71. J An, WW Carmichael. Use of a colorimetric protein phosphatase inhibition assay and enzyme linked immunosorbent assay for the study of microcystins and nodularins, Toxicon 32:1495–1507, 1994.
72. RR Stotts, AR Twardock, GD Koritz, WM Haschek, RK Manuel, WB Hollis, VR Beasley. Toxicokinetics of tritiated dihydromicrocystin-LR in swine. Toxicon 35:455–465, 1997.
73. BW Choi, M Namikoshi, F Sun, KL Rinehart, WW Carmichael, AM Kaup, WR Evans. Isolation of linear peptides related to the hepatotoxins nodularin and microcystins. Tetrahed Lett 34:7881–7884, 1993.
74. JRH Hoffman. Removal of microcystis toxins in water purification process. Water SA 2:58–60, 1976.
75. AM Keijola, K Himberg, AL Esala, K Sivonen, L Hiisvirta. Removal of cyanobacterial toxins in water treatment processes: laboratory and pilot-scale experiments. Toxicity Assess 3:643–656, 1988.
76. K Himberg, AM Keijola, L Hiisvirta, H Pysalo, K Sivonen. The effect of water treatment process on the removal hepatotoxins from microcystis and oscillatoria cyanobacterial: a laboratory study. Water Res 23:979–984, 1989.
77. BC Nicholson, J Rositano, MD Burch. Destruction of cyanobacterial peptide hepatotoxins by chlorine and chloramine. Water Res 28:1297–1303, 1994.
78. K Tsuji, T Watanuki, F Kondo, MF Watanabe, H Nakazawa, S Naito, M Suzuki, H Uchida, KI Harada. Stability of microcystins from cyanobacteria-IV. Effect of chlorination of decomposition. Toxicon 35:1033–1041, 1997.
79. J Rapala, K Lahti, K Sivonen, SI Niemelci. Biodegradability and adsorption on lake sediments of cyanobacterial hepatotoxins and anatoxin-a. Lett Appl Microbiol 19:423–428, 1994.
80. GJ Jones, DG Bourne, RL Blakeley, H Doelle. Degradation of the cyanobacterial hepatotoxin microcystin by aquatic bacteria. Nat Toxins 2:228–235, 1994.
81. DG Bourne, GJ Jones, RL Blakeley, A Jones, AP Negri, P Riddles. Enzymatic pathway for the

bacterial degradation of the cyanobacterial cyclic peptide toxin microcystin LR. Appl Environ Microbiol 62:4086–4094, 1996.
82. S Takenaka, MF Watanabe. Microcystin LR degradation by Pseudomonas aeruginosa alkaline protease. Chemosphere 34:749–757, 1997.
83. S Takenaka. Formation of 3-amino-2,6,8-trimethyl-10-phenydeca-4E,6E-dienoic acid from microcystin LR by treatment with various proteases, and its detection in mouse liver. Chemosphere 36: 2277–2282, 1998.
84. YS Yoo, WW Carmichael, RC Hoehn, SE Hrudey. Cyanobacterial (blue-green algal) toxins: A resource guide. Denver: AWWA Research Foundation, 1995, pp 145–158.
85. TW Lambert, CFB Holmes, SE Hrudey. Adsorption of microcystin-LR by activated carbon and removal in full scale water treatment. Water Res 30:1411–1422, 1996.
86. LA Lawton, BJPA Cornish, AWR Macdonald. Removal of cyanobacterial toxins (microcystins) and cyanobacterial cells from drinking water using domestic water filters. Water Res 32:633–638, 1998.
87. AM Warhurst, SL Raggett, GL McConnachie, SJT Pollard, V Chipofya, GA Codd. Adsorption of the cyanobacterial hepatotoxin microcystin-LR by a low-cost activated carbon from the seed husks of the pan-tropical tree, moringa oleifera. Sci Total Environ 207:207–211, 1997.
88. FS Chu, R Wedepohl. Algal toxins in drinking water? Research in Wisconsin. LakeLine, 14:41–42, 1994.
89. K Tsuji, S Naito, F Kondo, N Ishikawa, MF Watanabe, M Suzuki, KI Harada. Stability of microcystins from cyanobacteria-11. Effect of light on decomposition and isomerization. Environ Sci Technol 28:173–177, 1994.
90. K Tsuji, T Watanuki, F Kondo, MF Watanabe, S Suzuki, H Nakazawa, M Suzuki, H Uchida, K-I Harada. Stability of microcystins from cyanobacteria-11. Effect of UV light on decomposition and isomerization. Toxicon 33:1619–1631, 1995.
91. KI Harada, K Tsuji, MF Watanabe, F Kondo. Stability of microcystins from cyanobacteria-III. Effect of pH and temperature. Phycologia 35(Suppl 6):83–88, 1996.
92. H Twist, CA Codd. Degradation of the cyanobacterial hepatotoxin, nodularin under light and dark conditions. FEMS Microbiol Lett 151:83–88, 1977.
93. PKJ Robertson, LA Lawton, B Munch, J Rouzade. Destruction of cyanobacterial toxins by semiconductor photocatalysis. Chem Commun 4:393–394, 1997.
94. K Kaya, T Sano. A photodetoxification mechanism of the cyanobacterial hepatotoxin microcystin-LR by ultraviolet irradiation. Chem Res Toxicol 11:159–163, 1998.
95. F Kondo, H Matsumoto, S Yamada, N Ishikawa, E Ito, S Nagata, Y Ueno, M Suzuki, K Harada. Detection and identification of metabolites of microcystins formed in vivo in mouse and rat livers. Chem Res Toxicol 9:1355–1359, 1996.
96. DE Williams, SC Dawe, ML Kent, RJ Andersen, M Craig, CFB Holmes. Bioaccumulation and clearance of microcystins from salt water, mussels, mytilus edulis, and in vivo evidence for covalently bound microcystins in mussel tissues. Toxicon 35:1617–1625, 1997.
97. DE Williams, M Craig, SC Dawe, ML Kent, CFB Holmes, RJ Andersen. Evidence for a covalently bound form of microcystin-LR in salmon liver and dungeness crab larvae. Chem Res Toxicol 10: 463–469, 1997.
98. MF Watanabe, HD Park, F Kondo, KI Harada, H Hayashi, T Okino. Identification and estimate of microcystins in freshwater mussels. Nat Toxins 5:31–35, 1997.
99. EE Prepas, BG Kotak, LM Campbell, JC Evans, SE Hrudey, CFB Holmes. Accumulation and elimination of cyanobacterial hepatotoxins by the freshwater clam Anodonta grandis simpsoniana. Can J Fish Aquatic Sci 54:41–46, 1997.
100. M Liras, P Lindberg, H Nystrom, LA Annadotter, B Lawton, B Graf. Can ingested cyanobacteria be harmful to the signal crayfish (Pacifastacus leniusculus). Freshwater Biol 39:233–242, 1998.
101. S Hall, G Strichartz, (eds.) Marine toxins: origin, structure, and molecular pharmacology. ACS Symposium Series 418 ACS, 1990.
102. MM Wekell. Seafood toxins. J Assoc Off Anal Chem 73:113–117, 1990.
103. JJ Sullivan. Methods for analysis for DSP and PSP toxins in shellfish. J Shellfish Res 7:587–595, 1988.

104. MA Quilliam. Phycotoxins. J AOAC Int 81:142–151, 1998 (also see 1995–1997 Jan.–Feb. issues of J AOAC Int for reviews).
105. J Meriluoto. Chromatography of microcystins. Anal Chim Acta 352:277–298, 1997.
106. GK Poon, LJ Griggs, C Edwards, KA Beattie, GA Codd. Liquid chromatography-electrospray ionization mass spectrometry of cyanobacterial toxins. J Chromatogr 628:215–233, 1993.
107. LA Lawton, C Edwards, GA Codd. Extraction and high-performance liquid chromatographic method for the determination of microcystins in raw and treated waters. Analyst 119:1525–1530, 1994.
108. K Harada, H Murata, Z Qiang, M Suzuki, F Kondo. Mass spectrometric screening method for microcystins in cyanobacteria. Toxicon 34:701–710, 1996.
109. N Onyewuenyi, P Hawkins. Separation of toxic peptides (microcystins) in capillary electrophoresis, with the aid of organic mobile phase. J Chromatogr 749:271–277, 1996.
110. J Kiviranta, K Sivonen, SI Niemela, K Huovinen. Detection of toxicity of cyanobacteria by *Artemia salina* bioassay. Environ Toxicol Water Qual 6, 423–436, 1991.
111. FS Chu, X Huang, RD Wei. Enzyme-linked immunosorbent assay for microcystins in blue-green algal blooms. J Assoc Off Anal Chem 73:451–456, 1990.
112. KI Harada, K Matsuura, M Suzuki, H Oka, MF Watanabe, S Oishi, AM Dahlem, VR Beasley, WW Carmichael. Analysis and purification of toxic peptides from cyanobacteria by reversed phase high-pressure liquid chromatography. J Chromatogr 448:275–283, 1988.
113. K Tsuji, S Naito, F Kondo, MF Watanabe, S Suzuki, H Nakazawa, M Suzuki, T Shimada, K Harada. A clean-up method for analysts of trace amounts of microcystins in lake water. Toxicon 32:1251–1259, 1994.
114. Y Tsuchiya, M Watanabe. Disk type solid-phase extraction of microcystin in environmental water. Jpn J Toxicol Environ Health 43(3):190–196, 1997.
115. GJ Jones, TR Falconer, RM Wilkins. Persistence of cyclic peptide toxins in dried cyanobacterial crusts from Lake Mokoan, Australia. Environ Toxicol Water Qual 10:19–24, 1995.
116. C Rivasseau, S Martins, MC Hennion. Determination of some physicochemical parameters of microcystins (cyanobacterial toxins) and trace level analysis in environmental samples using liquid chromatography. J Chromatogr 799:155–169, 1998.
117. J Fastner, I Flieger, U Neumann. Optimised extraction of microcystins from field samples—a comparison of different solvents and procedures. Water Res 32:3177–3181, 1998.
118. W John, MW Raynor, B Rae. Micellar electrokinetic capillary chromatography of algal toxins. J High Res Chromatogr 20:34–38, 1997.
119. LR Lin, FS Chu. Kinetics of distribution of microcystin-LR in serum and liver cytosol of mice: an immunochemical analysis. J Agric Food Chem 42:1035–1040, 1994.
120. PF Solter, GK Wollenberg, X Huang, FS Chu, MT Runnegar. Prolonged sublethal exposure to the protein phosphatase inhibitor microcystin-LR results in multiple dose-dependent hepatotoxic effects. Toxicol Sci 44:87–96, 1998.
121. KI Harada Chemistry and detection of microcystins. In: MF Watnabe, KI Harada, WW Carmichael, H Fujiki, eds. Toxic microcystis. Boca Raton, FL: CRC Press, 1996, p 103.
122. C Rivasseau, MC Hennion, P Sandra. Determination of microcystins in cyanobacterial samples using microliquid chromatography. J Microcol Sep 8:542–551, 1996.
123. LA Lawton, C Edwards, KA Beattie, S Pleasance, GJ Dear, CA Codd. Isolation and characterization of microcystins from laboratory cultures and environmental samples of *Microcystis aeruginosa* and from an associated animal toxicosis. Nat Toxins 3:50–57, 1995.
124. TH Lee, YM Chen, HN Chou. First report of microcystins in Taiwan. Toxicon 36:247–255, 1998.
125. HW Siegelman, WH Adams, RD Stoner, DN Slatkin. In: EP Ragelis, ed. Seafood Toxins. ACS Symposium Series No. 262. Washington, DC: American Chemical Society, 1984, p 407.
126. JAO Meriluoto, JE Eriksson. Rapid analysis of peptide toxins in cyanobacteria. J Chromatogr 438: 93–99, 1988.
127. JAO Meriluoto, JE Eriksson, KI Harada, AM Dahlem, K Sovonen, WW Carmichael. Internal surface reversed-phase high performance liquid chromatographic separation of the cyanobacterial peptide toxins microcystin-LA, -LR, -YR, -RR and nodularin. J Chromatogr 509:390–395, 1990.

128. B Sedmak, G Kosi. Microcystins in Slovene freshwaters (Central Europe)—first report. Nat Toxins 5:64–73, 1997.
129. H Murata, H Shoji, M Oshikata, KI Harada, M Suzuki, F Kondo, H Goto. High-performance liquid chromatography with chemiluminescence detection of derivatized microcystins. J Chromatogr A 693:263–270, 1995.
130. RJ Wicks, PG Thiel. Environmental factors affecting the production of peptide toxins in floating scums of the cyanobacterium *Microcystis aeruginosa* in a hypertrophic African reservoir. Environ Sci Technol 24:1413–1418, 1990.
131. KP Bateman, P Thibault, DJ Douglas, RL White. Mass spectral analyses of microcystins from toxic cyanobacteria using on-line chromatographic and electrophoretic separations. J Chromatogr A. 712: 253–268, 1995.
132. M Shimizu, Y Iwasaki, S Yamada. New fluorogenic dienophile: Synthesis, reaction with vitamin D, vitamin A and microcystins, and application to fluorometric assays. Yakugaku Zasshi J Pharm Soc Jpn 115(8):584–602, 1995.
133. J Meriluoto, B Kincaid, MR Smyth, M Wasberg. Electrochemical detection of microcystins, cyanobacterial peptide hepatotoxins, following high-performance liquid chromatography. J Chromatogr 810:226–230, 1998.
134. M Erhard, H Vondohren, P Jungblut. Rapid typing and elucidation of new secondary metabolites of intact cyanobacteria using maldi-t of mass spectrometry. Nat Biotechnol 9:906–909, 1997.
135. MP Boland, MA Smillie, DZX Chen, CFB Holmes. A unified bioscreen for the detection of diarrhetic shellfish toxins and microcystins in marine and freshwater environments. Toxicon 31:1393–1405, 1993.
136. AV Humble, GM Gadd, GA Codd. Binding of copper and zinc to three cyanobacterial microcystins quantified by differential pulse polarography. Water Res 31:1679–1686, 1997.
137. T Battle, C Touchard, HJ Moulsdale, B Dowsett, GN Stacey. New cell substrates for in vitro evaluation of microcystin hepato-cytotoxicity. Toxicol in Vitro 11:557–567, 1997.
138. M Feuillade, G Jannpara, J Feuillade. Toxic compounds to artemia from blooms and isolates of the cyanobacterium planktothrix rubescens. Archiv fur Hydrobiol 138:175–186, 1996.
139. J Kiviranta, A Abdel-Hameed, K Sivonen, SI Niemela, G Carlberg. Toxicity of cyanobacteria to mosquito larve-screening of active compounds. Environ Toxicol Water Qual 8:63–68, 1993.
140. MJ Turell, LL Middlebrook. Mosquito inoculation—an alternative bioassay for toxins. Toxicon 26: 1089–1094, 1988.
141. LA Lawton, DL Campbell, KA Beattie, GA Codd. Use of a rapid bioluminescence assay for detecting cyanobacterial microcystin toxicity. Lett Appl Microbiol 11:205–207, 1990.
142. DL Campbell, LA Lawton, KA Beattie, GA Codd. 1994. Comparative assessment of the specificity of the brine shrimp and Microtox assays to hepatotoxic (microcystin-LR–containing) cyanobacteria. Environ Toxicol Water Qual 9:71–77, 1994.
143. K Lahti, J Ahtiainen, J Rapala, K Sivonen, SI Niemela. Assessment of rapid bioassays for detecting cyanobacterial toxicity. Lett Appl Microbiol 21:109–114, 1995.
144. L Volterra, M Bruno, PMB Gucci, E Pierdominici. Fast method for detecting toxic cyanophyte blooms. Environ Toxicol Water Qual 7:215–222, 1992.
145. Z Zhang, G Bai, S Deans-Zirattu, MF Browner, EYC Lee. Expression of the catalytic subunit of phosphorylase phosphatase (protein phosphatase-1) in Escherichia coli. J Biol Chem 267:1484–1490, 1992.
146. CJ Ward, KA Beattie, EYC Lee, GA Codd. Colorimetric protein phosphatase inhibition assay of laboratory strains and natural blooms of cyanobacteria—comparisons with high-performance liquid chromatographic analysis for microcystins. FEMS Microbiol Lett 153:465–473, 1997.
147. Y Sugiyama, K Fujimoto, II Ohtani, A Takai, M Isobe. Sensitive analysis of protein phosphatase inhibitors by the firefly bioluminescence system—application to pp1-gamma. Biosci Biotech Biochem 60:1260–1264, 1996.
148. CW Laidley, E Cohen, JE Casida. Protein phosphatase in neuroblastoma cells—[H^3]cantharidin binding site in relation to cytotoxicity. J Pharmacol Exp Ther 280:1152–1158, 1997.
149. FS Chu, X Huang, R Wei, WW Carmichael. Production and characterization of antibodies against microcystin. Appl Environ Microbiol 55:1928–1933, 1989.

150. R Kfir, E Johannsen, DP Botes. Monoclonal antibody specific for cyanoginosin-LA: preparation and characterization. Toxicon 24:543–552, 1986.
151. WP Brooks, GA Codd. Immunoassay of hepatotoxic cultures and water blooms of cyanobacteria using Microcystis aeruginosa peptide toxin polyclonal antibodies. Environ Technol Lett 9:1343–1348, 1988.
152. FS Chu, HP Lau, TS Fan, KS Zhang. Ethylene diamine modified bovine serum albumin as protein carrier in the production of antibody against mycotoxins. J Immunol Methods 55:73–78, 1982.
153. CM McDermott, R Feola, J Plude. Detection of cyanobacterial toxins (microcystins) in waters of northeastern Wisconsin by a new immunoassay technique. Toxicon 33:1433–1442, 1995.
154. S Nagata, H Soutome, T Tsutsumi, A Hasegawa, M Sekijima, M Sugamata, KI Harada, M Suganuma, Y Ueno. Novel monoclonal antibodies against microcystin and their protective activity for hepatotoxicity. Nat Toxins 3:78–86, 1995.
155. X Huang, FS Chu. Production and characterization of monoclonal antibodies against blue-green algal toxin microcystin. Abstract for American Society for Microbiology 1996 annual meeting, New Orleans, May 19–24, 1996.
156. S Saito, Y Nakano, K Kushida, M Shirai, K Harada, M Nakano. Cross-reactivity and neutralizing ability of monoclonal antibodies against microcystins. Microbiol Immunol 38:389–392, 1994.
157. B Liu, FY Yu, FS Chu. Anti-idiotype and anti-anti-idiotype antibodies generated from polyclonal antibodies against microcystin-LR. J Agric Food Chem 44:4037–4042, 1996.
158. X Huang, FS Chu. Production and characterization of anti-idiotype and anti-anti-idiotype antibodies against microcystins. ASM annual meeting, Atlanta, May 17–21, 1998.
159. T Tsutsumi, S Nagata, F Yoshida, Y Ueno. Anti-idiotype monoclonal antibodies against anti-microcystin antibody and their use in enzyme immunoassay. Toxicon 36:235–245, 1998.
160. T Tsutsumi, S Nagata, F Yoshida, Y Ueno. Development of sandwich immunoassay for microcystins, a natural contaminant in environmental water: an assay based on anti-immune complex antibodies. Jpn J Toxicol Environ Health 44:25, 1998.
161. S Nagata, T Tsutsumi, A Hasegawa, F Yoshida, Y Ueno, WF Watanabe. Enzyme immunoassay for direct determination of microcystins in environmental water. J AOAC Int 80:408–417, 1997.
162. L Shi, WW Carmichael, I Miller. Immuno-gold localization of hepatotoxins in cyanobacterial cells. Arch Microbiol 163:7–15, 1995.
163. T Yoshida, Y Makita, T Tsutsumi, S Nagata, F Tashiro, F Yoshida, M Sekijima, SI Tamura, T Harada, K Maita, and Y Ueno. Immunohistochemical localization of microcystin-LR in the liver of mice: a study on the pathogenesis of microcystin-LR induced hepatotoxicity. Toxicol Pathol 26: 411–418, 1998.
164. BH Liu, FY Yu, X Huang, FS Chu. Monitoring of microcystin-protein phosphatase adduct formation with immunochemical methods. Toxicon 38:619–632, 2000.
165. JR Lin, FS Chu. In vitro neutralization of the inhibitory effect of microcystin-LR to protein phosphatase 2A by antibody against the toxin. Toxicon 32:605–613, 1994.

30

Freshwater Hepatotoxins: Microcystin and Nodularin, Mechanisms of Toxicity and Effects on Health

Marcia Craig and Charles F.B. Holmes
University of Alberta, Edmonton, Alberta, Canada

I. INTRODUCTION

Cyanobacteria have been an intrinsic part of the Earth's ecosystem for over 3 billion years. One of the earliest organisms to harness the sun's energy by photosynthesis, these prokaryotes may have been vital in preparing the planet for oxygen-based life. In spite of their early contributions, they have lately been getting some bad press. Cyanobacteria are ubiquitous, growing in all depths of fresh, brackish, and marine waters and in most cases are an innocuous and vital part of the biosphere. However, planktonic forms periodically rise and proliferate in surface waters where they come into contact with wildlife and domestic animals, often resulting in illness and death.

Cyanobacteria were originally called blue-green algae, until it was discovered that they do not, as true algae do, have nucleated cells (1). These organisms began to be linked to animal deaths as early as the late nineteenth century. As the twentieth century progressed, the phenomenon known as a "bloom," or an exponential growth of cyanobacterial cells, became widely recognized as a potential problem for animal and human health. Cyanobacteria are ancient organisms, but the frequency of blooms and their connection to illness and death are perceived to be modern developments. Bloom development is dependent on nutrient-rich waters, frequently nurtured by activities associated with human civilization. As population densities rise, effluents of human activity increasingly find their way into natural and manmade water supplies. The eutrophication, or enrichment, of these waters promotes proliferation of cyanobacterial blooms and can reasonably be linked to the apparently growing incidence of intoxication. This biological threat has been addressed using many different approaches: the development of methods to control bloom formation and/or removal or inactivation of toxic elements, investigations of the nature of active bacterial substances and mechanisms of intoxication, and the use of cyanobacterial hepatotoxins as research tools.

Cyanobacteria come in toxic and nontoxic varieties. Over 40 species of known toxicity have been identified (2), but the factors affecting the production of toxins are poorly understood. About 50% of spontaneous cyanobacterial blooms are toxic. Freshwater blooms are the most obviously problematic, as the potential exists for contamination of drinking water supplies.

II. CYANOBACTERIAL HEPATOTOXINS: STRUCTURES AND PROPERTIES OF MICROCYSTIN AND NODULARIN

Cyanobacteria synthesize a plethora of substances that are, via different mechanisms, harmful to mammals, birds, and fish. Most extensively studied are the neurotoxins and hepatotoxins. Neurotoxins such as the anatoxins and saxitoxins (1–4) interrupt communications between neurons and muscle cells. (For a discussion of these molecules and a historical development, see ref. 1). This chapter focuses entirely on two families of hepatotoxins (reviewed earlier in refs. 1–5 and 119,124) that are similar in structure and in their physiological effects: the microcystins, produced by *Microcystis, Anabaena, Oscillatoria*, and *Nostoc* and the nodularins, produced by *Nodularia*.

Microcystins and nodularins are cyclic peptides containing a unique C_{20} amino acid, 3-amino-9-methoxy-2,6,8-trimethyl-10-phenyl-4,6-decadienoic acid, abbreviated Adda (6). Besides Adda, their molecular structures are very similar with respect to the sequence of amino acids except that microcystins are composed of seven residues, whereas nodularins have five residues (Figure 1).

A. Microcystins

The general structure of microcystin is cyclo(-D-Ala-X-D-MeAsp-Y-Adda-D-Glu-Mdha-) in which X and Y are variable amino acids, the composition of Adda is as described above, and Mdha stands for methyldehydroalanine. The two acidic amino acids D-MeAsp and D-Glu are connected by isolinkages.

Over 50 microcystins have been identified. All have the cyclic heptapeptide structure, but they may differ in the degree of methylation, configuration of Adda, or in amino acid composition at X and Y positions (5). Most microcystins are toxic, although there is a range of toxicity that can be related to structure. For instance, esterification of the D-glutamic acid and isomerization of Adda yield molecules with significantly reduced toxicity (5). Microcystins are named according to the variable residues in positions X and Y. One of the most active compounds, and one that is most frequently encountered in blooms, is microcystin-LR which has L-leucine and L-arginine in X and Y positions. When intraperitoneally injected in mice, microcystin-LR has an LD50 of 50–100 μg/kg body weight (7–9).

Substitutions in positions X and Y are common and lead to minor differences in toxicity compared to microcystin-LR. However, changes in amino acid composition can alter the hydrophobicity of the molecule, and thus modify its solubility and cellular permeability. Several microcystins containing methyldehydrobutyrine instead of the more common methyldehydroalanine were recently identified (10–12). It was noted that the methyl group of dehydrobutyrine did not significantly affect the toxicity of the molecule (11). Seven microcystins containing dehydroalanine or L-serine instead of N-methyldehydroalanine and glutamic acid and/or its methyl ester at the variable amino acid positions have been isolated (13).

B. Nodularin

The nodularin pentapeptide, cyclo (-D-MeAsp-Y-Adda-D-Glu-Mdhb-), is analogous to microcystin except that in nodularin there is an excision of -D-Ala-X- residues from the macrocycle and a methyldehydrobutyrine residue (Mdhb) replaces the methyldehydroalanine of microcystin (14). The first nodularin reported, isolated from brackish water *Nodularia spumigena*, had arginine in the Z (Figure 1) position (15). Nodularin-R is the most common nodularin encountered in cyanobacterial extracts and is generally referred to simply as nodularin in the literature. Subse-

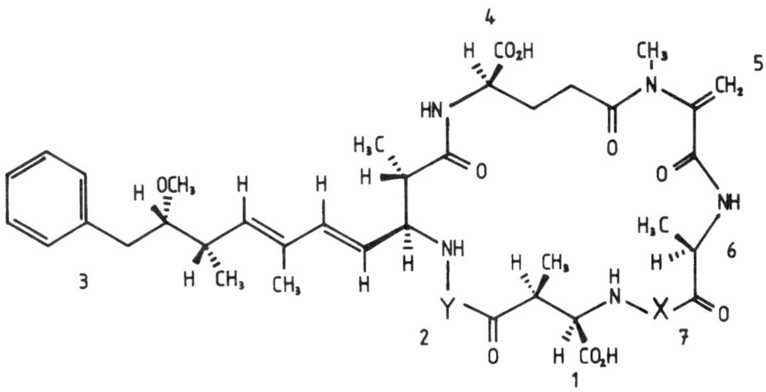

	AMINO ACID	
	X	Y
Microcystin-LR	Leu	Arg
Microcystin-LA	Leu	Ala
Microcystin-LL	Leu	Leu

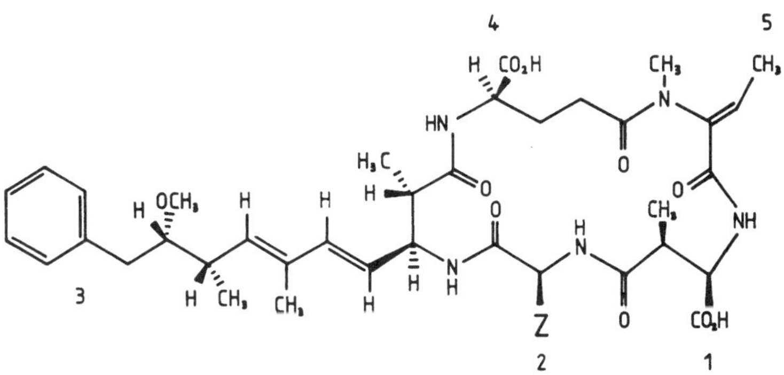

	SIDE CHAIN
	Z
Nodularin-R	Arg
Nodularin-V	Val

Figure 1 Primary sequences of selected microcystins (top panel) and nodularins (bottom panel). Residues for microcystin-LR are (1) β-linked D-erythro-β-methylaspartic acid (Masp), (2) L-arginine (Y), (3) β-(2S, 3S, 8S, 9S)-3-amino-9-methoxy-2,6,8-trimethyl-10-phenyldeca-4,6-dienoic acid (Adda), (4) γ-linked D-glutamic acid (Glu), (5) N-methyldehydroalanine (Mdha), (6) D-alanine, and (7) L-leucine (X). Residues for nodularin-V are (1) Masp, (2) L-valine (Z), (3) Adda, (4) D-Glu, and (5) N-methyldehydrobutyrine (Mdhb).

quently, the analogues nodularin-V and nodularin-I were isolated from a marine sponge (16). The investigators that identified nodularin-V called it motuporin, and it is often referred to by that name.

C. Toxicity of Microcystin and Nodularin

High doses of microcystin or nodularin cause blood to pool in the liver, producing hemorrhagic shock in affected animals. Chronic exposure to lower doses has been associated with tumor growth. In addition to their tendency to target the livers of exposed animals, microcystins and nodularins have a common molecular mechanism: the potent inhibition of serine/threonine protein phosphatases types 1 and 2A.

The sections that follow describe the biological ramifications of exposure to cyanobacterial hepatotoxins. Although significantly more research has been conducted using microcystins, in the instances where nodularins have been tested, it has been shown that the harmful effects of nodularins are similar to, and as potent as, those of microcystins (14,17,18).

D. Pathology of Microcystin Intoxication

Cyanobacterial intoxication, first reported in 1878 (19), involves lack of appetite, vomiting, diarrhea, stupor, convulsions, unconsciousness, and death. Reports of animal contacts with cyanobacteria are comprehensively reviewed elsewhere (20). A biochemical profile (21) of animals exposed to cyanobacteria was constructed using clinical data from sheep grazing the foreshore of Lake Mokoan in southeast Australia during cyanobacterial blooms in the summer and autumn seasons of 1991 and 1992. The toxicity of the blooms, dominated by *Microcystis aeruginosa*, was verified by intraperitoneal injections of cyanobacterial extracts in mice. Blood samples from the sheep were found to have high concentrations of bile acids and bilirubin and high activities of liver enzymes such as γ-glutamyl transferase and glutamate dehydrogenase compared to unexposed animals. The results indicated progressive liver damage and provided a convenient clinical profile for testing animals exposed to chronic, sublethal levels of toxin. Livers of several sheep were enlarged and engorged with fluid at necropsy.

Numerous studies of mammals injected with microcystins have shown that most of the toxin finds its way to the liver (22–27). In a typical experiment (27), 71% of injected tritium-labeled microcystin-LR was detected in the liver of mice 1 h after intraperitoneal administration. A single intraperitoneal injection of a lethal dose (50–100 μg/kg body weight) of microcystin-LR results in death within several hours due to liver hemorrhage and hepatic necrosis (9).

Microcystin is considerably less toxic when administered orally (26–28). In mice, LD50 values for microcystin-LR were 10.9 mg/kg when toxin was administered by oral gavage compared with 65.4 μg/kg by intraperitoneal injection (26). Although microcystin is less toxic when ingested, the frequency of animal deaths after drinking water contaminated by cyanobacteria indicates that, at sufficiently high doses, the peptide can pass through the intestinal tract into the bloodstream.

III. MECHANISMS OF MICROCYSTIN TOXICITY

A. Microcystin Uptake

Microcystins, particularly those with polar amino acids in the X and/or Y positions (for instance, microcystin-LR, microcystin-RR), are soluble in water. In general their polar nature prevents them from readily passing through cell membranes (28–30). This property may underlie the inefficacy of oral dosing. Organs other than liver do not accumulate significant quantities of microcystins in animals that have been treated with these toxins. This finding has been repeated many times in studies over the past decade and has led to the theory that microcystins are actively transported from the circulation into liver cells.

A number of agents which differ from each other chemically and biologically have been shown to protect animals from lethal doses of microcystin (9). These include the diazo dyes trypan red and trypan blue, silymarin, the antibiotic rifampin, and the immunosuppressant cyclosporin A. The uptake of [^{125}I]microcystin-YM in rat hepatocytes has been tested in the presence of rifampin (50 μM), cyclosporin A (5 μM), and the diazo dyes (20 μM) (9). It was concluded that microcystin toxicity is controlled by cellular uptake which was more or less inhibited by all of the chemoprotectants. The nature of the carrier itself, however, has not yet been identified.

B. Eukaryote Protein Receptors for Microcystin

Eukaryotes, from *Drosophila* to humans, produce proteins that have a particularly high affinity for microcystins and nodularins. The serine/threonine protein phosphatases, type 1 (PP1) and type 2A (PP2A), are enzymes implicated in a diverse array of intracellular processes such as glycogen metabolism, muscle contraction, cell cycle progression, and signal transduction (31). The catalytic subunits of PP1 and PP2A are inhibited by subnanomolar concentrations of cyanobacterial hepatotoxins. This is curious considering the vast evolutionary distance between higher eukaryotes and cyanobacteria. The fact that PP1 and PP2A are among the most highly conserved of known proteins (32) suggests that their relationship with the toxins might be ancient.

Potent inhibition of the catalytic subunits of PP1 and PP2A by microcystins and nodularins has been well characterized (18,33–35). The extreme sensitivities of the protein phosphatases to toxin inhibition has made it possible to measure microcystins and nodularins in quantities that are below the limits of detection in the mouse bioassay. When the high affinities between toxins and phosphatases were discovered, it was proposed that the inhibition of these enzymes provided a molecular basis for the harmful physiological effects of the toxins.

C. Structure Versus Activity of Microcystins and Nodularins

Structure-activity relationships for microcystins and nodularins were originally conducted with the mouse bioassay. Using this approach, chemically modified molecules and compounds isolated from natural sources were injected intraperitoneally into mice and the degree of toxicity was determined by the quantity required for a lethal dose to 50% of the animals. Some variants of microcystin and nodularin showed reduced toxicity in the mouse bioassay. With the discovery that compounds inactive in the mouse bioassay displayed a reduction in the ability to inhibit PP1 and PP2A (5,36,37), the highly sensitive protein phosphatase inhibition assay became a useful tool for evaluating potentially harmful compounds. As in the mouse bioassay, full activity in the protein phosphatase assay required the toxin to have an intact cyclic structure, a free D-Glu residue, and a 6(E) configuration of Adda.

D. Irreversible Toxin-Protein Bond

Evidence for the selective binding of microcystin-LR to PP1 and PP2A in animal and plant cell extracts was supported by experiments with other phosphatases that were unresponsive to the toxin (35). Furthermore, toxin binding to PP2A was prevented by okadaic acid, a polyether derived from a C_{38} fatty acid that potently inhibits PP2A. In addition, inhibition of PP1 by microcystin-LR was prevented by protein inhibitors of PP1, I-1, and I-2. Studies of microcystin-LR function in rat cytosol (38) indicated that the toxin formed complexes with components of the cytosol and could only be released under conditions of heat and protease activity. It was proposed that microcystin-LR was covalently bound to PP2A and that the site of complexation was the same as that of okadaic acid. Competition studies (39) supported the theory that okadaic acid and microcystin-LR bound PP2A in a mutually exclusive manner.

Evidence for the covalent nature of the bond between microcystin and PP1/2A accumulated over the next few years. Using [^{125}I]microcystin-YR and recombinant PP1, it was shown (40,41) that cysteine 273 of PP1 was linked covalently with the N-methydehydroalanine residue of microcystin in protein-toxin complexes. Mutation of cysteine 273 on PP1, or reduction of dehydroalanine on microcystin, abolished the formation of the covalent bond. Covalent binding between phosphatase and toxin was determined to be an event separate from inhibition of phosphatase activity (40,41).

An in vivo covalent association of microcystin with PP1 and PP2A was shown using antibodies specific to each phosphatase (41). After mice were treated with [^{125}I] microcystin-YR, liver extracts were mixed with antibodies raised against PP1 or PP2A. When immunoprecipitants were subjected to sodium dodecylsulfate–polyacrylamide gel electrophoresis (SDS-PAGE) and Western blotting, radioactivity was bound to the catalytic subunits of PP1 and PP2A.

Additional evidence for the importance of the glutamic and aspartic acid residues on microcystin for binding to protein phosphatases came from high-performance liquid chromatographic (HPLC) studies of microcystin-phosphatase complexes in our laboratory (42). Microcystin-LR, monomethylated at the free COOH group of glutamic acid or dimethylated at both glutamic and aspartic acids, was 100-fold or 1000-fold, respectively, less potent as an inhibitor of PP2A. As determined by reverse-phase HPLC, the methylated compounds showed no tendency to form covalent bonds with the phosphatase, supporting the idea that covalent interaction is dependent on initial binding of the toxin.

By now a mechanistic model of microcystin interaction with PP1/2A was coming into focus (Figure 2). The likely mechanism involved initial, fast (within minutes) binding of toxin to protein via microcystin residues Adda, glutamic acid, and, to a lesser extent, methylaspartic acid. This initial binding, concomitant with the inhibition of enzyme activity, was followed by a slower (over hours) covalent bond formation between cysteine 273 of PP1 (cysteine 266 of PP2A, according to sequence alignment with this homologous phosphatase) and the N-methyldehydroalanine group on microcystin.

E. Covalent Interactions with Protein Phosphatases: Differences Between Microcystins and Nodularins

In vitro studies (42) showed that covalent binding between PP1 or PP2A and microcystin-LL took place over several hours, although inhibition of phosphatase activity was complete within 10 min. Microcystin-LL has an IC50 = 0.2 nM for the inhibition of PP1 and PP2A, which is similar to that of microcystin-LR. Complex formation between protein and toxin was analyzed by reverse-phase HPLC. Although complexes of PP2A-microcystin-LR were distinguished as shoulders on parent protein peaks, PP2A-microcystin-LL was almost completely resolved from protein owing to the greater hydrophobicity of the microcystin-LL moiety. Experiments under similar conditions with nodularin-V (motuporin), an analogue of nodularin-R with an inhibitory potency equal to that of microcystin-LR, showed no sign of covalent complex formation after 96 h. Nodularins have an N-methyldehydrobutyrine group in a position corresponding to N-methyldehydroalanine on microcystin.

It was puzzling that microcystins and nodularins, equipotent inhibitors of PP1 and PP2A, were different in their chemical interactions with these proteins. The methyl group flanking the double bond of the dehydrobutyrine residue of nodularin is in a position to provide some struc-

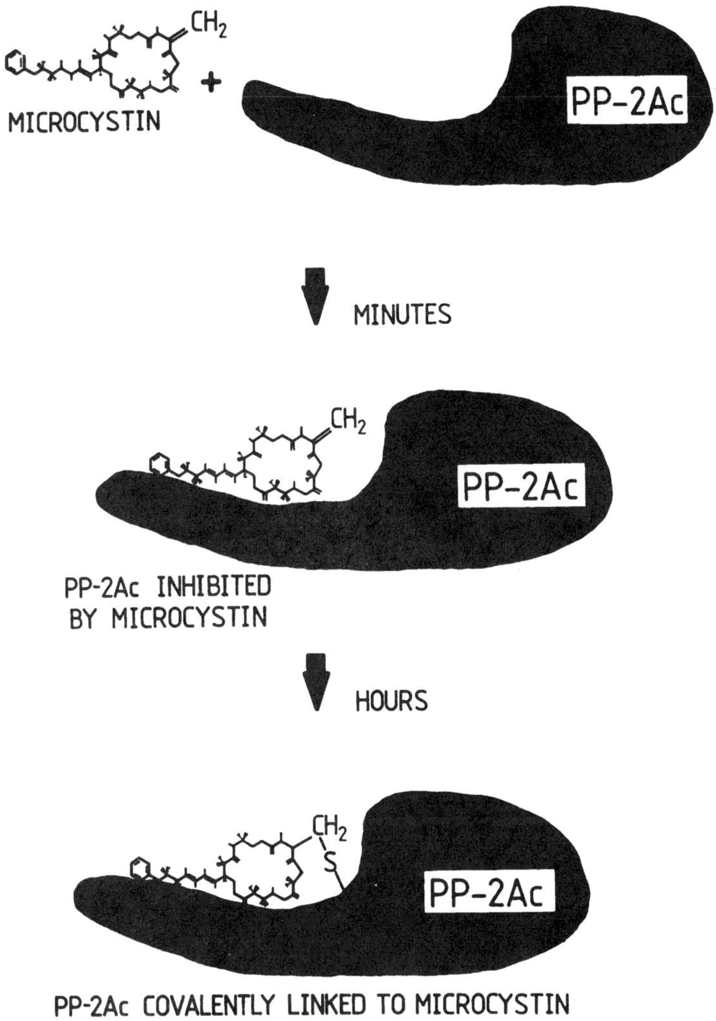

Figure 2 Proposed model scheme for outlining the time course by which microcystins interact with PP2A. This scheme also applies to the homologous enzyme PP1.

tural protection from interaction with the phosphatase, but it was difficult to understand the complete abrogation of covalent binding between nodularins and phosphatases. In 1995, determination of the solution structures of microcystin-LR and nodularin-V threw some light on the apparent paradox.

F. Structural Studies and Modeling of Toxin-Phosphatase Interactions

Nuclear magnetic resonance (NMR) solution structures of microcystin-LR and motuporin (nodularin-V) (43) showed that motuporin could potentially bind phosphatase in a way very similar to microcystin-LR. The residues essential to activity, namely, Adda, glutamic acid, and aspartic acid, could be superimposed to present a reactive surface to the protein. However, in this configuration, N-methyldehydroalanine of microcystin-LR and N-methyldehydrobutyrine of motuporin assumed quite different orientations. It was postulated that the apparent displacement of the dehydrobutyrine residue could account for the failure of nodularins to form covalent links with the phosphatases.

Microcystins in which the reactive double bond of dehydroalanine was reduced to dihydroalanine showed toxicity in the mouse bioassay similar to that of the parent compound (44). Studies with dihydromicrocystin-LA (42) indicated that the toxin retained full activity as an inhibitor of PP2A, although no covalent interaction between peptide and protein was observed. Other investigators (40) found that the IC50 value for reduced microcystin-YR, as an inhibitor of native PP1, was increased fivefold. In vivo effects of covalent complex formation between microcystin and phosphatase have not been determined.

The crystal structure of microcystin-LR bound to the catalytic subunit of recombinant PP1 was published in 1995 (45). It showed that the conformation of bound crystalline microcystin was similar to the solution structure of free microcystin, and it verified the presence of a covalent association between cysteine 273 of the protein and the dehydroalanine residue of the peptide (Figure 3). Using the structure coordinates of PP1 bound to microcystin-LR and the solution structure of motuporin (nodularin-V), a motuporin-protein complex was modeled with motuporin in a position corresponding to microcystin-LR bound to the protein (46). According to this model, the potentially reactive dehydrobutyrine group of motuporin was at least 8 Å from cysteine 273 of PP1, a distance not conducive to chemical bonding (Figure 4).

The striking difference between microcystins and nodularins, with respect to their ability to form covalent bonds with PP1 and PP2A, may underlie differences in their in vivo activities. Molecular modeling (46) predicts that the reactive dehydrobutyrine group of motuporin lies on the surface of the toxin-protein complex where it is theoretically available to react with other molecules in the cell. Although superficially both microcystins and nodularins appear to produce the same physiological degeneration, the effects of the toxin-protein complexes themselves have not been investigated. Microcystins and nodularins are well-characterized tumor promoters;

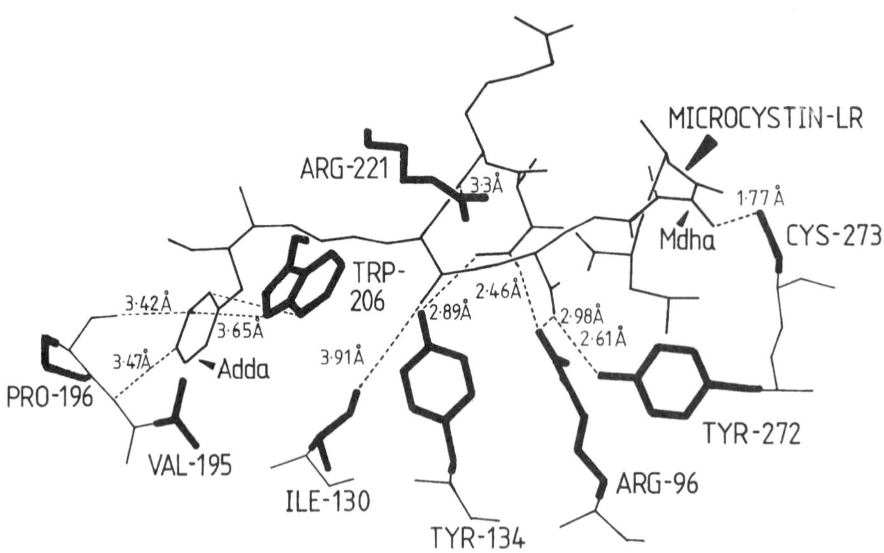

Figure 3 Essential interactions between microcystin-LR and PP1 catalytic subunit (PP1c), as deduced by x-ray crystallography. Principal residues from PP1c that interact with microcystin-LR are indicated. The hydrophobic Adda group on the toxin packs in a groove formed by a set of hydrophobic residues. The distances between key points of interaction are shown. Mdha is N-methyldehydroalanine on the toxin.

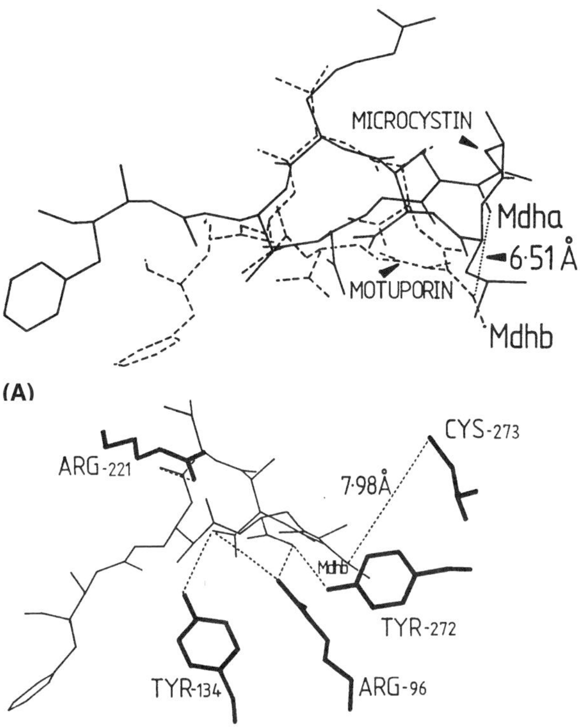

(B)

Figure 4 Panel A depicts superimposition of the NMR solution structure of nodularin-V on to microcystin-LR from the crystal structure of "toxin-bound" PP1 catalytic subunit (PP1c) (via Biosym Insight II). Panel B depicts possible PP1c contact residues in the hypothetical nodularin-V-bound PP1c complex. Mdhb denotes N-methyldehydrobutyrine in nodularin-V, whereas Mdha denotes N-methyldehydroalanine in microcystin-LR. Coordinates for PP1 and microcystin-LR were taken from (45).

however, some evidence for a role in tumor initiation by nodularins has been presented (47,56). These findings will be discussed further in Section IV.C.

IV. BIOLOGICAL EFFECTS OF CYANOBACTERIAL HEPATOTOXINS

A. Intoxication and Cell Architecture

Because PP1 and PP2A have high affinities for microcystins and nodularins (IC50 values in the 0.1–0.2 nM range), toxins that gain entry to cells are expected to be bound by available phosphatases. Thus, the biological effects of the toxins have reasonably been ascribed to their interactions with these enzymes. Histological symptoms of cyanobacterial intoxication are typically localized to hepatocytes; they include loss of normal cell-to-cell adhesion, disruption of intracellular architecture, intrahepatic hemorrhage, and finally cell death. Several investigators have observed a correlation between the breakdown of structural elements such as microfilaments, intermediate filaments, and microtubules and the hyperphosphorylation of intracellular proteins (48,49). The importance of phosphorylation in the regulation of cytoskeletal organization is well documented, but as yet the actual phosphorylation sites have not been identified.

In one study (50), immunofluorescence microscopy was used to compare microcystin-LR–induced cytoskeletal changes in cultured rat hepatocytes with changes in kidney epithelial cells and skin fibroblasts. It was noted that all cell types exhibited similar changes provided that nonhepatocytes were exposed to higher concentrations of toxin for longer periods. In a typical experiment, nonhepatocytes were incubated in 133 μM microcystin-LR for up to 24 h and hepatocytes were treated with 13.3 μM toxin for up to 1 h. In all cell types, microfilaments, intermediate filaments, and microtubules were observed to condense and collapse around the nucleus, although there were differences in the sequence of cytoskeletal events between hepatocytes and nonhepatocytes. The investigators speculated that the disparity might be due to differences in phosphorylated proteins between cell types.

B. Apoptosis

Microcystins have been implicated in apoptotic, or programmed, cell death (26,51,52). Characteristic features of apoptosis are condensation of chromatin, the appearance of apoptotic bodies, and fragmentation of DNA. Microcystin and nodularin at concentrations of 1 μM induced apoptosis in less than 2 h in freshly isolated hepatocytes from Atlantic salmon and rat (52). Apoptotic changes were determined by light and electron microscopy and were differentiated from cell necrosis by a lack of trypan blue uptake, cell surface alterations, and chromatin condensation typical of apoptosis. In addition, an apoptotic protease inhibitor (caspase) reduced the effects of nodularin.

C. Tumor Promotion and Initiation by Microcystin and Nodularin

Acute effects of microcystins and nodularins are well established and are similar for each class of toxin. That is, the lethal levels of the toxins are comparable, liver damage has been shown to be the principal cause of illness, and the inhibition of protein phosphatases is an important, probably primary, determinant of biological harm. Results of prolonged exposure to low levels of toxin have been harder to define. In the earliest studies (53), mice developed liver tumors after being supplied for up to 1 year with drinking water which contained sublethal doses of *Microcystis* extract. Later, tumor promotion was demonstrated on the skin of mice that drank water containing *Microcystis* extract after tumors were initiated with dimethylbenzanthracene (54). Tumor growth was also enhanced in rat livers when animals, pretreated with the initiator diethylnitrosamine, were injected intraperitoneally with microcystin-LR (55).

The properties of nodularin, with respect to structure, liver specificity, hepatotoxicity, and phosphatase inhibition, are similar to those of microcystin. In addition, nodularin has been shown to promote tumors in rat liver more effectively than microcystin (47,56). Tumor growth after exposure to nodularin, but not microcystin, was demonstrated in the absence of initiating chemicals (56). Tumor foci were measured in rat liver after animals received intraperitoneal injections of 200 mg/kg body weight of the carcinogen diethylnitrosoamine (DEN) followed by intraperitoneal injections of 25 μg/kg body weight microcystin or nodularin. Starting on the third week after DEN treatment, toxin was injected twice a week for 10 weeks over a period of 12 weeks. Both toxins increased the production of foci after treatment with DEN, with nodularin being about eight times as effective as microcystin-LR. In addition, a control group treated with nodularin alone had a higher volume of foci per liver than a group treated with DEN alone. The investigators attributed the difference in physiological effects between microcystin and nodularin to a difference in structure. They suggested that the smaller, five-membered ring of nodularin was more easily cell permeable that the seven-membered ring of microcystin. The significant difference in size might reasonably be a factor, but the difference in chemical interactions with the phosphatases—the fact that nodularin, unlike microcystin, does not form a covalent bond with PP1 or PP2A (42)—could be pertinent (see Section III.F).

However, recent experiments with microcystin in mouse liver demonstrated nodular formation in the absence of initiating chemicals (28). Microcystin-LR intraperitoneally injected at 20 μg/kg body weight, 100 times over 28 weeks, produced nodules up to 5 mm without pretreatment with initiator. The extent and type of injury was dependent on the level of dosage and the time over which exposure took place. Microcystin-LR treatment, up to 500 μg/kg, by intragastric intubation produced liver nodules only in aged mice. Tumor promotion and initiation by cyanobacterial hepatotoxins remains a subject for further study. Meanwhile, it is comforting that oral exposure has been confirmed repeatedly to be fairly innocuous.

D. Mutagenicity of Microcystin

Questions about the chronic toxicity of microcystin were addressed in early studies (57) in which it was reported that purified *Microcystis* was nonmutagenic based on the Ames *Salmonella typhimurium* mutagenicity test (58). Later, the Ames test and the *Bacillus subtilis* multigene sporulation test (120) were used (59) to look for mutations in cells grown in the presence of purified hepatotoxin and algal extracts of *Microcystis* previously found to be toxic in the mouse bioassay. In the sporulation test, 35 gene sites are surveyed for mutations compared to 1 site in the Ames test. In spite of the presence of high concentrations of hepatotoxin, up to 0.9 mg/mL, no mutagenicity was observed, although there were signs of cell toxicity at the highest dose in the sporulation assay. Hepatotoxins were also tested against human lymphocytes (59), and it was tentatively concluded that hepatotoxins are clastogenic, as an increased degree of chromosomal breakage was observed in exposed cells. Interpretation of the data was cautious owing to the presence of a high background of chromosomal breakage. The hepatotoxins were not identified in this study, so it is not possible to evaluate their hydrophobicity. However, as discussed above, the most common microcystins contain the polar residue arginine as one of the variable amino acids and are not generally cell permeable. Therefore, the absence of mutations detected in the bacterial cells treated with hepatotoxin could be a result of the inability of the toxins to enter the cells. It is possible that the human lymphocytes, which appeared to show effects from the toxins (59), were more permeable to hepatotoxins under the conditions of these experiments.

Recently, in experiments employing a hypermutable human cell line, RSa, a significant degree of mutagenicity was detected after cells were exposed to high concentrations of microcystin-LR (60). RSa cells exhibit hypermutability when mutation is evaluated as resistance to cell killing by drugs (e.g., ouabain). Cells were treated with toxin for 24 h and then tested for ouabain resistance and for base substitutions at K-*ras* codon 12 of genomic DNA. Mutagenic effects were observed with concentrations of microcystin-LR of 7.5–15.0 μg/mL. No effects on RSa cells, either in survival time or in ouabain resistance, were detected at microcystin concentrations less than 7.5 μg/mL (60). These conditions can be compared with those in which no mutagenic effects were seen in bacterial cells treated with 900 μg/mL purified hepatotoxin (59). Effects on hepatic cells, which apparently have uptake mechanisms for microcystin, would conceivably be observed at much lower concentrations. In mammals, it has been definitively shown that, whether ingested or injected, peptide toxin that enters the bloodstream accumulates in the liver. The risks of human exposure to hepatotoxins are discussed in Section VII.

Microcystins are sometimes referred to as being in the okadaic acid class of tumor promoters. Members of this class of molecules include okadaic acid, the polyether derived from a C_{38} fatty acid, and the spiro ketal calyculin A, as well as the polypeptide hepatotoxins microcystin and nodularin. Although diverse in structure, these molecules are all potent inhibitors of the serine/threonine protein phosphatases, PP1 and PP2A. Okadaic acid and calyculin A have been shown to promote tumor growth in mouse skin after initiation with dimethylbenzanthracene (DMBA) (62). In these experiments, the tumor promoter was applied directly on the skin rather than added to the animals' drinking water, as in experiments with microcystin (54). DNA from DMBA/okadaic acid–treated and DMBA/calyculin A–treated mice revealed the same mutation (62). In this case, the mutation was traced to the second nucleotide of codon 61 (CAA → CTA) in the c-Harvey-*ras* gene. The investigators concluded that the mutation was a direct effect of DMBA rather than a selective effect of specific tumor promoters. In the human cell line treated with microcystin (60), mutations were detected at K-*ras* codon 12 in the absence of an initiating compound. This suggests that microcystin can act as a mutagen and future work will be followed with interest.

E. DNA Damage by Microcystin

In recent studies (61,63), DNA damage was detected in mouse liver homogenates following administration of lethal, intraperitoneal doses of microcystin-LR. It is possible that the DNA damage was due to biochemical events related to cell necrosis, but the investigators claimed that DNA damage was measured before liver necrosis and leakage of hepatic enzymes could be detected (61). They proposed that nucleases responsible for early degradation of DNA may be involved.

Nonhepatic cells were also tested with microcystin-LR (61). Baby hamster kidney cells and mouse embryo erythrocytes that were incubated up to 3 h in 1 μg/mL toxin showed significant mortality with concomitant DNA damage. Again the DNA damage may have been a consequence of cell death by any means, but this was an interesting demonstration of the effects of microcystin on nonhepatic eukaryotes. These results are comparable to those of human lymphocytes referred to earlier (59) in Section IV.D that were affected by 0.9 μg/mL toxin. In contrast, human RSa cells (60) were affected only at microcystin concentrations of 7.5 μg/mL after 24 h. Signs of toxicity in bacteria (59) were observed only at hepatotoxin concentrations of 900 μg/mL.

In the absence of active transport mechanisms, permeability to microcystin may be affected by conditions of cell preparation. Bacterial cells, because of their need to survive in unprotected surroundings, would be expected to have more resistance than kidney cells or lymphocytes to exogenous substances. However, it is also possible that the bacteria were resistant for other reasons. The most obvious reason would be the absence in bacteria of PP1/2A–like enzymes. However, representatives of the PPP family of protein phosphatases, of which PP1 and PP2A are members, are being detected in growing numbers in prokaryotes (64). Many of these newly discovered enzymes have shown resistance to classic inhibitors such as microcystin and okadaic acid, but some have displayed moderate sensitivity to these compounds.

F. Protection Against Hepatotoxins

There is no known antidote to cyanobacterial poisoning. Although several compounds may afford some protection, they must be administered before toxin contact to be effective. Pretreatment of mice with protectants against microcystin uptake (9) was discussed earlier in Section III.A. Other work (65) showed that glutathione and cysteine may also afford some protection from microcystin. These sulfides react covalently with the dehydroalanine residue on the toxin. Glutathione and cysteine conjugates of microcystins-LR and -YR were severalfold less toxic than the uncomplexed toxins. Glutathione and cysteine complexes of microcystin may be less toxic because of steric hindrance of residues essential to protein binding. As mentioned before, the dehydroalanine group is not vital to the mechanism of inhibition.

Several other agents, administered intraperitoneally before the toxin, were tested for their ability to protect mice against lethal doses of microcysin-LR (63). High doses, 2 g/kg body weight, of glucose and mannose moderately increased survival time. It was proposed that their mechanism of protection may be dilution or inactivation of toxin in the peritoneal cavity. Glutathione was effective at 4 g/kg, but 250 mg/kg of N-acetylcysteine did not increase the survival rate of the mice. Dosages of glucose, mannitol, and glutathione were based on work done previously by another group (66). Glutathione protection may be rationalized by its demonstrated covalent binding to the dehydroalanine residue of microcystin-LR (65) and probable steric hindrance of toxin-phosphatase binding. N-acetylcysteine, smaller than glutathione, may provide less steric hindrance. LD50 values (i.v., mice) for microcystin-LR, microcystin-LR–glutathione and microcystin-LR–cysteine of 38, 630, and 267 μg/kg, respectively, have been reported (65). Thus, the smaller cysteine conjugate is more toxic than the glutathione conjugate. This agrees with the idea that the reducing sulfide compounds act sterically to hinder binding of the toxin to the phosphatase.

G. Cyanobacterial Toxins in Seafood

Hepatotoxins accumulate in the livers of exposed mammals and fish and toxins have been found in aquatic organisms such as sponges (16), crab larvae (67), mussels (68,69), and copepods (CFB Holmes and TLM McCready, unpublished data). Fish exposed to cyanobacteria undergo physiological symptoms similar to those of mammals; toxin targets the liver and fish sicken and die, apparently due to hepatocyte necrosis. Farmed fish are particularly sensitive to liver disease. Fish mortality from the so-called net-pen liver disease (NLD) has caused severe economic losses in farms raising Atlantic salmon on the west coasts of Canada and the northern United States. On autopsy, fish with NLD symptoms display liver injury and traces of microcystins not found in healthy fish (70). Suspecting that hepatotoxins from cyanobacteria cause NLD, investigators exposed healthy Atlantic salmon to microcystins and generated all the symptoms characteristic of NLD (71).

Because farmed fish are primarily raised for human consumption, the concern over net-pen liver disease relates directly to human health. After observing in preliminary work that Atlantic salmon appeared to tolerate higher levels of microcystin-LR than mice (70), this theory was tested by intraperitoneal administration of 33 mCi/mmol [^{3}H]dihydromicrocystin-LR, 1 mg/kg body weight, to Atlantic salmon (71). Salmon were sacrificed at several time points over 46 h and tissue distribution of radiolabel was estimated. The investigators concluded that Atlantic salmon metabolized the toxin much more rapidly than rats or mice. The maximum quantity of toxin detected, about 5% of the injected dose, was found in the fish liver 5 h after treatment compared to 50–70% measured in similar experiments with mice (22–24,27,72). In mice after 8 days, the liver and carcass retained 50 and 8% of incorporated radiolabel, respectively (72). In salmon after 46 h, the whole body incorporation of radiolabel was 12%. A small quantity of toxin was found in fish kidney 46 h postinjection, and the maximum quantity found in flesh was never more than one twentieth of the maximum liver load. A similar study (73) used oral gavage to introduce 247 mCi/mmol [^{3}H]microcystin-LR, 0.64 μg/kg body weight, into trout and found a significant proportion of toxin in flesh, although most was found in liver samples and none in kidney. The low level of toxin measured in this experiment, 0.3% (in liver and muscle) of injected toxin 24 h postgavage, suggests that in fish, as in mammals, oral administration is much less effective than intraperitoneal injection as a route to intoxication.

[^{3}H]dihydromicrocystin-LR, used in the experiments just described (71,73), was unable to react covalently with protein phosphatases, as the dehydroalanine residue was reduced to the unreactive dihydroalanine during the process of radioactive labeling. The absence of covalent interaction may either hasten or slow the process of microcystin degradation and elimination. Thus, the rates of transport, distribution, and clearance of dihydromicrocystin-LR found in these experiments might differ from those of unmodified dehydromicrocystin-LR.

The apparent resistance of fish to microcystin is supported by a recent cell study (52). Primary cultures of hepatocytes from Atlantic salmon were two to five times less sensitive to microcystin-LR and nodularin compared to cultures from rats. Cells were incubated for 90 mins in 1 μM (equivalent to 1 μg/mL for microcystin-LR) toxin, then stained and observed for necrotic and apototic changes by light and electron microscopy. Previous work in fish (70,71) suggested that the apparent resistance of Atlantic salmon to microcystin-LR was due to rapid clearance of the toxin, but inefficient uptake of toxin by salmon hepatocytes (52) also appears to be an important factor.

Human ingestion of shellfish and other filter feeders is a potential health hazard if the animals are collected during, or shortly after, a cyanobacterial bloom (74). In an effort to assess this risk, mussels were fed cyanobacteria that contained high levels of microcystin over a period of 3 days, a time period typical of a cyanobacterial bloom, then removed to fresh seawater, and sampled over a 2-month period (69). The Lemieux oxidation/gas chromatography/mass spectrometry (GC/MS) method (93–95) (see Section V.D for a discussion of the use of this method) was used to measure the total toxin burden. After an initially high level of microcystin, 337 μg/g mussel tissue, the levels of toxin were reduced to below detection limits, 10–20 ng, over 4 days. However, other investigators (75), using butanol/methanol/water extraction coupled with HPLC detection, found that toxin persisted in diminished but detectable amounts, 3–12 μg/mussel, for 2 months after exposure.

To our knowledge there are no documented illnesses or deaths of humans from the ingestion of seafood contaminated with hepatotoxins. According to current information, intoxication from the ingestion of seafood can probably be avoided by consuming animals that appear to be healthy and have not recently been associated with a cyanobacterial bloom. The risks of chronic, low-level consumption of cyanobacterial toxins are unknown. Furthermore, it is not known if toxin bound reversibly or irreversibly to protein phosphatase in seafood tissues is a threat to consumers.

V. ALGAL BLOOMS: PROLIFERATION AND DEGRADATION OF HEPATOTOXINS

A. Growth and Propagation of Cyanobacteria

Cyanobacteria are a normal component of both fresh and marine water sources. Species that produce hepatotoxins flourish mainly in fresh or brackish open waters but not in deep wells owing to their photosynthetic nature. They proliferate in seasons of warm temperatures and still air. Periods of exponential growth, or blooms, occur when nutrients, particularly nitrogen and phosphorus, are plentiful (1,20). Sewage, agricultural and industrial wastes, and other effluents induced by human activity provide fertile breeding conditions in exposed waters. Planktonic forms of cyanobacteria have the ability to rise and fall in the water to follow favorable conditions of light and nutrients. They usually propagate in the midst of water bodies such as lakes and reservoirs, but they are easily blown by prevailing winds. Thus, they may be moved near water-intake pipes of municipal reservoirs or to shorelines where animals are watered or where recreational activities take place. The potential variety of cyanobacterial species in blooms makes the possibility of toxicity unpredictable and best determined by laboratory tests.

B. Degradation of Hepatotoxins

The cyclic peptide hepatotoxins synthesized by cyanobacteria are harmless as long as they are confined to the cells that produce them unless the cells themselves are consumed, as is frequently the case when animals drink from infected waters. However, an increase in cell turnover is expected in a bloom situation, with concomitant leakage of toxins into the water from dying cells. Under natural conditions, maximum levels of toxin are measured when the bloom is fading and cells are dying. We have found that high levels of toxin associated with blooms fall over several days after the blooms fade (CFB Holmes and HA Luu, unpublished data). Another way in which toxins can be released in high quantities into the water is by treating blooms with copper sulfate or calcium hydroxide, substances which are effective in killing cyanobacterial cells (76,77). There is some question that these chemicals themselves have toxic properties in the concentrations required for efficient bloom elimination (78). Much of the expertise in dealing with cyanobacterial blooms has come from research done in Australia where there has been extensive experience with this problem due to the dry warm climate. In a study of microcystin degradation (79), the investigators report that water is often withheld from the reticulated supply for 7 days after chemical treatment to kill a bloom of cyanobacteria, as there is some evidence that toxins are degraded in this time. However, they conclude from this study that 7 days may not be sufficient to reduce toxin to safe levels (also see ref. 80). Most of the work on hepatotoxin degradation has been done with microcystin, but a profile of nodularin degradation has recently been published (123).

Early work with microcystins showed that these small cyclic peptides are extremely stable in distilled water (81,82). Furthermore, they prevail over the short term through wide ranges of pH and temperature and are only degraded completely under reflux in the presence of strong acid (83). We have noticed some degradation after exposure to high concentrations of chymotrypsin, an effect that was more pronounced with more hydrophobic microcystins, but it has been reported that they are resistant to enzymatic hydrolysis by trypsin, although no conditions were given (83). The half-life of microcystin-LR in natural water has been reported to be in the range of several days (77,80,84).

Since microcystins are stable in pure solutions, yet tend to disappear rapidly from natural water sources, several investigators have attempted to determine the factors responsible for depletion of the toxins. A recent review (83) discusses what has been achieved to date and provides a comprehensive list of references. The investigators propose five paths by which toxins may be dissipated and degraded: dilution, adsorption, thermal decomposition aided by temperature and pH, photolysis, and biological degradation. Work in this area is relevant to assessment of the risk associated with contaminated water and to the treatment of water for consumption. Some examples follow.

There is evidence that toxins are adsorbed by sediments present in native waters (83,85). Under purified conditions, a temperature of 40°C had a dramatic effect on microcystin-LR degradation at pH 1, a moderate effect at pH 9, but very little effect between pH 5 and 7 (86). A degradation metabolite detected was determined to be a linear peptide produced by hydrolysis at the methyldehydroalanine moiety. Although the degradation product was not tested for biological activity, linear peptide products of microcystin hydrolysis have previously been found to be inactive (87,88).

In the presence of cyanobacterial pigments, the Adda group of microcystin is isomerized by sunlight (89). Irradiation by sunlight, but not fluorescent light, over 15 days in the presence of pigments, particularly β-carotene, produced photolytic products which were identified as geometrical isomers involving the diene portion of the Adda group of microcystin (90). As discussed before, geometric isomers at the (E) C-6 double bond of Adda are nontoxic (5). Nontoxic 4(Z)-Adda- and 6(Z)-Adda-microcystins were also formed by ultraviolet (UV) irradiation (121).

A study of biological degradation of microcystins (80) had interesting implications for the stability of toxins under typical sample storage conditions. Water was collected from a lake contaminated with cyanobacteria and incubated at 4°C in the dark for several days. The release of 9 and 15% microcystin was observed after 3 and 7 days, respectively. Microcystins were separated by HPLC and identified by mass spectrometry. After 3 days, a portion of the lake water was filtered to remove cellular material. The filtrate was separated into two portions and stored for an additional 42 days: one portion at 4°C, the other at 20°C. The concentration of toxin remained almost constant up to 14 days at 4°C and then began to decrease more rapidly. Only 10% microcystin remained after 42 days at 4°C; no microcystin was detected after 42 days at 20°C. Similar release and degradation kinetics have been reported (77,79,82).

Considering the results reported above, it is probable that the eventual disappearance of microcystin, after its apparent stability for 14 days, was due to bacterial digestion. A similar lag phase was observed in microcystin depletion from an extract of *M. aeruginosa* added to natural waters (91). The investigators concluded that bacteria were the causative agents. They speculated that the lag was due to other carbon sources being exhausted before the peptides were attacked or to catabolite repressors present in the crude extract.

Two aspects of these results pertain to the management of drinking and recreational water supplies. First, since extremes of heat and pH are rarely sufficient in natural waters to ensure the degradation of cyanobacterial toxins after bloom kills, and the presence of metabolizing bacteria is bound to be variable, the only way to find out if toxins are above safe limits is laboratory testing. Second, samples to be tested for microcystin equivalents should be handled in a way that is appropriate to the type of measurement desired. Unfiltered samples from raw water may contain cyanobacterial cells whose contents may be released on freezing or on storage, increasing microcystin levels in the analysis. Raw water may contain bacterial cells from different species capable of degrading microcystins or of killing cyanobacteria. Thus, the effects of filtration, centrifugation, freeze/thaw, and storage procedures must be considered when results of the analysis are interpreted.

C. Metabolism and Clearance of Cyanobacterial Toxins: Problems of Quantitation

There have been numerous studies in which microcystins have been injected into mammals or fish and attempts made to measure in vivo toxin and its metabolites. In most of the early work, it was not known that the toxin might be covalently bound to a protein phosphatase; therefore methods were not tailored to detect toxin under these conditions. For instance, microcystins are famously heat-stable molecules and some published protocols reasonably employed heat denaturation followed by precipitation of protein components to isolate and partially purify in vivo toxin in liver extracts of treated animals. This procedure is likely to have precipitated any covalent protein complexes that may have been formed by toxin and phosphatase. In other procedures, monoclonal antibodies raised against microcystin-LR were used to separate microcystins from crude cell extracts without previously determining whether these antibodies had the ability to recognize the toxin bound to phosphatase.

Methanol extraction is frequently used to isolate microcystins, a method that would not be able to account for toxin bound covalently to protein (see Section V.D). In addition to extraction in methanol, samples are often transported and stored in this organic solvent. This is a good way to eliminate contaminating bacteria, but the spontaneous methylation of microcystins has been shown to take place under acidic conditions (86). Since methylation of acidic groups on microcystins reduces their toxic (and phosphatase inhibition) potential (5,37,42), caution must be observed if these conditions are present when toxins are to be identified by a toxicity (or phosphatase inhibition) assay.

Finally, tritiated microcystins (in which the reactive double bond of the dehydroalanine residue is reduced to dihydroalanine with sodium boro[^{3}H]hydride in order to insert the labeled hydrogen) have been employed many times to track toxin levels in the organs of treated animals, to identify toxin metabolites, and to measure clearance rates. Since tritiated dihydromicrocystin is unable to bind covalently to its natural protein receptors, toxin distribution, metabolism, and clearance rates may not be representative of native microcystin. In vitro covalent bond formation between microcystin and phosphatase takes place over a period of hours. The presence of toxin in the livers of animals exposed to microcystin has been demonstrated over a period of hours. Thus, in vivo covalent linkage of toxin to protein is a likely event resulting in some portion of administered toxin being bound irreversibly to protein phosphatase.

D. Measurement of Covalent Binding of Microcystin-LR In Vivo

As discussed previously, the fact that microcystins bind covalently to their intracellular phosphatase receptors must be considered in the interpretation of data from experiments in which animals have been fed or injected with these toxins.

In order to assess the quantity of toxin associated irreversibly with protein in vivo, one group administered ^{14}C-labeled microcystin-LR intraperitoneally to Atlantic salmon and measured the radioactivity in samples of liver after either digestion with potassium hydroxide (KOH) or extraction with methanol (92). The levels of radioactivity detected in digested samples were two to three times higher than levels detected in samples that had been extracted with methanol. ^{14}C-labeled microcystin-LR is fully capable of interacting covalently with PP1 and PP2A. Assuming that toxin bound covalently to protein would not be extracted with methanol, radioactivity measured after KOH digestion should reflect the total toxin burden, whereas radioactivity measured after methanol extraction would be expected to account for free, or reversibly bound, toxin. These results provided evidence for covalent binding of microcystin-LR in salmon liver.

In another attempt to measure microcystin bound covalently in vivo, salmon liver from fish treated with microcystin-LR was again processed by two different methods and the results were compared (67). In this experiment, half of the samples were treated by Lemieux oxidation and half by the methanol extraction procedure discussed above. In the Lemieux reaction, the Adda group of microcystin is modified and cleaved from the toxin molecule. It can then be separated by gas or liquid chromatography (LC) and identified by MS (93–95). Lemieux/GC/MS analysis theoretically measures the total burden of molecules that contain Adda, free or bound, in the sample. After 7 h, 31–34% of the total microcystin injected was detected in salmon liver by GC/MS after Lemieux oxidation, whereas only 1–8% was detected by protein phosphatase inhibition assay after methanol extraction. Only 24% of the total microcystin-LR present in the salmon liver was extractable with methanol. In a similar experiment using Dungeness crab larvae, 10,000-fold greater microcystin concentrations were measured in samples analyzed by Lemieux/GC/MS compared with analysis by the methanol/phosphatase inhibition assay (67). The investigators concluded that the significant difference in results between Lemieux/GC/MS and methanol extraction/phosphatase assay analyses was good evidence for the presence of covalently bound microcystin in vivo.

VI. DETECTION OF ELIMINATION OF HEPATOTOXINS IN RECREATIONAL AND DRINKING WATERS

A. Methods of Quantitation: Advantages and Limitations

Table 1 compares some of the most common methods of measuring microcystins and nodularins in recreational and drinking waters. This topic has recently been reviewed in ref. 96. As discussed below, the most accurate determinations are made when a combination of two or more techniques are utilized. Detection limits for purified standard samples are usually more favorable than limits encountered in raw samples. The mouse bioassay is the traditional method of detecting cyanobacterial toxins. Although unselective and insensitive, it has the advantage of being responsive to dangerous quantities of any biologically threatening substance.

The most sensitive way of detecting microcystins and nodularins is to use their potent ability to inhibit protein phosphatases. A radioactive assay which employs labeled protein substrates of PP1 and PP2A can detect subpicogram quantities of microcystin or nodularin, but obtaining pure, biological substances and labeling them is time consuming and costly. Recombinant PP1 efficiently dephosphorylates paranitrophenyl phosphate (*p*NPP), a small molecule whose dephosphorylated form can be detected by its optical density at 405 nm (*p*NPP is not a good substrate for the native enzyme PP1 (97) or for PP2A from some sources [98,99]). The ability of the catalytic subunit of recombinant PP1 to dephosphorylate *p*NPP has led to a convenient colorimetric phosphatase inhibition assay that can be used to detect phosphatase inhibitors at nanomolar concentrations.

When using phosphatase inhibition to detect microcystins and nodularins in native waters, where other substances that inhibit phosphatases may be present, it cannot be assumed that inhibition is due to cyanobacterial hepatotoxins. In an attempt to devise a method specifically to detect the cyclic peptide hepatotoxins, polyclonal antibodies raised against microcystin-LR

Table 1 Comparison of Detection Methods for Cyanobacterial Hepatotoxins

Method	Advantages	Disadvantages	Low detection range
Mouse bioassay	Detects all toxic substances	Nonselective, insensitive	1–200 μg
LC/MS	Selective, identification of toxin	Concentration of sample required, equipment cost	300–500 ng
LC/protein phosphatase assay	Selective, identification of toxin	Time consuming, equipment cost	300–500 ng
Protein phosphatase assay—radiolabeled	Selective, most sensitive method	Expensive, laborious, no identification of toxins	Sub picogram amounts
Protein phosphatase assay—colorimetric	Unselective, sensitive convenient	No identification of toxins, concentration of sample required	5–10 pg
Lemieux oxidation/ GC/MS	Specific to microcystin, nodularin	Laborious, equipment cost, nontoxic forms detected	10–20 ng
Immunoassay	Specific to microcystin, nodularin	Nontoxic forms detected	50–100 pg

LC, liquid chromatography; GC, gas chromatography; MS, mass spectrometry.

were used in an enzyme-linked immunosorbent assay (ELISA) (100). The antibodies bound to several microcystin and nodularin variants, including nontoxic forms. An arginine residue and an intact Adda group were essential for antibody specificity. The fact that nontoxic forms bind to the antibodies leads to false-positive results, lowering the resolution of this assay.

A combination of the colorimetric phosphatase inhibition assay, adapted from Takai and Mieskes, 1991 (122), and the ELISA (100), has been used to study structure-function relationships of several microcystin and nodularin congeners (101). It was found that the (E) configuration at the C-6 double bond of Adda was essential for antigenicity and that modification of the COOH of glutamic acid (producing a molecule with reduced toxicity) did not interfere with antibody recognition. Estimated IC50 values for microcystin-LR were reported as 3 ng/mL (3 μg/L) for the ELISA and 0.3 nM (0.3 μg/L) for the colorimetric assay (101). Thus, the sensitivities of these assays are in a range useful for detecting toxic levels of microcystin and nodularin in drinking water. Values of 0.5–1.5 μg microcystin-LR equivalents per liter have been suggested as safety guidelines by several experts in the field (2). In the combined protocol (101), false positives obtained by either the colorimetric assay or the ELISA were eliminated by the other, independent assay.

Reverse-phase LC separates hepatotoxins on the basis of their differing hydrophobicity. In one example of this technique (102), a combination of LC, protein phosphatase inhibition assay, and amino acid analysis was used to separate and identify components in the material from a cyanobacterial bloom. Individual toxins were identified by amino acid analyses of LC fractions that inhibited PP1. Several microcystins were characterized, including microcystin-LL, microcystin-LV, microcystin-LM, microcystin-LF, and microcystin-LZ (where Z represents an unknown hydrophobic amino acid). Although this combination of methods can theoretically identify toxins from any source, it is time consuming, expensive, and only practical when high concentrations of toxin are present, as in the material from a bloom.

Chemical methods are usually laborious and relatively insensitive. An example is the Lemieux oxidation (67,93–95) discussed in Section V.D. The Lemieux oxidation modifies and cleaves the Adda group from the toxin. Modified Adda can then be separated by chromatography and identified by mass spectrometry. This method relies on the unique nature of Adda to detect the presence of microcystin or nodularin. Detection limits for standard samples are 10–20 ng of toxin, but quantitation of field samples requires 5–20 μg for consistent results (D.E. Williams, personal communication). Because gas and liquid chromatographs and mass spectrometers are expensive items of equipment that require trained operators, they are valuable for research but impractical for routine water testing.

Since contaminated water is found in lakes, rivers, and reservoirs, field-testing kits would be of the greatest convenience. Such kits are under development in several laboratories and will likely employ the colorimetric protein phosphatase assay and/or immunoassay protocols.

The largest problem for routine monitoring of microcystins and nodularins in native waters is that they are typically present in quantities of 0.5–2 μg/L, which is near the detection limits of most methods. Thus, concentration of samples is often necessary in order to obtain consistently accurate estimates. This has the effect of concentrating other substances present in the samples which can interfere with measurement of the toxins.

B. Safe Limits for Hepatotoxins in Drinking Water

Contamination of drinking water sources by cyanobacterial hepatotoxins has escalated with increasing human populations. Effluents from sewage, agriculture, and industry encourage the growth of algal blooms in lakes and reservoirs, and traditional methods of water treatment often fail to inactivate or remove toxins produced by these blooms. Some of the difficulties in estimating a ''safe'' level for microcystin in drinking water have recently been reviewed (2). After studying the available data on cyanobacterial intoxication, the World Health Organization has recommended a guidance value for microcystin-LR of 1 μg/L (103). In Canada, the current recommended value is 1 μg/L, but the Health Ministry is considering a revision to 1.5 μg/L (M. Giddings, Health Canada, 1998, personal communication). Routine monitoring of drinking water for microcystin contamination is becoming widespread as awareness of the hazard grows. The suggested levels should be adequate to protect against acute intoxication, but they may need to be adjusted when the effects of chronic intake of low levels of hepatotoxins become known.

C. Cyanobacterial Bloom Management and Treatment of Drinking Water

Methods of dealing with cyanobacterial blooms and the treatment of drinking water supplies have been well reviewed (2,20,104). The most effective way of purifying contaminated waters is to use a combination of methods that can contend with the variable conditions expected in open water bodies. The best methods include microfiltration to remove cyanobacterial cells, adsorption of toxins with activated carbon, reverse osmosis, and ozonation. Besides being effective in removing hepatotoxins, these procedures do not disrupt the water ecology with chemical or biological treatments that can have unpredictable results. As discussed above, disruption of cyanobacterial cells by chemicals such as copper sulfate releases toxins into the water (see Section V.B). Widely used procedures such as coagulation, filtration by sand, and chlorination are generally ineffective for removing microcystins and nodularins (20). Procedures for filtering cyanobacterial cells are useful but, in cases of high algal content, filters quickly become clogged (104). When high quantities of toxins have been released owing to the death of a massive bloom, or to other conditions that have caused cell lysis, peptide toxins must be removed or otherwise inactivated.

Adsorption of contaminants by activated carbon, either powdered or granulated, is relatively inexpensive when compared to reverse osmosis and ozonation. However, treatment of water with activated carbon requires regular monitoring and replacement of saturated matrix. One group tested three types of domestic water filters, combinations of activated carbon, and ion-exchange resins (105). They concluded that none of the filters removed all of the cyanobacterial toxins, and they noted that filter performance had declined significantly when half of the recommended volume of water had passed through. It is generally recognized that activated carbon is effective in removing cyanobacterial toxins (20,106) but, because of the variable nature of raw water, it is difficult to standardize any carbon system with respect to the quantity of water that will exhaust it. True quality control can only be maintained by testing the treated product for toxin content.

We achieved 99.7% toxin removal, as measured by radiolabeled phosphatase inhibition assay, from 975 μg/L microcystin-LR equivalents in a crude cyanobacterial extract, by combined microfiltration and activated carbon treatment (C.F.B. Holmes and M. Craig, unpublished data). Toxin remaining after activated carbon treatment, 3 μg/L, was reduced to less than 0.3 μg/L—the approximate limit of detection of the phosphatase inhibition assay under these conditions—by reverse osmosis in about 1 h.

In a recent study (107), the elimination of microcystin-LR and microcystin-RR by reverse osmosis was examined. Several types of reverse osmosis membranes were tested with a crude methanolic extract of bloom material in tap water. Microcystin-LR and microcystin-RR were each present at 30–170 μg/L. A 96.7% and 99.9% removal of microcystin-LR and microcystin-RR, respectively, was achieved, as estimated by HPLC with a detection limit of 0.2 μg/L.

Given differences in membrane quality, pumping rates, and toxin concentrations, it is difficult to compare these experiments. It is clear that adsorption by activated carbon is a cost-efficient method for large-scale removal of microcystins and nodularins from crude extracts. Under field conditions where practical considerations and economics sometimes prevent conditions from being optimal, activated carbon combined with conventional treatment processes of coagulation, sedimentation, filtration, and chlorination generally removed more than 80% of the microcystin from raw water (106). Reverse osmosis is generally used for polishing water that has previously been treated by other methods; this yields a level of purification usually only available in research laboratories and medical facilities.

Methods that disrupt the Adda group or cleave the peptide ring, such as oxidative chemicals, ozonation (104), and UV irradiation (108,121), are effective in detoxifying microcystins and nodularins (see Section V.B), but they are expensive and not widely employed in water-purification plants.

The only practical systems in use have detection limits of approximately 0.2 μg/L. Oral ingestion of hepatotoxins below these limits is not an acute risk, but long-term effects of exposure to low levels of cyanobacterial hepatotoxins are not known. Hemodialysis patients are particularly at risk owing to the large volumes of water in direct contact with their blood (109,110) (see discussion in Section VII). Microcystins and nodularins can easily traverse the porous dialysis membrane and enter the blood of the patient whose system is already weakened by disease. Water used for dialysis is typically treated by several methods, including microfiltration, activated carbon, ion exchange, and reverse osmosis, and there is a growing awareness of the need for all of these systems to be fully functional. Now that sensitive tests are available, the need to regenerate or replace activated carbon and the integrity of osmotic membranes can be monitored.

VII. HUMAN HEALTH HAZARDS: RESULTS OF EXPOSURE TO CYANOBACTERIA

Incidents of human illness after exposure to cyanobacterial blooms have been documented since the late 1800s. Published reports of human injury incurred during recreational activities in contaminated waters include symptoms such as gastroenteritis, allergic and irritation reactions of skin, mouth, throat and lungs, headaches, and abdominal pain (2,20).

A 4-year study of two drinking water lakes in Japan detected the presence of microcystin in cell-free samples at levels of up to 4 μg/L in one lake and no microcystin in cell-free samples of the other lake (83,86). Values higher than 1 μg/L are rarely encountered in Canadian waters in the absence of blooms, but values of 5–15 μg/L have been measured during blooms, probably due to cell leakage (C.F.B. Holmes, unpublished data).

Chronic ingestion of drinking water containing cyanobacteria has been linked to liver disease (111,112). Epidemiological studies in China found a close correlation between the incidence of primary liver cancer and the drinking of pond or ditch waters. Using an ELISA method, microcystins were measured in water from several drinking water sources (112). Samples were frozen, so it is likely that there was some release of cell contents. The average content of microcystin in pond, ditch, and river waters was estimated to be 130 pg/mL (0.13 μg/L). The highest value measured was 1.5 μg/L, but samples from most water bodies were less than 0.5 μg/L. As discussed previously (in Section VI.A), antibodies used in ELISA methods are sensitive to nontoxic forms of hepatotoxins leading to false-positive results. The proportion of false-positive results that would be expected is not clear from the data presented here, but would be expected to increase measured values to some extent. Even without this consideration, microcystin concentrations in these drinking water sources in China appear to be within a range considered ''safe,'' according to current national and international guidelines (see Section VI.B).

Inhabitants of underdeveloped areas who are exposed to microcystins in their drinking water are often at risk of liver injury from other sources; for instance, from hepatitis B virus, aflatoxins in the diet, and poor nutrition. Thus, the effects of the hepatotoxins may be exacerbated by exposure to other disease-promoting elements (112). Some villages where water is provided by deep wells with no detectable microcystin have negligible incidences of hepatocellular carcinoma (111). The involvement of hepatotoxins in the high incidence of liver cancer in China is still under suspicion.

Until recently there were no reported deaths of humans due to cyanobacterial intoxication. But in 1996, 126 patients undergoing hemodialysis in Caruaru, Brazil, fell ill following the use of water from a reservoir with a massive growth of cyanobacteria (109,110,113). Sixty patients died (110). Their symptoms included myalgia, weakness, nausea, and tenderness around the liver and a range of neurological symptoms such as tinnitus, vertigo, headaches, deafness, blindness, and convulsions.

During dialysis, patients' blood and water are in free exchange across a porous membrane. In a normal 4-h dialysis session, the quantity of water that passes through the system is roughly 120 L. That is, in a normal week of three dialysis treatments, 360 L of water is in direct contact with a patient's blood. In the Brazilian incident, microcystins were detected in water from the reservoir, on dialysis filters, and in the liver tissue of patients who died. The concentration and pathological features of microcystin in the liver tissue were similar to those associated with severe liver damage in animals. Given the levels of microcystin found, and in the absence of other potential contaminants—chlorines and chloramines, agricultural compounds, trace elements, and heavy metals—it was concluded that the deaths were a consequence of high levels of microcystin in the water used for dialysis (109).

The incident in Caruaru, Brazil, occurred after a period of drought which left a water shortage in addition to the massive cyanobacterial bloom on the reservoir from which the water was taken. Municipal water goes to a treatment plant before being distributed. At the plant, the raw water is settled, filtered through layers of rock and sand, and chlorinated. In the dialysis clinic, the municipal water undergoes further purification, including filtration by sand and charcoal, ion-exchange resin columns (regenerated every 3 days), and finally a micropore filter. There was no reverse osmosis system in the clinic. In this instance, water used for dialysis of the stricken patients bypassed the treatment plant (110). When it was received by the dialysis unit, it was heavily chlorinated and then treated with the usual purification methods in the clinic. It was noted (109) that records at the dialysis center and at the city's water-treatment plant were poor and that water samples from the time of probable exposure were not available for testing.

VIII. MICROCYSTINS AND NODULARINS: TOOLS FOR SIGNAL TRANSDUCTION RESEARCH

PP1 and PP2A have been implicated in a multitude of cellular processes, but protein ligands that regulate these activities have been identified in only a modest number of cases. The strong attraction of PP1 and PP2A catalytic subunits for microcystins (IC50 = 0.1–0.2 nM) has lead to affinity methods for the purification of these enzymes and for isolating and characterizing regulatory subunits from biological systems (114–116).

Because microcystins and nodularins selectively inhibit only PP1 and PP2A, they are useful for abolishing the activity of these enzymes in crude preparations, allowing quantitation of other protein phosphatase activities (35). In combination, microcystin and okadaic acid can distinguish between PP1 and PP2A activities, as PP1 is less responsive than PP2A to okadaic acid, but both enzymes are equally responsive to microcystin.

In a comparison of native and recombinant enzymes (117), recombinant PP1 was found to be 5–100 times less sensitive than native PP1 to microcystin-LR. IC50 values for inhibition by microcystin-LR of α, β, and γ isoforms of recombinant PP1 were estimated to be 0.28 nM, 5.0 nM, and 0.28 nM, respectively, compared to 0.06 nM for native enzyme. This property can be used to differentiate recombinant and native enzymes. During prolonged storage, many properties of the native enzyme revert to those of recombinant protein. Native and recombinant enzymes differ in responses to substrates, inhibitors, and metal ions. Response to microcystin-LR is a convenient gauge of enzyme aging. In one interesting application, the selective inhibition of native/recombinant enzyme activity by microcystin-LR (along with other parameters) was used to assess a conformational change in PP1 from recombinant to native form after treatment with inhibitor-2, a mammalian protein inhibitor which may have the properties of a molecular chaperone (117).

IX. FUTURE WORK

The mechanisms of microcystin and nodularin intoxication are beginning to be understood, but much more needs to be learned about toxin uptake, metabolism and clearance, and the hazards of chronic exposure. The present suggested safe levels of microcystin may have to be lowered if it is shown definitively that low levels of exposure are linked to liver cancer. The importance of the covalent bond between phosphatases and microcystins is not known. The fact that nodularins do not react covalently with PP1 and PP2A may have biological implications that need to be investigated.

Water testing for the presence of hepatotoxins is presently confined to laboratories. The advantages of a field-testing kit cannot be overstated. Rural areas where cyanobacterial blooms are endemic suffer yearly economic losses due to the intoxication of farm animals. In many such situations, accurate quantitation of hepatotoxins is not warranted; the primary need is to have the ability to detect acute levels of toxin in order to prevent animals from drinking infected waters. Of the methods discussed, the phosphatase inhibition assay and the ELISA approach are the most likely to be developed for on-site testing.

In addition to the hepatotoxins, microcystins and nodularins, cyanobacteria produce a number of biologically active compounds, including neurotoxins, dermatotoxins, and gastrointestinal toxins (3). Antibacterial, antifungal, cytotoxic, molluscicidal, anti-inflammatory, and antiviral activities have also been demonstrated (118) using extracts obtained from samples of freshwater and terrestrial cyanobacteria. Microcystins and nodularins are routinely employed as research tools, and it is probable that other compounds from cyanobacteria will find uses in research or as pharmaceutical agents.

ACKNOWLEDGMENTS

The authors wish to acknowledge financial assistance from the Alberta Heritage Foundation for Medical Research (AHFMR) and the Medical Research Council of Canada. We are also indebted to Dr. J. Kuriyan (Rockefeller University) for making the structural coordinates for the crystal structure of protein phosphatase-1 available prior to publication and to Drs. R. Andersen and D. Williams (University of British Columbia) for much valuable discussion.

REFERENCES

1. WW Carmichael. The toxins of cyanobacteria. Sci Am 270(1):78–86, Jan, 1994.
2. IR Falconer. Algal toxins and human health. In: J Hrubec, ed. The Handbook of Environmental Chemistry. Vol 5. Part C. Quality and Treatment of Drinking water II. Berlin: Springer-Verlag, 1998, pp. 53–82.
3. K Sivonen. Cyanobacterial toxins and toxin production. Phycologia 35(Suppl):12–24, 1996.
4. PR Hunter. Cyanobacterial toxins and human health. J Appl Microbiol 84(Suppl):S35–S40, 1998.
5. KL Rinehart, M Namikoshi, BW Choi. Structure and biosynthesis of toxins from blue-green algae (cyanobacteria). J Appl Phycol 6:159–176, 1994.
6. DP Botes, PL Wessels, H Kruger, MTC Runnegar, S Santikarn, RJ Smith, JCJ Barna, DH Williams. Structural studies on cyanoginosin-LR, -YR, -YA and -YM: peptide toxins from *Microcystis aeruginosa*. J Chem Soc Perkin Trans 1:2747–2752, 1985.
7. IR Falconer, ARB Jackson, J Langely, MT Runnegar. Liver pathology in mice in poisoning by the blue-green alga *Microcystis aeruginosa*. Aust J Biol Sci 34:179–187, 1981.
8. JAO Meriluoto, A Sandstrom, JE Eriksson, G Ramaud, AG Craig, J Chattopadhyaya. Structure and toxicity of a peptide hepatotoxin from the cyanobacterium *Oscillatoria agardhii*. Toxicon 27:1021–1034, 1989.
9. M Runnegar, N Berndt, N Kaplowitz. Microcystin uptake and inhibition of protein phosphatases: effects of chemoprotectants and self-inhibition in relation to known hepatic transporters. Toxicol Appl Pharmacol 134:264–272, 1995.
10. KA Beattie, K Kaya, T Sano, GA Codd. Three dehydrobutyrine-containing microcystins from *Nostoc*. Phytochemistry 47:1289–1292, 1998.
11. T Sano, K Kaya. Two new (E)-2-amino-2-butenoic acid (Dhb)–containing microcystins isolated from *Oscillatoria agardhii*. Tetrahedron 54:463–470, 1998.
12. T Sano, KA Beattie, GA Codd, K Kaya. Two (Z)-dehydrobutyrine–containing microcystins from a hepatotoxic bloom of *Oscillatoria agardhii* from Soulseat Loch, Scotland. J Nat Prod 61:851–853, 1998.
13. M Namikoshi, M Yuan, K Sivonen, WW Carmichael, KL Rinehart, L Rouhiainen, F Sun, S Brittain, A Otsuki. Seven new microcystins possessing two L-glutamic acid units, isolated from *Anabaena* sp. Strain 186. Chem Res Toxicol 11:143–149, 1998.
14. JE Eriksson, JAO Meriluoto, HP Kujari, K Osterlund, K Fagerlund, L Hallbom. Preliminary characterization of a toxin isolated from the cyanobacterium *Nodularia spumigena*. Toxicon 26:161–166, 1988.
15. KL Rinehart, K Harada, M Namikoshi, C Chen, CA Harvis, MHG Munro, JW Blunt, PE Mulligan, VR Beasley, AM Dahlem, WW Carmichael. Nodularin, microcystin and the configuration of Adda. J Am Chem Soc 110:8557–8558, 1988.
16. ED De Silva, DE Williams, RJ Andersen, H Klix, CFB Holmes, TM Allen. Motuporin, a potent protein phosphatase inhibitor isolated from the Papua New Guinea sponge *Theonella swinhoei* Gray Tetrahedron Lett 33:1561–1564, 1992.
17. R Matsushima, S Yoshizawa, MF Watanabe, K-I Harada, M Furusawa, WW Carmichael, H Fujiki. In vitro and in vivo effects of protein phosphatase inhibitors, microcystins and nodularin, on mouse skin and fibroblasts. Biochem Biophys Res Commun 171:867–874, 1990.
18. S Yoshizawa, R Matsushima, MF Watanabe, K-I Harada, A Ichihara, WW Carmichael, H Fujiki. Inhibition of protein phosphatases by microcystins and nodularin associated with hepatotoxicity. J Cancer Res Clin Oncol 116:609–614, 1990.
19. G Francis. Poisonous Australian lake. Nature 18:11–12, 1878.
20. TW Lambert, CFB Holmes, SE Hrudey. Microcystin class of toxins: health effects and safety of drinking water supplies. Environ Rev 2:167–186, 1994.
21. CR Carbis. A biochemical profile for predicting the chronic exposure of sheep to *Microcystis aeruginosa*, an hepatotoxic species of blue-green algae. Res Vet. Sci 57:310–316, 1994.
22. NA Robinson, GA Miura, CF Matson, RE Dinterman, JG Pace. Characterization of chemically tritiated microcystin-LR and its distribution in mice. Toxicon 27:1035–1042, 1989.

23. NA Robinson, JG Pace, CF Matson, GA Miura, WB Lawrence. Tissue distribution, excretion and hepatic biotransformation of microcystin-LR in mice. J Pharmacol Exp Ther 256:176–182, 1991.
24. DM Toivola, JE Eriksson, DL Brautigan. Identification of protein phosphatase 2A as the primary target for microcystin-LR in rat liver homogenates. FEBS Lett 344:175–180, 1994.
25. RR Stotts, AR Twardock, WM Haschek, BW Choi, KL Rinehart, VR Beasley. Distribution of tritiated dihydromicrocystin in swine. Toxicon 35:937–953, 1997.
26. T Yoshida, Y Makita, S Nagata, T Tsutsumi, F Yoshida, M Sekijima, S Tamura, Y Ueno. Acute oral toxicity of microcystin-LR, a cyanobacterial hepatotoxin, in mice. Nat Toxins 5:91–95, 1997.
27. R Nishiwaki, T Ohta, E Sueoka, M Suganuma, K-I Harada, MF Watanabe, H Fujiki. Two significant aspects of microcystin-LR: specific binding and liver specificity. Cancer Lett 83:283–289, 1994.
28. E Ito, F Kondo, K Terao, K-I Harada. Neoplastic nodular formation in mouse liver induced by repeated intraperitoneal injections of microcystin-LR. Toxicon 35:1453–1457, 1997.
29. JE Eriksson, L Gronberg, S Nygard, JP Slotte, JAO Meriluoto. Hepatocellular uptake of ^{3}H-dihydromicrocystin-LR, a cyclic peptide toxin. Biochim Biophys Acta 1025:60–66, 1990.
30. MTC Runnegar, RG Gerdes, JR Falconer. The uptake of the cyanobacterial hepatotoxin microcystin by isolated rat hepatocytes. Toxicon 29:43–51, 1991.
31. D Barford, AK Das, M-P Egloff. The structure and mechanism of protein phosphatases: insights into catalysis and regulation. Annu Rev Biophys Biomol Struct 27:133–164, 1998.
32. P Cohen. The structure and regulation of protein phosphatases. Annu Rev Biochem 58:453–508, 1989.
33. JE Eriksson, D Toivola, JAO Meriluoto, H Karaki, Y-G Han, D Hartshorne. Hepatocyte deformation induced by cyanobacterial toxins reflects inhibition of protein phosphatases. Biochem Biophys Res Commun 173:1347–1353, 1990.
34. RE Honkanen, J Zwiller, RE Moore, SL Daily, BS Khatra, M Dukelow, AL Boynton. Characterisation of microcystin-LR, a potent inhibitor of type-1 and type-2A protein phosphatases. J Biol Chem 265:19401–19404, 1990.
35. C MacKintosh, KA Beattie, S Klumpp, P Cohen, GA Codd. Cyanobacterial microcystin-LR is a potent and specific inhibitor of protein phosphatase 1 and 2A from both mammals and higher plants. FEBS Lett 264:187–192, 1990.
36. R Nishiwaki-Matsushima, S Nishiwaki, T Ohta, S Yoshizawa, M Suganuma, K-I Harada, MF Watanabe, H Fujiki. Structure-function relationships of microcystins, liver tumor promotors, in interaction with protein phosphatase. Jpn J Cancer Res 82:993–996, 1991.
37. RR Stotts, M Namikoshi, WM Haschek, KL Rinehart, WW Carmichael, AM Dahlem, VR Beasley. Structural modifications imparting reduced toxicity in microcystins from *Microcystis* spp. Toxicon 31:783–789, 1993.
38. NA Robinson, CF Matson, JG Pace. Association of microcystin-LR and its biotransformation product with a hepatic-cytosolic protein. J Biochem Toxicol 6:171–180, 1991.
39. A Takai, K Sasaki, H Nagai, G Mieskes, M Isobe, K Isono, T Yasumoto. Inhibition of specific binding of okadaic acid to protein phosphatase 2A by microcystin-LR, calyculin-A and tautomycin: method of analysis of interactions of tight-binding ligands with target protein. Biochem J 306:657–665, 1995.
40. RW MacKintosh, KN Dalby, DG Campbell, PTW Cohen, P Cohen, C MacKintosh. The cyanobacterial toxin microcystin binds covalently to cysteine-273 on protein phosphatase 1. FEBS Lett 371: 236–240, 1995.
41. M Runnegar, N Berndt, S-M Kong, EYC Lee, L Zhang. In vivo and in vitro binding of microcystin to protein phosphatases 1 and 2A. Biochem Biophys Res Commun 216:162–169, 1995.
42. M Craig, HA Luu, TL McCready, D Williams, RJ Andersen, CFB Holmes. Molecular mechanisms underlying the interaction of motuporin and microcystins with type-1 and type-2 protein phosphatases. Biochem Cell Biol 74:569–578, 1996.
43. JR Bagu, FD Sonnichsen, D Williams, RJ Andersen, BD Sykes, CFB Holmes. Comparison of the solution structures of microcystin-LR and motuporin. Nat Struct Biol 2:114–116, 1995.
44. M Namikoshi, BW Choi, F Sun, KL Rinehart. Chemical characterization and toxicity of dihydro derivatives of nodularin and microcystin-LR, potent cyanobacterial cyclic peptide hepatotoxins. Chem Res Toxicol 6:151–158, 1993.

45. J Goldberg, H Huang, Y Kwon, P Greengard, AC Nairn, J Kuriyan. Three-dimensional structure of the catalytic subunit of protein serine/threonine phosphatase-1. Nature 376:745–753, 1995.
46. JR Bagu, BD Sykes, MM Craig, CFB Holmes. A molecular basis for different interactions of marine toxins with protein phosphatase-1. J Biol Chem 272:5087–5097, 1997.
47. H Fujiki, M Suganuma. Naturally-derived tumor promotors and inhibitors of carcinogenesis. J Toxicol Toxin Rev 15:129–156, 1996.
48. ML Wickstrom, SA Khan, WM Haschek, JF Wyman, JE Eriksson, DJ Schaeffer, VR Beasley. Alterations in microtubules, intermediate filaments and microfilaments induced by microcystin-LR in cultured cells. Toxicol Pathol 23:326–337, 1995.
49. S Ghosh, SA Khan, M Wickstrom, V Beasley. Effects of microcystin-LR on actin and the actin-associated proteins α-actinin and talin in hepatocytes. Nat Toxins 3:405–414, 1995.
50. SA Khan, ML Wickstrom, WM Haschek, DJ Schaeffer, S Ghosh, VR Beasley. Microcystin-LR and kinetics of cytoskeletal reorganization in hepatocytes, kidney cells and fibroblasts. Nat Toxins 4:206–214, 1996.
51. SA Khan, S Ghosh, M Wickstrom, LA Miller, R Hess, WM Haschek, VR Beasley. Comparative pathology of microcystin-LR in cultured hepatocytes, fibroblasts and renal epithelial cells. Nat Toxins 3:119–128, 1995.
52. KE Fladmark, MH Serres, NL Larsen, T Yasumoto, T Aune, SO Doskeland. Sensitive detection of apoptogenic toxins in suspension cultures of rat and salmon hepatocytes. Toxicon 36:1101–1114, 1998.
53. IR Falconer, JV Smith, ARB Jackson, A Jones, MT Runnegar. Oral toxicity of a bloom of the cyanobacterium *Microcystis aeruginosa* administered to mice over periods up to one year. J Toxicol Environ Health 24:291–305, 1988.
54. IR Falconer, TH Buckley. Tumor promotion by *Microcystis aeruginosa*. Med J Aust 150:351, 1989.
55. H Fujiki, R Matsushima, S Yoshizawa, M Suganuma, S Nishiwaki, T Ishikawa, WW Carmichael. Liver tumor promotion through the okadaic acid pathway, inhibition of protein phosphatases 1 and 2A (abst). 1991 AACR Meeting 157: 1991.
56. T Ohta, E Sueoka, N Iida, A Komori, M Suganuma, R Nishiwaki, M Tatematsu, S-J Kim, WW Carmichael, H Fujiki. Nodularin, a potent inhibitor of protein phosphatases 1 and 2A, is a new environmental carcinogen in male F344 rat liver. Cancer Res 54:6402–6406, 1994.
57. MTC Runnegar, IR Falconer. The in vivo and in vitro biological effects of the peptide hepatotoxin from the blue-green alga, *Microcystis aeruginosa*. SAJ Sci 78:363–366, 1982.
58. BN Ames, J McCann. Validation of the *Salmonella* test: a reply to Rinkus and Legston. Career Res 41:4192–4196, 1981.
59. WM Repavich, WC Sonzogni, JH Standridge, RE Wedepohl, LF Meisner. Cyanobacteria (blue-green algae) in Wisconsin waters: acute and chronic toxicity. Wat Res 24:225–231, 1990.
60. H Suzuki, MF Watanabe, Y Wu, T Sugita, K Kita, T Sato, X-L Wang, H Tanzawa, S Sekiya, N Suzuki. Mutagenicity of microcystin-LR in human RSa cells. Int J Mol Med 2:109–112, 1998.
61. PVL Rao, R Bhattacharya, MM Parida, AM Jana, ASB Bhaskar. Freshwater cyanobacterium *Microcystis aeruginosa* (UTEX 2385) induced DNA damage in vivo and in vitro. Environ Toxicol Pharmacol 5:1–6, 1998.
62. H Fujiki, M Suganuma, S Yoshizawa, H Kanazawa, T Sugimura, S Manam, SM Kahn, J Wei, S Hoshina, IB Weinstein. Codon 61 mutations in the c-Harvey-*ras* gene in mouse skin tumors induced by 7,12-dimethylbenz[a]anthracene plus okadaic acid class tumor promotors. Mol Carcinogen 2: 184–187, 1989.
63. PVL Rao, R Bhattacharya. The cyanobacterial toxin microcystin-LR induced DNA damage in mouse liver in vivo. Toxicology 114:29–36, 1996.
64. PJ Kennelly, M Potts. Life among the primitives: protein O-phosphatases in prokaryotes. Front Biosci, 4:372–385, 1999.
65. F Kondo, Y Ikai, H Oka, M Okumura, N Ishikawa, K-I Harada, K Matuura, H Murata, M Suzuki. Formation, characterization and toxicity of the glutathione and cysteine conjugates of toxic heptapeptide microcystins. Chem Res Toxicol 5:591–596, 1992.
66. SJ Hermansky, SJ Stohs, ZM Eldeen, VF Roche, KA Mereish. Evaluation of potential chemoprotectants against microcystin-LR hepatotoxicity in mice. J Appl Toxicol 11:65–74, 1991.

67. DE Williams, M Craig, S Dawe, ML Kent, CFB Holmes, RJ Andersen. Evidence for a covalently bound form of microcystin-LR in salmon liver and Dungeness crab larvae. Chem Res Toxicol 10: 463–469, 1997.
68. DZX Chen, MP Boland, MA Smillie, H Klix, C Ptak, RJ Andersen, CFB Holmes. Identification of protein phosphatase inhibitors of the microcystin class in the marine environment. Toxicon 31: 1407–1414, 1993.
69. DE Williams, SC Dawe, ML Kent, RJ Andersen, M Craig, CFB Holmes. Bioaccumulation and clearance of microcystins from salt water mussels, *Mytilus edulis*, and in vivo evidence for covalently bound microcystins in mussel tissues. Toxicon 35:1617–1625, 1997.
70. RJ Andersen, HA Luu, DZX Chen, CFB Holmes, ML Kent, M LeBlanc, FJR Taylor, DE Williams. Chemical and biological evidence links microcystin-LR to salmon 'netpen liver disease.' Toxicon 31:1315–1323, 1993.
71. DE Williams, ML Kent, RJ Andersen, H Klix, CFB Holmes. Tissue distribution and clearance of tritium-labeled dihydromicrocystin-LR epimers administered to Atlantic salmon via intraperitoneal injection. Toxicon 33:125–131, 1995.
72. NA Robinson, CF Matson, CF Miura, TG Lynch, JG Pace. Toxicokinetics of [^{3}H] microcystin-LR in mice. Fed Am Soc Exp Biol 4(3):2823, A753, 1990.
73. NR Bury, AD Newlands, FB Eddy, GA Codd. In vivo and in vitro intestinal transport of ^{3}H-microcystin-LR, a cyanobacterial toxin, in rainbow trout (*Oncorhynchus mykiss*). Aquat Toxicol 42:139–148, 1998.
74. IR Falconer, A Choice. Toxicity of edible mussels (*Mytilus edulis*) growing naturally in an estuary during a water bloom of the blue-green alga *Nodularia spumigena*. Environ Toxicol Wat Qual 7: 119–123, 1992.
75. JE Eriksson, JAO Meriluoto, T Lindholm. Accumulation of a peptide toxin from the cyanobacterium *Oscillatoria agardhii* in the freshwater mussel Anadonta cygnea. Hydrobiologia 183:211–216, 1989.
76. PR Hawkins, MTC Runnegar, ARB Jackson, IR Falconer. Severe hepatotoxicity caused by the tropical cyanobacterium (blue-green algae) *Cylindrospermopsis raciborskii* (Woloszynska) Seenaya and Subba Raju isolated from a domestic water supply reservoir. Appl Environ Microbiol 50:1292–1295, 1985.
77. SL Kenefick, SE Hrudey, HG Peterson, EE Prepas. Toxin release from *Microcystis aeruginosa* after chemical treatment. Wat Sci Technol 27:433–440, 1993.
78. P Prociv. Palm Island reconsidered. Was it copper poisoning? Aust NZ J Med 17:345–349, 1987.
79. GJ Jones, DG Bourne, RL Blakeley, H Doelle. Degradation of the cyanobacterial hepatotoxin microcystin by aquatic bacteria. Nat Toxins 2:228–235, 1994.
80. K Tsuji, S Setsuda, T Watanuki, F Kondo, H Nakazawa, M Suzuki, K-I Harada. Microcystin levels during 1992–95 for Lakes Sagami and Tsukui—Japan. Nat Toxins 4:189–194, 1996.
81. DP Botes, AA Tuinmann, PL Wessels, CC Viljoen, H Kruger, DH Williams, S Santikarn, RJ Smith, SJ Hammond. The structure of cyanoginosin-LA, a cyclic heptapeptide toxin from the cyanobacterium *Microcystis aeruginosa*. J Chem Soc Perkin Trans 1:2311–2318, 1984.
82. MF Watanabe, K Tsuji, Y Watanabe, K-I Harada, M Suzuki. Release of heptapeptide toxin (microcystin) during the decomposition process of *Microcystis aeruginosa*. Nat Toxins 1:48–53, 1992.
83. K-I Harada, K Tsuji. Persistence and decomposition of hepatotoxic microcystins produced by cyanobacteria in natural environment. J Toxicol Rev 17:385–403, 1998.
84. IT Cousins, DJ Bealing, HA James, A Sutton. Biodegradation of microcystin-LR by indigenous mixed bacterial populations. Wat Res 30:481–485, 1996.
85. J Rapala, K Lahti, K Sivonen, SI Niemela. Biodegradability and adsorption on lake sediments of cyanobacterial hepatotoxins and anatoxin-a. Lett Appl Microbiol 19:423–428, 1994.
86. K-I Harada, K Tsuji, MF Watanabe, F Kondo. Stability of microcystins from cyanobacteria-III effect of pH and temperature. Phycologia 35(Suppl 6):83–88, 1996.
87. BW Choi, M Namikoshi, F Sun, KL Rinehart, WW Carmichael, AM Kaup, WR Evans, VR Beasley. Isolation of linear peptides related to the hepatotoxins nodularin and microcystins. Tetrahedron Lett 34:7881–7884, 1993.

88. DG Bourne, GJ Jones, RL Blakeley, A Jones, AP Negri, P Riddles. Enzymatic pathway for the bacterial degradation of the cyanobacterial cyclic peptide toxin microcystin-LR. Appl Envir Microbiol 62:4086–4094, 1996.
89. K-I Harada, K Ogawa, K Matsuura, H Murata, M Suzuki, MF Watanabe, Y Itezono, N Nakayama. Structural determination of geometrical isomers of microcystins-LR and -RR from cyanobacteria by two dimensional NMR spectroscopic techniques. Chem Res Toxicol 3:473–481, 1990.
90. K Tsuji, S Naito, F Kondo, N Ishikawa, MF Matanabe, M Suzuki, K-I Harada. Stability of microcystins from cyanobacteria: effect of light on decomposition and isomerization. Environ Sci Technol 28:173–177, 1994.
91. GJ Jones, PT Orr. Release and degradation of microcystin following algicide treatment of a *Microcystis aeruginosa* bloom in a recreational lake, as determined by HPLC and protein phosphatase inhibition assay. Wat Res 28:871–876, 1994.
92. DE Williams, M Craig, S Dawe, ML Kent, RJ Andersen, CFB Holmes. ^{14}C-labeled microcystin-LR administered to Atlantic salmon via intraperitoneal injection provides in vivo evidence for covalent binding of microcystin-LR in salmon liver. Toxicon 35:985–989, 1997.
93. T Sano, K Nohara, F Shirai, K Kaya. A method for microdetection of total microcystin content in waterbloom of cyanobacteria (blue-green algae). Int J Environ Anal Chem 49:163–170, 1992.
94. Y Tanaka, S Takenaka, H Matsuo, S Kitamori, H Tokiwa. Levels of microcystins in Japanese lakes. Toxicol Environ Chem 39:21–27, 1993.
95. K-I Harada, H Murata, Z Qiang, M Suzuki, F Kondo. Mass spectrometric screening method for microcystins in cyanobacteria. Toxicon 34:701–710, 1996.
96. K-I Harada. Trace analysis of microcystins. Phycologia 35:36–41, 1996.
97. C MacKintosh. Assay and purification of protein (serine/threonine) phosphatases. In D.G. Hardie, ed. Protein Phosphorylation: A Practical Approach. Oxford, UK: IRL, 1993, pp 197–230.
98. C Ash, C MacKintosh, R MacKintosh, CR Fricker. Development of a colorimetric protein phosphorylation assay for detecting cyanobacterial toxins. Wat Sci Technol 31:47–49, 1995.
99. C Ash, C MacKintosh, R MacKintosh, CR Fricker. Use of a protein phosphatase inhibition test for the detection of cyanobacterial toxins in water. Wat Sci Technol 31:51–53, 1995.
100. FS Chu, X Huang, RD Wei. Enzyme-linked immunosorbent assay for microcystins in blue-green algal blooms. J Assoc Off Anal Chem 73:451–456, 1990.
101. J-S An, WW Carmichael. Use of a colorimetric protein phosphatase inhibition assay and enzyme linked immunosorbent assay for the study of microcystins and nodularins. Toxicon 32:1495–1507, 1994.
102. M Craig, TL McCready, HA Luu, MA Smillie, P Dubord, CFB Holmes. Identification and characterization of hydrophobic microcystins in Canadian freshwater cyanobacteria. Toxicon 31:1541–1549, 1993.
103. WHO Working Group Meeting on Chemical Substances in Drinking Water. Geneva, Report Section 5.2, 1997.
104. P Mouchet, V Bonnelye. Solving algae problems: French expertise and world-wide applications. J Wat SRT Aqua 47:125–141, 1998.
105. LA Lawton, BJPA Cornish, AWR Macdonald. Removal of cyanobacterial toxins (microcystins) and cyanobacterial cells from drinking water using domestic water filters. Wat Res 32:633–638, 1998.
106. TW Lambert, CFB Holmes, SE Hrudey. Adsorption of microcystin-LR by activated carbon and removal in full scale water treatment. Wat Res 30:1411–1422, 1996.
107. U Neumann, J Weckesser. Elimination of microcystin peptide toxins from water by reverse osmosis. Environ Toxicol Wat Qual 13:143–148, 1998.
108. K Kaya, T Sano. A photodetoxification mechanism of the cyanobacterial hepatotoxin microcystin-LR by ultraviolet irradiation. Chem Res Toxicol 11:159–163, 1998.
109. EM Jochimsen, WW Carmichael, J-S An, DM Cardo, ST Cookson, CEM Holmes, MB de C Antunes, DA de Melo Filho, TM Lyra, VST Barreto, SMFO Azevedo, WR Jarvis. Liver failure and death after exposure to microcystins at a hemodialysis center in Brazil. N Engl J Med 338:873–878, 1998.
110. S Pouria, A de Andrade, J Barbosa, RL Cavalcanti, VTS Barreto, CJ Ward, W Preiser, GK Poon,

GH Neild, GA Codd. Fatal microcystin intoxication in haemodialysis unit in Caruaru, Brazil. Lancet 352:21–26, 1998.
111. S-Z Yu. Primary prevention of hepatocellular carcinoma. J Gastroenterol Hepatol 10:674–682, 1995.
112. Y Ueno, S Nagata, T Tsutsumi, A Hasegawa, MF Watanabe, H-D Park, G-C Chen, G Chen, S-Z Yu. Detection of microcystins, a blue-green algal hepatotoxin, in drinking water sampled in Haimen and Fusui, endemic areas of primary liver cancer in China, by highly sensitive immunoassay. Carcinogen 17:1317–1321, 1996.
113. J Dunn. Algae kill dialysis patients in Brazil. BMJ 312:1183–1184, 1996.
114. S Nishiwaki, H Fujiki, M Suganuma, R Nishiwaki-Matsushima, T Sugimura. Rapid purification of protein phosphatase 2A from mouse brain by microcystin-affinity chromatography. FEBS Lett 279: 115–118, 1991.
115. G Moorhead, RW MacKintosh, N Morrice, T Gallagher, C MacKintosh. Purification of type 1 protein (serine/threonine) phosphatases by microcystin-Sepharose affinity chromatography. FEBS Lett 356:46–50, 1994.
116. M Campos, P Fadden, G Alms, Z Qian, TAJ Haystead. Identification of protein phosphatase-1-binding proteins by microcystin-biotin affinity chromatography. J Biol Chem 271:28478–28484, 1996.
117. DR Alessi, AJ Street, P Cohen, PTW Cohen. Inhibitor-2 functions like a chaperone to fold three expressed isoforms of mammalian protein phosphatase-1 into a conformation with the specificity and regulatory properties of the native enzyme. Eur J Biochem 213:1055–1066, 1993.
118. BS Falch, GM Konig, AD Wright, O Sticher, CK Angerhofer, JM Pezzuto, H Bachman. Biological activities of cyanobacteria: evaluation of extracts and pure compounds. Planta Med 61:321–328, 1995.
119. RM Dawson. The toxicology of microcystins. Toxicon 36:953–962, 1998.
120. LE Sacks, JT MacGregor. The *Bacillus subtilis* multigene sporulation test for detection of environmental mutagens. Chem Mut 9:165–181, 1984.
121. K Tsuji, S Naito, F Kondo, N Ishikawa, MF Matanabe, M Suzuki, K-I Harada, Stability of microcystins from cyanobacteria: effect of uv on decomposition and isomerization. Toxicon 33:1619–1631, 1995.
122. A Takai, G Mieskes. Inhibitory effect of okadaic acid on the ρ-nitrophenyl phosphate phosphatase activity of protein phosphatases. Biochem J 275:233–239, 1991.
123. H Twist, GA Codd. Degradation of the cyanobacterial hepatotoxin, nodularin, under light and dark conditions. FEMS Microbiol Lett 15:83–88, 1997.
124. GA Codd, CJ Ward, SG Bell. Cyanobacterial toxins: occurrence, modes of action, health effects and exposure routes. Arch Toxicol 19(Suppl):399–410, 1997.

31
Freshwater Hepatotoxins: Geographical Distribution of Toxic Cyanobacteria

Mariyo F. Watanabe
Rissho University, Magechi, Kumagaya City, Saitama, Japan

I. INTRODUCTION

Cyanobacteria (Cyanophyta, blue-green algae) are unique organisms that exhibit wide patterns of physiological and morphological adaptation under vast environmental conditions. One of their prominent features is that they appeared earlier than other photosynthetic organisms, a fact that was established by the finding of cyanobacteria in rocks as old as 3.5 billion years from the Warrawoona group in western Australia (1). It is said that they played an important role in the oxygenation of the atmosphere, as they were the first organisms able to carry out oxygenic photosynthesis and convert carbon dioxide into oxygen.

The ability of cyanobacteria to adapt to various environmental conditions allows them to survive under extreme conditions such as those found in hot springs, arctic regions, and deserts (2–4). Under the current circumstances of hydrobiological changes, cyanobacterial blooms proliferate in an extensive biomass in eutrophicated and hypereutrophicated lakes, reservoirs, and ponds, where certain kinds of species cannot survive owing to the deterioration of the water quality and physical environmental factors (5,6). Such cyanobacteria have been causing serious problems not only in domestic animals but also wildlife and in extremely rare cases in humans (7), although Dinophyta species, including Desmophyceae and Dinophyceae, are well known to produce toxins in marine environments. The accumulation of such toxins causes diarrheic shellfish poisoning (DSP) and paralytic shellfish poisoning (PSP) in humans (8). Cyanobacterial toxins found in freshwater used as a drinking water supply have been considered a possible hazard, and the accumulation of cyanobacterial toxins in mussels has already been pointed out (9,10). However, the most serious incident due to cyanobacterial toxins occurred when 50 patients died at a hemodialysis center in Caruara City, Brazil, in 1996 (11). This event revealed that waters from eutrophicated lakes should not be used as a supply of drinking water without giving a caution to and determining the hazards of cyanobacterial toxins beforehand.

Most toxic cyanobacteria can form water blooms on the surface of lakes, which offers sufficient nutrient conditions such as nitrogen and phosphorus (12). In particular, light wind contributes to the stagnation of water, and scum from cyanobacterial cells forms on surface water along downwind shores. Incidents of animal poisoning as a result of ingestion of such scum have been reported in many countries of the world (13). In this chapter, the occurrences of toxic cyanobacteria are reviewed with special reference to geographical distribution.

II. CYANOBACTERIAL SPECIES FORMING WATER BLOOMS AND ANIMAL POISONING

In a freshwater ecosystem, accelerating eutrophication leads to the development and persistence of water blooms of cyanobacteria. In such ecosystems, there are physical, chemical, and biotic conditions promoting cyanobacterial growth and bloom (5,14). Excessive nutrient loading due to human activities such as industrial drainage, agriculture, and household waste disposal is though to be the most important factor sustaining the conditions that enable the remarkable growth of cyanobacteria. However, this situation along with surface water temperature, strong light conditions, persistent water columns, and long water residence time can also be advantageous for other phyla of phytoplanktonic growth (Figure 1). In addition to the physical and chemical factors, the dominance of cyanobacteria in freshwater is supported by biological factors, including the biological characteristics of cyanobacteria themselves. Cyanobacteria possess gas vesicles inside the cells to regulate buoyancy. Because of this, they have an advantage by fitting the physiological conditions of the colonies or trichomes at a certain depth of lakes and attain a favorable growth and proliferation of biomass (15). The ability of cyanobacteria to store phosphorus internally as polyphosphate bodies under ambient phosphate conditions is also an advantage to survive under phosphorus-deficient conditions (16). Colony formation is also important to avoid sinking from the photosynthetic layer and to minimize the effects of grazing by zooplankton.

According to Pearl, cyanobacterial species can be grouped into four groups: unicellular picoplankton, colonial coccoid nannoplankton and microplankton, solitary filamentous microplankton, and colonial filamentous microplankton from morphological characteristics, such as size, shape, and aggregative behavior (17). However, from the more structural viewpoint of taxonomy, cyanobacteria can be divided into four groups according to the Komárek and Anagnostidis' classification system of cyanobacteria, which includes Chroococcales, Oscillatori-

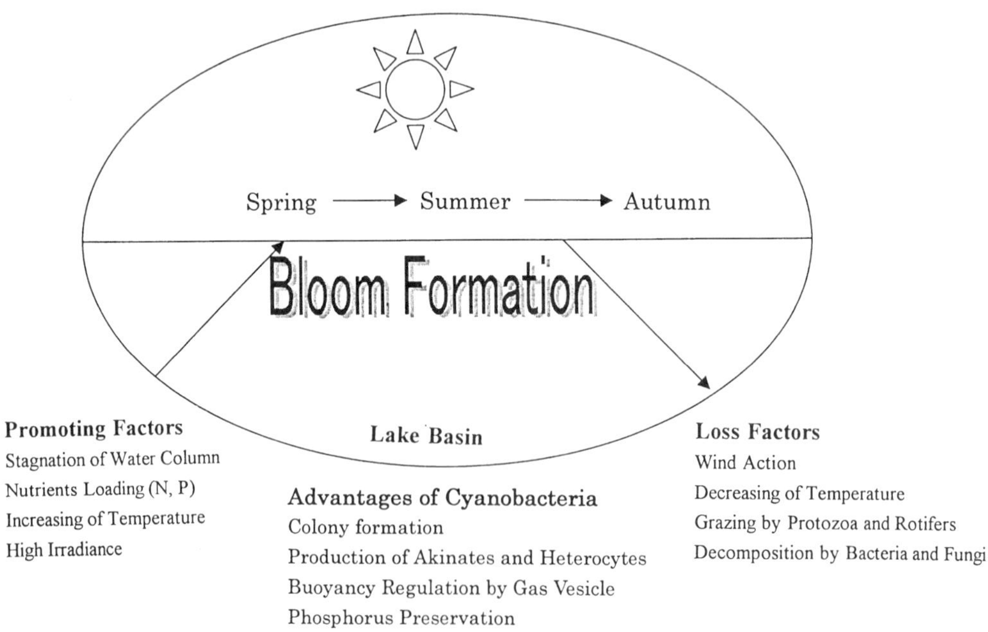

Figure 1 Schematic diagram of factors affecting the formation of cyanobacterial blooms.

ales, Nostocales, and Stigonematales (18–21). Chroococcales include species with coccoid cells, whose typical genera are *Microcystis*, *Chroococcus*, and *Merismopedia*. The other three groups include species possessing filaments or trichomes. Oscillatorialles has trichomes with only vegetative cells, being different from the other two groups, which have the ability to differentiate heterocysts and akinates. Those cells are able to survive during unfavorable growth periods. Although there exist about 80 genera in Chroococcales, 43 genera in Oscillatoriales, 32 genera in Nostcales, and 48 genera in Stigonematales, only a limited number of cyanobacterial species have caused animal toxicosis.

As mentioned above, among the plentiful number of species of cyanobacteria (more than 1500), only a limited number of them form buoyant aggregates of cells and filaments and water blooms, and an even more limited number of them are agents for animal poisoning. Table 1 shows the main species of cyanobacteria which form water blooms in fresh and brackish waters (22–49). There exist two types of water blooms: one type can be observed on the surface of water bodies; the other type can be observed at deeper water layers, usually in metalimnion or thermocline. Among species presented in Table 1, most cyanobacterial blooms, such as *Micro-*

Table 1 Main Species of Cyanobacteria Forming Water Blooms in Lakes

Species	Countries	Reference examples
Anabaena affinis Lemm.	United States, Japan	22,23
An. circinalis Rab.	Australia, Japan, Sweden	24, 25,26
An. crassa (Lem.)Kom.-Legn. & Cronb.	Japan	23
An. farciminiformis Cronb. et Kom.-Legn.	Sweden	27
An. flos-aquae Breb.	Finland, Sweden	22,27
An. lemmermannii P. Richt	Finland, Denmark	28,29
An. macrospora Kleb.	Japan	30
An. mendotae Trelease	Japan	26
An. planktonica Brunnth.	Japan	30
An. solitaria Kleb.	Sweden	25
An. spiroides Kleb.	Japan, Sweden	31,32
Anabaenopsis milleri Wolosz.	Greece	33
Aphanizomenon flos-aquae (L. Ralfs)	Japan, United States, Denmark	34,35,36
Apha. gracile Lemm.	Sweden	25
Apha. ovalisporum Forti	Israel	37
Cylindrospermopsis raciborskii (Wolosz.) Seena. & R. Raju	Brazil, Hungary	38,39
Gloeotrichia echinulata J.E. Smith	Denmark	36
Lyngbya limnetica Lemm.	Brazil	40
Microcystis aeruginosa (Kutz.) Kutz.	Many countries	41,42
M. viridis (A. Br.) Lemm.		43
M. wesenbergii (Kom.) Kom.		43
Nodularia spumigena Mert.	Australia, Finland,	44,45
Oscillatoria limnetica Lemm.	Sweden	25
Oscillatoria agardhii (Gom.) *(Planktothrix agardhii)*	Finland, Netherlands, Spain	46,47,48
O. mougeotii Bory ex Gom.	Japan	49
O. raciborskii Wolosz. *(Planktothrix raciborskii)*	Japan	22

cystis aeruginosa, Anabaena flos-aquae, and *A. circinalis*, are observed on the surface of lakes. However, several solitary filamentous species, such as *Oscillatoria agardhii* (*Planktothrix agardhii*), *O. mougeotii, Lyngbia limnetica*, and *Aphanizomenon flos-aquae*, can form the later type of blooms in mesotrophic and eutrophic lakes at a deeper layer where photon irradiance is 5–6% of that at the surface (35,49).

Cyanobacterial toxins resulting in animal poisoning are divided into hepatotoxins and neurotoxins. The toxic species are listed in Table 2. In spite of the wide occurrences of toxic cyanobacteria and the existence of a number of cyanobacterial species, a rather limited number of species are causative agents for animal poisoning. Since the first reports of animal deaths due to cyanobacteria by Francis in Lake Alexandrina in Australia in the 19th century (50), the incidents of animal poisoning have been reported from European, North American, and South African countries (Figure 2) (13,51,52). Wide occurrences of toxic cyanobacteria have been presented even in countries where there was no animal poisoning. In the United States, the presence of toxic blooms was reported in 27 states; in most cases associated with the loss of animals (7). The affected animals included domestic animals and pets as well as wildlife such as fish, birds, waterfowl, and mammals.

The toxic effects of freshwater cyanobacteria on humans have usually been the result of mismanagement of the utilization of lake waters. This is different from the cases of PSP by toxic marine phytoplankton. In a case of poisoning in Palm Island in Australia in the 1970s, over 130 individuals, mainly children, suffered from hepatoenteritis (53). This outbreak was attributed to the release of the toxin of *Cylindrospermopsis rachiborskii*, which resulted from the treatment of reservoir water by copper sulfate (54). It took 20 years to isolate the chemical structure of the toxin produced by *C. rachiborskii* and to determine its structure (55). Although several incidents of accidental poisoning have been reported in the United States and England,

Table 2 Toxins from Cyanobacteria

Toxins	Species
Neurotoxins	
Anatoxin-a	*Anabaena flos-aquae*
	An. circinalis
	An. lemmermanii
	An. planctonika
	An. solitaria
	Aphanizomenon flos-aqua
	Oscillatoria sp.
Anatoxin-a(S)	*Anabaena flos-aquae*
	A. lemmermanni
Saxitoxins	*Aphanizomenon flos-aquae*
	Anabaena circinalis
	Lyngbya wollei
Hepatotoxin	
Microcystins	*Microcystis aeruginosa*
	M. viridis
	Anabaena flos-aquae
	Nostoc sp.
	Oscillatoria agardhii
Nodularin	*Nodularia spumigena*
Cyrindrospermopsin	*Cylindrospermopsis rachiborskii*
	Umezakia natans

Figure 2 Countries and sites where the occurrences of toxic cyanobacteria were reported. ●: Animal toxicosis occurred. ▲: Occurrence of toxic bloom of cyanobacteria was reported.

involving people who accidentally came into contact with water containing cyanobacterial blooms (56), a more serious outbreak due to a cyanobacterial toxin caused a tragedy at a hemodialysis hospital in Brazil in 1996. Among 101 patients who were being treated with water supplied without proper management, about 50 patients died of acute liver failure (11). Microcystin-LR was detected in the livers of the victims. This accident warns about the hazard of cyanobacterial toxins when we use water supplied from lakes and reservoirs where the water blooms of cyanobacteria are observed.

III. OCCURRENCES OF TOXIC *MICROCYSTIS*

Microcystis is a typical genus that forms thick surface scum in eutrophic and hypereutrophic lakes. The species is made up of large colony aggregates which, in extreme cases, consist of over 1000 cells and play an important role in the biogeochemical cycle of organic matters. All colonies are composed of vegetative cells without cellular differentiation of heterocysts and akinates. The occurrence of *Microcystis* blooms is the most common, being distributed from subtropical to northern temperate regions (17).

Poisoning due to *Microcystis* involves acute hepatotoxicosis. The lethal toxicity of natural samples of *Microcystis* cells ranges from 11 to 360 mg/kg, with a lethal time of 30–120 min (57). This lethal time is clearly different from that of the natural samples dominated by *Anabaena flos-aquae* or *Aphanizomenon fos-aquae* (58). As a matter of fact, natural populations of *Microcystis* cells are composed of toxic and nontoxic species and strains and a low biomass of cyanobacterial cells. Therefore, it was estimated that isolated strains would have higher toxicity (59).

The toxins derived from *Microcystis* cells are called microcystins and constitute a large group of toxic compounds of cyanobacteria. Microcystins are known to be produced by species of *Anabaena, Oscillatoria*, and *Nostoc*. The toxins are cyclic heptapeptides and show a greater diversity in their chemical structure. The general structure is cyclo (-D-Ala-X-D-MeAsp-Z-Adda-D-Glu-Mdha-), including a unique C_{20} amino acid, 3-amino-9-methoxy-2,6,8-trimethyl-

10-phenyldeca-dienoic acid (Adda) (Figure 3). Until now, more than 50 variants have been reported from natural bloom samples and cultivated cells. Microcystin-LR is the most common member of the series. The frequent variations in structure are first found in the substitution of L–amino acids 2 and 4 (in a case of microcystin-LR, L-leucine and L-arginine). The next variation can be observed in the degree of methylation: β-methylasparatic acid (amino acid 3) replaced to D-aspartic acid and N-methyldehydroalanin (Mdha, amino acid 7) replaced to dehydroalanine. Table 3 shows the countries where the occurrences of toxic *Microcystis* and the variants have been reported (60–84). This implies that toxic *Microcystis* blooms are widely distributed in all the continents of the world. Microcystin-LR can be found in almost all bloom samples of *Microcystis* analyzed, although there exist some reports which described only new variants of microcystins. It was clearly shown that there existed toxic and nontoxic strains among those tentatively identified as small and large cell size groups of *M. aeruginosa* (85). However, the results of acute toxicity determined by mouse bioassay showed the high abundance of occurrences of toxic samples of *Microcystis*; for example, 13% in Australia, 50% in Japan, and 60% in Portugal (86–88). We can consider that natural blooms of *Microystis* populations consist of

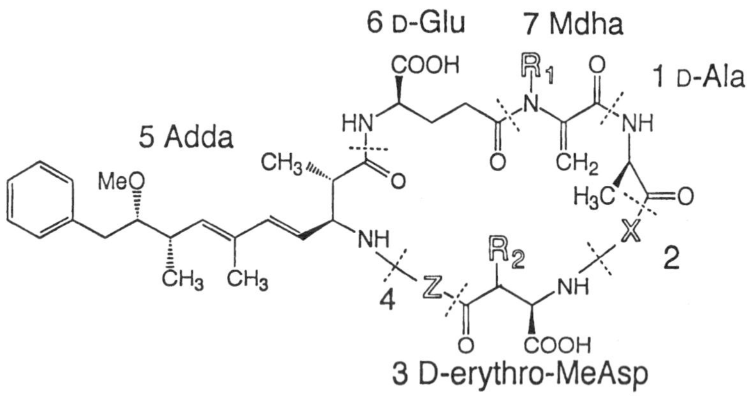

Microcystins	X	Z	R_1	R_2	MW
MC-LA	Leu	Ala	CH_3	CH_3	909
MC-LR	Leu	Arg	CH_3	CH_3	994
MC-FR	Phe	Arg	CH_3	CH_3	1028
MC-RR	Arg	Arg	CH_3	CH_3	1037
MC-YR	Tyr	Arg	CH_3	CH_3	1044
[D-Asp³]MC-LR	Leu	Arg	CH_3	H	980
[Dha⁷]MC-LR	Leu	Arg	H	CH_3	980
[D-Asp³, Dha⁷]MC-LR	Leu	Arg	H	H	966
[L-Ser⁷]MC-RR	Arg	Arg	H	CH_3	1041
MC-M(O)R	Met(O)	Arg	CH_3	CH_3	1028

Adda: 3-amino-9-methoxy-2,6,8-trimethyl-10-phenyldeca-4,6-dienoic acid
D-erythro-MeAsp: D-erythro-ß-methylasparatic acid
Mdha: N-methyldehydroalanine (Dha: dehydroalanine)

Figure 3 Structure of microcystins (MC). Typical variants of microcystins are also presented. X and Z represent variable L amino acids and R_1 and R_2 show the presence of CH_3.

Table 3 Occurrences of Toxic *Microcystis*

Country	Species	Bloom, strain	Microcystin	Reference
Australia	*M. aeruginosa*	Bloom	MC-LR	60
Brazil	*M. aeruginosa*	Bloom	MC-LR, MC-LF	61
Canada	*M. aeruginosa*	Bloom	MC-LL, MC-LV, MC-LM, MC-LF	62
	M. aeruginosa	Bloom	MC-LR	63
China	*M. aeruginosa*	Bloom	MC-LR	64
Denmark	*Microcystis* spp.	Bloom	MC-LR, -YR, -RR, [Dha^7]MC-RR, [DAsp^3]MC-RR, [Dha^7]MC-LR [DAsp^3]MC-LR	65
England	*M. aeruginosa*	Strain	MC-LR, MC-LY, MC-LW, MC-LF	66
Finland	*M. aeruginosa*	Bloom	[DAsp^3]MC-YR, [DAsp^3]MC-RR	67
	M. aeruginosa	Strain	[Dha^7]MC-LR	68
	Microcystis spp.	Bloom, Strain	MC-LR, -RR, -YR, [Dha^7]MC-FR, [D-Asp^3]MC-LR, [Dha^7]MC-LR, [DAsp^3, Dha^7]MC-LR, [DAsp^3]MC-RR, [Dha^7]MC-RR, [DAsp^3, Dha^7]MC-RR, [L-Ser^7]MC-RR, [Dha^7]MC-YR	69
France	*M. aeruginosa*	Bloom, Strain	MC-RR, MC-LR	70
Japan	*M. ae. M. viridis*	Strain	MC-LR, MC-YR, MC-RR	59
	M. ae. M. viridis	Strain	[6Z-Adda]MC-LR, [6Z-Adda]MC-RR,	71
	M. aeruginosa		MC-YR,[D-Asp^3]MC-LR	72
Germany	*Microcystis* sp.	Bloom	MC-LR, MC-YR, MC-RR	73
Ireland	*M. aeruginosa*	Bloom	Not identified	74[a]
Netherlands	*M. aeruginosa*	Bloom	Not identified	75
Norway	*M. aeruginosa*	Strain	MC-LR	76
Portugal	*M. aeruginosa*	Strain	MC-LR, -LA, -RR, -YR, -AR, [D-Asp^3]MC-LR	77
Russia	*M. aeruginosa*	Strain	[Dha^7]MC-YR, [Dha^7]MC-LR, [DAsp^3, Dha^7]MC-LR, [Dha^7]MC-RR, [DAsp^3, Dha^7]MC-RR	78
Slovenia	*Microcystis* spp.	Bloom	MC-RR, MC-LR	79
South Africa	*M. aeruginosa*	Strain	MC-LA	80
		Bloom, Strain	MC-LR, MC-YR, MC-YA, MC-YM	81
	M. aeruginosa	Bloom	MC-LR, -YR, -YA, -FA, -LA, -LAba	82
United States	*Microcystis* spp.	Bloom	MC-LR, -YR, -RR, -FR, -AR, -M(O)R, [DMAdda^5]MC-LR [Dha^7]MC-LR, [MSer^7]MC-LR, MC-WR [D-Glu-$OC_2H_3(CH_3)OH^6$]MC-LR	83
	Microcystis spp.	Bloom	[L-MeSer^7]MC-LR, [L-Me Lan^7]MC-LR, MC-(H_4)YR, MC-HilR	84

M(O), methionine S-oxide; Aba, aminoisobutyric acid; [DMAdda^5], 9-O-demethyl-Adda; Hil, homoisoleucine. The investigators did not identify microcystins in a bloom sample from a lake where *Microcystis* dominated.

a high percentage of toxic cells or toxic genetic groups, especially the typical type of *M. aeruginosa* even if nontoxic cells or groups are also constituents of the populations.

IV. SEASONAL CHANGES IN TOXIC *MICROCYSTIS*

Several morphological variations are often observed in bloom-forming *Microcystis* populations. In our study of Japanese lakes, the highest dominance was found in *M. aeruginosa*, and the rates of high relative abundance over 96.5% were estimated for 8 samples among 21 collected in 21 different bodies of water, whereas the highest abundance of *M. wesenbergii* was 71.0% and that of *M. viridis* was 79.6% (43).

Further, Komárek (42) recognized *M. ichthyoblabe* and *M. nobacekii* in *Microcystis* populations in Japanese lakes in addition to *M. aeruginosa, M. wesenbergii*, and *M. viridis*. However, we have often observed intermediate characteristics of *Microcystis* colonies which make their identification in natural populations difficult. Figure 4 shows a seasonal change of natural *Microcystis* populations accompanied by amounts of chlorophyll-a and microcystins in a pond in Tokyo (89). Colonies of *M. viridis* and *M. wesenbergii* are relatively easy to distinguish from *Microcystis* colonies, which show no clearly visible layer under a microscope. In the latter group, the colonies were divided into two groups depending on cell size: large and small cell groups (90). However, when the groups were compared with Komárek's classification (42), the *Microcystis* L corresponded to *M. aeruginosa*, and most colonies of *Microcystis* S were composed of those of *M. ichthyoblabe*.

A rapid growth of *Microcystis* spp. was observed from mid April to mid June (see Figure 4). At the later stage of the rapid growth, the highest amounts of total microcystins were estimated. Thereafter, the cell number of *Microcystis* fluctuated until October, whereas chlorophyll a concentration clearly decreased. The dominant species of *Microcystis* changed from *M. viridis* to *M. aeruginosa* S type (mostly *M. ichthyoblabe*). It is interesting that the contents of microcystins increased again in September and October followed by a small increase of chlorophyll a concentrations. This result suggested that the toxic problem of *Microcystis* was related to the growth stage of total *Microcystis* populations rather than to differences in species composition.

V. OCCURRENCES OF TOXIC *ANABAENA*

The toxins contained in *Anabaena* are both neurotoxins and hepatotoxins. However, it is still in question whether the microcystins contained in *Anabaena* cells actually caused the loss of animals, as the toxins were detected from almost isolated strains (91,92). Neurotoxins detected from *Anabaena* are anatoxin-a, anatoxin-a(S), and saxitoxins. Animal poisoning due to toxic *Anabaena* has occurred in relatively limited areas of the world, such as northern Europe, northern United States, Canada, and southeast regions of Australia. However, for example, animal poisoning caused by *An. circinalis* in the Dahring River in New South Wales in Australia resulted in enormous economic losses in 1992 (93).

A. Anatoxin A

Although a toxic strain of *Anabaena flos-aquae* isolated from a Canadian lake was first expected to contain several kinds of toxins (94), only two alkaloid neurotoxins, namely, anatoxin-a and anatoxin-a(S), and microcystin were characterized from the strain. Anatoxin-a is a secondary amine with a low molecular weight of 165 Da and a LD50 value of 200–250 μg/kg (Figure 5)

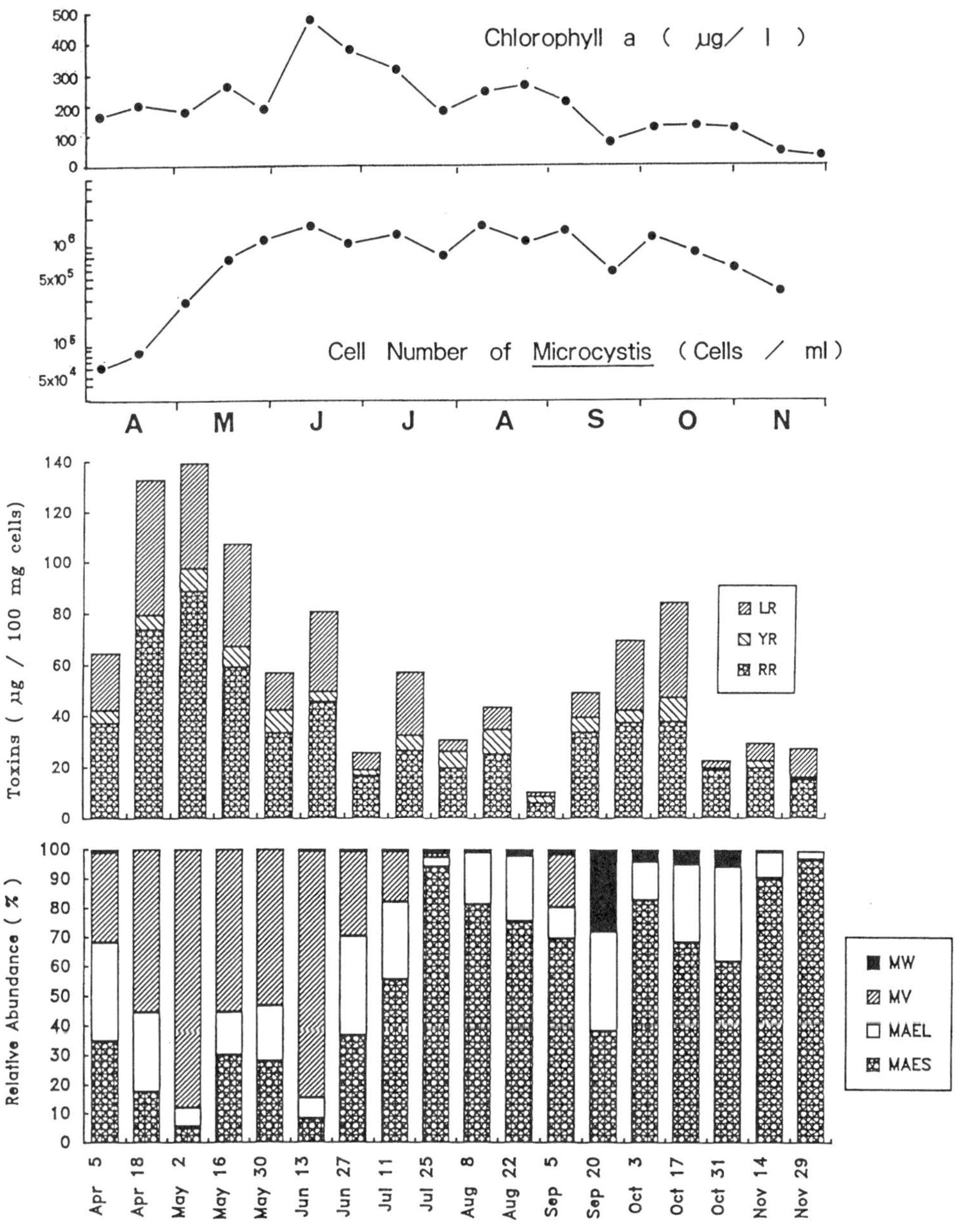

Figure 4 Seasonal changes in concentrations of cell number and chlorophyll a with compositions of microcystins and relative abundance in species composition observed in a pond in Tokyo. (From ref. 89.)

(95) whose structure was determined as 2-acetyl-9-azabicyclo (4-2-1)non-2-ene in the 1970s (96,97). The toxin is a potent postsynaptic depolarizing neuromuscular blocking agent that binds to the nicotinic-acetylcholine receptors and is not degraded by acetylchorineesterase (98).

Anatoxin-a was also reported to be contained in the toxic strains of *An. flos-aquae, An. lemmermannii, An. circinalis, Oscillatoria* sp., *Aphanizomenon* sp., and *Cylindrospermum* isolated from Finnish lakes (58). Bruno et al. detected anatoxin-a from *An. planktonica* forming

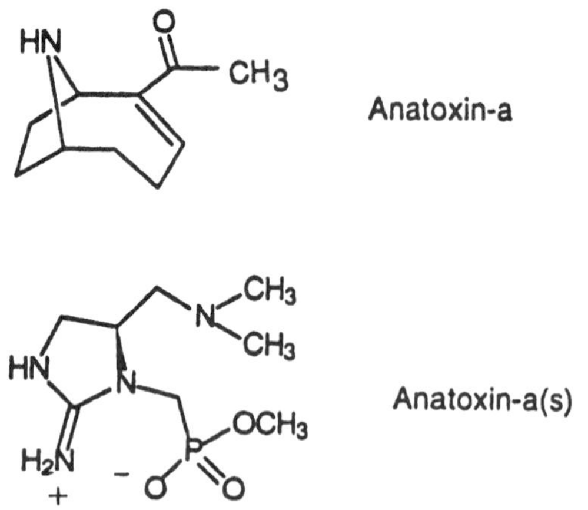

Figure 5 Structures of anatoxin-a and anatoxin-a(S).

a water bloom in a lake in Italy (99). Neurotoxic natural samples dominated by *An. flos-aquae* or *An. farciminiformis* were reported from Swedish lakes, but the components have not been clarified so far (27). From natural samples and isolated strains of *Anabaena* spp. in Japanese lakes, lower amounts of anatoxin-a were determined (100).

B. Anatoxin a(S)

Neurotoxic anatoxin-a(S) (see Figure 5) was distinguished from anatoxin-a by the fact that the former toxin induced hypersalivation, mucoid nasal discharge, and lacrimation in animals (101), although the toxins were contained in the same strain of *An. flos-aquae* NRC 525-17 isolated from a Canadian lake. Anatoxin-a(S) was characterized as a guanidiunium methyl phosphate with a molecular weight of 252 Da which binds to acetylcholineesterase and inhibits the enzyme from degrading acetylcholine (Figure 5) (98). Although animal poisoning of dogs and a few calves occurred only in Canada and the United States and were thought to be attributable to anatoxin-a(S) (98), a recent study revealed that anatoxin-a(S) was characterized from bloom-forming *An. lemmermanni* isolated from a lake in Denmark (29). In this case, the loss of birds, dogs, and fish was thought to be associated with the cyanobacterial species which formed the bloom in a Danish lake at that time.

C. Saxitoxins

The presence of neurotoxins in *Aph. flos-aquae* was first demonstrated by Sawyer et al. in 1968 in natural samples collected from lakes in New Hampshire in the United States (104). However, the structure was not determined until Ikawa et al. (105) confirmed the presence of saxitoxin and neosaxitoxin in *Aph. flos-aquae* in 1982. Saxitoxins and their analogues are potent neurotoxins which have been responsible for PSP in humans who consumed marine scallops, clams, and shellfish commercially produced (8). Saxitoxins and their derivatives are produced by marine dinoflagellates belonging to the genera *Alexandrium, Pyrodium*, and *Gymnodium* (see Chapter 2), which are taxonomically quite different from cyanobacteria.

The toxins are divided into three primary groups: carbamate, N-sulfocarbamoyl, and decarbomoyl compounds (see Figure 6) (106). The most toxic carbamate compounds, such as saxi-

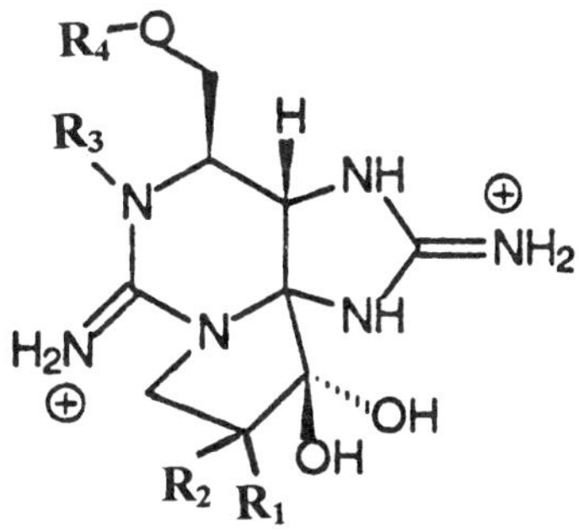

Toxin	R_1	R_2	R_3	R_4
STX	H	H	H	$CONH_2$
NEOSTX	H	H	OH	$CONH_2$
GTX_2	H	OSO_3^-	H	$CONH_2$
GTX_3	OSO_3^-	H	H	$CONH_2$
$dcGTX_2$	H	OSO_3^-	H	H
$dcGTX_3$	OSO_3^-	H	H	H

Figure 6 Structure of saxitoxins.

toxin, neosaxitoxin, and gonyautoxins, are thought to be responsible for the toxicity symptoms. Saxitoxin binds to nerve membrane and block the sodium channel of excitable membranes (107).

In the Murray-Dahring Basin in southeast Australia, animal deaths have been attributed to the occurrences of toxic cyanobacterial blooms and hepatotoxic and neurotoxic effects have been detected, but the neurotoxic substances have not been clarified (108). Recently, a wide range of saxitoxins were identified from *Anabaena circinalis* which formed severe blooms in

Table 4 Microcystins from *Anabaena* spp.

Species	Country	Microcystins	Reference
An. flos-aquae	Canada	[D-Asp^3]MC-LR, MC-LR	89
An. flos-aquae	Canada	[D-Asp^3]MC-HtyR	110
Anabaena sp.	Finland	[Dha^7]MC-HtyR, [D-Asp^3, Dha^7]MC-HtyR [L-Ser^7]MC-HtyR, [Dha^7]MC-HphR, MC-YR, [D-Asp^3]MC-HtyR, MC-FR	111
Anabaena sp.	Finland	[D-Asp^3, Dha^7]MC-LR, [Dha^7]MC-LR [D-Asp^3, Dha^7]MC-RR, [D-Asp^3]MC-RR [Dha^7]MC-RR, MC-RR	112
An. flos-aquae	Norway	[D-Glu(OMe)^6]MC-LR, [D-Asp^3, D-Glu(OMe)^6]MC-LR MC-LR, [D-Asp^3]MC-LR, MC-RR, [D-Asp^3]MC-RR	113
Anabaena sp.	Finland	[Dha^7]MC-E(OMe)E(OMe) [D-Asp^3, Dha^7]MC-E(OMe)E(OMe) [L-Ser^7]MC-E(OMe)E(OMe) [D-Asp^3, L-Ser^7]MC-E(OMe)E(OMe) [Dha^7]MC-EE(OMe)	114

Hty, homotyrosine; Hph, homophenylalanine; E(OMe), L-glutamic acid δ-methyl ester.

the surface waters, including saxitoxins (STXs), almost all gonyautoxins (GTXs), and decarbamoyl GTXs (109). The natural samples showed high abundances of GTX-2 and GTX-3, and the toxicities of STX equivalent ranged from 14.7 to 569 μg/g.

D. Microcystins

Microcystins were also detected from *Anabaena* species, although only the strains were isolated in Canadian and Finnish lakes. Since Krishnamurthy (91) identified microcystin-LR and [D-Asp^3]MC-LR in *An. flos-aquae*, more than 20 microcystins have been characterized (Table 4) (110–114). It has been shown that *Anabaena* species produced different variants of microcystins from the *Microcystis* species. The variants were composed of methyl ester derivatives of glutamic acid or homotyrosine as L-amino acid. Recently, Namikoshi et al. (114) characterized seven microcystins which posses L-glutamic acid and/or its methyl ester [E(OMe)] units as L-amino acids.

VI. OCCURRENCES OF TOXIC SPECIES OF *OSCILLATORIA*

Microcystins and anatoxin-a are toxins found in the *Oscillatoria* species. The occurrences of animal poisoning due to toxic *Oscillatoria* were first reported from England, where an incident involving the death of cattle occurred in a lake overwhelmingly dominated by *O. agardhii (Planktothrix agardhii)* (115). The toxic strains of *O. agardhii* were isolated from lakes in Norway and Finland. Two color types, red and green, are known among the strains of *O. agardhii* (116), but this does not relate to toxin production.

A toxin from *O. agardhii* was first identified as cyclo-(Ala-Arg-Asp-Arg-Adda-Glu-N-methyldehydroAla), namely, [D-Asp3]MC-RR, from a green strain by Meriluoto (117). Thereafter, from 13 strains of *O. agardhii* isolated from eight different Finnish lakes, Luukainen et al. (1993) isolated eight microcystins, six of which were identified as [D-Asp^3]MC-LR, [Dha^7]MC-LR, [D-Asp^3]MC-RR, [Dha^7]MC-RR, [d-Asp^3, Dha^7]MC-RR, and [D-Asp^3,Mser^7]MC-RR (118). Metalimnetic water blooms are often reported with occurrences of *O. agardhii* var. *ithothrix* in lakes whose surface water is not so turbid that the light cannot penetrate into the deeper water layers. The vertical distribution *O. agardhii* was described by the existence of a toxin peak in metalimnion in a lake in Finland (46).

Edwards et al. identified anatoxin-a from benthic *Oscillatoria* in a river whose waters had caused dog poisoning in Scotland (119). Homoanatoxin-a, an anatoxin homologue, was isolated from a strain of *Oscillatoria formosa* Boy in Norway by Skulberg et al. (120). Sedmak and Koji (1997) detected two peaks of microcystins and identified one of them as microcystin-YR in a natural bloom sample of *O. rubescens* in Slovenia (79)

VII. OCCURRENCES OF OTHER TOXIC SPECIES OF CYANOBACTERIA

Nodularia spumigena Mertens is the first species reported to have caused animal poisoning in Australia in the 19th century (50). This species develops heavy water blooms in brackish waters, and occurrences of toxic blooms have been reported in Australia, New Zealand, and the countries along the Baltic Sea and the coastal lakes along the North Sea (44,121,122). *Nodularia* toxin, named nodularin, is a cyclic pentapeptide hepatotoxin which has a novel amino acid, Adda,

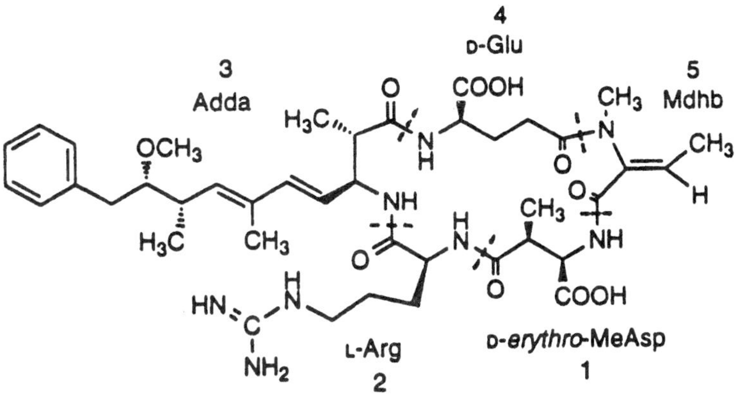

Figure 7 Structure of nodularin.

similarly to microcystin (Figure 7). Also similarly to microcystin, nodularin has caused death in mammals from liver hemorrhages. Both toxins inhibit protein phosphatases 1 and 2A in the liver (123), but it has been shown that nodularin is possibly a carcinogen (124). The accumulation of the toxin in the edible mussel *Mytilus edulis* was reported, and the lethal toxicity in the gut as high as 90 mg/kg to mice was estimated by intraperitonial injection when a high biomass of *N. spumigena* was observed (125). According to Lehitmäki et al. (1994), the temperature had a significant effect on growth and toxin production, whereas salinity had an effect on toxin production in one strain but no effect in another strain (126).

Cylindrospermopsis raciborskii (Wolosz.) Seeya and Subba Raju are cyanobacteria which caused hepatoenteritis in Palm Island in Australia, as described in Section II. The toxin, later named cylindrospermopsin, was released into surrounding water when the lake water was treated with copper sulfate. The toxin, identified as an alkaloid with a molecular weight of 415 Da (Figure 8), had effects in the liver, kidney, thymus, and heart (54,55). Terao et al. (1994) reported that the toxin inhibited protein synthesis (126). The dominance of *Cy. racirborskii* was also reported in one arm of Billings Reservoirs in São Paulo State in Brazil, where the species dominated from November to May with the highest density of 21,000 organisms per milliliter (38). In addition, the two cases of fish death occurring in the reservoir were supposed to be related to the presence of the species, although no data have been shown. Cylindrospermopsin was also isolated from a new species of cyanobacteria, *Umezakia natans*, whose trichomes extend vertically (127).

Apha. flos-aquae isolated in a lake in New Hampshire is the first species reported to contain saxitoxins in freshwater. By using the species, the biosynthesis of saxitoxin and neosaxitoxin was studied (128). The high possibility of neurotoxin production was shown in natural samples

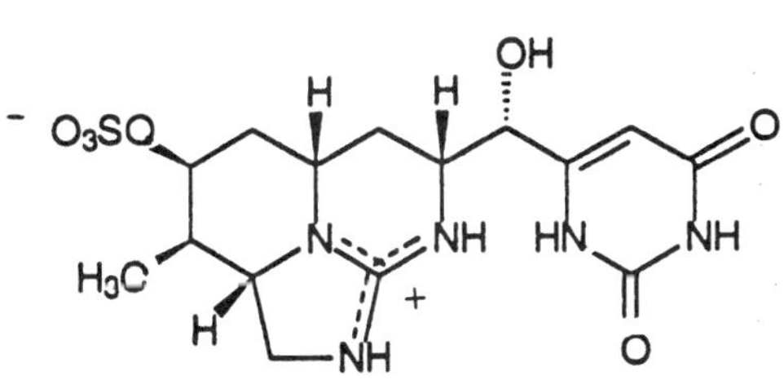

Figure 8 Structure of cylindrospermopsin.

dominated by *Apha. flos-aquae* in lakes in Finland (28), and Ekman-Ekebom et al. detected anatoxin-a from a natural bloom sample dominated by *Apha flos-aquae* in a lake (130).

Recently, the presence of PSP was reported from a mat-forming filamentous cyanobacterium, *Lyngbya wollei* (Farlow ex Gomont) comb. nov., which produced a very high-standing crop in lakes and reservoirs of the southeastern United States (131). The minimum lethal doses of natural samples were reported to be 150–1500 mg/kg of the lyophilized cells. Onodera et al. (132) identified six new saxitoxin analogues in addition to decarbamoylsaxitoxin and decarbamoylgonyautosins 2 and 3.

As described by Margalef (133), succession occupies a place in ecology similar to that of evolution in general biology. In an ecosystem, the process of action and reaction between the environment and the biological community brings about changes in the both components, and this causes succession. In a series of successions that occur in a lake, nutrient loading is considered to promote the growth of phytoplankton and the water plant, and the resulting accumulation of organic matter, which proceed to make a lake more shallow together with soil and sand. Even if such a process of eutrophication of a lake can be considered to be a natural phenomenon, recent nutrient loading has been accelerated by human activity. Blooms of cyanobacteria are commonly associated with nutrient-rich waters, which we have to use now and also in the future. Therefore, we must pay attention to the occurrences of toxic cyanobacteria when we use waters of lakes and reservoirs for drinking, keeping in mind the tragedy that happened in Brazil (11).

REFERENCES

1. SM Awramik. Ancient stromatolites and microbial mats. In: Y Cohen, RW Castenholz, HO Halvorsson eds. Microbial Mats: Stromatolites. New York: Liss, 1984, pp 1–22.
2. M Herdman. Akinates: structure and function. In: P Fay, CV Baalen, eds. The Cyanobacteria. Amsterdam: Elsevier, 1988, pp 227–250.
3. EPY Tang, R Tremblay, WF Vincent. Cyanobacterial dominance of polar freshwater ecosystems: Are high-latitude mat-formers adapted to low temperature? J Phycol 33:171–181, 1997.
4. J Komárek, J Lukavský. *Arthronema*, a new cyanophyte genus from Afro-Asian deserts. Arch Hydrobiol 80(Suppl):249–267, 1988.
5. HW Pearl. Nuisance phytoplankton blooms in coastal, estuarine, and inland waters. Limnol Oceanogr 33:823–847, 1988.
6. T Zohary, AM Pais-Madeira, R Robarts, KD Hambright. Interannual phytoplankton dynamics of a hypertrophic African lake. Arch Hydrobiol 136:105–126, 1996.
7. WW Carmichael, IR Falconer. Diseases related to freshwater blue-green algal toxins, and control measures. In: IR Falconer, ed. Algal Toxins in Seafood and Drinking Water. London: Academic Press, 1993, pp 187–209.
8. T Yasumoto, M Murata. Marine toxins. Chem Rev 93:1897–1909, 1993.
9. JE Eriksson, JAO Meriluoto, T Lindholm. Accumulation of a peptide toxin from the cyanobacterium *Oscillatoria agrdhii* in the freshwater mussel *Anadonta cygnea*. Hydrobiologia 183:211–216, 1989.
10. MF Watanabe, HD Park, F Kondo, K Harada, H Hayashi, T Okino. Identification of microcystins in freshwater mussels. Nat Toxins 5:31–35, 1997.
11. EM Jochimsen, WW Carmichael, JS An, DM Cardo, ST Cookson, CEM Holmes, et al. Liver failure and death after exposure to microcystins at a hemodialysis center in Brazil. New Engl J Med 338: 873–878, 1998.
12. CS Reynolds. Cyanobacterial water-blooms. Adv Bot Res 13:67–143, 1987.
13. WW Carmichael. A status report on planktonic cyanobacteria (blue-green alge) and their toxins. United States Environmental Protection Agency, EPA/600/R-92/079. Washington DC: Office of Research and development, 1992.

14. HW Pearl. A comparison of cyanobacterial bloom dynamics in freshwater, estuarine and marine environments. Phycologia 35(Suppl):25–35, 1996.
15. AE Walshby. Mechanisms of buoyancy regulation by planktonic cyanobacteria with gas vesicle. In: P Fay, CV Baalen, eds. The Cyanobacteria. Amsterdam: Elsevier, 1988, pp 377–392.
16. RD Simon. Inclusion bodies in the cyanobacteria: cyanophycin, polyphosphate, polyhedral bodies. In: P Fay, CV Baalen, eds. The Cyanobacteria. Amsterdam: Elsevier, 1988, pp 199–225.
17. HW Pearl. Growth and reproductive strategies of freshwater blue-green algae (Cyanobacteria). In: CD Sandgren, ed. Growth and Reproductive Strategies of Freshwater Phytoplankton. New York: Cambridge University Press, 1988, pp 261–312.
18. J Komárek, K Anagnostidis. Modern approach to the classification system of cyanophytes. 2-Chroococcales. Arch Hydrobiol 73(Suppl):157–226, 1986.
19. K Anagnostidis, J Komárek. Modern approach to the classification system of cyanophytes. 3-Oscillatoriales. Arch Hydrobiol 80(Suppl):327–472, 1988.
20. J Komárek, K Anagnostidis. Modern approach to the classification system of cyanophytes. 4-Nostocales. Arch Hydrobiol 82 (Suppl):247–345, 1989.
21. K Anagnostidis, J Komárek. Modern approach to the classification system of cyanophytes. 5-Stigonematales. Arch Hydrobiol Algol Stud 59:1–73, 1990.
22. AD Smith, JJ Gilbert. Spatial and temporal variability in filament length of a toxic cyanobacterium (*Anabaena affinis*). Freshwater Biol 33:1–11, 1995.
23. M Watanabe. Studies on planktonic blue-green algae 6. Bloom-forming species in Lake Biwa (Japan) in the summer of 1994. Bull Natl Sci Mus Tokyo B22:1–10, 1996.
24. V May. The occurrence of toxic cyanophyte blooms in Australia. In: WW Carmichael, ed. The Water Environment: Algal Toxins and Health. New York: Plenum Press, 1981, pp 127–142.
25. L Ramberg. Relations between planktic blue-green algal dynamics and environmental factors in four eutrophic Swedish lakes. Arch Hydrobiol 112:161–175, 1988.
26. M Watanabe, Y Niiyama. Freshwater algae from Lake Akan (5). Bull Natl Sci Mus Tokyo B16: 29–39, 1990 (in Japanese).
27. T Willén, R Mattsson. Water-blooming and toxin-producing cyanobacteria in Swedish fresh and brackish waters, 1981–1995. Hydrobiologia 353:181–192, 1997.
28. K Sivonen, SI Niemelä, RM Niemi, L Lepistö, TH Luona, LA Räsänen. Toxic cyanobacteria (blue-green algae) in Finnish fresh and coastal waters. Hydrobiologia 190:267–275, 1990.
29. P Henriksen. Detection of an anatoxin-a(s)–like anticholinesterase in natural blooms and cultures of cyanobacteria/blue-green algae from Danish lakes and in the stomach contents of poisoned birds. Toxicon 35:901–913, 1997.
30. M Watanabe. Studies on planktonic blue-green algae 4. Some *Anabaena* species with straight trichomes in Japan. Bull Natl Sci Mus Tokyo B18:123–137, 1992.
31. M Watanabe. The species of *Anabaena* from Hokkaido. J Jpn Bot 46:263–278, 1971.
32. PE Kellar, HW Pearl. Physiological adaptations in response to environmental stress during an N_2-fixing *Anabaena* bloom. Appl Environ Microbiol 40:587–595, 1980.
33. T Lanaras, CM Cook. Toxin extraction from an *Anabaenopsis milleri*-dominated bloom. Sci Total Environ 142:163–169, 1994.
34. S Hino, M Tada. Seasonal changes of nutrients, chlorophyll-a, and organic matter concentrations in highly eutrophic Lake Barato, Japan. Jpn J Limnol 46:268–278, 1985.
35. A Knopka. Metalimnetic cyanobacteria in hard-water lakes: Buoyancy regulation and physiological state. Limnol Oceanogr 34:1174–1184, 1989.
36. BA Jacobsen. Bloom formation of *Gloeotrichia echinulata* and *Aphanizomenon flos-aquae* in a shallow, eutrophic, Danish lake. Hydrobiologia 289:193–197, 1994.
37. U Pollingher, O Hadas, YZ Yacobi, T Zohary, T. Berman. *Aphanizomenon ovalisporum* (Forti) in Lake Kinneret, Israel. J Plankton Res 20:1321–1339, 1998.
38. RCR de Souza, MC Carvalho, AC Truzzi. *Cylindrospermopsis raciborskii* (Wolosz.) Seenaya and Subba Raju (Cyanophyceae) dominance and a contribution to the knowledge of Rio Pequeno arm, Billings Reservoir, Brazil. Environ Wat Quality 13:73–81, 1998.
39. M Présing, S Herodek, L Vörös, I Kóbor. Nitrogen fixation, ammonium and nitrate uptake during a bloom of *Cylindrospermopsis raciborskii* in Lake Balaton. Arch Hydrobiol 136:553–562, 1996.

40. CS Reynolds, JG Tundisi, K Hino. Observations on a metalimnetic *Lyngbya* population in a stably stratified tropical lake (Lagoa Carioca, Eastern Brasil). Arch Hydrobiol 97:7–17, 1983.
41. T Zohary, AM Pais Madeira. Structural, physical and chemical characteristics of *Microcystis aeruginosa* hyperscums from a hypertrophic lake. Freshwater Biol 23:339–352, 1990.
42. J Komárek. A review of water-bloom forming *Microcystis* species, with regard to populations from Japan. Arch Hydrobiol Algol Stud 64:115–127, 1991.
43. Y Watanabe, MF Watanabe, M. Watanabe. The distribution and relative abundance of bloom forming *Microcystis* species in several eutrophic waters. Jpn J Limnol 47:87–93, 1986.
44. K Sivonen, K Kononen, WW Carmichael, AM Dahlem, KL Rinehart, J Kiviranta, SI Niemelä. Occurrence of the hepatotoxic cyanobacterium *Nodularia spumigena* in the Baltic Sea and the structure of the toxin. Appl Environ Microbiol 55:1990–1995, 1989.
45. GJ Jones, SI Blackburn, NS Parker. A toxic bloom of *Nodularia spumigena* Mertens in Orielton Lagoon, Tasmania. Aust J Mar Freshwater Res 45:787–800, 1994.
46. T Lindholm, JE Eriksson, JAO Meriluoto. Toxic cyanobacteria and water quality problems-examples from a eutrophic lake on Åland, southwest Finland. Wat Res 23:481–486, 1989.
47. C Berger. Consistent blooming of *Oscillatoria agardhii* Gom. in shallow hypertrophic lakes. Verh Int Verein Limnol 22:910–916, 1984.
48. S Romo, R Miracle. Long-term periodicity of *Planktothrix agardhii*, *Pseudoanabaena galeata* and *Geitlerinema* sp. in a shallow hypertrophic lagoon, the Albufera of Valencia (Spain). Arch Hydrobiol 126:469–486, 1993.
49. MF Watanabe. Studies on the metalimnetic blue-green alga *Oscillatoria mougeotii* in a eutrophic lake with special reference to its population growth. Arch Hydrobiol 86:66–86, 1979.
50. G Francis. Poisonous Australian lake. Nature 18:11–12, 1878.
51. MF Watanabe. Occurrences of toxic cyanobacteria. In: MF Watanabe, KI Harada, H Fujiki eds., Water Bloom of Blue-green Algae and Their Toxins. Tokyo: University of Tokyo Press, 1994, pp 55–73 (in Japanese).
52. M Schwimmer, D Schwimmer. Medical aspects of phycology. In: DF Jackson ed., Algae, Man and the Environment. New York: Syracuse University Press, 1968, pp 279–358.
53. S Byth. Palm Island mystery disease. Med J Aust 2:40–42, 1980.
54. PR Hawkins, MTC Runnegar, ARB Jackson, IR Falconer. Severe hepatotoxicity caused by the tropical cyanobacterium (Blue-green alga) *Cylindrospermopsis raciborskii* (Woloszynska) Seenaya and Subba Raju isolated from a domestic water supply reservoir. Appl Environ Microbiol 50:1292–1295, 1985.
55. I Ohtani, RE Moore, MTC Runnegar. Cylindrospermopsin: a potent hepatotoxin from the blue-green alga *Cylindrospermopsis raciborskii*. J Am Chem Soc 114:7941–7942, 1992.
56. WH Billings. Water-associated human illness in northeast Pennsylvania and its suspected association with blue-green algae blooms. In: WW Carmichael, ed. The Water Environment: Algal Toxins and Health. New York: Plenum Press, 1981, pp 243–255.
57. MF Watanabe, K Harada, K Matsuura, S Oishi, Y Watanabe, M Suzuki. Heptapeptide toxins contained in natural samples of *Microcystis* species. Toxicity Assess 4:487–497, 1989.
58. K Sivonen, K Himberg, R Luukkainen, SI Niemelä, GK Poon, GA Codd. Preliminary characterization of neurotoxic blooms and strains from Finland. Toxicity Assess 4:339–352, 1989.
59. MF Watanabe, S Oishi, K Harada, K Matsuura, H Kawai, M Suzuki. Toxins contained in *Microcystis* species of cyanobacteria (Blue-green algae). Toxicon 26:1017–1025, 1988.
60. GJ Jones, PT Orr. Release and degradation of microcystin following algicide treatment a *Microcystis aeruginosa* bloom in a recreational lake, as determined by HPLC and protein phosphatase inhibition assay. Wat Res 28:871–876, 1994.
61. SMFO Azevedo, WR Evans, WW Carmichael, M Namikoshi. First report of microcystins from Brazilian isolate of the cyanobacterium *Microcystis aeruginosa*. J Appl Phycol 6:261–265, 1994.
62. M Craig, L McCready, HA Luu, MA Smillie, P Dubord, CFB Holmes. Identification and characterization of hydrophobic microcystins in Canadian freshwater cyanobacteria. Toxicon 31:1541–1549, 1993.
63. BG Kotak, AK-Y Lam, EE Prepas, SL Kenefick, SE Hrudey. Variability of the hepatotoxin microcystin-LR in hypereutrophic drinking water lakes. J Phycol 31:248–263, 1995.

64. WW Carmichael, JW He, J Eschedor, ZR He, YM Juan. Partial structural determination of hepatotoxic peptides from *Microcystis aeruginosa* (Cyanobacterium) collected in ponds of Central China. Toxicon 26:1213–1217, 1988.
65. P Henriksen, Ø Moestrup. Seasonal variations in microcystin contents of Danish Cyanobacteria. Nat Toxins 5:99–106, 1997.
66. LA Lawton, C Edwards, KA Beattie, S Pleasance, GJ Dear and GA Codd. Isolation and characterization of microcystins from laboratory cultures and environmental samples of *Microcystis aeruginosa* and from an associated animal toxicosis. Nat Toxins 3:50–57, 1995.
67. M Namikoshi, K Sivonen, WR Evans, F Sun, WW Carmichael, KL Rinehart. Isolation and structures of microcystins from a cyanobacterial water bloom (Finland). Toxicon 30:1473–1479, 1992.
68. J Kiviranta, M Namikoshi, K Sivonen, WR Evans, WW Carmichael, KL Rinehart. Structure determination and toxicity of a new microcystin from *Microcystis aeruginosa* strain 205. Toxicon 30: 1093–1098, 1992.
69. R Luukkainen, M Namikoshi, K Sivonen, KL Rinehart, SI Niemel. Isolation and identification of 12 microcystins from four strains and two bloom samples of *Microcyustis* spp.: structure of a new hepatotoxin. Toxicon 32:133–139, 1994.
70. C Vezie, L Brient, K Sivonen, G Bertru, JC Lefeuvre, M Salkinoja-Salonen. Variation of microcystin content of cyanobacterial blooms and isolated strains in Lake Grand-Lieu (France). Microb Ecol 35:126–135, 1998.
71. KI Harada, K Matsuura, M Suzuki, MF Watanabe, S Oishi, AM Dahlem, VR Beasley, WW Carmichael. Isolation and Characterization of the minor components associated with microcystins LR and RR in the cyanobacteria (blue-green algae). Toxicon 28:55–64, 1990.
72. M Shirai, A Ohtake, T Sano, S Matsumoto, T Sakamoto, A Sato, T Aida, K Harada, T Shimada, M Suzuki, M Nakano. Toxicity and toxins of natural blooms and isolated strains of Microcystis spp. (cyanobacteria) and improved procedure for purification of cultures. Appl Environ Microbiol 56:1241–1245, 1991.
73. C Jakobi, KL Rinehart, GA Codd, I Carmienke, J Weckesser. Occurrence of toxic water blooms containing microcystins in a German lake over a three year period. System Appl Microbiol 19: 249–254, 1996.
74. IR Scherlock, K James, FB Caudwell, C Mackintosh. First identification of microcystins in Irish lakes aided by a new derivatisation procedure for electrospray mass spectrometric analysis. Nat Toxins 5:247–254, 1997.
75. FI Kappers. Investigation of the presence of toxins produced by cyanobacteria (Blue-green algae) in the Netherlands. Sci Total Environ 18:359–361, 1981.
76. K Berg, WW Carmichael, OM Skulberg, C Benestad, B Underdal. Investigation of a toxic waterbloom of *Microcystis aeruginosa* (Cyanophyceae) in Lake Akerscvatn, Norway. Hydrobiologia 144: 97–103, 1987.
77. VM Vasconcelos, K Sivonen, WR Evans, WW Carmichael, M Namikoshi. Isolation and characterization of microcystins (heptapeptide hepatotoxins) from Portuguese strains of *Microcystis aeruginosa* Kutz. Emnd Elenkin. Arch Hydrobiol 134:295–305, 1995.
78. K Sivonen, M Namikoshi, WR Evans, BV Gromov, WW Carmichael, KL Rinehart. Isolation and structures of five microcystins from a Russian *Microcystis aeruginosa* strain CALU 972. Toxicon 30:1481–1485, 1992.
79. B Sedmak and G Kosi. Microcystins in Slovene freshwaters (Central Europe)—First report. Nat Toxins 5:64–73, 1997.
80. DP Botes, AA Tuinman, PL Wessels, CC Viljoen, H Kruger, DH Williams, S Santikarn RJ Smith, SJ Hammond. The structure of cyanoginosin-LA, a cyclic heptapeptide toxin from the cyanobacterium *Microcystis aeruginosa.* J Chem Soc Perkin Trans 1:2311–2318, 1984.
81. DP Botes, PL Wessels, H Kruger, MTC Runnegar, S Santikarn, RJ Smith, JCJ Barna, DH Williams. Structural studies on cyanoginosins-LR, -YR, -YA, and -YM, peptide toxins from *Microcystis aeruginosa.* J Chem Soc Perkin Trans 1:2747–2748, 1985.
82. WE Scott. Occurrence and significance of toxic cyanobacteria in South Africa. Wat Sci Technol 23:175–180, 1991.
83. M Namikoshi, KL Rinehart, R Sakai, RR Stotts, AM Dahlem, VR Beasley, WW Carmichael, WR

Evans. Identification of 12 hepatotoxins from a Homer Lake bloom of the cyanobacteria *Microcystis aeruginosa*, *Microcystis viridis*, *Microcystis wesenbergii*; nine new microcystins. J Org Chem 57: 866–872, 1992.

84. M Namikoshi, F Sun, BW Choi, KL Rinehart, WW Carmichael, WR Evans, VR Beasley. Seven more microcystins from Homer Lake cells: application of the general method for structure assignment of peptides containing α, β-dehydroamino acid unit(s). J Org Chem 60:3671–3679, 1995.
85. MF Watanabe, M Watanabe, T Kato, KI Harada and M Suzuki. Composition of cyclic peptide toxins among strains of *Microcystis aeruginosa* (blue-green algae, Cyanobacteria). Bot Mag Tokyo 104:49–57, 1991.
86. MF Watanabe, S Oishi. Toxicities of *Microcystis aeruginosa* collected from some lakes, reservoirs, ponds and a moat in Tokyo and adjacent regions. Jpn J Limnol 51:5–9, 1980.
87. PD Baker, AR Humpage. Toxicity associated with commonly occurring cyanobacteria in surface waters of the Murray-Darling Basin, Australia. Aust J Mar Freshwater Res 45:773–786, 1994.
88. VM Vasconcelos. Toxic cyanobacteria (blue-green algae) in Portuguese fresh waters. Arch Hydrobiol 130:439–451, 1994.
89. MF Watanabe, HD Park, M Watanabe. Composition of *Microcystis* and heptapeptide toxins. Verh Int Verein Limnol 25:2226–2229, 1994.
90. T Kato, MF Watanabe, M Watanabe. Allozyme divergence in *Microcystis* (Cyanophyceae) and its taxonomic inference. Arch Hydrobiol Algol Stud 64:129–140, 1991.
91. T Krishnamurthy, L Szafraniec, DF Hunt, J Shabanowitz, JR Yates, CR Hauer, WW Carmichael, O Skulberg, GA Codd, S Missler. Structural characterization of toxic cyclic peptides from blue-green algae by tandem mass spectrometry. Proc National Acad Sci USA 86:770–774, 1989.
92. M Namikoshi, K Sivonen, WR Evans, WW Carmichael, L Rouhiainen, R Luukkainen, KL Rinehart. Structures of three new homotyrosine-containing microcystins and a new homophenylalanine variant from *Anabaena* sp. strain 66. Chem Res Toxicol 5:661–666, 1992.
93. R Ressom, FS Soong, J Fitzgerald, L Turczymowicz, OE Saadi, D Roder, T Maynard, I Falconer. Health effects of toxic cyanobacteria (blue-green algae). National Health and Medical Research Council. Australian Government Publishing Service, 1994.
94. WW Carmichael, PR Gorham. Anatoxins from clones of *Anabaena flos-aquae* isolated from lakes of western Canada. Mitt Int Verein Limnol 21:285–295, 1978.
95. M Namikoshi, KL Rinehart. Bioactive compounds produced by cyanobacteria. J Indus Microbiol 17:373–384, 1996.
96. CS Huber. The crystal structure and absolute configulation 2,9-diacetyl-9-azabycyclo(4,2,1)non-2,3-ene. Acta Crystallogr B 28:2577–2582, 1972.
97. JP Delvin, OE Edwards, PR Gorham, MR Hunter, RK Pike, B Starvic. Anatoxin-a, a toxic alkaloid from *Anabaena flos-aquae*. Can J Chem 55:1367–1371, 1977.
98. WW Carmichael. The toxins of cyanobacteria. Sci Am Jan:64–72, 1994.
99. M Bruno, DA Barbini, E Pierdominic, AP Serse, A Ioppolo. Anatoxin-a and previously unknown toxin in *Anabaena planctonica* from blooms found in Lake Mulargia (Italy). Toxicon 32:369–373, 1994.
100. HD Park, MF Watanabe, KI Harada, H Nagai, M Suzuki, M Watanabe, H Hayashi. Hepatotoxin (microcystin) and neurotoxin (anatoxin-a) contained in natural blooms and strains of cyanobacteria from Japanese freshwaters. Nat Toxins 1:353–360, 1993.
101. WO Cook, VR Beasley, RA Lovell, AM Dahlem, SB Hooser, NA Mahmood, WW Carmichael. Consistent inhibition of peripheral chlolinesterases by neurotoxins from the freshwater cyanobacterium *Anabaena flos-aquae*: Studies of ducks, swine, mice and a steer. Environ Toxicol Chem 8: 915–922, 1989.
102. S Matsunaga, RE Moore, WP Niemczura, WW Carmichael. Anatoxin-a(S), a potent anticholinesterase from *Anabaena flos-aquae*. J Am Chem Soc 111:8021–8023, 1989.
103. NA Mahmood, WW Carmichael. Anticholinesterase poisonings in dogs from a cyanobacterial (blue-green algae) bloom dominated by *Anabaena flos-aquae*. Am J Vet Res 49:500–503, 1988.
104. PJ Sawyer, JH Gentile, JJ Sasner. Demonstration of a toxin from *Aphanizomenon flos-aquae* (L.) Ralfs. Can J Microbiol 14:1199–1204, 1968.

105. M Ikawa, K Wegener, TL Foxall, JJ Sasner Jr. Comparison of the toxins of the blue-green alga *Aphanizomenon flos-aquae* with the *Gonyaulax* toxins. Toxicon 20:747–752, 1982.
106. JJ Sullivan. Methods of analysis for algal toxins: dinoflagellate and diatom toxins. In: IR Falconer, ed. Algal Toxins in Seafood and Drinking Water. London: Academic Press, 1993, pp 29–48.
107. DG Baden, VL Trainer. Mode of action of toxins of seafood poisoning. In: IR Falconer, ed. Algal Toxins in Seafood and Drinking Water. London: Academic Press, 1993, pp 49–74.
108. PD Baker, AR Humpage. Toxicity associated with commonly occurring cyanobacteria in surface waters of the Murray-Darling Basin, Australia. Aust J Mar Freshwater Res 45:773–786, 1994.
109. AR Humpage, J Rositano, AH Bretag, R Brown, PD Baker, BC Nicholson, DA Steffensen. Paralytic shellfish poisons from Australian cyanobacterial blooms. Aust J Mar Freshwater Res 45:761–771, 1994.
110. KI Harada, K Ogawa, Y Kimura, H Murata, M Suzuki, PM Thorn, WR Evans, WW Carmichael. Microcystins from *Anabaena flos-aquae* NRC 525-17. Chem Res Toxicol 4:535–540, 1991.
111. M Namikoshi, K Sivonen, WR Evans, WW Carmichael, L Rouhiainen, R Luukkainen, KL Reinhart. Structures of three new homotyrosine-containing microcystins and a new homophenylalanine variant from *Anabaena* sp. strain 66. Chem Res Toxicol 5:661–666, 1992.
112. K Sivonen, M Namikoshi, WR Evans, WW Carmichael, F Sun, L Rouhiainen, R Luukkainen, KL Rinehart. Isolation and characterization of a variety of microcystins from seven strains of the cyanobacterial genus *Anabaena.* Appl Environ Microbiol 58:2495–2500, 1992.
113. K Sivonen, OM Skulberg, M Namikoshi, WR Evans, WW Carmichael, KL Rinehart. Two methyl ester derivates of microcystins, cyclic heptapeptide hepatotoxins, isolated from *Anabaena flos-aquae* strain CYA 83/1. Toxicon 30:1465–1471, 1992.
114. M Namikoshi, M Yuan, K Sivonen, WW Carmichael, KL Rinehart, L Rouhiainen, F Sun, S Brittain, A Otsuki. Seven new microcystins possessing two L-glutamic acis units, isolated from *Anabaena* sp. strain 186. Chem Res Toxicol 11:143–149, 1998.
115. CS Reynolds. Cattle deaths and blue-green algae: a possible instance from Cheshire, England. J Inst Water Eng Sci 34:74–76, 1983.
116. OM Skulberg, GA Codd, WW Carmichel. Toxic blue-green algal blooms in Europe: a growing problem. AMBIO 13:244–247, 1984.
117. JAO Meriluoto, A Sandström, JE Eriksson, G Remaud, AG Craig, J Chattopadhyaya. Structure and toxicity of a peptide hepatotoxin from the cyanobacterium *Oscillatoria agardhii.* Toxicon 27:1021–1034, 1989.
118. R Luukkainen, K Sivonen, M Namikoshi, M Färdig, KL Rinehart, SI Niemelä. Isolation and identification of eight microcystins from 13 *Oscillatoria agardhii* strains and structure of a new microcystin. Appl Environ Microbiol 59:2204–2209, 1993.
119. C Edwards, KA Beattie, CM Scrimgeour, GA Codd. Identification of anatoxin-a in benthic cyanobacteria (blue-green algae) and in associated dog poisonings at Loch Insh, Scotland. Toxicon 30: 1165–1175, 1992.
120. OM Skulberg, WW Carmichael, RA Andersen, S Matsunaga, RE Moore, R Skulberg. Investigations of a neurotoxic Oscillatorialean strain (Cyanophyceae) and its toxin. Isolation and characterization of homoanatoxin-a. Environ Toxicol Chem 11:321–329, 1992.
121. WW Carmichael. JT Eschedor, GML Patterson, RE Moore. Toxicity and partial structure of a hepatotoxic peptide produced by the cyanobacterium *Nodularia spumigena* Mertens emend. L575 from New Zealand. Appl Environ Microbiol 54:2257–2263, 1988.
122. T Heresztyn, BC Nicholson. Nodularin concentrations in Lakes Alexandrina and Albert, South Australia, during a bloom of the cyanobacterium (blue-green alga) *Nodularia spumigena* and degradation of the toxin. Environ Toxicol Water Qual 12:273–282, 1997.
123. R Matsushima, S Yoshizawa, MF Watanabe, KI Harada, M Fujisawa, WW Carmichael, H Fujiki. *In vitro* and *in vivo* effects of protein phosphatase inhibitors, microcystins and nodularin, on mouse skin and fibroblasts. Biochem Biophy Res Commun 171:867–874, 1990.
124. T Ohta, E Sueoka, N Iida, A Komori, M Suganuma, R Nishiwaki, M Tatematsu, SJ Kim, WW Carmichael, H Fujiki. Nodularin, a potent inhibitor of protein phosphatases 1 and 2A, is a new environmental carcinogen in male F344 rat liver. Cancer Res 54:6402–6406, 1994.
125. IR Falconer, A Choice. Toxicity of edible mussels (*Mytilus edulis*) growing naturally in an estuary

during a water bloom of the blue-green alga *Nodularia spumigena*. Environ Toxicol Wat Qual 7: 119–123, 1992.
126. J Lehtimäki, K Sivonen, R Luukkainen, SI Niemelä. The effects of incubation time, temperature, light, salinity, and phosphorus on growth and hepatotoxin production by *Nodularia* strains. Arch Hydrobiol 130:200–282, 1994.
127. K Terao, S Ohmori, K Igarashi, I Ohtani, MF Watanabe, KI Harada, E Ito, M Watanabe. Electron microscopic studies on experimental poisoning in mice induced by cylindrospermopsin isolated from blue-green alga *Umezakia natans*. Toxicon 32:833–843, 1994.
128. KI Harada, I Ohtani, K Iwamoto, M Suzuki, MF Watanabe, M Watanabe, K Terao. Isolation of cylindrospermopsin from a cyanobacterium *Umezakia natans* and its screening method. Toxicon 32:73–84, 1994.
129. Y Shimizu, M Norte, A Hori, A Genenah, M Kobayashi. Biosynthesis of saxitoxin analogues: the unexpected pathway. J Am Chem Soc 106:6433–6434, 1984.
130. M Ekman-Ekebom, M Kauppi, K Sivonen, M Niemi, L Lepistö. Toxic cyanobacteria in some Finnish lakes. Environ Toxicol Wat Qual 7:201–213, 1992.
131. WW Carmichael, WR Evans, QQ Yin, P Bell, EM Moczydlowski. Evidence for paralytic shellfish poisons in the freshwater cyanobacterium *Lyngbya wollei* (Farlow ex Gomont) comb. nov. Appl Environ Microbiol 63:3104–3110, 1997.
132. H Onodera, M Satake, Y Oshima, T Yasumoto, WW Carmichael. New saxitoxin analogues from the freshwater filamentous cyanobacterium *Lyngbya wollei*. Nat Toxins 5:146–151, 1997.
133. R Margalef. Perspectives in Ecological Theory. Chicago: University of Chicago Press, 1969, pp 26–50.

32

New Toxins on the Horizon

Kevin J. James, Alan G. Bishop, and Ambrose Furey
Cork Institute of Technology, Bishopstown, Cork, Ireland

I. INTRODUCTION

Poisoning incidents resulting from the consumption of seafood often go unreported unless there are a large number of individuals involved in a particular incident. Depending on the geographical location of these poisonings, full clinical and scientific investigations are not always undertaken owing to a lack of local resources or the inability to collect sufficient toxic material for chemical isolation studies. However, despite these difficulties, there have been many recent successes in the discovery of new seafood toxins, but they often involving protracted chemical analysis, multiple toxin isolation steps, and structural elucidation using very small amounts of isolated compounds. Other sources of information on potentially new seafood toxins arise from studies of the phytoplankton that have the potential to produce toxins, and this usually involves mass cultures to provide sufficient material for toxin isolation. In this chapter, we will discuss new toxins that are not detailed elsewhere in this text, and some lack detailed information, as they are from research activities that are currently in progress.

II. OKADAIC ACID–RELATED TOXINS

A. Okadaic Acid Isomers

Since the first reports of diarrheic shellfish poisoning (DSP) in Japan (1) in 1978, the illness is now recognised as a threat to public health throughout the world and has led to prolonged closures of shellfish culturing industries. Dinophysistoxin-1 (DTX-1) was identified as the causative toxin in Japan, and this originated from the dinoflagellate *Dinophysis fortii* (2). Although DTX-1 is rarely found in Europe, it was identified as the major toxin in shellfish from Norway (3). Most studies of DSP have implicated okadaic acid (OA) as the responsible toxin (4,5) arising from a variety of *Dinophysis* sp. (6). DTX-2, an isomer of okadaic acid (Figure 1a–c), was isolated from Irish mussels (7), and this was the predominant toxin during major DSP events in 1991 and 1994, when OA was also present at lower levels (8,9). DTX-2 was later identified in shellfish from Portugal (10) using fluorimetric liquid chromatographic analysis (LC-FL) and also in Spain, where detection with LC and mass spectrometry (MS) was recently used to demonstrate that DTX-2 was present in mussel samples that had been collected over several years (11). There was circumstantial evidence to implicate *D. acuta* as the source of DTX-2 from studies of shellfish toxicity in Ireland (9), where DTX-2 in shellfish was found soon after high

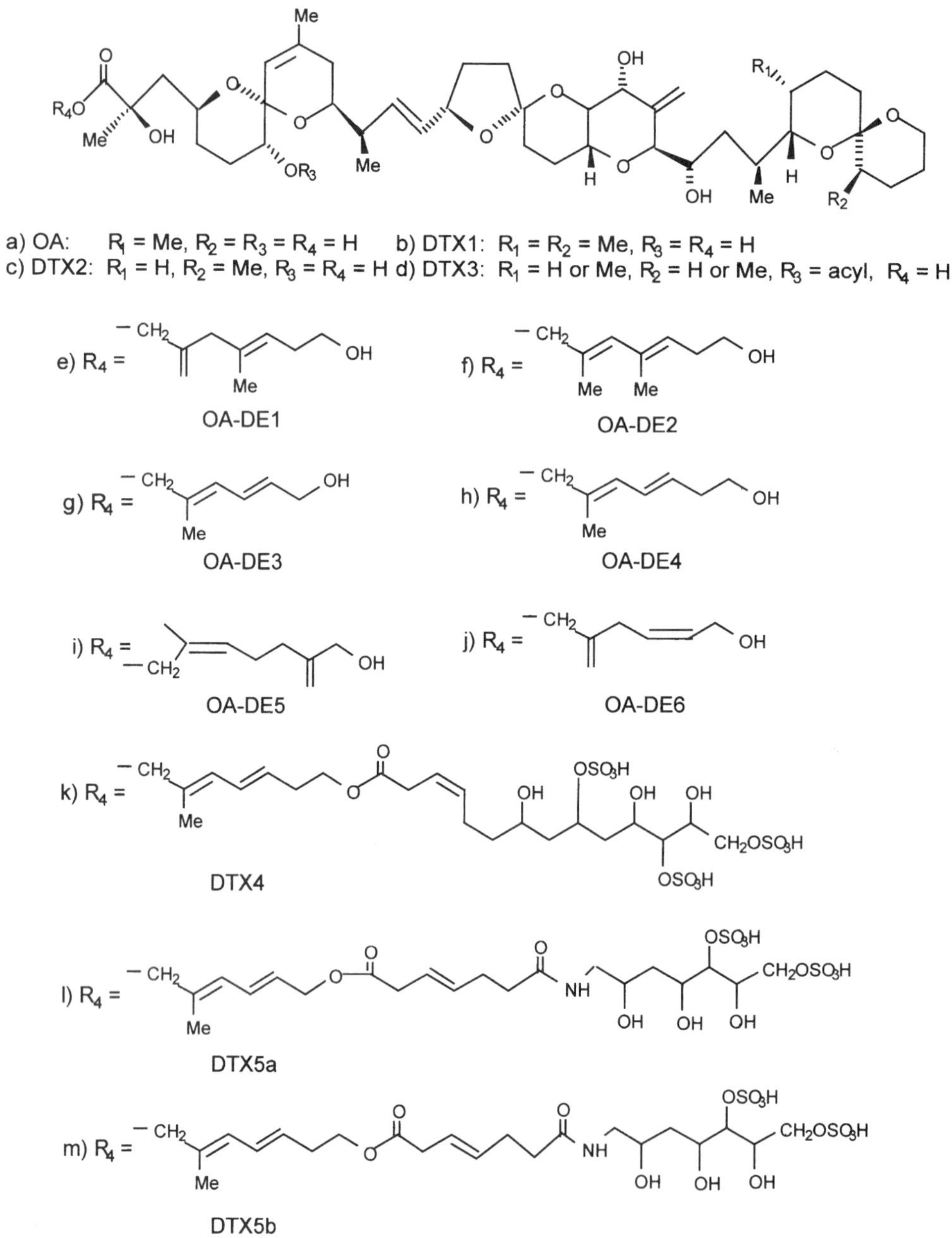

Figure 1 Structures of the acidic diarrheic shellfish poisoning (DSP) toxins (1a–1d) and the ester analogues of okadaic acid (1e–1m).

cell counts of this dinoflagellate were recorded in cultivation waters. Other evidence for the role of *D. acuta* was also obtained from toxin analysis of mixed phytoplankton populations in Spain (12).

D. acuta was confirmed as the progenitor of DTX-2 in Ireland by the analysis of unialgal samples (22–100 cells) collected manually from microscopic slides. LC-FL, LC-MS, and LC-mass spectrometry-tandem mass spectrometry (MS-MS) showed average toxin levels in cells of OA, 60 pg/cell, and DTX-2, 80 pg/cell (13). DTX-2 was also recently isolated from bulk phytoplankton samples (14) containing >60% *D. acuta*, which may prove to be a convenient source of this toxin.

Quilliam has produced evidence of the presence of other OA isomers in shellfish and phytoplankton (15). In fact, the certified reference material (MUS-2), prepared from shellfish and the cultured phytoplankton, *Prorocentrum lima*, not only contains OA and DTX-1 but also trace levels of three OA isomers, as shown by LC-MS analysis (16).

DTX-2B was isolated from shellfish, and LC-MS showed that it was an OA isomer (17,18). The intact protonated molecule of each DSP toxin analyte served as the precursor ion for collision-induced dissociation (CID) in the MS-MS experiments which were carried out by flow injection analysis (FIA). Figure 2 shows the positive-daughter mass spectra (range *m/z* 10–850) of the protonated molecules of OA, DTX-2, DTX-2B, and DTX-1 (19). Quilliam (15) had previously produced spectra for OA and DTX-1 using LC-MS-MS experiments with an atmospheric pressure ionization (API) source and an ion-spray interface. As expected, the spectrum of DTX-2 gave the same fragment ions as OA which was virtually indistinguishable from the spectrum of DTX-2B, thus showing that the latter was an isomer of OA and DTX-2. The main features of these spectra were the $[M + H]^+$ signal at *m/z* 805 and the sequential loss of

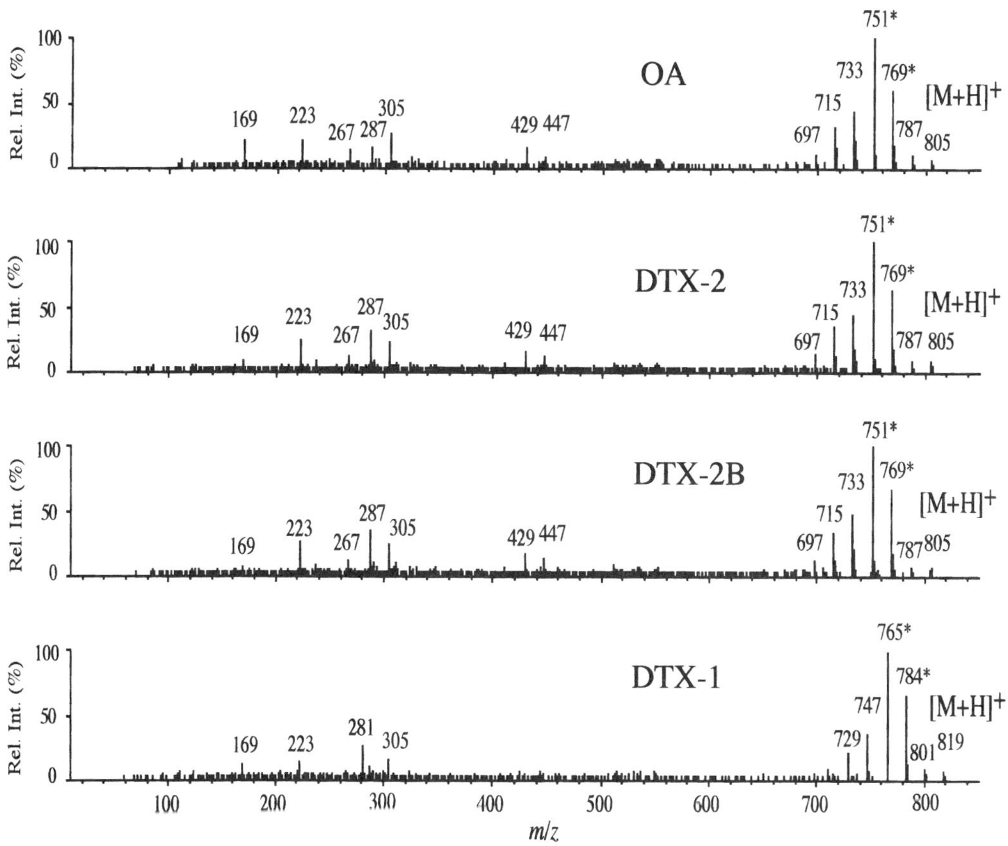

Figure 2 Positive-daughter mass spectra (range *m/z* 10–850), using collision-induced dissociation (CID), LC-MS-MS of the protonated molecules of OA, DTX-2, DTX-2B, and DTX-1. (From Ref. 19..)

water molecules producing signals due to $[M + H\text{-}nH_2O]^+$, with n = 1–6. These MS-MS spectra provided very useful structural information and fragment ions for confirmatory analysis in the selected reaction monitoring (SRM) mode, which was implemented using the parent-daughter ion combinations of *m/z* 805–751, *m/z* 805–769 for OA, DTX-2, and DTX-2B (19). However, using negative-ion FAB MS-MS, it was possible to distinguish OA from DTX-2, which produced spectral data identical to DTX-2B. Using this method, sequential fragmentation of the molecule occurred starting at the ether rings remote from the carboxylic acid group. Thus, OA gave ions *m/z* 803 $[M\text{-}H]^-$, 747 $[M\text{-}H\text{-}C_4H_8]^-$, and 703 $[M\text{-}H\text{-}C_5H_8O_2]^-$, whereas DTX-2 and DTX-2B both gave *m/z* 803 $[M\text{-}H]^-$, 737 $[M\text{-}H\text{-}C_5H_{11}]^-$, and 689 $[M\text{-}H\text{-}C_6H_{11}O_2]^-$ (T. Yasumoto and H. Naoki, unpublished data). These data suggest that DTX-2B may be an epimer of DTX-2.

DTX-2C, isolated in small amounts from *D. acuta*, was also shown to be an OA isomer by collision-induced dissociation (CID), as obtained by positive and negative MS-MS and implemented by FIA (20). Although DTX-2C was easily separated using LC-FL from the known toxins, OA, DTX-2, and DTX-1, as their 9-anthrylmethyl derivatives, the DTX-2B and DTX-2C derivatives were not resolved. Although the underivatized DTX-2C was difficult to separate from DTX-1, this did not present an analytical problem using selected ion-monitoring (SIM) LC-MS, as they produce different parent and fragment ions (Figure 3).

B. Okadaic Acid 1-Esters

The inability to produce viable cultures of *Dinophysis* sp. has greatly hampered toxinological studies on the production of DSP toxins by marine phytoplankton. However, DSP toxin–producing *Prorocentrum* sp. can be mass cultured, and a series of important studies, mainly from the groups of Yasumoto (Japan) and Wright (Canada), has led to the discovery of new toxins, some of which may play a role in the intracellular storage and transport of okadaic acid. In addition to these OA-related compounds, fast-acting macrocyclic imine toxins, prorocentrolides (21,22), have also been isolated from *Prorocentrum* sp. (see Section III.A). Several diol esters of OA have been isolated from cultures of *Prorocentrum* sp., and the structures of some of these esters which have C-7–C-9 unsaturated moieties are shown (Figure 1e–j). There have been several reports on the isolation of the OA diol ester, OA-DE1, from cultures of *P. maculosum* and *P. lima* (23–26). The isomeric diol ester, OA-DE2, was the major component in *P. lima* (26) and the C-7 and C-8 dienes, OA-DE3 and OA-DE4, were isolated from *P. maculosum* (previously described as *P. concavum*) (23). Biosynthetic studies on OA diol esters, using feeding experiments with various ^{13}C-labeled sodium acetates, led to enrichment in seven of the nine carbons in the side chain of OA-DE2. It was deduced that the side chain was derived from the condensation of two acetate units, with the branched methyl groups arising from the C-2 of acetate but leaving the remaining two carbons unattributed (26). There have been conflicting reports on the bioactivity of these compounds, but since it has previously been established that a free carboxylic acid was necessary for protein phosphatase inhibition (27), it is probable that hydrolysis to OA may account for this activity. Thus, in experiments to compare the effects of OA and OA-DE2 in rat myometrium, a level of activity for both compounds was observed but with the latter exhibiting a latent period before the appearance of a response (26).

A number of water-soluble OA derivatives, which were also produced by *P. lima* and *P. maculosum* cultures, have been structurally elucidated (28,29). These have been named DTX-4, DTX-5a, and DTX-5b (Figure 1k–l), and they are sulfated ester derivatives of OA-DE3 and OA-DE4. An aqueous methanol extract of *P. lima* was partitioned against hexane, diethyl ether, and 1-butanol and lipid-soluble DSP toxins were found mainly in the ether extract, whereas

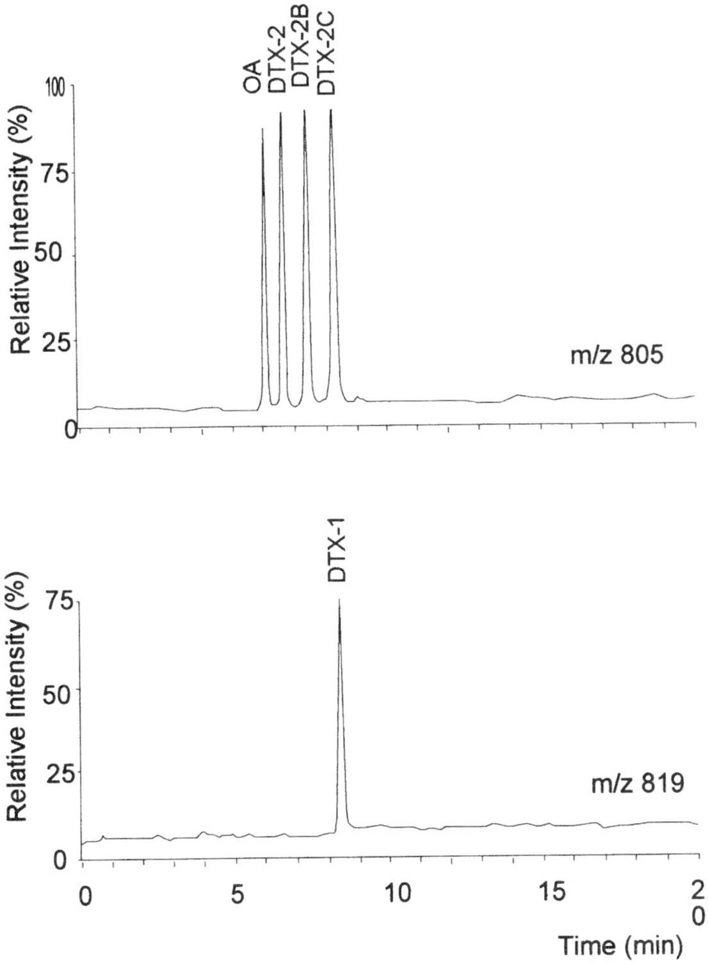

Figure 3 Chromatograms, using selected ion monitoring (SIM) LC-MS, of the acidic DSP toxins, OA, DTX-2, DTX-2B, DTX-2C, and DTX-1. (From Ref. 20.)

DTX-4 was the main toxin in the butanol phase. DTX-4 was the first polar DSP toxin to be isolated, and the structure of this compound shows that it is an ester derivative of OA-DE4 with three hydroxyl and three sulfate groups on a 14-carbon side chain (28). A report of high protein phosphatase inhibitory activity for DTX-4 (28) was later corrected (29), and some of the problems associated with measuring bioactivity of these compounds may be related to facile hydrolysis to OA.

In biosynthetic pulse labeling experiments using acetate, there was uniform enrichment of the OA skeleton and the side chain of DTX-4 (30). The subsequent discovery of two other sulfate derivatives of OA-DE3 and OA-DE4, DTX-5a and DTX-5b, respectively, led to the proposal that these bioinactive compounds are produced as protoxins that do not harm the host cell (29). Further, the excretion of these compounds from cells may be facilitated by the fact that they are water soluble and may be subsequently hydrolyzed to OA outside the cell. Experiments using esterases demonstrated the rapid hydrolysis of DTX-5b to OA, and it has been suggested that OA esters may play an important role in the storage within cells and intracellular transportation of OA-producing compounds (29). It was also demonstrated that it was important to destroy esterases by boiling prior to extraction of OA esters from cultures to prevent hydrolysis (31,32).

A specific method for DSP toxins in phytoplankton extracts using LC-MS has been developed by Quilliam (31,32), but it was also possible rapidly to determine DTX-4 and OA-DE4 in plankton extracts using LC-UV, as these compounds have a diene moiety. Using negative ion spray with an aqueous acetonitrile gradient, OA, DTX-1, and DTX-4 were determined directly in extracts of *P. lima* (Figure 4). Although the $[M\text{-}H]^-$ ion of DTX-4 was not observed at 1471.6, this toxin was readily detected by the $[M\text{-}3H]^{3-}$ and $[M\text{-}2H]^{2-}$ ions at 489.9 and 735.3, respectively. These studies also showed the presence of other compounds related to DTX-4 which contained additional hydroxyl and sulfate groups (Figure 4), and it is probable therefore that many more analogues of OA will be identified in the future.

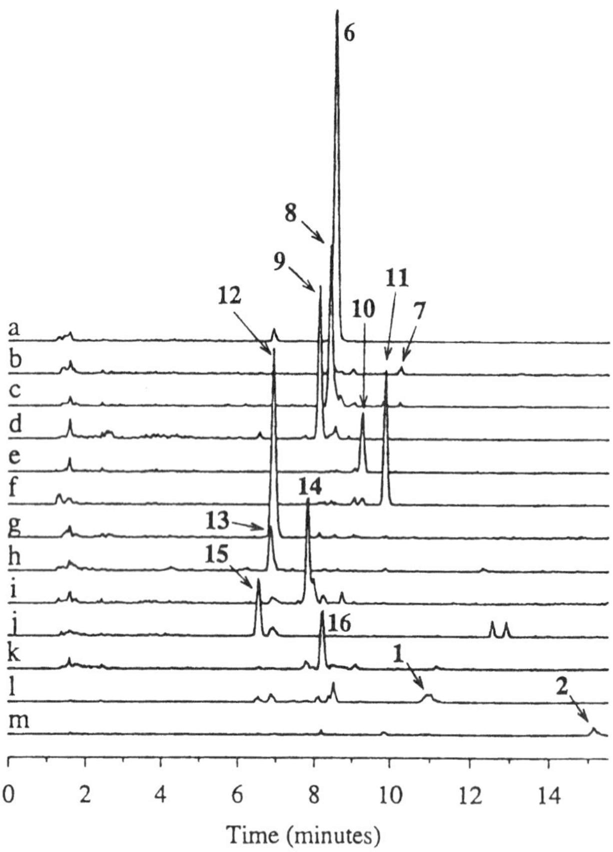

Figure 4 Negative ion-spray LC-MS analysis of an extract from *Prorocentrum lima*, analyzed using SIM of $[M\text{-}3H]^{3-}$ and $[M\text{-}2H]^{2-}$ ions. Peak identities (MW): a) 6 = DTX4 (1472.6; b) 7 = 6 + CH_2 (1486.6); c) 8 = 6 + O (1488.6); d) 9 = 6 + 2O (1504.6); e) 10 = 6 + 42u (1514.6); f) 11 = 6 + CH_2+ 2O (1518.6); g) 12 = 6 + SO_3 (1552.6); h) 13 = 6 + SO_3 + O (1568.6); i) 14 = 6 + 104u (1576.6); j) 15 = 6 + SO_3 + 2O (1584.6); k) 16 = 6 + CH_2 + SO_3 + 2O (1598.6); l) 1 = OA (804.5); m) 2 = DTX1 (818.5). Compounds 7, 11 and 16 are DTX1 analogues. (From ref. 32)

III. NITROGENOUS MACROCYCLIC SHELLFISH TOXINS

For some time, *Prorocentrum* sp. have been known (33) to produce toxins in addition to the lipophylic polyether toxins that were discussed in Section II.B. Prorocentrolide (21) was the first toxin to be isolated from a *Prorocentrum* sp. that was not related to OA. This is a macrocyclic compound with an imine moiety (Figure 5). Subsequently, there have been a number of recently characterized marine compounds that behave similarly to prorocentrolide in producing a rapid onset of symptoms in mouse bioassays and are thus referred to as ''fast-acting toxins'' (22). These include the pinnatoxins (34), gymnodimine (35), and the spirolides (36), but of these three examples, a phytoplanktonic source has only been confirmed in the case of gymnodimine. Although these macrocyclic compounds are structurally quite different, each contains a cyclic imine moiety which apparently is an important pharmacaphore, since the secondary amine reduction products of gymnodimine (37) and spirolide B (38) have greatly diminished bioactivities.

A. Prorocentrolides

As discussed in Section 1, *Prorocentrum* sp. produce OA, DTX-1, OA diol esters, and water-soluble OA sulfated esters. The first reported isolation of a fast-acting toxin was by Yasumoto's group, who obtained prorocentrolide from cultures of *P. lima* (21). A 1000-L culture of *P. lima* was extracted with acetone and methanol. After evaporation, the residue was partitioned between diethyl ether and water. The aqueous fraction was extracted with 1-butanol, and the components of this extract were chromatographed on silica, separated using gel permeation (Toyopearl HW-40), with another silica step, and reverse-phase chromatography on C_{18} silica with gradient elution (21). Prorocentrolide is structurally very different from the polyether toxins typically found in this phytoplankton and it contains two macrocyclic rings with an unusual cyclic imine moiety (Figure 5a).

However, as far back as 1984, Tindall et al. reported that polar and nonpolar toxins were present in Caribbean *P. concavum*, since after methanol extraction, subsequent distribution into

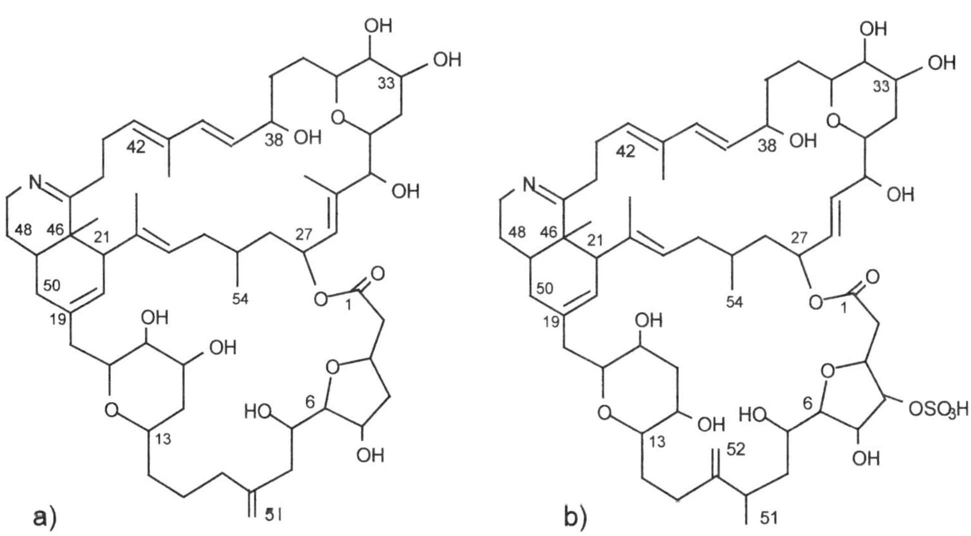

Figure 5 Prorocentrolide (a) and prorocentrolide B (b).

ether and butanol soluble fractions produced toxic responses in mouse bioassays (33). In addition, there were significant differences in the toxic responses, since the butanol, soluble fraction killed mice in minutes, whereas the nonpolar toxins in the ether fraction produced death in hours (33). Subsequently, chemical studies on the dinoflagellate *P. maculosum* Faust (formerly *P. concavum*) (39) were conducted in Canada (22) to identify the polar fast-acting toxin. Cells collected by centrifugation from 50-L cultures were extracted with methanol and the extract was brought to 70% aqueous methanol and washed with hexane. The methanol layer was diluted to 25% aqueous and extracted with diethyl ether and butanol. Prorocentrolide B was purified from the butanol fraction using gel permeation (Sephadex LH-20) and reverse-phase C_{18} chromatographic steps using open-column LC and semipreparative HPLC (22). The structure of prorocentrolide B was established by nuclear magnetic resonance (NMR) and MS-MS, which led to the determination of the relative stereochemistry of most of the ether ring systems. Molecular modeling demonstrated that there was considerable conformational flexibility within this molecule which limits the configuration information from Nuclear Overhauser Effect (NOE) measurements (22). The significant difference between these toxins is the presence of a sulfonate group in prorocentrolide B (Figure 5b).

B. Gymnodimine

The first reports of a major shellfish toxic event in North Island, New Zealand, occurred in the summer of 1992–1993 and epidemiological studies indicated that NSP was involved (40). Subsequently, a new toxin was isolated from shellfish (41) and named brevetoxin B_1 because of its structural similarity to the brevetoxins produced by the dinoflagellate *Gymnodinium breve* (42). The role played by *Gymnodinium* sp. in this toxic episode has been demonstrated by Mackenzie et al. (43), and in early 1994 *Gymnodimine* sp. were observed in Foveaux Strait, South Island and in Pakiri and Hawkes Bay, North Island (Figure 6). The difficulties associated with the taxonomy of this genus have also been highlighted (44).

As a consequence of the NSP outbreak, a nationwide shellfish biotoxin monitoring program was established in New Zealand. Early in 1994, monitoring for lipid-soluble toxins (NSP and DSP) produced high mouse bioassay values for oysters (*Tiostrea chilensis*) from Foveaux Strait which could not be attributed to the known toxins (45). Oyster samples (3 kg) were sent to Yasumoto's laboratory for toxin isolation. Acetone extracted the toxin which concentrated in the organic layer partitioning between ethyl acetate and water. Purification was achieved in multiple chromatographic steps using two C-8 reversed phases, DEAE cellulose, and gel permeation (Toyopearl HW-40), eluting with step gradients of aqueous methanol. The presence of toxin in eluent fractions was monitored using diode array ultraviolet (UV) detection and by mouse bioassays (46). The potent bioactive compound (2 mg), which was responsible for toxicity in mouse bioassays of shellfish extracts, was isolated and characterized using NMR (35). This new marine toxin ($C_{32}H_{45}NO_4$) is a pentacyclic compound with a 16-membered carbocyclic ring and an unusual fused ring containing an imine moiety (Figure 7). Interestingly, the ^{13}C NMR spectra measured in CD_3OD revealed only 30 carbon signals. The missing two signals were presumed to be due to carbons that were involved in imine-enamine tautomerism resulting in the exchange of protons with solvent deuterium during measurements. After changing to the aprotic solvent, pyridine-D_5, NMR measurements led to the observation of all 32 carbon signals (35).

The phytoplankton responsible for this toxicity in shellfish was soon suspected to be a *Gymnodinium* sp., and the toxin was therefore named gymnodimine (35). It was recognized that gymnodimine was related to the fast-acting toxins, prorocentrolide (Fig. 5) (21) and pinnatoxin (Figure 8) (34), which are also macrolides with a cyclic imine. The importance of this cyclic

Figure 6 Map of New Zealand showing locations from which contaminated shellfish were obtained.

imine moiety of gymnodimine to its bioactivity was shown by reduction to an amine, which dramatically reduced activity (37). The absolute stereochemistry of gymnodimine has also recently been established, and this was achieved by the x-ray structural analysis of a crystal of the p-bromobenzamide prepared from the reduced gymnodimine (37).

The presence of gymnodimine in shellfish caused considerable problems in New Zealand both for regulators and producers. High toxic responses in the mouse bioassays were obtained for most of 1994, and this led to prolonged closures and disruption of fisheries. However, experiments in which contaminated shellfish extracts were fed to rats produced no response, and the lack of epidemiological evidence of human toxicity in consumers of contaminated oysters led to a decision by the New Zealand regulatory agency not to use the presence of gymnodimine as a reason for imposing fishery closures. The mouse bioassay protocol was also modified by the incorporation of an additional ether extraction in the event of positive bioassays using acetone

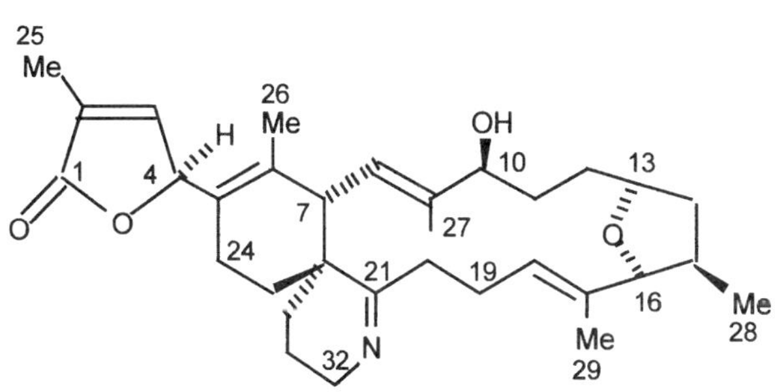

Figure 7 Gymnodimine.

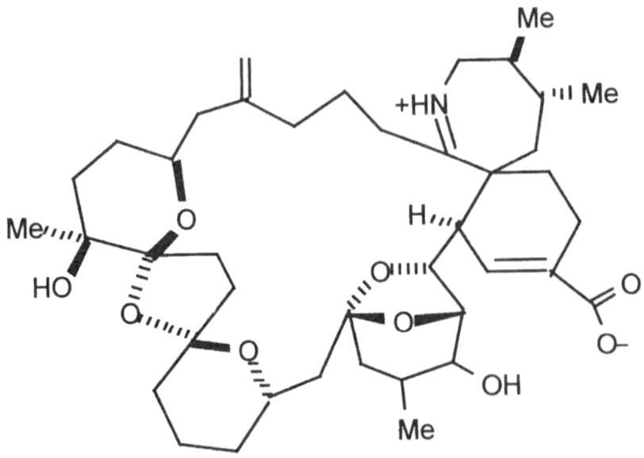

Figure 8 Pinnatoxin.

extraction of shellfish. This greatly reduced the number of positives in the bioassays with gymnodimine contaminated shellfish (45).

Large-scale culturing of a *Gymnodinium* sp. that strongly resembled *G. mikimotoi* was undertaken to confirm the source of gymnodimine. Thus, after isolation of this compound, and using NMR and LC-MS, it was shown to be identical to that obtained from oysters, *T. chilensis* (35,46). The observation that gymnodimine displayed potent ichthyotoxicity against a small freshwater fish at the level of 0.1 ppm led to the suggestion (35) that it might be responsible for massive fish kills that have been reported during blooms of *Gymnodinium* spp. in Japan (47).

C. Pinnatoxin A

Human toxicity resulting from the ingestion of the adductor muscle of the shellfish *Atrina pectinata (Pinna pectinata)* occurs frequently in the coastal regions of Japan and China (48). In China, human intoxications resulting from the ingestion of *Pinna attenuata* occurred in 1980 and 1989 (49). The symptoms reported in several thousand human illnesses have included diarrhea, paralysis, and convulsions and therefore indicated that a neurotoxin was responsible. A toxic extract, named pinnatoxin, was reported to be a Ca^{2+} channel activator (49). However, the identification of a specific toxin proved to be difficult, and it was not until recently that the responsible compound was isolated and structurally elucidated. The 75% ethanol extract from 45 kg of *Pinna muricata*, collected from Okinawa, Japan, was used for toxin isolation studies. A combination of gel permeation, ion exchange, and reverse-phase liquid chromatography was used to produce 3.5 mg of purified toxin that was named pinnatoxin A (34).

Pinnatoxin A, $C_{41}H_{61}NO_9$ (see Figure 8), is a polyether macrocyclic that contains iminium and carboxylate moieties together with a unique 6,7-spiro ring, a 5,6-bicyclo ring, and a 6,5,6-trispiroketal ring system involving 14 chiral centers (50). A biogenetic pathway for the production of this spiro system has been proposed (34) that includes a Diels-Alder reaction which is similar to the proposed biogenesis of prorocentrolide (21). The relative stereochemistry of pinnatoxin A was determined especially using NOESY and ROESY NMR experiments on pinnatoxin A and its methyl ester, produced by the reaction of the toxin with diazomethane (50).

D. Spirolides

A family of macrocyclic toxins were isolated from mussels (*Mytilus edulis*) and scallops (*Placopectena magellanicus*) which were collected from the eastern shore of Nova Scotia, Canada

(36). After the extraction of digestive glands with methanol, partitioning between hexane and chloroform gave a residue in the latter phase that was positive by mouse bioassay. The separation of the bioactive compounds was achieved using chromatography on silica and reverse-phase C_{18} silica followed by gel permeation on Sephadex LH-20. The progress of the isolation was monitored by mouse bioassay and final purification was achieved using reverse-phase HPLC to yield four compounds that were named spirolides A–D. Two of these toxins, spirolide B and D, were obtained in sufficient quantity to be structurally elucidated. These toxins were called spirolides, as they possess an unusual seven-membered cyclic imine moiety that was spiro-linked to a cyclohexene ring as well as a spiro-linked tricyclic ether ring system (Figure 9a, R = Me). Spirolide D differs from spirolide B only in that the latter has one less methyl group on the heptacyclic iminium ring (Figure 9a, R = H).

There are close structural similarities between the spirolides and pinnatoxin (34) discussed in the previous section. Pinnatoxin also contains these spiro-linked rings, one of which incorporates a heptacyclic imine, whereas gymnodimine (35) and prorocentolides (21,22) possess hexacyclic imine moieties. Although a specific phytoplankton source for spirolides has not been identified, a heterogeneous plankton sample collected at the same time as the spirolides were found in shellfish was also shown to contain these toxins based on HPLC and LC-MS data (36). It has not yet been established how the spirolides exert their toxic effects except that they have been shown to be weak activators of type L Ca^{2+} channels.

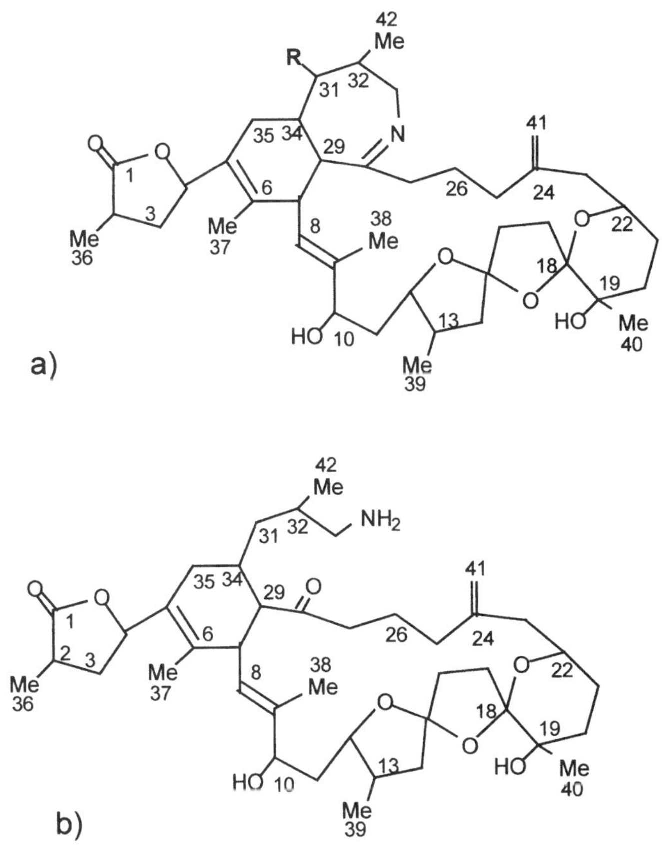

Figure 9 Spirolide B (a, R = H), spirolide D (a, R = Me), spirolide E (b), spirolide F (b, $\Delta_{2,3}$).

Further evidence that the cyclic iminium moiety is the pharmacaphore for these toxins in this section is the fact that the structurally different prorocentrolides induce identical and rapid symptoms in mice to those caused by the spirolides A–D but the minor structural variants, spirolide E (Figure 9b) and spirolide F (Figure 9b, $\Delta_{2,3}$), are not active (38). The latter were isolated from shellfish extracts and they differ from the toxic spirolides in that they are keto amines and do not have the characteristic heptacyclic iminium ring. To further substantiate the importance of the cyclic iminium moiety, spirolide B was reduced using sodium borohydride to produce the corresponding secondary amine. This compound failed to produce toxic symptoms in mice when administered at four times the equivalent spirolide B dose (38).

IV. PECTENOTOXIN-2 SECO ACIDS

Pectenotoxin-2 (PTX-2) (Figure 10) is a macrocyclic polyether toxin that, in addition to okadaic acid analogues, has been implicated in diarrheic shellfish poisoning. The dinoflagellate *Dinophysis fortii* has been shown to produce PTX-2, but a number of related compounds are found

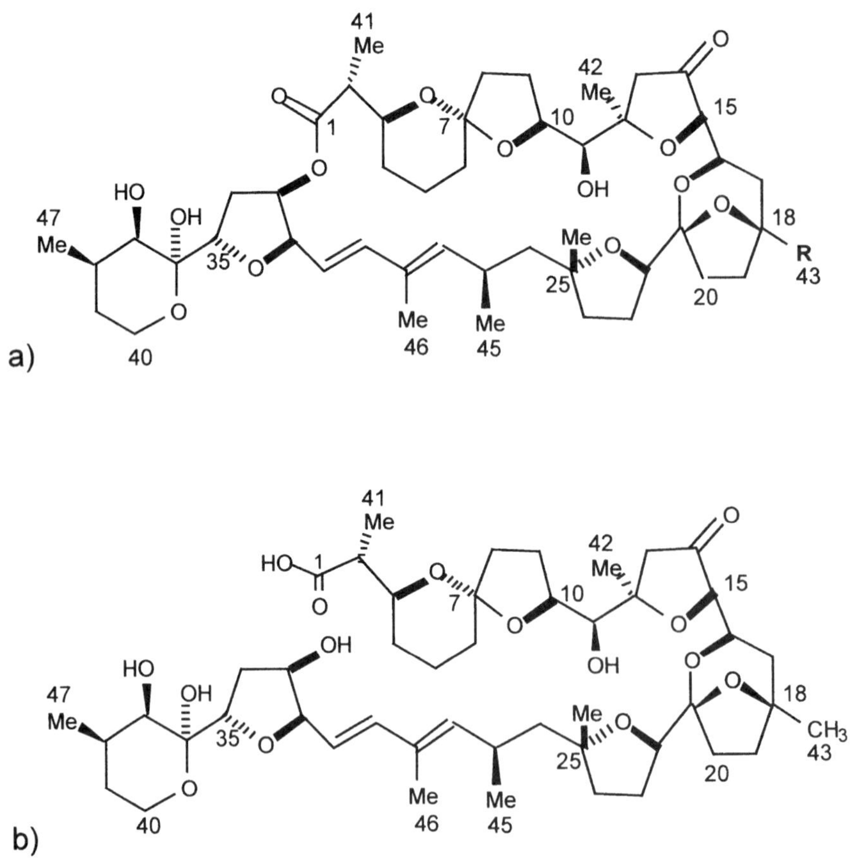

Figure 10 Pectenotoxins (PTXs): (a) PTX-1 (R = CH_2OH); PTX-2 (R = CH_3); PTX-3 (R = CHO); PTX-6 (R = COOH). Pectenotoxin-2 seco acids: (b) PTX-2SA (C-7, R); 7-*epi* PTX-2SA (C-7, S).

in shellfish owing to oxidation of the C43 methyl to alcohol (PTX-1), aldehyde (PTX-3), and carboxylic acid (PTX-6) moieties (24), and some of these are shown (Figure 10 a1–a4).

New pectenotoxin analogues have recently been identified in shellfish and phytoplankton collected in New Zealand and Ireland. Greenshell mussels (*Perna canaliculus*) that were exposed to a bloom which was predominantly *Dinophysis acuta* were harvested from the Marlborough Sounds, New Zealand (see Figure 6), in November 1996. Bulk phytoplankton was harvested near Union Hall off the coast of Ireland (Figure 11) using a large custom-built double-plankton net that was designed to concentrate *Dinophysis* spp. The biomass samples collected in September 1996 were predominantly (>60%) *D. acuta* and, by manually collecting cells from microscope slides, it was demonstrated that this dinoflagellate was the progenitor of DTX-2 in shellfish from Ireland (13). However, analysis of methanol extracts of phytoplankton using fluorimetric HPLC following ADAM (9-anthrylmethyl-diazomethane) derivatization also revealed the presence of unidentified acidic components that behaved similarly to the OA-type compounds during chromatography (Figure 12).

Isolation studies using extracts from New Zealand mussels were carried out using chromatography on alumina, silica, reverse-phase C_{18} silica, gel permeation (Toyopearl HW-40), and another C_{18} silica phase to produce two compounds that were identical to those that were isolated

Figure 11 Map of Ireland showing locations from which contaminated shellfish and toxic phytoplankton were obtained.

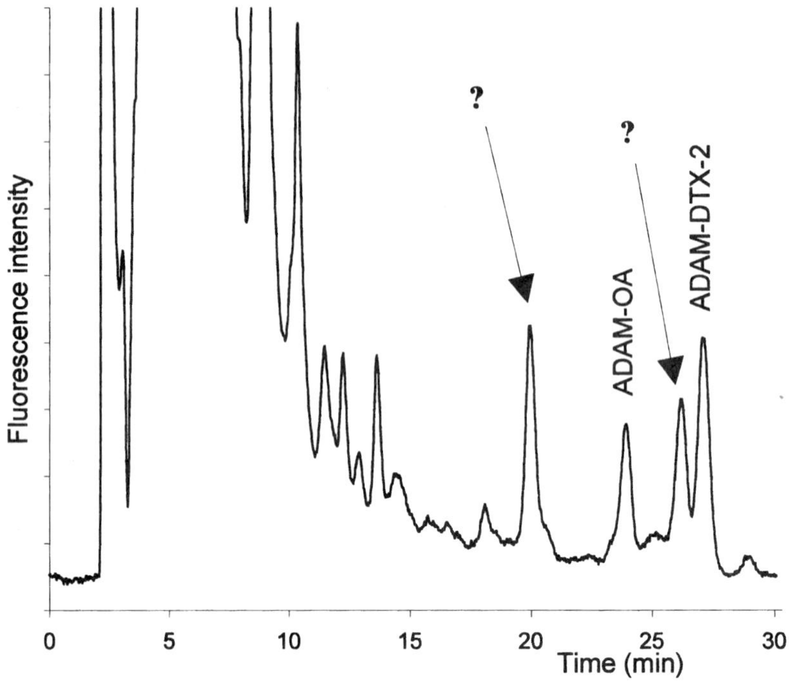

Figure 12 Chromatogram from the LC-FL analysis of an extract of phytoplankton (predominantly *D. acuta*), collected in Ireland, showing the 9-anthrylmethyl derivatives of OA, DTX-2, and other acidic compounds that were subsequently confirmed to be pectenotoxin-2 seco acids.

from phytoplankton in Ireland (51). Isolation from phytoplankton, which is a less complex matrix, required separations on silica, Sephadex LH-20, and two C_{18} silica phases. The progress of these separations was conveniently monitored using LC-UV as well as LC-FL following derivatization with ADAM. LC-MS studies on isolates were particularly informative, as these revealed the presence of three isomeric compounds corresponding to an $[M + H]^+$ at *m/z* 877 with fragmentation ions at *m/z* 859, 841, 823, 805, and 787 due to $[M + H - nH_2O]$ with n = 1–5 (52). The high-resolution FAB MS was consistent with the molecular formula, $C_{47}H_{72}O_{15}$, for these isomers (51). In addition, using LC-MS-MS, fragmentation ions were remarkably similar to those of standard pectenotoxin-2 (PTX-2), which has a parent ion $[M + H]^+$ at *m/z* 859 (52). From these data, it was apparent that these new compounds were related to PTX-2 except that the macrolactone ring had opened to produce a carboxylic acid. The structural elucidation of two of the new compounds was facilitated by the comparison of NMR data with that of PTX-2 whose relative structural configuration had recently been determined (53). From the ^{1}H-^{1}H COSY and TOCSY spectra, eight partial structures (H2–H6, H8–H11, H13, H15–H17, H19–H20, H22–H24, H26–H35, and H37–H40) were determined. ^{1}H connectivities in these partial structural structures were identical to those in PTX-2. Other NMR and negative-ion collision-induced dissociation (CID) MS-MS experiments were used to confirm the proposed structures (see Figure 10b1 and b2) for the new compounds, which have been named pectenotoxin-2 seco acid (PTX-2SA) and 7-*epi*-pectenotoxin-2 seco acid (7-*epi*-PTX-2SA) (51). There was evidence for the interconversion of these compounds during isolation and interconversion of pectenotoxins due to epimerization at C-7 has also been observed (54). However, since shellfish from Japan that were heavily contaminated with PTX-2 did not contain these seco acids, it is unlikely that they are artefacts of PTX-2 (51).

Further evidence of the worldwide distribution of these compounds in shellfish has recently been obtained. Suspected DSP-contaminated shellfish material that produced unidentified peaks using fluorimetric HPLC following derivatization of carboxylic acids were reexamined. It has been confirmed that shellfish from Croatia, Italy, China, and Ireland contained PTX-2SA and related compounds. In all of the shellfish samples that were found to contain PTX-2SAs, okadaic acid analogues were also present in similar concentrations (K.J. James, unpublished data). The implications for food quality and safety resulting from the presence of these compounds in shellfish have not yet been evaluated.

V. POLYCAVERNOSIDE A

The red alga, *Polycavernosa tsudai* (formerly *Gracilaria edulis*), had been eaten widely with no apparent health risk until 1991 when there were reports of human intoxication which occurred in Guam. Of the 13 confirmed reports of illness, three patients died and a variety of symptoms were evident, including diarrhea, hypersalivation, and muscle spasms (55). Two months after this incident, a bulk sample of *P. tsudai* (2.6 kg) was collected from the region where the suspected toxic alga had been harvested. After extraction with acetone and partitioning between water and dichloromethane, the lipophilic compounds were separated by chromatography using silica and several C_{18} silica phases and the eluates were monitored using mouse bioassays (56). Two compounds, with similar toxicities in mice (LD_{50}, i.p., 200–400 μg/kg), were isolated and named polycavernoside A and B, and were shown to be novel macrolide disaccharides (Figure 13). Although only 400 μg of the major component, polycavernoside A, was recovered, the planar structure was elucidated using NMR, and the proposed structure was supported by positive FAB MS (56). This trioxatridecane macrocycle is reminiscent of aplysiatoxins, which contain trioxadodecane (57). Attached to this macrocyclic ring is a glycoside residue containing two pyranosyl rings.

In view of the small amount of toxin present in this sample, it was speculated that the levels were probably much higher in the algae during the period of human intoxication. Toxic symptoms in laboratory animals were similar to those reported in affected humans. The occurrence of toxins in the algae is apparently transient and may have been responsible for previous fatal outbreaks of food poisoning caused by other *Gracilaria* sp. (58,59).

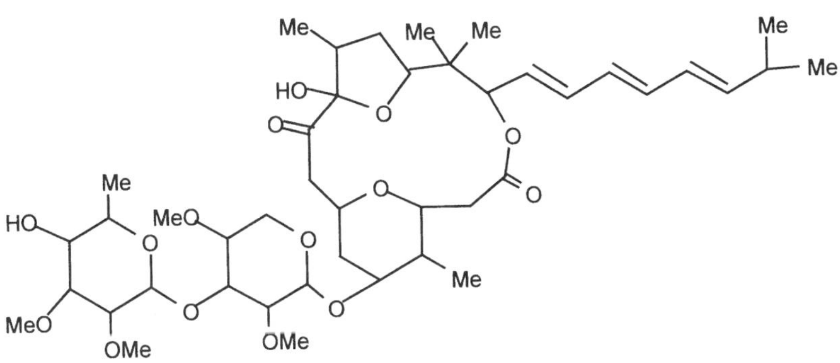

Figure 13 (a) Polycavernoside A and (b) polycavernoside B.

VI. AZASPIRACID POISONING TOXINS

A shellfish poisoning event occurred in The Netherlands in 1995 when at least eight severe gastrointestinal illnesses in humans were reported following the consumption of mussels (*Mytilus edulis*). These shellfish were traced to Killary Harbour, Ireland (Figure 11), where subsequent investigations revealed that a number of local intoxications had also occurred. Unfortunately, this toxicity was not uncovered during the DSP toxin–monitoring program in Ireland, which involved biweekly sampling in this region and utilized a recently modified rat bioassay. In subsequent chemical analysis of these shellfish, only low levels of OA and DTX-2 were detected. Toxicity persisted in shellfish from Killary Harbour for at least 8 months, which is an unusually long period for a toxic shellfish episode. In the DSP mouse test, a slow progressive paralysis was observed using extracts of mussels, and these neurotoxic symptoms were quite different from those typical of the DSP toxins.

Whole mussel meats (8 kg) collected from Killary Harbour in February, 1996, were extracted three times with acetone, which was evaporated and the residue was partitioned between hexane and 85% aqueous methanol. The predominant toxin was present in the methanol layer and purification was achieved by chromatography over silica, gel permeation (Toyopearl HW-40), weak cation exchange (CM-Toyopearl 650M), and a second gel permeation step. This isolation furnished 360 μg of purified toxin that was tentatively named killarytoxin-3 (KT-3) (60). Although this toxin seems stable in vivo, as it persists in shellfish for remarkably long periods, chemical studies on this compound were hampered by its instability in vitro. Thus, decomposition of toxin occurred in certain solvents, particularly chloroform, in slightly alkaline solution and during silica chromatography.

The ESI-MS spectrum of KT-3 gave an $[M + H]^+$ peak at *m/z* 842 and ^1H NMR showed that this toxin had few oxymethines or oxymethylenes, unlike the OA class of DSP toxins. When the structure of this new toxin was elucidated, it was apparent that it contained novel spiro five- and six-membered rings, with one of these containing nitrogen (Figure 14) (60). This toxin was renamed azaspiracid (61) and, although a number of nitrogen-containing dinoflagellate toxins have recently been discovered, such as pinnatoxin, prorocentrolide, gymnodimine, and

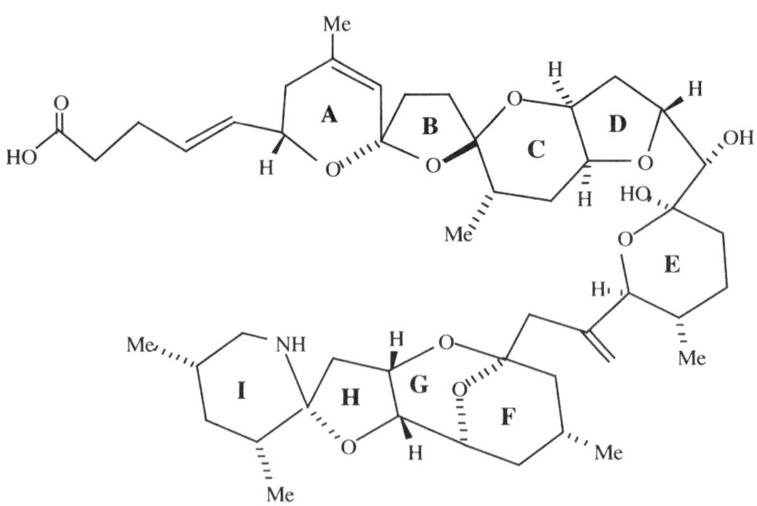

Figure 14 Azaspiracid

spirolides, these differ from azaspiracid in that they all contain an imine moiety and produce different toxic symptoms (see section 11.2).

In toxicity studies using azaspiracid, mice became progressively paralyzed with labored breathing and remained still in the corners of their cages (60). Diarrhea was not observed and, at low doses, mice died within 2–3 days, with a minimum lethal dose estimated at 200 μg/kg. Histopathological changes were observed in mice which were induced by both intraperitoneal and oral administration of azaspiracid. Marked necrosis of parenchymal cells in peripheral portions of lobuli in the pancreas and focal necrosis in the liver were also observed 20 h after administration of the toxin (62).

A second toxic event involving azaspiracid occurred in October 1997 on the northwest coast of Ireland (see Figure 11). At least 12 members of the small island community of Arranmore, County Donegal, became ill from eating mussels which were cultivated for the first time in this area. Once more, the period of toxicity in mussels was 8 months, and such a prolonged period of shellfish intoxication has serious implications for the economic viability of the shellfish industry. Two azaspiracid analogues have recently been isolated from Irish mussels, and they have been named azaspiracid-2 (AZ-2) and AZ-3. Structurally, they are methylazaspiracid (AZ-2) and desmethylazaspiracid (AZ-3), and although they usually occur in shellfish at lower levels than azaspiracid, they are more toxic in mouse (i.p.) bioassays (minimum lethal dose, 140 μg/kg). As the azaspiracids are structurally and toxicologically different from previously known toxins, a new shellfish toxic syndrome, named azaspiracid poisoning (AZP), has been declared (63).

The DSP mouse bioassay, which is typically used in shellfish biotoxin monitoring programs, produces a positive result when sufficiently large amounts of azaspiracids are present in shellfish. Paradoxically, shellfish contaminated with much lower levels of azaspiracids than were found in the two incidents discussed above led to much greater problems in Ireland in 1998. Thus, shellfish from the major cultivation areas in Ireland that had passed DSP mouse bioassays subsequently caused human intoxications in France and Italy with severe economic consequences for producers. Unlike previous DSP incidents, both mussels and oysters from the same region were similarly contaminated by AZP toxins, which is unusual considering the very different cultivation methods that were used.

The etiology of AZP is unknown and no unusual phytoplankton species were observed in any of the above toxic episodes. However, detailed studies were not undertaken until some time after human poisonings were reported and, because of the persistence of these toxins, the events that led to toxin accumulation in shellfish could have occurred much earlier. However, the pattern of methylation in azaspiracid (see Figure 14) is typical of toxins produced by dinoflagellates (J.L.C. Wright, personal communication).

The determination of azaspiracid in shellfish relies on the use of LC-MS and LC-MS-MS methods. The analyte was extracted from whole mussel meat with acetone and eluted from a reverse-phase LC column using acetonitrile-water, 85:15 (v/v), containing 0.03% trifluoroacetic acid. Using an ion-spray interface, in the positive ion mode, only the intact protonated molecule, $[M + H]^+$, was generated at *m/z* 842. This served as precursor ion for collision-induced dissociation (CID) and three product ions, $[M + H - nH_2O]^+$ with $n = 1 - 3$, were identified for the unambiguous toxin confirmation by selected reaction monitoring (SRM) LC-MS-MS analysis with a detection limit of 20 pg (64).

Using an ion-trap mass spectrometer, an LC-MS^3 method has been developed for the analysis of azaspiracids in shellfish (K.J. James et al., unpublished data). This highly specific and sensitive method is readily automated for shellfish monitoring, and Figure 15 shows the typical traces from a single analysis. The protonated molecules and the expected product ions

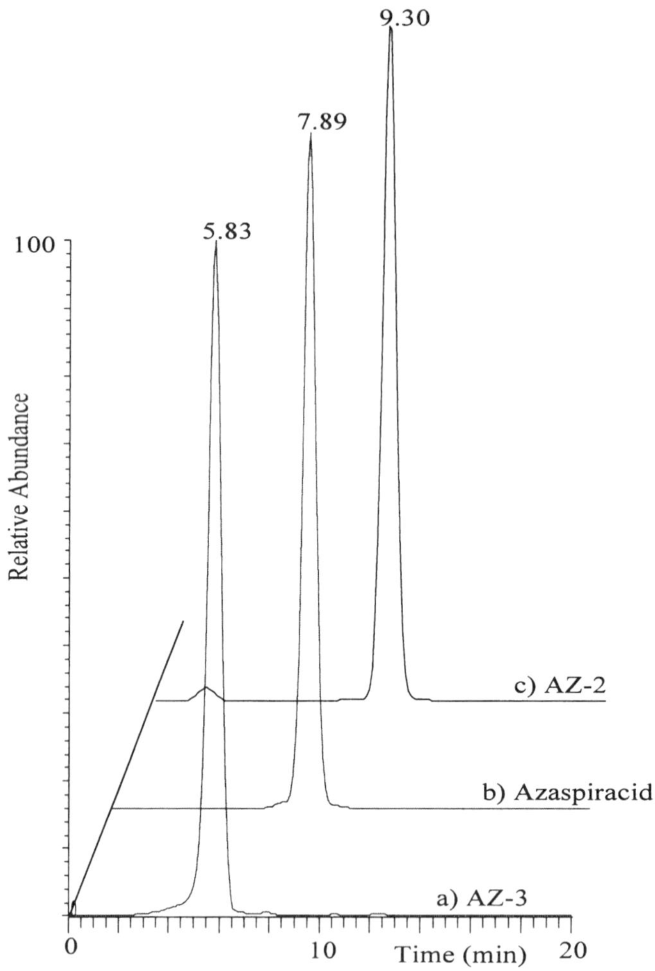

Figure 15 Chromatograms from the analysis of a toxic mussel extract (Donegal, 1997) using positive electrospray ion-trap LC-MS^3. (a) AZ-3: 5.83 min, *m/z* 828 → 810 → 792. (b) Azaspiracid: 7.89 min, *m/z* 842 → 824 → 806, c) AZ-2: 9.30 min, *m/z* 856 → 838 → 820. Relative abundance (a) 0.13, (b) 1.0, (c) 0.24. Chromatographic conditions: Luna C18(2) column (5 μm, 150 × 2 mm, Phenomenex); 30-mm guard column. Elution at 200 μL/min using acetonitrile/water (65/35) containing 0.05% TFA.

which are trapped and further fragmented are (1) AZ-3: *m/z* 828, 810, (2) azaspiracid: *m/z* 842, 824, (3) AZ-2: *m/z* 856, 838.

VII. NEW TOXINS NOT YET STRUCTURALLY ELUCIDATED

A. Carchatoxins

Ciguatera poisoning associated with the consumption of shark meat occurs occasionally, but a major intoxication episode in Madagascar in 1993 was attributed to a new toxin (65). A total of 188 individuals who had eaten meat from a single shark (*Carcharhinus leucas*) were hospitalized. Symptoms appeared 5–10 h after ingestion, and most patients had neurological symptoms, especially severe ataxia with frequent comas and seizures. Gastrointestinal disturbance was not present in most cases, and this clearly distinguished the intoxication from ciguatera poisoning. Also, unlike ciguatera, mortality was very high in this poisoning episode and at least 50 patients died. Using reverse-phase chromatography, two toxins were isolated in small quantities (30 μg each) and they were heat-stable lipophilic compounds that behaved chromatographically very different from ciguatoxin. However, there was insufficient material for structural elucidation of these toxins which have been named, carchatoxin-A and carchatoxin-B (65).

B. Cooliatoxin

A new toxin has been isolated from cultures of the benthic dinoflagellate *Coolia monotis*, and it was named cooliatoxin. After extracting mass cultures with methanol, an aqueous methanol medium was washed with hexane, and following fractionation between butanol and water, the toxin was predominantly present in the butanol phase. Purification was achieved using repeated chromatography on silica, gel permeation and reverse-phase HPLC. Cooliatoxin was found to be a potent cardiac stimulant in mice, but the major toxic symptom was acute respiratory distress. The action of cooliatoxin in vitro was an initial stimulation and a subsequent blocking of unmyelenated nerves (66).

Negative ion-spray MS produced a major signal at m/z 1061.5 $[M - H]^-$, whereas positive MS gave fragment ions that were attributed to water and sulfate loss. It was concluded that cooliatoxin was a polyether molecule containing one sulfate group. Yessotoxin is a disulfated polyether shellfish toxin whose absolute configuration has recently been determined (64). Since the $[M - H]^-$ ion of cooliatoxin corresponded closely to that of the monosulfated ion produced from the fragmentation of yessotoxin, and from similarities in toxicity data, these investigators concluded that cooliatoxin was probably a monosulfated yessotoxin (67).

REFERENCES

1. T Yasumoto, Y Oshima, M Yamaguchi. Occurrence of a new type of toxic shellfish poisoning in the Tohoku district. Bull Jpn Soc Sci Fish 44:1249–1255, 1978.
2. T Yasumoto, Y Oshima, W Sugawara, Y Fukuyo, H Oguri, T Igarashi, N Fujita. Identification of *Dinophysis fortii* as the causative organism of diarrhetic shellfish poisoning. Bull Jpn Soc Sci Fish 46:1405–1411, 1980.
3. J-S Lee, K Tangen, E Dahl, P Hovgaard, T Yasumoto. Diarrhetic shellfish toxins in Norwegian mussels. Nippon Suisa Gakkaishi 54:1953–1957, 1988.
4. HP van Egmond, T Aune, P Lassus, GJA Speijers, M Waldock. Paralytic and diarrhoeic shellfish poisons: occurrence in Europe, toxicity, analysis and regulation. J Nat Toxins 2:41–83, 1993.
5. DV Subba Rao, Y Pan, V Zitko, G Bugden, K Mackeigan. Diarrhetic shellfish poisoning (DSP) associated with a subsurface bloom of *Dinophysis norvegica* in Bedford Basin, eastern Canada. Marine Ecol-Prog Ser 97:117–126, 1993.
6. JS Lee, T Igarashi, S Fraga, E Dahl, P Hovgaard, T Yasumoto. Determination of diarrhetic shellfish toxins in various dinoflagellate species. J Appl Phycol 1:147–152, 1989.
7. T Hu, J Doyle, D Jackson, J Marr, E Nixon, S Pleasance, MA Quilliam, JA Water, JLC Wright. Isolation of a new diarrhetic shellfish poison from Irish mussels. J Chem Soc Chem Commun: 39–41, 1992.
8. EP Carmody, KJ James, SS Kelly, K Thomas. Complex shellfish toxin profiles in Irish mussels. In: P Lassus, G Arzul, E Erard, P Gentien, C Marcaillou LeBaut, eds. Harmful Marine Algal Blooms. Paris: Lavoisier, 1995, pp 273–278.

9. EP Carmody, KJ James, SS Kelly. Dinophysistoxin-2: the predominant diarrhetic shellfish toxin in Ireland. Toxicon 34:351–359, 1996.
10. P Vale, MA Sampayo. DTX-2 in Portuguese shellfish. In: T Yasumoto, Y Oshima, Y Fukuyo, eds. Harmful and Toxic Algal Blooms. Sendai, Japan: International Oceanographic Commission of UNESCO, 1996, pp 539–542.
11. A Gago-Martinez, JA Rodriguez-Vasquez, P Thibault, MA Quilliam. Simultaneous occurrence of diarrhetic and paralytic shellfish poisoning toxins in Spanish mussels in 1993. Nat Toxins 4:72–79, 1996.
12. J Blanco, M Fernandez, J Marino, B Reguera, A Miguez, J Maneiro, E Cacho, A Martinez. From *Dinophysis* spp. toxicity to DSP outbreaks: a preliminary model of toxin accumulation in mussels. In: P Lassus, G Arzul, E Erard, P Gentien, C Marcaillou LeBaut, eds. Harmful Marine Algal Blooms. Paris: Lavoisier, 1995, pp 777–782.
13. KJ James, AG Bishop, M Gillman, SS Kelly, C Roden, R Draisci, L Lucentini, L Giannetti, P Boria. Liquid chromatography with fluorimetric, mass spectrometric and tandem mass spectrometric detection for the investigation of the seafood-toxin-producing phytoplankton, *Dinophysis acuta.* J Chromatogr 777:213–221, 1997.
14. KJ James, AG Bishop, M Gillman, SS Kelly, C Roden, R Draisci, L Lucentini, L Giannetti. The diarrhoeic shellfish poisoning toxins of *Dinophysis acuta,* identification and isolation of dinophysistoxin-2 (DTX-2). In: B Reguera, J Blanco, ML Fernandez, T Wyatt, eds. Harmful Algae. Santiago de Compostela: Xunta de Galicia and International Oceanographic Commission of UNESCO, 1998, pp 489–492.
15. MA Quilliam. Analysis of diarrhetic shellfish poisoning toxins in shellfish tissue by liquid chromatography with fluorimetric and mass spectrometric detection. J AOAC Int 78:555–570, 1995.
16. MA Quilliam. Mussel tissue reference material for DSP toxins (MUS-2). Report, Institute of Marine Biosciences, NRC, Canada.
17. KJ James, EP Carmody, SS Kelly, R Draisci, L Lucentini, L Giannetti, R Staccini. A new constituent of diarrhetic shellfish toxins in Irish mussels. In: M Miraglia, E van Egmond, C Brera, J Gilbert, eds. Mycotoxins and Phycotoxins—Developments in Chemistry, Toxicology and Food Safety. Rome: 1998, pp 485–492.
18. KJ James, EP Carmody, M Gillman, SS Kelly, R Draisci, L Lucentini, L Giannetti. Identification of a new diarrhoetic toxin in shellfish by liquid chromatography with fluorimetric and mass spectrometric detection. Toxicon 35:973–978, 1997.
19. R Draisci, L Lucentini, L Giannetti, P Boria, KJ James, A Furey, M Gillman, SS Kelly. Determination of diarrhoetic shellfish toxins in mussels by micro high-performance liquid chromatography with tandem mass spectrometric detection. J AOAC Int 81:441–447, 1998.
20. R Draisci, L Giannetti, L Lucentini, C Marchiafava, KJ James, AG Bishop, BM Healy, SS Kelly. Isolation of a new okadaic acid analogue from phytoplankton implicated in diarrhoetic shellfish poisoning. J Chromatogr 798:137–145, 1998.
21. K Torigoe, M Murata, T Yasumoto, T Iwashita. Prorocentrolide, a toxic nitrogenous macrocycle from a marine dinoflagellate, *Prorocentrum lima.* J Am Chem Soc 110:7876–7877, 1988.
22. TM Hu, SW deFreitas, JM Curtis, Y Oshima, JA Walter, JLC Wright. Isolation and structure of prorocentrolide B, a fast-acting toxin from *Prorocentrum maculosum.* J Nat Prod 59:1010–1014, 1996.
23. TM Hu, J Marr, ASW deFreitas, MA Quilliam, JA Walter, JLC Wright, S Pleasance. New diol esters isolated from cultures of the dinoflagellates *Prorocentrum lima* and *Prorocentrum concavum.* J Nat Prod 55:1631–1637, 1992.
24. T Yasumoto, N Seino, Y Murakami, M Murata. Toxins produced by benthic dinoflagellates. Biol Bull 172:128–131, 1987.
25. T Hu, ASW deFreitas, J Doyle, D Jackson, J Marr, E Nixon, S Pleasance, MA Quilliam, JA Walter, JLC Wright. New DSP toxin derivatives isolated from toxic mussels and the dinoflagellates, *Prorocentrum lima* and *P. concavum.* In: TJ Smayda, Y Shimizu, eds. Toxic Phytoplankton Blooms in the Sea. Newport, RI: Elsevier, 1993, pp 507–512.
26. M Norte, A Padilla, LL Fernandez, ML Souto. Structural determination and biosynthetic origin of two ester derivatives of okadaic acid isolated from *Prorocentrum lima.* Tetrahedron 50:9175–9180, 1994.

27. S Nishiwaki, H Fujiki, M Suganuma, H Furuya-Suguri, R Matsushima, Y Iiada, M Ojika, K Yamada, D Uemera, T Yasumoto, FJ Schmitz, T Sugmura. Structure-activity relationship within a series of okadaic acid derivatives. Carcinogenesis 11:1837–1841, 1990.
28. T Hu, JM Curtis, JA Walter, JL Wright. Identification of DTX-4, a new water-soluble phosphatase inhibitor from the toxic dinoflagellate, *Prorocentrum lima.* J Chem Soc Chem Commun 597–599, 1995.
29. TM Hu, JM Curtis, JA Walter, JL Mclachlan, JLC Wright. Two new water-soluble DSP toxin derivatives from the dinoflagellate *Prorocentrum maculosum*: Possible storage and excretion products. Tetrahedron Lett 36:9273–9276, 1995.
30. J Needham, TM Hu, JL McLachlan, JA Walter, JLC Wright. Biosynthetic studies of the DSP toxin DTX-4 and an okadaic acid diol ester. J Chem Soc Chem Commun 1623–1624, 1995.
31. MA Quilliam. Liquid chromatography-mass spectrometry of seafood toxins. In: D Barcelo ed. Applications of LC-MS in Environmental Chemistry. Vol 59. Amsterdam: Elsevier, 1996, pp 415–444.
32. MA Quilliam, WR Hardstaff, N Ishida, JL McLachlan, AR Reeves, NW Ross, AJ Windust. Production of diarrhetic shellfish poisoning (DSP) toxins by *Prorocentrum lima* in culture and development of analytical methods. In: T Yasumoto, Y Oshima, Y. Fukuyo, eds. Harmful and Toxic Algal Blooms. Paris: Intergovernmental Oceanographic Commission of UNESCO, 1996, pp 289–292.
33. DR Tindall, RW Dickey, RD Carlson, G Morey-Gaines In: EP Ragelis, ed. Seafood Toxins. Washington, DC: American Chemical Society, 1984, pp 225–240.
34. D Uemera, T Chou, T Haino, A Nagatsu, S Fukuzawa, S-Z Zheng, H-S Chen. Pinnatoxin A: a toxic amphoteric macrocycle from the Okinawan bivalve, *Pinna muricata.* J Am Chem Soc 11:1155–1156, 1995.
35. T Seki, M Satake, L Mackenzie, HF Kaspar, T Yasumoto. Gymnodimine, a new marine toxin of unprecedented structure isolated from New Zealand oysters and the dinoflagellate, *Gymnodinium* sp. Tetrahedron Lett 36:7093–7096, 1995.
36. TM Hu, JM Curtis, Y Oshima, MA Quilliam, WM Watson-Wright, JLC Wright. Spirolides B and D, two novel macrocycles isolated from the digestive glands of shellfish. J Chem Soc Chem Commun 2159–2161, 1995.
37. M Stewart, JW Blunt, MHG Munro, WT Robinson, DJ Hannah. The absolute stereochemistry of the New Zealand shellfish toxin gymnodimine. Tetrahedron Lett 27:4889–4890, 1997.
38. TM Hu, JM Curtis, JA Walter, JLC Wright. Characterisation of biologically inactive spirolides E and F: identification of the spirolide pharmacaphore. Tetrahedron Lett 37:7671–7674, 1996.
39. MA Faust. Phycologia 32:410–418, 1993.
40. M Bates, M Baker, N Wilson, L Lane, S Handford. Epidemiological overview of the New Zealand shellfish toxicity outbreak. Royal Society of New Zealand Miscellaneous Series 24:35–40, 1993
41. H Ishida, A Nozawa, K Totoribe, N Muramatsu, H Nukaya, K Tsuji, K Yamaguchi, T Yasumoto, H Kaspar, N Berkett, T Kosuge. Brevetoxin B1, a new polyether marine toxin from the New Zealand shellfish, *Austrovenus stutchburyi.* Tetrahedron Lett 36:725–728, 1995.
42. YY Lin, M Risk, SM Ray, DV Engen, J Clardy, J Golik, JC James, K Nakanishi. Isolation and structure of brevetoxin B from the 'red tide' dinoflagellate *Ptychodiscus brevis* (*Gymnodinium breve*). J Am Chem Soc 103:6773–6775, 1981.
43. L Mackenzie, L Rhodes, D Till, FK Chang, H Kaspar, A Haywood, J Kapa, B Walker. A *Gymnodiminium* sp. bloom and the contamination of shellfish with lipid soluble toxins in New Zealand, Jan.–April 1993. In: P Lassus, G Arzul, E Erard, P Gentien, C Marcaillou LeBaut, eds. Harmful Marine Algal Blooms. Paris: Lavoisier, 1995, pp 777–782.
44. A Haywood, L Mackenzie, I Garthwaite, N Towers. *Gymnodinium breve* 'look-alikes': three gymnodinium isolates from New Zealand. In: T Yasumoto, Y Oshima, Y. Fukuyo, eds. Harmful and toxic algal blooms. Paris: Intergovernmental Oceanographic Commission of UNESCO, 1996, pp 227–230.
45. L Mackenzie, A Haywood, J Adamson, P Truman, D Till, T Seki, M Satake, T Yasumoto. Gymnodimine contamination of shellfish in New Zealand. In: T Yasumoto, Y. Oshima, Y. Fukuyo, eds. Harmful and toxic algal blooms. Paris: Intergovernmental Oceanographic Commission of UNESCO, 1996, pp 97–100.
46. T Seki, M Satake, L MacKenzie, HF Kaspar, T Yasumoto. Gymnodimine, a novel toxic imine isolated from the Foveaux Strait oysters and *Gymnodinium* sp. In: T Yasumoto, Y Oshima, Y Fukuyo, eds.

Harmful and toxic algal blooms. Paris: Intergovernmental Oceanographic Commission of UNESCO, 1996, pp 495–498.
47. T Okaichi. Red tide problems in the Seto Inland sea, Japan. In: T Okaichi, D Anderson, T Nemeto, eds. Proceedings of the First International Symposium on Red Tides. Amsterdam: Elsevier, 1988, pp 137–142.
48. T Otofuji, A Ogo, J Koishi, K Matsuo, H Tokiwa, T Yasumoto, K Nishihara, E Yamamoto, M Saisho, Y Kurihara, K Hayashida. Food Sanit Res 31:76–83, 1981.
49. SZ Zheng, FL Huang, SC Chen, XF Tan, JB Zou, J Peng, RW Xie. Chin J Mar Drugs 33:33–35, 1990.
50. T Chou, O Kamo, D Uemura. Relative stereochemistry of pinnatoxin A, a potent shellfish poison from Pinna muricata. Tetrahedron Lett 37:4023–4026, 1996.
51. M Daiguji, M Satake, KJ James, AG Bishop, L MacKenzie, H Naoki, T Yasumoto. Structures of new pectenotoxin analogs, pectenotoxin-2 seco acids and 7-*epi*-pectenotoxin-2 seco acid, isolated from a dinoflagellate and greenshell mussels. Chem Lett 653–654, 1998.
52. KJ James, AG Bishop, R Draisci, L Palleschi, C Marchiafava, E Ferretti, M Satake, T Yasumoto. Liquid chromatographic methods for the isolation and identification of new pectenotoxin-2 analogues from marine phytoplankton and shellfish. J Chromatogr 844:53–65, 1999.
53. K Sasaki, M Satake, T Yasumoto. Identification of the absolute configuration of pectenotoxin-6, a polyether macrolide compound, by NMR spectroscopic method using a chiral anisotropic reagent, phenylglycine methyl ester. Biosci Biotech Biochem 61:1783–1785, 1997.
54. K Sasaki, JLC Wright, T Yasumoto. J Org Chem 63:2475, 1998.
55. RL Haddock, OLT Cruz. Foodborne intoxication associated with seaweed. Lancet 338:195–196, 1991.
56. M Y-Yamashita, RL Haddock, T Yasumoto. Polycavernoside A: a novel glycosidic macrolide from the red alga, *Polycavernosa tsudai* (*Gracilaria edulis*). J Am Chem Soc 115:1147–1148, 1993.
57. Y Kato, PJ Scheuer. J Am Chem Soc 96:2245–2246, 1974.
58. S Aikawa, M Suzuki, K Ono. Rep Yamagata Prefectural Inst Public Health 13:81–82, 1981.
59. TJ Sonoda. J Food Hyg Soc Jpn 24:507–508, 1983.
60. M Satake, K Ofuji, KJ James, A Furey, T Yasumoto. New toxic event caused by Irish mussels. In: B Reguera, J Blanco, ML Fernandez, T Wyatt, eds. Harmful Algae. Vigo: Xunta de Galicia and International Oceanographic Commission of UNESCO, 1998, pp 468–469.
61. M Satake, K Ofuji, H Naoki, KJ James, A Furey, T McMahon, J Silke, T Yasumoto. Azaspiracid, a new marine toxin having unique spiro ring assemblies, isolated from Irish mussels, *Mytilus edulis*. J Am Chem Soc 120:9967–9968, 1998.
62. E Ito, K Terao, T MacMahon, T Yasumoto. Histopathological changes in mice caused by new toxin from Irish mussels. In: B Reguera, J Blanco, ML Fernandez, T Wyatt, eds. Harmful Algae. Vigo: Xunta de Galicia and International Oceanographic Commission of UNESCO, 1998, pp 588–589.
63. K Ofuji, M Satake, T McMahon, J Silke, KJ James, H Naoki, Y Oshima, T Yasumoto. Two analogs of azaspiracid isolated from mussels, Mytilus edulis, involved in human intoxications in Ireland. Nat Toxins 7:99–102, 1999.
64. R Draisci, L Palleschi, E Ferretti, A Furey, KJ James, M Satake, T Yasumoto. Development of a liquid chromatography–tandem mass spectrometry method for the identification of azaspiracid in shellfish. J. Chromatogr. 871:13–21, 2000.
65. P Boisier, G Ranaivoson, N Rasolofonirina, B Andriamahefazafy, J Roux, S Chanteau, M Satake, T Yasumoto. Fatal mass poisoning in Madagascar following ingestion of a shark (*Carcharhinus leucas*): clinical and epidemiological aspects and isolation of toxins. Toxicon 33:1359–1364, 1995.
66. MJ Holmes, RJ Lewis, A Jones, AW Wong Hoy. Cooliatoxin, the first toxin from *Coolia Monotis* (*Dinophyceae*). Nat Toxins 3:355–362, 1995.
67. H Takahashi, T Kusumi, Y Kan, M Satake, T Yasumoto. Determination of the absolute configuration of Yessotoxin, a polyether compound implicated in diarrhetic shellfish poisoning, by NMR spectroscopic method using a chiral anisotropic reagent, methoxy-(2-naphthyl)acetic acid. Tetrahedron Lett 37:7087–7090, 1996.

33
Marine Toxins as a Starting Point for Drugs

David J. Craik and Martin J. Scanlon
Centre for Drug Design and Development, The University of Queensland, St. Lucia, Brisbane, Australia

I. INTRODUCTION

A. Toxins as drug leads

It is clear from previous chapters that marine toxins have a very wide range of biological activities, and often these activities are induced by only minute amounts of toxin, that is, by their nature toxins are highly potent molecules. By definition, the biological activities exhibited by the toxins are harmful to the target organism, since the function of the toxin is either to protect the toxic species from attack by a predator or to immobilize potential prey. However, despite the generally harmful effects in their target organisms, toxins have a very great potential to be harnessed for favorable effects on nontarget organisms, and in particular in humans for therapeutic purposes. This apparent contradiction occurs because a molecule that is toxic at one concentration may well be useful when delivered at a lower and more controlled dose or to a different receptor site. One case of a dose-related differentiation of toxic versus useful effects applies to blood pressure–lowering agents. Perhaps the best known example is the drug Captopril (I), one of the highest selling medicines of the past decade. The lead design molecule for this drug was derived from a component of the venom of the South American pit viper where its severe hypotensive properties help to immobilize prey. Researchers at Bristol-Myers Squibb pharmaceuticals developed Captopril as an analogue of this molecule which, when given in appropriate dose, causes controlled lowering of blood pressure in humans. The molecule has been prescribed to millions of people as a treatment for hypertension. Its applicability to human use arises because it binds to an enzyme (angiotensin-converting enzyme) that is important in a cascade of events regulating blood pressure.

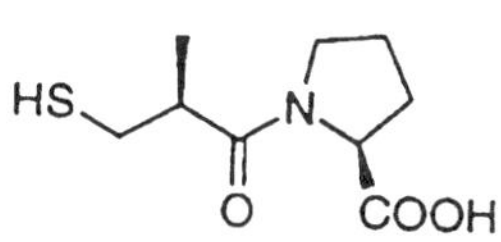

(I) Captopril

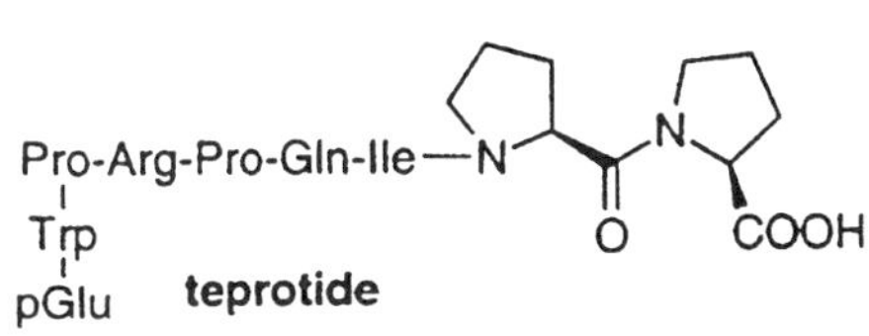

(II) toxic venom component

This example illustrates a case where a toxic molecule, (II), provided a lead for the development of an analogue, (I), which is a useful drug. Although it is possible that toxin components themselves may be directly useful as drugs, it is more often the case, as in the above example, that the toxic lead compound merely provides a design idea for the development of a drug molecule. This is because there are many factors, other than just potent biological activity, that are crucial in the development of drugs for human use even if the activity is at an important and relevant target site. These factors are addressed in Section II, but three particularly important factors which deserve mention at the outset are bioavailability via a desirable route (usually oral), lack of side effects, and stability.

Many toxins are delivered directly into the bloodstream of the target organism via a sting. Oral bioavailability is thus of little importance in toxin delivery, but it is often of paramount importance for human pharmaceuticals. Such bioavailability often requires modification of the structure of the toxin molecule to make it more suitable for the preferred delivery route. There are exceptions of course, and the conotoxin MVIIA, described in Section IV as a new treatment for pain is one—in this case, the lack of oral bioavailability in therapeutic applications is overcome by direct delivery of the toxin component into the spinal cord via an infusion pump. Similarly, the presence of side effects is of little consequence in toxin molecules. If the primary purpose is to disable prey or dissuade an attacking organism, then so long as the primary effect is exhibited, secondary effects at other binding sites are not important. In human therapeutic applications, however, such unwanted effects can dramatically reduce the applicability of the drug. Finally, for most modern human therapeutics, a degree of chemical stability and associated shelf life is necessary. This is often not the case for toxin molecules which are synthesized in vivo and used as needed. For all of these reasons and for others identified in Section II, it is usual for natural products to provide leads which are then further developed into drugs rather than be applied directly as drugs themselves.

B. Drug Development Process

The general scheme involved in the drug development process is illustrated in Figure 1. The starting point is the discovery of a novel bioactivity in a toxin or more usually in an extract of the venom, which may contain several components. The initial assay may be a broad screening assay that defines a general class of activity but does not provide any information on the mechanism of action. Once an active extract is discovered, the next step is separation and isolation of the active component(s). High-performance liquid chromatographic (HPLC) methods provide excellent separation of a wide range of chemical classes and may be applied after a crude chemical fractionation if the initial extract contains a large amount of extraneous material. In the case of venoms, this is often not necessary, and HPLC of venoms may be done directly without prior fractionation.

On isolation of the active component, the specific activity may be determined in an appropriate assay (discovery loop A in Figure 1). There are myriads of assays in widespread use in drug-screening programs, and these range from inhibition of enzyme activity, to effects on bacterial growth or viral replication, to functional effects in isolated tissues. In some cases, a crude classification of activity is obtained by examining effects in intact animals. For example, in much of the pioneering work on conotoxins done by Olivera and colleagues (1), injections of venom components into fish, mice, or other animals led to initial discoveries of activities. The activity in these cases was often described in terms of qualitative observed effects. For example, peptides which caused marked shaking of mice when injected intracutaneously were referred to as shaker peptides, those which caused crabs to rise on their hind legs in an aggressive display were dubbed King-Kong peptides, and those which caused juvenile mice to fall asleep were referred to as sleeper peptides. Subsequent assays are more specific; for example, involving the binding of isolated components to specific receptors on rat brain membrane preparations.

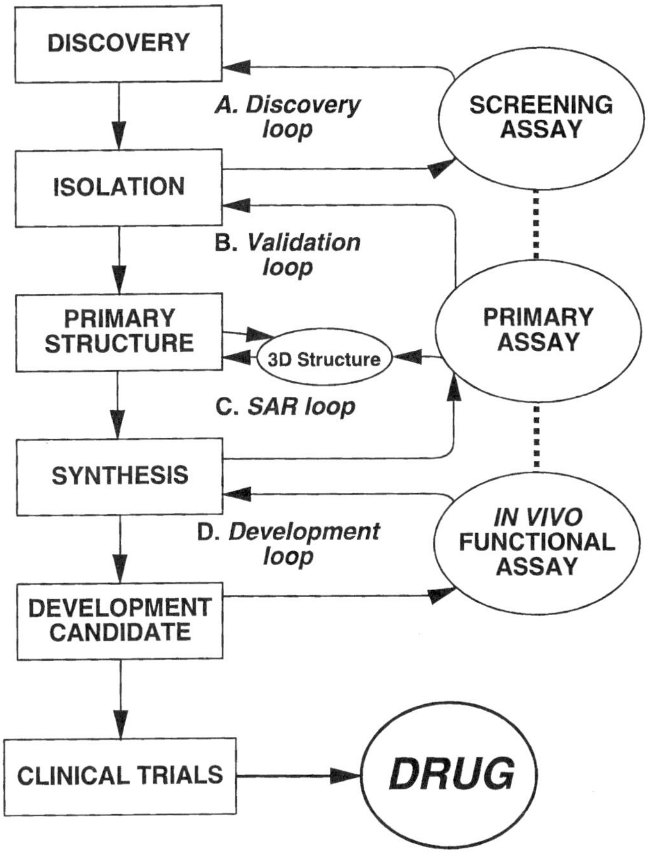

Figure 1 Schematic representation of the drug-design process and the central role of structural information.

Once an activity has been discovered and localized to a discrete chemical component, the next step is determination of the identity, or primary structure, of that component. In most cases, a general idea of the chemical class of the molecule may be known by analogy with related compounds or literature data. Chromatographic methods can then provide further insight into general features of the new compound, including its hydrophobicity and approximate molecular size. Mass spectrometry, however, provides the most valuable information at this early stage of identification and often allows the molecular weight to be established. The composition may be defined by elemental analysis for low molecular weight organic compounds or amino acid analysis for peptides. In the latter case, amino acid sequencing is now a routine procedure and allows the primary structure to be determined. If cysteine residues are present and connected via disulfide bonds, as is commonly the case in peptide toxins, their connectivity can be obtained by selective enzymatic fragmentation and sequencing or by partial reduction and mass spectroscopy analysis.

The methods described above for identification of the primary structure of the toxin require submicrogram quantities of material, and this is usually obtained from the native extract. However, for drug-design purposes, it is important to determine the three-dimensional structure, as it is the structural placement of functional groups within the molecule that determines receptor binding. Such three-dimensional (3D) structure determination may be done using either nuclear magnetic resonance (NMR) spectroscopy or x-ray crystallography. Both techniques require milligram quantities of material (sometimes up to 100 mg for x-ray crystallography if extensive crystallization trials are required), and hence the next step in the drug development process is the synthesis of suitable quantities of the toxin.

For nonpeptide toxins, organic synthesis may prove to be challenging and time consuming, particularly for larger molecules in which multiple chiral centers are present. Each novel structure presents new challenges, and there are few generalizations that can be made about the methods used. By contrast, for peptidic toxins, the methods of solid-phase synthesis are relatively standardized, and peptides consisting of less then 50 amino acids can now routinely be made using either manual or machine-assisted methods. Refolding of the peptide if disulfide bonds are present often proves to be the step where most innovation is required. In some cases, it may prove to be beneficial to use a suitable bacterial vector for the production of recombinant peptides.

Figure 1 emphasises that synthesis of the active toxin compound is a critical node in the developmental pathway. Not only does the synthesis open the possibility of 3D structure determination, it allows verification of the primary structure proposed on the basis of studies of native toxin, and it presents opportunities for modifying the structure with the ultimate aim of improving activity or the physicochemical/biological properties of the lead molecule.

The structure-verification aspect is illustrated in feedback loop B of Figure 1. Although modern spectroscopic methods for primary structure determination are very reliable, absolute certainty as to the structure is essential, and direct synthesis provides a way of detecting the presence of subtle but sometimes critical structural variations. For example, in the case of peptides, the presence of D–amino acids, glycosylation, or other posttranslational modifications may significantly affect biological activity but may be missed in the primary structure determination. If a proposed sequence is synthesized and it does not coelute on HPLC or display the same activity as the native sequence, such modifications should be suspected and probed for before structure-activity relationships (SARs) are developed.

Synthesis is also crucial in the development of SARs, as the ability to make analogues of the native sequence, either chemically or via molecular biology, is a prerequisite for such studies. Although it is not unusual for such SARs to be developed based on the presence of several related molecules of differing activities in a native toxin mixture, or in toxins from

Table 1 Top 10 Prescription Drugs in 1998[a]

Brand name	Manufacturer	Generic name	Therapeutic category[b]	Origin	MW
Premarin	Wyeth-Ayerst	Conjugated estrogens	Hormone replacement	Hormone	~342
Synthroid	Knoll	Levothyroxine	Hormone replacement	Hormone	798
Trimox	Apothecon	Amoxicillin	Antibiotic	Mold	365
Hydrocodone w/APAP[c]	Watson	Hyrdocodone w/APAP	Analgesic	Plant	494/151[c]
Prozac	Lilly	Fluoxetine	Antidepressant	De novo	345
Prilosec	Astra	Omeprazole	Antiulcer	De novo	345
Zithromax	Pfizer	Azithromycin	Antibiotic	Mold	785
Lipitor	Parke-Davis	Atorvastatin	Lipid reduction	De novo	1209
Norvasc	Pfizer	Amlodipine	Cardiovascular	De novo	567
Claritin	Schering	Loratidine	Antihistamine	De novo	382

[a] Based on US prescriptions.
[b] Refers to source of lead molecule.
[c] APAP, acetaminophen.

different species, synthesis is the only way of controlling the degree of structural diversity to be tested. The SAR feedback process is summarized by loop C in Figure 1.

The product of such a SAR study is a molecule that has better activity or physicochemical properties than the initial lead compound and is referred to as a development candidate. Occasionally, the initial lead compound may be taken forward as a development candidate; however, given the high costs of subsequent steps and the relatively low costs of SAR studies, it is usually important that the best candidate be thoroughly explored via SAR before proceeding to development.

The developmental process (loop D in Figure 1) involves extensive testing of the molecule both for efficacy and lack of toxicity in animal models. At this point, parallel development on formulating the active molecule in a way that maximizes its bioavailability and hence its effectiveness is undertaken. Clinical trials in humans are then done before the compound eventually reaches the market.

Table 2 Molecules with Potential Therapeutic Applications from Marine Organisms

Therapeutic application	Molecule	Source	Organism
Cardiovascular			
	Spongosine	Caribbean sponge	*Cryptotehia crypta*
	Anthopleurins	Sea anenomes	*Anthopleura* spp.
	Autonomoniums	Marine sponges	*Verongia fistularis*
	Anemonia toxins	Sea anenomes	*Anthopleura* spp.
	Doridosine	Nudibranch	*Anisodoris nobilis*
	Conotoxins	Marine snails	*Conus* spp.
	ShK	Sea anenomes	*Stichodactyla helianthus*
	Adenosine derivatives	Marine sponges	*Echinodictyum* spp.
Anti cancer			
	Didemnins	Caribbean tunicates	*Trididemnum* spp.
	Ara-C analogues	Caribbean sponge	*Cryptotehia crypta*
	Cembranoids	Soft corals	*Sinularia flexibilis*
	Dolastatin	Marine sponges	*Bugula neritina*
Muscle relaxants			
	5′-Deoxy-5-iodotubercidin	Red alga	*Hypnea valentiae*
	Flustramine A/B	Marine moss animals	*Flustra foliacae*
Anti-inflammatory			
	Flexibilide	Soft coral	*Sinularia flexibilis*
	Dendalone derivatives	Sponge	*Phyllospongia dendyi*
	Manoalide	Sponge	*Luffariella* spp.
Antibiotics			
	Acanthifolicin	Sponge	*Pandaros acanthifolium*
	Holothurins	Sea cucumbers	*Holothuroidae* spp.
Analgesia			
	Conotoxins	Marine snails	*Conus* spp.
Antiviral			
	Didemnins	Caribbean tunicates	*Trididemnum* spp.
	Spongouridine	Caribbean sponge	*Cryptotehia crypta*
CNS activity			
	Dactylyne	Sea hare	*Alysia dactylomela*

Source: Ref. 3.

C. Origins of Existing Drugs and Potential for Marine Toxin–based Drugs

Table 1 lists 10 of the top-selling prescription drugs in the United States in 1998 and identifies their origins. It is clear that a combination of rational design and discovery from natural sources was critical in the development of these compounds. It is notable that none of the top 10 drugs was derived from marine toxins. Products from marine sources are underrepresented even if the top 50 drugs are considered. This reflects the fact that the observation of biological effects from plants and animals has in the past reflected humankind's convenient access only to terrestrial species. With increasing exploration of marine resources, it is likely that new drugs will indeed be derived from the sea. Indeed, marine toxins are perhaps the most likely starting point for such discoveries, simply because they have been less explored than molecules from terrestrial sources and hence have a potential advantage in novelty of structure and pharmacology.

A hint as to the potential of marine toxins may be seen from the observation that two of the major conditions for which there are no effective cures at present are Alzheimer's disease and stroke. In both instances, compounds from marine sources have been identified in which there is at least a plausible mechanistic process by which drugs could be developed.

There have been several reviews over the last 20 years which have addressed the potential of marine sources of drugs (2–5) and a recent review that has more generally addressed the potential of toxins for drug discovery (6). The previous chapters of this book have extensively outlined the chemistry and pharmacology of a wide range of marine toxins. Table 2 summarizes many of the classes of activity and potential therapeutic applications described in these previous studies. Briefly, these include the full gamut of diseases and disorders, including, for example, anticancer compounds, antibiotics, and cardiovascular agents. In this chapter, we make no attempt to address all of these potential applications, but we will focus on selected examples of marine peptide toxins to illustrate the principles involved in their application to drug development.

Before describing these selected studies of marine toxins which show potential for the development of drugs, it is useful to describe briefly the factors which are important in drug molecules, in toxins, and in their respective target sites of action.

II. DRUGS, TOXINS, AND THEIR TARGETS

A. Drug Properties

The top 10 drugs listed in Table 1 share some common features relating to their chemistry. The majority have a molecular weight in the range of 300–500 and have water-lipid partition coefficients (log P) that favor their bioavailability. Although not essential features of drug molecules, these and certain other physicochemical considerations are important, as summarized below.

- Potency and selectivity for target site/disease
- Low toxicity
- Bioavailability
- Limited side effects
- Easy and economical synthesis
- Novelty

It is clear from this list that bioactivity alone is not sufficient for a molecule to be a useful drug candidate. Molecules which have some of the above properties may be useful lead compounds or may be useful receptor-binding probes and hence should not be underestimated, but ultimate success as a drug does require many features in addition to relevant biological activity.

B. Toxin Properties

As is clear from the preceding chapters, marine toxins have a diverse range of structures, sites of action, and physiological mechanisms. This diversity is detailed in Figure 2, which illustrates structures of a range of selected marine toxins that have potential therapeutic applications.

Consideration of Figure 2 suggests that many of the properties identified in Section A above as being desirable for drugs are in fact not present in toxins; e.g., in many cases, the toxins are complex molecules that would be extremely expensive to synthesize on the large scale required for pharmaceutical production; often they have poor oral bioavailability; and in the case of some of the larger peptides, they may be toxic and/or immunogenic. This reinforces the point made earlier that, in most cases, toxins should be regarded as starting points for drug design rather than ends in themselves.

As noted above, it is not the intention in this chapter to examine the pharmaceutical potential of each of the various classes of toxins described in this book, but rather via the use of selected examples to illustrate the principles involved in using toxins as starting points to drugs. With that aim we will focus on a class of toxins that offers perhaps the most potential—that of disulfide-rich peptides from marine venoms. In particular, much of the discussion will focus on the conotoxins.

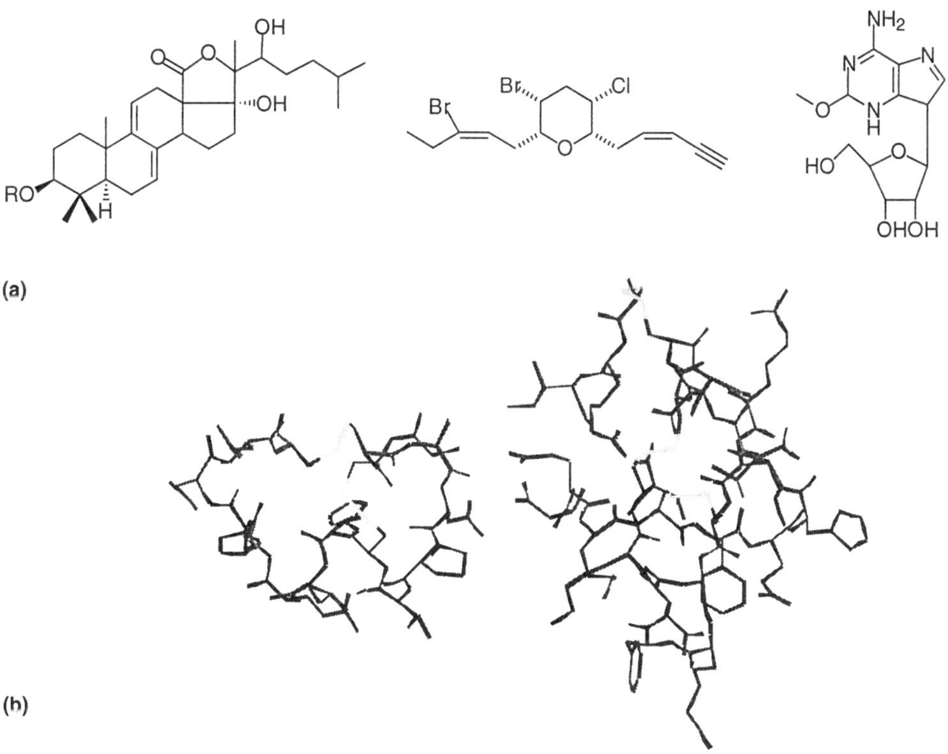

Figure 2 Structures of selected marine toxins from nonpeptide and peptidic families. (a) Nonpeptide toxins (b) peptide toxins.

The venoms of *Conus* snails typically contain complex mixtures of small disulfide-rich peptides (conotoxins) which interact specifically with a range of ion channels and receptors in the neuromuscular system. Targets currently identified include voltage-sensitive sodium (VSSC), calcium (VSCC), and potassium channels (VSKCs) and *N*-methyl-D-aspartate-glutamate (NMDA), vasopressin, and acetylcholine receptors (AChRs). Many of these conotoxins have been applied by researchers as pharmacological tools and biochemical probes for investigating the structure and function of receptors critical to the functioning of the neuromuscular system. Their high potency and specificity and convenient chemical synthesis of conotoxins also make them attractive leads in drug-design programs.

Conotoxins have been divided into a number of classes based on their pharmacological activities and cysteine frameworks. These include the α-, δ-, κ-, μ-, and ω-conotoxins and whose amino acid sequences are summarized in Table 3.

From the above sequences, it is clear that the different classes of conotoxins have a limited number of sequence motifs, or frameworks, based on the number and spacing of cysteine residues. The known frameworks are summarized in Figure 3, and they are commonly referred to as two-loop, three-loop, or four-loop framework based on the number of peptide backbone loops between cysteine residues.

Because the major physiological targets of the conotoxins are ion channels, it is valuable to outline here the therapeutic potential of the various channels.

C. Ion Channels as Drug Targets

Voltage-dependent ion channels are intrinsic membrane proteins that play an important role in fast communication in excitable cells. A short stretch of amino acids, the pore region, is the sole determinant of cation selectivity and also forms the binding site for many channel blockers. Attempts to characterize the structural basis for ion channel selectivity have been hampered by the fact that, like all integral membrane proteins, the voltage-gated ion channels are difficult to study by crystallographic or NMR methods. In fact, the only ion channel structure that has been published to date is that of a potassium channel from *Streptomyces lividans* (7). Consequently, most of the data describing ion channel structure have been obtained by indirect methods. For example, toxins that interact intimately with the pore region have been used as structural templates to deduce the spatial organization of this region of the ion channels. These models of pore structure are valuable for understanding the mechanisms of ion permeation, and ultimately may be useful for the rational design of drugs to modify ion channel function.

The various types of ion channels have some structural and functional similarities, but they also have some significant differences and each differs in potential drug-design applications (8). The various channels are examined briefly below.

Nicotinic Acetylcholine Receptors

The nicotinic acetylcholine receptors (nAChRs) are members of a gene superfamily of pentameric ligand-gated ion channels that are present in muscle and neuronal tissues. Activation of distinct subtypes of these presynaptic nAChRs by nicotinic agonists can selectively regulate the release of different neurotransmitters, including dopamine, norepinephrine, glutamate, and acetylcholine (9,10). Such receptors have been implicated in the pathophysiology of several neuropsychiatric disorders, including schizophrenia, Alzheimer's disease, Parkinson's disease, and Tourette's syndrome (9). Despite their importance, few of the nicotinic antagonists identified to date are highly selective between the multiple neuronal nAChR subtypes. Thus, the ability of the recently discovered conotoxin ImI to target homopentameric α_7 neuronal nAChRs with

Table 3 Major Classes of Conotoxins

α-conopeptides (two-loop framework peptides that inhibit nicotinic acetylcholine receptors)	
GI	E**CC**N-PA**C**GRHYS--**C**[a]
GIA	E**CC**N-PA**C**GRHYS--**C**GK[a]
GII	E**CC**H-PA**C**GKHFS--**C**[a]
MI	GR**CC**H-PA**C**GKNYS--**C**[a]
SI	I**CC**N-PA**C**GPKYS--**C**[a]
SIA	Y**CC**H-PA**C**GKNFD—**C**[a]
SII	G**CCC**NOA**C**GPBYG--**C**GTS**C**S
PnIA	G**CC**SLPP**C**AANNPDY**C**[a]
PnIB	G**CC**SLPP**C**ALSNPDY**C**[a]
ImI	G**CC**SDPR**C**AWR----**C**[a]
EI	RDO**CC**YHPT**C**NMSNPQI**C**[a]
MII	G**CC**SNPV**C**HLEHSNL**C**[a]
EpI	G**CC**SDPR**C**NMNNPDY(SO4)**C**[a]
AuIB	G**CC**SYPP**C**FATNPD-**C**
μ-conopeptides (three-loop framework that block sodium channels)	
GIIIA	RD**CC**TOOKK**C**KDRQ**C**KOQR**CC**A[a]
GIIIB	RD**CC**TOORK**C**KDRR**C**KOMK**CC**A[a]
GIIIC	RD**CC**TOOKK**C**KDRR**C**KOLK**CC**A[a]
PIIIA	RL**CC**GFOKS**C**RSRQ**C**KOHR**CC**[a]
ω-conopeptides (four-loop framework peptides that block calcium channels)	
GVIA	**C**KSOGSS**C**SOTSYN**CC**-RS**C**NOYTKR**C**Y
GVIB	**C**KSOGSS**C**SOTSYN**CC**-RS**C**NOYTKR**C**YG[a]
GVIC	**C**KSOGSS**C**SOTSYN**CC**-RS**C**NOYTKR**C**[a]
SVIA	**C**RSSGSO**C**GVTSI-**CC**GR-**C**--YRGK**C**T[a]
SVIB	**C**KLKGQS**C**RKTSYD**CC**SGS**C**GRS-GK**C**[a]
GVIIA	**C**KSOGTO**C**SRGMRD**CC**--S**C**LLYSNK**C**RRY[a]
GVIIB	**C**KSOGTO**C**SRGMRD**CC**T-S**C**LSYSNK**C**RRY[a]
MVIIA	**C**KGKGAK**C**SRLMYD**CC**TGS**C**RS--GK**C**[a]
MVIIB	**C**KGKGAS**C**HRTSYD**CC**TGS**C**NR--GK**C**[a]
MVIIC	**C**KGKGAP**C**RKTMYD**CC**SGS**C**GRR-GK**C**[a]
MVIID	**C**QGRGAS**C**RKTMYN**CC**SGS**C**NR--GR**C**[a]
TVIA	**C**LSOGSS**C**SOTSYN**CC**-RS**C**NOYSRK**C**Y[a]
δ-conopeptides (four-loop framework peptides that delay inactivation of sodium channels)	
TxVIA	W**C**KQSGEM**C**NLLDQN**CC**DGY-**C**IVLV**C**T
TxVIB	W**C**KQSGEM**C**NLLDQN**CC**DGY-**C**IVFV**C**T
GmVIA	VKP**C**RKEGQL**C**DPIFQN**CC**RGWN**C**-VLF**C**V
NgVIA	SK**C**FSOGTF**C**GIKOGL**CC**SVR-**C**FSLF**C**ISFE
PVIA	EA**C**YAPGTF**C**GIKOGL**CC**SEF-**C**LPGV**C**FG[a]
κ-conopeptides (four-loop framework peptides that block Shaker potassium channels)	
PVIIA	**C**RIONQK**C**FQHLDD**CC**RKK**C**NNRFNK**C**V

Amino acid sequences shown with cysteines (boldface) aligned within each structural framework.

[a] Processed carboxyl terminal; O = hydroxyproline residue, Z = γ-carboxyglutamic acid residue. Letter prefixes indicate conopeptides from the fish hunters *Conus magus* (M), *C. geographus* (G), *C. tulipa* (T), *C. striatus* (S), *C. purpurascens* (P), and *C. ermineus* (E); the mollusk hunters *C. textile* (Tx), *C. episcopatus* (Ep), *C. gloriamaris* (Gm), *C. nigropunctatus* (Ng), and *C. aulicus* (Au); and the worm hunter *C. imperialis* (Im).

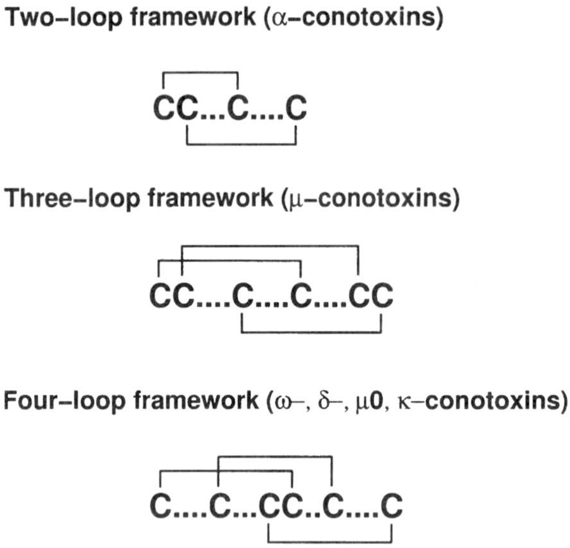

Figure 3 Disulfide-loop frameworks observed in conotoxins.

high specificity has considerable significance for both basic neuroscience and potential drug development. This toxin is thus chosen as an example to illustrate recent structural studies described in Section IV.

Sodium Channels

Sodium channels are composed of three separate subunits: α, β1, and β2, which are present in 1:1:1 stoichiometry. The β-subunits are integral membrane glycoproteins of ~30 kD each, whereas the α-subunit is a transmembrane glycoprotein of 260 kD. The α-subunit comprises four highly homologous domains, each of which contains six transmembrane helices. The pore region is formed by a short stretch of residues between the fifth and sixth transmembrane helices of the four domains. It is within the α-subunit that several binding sites (six or seven have been identified to date) for a diverse range of neurotoxins are located. Sodium channel–modulating agents have potential applications in the treatment of several conditions, including epilepsy, stroke, and cardiac arrhythmias. The tetrodotoxin (TTX)–resistant sodium channels are also an attractive target for the management of pain. Marine toxins that are capable of distinguishing between the different subtypes of sodium channels have provided insights into how specific Na^+ channel pathways could be modulated to control particular diseases. Furthermore, as discussed in Section IV, structural and functional data from closely related marine toxins have revealed some of the structural features that contribute to Na^+ channel subtype specificity.

Potassium Channels

K^+ channels represent the largest and most diverse family of ion channels and comprise four major functional classes: voltage-dependent K^+ channels, Ca^{2+}-activated K^+ channels, inwardly rectifying K^+ channels, and Na^+ activated K^+ channels. Regulation of these K^+ channels contributes to the control of potassium flow, cell volume, release of hormones and transmitters, and excitability of neurons and muscles. Such regulation contributes to signaling between neurons and mechanisms for cellular protection during stressful events such as ischemia. Reagents which modulate potassium channel activity have been suggested as potential antihypertensive and antianginal agents (11), as well as anti-ischemics (12), and immunosuppressants (13). Marine venoms have been found to contain a number of structurally diverse toxins that are capable of blocking K^+ channel currents (14,15). In Section IV, we describe how structural and functional data on these toxins and their mutants are lending insights into the structural requirements for K^+ channel binding and the factors governing channel subtype specificity.

Calcium Channels

Calcium channels are structurally related to sodium channels, with the main difference being the positioning and nature of the residues that line the selectivity filter in the pore of the channel. There are at least six pharmacologically distinct calcium channels: L-, N-, P-/Q-, T-, and R-type channels. Each of these channels contributes to a variety of cellular processes, such as neurotransmitter release, and reagents that modulate Ca^{2+} channel activity have widespread application as potential drugs. For example, L-type voltage-gated Ca^{2+} channels are a major therapeutic target for cardiovascular disorders, including hypertension, angina, selected arrhythmias, and peripheral vascular diseases. There has been widespread interest in one particular class of marine toxins, the ω-conotoxins, which are capable of selective modulation of specific subtypes of Ca^{2+} channels. Such activity has the potential for treatment of conditions such as pain and stroke, and, as discussed in detail in Section IV, one peptide (Ziconotide) is currently in clinical trials for these applications.

III. STRUCTURE-BASED DRUG DESIGN

As indicated in Section I, structure determination of potential drugs and their receptors plays a pivotal role in drug development. X-ray and NMR methods may both be applied here, but as most studies of marine toxins thus far have involved NMR, and this is our area of expertise, we will focus only on this method.

A. NMR Methods for the Structure Determination of Peptide Toxins

NMR spectroscopy is the preeminent method for structure determination of peptides and proteins in solution. The method relies on the measurement of a large number of distance restraints between pairs of protons. These restraints are used in a simulated annealing protocol to calculate a family of structures consistent with both the input restraints and with a force-field defining covalent geometry. The distance restraints are often supplemented with restraints on peptide backbone and side chain dihedral angles. The distance restraints are derived from NOESY spectra and the dihedral restraints from a combination of coupling constant and NOE data. Depending on the size of the protein being studied and the complexity of the spectra 2D, 3D, or 4D NMR methods may be required. The higher dimensional spectra (i.e., 3D or 4D) generally require uniform labeling of the protein with ^{15}N and/or ^{13}C isotopes so that spectral overlap may be resolved using the additional frequency dimensions associated with these NMR active nuclei as well as the usual proton chemical shift axis.

An assumption inherent in the NMR structure determination method is that the peptide or protein adopts predominantly a single conformation in solution. For linear peptides comprising fewer than approximately 30 amino acid residues, this is often not the case, with such small peptides being extremely flexible and adopting a myriad of conformations in solution. Thus, for these peptides, only qualitative conclusions can be drawn about solution conformations.

However, peptides which are cross linked by disulfide bonds are more restrained in their conformations and are very suitable for quantitative structure determination by NMR. As indicated above, the conotoxins are rich in disulfide bonds and are hence particularly amenable to conformational analysis by NMR.

An additional advantage of conotoxins is that their small size (generally less than 30 residues) means that spectral overlap is usually not a problem, and 2D rather than 3D or 4D NMR methods are sufficient for spectral assignment and structure determination. Because isotopic labeling is not required for such studies, it is in principle possible to determine structure from native peptides extracted from venom ducts; however, in practice, the amounts of material required (~1 mg) generally means that it is more convenient to synthesize the peptides.

To illustrate the principles involved in structure determination of the conotoxins, it is convenient to use as an example the data for conotoxin GS, a sodium channel blocking peptide whose structure was recently solved in our laboratory (16). The sequence of GS is given in Table 3.

For conotoxin GS, a series of 2D NMR spectra (DQF-COSY, E-COSY, TOCSY, and NOESY) was recorded. The TOCSY, DQF-COSY, and NOESY spectra were used to assign individual spectral peaks to specific protons in the molecule. The NOESY spectrum was then analyzed to derive distance restraints, whereas dihedral angle restraints were obtained from DQF-COSY and E-COSY spectra, respectively. From a series of TOCSY spectra recorded after dissolution of the sample in D_2O, those amide protons that were slowly exchanging were determined and used to deduce likely hydrogen bonds.

A set of 50 3D structures was calculated using 487 interproton distance restraints derived from 211 intraresidual, 114 sequential, 43 medium range ($|i-j| \leq 5$) and 119 long-range NOEs. These restraints were supplemented with 17 backbone and 8 side chain dihedral angle restraints from spin-spin coupling constants and 18 distance restraints defining 9 hydrogen bonds. The 20 structures with the lowest energies were chosen to represent the solution structure of conotoxin GS, and these are shown in Figure 4.

The structure consists of a β-hairpin and several turns cross linked by three disulfide bonds. The turns contain several cationic and polar residues that may be important for interacting with the sodium channel. Figure 4 also serves to illustrate one of the valuable features of the NMR method; that is, some indication of flexible regions of peptides can be discerned. In the case of GS, a region of apparent flexibility is present near the C-terminus of the molecule and in loop 4 of the structure (i.e., the loop in the backbone between the last two cysteine residues). This is just one example of how NMR provides more than just structural information for the drug-design process (17,18).

B. Additional Information Obtainable from NMR Studies

A complete determination of 3D structures is only one aspect of how NMR may be used to assist in drug-design programs associated with marine toxins. The main application of the structural studies is in the development of pharmacophore models to deduce likely receptor geometries and electrostatic characteristics. Other information relevant to drug design that can be obtained from NMR studies includes:

1. Determination of the correct folding of a series of closely related synthetic mutants
2. Determination of pK_a's of ionizable groups
3. Determination of internal flexibility in toxin molecules

In the first of these additional applications, NMR can be used as a quick screen to ensure that newly synthesized toxins within a given series (e.g., of alanine scan mutants or loop splice

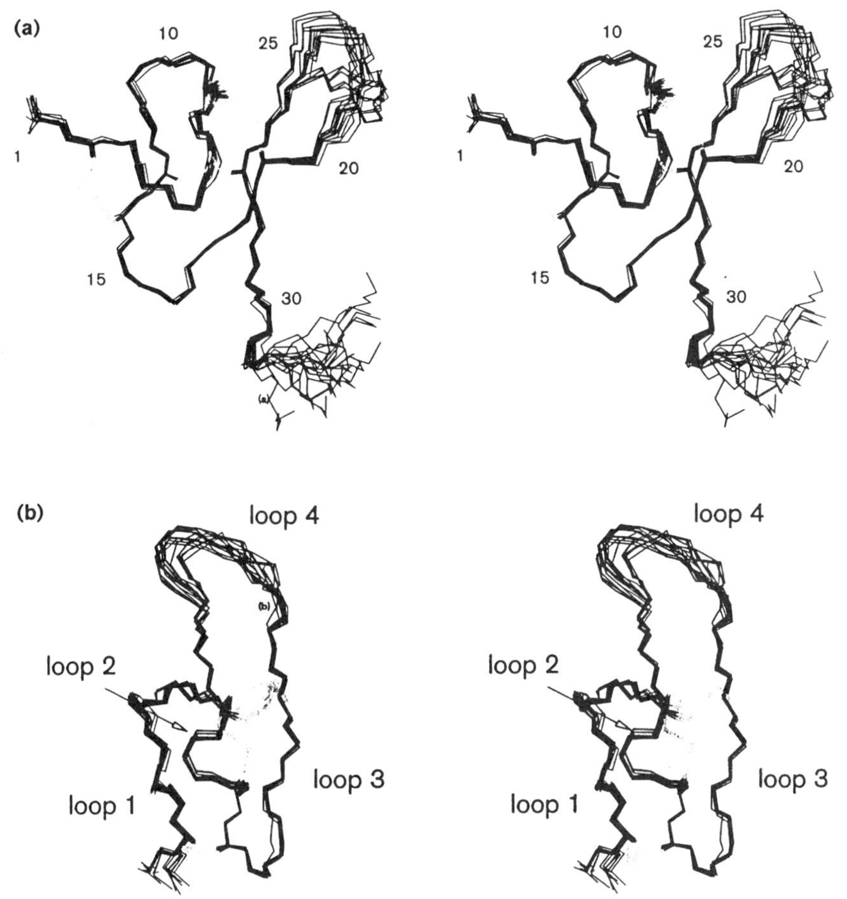

Figure 4 The three-dimensional structure of conotoxin GS. (**a**) Stereoview of 20 structures superimposed over the backbone (C, Cα, N) atoms of the well-defined region encompassing residues 2–20 and 25–31. The three disulfide bonds (2–14, 9–19, and 13–27) are shown in lighter shading. (**b**) Stereoview illustrating the orientation of the four loops in the structure. For clarity, the C-terminal residues 28–34 are not shown.

mutants) all fold in a similar manner. If the aim of studies of mutant peptides is to localize binding or functional activity to a particular residue or small region, it is important that all members of the series fold in the same way and do not undergo gross conformational changes associated with amino acid substitutions. The αH proton chemical shifts are a sensitive measure of the local environment within peptides and are readily measured in a single 2D NMR experiment. Thus, only a few minutes to hours of NMR data acquisition time is required to obtain an indication of correct folding.

NMR is also very useful for determining pK_a values of ionizable groups within peptide toxins, particularly for acidic residues like Asp and Glu which titrate in the pH range 4–6, which is convenient for NMR studies. His residue pK_a values may also be readily determined. A large change in pK_a value may indicate the presence of a salt bridge or other form of stabilization that might impact on the availability for receptor interactions of the amino acids involved.

An assumption inherent in many drug-design strategies is that the conformation determined in solution is the same as the biologically active form that binds to the receptor. This is expected to be the case for rigid molecules, and although many peptide toxins are extensively cross linked by disulfide bonds and hence expected to be rigid, surprises do occur. Thus, it is necessary to assess conformational flexibility of toxins that are leads in drug-design programs. Selective line broadening often gives clues to the presence of slow conformational interchanges, but NMR relaxation measurements are also extremely valuable for detecting a range of internal motions and now form an integral part of our structure-based design strategies.

C. Using Structures in Drug Design

To understand how a toxin exerts its function, it is necessary to elucidate both the structure of the toxin and the nature and spatial organization of the elements of the toxin that confer its specificity for a particular target. The small and highly constrained nature of many of these toxins makes them accessible to structure determination and many toxin structures have been determined by NMR spectroscopy and/or x-ray crystallography. However, the ion channels, which are targeted by many of these toxins, are membrane-associated proteins that have proved to be difficult to characterize by the conventional techniques of structural biology. Recently, the first structure of a potassium channel from *Streptomyces lividans* was solved by x-ray crystallography (7); however, to date there have been no reported structures of ion channel toxin complexes. Consequently, the binding capacity of different toxins has been probed by mutational analysis. By comparing this information with a knowledge of the 3D structures of the individual toxins, the topography that is critical for function may be determined.

IV. SELECTED EXAMPLES OF TOXIN-BASED DRUG DESIGN

In this section, we focus on describing marine toxins for which high-resolution 3D structures are available, as these provide candidate molecules for drug-design applications. Many of the drug development applications are currently in progress and as yet unpublished, but the potential of marine toxins as drugs is highlighted by the case study in this section, which describes clinical trials on conotoxin MVIIA, now referred to as Ziconotide. It is to be anticipated that similar success stories will arise in the coming years for many other toxins, including some of those described below.

A. Structural Families Observed in Marine Peptide Toxins

Table 4 summarizes some examples of marine peptide toxins for which 3D structures are currently available. Despite their structural and functional diversity, these peptide toxins share a number of similar features. They are all small proteins of less than 70 amino acids and they all possess a high proportion of cysteine residues that participate in disulfide bonds. The conotoxins are well represented in Table 4, reflecting a recent explosion in their structure determination. Over the last few years, our group alone has determined the structures of more than 30 conotoxins, and we are using these structures in several drug-design programs. From these studies, and from studies reported by colleagues in the literature, it has become clear that conotoxin structures fall into a limited number of families (see Table 3). However, as highlighted in Figure 5, toxins belonging to the same structural family, for example, conotoxins GS, PVIIA, and GVIA, may have totally different biological activities.

It is clear that marine toxins exhibit a diverse range of structures. However, it is worthy of note that the majority of these toxins may be thought of as ''mini-proteins'' that adopt well-defined solution structures with all the features of larger proteins. The defined presentation of amino acids on the surface of the frameworks in Figure 5 accounts for the specificity of their

Table 4 Structures of Selected Marine Toxins from Different Activity Classes

Channel target	Protein	PDB code	Source	Method	Reference
AChR	Erabutoxin A	5EBX	Sea snake	X-ray	52
	Erabutoxin B	3EBX	Sea snake	X-ray	53
		1ERA		NMR	54
	Conotoxin EpI	1A0M	Cone snail	X-ray	55
	Conotoxin ImI	1IMI	Cone snail	NMR	27,28
	Conotoxin GI	1NOT	Cone snail	X-ray	56
		IXGA		NMR	29
Sodium	Anthopleurin-A	1AHL	Sea anemone	NMR	32
	Anthopleurin-B	1APF	Sea anemone	NMR	31
	Conotoxin GS	1AG7	Cone snail	NMR	16
	Conotoxin GIIIA	1TCG	Cone snail	NMR	42,57
	Conotoxin GIIIB	1GIB	Cone snail	NMR	43
Calcium	Conotoxin MVIIA	1OMG	Cone snail	NMR	58
		1MVI		NMR	50
	Conotoxin GVIA	2CCO	Cone snail	NMR	59
		1OMC		NMR	60
Potassium	ShK	1ROO	Sea anemone	NMR	48
	BgK	1BGK	Sea anemone	NMR	47
	Conotoxin PVIIA	1AV3	Cone snail	NMR	45
		1KCP		NMR	46

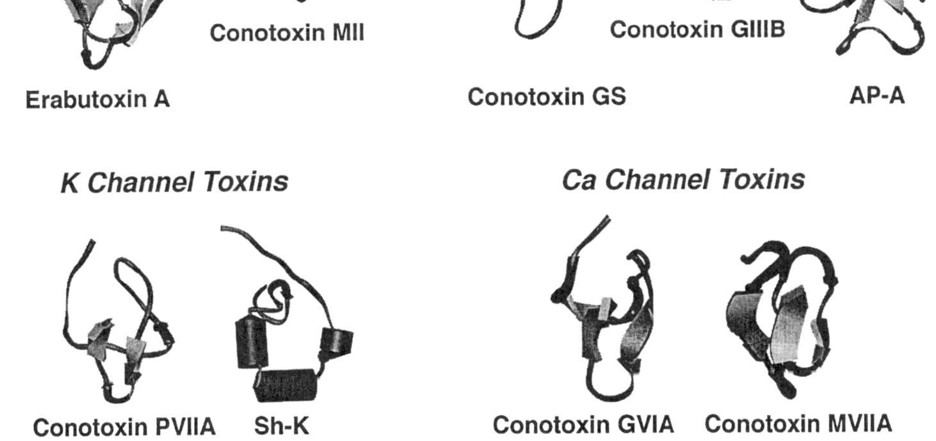

Figure 5 Ribbon diagrams which summarize the 3D structures of several classes of marine toxin.

binding interactions, and the small size of the molecules makes them valuable lead compounds in drug-design applications. In addition to the diversity of the natural marine toxins illustrated in Figure 5, examples will be provided where extra diversity has been engineered using chemical methods. In the following sections, we describe the structural findings in more detail. For convenience, the selected peptide structures are grouped according to their receptor targets.

B. Toxins that Block the Nicotinic Acetylcholine Receptor

α-Conotoxins are widespread in the venoms of cone snails and have been isolated from piscivorous, molluscivorous, and vermivorous species (19–26). These toxins are valuable ligands for probing structure-function relationships of various nAChR subtypes, as they are potent antagonists and exhibit marked selectivity between the peripheral and neuronal forms of the receptor. Typically, the α-conotoxins are 12–18 residues in length and are characterized by the presence of two conserved disulfide bonds and two loops in the peptide backbone between the cysteines. The number of amino in these two intracysteine loops varies, giving rise to the α3/5, α4/7, and α4/3 subclasses of α-conotoxins. Therefore, the α-conotoxins represent a family of similar toxins with different specificity for subtypes of the nAChR.

To illustrate potential applications of this class of conotoxins, we describe studies on conotoxin ImI, a recently discovered member of the family with selectivity for neuronal nAChRs, and on GI, the first and one of the smallest α-conotoxins to be discovered. The sequences of these peptides are given in Table 3.

Conotoxin IMI

α-Conotoxin ImI is derived from the venom of the vermivorous snail *Conus imperialis* and is the first identified peptide ligand that binds selectively to the neuronal α_7 homopentameric subtype of the nAChR. This receptor has emerged as a very important target in several disease states. The potential therapeutic value of a highly specific agonist is evident from findings that high doses of nicotine are beneficial to sufferers from cognitive and attention deficits, Parkinson's disease, Tourette's syndrome, and ulcerative colitis as well as schizophrenia. However, side effects arising from high doses of nicotine are often more detrimental than the disease state and alternative specific ligands are of much interest. It is important to understand the structural origin for the different selectivities of toxins that target the nAChR, so that highly selective agonists for the various receptor subtypes can be developed. Marine toxins such as ImI have the potential to define more precisely the physiological roles of the different nAChR subtypes as well as provide leads to the development of novel and selective ligands.

The structure of ImI has recently been reported (27,28) and mutational analyses have identified four residues that are crucial for activity. As shown in Figure 6, the functional residues are clustered on one face of the molecule. Furthermore, it has been found that α-conotoxins that are selective for neuronal nAChRs have similar structures and are different from those of α-conotoxins that are selective for muscular nAChRs. This is illustrated in Figure 6, which depicts backbone ribbons for the neuronal specific α-conotoxins ImI and EpI as well as the muscle-specific GI.

It has been suggested that these structural differences form the basis for selectivity of α-conotoxins for neuronal or muscular nAChR subtypes. Interestingly, the functionally essential side chains of ImI demonstrate high variability within the neuronal specific α-conotoxin family. It is this variability that is thought to form the basis for discrimination between different neuronal nAChR subtypes.

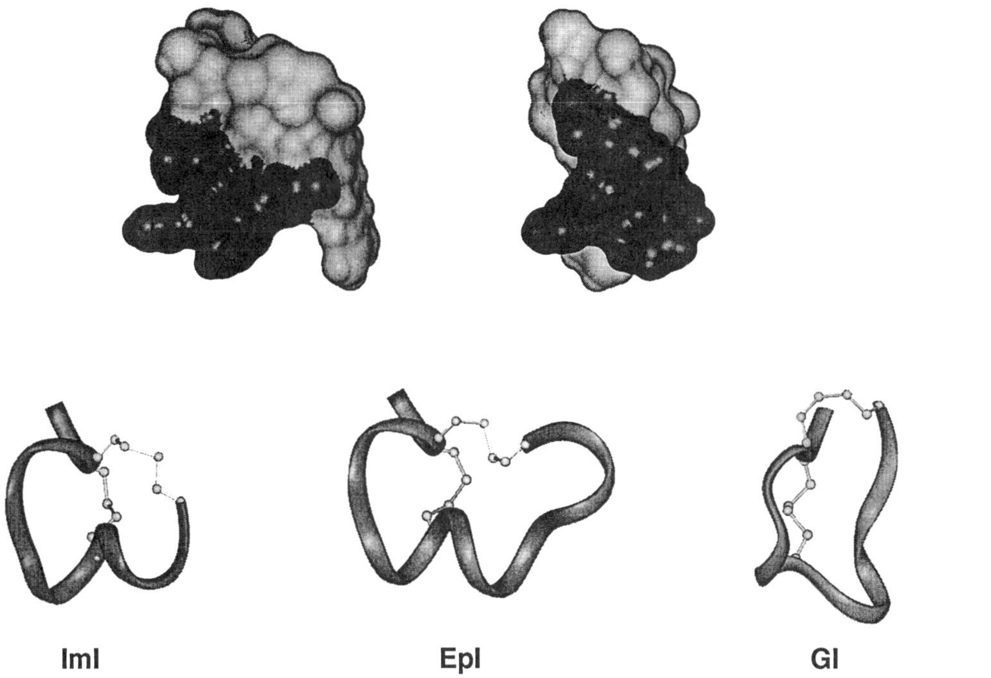

Figure 6 (**Top**) Connolly surfaces of ImI showing those residues that form the functional surface for interaction with the α7 neuronal nAChR in dark gray. The view shown on the right is in the same orientation as the ribbon diagram below, which has been rotated by 90° in the view on the left. (**Bottom**) Ribbon diagrams of ImI, EpI, and GI superimposed over the residues in the first loop.

Conotoxin GI

To further probe the structural requirements for activity within the α-conotoxin family, we have synthesized a series of nonnatural disulfide bond mutants of conotoxin G1. Our interest in GI is focused on its use as a model to explore conformational diversity resulting from disulfide bond engineering. As already noted, conotoxins are characterized by their particularly high content of cysteine, with the cysteine residues almost invariably being connected in pairs to form disulfide bonds. In peptide toxins, even more so than in larger proteins, these disulfide bonds have a crucial bearing on 3D structure and function. As the number of cysteine residues in a peptide increases the number of ways of connecting the cysteines in disulfide bonds increases dramatically, leading to a large number of potential isomers. It is interesting and highly significant that invariably only one of the possible isomers occurs naturally; that is, venoms do not normally contain different isomers of the same conotoxins with different connections of the disulfide bonds. However, using solid-phase chemical methods, it is possible selectively to produce each of the individual disulfide bond isomers. We have used this approach to synthesize all three possible disulfide bond isomers of the α-conotoxin G1 and have determined their structures (29). We refer to the three isomers as GI(2-7;3-13), GI(2-13;3-7), and GI(2-3;7-13), reflecting the respective disulfide bond connectivities between cysteine residues at positions 2, 3, 7, and 13 in the G1 sequence.

The structural findings may be summarized by noting that the native connectivity of the four constituent cysteine residues produces a significantly more stable and well-defined structure than either of the two alternative arrangements of the disulfide bonds (29). A single-solution conformation was detected for the native isomer, GI(2-7;3-13), which consists primarily of a distorted 3_{10} helix from residues 5–11. The two nonnative forms exhibit multiple conformations in solution, with the major populated forms being different in structure both from each other and with the native form. We concluded that the disulfide bonds in GI play a major role in determining both the structure and stability of the peptide. A trend for increased conformational flexibility was observed in the order GI(2-7;3-13) $<$ GI(2-13;3-7) $<$ GI(2-3;7-13).

Interest in making nonnative isomers arises because peptide analogues are widely regarded as valuable drug leads, and in recent years there has been much effort directed toward the development of peptide libraries. It has been of particular interest to develop methods to increase the surface variability of peptides, because the diversity of peptide libraries are, to some extent, limited by the use of the 20 natural amino acids. The study described above shows that the use of alternative disulfide bond connectivities provides another way of altering molecular conformations without modifying the sequence.

C. Sodium Channel–Binding Toxins

Sodium Channel Blockers from Sea Anemones

Anthopleurin-A (AP-A) and anthopleurin-B (AP-B) are 49–amino acid polypeptide toxins from the sea anemone *Anthopleura xanthogrammica* (30). They are members of a family of toxins that bind to the extracellular face of voltage-sensitive sodium channels (VSSCs) and retard channel inactivation. In the mammalian heart, this effect is manifest as an increase in the force of contraction. As a result there has been interest in exploiting the anthopleurins as lead compounds in the design of novel cardiac stimulants. AP-A, which differs from AP-B in 7 of its 49 residues, displays different selectivity compared with AP-B, acting preferentially on cardiac over neuronal sodium channels. Consequently, a knowledge of both the structure and binding determinants for each of the toxins may provide an insight into the basis for cardiac sodium channel subtype specificity. The structures of both AP-A and AP-B have been determined and are very similar (31,32). Each contains a four-stranded antiparallel β-sheet linked by three loops. The first loop in the structures is the largest and least well defined, but chemical modification studies have shown that several residues essential for activity are contained in this loop.

Two of the substitutions (L24 → F, T42 → N) are conservative and are unlikely to be important determinants of specificity. Two substitutions involve replacement of a neutral residue in AP-A with a basic one in AP-B (S12 → R, Q49 → K). Sodium channel binding studies have revealed that the differential binding of the two toxins is mediated, at least in part, by these residues (33). Of the remaining nonconserved residues, mutational studies have revealed that the only one that has a significant effect on binding affinities is the replacement of Val13 in AP-A with Pro in AP-B. The V13 → P mutation results in a significant change in the orientation of the first loop and is thought to affect the conformation of a cationic cluster that is an important determinant of both the binding affinity and channel isoform selectivity (34). As shown in Figure 7, of the seven residues that are different between AP-A and AP-B, only three are located on the face of the toxins that comprises the functional binding surface. The remainder, which have minimal effects on channel subtype specificity are located on the opposite face. This example highlights the way in which mutational analyses with toxins of known structure can lead to an understanding of the factors that govern binding.

Sodium Channel Conotoxins

The preceding example involves studies with toxins of similar structure that display differences in their specificity toward different subtypes of ion channels. However, there are also situations

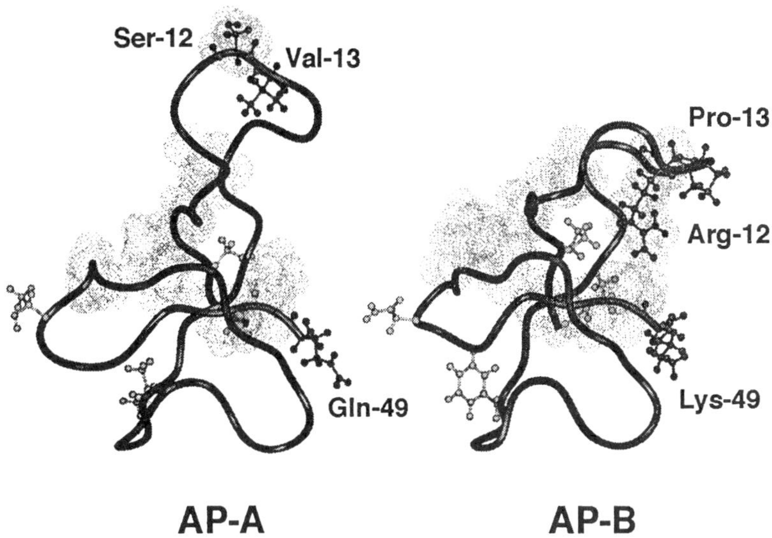

Figure 7 The solution structures of AP-A and AP-B, showing the backbone ribbon as well as the positions of the seven mutated residues. Those mutations that affect the binding specificity of the molecule are shown in ball-and-stick representation and colored dark gray. The noncritical substitutions are shown in light gray. The locations of those residues that are important for activity have been highlighted by the dotted surface.

in which toxins with completely different structures compete for binding at the same ion channel subtypes.

The piscivorous cone snail, *Conus geographus*, produces polypeptide neurotoxins that specifically inhibit skeletal muscle and eel electroplax sodium channels (35–37). These toxins, the μ-conotoxins and conotoxin GS, are attractive probes of sodium channel structure because of their high binding affinity and ability to discriminate between the skeletal muscle and neuronal and cardiac channel isoforms (36–39). It is remarkable that although these peptides belong to the same pharmacological class, they have different structural frameworks, as illustrated in Table 2.

The μ-conotoxins, a family of highly basic 22-residue polypeptides (GIIIA, GIIIB, and GIIIC), contain six cysteine residues which are paired in a 1–4, 2–5, 3–6 pattern to form three intramolecular disulfide bonds and a three-loop framework. Conotoxin GS has a strikingly different sequence and is 50% larger than the μ-conotoxins. This polypeptide contains six cysteine residues arranged in a similar 1–4, 2–5, 3–6 pattern (40); however, differences in the spacings between cysteine residues results in a four-loop framework rather than a three-loop framework. Despite the low sequence identity, conotoxin GS binds competitively with μ-conotoxin GIIIA, suggesting overlapping binding sites on the extracellular surface of skeletal muscle and eel electroplax sodium channels (36).

μ-Conotoxins 3D solution structures have been determined for μ-conotoxins GIIIA (41,42) and GIIIB (43) but not so far for GIIIC. The structures of GIIIA and GIIIB are similar and consist of a distorted 3_{10}-helix, a small β-hairpin, a cis-hydroxyproline, and several turns. The molecule is stabilized by three disulfide bonds, two of which connect the helix and the β-hairpin, forming a structural core with similarities to the CSαβ motif (44). This motif is common to several families of small proteins, including scorpion toxins and insect defensins. Other structural features include multiple arginine and lysine side chains that project into the solvent in a radial orientation relative to the core of the molecule. These cationic side chains form potential sites of interaction with anionic sites on sodium channels. Together, the structures of GIIIA and GIIIB provide a basis for further understanding of the structure-activity relationships of the μ-conotoxins and for their binding to skeletal muscle sodium channels.

Conotoxin GS Conotoxin GS (16) adopts a structure that consists of a compact, disulfide-bonded core from which several loops and the C-terminus project, as illustrated earlier in Figure 4. The main element of secondary structure is a double-stranded antiparallel β-sheet comprising residues 17–20 and 26–29 connected by a turn involving residues 21–25 to give a β-hairpin structure. A further peripheral β-strand involving residues 7–9 is almost perpendicular to the β-hairpin, with only Ser7 hydrogen bonded to the central β-strand forming an isolated β-bridge.

The availability of solution structures for several μ-conotoxins and conotoxin GS has facilitated the design of analogues that will be used to define the binding surface and to undertake complementary mutagenesis on the sodium channel to identify the interacting residues. These experiments with conotoxins may prove to be useful in modeling the outer vestibule of sodium channels as the peptide toxins from scorpions have been for potassium channels.

D. Potassium Channel Toxins

Conotoxin PVIIA

By comparison to other ion channels of eukaryotic cell membranes, voltage-sensitive potassium channels are relatively simple, and methodology has been developed for mapping their interactions with small peptide toxins. κ-PVIIA is a 27-residue polypeptide isolated from the venom of *Conus purpurascens*, and it is the first member of a new class of conotoxins that block potassium channels (14). PVIIA, therefore, is a valuable new probe of potassium channel structure. In recent studies, the solution structure and mode of channel binding of PVIIA have been determined (45,46), and this forms the basis for mapping the interacting residues at the conotoxin-ion channel interface.

The 3D structure of PVIIA resembles the triple-stranded β-sheet/cystine knot motif formed by a number of toxic and inhibitory peptides, including the ω-conotoxins and conotoxin GS, as described above. However, subtle structural differences, predominantly in loops 2 and 4, are observed between PVIIA and other conotoxins with similar structural frameworks. Electrophysiological binding data suggest that PVIIA blocks channel currents by binding in a voltage-sensitive manner to the external vestibule and occluding the pore. Comparison of structural and functional data for PVIIA with those for the well-characterized potassium channel blocker charybdotoxin has led to the suggestion that a lysine residue in PVIIA is responsible for physically occluding the pore of the potassium channel.

ShK

The venoms of a number of sea anemones have also been shown to contain a novel class of toxins that block potassium channels, including ShK from the sea anemone *Stichodactyla helianthus* (15). The toxins are basic molecules comprising 35–37 residues and contain three disulfide bonds. The main elements of secondary structure present in these toxins are two short stretches of α-helix, the remainder of the structure consisting of a series of extended loops and turns (47,48). Despite the relative lack of regular elements of secondary structure, the toxins are very well-defined in solution. Mutational analysis of ShK has identified a number of residues that are essential for high-affinity binding to the potassium channel (49). These residues (I7, R11, S20, K22, Y23, and F27) are clustered together on the surface of ShK where they form the potassium channel binding surface of the toxin.

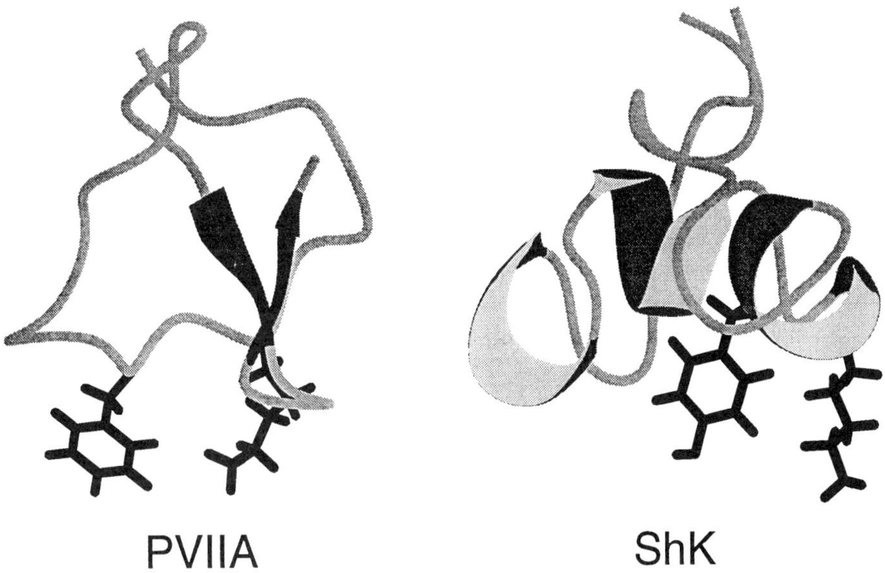

Figure 8 The solution structures of PVIIA and ShK. Despite the differences in size and secondary structure, each toxin contains a lysine and an aromatic residue in a similar orientation that are thought to be crucial for activity.

It is worthy of note that PVIIA and ShK, despite adopting vastly different structures, are both capable of targeting VSKCs. In both cases, a lysine residue, which occludes the pore of the channel, and an aromatic residue are thought to be major contributors to the functional binding surface of the toxins (Figure 8). In fact, potassium channel toxins from many phyla, both marine and terrestrial, all possess this functionally important dyad, consisting of a protruding lysine and a hydrophobic residue (47). It is thought that the dyad serves as a common anchor for all VSKCs, whereas other key functional residues surrounding the dyad determine the affinities and specificity toward the various VSKC subtypes.

E. Calcium Channel–Blocking Conotoxins

ω-Conotoxins

The ω-conotoxins have attracted much recent interest owing to their potential application in the treatment of chronic pain. They consist of a set of structurally related peptides that have a wide range of specificities for different subtypes of the VSCC. To understand their VSCC subtype differentiation, we studied the structure of two naturally occurring ω-conotoxins, MVIIA (specific to N-type VSCCs) and SVIB (specific to P/Q-type) and a synthetic hybrid, SNX-202, which has altered specificities to both VSCC subtypes (50). The secondary structures of the three peptides are almost identical, consisting of a triple-stranded β-sheet and several turns. The three-dimensional structures of SVIB and MVIIA are likewise quite similar, but some subtle differences are manifested as orientational differences between two key loops. A remarkable feature of the six cysteine/four-loop framework exemplified by the ω-conotoxins is the presence of a cystine knot within the structures. This motif consists of an embedded loop in the structure formed by two of the disulfide bonds and their connecting backbone segments. This loop is penetrated by the third disulfide bond in a remarkable example of Nature's engineering designs, which, as noted in previous sections, is also present in PVIIA and conotoxin GS.

From the above structural studies and a large number of other studies of molecules within this family, it is apparent that the ω-conotoxins form a consensus structure despite differences in sequence and VSCC subtype specificity. This indicates that the ω-conotoxin macrosites for the N/P/Q subfamily of VSCCs are related, with specificity for receptor targets being conferred by the positions of functional side chains on the surface of the peptides.

From the numerous studies that have characterized the relationships between structure and activity of the ω-conotoxins, it appears that their promise as useful drug targets is very likely to be fulfilled in the near future. This is illustrated in the following section, which provides a brief history (51) of the development of MVIIA from at first being just an interesting component of the venom of marine hunting snails to its development as Ziconotide, currently in clinical trial for the treatment of chronic pain and head trauma.

Case Study—Development of Ziconotide

The development of pharmaceutical products is an expensive (>US $300 million) and slow (>10 years) process, but there is still very great potential for individuals or small teams to make the initial breakthroughs that identify lead molecules and foster their development. In the case of the conotoxins, the field was pioneered by Baldomero Olivera in the early 1980s on his return to the Philippines after postdoctoral training in the United States in molecular biology. With a shortage of funds and resources to pursue molecular biology studies in the Philippines, Olivera turned to less expensive research and in particular to studies of the marine snails from the *Conus* genus. He had been aware from his youth of reports of the human fatalities from stings of one particular species, *C. geographus*, and began what he thought would be a short investigation to determine the active component of the venom. It soon became clear that the venom was an extremely complex mixture of a large number of peptides, many of which appeared to be inactive when injected intravenously into mice. Olivera subsequently returned to the United States to the University of Utah at Salt Lake City, and there his group made the major breakthrough they had been seeking when a number of purified venom components were injected directly into the central nervous system of mice. An amazing variety of responses were exhibited by different venoms ranging from, for example, sleeping, jumping, and scratching of the test animal, providing the first indication that the cone snail venoms were a virtual gold mine of peptides displaying neurological activity.

Some time after Olivera's work was published, a Californian start-up company, Neurex, became interested in the calcium channel blocking activity of some of the ω-conotoxins when they recruited George Miljanich from the University of California, San Francisco, and began a systematic investigation of structure-activity relationships of the ω-conotoxins. MVIIA proved to be a promising lead compound in animal models of pain; its activity resulting from an ability to block transmission of pain-related nerve impulses associated with N-type calcium channels in the spine. The selectivity of MVIIA for these channels, and not for P- and Q-type channels, made it useful, because pain could be blocked without affecting other vital functions of the nervous system. Neurex synthesized hundreds of variants in MVIIA in an attempt to find a more active molecule, but it appeared that Nature's evolutionary machinery had already produced an optimal molecule in MVIIA, or SNX-111, as Neurex referred to their synthetic derivative of it. The molecule, subsequently named Ziconotide, entered clinical trials and showed stunning efficacy for chronic pain patients.

Phase III clinical trials are now complete, and some indication of the value of the molecule can be seen from the recent decision of the Irish pharmaceutical company Elan to purchase

Neurex for $US 810 million, with Ziconotide a flagship molecule in the drug development pipeline. The molecule has some significant advantages over morphine in that it does not cause the constipation or respiratory suppression side effects of morphine, and unlike the opiates, it is effective against neuropathic pain, which is the result of nerve rather than tissue injury. Most significantly, Ziconotide shows no sign of the severe tolerance problems that develop in morphine-treated patients.

Ziconotide also has potential applications in other conditions associated with regulation of calcium channels. Thus, it shows promise in blocking cell death in the brain after head trauma and stroke. It prevents the calcium influxes that lead to apoptotic cell death, and that lead to excess glutamate release and so excitotoxic cell death. One of the difficulties with its application in these indications though was a marked lowering of blood pressure when given intravenously. This problem is managed in current phase III clinical trials by concurrent treatment with compensatory drugs. In the clinical trials for pain noted above, the hypotensive side effects were overcome by direct infusion of Ziconotide via an implanted pump and refilled via syringe.

These side effects illustrate one of the common features of drug development from natural toxins—that while their bioactivity makes them excellent starting points in drug design, the ultimate formulation may be based on analogues with improved bioavailability and/or reduced side effects. Second-generation Ziconotide products are likely to involve smaller molecules designed on the basis of structures of MVIIA-like peptides.

V. SUMMARY

Although there have been relatively few examples so far of marine toxins making it to the marketplace as drugs, this situation is clearly changing and many more examples of pharmaceuticals from the sea are anticipated over the next few years.

ACKNOWLEDGMENTS

D.J.C. is an Australian Research Council Professorial Fellow and M.J.S. is an ARC Australian Research Fellow.

REFERENCES

1. BM Olivera, WR Gray, R Zeikus, JM McIntosh, J Varga, J Rivier, V de Santos, LJ Cruz. Peptide neurotoxins from fish-hunting cone snails. Science 230:1338–1343, 1985.
2. PT Grant, AM Mackie. Drugs from the sea—fact or fantasy? Nature 267:786–788, 1977.
3. PN Kaul, P Daftari. Marine pharmacology: bioactive molecules from the sea. Annu Rev Pharmacol Toxicol 26:117–142, 1986.
4. DJ de Vries, PM Beart. Fishing for drugs from the sea: status and strategies. Trends Pharmacol Sci 16:275–279, 1995.
5. RS Norton. Structure and function of peptide and protein toxins from marine organisms. J Toxicol Toxin Rev 17:99–130, 1998.
6. AL Harvey, KN Bradley, SA Cochran, EG Rowan, JA Pratt, JA Quillfeldt, DA Jerusalinsky. What can toxins tell us for drug discovery? Toxicon 36:1635–1640, 1998.
7. DA Doyle, JM Cabral, RA Pfuetzner, A Kuo, JM Gulbis, SL Cohen, BT Chait, R MacKinnon. The structure of the potassium channel: molecular basis of K+ conduction and selectivity. Science 280: 69–77, 1998.

8. DJ Adams, PF Alewood, DJ Craik, R Drinkwater, RJ Lewis. Conotoxins and their potential therapeutic application. Drug Dev Res 46:219–234, 1999.
9. JM Kulak, TA Nguyen, BM Olivera, JM McIntosh. Alpha-conotoxin MII blocks nicotine-stimulated dopamine release in rat striatal synaptosomes. J Neurosci 17:5263–5270, 1997.
10. SA Kaiser, L Soliakov, SC Harvey, CW Luetje, S Wonnacott. Differential inhibition by alpha-conotoxin-MII of the nicotinic stimulation of [^{3}H] dopamine release from rat striatal synaptosomes and slices. J Neurochem 70:1069–1076, 1998.
11. LB Katz, EC Giardino, JJ Salata, JB Moore, Jr, R Falotico. RWJ 26629, a new potassium channel opener and vascular smooth muscle relaxant: a potential antihypertensive and antianginal agent. J Pharmacol Exp Ther 267:648–656, 1993.
12. GJ Grover. Protective effects of ATP sensitive potassium channel openers in models of myocardial ischaemia. Cardiovasc Res 28:778–782, 1994.
13. K Kalman, MW Pennington, MD Lanigan, A Nguyen, H Rauer, V Mahnir, K Paschetto, WR Kem, S Grissmer, GA Gutman, EP Christian, MD Cahalan, RS Norton, KG Chandy. ShK-Dap22, a potent Kvl.3-specific immunosuppressive polypeptide. J Biol Chem 273:32697–32707, 1998.
14. H Terlau, K-J Shon, M Grilley, M Stocker, W Stühmer, BM Olivera. Strategy for rapid immobilization of prey by a fish-hunting marine snail. Nature 381:148–151, 1996.
15. O Castaneda, V Sotolongo, AM Amor, R Stocklin, AJ Anderson, AL Harvey, A Engstrom, C Wernstedt, E Karlsson. Characterization of a potassium channel toxin from the Caribbean Sea anemone Stichodactyla helianthus. Toxicon 33:603–613, 1995.
16. JM Hill, PF Alewood, DJ Craik. Solution structure of the sodium channel antagonist conotoxin GS: a new molecular caliper for probing sodium channel geometry. Structure 5:571–583, 1997.
17. DJ Craik. NMR in Drug Design. Boca Raton, FL: CRC Press, 1996.
18. BJ Stockman. NMR spectroscopy as a tool for structure-based drug design. Prog NMR Spect 33: 109–151, 1998.
19. WR Gray, A Luque, BM Olivera, J Barrett, LJ Cruz. Peptide toxins from Conus geographus venom. J Biol Chem 256:4734–4740, 1981.
20. M McIntosh, LJ Cruz, MW Hunkapiller, WR Gray, BM Olivera. Isolation and structure of a peptide toxin from the marine snail Conus magus. Arch Biochem Biophys 218:329–334, 1982.
21. GC Zafaralla, C Ramilo, WR Gray, R Karlstrom, BM Olivera, LJ Cruz. Phylogenetic specificity of cholinergic ligands: alpha-conotoxin SI. Biochemistry 27:7102–7105, 1988.
22. RA Myers, GC Zafaralla, WR Gray, J Abbott, LJ Cruz, BM Olivera. alpha-Conotoxins, small peptide probes of nicotinic acetylcholine receptors. Biochemistry 30:9370–9377, 1991.
23. JM McIntosh, D Yoshikami, E Mahe, DB Nielsen, JE Rivier, WR Gray, BM Olivera. A nicotinic acetylcholine receptor ligand of unique specificity, alpha-conotoxin ImI. J Biol Chem 269:16733–16739, 1994.
24. M Fainzilber, A Hasson, R Oren, AL Burlingame, D Gordon, ME Spira, E Zlotkin. New mollusc-specific alpha-conotoxins block Aplysia neuronal acetylcholine receptors. Biochemistry 33:9523–9529, 1994.
25. JS Martinez, BM Olivera, WR Gray, AG Craig, DR Groebe, SN Abramson, JM McIntosh. alpha-Conotoxin EI, a new nicotinic acetylcholine receptor antagonist with novel selectivity. Biochemistry 34:14519–14526, 1995.
26. M Loughnan, T Bond, A Atkins, J Cuevas, DJ Adams, NM Broxton, BG Livett, JG Down, A Jones, PF Alewood, RJ Lewis. alpha-conotoxin EpI, a novel sulfated peptide from Conus episcopatus that selectively targets neuronal nicotinic acetylcholine receptors. J Biol Chem 273:15667–15674, 1998.
27. J Gehrmann, N Daly, PF Alewood, DJ Craik. The solution structure of alpha-conotoxin Im-I by 1H-NMR. J Med Chem 42:2364–2372, 1999.
28. IV Maslennikov, ZO Shenkarev, MN Zhmak, VT Ivanov, C Methfessel, VI Tsetlin, AS Arseniev. NMR spatial structure of alpha-conotoxin ImI reveals a common scaffold in snail and snake toxins recognizing neuronal nicotinic acetylcholine receptors. FEBS Lett 444:275–280, 1999.
29. J Gehrmann, PF Alewood, DJ Craik. Structure determination of the three disulfide bond isomers of

alpha-conotoxin GI: a model for the role of disulfide bonds in structural stability. J Mol Biol 278: 401–415, 1998.
30. S Shibata. Novel cardiostimulant polypeptides (anthopleurin-A, B and C) isolated from sea anemone. Jpn J Pharmacol 30:7P–9P, 1980.
31. SA Monks, PK Pallaghy, MJ Scanlon, RS Norton. Solution structure of the cardiostimulant polypeptide anthopleurin-B and comparison with anthopleurin-A (published erratum appears in Structure 1995; 3[10]:1126). Structure 3:791–803, 1995.
32. PK Pallaghy, MJ Scanlon, SA Monks, RS Norton. Three-dimensional structure in solution of the polypeptide cardiac stimulant anthopleurin-A. Biochemistry 34:3782–3794, 1995.
33. GR Benzinger, CL Drum, LQ Chen, RG Kallen, DA Hanck, D Hanck. Differences in the binding sites of two site-3 sodium channel toxins. Pflugers Arch 434:742–749, 1997.
34. GJ Kelso, CL Drum, DA Hanck, KM Blumenthal. Role for Pro-13 in directing high-affinity binding of anthopleurin B to the voltage-sensitive sodium channel. Biochemistry 35:14157–14164, 1996.
35. K Sato, Y Ishida, K Wakamatsu, R Kato, H Honda, Y Ohizumi, H Nakamura, M Ohya, JM Lancelin, D Kohda, et al. Active site of mu-conotoxin GIIIA, a peptide blocker of muscle sodium channels. J Biol Chem 266:16989–16991, 1991.
36. Y Yanagawa, T Abe, M Satake, S Odani, J Suzuki, K Ishikawa. A novel sodium channel inhibitor from Conus geographus: purification, structure, and pharmacological properties. Biochemistry 27: 6256–6262, 1988.
37. E Moczydlowski, BM Olivera, WR Gray, GR Strichartz. Discrimination of muscle and neuronal Na-channel subtypes by binding competition between [^{3}H] saxitoxin and mu-conotoxins. Proc Natl Acad Sci USA 83:5321–5325, 1986.
38. Y Ohizumi, S Minoshima, M Takahashi, A Kajiwara, H Nakamura, J Kobayashi. Geographutoxin II, a novel peptide inhibitor of Na channels of skeletal muscles and autonomic nerves. J Pharmacol Exp Ther 239:243–248, 1986.
39. L-Q Chen, M Chahine, RG Kallen, B RL, R Horn. Chimeric study of sodium channels from rat skeletal and cardiac muscle. FEBS Lett 309:253–257, 1992.
40. M Nakao, Y Nishiuchi, M Nakata, TX Watanabe, T Kimura, S Sakakibara. Synthesis and disulfide structure determination of conotoxin-GS a gamma-carboxyglutamic acid-containing neurotoxic peptide. Lett Pept Science 2:17–26, 1995.
41. JM Lancelin, D Kohda, S Tate, Y Yanagawa, T Abe, M Satake, F Inagaki. Tertiary structure of conotoxin GIIIA in aqueous solution. Biochemistry 30:6908–6916, 1991.
42. KH Ott, S Becker, RD Gordon, H Ruterjans. Solution structure of mu-conotoxin GIIIA analysed by 2D-NMR and distance geometry calculations. FEBS Lett 278:160–166, 1991.
43. JM Hill, PF Alewood, DJ Craik. Three-dimensional solution structure of mu-conotoxin GIIIB, a specific blocker of skeletal muscle sodium channels. Biochemistry 35:8824–8835, 1996.
44. B Cornet, JM Bonmatin, C Hetru, JA Hoffmann, M Ptak, F Vovelle. Refined three-dimensional solution structure of insect defensin A. Structure 3:435–448, 1995.
45. MJ Scanlon, D Naranjo, L Thomas, PF Alewood, RJ Lewis, DJ Craik. Solution structure and proposed binding mechanism of a novel potassium channel toxin kappa-conotoxin PVIIA. Structure 5: 1585–1597, 1997.
46. P Savarin, M Guenneugues, B Gilquin, H Lamthanh, S Gasparini, S Zinn-Justin, A Menez. Three-dimensional structure of kappa-conotoxin PVIIA, a novel potassium channel-blocking toxin from cone snails. Biochemistry 37:5407–5416, 1998.
47. M Dauplais, A Lecoq, J Song, J Cotton, N Jamin, B Gilquin, C Roumestand, C Vita, CLC de Medeiros, EG Rowan, AL Harvey, A Menez. On the convergent evolution of animal toxins. Conservation of a diad of functional residues in potassium channel-blocking toxins with unrelated structures. J Biol Chem 272:4302–4309, 1997.
48. JE Tudor, PK Pallaghy, MW Pennington, RS Norton. Solution structure of ShK toxin, a novel potassium channel inhibitor from a sea anemone. Nat Struct Biol 3.317–320, 1996.
49. MW Pennington, VM Mahnir, DS Krafte, I Zaydenberg, ME Byrnes, I Khaytin, K Crowley, WR Kem. Identification of three separate binding sites on SHK toxin, a potent inhibitor of voltage-depen-

dent potassium channels in human T- lymphocytes and rat brain. Biochem Biophys Res Commun 219:696–701, 1996.
50. KJ Nielsen, L Thomas, RJ Lewis, PF Alewood, DJ Craik. A consensus structure for omega-conotoxins with different selectivities for voltage-sensitive calcium channel subtypes: comparison of MVIIA, SVIB and SNX-202. J Mol Biol 263:297–310, 1996.
51. WA Wells. The snail companies. Neurex Corporation & Cognetix Inc. Chem Biol 5:R235–236, 1998.
52. PW Corfield, TJ Lee, BW Low. The crystal structure of erabutoxin a at 2.0-A resolution. J Biol Chem 264:9239–9242, 1989.
53. JL Smith, PW Corfield, WA Hendrickson, BW Low. Refinement at 1.4 A resolution of a model of erabutoxin b: treatment of ordered solvent and discrete disorder. Acta Crystallogr A 44:357–368, 1988.
54. H Hatanaka, M Oka, D Kohda, S Tate, A Suda, N Tamiya, F Inagaki. Tertiary structure of erabutoxin b in aqueous solution as elucidated by two-dimensional nuclear magnetic resonance. J Mol Biol 240: 155–166, 1994.
55. SH Hu, M Loughnan, R Miller, CM Weeks, RH Blessing, PF Alewood, RJ Lewis JL Martin. The 1.1 A resolution crystal structure of [Tyr15]EpI, a novel alpha-conotoxin from Conus episcopatus, solved by direct methods. Biochemistry 37:11425–11433, 1998.
56. LW Guddat, JA Martin, L Shan, AB Edmundson, WR Gray. Three-dimensional structure of the alpha-conotoxin GI at 1.2 A resolution. Biochemistry 35:11329–11335, 1996.
57. K Wakamatsu, D Kohda, H Hatanaka, JM Lancelin, Y Ishida, M Oya, H Nakamura, F Inagaki, K Sato. Structure-activity relationships of mu-conotoxin GIIIA: structure determination of active and inactive sodium channel blocker peptides by NMR and simulated annealing calculations. Biochemistry 31:12577–12584, 1992.
58. T Kohno, JI Kim, K Kobayashi, Y Kodera, T Maeda, K Sato. Three-dimensional structure in solution of the calcium channel blocker omega-conotoxin MVIIA. Biochemistry 34:10256–10265, 1995.
59. PK Pallaghy, BM Duggan, MW Pennington, RS Norton. Three-dimensional structure in solution of the calcium channel blocker omega-conotoxin. J Mol Biol 234:405–420, 1993.
60. JH Davis, EK Bradley, GP Miljanich, L Nadasdi, J Ramachandran, VJ Basus. Solution structure of omega-conotoxin GVIA using 2-D NMR spectroscopy and relaxation matrix analysis. Biochemistry 32:7396–7405, 1993.

34
Incidence of Marine Toxins on Industrial Activity

Juan M. Vieites and Francisco Leira Sanmartin
National Technical Center for Fish Products (CECOPESCA), Vigo, Spain

I. WORLD PRODUCTION OF BIVALVE MOLLUSKS AND OTHER RELATED SOCIOECONOMIC FACTS

This section deals with the most important facts regarding world production of bivalve mollusks, by species and countries. It also provides related socioeconomic information that will show the importance of the productive-extractive sector in the different countries, as well as its impact on the processing-canning sector, which greatly depends on the former.

First, it must be mentioned that over the last years important changes have occurred regarding bivalve mollusk–producing countries. New territories, little known so far, have appeared as leaders in the production and marketing rankings, taking the place of traditional producing world powers. This fact is not only the preserve of this sector but is also related to a new economic trend called economic globalization.

In recent years, aquaculture of bivalve molluscs has been increasingly important and, in some cases (e.g., mussels), has surpassed the extracting sector, which is suffering a decrease. There are data supporting this evolution toward culture or farming in different countries such as North Korea, Chile, Argentina, New Zealand, Australia, and the Netherlands. These countries have followed the way opened by pioneers such as Spain, where the culture industry (drawing together 69 companies) had a turnover of $140 million (U.S.) in 1996.

A. Scallops

The general prospect for the period 1991–1995 involves, as shown in Table 1, a growth of nearly 100% (from 846,800 tons in 1991 to 1,651,600 tons in 1995) in catches. This is mainly due to the sudden appearance of China in the list of producing countries (the increase in China's production accounts for 90% of the aforementioned total increase). Mention may also be made of the increase of market presentations and trade of frozen scallops. Specific analysis by geographical areas will be detailed as follows.

Asia–The Pacific

As mentioned above, the most significant fact is China being the major world producer of scallops (916,500 tons in 1995, reaching 55% of world production). China has become an exporting

Table 1 Yearly World Catches of Scallops by Major Landing Countries (in 1000 MT)

Country	1982	1983	1984	1985	1986	1987	1988	1989	1990	1991	1992	1993	1994	1995
Japan	176.4	213.3	209.2	369.4	226.8	249.6	297.8	341.6	421.7	367.9	401.5	465.3	470.3	502.7
United States	198.8	184.7	473.9	184.8	192.2	98.6	204.4	239.6	150.1	140.0	116.3	65.6	141.0	76.1
Canada	65.1	51.3	36.5	91.6	47.2	56.7	73.8	77.8	83.0	79.3	90.0	90.7	89.6	58.6
United Kingdom	18.2	23.3	22.8	14.5	18.3	17.6	21.5	14.3	13.8	17.3	18.8	20.1	18.7	18.5
Iceland	12.1	15.2	15.2	10.8	17.2	16.4	12.2	8.8	12.1	10.2	12.4	11.5	8.4	8.4
Australia	29.2	30.5	29.8	5.9	19.2	16.2	11.7	9.2	7.5	14.3	29.6	33.1	21.4	12.2
Peru	5.0	13.6	17.9	3.4	51.5	16.0	5.3	6.9	6.2	2.9	3.9	2.6	4.6	3.5
Norway	0.0	0.0	0.0	5.7	1.2	14.6	45.0	20.3	5.3	3.2	6.8	9.9	7.9	8.4
France	12.1	10.5	10.3	8.9	10.5	8.9	6.1	7.3	7.8	9.4	15.5	15.0	16.0	13.1
New Zealand	1.8	4.0	4.7	0.5	3.2	4.6	3.0	1.4	0.6	1.0	1.3	1.1	1.3	1.9
Former USSR	1.7	2.2	1.6	3.1	2.4	3.0	2.5	2.7	5.2	1.9	8.5	6.9	10.5	12.9
Faeroe Island	1.6	2.7	2.4	4.8	1.9	2.0	2.0	8.9	8.4	3.3	3.5	3.3	3.9	2.8
Chile	0.6	1.1	5.3	1.1	2.1	1.2	1.5	1.3	2.1	2.1	3.0	6.0	12.0	9.6
China	1.2	2.0	3.8	129.5	8.3	23.7	43.6	122.0	147.0	188.1	338.0	728.4	825.6	916.5
Others									6.6	5.9	2.0	2.3	3.1	6.4
Total	523.9	556.4	834.2	840.4	604.2	530.7	740.7	869.2	877.4	846.8	1051.1	1461.8	1634.3	1651.6

country as regards both volume (30,109 tons in 1995 as opposed to 3943 tons in 1991, which implies an increase of 665%) and value ($134,776,000 in 1995). In 1996, the United States imported 13,500 tons of fresh and frozen scallops from China, which accounts for 50% of American imports of this product. It has also made its way in the French frozen products market (2752 tons in 1996, the first supplier with 34% of total imports).

Japan is the second world producer of scallops (502,700 tons in 1995), although it hardly allocates a part of its production to external markets (only 1600 tons in 1996 exported to the United States out of total exports of 1918 tons). This is because Japan is the great consumer of this product, as shown by its import figures (750 tons of imported scallops for $3,550,000 in 1995).

Australia and New Zealand are among the first producing countries, although they have suffered significant drops in their catches scallops in recent years. Unlike China and Japan, these countries allocate a greater part of their production to exports. Australia allocated 3069 tons in 1995 (25% of its total production that year, 12,200 tons), mainly for Hong Kong, France, and New Zealand. That same year, New Zealand allocated 643 tons (35% of the total catches, 1900 tons), which were for France, the United States, and Australia.

American Continent

The United States is the third in the ranking of world producing countries with 76,100 tons scallops in 1996 (total production has decreased to 50% of the 1995 figures).

Canada is the fourth world producer of scallops with 58,600 tons in 1995. Its exports reached 7684 tons that year ($101,593,000), which were mainly for the United States (5900 tons in 1996) and France (about 1200 tons in 1996).

The rise in Chile's production of scallops (12,000 tons in 1994; 9600 tons in 1995) is also to be noted. The Chilean government is encouraging the creation and development of scallop farming companies (nearly 30 companies are working now, mainly in south Chile). The figures of foreign trade in this country increased in 1995 to 1131 tons of volume and $14,216,000 of value, mainly for Europe.

Europe

Production rates of scallops in Europe have not changed significantly over the last 5 year period. We must particularly stress the increase of production rates in France (16,000 tons in 1994 and 13,100 tons in 1995), Norway (162% of increase from 1991 to 1995) and the former Soviet Union (12,900 tons in 1995). The United Kingdom is still the main European producer (18,500 tons in 1995).

As for the foreign trade figures, Britain's exports of frozen scallops to France (357 tons in 1996), Belgium (251 tons), and Italy (574 tons) are highlighted. Total exports reached 3962 tons ($44,526,000) in the United Kingdom in 1995, 1712 tons in France, 2873 tons in Belgium, and 1333 tons in Denmark.

Spain is not among the leading scallop producers, although exports did rise to 120 tons for $518,000 in 1995.

B. Clams and Cockles

World production of clams and cockles has increased in 38% from 1991 to 1995. This has mainly been due to a dramatic rise in China's production (142%), which has made up for important decreases in the production rates of other countries such as Japan (−16%), Korea (−24%), and Thailand (−48%). These data are shown in Table 2.

As previously described for scallops, there are positive figures for clams and cockles in the trade of new market presentations. The frozen and canned products market has increased over the last years, and this trend may continue in the future.

Asia–The Pacific

A great number of leading producers of clams and cockles belong to this area, with China standing out over the rest (42% of the total world production). All the countries in this area entail 65%. China, with the species *Anadara granosa* and *Ruditapes philippinarum*, ranks in the first position as a producing country (900,900 tons in 1995). China's exports reached 88,132 tons that same year for $118,442,000. Its total figures of exporting value are slightly surpassed by South Korea (17,916 tons for $118,443,000 in 1995). The most important clam and cockle importing countries for China's foreign consignments are Japan (25,000 tons in 1996) and Europe.

Japan was the third leading producer of clams and cockles in 1995 (127,400 tons), although it is suffering a continuous fall in its catches. Japan ranks in first place among importing countries, which contrasts with its poor export figures (this could be due to the fact that it is the largest consumer of its own products).

The other important eastern power is South Korea, the fifth leading world producer of clams and cockles. Its catches are still decreasing (97,800 tons in 1995), allocating part of its production to exporting canned clams to the United States, Spain, and Japan, among others.

Thailand is another important producer of clams and cockles in this area (45,200 tons in 1995). It exported (mainly canned) 8% of its total production in 1994 to countries such as Canada (2000 tons), the United States (1700 tons), Italy (200 tons), and Japan (300 tons). Malaysia has faced a dramatic drop in its figures to 91,700 tons in 1994 (due to the overexploitation of resources) and recovered a production of 116,100 tons in 1995 (fourth place among producing countries); Indonesia recovered a production of 68,000 tons in 1995; Taiwan produced 20,800 tons that same year.

American Continent

The United States is the second leading world producer of clams and cockles country (404,400 tons in 1995). Production and import rates have shown no significant increase in recent years, which contrasts with consumer data on seafood in the United States (clams took seventh place in 1996, with 0.518 lb per capita). However, it seems that the aim of North American producers (United States and Canada) are seeking products with more added value and higher quality, as shown in the culture of giant clams, with a value of $12 per pound in the Far East market.

Canada is another important producer of clams and cockles (33,500 tons in 1995), but the culture is being seriously affected by the presence of marine toxins. Furthermore, Canada is a traditional importer of the canned product. Import figures in 1996 rose to 2805 tons, mainly coming from Thailand (1717 tons), the United States (708 tons), Indonesia (195 tons), and Malaysia (63 tons). As mentioned before, the Canadian clam and cockle industry is searching for new products with more added value.

Chile (35,400 tons in 1995) continues with the usual vicissitudes in production. Finally, Venezuela (34,500 tons in 1995) and Mexico (12,000 tons) have considerably increased their shares over the last decade and are starting to allocate some consignments of clams and cockles to foreign trade.

Table 2 Yearly World Clam and Cockle Landings (in Shell) by Major Producing Countries (in 1000 MT)

Country	1983	1984	1985	1986	1987	1988	1989	1990	1991	1992	1993	1994	1995
China	NA	146.4	173.9	192.0	220.4	237.7	264.4	291.3	371.3	510.7	707.2	844.4	900.9
United States	338.9	389.5	453.7	426.0	402.6	392.0	416.6	413.4	404.0	425.0	438.6	396.3	404.4
Japan	276.9	258.1	258.4	253.2	243.3	230.7	179.4	177.4	163.4	143.8	145.5	119.1	127.4
Korea	108.3	119.1	161.1	208.9	223.6	192.4	142.3	160.8	129.0	137.9	112.7	112.8	97.8
Thailand	48.4	67.1	103.1	114.8	142.4	122.6	104.4	100.7	86.2	89.4	63.1	64.2	45.2
Chile	30.6	37.7	43.3	51.1	51.6	66.5	52.0	36.2	50.7	59.0	39.8	31.3	35.4
Malaysia	42.3	67.3	46.8	46.0	45.7	34.9	39.8	36.3	46.9	66.2	100.0	91.7	116.1
United Kingdom	6.5	6.3	8.9	20.8	40.0	23.0	14.7	19.2	39.4	27.6	18.5	18.3	22.0
Italy	34.9	38.9	25.6	26.5	37.3	33.5	37.2	37.2	54.0	60.3	68.0	74.0	98.4
Indonesia	45.9	51.7	32.6	44.7	36.8	37.1	36.8	33.4	36.7	49.5	50.5	54.7	68.0
Canada	9.4	13.0	13.1	13.4	23.6	16.5	19.8	26.9	17.3	20.3	26.0	27.4	33.5
Venezuela	8.3	20.4	12.5	11.0	18.0	18.3	27.7	28.6	16.0	16.8	28.9	31.6	34.5
Taiwan	19.9	21.3	20.5	17.6	16.3	22.5	18.7	20.2	18.8	24.5	19.3	20.9	20.8
Mexico	7.9	7.6	8.0	8.7	15.0	20.5	23.3	31.7	31.2	29.4	12.4	15.6	12.0
Portugal	2.3	1.9	3.0	3.9	12.6	11.4	11.0	7.4	12.5	18.3	13.4	9.0	7.1
France	7.6	4.4	4.8	5.3	8.2	15.0	13.4	13.5	13.9	14.0	12.1	9.4	7.4
Spain	6.0	3.4	3.4	3.3	4.8	5.3	9.2	7.7	8.2	8.5	8.4	12.4	14.9
Others									47.2	89.8	110.0	87.0	88.2
Total	1083.7	1250.5	1346.6	1426.1	1489.8	1466.6	1459.1	1469.4	1546.7	1791.0	1967.1	2021.4	2134.0

Europe

The United Kingdom (22,000 tons in 1995) has completed the process of integration and development of marine culture techniques and has started to allocate part of its production of clams and cockles to foreign countries (2700 tons to Spain in 1996).

Italy, a traditional producer of clams and cockles, has reached 98,400 tons in catches in 1995 and also exports a substantial part of its share (2082 tons in 1995 for $3,713,000). In 1995, 12,100 tons of fresh clams and 7400 tons of frozen products were exported to Spain. Besides, Italy imports canned clams (although the figures have decreased by half in 1996 to a total of 2,622 tons), with Thailand (562 tons), Turkey (562 tons), Germany (415 tons), and the Netherlands (438 tons) being its main suppliers in 1996.

Portugal was the country that most suffered the poor production levels of 1994 (9000 tons as opposed to 13,400 in 1993) and its production has not recovered (7100 tons in 1995). In spite of this, it allocated about 2200 tons of clams and cockles to foreign trade with Spain.

France has also reduced its catches of clams and cockles in recent years (9400 tons in 1994 as opposed to 12,100 tons in 1993) and could not recover in 1995 (7400 tons). In spite of this, France has maintained its export figures (3100 tons of fresh and frozen products to Spain in 1996 and 33 tons of canned products to Italy that same year).

Spain, unlike its European neighbors, has maintained an upward line regarding its production of clams and cockles, reaching 8600 tons in 1994 and 14,900 tons in 1995 (allocating 285 tons for $782,000 to foreign trade).

C. Oysters

The most important fact, shown in Table 3, about the general increase in oyster catches in the world (1,338,800 tons in 1995 as opposed to 993,600 tons in 1991, which means a difference of 35%) is the dramatic rise in China's production, which has grown by 327% in its catches from 1991 to 1995. China has taken first place in the ranking of producing countries, reaching 28% (373,100 tons) of the world total in 1995.

Asia–The Pacific

As mentioned before, China's oyster production figures (373,100 tons in 1995) are significant, accounting for 28% of the world total. It also reached 2,230 tons of exports that year, which implies $10,556,000.

The second major Asian and third world producer of oysters in 1995 was Japan, with 227,300 tons. Japan exported 927 tons ($5,682,000) of oysters in 1995 (mainly for the United States and Hong Kong). These data contrast with its import figures (8646 tons), mainly destined for home consumption.

The largest oyster exporter, South Korea, reached 243,300 tons in catches in 1995, although these figures are still below usual production rates for the 1990s. However, the foreign trade figures are improving, reaching 8740 tons in fresh and canned oysters (for $66,857,000) and 4773 tons in 1996 of canned oysters. The main destinations are Japan, the United States (4500 tons in 1996), Canada (647 tons), and Hong Kong.

Other Asian producers are Taiwan (25,400 tons in 1995), the Philippines, and North Korea. The latter intends to open its borders to foreign capital to allow exploitation of 1 million hectares prepared for culture, which already reached 20,000 tons of oysters in 1996.

American Continent

The United States is the second in the ranking of countries producing oysters (239,100 tons in 1995), keeping an upward line from 1991 to 1995. It allocates 1154 tons to export (for

Table 3 Yearly World Catches of Oysters by Major Landing Countries (in 1000 MT)

Country	1982	1983	1984	1985	1986	1987	1988	1989	1990	1991	1992	1993	1994	1995
Korea	189.2	218.5	211.9	254.5	268.8	303.2	298.7	243.3	248.9	231.9	252.9	286.4	193.0	209.4
Japan	250.3	253.3	257.1	251.3	252.0	258.8	271.0	256.3	248.8	239.2	244.9	235.5	223.5	227.3
United States	315.2	307.1	284.3	260.5	234.3	217.6	167.7	158.4	148.7	169.1	194.7	206.2	219.1	239.1
France	96.3	109.1	112.1	139.6	136.0	138.2	137.8	140.4	153.8	142.3	135.7	144.9	149.0	147.5
Mexico	34.9	36.6	42.8	42.7	42.8	50.7	56.1	56.3	51.3	38.7	32.1	25.7	36.7	34.7
Taiwan	25.2	26.0	29.0	25.4	19.2	21.2	28.5	28.5	28.2	25.0	26.8	27.7	25.0	25.4
Philippines	19.0	11.5	14.8	15.5	16.7	16.3	15.9	17.3	18.6	16.5	16.1	18.5	13.1	12.5
New Zealand	11.1	11.9	10.4	10.8	7.8	5.8	8.1	9.7	5.3	6.7	3.0	2.9	2.6	3.9
Australia	8.0	7.8	7.2	8.0	6.8	7.8	9.4	8.4	7.1	8.1	8.3	9.7	9.5	9.2
Canada	NA	4.0	5.2	4.6	4.2	6.0	6.5	6.0	6.3	6.6	6.1	6.8	8.4	9.4
China	—	25.5	40.7	50.9	55.0	65.5	74.0	73.2	82.4	87.4	123.0	168.4	313.5	373.1
Others	17.9	13.6	16.0	14.1	16.6	12.0	20.0	50.3	8.4	22.1	24.5	65.5	45.5	47.3
Total	967.1	999.4	990.8	1026.3	1012.1	1109.9	1103.4	1048.1	1007.8	993.6	1079.4	1170.5	1220.9	1338.8

$6,614,000), importing 6600 tons in 1995 and 6400 tons in 1996. Its main supplier is South Korea, with 4500 tons that year.

The other great American producer of oysters, Mexico, reached 34,700 tons in 1995.

Europe

The major European producer of oysters is France (147,500 tons in 1995). This country has consolidated its position, acquired through the development of modern aquaculture techniques. France allocated 4435 tons to exports in 1996 (3816 tons in 1995 for $13,656,000). In 1995, the main destinations were Italy (more than 3000 tons), Benelux (500 tons), Germany, and Switzerland. France, in turn, also imports oysters, reaching 1355 tons in 1995 for $3,601,000.

The United Kingdom is a special case in the European framework. In spite of not being one of the major oyster producers, it has become an important exporter of this product (1355 tons in 1995), which is directly related to the development of culture in areas such as the Thames Estuary in the county of Kent. After several clean-up processes in the waters, 15 million Edulis and Pacific oysters were cultured in these waters in 1997.

D. Mussels

No doubt, mussels are the bivalve mollusks with the greatest importance and trade. This is especially relevant in Spain and particularly in Galicia, where production reached 200,000 tons in 1996. In this region, mussel culture employs 9000 people directly and 14,000 people indirectly (basically through the canning sector).

Unlike other species of bivalves (although the situation is changing regarding oysters and scallops), mussel production comes mainly from culture and not from catches of wild mussels. In 1995, 80% (1,022,000 tons) of the total production (1,264,000 tons) were raft-cultivated mussels and the remaining 20% (242,700 tons) wild mussels (see Tables 4–6 for the figures for cultivated and wild mussels and their total production by countries).

The reason for this feature lies in the expansion of mussel culture in different countries that have developed these techniques, becoming world producing powers. Another reason is the significant drop in the catches of wild mussels, partly due to overexploitation.

Wild mussels include different species such as the blue mussel (*Mytilus edulis*), which accounts for almost half of the total; Mediterranean mussel (*Mytilus galloprovincialis*), and green mussel (*Mytilus smaragdinus*). Cultured species include the ''Nei'' sea mussel (on which China bases its production), Korean mussel (typical of this country), greenshell from New Zealand (*Perna canaliculus*), and the above-mentioned blue, Mediterranean, and green mussels.

As reported by the FAO (Fisheries and Agriculture Organization) and shown in the statistics, world production of mussels has been around 1 million tons since 1990 (a figure that will apparently be surpassed by the next decade). Note the appearance of new competitors-producers, such as South Africa, which has increased its production rates (its annual producing capacity is estimated at about 90,000 tons) taking advantage of know-how transfer coming from Galicia.

The analysis of the data shows a slight decrease of world production in 1994 and 1995. This was mainly due to the poor campaigns in the major producing countries (China, Denmark, Italy, and South Korea in 1994 and China, Spain, the Netherlands, and Denmark in 1995).

According to the data reported by the FAO in 1995, two large mussel-producing areas can be distinguished. The first one includes China and other Asian countries (South Korea, Thailand, and the Philippines), which produce 46% (585,400 tons) of the world total. On the other hand lies Europe, which reaches 41% (520,600 tons) of the world production.

Table 4 Yearly World Mussel Production (Capture and Culture) by Major Producing Countries (in 1000 MT)

Country	1983	1984	1985	1986	1987	1988	1989	1990	1991	1992	1993	1994	1995
Spain	124.4	216.8	245.7	247.6	209.7	245.0	193.8	173.3	199.1	138.9	90.5	142.7	92.3
China	90.1	94.0	128.8	210.7	312.7	429.7	490.5	495.9	498.1	538.9	509.6	415.2	415.2
Netherlands	119.6	60.2	116.3	85.9	98.4	77.6	107.0	99.0	49.2	51.2	66.0	105.0	79.3
Denmark	66.5	81.5	84.1	105.0	93.6	72.5	75.6	93.3	125.7	136.3	136.7	129.3	107.4
Italy	69.0	66.3	76.0	78.6	85.4	85.4	100.5	110.3	100.7	104.7	108.3	92.7	116.4
France	60.5	52.9	61.0	66.3	61.0	54.9	59.1	62.0	62.9	64.0	73.0	71.9	75.0
Korea	51.2	36.6	58.0	43.1	32.6	27.4	16.3	19.7	24.2	21.5	60.8	45.7	82.0
Thailand	56.3	76.5	69.0	36.5	62.5	97.6	89.0	83.5	52.9	42.3	52.8	53.7	74.5
Germany	39.8	66.4	23.1	31.7	29.6	30.8	18.6	20.2	29.9	50.8	24.7	4.9	17.8
United States	14.5	13.0	15.8	20.7	21.5	35.7	24.4	17.9	17.5	22.4	16.4	12.5	13.6
Chile	13.6	17.1	17.2	21.6	18.6	21.9	21.5	19.5	15.5	20.3	18.8	20.3	17.7
New Zealand	7.6	9.8	10.9	15.6	17.0	18.0	21.5	24.0	43.6	46.5	47.0	47.0	62.7
Philippines	19.1	21.5	26.1	15.7	16.7	17.6	18.3	19.7	20.6	22.2	26.2	16.3	13.7
Ireland	5.7	12.6	10.4	10.6	14.9	16.0	12.3	18.6	19.3	13.4	17.8	13.0	15.6
Peru	5.3	9.4	4.7	8.7	9.4	9.1	12.8	16.5	3.8	7.8	6.0	7.2	11.2
Turkey	1.2	2.0	2.1	2.7	8.0	8.0	5.6	6.3	6.2	6.8	7.1	8.0	6.2
United Kingdom	5.8	4.3	5.8	6.3	4.9	6.8	8.9	5.5	6.4	8.9	12.5	14.4	15.3
Former USSR	14.9	11.0	4.9	1.0	4.8	1.1	1.4	2.5	1.8	1.2	1.4	1.6	2.2
Sweden	1.6	1.3	0.4	0.0	2.6	0.9	0.2	1.2	1.6	1.4	0.8	2.1	1.5
Australia	1.2	1.4	1.4	1.3	2.4	2.5	2.2	2.9	3.2	3.4	3.0	3.0	2.5
Canada	0.2	0.4	0.6	1.4	0.8	2.3	3.4	3.5	3.5	5.0	5.2	6.9	9.4
Total	774.3	861.0	941.1	1302.5	1071.6	1208.8	1240.2	1325.3	1304.1	1338.1	1312.6	1262.4	1264.0

Table 5 Yearly World Mussel Capture by Major Producing Countries (in 1000 MT)

Country	1984	1985	1986	1987	1988	1989	1990	1991	1992	1993	1994	1995
Denmark	81.5	84.1	87.5	85.7	72.5	75.6	93.3	125.8	136.3	136.7	129.3	107.4
Thailand	48.7	42.7	25.2	37.7	52.7	30.3	24.1	16.3	24.0	24.9	42.4	26.0
Italy	6.3	11.0	11.6	17.3	17.4	20.5	30.3	25.2	20.4	28.3	22.7	21.4
United States	11.1	13.8	18.1	19.0	33.7	23.0	16.5	17.2	21.7	15.7	11.7	12.7
Chile	17.0	17.2	19.8	17.1	19.4	18.9	17.1	12.4	16.5	15.7	16.3	11.8
Peru	9.4	4.7	8.7	9.4	9.1	12.8	16.5	3.9	7.8	6.0	7.2	11.2
France	0.1	0.1	2.6	0.7	1.2	6.5	9.9	6.8	4.8	3.7	6.7	10.8
Greece	0.4	0.4	1.2	0.9	0.2	0.1	—	0.3	8.2	2.9	2.3	10.8
United Kingdom	3.3	4.6	4.7	3.0	4.8	9.0	6.7	9.4	7.0	7.8	10.3	9.5
Korea	10.6	9.8	8.5	6.5	11.7	12.3	21.1	14.8	11.8	5.6	5.9	6.6
Turkey	2.0	2.1	2.7	8.0	8.3	5.3	6.3	6.3	6.8	7.1	8.0	6.0
Grand Total	206.0	202.8	204.2	219.2	243.0	228.7	257.7	253.4	276.7	261.2	277.6	242.7

Table 6 Yearly World Cultured Production of Mussels by Countries in 1000 MT

Country	1984	1985	1986	1987	1988	1989	1990	1991	1992	1993	1994	1995
China	136.6	128.9	210.7	312.7	429.7	490.5	495.9	498.2	538.9	509.6	415.2	415.2
Italy	60.0	65.0	67.0	68.1	85.4	80.0	80.0	75.0	84.3	80.0	70.0	95.0
Spain	230.0	245.7	247.0	245.5	243.0	193.0	173.6	195.2	138.9	90.5	142.7	92.3
Netherlands	60.1	116.3	85.9	98.4	77.6	107.0	98.8	49.3	51.2	66.0	105.0	79.3
Korea	26.0	48.2	34.6	26.1	15.7	7.9	9.8	9.5	9.7	55.2	39.8	75.4
France	52.8	54.9	58.0	55.7	49.6	50.0	61.8	61.9	59.2	70.0	64.4	64.2
New Zealand	9.8	10.9	15.6	17.7	24.6	24.0	24.0	43.6	46.5	47.0	45.0	62.5
Thailand	27.8	26.3	11.4	24.8	44.9	58.7	59.4	36.6	18.0	28.0	30.6	48.5
Germany	59.3	20.8	29.9	25.9	29.7	18.6	20.2	30.0	50.8	24.7	4.9	17.8
Ireland	13.7	10.4	10.9	14.9	12.6	13.6	15.3	19.3	13.4	17.8	13.0	15.6
Philippines	20.3	22.7	12.1	11.6	15.5	16.4	17.5	17.3	20.5	25.1	11.4	13.3
Greece	0.2	0.2	0.2	0.4	1.1	1.5	3.7	5.9	8.4	16.7	16.8	10.8
Canada	0.9	0.9	2.1	1.8	2.0	3.4	3.6	4.0	5.0	5.2	6.9	9.4
Grand Total	706.2	758.7	796.1	917.9	1047.6	1078.7	1080.0	1063.0	1061.4	1051.4	984.8	1022.4

Asia–The Pacific

China was the leading world producer of mussels in 1995 (415,200 tons), providing 33% of the world total. According to the data by the FAO, almost 100% of its production comes from culture, where China has developed and is still researching on new producing techniques. This country is also attempting to solve environmental and resources problems by means of specific programs oriented toward aquaculture. An interesting fact is that 54% (15,282,000 tons) of the country's total annual fishing production until September 1997 (28.3 million tons) comes from aquaculture. At present, the efforts made by the Chinese government are aimed at meeting international requirements as regards quality control and international standards. The Chinese foreign market has increased over the years, reaching exports of 9326 tons (for $15,266,000) in 1995, which contrasts with imports of 826 tons that year.

South Korea is another important producer of mussels in this area (82,000 tons in 1995 becoming the fifth world power). This country has developed mussel culture, which provides 91% of the total production and is being extensively developed. Its exports of fresh and frozen mussels in 1995 (139 tons) reached a value of $313,000. Its exports of canned products rose to 860 tons in 1996, which were mainly for Australia, Canada, and the United States.

Thailand has also recovered from the crises it faced in the early 1990s, reaching 74,500 tons of mussels in 1995. The percentage of cultured products in this country is much smaller that that of its Asian neighbors (65%). The data regarding foreign trade do not show any special focus on foreign markets.

The Philippines has decreased its production of mussels in recent years. It finished 1995 with 13,700 tons, of which 13,300 tons were cultured mussels (97%).

The study carried out in New Zealand revives the above-mentioned fact of searching for new market presentations. This country has strongly introduced itself into the market of frozen mussels in its varied forms. Moreover, it has dramatically increased (725%) its production over the last 15 years, which is mainly due to the efforts made to implement efficient culture systems. This has given rise to the fact that nearly 100% of the production comes from culture. Furthermore, mussels account for 11% of the global exports of the country (together with Pacific oysters). The main culture areas are Marlborough and Coromandel. Total production in 1996 rose to 66,000 tons (only 8% of fresh mussels were sold). Ninety percent of total production was exported (only 3% mussels are exported alive, as opposed to 75% of frozen mussels). Its main destinations were (65% of the total exports) the United States ($24,875,000 in 1996; $18,000,000 in ''half shell''), Japan ($11,400,000), Australia ($9,912,000), China, and South Korea.

American Continent

The United States produced of 13,600 tons of mussels in 1995, which means an increase relative to 1994 but not relative to previous years. The United States is traditionally a mussel-importing country (7024 tons in 1995) and hardly exports (994 tons in 1995). Its main suppliers are New Zealand, South Korea, and Japan.

Canada has vastly increased its production figures over the years, reaching 9400 tons of mussels in 1995. However, 1997 was not a good year, since there was a 3-month ban on mussel marketing (one of the most affected areas was the Vancouver coast). Note that 100% of the Canadian production comes from aquaculture.

The two most important South American producers of mussels in 1995 were Chile (17,700 tons) and Peru (11,200 tons). Chile is still suffering vicissitudes in its production, mainly because part of it comes from the capture of wild mussels (67%). Commercial species include Chorito (*Mytilus chilensis*), Cholga (*Aulacomya ater*), and Choro (*Mytilus chorus*), although only the first species is massively cultured. However, Chile is developing its culture techniques, which are technologically backward (both in equipment and knowledge), and it will be able to recover its production figures in the short term. The case of Peru is more significant because, with a production exclusively based on the capture of wild mussels, it has not stopped increasing its figures (as opposed to the situation in the rest of the world). Both countries hardly assign part of their production for foreign markets (Chile exported 295 tons in 1995); instead it is destined for the markets in neighboring countries, such as Argentina.

Europe

Europe, and particularly the European Union, is the second mussel-producing world power. Furthermore, if it were considered as a single territory (this concept will soon be considered as a reality), it would displace China from the first position in the world ranking (the production would be of about 540,760 tons in 1995). Conversely, the European Union is also the first world importer of mussels, as well as the major exporter (although in this case, it must be said that the main destinations of the product outside the borders of the member states are the Union partners). This is due to the fact that the flow with third countries is very weak (only 4% of the exchanges are not inside the Union) and that the balance of trade is lacking in terms of volume and value of exchanges (3300 tons for 7.3 million ECU in 1996).

After a detailed analysis of the countries, we find that the traditional major producer of mussels in the 1980s, Spain, has been surpassed over the years by some of its European partners, such as Italy. This country reached 116,400 tons in 1995 (82% from culture). Italy exported (mainly frozen) 2814 tons of mussels in 1995 (for $3,162,000), which were mainly for countries of the European Union (Spain and France are two of its main customers).

As regards Denmark (the second European producer in 1995), its 107,400 tons of mussels have a special meaning if we take into account that 100% comes from the catches of wild mussels. However, because of this type of production, which is more prone to pollution and other natural or environmental disasters, 1997 was a very poor year for the mussel industry. Pollution and high temperatures during the summer killed the coastal mussels, which can ruin the industry (both producing and processing industries). If this occurs, both imports (585 tons in 1995) and exports (9430 tons for $9,695,000) can be seriously affected. The two most important mussel areas are Limfjord (84 licenses with a share of 85 tons per week) and Waddenzee (6 licenses and an annual share that ranges from 5000 to 8000 tons), where we find most of the 300 direct jobs in the 15 companies of the Danish mussel industry.

Another important producer, the Netherlands (79,300 tons in 1995, almost 100% of which is cultivated) is the major world exporter (34,416 tons in 1995, for $69,993,000). The main foreign destinations for its products (both live and frozen) were its partners in Benelux (Belgium and Luxembourg received 27,500 tons), France (12,700 tons), and Germany (7,800 tons). The Netherlands, however, imported 11,360 tons in order to meet its internal and external needs, which makes it one of the largest importers.

France has improved its figures over the years, reaching 75,000 tons in 1995 (85% from culture). France is also the major importer of mussels in the world (25,861 tons in 1995 for $30,484,000 and 36,800 tons in 1996). Its most important suppliers were countries of the European Union, such as the Netherlands (14,500 tons in fresh mussels and 600 tons in frozen mussels), Spain (5000 tons in fresh mussels), Ireland (4500 tons in fresh mussels and 1400 tons in frozen mussels), and the United Kingdom (3600 tons of fresh mussels).

The United Kingdom reached a production of 15,300 tons of mussels in 1995 (9500 tons were wild mussels). The systematic development of culture in its waters has allowed this country to continue an upward line in recent years. In 1996, it reached 18,000 tons, including the expansion of mussel culture to new areas (Menai Straits and Poole Harbour, Morecambe Bay, River Exe, and Domoch Firth in Scotland). The main culture area is North Wales, with 7000 tons. Prospects say that the United Kingdom will soon surpass 20,000 tons in annual production. As for the trade figures, it has become a great exporter of the product (2535 tons in 1995 for $2,992,000). This balances its imports in volume (2082 tons that year) but not in value ($6,296,000).

Ireland has consolidated its mussel industry through the production of *M. edulis* by two different methods: extensive (11,000 tons in 1996) and intensive (6500 tons in 1996) culture. This means a global production of 17,500 tons (it exports approximately 90%). Its main destinations are France (6500 tons), Great Britain (1400 tons), Italy (600 tons), Spain (600 tons), and Germany (200 tons). Prospects point to a production over 20,000 tons by the year 2000, also searching for the marketing of products with more added value and for a more efficient organization of the exports and quality controls.

Germany produced 17,800 tons of mussels in 1995, 100% of which came from culture in areas such as Schleswig-Holstein (where there are eight licenses for aquaculturers) and Nieder Sachsen (four companies). These areas are threatened by their inclusion in Natural Parks, which limits their activity. Germany is also one of the main mussel importers (19,124 tons in 1995), which is due to the continuous increase in the consumption of sea products (it reached 176,780 tons in 1996, and there was an increase in the consumption of frozen products). It also reached 6172 tons (for $6,125,000) in exports, and its main suppliers are its partners in the European Union (the Netherlands, Spain, Italy, and Ireland, among others).

The most dramatic increase in mussel production within the European framework occurred in Greece. It reached 21,650 tons in 1995, which means an increase of 267% in the period 1991–1995. The cause for this growth is the development of techniques of mussel culture (in this case, the Mediterranean species, which means 50% of the total). Greece is also an important exporter of mussels (4869 tons in 1995 for $2,414,000).

Spanish (Galician) Production of Mussels

Spain was in the first position among mussel-producing countries for many years, which was because it was the first country to develop and implement culture techniques. However, as years went by and other countries imported these technologies, its figures have decreased and have been surpassed, which does not imply a specific loss, but it maintains a relevant position in the world context. A very important fact is that nearly 100% (more than 95%) of Spanish production comes from Galicia. Therefore, Galicia would become the second producing area in the world after China.

The mussel industry employs 9000 people directly (mainly culturers) and about 14,000 indirectly (in the processing-canning sector). Production value on a first-sale basis is between $70 million U.S. The Galician mussel sector is based on highly atomized family-based exploitations called "floating mussel beds." At present, about 3356 of these are productive, the majority being in the Ria de Arosa (2400), each one producing 50 tons per year.

The final destination of the production has changed over the years, and the figures for frozen mussels have increased to the detriment of fresh mussels. The consumption of canned mussels has leveled off. Destination of production could be summarized as follows:

33%: fresh
39%: canning industry
28%: frozen

Regarding the data of Spanish foreign trade, exports of live or fresh mussels amounted to 18,893 tons in 1995 (77% from Galicia). In 1996, there was an increase to 19,399 tons ($10 million U.S.), 75% of which came from Galicia. As for frozen mussels, year-on-year increase is very significant (rising from 699 tons in 1995 to 2773 tons in 1996). The marketing of canned mussels is almost entirely Galician.

The most common destinations of these exports are countries in the European Union, such as Italy (9100 tons of live or fresh mussels in 1996), France (the same volume), and Germany (600 tons).

As for imports, Spain imported 1200 tons of frozen mussels in 1996, mainly coming from New Zealand (997 tons for $2.3 million U.S.), Italy (122 tons), and France (53 tons). These figures mean a decrease relative to 1995, when imports rose to 1500 tons. Imports of fresh mussels are the most important in volume (3000 tons in 1996), the main suppliers being Italy (1072 tons), France (811 tons), and Ireland (796 tons). The figures for canned mussel imports are hardly important (650 tons in 1996).

All the previous data illustrate both the economic importance of bivalve mollusck and the negative effect that harmful algal blooms may produce in the economy of the productive regions (losses in North America and Japan have been estimated at more than $10 million). Although there is a lack of solid data, it is clear that the increase of toxic episodes and closure periods in the culture areas has become a serious problem for the economic sectors related to the culture and processing of bivalve mollusks. In addition to direct losses due to closure periods, the negative impact of toxic episodes in the population produces a significant decrease in the trade and consumption of bivalve molluscs. Furthermore, the increase in the incidence of toxic episodes worldwide has led to the implementation of costly monitoring programs to protect public health that must be taken into account. The total budget for the implementation of the Galician monitoring program is around $1,360,000 per year (16), which is an example of the increasing economic importance of harmful algal blooms in the production of bivalve mollusks.

II. REGULATIONS ON MARINE TOXINS

The increase of both toxic episodes and scientific knowledge associated with harmful algal blooms has raised an important issue concerning the differences in the regulations about marine toxins in the different countries. An additional effort by international organizations is desirable to solve these differences in order to satisfy the requirements of an equal worldwide trade of bivalve mollusks and protection of public health on a common basis.

At present, there are about 25 countries with specific regulations on marine phycotoxins (Tables 7 and 8), although many countries have general regulations on food or marine products, which indirectly cover the presence of toxins. Within this framework, the regulations show significant differences between countries, mainly centered on two aspects: analytical methods and tolerance limits, which will be detailed as follows.

A. Analytical Methods

In most countries, the official control of diarrheic shellfish poison (DSP) and paralytic shellfish poison (PSP) toxins is mainly based on mammalian bioassays. Other analytical options, such as the spectrophotometric assay for saritoxins (STX) and high-performance liquid chromatography (HPLC) for okadaic acid and gonyautoxins, are also included in the regulations of some of these countries (Austria, Sweden, and Korea, respectively) but mainly as complementary methods of

Table 7 International Regulations on PSP Toxins

Country	Toxin	Product	Tolerance limit	Reference method
Australia	STX	Shellfish	80 μg/100 g	Mouse bioassay
Austria	STX	Shellfish	40 μg/100 g	Spectrophotometric (mouse bioassay for confirmation)
Canada	PSP	Molluscs	80 μg/100 g	Mouse bioassay
European Union	PSP	Bivalve Mollusks	80 μg/100 g	Mouse bioassay (in association with a chemical method for detection of STX if necessary)
Guatemala	STX	Mollusks	400 MU/100 g	Mouse bioassay
Hong Kong	PSP	Shellfish	400 MU/100 g	Mouse bioassay
Japan	PSP	Bivalves	400 MU/100 g	Mouse bioassay
Korea	GTX	Bivalves	400 MU/100 g	Mouse bioassay HPLC
Norway	PSP	Mussels	40–80 μg/100 g	Mouse bioassay
Panama	PSP	Bivalves	400 MU/100 g	Mouse bioassay
Singapore	STX	Bivalves	80 μg/100 g	Mouse bioassay
Sweden	PSP	Mollusks	80 μg/100 g	Mouse bioassay
United States	PSP	Bivalves	80 μg/100 g	Mouse bioassay

STX, saxitoxin; PSP, paralytic shellfish poison; GTX, gonyautoxins; HPLC, high-performance liquid chromatography.

Table 8 International Regulations on DSP Toxins

Country	Toxin	Product	Tolerance limit	Reference method
Canada	DSP	Shellfish	0.2 μg/g	Mouse bioassay
European Union	DSP	Bivalve mollusks	Not detectable	Biological methods Mouse bioassay Observation period (h)
Belgium				24
Denmark				24
France				5
Italy				5
Luxembourg				24
Portugal				5
Spain				24
Galicia (Spain)				12
United Kingdom				5
Germany			40–60 μg/100 g	
Netherlands				Rat Assay
Sweden				Rat Assay HPLC (and mouse bioassay for confirmation)
Japan	DSP	Bivalves	5 MU/100 g	Mouse bioassay
Korea	DSP	Shellfish	5 MU/100 g	Mouse bioassay
Norway	DSP	Shellfish	5–7 MU/100 g	Mouse bioassay (observation period: 4 h)

DSP, diarrheic shellfish poison; HPLC, high-performance liquid chromatography.

mammalian bioassays. In spite of the increasing criticism about the use of animals in routine analysis, mammalian bioassays are considered as reference methods because of their wide covering range, which represents the main problem for the validation of alternative chemical methods (the lack of pure standards of several DSP and PSP toxins is an additional problem to solve). In any case, the covering range of mammalian bioassays is strongly influenced by different factors (e.g., extraction protocols, observation time), which show significant differences between countries, yielding different results. Furthermore, mammalian bioassays include substantially different techniques such as mouse or rat bioassays, and contradictory results may be obtained. These differences have become technical barriers for the international trade of bivalve mollusks, stressing the necessity of an additional effort to harmonize analytical criteria. In contrast to DSP and PSP, the situation of domoic acid is quite different, with HPLC as the reference method in all the countries where this toxin is being monitored.

B. Tolerance Limits

As occurs with analytical methods, the tolerance limits for marine toxins also show significant differences between countries (see Tables 7 and 8), which reflects the lack of scientific know-how for a solid risk assessment. The setting up of tolerance limits is mainly conditioned by the following factors:

- Toxicological data are very scarce to evaluate the real risk for consumers. Toxic profiles are quite different between countries. So, tolerance limits for PSP toxins may be generic (PSP toxins as a whole) or specific (STX in Guatemala, Australia, Austria, and Singapore; GTX in Korea). In addition, the products covered by the regulations are not the same (bivalve mollusks, mollusks, shellfish). In any case, all these differences should be removed to clarify the equivalencies in tolerance limits between countries.
- Formal criteria for the expression of tolerance limits change from MU (mouse units) to micrograms of toxins (per amount of product), which involves slight differences, as detailed in previous chapters.

Bearing in mind all these considerations, tolerance limits for PSP change from 40 μg STX/100 g to 80 μg (STX or PSP according to countries)/100 g. A tolerance limit of 400 MU (PSP, STX, or GTX)/100 g is also observed in the regulations of several countries, which corresponds to a value around 80 μg STX equivalents/100 g.

Tolerance limits for DSP toxins are also conditioned by the same factors previously described in the case of PSP. In general, the differences observed between countries (even within a common framework, as occurs in the European Union) have led to a serious criticism of the suitability of the criteria applied in each case.

In contrast to DSP and PSP, tolerance limits for domoic acid show a general agreement between countries (as occurs with analytical methods). Based on the Canadian model, most countries have set up a level of 20 μg domoic acid per gram of tissue to regulate the trade of bivalve mollusks.

In conclusion, we can state that the regulations on marine phycotoxins show significant differences between countries and, in many cases, are not based on solid scientific data. This situation involves an unequal consideration for public health and international trade of bivalve mollusks. So additional efforts at the international level are desirable in order to establish a solid basis for the legal consideration of marine phycotoxins.

III. DETOXIFICATION OF SHELLFISH

In recent years, the incidence of toxic episodes in cultured bivalve mollusks has significantly increased, becoming a serious problem for the economic sectors related to the culture and canning of bivalves. So researchers have been involved in the study of detoxification procedures to minimize the economic impact of marine toxins with limited success.

The transfer of toxic bivalves to zones free from toxic dinoflagellates was one method used for PSP detoxification. However, the variability in depuration rates according to species and the labor associated with bivalve relaying have prevented the application of this technique on a routine basis.

Ozone treatment has been used to detoxify PSP-containing bivalves exposed to blooms of *Alexandrium catenella, A. tamarense*, and *Gymnodinium breve* (2–4,7,25). Yet, no detoxifying effect of ozone in PSP-containing *Mya arenaria* has been shown (29). It has been suggested (5) that ozone could be effective only for bivalve mollusks recently contaminated with vegetative forms of dinoflagellates.

Thermal or saline stress have also been applied to toxic bivalves to accelerate the depuration rate in marine water, but this was only marginally effective (4,9). Other detoxification procedures, such as pH reduction (19) or chlorination were also unsuccessful for bivalve detoxification.

The application of high temperatures has been widely tested to decrease PSP toxicity in bivalve mollusks. Medcof et al. (17) and MacDonald (15) showed the effectiveness of pan frying to decrease PSP levels in toxic bivalves. Noguchi et al. (21,22) studied the sterilization effects on PSP toxicity in canned scallops (*Patinopecten yessoensis*) and shown a sharp decrease of toxic levels which continued moderately during storage. Similar results were obtained by Prakash et al. (24), which shown 90% detoxification during the sterilization process. Nagashima et al. (18) studied the kinetics of thermal degradation at 100, 110, and 120°C of PSP obtained from toxic bivalves. Mixtures of gonyautoxins 1 and 4 (GTX-1 and GTX-4) and gonyautoxins 2 and 3 (GTX-2 and GTX-3) show first order degradation kinetics, with GTX-1 and GTX-4 being more thermolabile. Although the results indicated a significant reduction of PSP toxicity by HPLC, thin-layer chromatography (TLC), and mouse bioassay, the heat treatments applied were more severe than the standard process used in the canning industry. Recently, Berenguer et al. (1) demonstrated the effectiveness of the canning process to decrease PSP toxicity in Mediterranean cockles (*Acanthocardia tuberculatum*) (toxin levels of 800 μg saxitoxin equivalents/100 g tissue) and yielding a product free from toxins. This work has brought about a change in the European legislation on PSP toxins (DOCE L 15/46), which authorizes the extraction of *A. tuberculatum* having toxin levels up to 300 μg saxitoxin equivalents (STX eq)/100 g tissue when the product is for the canning industry. The Canadian government has established similar regulations for clams (*Mya arenaria*) and mussels (*Mytilus edulis*) having toxin levels of up to 160 μg STX eq/100 g tissue. The current regulatory level for fresh bivalve mollusks in most countries is 80 μg STX eq/100 g tissue.

In spite of decreased PSP levels during canning, the mechanisms involved have not been fully described. Factors such as changes in toxin profile during heat treatment (11–13,20) and toxin transfer between different tissues of bivalves (23) demand a careful approach to determine the real loss of PSP in toxic bivalves. Vieites et al. (26) showed a significant reduction of PSP levels in canned mussel meat (over 50% of initial toxicity in all cases). Toxicity reduction was dependent on changes in toxin profile related to the packing media of mussels (pickled or brined). Detoxification percentages did not correspond to total toxin destruction, which was due to the transfer of toxins to cooking water and packing media of the canned product.

Leira et al. (14) studied the effect of the canning process on domoic acid levels of toxic scallops (*Pecten maximus*). These investigators found no significant destruction of domoic acid during the canning process (both in pickled or brined scallops), with recovery percentages close to 100%. As described for PSP, significant levels of domoic acid were found in the packing media of the canned product, which suggests the need to establish a surveillance protocol for packing media in canned bivalves. Industrial canning has also been ineffective in detoxifying DSP-containing mussels, showing only slight decreases in okadaic acid levels.

In conclusion, we can state that, nowadays, there are no completely safe industrial procedures to detoxify toxic bivalves (the only exception being canning for PSP-containing *A. tuberculata*). The canning process would also be helpful to decrease PSP levels in mussels, but there are many different factors, which influence the detoxifying effect. Further research needs to be made, but the general perception is that industrial detoxification of bivalves seems a quite remote possibility.

REFERENCES

1. JA Berenguer, L González, TM Jiménez, JB Legarda, JB Olmedo, PA Burdaspal. The effect of commercial processing on the paralytic shellfish poison (PSP) content of naturally-contaminated *Acanthocardia tuberculatum* L. Food Add Contam 10:217–230, 1993.
2. WJ Blogoslawski, FP Thurberg, MA Dawson. Ozone inactivation of a *Gymnodinium breve* toxin. Water Res 7:1701–1703, 1973.
3. WJ Blogoslawski, ME Stewart. Paralytic shellfish poison in Spisula solidissima: Anatomical location and ozone detoxification. Mar Biol 45:261–264.
4. WJ Blogoslawski, R Neve. Detoxification of shellfish. In: DL Taylor, HH Seliger, eds. Toxic Dinoflagellate Blooms. New York: Elsevier, 1979, p 473.
5. WJ Blogoslawski. Ozone depuration of bivalves containing PSP: Pitfalls and possibilities. J Shellfish Res 7:702–705, 1988.
6. B Dale, CM Yentsch. Red tide and paralytic shellfish poisoning. Oceanus 21:41–49, 1978.
7. MA Dawson, FP Thurberg, WJ Blogoslawski, JJ Sasner, M Ikawa. Inactivation of paralytic shellfish poison by ozone treatment. In: G Ruggieri, H Webber, eds. Proceedings of the Fourth Food-Drugs from the Sea Conference. Washington, DC: Marine Technology Society, 1976, p 152.
8. Resolution 96/77/EEC (DOCE n. L 15/46, 20/1/96)
9. ES Gilfillan, JW Hurst, SA Hansen, CP Leroyer, III. Final report to the New England Regional Commission, 1976.
10. GM Hallegraeff. Harmful algal blooms: A global overview. In: Manual on Harmful Marine Microalgae. GM Hallegraeff, DM Anderson, AD Cembella, eds. IOC Manuals and Guides No. 33, UNESCO, 1995.
11. S Hall, PB Reichardt. Cryptic paralytic shellfish toxins. In: EP Ragelis, ed. Seafood Toxins. Washington, DC: American Chemical Society, 1984, p 113.
12. T Harada, Y Oshima, T Yasumoto. Structures of two paralytic shellfish toxins, gonyautoxins V and VI, isolated from a tropical dinoflagellate, Pyrodinium bahamense var. compressa. Agric Biol Chem 46:1861, 1982.
13. M Kobayashi, Y Shimizu. Gonyautoxin VIII, a cryptic precursor of paralytic shellfish poisons. J Chem Soc Commun 827, 1981.
14. F Leira, JM Vieites, LM Botana, MR Vieytes. Domoic acid levels of naturally contaminated scallops as affected by canning. J Food Sci (in press).
15. EM MacDonald. The occurrence of paralytic shellfish poison in various species of shore animals along the Strait of Juan de Fuca in the State of Washington. MS thesis, University of Washington, 1970.
16. J Mariño, J Maneiro, J Blanco. In: B Reguera, J Blanco, ML Fernández, T Wyatt, eds. Harmful Algae. Xunta de Galicia (Regional Government of Galicia) and IOC-UNESCO, 1998.

17. JC Medcof, AH Leim, AB Needler, AWH Needler, J Gibbard, J Naubert. Paralytic shellfish poisoning on the Canadian Atlantic coast. Bull Fish Res Bd Can 75:1–32, 1947.
18. Y Nagashima, T Noguchi, M Tanaka, K Hashimoto. Thermal degradation of paralytic shellfish poison. J Food Sci 56:1572–75, 1991.
19. RA Neal. Fluctuation in level of paralytic shellfish toxin in four species of lamellibranch molluscs near Ketchikan, Alaska, 1963–1965. PhD dissertation, University of Washington, Seattle, 1967.
20. S Nishio, T Noguchi, Y Onoue, J Maruyama, K Hashimoto, H Seto. Isolation and properties of gonyautoxin-5, an extremely low-toxic component of paralytic shellfish poison. Bull Jpn Soc Sci Fish 48:959, 1982.
21. T Noguchi, Y Ueda, Y Onoue, M Kono, K Koyama, K Hashimoto, Y Seno, S Mishima. Reduction in toxicity of PSP infested scallops during canning process. Bull Jpn Soc Sci Fish 46:1273, 1980.
22. T Noguchi, Y Ueda, Y Onoue, M Kono, K Koyama, K Hashimoto, T Takeuchi, Y Seno, S Mishima. Reduction in toxicity of highly PSP-infested scallops during canning process and storage. Bull Jpn Soc Sci Fish 46:1339, 1980.
23. T Noguchi, Y Nagashima, J Maruyama, S Kamimura, K Hashimoto. Toxicity of the adductor muscle of markedly PSP-infested scallop Patinopecten yessoensis. Bull Jpn Soc Sci Fish 50:517, 1984.
24. A Prakash, JC Medcof, AD Tennant. Paralytic shellfish poisoning in Eastern Canada. Bull Fish Res Bd Can 177:1, 1971.
25. FP Thurberg. Inactivation of red-tide toxins by ozone treatment. In: WJ Blogoslawski, RG Rice, eds. Aquatic Application of Ozone. International Ozone Institute, 1975, p 50.
26. JM Vieites, LM Botana, MR Vieytes, F Leira. Canning process to diminish paralytic shellfish poison in naturally contaminated mussels (Mytilus galloprovincialis). J Food Prot (in press).
27. AW White, JL Martin, M Legresley, WJ Blogoslawski. Inability of ozonation to detoxify shellfish toxins in the soft-shell clams. In: In: DM Anderson, AW White, DG Baden, eds. Toxic Dinoflagellates, New York: Elsevier, 1985, pp 473–478.

35

Remote Sensing and Computerized Mapping for Development of Harmful Algal Bloom Prediction Methods

Ignacio Sordo, Joaquín A. Triñanes, José M. Cotos, and Carlos Hernández
University of Santiago de Compostela, Santiago de Compostela, Spain

I. INTRODUCTION

In the Rías Baixas, a series of marine rías located in Galicia (NW Spain; see Figure 1), the culture of mussels and other shellfish is a major local source of income; annual mussel production is approximately 200,000 ton (1). The risk of harmful algal blooms (HABs) is a recurrent threat. The Centro para o Control da Calidade do Medio Mariño (CCCMM, a marine environment quality control center belonging to the Galician government) carries out surveillance and research activities, including regular analysis of water and shellfish samples taken at stations adjoining the shellfish culture zones. As soon as toxin levels in mussel tissue exceed permitted limits, or toxic algae are detected in water samples, a temporary ban is put on shellfish harvesting in the affected areas (Figure 2). This efficiently prevents the marketing of shellfish containing harmful algae toxins, but results in serious economic losses to shellfish growers. The development and implementation of reliable methods for predicting the appearance of HABs would allow shellfish growers to harvest grown shellfish in advance, thus minimizing economic losses. In this chapter, we discuss, with particular reference to the situation in the Rias Baixas, some aspects of the possible roles of satellite remote sensing and geographical information systems in the development and/or implementation of HAB prediction methods (Figure 3).

II. RELEVANCE TO HAB PREDICTION

The environmental factors determining the sudden explosive proliferation of a minority algal population in a given area, or the activation of resting cysts, are complex and still poorly understood. The most immediate expected utility of satellite remote sensing is for early detection and monitoring of blooms in offshore areas (2) and prediction of the influx of these toxic algal populations into shellfish-growing areas as the result of mesoscale oceanic events. It may be worthwhile to point out that surveillance programs designed to detect such events by means of regular oceanographic cruises or flights would now be vastly more expensive than the use of satellite data.

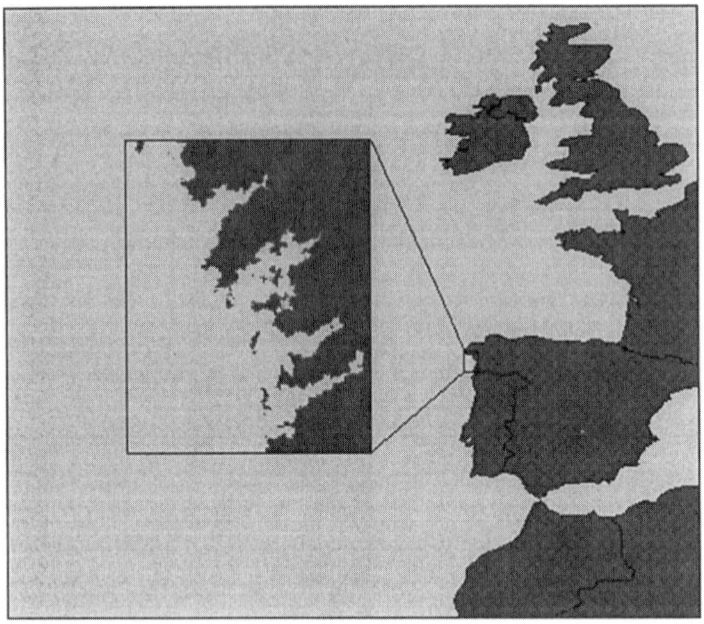

Figure 1 Location of the Galician Rías Baixas.

The two oceanic events that are currently thought to be chiefly responsible for the influx of toxic algal populations into shellfish-growing areas, at least in Galicia, are longshore currents and the relaxation or inversion of upwelling; it is possible that other events susceptible to remote sensing may also prove to be relevant as more data become available for correlation with observed bloom dynamics. A number of instances of longshore transport of HABs are documented in the literature. In several cases, it has been satellite images that indicated the origin of the

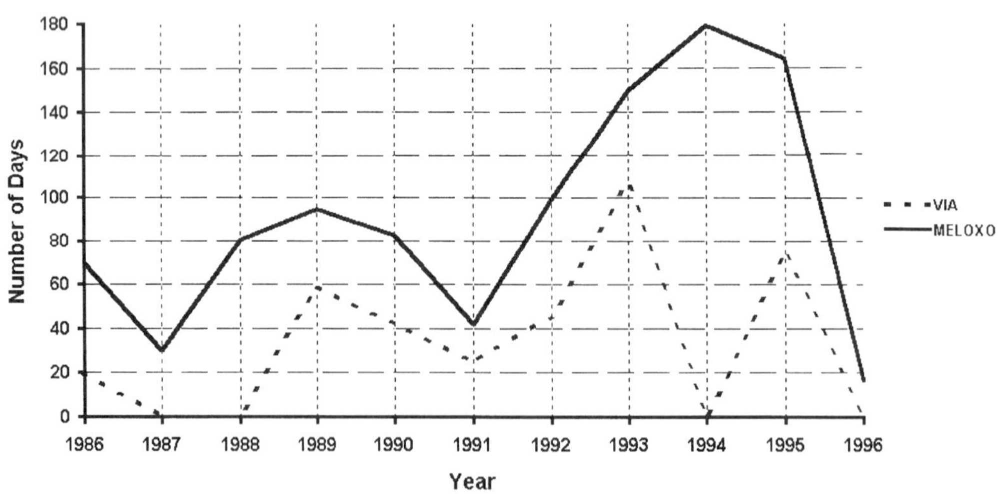

Figure 2 Total annual duration of HAB-provoked harvesting bans in two mussel-growing areas in the Galician Rías Baixas.

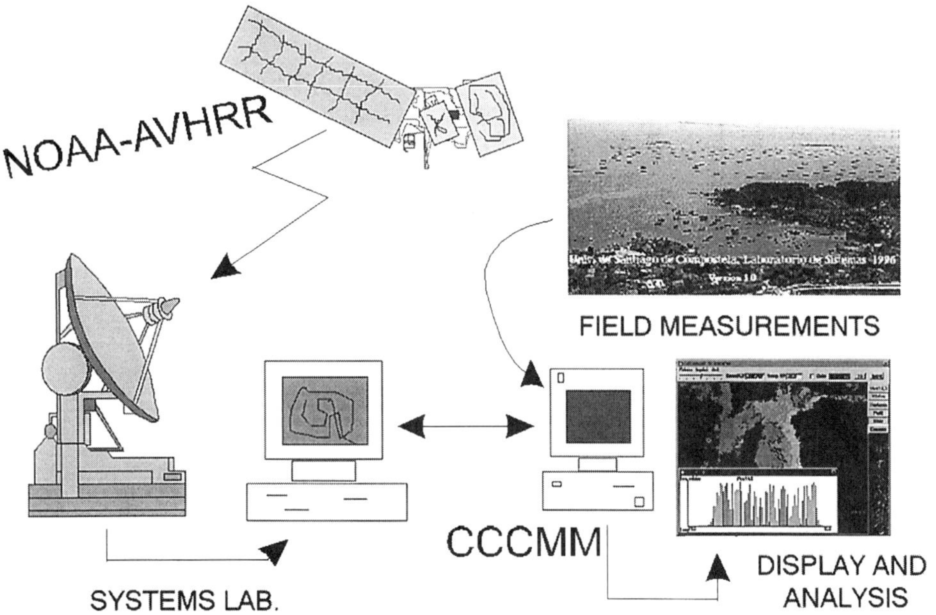

Figure 3 Operational relationship between the University of Santiago de Compostela's Systems Laboratory and the CCCMM center at Vilaxoán.

HAB: Dortch et al. (3) inferred that the presence of the dinoflagellate *Gymnodinium brevis* in the northern Gulf of Mexico was due to its transport from the west coast of Florida by currents that were visible in images provided by a National Oceanic and Atmospheric Administration (NOAA) satellite; and Tester et al. (4) similarly attributed the appearance of *G. brevis* off North Carolina to the shoreward movement of a filament of Gulf Stream water visible in sea surface temperature (SST) maps constructed from NOAA satellite data. In other cases, satellite data were not used, but the currents inferred would doubtless have been readily detectable in NOAA

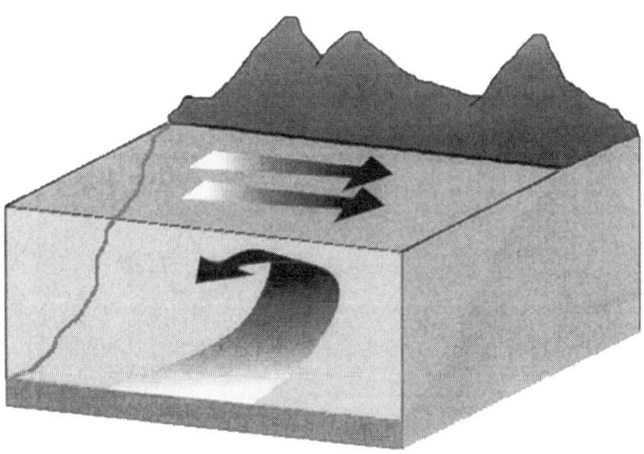

Figure 4 Oceanic upwelling is induced by steady longshore winds.

SST maps: Fraga (5) mentions that in 1993 the observation of *G. catenatum* at the edge of the Portuguese continental shelf off Figueira de Foz was followed 10 days later by a *G. catenatum* bloom in the Rias Baixas 200 km to the north. Satellite-assisted detection of currents from areas with known toxic algal populations toward shellfish-growing areas therefore holds out considerable promise for early identification of the risk of HAB.

Oceanic upwelling occurs off Galicia during the summer months when the prevailing northerly winds cause southward marine currents that are deflected offshore by the Coriolis force, the place of these surface water masses being taken by upwelling deep waters rich in

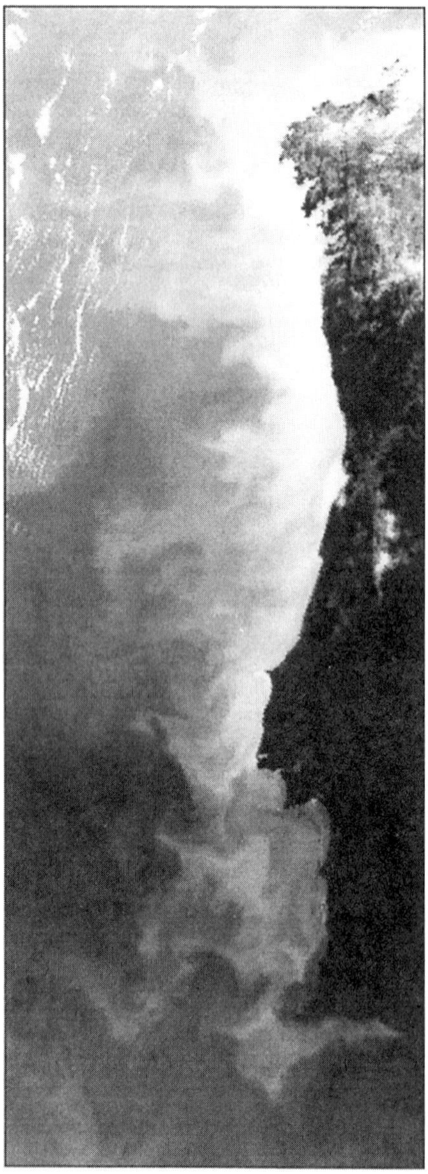

Figure 5 Intense coastal upwelling reflected in a NOAA SST map obtained in the summer of 1993 (light tones are colder).

Figure 6 A phytoplankton pigment map for the Atlantic off Spain constructed from CZCS data obtained in May 1982.

mineral nutrients (Figure 4). If the plankton that thrive at the interface between the cold and warm water masses include toxic species, then they may be carried into the rías by the inversion of the winds and currents that frequently occurs in late summer or early autumn. Da Silva and Moita (6) associated the relaxation of wind-induced upwelling with *G. catenatum* and *Dinophysis acuta* blooms during the summer of 1987; and Fraga et al. (7) attributed a *G. catenatum* paralytic shellfish poisoning (PSP) incident in the Ría de Vigo (the southernmost of the Rías Baixas) to a change in the wind that interrupted upwelling and caused the influx of dinoflagellate-laden offshore surface waters into the ría. Relaxation of upwelling is also known to alter the environmental structure of the waters inside the rías, which doubtless also influences HAB risk. Like longshore currents, upwelling fronts are readily detectable and quantifiable with the aid of data from satellite-borne sensors (Figure 5 shows a front visible in an SST map), and their remote surveillance, and that of the marine winds on which they depend, could therefore contribute substantially to the assessment of HAB risk.

III. SATELLITE PLATFORMS AND SENSORS

For many years, much of the remote sensing data relevant to the detection and study of mesoscale marine events and processes has come from the satellites of the NOAA TIROS (television infrared observation satellite) program and their successors the NOAA ATN (advanced TIROS-N) satellites (satellites of both series are usually referred to as NOAA-*n*; those currently in service are NOAA-12 and NOAA-14). Other satellites that will be mentioned in this chapter are those of the Landsat series, Nimbus-7 (which came out of service in 1986), the Japanese ADEOS (operational only between August 1996 and July 1997), the Indian IRS-P3 (launched in March 1996), and SeaStar (launched in August 1997). The characteristics that chiefly determine the utility of a satellite for a given remote sensing or surveillance function are the wavebands of the sensor(s) it carries, the spatial resolution of the images it records, and the period with which it passes over a target area of interest. The latter two factors are often inversely related: the longer the period between successive passes, the better the spatial resolution of the images obtained.

Table 1 SeaWiFS Wavebands

Band	Wavelength FWHM [μm]	Color	Measurement
1	0.402–0.422	Violet	Dissolved organic matter
2	0.433–0.453	Blue	Chlorophyll (blue absorption)
3	0.480–0.500	Blue/green	Chlorophyll
4	0.500–0.520	Green	Chlorophyll
5	0.545–0.565	Green/yellow	Chlorophyll
6	0.660–0.680	Red	Atmospheric aerosols
7	0.745–0.785	Near infrared	Atmospheric aerosols
8	0.845–0.885	Near infrared	Atmospheric aerosols

FWHM, full width at half maximum.

For direct satellite observation of existing HABs, the most useful wavebands are the higher frequency regions of the visible light spectrum. Nimbus-7 carried a visible light sensor, the Coastal Zone Colour Scanner (CZCS), that was specifically designed for studying phytoplankton populations; Figure 6 shows a phytoplankton pigment map for the Atlantic off Spain constructed from CZCS data obtained in May 1982 (note the higher concentrations of pigment in the colder

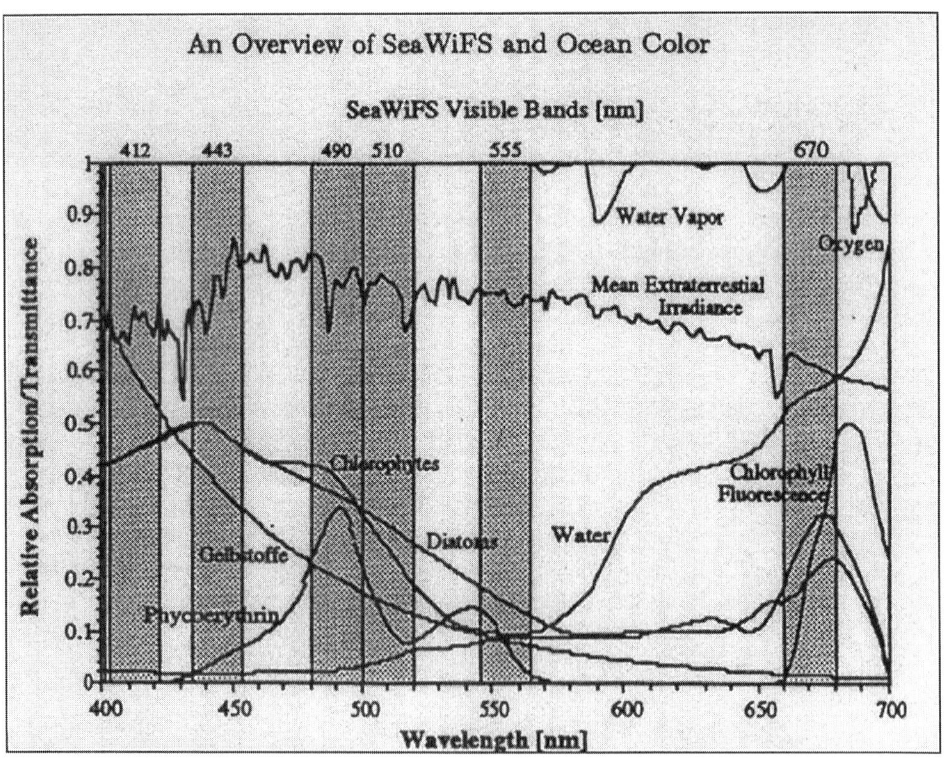

Figure 7 Spectral absorption characteristics of some common in-water optical constituents, mean extraterrestrial irradiance, and spectral transmittance of the atmospheric constituents oxygen and water vapor used to select the SeaWiFS visible bands. The bandwidths of the SeaWiFS bands are highlighted. (Figure 1 ref. 8, reprinted with permission.)

Table 2 AVHRR Wavebands

Band	Wavelengh FWHM [μm]
1	0.58–0.68
2	0.725–1.10
3	3.55–3.93
4	10.30–11.30
5	11.50–12.50

FWHM, full width at half maximum.

northern waters and in upwelling zones near the coast). Although Nimbus-7 ceased operations in 1986, its mention is relevant in relation to the formulae used to interpret data from visible light sensors on subsequent satellites (see Section VI). Since 1996, visible light data intended for oceanographic use have been obtained by ADEOS (now defunct) and IRS-P3, and have recently become available from the sea-viewing wide field-of-view sensor (SeaWiFS) on SeaStar. Since the IRS-P3 mission is concerned largely with systems development rather than the regular provision of data for applications, we shall refer here only to SeaWiFS, which comprises a series of sensors operating in the wavebands that are listed in Table 1 and shown, together with the absorption spectra of major seawater components, in Figure 7. SeaWiFS data have a spatial resolution of 1.1 km at nadir and the orbital period is 98.2 mins. Because of focusing on the visible part of the spectra, images at night are not valuable. Each day, one to

Figure 8 An SST map of part of the Rías Baixas region constructed using data from both Landsat and NOAA satellites.

three images are received by our antenna, with a time gap of about 1.5 hs between them. These figures are very appropriate for monitoring mesoscale phenomena in the open sea or on the continental shelf but are rather too coarse for detailed mapping of the Galician *rías*. For this later purpose, it is sometimes preferable to use data from the Landsat-borne Thematic Mapper (TM) sensor, which cover wavebands from the visible to the thermal infrared regions and have a spatial resolution of 30 or 120 m (depending on waveband); the drawback of Landsat data is its long sampling period: 16 days.

The mapping of oceanic thermal structure, which allows identification of currents, upwelling fronts, and other features (9–11), requires sensors measuring radiation in the thermal infrared region, such as the Landsat Thematic Mapper. More useful than the TM for oceanographic purposes is the advanced very high resolution radiometer (AVHRR) carried by NOAA ATN satellites, which senses in the wavebands listed in Table 2. Like SeaWiFS, AVHRR provides data with a spatial resolution of about 1.1 km, and about five to eight images are received by our antenna each day from NOAA-12 and NOAA-14.

Finally, it should be mentioned that data from a number of different sources (e.g., satellites, airborne sensors, oceanographic buoys) can often be combined to good purpose. Figure 8 shows an SST map of part of the Rias Baixas region that was constructed using data from both Landsat and NOAA satellites.

IV. DATA PROCESSING

As a remote sensing satellite circles the Earth, its sensors scan from side to side, building up images line by line (Figure 9); the band of terrain imaged in a single overpass of the satellite is called its swath. What the sensors measure and transmit to earth is the radiation they detect in the direction in which they are pointing at any given moment; they do not directly measure the

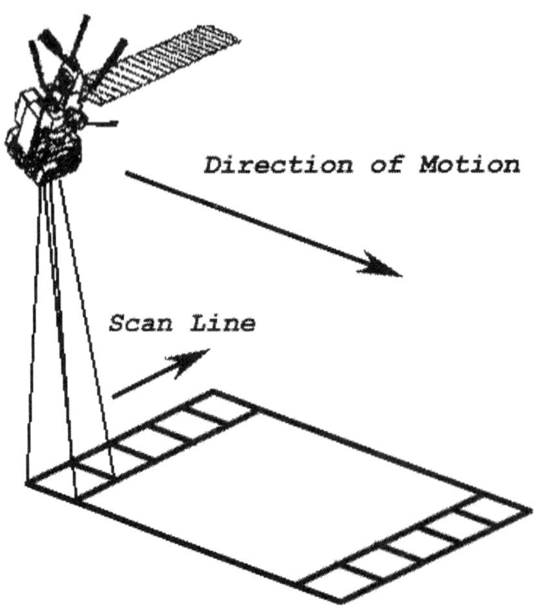

Figure 9 Remote sensing: scanning the swath from side to side.

variables of interest to most users, such as SST or plankton density. To determine the variables of interest, the raw satellite data must be subjected to several stages of data processing. Most of these stages depend on the variables to be determined, but one is common to all: adjustment of the raw radiance data in accordance with the calibration of the instrument so as to obtain true radiance values. Since modern satellite-borne sensors are not only calibrated before being launched but are also provided with mechanisms allowing their calibration to be continually updated, data adjustment is often preceded by recalibration using sensor measurements of certain reference objects.

For oceanographic purposes, another important stage of data processing is the masking of land and cloud. Cloud cover is a particular nuisance. Suppose, for example, that we have processed radiance data to obtain a map of ''sea surface temperature'' in a given region. The quotation marks are necessary here because certain areas of the map will in general correspond to land or cloud, not to sea (and the calculated temperature in these areas will be inaccurate because of having been calculated using parameters specific for the sea). Land areas can easily be masked out using knowledge of the shore line, but areas covered by cloud are not known a priori. Numerous methods have been proposed for identifying and masking cloud by analysis of SST and related data; they include the use of simple thresholds separating low-temperature regions (assumed to be cloudy) from higher temperature regions; adaptive thresholds (12); texture analysis and other statistical methods (13–15); and fuzzy logic (16,17). In our laboratory, we have developed a neural network (18) that after appropriate training identifies cloud areas when fed data, including SSTs, derived from all five AVHRR wavebands (Figure 10).

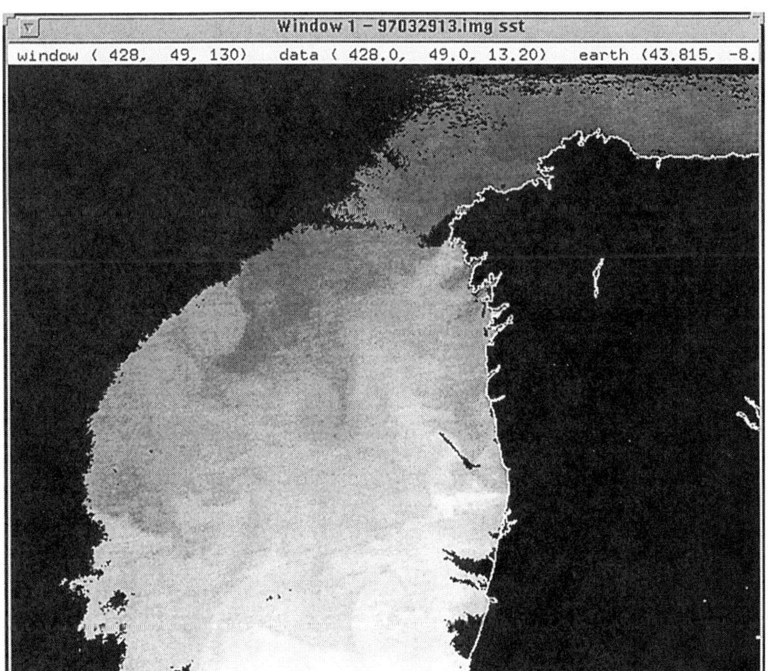

Figure 10 An SST map with cloud and land masked out.

V. HAB-RELATED INFORMATION OBTAINABLE FROM AVHRR DATA

A. Sea Surface Temperature

The radiation emitted by any object at a given wavelength is the product of its emissivity and the radiation that would be emitted at that wavelength by a perfect emitter (a "black body") with the same temperature as the object in question. Black body emission is given as a function of wavelength and temperature by Planck's law. Since the emissivity of the sea surface is near unity and can therefore be ignored, the radiance measured for a given sea area in each of the AVHRR wavebands allows Planck's law to be used to calculate an apparent temperature for that sea area (its *brightness temperature* for the waveband in question). However, none of these brightness temperatures is the real temperature of the target area, because the radiance detected by the AVHRR depends not only on the temperature of the sea but also on numerous other factors, notably the effect of the atmosphere. Figure 11 shows how the average percentage of radiance transmitted by the atmosphere varies with wavelength and hence why the best AVHRR wavebands for SST measurements are bands 3 (3.55–3.93 μm), 4 (10.3–11.3 μm), and 5 (11.5–12.5 μm); compare Table 2. However, even in these wavebands, transmittance is less than 100% and depends on spatial and temporal variation in atmospheric composition and thermal structure.

To correct for atmospheric effects, SSTs are calculated using empirical relationships between SST and the brightness temperatures in two or three bands. Specifically, the band 4 brightness temperature, T_4, is taken as the most reliable crude estimate of SST, and corrections are applied using the difference between T_4 and the brightness temperature for band 3 (T_3) or 5 (T_5), or the difference between T_3 and T_5. The three formulae that are commonly used to calculate the SST of a sea area viewed at an angle θ have the form

$$SST = A_0T_4 + A_1\Delta_{ij} + A_2\Delta_{ij}\Theta + A_3\Theta + A_4 \quad (1)$$

where $\Theta = \sec(\theta) - 1$, $\Delta_{ij} = T_i - T_j$ ($[ij]$ = [3,4], [4,5], or [3,5]), and the empirically determined parameters A_n depend on (i,j), on whether the radiance data are obtained during the daytime

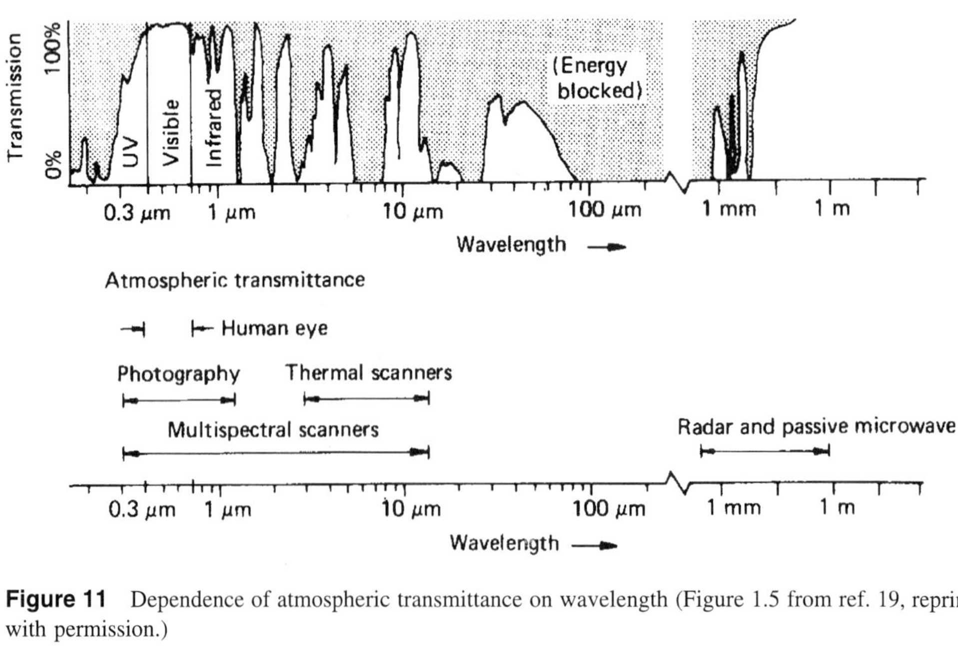

Figure 11 Dependence of atmospheric transmittance on wavelength (Figure 1.5 from ref. 19, reprinted with permission.)

or at night, and on which NOAA satellite from which the data are obtained. Ignoring the influence of viewing angle can lead to errors of up to 1°C at the edges of the swath (20); the error introduced by using globally optimized A_i that ignore spatial and temporal variation is less important.

The use of $\Delta_{ij} = T_3 - T_4$ (the "dual-window" method), $\Delta_{ij} = T_4 - T_5$ (the "split-window" method, so called because bands 4 and 5 adjoin each other; see Table 2) or $\Delta_{ij} = T_3 - T_5$ (the "triple-window" method) is determined largely by the time of day at which the radiance data are obtained, since in daytime reflected sunlight greatly increases the error of band 3 measurements. In our laboratory, we employ the split-window method during the daytime and the triple-window method at night. The resulting SSTs, which correspond to the conductive layer just below the sea-air interface (temperatures a few tens of centimeters deeper are typically 0.3–0.6°C warmer [21]), are accurate to within 0.3°C.

B. Suspended Particles Index

In the absence of radiance data for short wavelengths in the visible spectrum (the wavebands best reflecting plankton concentrations), between 1986 and 1996 formulae were developed to allow plankton blooms and other phenomena involving high suspended particles densities to be monitored using the shortest AVHRR wavebands, bands 1 and 2, which measure red and near infrared radiation, respectively (2,22). The physical basis for these formulae is the fact that suspended particles increase band 1 radiance but hardly affect band 2 radiance, and their concentration can therefore be estimated (roughly but nevertheless usefully) using essentially the difference between the two signals. Gower (2) defined a suspended particles index I_{SP} by

$$I_{SP} = G\left(\frac{N_1 - N_2}{C_{21}}\right) - AG\left(\frac{[p - 1024.5]}{1024}\right)^2 + O \tag{2}$$

where N_1 and N_2 are uncorrected radiance data for bands 1 and 2, respectively, C_{21} is the ratio between the reflectances of clear water in bands 1 and 2, A is an atmospheric correction, p is the digital position of the target area within the sensor scan line (the pixel number), G is a scale factor, and O is a baseline shift ensuring that even clear waters have positive I_{SP} values. Sea areas with high I_{SP} values (high particle densities) may correspond to river plumes (sediments carried out to sea), or to areas in which the sea bottom has been disturbed by storms; but in relatively calm waters unaffected by river currents, they may indicate high phytoplankton concentrations. In the I_{SP} map for March 30, 1997, shown in Figure 12, the light-shaded, high I_{SP} area off the Ría de Muros e Noia coincides with an area where fishing vessels reported red sea coloration presumably due to a *Mesodinium rubrum* bloom.

C. Surface Currents

As noted above (see Section II), HABs often seem to be initiated in Galician rías by harmful algae carried into the ría by marine currents. Marine currents can show up in series of satellite maps obtained at successive times: whatever variable the maps represent (e.g., SST, I_{SP}, plankton densities), if the values recorded in a given area in one map are shifted coherently in a particular direction in the next map in the series, then this must be due to their having been borne by a current. Such coherent shifts can be detected automatically by the *maximum cross-correlation method*, one variant of which we now describe (Figure 13).

First, a small area (for definiteness we consider a 15 × 15 pixel square) is defined around each pixel in the first map except for those in a border whose width is discussed below. Then,

Figure 12 An I_s map of the Atlantic off Spain for March 30, 1997.

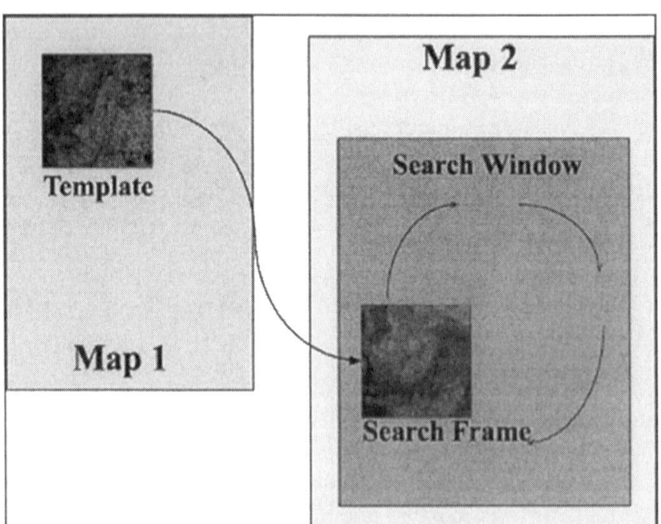

Figure 13 Calculation of currents: searching for maximum cross-correlation between a template in one map and the image in a search frame shifted around a larger search window in a second map of the same region acquired some time later.

for each such template area, a larger "search window" is defined around the same geographical area in the second map, and for each possible position of a 15 × 15 pixel search frame in the search window the cross correlation between the image in the template and the image in the search frame is calculated (cross correlation is a generalization of the familiar concept of correlation between two variables: the cross-correlation $\rho(I_1,I_2)$ between two $n \times n$ images $I_1(x,y)$ and $I_2(x,y)$ is defined by

$$\rho(I_1, I_2) = \mathrm{Cov}(I_1, I_2)/[\mathrm{Var}(I_1)\mathrm{Var}(I_2)]^{1/2} \quad (3)$$

where

$$\mathrm{Var}(I_i) = (1/n^2)\Sigma_x\Sigma_y[I_1(x,y) - \mu_1]^2 \quad (4)$$

and

$$\mathrm{Cov}(I_1,I_2) = (1/n^2)\Sigma_x\Sigma_y\ [I_1(x,y) - \mu_1][I_2(x,y) - \mu_2] \quad (5)$$

With μ_1 and μ_2 being the means of I_1 and I_2, respectively). Since a search frame area to which a current has borne the I values of the template area will have high cross correlation with the template area, the current at the center of the template area is provisionally estimated as defined by the vector from this point to the center of the most highly cross-correlated search frame area (and, of course, by the time interval between the two maps). However, the current so defined is only accepted as probably real if the associated cross correlation is high enough (in our laboratory we use a threshold of 0.7, which is rather higher than the thresholds afforded by significance test methods (23–25) but gives good results in practice); tentative currents failing

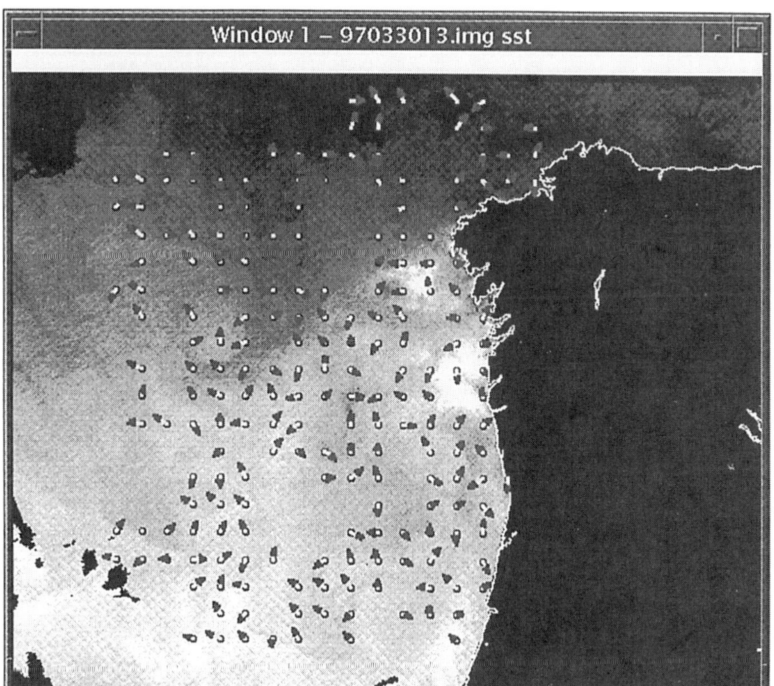

Figure 14 An SST map for March 30, 1997 and, superimposed, current vectors calculated from these data and those of another SST map acquired 9 h later.

this test are set to zero. Furthermore, to avoid detection of very localized currents that are irrelevant at the scale of interest, it is often advantageous to make currents passing the cross-correlation threshold test locally uniform by subjecting them to a median filter (26), which makes them identical to the median of the threshold-tested currents in the surrounding 15 × 15 pixel area.

The size of the search windows used in the second map must evidently depend on the strength of the currents that it is expected to detect. Specifically, the length L_2 of each side of the search window, in pixels, should be such that

$$L_2 = \text{Int}(L_1 + 2 \cdot S \cdot V_m \cdot \Delta t) \quad (6)$$

where L_1 is the length of each side of the template area (in pixels), V_m is the expected maximum current speed (km/h), Δt is the time interval between the two maps (in hours), S is the scale of the map (in pixels/km) and Int rounds to the nearest integer. V_m values should be specific for each zone of interest and time of year. The width of the border in which currents cannot be assigned in the first map (if the two maps cover exactly the same area) is $L_2/2$.

Figure 14 shows an SST map for March 30, 1997, together with the current vectors calculated from these data and those of another SST map acquired 9 h later. Note the northward current about 100 km from the northern Portuguese coast. Steady currents maintained throughout a certain season would be revealed by time-averaging the currents appearing in numerous successive maps like Figure 14.

VI. SEAWIFS PRODUCTS

Before transmission to Earth, SeaWiFS data are encoded using an encoding scheme that is changed fortnightly. The transmitted data are decoded using a key that is supplied by Orbital

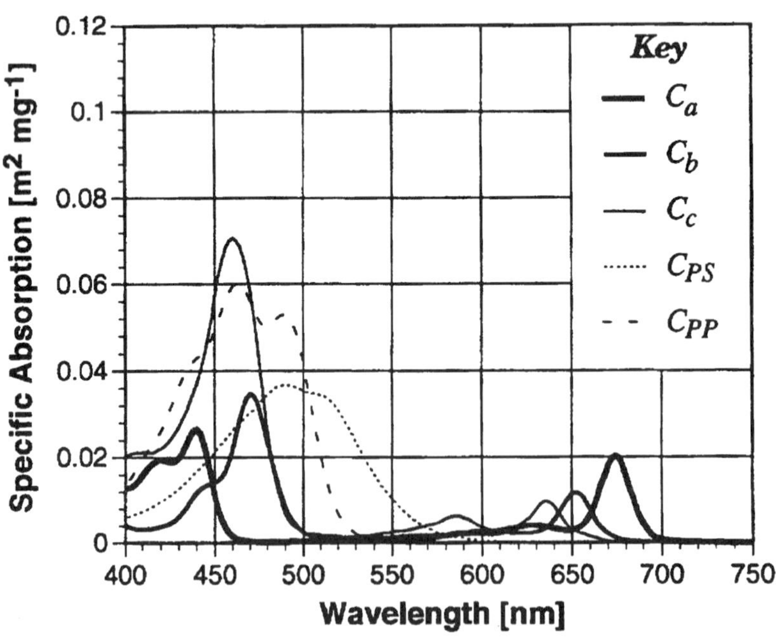

Figure 15 Absorption spectra of various plankton pigments. C_a, chlorophyll-a; C_b, chlorophyll-b; C_c, chlorophyll-c; C_{PS}, photosynthetic carotenoids; C_{PP}, photoprotectant-carotenoids. (Figure 1 from ref. 27, reprinted with permission)

Sciences Corporation (Dulles, Virginia), the owners of SeaStar in advance if the corresponding fee is paid or a fortnight later otherwise.

Correction for atmospheric effects is very much more important for SeaWiFS data than for AVHRR infrared data because up to 90% of the visible radiation received by the sensor originates in the atmosphere rather than at the sea surface. SeaDAS, the software package for processing SeaWiFS data, calculates corrected radiances using a database summarizing data on ozone concentration, surface wind speed, atmospheric pressure, and relative humidity over the past 40 years. The main quantities SeaDAS provides the user with are the normalized water-leaving radiances, L_{WM}, in bands 1–5, the atmospheric aerosol radiances, L_{a}, in bands 6–8, the aerosol optical thickness $\tau_{\mathrm{a}}(865)$ in band 8, the coefficient of diffuse attenuation at 490 nm (K_{490}), chlorophyll-a concentration, C_a (in mg/m^3), and the ''CZCS-type'' total pigment concentration $C_a + C_P$ (in mg/m^3; C_P is the concentration of phaeopigment). Here we shall concentrate on the last two.

Both C_a and $C_a + C_P$ are estimated from the ratio between the radiances measured in SeaWiFS bands 3 (centered on 490 nm) and 5 (centered on 555 nm) (28):

$$C_a = \exp\left(0.464 - 1.989 \ln\left[\frac{L_{\mathrm{WM}}(490)}{L_{\mathrm{WM}}(555)}\right]\right) \tag{7}$$

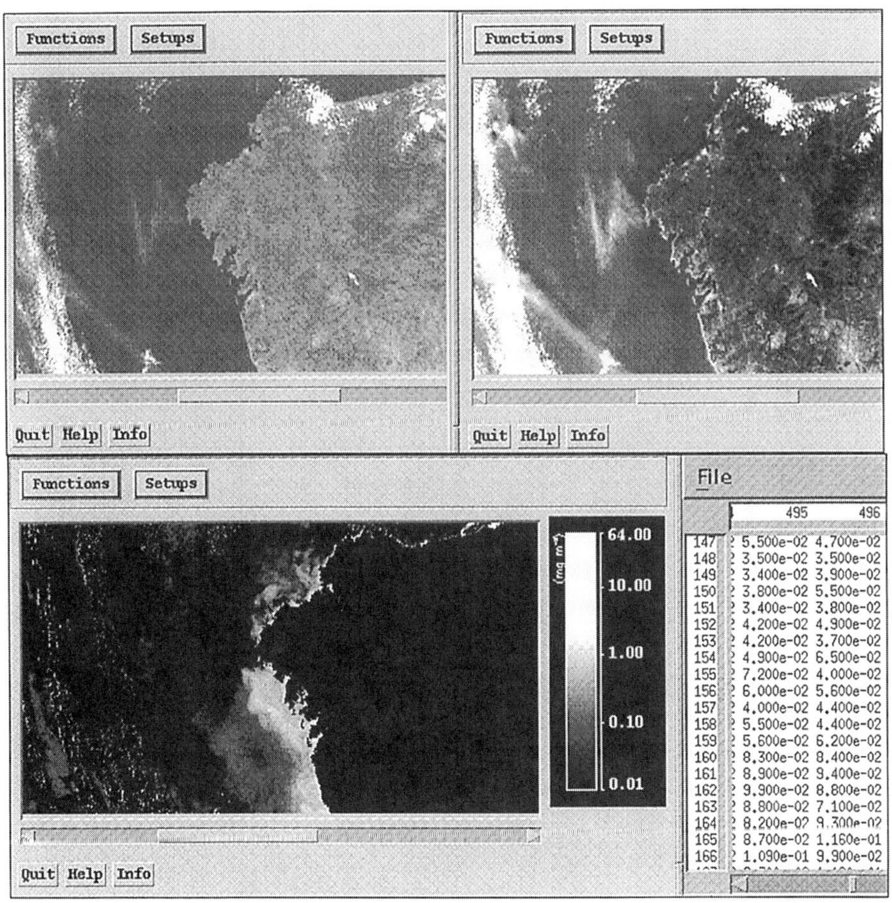

Figure 16 Maps of $L_{WM}(490)$ and $L_{WM}(555)$ off Galicia on October 29, 1997 (top), together with the C_a map derived using equation (7) (bottom).

$$C_a + C_P = \exp\left(0.696 - 2.085 \ln\left[\frac{L_{WM}(490)}{L_{WM}(555)}\right]\right) \tag{8}$$

Chlorophyll-a concentration is generally regarded as the best measure of viable phytoplankton biomass, but $C_a + C_P$ is also useful, partly because it provides continuity with earlier data obtained from the CZCS wavebands centered on 520 and 550 nm, so allowing joint analysis of CZCS and SeaWiFS data. Since SeaWiFS has more and narrower wavebands and better performance than the CZCS, better estimates of C_a and $C_a + C_P$ may be possible when enough sea-sample data are available for reliable parameterization of formulae with a sounder theoretical basis, but it is formulae 7 and 8 that are in general currently used (note, however, that whereas formula 8 appears to be very robust, the coefficients in formula 7 should preferably be optimized for each specific ocean region; and that more reliable formulae with a sounder theoretical basis are already available for estimating C_a and $C_a + C_P$ values less than 2 mg/m^3). The apparently paradoxical use of the waveband centered on 490 nm even though absorption by chlorophyll-a peaks in the 433- to 453-nm region (SeaWiFS band 2) and is practically zero at 490 nm (Figure 15) is explained by the high correlation between chlorophyll-a and carotenoid concentration, which in practice makes the $L_{WM}(490)/L_{WM}(555)$ ratio the more reliable predictor variable.

Figure 16 shows maps of $L_{WM}(490)$ and $L_{WM}(555)$ off Galicia on October 29, 1997, together with the derived C_a map. It may be hoped that such maps, and other information derivable from SeaWiFS measurements, such as data on primary production, dissolved organic matter concentration or suspended sediments, may assist understanding and eventually reliable prediction of HAB.

VII. GEOGRAPHICAL INFORMATION SYSTEMS FOR STUDYING HABs

The storage, interpretation, and joint analysis of the wealth of information with possible relevance to HABs that is now being acquired in particular shellfish production areas is enormously

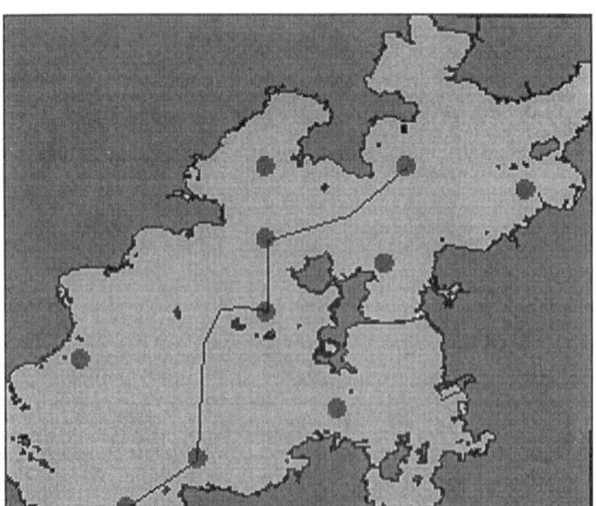

Figure 17 The 10 CCCMM monitoring stations in the Ría de Arousa, and the course of the section used for vertical profiling of marine variables in the ría.

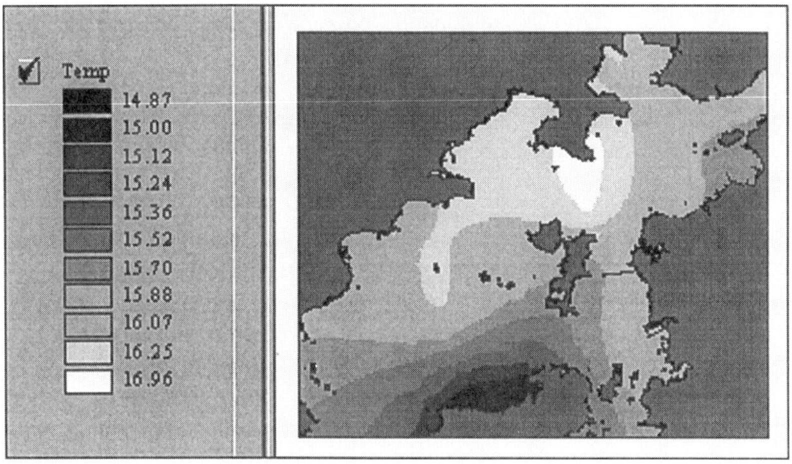

Figure 18 A surface temperature map of the Ría de Arousa for September 15, 1992, obtained by kriging data obtained at depths of 0–5 m at the 10 monitoring stations shown in Figure 17.

facilitated by incorporating it in automatic mapping systems relating it to local geography. A geographical information system (GIS) of this kind is currently being developed for the Rías Baixas. Even though its scope is currently limited to the data on marine physicochemical and biological variables that are measured at the CCCMM monitoring stations (e.g., temperature, salinity, dissolved oxygen, pH, chlorophyll concentrations, species-specific phytoplankton counts), it illustrates the capacity of such systems for condensing information in readily interpretable forms; in this case, maps of the CCCMM-monitored variables in horizontal and vertical sections of the Galician rías are constructed by kriging. For example, data obtained on September 15, 1992, at the 10 CCCMM monitoring stations in the Ría de Arousa (Figure 17), where about 80% of Spanish mussel production takes place, were used to construct the map of depthwise average sea temperature in the top 0–5 m shown in Figure 18 (similar maps can be produced for the 5- to 10-m and 10- to 15-m layers).

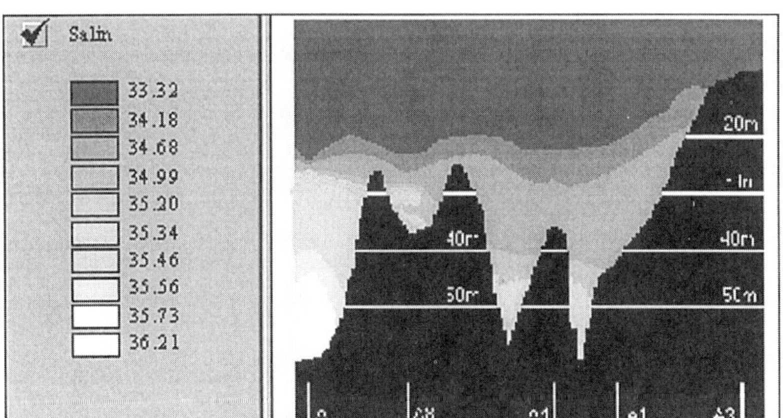

Figure 19 A salinity profile along the Ría de Arousa for January 2, 1996. The profile was constructed by kriging data collected at the CCCMM monitoring stations connected by a line in Figure 17.

The line joining some of the monitoring stations in Figure 17 indicates the bent section in which vertical profile maps are constructed. By way of example, Figure 19 shows a vertical profile of salinity for January 2, 1996.

Like the software for visualization and analysis of the satellite images supplied to the CCCMM by our laboratory, the HAB-oriented GIS has been developed within the Windows95 framework, and great effort has gone into maximizing its user-friendliness. It is envisaged that it will be readily adaptable for use in other areas as well as the Galician rías.

REFERENCES

1. B Reguera, MJ Campos, S Fraga, J Mariño, I Bravo. Oral Comunications in the CNEVA International Symposium on Marine Biotoxines, Paris, 1991.
2. JFR Gower. Red tide monitoring using AVHRR HRPT imagery from a local receiver. Remote Sens Environ 48:309–318, 1994.
3. Q Dortch, CA Moncreiff, W Mendenhall, ML Parsons, JS Franks, KW Hemphill. Spread of Gymnodinium Breve into Low Salinity Waters in the Northern Gulf of Mexico. VII International Conference on Harmful Algae, June, 1997, Vigo, Spain.
4. P Tester, RP Stumpf, FM Vukovich, PK Fowler, JT Turner. An expatriate red tide bloom: transport, distribution, and persistence. Limnol Oceanogr 36(5):1053–1061, 1991.
5. S Fraga. Harmful algal blooms in relation to wind induced coastal upwelling and river plumes. International Council for the Exploration of the Sea. Sess OCM 1993/L:38.
6. A Jorge da Silva, M Teresa Moita. Dynamic of toxic dinoflagellates during an upwelling event at the northwest coast off Spain. International Council for the Exploration of the Sea. Sess OCM 1993/L:66.
7. S Fraga, DM Anderson, I Bravo, B Reguera, KA Steindinger, CM Yentsch. Influence of upwelling relaxation on dinoflagellates and shellfish toxicity in Ría de Vigo, Spain. Estuarine, Coastal and Shelf Science 27:349–361, 1988.
8. SB Hooker, WE Esaias, GC Feldman, WW Gregg, CR McClain. An overview of SeaWiFS and Ocean Color. NASA Tech Memo 104566, Vol 1. SB Hooker and ER Firestone, eds. NASA Goddard Space Flight Center, Greenbelt, MD, 1992.
9. RJ Holyer, SH Peckinpaugh. Edge detection applied to satellite imagery of the oceans. IEEE Transact Geosci Remote Sens 27(1):46–56, 1989.
10. Cotos, JM Dinámica y Clasificación de Estructuras Oceánicas para Aplicación Operacional de Pesquerías Utilizando Teledetección e Ingeniería del Conocimiento. PhD dissertation, Dpto. Física Aplicada, Universidade de Santiago de Compostela, Spain.
11. R Lasker, J Peláez, RM Laurs. The use of satellite infrared imagery for describing ocean process in relation to spawning of the northern anchovy (Engraulis mordax). Remote Sens Environ 11:439–453, 1981.
12. JJ Simpson, JI Gobat. Improved cloud detection in GOES scenes over the oceans. Remote Sens Environ 52:79–94, 1995.
13. JJ Simpson, C Humphrey. An automated cloud screening algorithm for daytime advanced very high resolution radiometer imagery. J Geophys Res 95:13459–13481, 1990.
14. BA Eckstein, JJ Simpson. Cloud screening coastal zone color scanner images using channel 5. Int J Remote Sens 12:2359–2377, 1991.
15. TC Gallaudet, JJ Simpson. Automated cloud screening of AVHRR imagery using split-and-merge clustering. Remote Sens Environ 38:77–121, 1991.
16. JJ Simpson, RH Keller. An improved fuzzy logic segmentation of sea ice, clouds, and ocean in remotely sensed arctic imagery. Remote Sens Environ 54:290–312, 1995.
17. J Key, JA Maslanik, RG Barry. Cloud classification from satellite data using a fuzzy sets algorithm: a polar example. Int J Remote Sens 10:1823–1842, 1989.
18. JA Triñanes, J Torres, A Tobar, C Hernández. ''Clasificación de Imágenes Multiespectrales Mediante Redes Neuronales.'' Revista de Teledetección 3:46–50, 1994.

19. TM Lillesand, RW Kiefer. Remote Sensing and Image Interpretation. 2nd ed. New York: Wiley, 1987, pp 11.
20. PJ Minner. The numerical simulation of infrared satellite measurements over the greenland-iceland-norwegian Sea, SACLANTCEN SR-137, 1988.
21. CA Paulson, JJ Simpson. The temperature difference across the cool skin of the ocean. J Geophys Res 86:11044–11054, 1981.
22. RP Stumpf, MA Tyler. Satellite detection of bloom and pigment distribution in estuaries. Remote Sens Environ 24:385–404, 1988.
23. WJ Emery, AC Thomas, MJ Collins, WR Crawford, DL Mackas. An objective method for computing advective surface velocities from sequential infrared satellite images. J Geophys Res 91:C11:12865–12878, 1986.
24. R Tokmakian, PT Strub, J McClean-Padman. Evaluation of the maximum cross-correlation method of estimating sea surface velocities from sequential satellite images. J Atmo Ocean Technol 7:852–865, 1990.
25. OX Wu, D Pairman, SJ McNeill, EJ Barnes. Computing advective velocities from satellite images of sea surface temperature. IEEE Trans Geosci Remote Sens 30:(1):166–175, 1992.
26. JJ Simpson. Remote sensing in fisheries: a tool for better management in the utilization of a renewable resource. Can J Fish Aquat Sci 51:743–771, 1994.
27. RR Bidigare, ME Ondrusek, JH Morrow, DA Kiefer. In vivo absorption properties of algal pigments. SPIE Ocean Optics 1302:290–302, 1990.
28. A Aiken, GF Moore, CC Trees, SB Hooker, DK Clark. The SeaWiFS CZCS-Type Pigment Algorithm. NASA Tech Memo 104566, Vol 29, SB Hooker and ER Firestone, eds., NASA Goddard Space Flight Center, Greenbelt, MD, 1992.

Index